Abbreviations for Units

A	ampere	h	hour	N	newton
Å	angstrom (10^{-10} m)	Hz	hertz	nm	nanometer (10^{-9} m)
atm	atmosphere	in	inch	Pa	pascal
BTU	British thermal unit	J	joule	rad	radians
Bq	becquerel	K	kelvin	rev	revolution
C	coulomb	kg	kilogram	R	roentgen
°C	degree Celsius	km	kilometer	Sv	sievert
cal	calorie	keV	kilo-electron volt	s	second
Ci	curie	lb	pound	T	tesla
cm	centimeter	L	liter	u	unified mass unit
eV	electron volt	m	meter	V	volt
°F	degree Fahrenheit	MeV	mega-electron volt	W	watt
fm	femtometer, fermi (10^{-15} m)	mi	mile	Wb	weber
ft	foot	min	minute	y	year
G	gauss	mm	millimeter	μm	micrometer (10^{-6} m)
Gy	gray	mmHg	millimeters of mercury	μs	microsecond
g	gram	mol	mole	μC	microcoulomb
H	henry	ms	millisecond	Ω	ohm

Some Conversion Factors

Length

1 m = 39.37 in = 3.281 ft = 1.094 yard
1 m = 10^{15} fm = 10^{10} Å = 10^{9} nm
1 km = 0.6214 mi
1 mi = 5280 ft = 1.609 km
1 light-year = 1 $c \cdot$ y = 9.461×10^{15} m
1 in = 2.540 cm

Volume

1 L = 10^3 cm^3 = 10^{-3} m^3 = 1.057 qt

Time

1 h = 3600 s = 3.6 ks
1 y = 365.24 day = 3.156×10^7 s

Speed

1 km/h = 0.278 m/s = 0.6214 mi/h
1 ft/s = 0.3048 m/s = 0.6818 mi/h

Angle–angular speed

1 rev = 2π rad = 360°
1 rad = 57.30°
1 rev/min (rpm) = 0.1047 rad/s

Force–pressure

1 N = 10^5 dyn = 0.2248 lb
1 lb = 4.448 N
1 atm = 101.3 kPa = 1.013 bar = 760 mmHg = 14.70 lb/in^2

Mass

1 u = $[(10^{-3}\ \text{mol}^{-1})/N_A]$ kg = 1.661×10^{-27} kg
1 tonne = 10^3 kg = 1 Mg
1 kg = 2.205 lb

Energy–power

1 J = 10^7 erg = 0.7376 ft · lb = 9.869×10^{-3} L · atm
1 kW · h = 3.6 MJ
1 cal = 4.186 J
1 L · atm = 101.325 J = 24.22 cal
1 eV = 1.602×10^{-19} J
1 BTU = 778 ft · lb = 252 cal = 1054 J
1 horsepower = 550 ft · lb/s = 746 W

Thermal conductivity

1 W/(m · K) = 6.938 BTU · in/(h · ft^2 · °F)

Magnetic field

1 T = 10^4 G

Viscosity

1 Pa · s = 10 poise

College PHYSICS

Volume II

SECOND EDITION

Roger A. Freedman

Todd G. Ruskell

Philip R. Kesten

David L. Tauck

Vice President, STEM: *Ben Roberts*
Editorial Program Director: *Brooke Suchomel*
Program Manager: *Lori Stover*
Senior Development Editor: *Blythe Robbins*
Development Editor: *Meg Rosenburg*
Assistant Editor: *Kevin Davidson*
Marketing Manager: *Maureen Rachford*
Marketing Assistant: *Savannah DiMarco*
Content Director: *Natania Mlawer*
Media Editor: *Victoria Garvey*
Media Project Manager: *Daniel Comstock*
Associate Digital Marketing Specialist: *Cate McCaffery*
Physics Content Team Lead: *Josh Hebert*
Director, Content Management Enhancement: *Tracey Kuehn*
Managing Editor: *Lisa Kinne*
Senior Content Project Manager: *Kerry O'Shaughnessy*
Senior Workflow Project Manager: *Paul Rohloff*
Permissions Manager: *Jennifer MacMillan*
Photo Researcher: *Krystyna Borgen*
Director of Design, Content Management: *Diana Blume*
Interior Design: *Vicki Tomaselli*
Cover Design: *Lumina Datamatics, Inc.*
Art Manager: *Matthew McAdams*
Illustrations: *Precision Graphics*
Composition: *Lumina Datamatics, Inc.*
Printing and Binding: *LSC Communications*
Cover Art: *Quade Paul; Frank E. Fish*

Library of Congress Control Number: 2017955775

Volume II:
ISBN-13: 978-1-319-11511-1
ISBN-10: 1-319-11511-X

Printed in the United States of America
First printing

W. H. Freeman and Company
One New York Plaza
Suite 4500
New York, NY 10004-1562
www.macmillanlearning.com

ROGER:

To the memory of S/Sgt Ann Kazmierczak Freedman, WAC, and Pvt. Richard Freedman, AUS.

TODD:

To Susan and Allison, whose never-ending patience, love, and support made it possible. And to my parents, from whom I learned so much—especially my father, who so effectively demonstrates what it means to be an effective teacher both in and out of the classroom.

PHIL:

To my parents for instilling in me a love of learning, to my wife for her unconditional support, and to my children for letting their kooky dad infuse so much of their lives with science.

DAVE:

To my parents, Bill and Jean, for showing me how to lead a wonderful life, and to my sister and friends, teachers and students for helping me do it.

Preface

MOVE LEARNING FORWARD...

Bridge conceptual understanding and problem solving with resources to support active teaching and learning.

This new integrated learning system brings together a groundbreaking media program with an innovative text presentation of algebra-based physics. An experienced author team provides a unique set of expertise and perspectives to help students master concepts and succeed in developing problem-solving skills necessary for College Physics. Now available for the first time with SaplingPlus—an online learning platform that combines the heavily research-based FlipItPhysics Prelectures (derived from smartPhysics) with the robust Sapling homework system, in which every problem has targeted feedback, hints, and a fully worked and explained solution. This HTML5 platform gives students the ability to actively read with a fully interactive e-Book, watch Prelecture videos, and work or review problems with a mobile-accessible learning experience. Integration is available with Learning Management Systems to provide single sign-on and grade-sync capabilities compatible with the iClicker 2 and other classroom response systems to provide a seamless full course experience for you and your students.

Christine Pang

Roger Freedman This groundbreaking text boasts an exceptionally strong writing team that is uniquely qualified to write a college physics textbook. The *College Physics* author team is led by Roger Freedman, an accomplished textbook author of such best-selling titles as *Universe* (W. H. Freeman), *Investigating Astronomy* (W. H. Freeman), and *University Physics* (Pearson). Dr. Freedman is a lecturer in physics at the University of California, Santa Barbara. He was an undergraduate at the University of California campuses in San Diego and Los Angeles, and did his doctoral research in theoretical nuclear physics at Stanford University. He came to UCSB in 1981 after 3 years of teaching and doing research at the University of Washington. At UCSB, Dr. Freedman has taught in both the Department of physics and the College of Creative Studies, a branch of the university intended for highly gifted and motivated undergraduates. In recent years, he has helped to develop computer-based tools for learning introductory physics and astronomy and has been a pioneer in the use of classroom response systems and the "flipped" classroom model at UCSB. Roger holds a commercial pilot's license and was an early organizer of the San Diego Comic-Con, now the world's largest popular culture convention.

Mark Ramirez

Todd Ruskell As a Teaching Professor of Physics at the Colorado School of Mines, Todd Ruskell focuses on teaching at the introductory level and continually develops more effective ways to help students learn. One method used in large enrollment introductory courses is Studio Physics. This collaborative, hands-on environment helps students develop better intuition about, and conceptual models of, physical phenomena through an active learning approach. Dr. Ruskell brings his experience in improving students' conceptual understanding to the text, as well as a strong liberal arts perspective. Dr. Ruskell's love of physics began with a BA in physics from Lawrence University in Appleton, Wisconsin. He went on to receive an MS and a PhD in optical sciences from the University of Arizona. He has received awards for teaching excellence, including Colorado School of Mines' Alumni Teaching Award. Dr. Ruskell currently serves on the physics panel and advisory board for the NANSLO (North American Network of Science Labs Online) project.

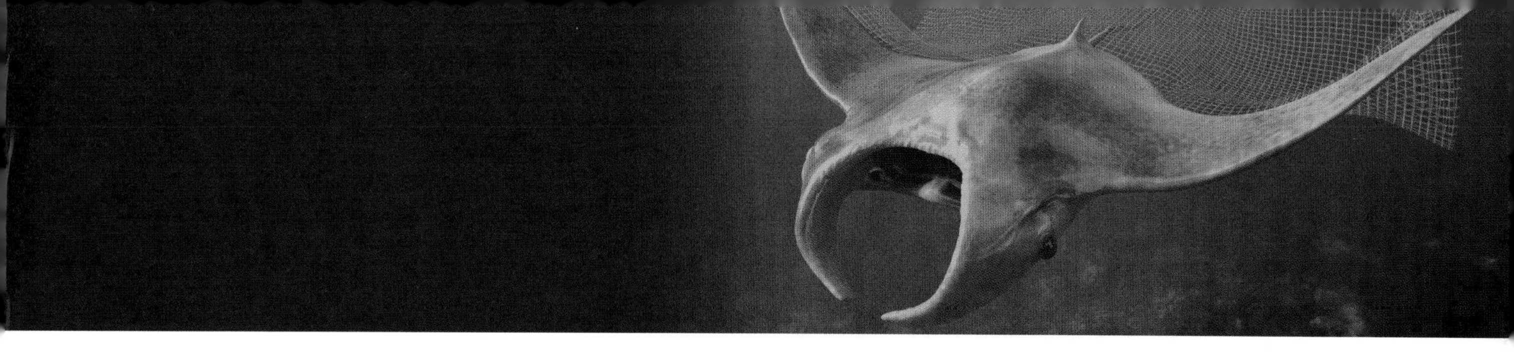

Build a conceptual foundation...before class

We place a high value on learning physics concepts prior to class time instruction and have created resources to make this as effective as possible for student learning and easy to implement for instructors. Groundbreaking Prelecture videos introduce students to physics topics and concepts as well as reinforce understanding with embedded questions ahead of class. In tandem, *College Physics*, Second Edition, provides an exceptional narrative and purposeful pedagogical tools focused on moving both conceptual learning and quantitative skill acquisition prior to class. A unique visual program and seamless blend of biological applications are interwoven throughout the book to provide relevance and interest for students taking algebra-based physics. By providing Prelectures with reading assignments, our goal is to jump-start student learning and allow for more productive class time.

Make classroom engagement meaningful

Instructor–student engagement can become more meaningful in class when students are aware of misconceptions and instructors have better insight into what students know before coming to class. Bridge questions (developed from research-based smartPhysics) provide a vehicle for students to demonstrate and communicate how well they understand the material that they learned before class. This invaluable instructor insight provides you with a way of identifying gaps in understanding and student misconceptions as you develop lectures to make the most efficient and meaningful use of class time. To further support an engaged classroom, Roger Freedman has refined an active classroom approach and shares in-class activities he has written to apply conceptual knowledge and engage students in the process of problem solving in class.

Develop problem-solving skills

Every effort in developing the print and digital materials for *College Physics*, Second Edition has been made to encourage students to develop a deep understanding of physics concepts and foster the reasoning and analytical skills necessary to solve problems. This goal motivated the student-centered pedagogy demonstrated in the worked examples and consistent Set Up—Solve It—Reflect problem-solving strategy found in *College Physics*, as well as the design and development of the Prelecture videos. Text problems incorporate conceptual questions, basic concepts, synthesis of multiple concepts, and life science applications. Paired with the Sapling homework platform, students are provided a tutorial experience with every problem in the system. Sapling adheres to the philosophy that every problem counts—therefore requiring that ALL problems have hints, targeted feedback, and a detailed solution—thus ensuring that students learn how to approach a problem, not just whether they answered correctly or incorrectly.

Philip Kesten

Philip Kesten, Associate Professor of Physics and Associate Vice Provost for Undergraduate Studies at Santa Clara University, holds a BS in physics from the Massachusetts Institute of Technology and received his PhD in high-energy particle physics from the University of Michigan. Since joining the Santa Clara faculty in 1990, Dr. Kesten has also served as Chair of Physics, Faculty Director of the ATOM and da Vinci Residential Learning Communities, and Director of the Ricard Memorial Observatory. He has received awards for teaching excellence and curriculum innovation, was Santa Clara's Faculty Development Professor for 2004–2005, and was named the California Professor of the Year in 2005 by the Carnegie Foundation for the Advancement of Education. Dr. Kesten is co-founder of Docutek (a SirsiDynix Company), an Internet software company, and has served as the Senior Editor for *Modern Dad,* a newsstand magazine.

Philip Kesten

David L. Tauck Unlike any other physics text on the market, *College Physics* includes a physiologist as a primary author. David Tauck, Associate Professor of Biology, holds both a BA in biology and an MA in Spanish from Middlebury College. He earned his PhD in physiology at Duke University and completed postdoctoral fellowships at Stanford University and Harvard University in anesthesia and neuroscience, respectively. Since joining the Santa Clara University faculty in 1987, he has served as Chair of the Biology Department, the College Committee on Rank and Tenure, and the Institutional Animal Care and Use Committee; he has also served as president of the local chapter of Phi Beta Kappa. Dr. Tauck currently serves as the Faculty Director in Residence of the da Vinci Residential Learning Community.

BUILDING A CONCEPTUAL FOUNDATION... BEFORE CLASS

Jump-start student learning with Prelectures, and free up class time for difficult concepts and problem-solving work.

Prelecture videos

Animated, narrated videos introduce core physics topics, laying the groundwork for conceptual understanding before students ever set foot in class. Each video is about 1–3 minutes long and is interspersed with conceptual questions. Each series can either be assigned in its entirety or divided into smaller, more tightly focused assignments. The full Prelecture activity is about 15 minutes long.

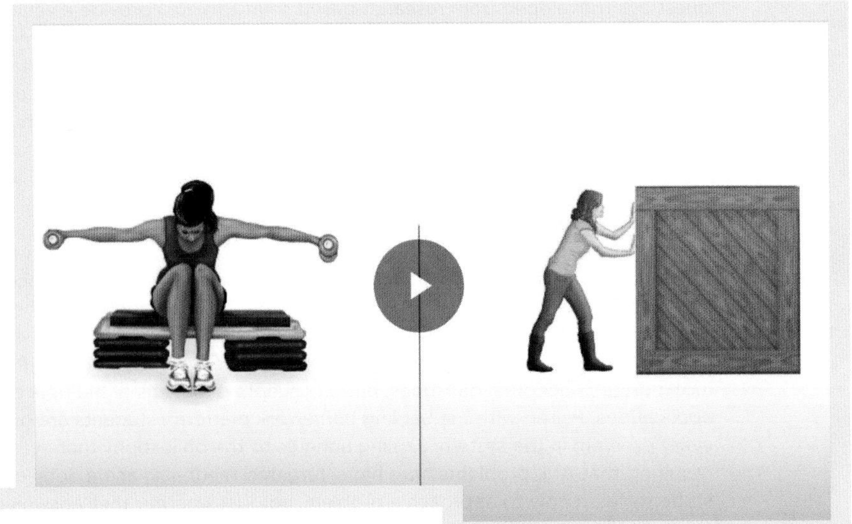

Angle between $\vec{F}$ and $\vec{d}$	Value of Work	Effect on Speed
$\theta = 90°$	$W = 0$	
$0° \leq \theta < 90°$	$W > 0$	will increase

Embedded questions reinforce student understanding along the way.

A box is pulled a distance d across the floor by a force F that makes an angle θ with the horizontal, as shown in the figure.

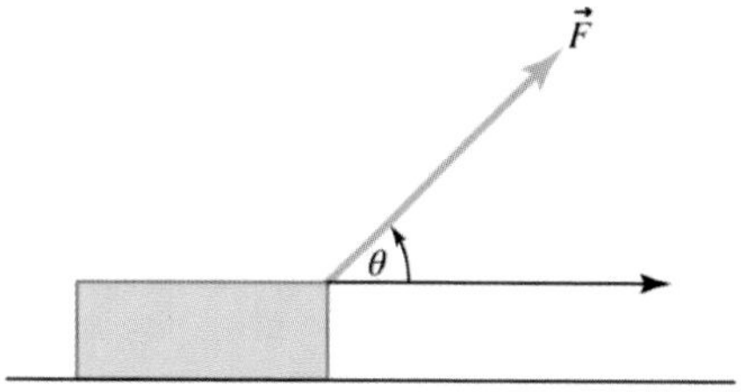

If the magnitude of the force was kept constant but the angle θ was increased toward 90°, the work done by the force in dragging the box would

- ◯ either increase or decrease, depending on what the initial angle θ was.
- ◯ increase.
- ◯ decrease.
- ◯ remain the same.
- ◯ either increase or decrease, depending on the magnitude of the force F.

$\vec{F}$

θ

Three objects of the same mass begin their motion at the same height. One object falls straight down, one slides down a low-friction inclined plane, and one swings in a circular arc on the end of a string. All three objects end at the same height.

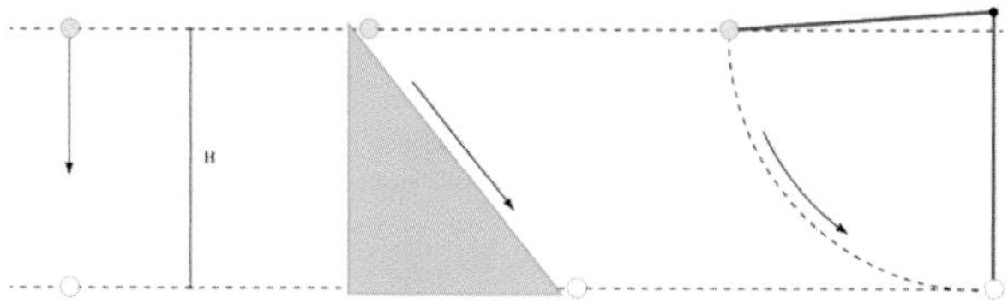

H

On which object does gravity do the most work?

- ◯ Gravity does equal work on all three objects.
- ◯ the object traversing the circular arc
- ◯ the object sliding down the low-friction incline
- ◯ the object in free fall

Briefly explain your choice.

Bridge questions

Multiple-choice and free-response questions review the content covered in the Prelectures and serve as a unique way for students to both demonstrate what they have learned and communicate misunderstandings or questions to an instructor prior to lecture.

A box sits on the horizontal bed of a truck that is accelerating to the left, as shown in the diagram. Static friction between the box and the truck keeps the box from sliding around as the truck accelerates.

a

μ_s

The work done on the box by the static friction force as the accelerating truck moves a distance D to the left is

- ◯ positive.
- ◯ negative.
- ◯ zero.
- ◯ dependent upon the speed of the truck.

Briefly explain your answer choice.

BUILDING A CONCEPTUAL FOUNDATION... BEFORE CLASS

Purposeful pedagogy helps students gain an accurate conceptual understanding of the most important physics concepts and avoid misconceptions.

Art and equations designed to teach

Visual narrative with word balloons

Average velocity of an object in linear motion

Displacement (change in position) of the object over a certain time interval: The object moves from x_1 to x_2, so $\Delta x = x_2 - x_1$.

$$v_{\text{average},x} = \frac{x_2 - x_1}{t_2 - t_1} = \frac{\Delta x}{\Delta t}$$

For both the displacement and the elapsed time, subtract the earlier value (x_1 or t_1) from the later value (x_2 or t_2).

Elapsed time for the time interval: The object is at x_1 at time t_1 and x_2 at time t_2, so the elapsed time is $\Delta t = t_2 - t_1$.

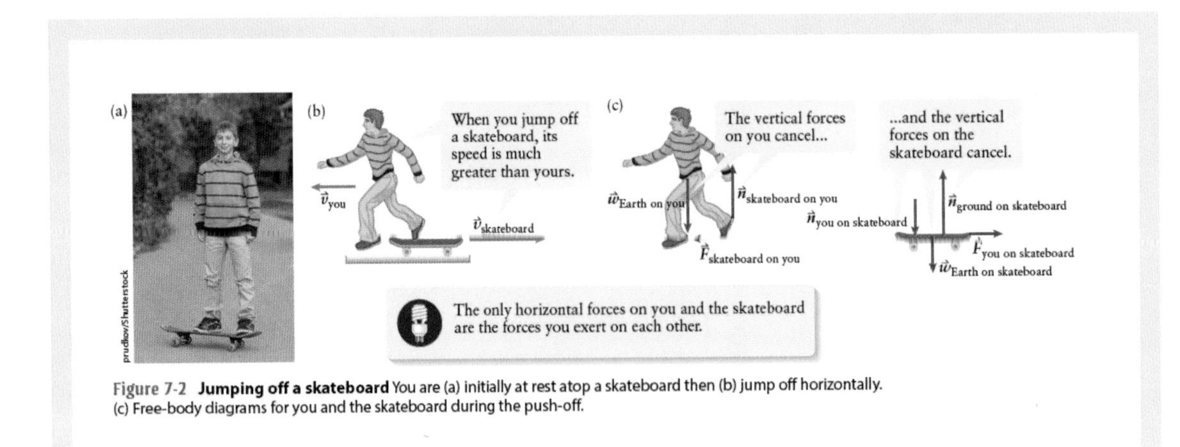

Figure 7-2 **Jumping off a skateboard** You are (a) initially at rest atop a skateboard then (b) jump off horizontally. (c) Free-body diagrams for you and the skateboard during the push-off.

Pedagogy emphasizing conceptual understanding

GOT THE CONCEPT? boxes help students think through the implications or connections of a physics concept.

GOT THE CONCEPT? 6-2 Slap Shot

A hockey player does work on a hockey puck in order to propel it from rest across the ice. When a constant force is applied over a certain distance, the puck leaves his stick at speed v. If instead he wants the puck to leave at speed $2v$, by what factor must he increase the distance over which he applies the same force? (a) $\sqrt{2}$; (b) 2; (c) $2\sqrt{2}$; (d) 4; (e) 8.

WATCH OUT! boxes draw attention to common student misconceptions and correct them immediately to promote a deeper understanding of physics.

WATCH OUT! The normal force exerted on an object is not always equal to the object's weight.

Since the maximum force of static friction is related to the normal force, it's important to remember that the normal force for an object on a surface is *not* always equal to the object's weight. (We cautioned you about this in Section 4-3). The situation shown in Figure 5-5 is an example of this caveat. Equation 5-4b shows that the normal force on this object is the object's weight multiplied by the cosine of the angle of the incline. Resist the temptation to always set the normal force equal to the weight!

TAKE-HOME MESSAGE. Every chapter section ends with a take-home message that is tied to outcome-based learning objectives listed at the beginning of each chapter and summarizes the main physics principles each student can take away.

TAKE-HOME MESSAGE FOR Section 6-3

✔ The net work done on an object (the sum of the work done on it by all forces) as it undergoes a displacement is equal to the change in the object's kinetic energy during that displacement.

✔ The formula for the kinetic energy of an object of mass m and speed v is $K = \frac{1}{2}mv^2$.

✔ The kinetic energy of an object is equal to the amount of work that was done to accelerate the object from rest to its present speed.

✔ The kinetic energy of an object is also equal to the amount of work the object can do in the process of coming to a halt from its present speed.

Deep integration of medical and biological examples to provide real-world relevance for life science majors

BIO-Medical icons point out biological applications.

BIO-Medical EXAMPLE 5-1 Friction in Joints

The wrist is made up of eight small bones called *carpals* that glide back and forth as you wave your hand from side to side. A thin layer of cartilage covers the surfaces of the carpals, making them smooth and slippery. In addition, the spaces between the bones contain synovial fluid, which provides lubrication. During a laboratory experiment, a physiologist applies a compression force to squeeze the bones together along their nearly planar bone surfaces. She then measures the force that must be applied parallel to the surface of contact to make them move. (Figure 5-7 shows the contact region between these two carpal bone surfaces.) When the compression force is 11.2 N, the minimum force required to move the bones is 0.135 N. What is the coefficient of static friction in the joint?

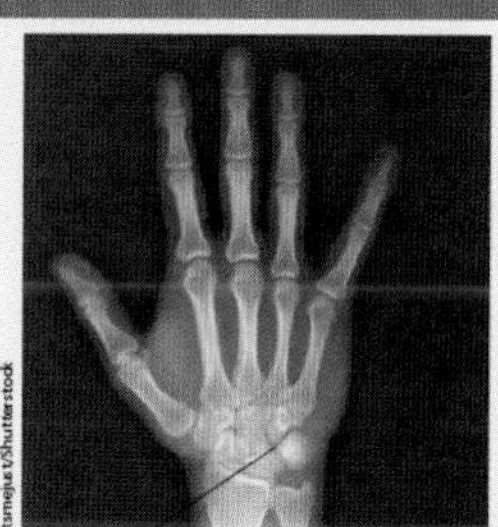

mage shows the carpal bones of the wrist.

BIO-Medical GOT THE CONCEPT? 8-9 Insects versus Birds

The maneuverability of a flying insect or a bird depends in large part on the contributions their wings make to the moments of inertia around their roll, pitch, and yaw axes (Figure 8-24). As much as 15% of the total body mass of a bird can be in its wings, while the wings of a typical insect are a considerably smaller fraction of its total mass. Which would you expect to be able to maneuver more quickly in flight: (a) a flying insect or (b) a bird?

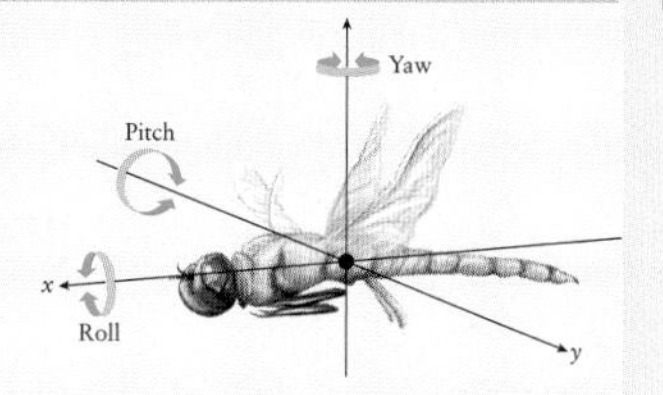

Figure 8-24 Rotation axes of an insect An insect can rotate in three ways: It can roll around its longitudinal x axis, pitch around its lateral y axis, or yaw around its z axis.

The electronic biological applications index on p. xxviii tells you where all of the BIO-Medical icons are in each chapter (by section) and how they relate to life science applications.

Life science examples are infused into problems throughout the text.

MAKE CLASSROOM ENGAGEMENT MEANINGFUL

Refine conceptual understanding in class with meaningful instructor–student engagement.

Bridge questions connect the Prelecture activity to the classroom experience, providing instructors with valuable insight into student understanding. With a better-prepared student audience, instructors can devote time to topics needing further explanation, or they can build on the knowledge students acquire before coming to class.

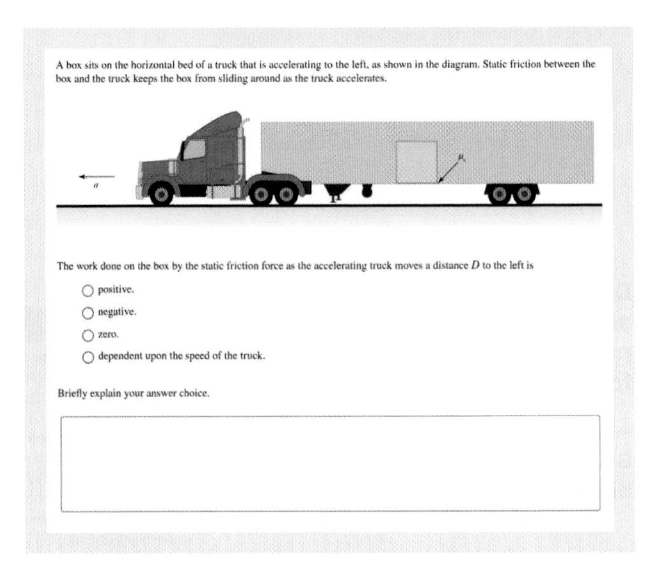

A box sits on the horizontal bed of a truck that is accelerating to the left, as shown in the diagram. Static friction between the box and the truck keeps the box from sliding around as the truck accelerates.

The work done on the box by the static friction force as the accelerating truck moves a distance D to the left is

- positive.
- negative.
- zero.
- dependent upon the speed of the truck.

Briefly explain your answer choice.

Activity 6-15. **Energy changes II [Accompanies Sections 6-7 and 6-8]**

A block is initially at point A at the bottom of an incline and is moving upward. *There is kinetic friction* between the block and the incline.

The block reaches a maximum height H at point C. Point B is halfway between points A and C. At point A, the gravitational potential energy is zero.

A, B, C, $\vec{v}_A$, $\vec{v}_B$, $\vec{v}_C = 0$, H, $H/2$

(a) Is mechanical energy conserved in this motion? *Explain* your reasoning.

In-class activities from Roger Freedman provide tough problems to tackle in class with annotations for instructors on learning objectives and student misconceptions. The activities can be used with iClicker Reef.

Integrated with iClicker
iClicker active learning simplified

In-class activities with iClicker

iClicker offers simple, flexible tools to help you give students a voice and facilitate active learning in the classroom. Students can participate with the devices they already bring to class using our iClicker Reef mobile app (which works with smartphones, tablets, or laptops) or iClicker remotes. We've now integrated iClicker with Macmillan's Sapling, and Roger Freedman has authored in-class activities that can be used with iClicker Reef. With Freedman, *College Physics*, Second Edition, and iClicker, it's easier than ever to promote engagement and synchronize student grades both in the classroom and at home.

Work 4

A box sits on the horizontal bed of a moving truck. Static friction between the box and the truck keeps the box from sliding around as the truck drives.

The work done on the box by the static frictional force as the truck moves a distance *D* is:

A. Positive
B. Negative
C. Zero

Work 4 Explanation

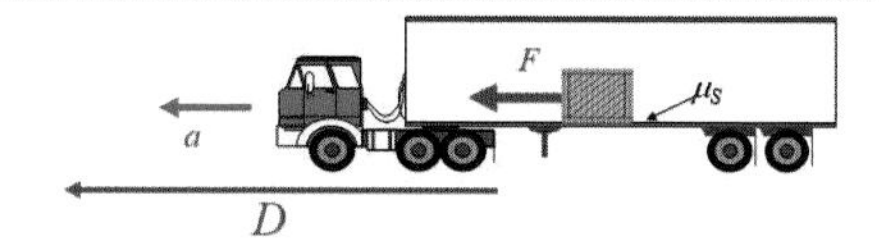

The work done on the box by the static frictional force as the truck moves a distance D is:

A) Positive B) Zero C) Negative

A. In the one example with the box it showed that if the change in *x* and force were in the same direction, the work would be positive.

B. It is zero because there is no movement.

: is negative because it is working in a negative direction.

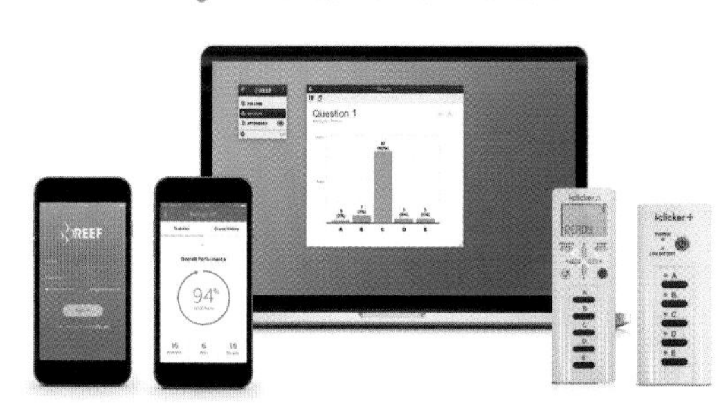

iClicker Reef access cards can also be packaged with Sapling or your Macmillan textbook at a significant savings for your students. To learn more, visit **iclicker.com** or talk to your Macmillan Learning representative.

DEVELOP
PROBLEM-SOLVING SKILLS

Encourage students to develop a deep understanding of physics concepts and foster the reasoning and analytical skills necessary to apply a consistent problem-solving strategy and a student-centered pedagogical framework.

Encourage strategic thinking

Set Up—Solve—Reflect problem-solving strategy

Worked examples mirror the approach scientists take to solve problems by developing reasoning and analysis skills.

BIO-Medical **EXAMPLE 5-7 Terminal Speed**

When diving straight down toward its prey, a peregrine falcon is acted on by two forces: a downward gravitational force, and a drag force of magnitude F_{drag} given by Equation 5-9 directed vertically upward (opposite to the direction of the falcon's motion through the air). As the falcon falls and its speed v increases, the value of F_{drag} also increases. When the drag force becomes equal in magnitude to the gravitational force, the net force on the falcon is zero and the falcon ceases to accelerate. It has reached its *terminal speed*, so it no longer speeds up nor does it slow down. Find the terminal speed of a female peregrine falcon of mass 1.2 kg, for which the value of c is $1.6 \times 10^{-3}\ N \cdot s^2/m^2$.

Set Up

The free-body diagram shows the two forces acting on the falcon. We use Equation 5-9 to find the value of the speed v at which the sum of these forces is zero, so that the acceleration is zero and the downward velocity is constant.

$$\sum \vec{F}_{ext\ on\ falcon} = \vec{F}_{drag\ on\ falcon} + \vec{w}_{falcon} = m\vec{a}_{falcon} = 0$$

Drag force for larger objects at faster speeds:

$$F_{drag} = cv^2 \qquad (5\text{-}9)$$

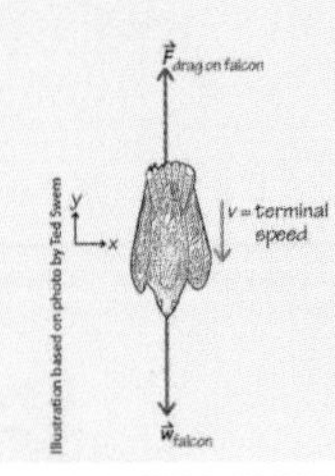

Solve

Write Newton's second law in component form and solve for the terminal speed v_{term}.

Newton's second law in component form applied to the falcon:

$$y: F_{drag\ on\ falcon} + (-w_{falcon}) = 0$$

At the terminal speed v_{term},

$$F_{drag\ on\ falcon} = cv_{term}^2 \text{ so}$$

$$cv_{term}^2 - w_{falcon} = 0$$

$$cv_{term}^2 = w_{falcon} = mg$$

$$v_{term}^2 = \frac{mg}{c}$$

$$v_{term} = \sqrt{\frac{mg}{c}}$$

Substitute the numerical values of m and c for the falcon.

Using $m = 1.2$ kg and $c = 1.6 \times 10^{-3}\ N \cdot s^2/m^2$,

$$v_{term} = \sqrt{\frac{(1.2\ kg)(9.80\ m/s^2)}{1.6 \times 10^{-3}\ N \cdot s^2/m^2}} = 86\ m/s = 310\ km/h = 190\ mi/h$$

Reflect

The high diving speeds attained by a peregrine falcon make it the fastest member of the animal kingdom.

The relationship $v_{term} = \sqrt{\frac{mg}{c}}$ explains the common notion that "heavier objects fall faster." A baseball and an iron ball of the same radius falling side by side in air have the same value of c (because they have the same shape and size), but the iron ball will have a greater terminal speed because its mass m is greater. So a heavier object *does* fall faster if we take the drag force into account. If the baseball and iron ball were dropped side by side in a vacuum, however, they would have the same acceleration of magnitude g and so would always have the same speed.

Set Up. The first step in each problem is to determine an overall approach and to gather the necessary pieces of information needed to solve it. These might include sketches, equations related to the physics, and concepts.

Solve. Rather than simply summarizing the mathematical manipulations required to move from first principles to the final answer, the authors show many intermediate steps in working out solutions to the sample problems, highlighting a crucial part of the problem-solving process.

Reflect. An important part of the process of solving a problem is to reflect on the meaning, implications, and validity of the answer. Is it physically reasonable? Do the units make sense? Is there a deeper or wider understanding that can be drawn from the result? The authors address these and related questions when appropriate.

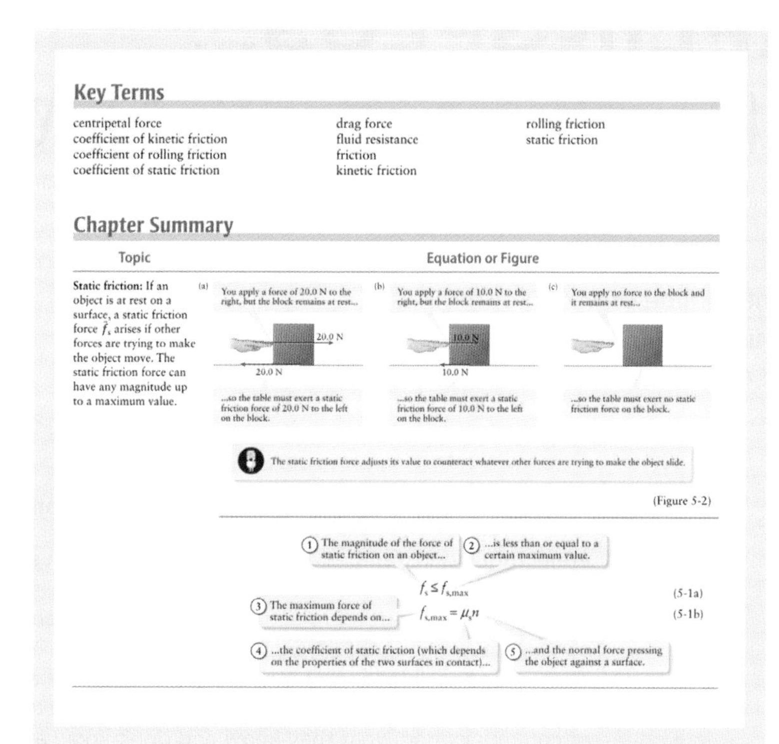

Key Terms

centripetal force
coefficient of kinetic friction
coefficient of rolling friction
coefficient of static friction
drag force
fluid resistance
friction
kinetic friction
rolling friction
static friction

Chapter Summary

Topic	Equation or Figure
Static friction: If an object is at rest on a surface, a static friction force $\vec{f}_s$ arises if other forces are trying to make the object move. The static friction force can have any magnitude up to a maximum value.	(a) You apply a force of 20.0 N to the right, but the block remains at rest... ...so the table must exert a static friction force of 20.0 N to the left on the block. (b) You apply a force of 10.0 N to the right, but the block remains at rest... ...so the table must exert a static friction force of 10.0 N to the left on the block. (c) You apply no force to the block and it remains at rest... ...so the table must exert no static friction force on the block. The static friction force adjusts its value to counteract whatever other forces are trying to make the object slide. (Figure 5-2)
	① The magnitude of the force of static friction on an object... ② ...is less than or equal to a certain maximum value. $f_s \le f_{s,max}$ (5-1a) ③ The maximum force of static friction depends on... $f_{s,max} = \mu_s n$ (5-1b) ④ ...the coefficient of static friction (which depends on the properties of the two surfaces in contact)... ⑤ ...and the normal force pressing the object against a surface.

Student-centered framework

Key terms/visual summary
Help students synthesize ideas after reading or class time.

End-of-chapter questions tagged with related worked examples. Students can go back and review worked examples while working problems.

47. •• An object of mass $3M$, moving in the $+x$ direction at speed v_0, breaks into two pieces of mass M and $2M$ as shown in **Figure 7-26**. If $\theta_1 = 45°$ and $\theta_2 = 30°$, determine the final velocities of the resulting pieces in terms of v_0. SSM Example 7-4

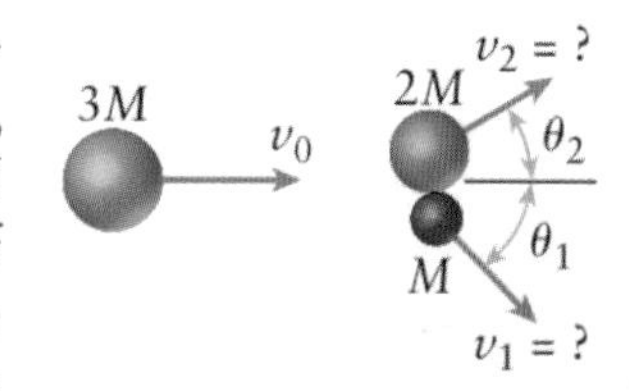

Figure 7-26 Problem 47

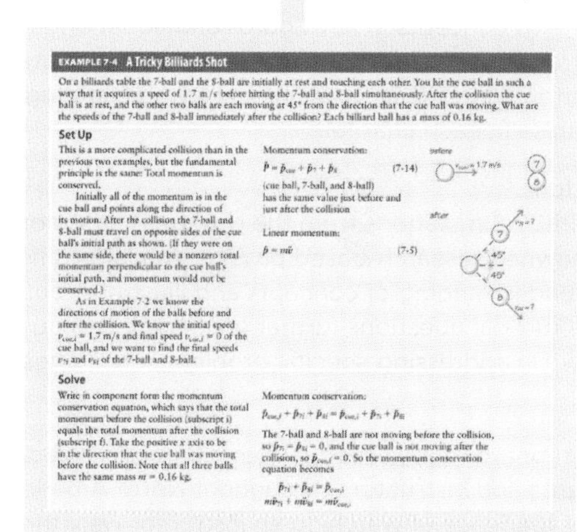

EXAMPLE 7-4 A Tricky Billiards Shot

On a billiards table the 7-ball and the 8-ball are initially at rest and touching each other. You hit the cue ball in such a way that it acquires a speed of 1.7 m/s before hitting the 7-ball and 8-ball simultaneously. After the collision the cue ball is at rest, and the other two balls are each moving at 45° from the direction that the cue ball was moving. What are the speeds of the 7-ball and 8-ball immediately after the collision? Each billiard ball has a mass of 0.16 kg.

Set Up

This is a more complicated collision than in the previous two examples, but the fundamental principle is the same: Total momentum is conserved.

Initially all of the momentum is in the cue ball and points along the direction of its motion. After the collision the 7-ball and 8-ball must travel on opposite sides of the cue ball's initial path as shown. (If they were on the same side, there would be a nonzero total momentum perpendicular to the cue ball's initial path, and momentum would not be conserved.)

As in Example 7-2 we know the directions of motion of the balls before and after the collision. We know the initial speed $v_{cue,i} = 1.7$ m/s and final speed $v_{cue,f} = 0$ of the cue ball, and we want to find the final speeds v_{7f} and v_{8f} of the 7-ball and 8-ball.

Momentum conservation:

$\vec{P} = \vec{p}_{cue} + \vec{p}_7 + \vec{p}_8$ (7-14)

(cue ball, 7-ball, and 8-ball) has the same value just before and just after the collision

Linear momentum:

$\vec{p} = m\vec{v}$ (7-5)

Solve

Write in component form the momentum conservation equation, which says that the total momentum before the collision (subscript i) equals the total momentum after the collision (subscript f). Take the positive x axis to be in the direction that the cue ball was moving before the collision. Note that all three balls have the same mass $m = 0.16$ kg.

Momentum conservation:

$\vec{p}_{cue,f} + \vec{p}_{7f} + \vec{p}_{8f} = \vec{p}_{cue,i} + \vec{p}_{7i} + \vec{p}_{8i}$

The 7-ball and 8-ball are not moving *before* the collision, so $\vec{p}_{7i} = \vec{p}_{8i} = 0$, and the cue ball is not moving after the collision, so $\vec{p}_{cue,f} = 0$. So the momentum conservation equation becomes

$\vec{p}_{7f} + \vec{p}_{8f} = \vec{p}_{cue,i}$
$m\vec{v}_{7f} + m\vec{v}_{8f} = m\vec{v}_{cue,i}$

Categorized end-of-chapter problems

- Conceptual questions
- Multiple-choice questions
- Estimation/numerical analysis
- Problems (organized by section)
- General problems

WITH SAPLINGPLUS...
EVERY PROBLEM COUNTS

This comprehensive and robust online platform combines innovative, high-quality teaching and learning features with Sapling Learning's acclaimed online Physics homework.

SaplingPlus for *College Physics,* Second Edition, features:

Prelecture videos/embedded questions/bridge questions
Developed based on research and principles that defined smartPhysics, animated and narrated Prelecture videos give students both a conceptual and quantitative understanding of core Physics topics. These videos are followed up by *bridge questions* that extend student learning to in-class engagement by giving students a means of communicating what they know and don't know and providing instructors with access to valuable insights to tailor class time.

Interactive e-Book
New! For the first time, the e-Book is also available through an app that allows students to read offline and have the book read aloud to them, in addition to the highlighting, note taking, and keyword search that you have come to expect.

Sapling Learning problems
Where every problem counts—with hints, targeted feedback, and detailed solutions.

Balloon art concept checks
Designed to guide students through the process of identifying important physics concepts in key figures and equations. Mirroring the visual narrative in the form of word balloons, these interactive questions reinforce key ideas from the text by highlighting important physics principles in each chapter.

PhET simulations
New HTML5 PhET Simulations from the University of Colorado at Boulder's renowned research-based physics simulations help students gain a visual understanding of concepts and illustrate cause-and-effect relationships. Tutorial questions further encourage this quantitative exploration, while addressing specific problem-solving needs.

P'Casts
250 total whiteboard mobile-ready videos. Carefully selected by physics students and instructors throughout North America to help simulate the experience of watching an instructor walk through the steps and explanation of Physics concepts while solving a problem.

Pocket worked examples
All worked examples from *College Physics* are available as a downloadable item for mobile devices.

ANATOMY OF A SAPLING PROBLEM

Hints
Clues attached to every problem encourage critical thinking by providing suggestions for completing the problem, without giving away the answer.

Hint

The potential energy in this scenario is *gravitational potential energy*, which depends on the height of the car above some reference point. In this case, take the bottom of the hill to be the reference point, so the potential energy at the bottom of the hill is zero.

For this question, use $g = 10\ m/s^2$ for the acceleration due to gravity.

Since the car is assumed to coast freely, without friction, you can make use of the principle of *energy conservation*.

A 2 kg toy car sits at the highest point of a 13 m high hill. The car is gently pushed forward until it begins to roll down the slope. Assuming the car coasts freely, without any friction or air resistance, how much kinetic energy (KE) and potential energy (PE) will it have at each of the indicated points? Complete the diagram by placing the correct label in each bin. Use $g = 10\ m/s^2$ for the acceleration due to gravity. Note, the diagram is not drawn to scale.

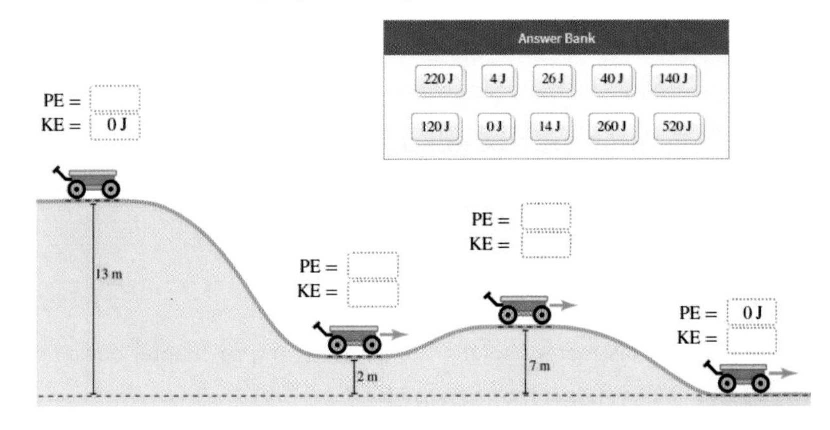

Targeted feedback
Each question includes wrong-answer specific feedback targeted to students' misconceptions.

Feedback

Remember that we are taking the bottom of the hill as our reference level. The potential energy at each point depends on the height h of that point relative to the bottom of the hill: $PE = mgh$.

A 3 kg toy car sits at the highest point of a 13 m high hill. The car is gently pushed forward until it begins to roll down the slope. Assuming the car coasts freely, without any friction or air resistance, how much kinetic energy (KE) and potential energy (PE) will it have at each of the indicated points? Complete the diagram by placing the correct label in each bin. Use $g = 10\ m/s^2$ for the acceleration due to gravity. Note, the diagram is not drawn to scale.

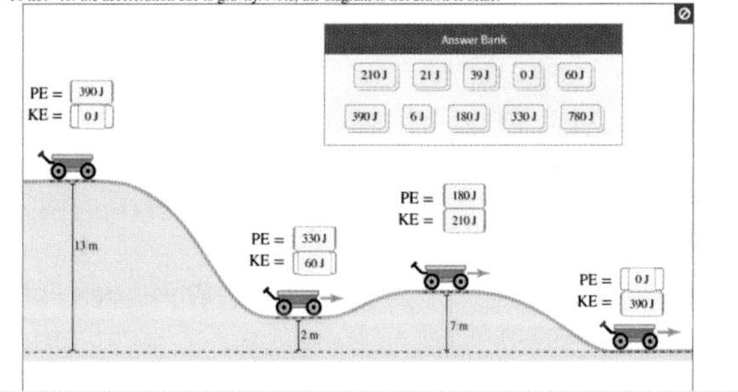

Detailed solutions
Fully worked solutions reinforce concepts and provide an in-product study guide for every problem in the Sapling Learning system.

Solution

The potential energy in this scenario is *gravitational potential energy*, which reference point. In this case, we take the bottom of the hill to be our reference At some height h above the bottom, the gravitational potential energy PE is g

$$PE = mgh$$

where m is the mass of the car (3 kg), and $g = 10\ m/s^2$ is the acceleration energy at each point. At the top of the hill, $h = 13$ m, so

$$PE_1 = (3\ kg) \times (10\ m/s^2) \times (13\ m) = 390\ J$$

In the valley, at the point where $h = 2$ m, the potential energy is

$$PE_2 = (3\ kg) \times (10\ m/s^2) \times (2\ m) = 60\ J$$

Similarly, the potential energy at the top of the smaller hill can be found

$$PE_3 = 210\ J$$

A 3 kg toy car sits at the highest point of a 13 m high hill. The car is gently pushed forward until it begins to roll down the slope. Assuming the car coasts freely, without any friction or air resistance, how much kinetic energy (KE) and potential energy (PE) will it have at each of the indicated points? Complete the diagram by placing the correct label in each bin. Use $g = 10\ m/s^2$ for the acceleration due to gravity. Note, the diagram is not drawn to scale.

Answer Bank

60 J | 0 J | 390 J | 39 J | 210 J | 180 J | 330 J | 780 J | 6 J | 21 J

PE = 390 J
KE = 0 J

The potential energy in this scenario is *gravitational potential energy*, which depends on the height of the car above some reference point. In this case, we take the bottom of the hill to be our reference point, so the potential energy at that point is zero. At some height h above the bottom, the gravitational potential energy PE is given by

$$PE = mgh$$

where m is the mass of the car (3 kg), and $g = 10\ m/s^2$ is the acceleration due to gravity. We can now calculate the potential energy at each point. At the top of the hill, $h = 13$ m, so

RESOURCES TO SUPPORT *COLLEGE PHYSICS*, SECOND EDITION, AND SAPLINGPLUS

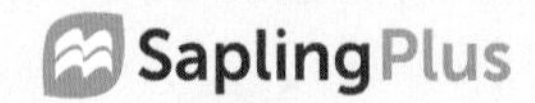

INSTRUCTOR RESOURCES

Housed in SaplingPlus

- iClicker questions
- Lecture slides
- Roger Freedman's in-class activities
- Test bank
- Instructor's solutions manual for end-of-chapter questions

STUDENT RESOURCES

Print supplement

- Student workbook problem-solving guide with solutions: ISBN: 1-319-16219-3

Housed in SaplingPlus

- Self-test concept checks
- Flash cards
- Interactive math tutorials
- Instructional videos
- PhET simulations
- Worked examples

PhysIndex—physical constants, conversion factors, and material data

Student support community

Course Solutions for Macmillan Physics

Custom Curriculum Solutions

SaplingPlus with e-Book

Textbooks

Available now for your Physics Lab...iOLab

The power of a lab in the palm of your hand

iOLab is a handheld data-gathering device that communicates wirelessly to its software and gives students a unique opportunity to see the concepts of physics in action. Students gain hands-on experience and watch their data graphed in real time. This can happen anywhere you have an iOLab device and a laptop: in the lab, in the classroom, in the dorm room, or in your basement. iOLab is flexible and makes it easy for instructors to design and implement virtually any experiment they want to assign their students or demonstrate in lecture.

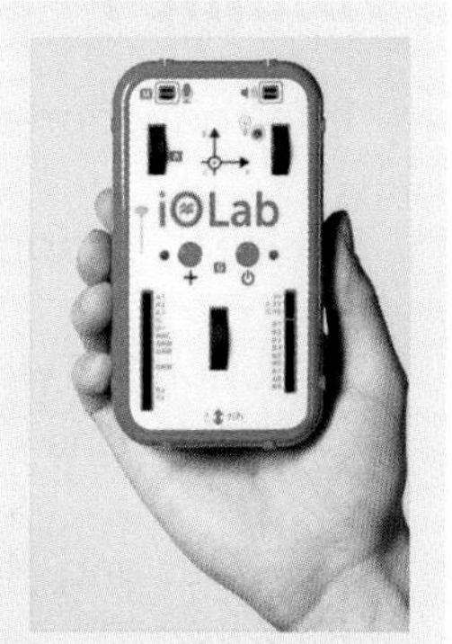

Reef In-Class Engagement

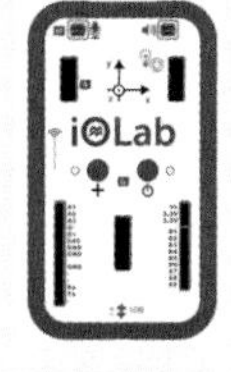

iOLab

Lab Solutions

BRIEF CONTENTS

Volume I

Volume II

Dear Students and Instructors,

Welcome to *College Physics!* We are excited to bring you this innovative text. No other college physics text presents material in quite this way. Our unique author team includes an experienced and highly successful textbook author, physicists who have spent years focusing on how students learn physics best, and even a biology professor who brings his perspective on what makes physics interesting to the students who take this course.

Our innovations, coupled with an engaging writing style, help students master concepts and succeed in developing and practicing the problem-solving skills they need to do well in this course. The visual impact of this text is something totally new: Word balloons throughout the text highlight key physics concepts in figures and equations so that students can learn the concepts in easy-to-manage pieces.

In addition, numerous features in the text are also designed to help students succeed in this course. Look for the **Chapter Goals** outlined at the beginning of each chapter; **Watch Out!** boxes that address misconceptions; **Got the Concept?** questions that test students' understanding of the material; **BIO-Medical** applications of physics in every chapter; **Take-Home Messages** that directly link to the chapter goals and help students focus on the important concepts presented in each chapter section; **Examples** that are broken down into key steps with an easy-to-follow *Set-Up*, *Solve*, *Reflect* structure; and a visual **Chapter Summary** at the end of each chapter, followed by **Questions and Problems**.

Our aim is to instill in students a deeper appreciation of physics—by showing how physics connects to their lives and their future careers, and by helping them succeed in the course. We hope you enjoy exploring and using this text!

Best Regards,

Roger, Todd, Phil, Dave

Contents

Volume II

Biological Applications

Unique and fully integrated physiological and biological applications are found throughout the text and are indicated by a **BIO-Medical** icon. Below is a list of select biological applications organized by chapter section for easy reference.

Chapter 1
Section 1-1: Hummingbirds and Physics
Section 1-3: Example 1-2 The World's Fastest Bird (speed)
Example 1-3 Hair Growth (unit conversion)

Chapter 2
Section 2-6: Example 2-10 Free Fall II: A Pronking Springbok (free fall)

Chapter 3
Section 3-3: Example 3-2 At What Angle Is Your Heart? (vector addition)
Section 3-8: Vestibular System (uniform circular motion)

Chapter 4
Section 4-2: Example 4-1 Small but Forceful (Newton's second law)
Section 4-5: Animal Propulsion (Newton's third law)

Chapter 5
Lubrication in Joints (static friction)
Section 5-2: Gecko Feet (static friction)
Example 5-1 Friction in Joints (static friction)
Section 5-5: Algal Spore (drag force)
Streamlined Animal Shapes (drag force)
Example 5-7 Terminal Speed (drag force)
Example 5-9 Making an Airplane Turn (apparent weight)

Chapter 6
Section 6-2: Example 6-2 Work Done by Actin (work)
Muscles and Doing Work (muscle function)
Section 6-5: Tendons (Hooke's Law)
Section 6-7: Arteries (energy transfer)
Muscles (potential energy)
Section 6-9: Exercise Rate (power)

Chapter 7
Section 7-1: Squid (momentum)
Section 7-7: Ballet Dancer (center of mass)

Chapter 8
Section 8-6: Lever Arms in Animal Anatomy (muscles)
Got the Concept? 8-8 Rotating the Human Jaw (moment of inertia)
Got the Concept? 8-9 Insects versus Birds (moment of inertia)
Section 8-9: Alligator Death Roll (angular momentum)
Got the Concept? 8-14 Danger! Falling Cat Zone (angular momentum)

Chapter 9
Section 9-1: Achilles Tendon (deformation)
Section 9-3: Example 9-1 Tensile Stress and Strain (sternoclavicular ligaments)
Section 9-4: Diving (compression)
Section 9-6: Arterial Blood Flow (elasticity)
Section 9-7: Example 9-4 Endothelial Cells and Shear (shear stress)
Section 9-8: Anterior Cruciate Ligament (stress)
Section 9-9: Example 9-6 Human ACL I: Maximum Force (tensile stress)
Example 9-7 Human ACL II: Breaking Strain (tensile strain)
Example 9-8 Human ACL III: The Point of No Return (tensile stress and strain)

Chapter 10
Section 10-5 Space Adaptation Sickness (weightlessness)
Muscle and Blood Loss (weightlessness)

Chapter 11
Section 11-5: Blood Pressure (gauge pressure)
Section 11-6: The Lungs (pressure differences)
Section 11-8: Fish Swim Bladder (density)
Example 11-10 Measuring Body Fat (Archimedes' principle)
Section 11-9: Measuring Blood Pressure (laminar versus turbulent flow)
Blood Flow (equation of continuity)
Example 11-12 How Many Capillaries? (equation of continuity)
Example 11-13 From Capillaries to the Vena Cavae (equation of continuity)
Section 11-10: Opercula (ram ventilation)
Section 11-11: Single-celled Alga *Chlamydomonas* (viscosity)
Blood Flow (Reynolds number)
Section 11-12: Gas Exchange (surface tension)

Chapter 12
Section 12-2: Got the Concept? 12-1 Heartbeats
Section 12-3: Intracellular Calcium Concentration (oscillations)
Section 12-4: Kangaroos and Dolphins (potential energy)
Section 12-5: Knee-jerk Reflex (physical pendulum)
Section 12-6: Example 12-7 Moment of Inertia of Swinging Leg (simple pendulum)
Section 12-7: Frog's Eardrum (underdamped oscillation)
Example 12-8 Oscillations of the Inner Ear (amplitude of oscillation)

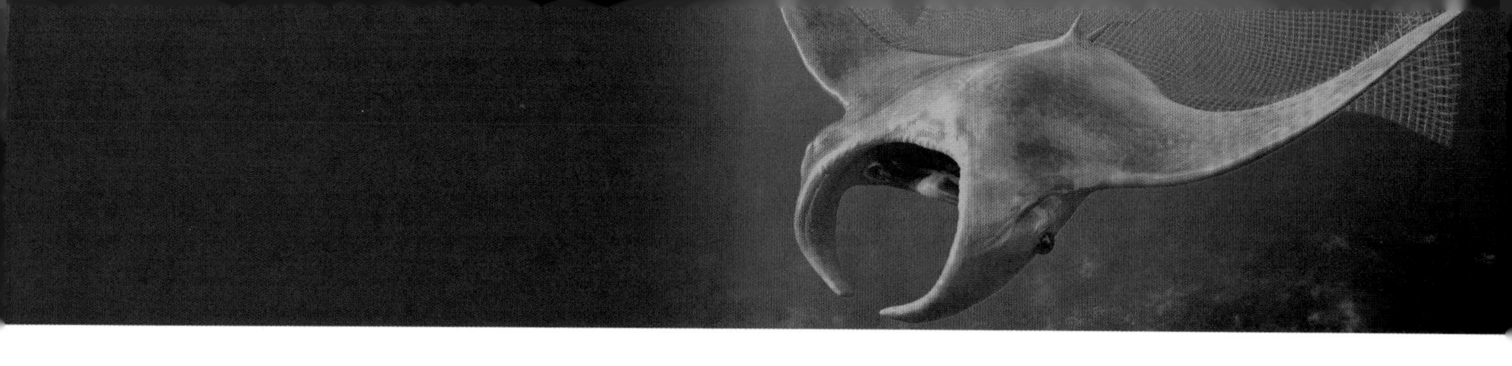

Acknowledgments

Bringing a textbook into its second edition requires the coordinated effort of an enormous number of talented professionals. We are grateful for the dedicated support of our in-house team at W. H. Freeman; thank you for sustaining our concept and developing this book into a beautiful new edition.

We especially want to thank Senior Development Editor Blythe Robbins and our development editor, Meg Rosenburg, for their dedication to this book. We would also like to thank our program manager, Lori Stover, for encouraging us and leading our team; Content Director Tania Mlawer, Media Editor Victoria Garvey, Media Project Manager Daniel Comstock, and Physics Content Team Lead Josh Hebert for their thoughtful contributions; and Assistant Editor Kevin Davidson, Permissions Manager Jennifer MacMillan, and Senior Content Project Manager Kerry O'Shaughnessy for their patience and attention to detail. Of course, none of this would be possible without the support of Editorial Program Director Brooke Suchomel and Vice President for STEM Ben Roberts. Special thanks also go to our talented marketing team, Marketing Manager Maureen Rachford, Associate Digital Marketing Specialist Cate McCaffery, and Marketing Assistant Savannah DiMarco for their insight and assistance.

Friends and Family

One of us (RAF) thanks his wife, Caroline Robillard, for her patience with the seemingly endless hours that went into preparing this textbook. I also thank my students at the University of California, Santa Barbara, for giving me the opportunity to test and refine new ideas for making physics more accessible.

One of us (TGR) thanks his wife, Susan, and daughter, Allison, for their limitless patience and understanding with the countless hours spent working on this book. I also thank my parents who showed me how to live a balanced life.

One of us (PRK) would like to acknowledge valuable and insightful conversations on physics and physics teaching with Richard Barber, John Birmingham, and J. Patrick Dishaw of Santa Clara University, and to offer these colleagues my gratitude. Finally, I offer my gratitude to my wife Kathy and my children Sam and Chloe for their unflagging support during the arduous process that led to the book you hold in your hands.

One of us (DLT) thanks his family and friends for accommodating my tight schedule during the years that writing this book consumed. I especially want to thank my parents, Bill and Jean, for their boundless encouragement and support, and for teaching me everything I've ever really needed to know—they've shown me how to live a good life, be happy, and age gently. I greatly appreciate my sister for encouraging me not to abandon a healthy lifestyle just to write a book. I also want to thank my nonbiological family, Holly and Geoff, for leading me to Sonoma County and for making Sebastopol feel like home.

Solutions Manual authors:
Bryan Armentrout, *Nova Southeastern University*
Michael Dunham, *State University of New York at Fredonia*
Andrew Ekpenyong, *Creighton University*

Accuracy checker:
Jose Lozano, *Bradley University*

Student Workbook authors:
Perry Hilburn, *Gannon University*
Seong-Gon Kim, *Mississippi State University*
Garrett Yoder, *Eastern Kentucky University*
Linda McDonald, *North Park University*
Avishek Kumar, *Arizona State University*
Michael Dunham, *SUNY Fredonia*

Test Bank authors:
Anna Harlick, *University of Calgary*
Debashis Dasgupta, *University of Wisconsin at Milwaukee*
Elizabeth Holden, *University of Wisconsin at Platteville*

iClicker Slides author:
Adam Lark, *Hamilton College*

Lecture Slides author:
Adam Lark, *Hamilton College*

Questions and Problems editors:
Jonathan Bratt
Michael Scott
Fredrick "Mike" DeArmond
Syed "Asif" Hassan

Second-Edition Reviewers

We would also like to thank the many colleagues who carefully reviewed chapters for us. Their insightful comments significantly improved our book.

Miah Muhammad Adel, *University of Arkansas at Pine Bluff*
Mikhail M. Agrest, *The Citadel*
Vasudeva Rao Aravind, *Clarion University of Pennsylvania*
Yiyan Bai, *Houston Community College*
E. C. Behrman, *Wichita State University*
Antonia Bennie-George, *Green River College*
Ken Bolland, *The Ohio State University*
Matthew Joseph Bradley, *Santa Rosa Junior College*
Matteo Broccio, *University of Pittsburgh*
Daniel J. Costantino, *The Pennsylvania State University*
Adam Davis, *Wayne State College*
Sharvil Desai, *The Ohio State University*

Eric Deyo, *Fort Hays State University*
Diana I. Driscoll, *Case Western Reserve University*
Davene Eyres, *North Seattle College*
William Falls, *Erie Community College*
Sambandamurthy Ganapathy, *SUNY Buffalo*
Vladimir Gasparyan, *California State University, Bakersfield*
Frank Gerlitz, *Washtenaw Community College*
Svetlana Gladycheva, *Towson University*
Romulus Godang, *University of South Alabama*
Javier Gomez, *The Citadel*
Rick Goulding, *Memorial University of Newfoundland*
Ania Harlick, *Memorial University of Newfoundland*
Erik Helgren, *California State University, East Bay*
Perry G. Hillburn, *Gannon University*
Zdeslav Hrepic, *Columbus State University*
Patrick Huth, *Community College of Allegheny County*
Matthew Jewell, *University of Wisconsin, Eau Claire*
Wafaa Khattou, *Valencia College*
Seong-Gon Kim, *Mississippi State University*
Patrick Koehn, *Eastern Michigan University*
Ameya S. Kolarkar, *Auburn University*
Maja Krcmar, *Grand Valley State University*
Elena Kuchina, *Thomas Nelson Community College*
Avishek Kumar, *Arizona State University*
Chunfei Li, *Clarion University of Pennsylvania*
Jose Lozano, *Bradley University*
Dan MacIsaac, *SUNY Buffalo State College*
Linda McDonald, *North Park University*
Francis Mensah, *Virginia Union University*
Victor Migenes, *Texas Southern University*
Krishna Mukherjee, *Slippery Rock University of Pennsylvania*
Rumiana Nikolova-Genov, *College of DuPage*
Moses Ntam, *Tuskegee University*
Martin O. Okafor, *Georgia Perimeter College*
Gabriela Petculescu, *University of Louisiana at Lafayette*
Yuriy Pinelis, *University of Houston, Downtown*
Sulakshana Plumley, *Community College of Allegheny County*
Lawrence Rees, *Brigham Young University*
David Richardson, *Northwest Missouri State University*
Carlos Roldan, *Central Piedmont Community College*
Jeffrey Sabby, *Southern Illinois University Edwardsville*
Arun Saha, *Albany State University*
Haiduke Sarafian, *Pennsylvania State University*
Katrin Schenk, *Randolph College*
Surajit Sen, *SUNY Buffalo*
Jerry Shakov, *Tulane University*
Douglas Sharman, *San Jose State University*
Ananda Shastri, *Minnesota State University, Moorhead*
Marllin L. Simon, *Auburn University*
Stanley J. Sobolewski, *Indiana University of Pennsylvania*
Erin C. Sutherland, *Kennesaw State University*
Sarah F. Tebbens, *Wright State University*
Fiorella Terenzi, *Florida International University*
Dmitri Tsybychev, *Stony Brook University*
Laura Whitlock, *Georgia Perimeter College*
Pushpa Wijesinghe, *Arizona State University*
Fengyuan Yang, *The Ohio State University*
Garett Yoder, *Eastern Kentucky University*
Jiang Yu, *Fitchburg State University*
Ulrich Zurcher, *Cleveland State University*

First-Edition Reviewers

We would also like to acknowledge the many reviewers who reviewed the first edition of this text, as their contributions live on in this edition.

Don Abernathy, *North Central Texas College*
Elise Adamson, *Wayland Baptist University*
Miah Muhammad Adel, *University of Arkansas, Pine Bluff*
Ricardo Alarcon, *Arizona State University*
Z. Altounian, *McGill University*
Abu Amin, *Riverland Community College*
Sanjeev Arora, *Fort Valley State University*
Llani Attygalle, *Bowling Green State University*
Yiyan Bai, *Houston Community College*
Michael Bates, *Moraine Valley Community College*
Luc Beaulieu, *Memorial University*
Jeff J. Bechtold, *Austin Community College*
David Bennum, *University of Nevada*
Satinder Bhagat, *University of Maryland*
Dan Boye, *Davidson College*
Jeff Bronson, *Blinn College*
Douglas Brumm, *Florida State College at Jacksonville*
Mark S. Bruno, *Gateway Community College*
Brian K. Bucklein, *Missouri Western State University*
Michaela Burkardt, *New Mexico State University*
Kris Byboth, *Blinn College*
Joel W. Cannon, *Washington & Jefferson College*
Kapila Clara Castoldi, *Oakland University*
Paola M. Cereghetti, *Lehigh University*
Hong Chen, *University of North Florida*
Zengjun Chen, *Tuskegee University*
Uma Choppali, *Dallas County Community College*
Todd Coleman, *Century College*
José D'Arruda, *University of North Carolina, Pembroke*
Tinanjan Datta, *Georgia Regents University*
Chad L. Davies, *Gordon College*
Brett DePaola, *Kansas State University*
Sandra Doty, *Ohio University*
James Dove, *Metro Community College*
Carl T. Drake, *Jackson State University*
Rodney Dunning, *Longwood University*
Vernessa M. Edwards, *Alabama A & M University*
Davene Eyres, *North Seattle Community College*
Hasan Fakhruddin, *Ball State University*
Paul Fields, *Pima Community College*
Lewis Ford, *Texas A & M University*
J.A. Forrest, *University of Waterloo*
Scott Freedman, *Philadelphia Academy Charter High School*
Tim French, *Harvard University*
James Friedrichsen III, *Austin Community College*
Sambandamurthy Ganapathy, *SUNY Buffalo*
J. William Gary, *University of California, Riverside*
L. Gasparov, *University of North Florida*
Vladimir Gasparyan, *California State University, Bakersfield*
Brian Geislinger, *Gadsden State Community College*
Oommen George, *San Jacinto College*
Anindita Ghosh, *Suffolk County Community College*
Alan I. Goldman, *Iowa State University*
Richard Goulding, *Memorial University of Newfoundland*
Morris C. Greenwood, *San Jacinto College Central*

Thomas P. Guella, *Worcester State University*
Alec Habig, *University of Minnesota, Duluth*
Edward Hamilton, *Gonzaga University*
C. A. Haselwandter, *University of Southern California*
Zvonko Hlousek, *California State University, Long Beach*
Micky Holcomb, *West Virginia University*
Kevin M. Hope, *University of Montevallo*
J. Johanna Hopp, *University of Wisconsin, Stout*
Leon Hsu, *University of Minnesota*
Olenka Hubickyj Cabot, *San Jose State University*
Richard Ignace, *East Tennessee State University*
Elizabeth Jeffery, *James Madison University*
Yong Joe, *Ball State University*
Darrin Eric Johnson, *University of Minnesota, Duluth*
David Kardelis, *Utah State University, College of Eastern Utah*
Agnes Kim, *Georgia State College*
Ju H. Kim, *University of North Dakota*
Seth T. King, *University of Wisconsin, La Crosse*
Kathleen Koenig, *University of Cincinnati*
Olga Korotkova, *University of Miami*
Minjoon Kouh, *Drew University*
Tatiana Krivosheev, *Clayton State University*
Michael Kruger, *University of Missouri, Kansas City*
Jessica C. Lair, *Eastern Kentucky University*
Josephine M. Lamela, *Middlesex County College*
Patrick M. Len, *Cuesta College*
Shelly R. Lesher, *University of Wisconsin, La Crosse*
Zhujun Li, *Richland College*
Bruce W. Liby, *Manhattan College*
David M. Lind, *Florida State University*
Jeff Loats, *Metropolitan State College of Denver*
Susannah E. Lomant, *Georgia Perimeter College*
Jia Grace Lu, *University of Southern California*
Mark Lucas, *Ohio University*
Lianxi Ma, *Blinn College*
Aklilu Maasho, *Dyersburg State Community College*
Ron MacTaylor, *Salem State University*
Eric Mandell, *Bowling Green State University*
Maxim Marienko, *Hofstra University*
Mark Matlin, *Bryn Mawr College*
Dan Mattern, *Butler Community College*
Mark E. Mattson, *James Madison University*
Jo Ann Merrell, *Saddleback College*
Michael R. Meyer, *Michigan Technological University*
Karie A. Meyers, *Pima Community College*
Andrew Meyertholen, *University of Redlands*
John H. Miller Jr., *University of Houston*
Ronald C. Miller, *University of Central Oklahoma*
Ronald Miller, *Texas State Technical College System*
Hector Mireles, *California State Polytechnic University, Pomona*
Ted Monchesky, *Dalhousie University*
Steven W. Moore, *California State University, Monterey Bay*
Mark Morgan-Tracy, *University of New Mexico*
Dennis Nemeschansky, *University of Southern California*
Terry F. O'Dwyer, *Nassau Community College*
John S. Ochab, *J. Sargeant Reynolds Community College*
Umesh C. Pandey, *Central New Mexico Community College*
Archie Paulson, *Madison Area Technical College*
Christian Poppeliers, *Augusta State University*
James R. Powell, *University of Texas, San Antonio*
Michael Pravica, *University of Nevada, Las Vegas*
Kenneth M. Purcell, *University of Southern Indiana*
Kenneth Ragan, *McGill University*
Milun Rakovic, *Grand Valley State University*
Jyothi Raman, *Oakland University*
Natarajan Ravi, *Spelman College*
Lou Reinisch, *Jacksonville State University*
Sandra J. Rhoades, *Kennesaw State University*
John Rollino, *Rutgers University, Newark*
Rodney Rossow, *Tarrant County College*
Larry Rowan, *University of North Carolina*
Michael Sampogna, *Pima Community College*
Tumer Sayman, *Eastern Michigan University*
Jim Scheidhauer, *DePaul University*
Paul Schmidt, *Ball State University*
Morton Seitelman, *Farmingdale State College*
Saeed Shadfar, *Oklahoma City University*
Weidian Shen, *Eastern Michigan University*
Jason Shulman, *Richard Stockton College of New Jersey*
Michael J. Shumila, *Mercer County Community College*
R. Seth Smith, *Francis Marion University*
Frank Somer, *Columbia College*
Chad Sosolik, *Clemson University*
Brian Steinkamp, *University of Southern Indiana*
Narasimhan Sujatha, *Wake Technical Community College*
Maxim Sukharev, *Arizona State University*
James H. Taylor, *University of Central Missouri*
Richard Taylor, *University of Oregon*
E. Tetteh-Lartey, *Blinn Community College*
Fiorella Terenzi, *Brevard Community College*
Gregory B. Thompson, *Adrian College*
Marshall Thomsen, *Eastern Michigan University*
Som Tyagi, *Drexel University*
Vijayalakshmi Varadarajan, *Des Moines Area Community College*
John Vasut, *Baylor University*
Dimitrios Vavylonis, *Lehigh University*
Kendra L. Wallis, *Eastfield College*
Laura Weinkauf, *Jacksonville State University*
Heather M. Whitney, *Wheaton College*
Capp Yess, *Morehead State University*
Chadwick Young, *Nicholls State University*
Yifu Zhu, *Florida International University*
Raymond L. Zich, *Illinois State University*

Electrostatics I: Electric Charge, Forces, and Fields

VisionDive/Shutterstock.com

What do you think?

A shark can detect the electric field produced by a small, electrically charged object. If a charged object 2 m from a shark's nose is moved to be only 1 m from the nose, the magnitude of the electric field that the shark detects will (a) decrease; (b) double; (c) increase by a factor of 4; (d) increase by a factor of 8; (e) increase by a factor of 16.

In this chapter, your goals are to:

- (16-1) Explain why studying electric phenomena is important.
- (16-2) Describe how objects acquire a net electric charge.
- (16-3) Recognize the differences between insulators, conductors, and semiconductors.
- (16-4) Use Coulomb's law to quantitatively describe the force that one charged particle exerts on another.
- (16-5) Explain the relationship between electric force and electric field.
- (16-6) Describe the connection between enclosed charge and electric flux described by Gauss's law.
- (16-7) Apply Gauss's law to symmetric situations and to the distribution of excess charge on conductors.

To master this chapter, you should review:

- (3-2, 3-3, and 3-4) How to add vectors and how to do vector calculations using components.
- (4-2) Newton's second law.

16-1 Electric forces and electric charges are all around you—and within you

You may think of *electricity* as the shock you get when you walk across a carpet and then touch a metal doorknob, or a commodity that you purchase from the power company. The reality is that electric phenomena are all around you, from the behavior of grains of pollen to the drama of a thunderstorm (**Figure 16-1**). As you read this sentence, your eye casts an image of the text onto your retina, and this image is transmitted to your brain along the optic nerve as a stream of electrical impulses. Electricity even explains why you don't fall through the floor: The molecules on the surface of the floor exert electric forces that repel the molecules on the underside of your shoes, and these forces are strong enough to balance Earth's downward gravitational force on you.

Electric phenomena are important because *all* ordinary matter is made of electrically charged objects. That's why we'll spend the next several chapters learning about electric

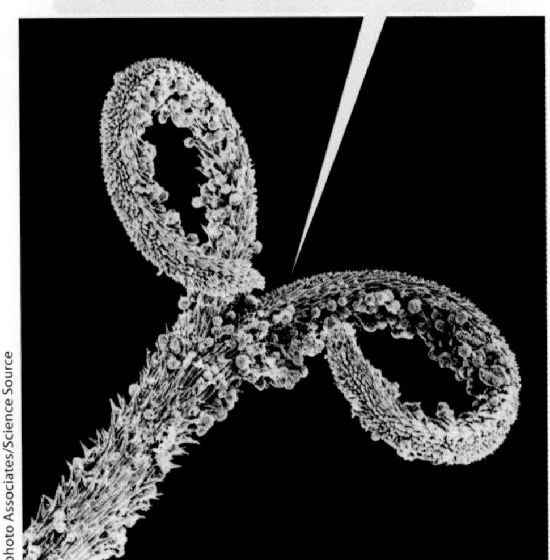

(a)
Grains of pollen wafting through the air carry a small amount of electric charge. This attracts them to the oppositely charged stigma of a flower and so aids in pollination.
Biophoto Associates/Science Source

(b)
The motion of air in a thundercloud causes positive and negative charges to separate and build up within the cloud. The release of these charges causes lightning.
KingWu/iStockphoto

Figure 16-1 Electric charges and forces Electrostatics—the physics that describes how opposite charges attract and like charges repel—plays an important role in (a) biology and (b) the weather.

TAKE-HOME MESSAGE FOR Section 16-1

✔ Electric phenomena are important because all ordinary matter is made of electrically charged objects.

✔ Electrostatics is the study of electric charges at rest and the forces that they exert on each other.

charges and their physics. The simplest place to begin our study of electricity is with *electrostatics*, the branch of physics that deals with electric charges at rest and how they interact with each other. In this chapter we'll first learn about the different kinds of electrical charge in nature, and describe the forces that two electrically charged objects exert on each other. We'll see how these forces manifest themselves in two important classes of materials called conductors and insulators. We'll go on to learn about the concept of an *electric field*, a powerful way of understanding how electric charges can exert forces on each other even over a distance. We'll conclude this chapter with a look at *Gauss's law*, an important tool for understanding electric fields.

16-2 Matter contains positive and negative electric charge

Figure 16-2 shows some simple experiments that demonstrate the nature of electric charge. If you hold two rubber rods next to each other, they neither attract nor repel each other. The same is true for two glass rods held next to each other, or for a rubber rod and a glass rod (**Figure 16-2a**). But if you rub the ends of each rubber rod with fur and each glass rod with silk, we find that the ends of the rubber rods repel each other and the ends of the glass rods repel each other, but the ends of the rubber and glass rods attract each other (**Figure 16-2b**). What's more, the end of the rubber rod attracts the piece of fur used to rub it, and the end of the glass rod attracts the piece of silk used to rub it (**Figure 16-2c**). In each case where there is a repulsion or an attraction, the repulsive or attractive force is stronger the closer the objects are held to each other.

Here's how we explain these experiments and others like them:

(1) *Matter has electric charge.* In addition to having mass, matter also possesses a property called **electric charge**. This charge comes in two forms, positive and negative. An object that has a net electric charge (more positive than negative, or more negative than positive) is **charged**. Most matter, however, contains equal amounts of positive and negative charge and is electrically **neutral**. In Figure 16-2a the rubber and glass rods are electrically neutral, as are the pieces of fur and silk.

(2) *Electric charge is a property of the constituents of atoms.* All ordinary matter is made up of atoms. Atoms are in turn composed of more fundamental particles called electrons, protons, and (except for the most common type of hydrogen atom) neutrons. Electrons carry a negative charge, while protons carry an equally large positive charge; neutrons are neutral. Protons and neutrons have about the same mass and are found in the dense nucleus at the center of an atom. A negatively charged electron has only about 0.05% the mass of a proton, and the electrons

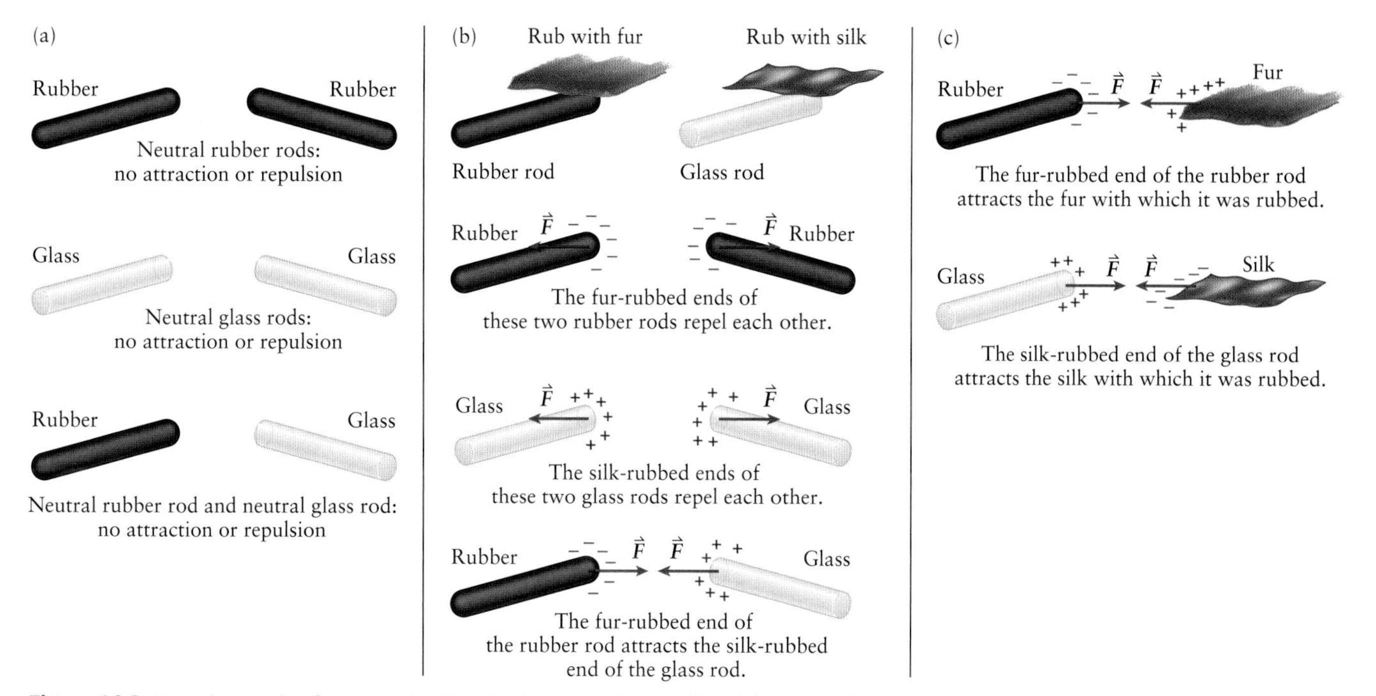

Figure 16-2 Experiments in electrostatics Electric charge can be transferred from one object to another by rubbing. Electrically charged objects exert forces on each other. Plus and minus signs indicate the net charge on each object (+ for positive, − for negative).

move around the massive, positively charged nucleus (**Figure 16-3**). A normal atom contains as many electrons as it does protons and so is electrically neutral.

(3) *Charge can be transferred between objects.* When dissimilar materials are rubbed together, negatively charged electrons can be transferred from one material to the other. As an example, consider the rubber rod and fur shown in Figure 16-2b. Initially the rod and fur are electrically neutral, so each has equal amounts of positive and negative charge. But when they are rubbed together, electrons are transferred from the fur to the rubber. This leaves the fur with a net positive charge (more positive charge than negative charge) and the rubber with a net negative charge. By contrast, when the glass rod and silk in Figure 16-2b (both of which are initially neutral) are rubbed together, electrons are transferred from the glass to the silk. So the silk ends up with a net negative charge and the glass rod with a net positive charge.

(4) *Electrically charged objects exert forces on each other.* Objects with electric charges of the same sign (both positive or both negative) repel each other, while objects with electric charges of opposite sign (one positive and the other negative) attract each other (Figure 16-2c). The force between charges is called the **electric force.** The magnitude of the force depends on the amount of charge on each object and on the distance between the objects. The force increases for a greater amount of charge, and it increases if the charged objects are brought closer to each other.

(5) *Charge is never created or destroyed.* In the experiments shown in Figure 16-2, charge is *transferred* between objects. Although charge moves between the rubber rod and the fur or between the glass rod and the silk, in each case the *total* charge of the two objects remains the same as before they were rubbed together. That is, charge is *conserved*; it is never created or destroyed. No one has ever observed a process in which charge is not conserved, so to the best of our knowledge the conservation of electric charge is an absolute law of nature.

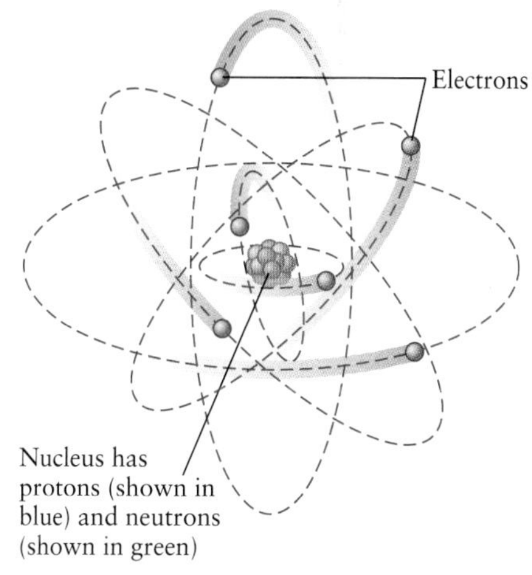

Figure 16-3 A simplified model of an atom Negatively charged electrons orbit the atom's nucleus, which contains most of the atom's mass. The nucleus contains two types of particles, positively charged protons and uncharged neutrons.

The character of the electric force explains the remarkable experiment shown in **Figure 16-4a**. A balloon is first rubbed against the girl's sleeve, causing electrons to transfer from her sleeve onto the rubber surface of the balloon (just like the rubber rod and fur in Figure 16-2b). When the negatively charged balloon is held next to the girl's head, electric forces attract the hair to the balloon—even though the hair itself is *neutral*. What happens is that the negatively charged electrons on the balloon repel the electrons in the atoms that comprise the hair, but also attract the positively charged nuclei of these atoms.

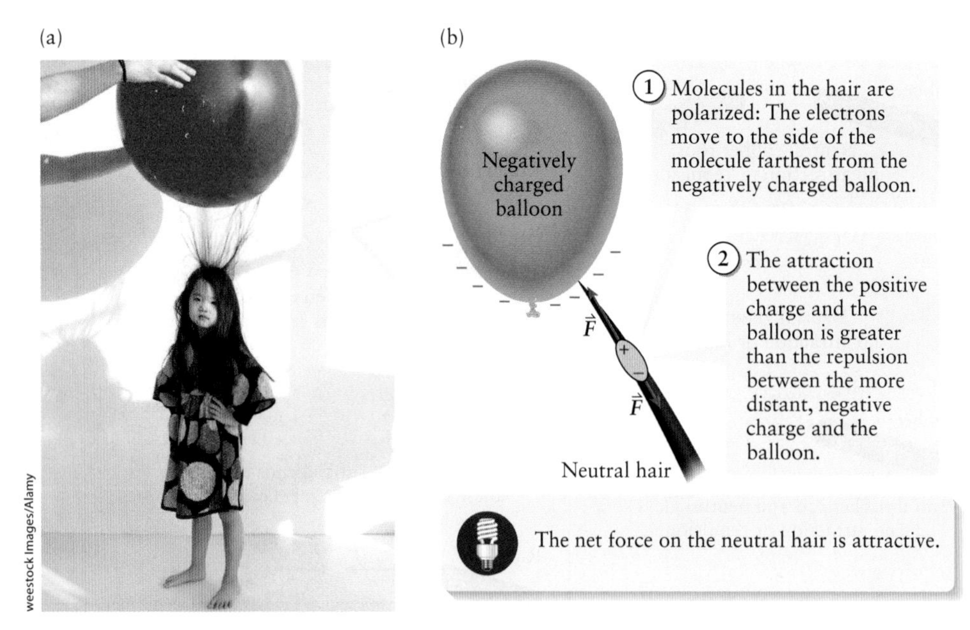

Figure 16-4 A hair-raising experiment (a) Rubbing a balloon on your sleeve gives the balloon an electric charge. A charged balloon can attract your hair and make it stand up, even though your hair has no net charge. (b) The nature of the electric force explains why this attraction happens.

The electrons and nuclei remain within their atoms but end up slightly displaced from each other (**Figure 16-4b**). We say that the hair becomes *polarized.* As we described above, the electric force is greater the closer charged objects are to each other. Figure 16-4b shows that the nuclei in the polarized hair are pulled slightly closer to the negatively charged balloon than the electrons in the hair. So the attractive force between the hair nuclei and the balloon is slightly greater than the repulsive force between the hair electrons and the balloon. The result is that the hair feels a net attractive force toward the balloon.

The unit of electric charge is the **coulomb** (C), named after the eighteenth-century French physicist Charles-Augustin de Coulomb who uncovered the fundamental law that governs the interaction of charges. (We'll discuss this law in Section 16-4.) As we mentioned above, electrons and protons have the same magnitude of electric charge; we call this magnitude e. Precise measurements show that

$$e = 1.60217657 \times 10^{-19}\ \text{C}$$

We use q or Q as the symbol for the charge of an object, so the charge on a proton is (to four significant figures) $q_{\text{proton}} = +e = +1.602 \times 10^{-19}$ C, and the charge on an electron is $q_{\text{electron}} = -e = -1.602 \times 10^{-19}$ C. No free particle has ever been detected with a charge smaller in magnitude than e. For that matter, no object has ever been found whose charge was not a multiple of e such as $+2e$, $-3e$, and so on. That is, charge is *quantized*: The charge of an object is always increased or decreased by an amount equal to an integer multiple of the fundamental charge e. It is impossible to add a fraction of a proton or electron to an object. (As we'll see in Chapter 28, there is strong evidence that protons and neutrons are composed of more fundamental particles called *quarks* that have charges of $+2e/3$ and $-e/3$. However, no isolated quarks have ever been observed outside of a larger particle such as a proton or neutron.)

While electrons and protons have opposite signs, it's completely arbitrary whether we choose electrons to have negative charge and protons to have positive charge or the other way around. The choice of sign is due to the American scientist and statesman Benjamin Franklin, who decided in the mid-eighteenth century that the sign of charge on a glass rod rubbed with silk (see Figure 16-2b) is positive. The electron and proton were not discovered until much later (1897 and 1919, respectively).

EXAMPLE 16-1 Electrons in a Raindrop

A water molecule is made up of two hydrogen atoms, each of which has one electron, and an oxygen atom, which has eight electrons. One water molecule has a mass of 2.99×10^{-26} kg. How many electrons are there in a single raindrop with a radius of 1.00 mm $= 1.00 \times 10^{-3}$ m, which has a mass of 4.19×10^{-6} kg? What is the total charge of all of these electrons?

Set Up

Each water molecule has 10 electrons (two from the hydrogen atoms and eight from the oxygen atom). So we need to determine the number of water molecules in the raindrop and multiply by 10 to find the number of electrons.

Charge of an electron:

$$q_{\text{electron}} = -e = -1.602 \times 10^{-19}\ \text{C}$$

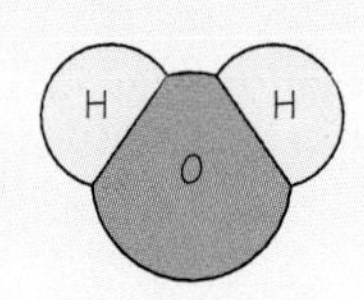

Solve

Determine the number of molecules in the raindrop, and from that determine the number of electrons.

The total number of molecules in the raindrop is the mass of the raindrop divided by the mass of a single water molecule:

$$\frac{\text{mass of raindrop}}{\text{mass of one water molecule}} = \frac{4.19 \times 10^{-6}\ \text{kg}}{2.99 \times 10^{-26}\ \text{kg/molecule}}$$
$$= 1.40 \times 10^{20}\ \text{molecules}$$

The number of electrons in the raindrop is

$$N = (1.40 \times 10^{20}\ \text{molecules})(10\ \text{electrons/molecule})$$
$$= 1.40 \times 10^{21}\ \text{electrons}$$

To find the charge of this number of electrons, multiply by the charge per electron.

Total charge of this number of electrons:

$$q = Nq_{\text{electron}} = (1.40 \times 10^{21}\ \text{electrons})(-1.602 \times 10^{-19}\ \text{C/electron})$$
$$= -224\ \text{C}$$

Reflect

How large a number is 1.40×10^{21}? The human body contains about 10^{14} cells. Even if you include all of the cells in all of the approximately 9 million (9×10^{6}) people in New York City, that's only about 9×10^{20} cells—still fewer than the number of electrons in a single raindrop.

The charge of -224 C is quite substantial, but remember that for every electron in a water molecule, there is also one proton (one in each hydrogen atom and eight in each oxygen atom). Since a proton carries exactly as much positive charge as an electron carries negative charge, our raindrop has zero *net* charge. In Section 16-4 we'll see how difficult it would be to separate the positive and negative charges of this raindrop.

GOT THE CONCEPT? 16-1 Transferring Charge

Consider the rubber rod and glass rod in Figure 16-2b. Compared to their masses before they are rubbed with fur and silk, respectively, what are their masses after being rubbed? (a) Both rods have more mass; (b) both rods have less mass; (c) the rubber rod has more mass and the glass rod has less mass; (d) the rubber rod has less mass and the glass rod has more mass; (e) the masses of both rods are unchanged.

TAKE-HOME MESSAGE FOR Section 16-2

✔ Electric charge is quantized. The smallest amount of charge that can be added to or removed from an object is equal to the fundamental charge $e = 1.602 \times 10^{-19}$ C.

✔ Charge can be neither created nor destroyed. However, charge can be transferred from one object to another, for example, by moving electrons between objects.

✔ All ordinary matter contains positive charge in the form of protons, which have a charge of $+e$, and negative charge in the form of electrons, which have a charge of $-e$. An object is electrically neutral if it contains equal amounts of positive and negative charge. If these amounts are not equal, the object has a net charge.

✔ Objects with a net charge exert electric forces on each other. These forces become weaker with increasing distance. If the objects have the same sign of charge (both positive or both negative), the forces are repulsive. If the objects have opposite signs of charge (one positive and one negative), the forces are attractive.

16-3 Charge can flow freely in a conductor but not in an insulator

All substances contain positive and negative charges. But how *mobile* those charges are depends on the specific material. The rubber, glass, silk, and fur shown in Figure 16-2 are all examples of **insulators**, substances in which charges are not able to move freely. All of the electrons in an insulator are bound tightly to the nuclei of atoms, and any excess charge added to an insulator tends to stay wherever it is placed. So when electrons get placed on one end of a rubber rod by rubbing it with fur, the excess electrons cannot redistribute themselves along the rod (**Figure 16-5a**). Thus there is no excess charge on the opposite end of the rod to attract the positively charged fur. (Due to a polarization effect like that shown in Figure 16-4, there is still a very weak attraction between the positively charged fur and the neutral, unrubbed end of the rubber rod.) Most nonmetals are insulators.

By contrast, a metal such as copper is an example of a **conductor**, a substance in which charges *can* move freely. In a copper atom the outermost, or valence, electron can easily be dislodged. (The valence electron is relatively far from the positively charged nucleus, and the many electrons closer to the nucleus tend to shield the valence electron from the charge of the nucleus.) As a result, electrons—including excess electrons added to the copper—can move between copper atoms relatively freely. This explains what happens when you rub one end of a copper rod with nylon (**Figure 16-5b**). Electrons are transferred from the nylon to the copper, giving the copper rod a net negative charge and the nylon a net positive charge. But unlike what we saw in Figure 16-5a, after one end of the copper rod has been rubbed, the nylon attracts *both* ends of the copper rod equally. That's because the excess electrons deposited on one end of the copper rod can easily move within the rod. When the positively charged nylon is brought close to *either* end of the copper rod, the excess electrons on the copper rush to that end. As a result, that end has a net negative charge and is attracted to the nylon.

Conductors and insulators are an essential part of all electric circuits, which are systems in which there is an ongoing current (a flow of charge) around a closed path. Electric circuits are at the heart of any device that uses a battery (such as a mobile phone, a flashlight, or an electric vehicle) or that you plug into a wall socket (such as a desktop computer, a toaster, or an electric fan). An electric circuit uses moving charges (typically electrons) to transfer energy along a conductor from one point in a circuit to another. In a flashlight, for example, electrons flowing through a copper wire carry energy from the battery (which is a repository of *electric potential energy*) to the light bulb, where the energy is converted into visible light. The electrons then return to the battery through a second copper wire to pick up more energy and repeat the process. Insulators play a crucial role in this process: Each conducting wire is clad in a sheath made of an insulator, which helps ensure that the electrons flow only along the length of the wires. (The visible part of an ordinary extension

(a)

The fur-rubbed end of the rubber rod attracts the fur with which it was rubbed...

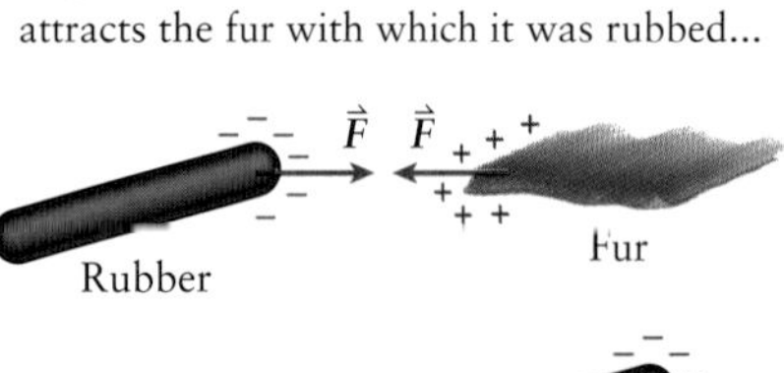

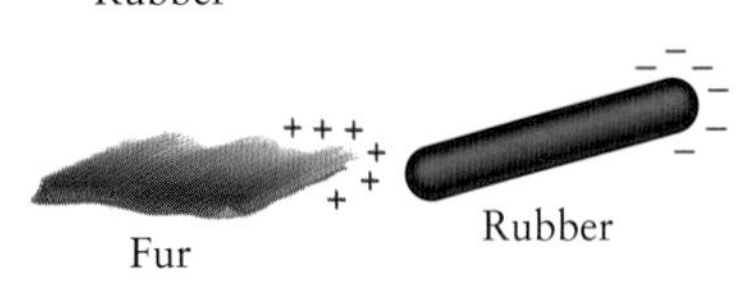

...but the end of the rubber rod that was not rubbed does not feel a strong attraction.

(b)

The nylon-rubbed end of the copper rod attracts the nylon with which it was rubbed...

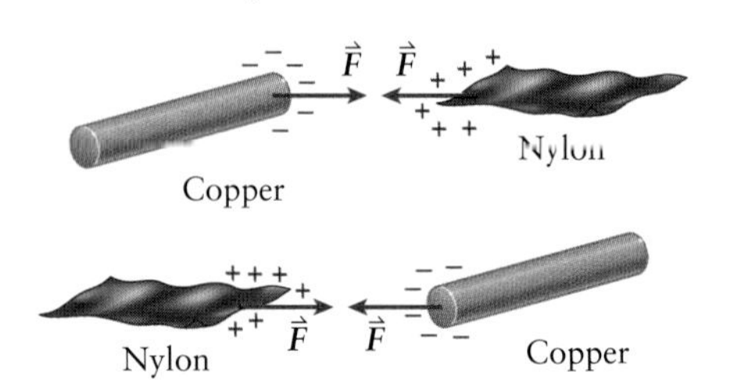

...and the end of the copper rod that was not rubbed also feels a strong attraction.

Figure 16-5 Insulators versus conductors (a) If you place excess charge at one location on an insulator, it remains at that location. (b) If you place excess charge at one location on a conductor, the excess charge can move freely through the conductor. (Due to a polarizing effect like that shown in Figure 16-4, there is still a very weak attraction between the positively charged fur and the neutral, unrubbed end of the rubber rod in part (a).)

cord or power cord is actually the insulating sheath. The copper wires are contained within the sheath.) We'll explore the idea of electric potential energy in detail in Chapter 17, and we'll devote Chapters 18 and 21 to exploring various important kinds of electric circuits.

Most metals are good electrical conductors *and* also good conductors of heat, with large values of thermal conductivity (see Section 14-7). That's because the flow of electric charge within a material and the flow of heat within that material both require that particles of the material be free to move. In Section 14-3 we learned that the temperature of a material is related to the kinetic energy of that material's particles. For heat to flow from one part of a material to another, the faster-moving particles in the high-temperature part of the material must transfer kinetic energy to the slower-moving particles in the low-temperature part of the object. So the presence of free electrons makes metals good thermal conductors as well as good conductors of electricity. Materials that are electrical insulators, in which the electrons are generally not free to move between atoms, also tend to be thermal insulators with low thermal conductivity.

Not all electric conductors are metals. In nearly all biological systems, electric charges are carried by *ions*—atoms with an excess or deficit of electrons. These ions are suspended in water, that is, in aqueous solution. Water is particularly good at holding ions in solution because the water molecule (H_2O), while electrically neutral, has slightly more negative charge at the oxygen atom of the molecule and slightly more positive charge at the hydrogen atoms. So when an ionic compound such as sodium chloride (NaCl) is dissolved in water, the positive ions (Na^+) are attracted to the negative (oxygen) end of H_2O molecules while the negative ions (Cl^-) are attracted to the positive (hydrogen) end. Because the H_2O molecules to which the ions are attached are free to move, the aqueous solutions which make up living things are conductors. It doesn't matter whether charge is carried by an electron or by an ion. As we'll see in Chapter 18, it also doesn't matter whether the moving charges are positive or negative: In either case there is a flow of charge.

GOT THE CONCEPT? 16-2 Sodium in Water

A normal atom of sodium (chemical symbol Na) has 11 electrons and is electrically neutral. But in the compound sodium chloride (NaCl) the sodium atom is actually an ion of charge $+e$. How does a sodium atom acquire this charge? (a) By annihilating one of its electrons; (b) by creating a new electron; (c) by transferring an electron to another atom or molecule; (d) by acquiring an electron from another atom or molecule; (e) by acquiring a proton from another atom or molecule.

There's an important third class of substances called **semiconductors**. These substances have electrical properties that are intermediate between those of insulators and conductors. A common example of a semiconductor is silicon. A silicon atom has four outer electrons, compared to just one for copper, which might suggest that it would be a good conductor. However, in pure silicon each of those electrons forms part of a chemical bond with a neighboring silicon atom, so the outer electrons have limited mobility. As a result, pure silicon conducts electricity far worse than a conductor such as copper, though still far better than an insulator such as rubber.

Semiconductors are of tremendous practical use in electric circuits. That's because it's possible to adjust their electrical properties by *doping*—that is, by adding a small amount of a second substance. Here's an example: If we take pieces of silicon that have been doped in different ways, we can arrange them so that charges will flow through the combination in one direction but will *not* flow in the opposite direction! Such a combination, called a *diode*, plays the same role in an electric circuit as the valves in the human heart, which allow blood to flow through the heart in one direction only. In Chapter 21 we'll look at some of the applications of semiconductors to modern technologies such as light-emitting diodes (LEDs), solar cells, and integrated circuits.

TAKE-HOME MESSAGE FOR Section 16-3

✔ Charges are free to move within a conductor but can move very little in an insulator.

✔ In a metal conductor such as copper, the moving charges are typically electrons. In biological systems ions in aqueous solution are the moving charges.

✔ Semiconductors have electrical properties intermediate between those of conductors and insulators.

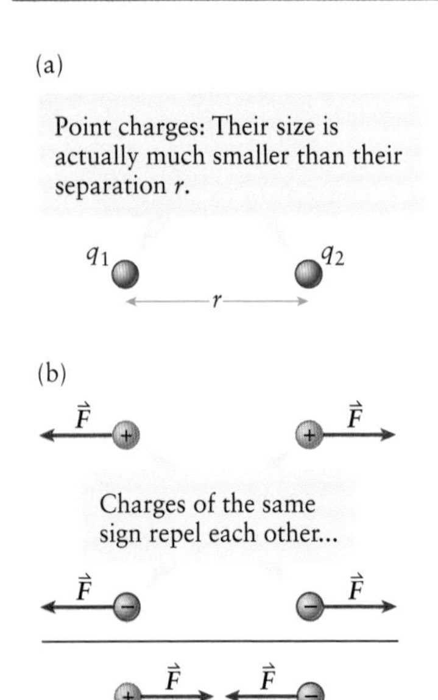

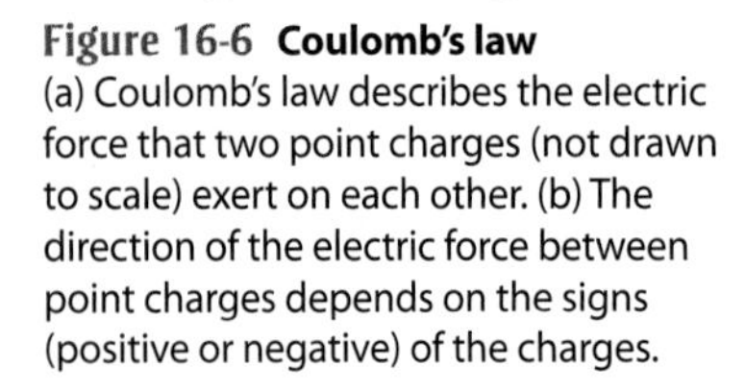

Figure 16-6 Coulomb's law
(a) Coulomb's law describes the electric force that two point charges (not drawn to scale) exert on each other. (b) The direction of the electric force between point charges depends on the signs (positive or negative) of the charges.

16-4 Coulomb's law describes the force between charged objects

We learned in Section 16-2 that electrically charged objects, or *charges* for short, exert forces on each other (see Figure 16-2). Careful measurements reveal that the force between two charges is directly proportional to the magnitude of each charge. The greater the magnitudes of the charges on the objects, the larger the force that acts on each charge. The force also depends on the distance between the charged objects; the closer the charges, the larger the force.

Coulomb's law summarizes the results of these measurements. Specifically, this law tells us about the force between two **point charges**, which are very small charged objects whose size is much smaller than the separation between them (**Figure 16-6a**). A point charge is an idealization, just like a massless rope or a frictionless incline, but it's a good description in many situations where charged objects interact with each other. Coulomb's law tells us the magnitude of the electric force that two point charges q_1 and q_2 separated by a distance r exert on each other:

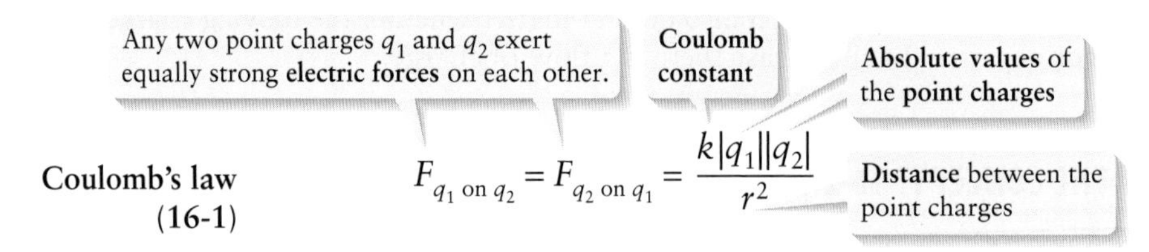

Coulomb's law (16-1)

$$F_{q_1 \text{ on } q_2} = F_{q_2 \text{ on } q_1} = \frac{k|q_1||q_2|}{r^2}$$

The value of the **Coulomb constant** k in Equation 16-1 is, to three significant figures,

$$k = 8.99 \times 10^9 \text{ N} \cdot \text{m}^2/\text{C}^2$$

Note that Equation 16-1 just tells you the *magnitude* of the electric force between two point charges. As **Figure 16-6b** shows, the *direction* of the electric force is such that charges of the same sign (both positive or both negative) repel each other, while charges of opposite sign (one positive and one negative) attract each other. **Figure 16-7** shows an application of this principle.

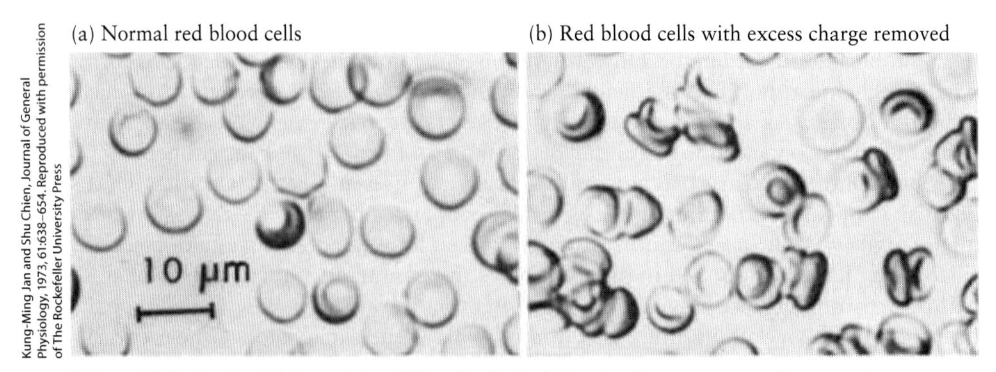

Kung-Ming Jan and Shu Chien, Journal of General Physiology, 1973, 61:638–654. Reproduced with permission of The Rockefeller University Press

Figure 16-7 Coulomb's law and red blood cells Red blood cells carry oxygen from your lungs to other parts of your body through the circulatory system. (a) Each red blood cell has a slight excess of electrons that gives it a net negative charge. Since all cells are negatively charged, they exert repulsive electric forces on each other that help keep the cells apart. (b) The red blood cells in this photo have been treated with an enzyme that removes the excess electrons. Without electric forces to keep them apart, the red blood cells tend to clump. If red blood cells in your body behaved like this, their flow through your circulatory system would be impeded and your body would be starved for oxygen.

You should notice the similarity between Coulomb's law and Newton's law of universal gravitation:

Gravitational constant (same for any two objects) **Masses** of the two objects

Any two objects (1 and 2) exert equally strong **gravitational forces** on each other.

$$F_{1 \text{ on } 2} = F_{2 \text{ on } 1} = \frac{Gm_1 m_2}{r^2}$$

Center-to-center distance between the two objects

The gravitational forces are attractive: $\vec{F}_{1 \text{ on } 2}$ pulls object 2 toward object 1 and $\vec{F}_{2 \text{ on } 1}$ pulls object 1 toward object 2.

Newton's law of universal gravitation (10-2)

In both Coulomb's law (Equation 16-1) and the law of universal gravitation (Equation 10-2), the force that one object exerts on the other is inversely proportional to the square of the distance between them. In addition, the gravitational force is proportional to the product of the masses of the two particles, while the electric force is proportional to the product of the charges of the two particles. As remarkable as these similarities are, there is one essential difference between the two laws: Newton's law of universal gravitation tells us that two objects with mass always attract each other, but Coulomb's law states that two charged objects can attract or repel each other, depending on the sign of the charges they carry.

The electric force also shares some properties with every other force we've discussed. The electric force is a vector and has a direction. The net electric force on an object is the vector sum of every separate electric force that acts on it. Furthermore, the electric forces that two charged objects exert on each other obey Newton's third law. When point objects with charges q_1 and q_2 interact, the force that object 1 exerts on object 2 is equal in magnitude to the force that object 2 exerts on object 1, but the two forces are in opposite directions (see Figure 16-6b).

EXAMPLE 16-2 Calculating Electric Force

(a) What is the electric force (magnitude and direction) between two electrons separated by a distance of 10.0 cm = 0.100 m? (b) Suppose you could remove all the electrons from a drop of water 1.00 mm in radius (see Example 16-1 in Section 16-2) and clump them into a ball 1.00 mm in radius. If this ball of electrons is 10.0 cm from the drop of water from which they were removed, what is the magnitude of the electric force between the drop of water and the ball of electrons?

Set Up

The two electrons repel because both have a negative charge $q = -e = -1.602 \times 10^{-19}$ C. From Example 16-1, the combined charge of all of the electrons in a water drop of this size is -224 C; the water drop was initially neutral, so the charge of the water drop after all of the electrons have been removed is $+224$ C. Since the ball of electrons and the water drop (with electrons removed) have opposite signs of charge, they attract each other.

We can use Coulomb's law (Equation 16-1) in both parts of this problem. That's because in both cases the charged objects are much smaller than the distance that separates them, so we can treat them as point charges. (Indeed, electrons are very small even compared to the dimensions of an atom.)

Coulomb's law:

$$F_{q_1 \text{ on } q_2} = F_{q_2 \text{ on } q_1} = \frac{k|q_1||q_2|}{r^2} \quad (16\text{-}1)$$

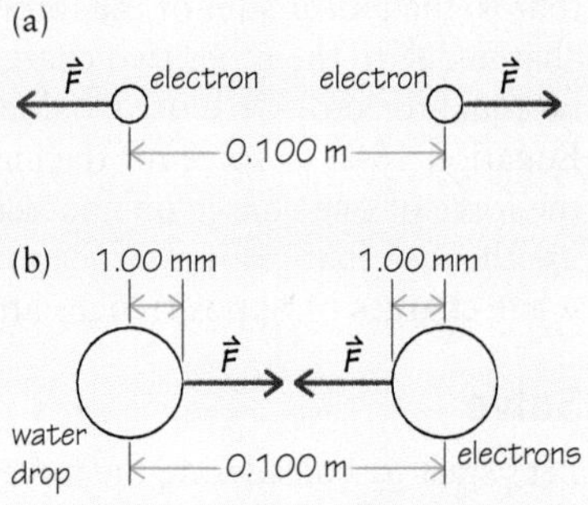

(a) $\vec{F}$ electron electron $\vec{F}$ 0.100 m

(b) 1.00 mm 1.00 mm $\vec{F}$ $\vec{F}$ water drop 0.100 m electrons

Solve

(a) Find the magnitude of the force that each electron exerts on the other.

For the two electrons the charges are

$$q_1 = q_2 = -e = -1.602 \times 10^{-19}\ \text{C}$$

The distance between the electrons is $r = 0.100$ m. From Equation 16-1 the magnitude of the force that each electron exerts on each other is

$$F = \frac{(8.99 \times 10^9\ \text{N}\cdot\text{m}^2/\text{C}^2)|-1.602 \times 10^{-19}\ \text{C}||-1.602 \times 10^{-19}\ \text{C}|}{(0.100\ \text{m})^2}$$

$$= 2.31 \times 10^{-26}\ \text{N}$$

(b) Find the magnitude of the force between the ball of electrons and the water drop from which they were extracted.

For the ball of electrons and the water drop with electrons removed, the charges are

$$q_1 = -224\ \text{C},\ q_2 = +224\ \text{C}$$

The distance between the two objects is $r = 0.100$ m. From Equation 16-1 the two objects exert forces on each other of magnitude

$$F = \frac{(8.99 \times 10^9\ \text{N}\cdot\text{m}^2/\text{C}^2)|-224\ \text{C}||+224\ \text{C}|}{(0.100\ \text{m})^2}$$

$$= 4.51 \times 10^{16}\ \text{N}$$

Reflect

The repulsive force between the two electrons is tiny because the electron charge is tiny. By contrast, the attractive force between the ball of electrons and the electron-free water drop is immense. To put this force into perspective, a solid cube of lead with a weight of 4.51×10^{16} N would be 7.4 *kilometers* on a side! This is the magnitude of force that you would have to exert to keep the electrons from flying back into the water drop. There is no known way to produce a force of this magnitude, which is why you'll never see an object with all of its electrons removed. It's relatively easy to remove a small fraction of an object's electrons, as for the fur and the glass rod in Figure 16-2b. But removing *all* of the electrons from a piece of fur or a glass rod is not a practical thing to do.

EXAMPLE 16-3 Three Charges in a Line

A particle with negative charge q is placed halfway between two identical particles, each of which carries the same positive charge: $Q_1 = Q_2 = +Q$. The distance between adjacent charges is d. If each of the three particles experiences a net electric force of zero, what is the magnitude of charge q in terms of Q?

Set Up

We want the net force on each point charge—that is, the *vector* sum of the forces on that charge due to the other two charges—to be equal to zero. We'll use Coulomb's law, Equation 16-1, to solve for the magnitude of the force of one charge on another. We'll also use the idea that charges of the same signs repel while charges of opposite signs attract.

Coulomb's law:

$$F_{q_1\ \text{on}\ q_2} = F_{q_2\ \text{on}\ q_1} = \frac{k|q_1||q_2|}{r^2} \qquad (16\text{-}1)$$

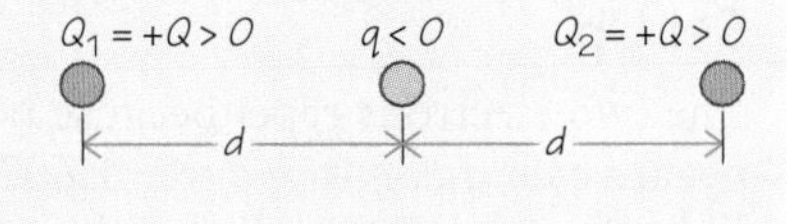

Solve

Let's start by considering the forces on positive charge Q_2. The other positive charge, Q_1, exerts a repulsive force $\vec{F}_{Q_1\ \text{on}\ Q_2}$ that pushes Q_2 away from Q_1, that is, to the right. The negative charge q exerts an attractive force $\vec{F}_{q\ \text{on}\ Q_2}$ that pulls Q_2 toward q, that is, to the left. In order for the net force on Q_2 to be zero, these two forces must have the same magnitude.

The net force on charge Q_2 must be zero:

$$\vec{F}_{Q_1\ \text{on}\ Q_2} + \vec{F}_{q\ \text{on}\ Q_2} = 0$$

For this to be true, $\vec{F}_{Q_1\ \text{on}\ Q_2}$ and $\vec{F}_{q\ \text{on}\ Q_2}$ must have the same magnitude:

$$F_{Q_1\ \text{on}\ Q_2} = F_{q\ \text{on}\ Q_2}$$

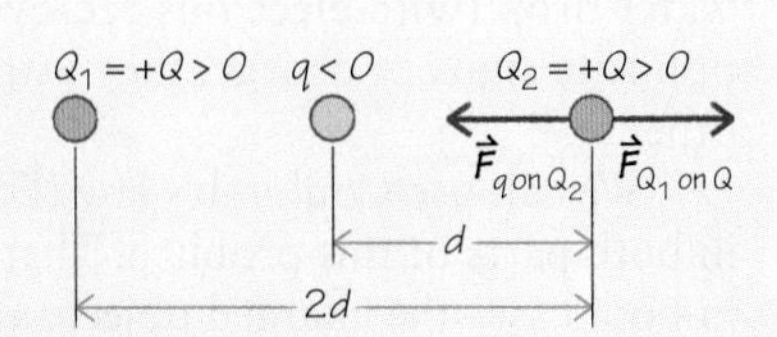

Use Coulomb's law (Equation 16-1) to find the magnitudes $F_{Q_1 \text{ on } Q_2}$ and $F_{q \text{ on } Q_2}$. Set these equal to each other and solve for the magnitude (absolute value) of q.

The distance between Q_1 and Q_2, each of which has a charge of magnitude Q, is $2d$. From Equation 16-1, the force of Q_1 on Q_2 has magnitude

$$F_{Q_1 \text{ on } Q_2} = \frac{k|Q_1||Q_2|}{(2d)^2} = \frac{kQ^2}{4d^2}$$

d is the distance between Q_1 and q, and also between Q_2 and q. Notice that Q_1 and Q_2 are separated by distance $2d$. Remember that q is negative, so we need to keep its absolute value.

The distance between q and Q_2 is d, so the magnitude of the force of q on Q_2 is

$$F_{q \text{ on } Q_2} = \frac{k|q||Q_2|}{d^2} = \frac{k|q|Q}{d^2}$$

In order for the forces to have the same magnitude,

$$\frac{kQ^2}{4d^2} = \frac{k|q|Q}{d^2}$$

Solve for the absolute value of q:

$$|q| = \frac{Q}{4}$$

Reflect

The value $|q| = Q/4$ satisfies the condition that there is zero net force on Q_2. Because Q_1 is twice as far from Q_2 as q, and because the electric force is inversely proportional to the square of the distance, the charge Q_1 must be $(2)^2 = 4$ times greater than the magnitude of q in order for the forces these charges exert on Q_2 to have the same magnitude.

You can see that since Q_1 and Q_2 have the same charge, the forces on Q_1 are the mirror images of those on Q_2 (a repulsive force to the *left* exerted by Q_2, which is a distance $2d$ from Q_1, and an attractive force to the *right* exerted by q, which is a distance d from Q_1). So the net force on Q_1 will be zero, too.

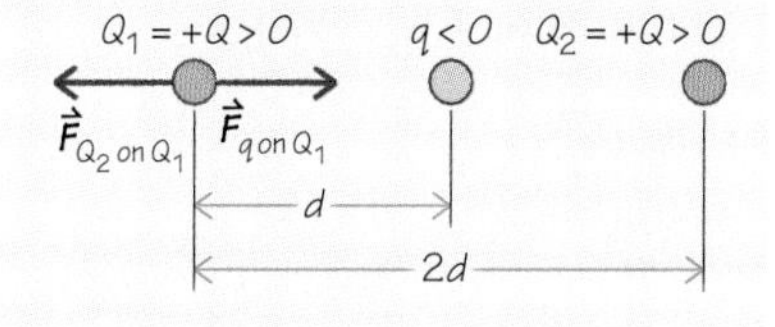

You can see that the net force on the negative charge q is also guaranteed to be zero. This charge is the same distance d from the two equal positive charges Q_1 and Q_2, so the force from Q_1 that pulls q to the left is just as great as the force from Q_2 that pulls q to the right.

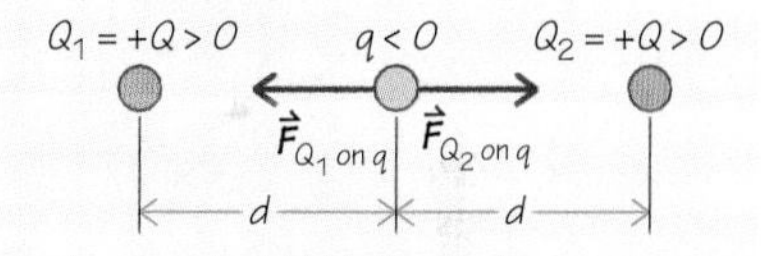

EXAMPLE 16-4 Three Charges in a Plane

Charges Q_1 and Q_3 are both positive and equal to 1.50×10^{-6} C; charge Q_2 is negative and equal to -1.50×10^{-6} C. Charges Q_1 and Q_2 are placed at a fixed position a distance $D = 6.00$ cm apart, and Q_3 is placed at a fixed position a distance $H = 4.00$ cm above the midpoint of the line that connects Q_1 and Q_2. Calculate the magnitude and direction of the force on Q_3 due to the other two charges.

Set Up

The net force on positive charge Q_3 is the vector sum of $\vec{F}_{1 \text{ on } 3}$, the *repulsive* force that the positive charge Q_1 exerts on Q_3, and $\vec{F}_{2 \text{ on } 3}$, the *attractive* force that the negative charge Q_2 exerts on Q_3. We'll use Coulomb's law to find the magnitude of each of these forces. We'll then add the two force vectors using components.

Coulomb's law:

$$F_{q_1 \text{ on } q_2} = F_{q_2 \text{ on } q_1} = \frac{k|q_1||q_2|}{r^2} \quad (16\text{-}1)$$

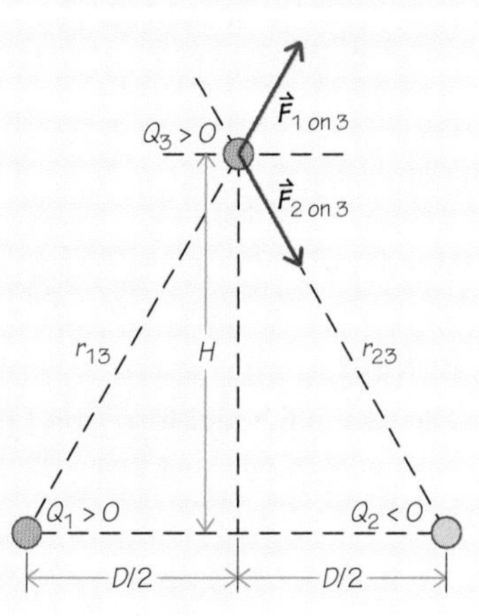

Solve

Use Equation 16-1 to find the magnitudes of the forces $\vec{F}_{1\text{ on }3}$ and $\vec{F}_{2\text{ on }3}$.

The figure above shows that the distance from Q_1 to Q_3 is the same as the distance from Q_2 to Q_3:

$$r_{13} = r_{23} = \sqrt{\left(\frac{D}{2}\right)^2 + H^2} = \sqrt{\left(\frac{6.00\text{ cm}}{2}\right)^2 + (4.00\text{ cm})^2}$$

$$= 5.00\text{ cm} = 5.00 \times 10^{-2}\text{ m}$$

All three charges have the same magnitude:

$$|Q_1| = |Q_2| = |Q_3| = 1.50 \times 10^{-6}\text{ C}$$

So from Equation 16-1, $\vec{F}_{1\text{ on }3}$ and $\vec{F}_{2\text{ on }3}$ have the same magnitude:

$$F_{1\text{ on }3} = \frac{k|Q_1||Q_3|}{r_{13}^2}$$

$$= \frac{(8.99 \times 10^9\text{ N}\cdot\text{m}^2/\text{C}^2)(1.50 \times 10^{-6}\text{ C})^2}{(5.00 \times 10^{-2}\text{ m})^2}$$

$$= 8.09\ N$$

$$F_{2\text{ on }3} = \frac{k|Q_2||Q_3|}{r_{23}^2} = F_{1\text{ on }3} = 8.09\text{ N}$$

Choose the positive x direction to be to the right and the positive y direction to be upward. Then find the x and y components of $\vec{F}_{1\text{ on }3}$ and $\vec{F}_{2\text{ on }3}$, and use these to calculate the components of the net force on Q_3.

The components of $\vec{F}_{1\text{ on }3}$ and $\vec{F}_{2\text{ on }3}$ are

$$F_{1\text{ on }3,x} = F_{1\text{ on }3}\cos\theta$$

$$F_{1\text{ on }3,y} = F_{1\text{ on }3}\sin\theta$$

$$F_{2\text{ on }3,x} = F_{2\text{ on }3}\cos\theta$$

$$F_{2\text{ on }3,y} = -F_{2\text{ on }3}\sin\theta$$

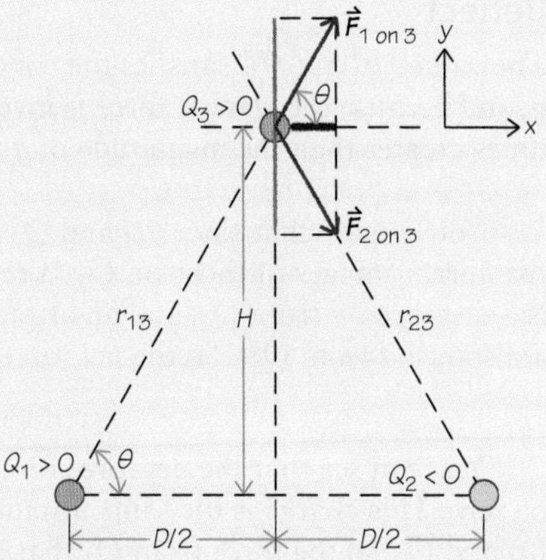

Since the magnitudes $F_{1\text{ on }3}$ and $F_{2\text{ on }3}$ are equal, the components of the net force on Q_3 are

$$F_{\text{net on }3,x} = F_{1\text{ on }3,x} + F_{2\text{ on }3,x}$$

$$= F_{1\text{ on }3}\cos\theta + F_{2\text{ on }3}\cos\theta$$

$$= 2F_{1\text{ on }3}\cos\theta$$

$$F_{\text{net on }3,y} = F_{1\text{ on }3}\sin\theta + (-F_{2\text{ on }3}\sin\theta)$$

$$= F_{1\text{ on }3}\sin\theta - F_{1\text{ on }3}\sin\theta = 0$$

From the figure,

$$\cos\theta = \frac{(D/2)}{r_{13}} = \frac{3.00\text{ cm}}{5.00\text{ cm}} = 0.600$$

So

$$F_{\text{net on }3,x} = 2(8.09\text{ N})(0.600) = 9.71\text{ N}$$

$$F_{\text{net on }3,y} = 0$$

So the net force on Q_3 is to the right and has magnitude 9.71 N.

Reflect

The two individual forces on Q_3 add to a net force that is neither directly away from Q_1 nor directly toward Q_2.

GOT THE CONCEPT? 16-3 Electric Force

In part (b) of Example 16-2 we imagined removing all of the electrons from a drop of water and moving them to a given distance from the water drop. We then calculated the electric force between the drop and the electrons. Suppose instead we removed only one-half of the electrons from the water drop then moved them to the same distance as in Example 16-2. Compared to the force we calculated in part (b) of Example 16-2, what would be the force between the electrons and the water drop in this case? (a) The same; (b) $1/\sqrt{2}$ as great; (c) 1/2 as great; (d) 1/4 as great; (e) 1/16 as great.

TAKE-HOME MESSAGE FOR Section 16-4

✔ Coulomb's law tells us the magnitude of the electric force that two point charges exert on each other. This magnitude is proportional to the product of the magnitudes of the two charges and inversely proportional to the square of the distance between them.

✔ The electric force between two point charges is repulsive if the two charges have the same sign (both positive or both negative) and attractive if the two charges have opposite signs (one positive and one negative).

16-5 The concept of electric field helps us visualize how charges exert forces at a distance

Most forces in our daily experience arise only when one object is in direct contact with another object, like the normal force that acts on your body when you sit in a chair or when you push directly on an object. Nevertheless, some forces, such as the gravitational force and the Coulomb force, appear to act even between objects separated by a distance. We can describe forces of this kind using the concept of a *field*. In this view every charged object modifies all of space by producing an electric field, which is strongest closest to the object but extends infinitely far away. A second charged object senses this change in space, interacting with the electric field and experiencing an electric force (**Figure 16-8**).

If we place a particle carrying charge q in an electric field $\vec{E}$, the force experienced by the particle is

If a particle with charge q is placed at a position where the **electric field** due to other charges is $\vec{E}$...

$$\vec{F} = q\vec{E}$$

...then the **electric force** on the particle is $\vec{F} = q\vec{E}$.

Electric field and electric force (16-2)

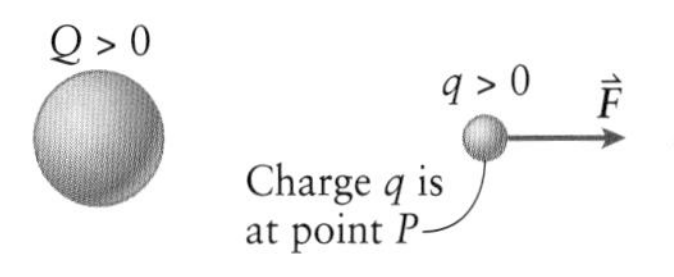

(1) To explain how charge Q exerts an electric force on charge q at a point P some distance from charge Q...

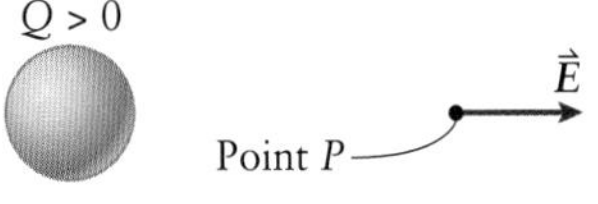

(2) ...we picture charge Q as producing an electric field at point P. This field is present whether or not there is any charge at point P.

(3) If we place a charge q at point P, it senses the electric field due to Q and experiences a force as a result.

(4) The force on charge q is in the same direction as the electric field if q is positive...

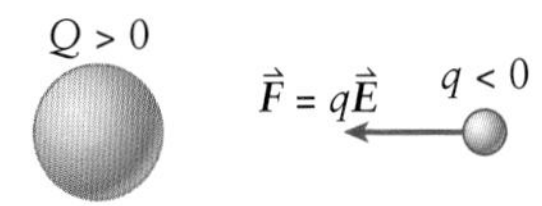

(5) ...and the force on charge q is in the direction opposite to the electric field if q is negative.

Figure 16-8 Electric field and electric force The electric field concept helps us visualize how a charge Q exerts an electric force on a second charge q.

An equivalent way to write Equation 16-2 is

$$\vec{E} = \frac{\vec{F}}{q} \tag{16-3}$$

Equation 16-3 tells us that we can interpret the electric field $\vec{E}$ at a certain point as the *electric force per charge* that acts on a charged object placed at that point. If an object with double the charge is placed at that point, it will experience double the force.

WATCH OUT! The direction of *force* versus the direction of *field*.

Although we can think of the electric field as the electric force per unit charge, you must be careful about the direction of the force. In particular, the direction of the electric force that a charge experiences depends on the *sign* of the charge as well as the *direction* of the electric field. Equation 16-2 tells us that a positive charge ($q > 0$) experiences a force in the same direction as the electric field $\vec{E}$, but a negative charge ($q < 0$) experiences a force in the direction opposite to $\vec{E}$ (Figure 16-8). So the electric field at a certain point is in the direction of the electric force that would be exerted on a *positive* charge placed at that point.

From Equation 16-3 we see that the SI units of electric field are newtons per coulomb, or N/C. If a 1-C charge experiences a 1-N force at a certain point in space, then at that point in space there is an electric field with a magnitude of 1 N/C.

Note that Equations 16-2 and 16-3 are strictly valid *only* if the particle of charge q is a *point* particle (that is, one with a very small size). That's because the value of $\vec{E}$ can be different at different places. If the charge q were spread over a large volume (say, a sphere 1 m in diameter), the value of $\vec{E}$ could be different at different points within that volume. In that case it wouldn't be clear which value of $\vec{E}$ to use in Equation 16-2 or 16-3. But if the charge q is within a point particle, it's clear what value of $\vec{E}$ to use: the value at the point where that particle is located.

Go to Interactive Exercise 16-1 for more practice dealing with electric fields.

Although this is the first time we've introduced the idea of a field, we've actually used the concept before. In Chapter 4 we found that we could express the gravitational force $\vec{w}$ on an object of mass m as $\vec{w} = m\vec{g}$, where $\vec{g}$ is the acceleration due to gravity. But we can also think of $\vec{g}$ as the *gravitational field* that Earth sets up in the space around it; $\vec{g}$ represents the value of that field vector (magnitude and direction) at a given point. An object of mass m placed at the point then experiences a gravitational force $\vec{w} = m\vec{g}$ (compare to Equation 16-2). Just as the concept of an electric field gives us a way to visualize how two charges can interact over a distance, the concept of a gravitational field helps us visualize how two objects with mass (Earth and the object of mass m) can interact without touching each other.

EXAMPLE 16-5 Determining Charge-to-Mass Ratio

When released from rest in a uniform electric field of magnitude 1.00×10^4 N/C, a certain charged particle travels 2.00 cm in 2.88×10^{-7} s in the direction of the field. You can ignore any nonelectric forces acting on the particle. (a) What is the *charge-to-mass ratio* of this particle (that is, the ratio of its charge q to its mass m)? (b) Other experiments show that the mass of this particle is 6.64×10^{-27} kg. What is the charge of this particle?

Set Up

The only force that acts on the particle is the electric force given by Equation 16-2. Since the particle accelerates in the direction of the electric field $\vec{E}$, the force on the particle must also be in the direction of $\vec{E}$. So the charge on the particle must be positive. Since $\vec{E}$ is uniform (it has the same value at all points), the force on the particle will be constant. Its acceleration

Electric field and electric force:

$$\vec{F} = q\vec{E} \tag{16-2}$$

Straight-line motion with constant acceleration:

$$x = x_0 + v_{0x}t + \frac{1}{2}a_x t^2 \tag{2-9}$$

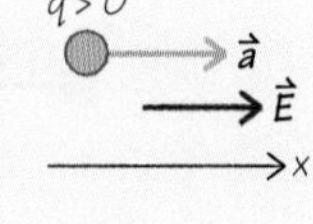

a_x will be constant as well, so we can use one of the constant-acceleration equations from Chapter 2 to determine a_x. We'll use this with Newton's second law and Equation 16-2 to learn what we can about this particle.

Newton's second law:

$$\sum \vec{F}_{\text{ext}} = m\vec{a} \quad (4\text{-}2)$$

Solve

(a) We are given that the particle travels 2.00 cm = 2.00×10^{-2} m in 2.88×10^{-7} s. Use this to determine the particle's constant acceleration.

Take the positive x direction to be the direction in which the particle moves. The particle begins at rest, so $v_{0x} = 0$. If we take the initial position of the particle to be $x_0 = 0$, then Equation 2-9 becomes

$$x = \frac{1}{2}a_x t^2$$

Solve for the acceleration:

$$a_x = \frac{2x}{t^2} = \frac{2(2.00 \times 10^{-2}\text{ m})}{(2.88 \times 10^{-7}\text{ s})^2} = 4.82 \times 10^{11}\text{ m/s}^2$$

Relate the acceleration to the net external (electric) force on the particle and solve for the charge-to-mass ratio.

The net force on the particle of charge q is the electric force in the x direction, which from Newton's second law is equal to the mass m of the particle multiplied by the acceleration a_x:

$$qE_x = ma_x$$

This says that the acceleration of a particle in an electric field depends on the particle's charge-to-mass ratio:

$$a_x = \frac{q}{m}E_x$$

In this example we know both a_x and E_x, so the charge-to-mass ratio is

$$\frac{q}{m} = \frac{a_x}{E_x} = \frac{4.82 \times 10^{11}\text{ m/s}^2}{1.00 \times 10^4\text{ N/C}}$$

Since 1 N = 1 kg · m/s², this is

$$\frac{q}{m} = 4.82 \times 10^7\text{ C/kg}$$

(b) Given the charge-to-mass ratio and the mass of the particle, determine the charge q.

The charge of the particle is

$$q = m\left(\frac{q}{m}\right) = (6.64 \times 10^{-27}\text{ kg})(4.82 \times 10^7\text{ C/kg})$$

$$= 3.20 \times 10^{-19}\text{ C}$$

The charge on a proton is $e = 1.60 \times 10^{-19}$ C; the charge on this particle is $2e$.

Reflect

The particle in this example has about four times the mass of a proton but only double the charge of a proton. For historical reasons it's known as an *alpha particle*; in fact it's the nucleus of a helium atom, which contains two protons (each with charge e) and two neutrons (each with nearly the same mass as a proton but with zero charge).

An important application of the force produced by an electric field is *electrophoresis*. Chemists use this technique to separate molecules of different kinds according to their charge and mass. In the simplest kind of electrophoresis, a small amount of a sample containing molecules of different kinds is placed on a strip of filter paper, and the paper is soaked with a solution that conducts electricity (**Figure 16-9**). An electric field of magnitude E is then applied along the length of the strip. Each molecule accelerates

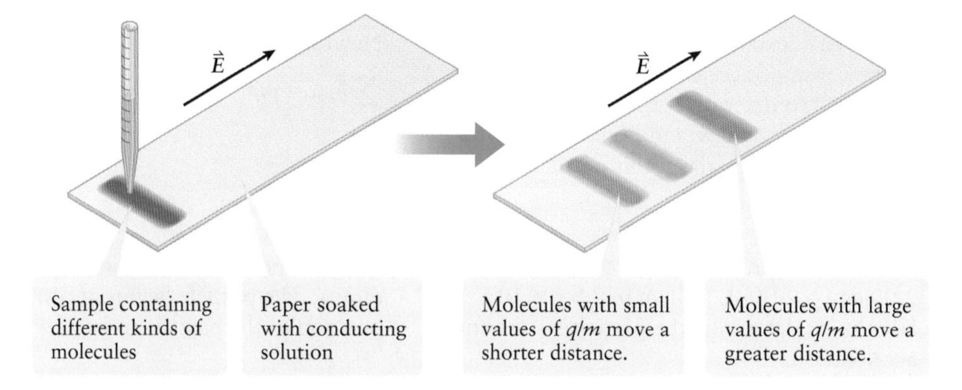

Figure 16-9 Paper electrophoresis This analytical technique used by chemists makes use of the electric field concept.

in response to the field, and that acceleration (taking into account fluid resistance on the molecules) is proportional to the charge-to-mass ratio q/m of the molecule. As a result, molecules with larger values of q/m move farther along the paper than do those with small values of q/m. The many applications of this specific technique, called *paper electrophoresis*, include analyzing currency to determine whether it is counterfeit (a forger's ink may have a different chemical composition than the ink used in legal currency) and looking for the presence of cancer antibodies or human immunodeficiency virus (HIV) in blood.

A different sort of electrophoresis is used for DNA profiling (also called genetic fingerprinting), which is an essential part of modern forensic science. A sample of human DNA is treated with an enzyme that breaks the long DNA strand into shorter segments. The sizes of these segments are characteristic of the person's genetic code, so measuring the segment sizes is a powerful technique for forensic identification. Unfortunately, the ratio of q/m for a segment of DNA is nearly the same for segments of any size, so paper electrophoresis isn't useful. Instead, a sample containing the DNA segments is placed in a special gel that is permeated by many microscopic pores. When an electric field is applied, all of the segments move in response, but the smaller segments move through the gel pores more easily than large ones. The result is that the DNA segments are spread out according to their size, allowing a genetic "fingerprint" to be made. Similar techniques are used in medical research for studying both DNA and proteins.

Electric Field of a Point Charge

Equation 16-2 tells us how a charge q responds to a given electric field $\vec{E}$. It also tells us how to determine the value of $\vec{E}$ at any point. As an example, **Figure 16-10** shows how we might determine the electric field around a positive point charge Q. We place a small positive charge q (which we call a *test charge*) at various locations around the charge Q and measure the force $\vec{F}$ on that small charge. The electric field at each location is given by Equation 16-3, $\vec{E} = \vec{F}/q$; since q is positive, the electric field is in the same direction as the force on the test charge. The Coulomb force exerted by Q repels a positive test charge q; thus, the direction of force at each location—and so the direction of the electric field at each location, shown by the blue vectors—is radially

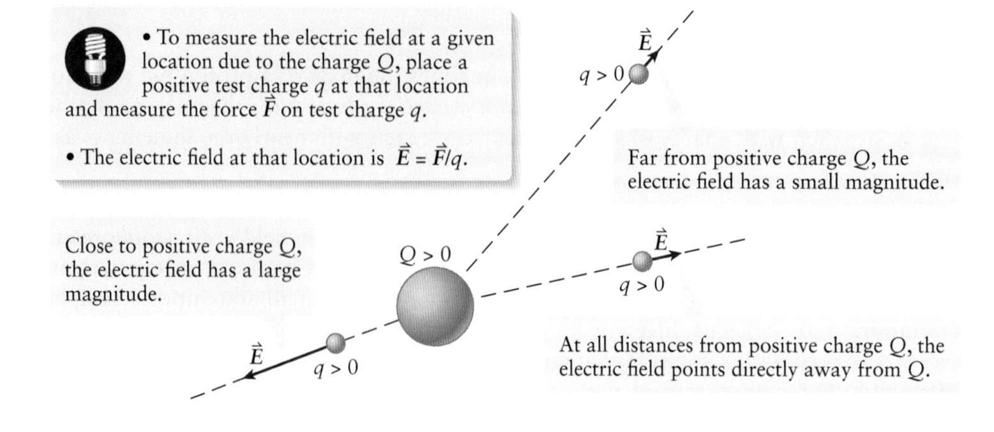

Figure 16-10 Mapping the electric field We can map out the electric field surrounding a point charge Q—in this case a positive charge—by placing a positive test charge $+q$ at various locations around Q.

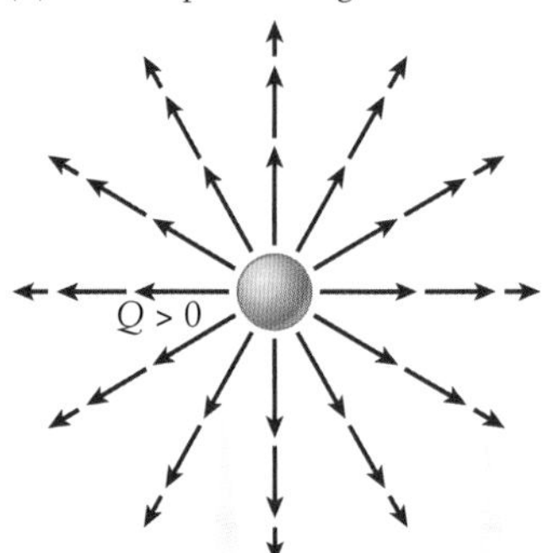
(a) Positive point charge: electric field vectors
Q > 0
Field due to a positive charge points away from the charge.
Field strength decreases with increasing distance from the charge.

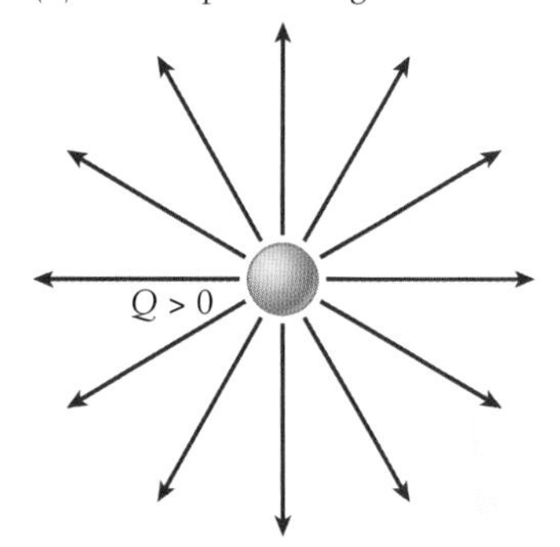
(b) Positive point charge: electric field lines
Q > 0
Field due to a positive charge points away from the charge.
Farther from the charge, where the field strength is weaker, field lines are farther apart.

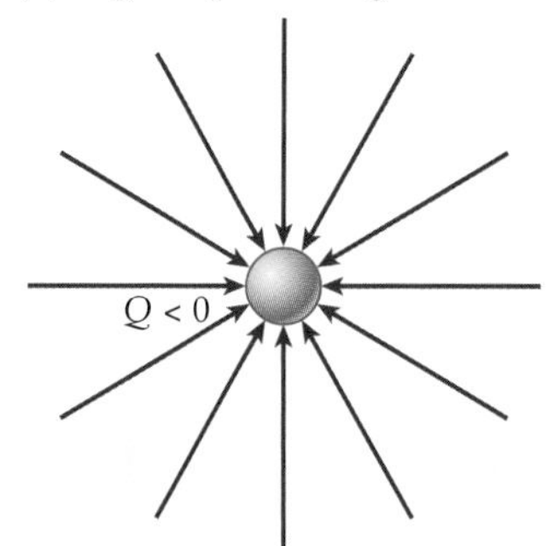
(c) Negative point charge: electric field lines
Q < 0
Field lines due to a negative charge point toward the charge.
Farther from the charge, where the field strength is weaker, field lines are farther apart.

Figure 16-11 Electric field and electric field lines The electric field around a charged object can be represented by (a) electric field vectors or (b), (c) electric field lines.

away from Q. As the lengths of the vectors show, the electric field magnitude decreases with increasing distance from Q. That's because the force between Q and q decreases with increasing distance in accordance with Coulomb's law (Section 16-4).

There are two ways we can depict the entire electric field around a positive charge Q. In **Figure 16-11a** we draw vectors to represent the electric field at a large number of points around Q. **Figure 16-11b** shows a simpler approach that's easier to draw: We connect adjacent vectors to form lines, called **electric field lines**. The direction of the field line passing through any point represents the direction of the field at that point. The magnitude of the electric field is shown by the *density* of the field lines—that is, how close they are to each other. Close to Q, where the field lines are close together, the field has a large magnitude. Far from Q, where the field lines are far apart, the magnitude of the electric field is smaller. If the charge Q is negative, a positive test charge q experiences an attractive force $\vec{F}$ toward Q, so the electric field $\vec{E} = \vec{F}/q$ due to charge Q is directed *toward* that charge (**Figure 16-11c**).

WATCH OUT! Electric fields are three-dimensional.

Figure 16-11 may give you the incorrect impression that the electric field and electric field lines of a point charge Q lie only on the plane of the page. Not so! The electric field completely surrounds the charge; it is three-dimensional. The field lines are arranged around the charge rather like the spines of a sea urchin (**Figure 16-12**).

NOAA

Figure 16-12 An electric field analogy The electric field lines of a point charge are arranged radially around the charge in three dimensions, much like the spines on a sea urchin.

We can use Coulomb's law, Equation 16-1, to write an expression for the *magnitude* of the electric field due to a point charge Q. If we place a test charge q a distance r from charge Q, Equation 16-1 tells us that the magnitude of the Coulomb force on q is $F = k|q||Q|/r^2$. From Equation 16-2 we can also write this force magnitude as $F = |q|E$, where E is the magnitude of the electric field due to charge Q at the position of test charge q. Setting these two expressions for F equal to each other, we get

$$|q|E = \frac{k|q||Q|}{r^2}$$

or

Magnitude of the electric field due to a point charge (16-4)

$$E = \frac{k|Q|}{r^2}$$

Magnitude of the electric field due to a point charge Q

Coulomb constant

Absolute value of charge Q

Distance between point charge Q and the location where the field is measured

The *magnitude* of the electric field due to a point charge Q decreases with increasing distance r from the charge. As Figure 16-11 shows the *direction* of the electric field depends on the sign of the charge Q. The field points radially outward from a positive point charge and radially inward toward a negative point charge.

Electric Field of an Arrangement of Charges

Go to Picture It 16-1 for more practice dealing with electric fields.

The results in Figure 16-11 and Equation 16-4 describe the electric field due to a single point charge (positive or negative). If several point charges Q_1, Q_2, Q_3, . . . are present, experiment shows that the *net* electric force $\vec{F}$ that these charges exert on a test charge q at any location is just the vector sum of the forces $\vec{F}_1$, $\vec{F}_2$, $\vec{F}_3$, . . . that these charges *individually* exert on q. If we use the symbol $\vec{E}$ for the net electric field produced at a given location by Q_1, Q_2, Q_3, . . . together, and the symbols $\vec{E}_1$, $\vec{E}_2$, $\vec{E}_3$, . . . for the electric fields that these charges produce individually, we can use Equation 16-2 to express this experimental result as

$$\vec{F} = \vec{F}_1 + \vec{F}_2 + \vec{F}_3 + \cdots \quad \text{or} \quad q\vec{E} = q\vec{E}_1 + q\vec{E}_2 + q\vec{E}_3 + \cdots$$

If we divide both sides of the second of these equations by q, we get

$$\vec{E} = \vec{E}_1 + \vec{E}_2 + \vec{E}_3 + \cdots \qquad (16\text{-}5)$$

In other words, *when there are two or more point charges present, the electric field at any point in space is the vector sum of the fields due to each charge separately*. The following examples illustrate how to use this principle.

EXAMPLE 16-6 Where *Is* the Electric Field Zero?

A point charge $Q_1 = +4.00$ nC (1 nC = 1 nanocoulomb = 10^{-9} C) is placed 0.500 m to the left of a point charge $Q_2 = +9.00$ nC. Find the position between the two point charges where the net electric field is zero.

Set Up

At any point between the two charges, the electric field $\vec{E}_1$ due to Q_1 points to the right (away from this positive charge) and the electric field $\vec{E}_2$ due to Q_2 points to the left (away from this positive charge). We want to find the point P where the total field $\vec{E} = \vec{E}_1 + \vec{E}_2$ equals zero. We'll use the symbol D for the 0.500-m distance between the two charges and x for the distance from Q_1 to point P: Then the distance from Q_2 to point P is $D - x$. Our goal is to find the value of x for which $\vec{E} = 0$.

Magnitude of the electric field due to a point charge Q:

$$E = \frac{k|Q|}{r^2} \qquad (16\text{-}4)$$

Total electric field:

$$\vec{E} = \vec{E}_1 + \vec{E}_2 \qquad (16\text{-}5)$$

Solve

If the net electric field at P is zero, then $\vec{E}_1$ (to the right) and $\vec{E}_2$ (to the left) must have equal magnitudes so that these two vectors cancel. Use Equation 16-4 to write this statement in equation form.

In order for the net electric field at P to be zero,

$$E_1 = E_2$$

The distance from Q_1 to P is x, and the distance from Q_2 to P is $D - x$, so from Equation 16-4

$$E_1 = \frac{k|Q_1|}{x^2} \quad \text{and} \quad E_2 = \frac{k|Q_2|}{(D-x)^2}$$

If these are equal to each other,

$$\frac{k|Q_1|}{x^2} = \frac{k|Q_2|}{(D-x)^2}$$

Solve this equation for x.

The factors of k cancel in the above equation, so

$$\frac{|Q_1|}{x^2} = \frac{|Q_2|}{(D-x)^2} \quad \text{or}$$

$$|Q_1|(D-x)^2 = |Q_2|x^2$$

Multiply out the quantity $(D - x)^2$ and rearrange:

$$|Q_1|(D^2 - 2Dx + x^2) = |Q_2|x^2$$

$$(|Q_2| - |Q_1|)x^2 + (2D|Q_1|)x + (-D^2|Q_1|) = 0$$

We can simplify this if we divide through by $|Q_1|$:

$$\left(\frac{|Q_2|}{|Q_1|} - 1\right)x^2 + 2Dx + (-D^2) = 0$$

This is a quadratic equation of the form $ax^2 + bx + c = 0$, with $a = |Q_2|/|Q_1| - 1 = |9.00 \text{ nC}|/|4.00 \text{ nC}| - 1 = 1.25$, $b = 2D = 2(0.500 \text{ m}) = 1.00 \text{ m}$, and $c = -D^2 = -(0.500 \text{ m})^2 = -0.250 \text{ m}^2$. The solutions are

$$x = \frac{-b \pm \sqrt{b^2 - 4ac}}{2a}$$

$$= \frac{-(1.00 \text{ m}) \pm \sqrt{(1.00 \text{ m})^2 - 4(1.25)(-0.250 \text{ m}^2)}}{2(1.25)}$$

$$= \frac{-(1.00 \text{ m}) \pm \sqrt{2.25 \text{ m}^2}}{2.50}$$

$$= \frac{-(1.00 \text{ m}) \pm (1.50 \text{ m})}{2.50}$$

$$= +0.200 \text{ m or } -1.00 \text{ m}$$

We want a positive value of x to correspond to a point to the right of Q_1, so the solution we want is $x = +0.200$ m. We conclude that point P is a distance $x = 0.200$ m to the right of charge Q_1 and a distance $D - x = 0.500 \text{ m} - 0.200 \text{ m} = 0.300$ m to the left of charge Q_2.

Reflect

Because charge Q_1 is smaller than Q_2, the location of the point where the net electric field is zero must be closer to Q_1 than to Q_2. That's just what we found. You can check the result $x = 0.200$ m by substituting this value into the above expressions for E_1 and E_2 and confirming that $E_1 = E_2$ for this value of x.

$\vec{E}_1$ $Q_1 = +4.00$ nC $Q_2 = +9.00$ nC $\vec{E}_2$ 1.00 m $D = 0.500$ m

But what's the significance of the second solution, $x = -1.00$ m? This refers to a point 1.00 m to the *left* of charge Q_1 and 1.00 m + 0.500 m = 1.50 m to the left of charge Q_2. Our calculation shows that at this point E_1 is equal to E_2. However, at this point $\vec{E}_1$ and $\vec{E}_2$ *both* point to the *left* (both fields point away from the positive charges that produce them). So at this point the electric fields $\vec{E}_1$ and $\vec{E}_2$ do *not* cancel, and the total field is not zero.

EXAMPLE 16-7 Field of an Electric Dipole

A combination of two point charges of the same magnitude but opposite signs is called an **electric dipole.** **Figure 16-13** shows an electric dipole made up of a point charge $+q$ and a point charge $-q$ separated by a distance $2d$. Derive expressions for the magnitude and direction of the net electric field due to these two charges at a point P a distance y along the midline of the dipole.

Figure 16-13 An electric dipole The field produced by an electric dipole at any point is the vector sum of the fields $\vec{E}_+$ and $\vec{E}_-$ caused by the positive charge $+q$ and the negative charge $-q$, respectively.

Set Up

The net field is the vector sum of the field $\vec{E}_+$ due to the charge $+q$ (which points away from $+q$) and the field $\vec{E}_-$ due to the charge $-q$ (which points toward $-q$). In Example 16-4 in Section 16-4 we used vector addition to find the net electric *force* exerted by two charges (one positive and one negative) on a third charge; here we use vector addition to find the net electric *field* due to the positive and negative charge. As in Example 16-4, we'll choose the positive x direction to be to the right and the positive y axis to be upward and add the two vectors using components.

Magnitude of the electric field due to a point charge Q:

$$E = \frac{k|Q|}{r^2} \qquad (16\text{-}4)$$

Total electric field:

$$\vec{E} = \vec{E}_1 + \vec{E}_2 \qquad (16\text{-}5)$$

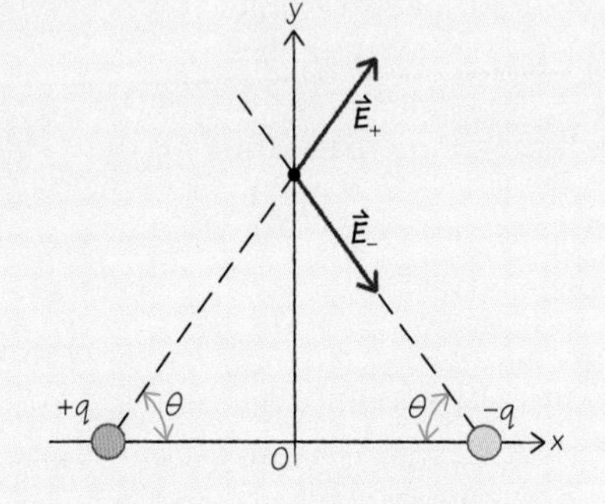

Solve

Use Equation 16-4 to find the magnitudes of the fields $\vec{E}_+$ and $\vec{E}_-$ at P.

The distance from $+q$ to point P is the same as the distance from $-q$ to P. Call this distance r:

$$r = \sqrt{y^2 + d^2}$$

Since $+q$ and $-q$ have the same magnitude (q, which is positive) and are the same distance from P, Equation 16-4 tells us that the fields that the two charges produce at P have the same magnitude:

$$E_+ = E_- = \frac{kq}{r^2} = \frac{kq}{y^2 + d^2}$$

Find the x and y components of $\vec{E}_+$ and $\vec{E}_-$, and use these to calculate the components of the net field at P.

The components of $\vec{E}_+$ and $\vec{E}_-$ are

$$E_{+,x} = E_+ \cos\theta$$
$$E_{+,y} = E_+ \sin\theta$$
$$E_{-,x} = E_- \cos\theta$$
$$E_{-,y} = -E_- \sin\theta$$

Since the magnitudes E_+ and E_- are equal, the components of the net field at P are

$$E_x = E_{+,x} + E_{-,x} = E_+ \cos\theta + E_- \cos\theta$$
$$= \frac{2kq}{y^2 + d^2} \cos\theta$$
$$E_y = E_{+,y} + E_{-,y} = E_+ \sin\theta + (-E_- \sin\theta)$$
$$= \frac{kq}{y^2 + d^2}\sin\theta - \frac{kq}{y^2 + d^2}\sin\theta = 0$$

From the figure

$$\cos\theta = \frac{d}{r} = \frac{d}{\sqrt{y^2 + d^2}}$$

So the components of the net electric field are

$$E_x = \frac{2kq}{(y^2 + d^2)}\frac{d}{\sqrt{y^2 + d^2}} = \frac{2kqd}{(y^2 + d^2)^{3/2}}$$

$$E_y = 0$$

The net electric field at point P is to the right and has magnitude $E = 2kqd/(y^2 + d^2)^{3/2}$.

Reflect

We can check our result by substituting $y = 0$ so that the point P is directly between the two charges and a distance d from each charge. Then $\vec{E}_+$ and $\vec{E}_-$ both point to the right, and the magnitude of the net electric field should be equal to the sum of the magnitudes of $\vec{E}_+$ and $\vec{E}_-$.

At $y = 0$, the net electric field has magnitude

$$E = \frac{2kqd}{(0 + d^2)^{3/2}} = \frac{2kqd}{d^3} = \frac{2kq}{d^2} = 2\left(\frac{kq}{d^2}\right)$$

This is just twice the magnitude of the field due to each individual charge:

$$E_+ = E_- = \frac{kq}{d^2}$$

Note that at points very far from the dipole, so that y is much greater than d, the magnitude of the field is inversely proportional to the *cube* of y: At double the distance, the field of a dipole is $(1/2)^3 = 1/8$ as great. This is a much more rapid decrease with distance than the field of a single point charge, for which E is inversely proportional to the *square* of the distance: At double the distance, the field of a point charge is $(1/2)^2 = 1/4$ as great. The dipole field decreases much more rapidly because the fields of $+q$ and $-q$ partially cancel each other.

If y is much greater than d, $y^2 + d^2$ is approximately equal to y^2. Then the magnitude of the net electric field due to the dipole is approximately

$$E_{\text{net}} = \frac{2kqd}{(y^2)^{3/2}} = \frac{2kqd}{y^3}$$

By using techniques like the ones we employed in Example 16-7, it's possible to calculate and map out the electric field at *all* points around an electric dipole. **Figure 16-14** shows the field lines. Note that as you move away from the dipole along its midline, the field lines become farther apart. This is a graphical way of showing that the magnitude of the field decreases with increasing distance, just as we found in Example 16-7.

GOT THE CONCEPT? 16-4 Electric Field I

The positive charge in the dipole shown in Figure 16-14 is attracted to the negative charge. To find the force of this attraction, the electric field $\vec{E}$ to use in Equation 16-2 is (a) the field due to the positive charge; (b) the field due to the negative charge; (c) the net field due to both charges; (d) none of these.

GOT THE CONCEPT? 16-5 Electric Field II

Suppose both of the charges in Figure 16-13 were negative and had the same magnitude. At point P in that figure, the net electric field due to these charges would (a) point to the left; (b) point to the right; (c) point straight up; (d) point straight down; (e) be zero.

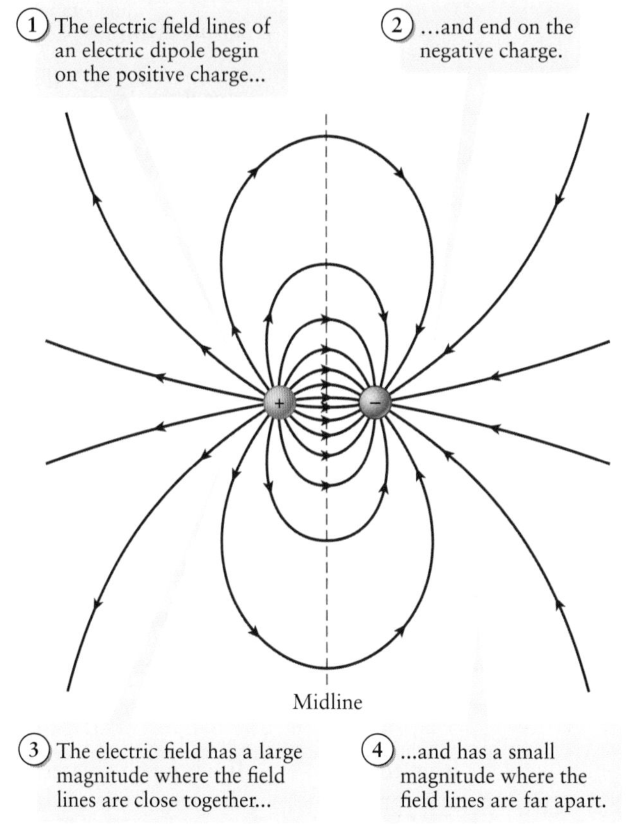

Figure 16-14 Field lines of an electric dipole At any point the electric field due to an electric dipole is the vector sum of the field due to the positive charge and the field due to the negative charge.

(1) The electric field lines of an electric dipole begin on the positive charge...

(2) ...and end on the negative charge.

Midline

(3) The electric field has a large magnitude where the field lines are close together...

(4) ...and has a small magnitude where the field lines are far apart.

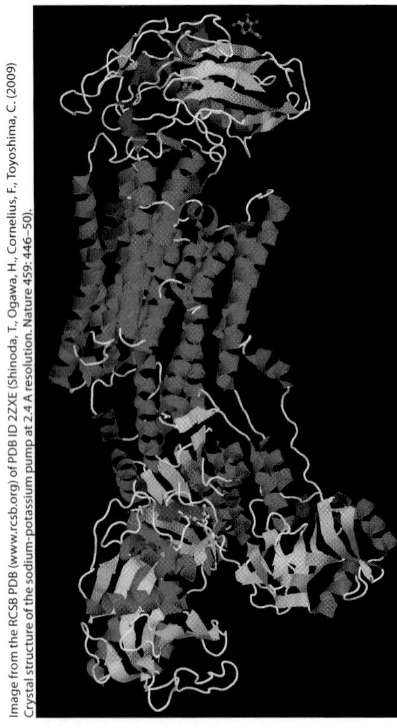

Figure 16-15 A protein molecule This protein molecule, a long chain of amino acids, coils on itself because different parts of the chain carry different amounts of electric charge. The electric forces between the parts pull the chain into the complex shape shown here. Understanding the details of protein folding requires knowing the electric field at any point due to the arrangement of charges along the chain. This very complicated problem can only be solved using a computer.

TAKE-HOME MESSAGE FOR Section 16-5

✔ Any charged object produces an electric field in the space around it. A second charged object responds to this electric field; this is the origin of the electric force that the first object exerts on the second.

✔ The electric field of a positive point charge points directly away from that charge; the electric field of a negative point charge points directly toward that charge.

✔ The net electric field due to two or more charges is the vector sum of the electric fields due to the individual charges.

16-6 Gauss's law gives us more insight into the electric field

Example 16-7 in the previous section shows how we can find the net electric field due to two point charges. In principle we can extend this approach to calculate the net electric field due to a collection of any number of point charges. If there are many such charges, however, the calculations can become very complex (**Figure 16-15**).

Happily there's an alternative and a much easier approach that we can use to find the electric field if the charges are arranged in a very *symmetrical* fashion—for example, uniformly distributed over a spherical volume. This approach uses a principle called *Gauss's law*, which is an alternative way to express Coulomb's law (Equation 16-1). We'll develop Gauss's law in this section and apply it to a variety of physical situations in the following section.

To understand Gauss's law, we first need to define a new quantity called *electric flux*. We'll do this by making an analogy between electric fields and the flow of water.

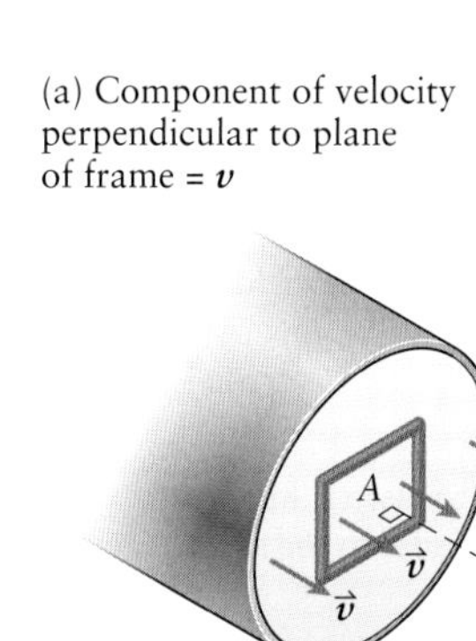

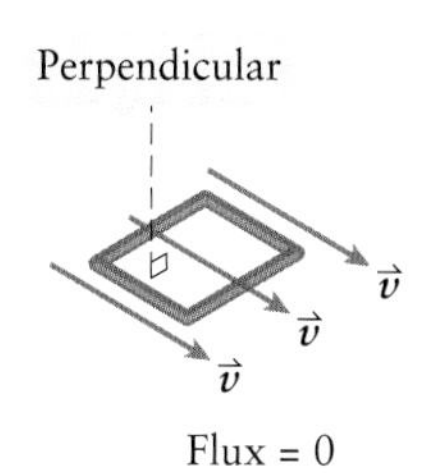

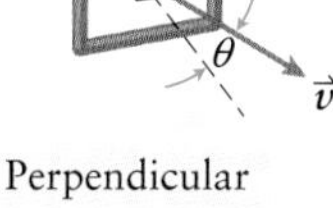

Figure 16-16 **Flux of water** The flux of water through a rectangular wire frame depends on the magnitude of the velocity $\vec{v}$ of the water, the area A of the wire frame, and the angle between the direction of $\vec{v}$ and the perpendicular to the plane of the frame.

Water Flux and Electric Flux

Figure 16-16 shows water flowing through a pipe. The vectors labeled $\vec{v}$ in Figure 16-16a represent the velocity of the water at each point. For simplicity we've assumed that the water velocity is the same everywhere. Now imagine that we place a rectangular wire frame of area A in the flow. We define the *flux* of water through this wire frame as the product of (i) the area A of the frame and (ii) the component of flow velocity $\vec{v}$ that's perpendicular to the plane of the frame. If we orient the frame so that it's face-on to the water flow as in Figure 16-16a, then $\vec{v}$ is perpendicular to the plane of the frame and the perpendicular component is just v. The flux is then equal to Av. If instead we orient the frame so that it is edge-on to the flow as in Figure 16-16b, the flow velocity $\vec{v}$ has no component perpendicular to the plane of the frame. In this case the flux is zero. If we orient the frame so that a line perpendicular to the frame is at an angle θ to the direction of $\vec{v}$ as in Figure 16-16c, the perpendicular component of $\vec{v}$ is $v \cos \theta$ and the flux is $A(v \cos \theta) = Av \cos \theta$.

Note that we've actually encountered the concept of flux before. In Section 11-9 we learned that if an incompressible fluid is in steady flow through a pipe of varying cross-sectional area A, the product of the area and the flow speed v has the same value at any two points 1 and 2 along the pipe: $A_1v_1 = A_2v_2$ (Equation 11-19). In the language we've just introduced, this says that the *flux* of the fluid through the entire pipe maintains the same value even if the cross-sectional area of the pipe changes.

Figure 16-17 shows how we extend the idea of flux to the electric field. Instead of a pipe carrying a fluid, let's look at a region of space where there is an electric field $\vec{E}$. We saw in Section 16-5 that the value of $\vec{E}$ can vary from point to point, so we consider a small enough region that we can treat $\vec{E}$ as having essentially the same value over that region. We then imagine a small rectangular area A that we can orient however we like. By analogy to the flux of water in Figure 16-16,

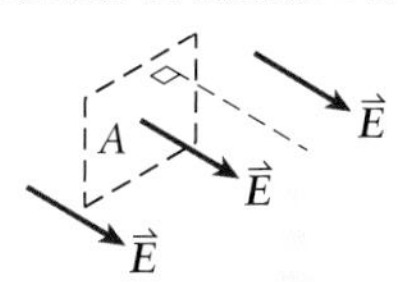

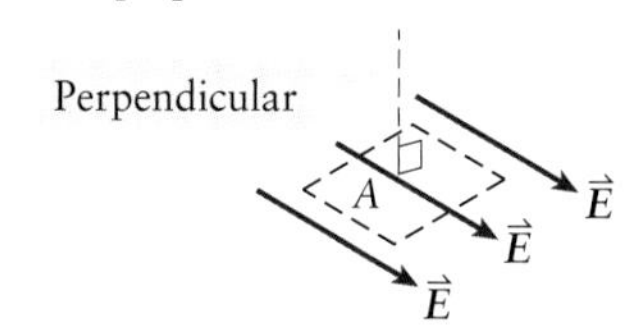

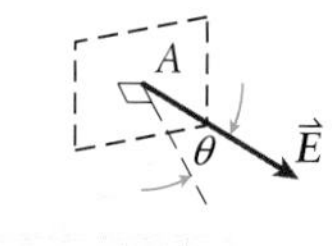

Figure 16-17 **Electric flux** The electric flux through a small rectangular surface depends on the magnitude of the electric field $\vec{E}$, the area A of the surface, and the angle between the direction of $\vec{E}$ and the perpendicular to the plane of the surface. (Compare Figure 16-16.)

we define the **electric flux** Φ (the uppercase Greek letter phi) through the area A as follows:

Electric flux (16-6)

Area of the surface | Magnitude of the electric field

Electric flux through a surface

$$\Phi = AE_{\perp} = AE\cos\theta$$

The component of electric field perpendicular to the surface | Angle between the electric field and the perpendicular to the surface

In Figure 16-17a we orient the area so that it is face-on to the electric field $\vec{E}$. Then the electric field is perpendicular to the surface and the angle $\theta = 0$. Therefore $E_{\perp}$ is equal to the field magnitude E and the electric flux through the surface is $\Phi = AE_{\perp} = AE$; alternatively, since $\theta = 0$, $\Phi = AE\cos 0 = AE(1) = AE$. In Figure 16-17b the area is edge-on to the electric field, so the electric field has zero component perpendicular to the area A. Then $E_{\perp} = 0$ and so $\Phi = AE_{\perp} = 0$. Equivalently, with this orientation $\theta = 90°$ and so $\Phi = AE\cos 90° = AE\,(0) = 0$. Finally, in Figure 16-17c the area is oriented at an angle θ between 0 and 90°. The component of the electric field perpendicular to the area is $E_{\perp} = E\cos\theta$, and the electric flux $\Phi = AE\cos\theta$ has a value between AE and zero.

Note that unlike the flow of water in Figure 16-16, there's nothing "flowing" through the area A in Figure 16-17. Unlike velocity $\vec{v}$ the electric field $\vec{E}$ does *not* represent motion of any kind. But we can still use the analogy between $\vec{v}$ and $\vec{E}$ as expressed by Figures 16-16 and 16-17 to define electric flux Φ in the manner given by Equation 16-6.

WATCH OUT! In calculating electric flux, the area is "imaginary."

In Figure 16-16 we measured the flux of water through an area outlined by a real, physical wire frame. But we really didn't need the frame: It was simply there to help us visualize the area in question. In defining electric flux as in Figure 16-17, we've done away with the wire frame entirely. You can think of the area A in Figure 16-17 and Equation 16-6 as "imaginary" in the sense that there's no physical object outlining the area.

Electric Flux Through a Closed Surface: Gauss's Law

Why is the idea of electric flux through a surface a useful one? To see the answer let's consider a *closed* surface—that is, one that encloses a volume. For example, let's find the electric flux through a spherical surface of radius r that is centered on and encloses a positive point charge q (**Figure 16-18a**). From Equation 16-4 the electric field $\vec{E}$ due to the point charge has the same magnitude $E = kq/r^2$ at every point on the surface because every point is the same distance r from the point charge. However, $\vec{E}$ points in different directions at different points on the spherical surface: straight upward at the top of the surface, to the left at the leftmost point on the surface, and so on. But if we look at a very small rectangular portion of the surface, then $\vec{E}$ points in essentially the same direction at every point on that rectangle. In fact, $\vec{E}$ is perpendicular to the rectangle of area ΔA, so $\theta = 0$ in Equation 16-6 and the flux through that rectangle is $(\Delta A)E = (\Delta A)(kq/r^2)$. If we now imagine that the entire spherical surface is made up of a very large number of such rectangles of area $\Delta A_1, \Delta A_2, \Delta A_3, \ldots, \Delta A_N$, the *total* electric flux through the spherical surface as a whole is just the sum of the fluxes through the individual rectangles:

$$\Phi = (\Delta A_1)\left(\frac{kq}{r^2}\right) + (\Delta A_2)\left(\frac{kq}{r^2}\right) + (\Delta A_3)\left(\frac{kq}{r^2}\right) + \cdots + (\Delta A_N)\left(\frac{kq}{r^2}\right)$$

(16-7)
$$= (\Delta A_1 + \Delta A_2 + \Delta A_3 + \cdots + \Delta A_N)\left(\frac{kq}{r^2}\right)$$

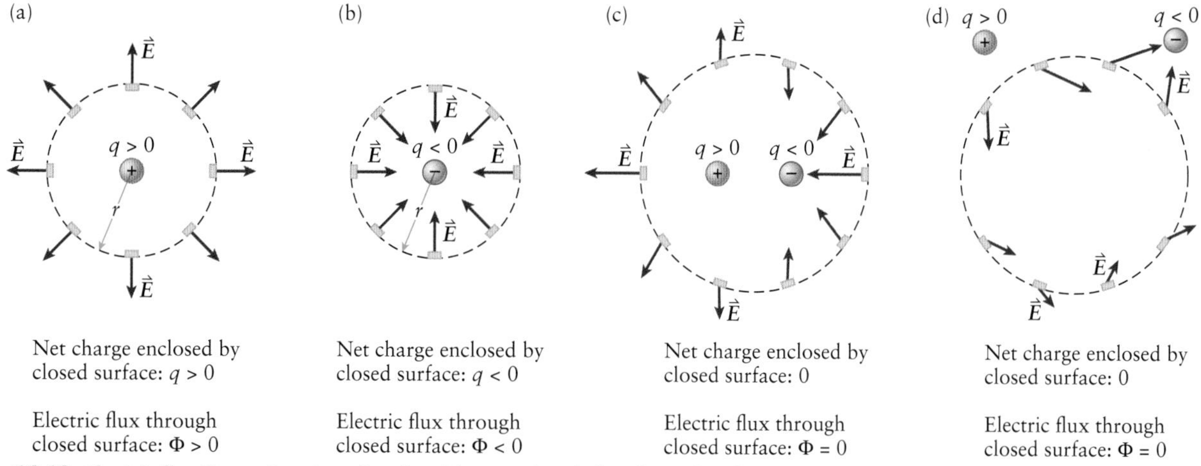

Figure 16-18 Electric flux through a closed surface The net electric flux through a closed spherical surface is (a) positive if a positive charge is enclosed by the surface, (b) negative if a negative charge is enclosed by the surface, (c) zero if equal amounts of positive and negative charge are enclosed by the surface, and (d) zero if no charge at all is enclosed by the surface.

In Equation 16-7 the quantity $\Delta A_1 + \Delta A_2 + \Delta A_3 + \cdots + \Delta A_N$ is the sum of the areas of all of the individual rectangles that make up the spherical surface, and so is equal to the total area A of the surface. The surface area of a sphere of radius r is $A = 4\pi r^2$, so Equation 16-7 becomes

See the Math Tutorial for more information on the surface area of geometric shapes.

$$\Phi = (4\pi r^2)\left(\frac{kq}{r^2}\right) = 4\pi kq \qquad (16\text{-}8)$$

Notice that the radius of the spherical surface surrounding the point charge q cancels out in this equation: The magnitude of the electric field decreases as the square of the sphere's radius r and the surface area increases at the same rate. So their product—the electric flux Φ—does not depend on the radius and is directly proportional to the charge q enclosed within the spherical surface.

Equation 16-8 also holds true if the charge q is negative (**Figure 16-18b**): The electric flux through the closed surface is *negative* in this case. The interpretation is that the flux is positive if the electric field points out of the closed surface, as in Figure 16-18a, but negative if the electric field points into the closed surface, as in Figure 16-18b.

What happens if we replace the single charge q with an electric dipole like that shown in Figure 16-14, with both a positive charge and a negative charge of equal magnitude? In **Figure 16-18c** the spherical surface is centered on the dipole and encloses both the positive and negative charges, so the *net* enclosed charge is zero. The figure shows that for each small rectangle on the surface where the electric field $\vec{E}$ points outward, giving a positive contribution to the flux, there is another small rectangle elsewhere on the surface where $\vec{E}$ points inward, giving an equally large negative contribution to the flux. These positive and negative contributions to the flux cancel, so the net electric flux through this surface—which encloses zero net charge—is zero. The same cancellation also happens for the spherical surface in **Figure 16-18d**, which does not enclose either charge. In this case as well, the net enclosed charge is zero and the net electric flux through the surface is zero.

Here's what we've concluded so far from Figure 16-18:

- Closed surfaces that enclose a point charge q, as in Figure 16-18a and Figure 16-18b, have a nonzero electric flux through them. This flux is proportional to the enclosed charge q, as in Equation 16-8.
- Closed surfaces that enclose zero net charge—either equal amounts of positive and negative charge, as in Figure 16-18c, or no charge at all, as in Figure 16-18d—have zero net electric flux through them. This is consistent with Equation 16-8, but with $q = 0$.

GOT THE CONCEPT? 16-6
Electric Flux I

? When a certain charged particle is placed at the center of a sphere, the net electric flux through the surface of the sphere is Φ_0. If the radius of the sphere is doubled, what would be the new flux through the sphere? (a) $\Phi_0/4$; (b) $\Phi_0/2$; (c) Φ_0; (d) $2\Phi_0$; (e) $4\Phi_0$.

We were able to draw these conclusions because the closed surfaces in the figure are spherical in shape and placed symmetrically with respect to the charges inside them. But it can be shown that the same conclusions hold true for a closed surface of *any* shape or placement. We can summarize these conclusions by rewriting Equation 16-8 with q replaced by q_{encl}, which represents the *net* charge enclosed by the closed surface:

Gauss's law (16-9)

$$\Phi = \frac{q_{encl}}{\varepsilon_0}$$

Electric flux through a closed surface

Net amount of charge enclosed within the surface

Permittivity of free space = $1/(4\pi k)$

This relationship is called **Gauss's law** after the scientist who first deduced it, the nineteenth-century German mathematician and physicist Carl Friedrich Gauss. In honor of Gauss a closed surface used to enclose charge in order to apply Gauss's law is referred to as a **Gaussian surface**. Equation 16-9 holds true for *any* Gaussian surface (**Figure 16-19**), no matter what its shape or size and no matter where the charges are located inside the surface. As we mentioned previously such surfaces are imaginary: The surface does not need to be made of any physical substance.

GOT THE CONCEPT? 16-7 Electric Flux II

When a certain charged particle is placed at the center of a sphere, the net electric flux through the surface of the sphere is Φ_0. If the sphere were elongated into a spheroid (such as a rugby ball or an American football) but still enclosed the same charge, what would be the new electric flux through the surface? (a) Less than Φ_0; (b) Φ_0; (c) more than Φ_0; (d) not enough information given to decide.

In Equation 16-9 we have replaced the combination $4\pi k$ that appears in Equation 16-8 with $1/\varepsilon_0$. Here ε_0, called the **permittivity of free space** for historic reasons, is equal to $1/(4\pi k) = 8.85 \times 10^{-12}\ C^2/(N \cdot m^2)$ to three significant figures. So we can state Gauss's law as

The net electric flux through any closed surface equals the net charge enclosed by that surface divided by the permittivity ε_0. Charges outside the surface have no effect on the net electric flux through the surface.

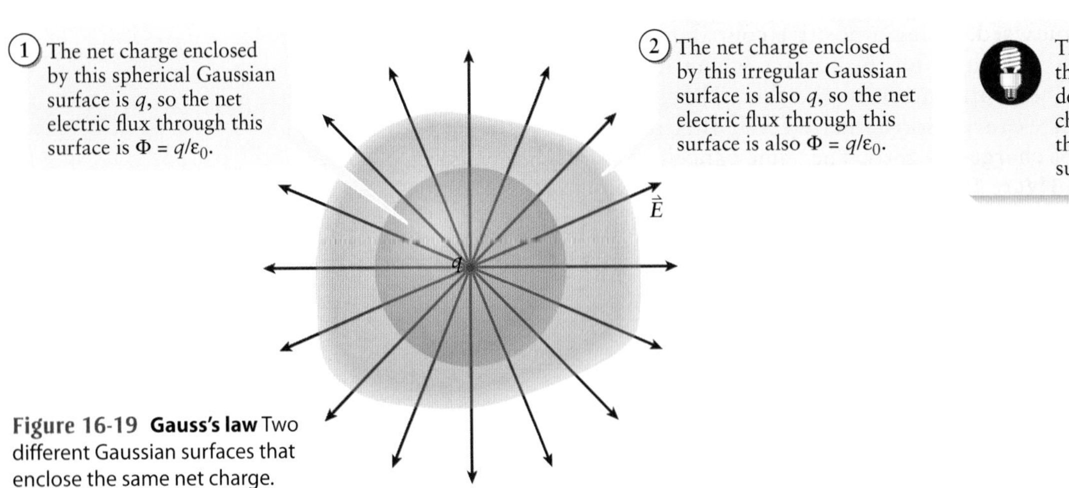

1 The net charge enclosed by this spherical Gaussian surface is q, so the net electric flux through this surface is $\Phi = q/\varepsilon_0$.
2 The net charge enclosed by this irregular Gaussian surface is also q, so the net electric flux through this surface is also $\Phi = q/\varepsilon_0$.
q
$\vec{E}$
The net electric flux through a Gaussian surface depends only on the net charge that it encloses, not the shape or size of the surface.

Figure 16-19 Gauss's law Two different Gaussian surfaces that enclose the same net charge.

This law holds true because of the $1/r^2$ character of the electric field due to a point charge (see the discussion following Equation 16-8).

We'll see in the following section how we can use Gauss's law to determine the electric field due to certain charge distributions and what Gauss's law tells us about how charges distribute themselves in a conductor.

GOT THE CONCEPT? 16-8 Electric Flux III

When a certain charged particle is placed at the center of a sphere, the net electric flux through the surface of the sphere is Φ_0. If a second, identical charged particle is placed outside the sphere, what would be the new electric flux through the surface? (a) Less than Φ_0; (b) Φ_0; (c) more than Φ_0; (d) not enough information given to decide.

TAKE-HOME MESSAGE FOR Section 16-6

✔ The electric flux through a surface is analogous to the flux of water through a wire frame. Electric flux depends on the magnitude of the electric field, the area of the surface, and the relative orientation of the field and the surface.

✔ For a closed surface (one that encloses a volume) an electric field $\vec{E}$ that points out of the closed surface makes a positive contribution to the electric flux. An electric field that points into the closed surface makes a negative contribution to the electric flux.

✔ Gauss's law states that the net electric flux through a closed surface is proportional to the amount of charge enclosed by the surface. If the net enclosed charge is zero, the net electric flux through the surface is zero.

16-7 In certain situations Gauss's law helps us to calculate the electric field and to determine how charge is distributed

Gauss's law tells us the value of the net electric flux through the closed Gaussian surfaces shown in Figures 16-18c and 16-18d. But that isn't enough to tell us the value of $\vec{E}$ at any *individual* point on these surfaces. Yet we *can* use Gauss's law to determine $\vec{E}$ in cases where the charge that produces the field is distributed in a particularly symmetric way. Let's look at an example.

Electric Field of a Spherical Charge Distribution

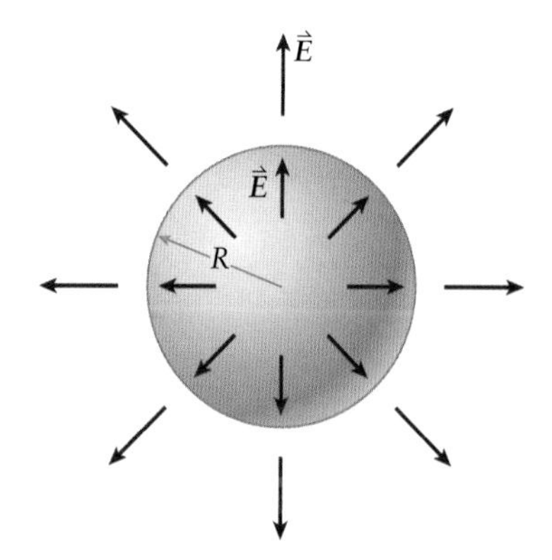

Charge Q is uniformly distributed throughout the volume of this sphere of radius R.

Figure 16-20 A uniformly charged sphere The spherical symmetry of this charge distribution tells us that the electric field $\vec{E}$ that it produces must be radial and that the magnitude of $\vec{E}$ can depend only on the distance from the center of the sphere.

In **Figure 16-20** charge Q is distributed uniformly throughout a spherical volume of radius R. (This could be a model of how electric charge is distributed over the volume of an atomic nucleus.) This distribution is *spherically symmetric*: You can rotate the sphere through any angle around its center and it looks exactly the same. Hence the electric field $\vec{E}$ caused by the charge distribution must also be spherically symmetric. This implies that the direction of the electric field must be either radially inward or radially outward, like the sea urchin spines shown in Figure 16-12. (If the field pointed in any other direction, the field lines would look different after rotating the sphere through some angle. The spherical symmetry says that's impossible.) So at any point $\vec{E}$ can have only a radial component E_r, which points either directly away from the center of the sphere ($E_r > 0$, as in Figure 16-20) or directly toward the center of the sphere ($E_r < 0$). This must be true for points inside the sphere as well as outside the sphere.

The spherical symmetry of the charge distribution in Figure 16-20 tells us something more: The value of E_r at a given point depends only on the radial distance r from the center of the sphere and not on the point's location around the sphere.

Given what spherical symmetry tells us about the electric field for the situation in Figure 16-20, what more can we learn by using Gauss's law? In **Figure 16-21a** we've

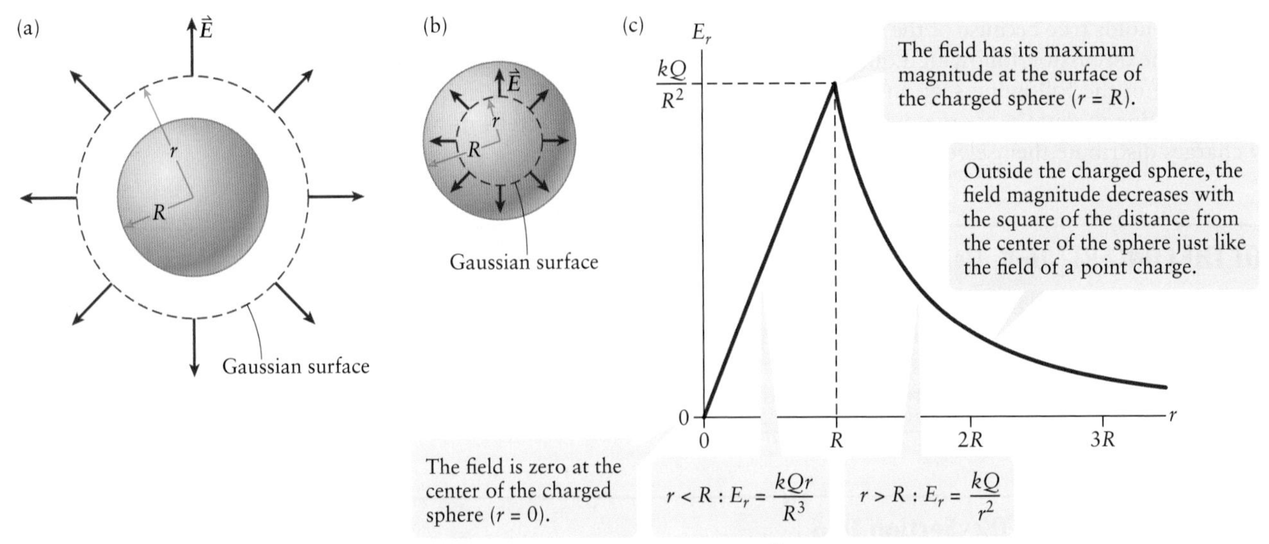

Figure 16-21 Finding the electric field of a uniformly charged sphere To find how $\vec{E}$ depends on distance from the center of the sphere of charge Q, we use (a) one Gaussian surface that encloses the entire sphere and (b) a second Gaussian surface that encloses only part of the sphere. (c) The radial electric field E_r as a function of the distance r from the center, graphed for the case $Q > 0$.

drawn a spherical Gaussian surface that's centered on the sphere and that encloses the entire sphere. This Gaussian surface has radius $r > R$ and surface area $A = 4\pi r^2$. Just like the spherical surface in Figure 16-18a or Figure 16-18b, the electric field is perpendicular to the surface at every point and has the same magnitude at every point. Using the same reasoning that we used for Equation 16-8, the net electric flux through this surface is just the area of the surface multiplied by the radial electric field:

$$\Phi = AE_r = 4\pi r^2 E_r \tag{16-10}$$

This flux is positive if the electric field points outward (so $E_r > 0$) and negative if the electric field points inward ($E_r < 0$). Since the entire charged sphere is enclosed within the Gaussian surface, the enclosed charge is $q_{\text{encl}} = Q$. Gauss's law, Equation 16-9, tell us that the flux is also equal to $q_{\text{encl}}/\varepsilon_0$. Equating Equations 16-9 and 16-10 yields

$$4\pi r^2 E_r = \frac{Q}{\varepsilon_0} \quad \text{and} \quad E_r = \frac{Q}{4\pi\varepsilon_0 r^2} = \frac{kQ}{r^2} \quad (r > R) \tag{16-11}$$

Equation 16-11 says that at points outside the sphere, the electric field is *exactly the same* as the field due to a point charge Q (Equation 16-4). The field is inversely proportional to the square of the distance r and is proportional to the charge Q on the sphere. The field points radially outward ($E_r > 0$) if the charge Q is positive and points radially inward ($E_r < 0$) if Q is negative. (Compare Figure 16-11.)

We can also use Gauss's law to tell us about the electric field at points *inside* the sphere. In **Figure 16-21b** we've drawn another spherical Gaussian surface of radius r and surface area $A = 4\pi r^2$ but with a radius r that's less than the radius R of the charged sphere. Again the net electric flux through this Gaussian surface is given by Equation 16-10. But now the enclosed charge is less than the total charge Q on the sphere. Since the charge is distributed uniformly, the fraction of charge that's enclosed by the Gaussian surface is just equal to the ratio of the volume within the Gaussian surface (a sphere of radius r, with volume $4\pi r^3/3$) to the total volume occupied by the charge Q (a sphere of radius R, with volume $4\pi R^3/3$). So Gauss's law applied to the Gaussian surface in Figure 16-21b tells us

$$4\pi r^2 E_r = \frac{q_{\text{encl}}}{\varepsilon_0} = \frac{1}{\varepsilon_0}\left[Q\left(\frac{4\pi r^3/3}{4\pi R^3/3}\right)\right] = \frac{Qr^3}{\varepsilon_0 R^3}$$

If we divide through by $4\pi r^2$, we get

$$E_r = \frac{Qr^3}{4\pi\varepsilon_0 r^2 R^3} = \frac{Qr}{4\pi\varepsilon_0 R^3} = \frac{kQr}{R^3} \quad (r < R) \tag{16-12}$$

Equation 16-12 tells us that the direction of the electric field is the same inside the charged sphere as outside: $E_r > 0$ (an outward field) if Q is positive and $E_r < 0$ (an inward field) if Q is negative. It also says that inside the sphere the magnitude of the field *increases* with increasing distance r from the center of the sphere. At the surface of the charged sphere, $r = R$, both Equation 16-11 and Equation 16-12 give the same result: $E_r = kQ/R^2$. At the very center of the sphere ($r = 0$), the electric field is zero. **Figure 16-21c** shows a graph of E_r as a function of r for the case of a positively charged sphere ($Q > 0$).

These conclusions about the electric field of a charged sphere would have been *very* difficult to obtain without using Gauss's law. (The alternative approach is to divide the charged sphere into a very large number of small segments, treating each segment as an individual point charge, and using vector addition to add the individual electric fields produced by all of the segments. That would take a lot of strenuous mathematics.) The relative ease with which we came to these conclusions shows the power of Gauss's law.

Electric Field of a Large, Flat, Charged Disk

Figure 16-22a shows a charge distribution with a different kind of symmetry: a uniformly charged plate or disk. (We'll see later that charged disks of this kind are found in an important device called a *capacitor*, used in many electric circuits.) This charge distribution has *rotational symmetry*, which means that it looks the same if you rotate it through any angle around an axis that passes vertically through the center of the disk. However, the electric field $\vec{E}$ due to this charge distribution can (and does) vary in a complicated way as you move from the center of the disk toward the edges.

To simplify the problem let's just consider what the electric field is like at points around the disk near the disk center. We imagine that the edges of the disk are so far away that we can regard them as being infinitely distant. Then our problem is that of finding the electric field due to an *infinite sheet* of charge (**Figure 16-22b**).

We'll use the symbol σ (the lowercase Greek letter "sigma") for the amount of charge per unit area on the sheet, also called the **surface charge density**. The sheet is uniformly charged, so σ has the same value everywhere on the sheet. The units of σ are coulombs per square meter (C/m^2).

Our infinite sheet of charge has *translational* symmetry: No matter which way or how far you move parallel to the disk, the charge distribution looks exactly the same. The same must therefore be true of the electric field produced by the charge distribution. So if the disk lies in the xy plane, the electric field $\vec{E}$ at any point cannot depend on the x or y coordinate of that point. The field can depend only on the z coordinate, which is the coordinate measured perpendicular to the plane in Figure 16-22.

Translational symmetry also tells us that $\vec{E}$ must be *perpendicular* to the plane of the sheet. This means that a point charge q placed in that field will feel a force $\vec{F} = q\vec{E}$ either directly toward or directly away from the sheet, depending on the sign of q. (If there were a component of $\vec{E}$ in some direction parallel to the plane of the sheet, a positive point charge would be pushed in that direction. This would only be the case if the point charge were repelled or attracted by one part of the sheet. But since the charge distribution is the same everywhere on the sheet, no part attracts or repels the positive charge more than any other part. So there can't be any component of $\vec{E}$ parallel to the sheet.) Thus $\vec{E}$ can depend only on the z coordinate and can have only a z component.

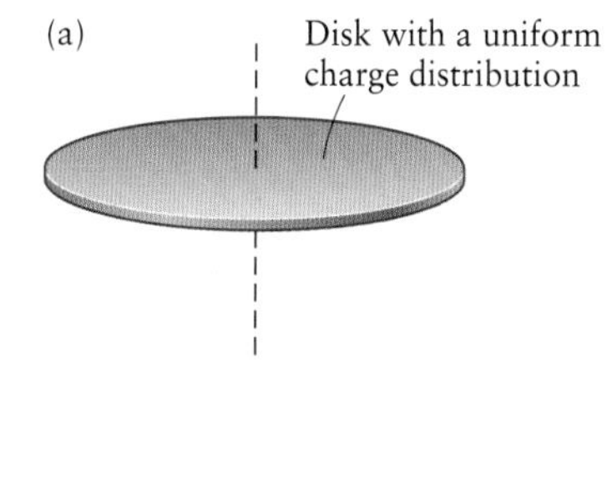

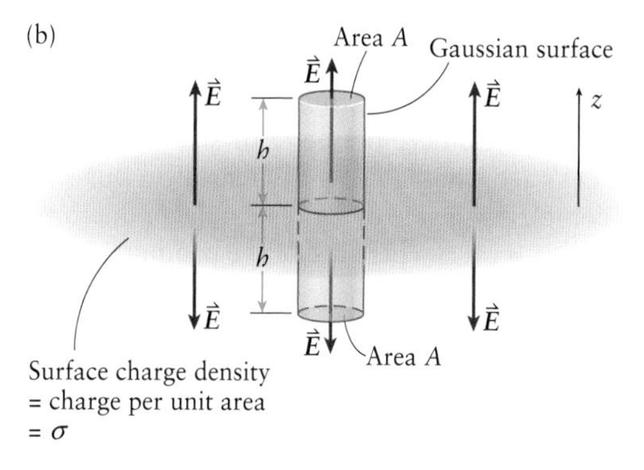

Figure 16-22 Finding the electric field of a uniformly charged disk (a) A uniformly charged disk; (b) to find the electric field of the disk at points close to the center of the disk, we can replace the disk by an infinitely large charged sheet.

The infinite sheet also has *reflection* symmetry: It looks the same if we flip the sheet upside down. The same must be true of the electric field. So if the field points upward above the sheet, it must point downward below the sheet, as we've indicated in Figure 16-22b.

To find the electric field at a certain distance h from the plane of the sheet, we'll use a cylindrical Gaussian surface as in Figure 16-22b. The top and bottom faces of the cylinder each have area A, and each is the same distance h from the sheet. The symmetries we have described tell us that the electric field $\vec{E}$ is perpendicular to the top and bottom faces and has the same value everywhere on each of these faces. Because $\vec{E}$ is parallel to the sides of the cylinder, there is *zero* electric flux through the sides: $\Phi_{\text{sides}} = 0$. The flux through the top and bottom faces of the cylinder, however, is not zero. If E_z is the z component of electric field a distance h above the infinite sheet, the flux through the top face of the cylinder is $\Phi_{\text{top}} = AE_z$. Note that this flux is positive if E_z is positive, which means that the electric field above the sheet is upward and points out of the Gaussian surface; this flux is negative if E_z is negative, which means that the electric field above the sheet is downward and points into the Gaussian surface. The reflection symmetry tells us that the flux through the bottom face is the same, so $\Phi_{\text{top}} + \Phi_{\text{bottom}} = 2\Phi_{\text{top}} = 2AE_z$. The net electric flux through the Gaussian surface in Figure 16-22b is therefore

$$\Phi = \Phi_{\text{sides}} + \Phi_{\text{top}} + \Phi_{\text{bottom}} = 0 + 2AE_z = 2AE_z \tag{16-13}$$

The area of the sheet enclosed within the Gaussian surface is A, so the amount of charge enclosed within the surface is the charge per unit area σ multiplied by the area A: $q_{\text{encl}} = \sigma A$. Using this and Equation 16-13 in Gauss's law, Equation 16-9, gives us an expression for the electric field component E_z a distance h above the sheet:

$$2AE_z = \frac{\sigma A}{\varepsilon_0} \quad \text{or} \quad E_z = \frac{\sigma}{2\varepsilon_0} \tag{16-14}$$

Equation 16-14 tells us that E_z has the same sign as the surface charge density σ. So the electric field $\vec{E}$ above the sheet points upward (away from the sheet) if the surface charge density is positive ($\sigma > 0$); $\vec{E}$ points downward (toward the sheet) if the surface charge density is negative ($\sigma < 0$). This agrees with the idea that electric fields point away from positive charges and toward negative charges.

Equation 16-14 also tells us that the value of E_z at a point a distance h above the sheet does *not* depend on h. (Note that h doesn't appear anywhere in this equation.) So for an infinite sheet of charge, the electric field is the same at all distances from the sheet.

These results are only approximate, since there's no such thing as a truly infinite sheet of charge. But they are valid for a charged disk at points that are relatively close to the disk, so the height h above the disk is small compared to the radius of the disk. We'll make use of Equation 16-14 in later chapters.

Excess Charge on Conductors

Gauss's law leads to a remarkable conclusion about a conductor to which we add excess charge so that the conductor has a net nonzero charge. Charges are free to move in a conductor, so if we add excess charges they will move in the conductor until they come to rest in equilibrium so that the net force on each added charge is zero. Because the electric force on a charge q is directly proportional to the electric field $\vec{E}$ (Equation 16-2), $\vec{E}$ inside the conductor must be *zero*. If it were not, excess charges inside the conductor would experience a force and be pushed to some new location. Note that this statement only applies *inside* the volume of the conductor. Outside the charged block of conductor the electric field need not be zero.

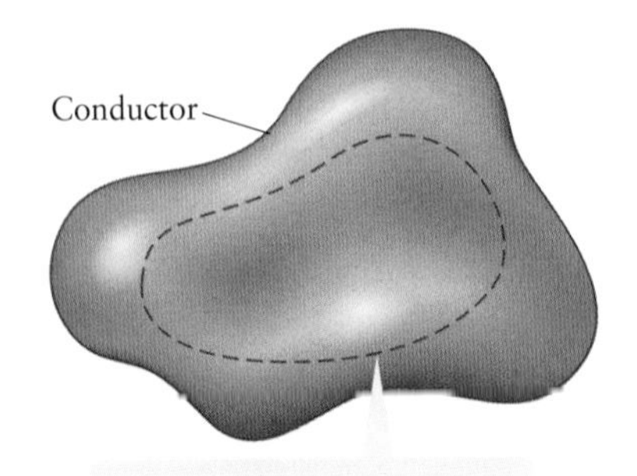

Figure 16-23 A charged conductor Gauss's law tells us that any excess charge on a conductor must reside on the surface of the conductor.

Let's see what Gauss's law tells us in this situation. Imagine a Gaussian surface that lies completely inside the volume of a conductor that carries excess charge (**Figure 16-23**). Since $\vec{E} = 0$ everywhere inside the conductor, the net electric flux Φ through this surface is zero. From Gauss's law, Equation 16-9, Φ is equal to the net charge q_{encl} enclosed by the surface divided by ε_0. So the net charge inside the Gaussian surface is zero. This holds true for any Gaussian surface, no matter how small, that lies entirely within the conductor. So there can be *no* excess charge within the volume of the conductor. Instead, *all of the excess charge on a conductor in equilibrium must reside on the surface.* That's why we depicted the charged copper rod in Figure 16-5b with its excess charge spread over the surface of the rod, not its interior.

Our conclusion that $\vec{E} = 0$ inside a conductor holds true *only* if all of the charges are at rest. We will see in later chapters that if an electric current is present inside a conductor, there must be a nonzero electric field to sustain the flow of charge.

EXAMPLE 16-8 Gauss's Law and Surface Charge

A conducting sphere of radius R_1 carries excess charge +7 C. This sphere is enclosed within a concentric spherical conducting shell of inner radius R_2 and outer radius R_3. The conducting shell carries a net charge +2 C. How much charge is on the inner surface of the conducting shell? How much is on the outer surface?

Set Up

In equilibrium the electric field inside the volume of the conducting shell must be zero. So if we imagine a spherical Gaussian surface that lies inside the shell, so that its radius r is intermediate between R_2 and R_3, the net electric flux through that surface will be zero. From Gauss's law, that means that the net charge enclosed by the Gaussian surface must be zero. The +2-C charge on the shell will arrange itself on the surfaces of the shell in order to make that happen.

Gauss's law:

$$\Phi = \frac{q_{\text{encl}}}{\varepsilon_0} \qquad (16\text{-}9)$$

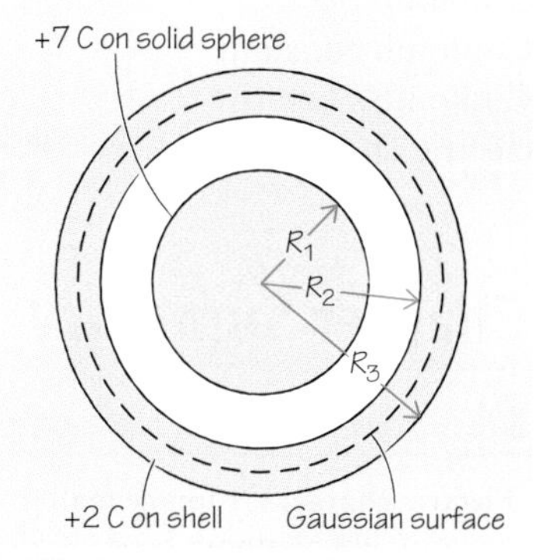

Solve

The charge enclosed within our Gaussian surface includes the charge of the central sphere ($q_{\text{sphere}} = +7$ C) and whatever charge q_{inner} is present on the inner surface of the shell. These charges must add to zero.

For the Gaussian surface

$$q_{\text{encl}} = q_{\text{sphere}} + q_{\text{inner}} = 0$$

So the charge on the inner surface of the shell equals the negative of the charge on the central sphere:

$$q_{\text{inner}} = -q_{\text{sphere}} = -7\text{ C}$$

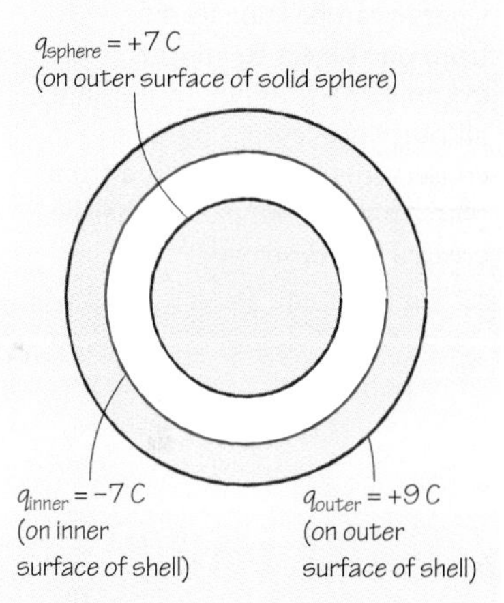

The total charge on the shell is $q_{\text{shell}} = +2$ C. Because all of the excess charge on a conductor resides on its surfaces, q_{shell} is the sum of the charge on the inner and outer surfaces.

$$q_{\text{shell}} = q_{\text{inner}} + q_{\text{outer}}$$

So

$$\begin{aligned} q_{\text{outer}} &= q_{\text{shell}} - q_{\text{inner}} \\ &= (+2\text{ C}) - (-7\text{ C}) \\ &= +9\text{ C} \end{aligned}$$

Reflect

While the shell as a whole carries a *positive* excess charge of +2 C, the inner surface of the shell has a *negative* charge. Here's why: When the positively charged sphere is placed inside the shell, some of the free electrons in the shell are drawn to the shell's inner surface, giving it a negative charge of −7 C. This leaves a deficit of electrons—a positive charge of +7 C—on the outside surface, which adds to the +2 C of excess charge to give the outer surface of the shell a charge of +9 C.

GOT THE CONCEPT? 16-9 Size of a Gaussian Surface

? In our derivation of Equation 16-13 for the electric field due to an infinite charged sheet, we used a Gaussian surface in the shape of a cylinder whose top and bottom faces had area A. If we used a larger cylinder with top and bottom faces of area $2A$, how would this change our result for the electric field? (a) E_z would be four times larger; (b) E_z would be twice as large; (c) E_z would be the same; (d) E_z would be one-half as great; (e) E_z would be one-fourth as great.

TAKE-HOME MESSAGE FOR Section 16-7

✔ Gauss's law can be used to calculate the electric field in certain cases where the charge distribution has a high degree of symmetry. These include a spherical distribution of charge and a very large, uniformly charged sheet.

✔ Gauss's law tells us that when excess charge added to a conductor is allowed to come to rest in equilibrium, the excess charge resides only on the surfaces of the conductor.

Key Terms

charged
conductor
coulomb
Coulomb constant
Coulomb's law
electric charge
electric dipole
electric field lines
electric flux
electric force
insulators
Gaussian surface
Gauss's law
neutral
permittivity of free space
point charges
semiconductors
surface charge density

Chapter Summary

Topic	Equation or Figure
Electric charge: Ordinary matter contains equal amounts of positive and negative charge. Charge can be transferred from one object to another, for example, by rubbing. In this and all other processes, charge is conserved: It can be moved from place to place but can neither be created nor destroyed.	(b) 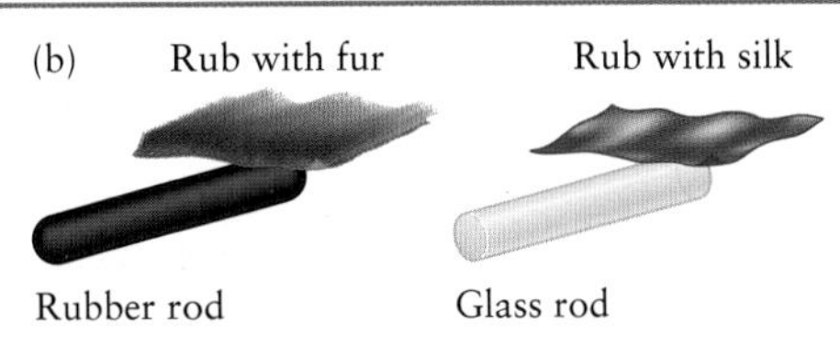  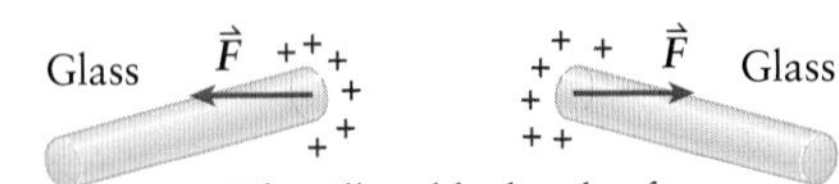 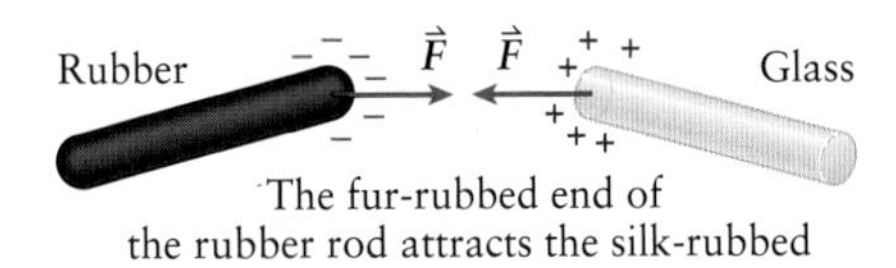 (Figure 16-2b)
Conductors, insulators, and semiconductors: In a conductor charges are free to move with ease; in an insulator charges can move very little. Semiconductors have properties intermediate between those of conductors and insulators.	(b) 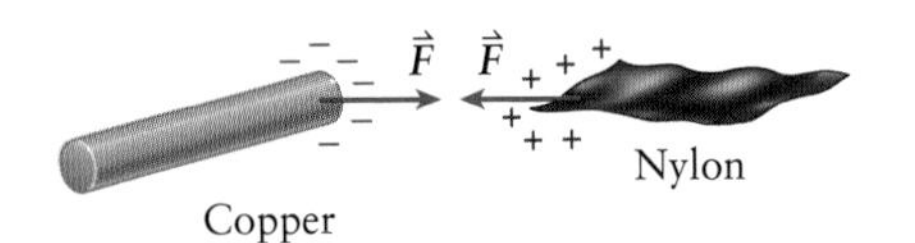 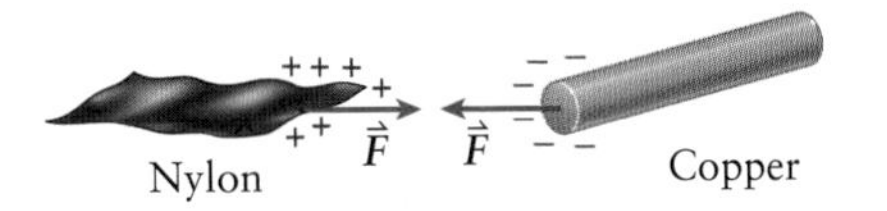 (Figure 16-5b)

Coulomb's law: The electric forces that two point charges exert on each other are proportional to the magnitudes of the charges and inversely proportional to the square of the distance between the charges. These forces are attractive if the two charges have opposite signs, and repulsive if the two charges have the same sign.

Any two point charges q_1 and q_2 exert equally strong **electric forces** on each other.

Coulomb constant

Absolute values of the **point charges**

$$F_{q_1 \text{ on } q_2} = F_{q_2 \text{ on } q_1} = \frac{k|q_1||q_2|}{r^2} \quad (16\text{-}1)$$

Distance between the point charges

Electric field: We can regard the interaction between charges as a two-step process: One charge sets up an electric field, and the other charge responds to that field. The electric field points away from a positive charge and toward a negative charge. The electric field of a single point charge is given by a simple equation; the field due to a combination of charges is the vector sum of the fields due to the individual charges.

If a particle with charge q is placed at a position where the **electric field** due to other charges is $\vec{E}$...

$$\vec{F} = q\vec{E} \quad (16\text{-}2)$$

...then the **electric force** on the particle is $\vec{F} = q\vec{E}$.

Coulomb constant

Absolute value of charge Q

Magnitude of the electric field due to a point charge Q

$$E = \frac{k|Q|}{r^2} \quad (16\text{-}4)$$

Distance between point charge Q and the location where the field is measured

Gauss's law: The electric flux through a surface equals the area of that surface multiplied by the component of electric field perpendicular to the surface. Gauss's law states that the net electric flux through a closed surface (one that encloses a volume) is proportional to the net charge enclosed within that volume. Gauss's law can be used to determine the electric field in situations where the charge distribution is highly symmetric. It also tells us that any excess charge on a conductor resides on the surfaces of the conductor.

Electric flux through a closed surface

Net amount of **charge enclosed** within the surface

$$\Phi = \frac{q_{\text{encl}}}{\varepsilon_0} \quad (16\text{-}9)$$

Permittivity of free space = $1/(4\pi k)$

(1) The net charge enclosed by this spherical Gaussian surface is q, so the net electric flux through this surface is $\Phi = q/\varepsilon_0$.

(2) The net charge enclosed by this irregular Gaussian surface is also q, so the net electric flux through this surface is also $\Phi = q/\varepsilon_0$.

$\vec{E}$

q

(Figure 16-19)

The net electric flux through a Gaussian surface depends only on the net charge that it encloses, not the shape or size of the surface.

Answer to What do you think? Question

(c) The magnitude of the electric field of a charged object measured at a distance r from the object is proportional to $1/r^2$ (see Section 16-5). If r is made one-half as great (decreased from $r = 2$ m to $r = 1$ m), the electric field magnitude changes by a factor of $1/(1/2)^2 = 1/(1/4) = 4$.

Answers to Got the Concept? Questions

16-1 (c) When the rubber rod is rubbed with fur, electrons are transferred to the rod from the fur. The electrons have mass, so the rod gains a little mass and the fur loses an equal amount of mass. By contrast, electrons are transferred from the glass rod to the silk when these two objects are rubbed together. As a result, the glass rod loses a little mass and the silk gains an equal amount of mass. The amount of mass involved is very small but is not zero.

16-2 (c) A sodium atom is neutral because it has 11 protons in its nucleus (each of charge $+e$) and 11 electrons (each of charge $-e$). If a sodium atom loses one of its 11 electrons by transferring it to another atom, it is left with 11 protons with a combined charge of $+11e$ and 10 electrons with a combined charge of $-10e$, and so it has a net positive charge of $+e$. In NaCl the electron is transferred to the chlorine atom, which then has a net negative charge of $-e$.

16-3 (d) Equation 16-1 tells us that the magnitude of the electric force between two charged objects is directly proportional to the product of the two charges. If we remove only half of the electrons from the neutral water drop, the ball of electrons has half the negative charge as before and the water drop has half the positive charge as before. So the electric force between the water drop and the ball of electrons will be $(\frac{1}{2}) \times (\frac{1}{2}) = \frac{1}{4}$ as great as before.

16-4 (b) Equation 16-2, $\vec{F} = q\vec{E}$, tells us the force $\vec{F}$ that acts on a charge q due to the electric field $\vec{E}$ produced by *other* charges. So to find the force that acts on the positive charge in Figure 16-14 due to the negative charge, in Equation 16-2 we let q be the positive charge (the charge that experiences the force) and let $\vec{E}$ be the electric field produced at the position of the positive charge by the negative charge (the other charge that exerts the force).

16-5 (d) If both charges were negative, the electric field at point P due to each charge would point directly toward that charge. So the electric field due to the right-hand charge would point down and to the right, while the electric field due to the left-hand charge would point down and to the left. Since point P is the same distance from each charge, and each charge has the same magnitude, the electric fields due to the two charges would have the same magnitude. The horizontal components of the two fields cancel (one points left and the other points right), so what remains is a net electric field that points straight down.

16-6 (c) The flux remains the same. According to Gauss's law, the net electric flux due to an enclosed charge does not depend on the size of the surface that encloses the charge.

16-7 (b) The flux remains the same. According to Gauss's law, the net electric flux due to an enclosed charge does not depend on the shape of the surface that encloses the charge.

16-8 (b) The flux remains the same. According to Gauss's law, the electric flux through a closed surface depends only on the charge enclosed *within* the surface. A charge outside the surface will cause positive flux on one part of the surface (where its electric field points out of the surface) and negative flux on another part of the surface (where its electric field points into the surface), but will have no net effect on the flux through the surface as a whole.

16-9 (c) The result does not depend on the size of the cylinder we choose. If we double the area, the flux through the top and bottom faces would double, but the amount of enclosed charge would double as well. In the derivation of Equation 16-14 these factors would cancel, so our result for E_z would be the same.

Questions and Problems

In a few problems you are given more data than you actually need; in a few other problems you are required to supply data from your general knowledge, outside sources, or informed estimate.

Interpret as significant all digits in numerical values that have trailing zeros and no decimal points.

For all problems use $\mathbf{g} = \mathbf{9.80\ m/s^2}$ for the free-fall acceleration due to gravity. Neglect friction and air resistance unless instructed to do otherwise.

•	Basic, single-concept problem
••	Intermediate-level problem; may require synthesis of concepts and multiple steps
•••	Challenging problem
SSM	*Solution is in Student Solutions Manual*
Example	*See worked example for a similar problem.*

Conceptual Questions

1. • How, if at all, would the physical universe be different if the proton were negatively charged and the electron were positively charged?

2. • How, if at all, would the physical universe be different if the proton's charge was very slightly larger in magnitude than the electron's charge?

3. • When an initially electrically neutral object acquires a net positive charge, does its mass increase or decrease? Why? SSM

4. • When you remove socks from a hot dryer, they tend to cling to everything. Two identical socks, however, usually repel. Why?

5. • Describe a set of experiments that might be used to determine if you have discovered a third type of charge other than positive and negative.

6. • How does a person become "charged" as he or she shuffles across a carpet, wearing cloth slippers, on a dry winter day?

7. • After combing your hair with a plastic comb, you find that when you bring the comb near a small bit of paper, the bit of paper moves toward the comb. Then, shortly after the paper touches the comb, it moves away from the comb. Explain these observations. SSM

8. • After combing your hair with a plastic comb, you find that when you bring the comb near an empty aluminum soft-drink can that is lying on its side on a nonconducting tabletop, the can rolls toward the comb. After being touched by the comb, the can is still attracted by the comb. Explain these observations.

9. • (a) A positively charged glass rod attracts a lighter object suspended by a thread. Does it follow that the object is negatively charged? (b) If, instead, the rod repels it, does it follow that the suspended object is positively charged?

10. • Some days it can be frustrating to attempt to demonstrate electrostatic phenomena for a physics class. An experiment that works beautifully one day may fail the next day if the weather has changed. Air-conditioning helps a lot while demonstrating the phenomena during the summer. Why?

11. • Discuss the similarities and differences between the gravitational and electric forces.

12. • Why is the gravitational force usually ignored in problems on the scale of particles such as electrons and protons?

13. • (a) What are the advantages of thinking of the force on a charge at a point P as being exerted by an electric field at P, rather than by other charges at other locations? (b) Is the convenience of the field as a calculation device worth inventing a new physical quantity? Or is there more to the field concept than that? SSM

14. • Do electric field lines point along the trajectory of positively charged particles? Why or why not?

15. • An electron and a proton are released in a region of space where the electric field is vertically downward. How do the electric forces acting on the electron and proton compare?

16. • Inside a uniform spherical charge distribution, why is it that as one moves out from the center, the electric field increases as r rather than decreases as $1/r^2$?

17. • Is the electric field $\vec{E}$ in Gauss's law only the electric field due to the charge inside the Gaussian surface, or is it the total electric field due to all charges both inside and outside the surface? Explain your answer. SSM

18. • If the net electric flux out of a closed surface is zero, does that mean the charge density must be zero everywhere inside the surface? Explain your answer.

Multiple-Choice Questions

19. • Electric charges of the opposite sign
   A. exert no force on each other.
   B. attract each other.
   C. repel each other.
   D. repel and attract each other.
   E. repel and attract each other depending on the magnitude of the charges.

20. • If two uncharged objects are rubbed together and one of them acquires a negative charge, then the other one
   A. remains uncharged.
   B. also acquires a negative charge.
   C. acquires a positive charge.
   D. acquires a positive charge equal to twice the negative charge.
   E. acquires a positive charge equal to half the negative charge.

21. • Metal sphere A has a charge of $-Q$. An identical metal sphere B has a charge of $+2Q$. The magnitude of the electric force on B due to A is F. The magnitude of the electric force on A due to B is
   A. $F/4$.
   B. $F/2$.
   C. F.
   D. $2F$.
   E. $4F$. SSM

22. • A balloon can be charged by rubbing it with your sleeve while holding it in your hand. You can conclude from this that the balloon is a(n)
   A. conductor.
   B. insulator.
   C. neutral object.
   D. Gaussian surface.
   E. semiconductor.

23. • A positively charged rod is brought near one end of an uncharged metal bar. The end of the metal bar farthest from the charged rod will be charged
   A. positively.
   B. negatively.
   C. neutral.
   D. twice as much as the end nearest the rod.
   E. none of the above ways. SSM

24. • A free positive charge released in an electric field will
   A. remain at rest.
   B. accelerate in the direction opposite to the electric field.
   C. accelerate in the direction perpendicular to the electric field.
   D. accelerate in the same direction as the electric field.
   E. accelerate in a circular path.

25. • Consider a point charge $+Q$ located outside a closed surface such as a sphere bound by the black circle in **Figure 16-24**. What is the net electric flux through the closed surface?
   A. $\frac{+Q}{\varepsilon_0}$
   B. $\frac{-Q}{\varepsilon_0}$
   C. 0
   D. $\frac{+2Q}{\varepsilon_0}$
   E. $\frac{-2Q}{\varepsilon_0}$

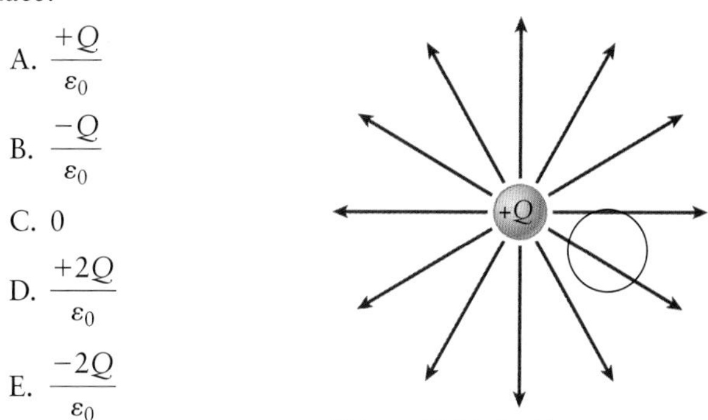

Figure 16-24 Problem 25

26. • If a charge is located at the center of a spherical volume and the electric flux through the surface of the sphere is Φ,

what would be the flux through the surface if the radius of the sphere were tripled?

A. 3Φ
B. 9Φ
C. Φ
D. Φ/3
E. Φ/9

27. • A point charge $+Q$ is at the center of a spherical conducting shell of inner radius R_1 and outer radius R_2, as shown in **Figure 16-25**. The charge on the inner surface of the shell is

A. $+Q$.
B. $-Q$.
C. 0.
D. $+Q/2$.
E. $-Q/2$. SSM

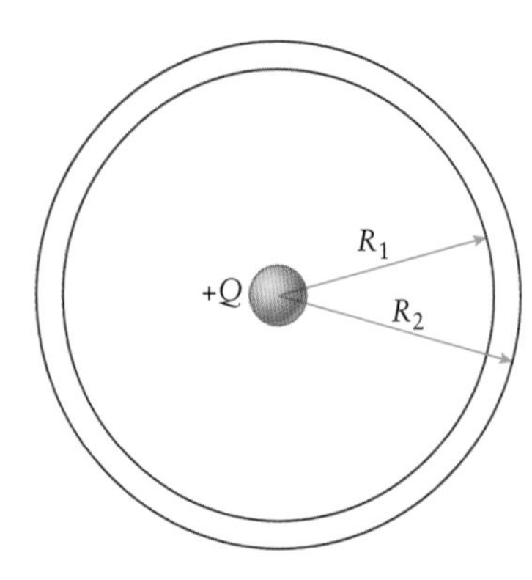

Figure 16-25 Problem 27

Estimation/Numerical Analysis

28. •• (a) Estimate the positive charge in you due to the protons in the molecules present in your body. (b) What is the net charge in your body?

29. • Estimate the amount of charge needed on both a comb and on a bit of tissue paper in order to generate an electric force of sufficient magnitude to support the weight of the paper.

30. • A cylindrical jar 10 cm in diameter and 15 cm tall is full of old copper pennies. Estimate the number of *valence electrons* in the jar.

31. • The *electrical breakdown* of an insulator occurs when an electric field becomes large enough to ionize the insulator, allowing electrical current to flow—lightning is an example of the electrical breakdown of air. The electrical breakdown of air occurs when the electric field reaches about 3×10^6 N/C. Estimate the total static charge on your body when you generate a static spark with your fingertip when getting out of your car.

32. • Suppose two people are standing near each other having a conversation. If both people carry the same amount of static charge, estimate how much charge each person would have to carry in order for the repulsive electric force to cancel the attractive gravitational force between them.

33. • Estimate the electric flux passing through the passenger window of your car when you become charged up after sliding into your vehicle.

34. • The magnitude of the repulsive force (F) between two $+2.5$-μC charges as a function of the distance of separation (r) is listed in the following table. Empirically derive a relationship between F and r by doing a curve fit to find the power n of r in the following formula:

$$F = \frac{kq_1q_2}{r^n}$$

Use a graphing calculator or spreadsheet.

r (m)	F (N)	r (m)	F (N)
0.003	5500	0.080	6
0.004	3000	0.100	5
0.005	2000	0.200	1
0.010	600	0.300	0.5
0.020	175	0.400	0.35
0.040	30	0.500	0.25
0.050	10	0.600	0.15

Problems

16-1 Electric forces and electric charges are all around you—and within you

16-2 Matter contains positive and negative electric charge

35. • The nucleus of a copper atom has 29 protons and 35 neutrons. What is the total charge of the nucleus? Example 16-1

36. • Five electrons are added to 1.00 C of positive charge. What is the net charge of the system? Example 16-1

37. •• How many coulombs of negative charge are there in 0.500 kg of water? SSM Example 16-1

38. • An ion has 17 protons, 18 neutrons, and 18 electrons. What is the net charge of the ion? Example 16-1

39. •• The charge per unit length on a glass rod is 0.00500 C/m. If the rod is 1 mm long, how many electrons have been removed from the glass rod? Example 16-1

40. • How many electrons must be transferred from an object to produce a charge of 1.60 C? Example 16-1

16-3 Charge can flow freely in a conductor but not in an insulator

41. • Suppose 2.00 C of positive charge are distributed evenly throughout a solid sphere of 1.27-cm radius. (a) What is the charge per unit volume for this situation? (b) Is the sphere insulating or conducting? How do you know?

42. • The maximum amount of charge that can be collected on a Van de Graaff generator's conducting sphere (30-cm diameter) is about 30 μC. Calculate the surface charge density, σ, of the sphere in C/m^2.

43. • **Biology** Most workers in nanotechnology are actively monitored for excess static charge buildup. The human body acts like an insulator as one walks across a carpet, collecting −50 nC per step. (a) What charge buildup will a worker in a manufacturing plant accumulate if she walks 25 steps? (b) How many electrons are present in that amount of charge? (c) If a delicate manufacturing process can be damaged by an electrical discharge of greater than 10^{12} electrons, what is the maximum number of steps that any worker should be allowed to take before touching the components? SSM

16-4 Coulomb's law describes the force between charged objects

44. • Two point charges are separated by a distance of 20.0 cm. The numerical value of one charge is twice that of the other. If each charge exerts a force of magnitude 45.0 N on the other, find the magnitude of the charges. Example 16-2

45. • The mass of an electron is 9.11×10^{-31} kg. How far apart would two electrons have to be in order for the electric force exerted by each on the other to be equal to the weight of an electron? SSM Example 16-2

46. • Charge A, $+5.00\ \mu C$, is positioned at the origin of a coordinate system. Charge B, $-3.00\ \mu C$, is fixed on the x axis at $x = 3.00$ m. (a) Determine the magnitude and direction of the force that charge B exerts on charge A. (b) What is the magnitude and direction of the force that charge A exerts on charge B? Example 16-2

47. • Point charge A with charge $q_A = +3.00\ \mu C$ is located at the origin. Point charge B with charge $q_B = -4.00\ \mu C$ is on the x axis at $x = 3.00$ m. Point charge C with charge $q_C = -2.00\ \mu C$ is on the x axis at $x = 6.00$ m. And point charge D with charge $q_D = +6.00\ \mu C$ is on the x axis at $x = 8.00$ m. What is the net electric force on point charge A due to the other three charges? Example 16-3

48. • A charge of $+3.00\ \mu C$ is located at the origin, and a second charge of $-2.00\ \mu C$ is located on the $x-y$ plane at the point (30.0 cm, 20.0 cm). Determine the electric force exerted by the $-2.00\ \mu C$ charge on the $3.00\ \mu C$ charge. Example 16-4

49. • A point charge with charge $q_1 = +5.00$ nC is fixed at the origin. A second point charge with charge $q_2 = -7.00$ nC is located on the x axis at $x = 5.00$ m. Where along the x axis will a third point charge of $q = +2.00$ nC charge need to be for the net electric force on it due to the two fixed charges to be equal to zero? SSM Example 16-3

50. • Two charges lie on the x axis, a -2.00-μC charge at the origin and a $+3.00$-μC charge at $x = 0.100$ m. At what position along the x axis, if any, should a third $+4.00$-μC charge be placed so that the net force on the third charge is equal to zero? Example 16-3

51. •• Point charge A with a charge of $+3.00\ \mu C$ is located at the origin. Point charge B with a charge of $+6.00\ \mu C$ is located on the x axis at $x = 7.00$ cm. And point charge C with a charge of $+2.00\ \mu C$ is located on the y axis at $y = 6.00$ cm. What is the net force (magnitude and direction) exerted on each charge by the others? SSM Example 16-4

52. • A charge q_1 equal to $0.600\ \mu C$ is at the origin, and a second charge q_2 equal to $0.800\ \mu C$ is on the x axis at 5.00 cm. (a) Find the force (magnitude and direction) that each charge exerts on the other. (b) How would your answer change if q_2 were $-0.800\ \mu C$? Example 16-2

53. •• A charge $Q_1 = +7.00\ \mu C$ is located on the y axis at $y = 4.00$ cm. A second charge $Q_2 = -4.00\ \mu C$ is located on the $x-y$ plane at the point (-3.00 cm, 4.00 cm). And a third charge $Q_3 = +6.00\ \mu C$ is located on the x axis at $x = -3.00$ cm. Determine the x and y components of the net electric force that acts on the third charge Q_3. Example 16-4

16-5 The concept of electric field helps us visualize how charges exert forces at a distance

54. • At point P in **Figure 16-26** the electric field is zero. (a) What are the signs of q_1 and q_2? (b) Describe their magnitudes. Example 16-6

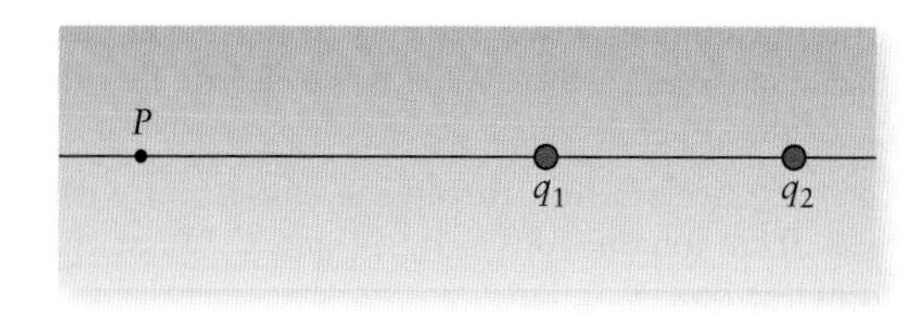

Figure 16-26 Problem 54

55. • Near the surface of Earth an electric field points radially downward and has a magnitude of approximately 100 N/C. What charge (magnitude and sign) would have to be placed on a penny that has a mass of 3.11 g to cause it to rise into the air with an upward acceleration of 0.190 m/s^2? Example 16-5

56. • Two charges are placed on the x axis, $+5.00\ \mu C$ at the origin and $-10.0\ \mu C$ at $x = 10.0$ cm. (a) Find the electric field on the x axis at $x = 6.00$ cm. (b) At what point(s) on the x axis is the electric field zero? Example 16-6

57. • In **Figure 16-27** the electric field at the origin is zero. If q_1 is 1.00×10^{-7} C, what is q_2? Example 16-6

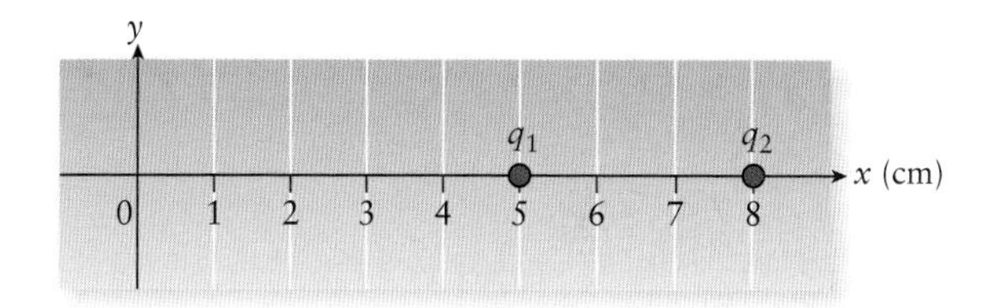

Figure 16-27 Problems 57 and 58

58. • In **Figure 16-27** if $q_1 = 1.00 \times 10^{-7}$ C and $q_2 = 2.00 \times 10^{-7}$ C, (a) what is the electric field $\vec{E}$ at the point $(x, y) =$ (0.00 cm, 3.00 cm)? (b) What is the force $\vec{F}$ acting on an electron at that position? Example 16-7

59. • In Figure 16-27 if $q_1 = 1.00 \times 10^{-7}$ C and $q_2 = 2.00 \times 10^{-7}$ C, (a) what is the electric field $\vec{E}$ at the point $(x, y) =$ (6.00 m, 3.00 m)? (b) What is the force $\vec{F}$ acting on a proton at that position? Example 16-7

60. •• In the Bohr model the hydrogen atom consists of an electron in a circular orbit of radius $a_0 = 5.29 \times 10^{-11}$ m around the nucleus. Using this model, and ignoring relativistic effects, what is the speed of the electron? Example 16-5

61. • **Biology** The cockroach *Periplaneta americana* can detect a static electric field of 8.0 kN/C using its long antennae. If we model the excess static charge on a cockroach as a point charge located at the end of each antenna, what magnitude of charge would each antenna possess in order for each antennae to experience a force of 3.0 μN from the external electric field? Example 16-5

16-6 Gauss's law gives us more insight into the electric field

62. • A rectangular area is rotated in a uniform electric field, from a position where the maximum electric flux goes through it to an orientation where only half the maximum flux goes through it. What is the angle of rotation?

63. • A point charge of 4.00×10^{-12} C is located at the center of a cubical Gaussian surface. What is the electric flux through each face of the cube?

64. • The net electric flux through a cubic box with sides that are 20.0 cm long is 4.80×10^3 N·m²/C. What charge is enclosed by the box?

65. •• A 10.0-cm-long uniformly charged plastic rod is sealed inside a plastic bag. The net electric flux through the bag is 7.50×10^5 N·m²/C. What is the linear charge density (charge per unit length) on the rod? SSM

66. •• **Figure 16-28** shows a prism-shaped object that is 40.0 cm high, 30.0 cm deep, and 80.0 cm long. The prism is immersed in a uniform electric field of 500 N/C directed parallel to the x axis. (a) Calculate the electric flux out of each of its five faces and (b) the net electric flux out of the entire closed surface. (c) If in addition to the given electric field the prism also enclosed a point charge of $-2.00\ \mu$C, qualitatively how would your answers above change, if at all?

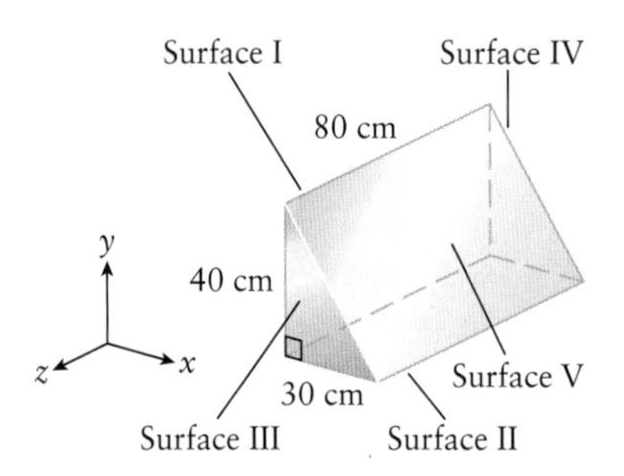

Figure 16-28 Problem 66

16-7 In certain situations Gauss's law helps us to calculate the electric field and to determine how charge is distributed

67. •• Use Gauss's law to find an expression for the electric field just outside the surface of a sphere carrying a uniform surface charge density σ (charge per unit area). SSM Example 16-8

68. • Determine the charge density for each of the following cases (assume that all densities are uniform): (a) a solid cylinder that has a length L, has a radius R, and carries a charge Q throughout its volume; (b) a flat plate (very thin) that has a width W, has a length L, and carries a charge Q on its surface area; (c) a solid sphere of radius R carrying a charge Q throughout its volume; and (d) a hollow sphere of radius R carrying a charge Q over its surface area.

69. • An electric field of magnitude 4.00×10^2 N/C exists at all points just outside the surface of a 2.00-cm-diameter steel ball bearing. Assuming the ball bearing is in electrostatic equilibrium, (a) what is the total charge on the ball? (b) What is the surface charge density on the ball? Example 16-8

70. •• Consider an infinite plane with a uniform charge distribution σ. (a) Use Gauss's law to find an expression for the electric field due to the plane. (b) What field would be created by two equal but oppositely charged parallel planes? Consider the region between the planes as well as the two regions outside the planes. The simplest way to express your answers is in terms of the surface charge density σ (the charge per unit area) on the plane.

71. •• A -3.20-μC charge sits in static equilibrium in the center of a conducting spherical shell that has an inner radius of 2.50 cm and an outer radius of 3.50 cm. The shell has a net charge of $-5.80\ \mu$C. Determine the charge on each surface of the shell and the electric field just outside the shell. Example 16-8

General Problems

72. •• A spherical party balloon on Earth's surface that is 25 cm in diameter contains helium at room temperature (20°C) and at a pressure of 1.3 atm. If one electron could be stripped from every helium atom in the balloon and removed to a satellite orbiting Earth 22,000 mi (32,187 km) above the planet, with what force would the balloon and the satellite attract each other when the satellite is directly above the balloon? Example 16-2

73. •• A plutonium-242 atom has a nucleus of 94 protons and 148 neutrons and has 94 electrons. The diameter of its nucleus is approximately 15×10^{-15} m. (a) Make a reasonable physical argument as to why we can treat the nucleus as a point charge for points outside of it. (b) Plutonium decays radioactively by emitting an *alpha particle* from its nucleus. The mass of the alpha particle is 6.6×10^{-27} kg, and the particle has two protons and two neutrons. If the alpha particle comes from the surface of the ^{242}Pu nucleus, what is its greatest acceleration? Example 16-2

74. • When a test charge of +5.00 nC is placed at a certain point, the force that acts on it has a magnitude of 0.0800 N and is directed northeast. (a) If the test charge were -2.00 nC instead, what force would act on it? (b) What is the electric field at the point in question? Example 16-5

75. •• **Biology** A red blood cell may carry an excess charge of about -2.5×10^{-12} C distributed uniformly over its surface. The cells, modeled as spheres, are approximately 7.5 μm in diameter and have a mass of 9.0×10^{-14} kg. (a) How many excess electrons does a typical red blood cell carry? (b) What is the surface charge density σ on the red blood cell? Express your answer in C/m² and in electrons/m². SSM Example 16-1

76. •• Three point charges are placed on the $x-y$ plane: a +50.0-nC charge at the origin, a -50.0-nC charge on the x axis at 10.0 cm, and a +150 nC charge at the point (10.0 cm, 8.00 cm). (a) Find the total electric force on the +150-nC charge due to the other two. (b) What is the electric field at the location of the +150-nC charge due to the presence of the other two charges? Example 16-4

77. •• Two small spheres each have a mass m of 0.100 g and are suspended as pendulums by light insulating strings from a common point, as shown in **Figure 16-29**. The spheres are given the same electric charge, and the two come to equilibrium when each string is at an angle of $\theta = 3.00°$ with the vertical. If each string is 1.00 m long, what is the magnitude of the charge on each sphere? Example 16-2

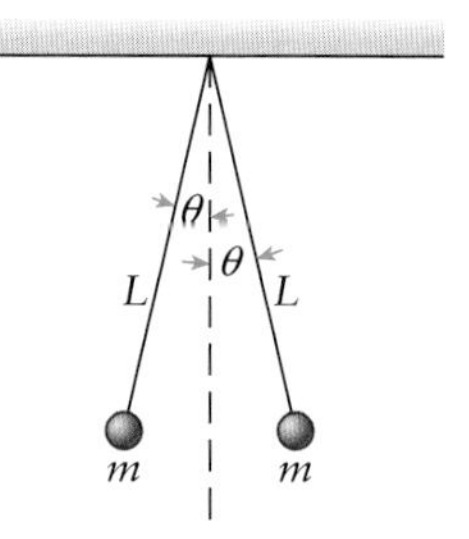

Figure 16-29 Problem 77

78. •• A small 1.00-g plastic ball that has a charge q of 1.00 C is suspended by a string that has a length L of 1.00 m in a uniform electric field, as shown in **Figure 16-30**. If the ball is in equilibrium when the string makes a 9.80° angle with the vertical as indicated by θ, what is the electric field strength? Example 16-5

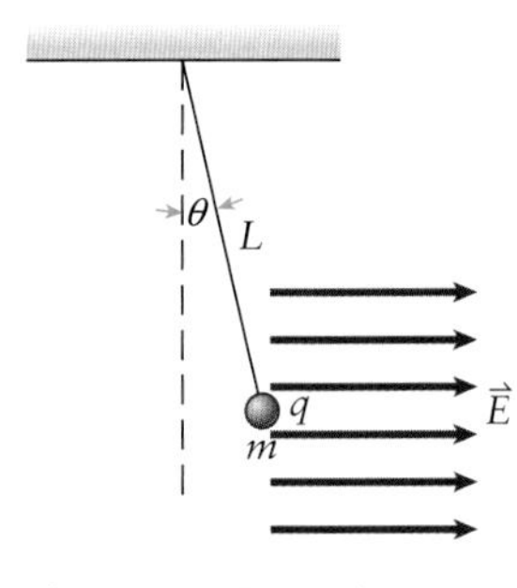

Figure 16-30 Problem 78

79. • **Biology** The 9-inch-long elephant nose fish in the Congo River generates a weak electric field around its body using an organ in its tail. When small prey (or even potential mates) swim within a few feet of the fish, they perturb the electric field. The change in the field is picked up by electric sensor cells in the skin of the elephant nose. These remarkable fish can detect changes in the electric field as small as 3.0 μN/C. (a) How much charge (modeled as a point charge) in the fish would be needed to produce such a change in the electric field at a distance of 75 cm? (b) How many electrons would be required to create the charge? SSM Example 16-1

80. ••• Three charges (q_A, q_B, and q_C) are placed at the vertices of the equilateral triangle that has sides of length s in **Figure 16-31**. Derive expressions for the electric field at (a) X (at the center of the triangle), (b) Y (at the midpoint of the side between q_B and q_C), and (c) Z (at the midpoint of the side between q_A and q_C). (d) Now use the following numerical values and calculate the electric field at those same points: s = 10.0 cm, q_A = +20.0 nC, q_B = −8.00 nC, and q_C = −10.0 nC. Example 16-7

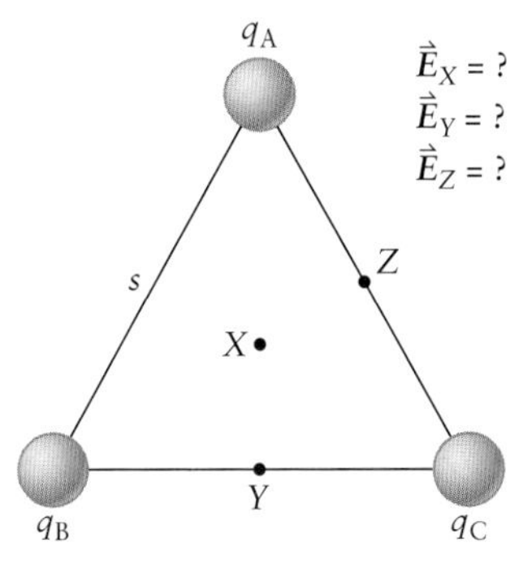

Figure 16-31 Problem 80

81. •• Calculate the electric field at the center of the hexagon shown in **Figure 16-32**. Assume the sides of the hexagon are all 5.00 cm long. SSM Example 16-7

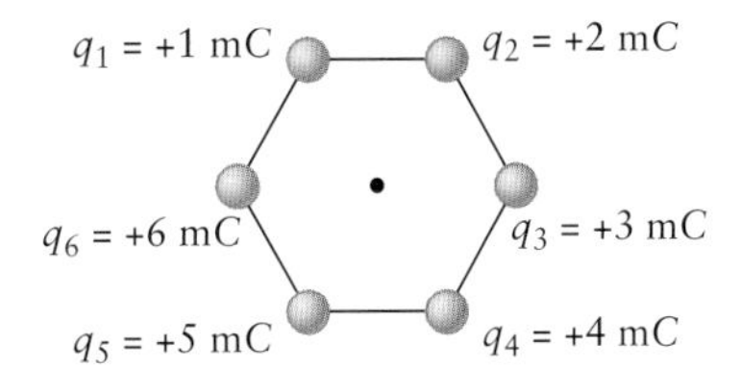

Figure 16-32 Problem 81

82. •• Electric fields up to 2.00×10^5 N/C have been measured inside of clouds during electrical storms. Neglect the drag force due to the air in the cloud and any collisions with air molecules. (a) What acceleration does the maximum electric field produce for protons in the cloud? Express your answer in SI units and as a fraction of g. (b) If the electric field remains constant, how far will the proton have to travel to reach 10% of the speed of light (3.00×10^8 m/s) if it started with negligible speed? (c) Can you neglect the effects of gravity? Explain your answer. Example 16-5

83. •• An electron with an initial speed of 5.00×10^5 m/s enters a region in which there is an electric field directed along its direction of motion. If the electron travels 5.00 cm in the field before being stopped, what are the magnitude and direction of the electric field? Example 16-5

84. •• An electron, released in a region where the electric field is uniform, is observed to have an acceleration of 3.00×10^{14} m/s^2 in the positive x direction. (a) Determine the electric field producing the acceleration. (b) Assuming the electron is released from rest, determine the time required for it to reach a speed of 11,200 m/s, the escape speed from Earth's surface. Example 16-5

85. •• **Chemistry** The iron atom (Fe) has 26 protons, 30 neutrons, and 26 electrons. The diameter of the atom is approximately 1.0×10^{-10} m, while the diameter of its nucleus is about 9.2×10^{-15} m. (You can reasonably model the nucleus as a uniform sphere of charge.) What are the magnitude and direction of the electric field that the nucleus produces (a) just outside the surface of the nucleus and (b) at the distance of the outermost electron? (c) What would be the magnitude and direction of the acceleration of the outermost electron due only to the nucleus, neglecting any force due to the other electrons? SSM Example 16-5

86. ••• An electron with kinetic energy K is traveling along the $+x$ axis, which is along the axis of a cathode-ray tube as shown in **Figure 16-33**. There is an electric field $E = 12.00 \times 10^4$ N/C pointed in the $+y$ direction between the deflection plates, which are 0.0600 m long and are separated by 0.0200 m. Determine the minimum initial kinetic energy the electron can have and still avoid colliding with one of the plates. Example 16-5

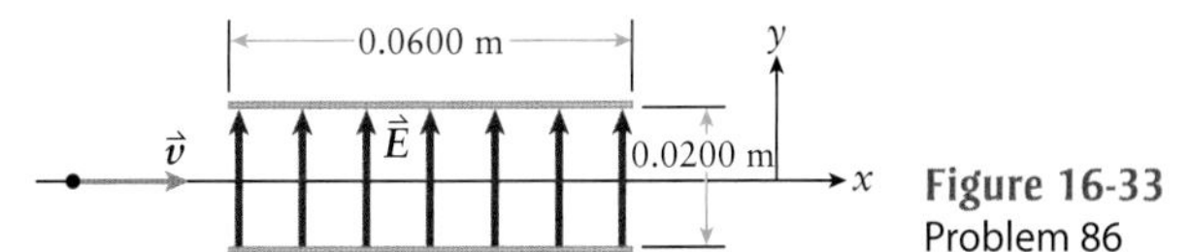

Figure 16-33 Problem 86

87. •• In the famous Millikan oil-drop experiment, tiny spherical droplets of oil are sprayed into a uniform vertical electric field. The drops get a very small charge (just a few electrons) due to friction with the atomizer as they are sprayed. The field is adjusted until the drop (which is viewed through a small telescope) is just balanced against gravity and therefore remains stationary. Using the measured value of the electric field, we can calculate the charge on the drop and from this calculate the charge e of the electron. In one apparatus the drops are 1.10 μm in diameter, and the oil has a density of 0.850 g/cm^3.(a) If the drops are negatively charged, which way should the electric field point to hold them stationary (up or down)? (b) Why? (c) If a certain drop contains four excess electrons, what magnitude electric field is needed to hold it stationary? (d) You measure a balancing field of 5183 N/C for another drop. How many excess electrons are on this drop? Example 16-5

88. •• **Biology** There is a naturally occurring vertical electric field near Earth's surface, pointing toward the ground. In fair weather conditions and in an open field, the strength of the electric field is 1.00×10^2 N/C. A spherical pollen grain, with a radius of 12 μm, is released from its parent plant by a light breeze, giving it a net charge of −0.80 fC (where 1 fC = 1×10^{-15} C). What is the ratio of the magnitudes of the electric

force to the gravitational force, $F_{\text{electric}}/F_{\text{gravitational}}$, acting on the pollen? Assume the volume mass density of the pollen is the same as water, 1.00×10^3 kg/m^3, which is a primary constituent of pollen. Example 16-5

89. •• Two hollow, concentric, spherical shells are covered with charge (**Figure 16-34**). The inner sphere has a radius R_i and a surface charge density of $+\sigma_i$, while the outer sphere has a radius R_o and a surface charge density of $-\sigma_o$. Derive an expression for the electric field in the following three radial regions: (a) $r < R_i$, (b) $R_i < r < R_o$, and (c) $r > R_o$. SSM Example 16-8

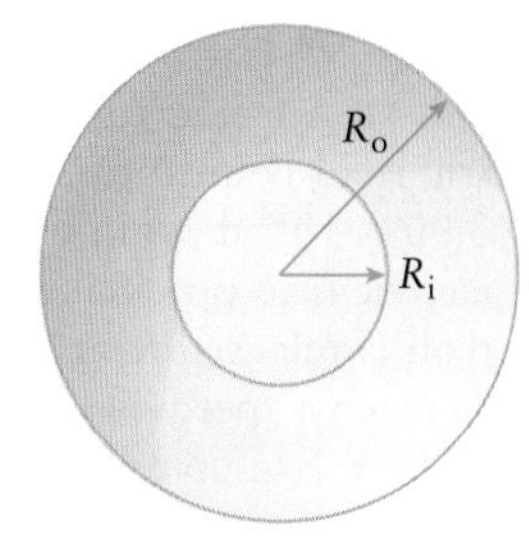

Figure 16-34 Problem 89

90. •• A charge $q_1 = +2q$ is at the origin, and a charge $q_2 = -q$ is on the x axis at $x = a$. Find expressions for the total electric field on the x axis in each of the regions (a) $x < 0$; (b) $0 < x < a$; and (c) $a < x$. (d) Determine all points on the x axis where the electric field is zero. (e) Use a graphing calculator or spreadsheet to make a plot of E_x versus x for all points on the x axis, and (f) qualitatively discuss what happens for $-\infty < x < \infty$. Example 16-6

91. • Two point charges are enclosed by a spherical conducting shell that has an inner and outer radius of 13.0 and 15.2 cm, respectively. One point charge has a charge of $q_1 = 8.30\ \mu\text{C}$, while the second point charge has an unknown charge q_2. The conducting shell is known to have a net electric charge of $-2.40\ \mu\text{C}$, but measurements find that the charge on the outer shell is $+3.70\ \mu\text{C}$. Determine the charge q_2 of the second point charge. Example 16-8

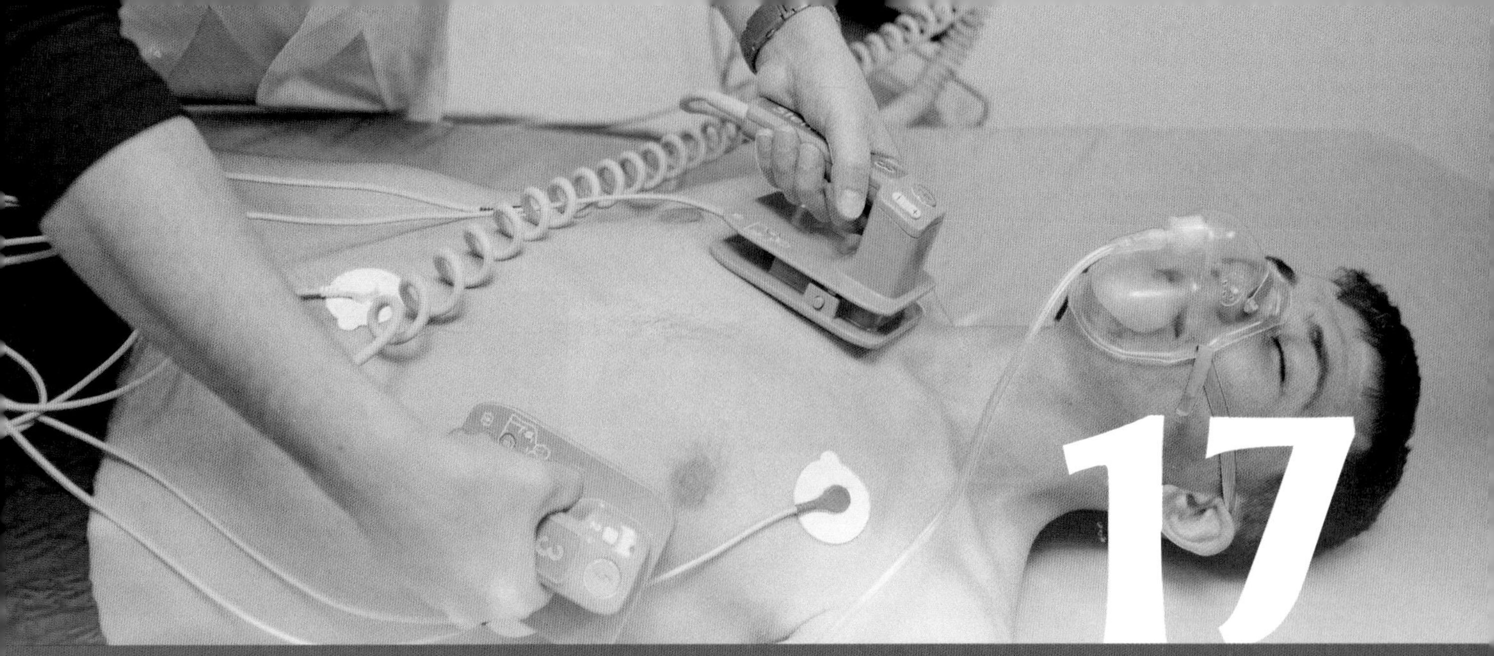

Electrostatics II: Electric Potential Energy and Electric Potential

muratseyit/E+/Getty Images

What do you think?

A cardiac defibrillator works by delivering an electric shock to a malfunctioning heart to jump-start it back into its normal rhythm. Electric energy is stored for this purpose in the defibrillator by separating positive charge $+q$ from an equal amount of negative charge $-q$. To release this energy to the heart, conducting paddles are placed on the patient's chest, and charge $-q$ in the form of electrons is allowed to flow through the paddles and patient until it reaches the stored charge $+q$ within the defibrillator. If the value q of the charge stored in a defibrillator is doubled, by what factor will the stored energy increase? (a) $\sqrt{2}$; (b) 2; (c) 4; (d) 8; (e) 16.

In this chapter, your goals are to:

- (17-1) Explain the significance of energy in electrostatics.
- (17-2) Discuss how the work done on a charged particle by the electric field relates to changes in electric potential energy.
- (17-3) Explain the difference between electric potential and electric potential energy.
- (17-4) Recognize why equipotential surfaces are perpendicular to electric field lines.
- (17-5) Explain what is meant by capacitance and describe how the capacitance of a parallel-plate capacitor depends on the size of the plates and their separation.
- (17-6) Calculate the electric energy stored in a capacitor.
- (17-7) Explain how to treat capacitors attached in series and in parallel as a single equivalent capacitance.
- (17-8) Describe how the capacitance of a capacitor increases when an insulating material other than a vacuum is placed between the plates of the capacitor.

To master this chapter, you should review:

- (6-2) The work done by a force on a moving object.
- (6-6) The relationship between a conservative force and the potential energy associated with that force.
- (10-3) The generalized expression for gravitational potential energy.
- (16-5, 16-7) The electric field due to a point charge and to a sheet of charge.

17-1 Electric energy is important in nature, technology, and biological systems

Everyone is familiar with the idea that *electricity* implies *energy*. The energy to run your computer, lighting, and television is delivered to your home in the form of electricity. A bolt of lightning releases so much energy that it heats air to a temperature of 30,000°C, causing the air to glow with the characteristic light that we call a lightning flash (see Figure 16-1b). A defibrillator saves lives by delivering a sharp punch of electric energy to a malfunctioning heart (see the photo that opens this chapter). When you

Figure 17-1 Electric energy Sources of electric energy in (a) technology and (b) nature.

(a)

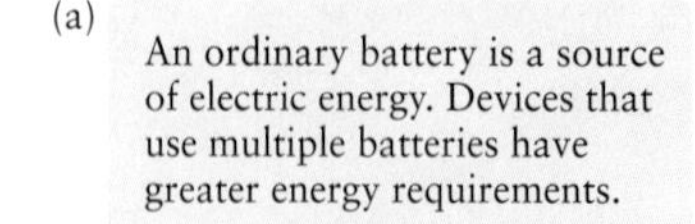

(b)

purchase an ordinary electric battery, you are really paying for the electric energy that the battery can deliver to whatever electric circuit you plug it into (**Figure 17-1a**). Some specialized species of fish such as electric rays (**Figure 17-1b**) are equipped with organic batteries; they can deliver an intense burst of electric energy to stun their prey or ward off predators.

An electric ray (order *Torpediniformes*) is equipped with a large number of organic batteries that work together to deliver intense electric shocks.
NHPA/NHPA/Superstock

In this chapter we'll make clear what is meant by electric potential energy. We'll gain insight into this new kind of energy by considering the similarities to and differences from gravitational potential energy. Just as there is a change in gravitational potential energy when a massive object moves up or down in the presence of Earth's gravitational field, there is a change in electric potential energy when a charged object moves along with or opposite to an electric field. We'll go on to introduce the useful concepts of electric potential, or electric potential energy per unit charge, and voltage, which is the difference in the value of electric potential at two positions. We'll also learn about equipotential surfaces, which are surfaces on which the electric potential has the same value at every point.

An important device for storing electric energy is a capacitor. In its simplest form a capacitor is just two pieces of metal, called capacitor plates, placed close to each other but not in contact. We'll examine the key properties of capacitors, including how to combine two or more of them for even greater energy storage. We'll conclude the chapter by seeing how capacitors can be made even more effective by inserting an insulating material between the plates. In Chapter 18 we'll learn how batteries like those shown in Figure 17-1a provide electric energy to the components of an electric circuit.

TAKE-HOME MESSAGE FOR Section 17-1

✔ Electric potential energy associated with a point charge is analogous to the gravitational potential energy associated with a massive object.

✔ Electric potential is electric potential energy per charge.

✔ A capacitor is a device for storing electric energy.

17-2 Electric potential energy changes when a charge moves in an electric field

We learned in Section 6-6 that Earth's gravitational force is a *conservative* force: As an object moves from one position to another, the work done on the object by the gravitational force depends only on where the object starts and where it ends up, not on how it gets there. Hence we can express the work done by gravity W_{grav} in terms of a change in the gravitational potential energy U_{grav} of the system of Earth and object:

$$W_{grav} = -\Delta U_{grav} \quad (6\text{-}16)$$

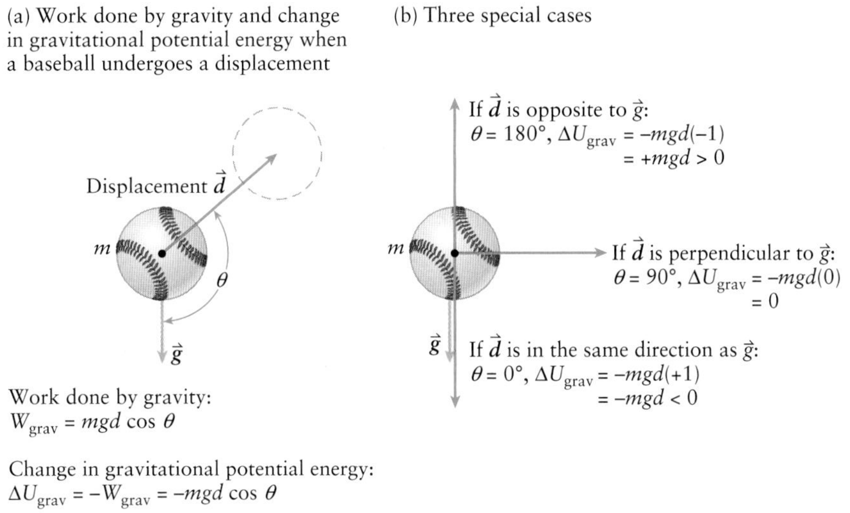

(a) Work done by gravity and change in gravitational potential energy when a baseball undergoes a displacement

Displacement $\vec{d}$

m

θ

$\vec{g}$

Work done by gravity:
$W_{grav} = mgd \cos \theta$

Change in gravitational potential energy:
$\Delta U_{grav} = -W_{grav} = -mgd \cos \theta$

(b) Three special cases

If $\vec{d}$ is opposite to $\vec{g}$:
$\theta = 180°$, $\Delta U_{grav} = -mgd(-1)$
$= +mgd > 0$

m

If $\vec{d}$ is perpendicular to $\vec{g}$:
$\theta = 90°$, $\Delta U_{grav} = -mgd(0)$
$= 0$

$\vec{g}$

If $\vec{d}$ is in the same direction as $\vec{g}$:
$\theta = 0°$, $\Delta U_{grav} = -mgd(+1)$
$= -mgd < 0$

Figure 17-2 Change in gravitational potential energy When an object of mass m moves in the presence of gravity, the gravitational force can do work on it and there can be a change in gravitational potential energy.

The minus sign means that if the object moves downward so that gravity does positive work (the gravitational force and the object's displacement are in the same direction), there is a negative change in gravitational potential energy (U_{grav} decreases). If the object moves upward so that gravity does negative work (force and displacement are in opposite directions), there is a positive change in gravitational potential energy (U_{grav} increases).

To be specific, suppose an object of mass m moves in a region where the acceleration $\vec{g}$ due to Earth's gravity is the same at all positions, so the gravitational force $\vec{F} = m\vec{g}$ on the object is uniform. The object's displacement is $\vec{d}$, and that displacement is at an angle θ to the direction of the force $\vec{F} = m\vec{g}$ and so at the same angle θ to the direction of $\vec{g}$ (**Figure 17-2a**). Then the work done by gravity is $W = Fd \cos \theta$ (Equation 6-2), and the change in gravitational potential energy is

See the Math Tutorial for more information on trigonometry.

$$\Delta U_{grav} = -W_{grav} = -mgd \cos \theta \tag{17-1}$$

Figure 17-2b shows three special cases. If the object moves a distance d straight upward, opposite to the direction of the gravitational force, $\theta = 180°$ and $\cos \theta = \cos 180° = -1$. In this case gravity does negative work on the object, and the gravitational potential energy increases: $\Delta U_{grav} = -mgd(-1) = +mgd$. If the object moves a distance d straight downward, in the direction of the gravitational force, $\theta = 0$ and $\cos \theta = \cos 0° = 1$. Then gravity does positive work on the object and the gravitational potential energy decreases: $\Delta U_{grav} = -mgd(+1) = -mgd$. Gravity does zero work, and there is no change in gravitational potential energy, if the object moves horizontally: Then $\theta = 90°$ and $\Delta U_{grav} = -mgd \cos 90° = 0$. These results are the same ones that we found in Section 6-6.

Like gravity, the electric force due to the interaction between charged objects is a conservative force. You can see this easily for the special case in which an object moves in a region of space where there is a *uniform* electric field $\vec{E}$—that is, where the value of $\vec{E}$ is the same at all positions (**Figure 17-3**). The electric force on an object of charge q is the same everywhere in this region and equal to $\vec{F} = q\vec{E}$. That's exactly like the situation in Figure 17-2: In a region where $\vec{g}$ is the same at all positions, the gravitational force on an object of mass m is the same everywhere and equal to $\vec{F} = m\vec{g}$. This uniform gravitational force is conservative, so the force $\vec{F} = q\vec{E}$ due to a uniform electric field must be conservative as well. Hence we can express the work done by the electric force in terms of a change in *electric* potential energy.

Electric Potential Energy in a Uniform Field

This analogy between gravitational force and electric force tells us how to write the change in electric potential energy for an object of charge q that undergoes a displacement $\vec{d}$ in the presence of a uniform electric field $\vec{E}$ (**Figure 17-3a**). Following the same steps that we used to find Equation 17-1 above, you can see that if θ is the angle

Go to Picture It 17-1 for more practice dealing with electric potential energy.

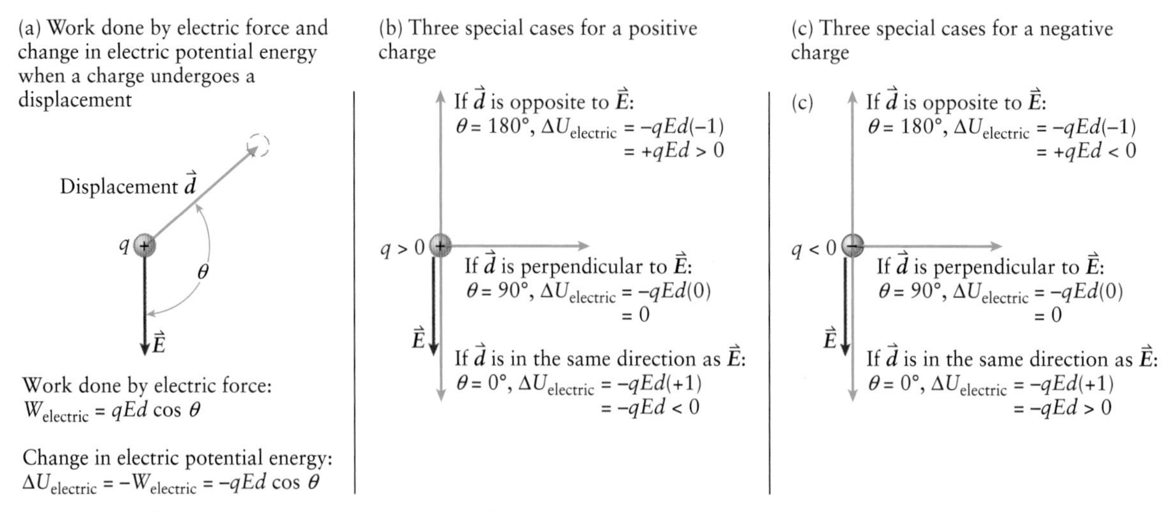

Figure 17-3 Change in electric potential energy When an object of charge q moves in the presence of a uniform electric field $\vec{E}$, the electric force can do work on it and there can be a change in electric potential energy (compare Figure 17-2).

between the directions of $\vec{d}$ and $\vec{E}$, then the work done on the charge by the electric force $\vec{F} = q\vec{E}$ is $W_{electric} = qEd \cos\theta$. The change in **electric potential energy** equals the negative of the electric work done on the charge:

Electric potential energy for a charge in a uniform electric field (17-2)

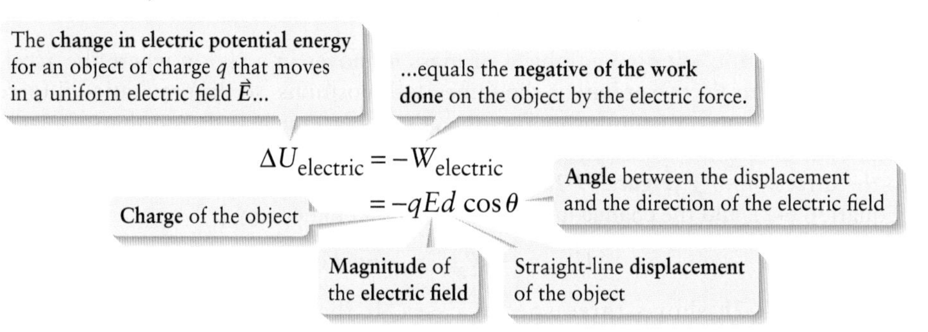

Just as for the gravitational case, Equation 17-2 tells us how much the electric potential energy *changes* when the charge is displaced as shown in Figure 17-3a. Note also that the electric potential energy is a shared property of the charge q and the other charges that produce the electric field $\vec{E}$.

Suppose q is positive, so the force $\vec{F} = q\vec{E}$ on the charged object is in the *same* direction as the uniform electric field $\vec{E}$ (**Figure 17-3b**). If this positive charge moves a distance d opposite to $\vec{E}$, then $\theta = 180°$ and $\cos\theta = -1$. In this case the electric force does negative work on the charge, $\Delta U_{electric} = (-qEd)(-1) = +qEd$ is positive, and the electric potential energy of the system increases. That's exactly what happens to *gravitational* potential energy when a massive object moves upward from Earth's surface, in the direction opposite to the gravitational force (Figure 17-2b). If the positive charge q instead moves in the same direction as the electric field, and so in the same direction as the electric force, then $\theta = 0$ and $\cos\theta = +1$. Then the electric force does positive work on the object, so $\Delta U_{electric} = (-qEd)(+1) = -qEd$ is negative and the electric potential energy of the system decreases. That's the same thing that happens when a massive object falls toward Earth's surface, in the same direction as the gravitational force on the object: Gravitational potential energy decreases (Figure 17-2b).

For an object with *negative* charge ($q < 0$), the force $\vec{F} = q\vec{E}$ on this object is in the direction *opposite* to the electric field $\vec{E}$ (**Figure 17-3c**). If such a negative charge moves a distance d opposite to the direction of $\vec{E}$ so that $\theta = 180°$ and $\cos\theta = -1$ in Equation 17-2, the electric force does *positive* work on the charge because the force and displacement are in the same direction. Then, because $q < 0$, $\Delta U_{electric} = +qEd$ is *negative*, and the electric potential energy of the system decreases. In the same way, if a negative charge moves a distance d in the direction of $\vec{E}$ so that $\theta = 0$ and $\cos\theta = +1$ in Equation 17-2, the electric force does *negative* work on the object (the electric force is opposite to E and so opposite the displacement). The electric potential energy change in this case is $\Delta U_{\text{electric}} = (-qEd)(+1) = -qEd$, which is positive because q is negative. So the electric potential energy of the system increases. These results are exactly opposite to what happens for a positive charge that moves a distance d in the direction of the electric field.

EXAMPLE 17-1 Electric Potential Energy Difference in a Uniform Field

An electron (charge $q = -e = -1.60 \times 10^{-19}$ C, mass $m = 9.11 \times 10^{-31}$ kg) is released from rest in a uniform electric field. The field points in the positive x direction and has magnitude 2.00×10^2 N/C. Find the speed of the electron after it has moved 0.300 m.

Set Up

The only force acting on the electron of charge q is the conservative electric force, so mechanical energy is conserved (Equation 6-23). We'll find the change in electric potential energy using Equation 17-2 and from this find the change in kinetic energy of the electron. The electron's initial kinetic energy is zero because it starts at rest; once we know the final kinetic energy, we can determine the electron's final speed.

The electron has a negative charge, so the force $\vec{F} = q\vec{E}$ is directed opposite to the electric field $\vec{E}$, and the electron will move in the direction opposite to $\vec{E}$ when released from rest. So the angle in Equation 17-2 between electric field and displacement will be $\theta = 180°$.

Conservation of mechanical energy:

$$K_{\text{i}} + U_{\text{electric,i}} = K_{\text{f}} + U_{\text{electric,f}} \qquad (6\text{-}23)$$

Electric potential energy change for a charge in a uniform electric field:

$$\Delta U_{\text{electric}} = -qEd\cos\theta \qquad (17\text{-}2)$$

Kinetic energy:

$$K = \frac{1}{2}mv^2 \qquad (6\text{-}8)$$

$E = 2.00 \times 10^2$ N/C

$d = 0.300$ m

$-e$

$+x$

Solve

Use Equation 17-2 to solve for the change in electric potential energy.

The change in electric potential energy equals the difference between the final and initial electric potential energy:

$$\Delta U_{\text{electric}} = U_{\text{electric,f}} - U_{\text{electric,i}}$$

From Equation 17-2,

$$\begin{aligned}\Delta U_{\text{electric}} &= U_{\text{electric,f}} - U_{\text{electric,i}} \\ &= -qEd\cos\theta \\ &= -(-1.60 \times 10^{-19}\ \text{C})(2.00 \times 10^2\ \text{N/C})(0.300\ \text{m})\cos 180° \\ &= -(-9.60 \times 10^{-18}\ \text{N}\cdot\text{m})(-1) \\ &= -9.60 \times 10^{-18}\ \text{N}\cdot\text{m} = -9.60 \times 10^{-18}\ \text{J}\end{aligned}$$

Find the final kinetic energy and the final speed of the electron.

The initial kinetic energy of the electron is $K_{\text{i}} = 0$. Solve the energy conservation equation for the final kinetic energy of the electron:

$$\begin{aligned}K_{\text{f}} &= K_{\text{i}} + U_{\text{electric,i}} - U_{\text{electric,f}} = 0 - (U_{\text{electric,f}} - U_{\text{electric,i}}) \\ &= 0 - (-9.60 \times 10^{-18}\ \text{J}) = 9.60 \times 10^{-18}\ \text{J}\end{aligned}$$

Finally, solve for the final speed of the electron:

$$K_{\text{f}} = \frac{1}{2}mv_{\text{f}}^2 \text{ so}$$

$$v_{\text{f}} = \sqrt{\frac{2K_{\text{f}}}{m}} = \sqrt{\frac{2(9.60 \times 10^{-18}\ \text{J})}{9.11 \times 10^{-31}\ \text{kg}}} = 4.59 \times 10^6\ \text{m/s}$$

Reflect

The electric potential energy decreases when we release the electron in the electric field, in the same way that the gravitational potential energy decreases when you release a ball in the presence of Earth's gravity. In both situations, the kinetic energy increases by the same amount that the potential energy decreases.

You should verify that the force on the electron is very small, only 3.2×10^{-17} N in magnitude. The electron has only a tiny mass, however, and our results show that it acquires a ferocious speed (about 1.5% of the speed of light) after moving just 0.300 m.

Electric Potential Energy of Point Charges

Electric potential energy is particularly important for the case of two point charges interacting with each other. An example from chemistry is the dissociation energy of an ionic compound such as sodium chloride (NaCl)—that is, the energy that must be given to the molecule to break it into its component atoms. The dissociation energy is determined in large part by the change in electric potential energy required to separate the positive sodium ion (Na^+) and the negative chloride ion (Cl^-), both of which behave much like point charges, as well as by the change in electric potential energy required to move an electron from the Cl^- ion to the Na^+ ion to make the atoms neutral.

Just as for a charge moving in a uniform electric field, we can use our knowledge of gravitation to find an expression for the electric potential energy of two point charges q_1 and q_2 separated by a distance r. In Section 10-2 we saw the following expression for the attractive *gravitational* force between two objects with *masses* m_1 and m_2 separated by a distance r:

Newton's law of universal gravitation (10-2)

$$F_{1 \text{ on } 2} = F_{2 \text{ on } 1} = \frac{Gm_1m_2}{r^2}$$

Gravitational constant (same for any two objects)

Masses of the two objects

Any two objects (1 and 2) exert equally strong **gravitational forces** on each other.

Center-to-center distance between the two objects

The gravitational forces are attractive: $\vec{F}_{1 \text{ on } 2}$ pulls object 2 toward object 1 and $\vec{F}_{2 \text{ on } 1}$ pulls object 1 toward object 2.

The gravitational potential energy of these two objects is given by Equation 10-4 in Section 10-3:

Gravitational potential energy (10-4)

$$U_{\text{grav}} = -\frac{Gm_1m_2}{r}$$

Gravitational constant (same for any two objects)

Masses of the two objects

Gravitational potential energy of a system of two objects (1 and 2)

Center-to-center distance between the two objects

The gravitational potential energy is zero when the two objects are infinitely far apart. If the objects are brought closer together (so r is made smaller), U_{grav} decreases (it becomes more negative).

Figure 17-4a graphs the gravitational potential energy given by Equation 10-4. We choose the point at which potential energy U_{grav} is zero to be where the two objects are infinitely far apart, so $r \to \infty$. The gravitational potential energy is negative for any finite value of r and increases—that is, becomes less negative—as the objects move farther apart. That's because the change in gravitational potential energy is the negative of the work done by the gravitational force: $\Delta U_{\text{grav}} = -W_{\text{grav}}$ (see Equation 6-16 at the beginning of this section). If we hold object m_1 stationary and move object m_2 farther away, increasing the distance r, the attractive gravitational force does negative work. Then $W_{\text{grav}} < 0$, so $\Delta U_{\text{grav}} > 0$ and the gravitational potential energy increases, as Figure 17-4a shows.

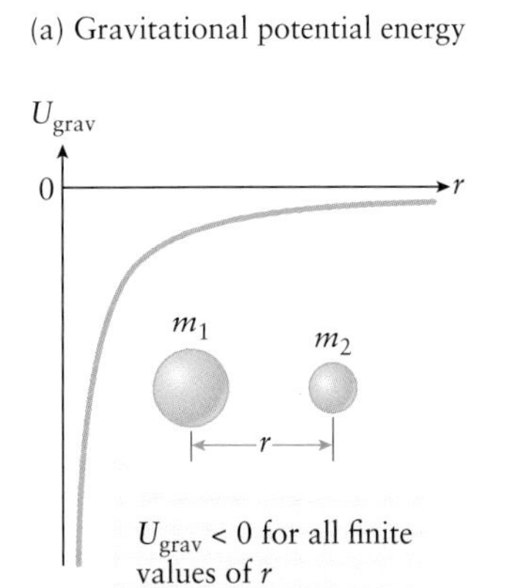

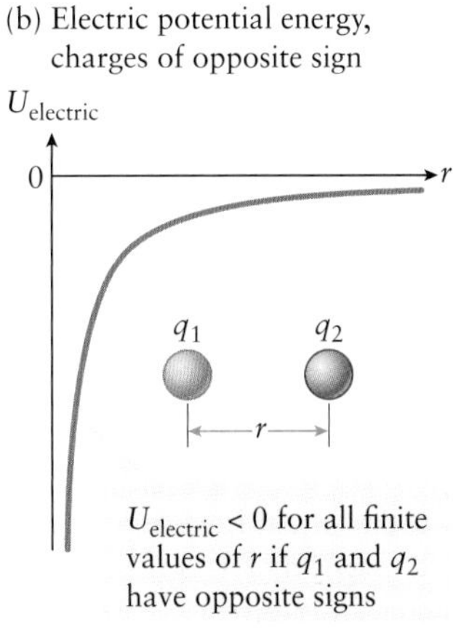

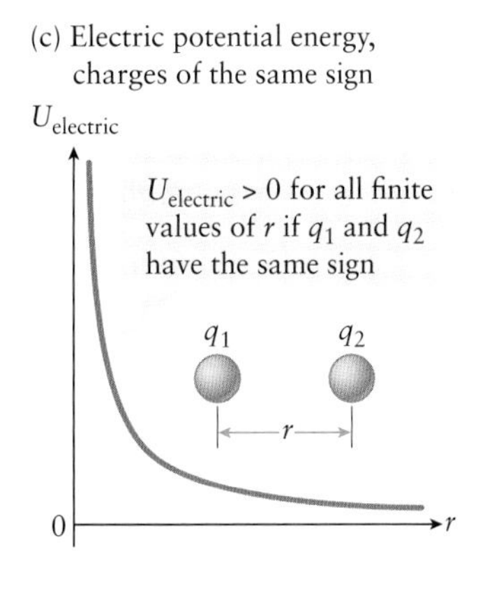

Figure 17-4 Potential energy for two masses and for two charges The electric potential energy $U_{electric}$ for two point charges is similar to the gravitational potential energy for two masses. The sign of $U_{electric}$ depends on the signs of the two charges.

Now compare Equation 10-2 to Coulomb's law for the electric force between two point charges, Equation 16-1:

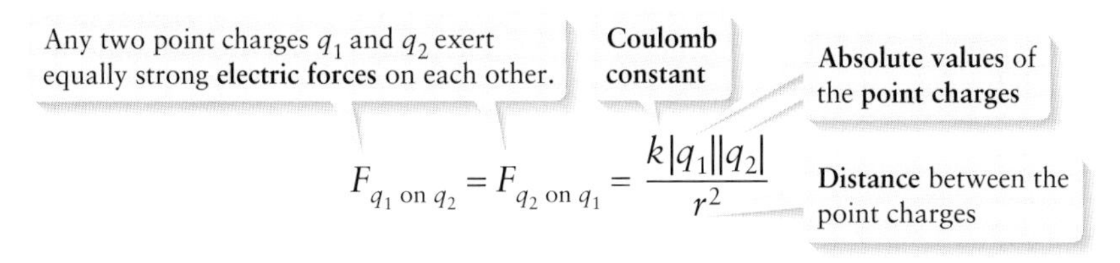

$$F_{q_1 \text{ on } q_2} = F_{q_2 \text{ on } q_1} = \frac{k|q_1||q_2|}{r^2}$$

Coulomb's law (16-1)

If we replace m and G in Equation 10-2 with q and k, we see that Equation 16-1 is identical to Equation 10-2 for the gravitational force, but with an important difference: The electric force is *attractive* (like the gravitational force) if the two charges q_1 and q_2 have opposite signs (one negative and the other positive) but *repulsive* if q_1 and q_2 have the same sign (either both positive or both negative). Here's an expression for the electric potential energy of two point charges that accounts for both of these possibilities:

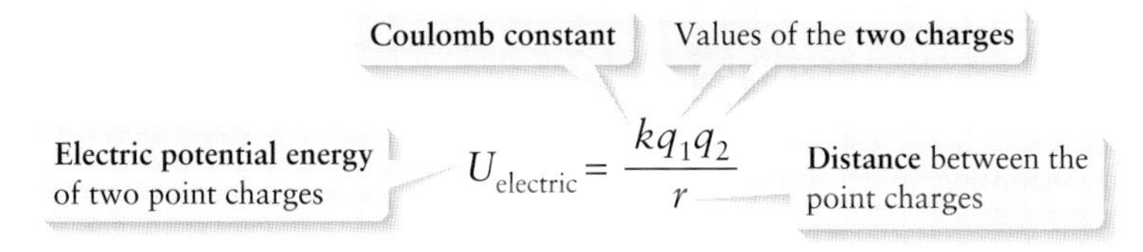

$$U_{electric} = \frac{kq_1q_2}{r}$$

Electric potential energy of two point charges (17-3)

This expression is very similar to Equation 10-4 for gravitational potential energy. Equation 17-3 shows that $U_{electric}$ is inversely proportional to the distance r, so the electric potential energy is zero when the two charges are infinitely far apart ($r \rightarrow \infty$). But unlike gravitational potential energy, $U_{electric}$ can be either negative or positive depending on the signs of the two charges. If q_1 and q_2 have different signs (one positive and one negative) so that the two charges attract each other, then $U_{electric} < 0$ for any finite distance r between the charges (**Figure 17-4b**). This makes sense: If we hold charge q_1 at rest and move charge q_2 farther away, the attractive electric force on q_2 does negative work (the force on q_2 is opposite to the displacement of q_2). In this case the change in electric potential energy is positive, and the electric potential energy increases (becomes less negative) with increasing distance r, just like the gravitational potential energy of two masses. But if q_1 and q_2 have the same sign (both positive or both negative) so that the two charges repel each other, then $U_{electric} > 0$ for any finite distance r between the charges (**Figure 17-4c**). In this case when q_2 moves away from q_1, the repulsive electric force on q_2 does positive work (the force on q_2 is in the same direction as the displacement of q_2). Then the change in electric potential energy is negative, and the electric potential energy *decreases* (becomes less positive) with increasing distance r.

Here's a useful way to interpret U_{electric} as given by Equation 17-3:

The electric potential energy of a pair of point charges equals the amount of work you would have to do to bring the charges to their current positions from infinitely far away.

If the two charges have opposite signs as in Figure 17-4b, you would have to do *negative* work to oppose the attractive electric force that pulls the two charges together. (If you didn't exert a force to do this work, the electric attraction would make the charges crash into each other instead of stopping a distance r apart.) So in this case the electric potential energy of the two charges is negative, as Figure 17-4b shows. By contrast, if the two charges have the same sign as in Figure 17-4c, you would have to do positive work to push the two charges together to overcome the repulsive electric force that pushes the two charges apart. So the electric potential energy of these two charges is positive, as depicted in Figure 17-4c.

The same idea holds for an assemblage of three or more point charges. To put charges q_1, q_2, and q_3 into proximity to each other, you would first have to do work to bring q_1 and q_2 from infinity to a distance r_{12} from each other. If you then brought in the third charge q_3 while keeping the other two charges stationary, you would have to do additional work against the electric force that q_1 exerts on q_3 *and* against the electric force that q_2 exerts on q_3. If q_3 ends up a distance r_{13} from q_1 and a distance r_{23} from q_2, the electric potential energy of the assemblage is the total amount of work that you did:

$$U_{\text{electric}} = \frac{kq_1q_2}{r_{12}} + \frac{kq_1q_3}{r_{13}} + \frac{kq_2q_3}{r_{23}} \quad (17\text{-}4)$$

(electric potential energy of three charges)

Go to Interactive Exercise 17-1 for more practice dealing with electric potential energy.

The total electric potential energy of a system of three charges is the sum of three terms, one for each pair of charges in the system (q_1 and q_2, q_1 and q_3, and q_2 and q_3). Each such term is the same as Equation 17-3. You can easily extend this idea to an assemblage made up of any number of point charges.

The following examples show how to do calculations using Equations 17-3 and 17-4.

EXAMPLE 17-2 Electric Potential Energy and Nuclear Fission

When a nucleus of uranium-235 (92 protons and 143 neutrons) absorbs an additional neutron, it undergoes a process called *nuclear fission* in which it breaks into two smaller nuclei. One possible fission is for the uranium nucleus to divide into two palladium nuclei, each of which has 46 protons and is 5.9×10^{-15} m in radius. The palladium nuclei then fly apart due to their electric repulsion. If we assume that the two palladium nuclei begin at rest and are just touching each other, what is their combined kinetic energy when they are very far apart?

Set Up

We can treat the two spherical nuclei as though they were point charges of $q = +46e$ located at the centers of the two nuclei. (To motivate this, recall from Section 16-7 that the electric field outside a sphere of charge Q is the same as that of a point charge Q located at the sphere's center.) Equation 17-3 then tells us the electric potential energy of the two palladium nuclei when they begin at rest. We'll then use energy conservation to find the combined kinetic energy when the palladium nuclei are very far apart.

Electric potential energy of two point charges:

$$U_{\text{electric}} = \frac{kq_1q_2}{r} \quad (17\text{-}3)$$

Conservation of mechanical energy:

$$K_{\text{i}} + U_{\text{electric,i}} = K_{\text{f}} + U_{\text{electric,f}} \quad (6\text{-}23)$$

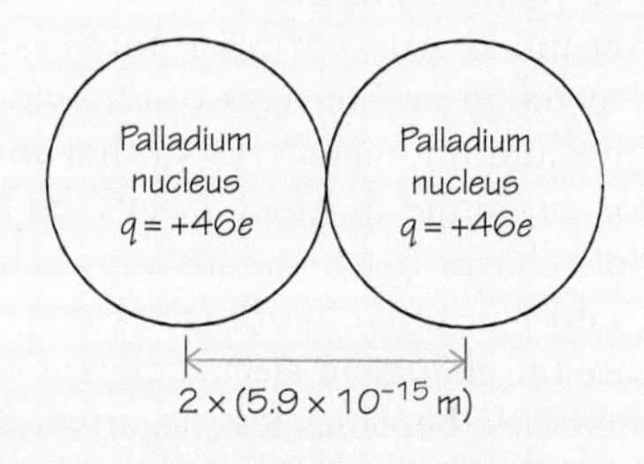

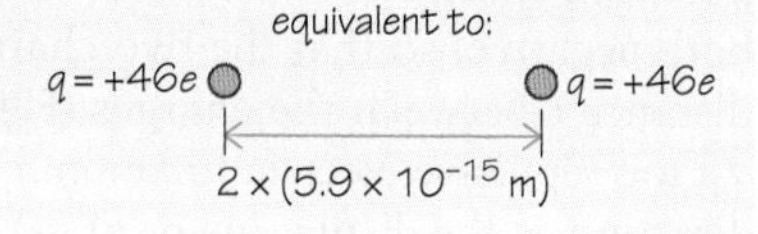

Solve

Use Equation 17-3 to solve for the initial electric potential energy when the two nuclei are just touching.

Each nucleus has charge $+46e$:

$q_1 = q_2 = +46(1.60 \times 10^{-19}\ \text{C}) = +7.36 \times 10^{-18}\ \text{C}$

The separation between the charges is twice the radius of either nucleus:

$r = 2(5.9 \times 10^{-15}\ \text{m})$

The electric potential energy is

$$U_{\text{electric,i}} = \frac{kq_1q_2}{r} = \frac{(8.99 \times 10^9\ \text{N}\cdot\text{m}^2/\text{C}^2)(7.36 \times 10^{-18}\ \text{C})^2}{2(5.9 \times 10^{-15}\ \text{m})}$$
$$= 4.1 \times 10^{-11}\ \text{N}\cdot\text{m} = 4.1 \times 10^{-11}\ \text{J}$$

Use energy conservation to find the combined final kinetic energy of the two palladium nuclei.

The palladium nuclei begin at rest, so the initial kinetic energy is zero:

$K_i = 0$

The palladium nuclei end up very far apart, so their separation is essentially infinite ($r \to \infty$). The final electric potential energy is therefore zero:

$U_{\text{electric,f}} = 0$

From the energy conservation equation, the final kinetic energy is

$$K_f = K_i + U_{\text{electric,i}} - U_{\text{electric,f}}$$
$$= 0 + 4.1 \times 10^{-11}\ \text{J} - 0 = 4.1 \times 10^{-11}\ \text{J}$$

Reflect

All of the initial electric potential energy is converted into kinetic energy of the palladium nuclei. The energy released by the fission of a single uranium-235 nucleus, 4.1×10^{-11} J, is very small. But imagine that you could get 1.0 kg of uranium-235, which contains about 2.6×10^{24} uranium atoms, to undergo fission at once. (This is the principle of a *fission bomb.*) The released energy would be $(2.6 \times 10^{24}) \times (4.1 \times 10^{-11}\ \text{J}) = 1.1 \times 10^{14}$ J, equivalent to the energy given off by exploding 26,000 tons of TNT. This suggests the terrifying amount of energy that can be released by a fission weapon.

EXAMPLE 17-3 Electric Potential Energy of Three Charges

A particle with charge $q_1 = +4.30\ \mu\text{C}$ is located at $x = 0$, $y = 0$. A second particle with charge $q_2 = -9.80\ \mu\text{C}$ is located at $x = 0$, $y = 4.00$ cm, and a third particle with charge $q_3 = +5.00\ \mu\text{C}$ is located at position $x = 3.00$ cm, $y = 0$. (Note that $1\ \mu\text{C} = 1$ microcoulomb $= 10^{-6}$ C.) What is the total electric potential energy of these three charges?

Set Up

We'll use Equation 17-4 to find the value of U_{electric}. We're given the values of the three charges; we'll use the positions of the charges to find the distances r_{12}, r_{13}, and r_{23} between them.

Electric potential energy of three charges:

$$U_{\text{electric}} = \frac{kq_1q_2}{r_{12}} + \frac{kq_1q_3}{r_{13}} + \frac{kq_2q_3}{r_{23}} \quad (17\text{-}4)$$

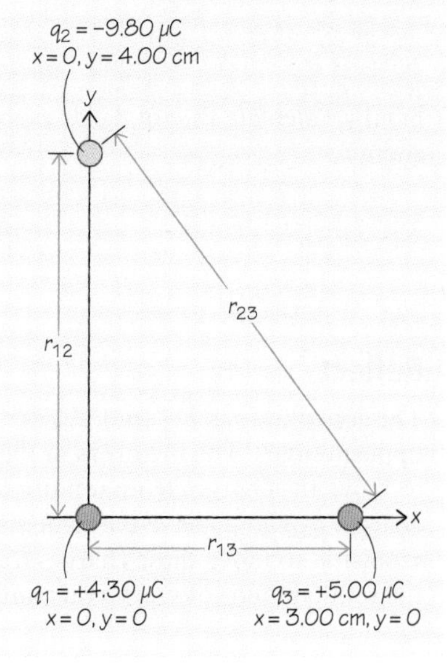

Solve

The figure above shows that charges 1 and 2 are $r_{12} = 4.00$ cm apart, and that charges 1 and 3 are $r_{13} = 3.00$ cm apart. The distance r_{23} between charges 2 and 3 is the hypotenuse of a right triangle of sides r_{12} and r_{13}.

From the figure above,

$$r_{12} = 4.00 \text{ cm} = 4.00 \times 10^{-2} \text{ m}$$
$$r_{13} = 3.00 \text{ cm} = 3.00 \times 10^{-2} \text{ m}$$

From the Pythagorean theorem,

$$\begin{aligned} r_{23} &= \sqrt{r_{12}^2 + r_{13}^2} \\ &= \sqrt{(4.00 \times 10^{-2} \text{ m})^2 + (3.00 \times 10^{-2} \text{ m})^2} \\ &= 5.00 \times 10^{-2} \text{ m} \end{aligned}$$

Calculate each term in the expression for electric potential energy, Equation 17-4.

We can write Equation 17-4 as

$$U_{\text{electric}} = U_{12} + U_{13} + U_{23}$$

Each term on the right-hand side of this equation represents the contribution to the electric potential energy due to a specific pair of point charges. The contribution due to the interaction between charges q_1 and q_2 is

$$\begin{aligned} U_{12} &= \frac{kq_1q_2}{r_{12}} \\ &= \frac{(8.99 \times 10^9 \text{ N} \cdot \text{m}^2/\text{C}^2)(+4.30 \times 10^{-6} \text{ C})(-9.80 \times 10^{-6} \text{ C})}{4.00 \times 10^{-2} \text{ m}} \\ &= -9.47 \text{ J} \end{aligned}$$

The contribution due to the interaction between charges q_1 and q_3 is

$$\begin{aligned} U_{13} &= \frac{kq_1q_3}{r_{13}} \\ &= \frac{(8.99 \times 10^9 \text{ N} \cdot \text{m}^2/\text{C}^2)(+4.30 \times 10^{-6} \text{ C})(+5.00 \times 10^{-6} \text{ C})}{3.00 \times 10^{-2} \text{ m}} \\ &= +6.44 \text{ J} \end{aligned}$$

The contribution due to the interaction between charges q_2 and q_3 is

$$\begin{aligned} U_{23} &= \frac{kq_2q_3}{r_{23}} \\ &= \frac{(8.99 \times 10^9 \text{ N} \cdot \text{m}^2/\text{C}^2)(-9.80 \times 10^{-6} \text{ C})(+5.00 \times 10^{-6} \text{ C})}{5.00 \times 10^{-2} \text{ m}} \\ &= -8.81 \text{ J} \end{aligned}$$

Finally, calculate the total electric potential energy.

The total electric potential energy is the sum of the three terms calculated above:

$$\begin{aligned} U_{\text{electric}} &= U_{12} + U_{13} + U_{23} \\ &= (-9.47 \text{ J}) + (+6.44 \text{ J}) + (-8.81 \text{ J}) \\ &= -11.84 \text{ J} \end{aligned}$$

Reflect

The contributions U_{12} and U_{23} are negative because these pairs of charges (q_1 and q_2 for U_{12}, q_2 and q_3 for U_{23}) have opposite signs. The members of these pairs attract each other, so you must do negative work against each attractive force to move the charges from infinity to the positions shown in the figure. The contribution U_{13}, however, is positive because charges q_1 and q_3 have the same sign (both positive). These two charges repel, so you must do positive work against the repulsion to move these charges from infinity to their positions. The total potential energy is negative, which shows that the total amount of work you would do to move all three charges from infinity is negative.

GOT THE CONCEPT? 17-1 Electric Potential Energy

? Suppose you reversed the signs of the three point charges in Example 17-3 so that $q_1 = -4.30\ \mu C$, $q_2 = +9.80\ \mu C$, and $q_3 = -5.00\ \mu C$. If the positions of the charges remain unchanged, what would be the total electric potential energy of this assemblage? (a) More negative than -11.84 J; (b) -11.84 J; (c) between -11.84 J and $+11.84$ J; (d) $+11.84$ J; (e) more positive than $+11.84$ J.

TAKE-HOME MESSAGE FOR Section 17-2

✔ The electric potential energy associated with a point charge can change if the charge changes position in an electric field.

✔ The electric potential energy of a pair of point charges equals the amount of work you would have to do to move those charges from infinity to their present positions.

✔ The electric potential energy of an assemblage of three or more point charges is the sum of the electric potential energies for each pair of charges in the assemblage.

17-3 Electric potential equals electric potential energy per charge

Our discussion in Section 17-2 shows that if a point charge q changes position, the potential energy change $\Delta U_{\text{electric}}$ depends on both the magnitude and the sign (positive or negative) of q (see Figure 17-3). We can simplify things by considering the potential energy *per charge*—that is, the electric potential energy for a charge at a given position divided by the value of that charge. We call this quantity the **electric potential** V:

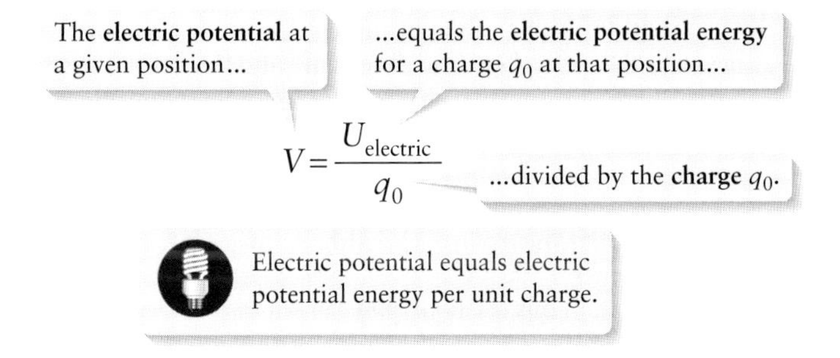

Electric potential related to electric potential energy (17-5)

We call the charge q_0 the **test charge**: Its charge has such a small magnitude that it doesn't affect the other charges that create the electric field in which q_0 moves. Because we divide out the value of q_0, the value of the potential V at a given position does *not* depend on the value of the point charge q_0 that we place there. Instead, V is determined by the other charges that produce the electric field at the position where we place the test charge.

Note that electric potential V has the same relationship to electric potential energy U_{electric} as electric field $\vec{E}$ has to electric force $\vec{F}$: V is electric potential energy per charge, just as $\vec{E}$ is electric force per charge. Like electric potential energy, but unlike electric field, potential V is a *scalar* quantity.

For electric potential energy, the value of U_{electric} for a charge at a given position is not as important as the potential energy *difference* $\Delta U_{\text{electric}}$ when the charge moves from a point a to a different point b. The same is true for electric potential. From Equation 17-5, the **electric potential difference** ΔV between two points is

Electric potential difference related to electric potential energy difference (17-6)

The **difference in electric potential** between two positions...

...equals the **change in electric potential energy** for a charge q_0 moved between these two positions...

$$\Delta V = \frac{\Delta U_{\text{electric}}}{q_0}$$

...divided by the **charge** q_0.

Electric potential difference equals electric potential energy difference per unit charge.

The SI unit of electric potential and electric potential difference is the **volt** (V), named after the Italian scientist Alessandro Volta. For example, a common AA or AAA flashlight battery has "1.5 volts" written on its side. This means that the electric potential at the positive terminal of the battery (labeled +) is 1.5 V greater than the electric potential at the negative terminal of the battery (labeled −). In other words, 1.5 V is the electric potential *difference* between the terminals of the battery. In Chapter 18 we'll see how this electric potential difference causes electric charge to flow when a battery is included in an electric circuit. (Note that Volta invented the electric battery in 1800.) Equations 17-5 and 17-6 show that 1 volt is equal to 1 joule per coulomb, and we know that 1 joule is equal to 1 newton multiplied by 1 meter. So

$$1\ \text{V} = 1\,\frac{\text{J}}{\text{C}} = 1\,\frac{\text{N}\cdot\text{m}}{\text{C}}$$

This means that if the electric potential at point b is 1 V higher than at point a, it takes 1 J of work to move a charge of +1 C from a to b. If a charge of +1 C at point b is released from rest and moves to point a, the electric potential that the charge experiences decreases by 1 V, the electric potential energy associated with the charge decreases by 1 J, and the charge acquires 1 J of kinetic energy.

We saw in Chapter 16 that the SI units of electric field are newtons per coulomb, or N/C. Because 1 V = 1 N·m/C, it follows that 1 V/m = 1 N/C. So an equally good set of units for electric field is V/m (volts per meter).

We'll often abbreviate the terms "electric potential" and "electric potential difference" as simply "potential" and "potential difference," respectively. Since the unit of electric potential is the volt, it's also common to refer to electric potential difference as **voltage**.

WATCH OUT! Don't confuse the symbol *V* for electric potential difference and the abbreviation V for volts.

In this chapter and others you'll see mathematical statements such as "V = 10 V." Such statements are perfectly legal because V and V are different symbols. V (italicized) represents the electric potential at a point in space. V (not italicized) is an abbreviation for the unit of electric potential, the volt. In this textbook you can tell the difference by noticing which font is used and by paying close attention to the context in which V and V are used. But it's pretty hard to handwrite something in italics, so make sure you know the distinction between V and V in your own work on your homework assignments and exams.

WATCH OUT! Electric potential and electric potential energy are not the same.

Remember that electric potential V and electric potential energy U_{electric} are related but different scalar quantities. Electric potential energy U_{electric} refers to the energy associated with a particular amount of charge at a given position. Electric potential V is the energy associated with a *unit* charge at that position.

Electric Potential in a Uniform Electric Field

We can use Equation 17-6 to find the electric potential difference between two points a and b in a uniform electric field (**Figure 17-5**). From Equation 17-2 the change or difference in electric potential energy when a charge q_0 is moved from a to b is

$\Delta U_{\text{electric}} = -q_0 Ed \cos\theta$. Equation 17-6 tells us that the electric potential difference between a and b is $\Delta U_{\text{electric}}$ divided by q_0:

$$\Delta V = \frac{\Delta U_{\text{electric}}}{q_0} = \frac{-q_0 Ed \cos\theta}{q_0}$$
$$= -Ed\cos\theta \qquad (17\text{-}7)$$

Equation 17-7 shows that ΔV is positive if $\cos\theta$ is negative, which is the case if the angle θ is greater than 90° as depicted in Figure 17-5. In other words, *if you move in a direction opposite to the electric field, the electric potential increases.* If the angle θ is less than 90°, $\cos\theta$ is positive and ΔV is negative; *if you move in the direction of the electric field, the electric potential decreases.*

We know that the electric force on a positive charge is in the direction of the electric field $\vec{E}$, and the electric force on a negative charge is in the direction opposite to $\vec{E}$. So our observations about how electric potential changes with position tell us that

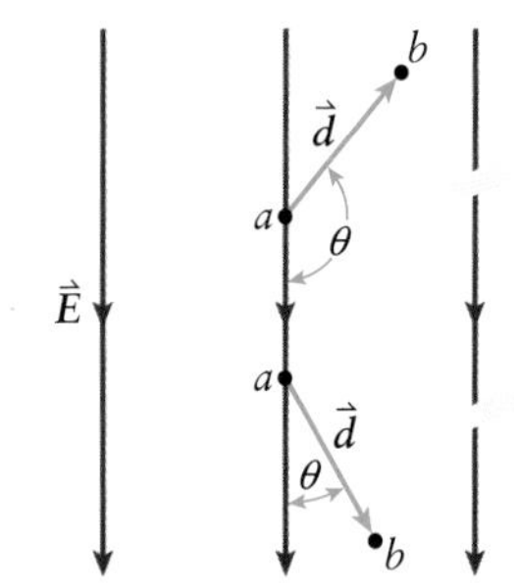

If you move in a direction opposite to the electric field ($\theta > 90°$), the electric potential increases ($\Delta V > 0$).

If you move in the direction of the electric field ($\theta < 90°$), the electric potential decreases ($\Delta V < 0$).

If you move from point a to point b in a uniform electric field $\vec{E}$, the change in electric potential is $\Delta V = -Ed\cos\theta$.

Figure 17-5 Change in electric potential Calculating the difference in electric potential between two points in a uniform electric field $\vec{E}$.

If an object has positive charge, the electric force on that object pushes it toward a region of lower electric potential. If an object has negative charge, the electric force on that object pushes it toward a region of higher electric potential.

As we'll see below, these observations hold true whether or not the electric field is uniform.

Comparing Equations 17-2 and 17-7 shows that the right-hand side of Equation 17-7 equals the negative of the work done by the electric field ($q_0 Ed \cos\theta$) divided by the charge q_0. So another way to think of the potential difference between two points is as the negative of the work done by the electric field per unit charge when a charged object is moved between those points. Alternatively, the potential difference between two points equals the work that *you* must do per unit charge against the electric force to move a charged object between those points. So if you take a charge of +1 C that is at rest at point a and move it to point b where it is again at rest, and the potential at b is 1 V higher than the potential at a, the electric field does −1 J of work on the charge and you do +1 J of work on that charge.

Go to Picture It 17-2 for more practice dealing with electric potential.

EXAMPLE 17-4 Electric Potential Difference in a Uniform Field I

A uniform electric field points in the positive x direction and has magnitude 2.00×10^2 V/m. Points a and b are both in this field: Point b is a distance 0.300 m from a in the negative x direction. Determine the electric potential difference $V_b - V_a$ between points a and b.

Set Up

We apply Equation 17-7 to the path shown in the figure. Note that the magnitude of the displacement is d = 0.300 m, and the angle between the electric field and the displacement is $\theta = 180°$. Note also that the potential difference ΔV equals the potential at the *end* of the displacement $\vec{d}$ (that is, at point b) minus the potential at the *beginning* of the displacement (that is, at point a).

Potential difference between two points in a uniform electric field:

$$\Delta V = V_b - V_a = -Ed\cos\theta \qquad (17\text{-}7)$$

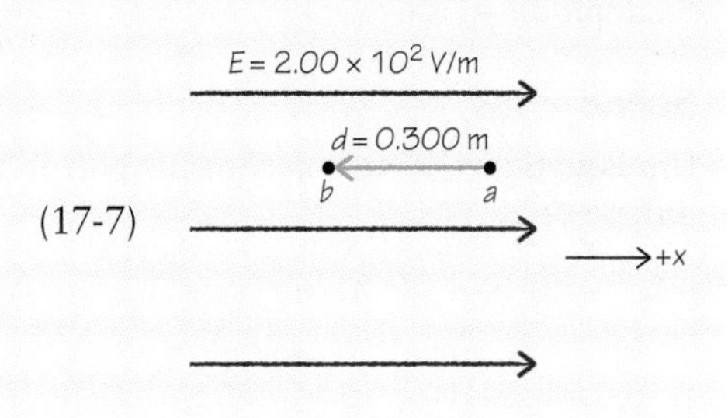

Solve

Use Equation 17-7 to solve for the potential difference.

Calculate the potential difference from the electric field magnitude E, the displacement d, and the angle θ:

$$\Delta V = V_b - V_a = -Ed\cos\theta$$
$$= -(2.00 \times 10^2\ \text{V/m})(0.300\ \text{m})\cos 180°$$
$$= -(60.0\ \text{V})(-1)$$
$$= +60.0\ \text{V}$$

Reflect

We can check our result by comparing with Example 17-1 in Section 17-2, where we considered the change in electric potential energy $\Delta U_{\text{electric}}$ for an electron that undergoes the same displacement in this same electric field. Using Equation 17-6 we find the same value of $\Delta U_{\text{electric}}$ as in Example 17-1.

The positive value of $\Delta V = V_b - V_a$ means that point b is at a higher potential than point a. This agrees with our observation above that if you travel opposite to the direction of the electric field, the electric potential increases. The value $E = 2.00 \times 10^2$ V/m means that the electric potential increases by 2.00×10^2 V for every meter that you travel opposite to the direction of $\vec{E}$.

If a charge $q_0 = -1.60 \times 10^{-19}$ C travels from a to b, the charge in electric potential energy is given by Equation 17-6:

$$\Delta V = \frac{\Delta U_{\text{electric}}}{q_0}$$

so

$$\Delta U_{\text{electric}} = q_0 \Delta V = (-1.60 \times 10^{-19}\ \text{C})(+\ 60.0\ \text{V})$$
$$= -9.60 \times 10^{-18}\ \text{V} \cdot \text{C} = -9.60 \times 10^{-18}\ \text{J}$$

(Recall from above that 1 V = 1 J/C.)

EXAMPLE 17-5 Electric Potential Difference in a Uniform Field II

Determine the electric potential difference $\Delta V = V_c - V_a$ between points a and c in the uniform electric field of Example 17-4. Point c is a distance 0.500 m from point a, and the straight-line path from a to c makes an angle of 126.9° with respect to the electric field.

Set Up

Again we'll use Equation 17-7 to calculate the potential difference between the two points. Since the displacement from point a to point c points generally opposite to the direction of the electric field, we expect that $\Delta V = V_c - V_a$ will be positive, just like $\Delta V = V_b - V_a$ in Example 17-4.

Potential difference between two points in a uniform electric field:

$$\Delta V = V_c - V_a = -Ed\cos\theta \qquad (17\text{-}7)$$

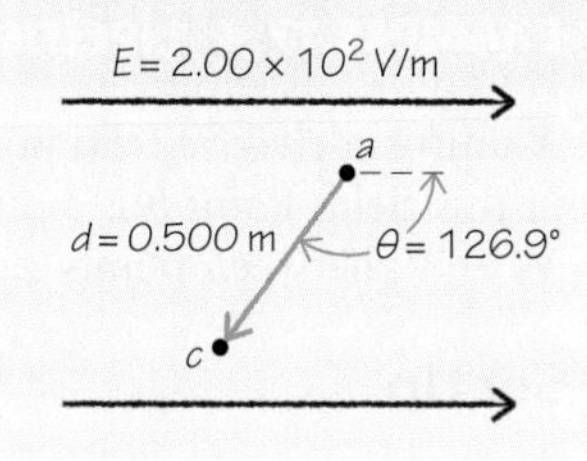

Solve

Use Equation 17-7 to solve for the potential difference.

Calculate the potential difference from the electric field magnitude E, the displacement d, and the angle θ:

$$\Delta V = V_c - V_a = -Ed\cos\theta$$
$$= -(2.00 \times 10^2\ \text{V/m})(0.500\ \text{m})\cos 126.9°$$
$$= -(2.00 \times 10^2\ \text{V/m})(0.500\ \text{m})(-0.600)$$
$$= +60.0\ \text{V}$$

Reflect

The potential difference between points a and c is the *same* as the potential difference between points a and b in Example 17-4. Equation 17-7 tells us why this should be: The potential difference $\Delta V = -Ed \cos \theta$ involves the magnitude E of the electric field multiplied by $d \cos \theta$, which is the component of the displacement $\vec{d}$ in the direction of $\vec{E}$. In both examples the electric field magnitude has the same value (2.00×10^2 V/m), as does the component of displacement in the direction of the electric field (-0.300 m).

Another way to come to this same conclusion is to recognize that the displacement from a to c can be broken down into two parts: a displacement in the negative x direction from a to b, followed by a displacement from b to c in the y direction. The potential difference for the displacement from a to b is $V_b - V_a = +60.0$ V as we calculated in Example 17-4; the potential difference for the displacement from b to c is $V_c - V_b = -Ed \cos 90° = 0$, since that displacement is perpendicular to the electric field. So the net potential difference between a and c is $V_c - V_a = (V_c - V_b) + (V_b - V_a) = +60.0 \text{ V} + 0 \text{ V} = +60.0$ V.

EXAMPLE 17-6 Transmission Electron Microscope

A *transmission electron microscope* forms an image by sending a beam of fast-moving electrons rather than a beam of light through a thin sample. As we will see in Chapter 26, such fast-moving electrons behave very much like a light wave. If the electrons have sufficiently high energy, the image that they form can show much finer detail than even the best optical microscope. The electrons are emitted from a heated metal filament and are then accelerated toward a second piece of metal called the *anode* that is at a potential 2.50 kV (1 kV = 1 kilovolt = 10^3 V) higher than that of the filament. If the electrons leave the filament initially at rest, how fast are the electrons traveling when they pass the anode?

Set Up

As the electrons move through the potential difference between the filament and the anode, the electric potential energy will change in accordance with Equation 17-6. Each electron has a negative charge $q_0 = -e$, so an *increase* in electric potential ($\Delta V > 0$) means a *decrease* in electric potential energy ($\Delta U_{\text{electric}} < 0$). The total mechanical energy (kinetic energy plus potential energy) is conserved as the electron moves because there are no forces acting on it other than the conservative electric force, so the electron kinetic energy will increase as the potential energy decreases.

Electric potential difference related to electric potential energy difference:

$$\Delta V = \frac{\Delta U_{\text{electric}}}{q_0} \qquad (17\text{-}6)$$

Mechanical energy is conserved:

$$E = K + U_{\text{electric}} = \text{constant}$$

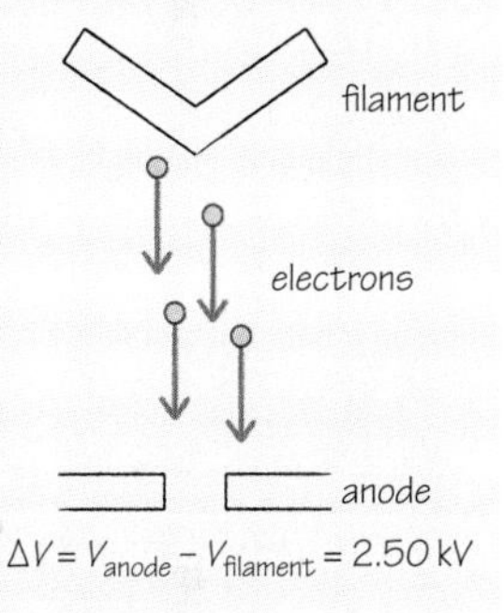

Solve

Solve Equation 17-6 for the change in electric potential energy as the electron moves from the filament to the anode, which is at a potential 2.50 kV higher than the filament.

From Equation 17-6,

$$\Delta U_{\text{electric}} = q_0 \Delta V$$

The electron has change $q_0 = -e$, so

$$\begin{aligned}\Delta U_{\text{electric}} &= -e\Delta V \\ &= -(1.60 \times 10^{-19} \text{ C})(+2.50 \times 10^3 \text{ V}) \\ &= -4.00 \times 10^{-16} \text{ J}\end{aligned}$$

(Recall that 1 V = 1 J/C, so 1 J = 1 C·V.)

The conservation of mechanical energy tells us that the change in the kinetic energy of the electron is equal to the negative of the change in the electric potential energy.

Mechanical energy is conserved:

$$E = K + U_{\text{electric}} = \text{constant}$$

So the *change* in mechanical energy is zero:

$$\Delta E = \Delta K + \Delta U_{\text{electric}} = 0$$

The change in the kinetic energy of the electron is

$$\begin{aligned}\Delta K &= -\Delta U_{\text{electric}} = -(-4.00 \times 10^{-16} \text{ J}) \\ &= +4.00 \times 10^{16} \text{ J}\end{aligned}$$

Each electron begins with zero kinetic energy, so the change in its kinetic energy is equal to its final kinetic energy as it reaches the anode.

Since the electron begins with zero kinetic energy at the filament, the change in its kinetic energy is

$$\Delta K = +4.00 \times 10^{-16}\ \text{J} = K_{\text{anode}} - K_{\text{filament}} = K_{\text{anode}}$$

Use this to find the speed of the electron at the anode:

$$K_{\text{anode}} = \frac{1}{2} m_{\text{electron}} v_{\text{anode}}^2, \text{ so}$$

$$v_{\text{anode}} = \sqrt{\frac{2K_{\text{anode}}}{m_{\text{electron}}}} = \sqrt{\frac{2(4.00 \times 10^{-16}\ \text{J})}{9.11 \times 10^{-31}\ \text{kg}}}$$

$$= 2.96 \times 10^7\ \text{m/s}$$

Reflect

The electrons are accelerated to nearly one-tenth of the speed of light ($c = 3.00 \times 10^8$ m/s).

Example 17-6 is just one of many situations in which an object with a charge of $-e$ (such as an electron) or $+e$ (such as a proton) moves through a potential difference. A common unit for the potential energy change in such situations is the **electron volt** (eV), which is equal to the magnitude e of the charge on the electron multiplied by 1 volt. Since $e = 1.60 \times 10^{-19}$ C, it follows that

$$1\ \text{eV} = (1.60 \times 10^{-19}\ \text{C})(1\ \text{V}) = 1.60 \times 10^{-19}\ \text{C} \cdot \text{V} = 1.60 \times 10^{-19}\ \text{J}$$

In Example 17-6, the electron moves through a potential difference of $+2.50 \times 10^3$ V, so the change in electric potential energy is -2.50×10^3 eV, and the kinetic energy that the electron acquires is $+2.50 \times 10^3$ eV. We also use the abbreviations 1 keV = 10^3 eV, 1 MeV = 10^6 eV, and 1 GeV = 10^9 eV. (The largest particle accelerator in the world, the Large Hadron Collider at the European Organization for Nuclear Research CERN near Geneva, Switzerland, accelerates protons to a kinetic energy of 7×10^{12} eV = 7 TeV. This is equivalent to making the protons pass through a potential difference of 7×10^{12}, or 7 trillion, volts.) In later chapters we'll see that the electron volt is a useful unit for expressing energies on the atomic or nuclear scale.

Electric Potential Due to a Point Charge

Note that Equation 17-7 is useful only in the case of a uniform electric field. Another important case is the electric potential due to a point charge Q. We know from Equation 17-3 that if we place a test charge q_0 a distance r from a point charge Q, the electric potential energy of the system of two charges is

$$U_{\text{electric}} = \frac{kq_0Q}{r}$$

Equation 17-5 tells us that to find the electric potential due to the point charge Q, we must divide U_{electric} by the value of the test charge q_0:

Electric potential due to a point charge (17-8)

Electric potential due to a point charge Q

Coulomb constant

$$V = \frac{kQ}{r}$$

Value of the point charge

Distance from the point charge Q to the location where the potential is measured

Equation 17-8 says that all points that are the same distance r from a point charge have the same electric potential due to that charge. It also says that if $Q > 0$, the electric potential due to the charge is positive and decreases (becomes less positive) as you move farther away from the charge so that r increases. If $Q < 0$, the electric potential due to the charge is negative and decreases (becomes more negative) as you move closer to the charge. For either sign of Q, the electric potential goes to zero at an infinite distance from the point charge.

These observations about Equation 17-8 are consistent with our previous statements about electric potential and electric force. A positive test charge q_0 placed near a positive charge Q feels an electric force that pushes it farther away from Q, toward regions where the potential V due to Q is lower (less positive). If instead that positive test charge is placed near a negative charge Q, the test charge feels an electric force that pulls it toward Q—again toward regions where the potential V is lower (in this case more negative).

WATCH OUT! Don't confuse the formulas for electric potential and electric field due to a point charge.

Be sure that you recognize the differences between Equation 17-8, $V = kQ/r$, and Equation 16-4 for the magnitude of the electric field due to a point charge Q, $E = k|Q|/r^2$. Equation 17-8 says that the potential V due to a point charge is inversely proportional to r and can be positive or negative, depending on the sign of Q. By contrast, Equation 16-4 tells us that the magnitude E of the field due to a point charge is inversely proportional to the *square* of r. Furthermore, because E is the magnitude of a vector, it is always positive (it is proportional to the absolute value of Q).

If there is not a single point charge but an assemblage of charges, the total electric potential at a given position due to these charges is the sum of the individual potentials. For example, for the case of three charges Q_1, Q_2, and Q_3, the potential at a point that is a distance r_1 from the first charge, a distance r_2 from the second charge, and a distance r_3 from the third charge is

$$V = \frac{kQ_1}{r_1} + \frac{kQ_2}{r_2} + \frac{kQ_3}{r_3} \qquad (17\text{-}9)$$

(electric potential of three charges)

GOT THE CONCEPT? 17-2 Electric Potential Difference

Four point charges are each moved from one position to another position where the electric potential has a different value. Rank the four charges in order of the electric potential energy change that takes place, from most positive to most negative. (a) A +0.0010 C charge that moves through a potential increase of 5.0 V; (b) a +0.0020 C charge that moves through a potential decrease of 2.0 V; (c) a −0.0015 C charge that moves through a potential decrease of 4.0 V; (d) a −0.0010 C charge that moves through a potential increase of 2.5 V.

TAKE-HOME MESSAGE FOR Section 17-3

✔ The electric potential at a certain position equals the electric potential energy per unit charge for a test charge at that position. Electric potential is a scalar, not a vector, quantity. Electric potential decreases as you move in the direction of the electric field.

✔ The electric potential due to a point charge is positive if the charge is positive and negative if the charge is negative. As you move farther from an isolated point charge, the electric potential becomes closer to zero.

✔ The electric potential due to a collection of charges is the sum of the potentials due to the individual charges.

✔ The electric potential difference between two points equals the difference in electric potential energy per unit charge between those points.

17-4 The electric potential has the same value everywhere on an equipotential surface

In Examples 17-4 and 17-5 in the previous section, we looked at the potential differences between two pairs of points in an electric field, a and b in Example 17-4 and a and c in Example 17-5. Although the electric field is the same in both examples, the distance from a to c in Example 17-5 is clearly longer than the distance from a to b in Example 17-4. So you might have expected that the potential difference $V_c - V_a$ in Example 17-5 would be greater in magnitude than the potential difference $V_b - V_a$ in Example 17-4. In fact, we found that the potential differences were equal, which tells us that the potential is the *same* at points b and c. You can see why from Equation 17-7, $\Delta V = -Ed \cos \theta$. A displacement $\vec{d}$ from point b to point c is perpendicular to the electric field $\vec{E}$, so the angle in Equation 17-7 is $\theta = 90°$, and $\cos \theta = \cos 90° = 0$. Thus the potential difference ΔV between points b and c must be zero.

In general, the electric potential will be the same at any two points that lie along a curve perpendicular to electric field lines. Such a curve is called an *equipotential curve* or simply an **equipotential**. For the case of a uniform electric field, the electric potential has the same value anywhere on a plane that's perpendicular to the electric field. Such a plane is an example of an **equipotential surface**, one on which the electric potential has the same value at all points (**Figure 17-6**). No work is required to move a charge from one point to another along any path on an equipotential surface. As we described in Section 17-3, the electric potential decreases as you move in the direction of the electric field $\vec{E}$, so the value of the potential V is lower for equipotential surfaces that are "downstream" in the electric field than on surfaces that are "upstream."

Figure 17-7 shows both electric field lines and equipotentials for the case of a *nonuniform* electric field. The equipotential surfaces are perpendicular at *all* points to the field lines, just as for the case of a uniform field in Figure 17-6. But since the electric field lines are not parallel lines, the equipotential surfaces are not flat planes. In general, *any* surface that is everywhere perpendicular to the electric field is an equipotential surface, and the value of the potential V on an equipotential surface is lower the farther "downstream" in the electric field that surface is.

In Figure 17-7 points e and f are both on the equipotential for which $V = +50$ V, and points s and t are both on the equipotential for which $V = 0$ V. So the potential difference between points e and s is the same as the potential difference between points f and t:

$$V_e - V_s = V_f - V_t = (+50 \text{ V}) - (0 \text{ V}) = +50 \text{ V}$$

However, the distance between points e and s is less than that between points f and t. Equation 17-7 tells us that the potential difference between two points is

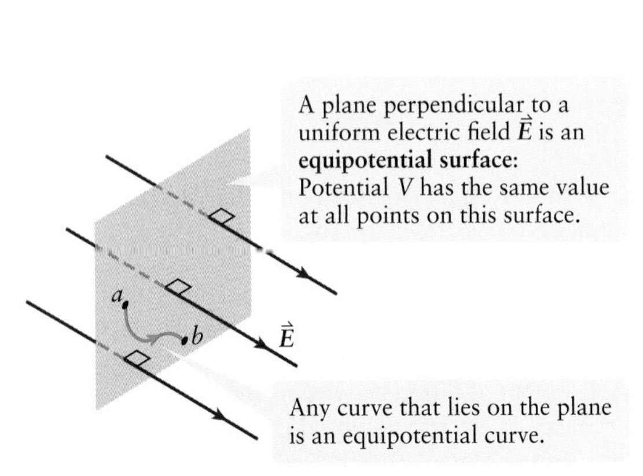
A plane perpendicular to a uniform electric field $\vec{E}$ is an **equipotential surface**: Potential V has the same value at all points on this surface.
a
b
$\vec{E}$
Any curve that lies on the plane is an equipotential curve.

Figure 17-6 Equipotential surfaces I In a uniform electric field equipotential surfaces are planes perpendicular to the electric field lines.

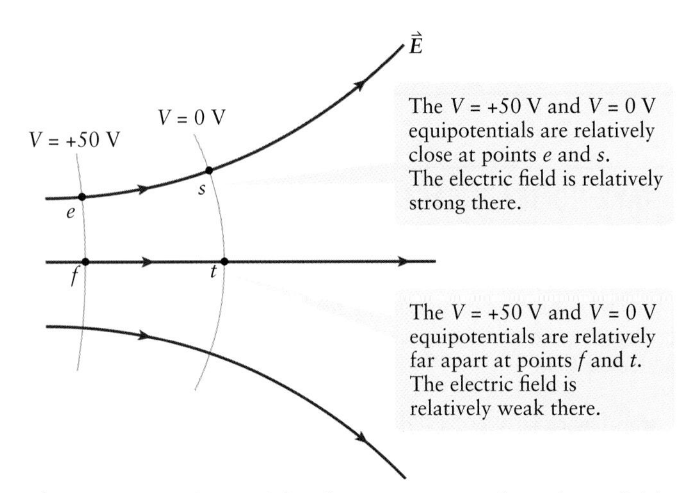
$\vec{E}$
V = +50 V
V = 0 V
e
s
f
t
The V = +50 V and V = 0 V equipotentials are relatively close at points e and s. The electric field is relatively strong there.
The V = +50 V and V = 0 V equipotentials are relatively far apart at points f and t. The electric field is relatively weak there.

Figure 17-7 Equipotential surfaces II In a nonuniform electric field equipotential surfaces are curved surfaces (seen here from the side) that are everywhere perpendicular to the electric field lines.

proportional to the distance d between the points and the electric field magnitude E. (This equation is strictly valid only for a uniform field, but is still approximately true for a nonuniform field.) So the electric field must be greater between the points e and s that are closer together, and less between the points f and t that are farther apart. This is an example of a general rule:

> *Where two adjacent equipotential surfaces are close together, the electric field is relatively strong. Where these surfaces are far apart, the electric field is relatively weak.*

Figure 17-8 shows an application of this idea. For a positive point charge (**Figure 17-8a**) or a negative point charge (**Figure 17-8b**), the electric field points radially outward or inward. The equipotential surfaces are everywhere perpendicular to the field lines, so they are spheres centered on the point charge. The radial distance from one spherical equipotential surface to the next is the same no matter where you are around the sphere, so the electric field magnitude is the same at all points a given distance from the point charge. This agrees with Equation 16-4 for the field magnitude E due to a point charge, $E = k|Q|/r^2$; the value of E depends only on the distance from the charge, not on where you are around the charge. For an electric dipole, however, the situation is different (**Figure 17-8c**). The field lines point from the positive charge to the negative one, and the equipotential surfaces are neither spherical nor centered on the point charges. (They are, however, everywhere perpendicular to the field lines.) Adjacent equipotential surfaces are close together between the two charges because the electric field is strong there. To the left of the left-hand charge or to the right of the right-hand charge, the electric field is relatively weak and adjacent equipotential surfaces are farther apart.

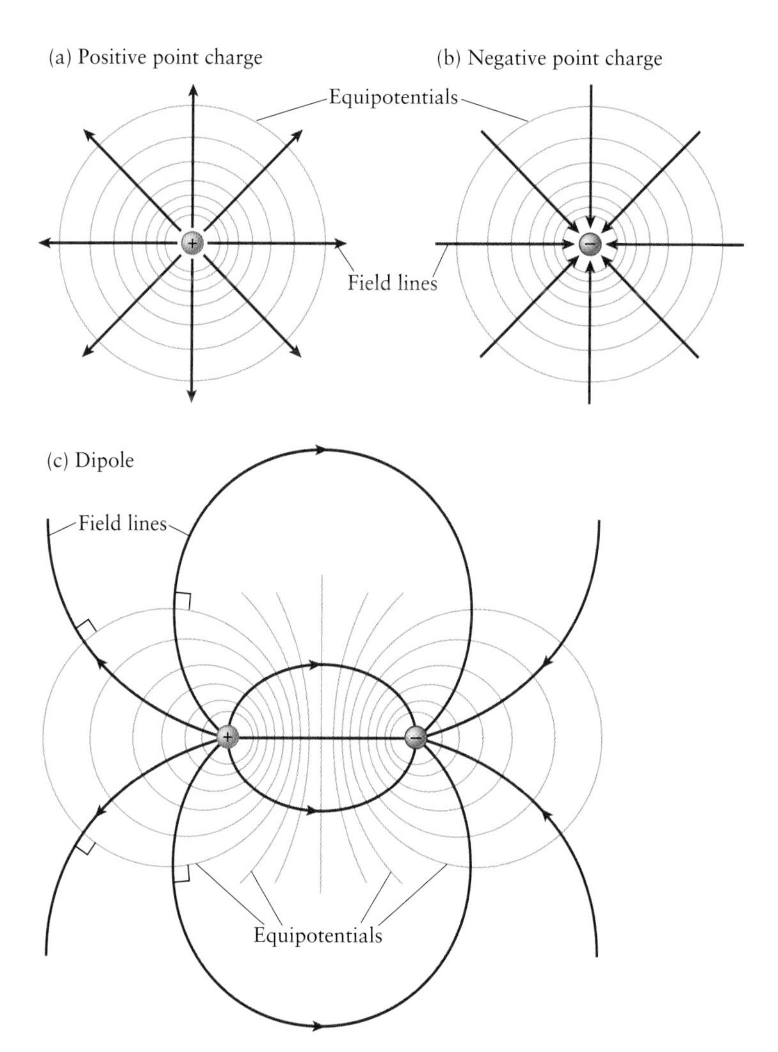

Figure 17-8 Equipotential surfaces III (a), (b) The equipotential surfaces for a positive or negative point charge are spheres centered on the point charge. (You can see this from Equation 17-8, $V = kQ/r$. This says that the potential V has the same value at all points that are the same distance from a point charge Q—that is, at all points on a sphere of radius r centered on the charge.) **(c)** For a dipole (point charges $+Q$ and $-Q$) the equipotentials are more complicated.

The equipotential concept is helpful for understanding the electric field around a charged conductor. We learned in Section 16-7 that if we put excess charge on a conductor and allow the individual excess charges to move to their equilibrium positions, all of the excess charge will end up on the conductor's surface and the electric field $\vec{E}$ will be zero everywhere inside the conductor. Since $\vec{E}$ is uniform inside the conductor (it has the same value at all points), we can use Equation 17-7 to calculate the potential difference between two points inside the conductor separated by a distance d: $\Delta V = -Ed\cos\theta = 0$ because $E = 0$. In other words, *the electric potential has the same value everywhere inside a conductor in equilibrium*. We say that a conductor in equilibrium is an **equipotential volume**. That's why we can make statements like "This conductor is at a potential of +20 V" or "The conductor at the positive end of a AA battery is at a potential 1.5 V higher than the conductor at the negative end"—the value of potential is the same everywhere throughout the volume of the conductor.

The electric field *outside* a charged conductor is *not* zero. If the excess charge on the conductor is positive, the outside electric field will point away from the surface of the conductor; if the excess charge is negative, the outside electric field will point toward the surface. Because the surface of the conductor is part of the equipotential volume, it is itself an equipotential surface. Because field lines and equipotential surfaces are always perpendicular, we conclude that *the electric field just outside a conductor in equilibrium must be perpendicular to the surface of the conductor*. Since the electric field always points toward lower electric potential, we can also conclude that *a positively charged conductor is at a higher electric potential than an adjacent negatively charged conductor*. We'll use these ideas about conductors in the following section to help us understand an important device called a *capacitor*.

GOT THE CONCEPT? 17-3 Equipotentials and Electric Field

? The orange vertical lines in **Figure 17-9** represent equipotential curves in some region of space. Which statement is correct about the electric field $\vec{E}$ at point A compared to the electric field at point B? (a) At A the field $\vec{E}$ points to the right and has a greater magnitude than at B; (b) at A the field $\vec{E}$ points to the right and has a smaller magnitude than at B; (c) at A the field $\vec{E}$ points to the left and has a greater magnitude than at B; (d) at A the field $\vec{E}$ points to the left and has a smaller magnitude than at B; (e) none of these.

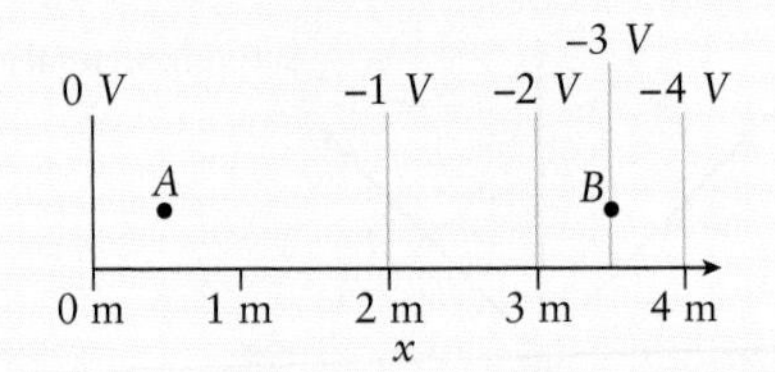

Figure 17-9 **What equipotentials tell you about $\vec{E}$** What can you infer from this figure about the electric field at A and B?

TAKE-HOME MESSAGE FOR Section 17-4

✔ The potential is the same everywhere along an equipotential curve or equipotential surface.

✔ Any curve or surface that is perpendicular to the electric field lines is an equipotential.

✔ Where equipotential surfaces are close, the electric field is relatively strong.

✔ A conductor in equilibrium has the same potential throughout its volume. The electric field at the surface of a conductor is perpendicular to the surface.

17-5 A capacitor stores equal amounts of positive and negative charge

The surface of every cell in your body is a *membrane* composed of a phospholipid bilayer that separates the fluid inside the cell and the fluid outside the cell. Negative charge accumulates on the membrane's interior surface, and this attracts positive charge onto the exterior surface. The result is a potential difference between the inner and outer surfaces of the membrane, and an electric field within the membrane that points from the outside in. This field helps drive essential ions through apertures in the membrane. They are also the source of the electrical signal used by the specialized cells called neurons to code, process, and transmit information.

A system or device that can store positive and negative charge like a cell membrane is called a **capacitor.** In technological applications a capacitor uses two pieces of metal called **plates.** One plate holds a positive charge q, and the other carries a negative charge $-q$. (Note that the *net* charge on the capacitor is zero.) It takes work to separate the positive and negative charges against the electric forces that attract them to each other, and this work goes into increasing the electric potential energy of the system of charges. So a capacitor is a device for storing electric potential energy.

We often need to draw capacitors in diagrams that represent electric circuits. The standard symbol for a capacitor in the diagram of a circuit is

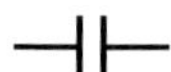

The two closely spaced parallel lines represent the two plates of the capacitor, and the straight horizontal lines represent wires that can connect the capacitor to an electric circuit.

The simplest geometry for a capacitor is two large parallel plates, one with charge $+q$ and the other with charge $-q$. **Figure 17-10a** shows these two plates separated from each other. We saw in Section 16-7 that close to a large charged plate and far from its edges, the electric field due to that plate is uniform, is perpendicular to the plane of the plate, and has magnitude $E = \sigma/(2\varepsilon_0)$ (from Equation 16-14). In this expression σ is the charge per unit area, equal to the charge q divided by the area A of the plate: $\sigma = q/A$. Since each plate has the same magnitude of charge, the field $\vec{E}_+$ due to the charge on the positive plate has the same magnitude as the field $\vec{E}_-$ due to the charge on the negative plate.

For a capacitor the *net* electric field is the vector sum of $\vec{E}_+$ and $\vec{E}_-$: $\vec{E} = \vec{E}_+ + \vec{E}_-$. **Figure 17-10b** shows the two plates moved into position to form a **parallel-plate capacitor.** Because the magnitude of each field does not depend on the distance from the plate that generates the field, $\vec{E}_+$ cancels $\vec{E}_-$ in the region above the upper plate and below the lower plate. Between the plates the fields $\vec{E}_+$ and $\vec{E}_-$ have the same magnitude and point in the same direction, so the net field between the plates has twice the magnitude of the field due to either plate by itself:

$$E = 2\left(\frac{\sigma}{2\varepsilon_0}\right) = \frac{\sigma}{\varepsilon_0} = \frac{q}{\varepsilon_0 A} \tag{17-10}$$

(electric field in a parallel-plate capacitor)

(a) Two large plates (viewed from the side) carry charge $+q$ and $-q$. Close to the plates the electric fields are nearly constant, so we represent them by straight, parallel field lines.

+ + + + + + + +
+q
$\vec{E}_+$

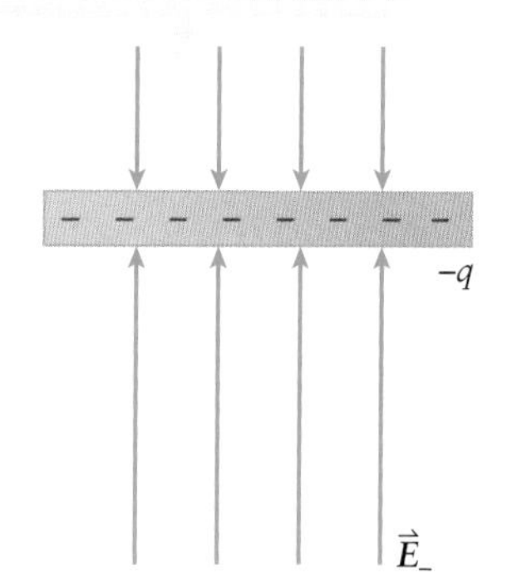

(b) Because the fields are nearly constant, the magnitude of the field due to each plate is the same at any point near the plates...

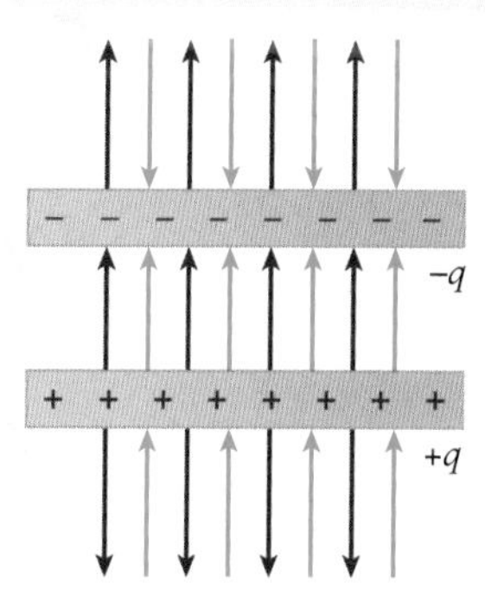

(c) ...so when the plates are placed close together, the fields cancel in the region outside the plates. The fields add in between the plates.

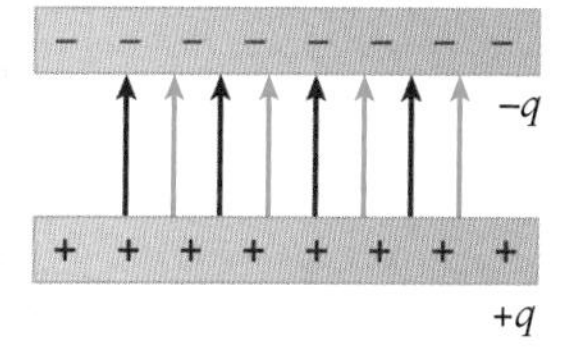

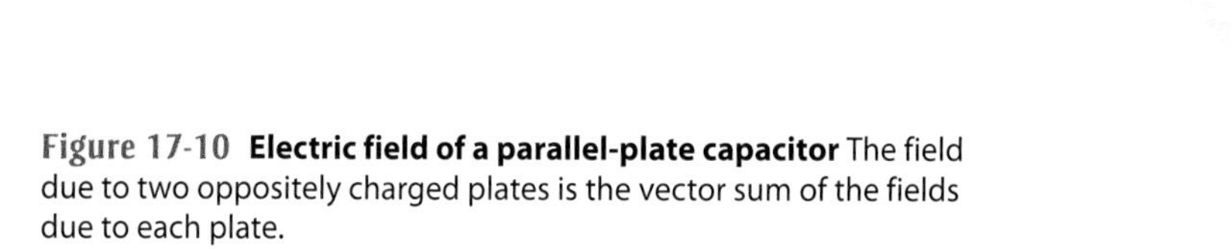

Figure 17-10 Electric field of a parallel-plate capacitor The field due to two oppositely charged plates is the vector sum of the fields due to each plate.

Note that the electric field points from the positive plate to the negative plate, just as we described in our discussion of conductors in Section 17-4.

(Equation 17-10 and the claim that $\vec{E}_+$ and $\vec{E}_-$ cancel outside the capacitor are strictly true only if the plates are large and close together, and when we consider points in space far from the edges of the capacitor. This idealization is physically reasonable for the kinds of problems we will encounter.)

We can substitute Equation 17-10 into Equation 17-7 to calculate the *potential difference* between the plates of a parallel-plate capacitor. The electric field points from the positive plate to the negative plate. If we let d be the distance between the plates, and we travel from the negative plate to the positive plate, the angle between the electric field and the displacement is $\theta = 180°$. The potential difference is therefore

$$\begin{aligned} V = V_+ - V_- &= -Ed\cos\theta = -Ed(-1) \\ &= Ed = \left(\frac{q}{\varepsilon_0 A}\right)d \\ &= \frac{qd}{\varepsilon_0 A} \end{aligned} \quad (17\text{-}11)$$

WATCH OUT! The symbol *V* sometimes stands for potential, sometimes for potential difference.

Previously in this chapter we've used V to denote the electric potential at a given point and ΔV to denote the difference in electric potential between two points. In Equation 17-11, and throughout our discussion of capacitors, we'll follow common practice and use V as the symbol for the potential difference, or voltage, between the two capacitor plates. Just remember "for capacitors, V means voltage." We'll frequently refer to V as the "voltage *across* the capacitor."

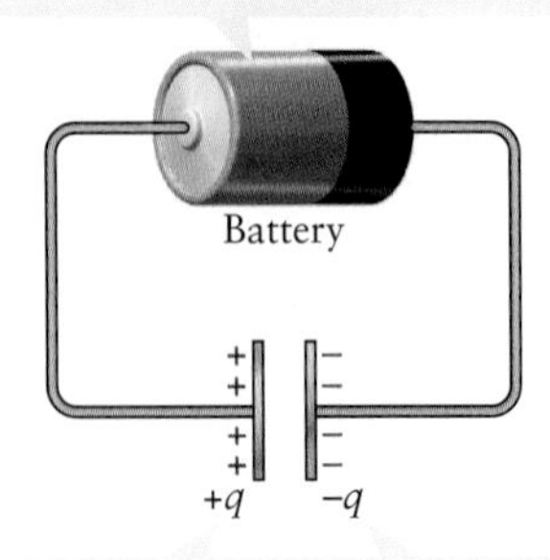
When this battery is connected to the plates of a capacitor, it creates a potential difference V between the plates...
Battery
+q
−q
...which drives electrons from one plate, through the battery, and onto the other plate. The plate that lost electrons ends up with charge +q, and the other plate ends up with charge −q.

Figure 17-11 Charging a capacitor The charges that appear on the two capacitor plates have the same magnitude but opposite signs.

For a capacitor of plate area A and plate separation d, the potential difference V between the plates is proportional to the magnitude of the charge q on each plate.

We can rewrite Equation 17-11 as an expression for the amount of charge on each plate. To charge an initially uncharged capacitor, we apply a voltage V between its plates as in **Figure 17-11**. Electrons are driven to one of the plates, giving it a charge $-q$; the plate from which the electrons were taken is left with a charge $+q$. (As we mentioned previously the net charge on the capacitor remains zero.) Equation 17-11 tells us that the magnitude q of the charge that each plate acquires is proportional to the applied voltage V:

$$q = \frac{\varepsilon_0 A}{d}V \quad (17\text{-}12)$$

(charge on a parallel-plate capacitor)

The quantity $\varepsilon_0 A/d$ in Equation 17-12 is called the **capacitance** of the capacitor. We use the symbol C for capacitance:

Capacitance of a parallel-plate capacitor (17-13)

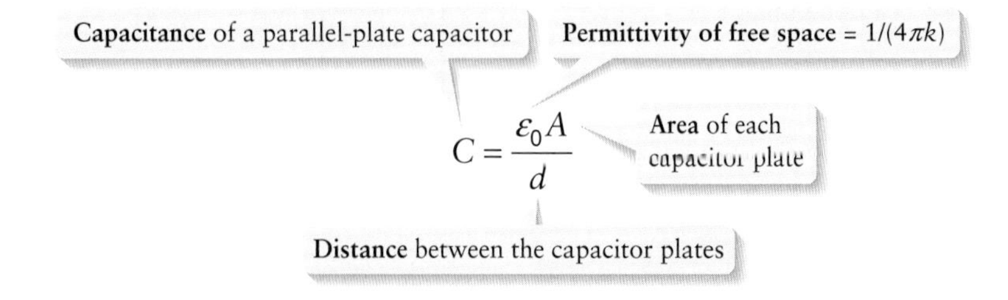
Capacitance of a parallel-plate capacitor
Permittivity of free space = 1/(4πk)
C = ε0A/d
Area of each capacitor plate
Distance between the capacitor plates

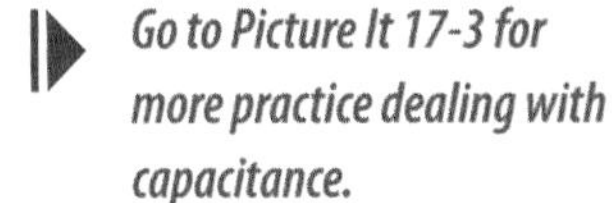
Go to Picture It 17-3 for more practice dealing with capacitance.

For a parallel-plate capacitor, C depends only on the area A of the plates, the distance d between them, and the material between them. We've assumed that the plates are separated by vacuum; in Section 17-8 we'll explore what happens if the space between the capacitor plates is filled with a different material.

Capacitance tells us the amount of charge that can be stored on a capacitor held at a given voltage. From Equations 17-12 and 17-13,

A capacitor carries a charge $+q$ on its positive plate and a charge $-q$ on its negative plate.

The magnitude of q is directly proportional to V, the **voltage** (potential difference) between the plates.

$$q = CV$$

The constant of proportionality between charge q and voltage V is the **capacitance** C of the capacitor.

Charge, voltage, and capacitance for a capacitor
(17-14)

The greater the capacitance, the more charge is present for a given voltage. Note that while Equations 17-12 and 17-13 are valid for a parallel-plate capacitor only, Equation 17-14 is valid for capacitors of *any* geometry. In the problems at the end of this chapter, you'll analyze some other simple types of capacitor.

Equation 17-14 shows that the unit of capacitance is the coulomb per volt, also known as the **farad** (symbol F): $1\text{ F} = 1\text{ C/V}$. The farad is named for the nineteenth-century English physicist Michael Faraday. The capacitors used in consumer electronics are typically in the range of 10 pF (10×10^{-12} F) to 10,000 μF ($10{,}000 \times 10^{-6}$ F); 1 F is an extremely large capacitance. Using $1\text{ F} = 1\text{ C/V}$, $1\text{ V} = 1\text{ J/C}$, and $1\text{ J} = 1\text{ N}\cdot\text{m}$, we see that $1\text{ F} = (1\text{ C})/(1\text{ J/C}) = 1\text{ C}^2/\text{J} = 1\text{ C}^2/(\text{N}\cdot\text{m})$. We can express the value of the constant ε_0 (the permittivity of free space) as

$$\varepsilon_0 = 8.85 \times 10^{-12}\,\frac{\text{C}^2}{\text{N}\cdot\text{m}^2} = 8.85 \times 10^{-12}\,\frac{\text{F}}{\text{m}}$$

These units are useful in calculations of capacitance, as we'll see below.

An everyday application of Equation 17-13 is the touchscreen on a mobile device such as a smartphone or tablet (**Figure 17-12**). Behind the device's glass screen is a layer of a special transparent conductor called indium tin oxide (ITO), which is actually a solid mixture of indium oxide, In_2O_3, and tin oxide, SnO_2. When you touch your finger to the screen, the conducting ITO layer acts as one plate of a capacitor and your finger—which is also a conductor—acts as the other plate. Sensor circuits in the mobile device detect where on the screen a capacitance appears due to this capacitor, which is how the device "knows" where on the screen it has been touched. (To verify that the object touching the screen has to be a conductor, try using the rubber eraser on a pencil to touch the screen of a mobile device. Because rubber is an insulator, not a conductor, the screen won't respond.) The sensor circuits are adjusted so that they will register only if the capacitance C is above a certain minimum value. That explains why you have to physically touch the screen: According to Equation 17-13, C increases as the distance d between the plates decreases, so the capacitance is largest when your finger is touching the screen and so closest to the ITO layer. Your finger still acts as a capacitor plate if you hold it a slight distance away from the screen, but the capacitance is now too low to trigger the sensor circuits.

iStockphoto/Thinkstock

Figure 17-12 Capacitive touchscreens Mobile devices such as these use the physics of capacitors to determine where your finger touches the screen.

EXAMPLE 17-7 A Parallel-Plate Capacitor

Two square, parallel conducting plates each have dimensions 5.00 cm by 5.00 cm and are placed 0.100 mm apart. Determine the capacitance of this configuration.

Set Up

We'll use Equation 17-13 to determine the value of C for this capacitor.

Capacitance of a parallel-plate capacitor:

$$C = \frac{\varepsilon_0 A}{d} \qquad (17\text{-}13)$$

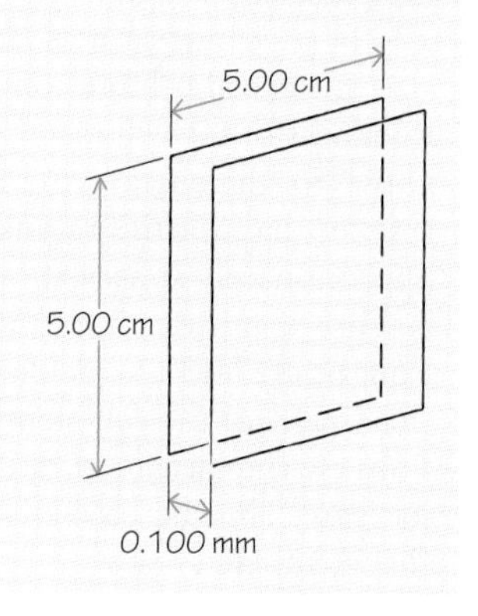

Solve

The area of each plate is the product of the length of the two sides, and the plate separation is given. We need to convert all dimensions into SI units.

Area $A = (5.00 \times 10^{-2}\text{ m})(5.00 \times 10^{-2}\text{ m}) = 2.50 \times 10^{-3}\text{ m}^2$

Plate separation $d = 1.00 \times 10^{-4}\text{ m}$

Using Equation 17-13,

$$C = \frac{(8.85 \times 10^{-12}\text{ F/m})(2.50 \times 10^{-3}\text{ m}^2)}{1.00 \times 10^{-4}\text{ m}} = 2.21 \times 10^{-10}\text{ F}$$

$$= 221\text{ pF}$$

(Recall that the prefix "p," or "pico–" represents 10^{-12}.)

Reflect

Capacitors with capacitances of a few hundred picofarads are found in calculators, cell phones, and portable music players.

To make a 1-F capacitor with the same separation $d = 0.100$ mm between plates, we would have to increase the plate area by a factor of $(1\text{ F})/(2.21 \times 10^{-10}\text{ F}) = 4.52 \times 10^9$. The length of each side would have to be increased by a factor of the square root of 4.52×10^9, or 6.73×10^4. The sides of the plates would be $(6.73 \times 10^4)(5.00 \times 10^{-2}\text{ m}) = 3.36 \times 10^3$ m, or 3.36 *kilometers* (about 2 miles). One farad is a *very* large capacitance.

BIO-Medical

EXAMPLE 17-8 Insulin Release

The hormone insulin minimizes variations in blood glucose levels. Pancreatic beta cells ("β cells") synthesize insulin and store it in *vesicles*, bubble-like organelles approximately 150 nm in radius within the cytoplasm of the cells. To release insulin, vesicles fuse with the membrane of the β cell. This increases the surface area of the β cell by the surface area of the fused vesicles (**Figure 17-13**). The thickness of the cell membrane does not change, so the increase in surface area increases the capacitance of the β cell membrane. Experiment shows that the membrane capacitance increases at 1.6×10^{-13} F/s during insulin release. If the membrane capacitance is approximately 1 μF per square centimeter of surface area, estimate the number of vesicles that fuse with the cell membrane per second during insulin release.

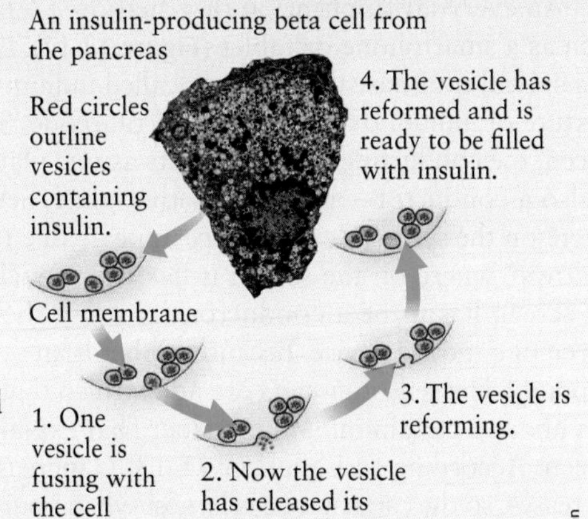

Figure 17-13 Changing membrane, changing capacitance To study this important process in a pancreatic β cell, scientists monitor changes in the capacitance of the cell membrane.

Set Up

Although the cell membrane is not flat like the parallel-plate capacitor shown in Figure 17-10, we can treat it as such because the size of the cell is large compared to the thickness of the cell membrane. (It's like being above Earth's surface at a height that's much smaller than Earth's radius. While Earth is approximately spherical, when seen from such a short distance, it appears to be flat.) We'll use Equation 17-13 to calculate the rate at which the area A of the cell membrane must increase to cause the measured rate of capacitance increase. This increase in area comes from fusing vesicles.

Capacitance of a parallel-plate capacitor:

$$C = \frac{\varepsilon_0 A}{d} \qquad (17\text{-}13)$$

Surface area of a sphere of radius r:

$$A_{\text{sphere}} = 4\pi r^2$$

Solve

Find the rate at which the membrane area must increase to cause this rate of capacitance increase.

Rate of change of the capacitance of the membrane:

$$\frac{\Delta C_m}{\Delta t} = 1.6 \times 10^{-13}\ \text{F/s}$$

Equation 17-13 tells us that the capacitance C_m is directly proportional to the membrane surface area A_m. We are told that the capacitance per unit area is approximately 1 μF/cm^2, so the change in capacitance ΔC_m that corresponds to a given change in area ΔA_m is

$$\Delta C_m = \left(1\ \frac{\mu\text{F}}{\text{cm}^2}\right)\Delta A_m$$

Since the capacitance increases by 1.6×10^{-13} F in 1 s, the increase in area of the membrane in 1 s must be

$$\Delta A_m = \frac{\Delta C_m}{1\ \mu\text{F/cm}^2}$$
$$= (1.6 \times 10^{-13}\ \text{F})\left(1\ \frac{\text{cm}^2}{\mu\text{F}}\right)\left(\frac{1\ \mu\text{F}}{10^{-6}\ \text{F}}\right)\left(\frac{1\ \text{m}}{100\ \text{cm}}\right)^2$$
$$= 1.6 \times 10^{-11}\ \text{m}^2$$

(We used 1 μF = 1 microfarad = 10^{-6} F and 1 m = 100 cm.)

Now we can calculate how many vesicles must fuse with the membrane per second to cause this increase in area.

The surface area of a single vesicle is

$$A_{\text{vesicle}} = 4\pi r_{\text{vesicle}}^2 = 4\pi(150\ \text{nm})^2 = 4\pi(150 \times 10^{-9}\ \text{m})^2$$
$$= 2.8 \times 10^{-13}\ \text{m}^2$$

The number of vesicles that must add their area to the membrane in 1 s to cause an area increase $\Delta A_m = 1.6 \times 10^{-11}\ \text{m}^2$ is

$$\frac{\Delta A_m}{A_{\text{vesicle}}} = \frac{1.6 \times 10^{-11}\ \text{m}^2}{2.8 \times 10^{-13}\ \text{m}^2/\text{vesicle}} = 57\ \text{vesicles}$$

Our final answer should have just one significant figure, since we were given the capacitance per unit area 1 μF/cm^2 to just one significant figure. So our final answer is that about 60 vesicles fuse with the membrane wall per second.

Reflect

The ability to detect small changes in capacitance makes it possible to study the molecular mechanisms of hormone release from single cells and to verify previous biochemical measurements of insulin release from β cells. After insulin release, new vesicles form by pinching off from the cell membrane and are refilled with insulin. By recycling the vesicle membrane, β cells maintain their size over the long term.

GOT THE CONCEPT? 17-4 Capacitors

A parallel-plate capacitor has a potential difference V between its plates. If the potential difference is increased to $2V$, what effect does this have on the capacitance C?

(a) C increases by a factor of 4; (b) C increases by a factor of 2; (c) C becomes $\frac{1}{2}$ as great; (d) C becomes $\frac{1}{4}$ as great; (e) C is unchanged.

TAKE-HOME MESSAGE FOR Section 17-5

✔ A capacitor has two plates, one with charge $+q$ and the other with charge $-q$.

✔ For a given capacitor the potential difference between the plates is proportional to the magnitude of the charge q on each plate. The proportionality constant, called the capacitance, depends on the geometry of the plates and on the material in the space between the plates.

17-6 A capacitor is a storehouse of electric potential energy

In electric circuits, capacitors are most useful because of their ability to store electric potential energy. An applied voltage V such as that supplied by the battery in Figure 17-11 charges a capacitor by effectively pulling negative charges from one of the plates and depositing them on the other. To move the negative charges away from the positive charges requires work, and it is this work that results in electric potential energy being stored in the capacitor. At a later time the potential energy can be transferred by charge leaving the capacitor and passed on to other parts of the circuit. (That's what happens in the electronic flash unit in a camera or mobile phone. The device's battery charges a capacitor, and the energy stored in the charged capacitor is used to produce a short, intense burst of light.)

Let's see how to calculate the amount of electric potential energy stored in a capacitor that has charge $+q$ and $-q$ on its positive and negative plates, respectively. This is equal to the amount of electric potential energy that's added to the capacitor if we start with both plates uncharged (which we can regard as a state of zero potential energy) and move charge $+q$ from the first plate to the second one, leaving charge $-q$ on the first plate. If the potential difference between the plates had a constant value ΔV, Equation 17-6 tells us that the change in electric potential energy $\Delta U_{\text{electric}}$ in this process would be

$$\Delta V = \frac{\Delta U_{\text{electric}}}{q} \quad \text{so} \quad \Delta U_{\text{electric}} = q\,\Delta V$$

However, the potential difference between the plates (that is, the voltage across the capacitor) does *not* stay constant as we transfer charge from one plate to the other! Equation 17-14 tells us that the potential difference between the plates is proportional to the amount of charge on the positive plate. So as we transfer more charge from one plate to the other, the potential difference across which the charge must move increases from zero (its starting value) to a final value V given by Equation 17-14: $q = CV$, so $V = q/C$. To correctly calculate the amount of potential energy stored in the capacitor when it is charged, we have to replace ΔV in Equation 17-6 by the *average* value of the potential difference during the charging process. Because potential difference increases in direct proportion to the charge, this average value is just the average of the starting potential difference and the final potential difference V:

$$\begin{aligned}\Delta U_{\text{electric}} &= q\,\Delta V_{\text{average}}\\ &= q\left[\frac{(\text{starting potential difference}) + (\text{final potential difference})}{2}\right]\\ &= q\left(\frac{0+V}{2}\right) = \frac{1}{2}qV \qquad (17\text{-}15)\end{aligned}$$

Equation 17-14 tells us that $q = CV$ and $V = q/C$, so we can also write Equation 17-15 as

$$\Delta U_{\text{electric}} = \frac{1}{2}(CV)V = \frac{1}{2}CV^2 \quad \text{or} \quad \Delta U_{\text{electric}} = \frac{1}{2}q\left(\frac{q}{C}\right) = \frac{q^2}{2C} \qquad (17\text{-}16)$$

If we say that the electric potential energy of the initial uncharged capacitor was zero, the final potential energy U_{electric} is just equal to the increase in potential energy $\Delta U_{\text{electric}}$ given by Equation 17-15 or Equation 17-16. So we can write the potential energy stored in the capacitor in three ways:

Electric potential energy stored in a capacitor (17-17)

The **electric potential energy** stored in a charged capacitor...

$$U_{\text{electric}} = \frac{1}{2}qV = \frac{1}{2}CV^2 = \frac{q^2}{2C}$$

...can be expressed in three ways in terms of the **charge** q, **potential difference** V, and **capacitance** C.

Equation 17-17 says that the energy stored in a charged capacitor is proportional to the *square* of the charge q or, equivalently, to the square of the potential difference (voltage) V across the capacitor. These results are true for capacitors of all kinds, not just the simple parallel-plate capacitor that we discussed in Section 17-5.

The last of the three expressions for electric potential energy in Equation 17-17, $U_{electric} = q^2/(2C)$, is very similar to the equation for *spring* potential energy that we learned in Section 6-6:

Spring potential energy of a stretched or compressed spring

Spring constant of the spring

$$U_{spring} = \frac{1}{2}kx^2$$

Spring potential energy (6-19)

Extension of the spring ($x > 0$ if spring is stretched, $x < 0$ if spring is compressed)

The only differences between $U_{electric} = q^2/(2C)$ from Equation 17-17 and the expression in Equation 6-19 is that the spring displacement x is replaced by the charge q, and the spring constant k is replaced by the reciprocal of the capacitance, $1/C$. This similarity isn't surprising. To add potential energy to a spring by stretching it, you have to pull against the force of magnitude $F = kx$ that the spring exerts on you. The greater the distance that the spring is already stretched, the more force it exerts and the harder it is to stretch it further. In exactly the same way, to add potential energy to a capacitor by increasing the magnitude of charge on the two plates, you have to transfer charge against a voltage $V = (1/C)q$. The greater the charge that is already on the plates, the greater the voltage and the harder it is to increase q.

BIO-Medical EXAMPLE 17-9 A Defibrillator Capacitor

A defibrillator, like the one shown in the photograph that opens this chapter, is essentially a capacitor that is charged by a high-voltage source and then delivers the stored energy to a patient's heart. (a) How much charge does the 80.0-μF capacitor in a certain defibrillator store when it is fully charged by applying 2.50 kV? (b) How much energy can this defibrillator deliver?

Set Up

We're given the capacitance $C = 80.0\ \mu\text{F}$ (recall $1\ \mu\text{F} = 1$ micro farad $= 10^{-6}$ F) and the potential difference $V = 2.50$ kV (recall 1 kV = 1 kilovolt = 10^3 V) between the capacitor plates. We'll use Equation 17-14 to determine the magnitude q of the charge on each capacitor plate and Equation 17-17 to find the potential energy $U_{electric}$ stored in the charged capacitor. If we assume that no energy is lost in the process of being transferred to the patient, this is equal to the energy that the defibrillator delivers.

Charge, voltage, and capacitance for a capacitor:

$$q = CV \tag{17-14}$$

Electric potential energy stored in a capacitor:

$$U_{electric} = \frac{1}{2}qV = \frac{1}{2}CV^2 = \frac{q^2}{2C} \tag{17-17}$$

Solve

(a) Substitute the given values of C and V into Equation 17-14 to solve for the charge q.

Charge on the capacitor plates:

$$\begin{aligned} q &= CV = (80.0\ \mu\text{F})(2.50\ \text{kV}) \\ &= (80.0 \times 10^{-6}\ \text{F})(2.50 \times 10^3\ \text{V}) \\ &= 0.200\ \text{F} \cdot \text{V} = 0.200\ \text{C} \end{aligned}$$

(Recall that 1 F = 1 C/V, so 1 F · V = 1 C.)

(b) Since the values of C and V are given, let's use the second of the three relationships in Equation 17-17 to find the stored electric potential energy.

Energy stored in the capacitor:

$$\begin{aligned} U_{electric} &= \frac{1}{2}CV^2 \\ &= \frac{1}{2}(80.0 \times 10^{-6}\ \text{F})(2.50 \times 10^3\ \text{V})^2 \\ &= 2.50 \times 10^2\ \text{F} \cdot \text{V}^2 \\ &= 2.50 \times 10^2\ \text{J} \end{aligned}$$

(Recall that 1 V = 1 J/C and 1 F = 1 C/V, so $1\ \text{F} \cdot \text{V}^2 = 1\ [\text{C/V}]\text{V}^2 = 1\ \text{C} \cdot \text{V} = 1\ \text{C} \cdot [\text{J/C}] = 1\ \text{J}$.)

Reflect

The American Heart Association recommends that a defibrillator shock should deliver between 40 and 360 J in order to be effective, so our numerical result is in the recommended range.

We can double-check our answers by using the other two relationships in Equation 17-17. Happily, by using the value of q that we calculated in part (a), we get the same result for U_{electric} in part (b) as we found above.

Remember that in solving any problem, you should *always* take advantage of alternative ways to find the answer in order to check your results.

One alternative way to calculate the energy stored in the capacitor:

$$
\begin{aligned}
U_{\text{electric}} &= \frac{1}{2}qV \\
&= \frac{1}{2}(0.200\ \text{C})(2.50 \times 10^{3}\ \text{V}) \\
&= 2.50 \times 10^{2}\ \text{C}\cdot\text{V} \\
&= 2.50 \times 10^{2}\ \text{J}
\end{aligned}
$$

Another alternative way to calculate the energy stored in the capacitor:

$$
\begin{aligned}
U_{\text{electric}} &= \frac{q^2}{2C} \\
&= \frac{(0.200\ \text{C})^2}{2(80.0 \times 10^{-6}\ \text{F})} \\
&= 2.50 \times 10^{2}\ \frac{\text{C}^2}{\text{F}} \\
&= 2.50 \times 10^{2}\ \text{J}
\end{aligned}
$$

(Recall that 1 V = 1 J/C and 1 F = 1 C/V. You should verify that 1 C^2/F = 1 J.)

GOT THE CONCEPT? 17-5 Spreading the Plates of a Capacitor I

Suppose you increase the distance between the plates of a charged parallel-plate capacitor without changing the amount of charge stored on the plates. What will happen to the energy stored in the capacitor? (a) It will decrease; (b) it will remain the same; (c) it will increase; (d) not enough information given to decide.

GOT THE CONCEPT? 17-6 Spreading the Plates of a Capacitor II

Suppose you increase the distance between the plates of a parallel-plate capacitor while holding the potential difference between the plates constant. (You could do this by keeping the plates connected to a battery, as in Figure 17-11.) What will happen to the energy stored in the capacitor? (a) It will decrease; (b) it will remain the same; (c) it will increase; (d) not enough information given to decide.

TAKE-HOME MESSAGE FOR Section 17-6

✔ A charged capacitor stores electric potential energy. The amount of energy stored is proportional to the square of the magnitude of the charge on each plate and also proportional to the square of the potential difference between the plates.

17-7 Capacitors can be combined in series or in parallel

In both biological systems and electric circuits, it is not uncommon for more than one capacitor to be connected together in some way. The net result is that the capacitor combination behaves as though it were a *single* capacitor, with an **equivalent capacitance** that depends on the properties of the individual capacitors actually present. (An analogy is lifting a heavy conference table to move it across the room. Four people of normal strength could do the job, or it could be done by a single weightlifter. We would say that the four people together have an "equivalent strength" comparable to that of the weightlifter.) Our goal in this section is to find the equivalent capacitance in different situations.

Capacitors in Series

Figure 17-14 shows three initially uncharged capacitors with capacitances C_1, C_2, and C_3 that are connected end to end. The capacitors become charged when the combination is connected to a battery of voltage V as shown in the figure. Note that the negative plate of one capacitor is connected to the positive plate of the next capacitor in the combination. Capacitors connected in this way are said to be in **series**.

What is the equivalent capacitance of this series combination? To answer this question let's first determine the charges q_1, q_2, and q_3 on the individual capacitors and the voltages V_1, V_2, and V_3 across the individual capacitors. For each capacitor the magnitudes of the charge on the positive and negative plates must be equal. So as the battery draws negative charge from the left-hand plate of C_1, whatever positive charge $+q$ that plate acquires must be balanced by charge $-q$ on the right plate. The charging of the right plate of C_1 occurs as negative charge is drawn from the left plate of C_2, and because this whole section is initially uncharged, the left plate of C_2 acquires charge $+q$. If we apply the same reasoning to C_3, we see that all three capacitors acquire the *same* charge:

$$q_1 = q_2 = q_3 = q \quad \text{(capacitors in series)} \tag{17-18}$$

Thus the series combination of three capacitors is equivalent to a single capacitor with charge q. You can think of the charge on the negative plate of C_1 as canceling the charge on the positive plate of C_2 and likewise for the charges on the negative plate of C_2 and the positive plate of C_3.

The charges and voltages for each capacitor are also given by Equation 17-14, $q = CV$. If we substitute this into Equation 17-18, we get

$$q = C_1V_1 = C_2V_2 = C_3V_3 \quad \text{(capacitors in series)}$$

which we can rewrite as expressions for the voltages across the individual capacitors:

$$V_1 = \frac{q}{C_1} \quad V_2 = \frac{q}{C_2} \quad V_3 = \frac{q}{C_3} \quad \text{(capacitors in series)} \tag{17-19}$$

Now, the voltage V of the battery equals the voltage across the combination of three capacitors. This is just the sum of the voltages across the individual capacitors:

$$V = V_1 + V_2 + V_3 \quad \text{(capacitors in series)} \tag{17-20}$$

A battery is connected to three capacitors in series.

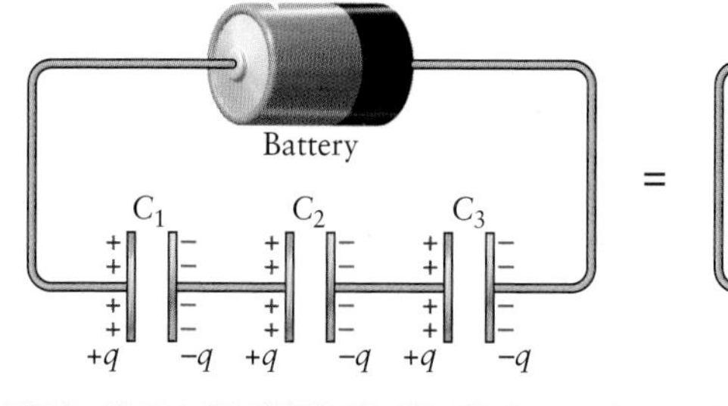

=

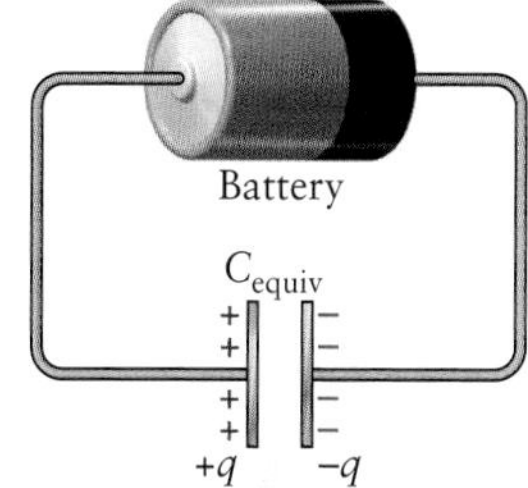

• The magnitude q of the charge on each capacitor must be the same.
• The voltage across each capacitor does not have to be the same.
• The sum $V_1 + V_2 + V_3$ of the individual capacitor voltages equals the battery voltage V.

The three capacitors are equivalent to a single capacitor C_{equiv}.

Figure 17-14 Capacitors in series What is the equivalent capacitance of this series combination?

If we substitute Equations 17-19 into Equation 17-20, we get a relationship between the charge q on each capacitor and the voltage V across the combination—that is, between the charge q on the equivalent capacitor and the voltage V across the equivalent capacitor:

(17-21)
$$V = \frac{q}{C_1} + \frac{q}{C_2} + \frac{q}{C_3} = q\left(\frac{1}{C_1} + \frac{1}{C_2} + \frac{1}{C_3}\right)$$
(capacitors in series)

For the equivalent capacitor alone, of capacitance C_{equiv}, Equation 17-14 says that $q = C_{equiv}V$ or $V = q/C_{equiv}$. If we compare this to Equation 17-21, we see that

If capacitors are in series, the reciprocal of the equivalent capacitance is the sum of the reciprocals of the individual capacitances.

Equivalent capacitance of capacitors in series (17-22)

$$\frac{1}{C_{equiv}} = \frac{1}{C_1} + \frac{1}{C_2} + \frac{1}{C_3}$$

Equivalent capacitance of capacitors in series

Capacitances of the individual capacitors

If there are more than three capacitors in series, the same rule given in Equation 17-22 applies. The equivalent capacitance of a series combination is always *less* than the smallest capacitance of any of the individual capacitors.

Capacitors in Parallel

Figure 17-15 shows an alternative way to connect the three initially uncharged capacitors C_1, C_2, and C_3 to a battery with voltage V. Capacitors connected in this way are said to be in **parallel**. Notice the difference in the arrangement of the capacitors in parallel (Figure 17-15) compared to capacitors in series (Figure 17-14). In a series arrangement, the right-hand plate of one capacitor is connected to the left-hand plate of the capacitor to its right. In a parallel arrangement, all of the right-hand plates are connected, and all of the left-hand plates are connected.

To find the equivalent capacitance of capacitors in parallel, first note that all of the right-hand capacitor plates are connected to one terminal of the battery and all of the left-hand plates are connected to the other terminal. So each of the voltages V_1, V_2, and V_3 across the individual capacitors is equal to the voltage V across the battery:

(17-23)
$$V = V_1 = V_2 = V_3$$
(capacitors in parallel)

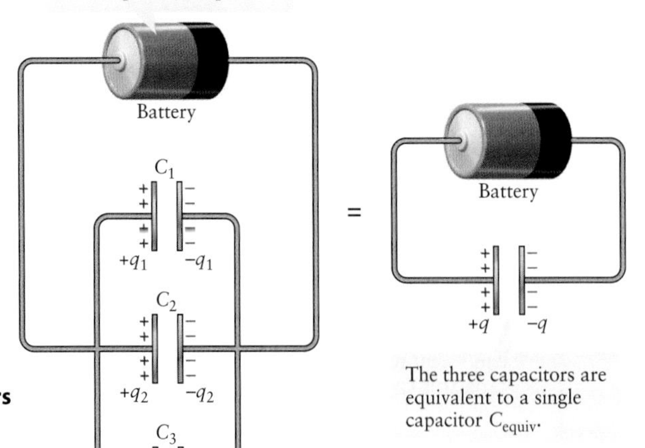

• The voltage V across each capacitor must be the same.
• The charges on each capacitor do not have to be the same.
• The sum $q_1 + q_2 + q_3$ of the individual capacitor charges equals the charge q on the equivalent capacitor.

Figure 17-15 Capacitors in parallel What is the equivalent capacitance of this parallel combination?

Equation 17-23 coupled with Equation 17-14, $q = CV$, then tells us the charges q_1, q_2, and q_3 on the individual capacitors:

$$q_1 = C_1V \quad q_2 = C_2V \quad q_3 = C_3V \qquad \text{(17-24)}$$
(capacitors in parallel)

The *total* charge acquired by all three capacitors is the sum of the charges q_1, q_2, and q_3. From Equation 17-24,

$$\begin{aligned} q &= q_1 + q_2 + q_3 = C_1V + C_2V + C_3V \\ &= (C_1 + C_2 + C_3)V \end{aligned} \qquad \text{(17-25)}$$
(capacitors in parallel)

For the equivalent capacitor of capacitance C_{equiv} that corresponds to the three capacitors in parallel, Equation 17-14 says that $q = C_{equiv}V$. Comparing this to Equation 17-25 shows that

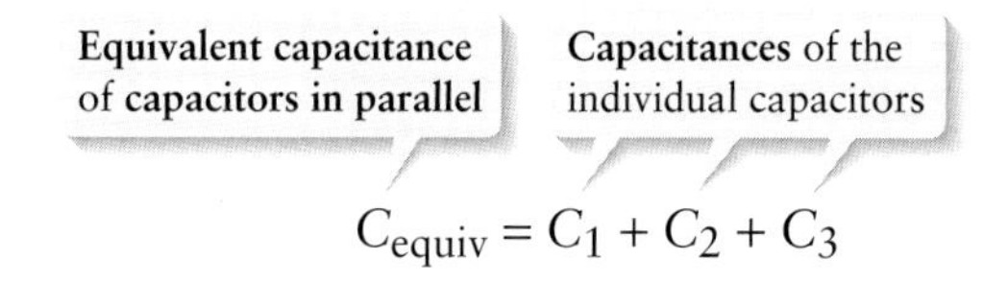

$$C_{equiv} = C_1 + C_2 + C_3$$

Equivalent capacitance of capacitors in parallel (17-26)

If capacitors are in parallel, the equivalent capacitance is the sum of the individual capacitances.

If there are more than three capacitors in parallel, the same rule given in Equation 17-26 applies. The equivalent capacitance of a parallel combination is always *greater* than the smallest capacitance of any of the individual capacitors.

Example 17-10 shows how to do calculations with capacitors in series and in parallel. (We'll see a biological application of these calculations in the following section.) Many real networks of capacitors are more complex: They have a mixture of series and parallel combinations. To find the equivalent capacitance of such a network, we identify any small grouping of capacitors that are either entirely in series or entirely in parallel, find the equivalent capacitance of each group, and then combine them in larger and larger groupings, using the series and parallel rules, until we have accounted for all of the capacitors in the network. Example 17-11 illustrates how to do this.

WATCH OUT! Capacitors in series and parallel have different properties.

Here's a summary of the differences between series and parallel combinations of capacitors. In a *series* combination, each capacitor has the same charge, but there are different voltages across capacitors with different capacitances. In a *parallel* combination, there is the same voltage across each capacitor, but there are different charges on capacitors with different capacitances.

EXAMPLE 17-10 Two Capacitors in Series or in Parallel

(a) If two capacitors are connected in series, find their equivalent capacitance for the case when the capacitors have different capacitances $C_1 = 2.00\ \mu F$ and $C_2 = 4.00\ \mu F$ and for the case where both have the same capacitance $C_1 = C_2 = 2.00\ \mu F$. (b) Repeat part (a) for the two capacitors connected in parallel.

Set Up

We'll use Equations 17-22 and 17-26 to find the equivalent capacitance C_{equiv} in each case. Since there are only two capacitors in each combination, we drop the C_3 term from these equations.

Two capacitors in series:

$$\frac{1}{C_{equiv}} = \frac{1}{C_1} + \frac{1}{C_2} \qquad \text{(17-22)}$$

Two capacitors in parallel:

$$C_{equiv} = C_1 + C_2 \qquad \text{(17-26)}$$

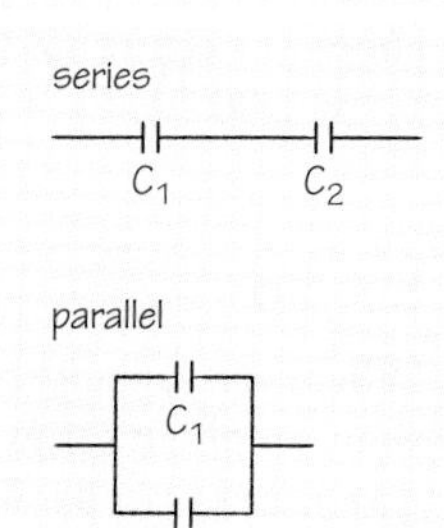

Solve

(a) We first apply Equation 17-22 to the case where $C_1 = 2.00\ \mu F$ and $C_2 = 4.00\ \mu F$.

With capacitors $C_1 = 2.00\ \mu F$ and $C_2 = 4.00\ \mu F$ in series, Equation 17-22 becomes

$$\frac{1}{C_{equiv}} = \frac{1}{2.00\ \mu F} + \frac{1}{4.00\ \mu F}$$

$$= 0.500\ \mu F^{-1} + 0.250\ \mu F^{-1} = 0.750\ \mu F^{-1}$$

To find C_{equiv} take the reciprocal of $1/C_{equiv}$:

$$C_{equiv} = \frac{1}{0.750\ \mu F^{-1}} = 1.33\ \mu F$$

Note that C_{equiv} is less than either C_1 or C_2.

Now repeat the calculation for the case where $C_1 = C_2 = 2.00\ \mu F$.

Follow the same steps but with both capacitances equal to $2.00\ \mu F$:

$$\frac{1}{C_{equiv}} = \frac{1}{2.00\ \mu F} + \frac{1}{2.00\ \mu F}$$

$$= 0.500\ \mu F^{-1} + 0.500\ \mu F^{-1} = 1.00\ \mu F^{-1}$$

$$C_{equiv} = \frac{1}{1.00\ \mu F^{-1}} = 1.00\ \mu F$$

For this case of identical capacitors in series, C_{equiv} is exactly one-half of $C_1 = C_2 = 2.00\ \mu F$.

(b) Apply Equation 17-26 to the case where $C_1 = 2.00\ \mu F$ and $C_2 = 4.00\ \mu F$.

With capacitors $C_1 = 2.00\ \mu F$ and $C_2 = 4.00\ \mu F$ in parallel, Equation 17-26 becomes

$$C_{equiv} = 2.00\ \mu F + 4.00\ \mu F = 6.00\ \mu F$$

Note that C_{equiv} is greater than either C_1 or C_2.

Now repeat the calculation for the case where $C_1 = C_2 = 2.00\ \mu F$.

Follow the same steps but with both capacitances equal to $2.00\ \mu F$:

$$C_{equiv} = 2.00\ \mu F + 2.00\ \mu F = 4.00\ \mu F$$

For this case of identical capacitors in parallel, C_{equiv} is exactly twice as great as $C_1 = C_2 = 2.00\ \mu F$.

Reflect

Our results illustrate the following general results: Connecting capacitors in series reduces the capacitance, while connecting them in parallel increases the capacitance.

EXAMPLE 17-11 Multiple Capacitors

Find the equivalent capacitance of the three capacitors shown in **Figure 17-16.** The individual capacitances are $C_1 = 1.00\ \mu F$, $C_2 = 2.00\ \mu F$, and $C_3 = 6.00\ \mu F$.

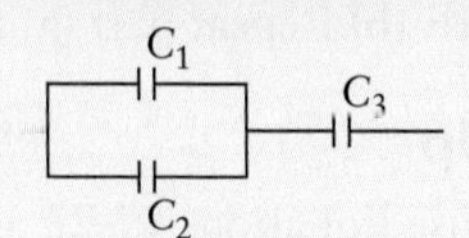

Figure 17-16 A capacitor network These three capacitors are neither all in series nor all in parallel.

Set Up

Whenever capacitors are combined in ways other than purely in series or purely in parallel, we look for groupings of capacitors that *are* either in parallel or in series, and then we combine the groups one at a time.

Two capacitors in series:

$$\frac{1}{C_{equiv}} = \frac{1}{C_1} + \frac{1}{C_2} \quad (17\text{-}22)$$

Replace the parallel capacitors C_1 and C_2 by their equivalent capacitor C_{12}:

C_{12} C_3

Then find the equivalent capacitance of C_{12} and C_3 in series.

Notice in Figure 17-16 that C_1 and C_2 are in parallel because their two right plates are directly connected, as are their two left plates. We can therefore use Equation 17-26 to find C_{12} (the equivalent capacitance of the combination of C_1 and C_2) by using our relationship for capacitors in parallel.

Capacitor C_3 is in series with C_{12}, so we can find their combined capacitance by using Equation 17-22. This result is C_{123}, the equivalent capacitance of all three capacitors.

Two capacitors in parallel:

$$C_{equiv} = C_1 + C_2 \qquad (17\text{-}26)$$

Solve

First find the equivalent capacitance of the parallel capacitors C_1 and C_2.

For $C_1 = 1.00\ \mu F$ and $C_2 = 2.00\ \mu F$ in parallel, the equivalent capacitance C_{12} is given by Equation 17-26:

$$C_{12} = C_1 + C_2 = 1.00\ \mu F + 2.00\ \mu F = 3.00\ \mu F$$

Then find C_{123}, the equivalent capacitance of the series capacitors C_{12} and C_3. This is the equivalent capacitance of the entire network of C_1, C_2, and C_3.

Since $C_{12} = 3.00\ \mu F$ and $C_3 = 6.00\ \mu F$ are in series, their equivalent capacitance C_{123} is given by Equation 17-22:

$$\frac{1}{C_{123}} = \frac{1}{C_{12}} + \frac{1}{C_3} = \frac{1}{3.00\ \mu F} + \frac{1}{6.00\ \mu F}$$

$$= 0.333\ \mu F^{-1} + 0.167\ \mu F^{-1} = 0.500\ \mu F^{-1}$$

Take the reciprocal of this to find C_{123}:

$$C_{123} = \frac{1}{0.500\ \mu F^{-1}} = 2.00\ \mu F$$

Reflect

As we described previously, when capacitors are connected in parallel, the equivalent capacitance is always greater than the greatest individual capacitor. That's why C_{12} is greater than either C_1 or C_2. When capacitors are connected in series, the equivalent capacitance is always less than the least individual capacitor, which is why C_{123} is less than either C_{12} or C_3.

GOT THE CONCEPT? 17-7 Energy in a Capacitor Combination

? Capacitor 1 has capacitance $C_1 = 1.0\ \mu F$, and capacitor 2 has capacitance $C_2 = 2.0\ \mu F$. You connect the initially uncharged capacitors to each other then connect the capacitor combination to a battery. Which capacitor stores the greater amount of electric potential energy if the two capacitors are connected in series? If they are connected in parallel? (a) Capacitor 1 for both the series and parallel cases; (b) capacitor 2 for both the series and parallel cases; (c) capacitor 1 for the series case, capacitor 2 for the parallel case; (d) capacitor 2 for the series case, capacitor 1 for the parallel case; (e) not enough information given to decide.

TAKE-HOME MESSAGE FOR Section 17-7

✔ Whenever two or more capacitors are connected, the equivalent capacitance is the capacitance of a single capacitor that is the equivalent of the combined capacitors.

✔ Capacitors are connected in series when they are connected one after another. In a series combination, all capacitors have the same charge, and the equivalent capacitance is less than that of any of the individual capacitors.

✔ Capacitors are connected in parallel when all of their right-hand plates are directly connected to each other and all of their left-hand plates are directly connected to each other. In a parallel combination, the voltage is the same across all capacitors, and the equivalent capacitance is greater than that of any of the individual capacitors.

17-8 Placing a dielectric between the plates of a capacitor increases the capacitance

So far in our discussion of capacitors, we've assumed that there is only vacuum between the capacitor plates. In most situations, however, the two plates are separated by a layer of a **dielectric**, a material that is both an insulator and *polarizable*—that is, in which there is a separation of positive and negative charge within the material when it's exposed to an electric field. Dielectrics used in commercial capacitors include glass, ceramics, and plastics such as polystyrene. These dielectrics not only help to keep the positive and negative plates from touching each other but also increase the capacitance of the capacitor. In this section we'll see how dielectrics make this possible.

Let's consider an isolated parallel-plate capacitor with charges $+q$ and $-q$ on its plates and with vacuum in the space between its plates (**Figure 17-17a**). The charge creates a uniform electric field $\vec{E}_0$ between the plates. We now insert a dielectric material that fills the space between the plates of this capacitor. What happens then depends on the kind of molecules that make up the dielectric.

If the molecules are *polar*—that is, if one end of the molecule has a positive charge and the other end has a negative charge of the same magnitude—the molecules will orient themselves so that their positive ends are pointed toward the negatively charged plate and their negative ends toward the positively charged plate, as in **Figure 17-17b**. (The most common molecule in your body, the water molecule H_2O, is a polar molecule. Others include ammonia, NH_3, and sucrose, $C_{12}H_{22}O_{11}$.)

If the molecules are not polar, so there normally is no separation of charge within the molecule, the electric field between the plates of the capacitor will *induce* a slight separation of the positive and negative charges in each molecule. (We described this

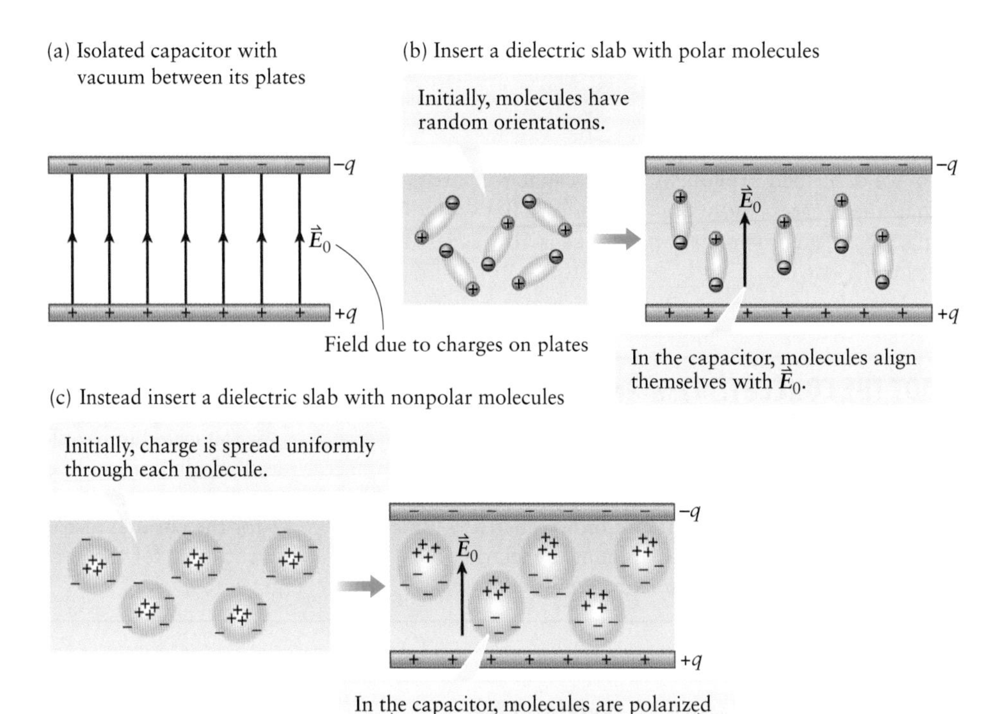

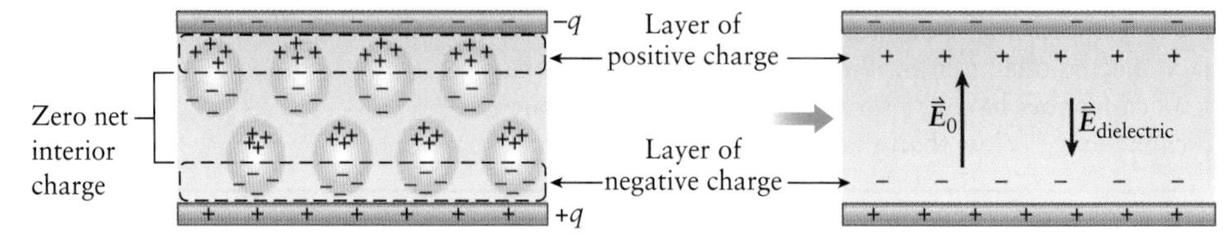

Figure 17-17 A dielectric in a parallel-plate capacitor A dielectric slab inserted between the plates of an isolated capacitor reduces the electric field between the plates.

process in Section 16-2.) The molecules will then orient themselves in the same manner as polar molecules would (**Figure 17-17c**). In either case we say that the dielectric becomes *polarized* once it has been inserted between the plates of the capacitor.

In the interior of the polarized dielectric in Figures 17-17b or 17-17c, there are as many positive ends of molecules as there are negative ends, so the interior of the dielectric is neutral. But there *is* a net charge at the surfaces of the dielectric: There is a layer of positive charge on the surface next to the negatively charged plate, and there is a layer of negative charge on the surface next to the positively charged plate. (The dielectric as a whole is neutral, so these two layers have the same magnitude of charge.) These two layers of charge create an electric field $\vec{E}_{\text{dielectric}}$ that points opposite to the electric field $\vec{E}_0$ created by the charges on the plates of the capacitor (**Figure 17-17d**). Hence $\vec{E}_{\text{dielectric}}$ partially cancels $\vec{E}_0$, and the *net* electric field $\vec{E} = \vec{E}_0 + \vec{E}_{\text{dielectric}}$ between the plates is less than $\vec{E}_0$. So the effect of the dielectric is to *reduce* the electric field between the plates. We can express this as

Electric field in an isolated parallel-plate capacitor **with dielectric** between the plates

Electric field in that same capacitor with **no dielectric** between the plates

$$E = \frac{E_0}{\kappa}$$

Dielectric constant; note that $\kappa \geq 1$

For an isolated charged capacitor, a dielectric reduces the electric field between the plates.

Electric field in an isolated parallel-plate capacitor with a dielectric (17-27)

The quantity κ (the Greek letter "kappa") is the **dielectric constant** of the material. The greater the value of κ, the more the dielectric is polarized when it is placed between the plates of a charged capacitor, so the greater the magnitude of the field $\vec{E}_{\text{dielectric}}$ due to the dielectric and the smaller the magnitude E of the net field. **Table 17-1** lists values of the dielectric constant for a variety of materials.

TABLE 17-1 Dielectric Constants (at 20°C and 1 atm)

Material	κ
vacuum	1
air	1.00058
lipid	2.2
paraffin	2.2
paper	2.7
ceramic (porcelain)	5.8
water	80

What does Equation 17-27 tell us about how the dielectric affects the capacitance of the capacitor? We saw in Section 17-5 that the potential difference V between the positive and negative plates is proportional to the magnitude E of the electric field between the plates (see Equation 17-11, which states that $V = Ed$). Equation 17-27 says that the dielectric reduces the value of the electric field by a factor $1/\kappa$, so V is also reduced by a factor $1/\kappa$. Now, Equation 17-14 tells us that the capacitance C is related to potential difference V and the magnitude q of the charges on the plates by $q = CV$, or $C = q/V$. While the dielectric reduces V by a factor $1/\kappa$, it has no effect on the value of q. (Since we specified that the capacitor is isolated, no charge can leave either plate.) Because the capacitance $C = q/V$ is inversely proportional to the potential difference V, which is reduced by a factor $1/\kappa$, it follows that C *increases* by a factor of $1/(1/\kappa) = \kappa$. If C_0 is the capacitance without the dielectric, the capacitance with the dielectric is

Capacitance of a parallel-plate capacitor **with dielectric** between the plates

Capacitance of that same capacitor with **no dielectric** between the plates

$$C = \kappa C_0$$

A dielectric increases the capacitance of a capacitor.

Dielectric constant; note that $\kappa \geq 1$

Capacitance of a parallel-plate capacitor with a dielectric (17-28)

As an example, Table 17-1 tells us that the dielectric constant of porcelain is $\kappa = 5.8$. From Equation 17-28 the capacitance of a capacitor with porcelain filling the space between its plates is 5.8 times greater than an identical capacitor with vacuum between its plates. That's fundamentally because the electric field in the porcelain-filled capacitor is only $1/5.8 = 0.17$ as great as for an identical vacuum capacitor carrying the same charge (Equation 17-27).

Example 17-12 shows how to apply these ideas to the membrane of a cell, which we can regard as a capacitor with multiple layers of dielectric between its plates.

BIO-Medical EXAMPLE 17-12 Cell Membrane Capacitance

In all cells, it is easier for positive potassium ions (K^+) to flow out of the cell than it is for negative ions. As a result, there is negative charge on the inside of the cell membrane and positive charge on the outside of the membrane, much like a capacitor. Consider a typical cell with a membrane of thickness 7.60 nm = 7.60×10^{-9} m. (a) Find the capacitance of a square patch of membrane 1.00 μm = 1.00×10^{-6} m on a side, assuming that the membrane has dielectric constant $\kappa = 1$. (b) The actual structure of the membrane is a layer of lipid surrounded by layers of a polarized aqueous solution and layers of water (**Figure 17-18**). Taking account of this structure, find the capacitance of a square patch of membrane 1.00 μm on a side.

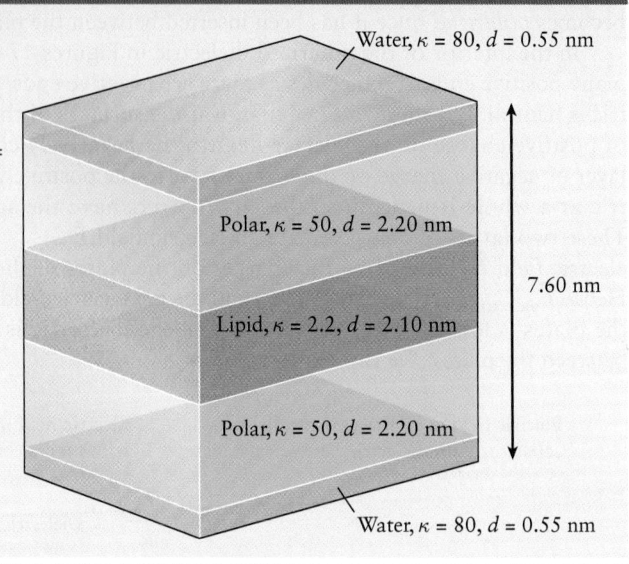

Figure 17-18 Membrane capacitance A membrane acts as a capacitor filled with a dielectric. What is the capacitance of a membrane that consists of a layer of lipid surrounded by layers of a polarized aqueous solution and water?

Set Up

A dielectric constant $\kappa = 1$ corresponds to vacuum, so in part (a) we can find the capacitance C_0 of the membrane under the assumption $\kappa = 1$ by using Equation 17-13 from Section 17-5.

In part (b) we must deal with the more complicated situation shown in Figure 17-18. We can't simply use Equation 17-28 to find the capacitance because there's not a single value of κ for the arrangement of layers that make up the membrane. Instead we'll imagine that there's a very thin conducting sheet separating each layer from the next. If there are charges $+q$ and $-q$ on the outer and inner surfaces of the multilayer membrane, there will also be charges $+q$ and $-q$ on the surfaces of these conducting sheets. This is exactly like the situation with capacitors in series (Section 17-7). So we'll treat each of the five layers in Figure 17-18 as an individual capacitor, then use Equation 17-22 to find the capacitance of the combination.

Parallel-plate capacitor with vacuum between the plates:

$$C_0 = \frac{\varepsilon_0 A}{d} \qquad (17\text{-}13)$$

Parallel-plate capacitor with dielectric:

$$C = \kappa C_0 \qquad (17\text{-}28)$$

Five capacitors in series:

$$\frac{1}{C_{\text{equiv}}} = \frac{1}{C_1} + \frac{1}{C_2} + \frac{1}{C_3} + \frac{1}{C_4} + \frac{1}{C_5} \qquad (17\text{-}22)$$

Solve

(a) First use Equation 17-13 to find the capacitance, assuming that the membrane has $\kappa = 1$ (equivalent to vacuum).

The area A for the capacitor of interest is a square 1.00 μm on a side:

$$A = (1.00\ \mu\text{m})^2 = (1.00 \times 10^{-6}\ \text{m})^2 = 1.00 \times 10^{-12}\ \text{m}^2$$

A parallel-plate capacitor with this area and with $d = 7.60 \times 10^{-9}$ m has capacitance

$$C_0 = \frac{\varepsilon_0 A}{d} = \frac{(8.85 \times 10^{-12}\ \text{F/m})(1.00 \times 10^{-12}\ \text{m}^2)}{7.60 \times 10^{-9}\ \text{m}} = 1.16 \times 10^{-15}\ \text{F}$$

(b) Calculate the capacitance of each of the five layers separately using Equations 17-13 and 17-28.

The capacitance of the first layer of water is $\kappa_1 = 80$ times the capacitance of a vacuum capacitor of thickness $d_1 = 0.55$ nm:

$$C_1 = \kappa_1\left(\frac{\varepsilon_0 A}{d_1}\right) = (80)\frac{(8.85 \times 10^{-12}\ \text{F/m})(1.00 \times 10^{-12}\ \text{m}^2)}{0.55 \times 10^{-9}\ \text{m}} = 1.3 \times 10^{-12}\ \text{F}$$

Similarly, for the second layer of polarized aqueous solution with $\kappa_2 = 50$ and thickness $d_2 = 2.20$ nm, the capacitance is

$$C_2 = \kappa_2\left(\frac{\varepsilon_0 A}{d_2}\right)$$
$$= (50)\frac{(8.85 \times 10^{-12}\ \text{F/m})(1.00 \times 10^{-12}\ \text{m}^2)}{2.20 \times 10^{-9}\ \text{m}}$$
$$= 2.0 \times 10^{-13}\ \text{F}$$

The capacitance of the third lipid layer with $\kappa_3 = 2.2$ and thickness $d_3 = 2.10$ nm is

$$C_3 = \kappa_3\left(\frac{\varepsilon_0 A}{d_3}\right)$$
$$= (2.2)\frac{(8.85 \times 10^{-12}\ \text{F/m})(1.00 \times 10^{-12}\ \text{m}^2)}{2.10 \times 10^{-9}\ \text{m}}$$
$$= 9.3 \times 10^{-15}\ \text{F}$$

The fourth layer of polarized aqueous solution is identical to the second layer, and the fifth water layer is identical to the first layer. So

$$C_4 = C_2 = 2.0 \times 10^{-13}\ \text{F}$$
$$C_5 = C_1 = 1.3 \times 10^{-12}\ \text{F}$$

Calculate the net capacitance of the five layers in series using Equation 17-22.

The equivalent capacitance C_{equiv} of the five-layer stack is given by

$$\frac{1}{C_{\text{equiv}}} = \frac{1}{C_1} + \frac{1}{C_2} + \frac{1}{C_3} + \frac{1}{C_4} + \frac{1}{C_5}$$

Since $C_4 = C_2$ and $C_5 = C_1$, this is

$$\frac{1}{C_{\text{equiv}}} = \frac{1}{C_1} + \frac{1}{C_2} + \frac{1}{C_3} + \frac{1}{C_2} + \frac{1}{C_1} = \frac{2}{C_1} + \frac{2}{C_2} + \frac{1}{C_3}$$
$$= \frac{2}{1.3 \times 10^{-12}\ \text{F}} + \frac{2}{2.0 \times 10^{-13}\ \text{F}} + \frac{1}{9.3 \times 10^{-15}\ \text{F}}$$
$$= 1.2 \times 10^{14}\ \text{F}^{-1}$$

The reciprocal of this is C_{equiv}:

$$C_{\text{equiv}} = \frac{1}{1.2 \times 10^{14}\ \text{F}^{-1}} = 8.4 \times 10^{-15}\ \text{F}$$

Reflect

The equivalent capacitance of the stack of five layers is less than that of any individual layer. This is just what we saw in Section 17-7 for capacitors in series.

The capacitance $C_{\text{equiv}} = 8.4 \times 10^{-15}$ F is greater than the capacitance $C_0 = 1.16 \times 10^{-15}$ F calculated assuming $\kappa = 1$ by a factor $C_{\text{equiv}}/C_0 = (8.4 \times 10^{-15}\ \text{F})/(1.16 \times 10^{-15}\ \text{F}) = 7.2$. So the effective dielectric constant of the membrane is 7.2, intermediate between the largest ($\kappa_1 = 80$) and smallest ($\kappa_3 = 2.2$) values of dielectric constant for the individual layers.

GOT THE CONCEPT? 17-8 Inserting a Dielectric into a Capacitor

? An isolated parallel-plate capacitor (one that is not connected to anything else) has a vacuum between its plates and has charges $+q$ and $-q$ on its plates. If you insert a slab of dielectric with dielectric constant κ that fills the space between the plates, the capacitance increases. What happens to the energy stored in the capacitor as you insert the dielectric slab? Will you have to push the slab into the capacitor, or will you feel the capacitor pulling the slab in? (a) Stored energy increases, you will have to push the slab in; (b) stored energy increases, the slab will be pulled in; (c) stored energy decreases, you will have to push the slab in; (d) stored energy decreases, the slab will be pulled in; (e) not enough information given to decide.

TAKE-HOME MESSAGE FOR Section 17-8

✔ When a dielectric material is placed in an electric field $\vec{E}_0$, it becomes polarized. This produces an additional electric field that partially cancels $\vec{E}_0$ and so reduces the net field in the dielectric. The dielectric constant is a measure of how much the field is reduced.

✔ If a dielectric material fills the space between the plates of a capacitor, the capacitance is greater than if there is vacuum between the plates.

Key Terms

capacitance
capacitor
dielectric
dielectric constant
electric potential
electric potential difference
electric potential energy
electron volt
equipotential
equipotential surface
equipotential volume
equivalent capacitance
farad
parallel (capacitors)
parallel-plate capacitor
plates
series (capacitors)
test charge
volt
voltage

Chapter Summary

Topic	Equation or Figure
Electric potential energy: Like the gravitational force, the electric force is a conservative force and has an associated potential energy. The change in electric potential energy when a charged object moves from one point to another depends on the sign of the object's charge. The sign of the electric potential energy of two point charges depends on whether the two charges have the same sign or different signs.	The **change in electric potential energy** for an object of charge q that moves in a uniform electric field $\vec{E}$... ...equals the **negative of the work done** on the object by the electric force. $\Delta U_{\text{electric}} = -W_{\text{electric}} = -qEd\cos\theta$ (17-2). **Charge** of the object; **Angle** between the displacement and the direction of the electric field; **Magnitude of the electric field**; Straight-line **displacement** of the object. $U_{\text{electric}} = \dfrac{kq_1q_2}{r}$ (17-3). **Coulomb constant**; Values of the **two charges**; **Electric potential energy** of two point charges; **Distance** between the point charges
Electric potential: The electric potential at a given position equals the electric potential energy for a point charge q_0 at that position, divided by the value of q_0. If you move in a direction opposite to the electric field, the electric potential increases; if you move in the direction of the electric field, the electric potential decreases. The electric potential due to a point charge is inversely proportional to the distance from the charge. Voltage is the	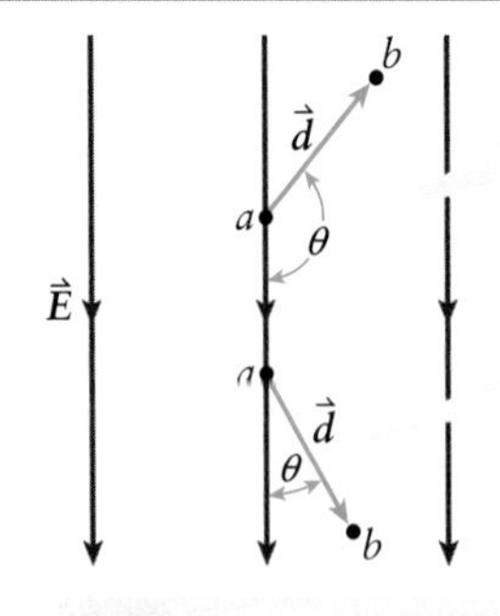 If you move in a direction opposite to the electric field ($\theta > 90°$), the electric potential increases ($\Delta V > 0$). If you move in the direction of the electric field ($\theta < 90°$), the electric potential decreases ($\Delta V < 0$). (Figure 17-5)  If you move from point a to point b in a uniform electric field $\vec{E}$, the change in electric potential is $\Delta V = -Ed\cos\theta$.

difference in electric potential between two locations.

Electric potential due to a point charge Q

Coulomb constant

$$V = \frac{kQ}{r} \quad (17\text{-}8)$$

Value of the point **charge**

Distance from the point charge Q to the location where the potential is measured

Equipotentials: The electric potential has the same value everywhere on an equipotential surface. An equipotential surface is everywhere perpendicular to the electric field. The electric potential has the same value throughout the volume of a conductor in equilibrium.

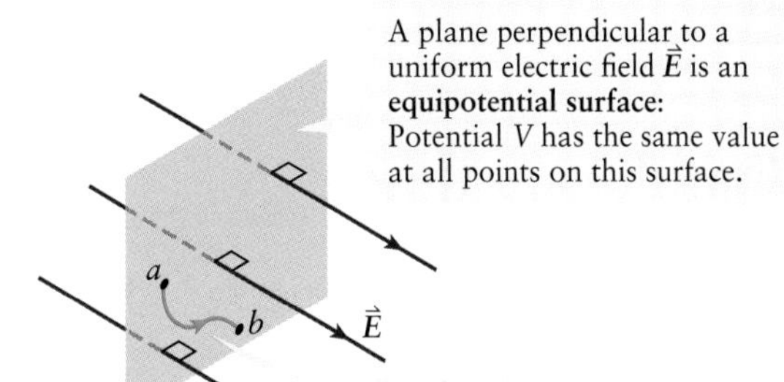

A plane perpendicular to a uniform electric field $\vec{E}$ is an **equipotential surface:** Potential V has the same value at all points on this surface.
a
b
$\vec{E}$
Any curve that lies on the plane is an equipotential curve.

(Figure 17-6)

Capacitors: A capacitor has two conducting plates that store charges $+q$ and $-q$, of the same magnitude but opposite sign. The potential difference V between the plates is proportional to the charge q; the proportionality constant, the capacitance C, depends only on the geometry of the capacitor and the substance between the plates.

A capacitor carries a charge $+q$ on its positive plate and a charge $-q$ on its negative plate.

The magnitude of q is directly proportional to V, the **voltage** (potential difference) between the plates.

$$q = CV \quad (17\text{-}14)$$

The constant of proportionality between charge q and voltage V is the **capacitance** C of the capacitor.

Energy stored in a capacitor: A charged capacitor stores electric potential energy. The stored energy can be expressed by any two of the three quantities: capacitance C, charge q, and potential difference V.

The **electric potential energy** stored in a charged capacitor...

$$U_{\text{electric}} = \frac{1}{2}qV = \frac{1}{2}CV^2 = \frac{q^2}{2C} \quad (17\text{-}17)$$

...can be expressed in three ways in terms of the **charge** q, **potential difference** V, and **capacitance** C.

Capacitors in series and parallel: A collection of capacitors in a circuit behaves as though it were a single capacitor, with an equivalent capacitance C_{equiv}. The equivalent capacitance is different for capacitors in series (with the positive plate of one capacitor connected to the negative plate of another) than for capacitors in parallel (with positive plates connected to positive plates and negative plates to negative plates).

If capacitors are in series, the reciprocal of the equivalent capacitance is the sum of the reciprocals of the individual capacitances.

$$\frac{1}{C_{\text{equiv}}} = \frac{1}{C_1} + \frac{1}{C_2} + \frac{1}{C_3} \quad (17\text{-}22)$$

Equivalent capacitance of capacitors in series

Capacitances of the individual capacitors

Equivalent capacitance of capacitors in parallel

Capacitances of the individual capacitors

$$C_{\text{equiv}} = C_1 + C_2 + C_3 \quad (17\text{-}26)$$

If capacitors are in parallel, the equivalent capacitance is the sum of the individual capacitances.

Dielectrics: A dielectric is an insulator that becomes polarized when placed in an electric field. When the space between the plates of a capacitor is filled with a dielectric, the electric field between the plates decreases (for a given charge q on the plates) and the capacitance increases.

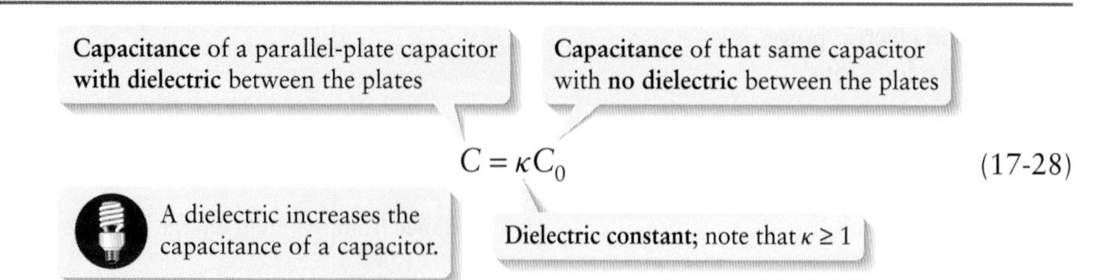

$$C = \kappa C_0 \tag{17-28}$$

Answer to What do you think? Question

(c) The energy stored by separating charge $+q$ from charge $-q$ in a capacitor is proportional to q^2 (see Section 17-6, especially Equation 17-17). So doubling the value of q increases the stored energy by a factor of $2^2 = 4$.

Answers to Got the Concept? Questions

17-1 (b) Although q_1, q_2, and q_3 all have reversed sign, the signs of the *products* q_1q_2, q_1q_3, and q_2q_3 in Equation 17-4 remain the same. So if you change all of the positive charges in an assemblage to negative and vice versa, there is no effect on the electric potential energy of the assemblage.

17-2 (c), (a), (d), (b) From Equation 17-6, the potential energy change $\Delta U_{\text{electric}}$ for a point charge q_0 is related to the potential change ΔV by $\Delta V = \Delta U_{\text{electric}}/q_0$, or $\Delta U_{\text{electric}} = q_0\,\Delta V$. For the four cases we have (a) $\Delta U_{\text{electric}} = (+0.0010\text{ C})(5.0\text{ V}) = +0.0050\text{ C}\cdot\text{V} = +0.0050\text{ J}$; (b) $\Delta U_{\text{electric}} = (+0.0020\text{ C})(-2.0\text{ V}) = -0.0040\text{ J}$; (c) $\Delta U_{\text{electric}} = (-0.0015\text{ C})(-4.0\text{ V}) = +0.0060\text{ J}$; (d) $\Delta U_{\text{electric}} = (-0.0010\text{ C})(+2.5\text{ V}) = -0.0025\text{ J}$.

17-3 (b) In the vicinity of A, the potential decreases by 1 V (from $V = 0$ V to $V = -1$ V) in the 2-m distance between $x = 0$ and $x = 2$ m. The electric field $\vec{E}$ points in the direction of decreasing potential, which is to the right (the positive x direction) at both points. In the vicinity of B, the equipotentials are much closer together; the potential decreases by 1 V (from $V = -2$ V to $V = -3$ V) in the 0.5-m distance from $x = 3$ m to $x = 3.5$ m. Where equipotentials are farther apart, as at A, the electric field has a smaller magnitude.

17-4 (e) Equation 17-13 shows that the capacitance of a parallel-plate capacitor depends on its geometry only. It does not depend on the amount of charge on the plates or on the potential difference between the plates.

17-5 (c) The stored energy increases. The positive and negative charges on the two plates attract each other, so to pull the plates apart you must do positive work on the system. The work you do increases the electric potential energy stored in the capacitor. To verify this conclusion note from Equation 17-13 that if you increase the spacing d between the plates, the capacitance $C = \varepsilon_0 A/d$ will decrease. Equation 17-17, $U_{\text{electric}} = q^2/2C$, then tells us that if the charge q remains constant and the capacitance C decreases, the stored energy U_{electric} will increase.

17-6 (a) The stored energy decreases. Equation 17-13 tells us that as you increase the spacing d between the plates, the capacitance $C = \varepsilon_0 A/d$ decreases. Equation 17-17, $U_{\text{electric}} = (1/2)CV^2$, then tells us that if the potential difference V remains constant and C decreases, the stored energy must decrease. Note that in the situation of Got the Concept? 17-5, the stored energy *increases* when the plates are pulled apart. Why is the answer different here? The explanation is that in Got the Concept? 17-5 the charge q remained constant. In the current situation, however, the charge q has to change in accordance with Equation 17-14, $q = CV$: The charge q decreases as the capacitance C decreases. (Electrons flow from the negative plate through the battery and onto the positive plate, so the magnitude of charge on both plates decreases.) Equation 17-17 also tells us that we can express the stored energy as $U_{\text{electric}} = (1/2)qV$, so the stored energy will decrease if q decreases while V remains the same.

17-7 (c) Equation 17-17 gives us two useful expressions for the electric potential energy stored in a capacitor with capacitance C: $U_{\text{electric}} = (1/2)CV^2$ (which says that for a given voltage V the stored energy is proportional to C) and $U_{\text{electric}} = q^2/2C$ (which says that for a given charge q the stored energy is inversely proportional to C). In a series combination both capacitors carry the same charge q, so $U_{\text{electric}} = q^2/2C$ tells us that the capacitor with the smaller capacitance ($C_1 = 1.0\ \mu\text{F}$) stores the greater amount of energy. In a parallel combination, the voltage V is the same for both capacitors, so $U_{\text{electric}} = (1/2)CV^2$ tells us that the capacitor with the greater capacitance ($C_2 = 2.0\ \mu\text{F}$) stores the greater amount of energy. We can draw these conclusions without knowing the specific values of q for the series case or V for the parallel case.

17-8 (d) Because the capacitor is isolated, the charge q does not change when the dielectric slab is inserted. So among the expressions for stored energy given in Equation 17-17, the one to use is $U_{\text{electric}} = q^2/2C$. Since q remains constant but capacitance C increases as you insert the slab, it follows that U_{electric} decreases. Just as a ball is pulled downward by the gravitational force, in the direction that decreases gravitational potential energy, the slab must be pulled into the capacitor by an electric force because motion in that direction decreases electric potential energy U_{electric}.

Questions and Problems

In a few problems you are given more data than you actually need; in a few other problems you are required to supply data from your general knowledge, outside sources, or informed estimate.

Interpret as significant all digits in numerical values that have trailing zeros and no decimal points. For all problems use $g = 9.80 \text{ m/s}^2$ for the free-fall acceleration due to gravity. Neglect friction and air resistance unless instructed to do otherwise.

•	Basic, single-concept problem
••	Intermediate-level problem; may require synthesis of concepts and multiple steps
•••	Challenging problem
SSM	*Solution is in Student Solutions Manual*
Example	*See worked example for a similar problem*

Conceptual Questions

1. • What is the difference between electric potential and electric field?

2. • What is the difference between electric potential and electric potential energy?

3. • Explain why electric potential requires the existence of only one charge, but a finite electric potential energy requires the existence of two charges.

4. • An electron is released from rest in an electric field. Will it accelerate in the direction of increasing or decreasing potential? Why?

5. • Does it make sense to say that the voltage at some point in space is 10.3 V? Explain your answer. SSM

6. •• Explain why an electron will accelerate toward a region of lower electric potential energy but higher electric potential.

7. • (a) If the electric potential throughout some region of space is zero, does it necessarily follow that the electric field is zero? (b) If the electric field throughout a region is zero, does it necessarily follow that the electric potential is zero? SSM

8. • Discuss how a topographical map showing various elevations around a mountain is analogous to the equipotential lines surrounding a charged object.

9. • How much work is required to move a charge from one end of an equipotential path to the other? Explain your answer.

10. • Explain why capacitance depends neither on the stored charge Q nor on the potential difference V between the plates of a capacitor.

11. • Describe three methods by which you might increase the capacitance of a parallel-plate capacitor.

12. • If the voltage across a capacitor is doubled, by how much does the stored energy change?

13. •• You charge a capacitor and then remove it from the battery. The capacitor consists of large movable plates with air between them. You pull the plates a bit farther apart. What happens to the stored energy? SSM

14. •• The capacitance of several capacitors in series is less than any of the individual capacitances. What, then, is the advantage of having several capacitors in series?

15. • What is the advantage to arranging several capacitors in parallel?

16. • Which way of connecting (series or parallel) three identical capacitors to a battery would store more energy?

17. • Qualitatively explain why the equivalent capacitance of a parallel combination of identical capacitors is larger than the individual capacitances.

18. • Does inserting a dielectric into a capacitor increase or decrease the energy stored in the capacitor? Explain your answer.

19. • What are the benefits, if any, of filling a capacitor with a dielectric other than air?

20. •• Capacitors A and B are identical except that the region between the plates of capacitor A is filled with a dielectric. As shown in **Figure 17-19**, the plates of these capacitors are maintained at the same potential difference by a battery. Is the electric field magnitude in the region between the plates of capacitor A smaller, the same, or larger than the field in the region between the plates of capacitor B? Explain your answer.

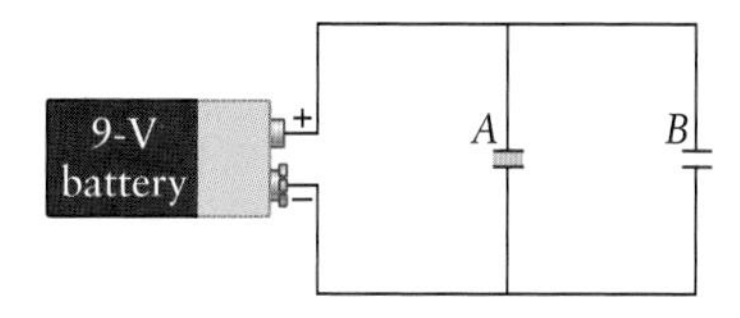

Figure 17-19 Problem 20

Multiple-Choice Questions

21. •• For a positive charge moving in the direction of the electric field,
    A. its potential energy increases and its electric potential increases.
    B. its potential energy increases and its electric potential decreases.
    C. its potential energy decreases and its electric potential increases.
    D. its potential energy decreases and its electric potential decreases.
    E. its potential energy and its electric potential remain constant. SSM

22. •• If a negative charge is released in a uniform electric field, it will move
    A. in the direction of the electric field.
    B. from high potential to low potential.
    C. from low potential to high potential.
    D. in a direction perpendicular to the electric field.
    E. in circular motion.

23. • An equipotential surface must be
    A. parallel to the electric field at every point.
    B. equal to the electric field at every point.
    C. perpendicular to the electric field at every point.
    D. tangent to the electric field at every point.
    E. equal to the inverse of the electric field at every point.

24. • A positive charge is moved from one point to another point along an equipotential surface. The work required to move the charge
    A. is positive.
    B. is negative.
    C. is zero.
    D. depends on the sign of the potential.
    E. depends on the magnitude of the potential.

25. • The electric potential at a point equidistant from two particles that have charges $+Q$ and $-Q$ is
   A. larger than zero.
   B. smaller than zero.
   C. equal to zero.
   D. equal to the average of the two distances times the charges.
   E. equal to the net electric field.

26. • Four point charges of equal magnitude but differing signs are arranged at the corners of a square (**Figure 17-20**). The electric field E and the potential V at the center of the square are
   A. $E = 0$; $V \neq 0$.
   B. $E = 0$; $V = 0$.
   C. $E \neq 0$; $V \neq 0$.
   D. $E \neq 0$; $V = 0$.
   E. $E = 2V^{1/2}$.

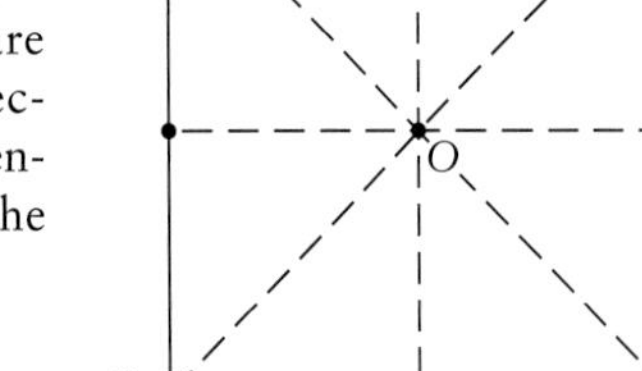

Figure 17-20 Problems 26 and 40

27. •• An isolated parallel-plate capacitor carries a charge Q. If the separation between the plates is doubled, the electrical energy stored in the capacitor will be
   A. halved.
   B. doubled.
   C. unchanged.
   D. quadrupled.
   E. quartered. SSM

28. •• A parallel-plate capacitor is connected to a battery that maintains a constant potential difference across the plates. If the separation between the plates is doubled, the electrical energy stored in the capacitor will be
   A. halved.
   B. doubled.
   C. unchanged.
   D. quadrupled.
   E. quartered.

29. • Capacitors connected in series have the same
   A. charge.
   B. voltage.
   C. dielectric.
   D. surface area.
   E. separation.

30. • Capacitors connected in parallel have the same
   A. charge.
   B. voltage.
   C. dielectric.
   D. surface area.
   E. separation.

Estimation/Numerical Analysis

31. • A pollen grain has a maximum electric charge limit it can carry before the electric field it generates exceeds the dielectric limit for air of 3×10^6 N/C. Estimate the maximum voltage at the surface of a pollen grain that is carrying its maximum electric charge, assuming $V = 0$ at infinity.

32. • Estimate the amount of energy released in a typical "finger-to-door knob" spark.

33. • Estimate the amount of energy released in a cloud-to-ground lightning strike.

34. • If a source charge is in the microcoulomb range, how far from the charge should you be to have an electric potential in the millivolt range as compared to the potential a long way from the charge?

35. • Estimate the number of 100 μF capacitors, connected in parallel, that would provide enough energy to get an electric car moving at 20 m/s (the mass of a 1 μF capacitor equals 0.005 g).

Problems

17-1 Electric energy is important in nature, technology, and biological systems

17-2 Electric potential energy changes when a charge moves in an electric field

36. • A point charge q_0 that has a charge of 0.500 μC is at the origin. (a) A second particle q that has a charge of 1.00 μC and a mass of 0.0800 g is placed at $x = 0.800$ m. What is the potential energy of this system of charges? (b) If the particle with charge q is released from rest, what will its speed be when it reaches $x = 2.00$ m? Example 17-2

37. • A uniform electric field of 2.00 kN/C points in the $+x$ direction. (a) What is the change in potential energy $U_{\text{electric},b} - U_{\text{electric},a}$ of a +2.00 nC test charge as it is moved from point a at $x = -30.0$ cm to point b at $x = +50.0$ cm? (b) The same test charge is released from rest at point a. What is its kinetic energy when it passes through point b? (c) If a negative charge instead of a positive charge were used in this problem, qualitatively how would your answers change? SSM Example 17-1

38. •• **Biology** Two red blood cells each have a mass of 9.0×10^{-14} kg and carry a negative charge spread uniformly over their surfaces. The repulsion arising from the excess charge prevents the cells from clumping together. One cell carries −2.50 pC of charge and the other −3.10 pC, and each cell can be modeled as a sphere 7.5 μm in diameter. (a) What speed would they need when very far away from each other to get close enough to just touch? Ignore viscous drag from the surrounding liquid. (b) What is the magnitude of the maximum acceleration of each cell in part (a)? Example 17-2

39. • Three charges lie on the x axis. Charge $q_1 = +2.20$ μC is at $x = -30.0$ cm, charge $q_2 = -3.10$ μC is at the origin, and charge $q_3 = +1.70$ μC is at $x = 25.0$ cm. Calculate the potential energy of the system of charges. Example 17-3

40. • Calculate the potential energy of the system of charges in Figure 17-20, if the magnitude of each charge is $|Q| = 4.40$ μC and the length of each side of the large square connecting each charge is 60.0 cm. Example 17-3

17-3 Electric potential equals electric potential energy per charge

41. • A uniform electric field of magnitude 28.0 V/m makes an angle of 30.0° with a displacement of length 10.0 m. What is the potential difference of the final position relative to the initial position of this displacement? Example 17-5

42. • At a certain point in space, there is a potential of 800 V relative to zero. What is the potential energy of the system when a +1.0 μC charge is placed at that point in space? Example 17-6

43. • How much work is required to move a 2.0-C positive charge from the negative terminal of a 9.0-V battery to the positive terminal? SSM Example 17-6

44. • **Biology** A potential difference exists between the inner and outer surfaces of the membrane of a cell. The inner surface is negative relative to the outer surface. If 1.5×10^{-20} J of work is required to eject a positive sodium ion (Na^+) from the interior of the cell, what is the potential difference between the inner and outer surfaces of the cell? Example 17-6

45. • **Chemistry** What is the electric potential due to the nucleus of hydrogen at a distance of 5.00×10^{-11} m? Assume the potential is equal to zero as $r \to \infty$.

46. • (a) What is the electric potential due to a point charge of $+2.00\ \mu C$ at a distance of 0.500 cm? (b) How will the answer change if the charge has a value of $-2.00\ \mu C$? Assume the potential is equal to zero as $r \to \infty$.

47. • The electric potential has a value of -200 V at a distance of 1.25 m from a point charge. What is the value of that charge? Assume the potential is equal to zero as $r \to \infty$.

48. • At point P in **Figure 17-21** the electric potential is zero. (As usual, we take the potential to be zero at infinite distance.) (a) What can you say about the two charges? (b) Are there any other points of zero potential on the line connecting P and the two charges?

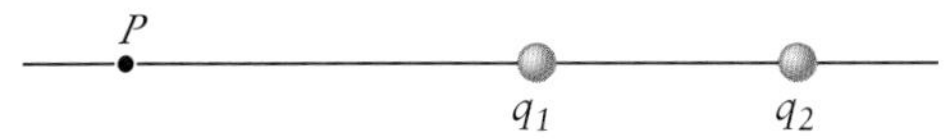

Figure 17-21 Problem 48

49. • Two point charges are placed on the x axis: $+0.500\ \mu C$ at $x = 0$ and $-0.200\ \mu C$ at $x = 10.0$ cm. At what point(s), if any, on the x axis is the electric potential equal to zero? SSM

50. • A charge of $+2.00\ \mu C$ is at the origin and a charge of $-3.00\ \mu C$ is on the y axis at $y = 40.0$ cm. (a) What is the potential at point a, which is on the x axis at $x = 40.0$ cm? (b) What is the potential difference $V_b - V_a$ when point b is at (40.0 cm, 30.0 cm)? (c) How much work is required to move an electron at rest from point a to rest at point b?

51. •• Calculate the electric potential at the origin O due to the point charges in **Figure 17-22**.

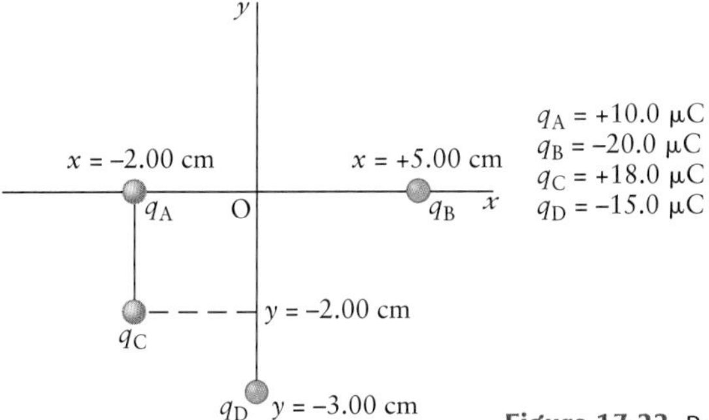

Figure 17-22 Problem 51

52. • In the Bohr model of the hydrogen atom, an electron in the lowest energy state moves around the nucleus at a speed of 2.19×10^6 m/s at a distance of 0.529×10^{-10} m from the nucleus. (a) What is the electric potential due to the hydrogen nucleus at this distance? (b) How much energy is required to ionize a hydrogen atom, whose electron is in this lowest energy state? Assume the electric potential goes to zero as $r \to \infty$.

53. • As shown in **Figure 17-23**, two large parallel plates, which are aligned along the y axis, are separated by a distance $d = 30.0$ cm and are at different electric potentials. The center of each plate has a small opening that lies on the x axis. A proton, traveling on the x axis, passes through the first plate with a speed of 2.50×10^5 m/s, and then leaves through the second plate with a speed of 7.80×10^5 m/s. Calculate the potential difference, $V_2 - V_1$, between the two plates. Note that a positive potential difference indicates the second plate is at a higher potential than the first plate. Example 17-4

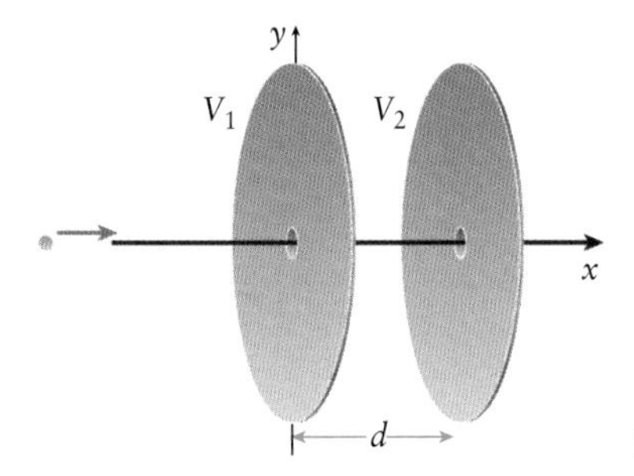

Figure 17-23 Problem 53

17-4 The electric potential has the same value everywhere on an equipotential surface

54. • Electric field lines for a system of two point charges are shown in **Figure 17-24**. Reproduce the figure and draw on it some equipotential lines for the system.

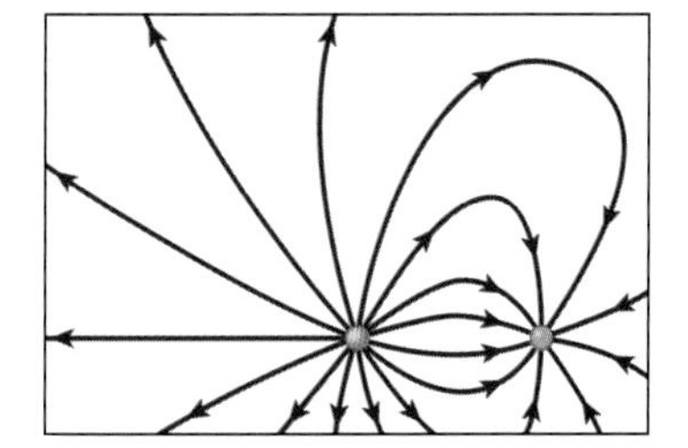

Figure 17-24 Problem 54

55. • In **Figure 17-25**, equipotential lines are shown at 1-m intervals. What is the electric field at (a) point A and (b) point B?

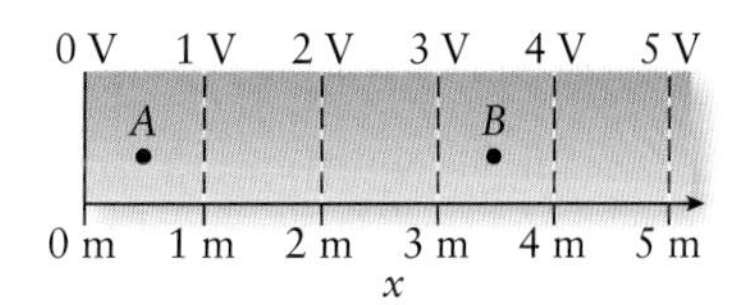

Figure 17-25 Problem 55

56. • Equipotential lines for some region of space are shown in **Figure 17-26**. What is the approximate electric field at (a) point A and (b) point B?

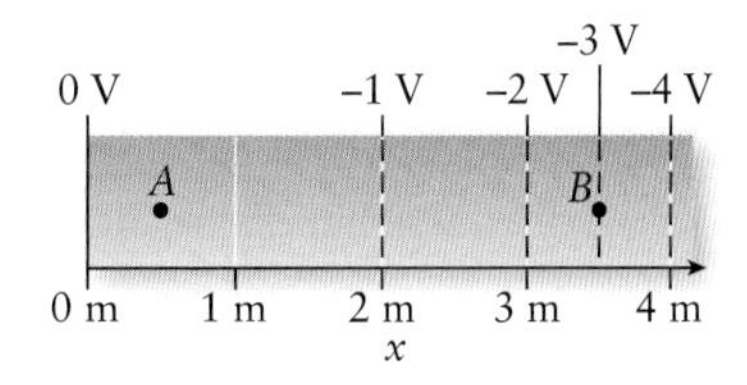

Figure 17-26 Problem 56

57. •• Draw the equipotential lines and electric field lines surrounding (a) two positive charges; (b) two negative charges (**Figure 17-27**). SSM

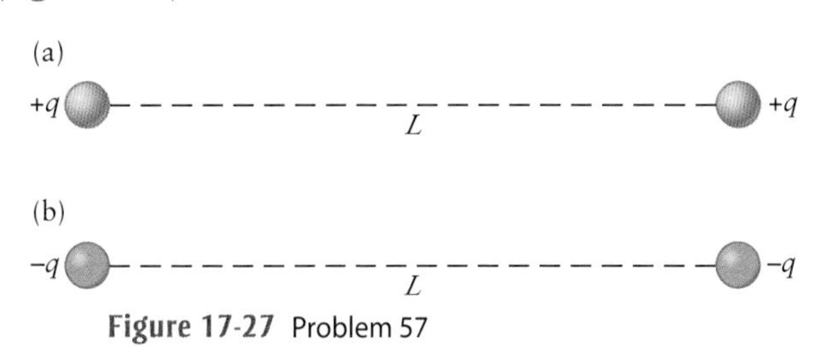

Figure 17-27 Problem 57

58. •• Draw the equipotential lines and electric field lines surrounding a dipole ($+q$ is a distance L from $-q$) (**Figure 17-28**).

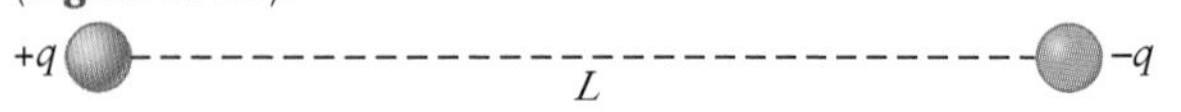

Figure 17-28 Problem 58

59. •• Draw the electric field lines and the electric equipotential lines for the charge distribution in **Figure 17-29**.

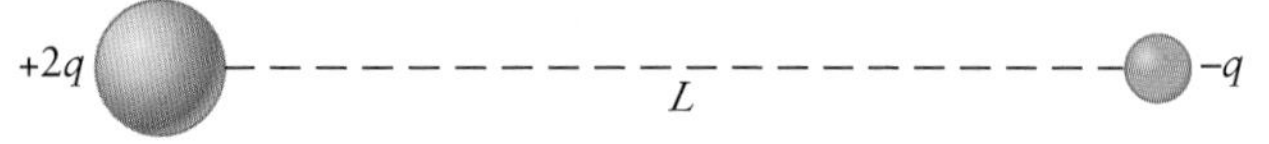

Figure 17-29 Problem 59

17-5 A capacitor stores equal amounts of positive and negative charge

60. • Using a single 10.0-V battery, what capacitance do you need to store 10.0 μC of charge?

61. • A 2.00-μF capacitor is connected to a 12.0-V battery. What is the magnitude of the charge on each plate of the capacitor?

62. • A parallel-plate capacitor has a plate separation of 1.00 mm. If the material between the plates is air, what plate area is required to provide a capacitance of 2.00 pF? Example 17-7

63. • A parallel-plate capacitor has square plates that have edge lengths equal to 1.00×10^2 cm and are separated by 1.00 mm. What is the capacitance of this device? Example 17-7

64. • An air-filled parallel-plate capacitor has plates measuring 10.0 cm × 10.0 cm and a plate separation of 1.00 mm. If you want to construct a parallel-plate capacitor of the same capacitance but with plates measuring 5.00 cm × 5.00 cm, what plate separation do you need? Example 17-7

65. • A parallel-plate capacitor has square plates that have edge length equal to 1.00 m. If the material between the plates is air, what separation distance is required to provide a capacitance of 8850 pF? SSM Example 17-7

17-6 A capacitor is a storehouse of electric potential energy

66. • Using a single 10.0-V battery, what capacitance do you need to store 1.00×10^{-4} J of electric potential energy? Example 17-9

67. • A parallel-plate capacitor has square plates that have edge length equal to 1.00×10^2 cm and are separated by 1.00 mm. It is connected to a battery and is charged to 12.0 V. How much energy is stored in the capacitor? Example 17-9

68. • You charge a 2.00-μF capacitor to 50.0 V. How much additional energy must you add to charge it to 100 V? Example 17-9

69. • A capacitor has a capacitance of 80.0 μF. If you want to store 160 J of electric energy in this capacitor, what potential difference do you need to apply to the plates? Example 17-9

70. •• (a) You want to store 1.00×10^{-5} C of charge on a capacitor, but you only have a 100-V voltage source with which to charge it. What must be the value of the capacitance? (b) You want to store 1.00×10^{-3} J of energy on a capacitor, and you only have a 100-V voltage source with which to charge it. What must be the value of the capacitance? Example 17-9

71. •**Medical** A defibrillator containing a 20.0-μF capacitor is used to shock the heart of a patient by holding it to the patient's chest. Just prior to discharging, the capacitor has a voltage of 10.0 kV across its plates. How much energy is released into the patient, assuming no energy losses? SSM Example 17-9

17-7 Capacitors can be combined in series or in parallel

72. •• How should four 1.0-pF capacitors be connected to have a total capacitance of 0.75 pF? Example 17-11

73. • Three capacitors have capacitances 10.0 μF, 15.0 μF, and 30.0 μF. What is their effective capacitance if the three are connected (a) in parallel and (b) in series? Example 17-10

74. •• A series circuit consists of a 0.50-μF capacitor, a 0.10-μF capacitor, and a 220-V battery. Determine the charge on each of the capacitors. Example 17-10

75. • Two capacitors provide an equivalent capacitance of 8.00 μF when connected in parallel and 2.00 μF when connected in series. What is the capacitance of each capacitor? SSM Example 17-10

76. •• A 2.00-μF capacitor is first charged by being connected across a 6.00-V battery. It is then disconnected from the battery and connected across an uncharged 4.00-μF capacitor. Calculate the final charge on each of the capacitors. Example 17-10

77. • A 0.0500-μF capacitor and a 0.100-μF capacitor are connected in parallel across a 220-V battery. Determine the charge on each of the capacitors. Example 17-10

78. • What is the equivalent capacitance of the network of three capacitors shown in **Figure 17-30**? Example 17-11

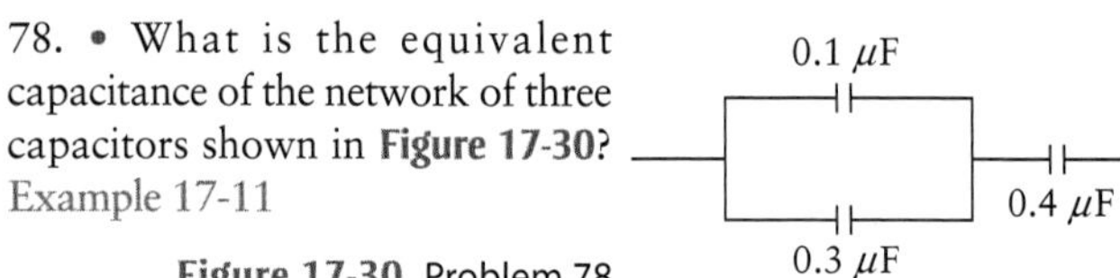

Figure 17-30 Problem 78

79. • Calculate the equivalent capacitance between a and b for the combination of capacitors shown in **Figure 17-31**. Example 17-11

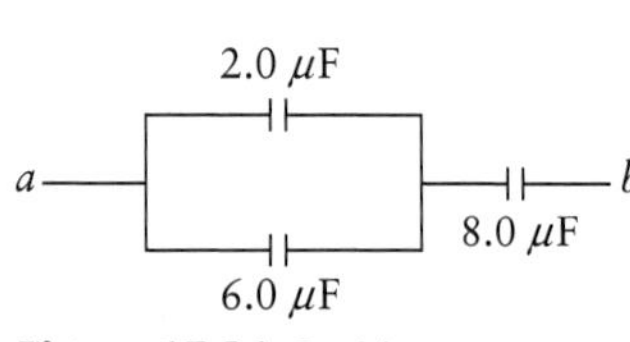

Figure 17-31 Problem 79

80. • A 10.0-μF capacitor, a 40.0-μF capacitor, and a 100.0-μF capacitor are connected in parallel across a 12.0-V battery. (a) What is the equivalent capacitance of the combination? (b) What is the charge on each capacitor? (c) What is the potential difference across each capacitor? Example 17-10

81. • A 10.0-μF capacitor, a 40.0-μF capacitor, and a 100.0-μF capacitor are connected in series across a 12.0-V battery. (a) What is the equivalent capacitance of the combination? (b) What is the charge on each capacitor? (c) What is the potential difference across each capacitor? SSM Example 17-10

82. •• For the capacitor network shown in **Figure 17-32**, the potential difference across ab is 75.0 V. How much charge and how much energy are stored in this system? Example 17-11

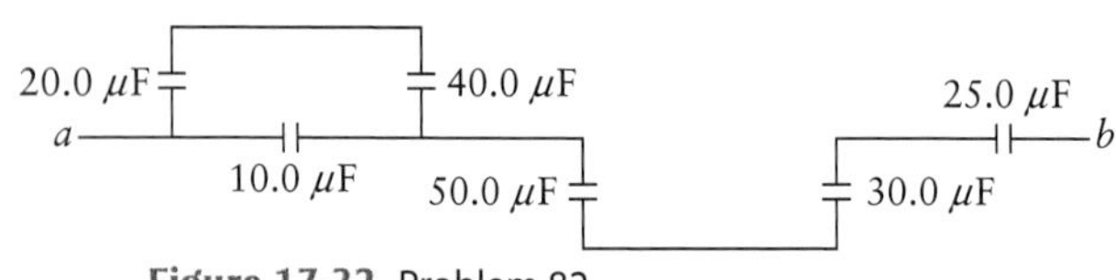

Figure 17-32 Problem 82

17-8 Placing a dielectric between the plates of a capacitor increases the capacitance

83. • What is the dielectric constant of the material that fills the gap between a parallel-plate capacitor with plate area of 20.0 cm^2 and plate separation of 1.00 mm if the capacitance is measured to be 0.0142 μF? Example 17-12

84. • A parallel-plate capacitor has plates of 1.00 cm by 2.00 cm. The plates are separated by a 1.00-mm-thick piece of paper. What is the capacitance of this capacitor? The dielectric constant for paper is 2.7. Example 17-12

85. •• A 2800-pF air-filled capacitor is connected to a 16-V battery. If you now insert a ceramic dielectric material (k = 5.8) that fills the space between the plates, how much charge will flow from the battery? SSM Example 17-12

86. •• A parallel-plate capacitor has square plates 1.00×10^2 cm on a side that are separated by 1.00 mm. It is connected to a battery and charged to 12.0 V. How much energy is stored in the capacitor if a ceramic dielectric material (κ is 5.8) fills the space between the plates? Example 17-12

87. ••• (a) Determine the capacitance of the parallel-plate capacitor shown in **Figure 17-33**. The dielectric with constant κ_1 fills up one-quarter of the area, but the full separation of the plates. The materials with constants κ_2 and κ_3 fill the other three-quarters of the area, and divide the separation of the plates in half. (b) What happens to the capacitance if the material with constant κ_3 is replaced by air? Example 17-12

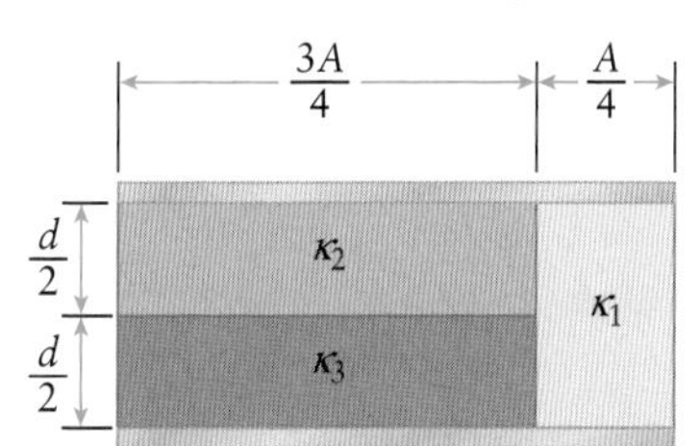

Figure 17-33 Problem 87

88. •• A parallel-plate capacitor that has a plate separation of 0.50 cm is filled halfway with a slab of dielectric material (κ is 5.0) (**Figure 17-34**). If the plates are 1.25 cm by 1.25 cm in area, what is the capacitance of this capacitor? Example 17-12

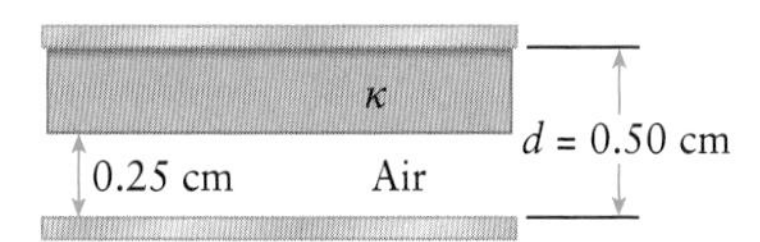

Figure 17-34 Problem 88

General Problems

89. •• A parallel-plate capacitor has a plate separation of 1.5 mm and is charged to 600 V. If an electron leaves the negative plate, starting from rest, how fast is it going when it hits the positive plate? Example 17-6

90. ••• As shown in **Figure 17-35**, three particles, each with charge q, are at different corners of a rhombus with sides of length a and with one diagonal of length a and the other of length b. (a) What is the electric potential energy of the charge distribution? (b) What is the electric potential at the vacant corner of the rhombus? (c) How much work by an external agent is required to bring a fourth particle, also of charge q, from rest at infinity to rest at the vacant corner? (d) What is the total electric potential energy of the four charges? SSM Example 17-3

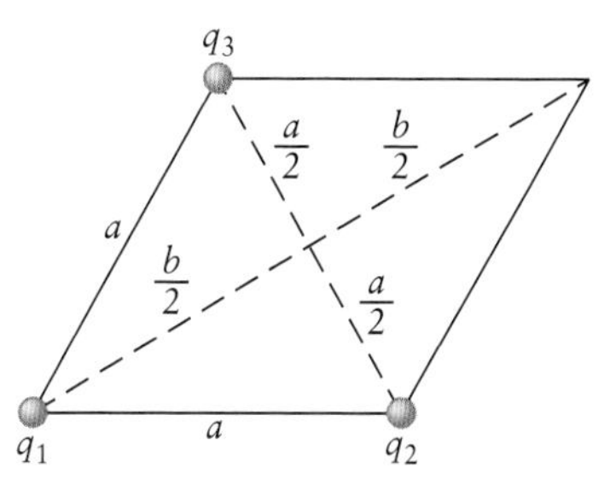

Figure 17-35 Problem 90

91. •• A lightning bolt transfers 20 C of charge to Earth through an average potential difference of 30 MV. (a) How much energy is dissipated in the bolt? (b) What mass of water at 100°C could this energy turn into steam? Example 17-6

92. • In 2004, physicists at the SLAC National Accelerator Laboratory fired electrons toward each other at very high speeds so that they came within 1.0×10^{-15} m of each other (approximately the diameter of a proton). (a) What was the electric force on each electron at closest approach? (b) Would you be able to feel a force of this magnitude acting on you? (c) What kinetic energy must each electron have had when they were far apart to be able to get this close? Example 17-2

93. • **Biology** Potassium ions (K^+) move across a 8.0-nm-thick cell membrane from the inside to the outside. The potential inside the cell is −70.0 mV, and the potential outside is zero. (a) What is the change in the electrical potential energy of the potassium ions as they move across the membrane? Does their potential energy increase or decrease? Example 17-6

94. • Calculate the equivalent capacitance of the combination in **Figure 17-36**. SSM Example 17-11

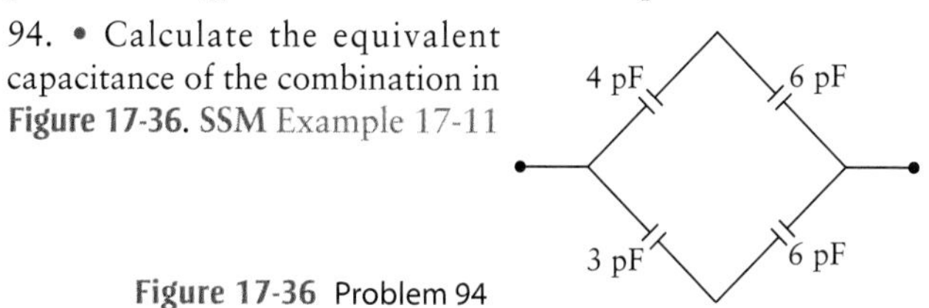

Figure 17-36 Problem 94

95. •• Suppose you are supplied with five identical capacitors (each with a capacitance of 10.0 μF). Determine all of the unique combinations that use all five capacitors and the equivalent capacitance of each combination. Example 17-11

96. •• Determine the equivalent capacitance of the combination in **Figure 17-37**. Example 17-11

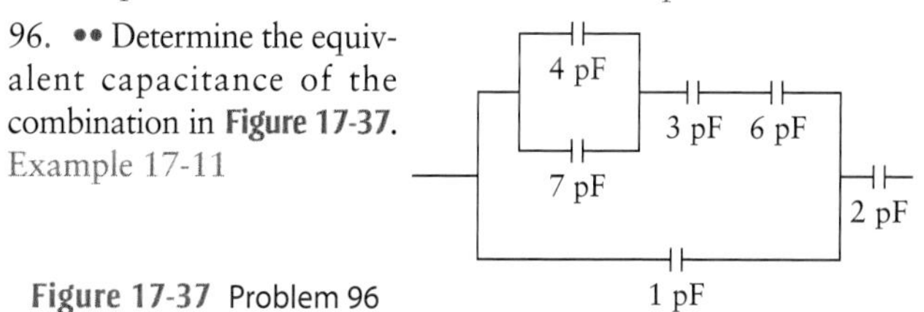

Figure 17-37 Problem 96

97. • The arrangement of four capacitors in **Figure 17-38** has an equivalent capacitance of 8.00 μF. Calculate the value of C_x. Example 17-11

Figure 17-38 Problem 97

98. •• Calculate the charge stored on each capacitor in the circuit shown in **Figure 17-39**. Example 17-10

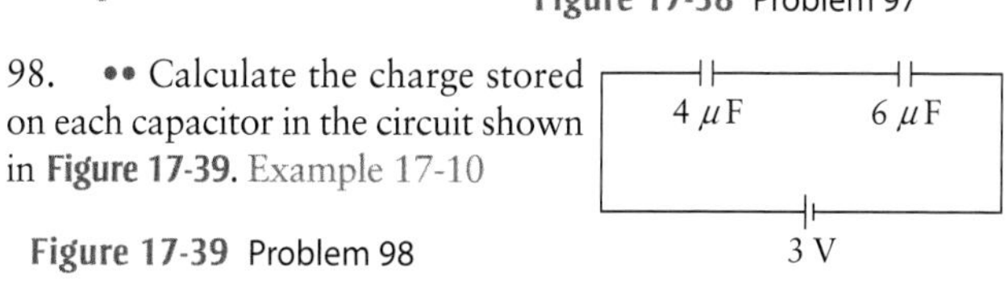

Figure 17-39 Problem 98

99. •• A parallel-plate capacitor is made by sandwiching 0.100-mm sheets of paper (dielectric constant 2.7) between

three sheets of aluminum foil (A, B, and C in **Figure 17-40**) and rolling the layers into a cylinder. A capacitor that has an area of 10 m^2 is fabricated this way. What is the capacitance of this capacitor? Example 17-12

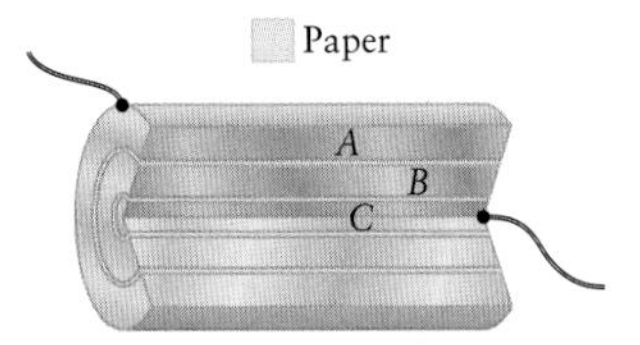

Figure 17-40 Problem 99

100. •• A parallel-plate capacitor with a capacitance of 5.00 μF is fully charged with a 12.0-V battery. The battery is then removed. How much work is required to triple the separation between the plates? Example 17-9

101. •• An air-filled parallel-plate capacitor is connected to a battery with a voltage V. (a) The plates are pulled apart, doubling the gap width, while they remain connected to the battery. By what factor does the potential energy of the capacitor change? (b) If the capacitor is removed from the battery, what happens to the stored potential energy when the gap width is doubled? Explain your answer. Example 17-9

102. •• An air-filled parallel-plate capacitor is attached to a battery with a voltage V. While attached to the battery, the area of the plates is doubled and the separation of the plates is halved. During this process, what happens to (a) the capacitance, (b) the charge on the positive plate, (c) the potential across the plates, and (d) the potential energy stored in the capacitor? (e) How would your answers change if, once the capacitor was charged, it was disconnected from the battery while the area and separation were changed as above? Example 17-7

103. •• A honey bee of mass 130 mg has accumulated a static charge of +1.8 pC. The bee is returning to her hive by following the path shown in **Figure 17-41**. Because Earth has a naturally occurring electric field near ground level of around 100 V/m pointing vertically downward, the bee experiences an electric force as she flies. (a) What is the change in the bee's electric potential energy, $\Delta U_{electric}$, as she flies from point A to point B? (b) Compute the ratio of the bee's change in electric potential energy to her change in gravitational potential energy, $\Delta U_{electric}/\Delta U_{grav}$. Example 17-4.

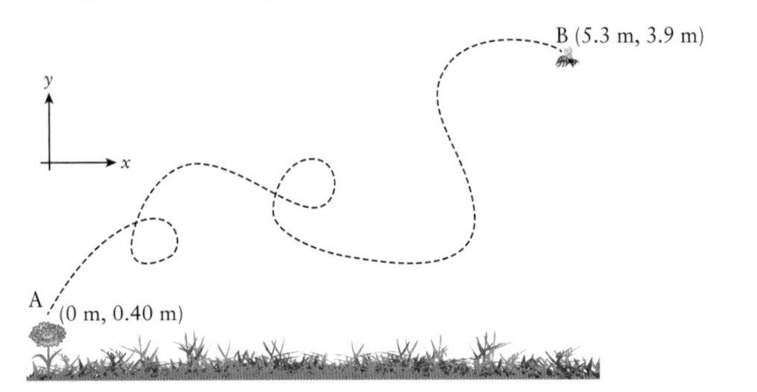

Figure 17-41 Problem 103

104. ••• A parallel-plate capacitor has area A and separation d. (a) What is its new capacitance if a *conducting* slab of thickness $d' < d$ is inserted between, and parallel to, the plates as shown in **Figure 17-42**? (b) Does your answer depend on where the slab is positioned vertically between the plates? SSM Example 17-12

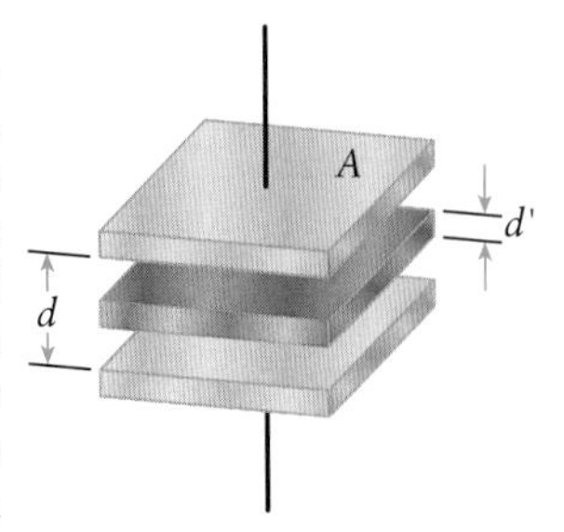

Figure 17-42 Problem 104

105. ••• Three 0.18-μF capacitors are connected in parallel across a 12-V battery (**Figure 17-43**). The battery is then disconnected. Next, one capacitor is carefully disconnected so that it doesn't lose any charge and is reconnected with its positively charged and negatively charged sides reversed. (a) What is the potential difference across the capacitors now? (b) By how much has the stored energy of the combination of capacitors changed in the process? Example 17-9

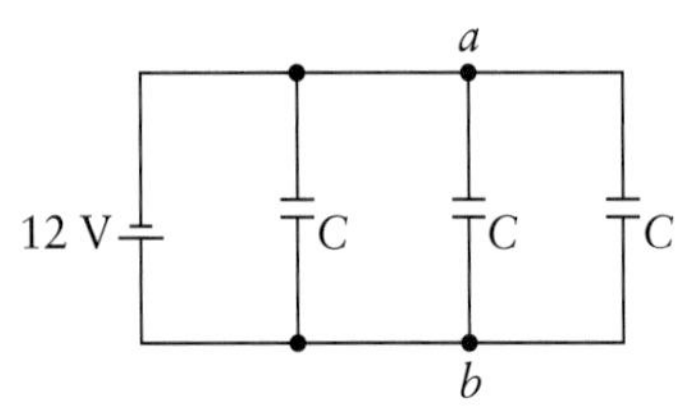

Figure 17-43 Problem 105

106. •• (a) Calculate the charge and energy stored on the 25-μF capacitor when the switch S is placed at position A in **Figure 17-44**. (b) Repeat for both the 25-μF and the 20-μF capacitors after the switch is then placed at position B. Example 17-10

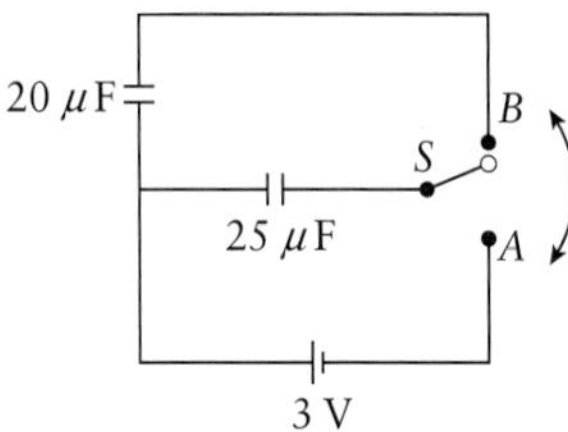

Figure 17-44 Problem 106

107. •• **Figure 17-45** shows equipotential curves for 30 V, 10 V, and −10 V. A proton follows the path shown in the figure. If the proton's speed at point A is 80 km/s, what is its speed at point B? Example 17-6

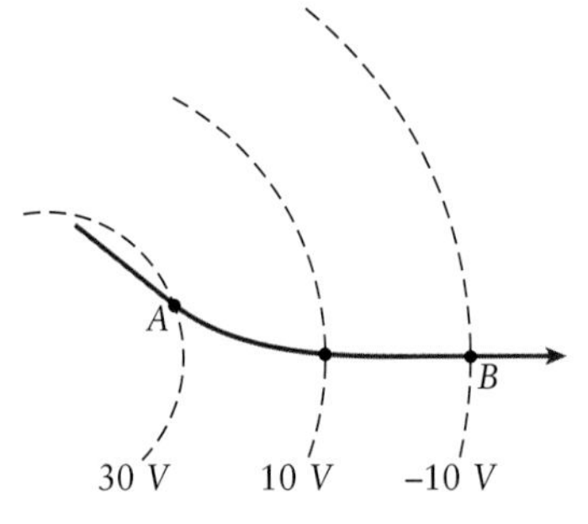

Figure 17-45 Problem 107

108. ••• A parallel-plate, air-filled capacitor has a charge of 20.0 μC and a gap width of 0.100 mm. The potential difference between the plates is 200 V. (a) What is the electric field between the plates? (b) What is the surface charge density on the positive plate? (c) If the plates are moved closer together while the charge remains constant, how do the electric field, surface charge density, and potential difference change, if at all? Explain your answers. Example 17-7

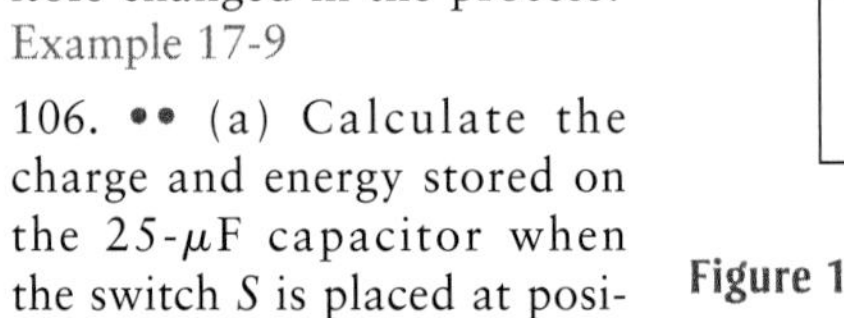

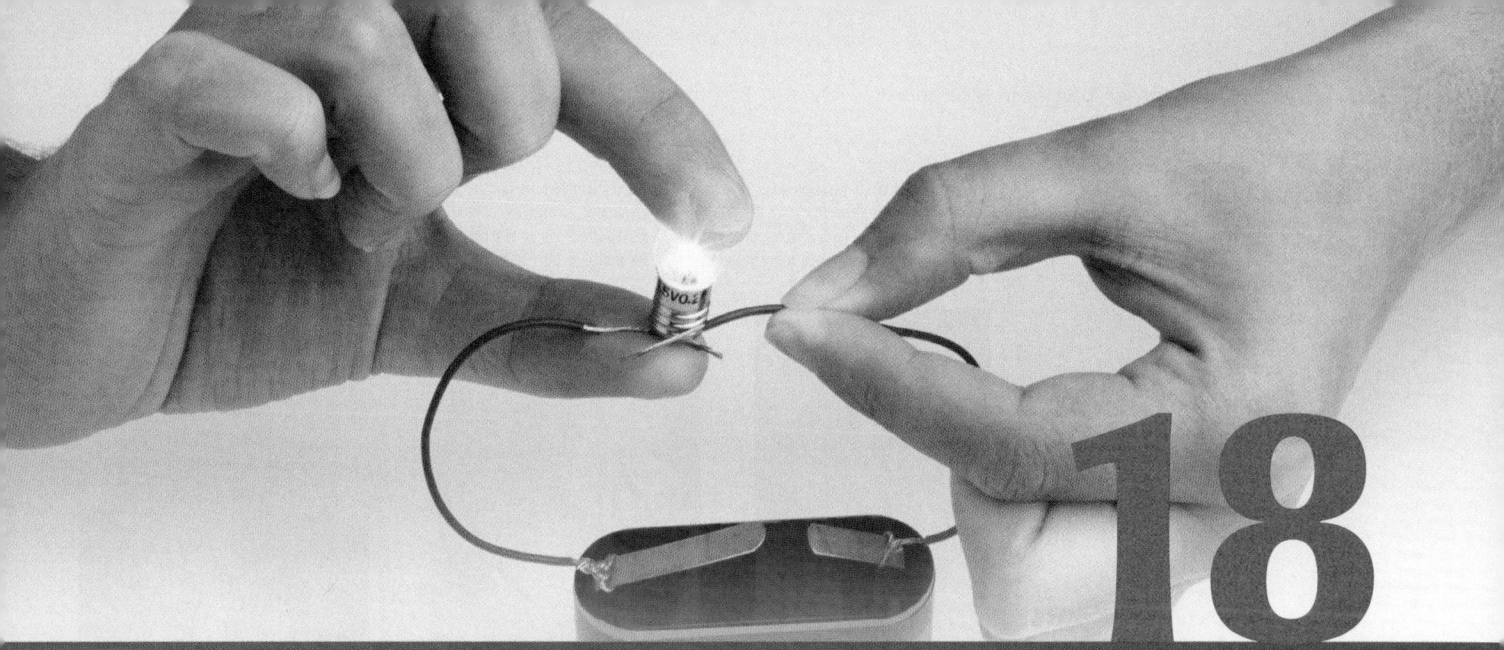

Electric Charges in Motion

Dave King/Dorling Kindersley/Getty Images

In this chapter, your goals are to:

- (18-1) Recognize why moving electric charges are important.
- (18-2) Explain the meaning of current and drift speed, and the difference between direct and alternating current.
- (18-3) Describe the relationships among voltage, current, resistance, and resistivity for charges moving in a wire.
- (18-4) Calculate the resistance of a resistor and the current that a given voltage produces in that resistor.
- (18-5) Discuss Kirchhoff's rules and how to apply them to single-loop and multiloop circuits.
- (18-6) Calculate the power into or out of a circuit element.
- (18-7) Explain what happens when a capacitor in series with a resistor is charged or discharged.

To master this chapter, you should review:

- (5-5) How fluids exert drag forces on objects moving through them.
- (6-9) How power is the rate at which energy is transferred.
- (12-3) How to describe the position of an object oscillating with simple harmonic motion.
- (16-3) How the loosely bound electrons in metals allow them to conduct electricity.
- (16-7) Why the electric field outside a uniformly charged sphere is the same as if all the charges were located at the center of the sphere.
- (17-3) How electric field and potential difference are related, and how potential difference and electric potential energy difference are related.
- (17-5, 17-6, 17-7) The relationship among charge, voltage, and capacitance; the energy stored in a capacitor; and the equivalent capacitance of capacitors attached in series and parallel.

What do you think?

The simplest of all electric circuits is a battery connected to a light bulb, such as the one you find inside an ordinary flashlight. The battery causes electrons to move through the circuit. As electrons pass through the light bulb, what is the principal kind of energy that they transfer to the light bulb to make it shine? (a) Kinetic energy; (b) electric potential energy; (c) both kinetic energy and electric potential energy; (d) neither kinetic energy nor electric potential energy; (e) answer depends on the details of how the light bulb is constructed.

18-1 Life on Earth and our technological society are only possible because of charges in motion

In the previous two chapters we've investigated electric force, electric field, electric potential energy, and electric potential. In our investigations we considered electric charges that were fixed in place. But in many important situations in nature and

(a)
Wires and cables are made of conductors that allow moving charges to flow along their length.
iStockphoto/Thinkstock

(b)
The current in a mobile device is provided by a battery (a source of emf).
Daniel M Ernst/Shutterstock.com

(c)
A camera's electronic flash uses energy stored in a capacitor to produce a burst of light.
Henrik Sorensen/The Image Bank/Getty Images

Figure 18-1 Charges in motion (a) An essential part of any electric circuit. (b) and (c) Two common devices that use currents (moving electric charges).

technology electric charges are in *motion*. You are able to read these words thanks to electric charges that travel along the optic nerve from your eye to your brain, transmitting the image of those words in the form of a coded electrical signal. If you are reading these words after sunset, you are either looking at a printed page illuminated by a light bulb that's powered by moving electric charges or else reading them on a tablet or other electronic device that operates using complex electric circuitry.

In this chapter we'll look at the basic physics of electric charges in motion. We'll introduce the idea of *current*, which measures the rate at which charges move through a conductor (**Figure 18-1a**). Ordinary conductors have *resistance* to the flow of charge, so it's necessary to set up a voltage between the ends of a conductor in order to produce a current. This is the role of a *source of emf*, the most common example of which is an ordinary battery (**Figure 18-1b**). We'll analyze simple circuits that include a source of emf and one or more *resistors* (circuit elements that have resistance), and see how to treat combinations of resistors.

Fundamentally, an electric circuit is used to transfer energy from one place to another (for instance, from a battery to a light bulb, as in the photograph that opens this chapter). We'll see how to describe the *power*, or rate of energy transfer, associated with any circuit element. Finally, we'll study circuits that include a capacitor, which make it possible to deliver energy in a quick burst (**Figure 18-1c**).

TAKE-HOME MESSAGE FOR Section 18-1

✔ Technological devices and biological systems depend on electric charges in motion.

✔ An electric circuit is fundamentally a means of transferring energy.

18-2 Electric current equals the rate at which charge flows

Figure 18-2 shows a common situation in which electric charges are in motion. What sets charges into motion is the **battery**, also known as an *electrochemical cell*. Inside the cell are two different substances that, due to their different chemical properties, undergo a chemical reaction so that each substance ends up with an excess or deficit of electrons. In an ordinary alkaline battery, like an AA or D cell, the two substances are zinc and manganese dioxide. The zinc ends up with an electron excess, while the manganese dioxide ends up with an electron deficit.

The two terminals of the battery are each connected to one of these substances. Electrons can flow between each substance and the metal terminal to which it is attached. As a result, the terminal attached to the substance with an electron excess also has an electron excess and is called the *negative* terminal. The other terminal, attached to the substance with an electron deficit, also has an electron deficit and is called the *positive* terminal. These terminals are marked on the battery (and in Figure 18-2) by a minus sign and a plus sign, respectively. Because of this charge imbalance, the positive terminal is at a higher electric potential than the negative terminal. The potential difference, or voltage, between the terminals depends on the two substances within the battery. For zinc and manganese dioxide, the voltage is 1.5 V, a number that you will see written on the case of any alkaline battery.

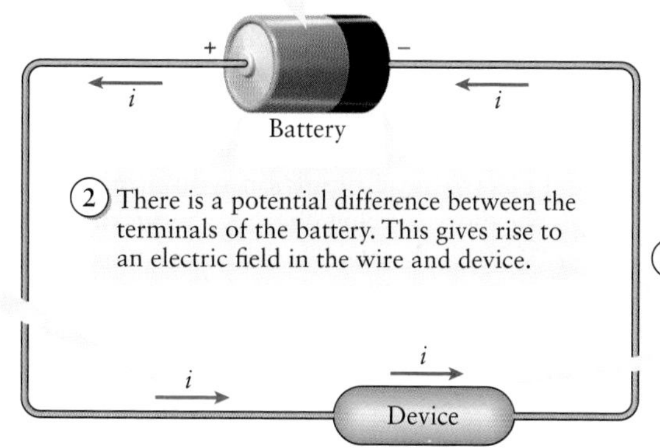

Figure 18-2 Current in a circuit Charge flows in a wire loop due to the potential difference (voltage) supplied by a battery.

We know from Section 17-3 that charges tend to flow between two points when there is a potential difference between those points: The electric force pushes positive charges from high to low potential and negative charges from low to high potential. If a battery is isolated and not connected to anything, there is no way for charge to flow outside the battery from one terminal to another. But charge *can* flow if a metallic wire is used to connect the two terminals to each other as in Figure 18-2. (As we saw in Section 16-3, metals are especially good electrical conductors because one or more of the electrons associated with a metal atom are only weakly bound to that atom. As a result, these loosely bound electrons can move with relative freedom within the metal.) So the battery and wire form a complete loop or **circuit**, and charge flows continuously through the circuit from one terminal of the battery to the other.

Here's another way to see why charges move in the wire. The potential difference between the battery terminals means that there is an electric *field* $\vec{E}$ within the wire. This field points along the length of the wire in the direction from the positive terminal to the negative terminal. A mobile charge with charge q within the wire will feel a force $\vec{F} = q\vec{E}$ from the electric field, and this force pushes the charges through the wire. This force is necessary because mobile charges within the wire collide very frequently with the atoms that make up the wire. The net effect of the collisions is to slow or retard the motion of the mobile charges through the wire, rather like how an algal spore moving through water is retarded by fluid resistance (see Section 5-5). The speed at which charge flows is determined by the balance between the retarding force due to collisions and the forward electric force $\vec{F} = q\vec{E}$.

Later in this chapter we'll see how the flow of charge delivers energy to a device (like a light bulb) that's part of the circuit. For now, however, let's concentrate on the properties of the charge flow itself.

Current

The **current** in a circuit like that in Figure 18-2 equals the rate at which charge flows past any point in the circuit. In particular, if an amount of charge Δq moves past a certain point in a time Δt, the current i is

See the Math Tutorial for more information on direct and inverse proportions.

Current in a circuit

Amount of **charge** that flows past a certain point in the circuit

$$i = \frac{\Delta q}{\Delta t}$$

Time required for that amount of charge to flow past that point

Definition of current (18-1)

The SI unit of current is the **ampere**, named after the French scientist and mathematician André Ampère. One ampere (abbreviated amp or A) is equivalent to one coulomb of charge passing a given point per second: 1 A = 1 C/s.

In much the same way as we treated fluid flow in Chapter 11, we'll begin by limiting our discussion of currents to a *steady* flow of charge. Then the current i has the same value at all *times*. In addition, the current has the same value at all *points* in a simple circuit like that in Figure 18-2 in which charges move around a single loop. The moving charges cannot "pile up" or accumulate at any point in the circuit (if they did, their mutual electrostatic repulsion would make them spread apart again). There is also no way for more moving charges to join the flow or for charges to leave the flow. So the value of the current is the same everywhere in the circuit.

Note that current is *not* a vector quantity. For example, in Figure 18-2 the current is to the left in the upper part of the circuit, downward in the left-hand part of the circuit, to the right in the lower part of the circuit, and upward in the right-hand part of the circuit. So there is no single vector that describes the direction of current in every part of the circuit. Instead, we simply say that the current in Figure 18-2 is counterclockwise around the circuit. If we reversed the battery so that the positive terminal was on the right rather than on the left, the current would be clockwise.

WATCH OUT! The direction of current is chosen to be the direction in which *positive* charges would flow.

In ordinary wires and in common electric devices such as light bulbs and toasters, the charges that are free to move are negatively charged electrons. In an electric circuit such as that shown in Figure 18-2, electrons flow through the wire from low potential to high potential and so from the negative terminal of the battery toward the positive terminal. But in Figure 18-2 we show the current flowing through the wire from the battery's *positive* terminal to its *negative* terminal. That's because the convention is that current flows in the direction that positively charged objects would move, regardless of whether the moving objects are positively or negatively charged. We'll use this convention throughout this book. (This convention is often attributed to the eighteenth-century American scientist and statesman Benjamin Franklin. The discovery that the moving charges in wires are negatively charged came decades after Franklin's death.) Note that in most biological systems such as neurons and muscle cells, the moving charges actually *are* positive (they are positive ions, atoms that have lost one or more electrons each).

Because an electron has a very small charge ($q = -e = -1.60 \times 10^{-19}$ C), typical currents involve the flow of a very large number of electrons. For example, suppose the current i in the circuit shown in Figure 18-2 is 1.00 A or 1.00 C/s. (That's roughly the current in a large flashlight.) Then the rate at which electrons move past any specific point on the wire is

$$1.00 \text{ A} = 1.00\frac{\text{C}}{\text{s}} = \left(1.00\frac{\text{C}}{\text{s}}\right)\left(\frac{1 \text{ electron}}{1.60 \times 10^{-19} \text{ C}}\right) = 6.25 \times 10^{18}\,\frac{\text{electron}}{\text{s}}$$

(Here we're considering the *magnitude* of the current, so we're ignoring the negative sign on the charge of the electron.) This result says that if we were to pass an imaginary plane through the cross section of the wire, we would find about 6×10^{18} electrons crossing that plane per second.

EXAMPLE 18-1 Charging a Sphere

A large, hollow metal sphere is electrically isolated from its surroundings, except for a wire that can carry a current to charge the sphere. Initially the sphere is uncharged and is at electrical potential zero. If the sphere has a radius of 0.150 m and the current is a steady 5.00 μA = 5.00×10^{-6} A, how long does it take for the sphere to attain a potential of 4.00×10^{5} V?

Set Up

The electric potential at the surface of a charged sphere of radius R depends on the amount of excess charge Q on the sphere. In Section 16-7 we learned that the electric *field* outside a uniformly charged sphere is identical to that of a particle with the same charge, located at the center of the sphere. So the electric *potential* at a point just outside the sphere must be given by Equation 17-8, the equation for the potential of a charged particle with total charge Q located at the center of the sphere. We'll rearrange this equation to solve for the amount of charge required to generate the given electric potential. Then we'll find the time required for this charge to reach the sphere.

Electric potential of a charged sphere:

$$V = \frac{kQ}{R} \qquad (17\text{-}8)$$

Definition of current:

$$i = \frac{\Delta q}{\Delta t} \qquad (18\text{-}1)$$

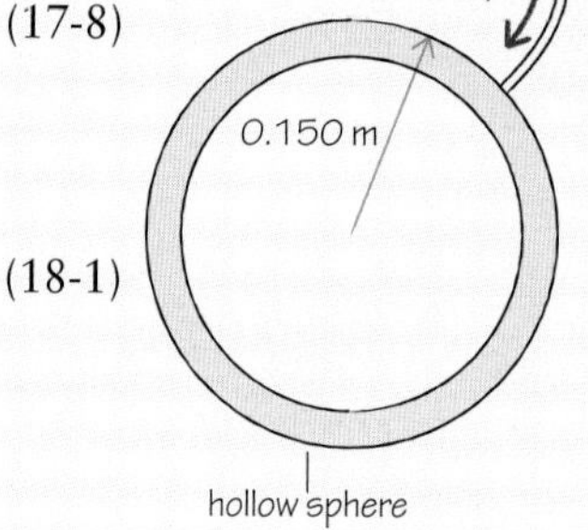

Solve

Use Equation 17-8 to solve for the final charge on the sphere.

From Equation 17-8,

$$Q = \frac{VR}{k} = \frac{(4.00 \times 10^5\ \text{V})(0.150\ \text{m})}{(8.99 \times 10^9\ \text{N}\cdot\text{m}^2/\text{C}^2)}$$

$$= 6.67 \times 10^{-6}\,\frac{\text{V}\cdot\text{C}^2}{\text{N}\cdot\text{m}}$$

Since $1\ \text{V} = 1\ \text{J/C} = 1\ \text{N}\cdot\text{m/C}$, this becomes

$$Q = \left(6.67 \times 10^{-6}\,\frac{\text{V}\cdot\text{C}^2}{\text{N}\cdot\text{m}}\right)\left(\frac{1\ \text{N}\cdot\text{m/C}}{1\ \text{V}}\right)$$

$$= 6.67 \times 10^{-6}\ \text{C} = 6.67\ \mu\text{C}$$

Our result $Q = 6.67\ \mu\text{C}$ is the amount of charge Δq that must flow onto the sphere in a time Δt. Use Equation 18-1 to solve for Δt. Recall that $1\ \text{A} = 1\ \text{C/s}$.

From Equation 18-1,

$$\Delta t = \frac{\Delta q}{i} = \frac{6.67 \times 10^{-6}\ \text{C}}{5.00 \times 10^{-6}\ \text{A}} = \frac{6.67 \times 10^{-6}\ \text{C}}{5.00 \times 10^{-6}\ \text{C/s}}$$

$$= 1.33\ \text{s}$$

Reflect

A charged sphere like this is used in a Van de Graaff generator, which you may have seen demonstrated in your physics class. If you have, you know that when the generator is turned on to charge the sphere, the sphere can begin to throw off sparks within a second or so. So a result for Δt on the order of 1 s is reasonable.

Drift Speed

The value of the current i tells us what quantity of charge flows past a given point in a circuit per second. Let's see how to relate this to the **drift speed** v_{drift}, which is the average speed at which **mobile charges**—those that are free to move throughout the conducting material in the circuit—move ("drift") through the circuit.

Figure 18-3 shows a wire that has a cross-sectional area A and carries a current i. Charges are moving ("drifting") through the green region at an average speed of v_{drift}. At any time the total moving charge in that region is Δq. Note that Δq is the product of n (the number of mobile charges per volume), the volume $A\Delta x$ of the green region (a cylinder of area A and length Δx), and the amount of charge e on each mobile charge:

$$\Delta q = n(\text{Volume})e = n(A\Delta x)e \qquad (18\text{-}2)$$

If the mobile charges are electrons, each has a charge $-e$ rather than e. But since we take the direction of current to be the direction in which positive charges would flow, we'll use the convenient fiction that the charge is positive. The time required for this

Figure 18-3 Drift speed The current i in a wire is proportional to the speed v_{drift} at which moving charges drift through the wire.

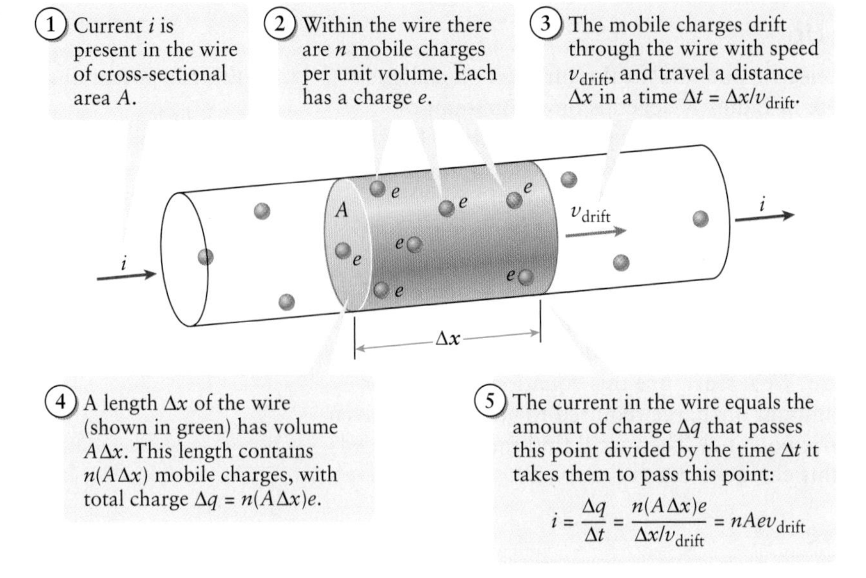

volume of charge to drift the distance Δx along the length of the wire is $\Delta t = \Delta x/v_{drift}$ (time equals distance divided by speed). So from Equations 18-1 and 18-2, the current in the wire is

$$i = \frac{\Delta q}{\Delta t} = \frac{n(A\Delta x)e}{\Delta x/v_{drift}}$$

or, simplifying,

Current and drift speed (18-3)

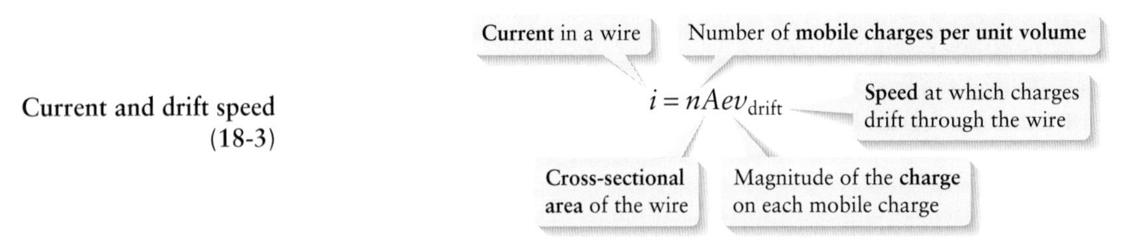

The number of moving charges per volume (n) is different for different materials. Equation 18-3 tells us that for a wire made of a given material and with a given cross-sectional area A, the current i is directly proportional to the drift speed.

WATCH OUT! The drift speed is not the same as the speed at which individual charges move.

Even if there is no current through a wire, the mobile charges within that wire are still in motion. However, their motions are in random directions, so there is no *net* motion of charge. It's rather like a swarm of bees flying around a hive: Individual bees move in different directions, but the swarm as a whole stays in the same place. If the swarm moves to a different hive, the motions of individual bees will be different but the swarm as a whole will "drift" together to the new hive. In the same way, if a current is present in the wire in Figure 18-3, the individual charges within the green volume can be moving at different speeds and in different directions, but this "swarm" of charge drifts along the wire at speed v_{drift}.

EXAMPLE 18-2 Electron Drift Speed in a Flashlight

In a large flashlight the distance from the on–off switch to the light bulb is 10.0 cm. How long does it takes electrons to drift this distance if the flashlight wires are made of copper, are 0.512 mm in radius, and carry a current of 1.00 A? There are 8.49×10^{28} atoms in 1 m^3 of copper, and one electron per copper atom can move freely through the metal.

Set Up

We'll use Equation 18-3 to determine the drift speed from the information given about the number of mobile electrons, the radius of the wire, and the current. From this we'll be able to find the drift time by using the familiar relationship between speed, distance, and time.

Current and drift speed:

$$i = nAev_{\text{drift}} \quad (18\text{-}3)$$

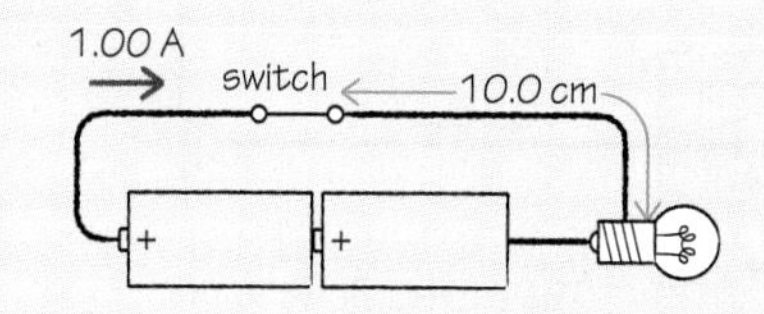

Solve

Calculate the drift speed of the electrons in the wire.

Solve Equation 18-3 for the drift speed:

$$v_{\text{drift}} = \frac{i}{nAe}$$

There is one mobile electron per atom and 8.49×10^{28} atoms/m^3, so the value of n is 8.49×10^{28} electrons/m^3. The wire has a circular cross section with radius $r = 0.512$ mm $= 0.512 \times 10^{-3}$ m, so its cross-sectional area is

$$A = \pi r^2 = \pi(0.512 \times 10^{-3}\text{ m})^2 = 8.24 \times 10^{-7}\text{ m}^2$$

The magnitude of the charge per electron is $e = 1.60 \times 10^{-19}$ C/electron, and the current is $i = 1.00$ A $= 1.00$ C/s. So the drift speed is

$$v_{\text{drift}} = \frac{i}{nAe}$$

$$= \frac{1.00\text{ C/s}}{(8.49 \times 10^{28}\text{ electrons/m}^3)(8.24 \times 10^{-7}\text{ m}^2)(1.60 \times 10^{-19}\text{ C/electron})}$$

$$= 8.94 \times 10^{-5}\text{ m/s} = 0.0894\text{ mm/s}$$

Then calculate how long it takes electrons to drift the distance from switch to light bulb.

At this drift speed the time it takes electrons to travel a distance $d = 10.0$ cm $= 0.100$ m from the flashlight on–off switch to the light bulb is

$$t = \frac{d}{v_{\text{drift}}} = \frac{0.100\text{ m}}{8.94 \times 10^{-5}\text{ m/s}} = 1.12 \times 10^3\text{ s}$$

$$= (1.12 \times 10^3\text{ s})\left(\frac{1\text{ min}}{60\text{ s}}\right) = 18.6\text{ min}$$

Reflect

The phrase "a snail's pace" refers to something that moves very slowly. But an ordinary snail moves at about *twice* the drift speed of electrons in this wire. The drift speed is so slow because electrons in the wire are continually colliding with the copper atoms, which slows their progress tremendously. (More sophisticated physics shows that an electron in copper moves in *random* motion at an average speed of about 10^6 m/s, about 10^{10} times faster than the drift speed. In the analogy we made earlier between electrons and a swarm of bees, you should think of the electrons in this wire as *very* fast-moving bees within a swarm that's drifting very, very slowly.)

At their slower-than-a-snail's pace, it takes electrons more than a quarter of an hour to travel from the on–off switch to the light bulb. Why, then, does the flashlight turn on immediately when you move the switch to the on position?

The explanation is that the wire is full of mobile electrons, and these electrons drift in response to the electric field in the wire. With the switch in the off position, there is no electric field and so no drift. An electric field is set up only when the switch is put in the on position, making a complete circuit like that shown in Figure 18-2. Changes in the electric field propagate through the wire at close to the speed of light (3×10^8 m/s), so the field is set up throughout the circuit, and the electrons begin to drift, in a tiny fraction of a second. That's why the light comes on nearly instantaneously.

Direct Current and Alternating Current

In the circuit shown in Figure 18-2 the current always flows around the circuit in the same direction, as shown by the arrows labeled i. Current of this kind is called **direct current** or **dc** for short. You'll find direct current in any device that's powered by a battery, such as a flashlight, a television remote control, or a mobile phone. That's because the potential difference between the two terminals of the battery that powers the circuit always has the same sign: The positive terminal is always at a higher potential than the negative terminal.

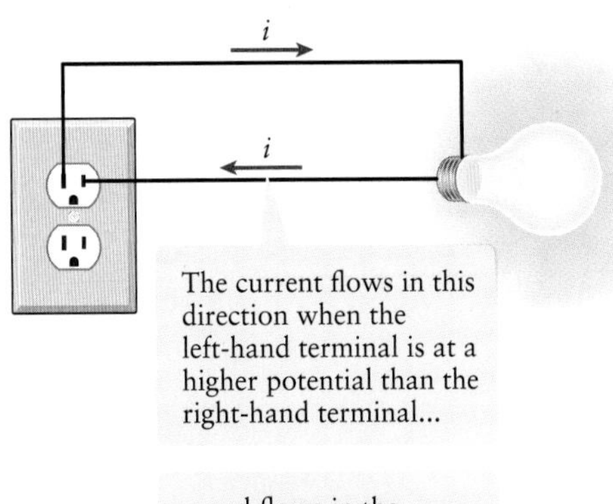
i
i
The current flows in this direction when the left-hand terminal is at a higher potential than the right-hand terminal...

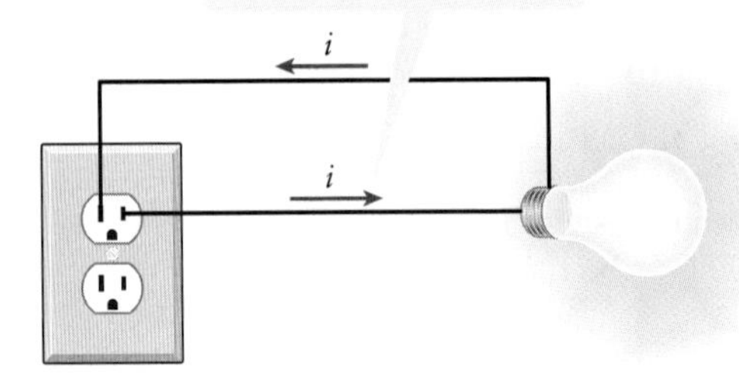
...and flows in the opposite direction when the left-hand terminal is at a lower potential than the right-hand terminal.
i
i

The alternating voltage of the wall socket produces an **alternating current.**

Figure 18-4 Alternating current The potential difference between the terminals of a wall socket varies sinusoidally. As a result, the current in a circuit that contains this socket alternates direction.

Something very different happens in a toaster or a table lamp that you plug into a wall socket. The potential difference between the two terminals in a wall socket is not constant but instead oscillates or *alternates*. At one instant the left-hand terminal is at a higher potential than the right-hand terminal; a short time later the left-hand terminal is the one at the lower potential, and a short time after that the left-hand terminal is again at the higher potential. As a result, the current in a device plugged into a wall socket alternates direction (**Figure 18-4**). This is called **alternating current** or **ac** for short. (The third terminal found in most wall sockets, called the *ground*, remains at a constant potential and is not directly involved in producing the current.)

The potential difference or voltage between the two terminals of a wall socket varies with time in a sinusoidal way, just as does the position of an object in simple harmonic motion (Section 12-3). We can write this voltage as

$$V(t) = V_0 \sin \omega t = V_0 \sin 2\pi f t \tag{18-4}$$

This expression says that the potential difference between the left-hand and right-hand terminals in a wall socket varies between $+V_0$ and $-V_0$. In Equation 18-4 f is the frequency at which the voltage oscillates, and $\omega = 2\pi f$ is the corresponding angular frequency. In North America, Central America, and much of South America, the frequency used is $f = 60$ Hz; in most of the rest of the world, $f = 50$ Hz is used. The current in a circuit driven by such a wall socket oscillates with the same frequency. So in an ac circuit electrons just oscillate back and forth around an equilibrium position. That's very different from a dc circuit, in which electrons plod slowly in the same direction around the circuit.

If $f = 60$ Hz, the *period* of oscillation of the current is $T = 1/f = 1/(60 \text{ Hz}) = 1/60$ s. This means that the current in the circuit shown in Figure 18-4 moves in one direction for $(1/2) \times (1/60)$ s $= 1/120$ s, then moves in the opposite direction for $1/120$ s, and so on. Since the drift speed of electrons in a wire is typically very slow (see Example 18-2), electrons can move only a very short distance (typically around 10^{-6} m or less) in $1/120$ s.

In Chapter 21 we'll study alternating current in detail, learn how an alternating voltage is generated, and see why it's used instead of direct current (dc) for wall sockets. For the remainder of this chapter, we'll concentrate exclusively on dc circuits.

GOT THE CONCEPT? 18-1 Current in a Wire

Suppose that part of the wire in Example 18-2 were only half the thickness of the rest of the wire, with a radius of 0.256 mm instead of 0.512 mm. Compared to the rest of the wire, would the current in the thinner part be (a) four times greater; (b) twice as great; (c) the same; (d) one-half as great; or (e) one-fourth as great?

GOT THE CONCEPT? 18-2 Drift Speed in a Wire

Suppose that part of the wire in Example 18-2 were only half the thickness of the rest of the wire, with a radius of 0.256 mm instead of 0.512 mm. Compared to the rest of the wire, would the drift speed in the thinner part be (a) four times faster; (b) twice as fast; (c) the same; (d) one-half as fast; or (e) one-fourth as fast?

TAKE-HOME MESSAGE FOR Section 18-2

✔ An electric field applied in a wire results in an electric current (a net flow of charge in the wire).

✔ The SI unit of current is the ampere (A): 1 A = 1 C/s.

✔ Current has a magnitude and a direction, but it is not a vector.

✔ By convention, the direction assigned to a current is the direction in which positive charge carriers would move. In typical metals like those used in wires, it is the negative electrons that actually move and carry current.

✔ Direct current (dc) always travels the same direction around a circuit. Alternating current (ac) continually changes direction.

18-3 The resistance to current through an object depends on the object's resistivity and dimensions

We saw in Section 18-2 that current exists in a wire only if there is a potential difference between the ends of the wire. This gives rise to an electric field inside the wire, and this field exerts a force that causes mobile charges to move. However, due to collisions between the charges and the atoms of the wire, charges drift through the wire at a relatively slow speed. To better understand current we need to answer three questions:

(1) How is the electric field in a current-carrying wire related to the potential difference between the ends of the wire?
(2) How is the resulting current related to the electric field in the wire?
(3) How is the resulting current related to the potential difference between the ends of the wire?

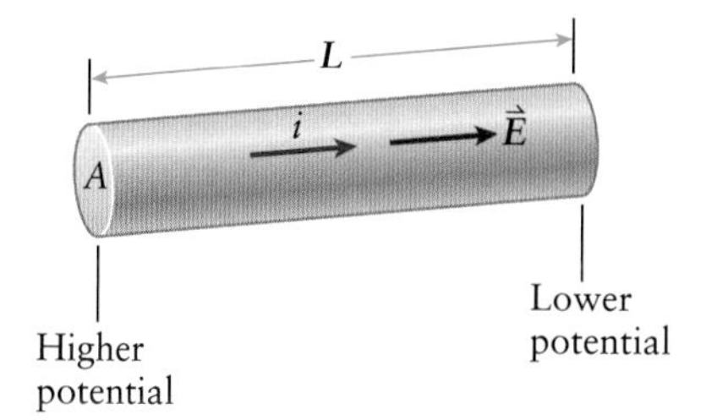

Figure 18-5 Inside a current-carrying wire The electric field in a current-carrying wire points from the higher-potential end toward the lower-potential end. The current is in the same direction.

To answer the first question, consider a straight wire of uniform cross-sectional area A and length L (**Figure 18-5**). If there is a potential difference between the ends of the wire, the electric field $\vec{E}$ points along the length of the wire from the high-potential end toward the low-potential end. Since the wire is uniform, we expect that the magnitude E of the field will be uniform as well. From Chapter 17 the potential difference V is just equal to the field magnitude multiplied by the length of the wire:

$$V = EL \quad \text{or} \quad E = \frac{V}{L} \tag{18-5}$$

This is the same as Equation 17-7, $\Delta V = -Ed \cos \theta$. We've changed the symbol for potential difference from ΔV to V, replaced the distance d by L, and used $\theta = 180°$ for the angle between the direction of $\vec{E}$ and the direction that we imagine traveling along the length of the wire to measure the potential difference. With this choice the potential difference V is positive. Remember that potential always increases as we travel opposite to $\vec{E}$ and decreases as we travel in the direction of $\vec{E}$. Remember also that the direction of current, which we choose to be the direction in which positive charges would move, is in the direction of $\vec{E}$ and so from high potential to low potential. For example, if there is a 1.5-V potential difference ($V = 1.5$ V) between the ends of a wire 10 cm in length ($L = 10$ cm $= 0.10$ m), Equation 18-5 tells us that the electric field inside the wire has magnitude $E = V/L = (1.5\text{ V})/(0.10\text{ m}) = 15\text{ V/m}$.

For the answer to the second question, we must turn to experiments on current-carrying wires. For many materials (including common conductors such as copper), experiments show that the current i that arises in a wire when an electric field is present inside the wire is directly proportional to both the electric field magnitude E and the cross-sectional area A of the wire. We can write this relationship as

$$i = \frac{EA}{\rho} \tag{18-6}$$

In Equation 18-6 the quantity ρ (the Greek letter "rho") is called the **resistivity**. The value of ρ depends on the material of which the wire is made and tells us how well or poorly this material inhibits the flow of electric charge. Equation 18-6 says that for a given cross-sectional area A and a given electric field magnitude E, a *greater* value of the resistivity means a *smaller* amount of current i. (It's unfortunate that the same Greek letter that we use as the symbol for density is also used as the symbol for resistivity. We promise never to use density and resistivity in the same equation—and you shouldn't, either!)

The units of resistivity are volt-meters per ampere, or V · m/A. For reasons that we will see below, the unit V/A (volts per ampere) appears very often and is given its own name, the **ohm** (symbol Ω, the uppercase Greek letter "omega"): 1 Ω = 1 V/A. In terms of this we can write the units of resistivity as ohm-meters or Ω · m.

TABLE 18-1 Resistivity of Some Conductors and Insulators at 21°C

Conductor	ρ ($\Omega \cdot$ m)
aluminum	2.733×10^{-8}
copper	1.725×10^{-8}
gold	2.271×10^{-8}
iron	9.98×10^{-8}
nichrome	150×10^{-8}
nickel	7.2×10^{-8}
silver	1.629×10^{-8}
titanium	43.1×10^{-8}
tungsten	5.4×10^{-8}

Insulator	ρ ($\Omega \cdot$ m)
glass	10^{12}
hard rubber	10^{13}
fused quartz	7.5×10^{17}

For copper, which is a good conductor of electricity, the resistivity is very low: $\rho = 1.725 \times 10^{-8}\ \Omega \cdot \text{m}$ at 21°C. For hard rubber, which is a very poor conductor (and a good insulator), $\rho = 10^{13}\ \Omega \cdot \text{m}$ at 21°C. The explanation for the huge ratio between these two values of resistivity is that copper has many mobile electrons per cubic meter, while hard rubber has hardly any. The value of resistivity also depends on temperature. In metals atoms move more rapidly and are likely to be less well organized at higher temperatures compared to lower temperatures. As a result, moving charges suffer more collisions with the atoms in a metal when the temperature is higher. Consequently the resistivity of most metals increases as temperature increases. In other materials, such as ceramics, resistivity *decreases* with increasing temperature. **Table 18-1** lists the values of resistivity for a variety of substances at 21°C.

In practice, it's more useful to know the answer to the third question: how the current i in a straight wire is related to the potential difference V between the ends of the wire, rather than the electric field E inside the wire. (That's because electrical meters measure potential difference directly, not electric field.) To answer this substitute the expression $E = V/L$ from Equation 18-5 into Equation 18-6:

$$i = \frac{EA}{\rho} = \left(\frac{V}{L}\right)\left(\frac{A}{\rho}\right) = V\left(\frac{A}{\rho L}\right) \quad \text{or} \quad V = i\left(\frac{\rho L}{A}\right) \tag{18-7}$$

We define the **resistance** R of a wire as

Definition of resistance (18-8)

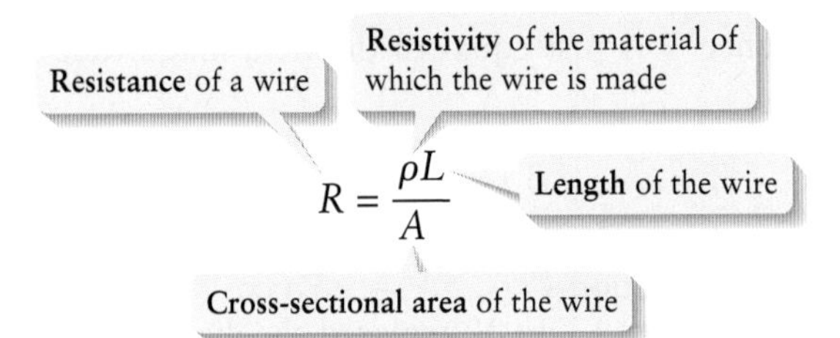

In terms of this new quantity, we can rewrite Equation 18-7 as

Relationship among potential difference, current, and resistance (18-9)

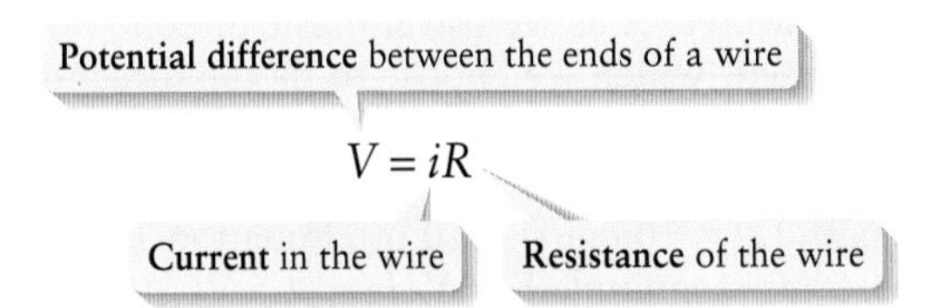

Equation 18-9 says the potential difference, or voltage, V required to produce a current i in a wire is proportional to i and to the resistance of the wire. Equation 18-8 further tells us that the resistance of a wire depends on the resistivity of the material of which the wire is made, the length of the wire, and the cross-sectional area of the wire. For a given material, the resistance is greater for a wire that is long (large L) and thin (small A) than for a wire that is short (small L) and thick (large A). You can see an example by looking inside the slots of an ordinary kitchen toaster. The wire that makes up the toaster's heating coils is very thin and if uncoiled would be very long. Hence the coils have a high resistance. The resistance is made even higher by making the wire out of nichrome, which has a resistivity about 10^2 times greater than that of copper (see Table 18-1).

Since resistivity ρ has units of ohm-meters ($\Omega \cdot$ m), length L has units of meters, and area A has units of meters squared, Equation 18-8 says that the units of resistance R are ohms. This agrees with Equation 18-9, since $1\ \Omega = 1\ \text{V/A}$: To produce a 1-A current in a wire with a resistance of $1\ \Omega = 1\ \text{V/A}$, a voltage of 1 V is required.

An ohm is a relatively small resistance. The heating coils of a toaster have a resistance R of about 10 to 20 Ω. A typical **resistor**—a circuit component intended to add resistance to the flow of current—such as you will find for sale in an electronics store will likely have a resistance in the range from 1 kΩ (1 kΩ = 1 kilohm = $10^3\ \Omega$) to

$10\ \text{M}\Omega (1\ \text{M}\Omega = 1\ \text{megohm} = 10^6\ \Omega)$. The resistance of a human body measured on the skin can be more than $0.5\ \text{M}\Omega$. The standard symbol for a resistor is a jagged line:

Equation 18-9 is often referred to as "Ohm's law," after the nineteenth-century Bavarian physicist Georg Ohm, whose pioneering experiments increased our understanding of electric current (and for whom the ohm is named). By itself it suggests that voltage and current are directly proportional to each other. This proportionality holds true if the resistance R remains constant as the voltage is changed, and this is in fact the case for many conducting materials over a wide range of voltages. Materials that have this property are referred to as *ohmic*. However, for many materials (including those used in a variety of electronic devices) the value of the resistance changes as the potential difference across the material changes. We can still use Equation 18-9 for such *nonohmic* materials, provided we keep in mind that R is not a constant.

EXAMPLE 18-3 Stretching a Wire

A 10.0-m-long wire has a radius of 2.00 mm and a resistance of 50.0 Ω. If the wire is stretched to 10.0 times its original length, what will be its new resistance?

Set Up

Equation 18-8 tells us that resistance of the wire depends on its resistivity (which is a property of the material of which the wire is made, and doesn't change if the dimensions change). It also depends on the length and cross-sectional area, both of which change when the wire is stretched. To find the new cross-sectional area, we'll use the idea that the volume of the wire (the product of its length and cross-sectional area) does not change as it's stretched.

Definition of resistance:

$$R = \frac{\rho L}{A} \qquad (18\text{-}8)$$

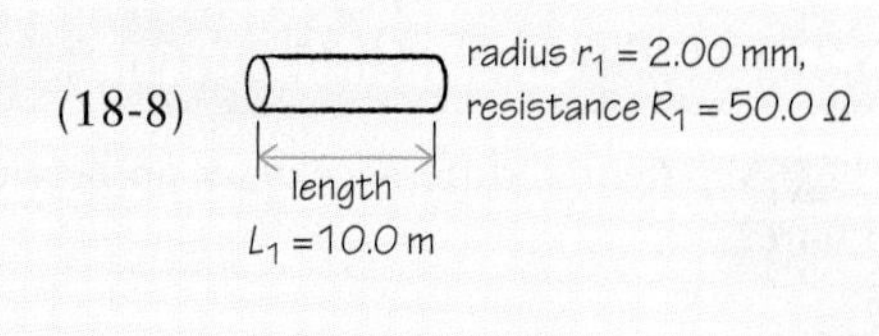

Solve

Find the cross-sectional area of the wire after it has been stretched.

The original cross-sectional area A_1 of the wire of radius $r_1 = 2.00\ \text{mm} = 2.00 \times 10^{-3}\ \text{m}$ is

$$A_1 = \pi r_1^2 = \pi(2.00 \times 10^{-3}\ \text{m})^2 = 1.26 \times 10^{-5}\ \text{m}^2$$

The volume of the wire is A_1L_1, where $L_1 = 10.0$ m is the wire's initial length. If we stretch the wire to a new length $L_2 = 10.0L_1 = 1.00 \times 10^2$ m, the cross-sectional area will change to a new value A_2, but the volume will be the same:

$$A_2L_2 = A_1L_1$$

$$A_2 = \frac{A_1L_1}{L_2} = \frac{(1.26 \times 10^{-5}\ \text{m}^2)(10.0\ \text{m})}{1.00 \times 10^2\ \text{m}} = 1.26 \times 10^{-6}\ \text{m}^2$$

The new cross-sectional area is 1/10.0 of the initial area.

Calculate the new resistance of the wire.

The initial resistance of the wire (before being stretched) is

$$R_1 = \frac{\rho L_1}{A_1} = 50.0\ \Omega$$

The resistance of the wire after being stretched is

$$R_2 = \frac{\rho L_2}{A_2}$$

If we take the ratio of the two resistances, the unknown value of resistivity will cancel out:

$$\frac{R_2}{R_1} = \frac{\rho L_2/A_2}{\rho L_1/A_1} = \frac{L_2 A_1}{L_1 A_2}$$

$$= \left(\frac{1.00 \times 10^2 \text{ m}}{10.0 \text{ m}}\right)\left(\frac{1.26 \times 10^{-5} \text{ m}^2}{1.26 \times 10^{-6} \text{ m}^2}\right)$$

$$= (10.0)(10.0) = 1.00 \times 10^2$$

The length has increased by a factor of 10.0, and the cross-sectional area has decreased by a factor of 10.0, so the new value of resistivity is greater than the old value by a factor of $(10.0)^2 = 1.00 \times 10^2$:

$$R_2 = 1.00 \times 10^2\ R_1 = (1.00 \times 10^2)(50.0\ \Omega)$$

$$= 5.00 \times 10^3\ \Omega = 5.00\ \text{k}\Omega$$

Reflect

Stretching the wire makes it longer and thinner, and both of these changes make the resistance of the wire increase. Hence the stretched wire in this problem has a much higher resistance than the wire had initially.

EXAMPLE 18-4 Calculating Current

If a 12.0-V potential difference is set up between the ends of each wire in Example 18-3, how much current will flow in each wire?

Set Up

We know the resistance of each wire from Example 18-3, so we can use Equation 18-9 to solve for the current i in each wire.

Relationship between potential difference, current, and resistance:

$$V = iR \qquad (18\text{-}9)$$

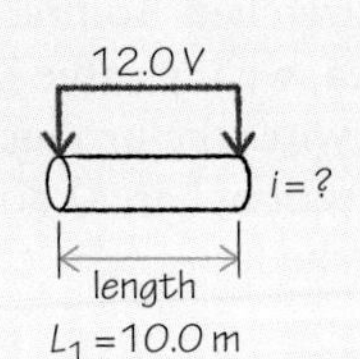

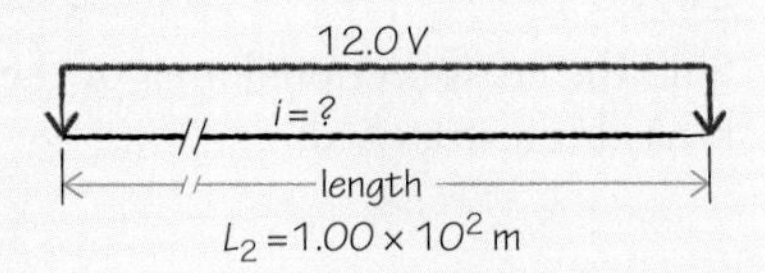

Solve

Rewrite Equation 18-9 as an expression for the current in terms of the voltage and resistance. Then solve for the current in each case.

From Equation 18-9,

$$i = \frac{V}{R}$$

For the wire before it is stretched, $R = R_1 = 50.0\ \Omega$. The current that results from a 12.0-V potential difference between the ends of this wire is

$$i_1 = \frac{V}{R_1} = \frac{12.0\text{ V}}{50.0\ \Omega} = 0.240\frac{\text{V}}{\Omega}$$

Recall that $1\ \Omega = 1$ V/A, so $1\text{ A} = 1\text{ V}/\Omega$. So the current in the wire before it is stretched is

$$i_1 = 0.240\text{ A}$$

After the wire is stretched the resistance is $R = R_2 = 5.00 \times 10^3\ \Omega$. A 12.0-V potential difference between the ends of this wire produces a current

$$i_2 = \frac{V}{R_2} = \frac{12.0\text{ V}}{5.00 \times 10^3\ \Omega} = 2.40 \times 10^{-3}\text{ A} = 2.40\text{ mA}$$

Reflect

The stretched wire has a greater resistance than the original wire, so the same potential difference produces a smaller current. Note that the milliampere (1 mA = 10^{-3} A) is a commonly used unit of current, as are the microampere (1 μA = 10^{-6} A), the nanoampere (1 nA = 10^{-9} A), and the picoampere (1 pA = 10^{-12} A).

GOT THE CONCEPT? 18-3 Is It Ohmic?

? A wire is made of a certain conducting material. You apply different potential differences V across the wire and measure the current i that results. Your results are $V = 2.0$ V, $i = 0.15$ A; $V = 4.0$ V, $i = 0.28$ A; $V = 8.0$ V, $i = 0.50$ A. Is the material of which the wire is made (a) ohmic or (b) nonohmic?

TAKE-HOME MESSAGE FOR Section 18-3

✔ Resistivity ρ is a measure of how well or poorly a particular material inhibits an electric current. The SI units of resistivity are volt-meters per ampere or ohm-meters.

✔ The value of resistivity differs from material to material. The resistivity is low for conductors and high for insulators.

✔ The resistance R of an object is a measure of the current through the object for a given potential difference between its ends. The value of R depends on the object's shape and on the resistivity of the material from which it is made. A long, narrow wire has a much higher resistance than a short, thick one.

18-4 Resistance is important in both technology and physiology

Electrical resistance can be found in every technological device, from the wires in an automobile ignition system to the resistive elements (resistors) in the circuits of a computer or mobile phone. In electronic devices, resistors are often small and cylindrical with colored bands (**Figure 18-6**). Other resistors look more like tiny, black cubes with the number of ohms etched on one face.

What purpose does a resistor serve? As we have seen, a potential difference V between two points in a conducting material causes a current i. But according to Equation 18-9, $V = iR$, the amount of current that the potential difference produces depends on the resistance of the conducting material. The greater the resistance R, the smaller the current i. Stated another way, the resistance allows us to control the current due to any particular applied voltage.

The circuit elements with colored bands are resistors. The particular colors on each resistor indicate the value of its resistance.

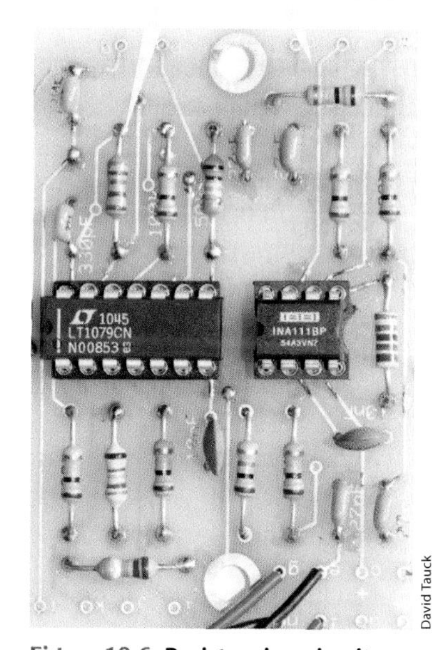

David Tauck

Figure 18-6 Resistors in a circuit A typical electronic circuit is likely to contain many resistors.

BIO-Medical

An important application of this idea takes place in every one of the millions of cells in your body. In order for a cell to live, there must be a higher concentration of positively charged potassium ions (K^+) inside the cell than in the surrounding fluid. This difference in concentration means that K^+ ions tend to leak out of the cell through pathways called *potassium channels* (**Figure 18-7**). The flow of K^+ ions constitutes a current, and each potassium channel acts like a resistor. For a given potential difference between the interior and exterior of the cell, the amount of current that flows through these channels depends on their resistance. So the flow of potassium through the membranes of your cells is determined by the electrical resistance of the membrane channels.

Why doesn't the flow of K^+ continue until there is equal concentration of these ions inside and outside the cell? Figure 18-7 shows the reason: As positive ions leave the cell, the cell interior is left with a net negative charge and a lower potential than the outside of the cell. This potential difference, called the *membrane potential*, gives rise to an electric field across the membrane that points from the outside to the inside of the cell. This field opposes additional flow of K^+ ions out of the cell. There is still some random flow of K^+ ions in and out of the cell, but the *net* flow stops when the membrane potential reaches an equilibrium value.

As the following example shows, we can use Equation 18-9 to determine the resistance of the membrane to potassium ion flow.

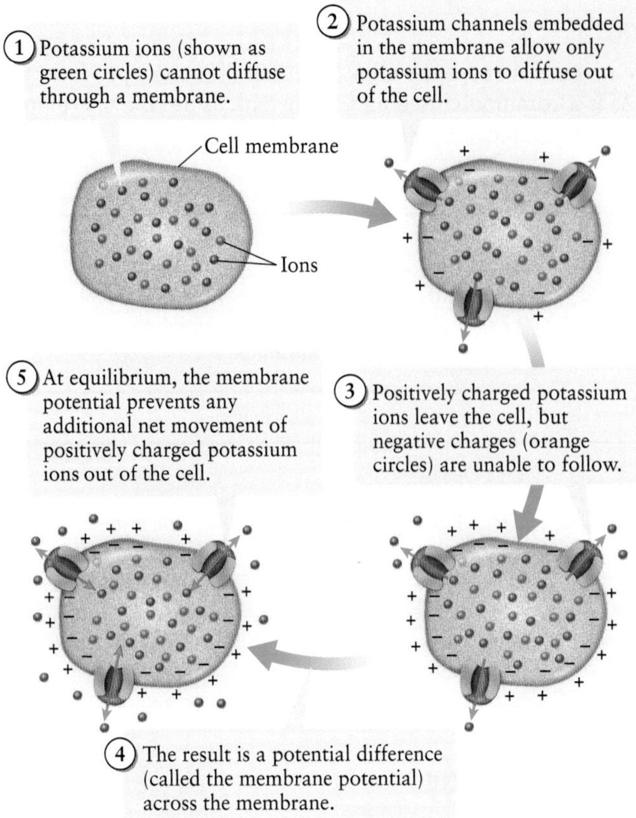

1 Potassium ions (shown as green circles) cannot diffuse through a membrane.

2 Potassium channels embedded in the membrane allow only potassium ions to diffuse out of the cell.

Cell membrane

Ions

3 Positively charged potassium ions leave the cell, but negative charges (orange circles) are unable to follow.

4 The result is a potential difference (called the membrane potential) across the membrane.

5 At equilibrium, the membrane potential prevents any additional net movement of positively charged potassium ions out of the cell.

Figure 18-7 Membrane potential The concentration of positively changed potassium ions (K^+) is always higher inside cells than outside, which causes K^+ ions to leak out. This flow of K^+ ions is a current; the channels that allow K^+ ions to leak out of the cell behave like resistors.

BIO-Medical

EXAMPLE 18-5 Resistance of a Potassium Channel

Using an instrument called a patch clamp, scientists are able to control the electrical potential difference across a tiny patch of cell membrane and measure the current through individual potassium channels. In one experiment a voltage of 0.120 V was applied across a patch of membrane, and as a result K^+ ions carried 6.60 pA of current (1 pA = 1 picoampere = 10^{-12} A). What is the resistance of this K^+ channel?

Set Up

Whether the mobile charges are negative electrons or positive ions, we can use Equation 18-9 to relate potential difference (voltage), current, and resistance.

Relationship among potential difference, current, and resistance:

$$V = iR \qquad (18\text{-}9)$$

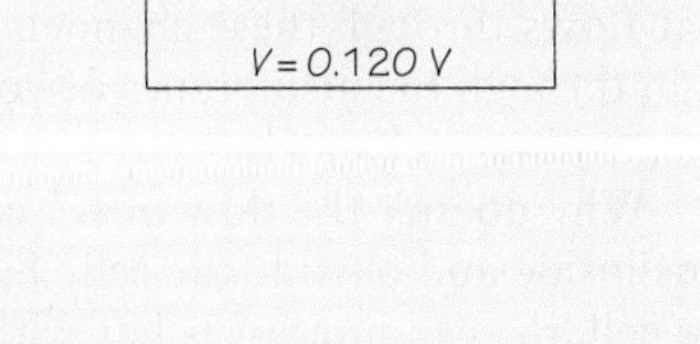

V = 0

K^+

membrane

K^+

V = 0.120 V

Solve

Rewrite Equation 18-9 to solve for the resistance.

From Equation 18-9,

$$R = \frac{V}{i}$$

The current is $i = 6.60 \text{ pA} = 6.60 \times 10^{-12}$ A, so

$$R = \frac{0.120 \text{ V}}{6.60 \times 10^{-12} \text{ A}} = 1.82 \times 10^{10} \ \Omega$$

Reflect

This is an immensely high resistance compared to the values found in circuits like that shown in Figure 18-6. Yet it is very characteristic of the channels in cell membranes, which typically have resistances in the range from 10^9 to 10^{11} Ω. A potassium channel is only about 10^{-9} m wide—so small that K^+ ions must pass through it in single file—so it's not surprising that the resistance is so high.

Although the current through a single potassium channel is very small, the electric field required to produce that current is tremendous. The thickness of the cell membrane is only about 7.5 nm = 7.5×10^{-9} m. Can you use Equation 18-5 to show that a potential difference of 0.120 V across this distance corresponds to an electric field magnitude of 1.6×10^7 V/m? (By comparison, the electric field inside the wires of a flashlight is only about 10 V/m.)

Membrane potential plays a crucial role in the nervous system. When a nerve signal propagates along an axon—a long, cable-like fiber that extends from a nerve cell—it does so in the form of a *variation* in the membrane potential of the axon. This variation, which travels along the length of the axon, is called an *action potential.* At rest the baseline membrane potential is lower on the inside compared to the outside surface of a cell. But if the potential on the interior of the membrane becomes just a little less negative, sodium ion (Na^+) channels in the membrane open. Because the concentration of Na^+ is always higher outside of cells compared to inside, Na^+ ions rush into the axon. This flow is so great that the interior surface of the axon membrane ends up with a positive charge and the exterior with a negative charge, just the reverse of what's shown in Figure 18-7. This inversion of charge is the action potential. The Na^+ channels then start closing as potassium (K^+) channels begin to open, allowing K^+ ions to flow out of the axon and restore the resting polarity of the cell membrane. You can think of an action potential as a momentary voltage "flip" that propagates along the length of the axon like a wave pulse along a stretched string.

The speed with which an action potential can propagate along an axon depends in part on the resistance of the axon to current flowing along its length. As Equation 18-8 tells us, this longitudinal resistance depends on both the length and diameter of the axon. The larger the diameter of the axon, for example, the lower the longitudinal resistance and the faster the electrical signal propagates. (In mammals the signal propagates at about 10 m/s in axons of 3 μm diameter, but at about 50 m/s in axons of 10 μm diameter.) Some axons, such as the giant ones that mediate the escape reflex in squid, have exceptionally large diameters.

BIO-Medical EXAMPLE 18-6 Giant Axons in Squid

Running along each side of the back of a squid is a tube-like structure that can be as large as 1.50 mm in diameter in some species. Originally thought to be blood vessels, these structures are actually giant axons that are part of the squid's nervous system. Although the vast majority of axons in the squid range in diameter from about 10.0 to 50.0 μm, as the squid develops, axons from about 30,000 neurons fuse together to form these giant axons. Compare the resistance of a giant axon to current along its length to the resistance of an axon with the same length but a more typical diameter of 15.0 μm.

Set Up

Equation 18-8 tells us the resistance of an axon in terms of its resistivity, length, and cross-sectional area. We'll use this to express the ratio of the resistance of a giant axon to that of an ordinary axon of the same composition (and hence same resistivity) and length but of smaller cross-sectional area.

Definition of resistance:

$$R = \frac{\rho L}{A} \quad (18\text{-}8)$$

brain

giant axon

Solve

Rewrite Equation 18-8 in terms of the diameter of an axon.

An axon has a circular cross section of radius r, equal to one-half its diameter d, so its cross-sectional area is

$$A = \pi r^2 = \pi\left(\frac{d}{2}\right)^2 = \frac{\pi d^2}{4}$$

If we substitute this into Equation 18-8, we get an expression for the resistance of an axon of length L and diameter d:

$$R = \frac{\rho L}{(\pi d^2/4)} = \frac{4\rho L}{\pi d^2}$$

Write the ratio of the resistance of a giant axon to that of a typical axon.

Let d_{giant} be the diameter of a giant axon and $d_{typical}$ be the diameter of a typical axon. If the two axons have the same length L and contain the same material of resistivity ρ, their resistances are

$$R_{giant} = \frac{4\rho L}{\pi d_{giant}^2} \quad \text{and} \quad R_{typical} = \frac{4\rho L}{\pi d_{typical}^2}$$

The ratio of these two resistances is

$$\frac{R_{giant}}{R_{typical}} = R_{giant} \times \frac{1}{R_{typical}} = \frac{4\rho L}{\pi d_{giant}^2} \times \frac{\pi d_{typical}^2}{4\rho L}$$

The factors of 4, ρ, L, and π cancel, so

$$\frac{R_{giant}}{R_{typical}} = \frac{d_{typical}^2}{d_{giant}^2} = \left(\frac{d_{typical}}{d_{giant}}\right)^2$$

We are given $d_{giant} = 1.50$ mm $= 1.50 \times 10^{-3}$ m and $d_{typical} = 15.0\ \mu$m $=$ 15.0×10^{-6} m $= 1.50 \times 10^{-5}$ m. So the ratio of resistances of the two axons is

Substitute the numerical values of the diameters of the two axons.

$$\frac{R_{giant}}{R_{typical}} = \left(\frac{1.50 \times 10^{-5}\ \text{m}}{1.50 \times 10^{-3}\ \text{m}}\right)^2 = (1.00 \times 10^{-2})^2$$

$$= 1.00 \times 10^{-4} = \frac{1}{1.00 \times 10^4}$$

The resistance of the giant axon is one ten-thousandth as great as that of a typical axon.

Reflect

When a squid recognizes danger, it sends electrical signals (action potentials) along the giant axons to trigger muscle contraction. Because the longitudinal resistance along an axon determines the speed at which nerve signals propagate, signals get to the muscle as much as 10,000 times more quickly through the giant axon than they would if the squid's nervous system was comprised entirely of ordinary nerve cells with small-diameter axons.

GOT THE CONCEPT? 18-4 Resistance and Diameter

Two copper wires have the same length, but one has four times the diameter of the other. Compared to the wire that has the smaller diameter, the wire that has the larger diameter has a resistance that is (a) 16 times greater; (b) 4 times greater; (c) the same; (d) 1/4 as great; (e) 1/16 as great.

TAKE-HOME MESSAGE FOR Section 18-4

✔ Electrical resistance is important in technology and in the physiology of cells.

✔ For a given applied voltage, the greater the resistance in an electrical system, the smaller the current.

18-5 Kirchhoff's rules help us to analyze simple electric circuits

Many simple electric circuits are made up of a source of electric potential difference and one or more resistors. An example is a light like that shown in the photograph that opens this chapter: The batteries provide the potential difference, and the filament of the light bulb acts as a resistor. We'll refer to batteries and resistors collectively as **circuit elements.** (Later in this chapter we'll consider circuits that also include capacitors as circuit elements.)

In this section we'll see how to analyze circuits made up of a battery and one or more resistors. In particular, we'll see how to determine the voltage across each circuit element as well as the current through each circuit element.

A Single-Loop Circuit and Kirchhoff's Loop Rule

Figure 18-8a shows a battery connected to a single resistor of resistance R. In fact, there are *two* resistors in this circuit; the other one, which we label r, is the **internal resistance** of the battery itself. This reflects the resistance that mobile charges encounter as they pass through the battery. So we can think of the battery as having two components, its **emf** (pronounced "ee-em-eff")—the aspect of the battery that causes charges in the circuit to move—and its internal resistance. Although we draw these as separate entities in Figure 18-8a, they cannot in fact be separated. When we say that a D or AA cell is a "1.5-volt battery" or that the battery in an automobile is "a 12-volt battery," we're actually stating the value of the battery's emf. We use the symbol ε, an uppercase script "e," for emf. So $\varepsilon = 1.5$ V for a D or AA cell and $\varepsilon = 12$ V for a standard automotive battery. (Note that the term "emf" comes from the older term "electromotive force." Although an emf is what pushes mobile charges through the circuit, it has units of volts, not newtons. So it's not accurate to call an emf a "force.") We'll often refer to a battery as a **source of emf.** In later chapters we'll encounter other sources of emf.

The symbol in Figure 18-8a for a source of constant emf is similar to the symbol for a capacitor (see Section 17-5), but with two parallel lines of unequal length:

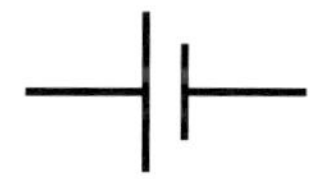

The longer of the two parallel lines represents the positive terminal of the source. Since we regard current as flowing in the direction that positive charges would move, such a source of emf causes current to flow out of its positive terminal and into its negative terminal (see Figure 18-2).

We call the circuit in Figure 18-8a a **single-loop circuit** because there's only a single path that moving charges can follow around the circuit. Think about what happens to such a charge as it travels through this circuit in the direction of the current i. (We'll continue to use the convenient fiction that these are positive charges.) As the charge passes through the source of emf (the battery) from the negative terminal to the positive terminal, it experiences an *increase* in electric potential (a *voltage rise*) of ε. When it moves through the internal resistance, however, the charge experiences a *decrease* in electric potential (a *voltage drop*) of ir. (Remember that the current moves in the direction of the electric field, and the electric field points in the direction from high potential to low potential. So the potential decreases as you move with the current through a resistor.) The charge experiences an additional decrease in potential (voltage drop) of iR when it moves through the resistor of resistance R.

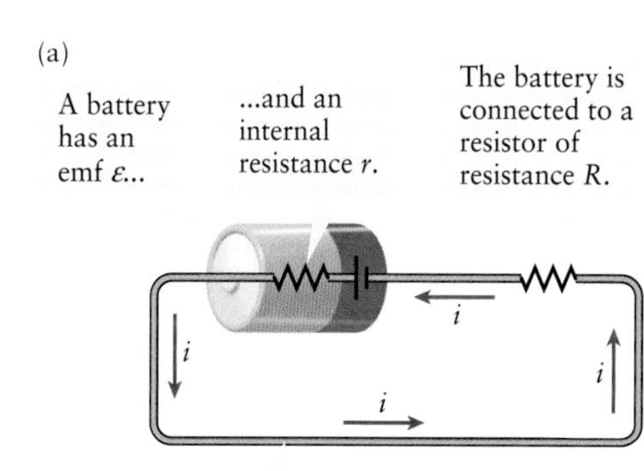

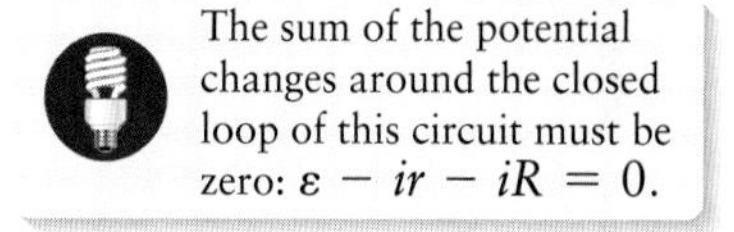

The sum of the potential changes around the closed loop of this circuit must be zero: ε − ir − iR = 0.

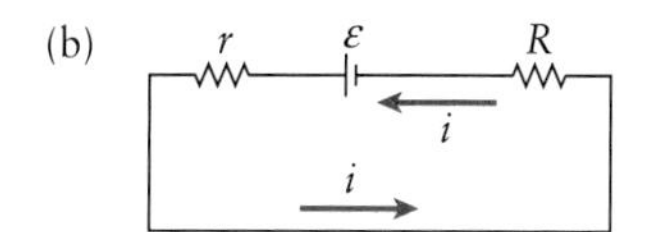

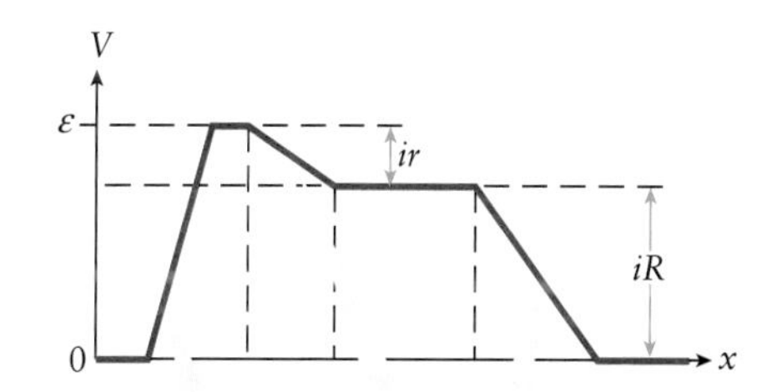

Figure 18-8 Kirchhoff's loop rule (a) A circuit made up of a battery (which has an emf ε and an internal resistance r) and a resistor R. (b) The changes in electric potential around the circuit.

There's also resistance in the wires that connect the battery and the resistor. However, the resistance of the connecting wires is generally quite small compared to the values of r and R. So we'll ignore the wire resistance and assume that a moving charge experiences no change in potential as it traverses the wires.

The current in Figure 18-8a does not change with time. So when the charge finishes a trip around the circuit and returns to where it started, the value of the potential it experiences at the starting point must have the same value as when the charge left that point. In other words, the *net change* in potential for a round trip around the loop must be *zero* (**Figure 18-8b**). This idea was first proposed by the Prussian physicist Gustav Kirchhoff in the mid-nineteenth century and is called **Kirchhoff's loop rule:**

> *The sum of the changes in potential around a closed loop in a circuit must equal zero.*

In equation form, we can write Kirchhoff's loop rule for the circuit in Figure 18-8a as

$$\varepsilon - ir - iR = 0 \tag{18-10}$$

We can rearrange Equation 18-10 to solve for the current in the circuit:

$$\varepsilon = ir + iR = i(r + R)$$

$$i = \frac{\varepsilon}{r + R}$$

To give a specific numerical example, suppose $\varepsilon = 12.0$ V, $r = 1.00\ \Omega$, and $R = 19.0\ \Omega$. Then the current is

$$i = \frac{12.0\ \text{V}}{1.00\ \Omega + 19.0\ \Omega} = \frac{12.0\ \text{V}}{20.0\ \Omega} = 0.600\ \text{A}$$

(Recall that 1 A = 1 V/Ω.) You can see that if the battery had a greater internal resistance r, the current would be smaller. As an example, the internal resistance r of a disposable battery in a flashlight or television remote control increases as the battery is used. Eventually r becomes so great that the current becomes too small to make the device work, which means that it's time to replace the battery. The emf ε of the used battery is almost the same as when it was new; it's the internal resistance that makes the battery no longer useful.

Note that for this numerical example, the voltage drop across the internal resistance is $ir = (0.600\ \text{A})(1.00\ \Omega) = 0.600$ V, and the voltage drop across the 19.0-Ω resistor is $iR = (0.600\ \text{A})(19.0\ \Omega) = 11.4$ V. The sum of the potential changes is zero, just as Kirchhoff's loop rule says it must be:

$$\varepsilon - ir - iR = 12.0\ \text{V} - 0.600\ \text{V} - 11.4\ \text{V} = 0$$

For this single-loop circuit, the *current* $i = 0.600$ A is the same through the internal resistance r as through the resistor R. That's because no charges can appear or disappear at any place in the circuit (see Section 18-2). However, the *voltage drops* are different for r and R because the values of resistance are different.

WATCH OUT! The voltage across a battery in a circuit is less than the emf.

The example we have just given shows that when a battery is in a circuit, the voltage V across the battery (that is, the potential difference between its terminals) is *not* equal to the emf ε. Rather, the voltage across the battery is equal to the emf *minus* the potential drop ir across the internal resistance of the battery. For the battery with emf ε = 12.0 V described above, the voltage is $V = \varepsilon - ir = 12.0\ \text{V} - (0.600\ \text{A})(1.00\ \Omega) = 12.0\ \text{V} - 0.600\ \text{V} = 11.4$ V. The only time the voltage $V = \varepsilon - ir$ across a battery is equal to the emf is when $i = 0$: that is, when the battery is disconnected from the circuit, so there is no current through the battery. So a 1.5-V battery has a 1.5-V potential difference between its terminals only when the battery isn't connected to anything!

Resistors in Series

An important application of Kirchhoff's loop rule is to resistors in **series**. That is, the resistors are connected end to end. (We used the same nomenclature for capacitors in series in Section 17-7.) **Figure 18-9a** shows a circuit that has three resistors R_1, R_2, and R_3 in series connected to a source of emf such as a battery with voltage V. This voltage includes the emf and internal resistance of the battery. Imagine we replace the three separate resistors with an **equivalent resistance**—that is, a single resistor of a resistance R_{equiv} that gives the same current as the series combination, as shown in **Figure 18-9b**. What is the equivalent resistance in terms of R_1, R_2, and R_3?

(a)

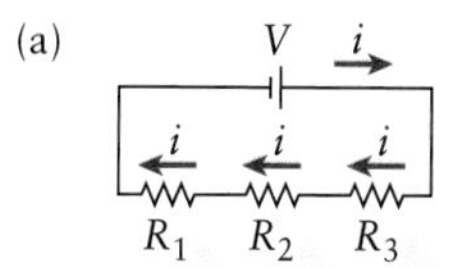

The current through each resistor in series is the same.

(b)

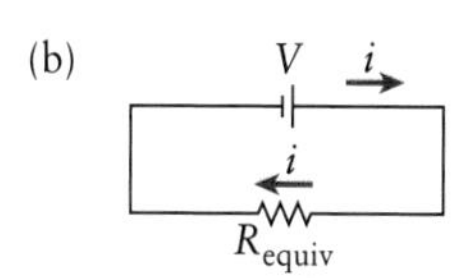

• The current i through R_{equiv} is the same as through each of the resistors in series.
• The voltage drop V across R_{equiv} is the same as across the combination of resistors in series.

Figure 18-9 Resistors in series
(a) A circuit contains three resistors connected in series to a source of emf. (b) The three resistors have been replaced by a single, equivalent resistor.

For R_{equiv} to have the same effect in the circuit as R_1, R_2, and R_3 together, it must give rise to the same current i and the same voltage drop. As we discussed above, the voltage drop across a resistance R that carries current i is iR. The same current is present through each of the three resistors shown in Figure 18-9a, and the total voltage drop across the three resistors in series is the sum of the individual voltage drops: $iR_1 + iR_2 + iR_3$. The voltage drop through the equivalent resistance is iR_{equiv}, so

$$iR_{equiv} = iR_1 + iR_2 + iR_3 = i(R_1 + R_2 + R_3)$$
$$R_{equiv} = R_1 + R_2 + R_3$$
(resistors in series)

For example, if the three resistors have resistances $R_1 = 25.0\ \Omega$, $R_2 = 12.0\ \Omega$, and $R_3 = 36.0\ \Omega$, they are equivalent to a single resistor with resistance $R_{equiv} = 25.0\ \Omega + 12.0\ \Omega + 36.0\ \Omega = 73.0\ \Omega$. This means that if we replace the three resistors by a single 73.0-Ω resistor, the current in the circuit will be exactly the same. The net voltage drop is also the same. If the current in our example is $i = 1.00$ A, the voltage drops across the individual resistors are $V_1 = iR_1 = (1.00\ \text{A})(25.0\ \Omega) = 25.0$ V, $V_2 = iR_2 = (1.00\ \text{A})(12.0\ \Omega) = 12.0$ V, and $V_3 = iR_3 = (1.00\ \text{A})(36.0\ \Omega) = 36.0$ V, and the net voltage drop is 25.0 V + 12.0 V + 36.0 V = 73.0 V. The voltage drop across the equivalent resistance is $iR_{equiv} = (1.00\ \text{A})(73.0\ \Omega) = 73.0$ V, the same as for the three resistors in series.

In general, if there are N resistors arranged in series the equivalent resistance is

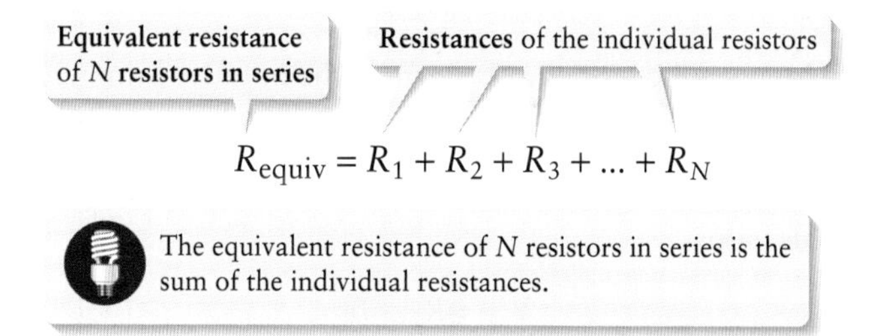

$$R_{equiv} = R_1 + R_2 + R_3 + \ldots + R_N \quad (18\text{-}11)$$

Equivalent resistance of resistors in series (18-11)

Equation 18-11 tells us that by combining resistors in series we create a circuit with a higher equivalent resistance than that of any of the individual resistors. The numerical example above illustrates this: with $R_1 = 25.0\ \Omega$, $R_2 = 12.0\ \Omega$, and $R_3 = 36.0\ \Omega$ in series, $R_{equiv} = 73.0\ \Omega$ is greater than any of R_1, R_2, and R_3.

Our analysis tells us that for resistors in series, the *current* is the same through each resistor, but the *voltage drop* is different for different resistors. As we will see, this is not the case if the resistors are in an arrangement other than series.

Kirchhoff's Junction Rule and Resistors in Parallel

Figure 18-10 shows a circuit with two resistors R_1 and R_2 connected in **parallel** to a battery. (As in Figure 18-9, the potential V includes both the emf and the internal resistance of the battery.) Unlike the single-loop circuits shown in Figures 18-8 and 18-9, this is a **multiloop circuit:** There is more than one pathway that a moving charge can take from the positive terminal of the battery through the circuit to the

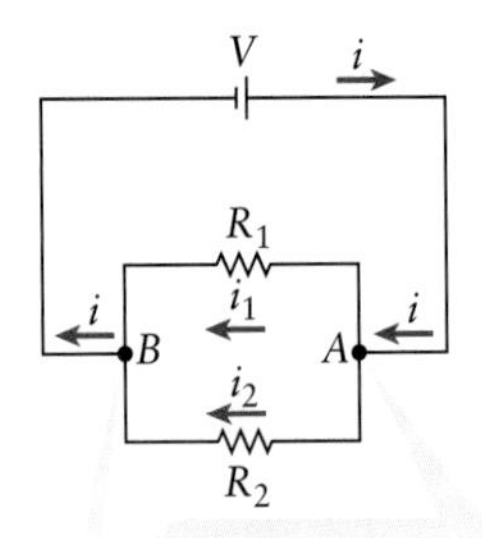

① At junction A current i splits into current i_1 and current i_2.

② At junction B current i_1 and current i_2 recombine to current i.

Figure 18-10 Kirchhoff's junction rule At the circuit junctions A and B, the net current into the junction must equal the net current out of that junction.

negative terminal. In particular, the circuit in Figure 18-10 has two **junctions** at A and B where the current either breaks into two currents (as at A) or comes together (as at B). What is the relationship among the current i that passes through the battery, the current i_1 that passes through resistor R_1, and the current i_2 that passes through resistor R_2?

One condition on i, i_1, and i_2 is that charge can neither be created nor destroyed, nor can it pile up anywhere in the circuit. This means that the rate at which charge *arrives* at a junction must be equal to the rate at which charge *leaves* that junction. This is our second rule of electric circuits. Like the loop rule, this was also proposed by Kirchhoff and is called **Kirchhoff's junction rule:**

The sum of the currents flowing into a junction equals the sum of the currents flowing out of it.

Let's apply this rule to the junctions shown in Figure 18-10. Current i flows into junction A, and currents i_1 and i_2 flow out of it. So the junction rule tells us that i (the sole current flowing into junction A) must equal $i_1 + i_2$ (the sum of the currents flowing out of A). At junction B the sum of the currents flowing in is $i_1 + i_2$ and the sole current flowing out is i. So by analyzing either junction we can conclude that

$$i = i_1 + i_2 \tag{18-12}$$

The current divides itself into i_1 and i_2 when it reaches junction A, with no extra current being added and no current being lost. These currents rejoin at junction B.

Equation 18-12 by itself doesn't tell us how much of current i takes the branch through resistor R_1 (as current i_1) and how much takes the branch through resistor R_2 (as current i_2). To determine this, let's apply the *loop* rule to two different loops through the circuit in Figure 18-10. First consider a loop that starts at the negative terminal of the battery, passes through the battery to the positive terminal (voltage rise V), then follows the path of current i_1 through resistor R_1 (voltage drop i_1R_1), and returns to the negative terminal of the battery. (As before, we'll ignore the resistance of the wires, so there is no voltage drop as a charge traverses the wires.) From the loop rule, the net change in electric potential for this loop is zero:

$$V - i_1R_1 = 0 \tag{18-13}$$

The second loop we'll consider also starts at the negative terminal of the battery and passes through the battery to the positive terminal (voltage rise V) but then follows the path of current i_2 through resistor R_2 (voltage drop i_2R_2) before returning to the battery's negative terminal. The loop rule says that the net change in electric potential is also zero for this loop:

$$V - i_2R_2 = 0 \tag{18-14}$$

If you compare Equations 18-13 and 18-14, you'll see that these equations can both be true only if

$$i_1R_1 = i_2R_2 \tag{18-15}$$

In other words, for resistors in parallel the *voltage drop* must be the same for each resistor. However, the *currents* are different for different resistors in parallel: If R_1 is less than R_2, the current will be greater in R_1 (with the smaller resistance) and smaller in R_2 (with the greater resistance). Compare this to resistors in series, for which the current is the same but the voltage drops are different for different resistors.

As an illustration, suppose $V = 12.0$ V, $R_1 = 3.00\ \Omega$, and $R_2 = 6.00\ \Omega$. From Equation 18-13 the current i_1 through resistor R_1 is given by

$$V - i_1R_1 = 0 \quad \text{so} \quad i_1R_1 = V \quad \text{and} \quad i_1 = \frac{V}{R_1} = \frac{12.0\text{ V}}{3.00\ \Omega} = 4.00\text{ A}$$

We can then find the current i_2 using Equation 18-15:

$$i_1R_1 = i_2R_2 \quad \text{and} \quad i_2 = \frac{i_1R_1}{R_2} = \frac{(4.00\ \text{A})(3.00\ \Omega)}{6.00\ \Omega} = 2.00\ \text{A}$$

Current $i_1 = 4.00$ A is twice as great as current $i_2 = 2.00$ A because resistance $R_1 = 3.00\ \Omega$ is half as great as resistance $R_2 = 6.00\ \Omega$. Note that the total current i that passes through the battery is $i = i_1 + i_2 = 4.00\ \text{A} + 2.00\ \text{A} = 6.00\ \text{A}$.

We can now find the equivalent resistance of a set of resistors in parallel. **Figure 18-11a** shows three resistors R_1, R_2, and R_3 connected in parallel to a battery with voltage V (including the emf and internal resistance). The currents through these three resistors are i_1, i_2, and i_3, respectively. **Figure 18-11b** shows the three resistors replaced by an equivalent resistance R_{equiv}. As we did for resistors in series, we'll determine R_{equiv} by demanding that the net current i and the voltage drop be the same for the actual set of resistors in Figure 18-11a and the equivalent resistance in Figure 18-11b. From our discussion above we see that the voltage drop through each of the resistors in Figure 18-11a is equal to V:

$$V = i_1R_1, \quad V = i_2R_2, \quad V = i_3R_3$$

If we divide each of these equations by R_1, R_2, and R_3, respectively, we get expressions for the current through each resistor:

$$i_1 = \frac{V}{R_1}, \quad i_2 = \frac{V}{R_2}, \quad i_3 = \frac{V}{R_3} \tag{18-16}$$

The junction rule tells us that the total current that passes through the battery is the sum of the currents through the individual resistors: $i = i_1 + i_2 + i_3$. From Equations 18-16 we can write this as

$$i = \frac{V}{R_1} + \frac{V}{R_2} + \frac{V}{R_3} = V\left(\frac{1}{R_1} + \frac{1}{R_2} + \frac{1}{R_3}\right) \tag{18-17}$$

The total current must be the same for the circuit in Figure 18-11b with the equivalent resistance. If we apply the loop theorem to this circuit, we get $V - iR_{\text{equiv}} = 0$, so $iR_{\text{equiv}} = V$ or

$$i = \frac{V}{R_{\text{equiv}}} = V\left(\frac{1}{R_{\text{equiv}}}\right) \tag{18-18}$$

Since Equations 18-17 and 18-18 are both expressions for the total current i, they can both be valid only if the right-hand sides of these equations are equal to each other:

$$\frac{1}{R_{\text{equiv}}} = \frac{1}{R_1} + \frac{1}{R_2} + \frac{1}{R_3} \tag{18-19}$$

(resistors in parallel)

As an illustration, suppose the three resistors in Figure 18-11a are $R_1 = 25.0\ \Omega$, $R_2 = 12.0\ \Omega$, and $R_3 = 36.0\ \Omega$. According to Equation 18-19 they are equivalent to a single resistor with resistance R_{equiv} given by

$$\frac{1}{R_{\text{equiv}}} = \frac{1}{25.0\ \Omega} + \frac{1}{12.0\ \Omega} + \frac{1}{36.0\ \Omega} = 0.151\ \Omega^{-1}$$

$$R_{\text{equiv}} = \frac{1}{0.151\ \Omega^{-1}} = 6.62\ \Omega$$

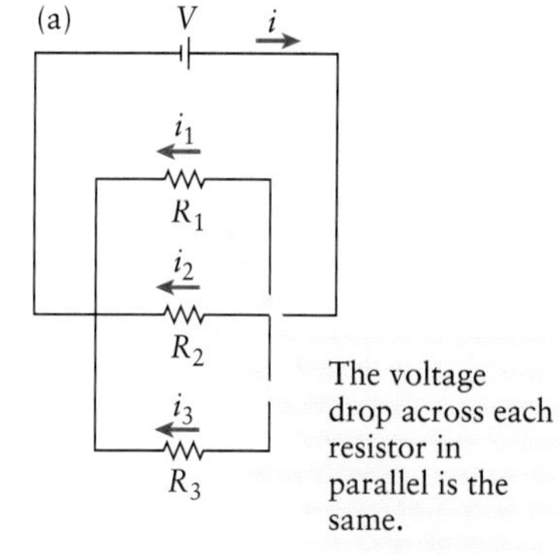

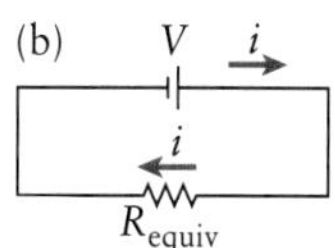

• The current i through R_{equiv} is the same as through the combination of resistors in parallel.
• The voltage drop V across R_{equiv} is the same as across each of the resistors in parallel.

Figure 18-11 Resistors in parallel (a) A circuit contains three resistors connected in parallel to a source of emf. (b) The three resistors have been replaced by a single, equivalent resistor.

Go to Picture It 18-1 for more practice dealing with resistors in combination.

If the net current through the battery is 1.00 A, the voltage drop across the equivalent resistance is $iR_{equiv} = (1.00\text{ A})(6.62\ \Omega) = 6.62$ V. The voltage drop across each of the individual resistors in parallel must be the same, so $6.62\text{ V} = i_1R_1 = i_2R_2 = i_3R_3$. You can use these relationships to show that $i_1 = 0.265$ A, $i_2 = 0.551$ A, and $i_3 = 0.184$ A (note that the current is greatest through resistor R_2, which has the smallest of the three resistances). The net current through the battery is $i = i_1 + i_2 + i_3 = 1.00$ A, just as for the battery connected to the equivalent resistance.

If we have N resistors in parallel, Equation 18-19 becomes

The reciprocal of the equivalent resistance of N resistors in parallel is the sum of the reciprocals of the individual resistances.

Equivalent resistance of resistors in parallel (18-20)

$$\frac{1}{R_{equiv}} = \frac{1}{R_1} + \frac{1}{R_2} + \frac{1}{R_3} + \ldots + \frac{1}{R_N}$$

Equivalent resistance of N resistors in parallel

Resistances of the individual resistors

Equation 18-20 tells us that by combining resistors in parallel, we create a circuit with a smaller equivalent resistance than any of the individual resistors. The numerical example above illustrates this: With $R_1 = 25.0\ \Omega$, $R_2 = 12.0\ \Omega$, and $R_3 = 36.0\ \Omega$ in parallel, the equivalent resistance $R_{equiv} = 6.62\ \Omega$ is less than any of R_1, R_2, and R_3.

WATCH OUT! Resistors do not combine in the same way as capacitors.

Be careful to distinguish between the rules for equivalent *resistance* that we've developed in this section and the rules for equivalent *capacitance* that we found in Section 17-7. For resistors in series, we find the equivalent resistance by adding the individual resistances ($R_{equiv} = R_1 + R_2 + \cdots$); for resistors in parallel, we find the reciprocal of the equivalent resistance by adding the reciprocals of the individual resistances ($1/R_{equiv} = 1/R_1 + 1/R_2 + \cdots$). These rules are reversed for capacitors. For capacitors in series, we find the reciprocal of the equivalent capacitance by adding the reciprocals of the individual capacitances ($1/C_{equiv} = 1/C_1 + 1/C_2 + \cdots$); for capacitors in parallel, we find the equivalent capacitance by adding the individual capacitances ($C_{equiv} = C_1 + C_2 + \cdots$).

The following examples illustrate some applications of the relationships we've developed in this section. In the second example, we'll see how to analyze resistors arranged in a combination that is neither purely series nor purely parallel.

BIO-Medical EXAMPLE 18-7 Giant Axons in Squid Revisited

As we saw in Example 18-6 (Section 18-4), in a squid axons of approximately 30,000 nerve cells fuse together to form each giant axon. A typical axon has a diameter of 15.0 μm, is 10.0 cm long, and has a resistivity of about 3100 $\Omega \cdot$ m. Find the resistance of a giant squid axon by considering it as 30,000 separate axons in parallel.

Set Up

We can find the resistance of a typical axon by using Equation 18-8. Each of these individual axons acts as a separate conducting path that mobile charges can follow between a point of high potential and low potential. That's just like the three resistors in parallel shown in Figure 18-11a, so we can use Equation 18-20 to determine the equivalent resistance of the giant axon as a whole in terms of the resistances of the individual axons.

Definition of resistance:

$$R = \frac{\rho L}{A} \quad (18\text{-}8)$$

Equivalent resistance of resistors in parallel:

$$\frac{1}{R_{equiv}} = \frac{1}{R_1} + \frac{1}{R_2} + \frac{1}{R_3} + \cdots + \frac{1}{R_N} \quad (18\text{-}20)$$

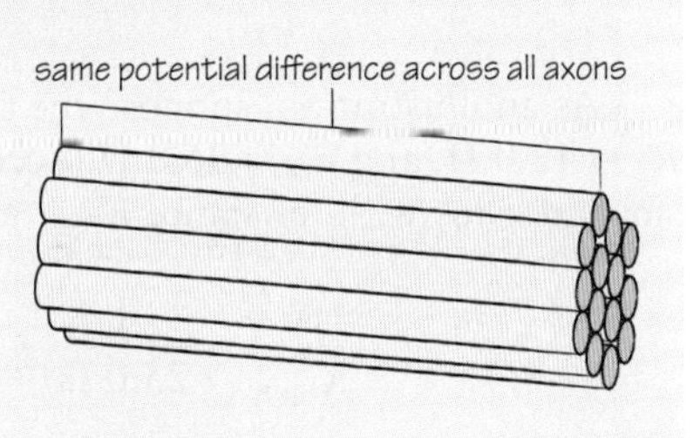

Solve

Calculate the resistance of an individual axon.

Each individual axon has resistivity $\rho = 3100\ \Omega \cdot \text{m}$, length $L = 10.0\ \text{cm} = 0.100\ \text{m}$, and diameter $15.0\ \mu\text{m} = 15.0 \times 10^{-6}\ \text{m}$. The radius r is one-half of the diameter:

$$r = \frac{1}{2}(15.0 \times 10^{-6}\ \text{m}) = 7.50 \times 10^{-6}\ \text{m}$$

If the axon has a circular cross section, its cross-sectional area is $A = \pi r^2$. So from Equation 18-8 the resistance is

$$R = \frac{\rho L}{A} = \frac{\rho L}{\pi r^2} = \frac{(3100\ \Omega \cdot \text{m})(0.100\ \text{m})}{\pi(7.50 \times 10^{-6}\ \text{m})^2} = 1.75 \times 10^{12}\ \Omega$$

This resistance is very large because the axon is long and thin, and because the axon resistivity is much higher than that of metallic conductors (see Table 18-1).

Calculate the resistance of the giant axon, as if it were 30,000 individual axons in parallel.

If N axons are in parallel and each has the same resistance R, there are N identical terms on the right-hand side of Equation 18-20. So

$$\frac{1}{R_{\text{equiv}}} = \frac{1}{R_1} + \frac{1}{R_2} + \frac{1}{R_3} + \cdots + \frac{1}{R_N} = \frac{N}{R}$$

Take the reciprocal of both sides:

$$R_{\text{equiv}} = \frac{R}{N}$$

There are $N = 30{,}000$ individual axons, each of which has resistance $R = 1.75 \times 10^{12}\ \Omega$, so the equivalent resistance of the giant axon is

$$R_{\text{equiv}} = \frac{1.75 \times 10^{12}\ \Omega}{30{,}000} = 5.85 \times 10^{7}\ \Omega$$

Reflect

In Example 18-6 we found that the resistance of a giant squid axon is about 10,000 times smaller than that of a normal axon. But in this example we've found that the equivalent resistance of a giant axon is 30,000 times smaller than that of an individual axon. Why do our answers differ by a factor of 3?

The explanation is that in this example we made the implicit assumption that the total cross-sectional area of the giant axon is 30,000 times the cross-sectional area of a small axon. However, in an actual squid the cross-sectional area of the giant axon with a diameter of 1.50 mm is smaller than the total cross-sectional area of 30,000 small axons each of diameter 15.0 μm. Although about 30,000 cells contribute to the formation of the giant axon, the process by which the giant axon forms is *not* an actual fusing together of 30,000 fully formed smaller axons. Nevertheless, it's encouraging that our calculations in Example 18-6 and this problem are pretty close.

EXAMPLE 18-8 Resistors in Combination

Figure 18-12 shows two different combinations of three identical resistors, each with resistance R. Find the equivalent resistance of the combination in (a) Figure 18-12a and (b) Figure 18-12b.

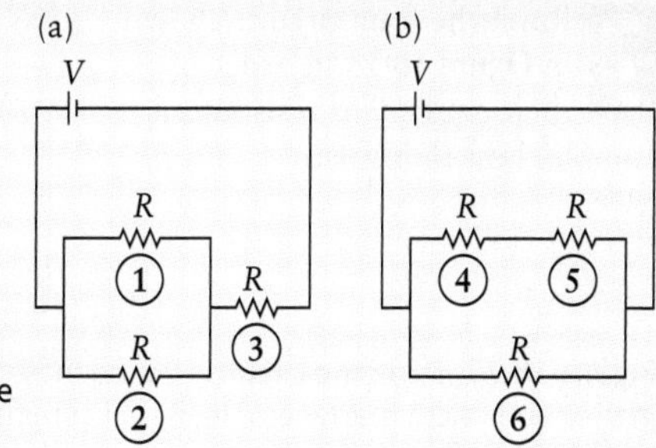

Figure 18-12 Two combinations of three identical resistors What is the equivalent resistance of each combination?

Set Up

Neither combination in Figure 18-12 is a simple series or parallel arrangement of resistors. But in Figure 18-12a resistors 1 and 2 are in parallel with each other, and that combination is in series with resistor 3. Similarly, in Figure 18-12b resistors 4 and 5 are in series with each other, and that combination is in parallel with resistor 6. So we can use Equations 18-11 and 18-20 together to find the equivalent resistances of both arrangements of resistors.

Equivalent resistance of resistors in series:

$$R_{equiv} = R_1 + R_2 + R_3 + \cdots + R_N \quad (18\text{-}11)$$

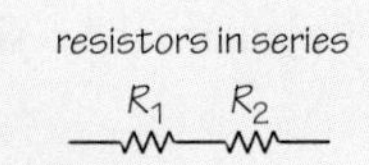

Equivalent resistance of resistors in parallel:

$$\frac{1}{R_{equiv}} = \frac{1}{R_1} + \frac{1}{R_2} + \frac{1}{R_3} + \cdots + \frac{1}{R_N} \quad (18\text{-}20)$$

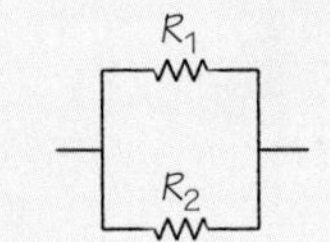

Solve

(a) For the arrangement in Figure 18-12a, first find the equivalent resistance of resistors 1 and 2.

Resistors 1 and 2 in Figure 18-12a are in parallel, so their equivalent resistance R_{12} is given by Equation 18-20:

$$\frac{1}{R_{12}} = \frac{1}{R} + \frac{1}{R} = \frac{2}{R}$$

$$R_{12} = \frac{R}{2}$$

The combination of resistors 1 and 2 is in series with resistor 3. This tells us the overall equivalent resistance.

Equivalent resistor R_{12} is in series with resistor 3. The equivalent resistance R_{123} of the entire combination is given by Equation 18-11:

$$R_{123} = R_{12} + R = \frac{R}{2} + R = \frac{3R}{2}$$

(b) For the arrangement in Figure 18-12b, first find the equivalent resistance of resistors 4 and 5.

Resistors 4 and 5 in Figure 18-12b are in series, so their equivalent resistance R_{45} is given by Equation 18-11:

$$R_{45} = R + R = 2R$$

The combination of resistors 4 and 5 is in parallel with resistor 6. This tells us the overall equivalent resistance.

Equivalent resistor R_{45} is in parallel with resistor 6. The equivalent resistance R_{456} of the entire combination is given by Equation 18-20:

$$\frac{1}{R_{456}} = \frac{1}{R_{45}} + \frac{1}{R} = \frac{1}{2R} + \frac{1}{R} = \frac{3}{2R}$$

$$R_{456} = \frac{2R}{3}$$

Reflect

If we had more than three resistors, or if the resistors had different values, we could create a large number of combinations and equivalent resistances.

GOT THE CONCEPT? 18-5 Combinations of Resistors I

? Rank the four circuits shown in **Figure 18-13** in order of their equivalent resistance, from highest to lowest.

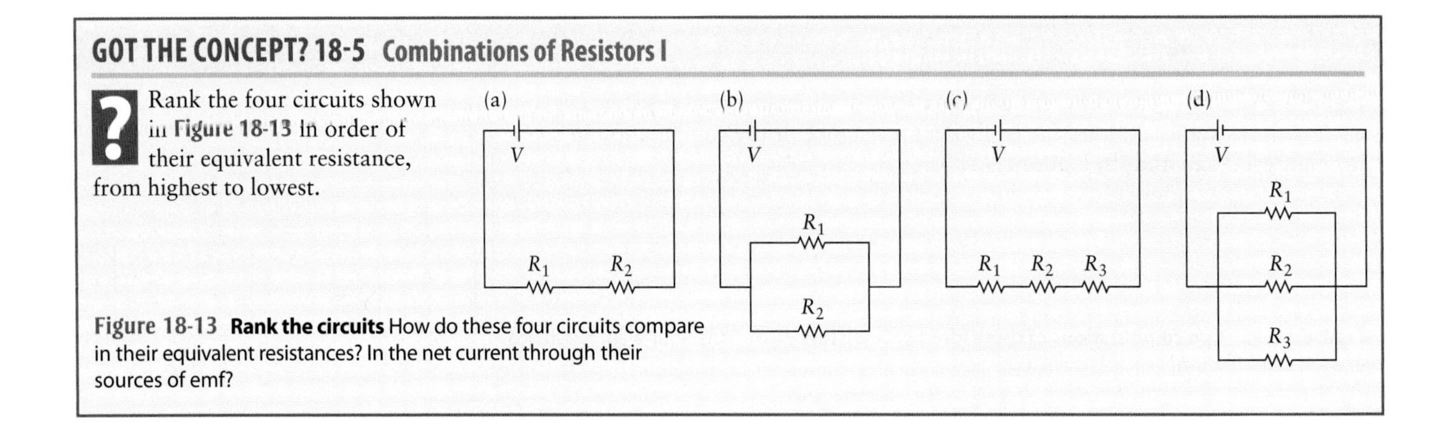

Figure 18-13 Rank the circuits How do these four circuits compare in their equivalent resistances? In the net current through their sources of emf?

GOT THE CONCEPT? 18-6 Combinations of Resistors II

Rank the four circuits shown in Figure 18-13 in order of the current through the source of emf V, from highest to lowest.

TAKE-HOME MESSAGE FOR Section 18-5

✔ In traversing a battery from its negative terminal to its positive terminal, the potential increases (a voltage rise) by an amount that depends on the battery's emf, its internal resistance, and the current i. In traversing a resistor of resistance R in the direction of the current, the potential drops by iR (a voltage drop).

✔ Kirchhoff's loop rule says that the sum of potential changes around a closed loop in a circuit is zero.

✔ Kirchhoff's junction rule says that the net current into a circuit junction equals the net current out of that junction.

✔ If resistors are connected in series, the current is the same in each resistor but the voltage drops are different across different resistors. The equivalent resistance equals the sum of the individual resistances.

✔ If resistors are connected in parallel, the voltage drop is the same for each resistor but the currents are different for different resistors. The reciprocal of the equivalent resistance equals the sum of the reciprocals of the individual resistances.

18-6 The rate at which energy is produced or taken in by a circuit element depends on current and voltage

The aspects of electric circuits that we've concentrated on so far are voltage, current, and resistance. But an electric circuit is fundamentally a way to transfer *energy* from one place to another, such as from a battery to a flashlight bulb (where the energy is converted into visible light) or from a wall socket to a toaster (where the energy is used to heat your morning bread or bagel). In most applications what's of interest is the *rate* at which energy is transferred into or out of a circuit element. For example, in order for a toaster to be useful, it must heat the bread rapidly enough that it becomes toast in a minute or so, not an hour.

As we learned in Section 6-9, **power** is the rate at which energy is transferred into or out of an object. The unit of power is the joule per second, or watt (abbreviated W): 1 W = 1 J/s. You can see the importance of power in electric circuits from the numbers that are used to describe various electric devices: An amplifier for a home audio system is rated by its power output (perhaps 75 to 100 W), and any light bulb is stamped with the power that must be supplied to it for normal operation (say, 13 or 60 W).

In this section we'll see how to calculate the power *output* of a source of emf such as a battery, which is fundamentally a source of electric potential energy. We'll also see how to calculate the power *input* of a resistor, which absorbs electric potential energy and converts it into other forms of energy.

Power in a Circuit Element

The key to understanding energy and power in electric circuits is that there is a potential difference across each circuit element, which means that a change takes place in electric potential energy when a moving charge traverses a circuit element. Remember from Section 17-3 the relationship between electric potential difference and electric potential energy difference:

The **difference in electric potential** between two positions...

...equals the **change in electric potential energy** for a charge q_0 moved between these two positions...

$$\Delta V = \frac{\Delta U_{\text{electric}}}{q_0}$$

...divided by the **charge** q_0.

Electric potential difference related to electric potential energy difference (17-6)

Electric potential difference equals electric potential energy difference per unit charge.

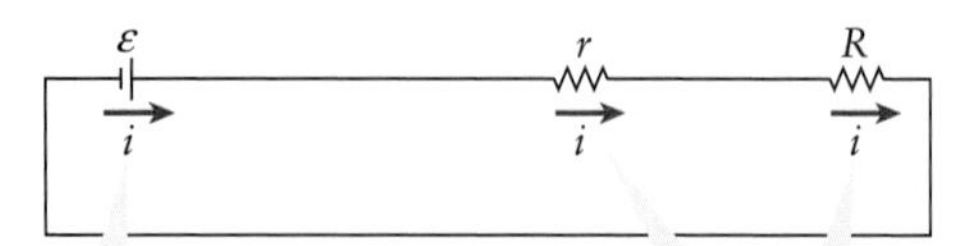

Charges moving through the emf undergo a potential change of $+\mathcal{E}$, so electric potential energy increases; energy is extracted from the emf.

Charges undergo a potential change of $-ir$ moving through the internal resistance and $-iR$ moving through the resistor. Electric potential energy decreases; energy goes into the resistances.

Figure 18-14 Energy and power in a single-loop circuit Energy flows from the emf into the moving charges; energy flows from the moving charges into the internal resistance and the resistor.

Let's rewrite this equation for the case in which a small quantity Δq of charge moves from one end of a circuit element to the other. If the potential difference between the ends of the element is ΔV, then Equation 17-6 becomes

$$\Delta V = \frac{\Delta U_{\text{electric}}}{\Delta q} \quad \text{or} \quad \Delta U_{\text{electric}} = (\Delta q)\Delta V \tag{18-21}$$

We'll continue to use the idea that current is in the direction in which positive charges would flow, so the moving charge is positive: $\Delta q > 0$. Then Equation 18-21 tells us that there is an *increase* in electric potential energy ($\Delta U_{\text{electric}} > 0$) if the charge Δq traverses a circuit element from low potential to high potential, so $\Delta V > 0$. That's the case for a charge that travels through the source of emf in **Figure 18-14** from the negative terminal to the positive terminal. Since energy is conserved, it must be that the amount of energy extracted from the source is equal to the increase in electric potential energy.

There is a *decrease* in electric potential energy ($\Delta U_{\text{electric}} < 0$) if the charge Δq traverses a circuit element from high potential to low potential, so $\Delta V < 0$. That's what happens when charge Δq travels through the resistor R in Figure 18-14 in the direction of the current. The lost electric potential energy is deposited into the resistor, which causes an increase in the resistor's temperature. If the resistor is the filament of a conventional flashlight bulb, the increased temperature causes the filament to emit radiation, some of which is in the form of visible light (see Section 14-7).

WATCH OUT! Potential energy may change as charges move around a circuit, but current does not.

It's a common misconception that current is "used up" when it passes through a resistor and that current is "added to" when it passes through a source of emf. In fact, there is *no* change in the current as it passes through either a resistor or a source: Moving charges leave any circuit element at the same rate as they enter the element. All that changes is the electric potential energy associated with the charges.

With a current i in a circuit, the power P for each circuit element is just the rate at which electric potential energy changes in that element. This equals the change in electric potential energy $\Delta U_{\text{electric}}$ for a charge Δq that enters the element, as given by Equation 18-21, divided by the time Δt that it takes each new bit of charge Δq to enter the element:

$$P = \frac{\Delta U_{\text{electric}}}{\Delta t} = \frac{\Delta q\,\Delta V}{\Delta t} = \left(\frac{\Delta q}{\Delta t}\right)\Delta V = (i)\Delta V \tag{18-22}$$

In the last part of Equation 18-22, we've used the idea that the ratio $\Delta q/\Delta t$ equals the rate at which charge enters the circuit element—in other words, this ratio equals the current i through the element (see Equation 18-1).

Equation 18-22 states that the power can be positive or negative depending on the sign of the potential difference ΔV between the ends of the circuit element. Instead of worrying about these signs, we'll replace ΔV in Equation 18-22 with the voltage V across the circuit element, which we'll regard as the absolute value of the potential difference between the ends of the element. Then P is always positive, and we can write

Power produced by or transferred into a circuit element

$$P = iV$$

Current through the circuit element

Voltage (absolute value) across the circuit element

Power for a circuit element (18-23)

Source of emf: Power flows out of the source and into the moving charges.
Resistor: Power flows out of the charges and into the resistor.

Equation 18-23 says that the power that flows into or out of any circuit element is equal to the product of the current through the element multiplied by the voltage across the element. For a given voltage V, each small amount of charge Δq that traverses the circuit element transfers the same amount of energy into or out of the element; the more charge that traverses the element per unit time and so the greater the current i, the greater the *rate* of energy transfer P.

WATCH OUT! The units of power take time into account.

The units of power sometimes cause confusion because it seems that watts require an additional time measurement. For example, to find the power in watts that goes into a flashlight bulb, a student might ask, "Find the power for what amount of time?" That's not a sensible question: Time is already included in the units of power, since 1 watt is equal to 1 joule *per second*. When a light bulb is rated at 120 W, that means it requires 120 J of energy every second to operate.

Note that the power company charges its customers based not on how much *power* they use at a given time but on the total amount of *energy* that they use. Since power is energy per time, the units of energy are the units of power multiplied by the units of time. That's why the power company bills in terms of kilowatt-hours (kWh): 1 kWh = 1000 watt-hours, or the amount of energy it takes to run a device that uses 1000 W of power (typical for a microwave oven) for one hour. Since 1 h = 3600 s, 1 kWh equals $(1000 \text{ W})(3600 \text{ s}) = 3.6 \times 10^6 \text{ W}\cdot\text{s} = 3.6 \times 10^6 \text{ J}$.

If the circuit element is a resistor with resistance R, Equation 18-9 tells us that the voltage V across the resistor is equal to the product of the current and the resistance: $V = iR$, or equivalently $i = V/R$. We can use these two expressions to rewrite Equation 18-23 in two equivalent forms for the special case of a resistor:

$$P = i(iR) \quad \text{or} \quad P = \left(\frac{V}{R}\right)V$$

We can simplify these to

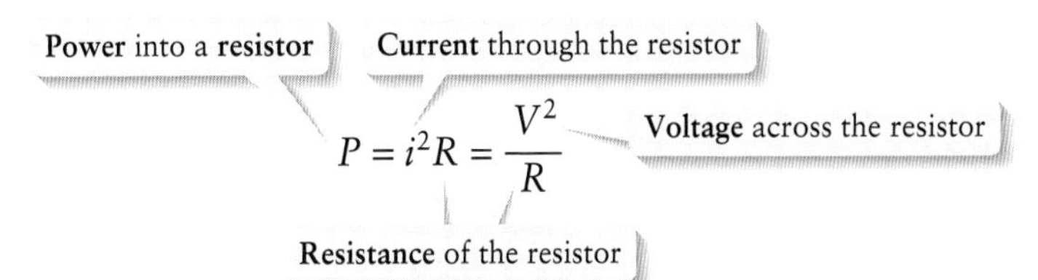

Power for a resistor (18-24)

The expression $P = i^2R$ is useful if we know the current through a resistor of known resistance, while $P = V^2/R$ is useful if we know the voltage across that resistor. The following examples show how to use Equations 18-23 and 18-24.

EXAMPLE 18-9 Power in a Single-Loop Circuit

A battery with emf 12.0 V and internal resistance 1.00 Ω is connected to a resistor with resistance 19.0 Ω. Find (a) the rate at which energy is supplied by the emf, (b) the rate at which energy flows into the internal resistance, and (c) the rate at which energy flows into the resistor.

Set Up

We can find the current in the circuit using Kirchhoff's loop rule (the sum of the changes in potential around a closed loop in a circuit must equal zero). Equation 18-9 then tells us the voltage across either resistance. We'll use Equation 18-23 to determine the power for each element of the circuit, and check our results for the two resistances using Equations 18-24.

Relationship among potential difference, current, and resistance:

$$V = iR \qquad (18\text{-}9)$$

Power for a circuit element:

$$P = iV \qquad (18\text{-}23)$$

Power for a resistor:

$$P = i^2R = \frac{V^2}{R} \qquad (18\text{-}24)$$

$\varepsilon = 12.0\ V$ $r = 1.00\ \Omega$ $R = 19.0\ \Omega$ a i

Solve

(a) First apply Kirchhoff's loop rule to determine the current i. Start at the point a in the circuit and go around in the direction of the current.

There is a voltage rise of $\varepsilon = 12.0$ V going across the emf, a voltage drop of ir going across the internal resistance $r = 1.00\ \Omega$, and a voltage drop of iR going across the resistor with resistance $R = 19.0\ \Omega$. The sum of the potential changes around the circuit is zero:

$$+\varepsilon + (-ir) + (-iR) = 0$$

Rearrange this to solve for the current i:

$\varepsilon - i(r + R) = 0$, so $i(r + R) = \varepsilon$ and

$$i = \frac{\varepsilon}{r + R} = \frac{12.0\ \text{V}}{1.00\ \Omega + 19.0\ \Omega} = \frac{12.0\ \text{V}}{20.0\ \Omega} = 0.600\ \text{A}$$

(Recall that 1 A = 1 V/Ω.)

Use Equation 18-23 to find the power extracted from the emf.

The voltage across the emf itself is $\varepsilon = 12.0$ V. The rate at which energy is extracted from the emf is

$$P_{\text{emf}} = i\varepsilon = (0.600\ \text{A})(12.0\ \text{V}) = 7.20\ \text{A}\cdot\text{V}$$

Since 1 A = 1 C/s, 1 V = 1 J/C, and 1 W = 1 J/s,

$$P_{\text{emf}} = 7.20\frac{\text{C}}{\text{s}}\cdot\frac{\text{J}}{\text{C}} = 7.20\frac{\text{J}}{\text{s}} = 7.20\ \text{W}$$

(b) First use Equation 18-9 to find the potential difference across the internal resistance $r = 1.00\ \Omega$. Then find the power in the internal resistance using Equation 18-23.

From Equation 18-9 the voltage across the internal resistance is

$$V_r = ir = (0.600\ \text{A})(1.00\ \Omega) = 0.600\ \text{V}$$

From Equation 18-23 the rate at which energy flows into the internal resistance is

$$P_r = iV_r = (0.600\ \text{A})(0.600\ \text{V}) = 0.360\ \text{W}$$

This power into the internal resistance goes into heating the battery.

(c) Repeat part (b) for the resistor of resistance $R = 19.0\ \Omega$.

The voltage across the resistor is

$$V_R = iR = (0.600\ \text{A})(19.0\ \Omega) = 11.4\ \text{V}$$

The rate at which energy flows into the resistor is

$$P_R = iV_R = (0.600\ \text{A})(11.4\ \text{V}) = 6.84\ \text{W}$$

Reflect

Note that the rate at which energy is extracted from the source of emf is equal to the *net* rate at which energy flows into the internal resistance and the resistor. This is equivalent to saying that energy is conserved in the circuit.

The net rate of energy flow into the two resistances is

$$P_r + P_R = 0.360\ \text{W} + 6.84\ \text{W} = 7.20\ \text{W}$$

This is the same as the rate at which energy flows out of the source of emf, $P_{emf} = 7.20$ W.

We can check our result $P_R = 6.84$ W for the power into the resistor by showing that we get the same results using Equations 18-24. Can you use the same approach to check the result $P_r = 0.360$ W for the internal resistance?

From the first of Equations 18-24, the power into the 19.0-Ω resistor is

$$P_R = i^2R = (0.600\ \text{A})^2(19.0\ \Omega) = 6.84\ \text{W}$$

(Note that $1\ \text{A}^2 \cdot \Omega = 1\ \text{A} \cdot \text{V} = 1$ W.)

To use the second of Equations 18-24, use the voltage $V_R = 11.4$ V across the resistor:

$$P_R = \frac{V_R^2}{R} = \frac{(11.4\ \text{V})^2}{19.0\ \Omega} = 6.84\ \text{W}$$

(Note that $1\ \text{V}^2/\Omega = 1\ \text{V} \cdot (\text{V}/\Omega) = 1\ \text{V} \cdot \text{A} = 1$ W.)

This agrees with the result for P_R found above.

EXAMPLE 18-10 Power in Series and Parallel Circuits

Two resistors, one with $R_1 = 2.00\ \Omega$ and one with $R_2 = 3.00\ \Omega$, are both connected to a battery with emf $\varepsilon = 12.0$ V and negligible internal resistance. Find the power delivered by the battery and the power absorbed by each resistor if the resistors are connected to the battery (a) in series and (b) in parallel.

Set Up

We'll use the same tools as in the previous example, plus Equations 18-11 and 18-20 for the equivalent resistance of resistors in series and in parallel. In each case we'll use the equivalent resistance to find the current through the battery, then use Equation 18-23 to determine the power provided by the battery. If the resistors are in series, the current through each is the same as the current through the battery, so we'll use the first of Equations 18-24 ($P = i^2/R$) to determine the power into each resistor. For resistors in parallel the voltage is the same across each resistor, so in that case we'll find the power into each resistor using the second of Equations 18-24 ($P = V^2/R$).

Relationship among potential difference, current, and resistance:

$$V = iR \quad (18\text{-}9)$$

Equivalent resistance of two resistors in series:

$$R_{equiv} = R_1 + R_2 \quad (18\text{-}11)$$

Equivalent resistance of two resistors in parallel:

$$\frac{1}{R_{equiv}} = \frac{1}{R_1} + \frac{1}{R_2} \quad (18\text{-}20)$$

Power for a circuit element:

$$P = iV \quad (18\text{-}23)$$

Power for a resistor:

$$P = i^2R = \frac{V^2}{R} \quad (18\text{-}24)$$

series: ε = 12.0 V, R_1 = 2.00 Ω, R_2 = 3.00 Ω

parallel: ε = 12.0 V, R_1 = 2.00 Ω, R_2 = 3.00 Ω

Solve

(a) The two resistors in series are equivalent to a single resistor R_{equiv}. This is connected directly to the terminals of the battery, across which the voltage is ε (we're told to ignore the internal resistance). So the voltage across R_{equiv} is also equal to ε, which tells us the current through the circuit. Equation 18-23 then tells us the power delivered by the battery.

From Equation 18-11 the equivalent resistance of the two resistors in series is

$$R_{equiv} = R_1 + R_2 = 2.00\ \Omega + 3.00\ \Omega = 5.00\ \Omega$$

The voltage drop across this equivalent resistance is $V = iR_{equiv}$ from Equation 18-9, which is also equal to the emf $\varepsilon = 12.0$ V of the battery. So the current through the equivalent resistance is given by

$$\varepsilon = iR_{equiv} \quad \text{or} \quad i = \frac{\varepsilon}{R_{equiv}} = \frac{12.0\ \text{V}}{5.00\ \Omega} = 2.40\ \text{A}$$

This is also the current through the battery. Since the voltage across the battery is equal to $\varepsilon = 12.0$ V, the power delivered by the battery is

$$P_{battery} = i\varepsilon = (2.40\text{ A})(12.0\text{ V}) = 28.8\text{ W}$$

The current $i = 2.40$ A is the same through both resistors in this series circuit. Use this to calculate the power that goes into each resistor.

Using the first of Equations 18-24, the power into resistor $R_1 = 2.00\ \Omega$ is

$$P_1 = i^2R_1 = (2.40\text{ A})^2(2.00\ \Omega) = 11.5\text{ W}$$

The power into resistor $R_2 = 3.00\ \Omega$ is

$$P_2 = i^2R_2 = (2.40\text{ A})^2(3.00\ \Omega) = 17.3\text{ W}$$

The net power into the two resistors is equal to the power supplied by the battery, as it should be:

$$P_1 + P_2 = 11.5\text{ W} + 17.3\text{ W} = 28.8\text{ W} = P_{battery}$$

(b) Follow the same steps as in part (a) to find the equivalent resistance of the two resistors in parallel, the current through the battery connected to that parallel arrangement, and the power delivered by the battery in this situation.

From Equation 18-20 the equivalent resistance of the two resistors in parallel is given by

$$\frac{1}{R_{equiv}} = \frac{1}{R_1} + \frac{1}{R_2} = \frac{1}{2.00\ \Omega} + \frac{1}{3.00\ \Omega} = 0.833\ \Omega^{-1}$$

$$R_{equiv} = \frac{1}{0.833\ \Omega^{-1}} = 1.20\ \Omega$$

Equation 18-9 tells us that the voltage drop across this equivalent resistance is $V = iR_{equiv}$; since this equivalent resistance is connected to the terminals of the battery, the voltage drop is also equal to the battery emf $\varepsilon = 12.0$ V. So the current through the equivalent resistance and through the battery is given by

$$\varepsilon = iR_{equiv} \text{ or } i = \frac{\varepsilon}{R_{equiv}} = \frac{12.0\text{ V}}{1.20\ \Omega} = 10.0\text{ A}$$

From Equation 18-23 the power delivered by the battery is

$$P_{battery} = i\varepsilon = (10.0\text{ A})(12.0\text{ V}) = 1.20 \times 10^2\text{ W}$$

Note that this is more than four times as much power as the same battery delivers to the same resistors in series.

The voltage is the same across resistors in parallel. Each resistor is effectively connected directly to the terminals of the battery, so the voltage across each resistor is $V = \varepsilon = 12.0$ V. Use this to calculate the power that goes into each resistor.

Using the second of Equations 18-24, we find that the power into resistor $R_1 = 2.00\ \Omega$ is

$$P_1 = \frac{V^2}{R_1} = \frac{\varepsilon^2}{R_1} = \frac{(12.0\text{ V})^2}{2.00\ \Omega} = 72.0\text{ W}$$

The power into resistor $R_2 = 3.00\ \Omega$ is

$$P_2 = \frac{V^2}{R_2} = \frac{\varepsilon^2}{R_2} = \frac{(12.0\text{ V})^2}{3.00\ \Omega} = 48.0\text{ W}$$

As for the series case, the net power into the two resistors is equal to the power supplied by the battery:

$$P_1 + P_2 = 72.0\text{ W} + 48.0\text{ W} = 1.20 \times 10^2\text{ W} = P_{battery}$$

Reflect

Although the same battery and same resistors are used in both circuits, the power provided by the battery is *much* different in the two circuits. That's because the current through the battery is different for the two circuits: $i = 2.40$ A for the series circuit, $i = 10.0$ A for the parallel circuit.

Reflect

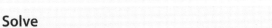

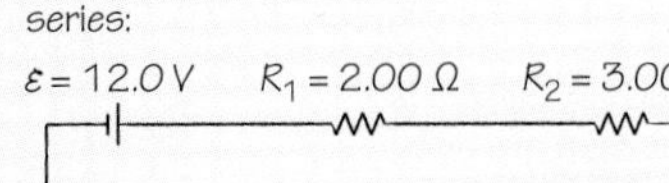

What's more, the power into each resistor is very different in the two circuits, because the voltage and current for each resistor are greater for the parallel circuit than for the series circuit. If the resistors are light bulbs that take in power and use it to produce light, the lights will glow brighter in the parallel circuit than in the series circuit.

Use Equation 18-9 to find the voltage across each resistor in the series circuit:

$$V_1 = iR_1 = (2.40\text{ A})(2.00\ \Omega) = 4.80\text{ V}$$

$$V_2 = iR_2 = (2.40\text{ A})(3.00\ \Omega) = 7.20\text{ V}$$

(compared to $V = \varepsilon = 12.0$ V for each resistor in the parallel circuit)

Use Equation 18-9 to find the current through each resistor in the parallel circuit:

$$i_1 = \frac{\varepsilon}{R_1} = \frac{12.0\text{ V}}{2.00\ \Omega} = 6.00\text{ A}$$

$$i_1 = \frac{\varepsilon}{R_2} = \frac{12.0\text{ V}}{3.00\ \Omega} = 4.00\text{ A}$$

(compared to $i = 2.40$ A for each resistor in the series circuit)

Note that the current is the same for both resistors in the series circuit, so more power goes into the resistor with the greater resistance ($P_2 = 17.3$ W into $R_2 = 3.00\ \Omega$ versus $P_1 = 11.5$ W into $R_1 = 2.00\ \Omega$). That follows from the relationship $P = i^2R$ for resistors. By contrast, the voltage is the same for the two resistors in the parallel circuit, so more power goes into the resistor with the smaller resistance ($P_1 = 72.0$ W into $R_1 = 2.00\ \Omega$ versus $P_2 = 48.0$ W into $R_2 = 3.00\ \Omega$). That agrees with the relationship $P = V^2/R$ for resistors. So if the resistors are light bulbs, the light bulb with $R_2 = 3.00\ \Omega$ is the brighter one in the series circuit, but the bulb with $R_1 = 2.00\ \Omega$ is the brighter one in the parallel circuit! When comparing the power in resistors, choose wisely among the relationships $P = iV$, $P = i^2R$, and $P = V^2/R$.

We've seen how to apply Equation 18-23, $P = iV$, and Equations 18-24, $P = i^2R = V^2/R$, to dc circuits in which there is a steady current that does not vary with time. But these same equations also apply to circuits in which the current is *not* constant and varies. In the following section we'll investigate the time-varying current in a circuit that includes both a resistor and a capacitor.

GOT THE CONCEPT? 18-7 Batteries in Series

Consider a resistor R connected to a single source of emf ε and negligible internal resistance (**Figure 18-15a**). If the same resistor is instead connected to two sources of emf ε in series, each with negligible internal resistance (**Figure 18-15b**), by what factor does the power into the resistor increase? (a) $\sqrt{2}$; (b) 2; (c) 4; (d) 8; (e) 16.

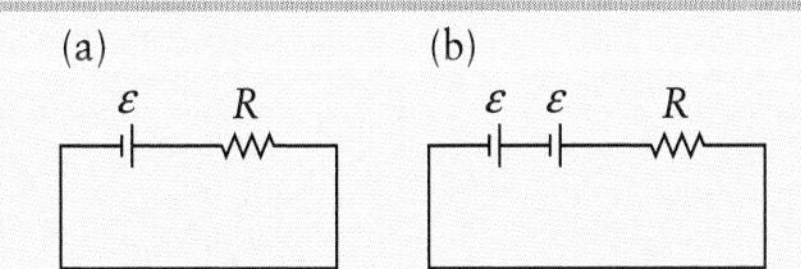

Figure 18-15 One emf or two How does adding a second source of emf affect the power delivered to the resistor?

GOT THE CONCEPT? 18-8 Ranking the Power

Five identical resistors, *A*, *B*, *C*, *D*, and *E*, are connected to a source of emf as shown in **Figure 18-16**. The source has negligible internal resistance. Rank the five resistors in order of the amount of power that goes into each resistor, from greatest to smallest. If the power in two resistors is the same, indicate as such.

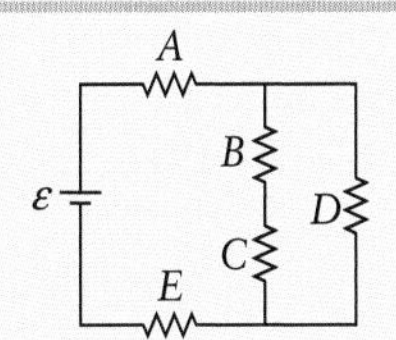

Figure 18-16 Rank the resistors How do these five identical resistors compare in the power that they absorb?

TAKE-HOME MESSAGE FOR Section 18-6

✔ Power is the rate at which energy is transferred into or out of a system. The unit of power is the watt (1 W = 1 J/s).

✔ The power into or out of any circuit element equals the current through the element multiplied by the voltage across the element.

✔ The power into a resistor is proportional to the square of the current through the resistor or, equivalently, the square of the voltage across the resistor.

(a) The position of this switch determines whether the source of emf ε is part of the circuit.

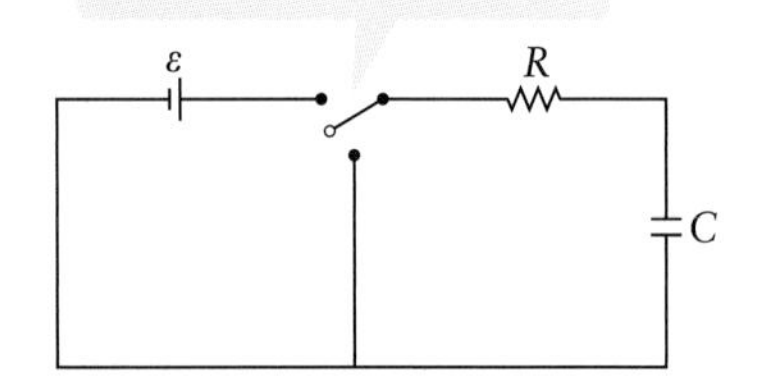

(b) With the switch in the up position, the emf causes charge to flow. This current charges the capacitor, so q increases.

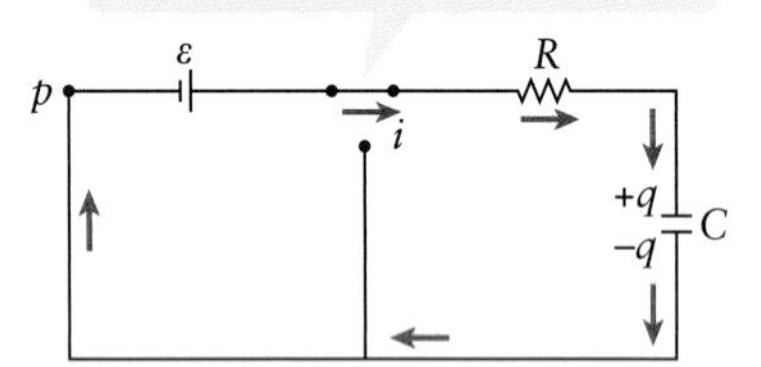

Figure 18-17 Charging a series *RC* circuit (a) In this circuit a resistor and an initially uncharged capacitor are connected in series to a source of emf. (b) The switch is moved to the up position at $t = 0$, beginning the charging process.

18-7 A circuit containing a resistor and capacitor has a current that varies with time

In an ordinary dc circuit the battery provides a steady current and delivers energy to the other circuit elements at a steady rate. In some circuits, however, what's required is a short burst of energy. That's the case for the electronic flash in a camera or mobile phone (Figure 18-1c): Energy has to be delivered to the flash lamp in a very brief time interval to produce a short-duration, high-intensity flash.

The simplest way to deliver a short burst of electric potential energy to a circuit element is by using a *capacitor* as an energy source. (You should review our discussion of capacitors in Sections 17-5 and 17-6.) We'll first examine a circuit in which a battery is used to charge a capacitor, then see what happens when the charged capacitor is discharged and its stored energy is delivered to a resistor.

A Series *RC* Circuit: Charging the Capacitor

Figure 18-17a shows a **series *RC* circuit**, with a resistor of resistance R and a capacitor of capacitance C connected in series to a battery of emf ε. (We will assume that the internal resistance of the battery is so small compared to R that it can be ignored.)

Initially the capacitor in Figure 18-17a is uncharged. If the switch in this circuit is moved to the up position (**Figure 18-17b**), the circuit is completed and the battery will cause charge to begin to flow. We'll call $t = 0$ the time when the switch is thrown. As time passes, positive charge $+q$ builds up on the upper capacitor plate and an equal amount of negative charge $-q$ builds up on the lower capacitor plate. (The two charges must be of equal magnitude since the current is the same throughout the circuit. Hence positive charge leaves the initially uncharged lower plate at the same rate that it arrives at the initially uncharged upper plate, leaving as much negative charge on the lower plate as there is positive charge on the upper plate.) So the circuit in Figure 18-17b is a *charging* series RC circuit.

Let's apply Kirchhoff's loop rule to the circuit shown in Figure 18-17b. In Section 17-5 we found that the magnitude of the charge q on the capacitor plates is proportional to the voltage V across the capacitor:

Charge, voltage, and capacitance for a capacitor (17-14)

A **capacitor** carries a **charge** $+q$ on its positive plate and a charge $-q$ on its negative plate.

The magnitude of q is directly proportional to V, the **voltage** (potential difference) between the plates.

$$q = CV$$

The constant of proportionality between charge q and voltage V is the **capacitance** C of the capacitor.

We can rewrite Equation 17-14 as $V = q/C$. If we start at point p in Figure 18-17b and move clockwise around the circuit, we encounter a voltage rise of $+\varepsilon$ as we cross the source of emf, a voltage drop $-iR$ as we cross the resistor, and a voltage drop $-q/C$ as we cross the capacitor from the positive plate to the negative plate (which is at lower potential). Kirchhoff's loop rule tells us that the sum of these voltages must be zero:

$$\varepsilon + (-iR) + \left(-\frac{q}{C}\right) = 0 \qquad (18\text{-}25)$$

If we solve Equation 18-25 for the current i in the circuit, we get

$$i = \frac{\varepsilon}{R} - \frac{q}{RC} \qquad (18\text{-}26)$$

Equation 18-26 tells us that the current in the circuit of Figure 18-17b *cannot* be constant. As charge builds up on the capacitor, the value of q increases. Hence the right-hand side of Equation 18-26 decreases with time, so the current i must decrease with time as well. What's happening is that the voltage $V = q/C$ across the capacitor increases as the charge increases. This voltage opposes that of the source of emf, so the current decreases. The current is what's causing the charge to increase, so q increases at an ever-slower rate as time goes by.

We'd like to know the capacitor charge q and the current i as functions of time. Solving Equation 18-26 for these functions is a problem in calculus that's beyond our scope. Instead we'll present the solutions and see that they make sense:

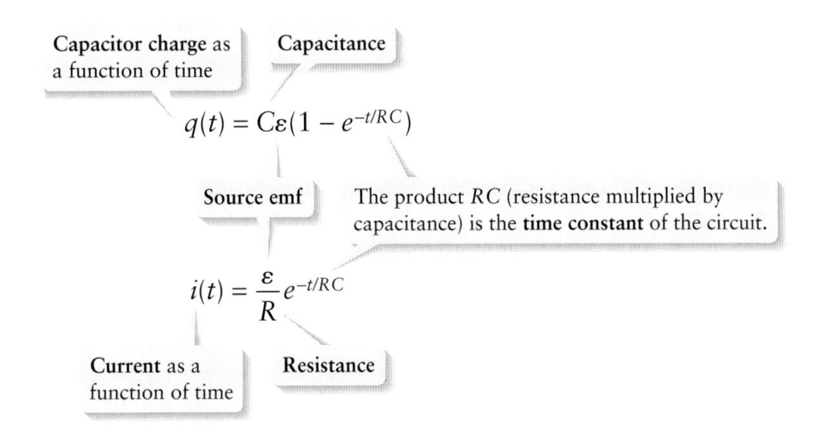

Capacitor charge and current in a charging series RC circuit (18-27)

In these equations, $t = 0$ is when the switch in Figure 18-17b is moved to the up position.

Figure 18-18 graphs the charge q and current i as given by Equations 18-27. These equations involve the **exponential function**, the irrational number $e = 2.71828\ldots$ raised to a power. (Note that e is *not* the same as the magnitude of the charge on the electron, for which we unfortunately use the same symbol.) The number e has special properties: If a population has N_0 members and grows at a rate r per year (for example, if the population grows by 2% per year, $r = 0.02$), then after t years the population will be $N(t) = N_0 e^{rt}$. In Equations 18-27 the exponential function has a *negative* exponent, which means that this function decreases with time. In particular, $e^{-t/RC}$ decreases by a factor of $1/e = 0.36787\ldots$ every time the quantity t/RC increases by 1—that is, whenever time t increases by RC. At $t = 0$ (when the switch in Figure 18-17 is closed and charge begins to flow), $e^{-t/RC} = e^0 = 1$; at $t = RC$, $e^{-t/RC} = e^{-1} = 1/e = 0.368$ to three significant figures; at $t = 2RC$, $e^{-t/RC} = e^{-2} = 1/e^2 = 0.135$; and so on.

See the Math Tutorial for more information on exponents and logarithms.

The quantity RC is called the **time constant** of a series RC circuit. (Note that the product RC has units of ohms times farads. Since $1\ \Omega = 1$ V/A, 1 F $= 1$ C/V, and 1 A $= 1$ C/s, it follows that $1\ \Omega \cdot \text{F} = 1$ (V/A)(C/V) $= 1$ C/A $= 1$ s. So the quantity RC does indeed have units of time.) The smaller the time constant, the more rapidly

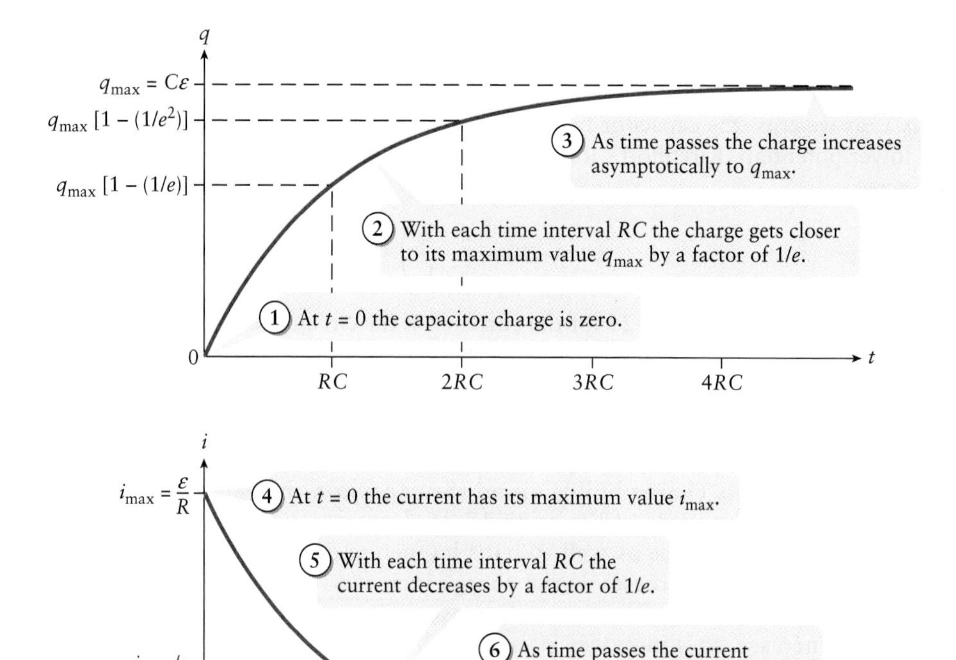

Figure 18-18 Charge and current in a charging series *RC* circuit Capacitor charge q starts at zero and approaches a maximum value; current i starts at a maximum value and approaches zero.

With the switch in the down position, the voltage across the capacitor causes a current. This current discharges the capacitor, so q decreases.

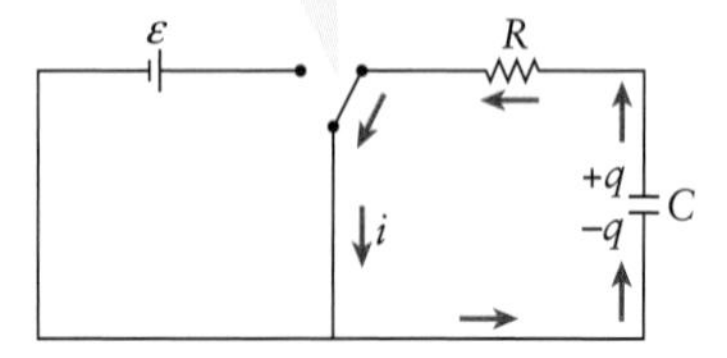

Figure 18-19 Discharging a series *RC* circuit When the switch in the circuit of Figure 18-17 is moved to the down position, the discharging process begins.

the charge q approaches its maximum value q_{max} and the more rapidly the current decreases to its final value of zero. This makes sense: A smaller resistance R (and so a smaller value of RC) allows a greater current, which means that the capacitor can accumulate charge at a faster rate. A smaller capacitance C also decreases the time constant because the capacitor can store less charge for a given voltage and so can be charged more rapidly.

At $t = 0$ the capacitor is uncharged, so there is zero voltage across the capacitor. As a result, at $t = 0$ the voltage iR across the resistor equals the voltage ε across the source, so $\varepsilon = iR$ and $i = i_{max} = \varepsilon/R$. After a very long time the current has dropped to zero, so the voltage iR across the resistor is zero. Then the voltage q/C across the capacitor equals the voltage ε across the source, so $\varepsilon = q/C$ and $q = q_{max} = C\varepsilon$.

A Series *RC* Circuit: Discharging the Capacitor

Some time after the switch in Figure 18-17b was thrown to the up position, the capacitor is charged to a charge q_{max}. (This may be less than $C\varepsilon$, depending on how long the source has had to charge the capacitor.) We now throw the switch to the down position, as in **Figure 18-19**, and restart our clock so that $t = 0$ is the time when the switch is moved to the new position. The emf is no longer part of the circuit, so the positive charge on the upper plate of the capacitor is free to move counterclockwise around the circuit to cancel the negative charge on the lower plate. So the value of the capacitor charge q decreases, and the electric potential energy stored in the capacitor is transferred into the resistor. We call this a *discharging* series RC circuit.

Kirchhoff's loop rule for the discharging circuit is now the same as Equation 18-25, but with the emf ε removed:

$$(-iR) + \left(-\frac{q}{C}\right) = 0 \qquad (18\text{-}28)$$

Solving Equation 18-28 for the current i in the circuit gives

$$i = -\frac{q}{RC} \qquad (18\text{-}29)$$

The minus sign in Equation 18-29 means that the current flows in the direction opposite to that in the charging RC circuit of Figure 18-17b. (You can see this in Figure 18-19.) As the capacitor discharges and the charge q decreases, the current i will decrease in magnitude. The charge and current as functions of time are

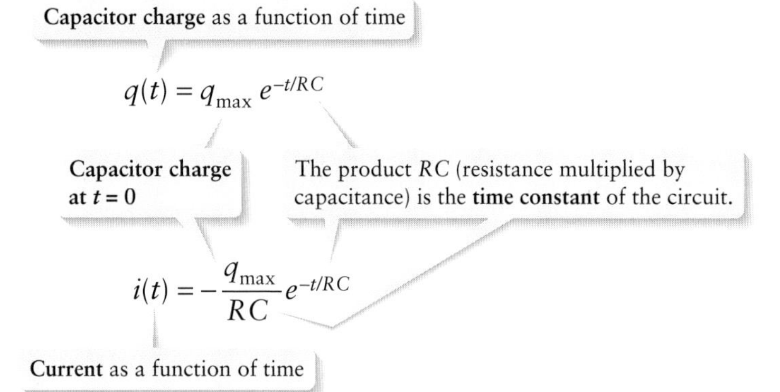

Capacitor charge and current in a discharging series RC circuit (18-30)

In these equations $t = 0$ is when the switch in Figure 18-19 is moved to the down position.

The graphs in **Figure 18-20** show the charge q and current i as given by Equations 18-30. Both functions are proportional to $e^{-t/RC}$, so both decrease by a factor of $1/e$ whenever time t increases by one time constant RC. So the value of RC determines how rapidly the capacitor charges *and* how rapidly it discharges.

Go to Interactive Exercises 18-1 and 18-2 for more practice dealing with RC circuits.

Go to Interactive Exercise 18-3 for more practice dealing with time constants.

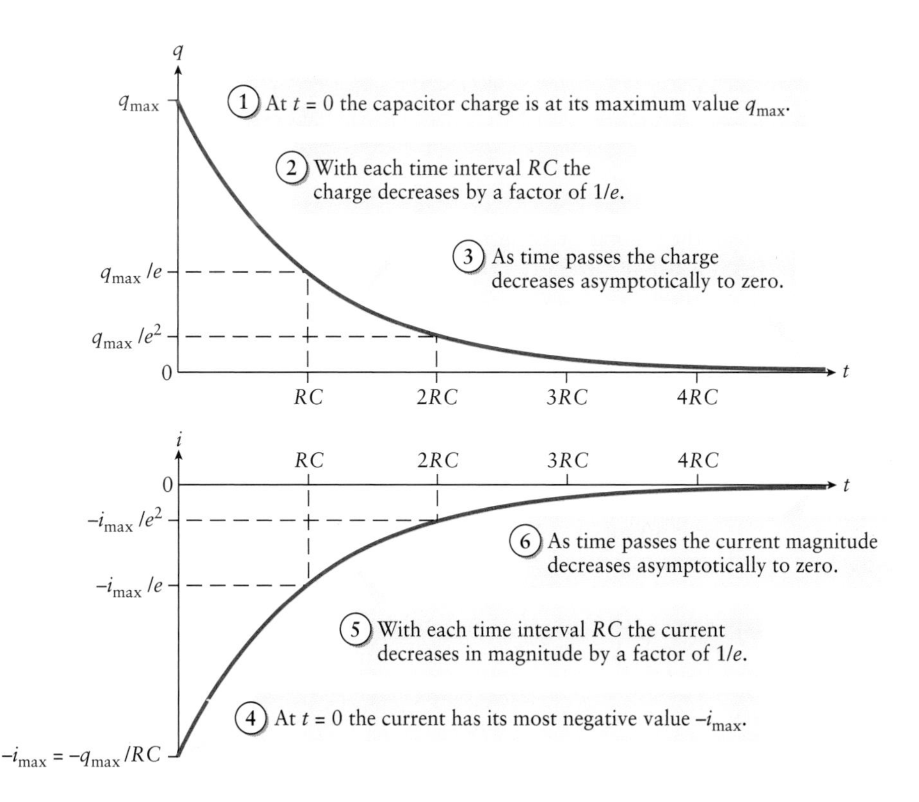

Figure 18-20 Charge and current in a discharging series RC circuit Capacitor charge q starts at a maximum value and approaches zero; current i starts at a maximum (negative) value and approaches zero.

EXAMPLE 18-11 A Charging Series *RC* Circuit

A 10.0-MΩ resistor is connected in series with a 5.00-μF capacitor. When a switch is thrown these circuit elements are connected to a 24.0-V battery of negligible internal resistance. The capacitor is initially uncharged. (a) What is the current in the circuit immediately after the switch is moved so that charging begins? (b) What is the charge on the capacitor once it is fully charged? (c) Find the capacitor charge, current, power provided by the battery, power taken in by the resistor, and power taken in by the capacitor at $t = 50.0$ s. (d) When the capacitor is fully charged, find the total energy that has been delivered by the battery and the total energy that has been delivered to the capacitor.

Set Up

We are given $R = 10.0\ \text{M}\Omega = 10.0 \times 10^6\ \Omega$, $C = 5.00\ \mu\text{F} = 5.00 \times 10^{-6}$ F, and $\varepsilon = 24.0$ V. Equations 18-27 tell us the capacitor charge and current at any time, including at $t = 0$ (when the switch is first closed) and $t \to \infty$ (long after the switch is closed, so the capacitor is fully charged). We'll use Equations 18-23 and 18-24 to find the power out of the battery and into the resistor and capacitor. (Equation 17-14 will help us in this.) In order to charge the capacitor to its maximum charge q_{max}, the total charge that must pass through the battery is q_{max}; we'll use this and Equation 17-6 to find the total energy delivered by the battery. Equation 17-17 tells us the total energy that is stored in the charged capacitor.

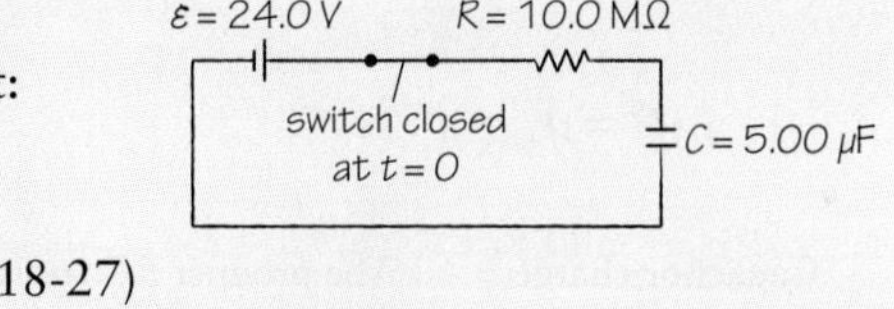

Capacitor charge and current in a charging series *RC* circuit:

$$q(t) = C\varepsilon(1 - e^{-t/RC})$$
$$i(t) = \frac{\varepsilon}{R}e^{-t/RC} \qquad (18\text{-}27)$$

Power for a circuit element:

$$P = iV \qquad (18\text{-}23)$$

Power for a resistor:

$$P = i^2R = \frac{V^2}{R} \qquad (18\text{-}24)$$

Charge, voltage, and capacitance for a capacitor:

$$q = CV \qquad (17\text{-}14)$$

Electric potential difference related to electric potential energy difference:

$$\Delta V = \frac{\Delta U_{\text{electric}}}{q_0} \qquad (17\text{-}6)$$

Electric potential energy stored in a capacitor:

$$U_{\text{electric}} = \frac{1}{2}qV = \frac{1}{2}CV^2 = \frac{q^2}{2C} \qquad (17\text{-}17)$$

Solve

(a) Find the current at $t = 0$.

From the second of Equations 18-27, the current when the switch is first closed at $t = 0$ is

$$i(0) = \frac{\varepsilon}{R}e^{-(0)/RC} = \frac{\varepsilon}{R}e^0$$

Since any number raised to the power 0 equals 1, we have $e^0 = 1$ and

$$i(0) = i_{\text{max}} = \frac{\varepsilon}{R} = \frac{24.0\ \text{V}}{10.0 \times 10^6\ \Omega}$$
$$= 2.40 \times 10^{-6}\ \text{A} = 2.40\ \mu\text{A}$$

(b) Find the capacitor charge long after the switch is closed ($t \to \infty$).

The first of Equations 18-27 tells us the capacitor charge $q(t)$. As $t \to \infty$, the exponent $-t/RC \to -\infty$. Any number raised to the power $-\infty$ is zero, so

$$e^{-t/RC} \to 0$$

$$q(t) = C\varepsilon(1 - e^{-t/RC}) \to q_{\text{max}} = C\varepsilon(1 - 0) = C\varepsilon$$
$$= (5.00 \times 10^{-6}\ \text{F})(24.0\ \text{V})$$
$$= 1.20 \times 10^{-4}\ \text{C} = 0.120\ \text{mC}$$

(c) The time constant for this circuit is $RC =$ 50.0 s, so we are actually being asked about the behavior of the circuit one time constant after the switch is closed. Use this to find charge q, current i, and the power out of or into each circuit element.

The time constant for this circuit is

$$RC = (10.0 \times 10^6\ \Omega)(5.00 \times 10^{-6}\ \text{F}) = 50.0\ \text{s}$$

so at $t = 50.0$ s, $t/RC = (50.0\ \text{s})/(50.0\ \text{s}) = 1.00$

From Equations 18-27,

$$q = C\varepsilon(1 - e^{-1.00}) = (5.00 \times 10^{-6}\ \text{F})(24.0\ \text{V})(1 - 0.368)$$
$$= 7.58 \times 10^{-5}\ \text{C} = 0.632 q_{max}$$
$$i = \frac{\varepsilon}{R} e^{-1.00} = \frac{24.0\ \text{V}}{10.0 \times 10^6\ \Omega}(0.368)$$
$$= 8.83 \times 10^{-7}\ \text{A} = 0.368 i_{max}$$

The voltage across the battery is $\varepsilon = 24.0$ V, so from Equation 18-23 the power out of the battery is

$$P_{battery} = i\varepsilon = (8.83 \times 10^{-7}\ \text{A})(24.0\ \text{V})$$
$$= 2.12 \times 10^{-5}\ \text{W} = 21.2\ \mu\text{W}$$

From the first of Equations 18-24, the power into the resistor is

$$P_R = i^2 R = (8.83 \times 10^{-7}\ \text{A})^2(10.0 \times 10^6\ \Omega)$$
$$= 7.80 \times 10^{-6}\ \text{W} = 7.80\ \mu\text{W}$$

Equation 17-14, $q = CV$, tells us that the voltage across the capacitor is $V = q/C$. Combining this with Equation 18-23 gives the power into the capacitor:

$$P_C = i\left(\frac{q}{C}\right) = (8.83 \times 10^{-7}\text{A})\left(\frac{7.58 \times 10^{-5}\ \text{C}}{5.00 \times 10^{-6}\ \text{F}}\right)$$
$$= 1.34 \times 10^{-5}\ \text{W} = 13.4\ \mu\text{W}$$

Note that the net power into the resistor and capacitor combined equals the power out of the battery:

$$P_R + P_C = 7.80\ \mu\text{W} + 13.4\ \mu\text{W} = 21.2\ \mu\text{W} = P_{battery}$$

(d) Use the maximum charge stored by the capacitor, which is the total charge moved across the battery, and Equation 17-6 to calculate the change in electric potential energy imparted by the battery. Use Equations 17-7 to calculate the electric potential energy stored in the capacitor.

Long after the switch is closed, the total amount of charge that has passed through the battery and to the positive capacitor plate is $q_{max} = 1.20 \times 10^{-4}$ C. From Equation 17-6 the potential energy change that was imparted by moving this charge across the 24.0-V emf of the battery is

$$\Delta U_{battery} = q_{max}\,\varepsilon = (1.20 \times 10^{-4}\ \text{C})(24.0\ \text{V})$$
$$= 2.88 \times 10^{-3}\ \text{J} = 2.88\ \text{mJ}$$

The last of Equations 17-17 tells us the amount of energy that went into the capacitor to store charge q_{max} there:

$$U_C = \frac{q_{max}^2}{2C} = \frac{(1.20 \times 10^{-4}\ \text{C})^2}{2(5.00 \times 10^{-6}\ \text{F})}$$
$$= 1.44 \times 10^{-3}\ \text{J} = 1.44\ \text{mJ}$$

So exactly one-half of the energy taken from the battery goes into the capacitor: $U_C = (1/2)\Delta U_{battery}$.

Reflect

Our results for charge q and current i in part (c) agree with Figure 18-18: After one time constant the capacitor charge has reached $[1 - (1/e)] = 0.632 = 63.2\%$ of its fully charged value q_{max}, and the current has decreased to $1/e = 0.368 = 36.8\%$ of its initial value i_{max}. Can you show that after five time constants ($t = 5RC = 250$ s) the charge will have reached 99.3% of q_{max} and the current will have decreased to just 0.674% of i_{max}?

The power calculations in part (c) show that all of the power extracted from the battery is accounted for: Part of the energy extracted from the battery goes into the resistor, and the rest goes into adding to the electric potential energy stored in the capacitor. We've shown this for a specific instant, but it's true at *all* times during the charging process.

These results from part (c) also help us understand our calculations in part (d): Only one-half of the energy extracted from the battery goes into the capacitor, so the other half must have gone into the resistor. This is a general result for charging *any* series RC circuit.

The physics of capacitors in circuits helps explain what happens in the propagation of action potentials along axons, which we discussed in Section 18-4. All biological membranes, including the axonal membrane, act like a capacitor. Under resting conditions the inner surface of the membrane is negatively charged and the outer surface is positively charged (see Figure 18-7). As we described in Section 18-4, as an action potential propagates along an axon, the resting charge distribution reverses: At the peak of an action potential, the inner surface of the membrane is positively charged and the outer surface is negatively charged. When the membrane returns to its initial charge state, it's analogous to a capacitor discharging as in the circuit of Figure 18-19. The value of C is the capacitance of a segment of axon membrane, and the value of R is the resistance of that segment to charge flowing through channels in the membrane. (This is different from the resistance we discussed in Section 18-4, especially Example 18-6. There we considered the *longitudinal* resistance for current flowing along the length of the axon; the value of R we're considering here is the *transverse* resistance to charges flowing across the axon membrane, perpendicular to the axon's length.)

The value of the time constant RC determines the rate of discharge: The smaller the value of RC, the faster the capacitance of the membrane discharges and the more rapidly the signal propagates along the axon. We saw in Section 18-4 that some invertebrates such as squid have large-diameter axons to allow action potentials to propagate quickly from the brain to the muscles that mediate escape reflexes. Another way to make signals propagate more rapidly along an axon is to decrease the capacitance C of the membrane, thus reducing the value of RC. In vertebrates special cells wrap themselves around some axons, forming a thick insulating layer called *myelin* (**Figure 18-21**). The myelin acts as a capacitor between the surface of the axon and its environment. This additional capacitance is in series with the capacitance of the axon's cell membrane and so decreases the effective capacitance of the axon. (Recall from Section 17-7 that when capacitors are placed in series, the effective capacitance is less than that of any individual capacitor.)

Only certain sections of the axon are coated with myelin. The action potential in the nonmyelinated gaps, or *nodes*, produces a signal that propagates rapidly along the myelinated regions of the axon. The signal weakens in strength in the myelinated region where no action potential is generated, but is reinvigorated when it reaches the next node.

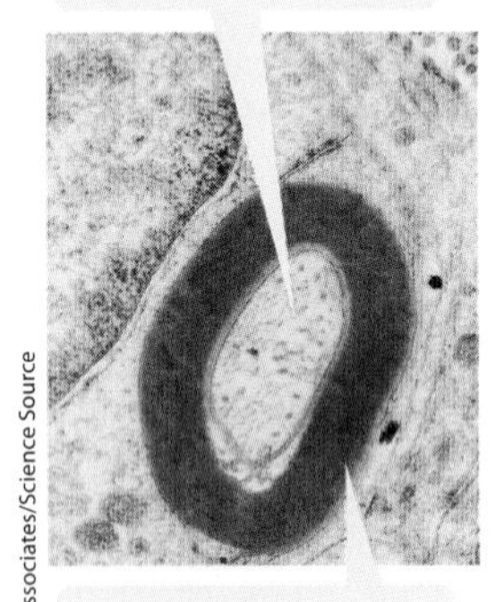

Figure 18-21 A myelinated axon In vertebrates special cells wrap themselves around some axons, forming a thick insulating layer called myelin. This decreases the effective capacitance of the axon and increases the speed at which nerve signals can propagate along the length of the axon.

GOT THE CONCEPT? 18-9 Discharging an *RC* Circuit

? The switch in the circuit of Figure 18-19 is moved to the down position, causing the capacitor to discharge. Suppose that the resistor is a light bulb and that its brightness is proportional to the power into the resistor. Compared to the brightness of the bulb when the switch is moved to the down position at $t = 0$, the brightness of the bulb at $t = RC$ is (a) the same; (b) $1 - (1/e^2) = 0.865$ as great; (c) $1 - 1/e = 0.632$ as great; (d) $1/e = 0.368$ as great; (e) $1/e^2 = 0.135$ as great.

TAKE-HOME MESSAGE FOR Section 18-7

✔ A circuit containing a resistor and a capacitor connected in series is a series RC circuit.

✔ Connecting the resistor and an uncharged capacitor to a source of emf causes a current that charges the capacitor. Switching the circuit to remove the source of emf causes the capacitor to discharge through the resistor. In either case an exponential function describes how the current and the capacitor charge vary with time.

✔ The rate of charge or discharge in a series RC circuit depends on the time constant, equal to the product RC. In a time interval RC the current decreases by a factor of $1/e$.

Key Terms

ac (alternating current)
alternating current (ac)
ampere
battery
circuit
circuit element
current
dc (direct current)
direct current (dc)
drift speed
emf
equivalent resistance
exponential function
internal resistance
junction
Kirchhoff's junction rule
Kirchhoff's loop rule
mobile charge
multiloop circuit
ohm
parallel (resistors)
power
resistance
resistivity
resistor
series (resistors)
series RC circuit
single-loop circuit
source of emf
time constant

Chapter Summary

Topic	Equation or Figure	
Current: Electric charge flows around a circuit in response to an electric potential difference, such as that provided by a battery (a source of emf). Current is the rate of charge flow and is related to the speed at which mobile charges drift through the circuit. In ordinary metals the mobile charges are negatively charged electrons, but it's conventional to take the direction of the current to be the direction in which positive charges would flow.	$i = \frac{\Delta q}{\Delta t}$ — **Current** in a circuit; Amount of **charge** that flows past a certain point in the circuit; **Time** required for that amount of charge to flow past that point	(18-1)
	$i = nAev_{drift}$ — **Current** in a wire; Number of **mobile charges per unit volume**; **Speed** at which charges drift through the wire; **Cross-sectional area** of the wire; Magnitude of the **charge** on each mobile charge	(18-3)
Resistance: Resistivity is a measure of how difficult it is for charges to flow through a material. The resistance of a wire or circuit element depends on the dimensions of the object as well as the resistivity of the material of which the object is made. Resistance and resistivity are not constants but vary with temperature. A resistor is a circuit element whose most important property is its resistance.	$R = \frac{\rho L}{A}$ — **Resistance** of a wire; **Resistivity** of the material of which the wire is made; **Length** of the wire; **Cross-sectional area** of the wire	(18-8)
Voltage, current, and resistance: In order for current to exist in a conductor with resistance, there must be a potential difference (voltage) between the ends of the conductor. For a given resistance a greater voltage produces a greater current.	$V = iR$ — **Potential difference** between the ends of a wire; **Current** in the wire; **Resistance** of the wire	(18-9)

Rules for circuits: Kirchhoff's loop rule states that the sum of the changes in electric potential around a closed loop in a circuit must equal zero. Kirchhoff's junction rule states that the sum of the currents into a junction equals the sum of the currents out of it. These rules allow us to analyze the currents and voltages in electric circuits.

(b)

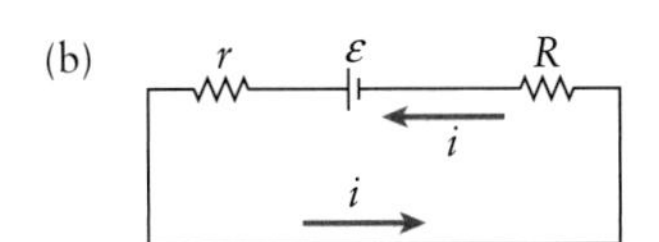

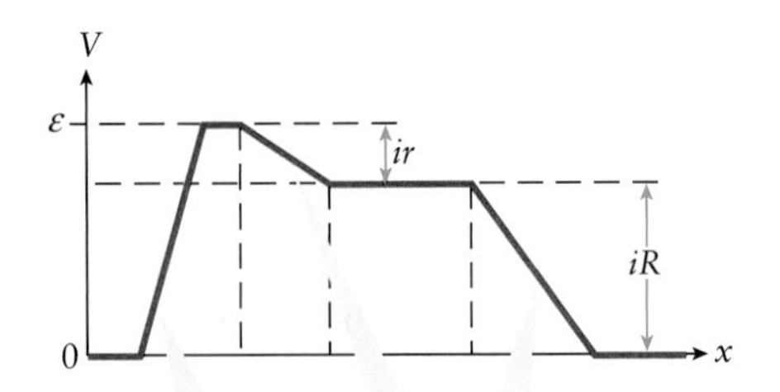

The potential increases by ε from the negative terminal to the positive terminal of the emf.

The potential drops by ir across the internal resistance and drops by iR across the resistor.

(Figure 18-8b)

Resistors in series and parallel: A collection of resistors in a circuit behaves as though it were a single resistor, with an equivalent resistance R_{equiv}. The equivalent resistance is different for resistors in series (all of which carry the same current) than for resistors in parallel (all of which have the same voltage).

Equivalent resistance of N resistors in series

Resistances of the individual resistors

$$R_{\text{equiv}} = R_1 + R_2 + R_3 + \ldots + R_N \tag{18-11}$$

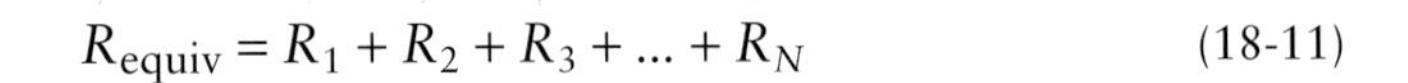

The equivalent resistance of N resistors in series is the sum of the individual resistances.

The reciprocal of the equivalent resistance of N resistors in parallel is the sum of the reciprocals of the individual resistances.

$$\frac{1}{R_{\text{equiv}}} = \frac{1}{R_1} + \frac{1}{R_2} + \frac{1}{R_3} + \ldots + \frac{1}{R_N} \tag{18-20}$$

Equivalent resistance of N resistors in parallel

Resistances of the individual resistors

Power in circuits: Power is the rate of energy transfer. In an electric circuit, energy flows from a source of emf into the other circuit elements. The power into a resistor can be expressed in terms of the resistance and either the current through the resistor or the voltage across the resistor.

Power produced by or transferred into a circuit element

$$P = iV \tag{18-23}$$

Current through the circuit element

Voltage (absolute value) across the circuit element

Source of emf: Power flows out of the source and into the moving charges.
Resistor: Power flows out of the charges and into the resistor.

Power into a **resistor**

Current through the resistor

$$P = i^2R = \frac{V^2}{R} \tag{18-24}$$

Voltage across the resistor

Resistance of the resistor

Series *RC* circuits: If a resistor R and capacitor C are connected in series to a source of emf, the capacitor charge increases toward a maximum value. The current in the circuit (which carries the charge to the capacitor) begins with a large value and gradually decreases to zero. If the source is taken out of the circuit, the capacitor discharges. The current in the circuit now carries the charge away from the capacitor and again gradually decreases to zero. The rate of charging or discharging depends on the time constant of the circuit, equal to the product RC.

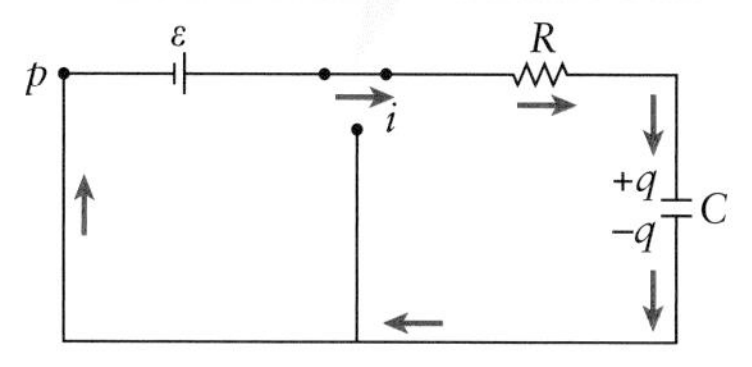
(b)
With the switch in the up position, the emf causes charge to flow. This current charges the capacitor, so q increases.
ε
R
p
i
+q
−q
C

(Figure 18-17b)

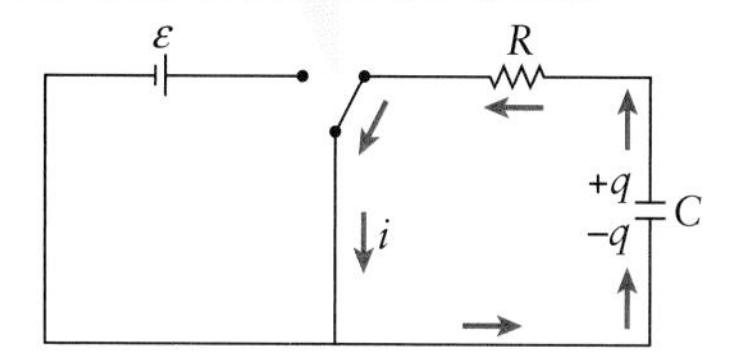
With the switch in the down position, the voltage across the capacitor causes a current. This current discharges the capacitor, so q decreases.
ε
R
i
+q
−q
C

(Figure 18-19)

Answer to What do you think? Question

(b) When electrons pass through the battery, they gain electric potential energy. They lose most of this potential energy as they pass through the filament of the light bulb, and this lost energy heats the filament and causes it to glow. The electrons lose the rest of the potential energy as they pass through the wires that connect the battery and light bulb. If electrons lost *kinetic* energy anywhere in the circuit, they would slow down and create an electron "traffic jam." The electric repulsion between electrons prevents such "jams" from happening, so the electrons must maintain the same average speed and average kinetic energy as they travel around the circuit.

Answers to Got the Concept? Questions

18-1 (c) The current i in a simple circuit (in which the moving charges travel around a single loop) has the same value at all points in the circuit. This is true whether the thickness of the wire varies or is the same at all points.

18-2 (a) As we saw in the answer to the previous Got the Concept? question, the current i has the same value at all points in the circuit. Equation 18-3 shows that for this to be true the product Av_{drift} (the cross-sectional area of the wire multiplied by the drift speed) must also have the same value at all points. (The quantity n, the number of mobile charges per unit volume, depends only on what the wire is made of, not on its dimensions.) The thinner wire has a radius r that is half as great and a cross-sectional area $A = \pi r^2$ that is one-quarter as great as the rest of the wire. So the drift speed of the electrons must be four times greater in the thinner part of the wire.

18-3 (b) For an ohmic material the current and voltage are directly proportional, so doubling the voltage should cause a doubling of current. That is *not* the case for this material: Doubling the voltage from 2.0 to 4.0 V increases the current by a factor of only (0.28 A)/(0.15 A) = 1.9, and doubling the voltage again from 4.0 to 8.0 V increases the current by a factor of only (0.50 A)/(0.28 A) = 1.8. So this material is nonohmic.

18-4 (e) As in Example 18-6 the resistance of wire of resistivity ρ, length L, and diameter d is $R = 4\rho L/\pi d^2$. This says that the resistance is inversely proportional to the square of the diameter. So if the diameter is increased by a factor of 4 while leaving the resistivity and length the same, the resistance will change by a factor of $1/4^2 = 1/16$.

18-5 (c), (a), (b), (d) Combining resistors in series results in a circuit with larger equivalent resistance; combining them in parallel results in a smaller equivalent resistance. Therefore, circuits (a) and (c) have higher resistance than circuits (b) and (d). For the same reason the equivalent resistance of (c) with three resistors in series is higher than that of circuit (a) with two resistors in series, and the equivalent resistance of circuit (b) with two resistors in parallel is higher than that of circuit (d) with three resistors in parallel. So in order of equivalent resistance from highest to lowest, the circuits are (c), (a), (b), and (d). This conclusion doesn't depend on how the value of R_3 compares to the values of R_1 or R_2: Any additional resistance in series increases the equivalent resistance, and any additional resistance in parallel decreases the equivalent resistance.

18-6 (d), (b), (a), (c) The larger the equivalent resistance the smaller the current for a fixed voltage. So this ranking of the circuits in order of decreasing current is just the opposite of the ranking in order of decreasing equivalent resistance in Got the Concept? 18-5.

18-7 (c) Placing two sources of emf in series is equivalent to increasing the emf from ε to 2ε. This doubles the voltage V across the resistor, which from $V = iR$ (Equation 18-9) means that the current i through the resistor doubles as well. From Equations 18-24, $P = i^2R = V^2/R$, the power into the resistor is proportional to either the square of the current i or the square of the resistor voltage V. So the power into the resistor increases by a factor of $2^2 = 4$. A device such as a flashlight or television remote control that uses two batteries uses four times as much power as one with a single battery.

18-8 $P_A = P_E > P_D > P_B = P_C$ The full current that passes through the source also passes through resistors A and E. This current divides up in order to pass through two branches—one with resistors B and C, one with resistor D—so only a fraction of the full current passes through resistors B, C, and D. Since the power into a resistor is given by $P = i^2R$ (the first of Equations 18-24), it follows that the same power goes into A and E ($P_A = P_E$) and that this is greater than the power into any of the other three resistors. The voltage across the branch with resistors B and C is the same as that across the branch with resistor D (the branches are in parallel). All of this voltage is across resistor D, but only half of this voltage is across resistor B and half across resistor C. From the second of Equations 18-24, $P = V^2/R$, this means that there is more power into D than into B or C but equal amounts into B and C (so $P_D > P_B = P_C$). So $P_A = P_E > P_D > P_B = P_C$. If the resistors are light bulbs, their ranking by brightness will be the same as this ranking by power.

18-9 (e) The second of Equations 18-30 shows that the current in a discharging series RC circuit is proportional to $e^{-t/RC}$, so at $t = RC$ the current is $e^{-1} = 1/e = 0.368$ as great as its value at $t = 0$ (when $e^{-t/RC} = e^0 = 1$). The power $P = i^2R$ (Equations 18-24) into a resistor is proportional to the square of i, and i is $1/e$ as great as at $t = 0$, so the power is $1/e^2 = (0.368)^2 = 0.135$ as great as at $t = 0$.

Questions and Problems

In a few problems you are given more data than you actually need; in a few other problems you are required to supply data from your general knowledge, outside sources, or informed estimate.

Interpret as significant all digits in numerical values that have trailing zeros and no decimal points.

For all problems use $g = 9.80\ \text{m/s}^2$ for the free-fall acceleration due to gravity. Neglect friction and air resistance unless instructed to do otherwise.

•	Basic, single-concept problem
••	Intermediate-level problem, may require synthesis of concepts and multiple steps
•••	Challenging problem
SSM	*Solution is in Student Solutions Manual*
Example	*See worked example for a similar problem.*

Conceptual Questions

1. • We distinguish the direction of current in a circuit. Why don't we consider it a vector quantity?

2. • Under ordinary conditions the drift speed of electrons in a metal is around 10^{-4} m/s or less. Why doesn't it take a long time for a light bulb to come on when you flip the wall switch that is several meters away?

3. • We justified a number of electrostatic phenomena by the argument that there can be no electric field in a conductor. Now we say that the current in a conductor is driven by a potential difference and thus there is an electric field in the conductor. Is this statement a contradiction? SSM

4. • Two wires, A and B, have the same physical dimensions but are made of different materials. If A has twice the resistance of B, how do their resistivities compare?

5. • **Biology** When a bird lands with both feet on a high voltage wire, will the bird be electrocuted? Explain your answer.

6. •• Many ordinary strings of holiday lights contain about 50 bulbs connected in parallel across a 110 V line. Sixty years ago a string of 50 bulbs would be connected in series across the line. What would happen if you could put one of the old-style bulbs into a modern holiday light set? (The light sockets are made differently to prevent this.) SSM

7. • **Biology** Explain how an action potential is generated.

8. • If the only voltage source you have is 36 V, how could you light some 6-V light bulbs without burning them out?

9. • An ammeter measures the current through a particular circuit element. (a) How should it be connected with that element, in parallel or in series? (b) Should an ammeter have a very large or a very small resistance? Why?

10. • For a given source of constant voltage, will more heat develop in a large external resistance connected across it or a small one? SSM

11. • The average drift velocity of electrons in a wire carrying a steady current is constant even though the electric field within the wire is doing work on the electrons. What happens to this energy?

12. • Is current dissipated when it passes through a resistor? Explain your answer.

13. • Give a simple physical explanation for why the charge on a capacitor in an RC circuit can't be changed instantaneously.

14. • (a) Does the time required to fully charge a capacitor through a given resistor with a battery depend on the voltage of the battery? (b) Does it depend on the total amount of charge to be placed on the capacitor? Explain your answers.

Multiple-Choice Questions

15. • If a current-carrying wire has a cross-sectional area that gradually becomes smaller along the length of the wire, the drift velocity
    A. increases along the length of the wire.
    B. decreases along the length of the wire.
    C. remains the same along the length of the wire.
    D. increases along the length of the wire only if the resistivity increases too.
    E. decreases along the length of the wire only if the resistivity decreases too. SSM

16. • What causes an electric shock?
    A. current
    B. voltage
    C. both current and voltage
    D. resistance and current
    E. resistance and voltage

17. • Two copper wires have the same length, but one has twice the diameter of the other. Compared to the one that has the smaller diameter, the one that has the larger diameter has a resistance that is
   A. larger by a factor of 2.
   B. larger by a factor of 4.
   C. the same.
   D. smaller by a factor of 1/2.
   E. smaller by a factor of 1/4.

18. • When a thin wire is connected across a voltage of 1 V, the current is 1 A. If we connect the same wire across a voltage of 2 V, the current is
   A. (1/4) A.
   B. (1/2) A.
   C. 1 A.
   D. 2 A.
   E. 4 A. SSM

19. • When a wire that has a large diameter and a length L is connected across the terminals of an automobile battery, the current is 40 A. If we cut the wire to half of its original length and connect one piece that has a length $L/2$ across the terminals of the same battery, the current will be
   A. 10 A.
   B. 20 A.
   C. 40 A.
   D. 80 A.
   E. 160 A.

20. • A charge flows from the positive terminal of a 6-V battery, through a light bulb, and through the battery back to the positive terminal. The total voltage change experienced by the charge is
   A. 1 V.
   B. 6 V.
   C. 0 V.
   D. −1 V.
   E. −6 V.

21. • When a second light bulb is added in series to a circuit with a single light bulb, the resistance of the circuit
   A. increases.
   B. decreases.
   C. remains the same.
   D. doubles.
   E. triples.

22. • When a light bulb is added in parallel to a circuit with a single light bulb, the resistance of the circuit
   A. increases.
   B. decreases.
   C. remains the same.
   D. doubles.
   E. triples. SSM

23. • If we use a 2-V battery instead of a 1-V battery to charge the capacitor shown in **Figure 18-22**, the time constant will
   A. be four times greater.
   B. double.
   C. remain the same.
   D. be half as much.
   E. be four times less.

Figure 18-22 Problem 23

Estimation/Numerical Analysis

24. • Estimate the amount of electric current that is drawn by three appliances you use on a daily basis.

25. • Estimate the current in a high voltage transmission line such as you might see in rural areas, sometimes close to remote interstate highways.

26. • Estimate the resistance of the coils in your toaster.

27. •• Estimate the electric energy usage of a citizen of a developed country such as Canada or the United States compared to a citizen of a developing nation.

28. • Estimate the current provided by the battery in a cellular phone.

29. • Estimate the value of resistors and capacitors that are used in a common flash attachment of a disposable camera. SSM

30. The current through a resistor R of 25 Ω is measured at time intervals of 500 μs after closing the switch that connects it to a 5-V battery through a capacitor C with an unknown value (**Figure 18-23**). The magnitude of the current through the resistor is shown in the table below. Graph the current as a function of time and determine the capacitance.

Figure 18-23 Problem 30

Time (s)	Current (A)
0	0.200
5.00×10^{-4}	0.109
1.00×10^{-3}	0.060
1.50×10^{-3}	0.032
2.00×10^{-3}	0.018
2.50×10^{-3}	0.010
3.00×10^{-3}	0.005
3.50×10^{-3}	0.003

31. Estimate the resistance of your phone charger's cable.

Problems

18-1 Life on Earth and our technological society are only possible because of charges in motion
18-2 Electric current equals the rate at which charge flows

32. • A steady current of 35 mA exists in a wire. How many electrons pass any given point in the wire per second? Example 18-1

33. • A light bulb requires a current of 0.50 A to emit a normal amount of light. If the light is left on for 1.0 h, how many electrons pass through the bulb? Example 18-1

34. • A synchrotron radiation facility creates a circular electron beam with a current of 487 mA when the electrons have a speed approximately equal to the speed of light. How many electrons pass a given point in the accelerator per hour? SSM Example 18-1

35. •• A copper wire that has a diameter of 2.00 mm carries a current of 10.0 A. Assuming that each copper atom contributes one free electron to the metal, find the drift speed of the electrons in the wire. The molar mass of copper is 63.5 g/mol, and the density of copper is 8.95 g/cm^3. Example 18-2

36. • Aluminum wiring can be used as an alternative to copper wiring. Suppose a 40-ft run of 12-gauge aluminum wire (2.05232 mm in diameter) carrying a 7.0-A current is used in a circuit. How long would it take electrons to travel the length of this run and back? There are 6.03×10^{28} atoms in 1 m^3 of aluminum, with each atom contributing three free electrons to the metal. Example 18-2

37. •• **Biology** Doubly charged calcium ions, Ca^{2+}, are released into your bloodstream and flow through your

capillaries. The diameter of a capillary is about 5 μm, calcium ion concentrations are approximately 1 mol/m^3, and typical electric current flow due to the Ca^{2+} ions is around 1.2 nA. Calculate the drift speed of the calcium ions. Example 18-2

38. • **Biology** 2.00×10^{-3} mol of potassium ions pass through a cell membrane in 4.00×10^{-2} s. (a) Calculate the electric current and (b) describe its direction relative to the motion of the potassium ions. Example 18-1

39. • **Biology** Cell membranes contain channels that allow ions to cross the phospholipid bilayer. A particular K^+ channel carries a current of 1.9 pA. How many K^+ ions pass through it in 1.0 ms? Example 18-1

18-3 The resistance to current through an object depends on the object's resistivity and dimensions

40. • If a copper wire has a resistance of 2.00 Ω and a diameter of 1.00 mm, how long is it? Example 18-3

41. • Calculate the resistance of a piece of copper wire that is 1.00 m long and has a diameter of 1.00 mm. Example 18-3

42. • The resistance ratio of two conductors that have equal cross-sectional areas and equal lengths is 1:3. What is the ratio of the resistivities of the materials from which they are made? Example 18-3

43. • An 8.00-m-long length of wire has a resistance of 4.00 Ω. The wire is uniformly stretched to a length of 16.0 m. Find the resistance of the wire after it has been stretched. SSM Example 18-3

44. • When 120 V is applied to the filament of a 75-W light bulb, the current drawn is 0.63 A. When a potential difference of 3.0 V is applied to the same filament, the current is 0.086 A. Is the filament made of an ohmic material? Explain your answer. Example 18-4

45. • A power transmission line is made of copper that is 1.80 cm in diameter. What is the resistance of 1 mi of the line? Example 18-3

18-4 Resistance is important in both technology and physiology

46. •• There is a current of 112 pA when a certain potential is applied across a certain resistor. When that same potential is applied across a resistor made of the identical material but 25 times longer, the current is 0.044 pA. Compare the effective diameters of the two resistors. Example 18-6

47. • A certain flexible conducting wire changes shape as environmental variables, such as temperature, change. If the diameter of the wire increases by 25% while the length decreases by 12%, by what factor does the resistance of the wire change? Example 18-3

48. • **Biology** Cell membranes contain channels that allow K^+ ions to leak out. Consider a channel that has a diameter of 1.0 nm and a length of 10 nm. If the channel has a resistance of 18 GΩ, what is the resistivity of the solution in the channel? Example 18-6

49. • If a light bulb draws a current of 1.0 A when connected to a 12-V circuit, what is the resistance of its filament? Example 18-4

50. • If a flashlight bulb has a resistance of 12.0 Ω, how much current will the bulb draw when it is connected to a 6.0-V circuit? Example 18-4

51. • A light bulb has a resistance of 8.0 Ω and a current of 0.5 A. At what voltage is it operating? Example 18-4

18-5 Kirchhoff's rules help us to analyze simple electric circuits

52. • Calculate the voltage difference $V_A - V_B$ in each of the situations shown in **Figure 18-24**. Example 18-8

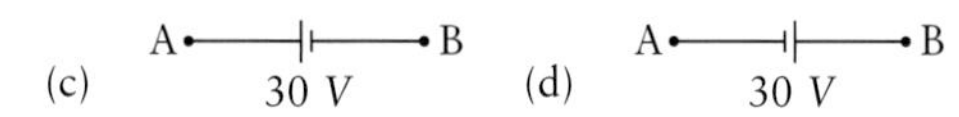

Figure 18-24 Problem 52

53. • An 18.0-Ω resistor and a 6.00-Ω resistor are connected in series. What is their equivalent resistance? Example 18-8

54. • An 18.0-Ω resistor and a 6.00-Ω resistor are connected in parallel. What is their equivalent resistance? Example 18-8

55. • A 9.00-Ω resistor and a 3.00-Ω resistor are connected in series across a 9.00-V battery. Find (a) the current through, and (b) the voltage drop across, each resistor. SSM Example 18-8

56. • A 9.00-Ω resistor and a 3.00-Ω resistor are connected in parallel across a 9.00-V battery. Find (a) the current through, and (b) the voltage drop across, each resistor. Example 18-8

57. • A potential difference of 3.6 V is applied between points a and b in **Figure 18-25**. Find (a) the current in each of the resistors and (b) the total current the three resistors draw from the power source. Example 18-8

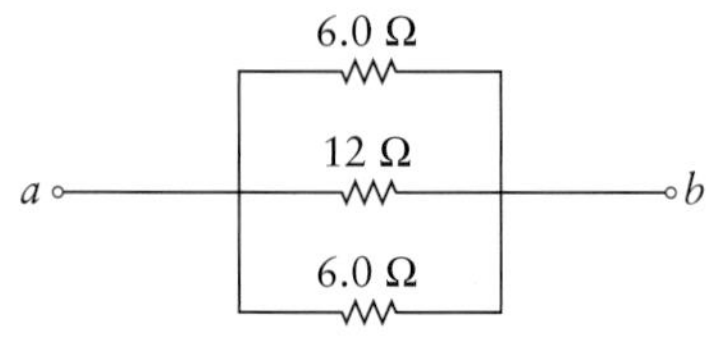

Figure 18-25 Problem 57

58. • The four resistors in **Figure 18-26** have an equivalent resistance of 8 Ω. Calculate the value of R_x. Example 18-8

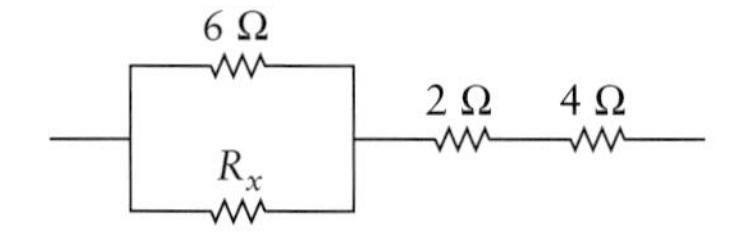

Figure 18-26 Problem 58

59. • Find the equivalent resistance of the combination of resistors shown in **Figure 18-27**. SSM Example 18-8

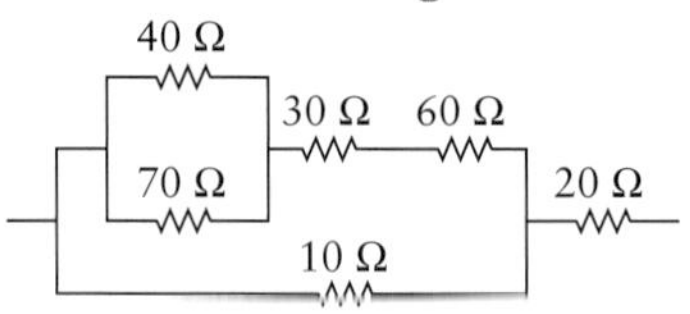

Figure 18-27 Problem 59

60. •• A metal wire of resistance 48 Ω is cut into four equal pieces that are then connected side by side to form a new wire which is one-quarter of the original length. What is the resistance of the new wire? Example 18-8

61. •• Two resistors A and B are connected in series to a 6.0-V battery; the voltage across resistor A is 4.0 V. When A and B are connected in parallel across a 6.0-V battery, the current

through B is 2.0 A. What are the resistances of A and B? The batteries have negligible internal resistance. Example 18-8

62. •• A potential difference of 7.50 V is applied between points a and c in **Figure 18-28**. (a) Find the difference in potential between points b and c. (b) Is the current through the 60-Ω resistor larger or smaller than that through the 35-Ω resistor? Why? Example 18-8

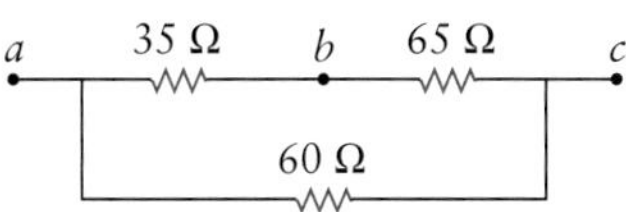

Figure 18-28 Problem 62

18-6 The rate at which energy is produced or taken in by a circuit element depends on current and voltage

63. • A heater is rated at 1500 W. How much current does it draw when it is connected to a 120-V voltage source? SSM Example 18-9

64. • A 4.0-Ω resistor is connected to a 12-V voltage source. What is the power dissipated by the resistor? Example 18-9

65. • How much power is used by a 12-Ω coffeemaker that draws a current of 15 A? Example 18-9

66. •• When connected in parallel across a 120-V source, two light bulbs consume 60 and 120 W, respectively. What powers do the light bulbs consume if instead they are connected in series across the same source? Assume the resistance of each light bulb is constant. Example 18-10

67. • A 40-Ω transmission line carries 1200 A. Calculate the rate at which electrical energy is converted to thermal energy due to resistance. Example 18-9

68. • A stereo speaker has a resistance of 8.0 Ω. If the power output is 40 W, calculate the current passing through the speaker wires. Example 18-9

69. •• Three resistors are connected to a battery with voltage $V_b = 9.00$ V, as shown in **Figure 18-29**. The power dissipated by each resistor is 1.50 W. What is the resistance of each resistor? Example 18-10

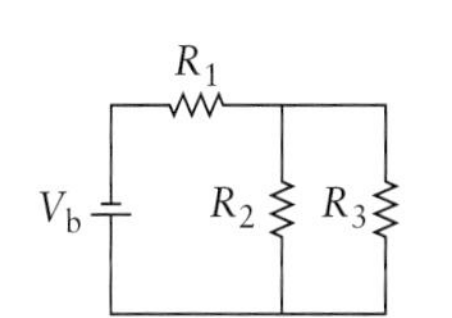

Figure 18-29 Problem 69

70. • A 10.0-V battery is connected in series with two resistors $R_1 = 10.0$ Ω and $R_2 = 40.0$ Ω (**Figure 18-30**). Calculate the power dissipated in each of the resistors. Example 18-10

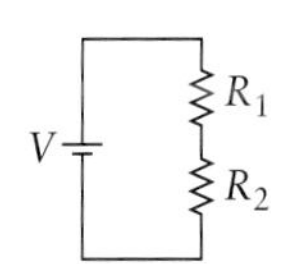

Figure 18-30 Problem 70

71. •• A battery with voltage $V_b = 12.0$ V is connected to resistors $R_1 = 8.00$ Ω, $R_2 = 10.0$ Ω, and $R_3 = 12.0$ Ω as shown in **Figure 18-31**. Calculate the total power provided by the battery and the power dissipated by each of the three resistors. Example 18-10

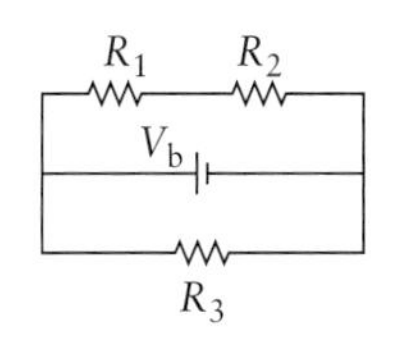

Figure 18-31 Problem 71

18-7 A circuit containing a resistor and capacitor has a current that varies with time

72. • A 4.00-MΩ resistor and a 3.00-μF capacitor are connected in series with a power supply. What is the time constant for the circuit? SSM Example 18-11

73. • A capacitor of 20.0 μF and a resistor of 1.00×10^2 Ω are quickly connected in series to a battery of 6.00 V. What is the charge on the capacitor 0.00100 s after the connection is made? Example 18-11

74. •• A 10.0-μF capacitor has an initial charge of 80.0 μC. If a 25.0-Ω resistor is connected across it, (a) what is the initial current in the resistor? (b) What is the time constant of the circuit? Example 18-11

75. •• A 12.5-μF capacitor is charged to 50.0 V, then discharged through a 75.0-Ω resistor. (a) How long after discharge begins will the capacitor have lost 90.0% of its initial (i) charge and (ii) energy? (b) What is the current through the resistor at both times in part (a)? Example 18-11

General Problems

76. • If your local power company charges \$0.11 per kW · h, what would it cost to run a 1500-W heater continuously during an 8.0-h night? SSM

77. •• A house is heated by a 24-kW electric furnace. The local power company charges \$0.10 per kW · h and the heating bill for January is \$218. How long must the furnace have been running on an average January day?

78. •• **Biology** A single ion channel is selectively permeable to K^+ and has a resistance of 1.0 GΩ. During an experiment the channel is open for approximately 1.0 ms while the voltage across the channel is maintained at +80.0 mV with a patch clamp. How many K^+ ions travel through the channel? Example 18-5

79. •• How much power is dissipated in each resistor shown in **Figure 18-33**? Example 18-10

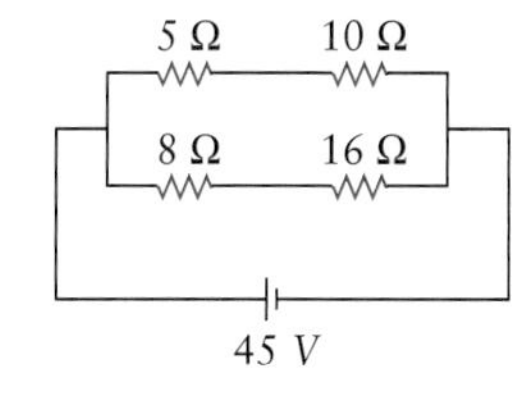

Figure 18-33 Problem 79

80. • Determine the current through each resistor in **Figure 18-34**. Example 18-8

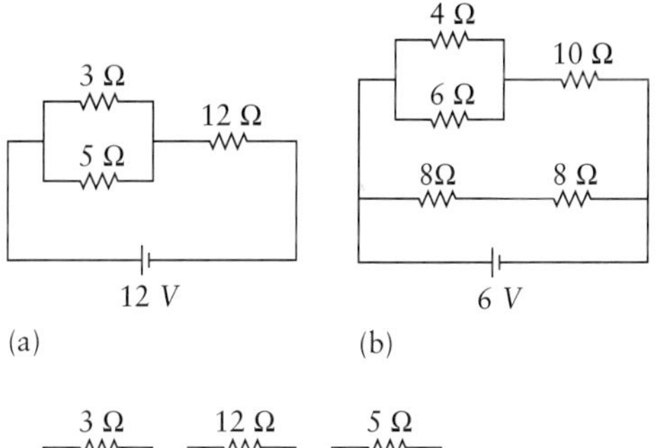

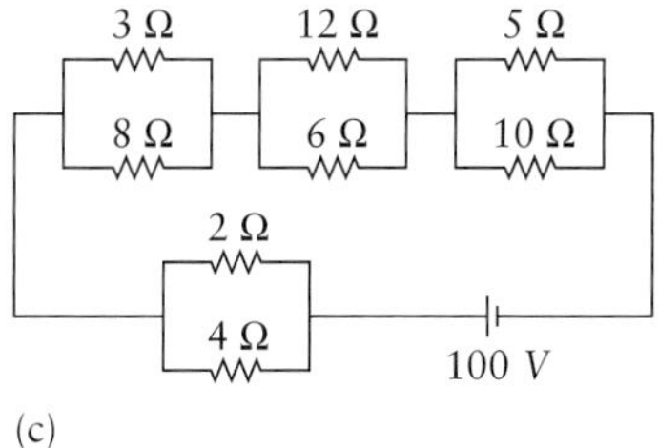

Figure 18-34 Problem 80

81. ••• An electric heater consists of a resistor connected across 110 V. It is used to heat 200.0 g of water from 20.0°C to 90.0°C in 2.70 min. (a) If 90% of the energy drawn from the power source goes into heating the water, what is the resistance of the heater? (b) How much longer will it take to heat your water if you have to power the water heater with your 12.0-V car battery? Example 18-9

82. ••• Lightning bolts can carry as much as 30 C of charge and can travel between a cloud and the ground in around 100 μs. Potential differences have been measured as high as 400 million volts. (a) What is the average current in such a lightning strike? How does it compare with typical household currents? (b) What is the resistance of the air during such a strike? (c) How much energy is transferred during such a strike? (d) What mass of water at 100°C could the lightning bolt evaporate? Example 18-9

83. •• In **Figure 18-35**, a potential difference of 5.00 V is applied between points a and b. Determine (a) the equivalent total resistance, (b) the current in each resistor, and (c) the power dissipated in each resistor. Example 18-10

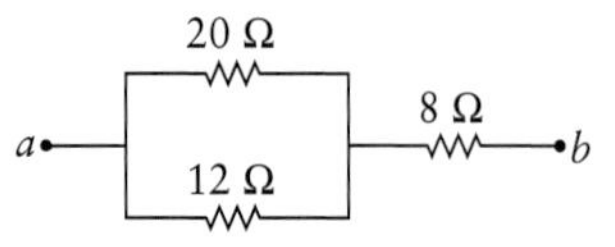

Figure 18-35 Problem 83

84. •• **Biology** Giant electric eels can deliver a voltage shock of 500 V and up to 1.0 A of current for a brief time. A snorkeler in salt water has a body resistance of about 600 Ω. A current of about 500 mA can cause heart fibrillation and death if it lasts too long. (a) What is the maximum power a giant electric eel can deliver to its prey? (b) If the snorkeler is struck by the eel, what current will pass through her body? Is this large enough to be dangerous? (c) What power does the snorkeler receive from the eel? Example 18-10

85. • **Biology** Most of the resistance of the human body comes from the skin, as the interior of the body contains aqueous solutions that are good electrical conductors. For dry skin, the resistance between a person's hands is typically 500 kΩ. The skin is on average about 2.0 mm thick. We can model the body between the hands as a cylinder 1.6 m long and 14 cm in diameter with the skin wrapped around it. What is the resistivity of the skin? Example 18-11

86. • A capacitor has a value of 160 μF. What is the resistance of the charging circuit if it takes 10.0 s to charge the capacitor to 80.0% of its maximum charge? Example 18-11

87. •• (a) How much time (in terms of the time constant RC) will it take before a discharging capacitor holds only 50.0% of the original charge? (b) What percentage of the original charge will the capacitor hold at a time that is 3 times the time constant? Example 18-11

88. •• A 3.0-μF capacitor is put across a 12-V battery. After it is fully charged the capacitor is disconnected and placed in series through an open switch with a 2.00×10^2-Ω resistor. (a) Determine the charge on the capacitor before it is discharged. (b) What is the initial current through the resistor when the switch is closed? (c) At what time will the current reach 37% of its initial value? Example 18-11

89. •• You need a 75-Ω resistor, but you have only a box of 50-Ω resistors on hand. (a) How can you make a 75-Ω resistor using the resistors you have on hand? (b) How could you use your 50-Ω resistors to make a 60-Ω resistor? Example 18-8

90. •• A probe designed to measure the voltage across a circuit element is connected across that element. The probe should change the resistance or capacitance of the circuit as little as possible. (a) When it is connected across a circuit element, is the probe in series or in parallel with that element? (b) Should a probe connected in this way have (i) a very large or a very small resistance? (ii) A very large or a very small capacitance? Explain your reasoning using the properties of series and parallel connections.

91. •• Batteries release their energy rather slowly and are very damaging environmentally. Capacitors would be much cleaner for the environment and can be quickly recharged. Unfortunately they don't store much energy. (a) A new 1.5-V AAA battery has a "capacity" (*not* capacitance) of 1250 mA · h. What does this "capacity" actually represent? Express it in standard SI units. (b) How many joules of energy can be stored in the AAA battery? (c) At a steady current of 400 mA, how many hours will the AAA battery last? (d) How much energy can be stored in a typical 10-μF capacitor charged to a potential of 1.5 V? How does that compare to the energy stored in the AAA battery?

92. •• An *ultracapacitor* is a very high-capacitance device designed so that the spacing of the plates is around 1000 times smaller than in ordinary capacitors. The plates also contain millions of microscopic nanotubes, which increase their effective area 100,000 times. (a) If the plate separation and effective area of a 10-μF capacitor are changed as described above to make an ultracapacitor, what is its new capacitance? (b) How much energy does the ultracapacitor store if charged to 1.5 V? Compare this to the energy stored in an ordinary 10-μF capacitor at 1.5 V. (c) Compare the energy stored in the ultracapacitor to that of the AAA battery in the previous problem. (d) If the ultracapacitor is to take 1.0 min to decrease to $1/e$ of its initial maximum charge, through what resistance must it discharge? Example 18-11

Magnetism

Peter Grant/Getty Images

In this chapter, your goals are to:

- (19-1) Recognize that magnetic forces can act over large distances.
- (19-2) Recognize that magnetism is fundamentally an interaction between moving electric charges.
- (19-3) Calculate the magnitude and direction of the magnetic force on a charged particle.
- (19-4) Describe how a mass spectrometer uses magnetic fields to sort ions according to their mass.
- (19-5) Calculate the magnetic force on a current-carrying wire.
- (19-6) Explain why a current loop in a uniform magnetic field experiences a net torque but zero net force.
- (19-7) Describe the principle of Ampère's law and how to use it.
- (19-8) Calculate the magnetic force that parallel current-carrying wires exert on each other.

To master this chapter, you should review:

- (5-6) Uniform circular motion.
- (8-6) The torque generated by a force.
- (16-5) The properties of electric field lines and the nature of electric dipoles.
- (18-2) The relationship between the current in a wire and the drift speed of charges in the wire.
- (18-6) Power into a current-carrying resistor.

What do you think?

The aurora borealis ("northern lights") is caused by fast-moving, electrically charged subatomic particles ejected from the Sun. Earth's magnetic field exerts a magnetic force on these particles that steers them toward our planet's north magnetic pole. When these particles enter Earth's upper atmosphere, they collide with the atoms there and cause the atoms to emit an eerie glow. (A similar effect in the southern hemisphere is called the aurora australis.) In what direction are these subatomic particles moving when the magnetic force on them is strongest? (a) In the same direction as the magnetic field; (b) opposite to the magnetic field; (c) perpendicular to the magnetic field; (d) either (a) or (b); (e) the magnetic force doesn't depend on the direction of motion.

19-1 Magnetic forces, like electric forces, act at a distance

If you rub a balloon on your head, then move the balloon a few centimeters away, the hairs on your head will stand up (see Figure 16-4). This is a result of *electric* forces. The rubbing transfers electrons between your hair and the balloon, giving one a net negative charge and the other a net positive charge; these attract each other, even over a distance, and this makes your hair stand up. The effect is rather feeble, however, and only the finest hairs respond noticeably to these electric forces.

Figure 19-1a shows another force that acts at a distance like the electric force but can be *much* stronger than that between the balloon and your hair. Certain objects called **magnets**, made of one of a handful of special materials such as iron, cobalt, and nickel, can exert strong **magnetic forces** on other magnets. (The name *magnet* comes from the region of ancient Greece known as Magnesia, where these objects were

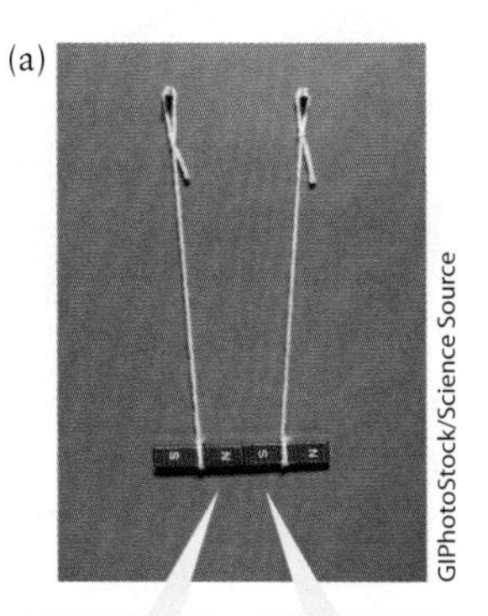

GIPhotoStock/Science Source
The north pole of the left-hand bar magnet...
...attracts the south pole of the right-hand bar magnet.

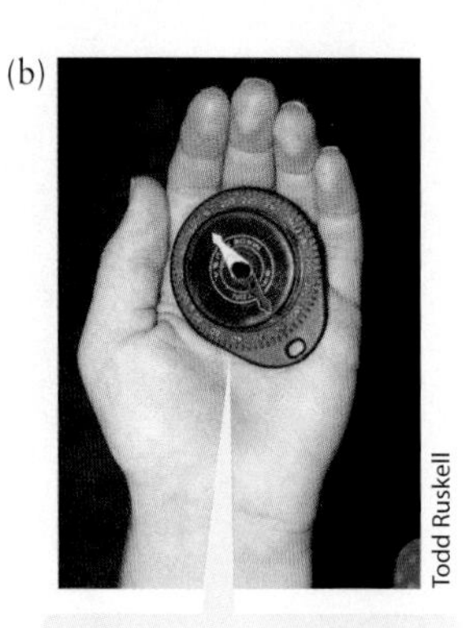

Todd Ruskell
The magnetized compass needle aligns with Earth's magnetic field.

Cordelia Molloy/Science Source
The magnetic field required to pick up pieces of iron at a scrap yard comes from an electromagnet—a device that generates a magnetic field when there is current through its wires.

Figure 19-1 Magnetic forces (a) The force between two bar magnets. (b) A compass needle interacting with Earth's magnetic field. (c) A large electromagnet.

discovered more than 2500 years ago.) Just as we use the umbrella term "electricity" to refer to interactions between electrically charged objects, we use the term "**magnetism**" to describe the interactions between magnets. One application of magnetic interactions is the compass (**Figure 19-1b**). Earth's core acts like a giant magnet and produces a *magnetic field* in the surrounding space. (We'll see that magnetic fields are analogous to electric fields, but with important differences.) The magnetic needle of a compass aligns with Earth's field and points north. Earth's magnetic field is relatively weak; the field produced by the electromagnet in **Figure 19-1c** is thousands of times stronger.

Our goal in this chapter is to understand magnetism. We begin in the next section by investigating the properties of magnets and realizing that they are created by moving charges. From there we develop a full understanding of magnetic forces, which requires us to understand two things: (1) the nature of the magnetic field that moving charges *produce* and (2) how moving charges *respond* to the magnetic fields produced by other moving charges. It turns out to be easiest to look at the second of these first. We'll analyze the magnetic force that acts on a moving charge placed in a magnetic field, as well as the magnetic force on a wire that carries a current (a collection of charges moving in the wire) and is placed in a magnetic field. Later in the chapter we'll see how to calculate the magnetic field produced by a given collection of charges in motion.

TAKE-HOME MESSAGE FOR Section 19-1

✔ Magnetic forces can act over large distances, much like electric forces.

19-2 Magnetism is an interaction between moving charges

Like the electric force, magnetic forces become stronger as the objects are moved closer together. However, the magnetic force is not simply a form of the electric force. The attracting magnets in Figure 19-1a are *not* electrically charged: Their atoms are made of positively charged nuclei and negatively charged electrons, but their net charges are zero. Any electric forces between these magnets are very weak and are not responsible for the strong attraction shown in Figure 19-1a.

As we did for the electric force, we can explain how the magnetic force acts at a distance by invoking the idea of a field. Just as an electrically charged object sets up an electric field in the space around it, a magnet sets up a **magnetic field** in the space around it (**Figure 19-2a**). A second magnet placed in this field experiences a magnetic force that depends on the magnitude and direction of the field. We use the symbol $\vec{B}$ for magnetic field. Unlike the electric field $\vec{E}$ of a point charge, which points either directly away from or directly toward the charge, the magnetic field $\vec{B}$ of a magnet points away from one end of a magnet toward the other end. These two ends are called the **magnetic poles** of the magnet, one of which is called the *north pole* and the other the *south pole*. The meaning of the names is that Earth itself acts like a giant magnet. If allowed to swing freely, a magnet will orient itself so that its north pole is toward Earth's magnetic north pole and its south pole is toward Earth's magnetic south pole (**Figure 19-2b**).

By convention we choose the direction of the magnetic field $\vec{B}$ produced by a magnet to be such that $\vec{B}$ points away from the magnet's north pole and toward the magnet's south pole. This should remind you of an electric dipole (see Section 16-5) made up of a positive charge $+q$ and a negative charge $-q$, for which the electric field

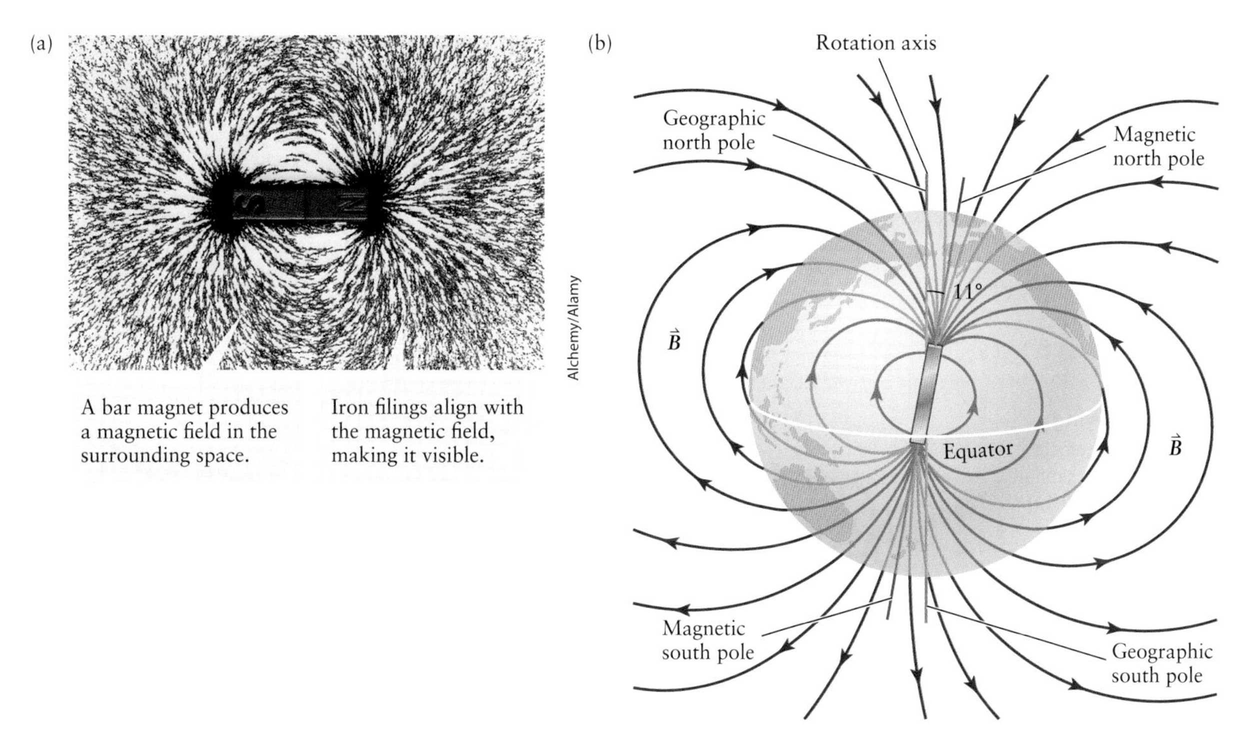

Figure 19-2 The magnetic fields of a bar magnet and Earth (a) The magnetic field of a bar magnet points out of its north pole (colored red) and into its south pole (colored black). (b) Earth's magnetic field has a similar pattern. Although Earth's field is produced in a different way—by electric currents in the liquid portion of our planet's interior—the field is much the same as if there were a giant bar magnet inside Earth. This "bar magnet" is tipped by about 11° from Earth's rotation axis, which is why the magnetic north and south poles are not at the same locations as the true, or geographic, poles. A compass needle points toward the north magnetic pole, not the true north pole.

$\vec{E}$ points away from the positively charged end and toward the negatively charged end. Indeed, a magnet like that shown in Figure 19-2a is often called a **magnetic dipole**.

The magnetic forces between magnets can be either attractive or repulsive. If the north pole of one magnet is close to the south pole of a second magnet, the force attracts the two magnets toward each other (**Figure 19-3a**); if the north poles are close together or the south poles are close together, the force makes the two magnets repel each other (**Figure 19-3b**). This is very different from the behavior of electrically charged objects but analogous to the way in which electric dipoles interact. However, there is an important distinction between electric and magnetic dipoles. You can take an electric dipole apart by separating its component positive and negative charges, but you *cannot* separate the north and south poles of a magnet. If you cut a magnet in half, each half has a north pole and a south pole (**Figure 19-3c**). The same is true no matter how many small pieces you cut a magnet into; a very small piece produces only a weak magnetic field, but is still a magnetic dipole with both a north pole and a south pole. (Physicists have speculated about the existence of *magnetic monopoles*, particles that have the properties of an isolated north or south magnetic pole. Many experiments have been performed to look for evidence of a magnetic monopole. None has been found.)

What is it about magnets that causes them to produce and respond to magnetic fields? And why is it impossible to separate their north and south poles? The answers to these questions were revealed by a set of crucial experiments in the nineteenth century. **Figure 19-4a** shows a version of one of these experiments. A coil of copper wire loops through a flat piece of clear plastic, and iron filings are spread over the plastic. Since copper is not magnetic, the filings lie wherever they were placed. But when there is an electric current through the coil, the filings line up just as they do around the magnet in Figure 19-2. This demonstrates that the moving charges that make up the current in the coil *produce* a magnetic field. The field points out of the coil at one end, which is the "north pole" of the coil, and points into the other end, which is the coil's "south pole."

We can verify that the coil produces a magnetic field by carrying out the experiment shown in **Figure 19-4b**, in which we've replaced one of the two magnets from Figure 19-3a with a coil. When the current in the coil is turned on, the coil and the magnet are attracted to each other. This can happen only if the current-carrying coil is

(a) These magnets attract.

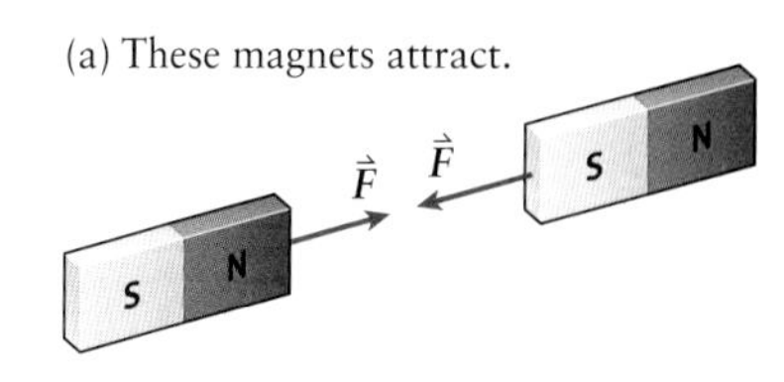

(b) These magnets repel.

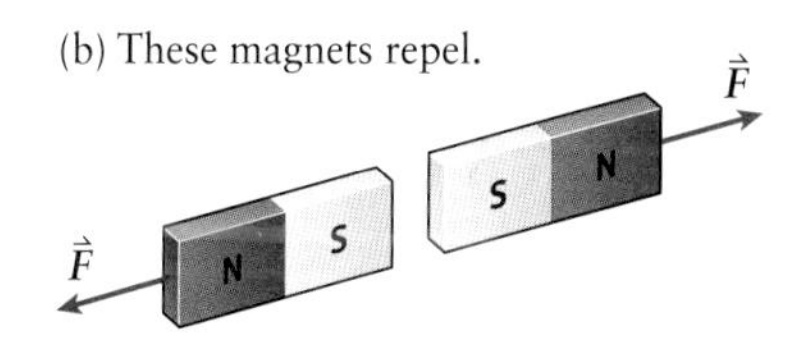

(c) If we cut a magnet in half, each half has a north and south pole.

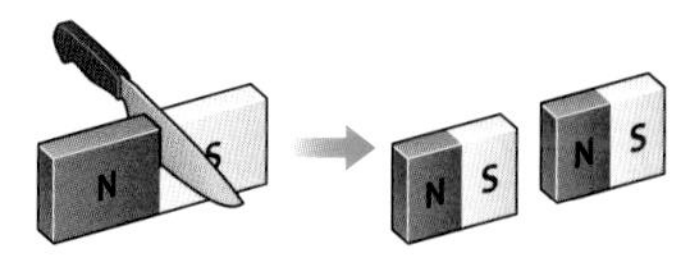

Figure 19-3 Bar magnets and magnetic poles (a) The opposite poles of these magnets attract. (b) Like poles of these magnets repel. (c) Every magnet has both a north and south pole; they cannot be separated.

(a) The magnetic field created by current in a coil.

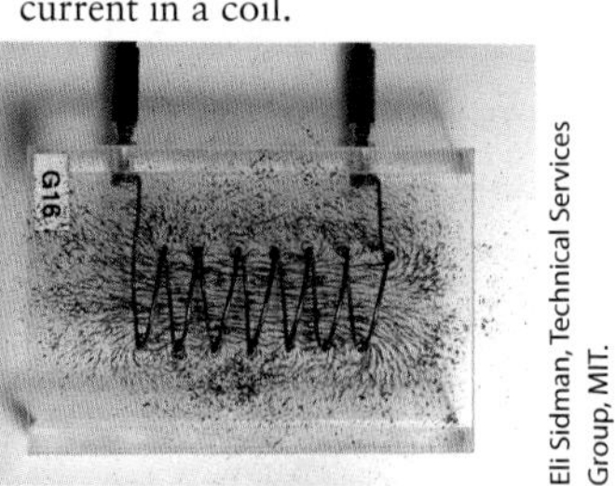

Eli Sidman, Technical Services Group, MIT.

(b) This coil and magnet attract.

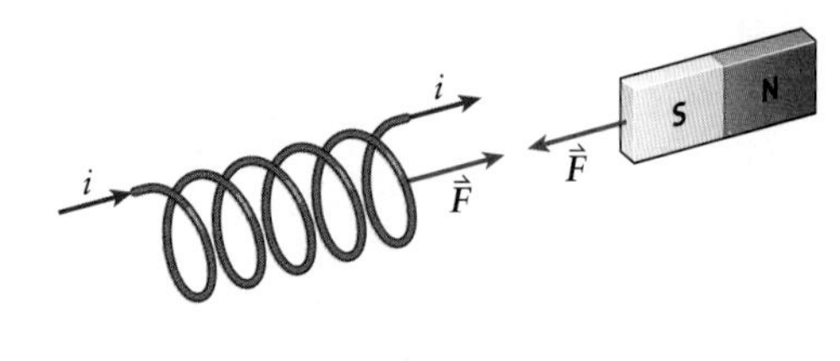

(c) This coil and magnet repel.

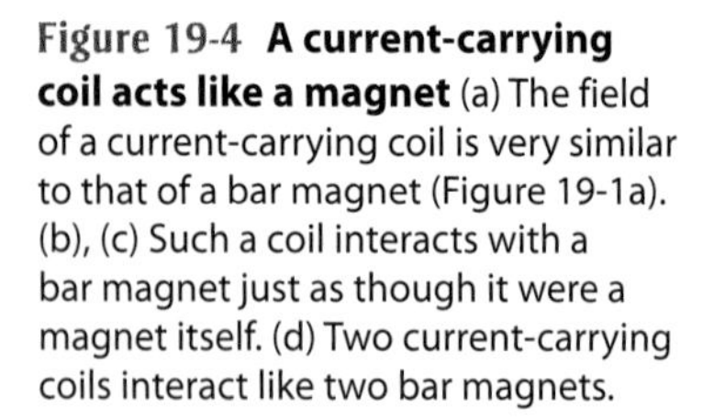

(d) These coils attract, just like two magnets.

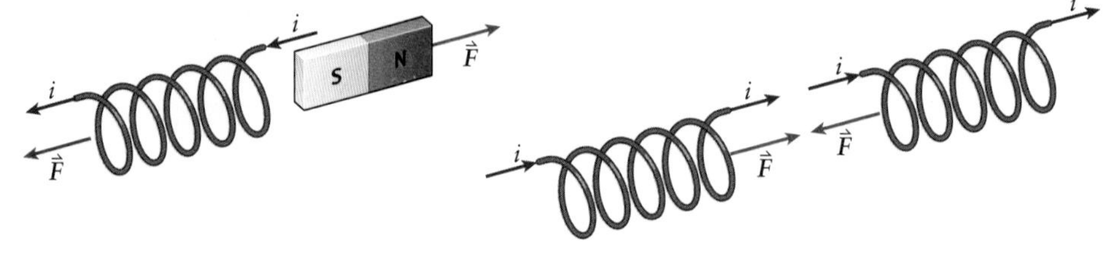

Figure 19-4 A current-carrying coil acts like a magnet (a) The field of a current-carrying coil is very similar to that of a bar magnet (Figure 19-1a). (b), (c) Such a coil interacts with a bar magnet just as though it were a magnet itself. (d) Two current-carrying coils interact like two bar magnets.

indeed producing a magnetic field, and the magnet is responding to that field. If we flip the coil over, as shown in **Figure 19-4c**, the coil and the magnet now repel each other. This is just what would happen if the coil were an iron magnet that we flipped over to interchange the north and south poles (see Figures 19-3a and 19-3b).

In order for the coil in Figures 19-4b and 19-4c to be attracted to or repelled from the magnet, it must also be true that the moving charges that make up the current in the coil respond to the magnetic field of the magnet. In other words, a current-carrying coil and a magnet are fundamentally the same, in that both objects *produce* as well as *respond to* magnetic fields. **Figure 19-4d** shows an experiment that verifies this: Two current-carrying coils with the same orientation attract each other, just as do two magnets with the same orientation.

The series of experiments shown in Figure 19-4 suggests that *magnetism is an interaction between charges in motion.* In order for two objects to exert magnetic forces on each other, there must be moving charges in both objects. We use the umbrella term "**electromagnetism**" to include both electricity and magnetism, since both involve interactions between charges. The distinction between electricity and magnetism is that two charges exert *electric* forces on each other whether or not the charges are moving, while two charges exert *magnetic* forces on each other only if both charges are in motion. (The objects in the experiments shown in Figure 19-4 are all electrically neutral. So there are no electric forces in those experiments, only magnetic forces.)

You may be wondering if magnetism must always involve charges in motion, since the ordinary magnets shown in Figures 19-1, 19-2, and 19-3 are not attached to sources of emf and so do not carry currents. The answer is that there *are* charges in motion inside an ordinary iron magnet: The moving charges are the electrons within the atoms of the magnet, and their motions within the atom are like that of electrons in the circular coils of Figure 19-4. Each individual atom in a magnet produces only a tiny magnetic field. But because a large fraction of the atoms that make up the magnet are oriented so that their electron motions are in the same direction as in the surrounding atoms, their magnetic fields add together to make a substantial total field. This explains why cutting a magnet into pieces doesn't leave you with a separate north pole and south pole: Each piece is simply a smaller version of the original magnet.

GOT THE CONCEPT? 19-1 Electric and Magnetic Forces

? Two identical objects each have a positive charge Q. One of the objects is held in place, while the other object is in motion a short distance from the first object. What kinds of forces do the two objects exert on each other?

(a) Both electric and magnetic forces; (b) electric forces but not magnetic forces; (c) magnetic forces but not electric forces; (d) neither electric forces nor magnetic forces; (e) answer depends on how the second object is moving.

TAKE-HOME MESSAGE FOR Section 19-2

✔ Magnetism is an interaction between moving charged particles. A moving charged particle produces a magnetic field, and a moving charged particle in a magnetic field experiences a force. A stationary charged particle neither produces nor responds to a magnetic field.

✔ In a magnet, charges are in motion at the atomic level to produce a magnetic field.

19-3 A moving point charge can experience a magnetic force

We'll begin our study of magnetic forces by considering the force on a single charged particle moving in a magnetic field $\vec{B}$. For example, this could be an electron moving in a current-carrying wire in the vicinity of a magnet, or a moving proton in Earth's upper atmosphere that is acted on by our planet's magnetic field. Here's what experiments tell us about the magnetic force on such a moving charged particle:

(1) A charged particle can experience a force when placed in a magnetic field, but *only when it is moving.*
(2) The magnetic force depends on the direction of the charged particle's velocity relative to the direction of the magnetic field. The charged particle does not experience a magnetic force when its velocity is parallel to or opposite to the magnetic field.
(3) The magnitude of the force is proportional to the charge on the particle, to the magnitude of the magnetic field, and to the speed of the particle.
(4) The direction of the force is perpendicular to both the direction of motion of the charged particle and the direction of the magnetic field.
(5) The direction of the force depends on the sign of the charge.

Figure 19-5a shows the magnetic force on a particle with positive charge q moving with velocity $\vec{v}$ in the presence of a magnetic field $\vec{B}$. If θ is the angle

(a) Magnetic force on a positive charge

Your right hand helps determine the direction of the magnetic force.

$\vec{F}$ $\vec{B}$ θ $q > 0$ $\vec{v}$

The force $\vec{F}$ is perpendicular to both the velocity $\vec{v}$ and the magnetic field $\vec{B}$.

(b) Magnetic force on a negative charge

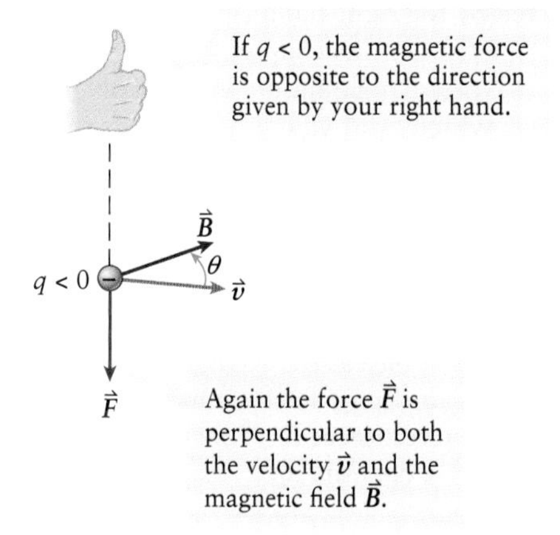

If $q < 0$, the magnetic force is opposite to the direction given by your right hand.
$\vec{B}$
θ
$q < 0$
$\vec{v}$
$\vec{F}$
Again the force $\vec{F}$ is perpendicular to both the velocity $\vec{v}$ and the magnetic field $\vec{B}$.

(c) Magnetic force on a positive charge, $\theta = 90°$

For a given speed v, the magnetic force is greatest if the velocity $\vec{v}$ is perpendicular to the magnetic field $\vec{B}$.

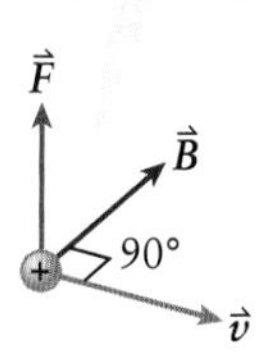

(d) Magnetic force on a positive charge, $\theta = 0$ or 180°

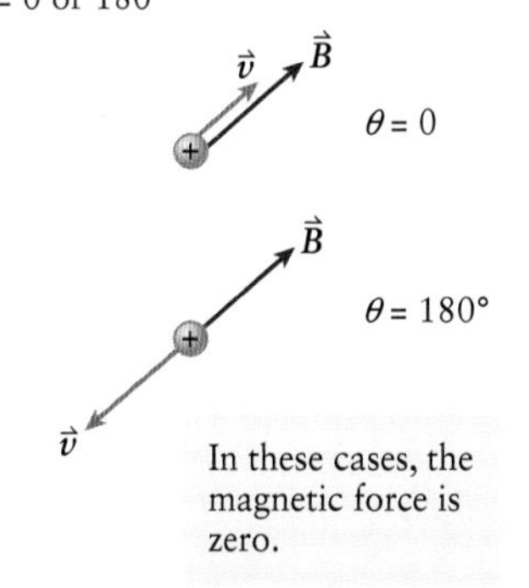

Figure 19-5 Magnetic force on a moving charged particle The direction of the magnetic force on a particle with charge q depends on whether (a) q is positive or (b) q is negative. The magnitude of the force $F = |q|vB \sin\theta$ goes from (c) a maximum value if $\theta = 90°$ to (d) zero if $\theta = 0$ or $\theta = 180°$.

between the velocity of the charged particle and the magnetic field, the magnitude of the magnetic force $\vec{F}$ is

Magnitude of magnetic force on a moving charged particle (19-1)

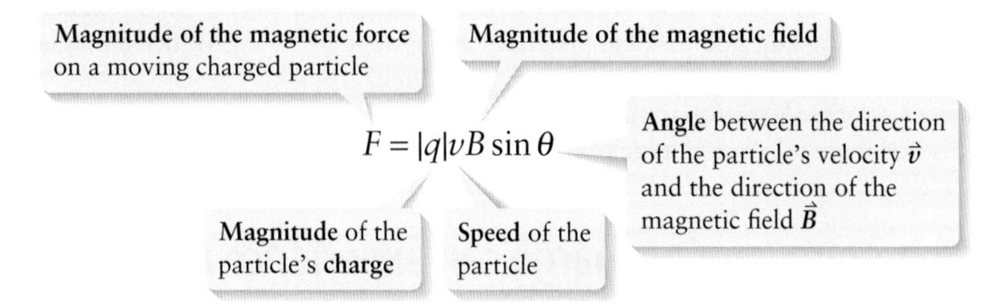

$$F = |q|vB \sin\theta$$

The SI units of magnetic field are **tesla** (T). In Equation 19-1, if charge q is in coulombs (C), speed v is in meters per second (m/s), and magnetic field magnitude B is in tesla (T), the force F is in newtons (N). So 1 T is the same as $1\ \text{N}\cdot\text{s}/(\text{C}\cdot\text{m})$ or $1\ \text{N}/(\text{A}\cdot\text{m})$. (Recall that one ampere equals one coulomb per second: $1\ \text{A} = 1\ \text{C/s}$.) The strongest magnetic field most of us will experience directly is the roughly 2-T field inside a magnetic resonance imaging (MRI) scanner. Other magnetic fields in our environment are much weaker: The strength of Earth's magnetic field is about 5×10^{-5} T, the electrical impulses that drive the contraction and relaxation of your heart produce magnetic fields of about 5×10^{-11} T, and the magnetic fields in your brain are on the order of 10^{-13} T. An alternative unit for magnetic field is the *gauss* (G): $1\ \text{G} = 10^{-4}$ T.

Here's a **right-hand rule** for determining the direction of the magnetic force on a moving charge. First extend the fingers of your right hand along the direction of the velocity $\vec{v}$ and orient your hand so that your palm is facing in the direction that the magnetic field $\vec{B}$ points. Then extend your right thumb so that it is perpendicular to the four fingers of your right hand and swivel your right hand so that the fingers now point in the direction of $\vec{B}$. If the particle has a positive charge, so $q > 0$, your thumb points in the direction of the force $\vec{F}$. If the particle has a negative charge, so $q < 0$ (**Figure 19-5b**), the force $\vec{F}$ points in the direction opposite to that given by the right-hand rule. In either case, the force $\vec{F}$ is perpendicular to the velocity $\vec{v}$ *and* perpendicular to the magnetic field $\vec{B}$. If we draw the vectors $\vec{v}$ and $\vec{B}$ with their tails together, they form a plane; the force $\vec{F}$ is perpendicular to that plane, with a direction given by the right-hand rule. (We will encounter a number of other right-hand rules for magnetism in this chapter.)

Note that Equation 19-1 involves the magnitude or absolute value of the charge, $|q|$, which is always positive. So the magnitude F of the magnetic force does *not* depend on whether the charge is positive or negative.

WATCH OUT! The magnetic force on a moving charged particle is never in the same direction as the magnetic field.

From Chapter 16 you're used to the idea that the electric force on a particle with charge q placed in an electric field $\vec{E}$ is $\vec{F}_{\text{electric}} = q\vec{E}$. This force is in the same direction as $\vec{E}$ if q is positive; if q is negative, it is in the direction opposite to $\vec{E}$. Figures 19-5a and 19-5b show that the magnetic force on a moving charged particle is very different: This force is *perpendicular* to the direction of the magnetic field. Electric and magnetic forces are quite different from each other!

Go to Interactive Exercises 19-1 and 19-2 for more practice dealing with magnetic forces.

Equation 19-1 tells us that the magnitude of the magnetic force on a moving charged particle depends on the angle θ between the directions of $\vec{v}$ and $\vec{B}$. For a given particle speed v, the magnetic force has its greatest magnitude if the velocity $\vec{v}$ is perpendicular to the magnetic field $\vec{B}$; then $\theta = 90°$, $\sin\theta = 1$, and $F = |q|vB$ (**Figure 19-5c**). If $\theta = 0$ or $\theta = 180°$, so the particle is moving either in the same direction as $\vec{B}$ or in the direction opposite to $\vec{B}$, $\sin\theta = 0$ and the force on the particle is *zero*. Thus a charged particle moving in the direction of the magnetic field, or in the direction opposite to the magnetic field, experiences no magnetic force (**Figure 19-5d**).

We've shown the vectors $\vec{v}$, $\vec{B}$, and $\vec{F}$ in perspective in Figure 19-5. Drawing in perspective isn't easy to do, so we'll often use a simple convention for drawing vectors that are pointed either into or out of the pages of this book (**Figure 19-6**). Think of a vector as an arrow, with a sharp point at its head and feathers at its tail. If a vector is directed perpendicular to the plane of the page and pointed toward you—that is, *out of* the page—we'll depict it with the symbol ⊙. The dot in the center of this symbol represents the sharp point of the arrowhead. If a vector is directed perpendicular to the plane of the page and pointed away from you—that is, *into* the page—we'll depict it with the symbol ⊙. The "X" in the center of this symbol represents the feathers on the tail of the arrow.

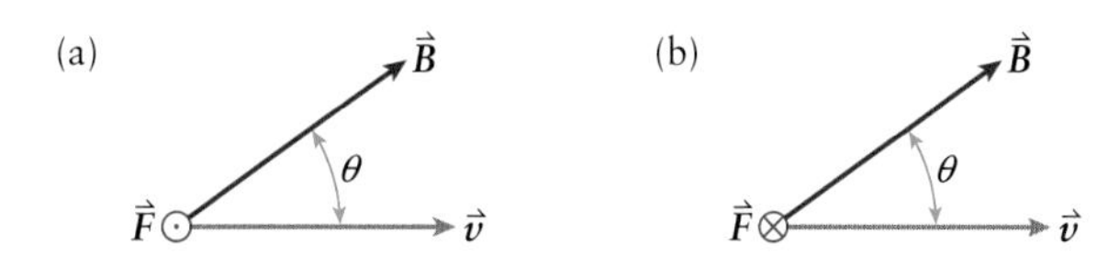

If $q > 0$, the force vector in this situation points out of the page (as indicated by the circle with a dot).

If $q < 0$, the force vector in this situation points into the page (as indicated by the circle with an X).

Figure 19-6 Vectors out of the page and into the page We use the symbols ⊙ and ⊗ to denote vectors that point out of the page and into the page, respectively.

EXAMPLE 19-1 Magnetic Forces on a Proton and an Electron

At a location near our planet's equator, the direction of Earth's magnetic field is horizontal (that is, parallel to the ground) and due north, and the magnitude of the field is 2.5×10^{-5} T. Find the direction and magnitude of the magnetic force on a particle moving at 1.0×10^4 m/s if the particle is (a) a proton moving horizontally and due east, (b) an electron moving horizontally and due east, and (c) a proton moving horizontally in a direction 25° east of north. Recall that a proton has charge $e = 1.60 \times 10^{-19}$ C and an electron has charge $-e$.

Set Up

In each case we'll use Equation 19-1 to find the magnitude of the force on the moving proton or electron. The right-hand rule (Figure 19-5) will tell us the direction of the magnetic force.

Magnetic force on a moving charged particle:

$$F = |q|vB \sin\theta \qquad (19\text{-}1)$$

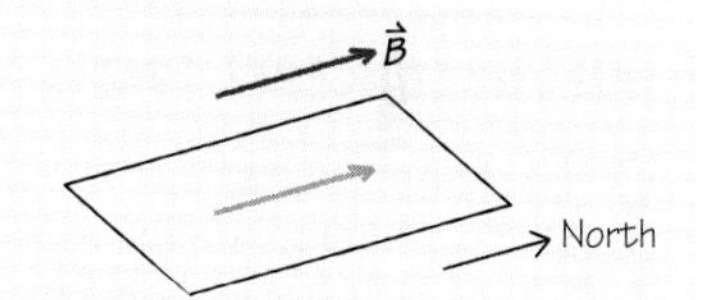

Solve

(a) The velocity vector $\vec{v}$ of the proton is perpendicular to the magnetic field vector $\vec{B}$, so in Equation 19-1 the angle $\theta = 90°$ and $\sin\theta = 1$. The vectors $\vec{v}$ and $\vec{B}$ both lie in a horizontal plane; the magnetic force vector $\vec{F}$ must be perpendicular to this plane, so $\vec{F}$ is vertical. The right-hand rule tells us that the force points vertically upward.

Charge on the proton:

$$q = e = 1.60 \times 10^{-19}\ \text{C}$$

The proton velocity vector is perpendicular to the magnetic field vector ($\theta = 90°$), so the magnitude of the magnetic force on the proton is

$$\begin{aligned} F &= |q|vB \sin\theta = evB \sin\theta \\ &= (1.60 \times 10^{-19}\ \text{C})(1.0 \times 10^4\ \text{m/s})(2.5 \times 10^{-5}\ \text{T}) \sin 90° \\ &= 4.0 \times 10^{-20}\ \text{N} \end{aligned}$$

(b) The electron has the same magnitude of charge as the proton and the same velocity, so it experiences the same magnitude of magnetic force. But its charge is negative, so the direction of the magnetic force $\vec{F}$ on the electron is *opposite* to the direction given by the right-hand rule. Hence the force $\vec{F}$ points vertically downward.

Charge on the electron:

$$q = -e = -1.60 \times 10^{-19}\ \text{C}$$

The electron velocity vector is perpendicular to the magnetic field vector ($\theta = 90°$), so the magnitude of the magnetic force on the electron is

$$\begin{aligned} F &= |q|vB \sin\theta = evB \sin\theta \\ &= (1.60 \times 10^{-19}\ \text{C})(1.0 \times 10^4\ \text{m/s})(2.5 \times 10^{-5}\ \text{T}) \sin 90° \\ &= 4.0 \times 10^{-20}\ \text{N} \end{aligned}$$

(c) The proton velocity $\vec{v}$ and the magnetic field $\vec{B}$ again both lie in a horizontal plane, but now the angle between these vectors is $\theta = 25°$. So the magnitude of the magnetic force is less than in part (a). The direction of the force is the same as in part (a), however.

The proton velocity vector is at an angle $\theta = 25°$ to the magnetic field vector, so the magnitude of the magnetic force on the proton is

$$F = |q|vB \sin\theta = evB \sin\theta$$
$$= (1.60 \times 10^{-19}\ \text{C})(1.0 \times 10^4\ \text{m/s}) \times (2.5 \times 10^{-5}\ \text{T}) \sin 25°$$
$$= 1.7 \times 10^{-20}\ \text{N}$$

This is smaller than in part (a) because $\sin 25° = 0.42$ compared to $\sin 90° = 1$.

Reflect

This example illustrates how the magnetic force on a moving charged particle depends on the direction in which the particle is moving. Note that the force magnitudes in parts (a), (b), and (c) are very small because a single electron or proton carries very little charge. These particles also have very small mass, however (1.67×10^{-27} kg for the proton and 9.11×10^{-31} kg for the electron), so the resulting *accelerations* are tremendous. Can you use Newton's second law to show that the proton in part (a) and the electron in part (b) have accelerations of $2.4 \times 10^7\ \text{m/s}^2$ and $4.4 \times 10^{10}\ \text{m/s}^2$, respectively?

GOT THE CONCEPT? 19-2 A Proton in a Magnetic Field

A proton is fired into a region of uniform magnetic field pointing into the page (**Figure 19-7**). The proton's initial velocity is shown by the blue vector, which lies in the plane of the page. In which direction does the proton's trajectory bend? (a) Toward the top of the figure; (b) toward the bottom of the figure; (c) out of the figure; (d) into the figure; (e) answer depends on the speed of the proton.

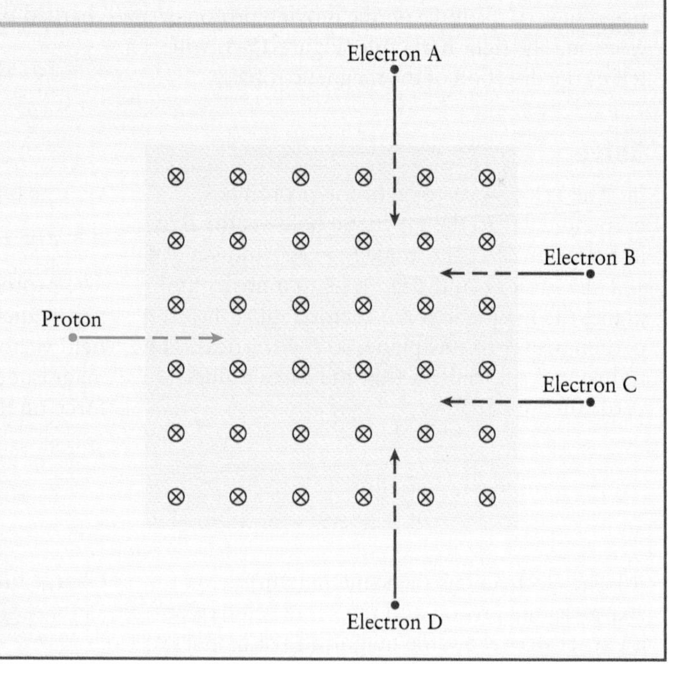

Figure 19-7 Charged particles in a magnetic field How will each particle be deflected when it enters this region of uniform magnetic field?

GOT THE CONCEPT? 19-3 Electrons in a Magnetic Field

Four electrons, A, B, C, and D, are fired into a region of uniform magnetic field pointing into the page (Figure 19-7). The initial velocities of the electrons are shown by the red vectors, which lie in the plane of the page. Each electron will follow a trajectory that is bent by the magnetic force. For which electron could that trajectory lead to the point on the left-hand side where the proton enters the field region? (There may be more than one correct answer.) (a) Electron A; (b) electron B; (c) electron C; (d) electron D; (e) none of the electrons.

TAKE-HOME MESSAGE FOR Section 19-3

✔ A charged particle moving in a magnetic field can experience a magnetic force. The magnitude of this force depends on the angle θ between the particle's velocity $\vec{v}$ and the direction of the magnetic field $\vec{B}$. The force is maximum if $\theta = 90°$ ($\vec{v}$ and $\vec{B}$ are perpendicular) and zero if $\theta = 0$ or $180°$ ($\vec{v}$ and $\vec{B}$ are either in the same direction or opposite directions).

✔ The magnitude of the magnetic force on a charged particle is proportional to the amount of charge, to the speed of the charged particle, and to the magnitude of the magnetic field.

✔ The direction of the magnetic force is perpendicular to both the particle's velocity $\vec{v}$ and the direction of the magnetic field $\vec{B}$ and depends on the sign of the charge. A right-hand rule helps tell us the force direction.

✔ The SI unit of magnetic field is the tesla (T).

19-4 A mass spectrometer uses magnetic forces to differentiate atoms of different masses

An important application of the magnetic force on a moving charged particle is the **mass spectrometer**, a device used by chemists to determine the masses of individual atoms and molecules. **Figure 19-8** shows a simplified version of one type of mass spectrometer. A sample to be analyzed is first vaporized, and its atoms and molecules are ionized by removing one or more electrons so all of the ions have a positive charge. A beam of the moving ions then passes through a *velocity selector*, a region where there is *both* an electric field $\vec{E}$ and a magnetic field $\vec{B}$. The fields $\vec{E}$ and $\vec{B}$ are oriented perpendicular to the ion beam as well as perpendicular to each other. You should be able to show in Figure 19-8 that in this region the electric force on a positive ion acts to the left while the magnetic force acts to the right. An ion will pass through this region without deflection only if these two forces cancel so that the *net* force on the ion is zero. For an ion of positive charge q, the electric force has magnitude $F_{electric} = qE$, and the magnetic force has magnitude $F_{magnetic} = qvB \sin 90° = qvB$ (the ion velocity and the magnetic field are perpendicular, so $\theta = 90°$ in Equation 19-1). So the condition for an ion to continue through this region without being deflected is

$$qE = qvB \quad \text{or} \quad v = \frac{E}{B} \tag{19-2}$$

(speed of ions emerging from a velocity selector)

If an ion has a speed different from $v = E/B$, it will be deflected out of the beam. This is why we call this arrangement of fields a velocity selector: After passing through the $\vec{E}$ and $\vec{B}$ fields, the only ions that remain in the beam are those with a speed given by Equation 19-2.

As Figure 19-8 shows, the ion beam then passes into a second region with a magnetic field as before but no electric field. The beam deflects to the right as a result of the magnetic force and continues to deflect. That's because as the direction of the velocity vector changes, the direction of the magnetic force also changes in order to remain perpendicular to the ion velocity $\vec{v}$. Because $\vec{v}$ and $\vec{B}$ remain perpendicular, the angle θ in Equation 19-1 remains equal to 90° as the ions deflect and the magnitude of the magnetic force remains the same, $F_{magnetic} = qvB \sin 90° = qvB$. You probably recognize that what's going on here is exactly the situation that we described in Section 5-6. Because an ion is acted on by a force of constant magnitude that points perpendicular to the ion's velocity, the ion moves at a constant speed in a circular path—that is, in uniform circular motion. We can use Newton's second law to find the radius r of the circular path followed by an ion of charge q and mass m that moves with speed v. The

(3) In this region ions follow a circular path under the influence of a magnetic field. For ions with the same charge, the trajectory depends on the ion mass.
$\odot\vec{B}$
Aperture
Low-mass ion
High-mass ion
Detector
$\vec{v}$
$\vec{E}$
$\vec{B}\odot$
(2) Only ions with speed $v = E/B$ pass through the aperture: Slow ions are deflected to the left of the aperture, fast ions to the right.
(1) Positive ions move upward through the velocity selector. A uniform magnetic field points out of the plane of the figure.

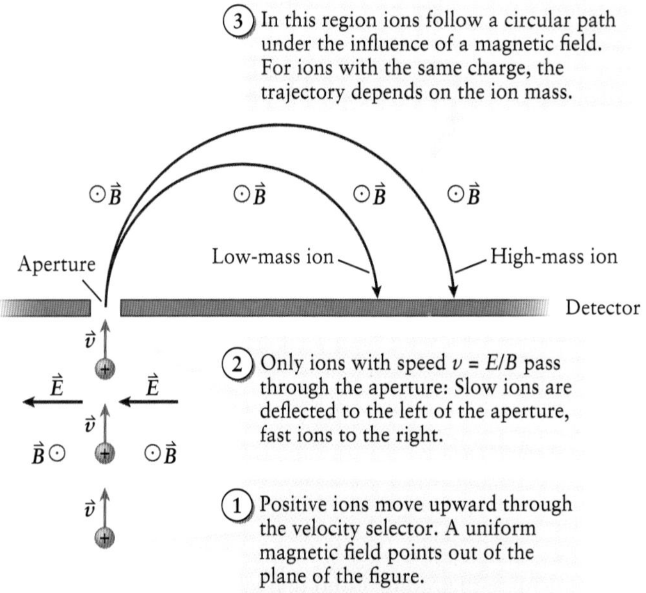

Figure 19-8 **A mass spectrometer** Electric and magnetic forces are used in this device to separate different ions according to their mass.

acceleration in uniform circular motion has magnitude $a = v^2/r$, and the net force on the ion has magnitude qvB, so

(19-3) $$qvB = \frac{mv^2}{r} \quad \text{or} \quad r = \frac{mv^2}{qvB} = \frac{mv}{qB}$$

(radius of circular path followed by charged particles in a uniform magnetic field)

Equation 19-3 is the key to understanding how the mass spectrometer works. All of the ions enter the aperture in Figure 19-8 moving at the same speed v (thanks to the velocity selector), and all are exposed to the same magnetic field of the same magnitude B. But ions of different mass m (and the same charge q) will follow paths with different radii and so will land at different locations on the detector in Figure 19-8. By counting how many ions land at each location, we can learn about the composition of the sample being analyzed. Example 19-2 illustrates how this works.

EXAMPLE 19-2 Measuring Isotopes with a Mass Spectrometer

Most oxygen atoms are the isotope ^{16}O ("oxygen-16"), which contains eight electrons, eight protons, and eight neutrons. The mass of this atom is 16.0 u, where 1 u = 1 atomic mass unit = 1.66×10^{-27} kg. The second most common isotope is ^{18}O ("oxygen-18"), which has two additional neutrons and so is more massive than an atom of ^{16}O: Each atom has a mass of 18.0 u. To determine the relative abundances of ^{16}O and ^{18}O, you send a beam of singly ionized oxygen atoms ($^{16}O^+$ and $^{18}O^+$, each with a charge $+e = +1.60 \times 10^{-19}$ C) through a mass spectrometer like that shown in Figure 19-8. The magnetic field strength in both parts of the spectrometer is 0.0800 T, and the electric field strength in the velocity selector is 4.00×10^3 V/m. (a) What is the speed of the beam that emerges from the velocity selector? (b) How far apart are the points where the ^{16}O and ^{18}O ions land?

Set Up

We use Equation 19-2 to find the speed v of ions that pass undeflected through the velocity selector and Equation 19-3 to find the radius of the circular path that each isotope follows. The distance from where each ion enters the region of uniform magnetic field to where it strikes the detector is the diameter (twice the radius) of the circular path.

Speed of ions emerging from a velocity selector:

$$v = \frac{E}{B} \qquad (19\text{-}2)$$

Radius of circular path followed by charged particles in a uniform magnetic field:

$$r = \frac{mv}{qB} \qquad (19\text{-}3)$$

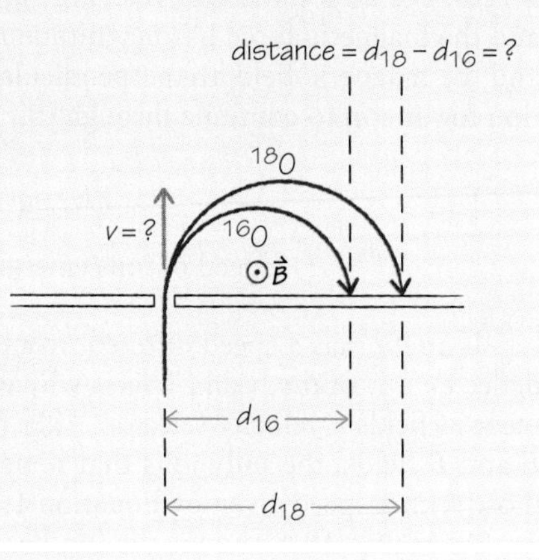

Solve

(a) Calculate the speed of ions that emerge from the velocity selector. Note that this speed does not depend on the charge or mass of the ion, so the $^{16}O^+$ and $^{18}O^+$ ions both emerge with this speed.

From Equation 19-2,

$$v = \frac{E}{B} = \frac{4.00 \times 10^3 \text{ V/m}}{0.0800 \text{ T}} = 5.00 \times 10^4 \frac{\text{V}}{\text{T} \cdot \text{m}}$$

From above, 1 T = 1 N·s/(C·m), so 1 T·m = 1 N·s/C. We also recall from Chapter 17 that 1 V = 1 J/C = 1 N·m/C. So we can write the speed as

$$v = 5.00 \times 10^4 \frac{\text{V}}{\text{T} \cdot \text{m}} = 5.00 \times 10^4 \left(\frac{\text{N} \cdot \text{m}}{\text{C}}\right)\left(\frac{\text{C}}{\text{N} \cdot \text{s}}\right)$$
$$= 5.00 \times 10^4 \text{ m/s}$$

(b) Calculate the radius r and diameter d of the circular path followed by each isotope.

The masses of the two atoms are

For ^{16}O: $m_{16} = (16.0 \text{ u})(1.66 \times 10^{-27} \text{ kg/u}) = 2.66 \times 10^{-26}$ kg

For ^{18}O: $m_{18} = (18.0 \text{ u})(1.66 \times 10^{-27} \text{ kg/u}) = 2.99 \times 10^{-26}$ kg

The mass of each positive ion ($^{16}O^+$ and $^{18}O^+$) is slightly less than the mass of the neutral atom; the difference is the mass of one electron, which is 9.11×10^{-31} kg $= 0.0000911 \times 10^{-26}$ kg. This difference is so small that we can ignore it.

For ^{16}O,

$$r_{16} = \frac{m_{16}v}{qB} = \frac{(2.66 \times 10^{-26}\ \text{kg})(5.00 \times 10^4\ \text{m/s})}{(1.60 \times 10^{-19}\ \text{C})(0.0800\ \text{T})}$$

$$= 0.104 \frac{\text{kg} \cdot \text{m}}{\text{T} \cdot \text{C} \cdot \text{s}}$$

Since $1\ \text{T} = 1\ \text{N} \cdot \text{s}/(\text{C} \cdot \text{m})$ and $1\ \text{N} = 1\ \text{kg} \cdot \text{m/s}^2$, it follows that $1\ \text{T} = \text{kg}/(\text{C} \cdot \text{s})$ and $1\ \text{T} \cdot \text{C} \cdot \text{s} = 1$ kg. So for ^{16}O we have

$$r_{16} = 0.104 \frac{\text{kg} \cdot \text{m}}{\text{kg}} = 0.104\ \text{m}$$

$$d_{16} = 2r_{16} = 0.208\ \text{m} = 20.8\ \text{cm}$$

For ^{18}O,

$$r_{18} = \frac{m_{18}v}{qB} = \frac{(2.99 \times 10^{-26}\ \text{kg})(5.00 \times 10^4\ \text{m/s})}{(1.60 \times 10^{-19}\ \text{C})(0.0800\ \text{T})}$$

$$= 0.117\ \text{m}$$

$$d_{18} = 2r_{18} = 0.234\ \text{m} = 23.4\ \text{cm}$$

The distance between the positions where the ^{16}O and ^{18}O ions land is the difference between the two diameters.

The distance between where the ^{16}O and ^{18}O ions land is

$$d_{18} - d_{16} = 23.4\ \text{cm} - 20.8\ \text{cm} = 2.6\ \text{cm}$$

This is a substantial distance, so the mass spectrometer does a good job of separating the two isotopes.

Reflect

Experiments of this kind show that, on average, 99.8% of oxygen atoms are ^{16}O and 0.2% are ^{18}O. (Making up a small fraction of a percent is a third isotope, ^{17}O.)

Measurements of the ratio of ^{18}O to ^{16}O are important to the science of *paleoclimatology*, the study of Earth's ancient climate. One way to determine the average temperature of the planet in the distant past is to examine ancient ice deposits in Greenland and Antarctica. These deposits endure for hundreds of thousands of years; deposits near the surface are more recent, while deeper deposits are older. The ice comes from ocean water that evaporated closer to the equator and then fell as snow in the far north or far south. Each molecule of water is made up of two hydrogen atoms and one oxygen atom, which could be ^{16}O or ^{18}O. A water molecule can more easily evaporate if it contains lighter ^{16}O than if it contains heavier ^{18}O, so the water that evaporated and fell on Greenland and Antarctica as snow contains an even smaller percentage of ^{18}O than ocean water. This deficiency becomes even more pronounced for colder climates. It has been shown that a decrease of one part per million of ^{18}O in ice indicates a 1.5°C drop in sea-level air temperature at the time it originally evaporated from the oceans.

Using mass spectrometers to analyze ancient ice from Greenland and Antarctica, paleoclimatologists have been able to determine the variation in Earth's average temperature over the past 160,000 years. They have also analyzed the amount of atmospheric carbon dioxide (CO_2) that was trapped in the ice as it froze. An important result of these studies is that higher levels of atmospheric CO_2 have gone hand-in-hand with elevated temperatures for the last 160,000 years, which is just what we would expect from our discussion of global warming in Section 14-7. In the same way, the tremendous increase in CO_2 levels in the past century due to burning fossil fuels has gone hand-in-hand with recent dramatic increases in our planet's average temperature.

GOT THE CONCEPT? 19-4 Ions in a Mass Spectrometer

? When passed through the mass spectrometer of Example 19-2, which of the following ions would follow nearly the same path as a $^{16}O^+$ ion? Assume that all ions are moving at the speed given by Equation 19-2. (a) A doubly charged oxygen-16 ion ($^{16}O^{2+}$); (b) a singly charged sulfur-32 ion ($^{32}S^+$); (c) a doubly charged sulfur-32 ion ($^{32}S^{2+}$); (d) more than one of these; (e) none of these.

TAKE-HOME MESSAGE FOR Section 19-4

✔ In a mass spectrometer, a magnetic field causes ions of the same speed and charge but different masses to follow different paths.

✔ In a velocity selector, the only charged objects that pass through undeflected are those traveling at the speed for which the electric and magnetic forces have equal magnitudes but opposite directions.

19-5 Magnetic fields exert forces on current-carrying wires

The magnetic force that we've described may seem to apply only in certain very special circumstances. But in fact, you use magnetic forces whenever you listen to recorded music through earbuds (**Figure 19-9a**). Within each earbud is a small but powerful magnet adjacent to a flexible plastic cone with a coil of wire attached to it (**Figure 19-9b**). This coil is connected through the earbud wires to your music player, which sends the musical signal to the coil in the form of a varying electric current. Charges within the coil are thus set into motion, and these moving charges experience magnetic forces exerted by the magnet. These forces pull on the coil that contains the charges and on the plastic cone to which the coil is attached. We learned in Section 19-3 that the direction of magnetic force depends on the direction in which charges move. So the force on the charges, coil, and plastic cone reverses whenever the current in the coil changes direction. The result is that the plastic cone oscillates back and forth in response to the signal coming from your player. This oscillation pushes on the surrounding air, producing a sound wave—the sound of music—that travels to your eardrum.

Let's see how to find the magnitude of the magnetic force on a current-carrying wire, such as a segment of the coil in Figure 19-9b. **Figure 19-10** shows a straight wire of length ℓ that carries a current i. A magnetic field $\vec{B}$ points at an angle θ to the direction of the current and has the same value along the entire length of the wire. We learned in Section 18-2 that the current in the wire is related to the speed at which individual charges drift through the wire:

Current and drift speed (18-3)

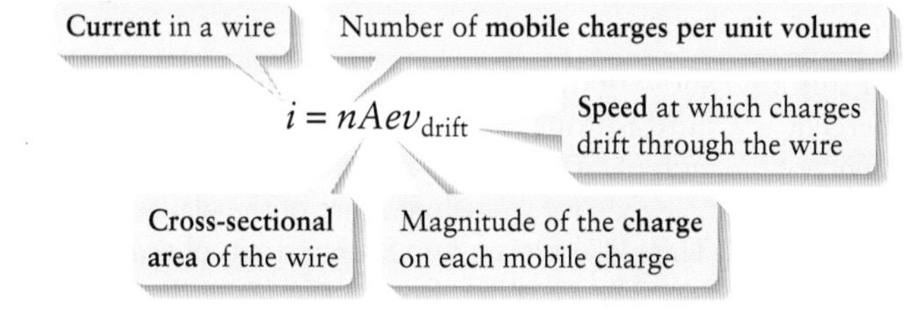

(a) (b)

Jupiterimages/Thinkstock

Christina Micek

Front cover

Transparent plastic cone

Coil (attached to plastic cone)

Magnet (plastic cone is attached to this at its edges)

Figure 19-9 Earbud physics (a) Earbuds and other speakers use magnetic forces to produce sound. (b) Internal construction of an earbud.

The force $\vec{F}$ on a current-carrying wire in a magnetic field $\vec{B}$ is perpendicular to both $\vec{B}$ and the length of the wire. The direction is given by a right-hand rule.

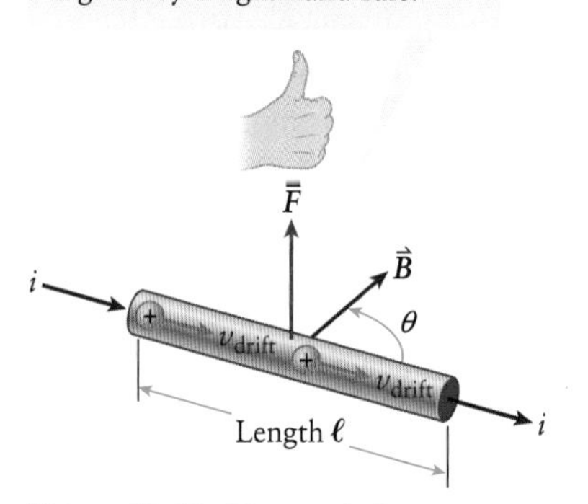

Figure 19-10 Magnetic force on a current-carrying wire The force that a magnetic field exerts on this wire is the sum of the forces on all the moving charges within the wire.

From Equation 19-1, the magnetic force on an individual charge e moving through the wire at speed v_{drift} has magnitude $ev_{\text{drift}}B \sin \theta$. Each such moving charge feels a force of the same magnitude and in the same direction. The total number of such moving charges in the wire shown in Figure 19-10 is n (the number of moving charges per unit volume) multiplied by the volume V of the wire, where $V = A\ell$ (the cross-sectional area of the wire multiplied by its length). So the magnitude of the *net* magnetic force on all of the moving charges in the wire is

$$F = (nA\ell)\,(ev_{\text{drift}}B \sin \theta)$$

If we rearrange the terms in this equation, we get

$$F = (nAev_{\text{drift}})\ell B \sin \theta \qquad (19\text{-}4)$$

The quantity in parentheses in Equation 19-4 is just the current i as given by Equation 18-3. So the magnitude of the magnetic force on a current-carrying wire is

Magnitude of magnetic force on a current-carrying wire (19-5)

$$F = i\ell B \sin \theta$$

- F: **Magnitude of the magnetic force** on a current-carrying wire
- B: **Magnitude of the magnetic field** (assumed uniform over the length of the wire)
- i: **Current** in the wire
- ℓ: **Length** of the wire
- θ: **Angle** between the direction of the current and the direction of the magnetic field $\vec{B}$

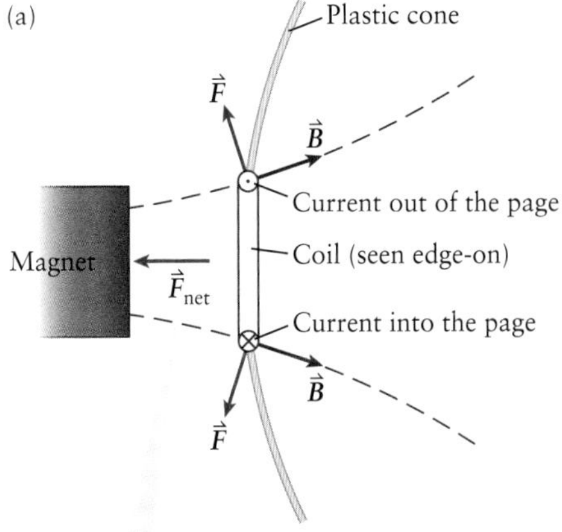

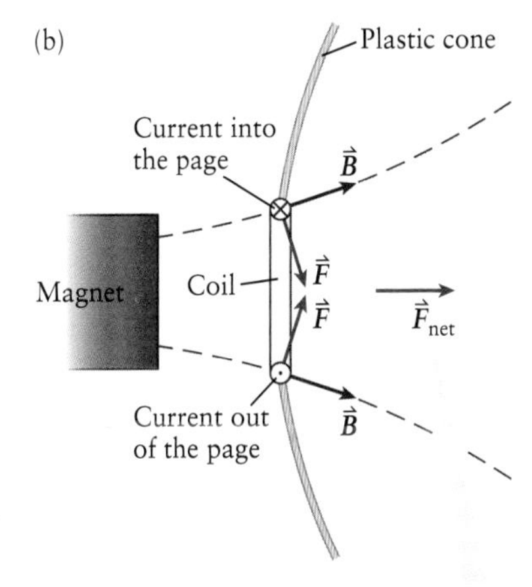

Figure 19-11 Magnetic forces in an earbud (a) The magnetic field produced by an earbud magnet is not uniform, so the field exerts a net force on the coil and the plastic cone to which it is attached. (b) Reversing the direction of the current reverses the direction of the net force.

Just as for the magnetic force on a moving charged particle, a right-hand rule tells you the direction of the magnetic force on a current-carrying wire. Extend the fingers of your right hand in the direction of the current and orient your hand so that the palm is facing the direction that the magnetic field vector $\vec{B}$ points. With your right thumb extended, swivel your right hand so that the fingers now point in the direction of $\vec{B}$. Your outstretched thumb points in the direction of the magnetic force exerted on the wire (Figure 19-10).

Equation 19-5 says that there is *no* magnetic force on the wire if the magnetic field $\vec{B}$ points either in the same direction as the current ($\theta = 0$) or in the direction opposite to the current ($\theta = 180°$). In either case, $\sin \theta = 0$ and $F = 0$. For a given magnetic field of magnitude B, the force is greatest if $\vec{B}$ points perpendicular to the current so $\theta = 90°$ and $\sin \theta = \sin 90° = 1$. This idea is used in the design of the earbuds shown in Figure 19-9. The wires in each earbud are curved into a circular coil, not straight as in Figure 19-10. But we can treat the coil as being made up of many short segments, each of which is effectively a straight piece of wire. As **Figure 19-11a** shows, the magnetic field from the magnet in each earbud is perpendicular to each such segment. The forces on different segments are in different directions, but the vector sum of these forces is toward the magnet. If the current is in the reverse direction, as in **Figure 19-11b**, all of the individual forces also reverse direction and the vector sum of the forces is away from the magnet. The current from the music player continually reverses direction, so the coil and attached diaphragm oscillate back and forth, producing a sound wave.

In the following section we'll see how to use Equation 19-5 to help us understand how electric motors work. For now, let's see how to use a magnetic field to levitate a current-carrying wire.

EXAMPLE 19-3 Magnetic Levitation

You set up a uniform horizontal magnetic field that points from south to north and has magnitude 2.00×10^{-2} T. (This is about 400 times stronger than Earth's magnetic field, but easily achievable with common magnets.) You want to place a straight copper wire of diameter 0.812 mm in this field, then run enough current through the wire so that the magnetic force will make the wire "float" in midair. This is called *magnetic levitation*. What minimum current is required to make this happen? The density of copper is 8.96×10^{3} kg/m^3.

Set Up

In order to make the wire "float," there must be an upward magnetic force on the wire that just balances the downward gravitational force. We know from Equation 19-5 that to maximize the magnetic force the current direction should be perpendicular to the magnetic field $\vec{B}$. The right-hand rule then shows that the current should flow from west to east so that the magnetic force is directed upward. We're not given the mass of the wire, but we can express the mass (and hence the gravitational force on the wire) in terms of its density. We're also not given the length of the wire; as we'll see, this will cancel out of the calculation.

Magnetic force on a current-carrying wire:

$$F = i\ell B \sin\theta \qquad (19\text{-}5)$$

Definition of density:

$$\rho = \frac{m}{V} \qquad (11\text{-}1)$$

Volume of a cylindrical wire of cross-sectional area A and length ℓ:

$$V = A\ell$$

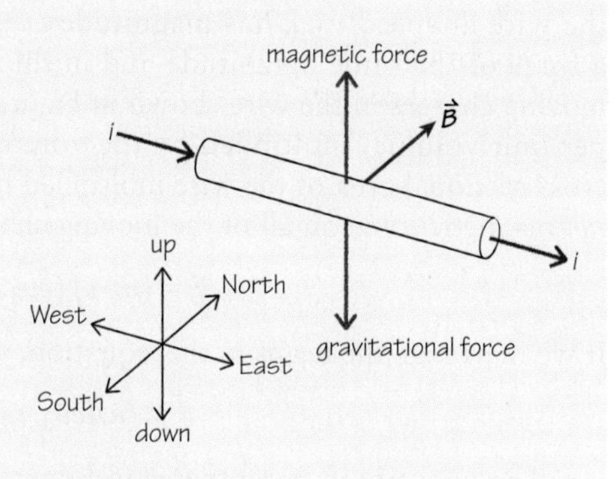

Solve

Write expressions for the two forces (magnetic and gravitational) that act on the wire.

Assume the wire has length ℓ. Since the current flows in a direction perpendicular to the magnetic field, the angle θ in Equation 19-5 is 90°. So the magnitude of the upward magnetic force is

$$F = i\ell B \sin 90° = i\ell B$$

The magnitude of the downward gravitational force on the wire of mass m is

$$w = mg$$

From Equation 11-1, the mass equals the density of copper multiplied by the volume of the wire:

$$m = \rho V = \rho A\ell$$

The cross-sectional area A of the wire is that of a circle of radius r:

$$A = \pi r^2 \quad \text{so} \quad m = \rho A\ell = \rho(\pi r^2)\ell$$

So the magnitude of the gravitational force is

$$w = mg = \rho(\pi r^2)\ell g$$

If the wire is floating in equilibrium, the upward magnetic force must balance the downward gravitational force. Use this to solve for the required current i.

In equilibrium the net vertical force on the wire is zero:

$$F - w = 0, \text{ so } F = w \text{ and } i\ell B = \rho(\pi r^2)\ell g$$

The length ℓ of the wire cancels out of this equation (both the magnetic force and the gravitational force are proportional to ℓ):

$$iB = \rho(\pi r^2)g$$

Solve for the current i:

$$i = \frac{\rho(\pi r^2)g}{B}$$

$$= \frac{(8.96 \times 10^3\ \text{kg/m}^3)(\pi)(0.406 \times 10^{-3}\ \text{m})^2(9.80\ \text{m/s}^2)}{2.00 \times 10^{-2}\ \text{T}}$$

$$= 2.27\ \frac{\text{kg}}{\text{T}\cdot\text{s}^2} = 2.27\ \text{A}$$

[Check on units: We know from Section 19-3 that $1\ \text{T} = 1\ \text{N/(A}\cdot\text{m)}$, and we also know that $1\ \text{N} = 1\ \text{kg}\cdot\text{m/s}^2$. Therefore, $1\ \text{T} = 1\ \text{kg/(A}\cdot\text{s}^2)$, so $1\ \text{kg/(T}\cdot\text{s}^2) = 1\ \text{A}$.]

Reflect

A current of 2.27 A is relatively small, so this experiment in magnetic levitation is not too difficult to perform. Note that the required current i is inversely proportional to the magnitude B of the magnetic field. You can see that if you tried to make a wire "float" using Earth's magnetic field, which is about 1/400 as strong as the field used here, you would need to use an immense current of $400 \times 2.27\ \text{A} = 909\ \text{A}$. That's not practical because a current of that magnitude would cause the wire in this example to melt! (Recall from Equation 18-24 in Section 18-6 that the power into a resistor with resistance R that carries current i is $P = i^2R$. The wire in this example has a small cross-sectional area, so its resistance R will be fairly large. The power delivered to the wire by a 909-A current will quickly increase its temperature to above the melting point of copper, 1085°C.)

A practical application of magnetic levitation is train design. By using magnetic forces to make a train float just above the track, the rolling friction between the wheels and the track can be completely eliminated and very high speeds achieved. (Magnetic forces are also used to propel the train forward.) A train of this type in commercial operation in Shanghai, China, reaches a top speed of 431 km/h (268 mi/h). Such train lines require special magnets and wires capable of sustaining very high currents.

GOT THE CONCEPT? 19-5 Direction of Magnetic Force on a Wire

A long, straight wire carrying a current can be placed in various orientations with respect to a constant magnetic field as shown in **Figure 19-12a, b, c, d, e,** and **f.** What is the direction of the force in each case? Give your answers in terms of the positive and negative x, y, and z directions. If the force is zero, say so. (In each case the positive z direction is out of the plane of the figures.)

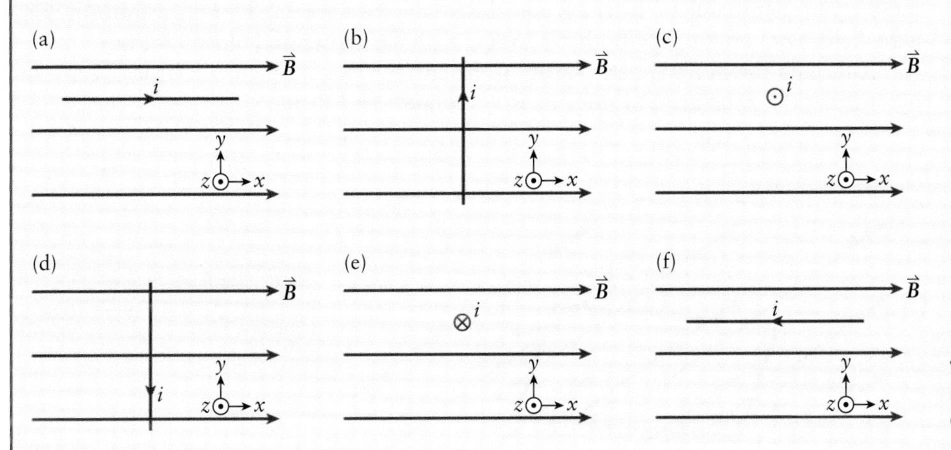

Figure 19-12 Which way is the force? What is the direction of the magnetic force on the wire in each case?

TAKE-HOME MESSAGE FOR Section 19-5

✔ A current-carrying wire in a magnetic field experiences a magnetic force that is perpendicular to both the current direction and the magnetic field direction.

✔ The magnitude of this magnetic force is maximum if the current and magnetic field directions are perpendicular, and zero if the wire axis is along the magnetic field.

19-6 A magnetic field can exert a torque on a current loop

A common everyday application of magnetic forces on current-carrying wires is an electric motor. An electric motor is at the heart of a kitchen blender, a vacuum cleaner, the starter for an internal-combustion automobile, and many other devices (**Figure 19-13a**). Inside any electric motor you'll find a rotating portion (called the *rotor*) that's wrapped with coils of wire (**Figure 19-13b**), as well as magnets in the stationary part of the motor. When you turn the motor on, causing a current through the coils, the magnets exert forces on the current-carrying wire of the coils. The net effect is

(a)

iStockphoto/Thinkstock

(b)

iStockphoto/Thinkstock

Figure 19-13 Electric motors and magnetic torque (a) This fan uses an electric motor to rotate the fan blades. (b) The key components of any electric motor are current-carrying coils and a source of magnetic field. The magnetic field exerts forces on the current-carrying wires of the coils, and these produce a torque that makes the rotating part of the motor spin.

that there is a magnetic *torque* that makes the rotor spin. Let's look at a simplified version of this process to see how such a magnetic torque arises.

Figure 19-14a shows a straight wire bent into a single rectangular loop of wire with sides of length L and W. The loop carries a current i provided by a source of emf (not shown), so we call it a **current loop**. This loop is immersed in a uniform magnetic field of magnitude B and is free to rotate around an axis (shown in green). You should apply the right-hand rule for magnetic forces on a current-carrying wire (see Section 19-5) to each side of this loop. You'll see that the left-hand side of the loop experiences a force to the left, the right-hand side feels a force to the right, the top of the loop feels an upward force, and the bottom of the loop feels a downward force, as Figure 19-14a shows. Each segment of the loop carries the same current and is in the same magnetic field. Since the left- and right-hand sides are the same length and at the same angle to the magnetic field, the forces on these two sides have the same magnitude but opposite directions. Thus these forces cancel, and there is zero net force to the left or the right. The forces on the top and bottom segments of the wire also cancel for the same reason. So the net *force* on this rectangular loop is zero. (You should be able to convince yourself that the same would be true if we reversed the direction of the current around the loop or the direction of the magnetic field, or if we changed the angle between the direction of the field and the plane of the loop.) It turns out that there is zero net magnetic force on *any* closed current-carrying loop in a uniform magnetic field, not just loops with the rectangular shape shown in Figure 19-14a.

Although the net magnetic force on the current loop in Figure 19-14a is zero, the net magnetic *torque* around the axis is not. **Figure 19-14b** is a side view of the current

WATCH OUT! A current-carrying coil can feel a net magnetic force if the $\vec{B}$ field is not uniform.

In Figure 19-14a we assumed that the magnetic field is uniform (its magnitude and direction are the same at all points) and found that the net magnetic force on the current loop is zero. But if the magnetic field magnitude were greater at the left-hand side of the loop than at the right-hand side, then the left-hand side would experience a greater magnetic force. This would result in a net force on the loop to the left. In this chapter we'll restrict our discussion to the simple case in which the field is uniform.

(a) Perspective view

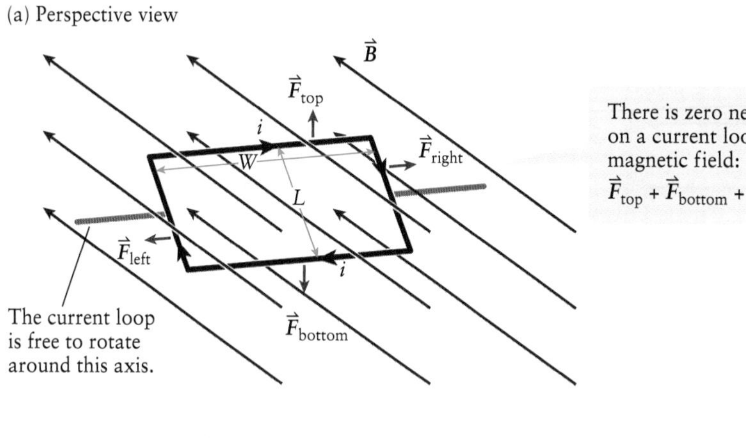

There is zero net magnetic force on a current loop in a uniform magnetic field:
$\vec{F}_{top} + \vec{F}_{bottom} + \vec{F}_{left} + \vec{F}_{right} = 0$

(b) Side view

$\vec{F}_{top}$ ϕ $\vec{B}$ i ϕ ϕ $\vec{F}_{bottom}$

There can be a nonzero net magnetic torque on a current loop in a uniform magnetic field.

Normal to the plane of the current loop

The torque magnitude depends on the angle ϕ.

Figure 19-14 A current loop in a uniform magnetic field (a) The magnetic forces on the sides of the current loop. (b) These forces can give rise to a net magnetic torque.

loop; the axis shown in green in Figure 19-14a is perpendicular to the plane of this figure and passes through the center of the loop. The forces on the near and far sides of the loop, which point out of and into the page, respectively, are directed parallel to this axis and so produce no torque. However, the forces that act on the top and bottom sides of the loop both give rise to a torque. For the situation shown in Figure 19-14b, both of these forces tend to make the loop rotate clockwise around the axis, so the net torque is clockwise.

Let's calculate the magnitude of the magnetic torque on the loop in Figure 19-14b. First note that the magnetic field is perpendicular to the current in both the top side of the loop and the bottom side of the loop. So in Equation 19-5 for the magnetic force on a current-carrying wire, the angle θ equals 90°. Both the top and bottom sides of the loop have length W (see Figure 19-14a), so the magnitude of the magnetic force on each of these sides is

$$F = iWB \sin 90° = iWB \qquad (19\text{-}6)$$

To find the torque around the axis that each of these forces produces, we use Equation 8-18:

$\sqrt{x}$ *See the Math Tutorial for more information on trigonometry.*

Magnitude of torque (8-18)

$$\tau = rF \sin \phi$$

Magnitude of the torque produced by a force acting on an object

Magnitude of the force

Distance from the rotation axis of the object to where the force is applied

Angle between the vector $\vec{r}$ (from the rotation axis to where the force is applied) and the force vector $\vec{F}$

In Figure 19-14b the distance r is one-half of the length L of the near or far side of the loop, and ϕ is the angle between the direction of the near or far side of the current loop and the force exerted on either the top or bottom of the loop. This angle phi is also equal to the angle between the direction of the magnetic field and an imaginary line that's perpendicular to the plane of the current loop. We call this imaginary line the **normal** to the plane of the loop. From Equations 8-18 and 19-6, the magnitude of the torque due to the force on either the top or bottom side of the loop is

$$\tau_{\text{one side}} = \left(\frac{L}{2}\right)F \sin \phi = \left(\frac{L}{2}\right)(iWB) \sin \phi = \frac{1}{2}i\,(LW)B \sin \phi \qquad (19\text{-}7)$$

The product LW (length times width) in Equation 19-7 is just the area of the current loop: $A = LW$. As we mentioned above, the torque from the top of the loop and the torque from the bottom of the loop both act in the same direction, so the *net* torque on the loop is just double the torque on one side given by Equation 19-7. Using $A = LW$, we have

$$\tau = 2\tau_{\text{one side}} = 2\left(\frac{1}{2}\right)i\,(LW)B \sin \phi = 2\left(\frac{1}{2}\right) iAB \sin \phi$$

or

Magnitude of magnetic torque on a current loop (19-8)

$$\tau = iAB \sin \phi$$

Magnitude of the magnetic torque on a current loop

Magnitude of the magnetic field

Current in the loop

Area of the loop

Angle between the normal to the plane of the loop and the direction of the magnetic field $\vec{B}$

Although we've assumed a rectangular loop, Equation 19-8 applies to a current loop of area A of *any* shape.

Typically the rotating coil in an electric motor is not just a single loop of wire but has many *turns* (equivalent to many single coils stacked on top of each other). If a coil has N turns, the magnetic torque on it is N times greater than the value given by Equation 19-8.

EXAMPLE 19-4 Angular Acceleration of a Current Loop

A length of copper wire is formed into a square loop with 50 turns. The loop is free to turn about a frictionless axis that lies in the plane of the loop and passes through its center. Each side of the loop is 2.00 cm long, and the moment of inertia of the loop about the axis of rotation is 4.00×10^{-6} kg · m^2. The loop lies in a region where there is a uniform magnetic field of magnitude 1.50×10^{-2} T that is perpendicular to the rotation axis of the loop. The current in the loop is 0.500 A. Find the angular acceleration of the loop (magnitude and direction) (a) when the loop is released from rest from the orientation shown in **Figure 19-15**, (b) after the loop has rotated 90°, and (c) after the loop has rotated 180°.

Figure 19-15 A square current loop in a magnetic field When this square current loop is released from rest, how will it begin to rotate?

Set Up

For each orientation we'll use Equation 19-8 to find the magnitude of the magnetic torque on the loop. There is no other torque on the loop (there is no friction, and gravity exerts zero torque since the center of mass of the loop lies on the rotation axis). So the magnetic torque is also the net torque, and we can use Equation 8-20 to find the magnitude of the angular acceleration. We'll find the direction of the angular acceleration by looking at the directions of the magnetic forces on the individual sides of the loop. Figure 19-14 shows that forces on the sides parallel to the rotation axis tend to affect rotation. The forces on the other sides cause no torque (compare Figure 19-14a).

Magnitude of magnetic torque on a current loop:

$$\tau = iAB \sin \phi \qquad (19\text{-}8)$$

Newton's second law for rotational motion:

$$\sum \tau_{\text{ext}, z} = I\alpha_z \qquad (8\text{-}20)$$

Solve

(a) First find the direction of the magnetic torque and hence the direction of the angular acceleration.

The right-hand rule for the magnetic force on a current-carrying wire shows that the forces on the wire tend to cause a clockwise rotation. So the angular acceleration is clockwise, and when the loop is released it will begin to rotate in the clockwise direction.

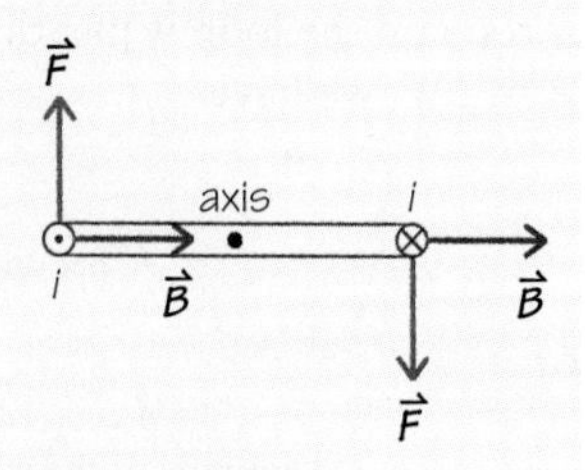

Find the magnitude of the angular acceleration.

The normal to the loop is perpendicular to the magnetic field, so in Equation 19-8 $\sin \phi = \sin 90° = 1$. The loop is square, so the cross-sectional area of the loop is just the square of the length of one side (2.00 cm = 0.0200 m). Since there are 50 turns, the total torque is $N = 50$ times greater than that given by Equation 19-8:

$$\begin{aligned}\tau &= NiAB \sin \phi \\ &= (50)(0.500 \text{ A})(0.0200 \text{ m})^2(1.50 \times 10^{-2} \text{ T})(1) \\ &= 1.50 \times 10^{-4} \text{ T} \cdot \text{A} \cdot \text{m}^2 = 1.50 \times 10^{-4} \text{ N} \cdot \text{m}\end{aligned}$$

(Recall that 1 T = 1 N/(A · m), so 1 T · A · m^2 = 1 N · m.)

This is the net torque on the loop, so from Equation 8-20 the magnitude of the loop's angular acceleration is

$$\alpha = \frac{\tau}{I} = \frac{1.50 \times 10^{-4}\ \text{N}\cdot\text{m}}{4.00 \times 10^{-6}\ \text{kg}\cdot\text{m}^2}$$
$$= 37.5\ \text{rad/s}^2$$

(If the torque is in N · m and the moment of inertia is in kg · m^2, the angular acceleration in Equation 8-20 is in rad/s^2.)

When released, the loop will begin to rotate in the clockwise direction, as shown above. Note that as the loop rotates, the angle ϕ and the value of sin ϕ will decrease, so the torque and angular acceleration will decrease: This is *not* a situation with constant angular acceleration.

(b) Find the angular acceleration when the loop has rotated 90° from its initial orientation.

When the loop has rotated 90°, the normal to the loop is in the same direction as the magnetic field, so $\phi = 0$ and sin $\phi = 0$. From Equation 19-8 it follows that there is zero torque on the loop at this point in its rotation. We can also see this using the right-hand rule for the magnetic forces on the wires of the loop: These forces do not exert any torque around the axis. Therefore, this orientation represents an equilibrium position for the loop, and the angular acceleration is zero. Note that the loop will be in motion as it passes through this position (there has been an angular acceleration ever since the loop was released). As a result, the loop doesn't stop at this position but keeps on rotating.

(c) Find the angular acceleration when the loop has rotated by 180° from its initial orientation.

At the 180° position, the normal to the loop is again perpendicular to the direction of the magnetic field, just as in part (a). So again the angle $\phi = 90°$, and again the magnitude of the angular acceleration is $\alpha = 37.5\ \text{rad/s}^2$. However, since the loop has been flipped over relative to its original orientation, the directions of the forces are reversed. So the torque and angular acceleration are now *counterclockwise*. In fact, the angular acceleration has been increasingly counterclockwise ever since the loop moved past the position in (b). Since passing that position, the loop has been rotating in a clockwise direction but has been slowing down due to the counterclockwise angular acceleration.

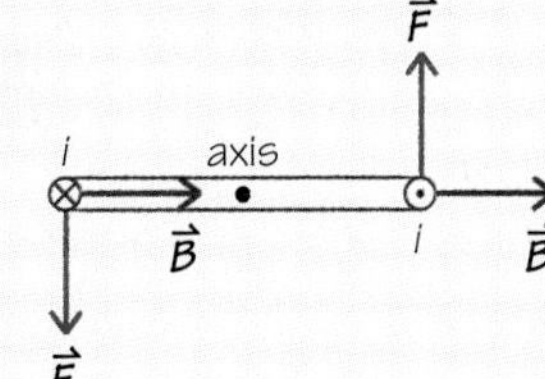

Reflect

The motion of the loop in this example should remind you of the motion of a pendulum (Section 12-5). When displaced from equilibrium and released, the pendulum will swing toward its equilibrium orientation (hanging straight down) but overshoot that equilibrium and swing to the other side of equilibrium. If there is no friction, the pendulum will keep swinging back and forth indefinitely. The same is true for this current loop: If the axis on which it rotates is frictionless, it will oscillate back and forth between the orientation in part (a) and the orientation in part (c).

In an electric motor we want the coil to continue rotating in the same direction, not oscillate back and forth like the coil in Example 19-4. To make this happen, the connection between the coil and the source of emf is arranged so that when the coil is at its equilibrium position (as in part (b) of Example 19-4), the direction of the current *reverses*. As a result, when the coil moves past this equilibrium position the torque is in the same direction as the rotation, and the rotation continues to speed up. The current reverses again after another half-rotation, so the torque is always in the same direction.

With this arrangement the coil would continue to gain rotational speed without limit if there were no other torques acting on it. In practice there are other torques that oppose the rotation, and the rotational speed reaches an upper limit. That's the case for the electric fan in Figure 19-13a: Air resistance on the fan blades increases as the fan turns faster, and the fan speed stabilizes when the torque due to air resistance just balances the torque of the electric motor.

GOT THE CONCEPT? 19-6 Magnetic Torque

Suppose the number of turns in the current loop of Example 19-4 were increased from 50 to 100. Would this make the maximum angular acceleration of the loop (a) four times greater, (b) twice as great, (c) 1/2 as great, (d) 1/4 as great, or (e) none of these?

TAKE-HOME MESSAGE FOR Section 19-6

✔ There is zero net force on a current loop in a uniform magnetic field.

✔ A uniform magnetic field can exert a torque on a current loop. The torque is maximum if the normal to the plane of the loop is perpendicular to the magnetic field direction.

19-7 Ampère's law describes the magnetic field created by current-carrying wires

So far in our discussion of magnetism, we've looked at the forces and torques that a magnetic field exerts on a current-carrying wire. But current-carrying wires also *produce* magnetic fields (see Figure 19-1c and Figure 19-4). As we described in Section 19-2, all magnetic fields are produced by electric charges in motion, whether it's a current in the coils of the electromagnet shown in Figure 19-1c, electrons in motion within the atoms of an iron bar magnet, or electric currents in the human brain as shown in **Figure 19-16**. To complete our understanding of magnetic forces, we need to be able to calculate the magnetic field produced by an arrangement of charges in motion. This is much like Chapter 16, where we needed to learn how to calculate the electric field due to an arrangement of charges in order to complete our understanding of electric forces.

We saw in Section 16-5 that the electric field that is due to a charge distribution is just the vector sum of the electric fields due to all of the charges in the distribution. These calculations can be rather challenging unless the charge distribution is very simple. When we look at the analogous problem of finding the *magnetic* field due to a distribution of moving charges, we find that the problem is even more complicated. That's because the magnetic field due to even a single moving charged particle is itself rather complex: The field does not point directly away from or toward the moving charge, but in a direction perpendicular to the velocity vector. Rather than looking at how to do calculations of this kind, we'll look at the result for just one important situation, the magnetic field due to a long, straight, current-carrying wire. We'll then see an alternative approach for calculating a magnetic field using *Ampère's law*. This law is a very powerful one, but like Gauss's law for electric fields (Sections 16-6 and 16-7), it can be used for field calculations only in certain simple situations. We'll conclude the section with a look at the magnetic field produced by a current loop, as well as some of the applications of this field.

Electric currents in the human brain generate weak magnetic fields. The colors in this magnetoencephalogram represent the strength of the magnetic field produced in different regions.

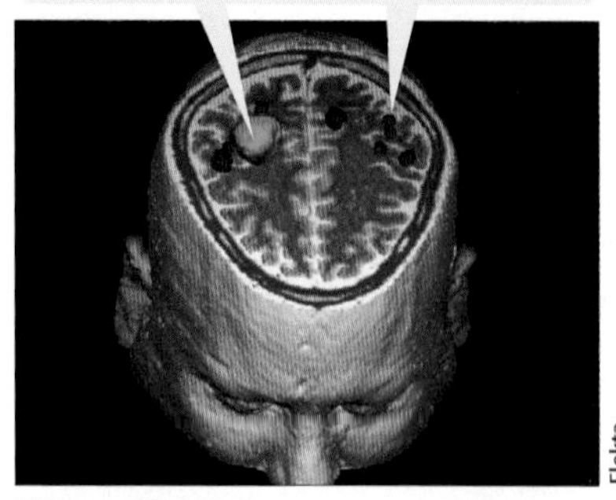

Elekta

Figure 19-16 Currents as sources of magnetic field Electric currents in biological systems produce magnetic fields.

A Long, Straight Wire and Ampère's Law

Consider a very long, straight wire that carries a constant current i. Experiment and calculation both show that the magnetic field due to this current has the properties shown in **Figure 19-17**. Note that the wire and the magnetic field that it produces have

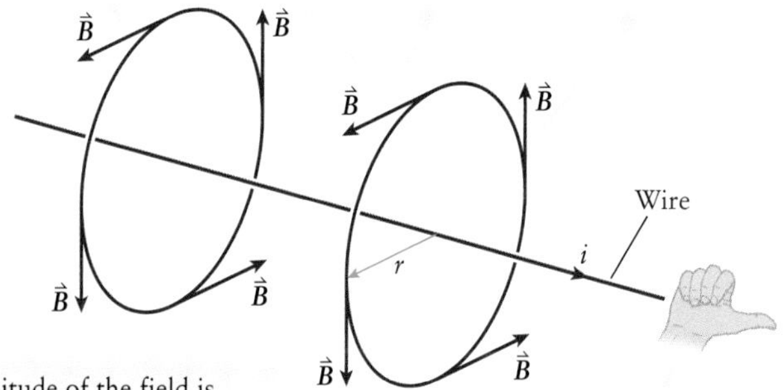

(1) The magnetic field lines due to a long, straight, current-carrying wire are circles that lie in planes perpendicular to the axis of the wire.
(2) The magnetic field vectors are tangential to the circular magnetic field lines and perpendicular to the axis of the wire.
$\vec{B}$
Wire
i
r
(3) The magnitude of the field is proportional to the current i and inversely proportional to the distance r from the axis of the wire.
(4) Right-hand rule for the field direction: If you point your right thumb in the direction of the current and curl your fingers, the magnetic field curls around the field lines in the direction of the curled fingers of your right hand.

Figure 19-17 Magnetic field of a long, straight, current-carrying wire The magnitude of this field is given by Equation 19-9.

cylindrical symmetry: The wire and the field pattern look exactly the same if you rotate the wire around its length. The magnitude of the field is given by

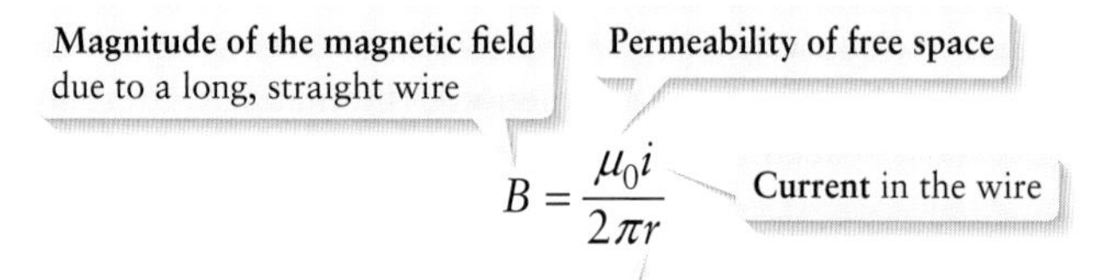

$$B = \frac{\mu_0 i}{2\pi r}$$

Magnitude of magnetic field due to a long, straight wire (19-9)

Equation 19-9 and the field properties shown in Figure 19-17 are strictly correct only if the wire is infinitely long. But they are very good approximations if the distance r is small compared to the length of the wire.

The constant μ_0 in Equation 19-9, called the **permeability of free space** for historic reasons, plays a role in magnetism that's comparable to the role of the constant ε_0, the *permittivity* of free space, in electricity (see Section 16-6). Its value is *exactly* $\mu_0 = 4\pi \times 10^{-7}\ \mathrm{T \cdot m/A}$. The value is known exactly because of the way that the tesla is defined: At a distance of 1 m from a long, straight wire carrying a current of 1 A, the magnetic field strength is exactly

$$B = \frac{\mu_0 i}{2\pi r} = \frac{(4\pi \times 10^{-7}\ \mathrm{T \cdot m/A})(1\ \mathrm{A})}{2\pi(1\ \mathrm{m})} = 2 \times 10^{-7}\ \mathrm{T}$$

WATCH OUT! Remember that vectors are straight, never curved.

When drawing the magnetic field around a long, straight wire, it may be tempting to draw curved arrows to represent how the magnetic field lines curl around the wire. But a vector always denotes a *single* direction and so *cannot* be curved. At any point along a magnetic field line, the direction of the field is always along the *tangent* to the field line, as shown in Figure 19-17.

The magnetic field around current-carrying wires with other geometries is much more complicated. As an example, **Figure 19-18** shows some of the magnetic field lines for a straight helical coil of wire. Such a coil is called a **solenoid**. Close to an

Figure 19-18 Magnetic field of a solenoid Compare this illustration to the photograph of a solenoid and its field in Figure 19-4a.

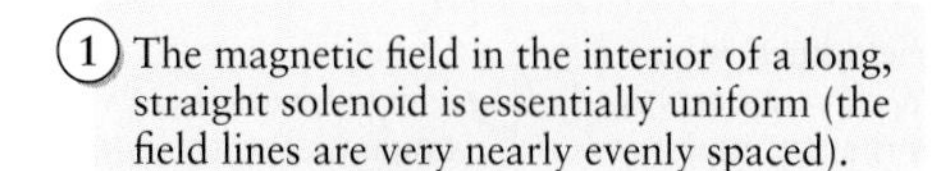

(1) The magnetic field in the interior of a long, straight solenoid is essentially uniform (the field lines are very nearly evenly spaced).

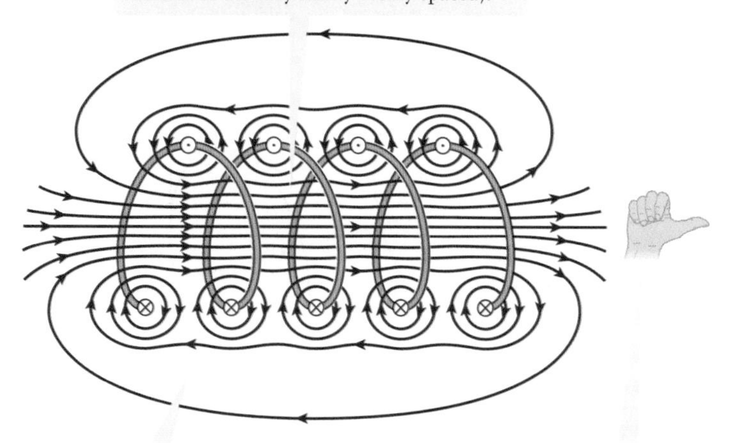

(2) The magnetic field outside the solenoid is very weak (the field lines are far apart).

(3) Right-hand rule for the field direction: If you curl the fingers of your right hand around the solenoid in the direction of the current, the magnetic field inside the solenoid points in the direction of your right thumb.

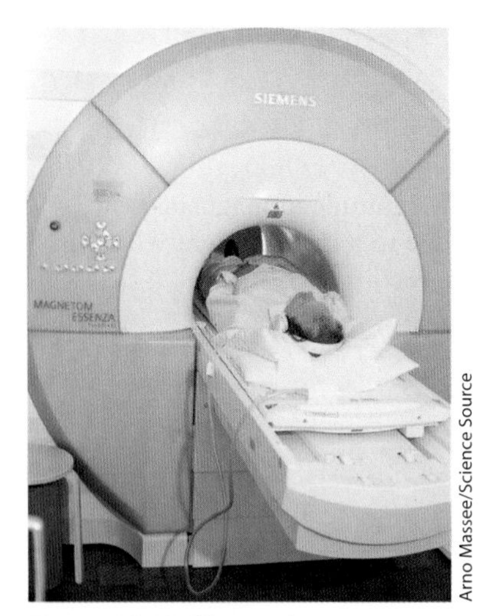

Arno Massee/Science Source

Figure 19-19 Magnetic resonance imaging (MRI) The medical imaging technique known as MRI requires that the patient be immersed in a strong, uniform magnetic field. In this MRI device this is done by having the patient lie inside a solenoid.

individual wire, the field lines resemble those around the long, straight wire shown in Figure 19-17. In the space outside the solenoid, the magnetic field is very weak, as you can see from the large spacing between adjacent field lines. (Recall from Section 16-5 that the same is true for electric fields: Where field lines are far apart, the field magnitude is small.) But in the interior of the solenoid, the magnetic field lines are close together and nearly evenly spaced, indicating that the magnetic field there is strong and nearly uniform.

BIO-Medical This property of solenoids explains why a conventional magnetic resonance imaging (MRI) scanner is in the form of a long tube inside which the patient lies (**Figure 19-19**). This tube is actually the interior of a solenoid like that shown in Figure 19-18, so the patient is bathed in a strong, uniform magnetic field—which is just what MRI requires. (In Chapter 27 we'll learn more about the physics of MRI.)

Can we use Equation 19-9 to calculate the field inside a solenoid? Not directly, no, because the wires that make up the solenoid are not straight. But we can use Equation 19-9 to illustrate a useful principle about magnetic fields and their sources, and then use that principle to determine the solenoid field. Imagine that we draw a circle of radius r around a long, straight wire as shown in **Figure 19-20**. Imagine further that we break the circle into a number of segments of length $\Delta\ell$. If $\Delta\ell$ is sufficiently small, we can treat each segment as being straight. Then for each segment take the component $B_{\|}$ of magnetic field parallel to that segment and multiply it by the segment length $\Delta\ell$. If we add up the values of these products for every segment in the circle, the result is a quantity called the **circulation** of the magnetic field around the circle:

$$\text{Circulation} = \sum B_{\|}\Delta\ell \tag{19-10}$$

For the circle shown in Figure 19-20, Equation 19-9 tells us that $B_{\|}$ is equal to $\mu_0 i/(2\pi r)$ at every point around the circle. That's because the magnetic field is everywhere tangent to the circle and so parallel to a short segment of length on the circle. Therefore, the circulation of the magnetic field as defined by Equation 19-10 is

$$\text{Circulation} = \sum\left(\frac{\mu_0 i}{2\pi r}\right)\Delta\ell = \frac{\mu_0 i}{2\pi r}\sum\Delta\ell \tag{19-11}$$

In Equation 19-11 we've taken the quantity $\mu_0 i/(2\pi r)$ outside the sum because it has the same value everywhere around the circle and so in all terms of the sum. The quantity $\sum\Delta\ell$ is the sum of the lengths of all of the segments that make up the

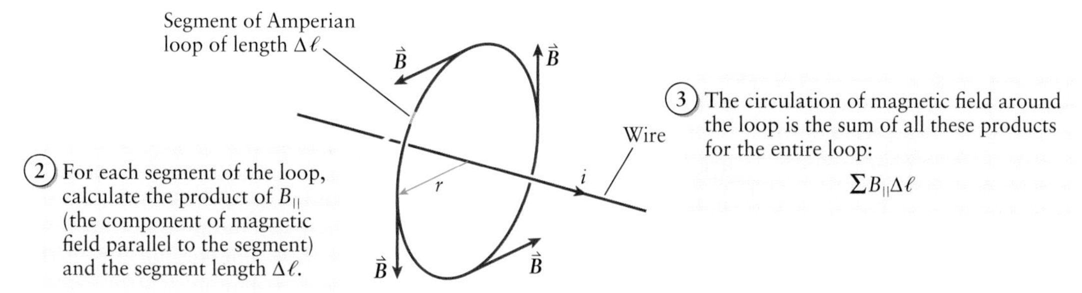

Figure 19-20 Ampère's law and circulation Ampère's law states that the circulation of the magnetic field around a closed path (called an *Amperian loop*) is proportional to the current that passes through that path.

circle—that is, the circumference of the circle, which is equal to $2\pi r$. So the circulation in Equation 19-11 is equal to

$$\text{Circulation} = \frac{\mu_0 i}{2\pi r}(2\pi r) = \mu_0 i \tag{19-12}$$

The circulation as given by Equation 19-12 does *not* depend on the radius r of the circle. (If the circle is larger, the magnetic field has a smaller magnitude, but it takes more segments of length $\Delta\ell$ to go all the way around the circle.) Remarkably, the same result holds true even if we draw a noncircular path around the wire: The circulation around *any* path that encloses the wire is equal to $\mu_0 i$. This is an example of **Ampère's law**, which was discovered by the French physicist André-Marie Ampère (pronounced "ahm-pair") in 1826:

Circulation of magnetic field around an Amperian loop

Permeability of free space

$$\sum B_{\|}\Delta\ell = \mu_0 i_{\text{through}}$$

Current through the Amperian loop

Ampère's law (19-13)

An **Amperian loop** is simply a closed path in space; the circle in Figure 19-20 is an example. The subscript "through" reminds us that the right-hand side of Equation 19-13 should include only current that passes through the interior of the Amperian loop.

It can be shown that Ampère's law is true for *any* magnetic field and *any* Amperian loop, no matter what the geometry of the current that produces the magnetic field. (The proof is beyond our scope.) However, it's only in certain very symmetric situations that Ampère's law helps us calculate the value of magnetic field. (This is much like Gauss's law for electric fields. We saw in Sections 16-6 and 16-7 that Gauss's law is always true but is useful for electric field calculations only in certain situations.) Happily, one such situation is the field of a solenoid.

Using Ampère's Law: Magnetic Field of a Solenoid

As Figure 19-18 shows, inside the solenoid the fields from each loop add to create a total field that runs generally parallel to the central axis, particularly at points close to the axis and far from the ends. Outside the solenoid the fields nearly cancel, especially far from the ends. The fields outside the solenoid cancel more completely when the length of the solenoid is large compared to its diameter and when the coils of the solenoid are tightly wound. If there are N turns or *windings* of wire in the solenoid, the *winding density* is $n = N/L$. (The units of n are windings per meter.) Let's see how to use Ampère's law to find the magnetic field of a long, narrow, tightly wound (ideal) solenoid of length L, diameter D, and winding density n carrying a current i. In particular, we'll find the field inside the coil and far from the ends, where we expect the field to be relatively uniform.

To use Ampère's law we must first select an Amperian loop *through* which current flows and *around* which we can calculate the magnetic field. The point at which we want to determine the field—in this case, a point inside the solenoid—must lie on the Amperian loop. **Figure 19-21** shows our choice of Amperian loop on a cut-away view of a section of the solenoid. As in Figure 19-18, the current is shown coming out of the page along the top part of the windings and going into the page along the bottom. We picked a rectangle as the Amperian loop and positioned it so that the bottom of the loop is parallel to the uniform field inside the solenoid. The left-hand side of Ampère's law, Equation 19-13, can then be written as

$$\sum B_{\parallel}\Delta\ell = B_{\parallel 1}\ell_1 + B_{\parallel 2}\ell_2 + B_{\parallel 3}\ell_3 + B_{\parallel 4}\ell_4 \tag{19-14}$$

Side 4 is in the region of zero magnetic field outside the solenoid, so $B_{\parallel 4} = 0$ and $B_{\parallel 4}\ell_4 = 0$. Side 1 runs partially through this region of zero field, so $B_{\parallel 1} = 0$ there; inside the solenoid the field is nonzero but is perpendicular to side 1, so $B_{\parallel 1} = 0$ there as well. The same is true for side 3, so in Equation 19-14 $B_{\parallel 1}\ell_1 = B_{\parallel 3}\ell_3 = 0$. The only side of the Amperian loop that makes a nonzero contribution to the path integral is side 2. Here the magnetic field of magnitude B is parallel to the side of length ℓ_2, so $B_{\parallel 2} = B$ and Equation 19-14 becomes

$$\sum B_{\parallel}\Delta\ell = B\ell_2 \tag{19-15}$$

On the right-hand side of the Ampère's law equation, we need to evaluate i_{through}, the current that passes through the loop. This quantity is *not* just the current in the solenoid i, because every winding of the coil that passes through the loop brings a contribution i to i_{through}. The length of solenoid enclosed by the loop is ℓ_2, and there are n windings per meter, so there are $n\ell_2$ windings that pass through the loop. Each winding carries current i, so the total current through the loop is $n\ell_2$ multiplied by i:

$$i_{\text{through}} = n\ell_2 i \tag{19-16}$$

We can now substitute Equations 19-15 and 19-16 into the two sides of Equation 19-13 for Ampère's law:

$$B\ell_2 = \mu_0(n\ell_2 i)$$

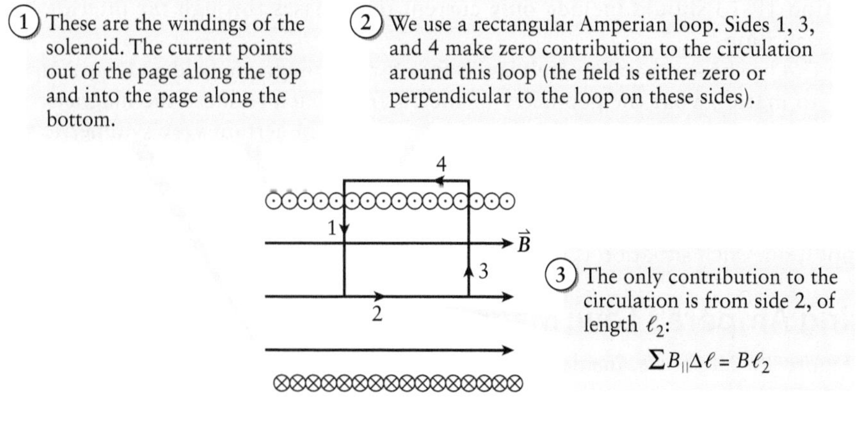

(1) These are the windings of the solenoid. The current points out of the page along the top and into the page along the bottom.

(2) We use a rectangular Amperian loop. Sides 1, 3, and 4 make zero contribution to the circulation around this loop (the field is either zero or perpendicular to the loop on these sides).

4

1

$\vec{B}$

3

2

(3) The only contribution to the circulation is from side 2, of length ℓ_2:

$\Sigma B_{\parallel}\Delta\ell = B\ell_2$

Figure 19-21 Ampère's law and the field of a solenoid The Amperian loop shown here helps us to calculate the magnetic field inside a long, straight solenoid.

The length ℓ_2 of the Amperian loop cancels, and we are left with

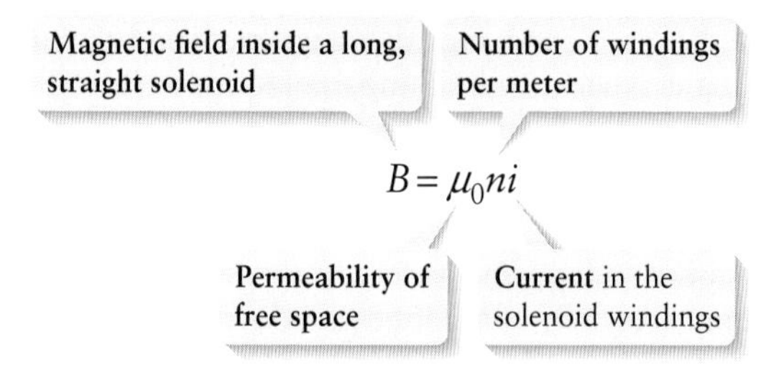

Magnetic field inside a long, straight solenoid (19-17)

In deriving Equation 19-17 we've assumed that the field inside the solenoid is perfectly uniform and parallel to the solenoid axis and that the field outside the solenoid is exactly zero. These assumptions are strictly valid only for an infinitely long solenoid. But Equation 19-17 is a good approximation at points near the middle of any solenoid, especially one whose length is large compared to its diameter.

Note that Ampère's law also tells us the *direction* of the magnetic field inside the solenoid. Just as for the case of a long, straight wire (Figure 19-17), point your right thumb in the direction of the current through the Amperian loop in Figure 19-21. This current points out of the page, so you should point your thumb in that direction. If you curl the fingers of your right hand, they will curl in a counterclockwise direction: from left to right below the enclosed windings, and from right to left above the enclosed windings. That's consistent with the direction of the magnetic field inside the solenoid, which is from left to right.

Equation 19-17 tells us that the more windings that can be packed into a length of the solenoid, the greater the field magnitude inside the solenoid. A typical solenoid used for electronic applications is therefore more likely to look like the one in **Figure 19-22**, with many layers of tightly packed windings, than the one shown in Figure 19-18.

In the following example we use Ampère's law to find the magnetic field due to a rather different distribution of current.

Paul Silverman/Fundamental Photographs

Figure 19-22 A real-life solenoid When the current to this solenoid is turned on, the iron rod that sticks out from its end experiences a strong magnetic force and is pulled into the solenoid. Such a device can be used to unlock a security door, as well as many other applications.

EXAMPLE 19-5 Magnetic Field Due to a Coaxial Cable

A coaxial cable consists of a solid conductor of radius R_1 surrounded by insulation, which in turn is surrounded by a thin conducting shell of radius R_2 made of either fine wire mesh or a thin metallic foil (**Figure 19-23**). The combination is enclosed in an outer layer of insulation. The inner conductor carries current in one direction, and the outer conductor carries it back in the opposite direction. For a coaxial cable that carries a constant current i, find expressions for the magnetic field (a) inside the inner conductor at a distance $r < R_1$ from its central axis, (b) in the space between the two conductors, and (c) outside the coaxial cable. Assume that the moving charge in the inner conductor is distributed uniformly over the volume of the conductor.

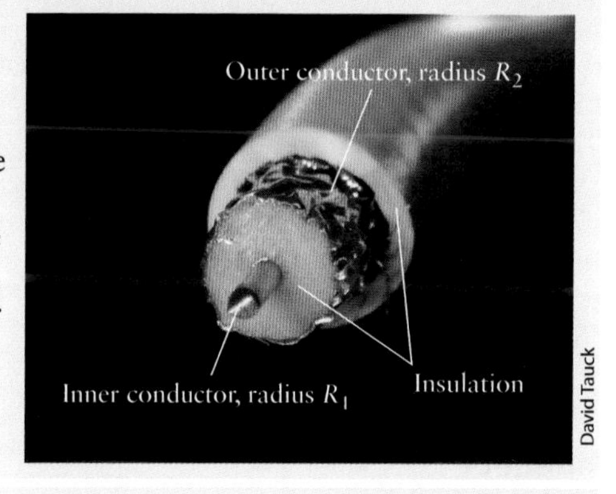

David Tauck

Figure 19-23 A coaxial cable Equal amounts of current flow in opposite directions in the inner and outer conductors of this cable.

Set Up

Both the inner conductor separately and the coaxial cable as a whole have the same cylindrical symmetry as a long, straight wire (Figure 19-17). So we expect that the field lines are circles concentric with the axis of the cable. Just as for the long, straight wire, this means that it's natural to choose circular paths concentric with the cable axis as the Amperian loops. To find the field in the three regions, we'll choose the radius r of the Amperian loop to be less than R_1 in part (a), between R_1 and R_2 in part (b), and greater than R_2 in part (c).

Ampère's law:

$$\sum B_{\parallel}\Delta\ell = \mu_0 i_{\text{through}} \quad (19\text{-}13)$$

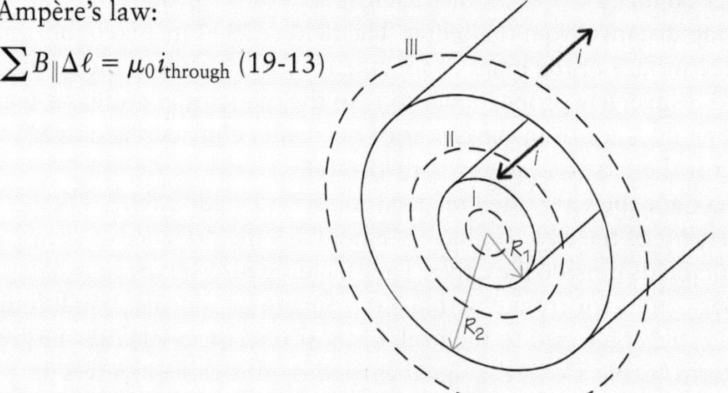

Dashed circles labeled I, II, and III: Amperian loops for parts (a), (b), and (c), respectively

Solve

(a) Find the field inside the inner conductor by using an Amperian loop of radius $r < R_1$.

Inside the inner conductor the magnetic field has magnitude B_{inner} and points tangent to the Amperian loop, so $B_{\parallel} = B_{inner}$. The left-hand side of the Ampère's law equation is

$$\sum B_{\parallel}\Delta\ell = B_{inner}\sum \Delta\ell = B_{inner}(2\pi r)$$

The Amperian loop encloses area πr^2, which is less than the cross-sectional area πR_1^2 of the inner conductor. The current through the loop is therefore a fraction $(\pi r^2)/(\pi R_1^2)$ of the total current i in the inner conductor:

$$i_{through} = i\left(\frac{\pi r^2}{\pi R_1^2}\right) = i\frac{r^2}{R_1^2}$$

Insert these into Equation 19-13 and solve for B_{inner}:

$$B_{inner}(2\pi r) = \mu_0 i\frac{r^2}{R_1^2}$$

$$B_{inner} = \mu_0 i\frac{r^2}{2\pi r R_1^2} = \frac{\mu_0 i r}{2\pi R_1^2}$$

(b) Find the field between the conductors by using an Amperian loop of radius r, where $R_1 < r < R_2$.

Between the two conductors the magnetic field of magnitude $B_{between}$ also points tangent to the Amperian loop, so as in part (a) the left-hand side of Equation 19-13 is

$$\sum B_{\parallel}\Delta\ell = B_{between}(2\pi r)$$

The Amperian loop encloses the entire inner conductor, so $i_{through} = i$. Insert these into Equation 19-13 and solve for $B_{between}$:

$$B_{between}(2\pi r) = \mu_0 i$$

$$B_{between} = \frac{\mu_0 i}{2\pi r}$$

(c) Find the field outside the cable by using an Amperian loop of radius $r > R_2$.

Just as in parts (a) and (b), outside the outer conductor the magnetic field of magnitude B_{outer} points tangent to the Amperian loop, so

$$\sum B_{\parallel}\Delta\ell = B_{outer}(2\pi r)$$

The Amperian loop encloses both conductors, each of which carries current i. Since the currents flow in opposite directions, the *net* current through the loop is $i_{through} = 0$. So Equation 19-13 tells us

$$B_{outer}(2\pi r) = \mu_0(0)$$

$$B_{outer} = 0$$

Reflect

Our result from (a) says that the magnetic field is zero at the center of the inner conductor ($r = 0$), then increases in direct proportion to r with increasing distance from the center. The field reaches its maximum value at the outer surface of the inner conductor ($r = R_1$). Between the conductors the field is inversely proportional to r, so the magnitude decreases with increasing distance from the center of the cable. Outside the outer conductor there is *zero* magnetic field.

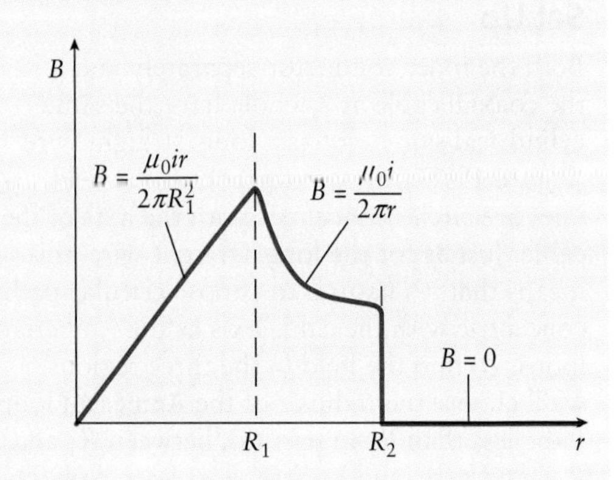

Coaxial cables are often referred to as "shielded" cables. The arrangement of the two conductors eliminates the presence of stray magnetic fields outside the cable. The shielding also serves to isolate the inner conductor from external electromagnetic signals. You'll find a coaxial cable connected to the back of most television sets (it's the "cable" in the term "cable TV"); the signal carried by this cable involves an alternating current and hence a varying magnetic field rather than a steady one, but the shielding principle is the same.

Magnetic Field of a Current Loop

An important special case for which Ampère's law is *not* helpful is the magnetic field produced by a current loop (a current-carrying wire bent into a circle). **Figure 19-24** shows some of the magnetic field lines for such a loop.

Unlike the case for a long, straight wire (Figure 19-17) or for a coaxial cable (Example 19-5), the magnetic field does *not* have the same magnitude at all points on a field line: The magnitude B is greater where the lines are closer together. So choosing an Amperian loop that coincides with a field line will not give us a simple equation for the magnitude B, as was the case in Example 19-5. To calculate the magnetic field at a given point in this situation, it's necessary to find the contribution to the field due to each short segment of the loop, then add those contributions using vector arithmetic. Such a calculation is beyond our scope. Here's the result for the magnitude of the magnetic field along the axis of the current loop, labeled y in Figure 19-24:

$$B = \frac{\mu_0 i}{2} \frac{R^2}{(R^2 + y^2)^{3/2}} \tag{19-18}$$

In Equation 19-18 i is the current in the loop, R is the radius of the loop, and y is the coordinate along the y axis, where $y = 0$ represents the plane of the loop. If we substitute $y = 0$ into Equation 19-18, we get the field magnitude at the very center of the loop. At points far from the loop, so y is much greater than R, we can replace $R^2 + y^2$ with y^2 to good approximation. Equation 19-18 then becomes

$$B = \frac{\mu_0 i}{2} \frac{R^2}{(y^2)^{3/2}} = \frac{\mu_0 i}{2} \frac{R^2}{|y|^3} \tag{19-19}$$

(We've added the absolute value signs because y can be positive or negative, but the field magnitude B must be positive.) So at large distances from a current loop, the magnetic field is inversely proportional to the *cube* of the distance from the loop.

This result is reminiscent of the *electric* field of an electric dipole (a combination of a positive charge q and a negative charge $-q$): We found in Example 16-7 (Section 16-5)

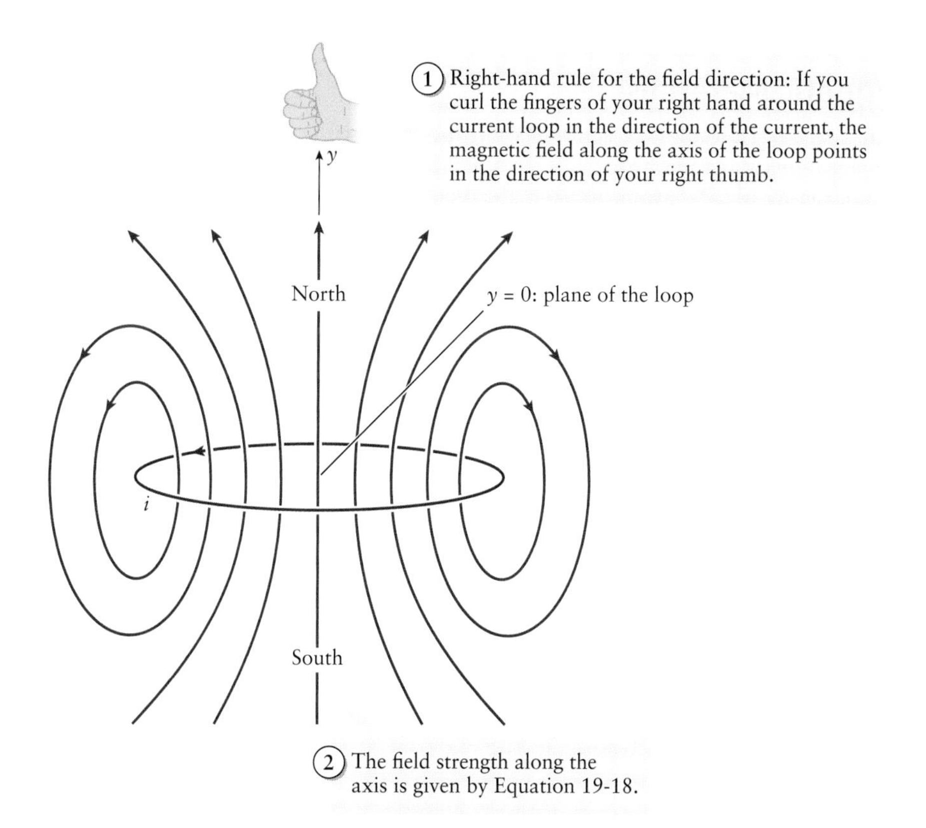

(1) Right-hand rule for the field direction: If you curl the fingers of your right hand around the current loop in the direction of the current, the magnetic field along the axis of the loop points in the direction of your right thumb.
y
North
y = 0: plane of the loop
i
South
(2) The field strength along the axis is given by Equation 19-18.

Figure 19-24 Magnetic field due to a current loop This magnetic field pattern is similar to the electric field pattern of an electric dipole (Figure 16-14).

that at large distances from an electric dipole, the electric field due to that dipole is inversely proportional to the cube of the distance. The overall magnetic field pattern of a current loop also has some similarities to the electric field pattern of an electric dipole. (Compare Figure 19-24 and Figure 16-14; if you rotate Figure 16-14 clockwise 90°, the similarities will be more evident.)

In light of these similarities, we use the term *magnetic dipole* to refer to a current loop. The two "poles" of a magnetic dipole are the points just above and just below the loop, as Figure 19-24 shows. We call these poles north and south by analogy to the poles of a bar magnet: The magnetic field points away from the current loop at its north pole and points toward the current loop at its south pole, just as for a bar magnet (see Section 19-2).

Note that unlike the two charges that make up an electric dipole, the north and south poles of a current loop can never be separated: The current loop must always have two sides! As we discussed in Section 19-2, a permanent magnet such as a bar magnet acts as a magnetic dipole. That's because a permanent magnet is really just a collection of *atomic* current loops, each the result of electron motions within the atom. Their combined effect is the same as electrons moving around the circular loop of wire in Figure 19-24. The poles of such a magnet can no more be separated than can the two sides of a current loop.

Earth's magnetic field is nearly that of a dipole, with the axis of the field tilted about 11° from Earth's rotation axis (Figure 19-2). The magnetic field is produced because molten material in the outer regions of Earth's core is in a state of continuous motion, and this motion gives rise to electric currents that generate the field. The photo that opens this chapter illustrates one dynamic consequence of our planet having a magnetic field.

Anyone who uses a compass to navigate makes use of Earth's magnetic field, which points generally from south to north. Other living organisms also take advantage of Earth's field to guide them from location to location. Pigeons, honeybees, and sea turtles, among others, rely to some extent on an internal magnetic compass to navigate. Sea turtles, for example, have been observed to travel hundreds of kilometers and still find their way back to their nesting sites along relatively direct paths. Yet when the turtles are transported away from their nests after a magnet has been attached to their heads, they take wildly circuitous routes back to the nesting site. The field of the attached magnet clearly disrupts the turtles' ability to determine their position using Earth's magnetic field.

Magnetic Materials

While there are circulating electrons within every kind of atom, not all materials have the same magnetic properties. In some materials there is a net rotation of electron charge within the atom, so each atom behaves like a current loop. In most cases these atomic current loops are randomly oriented, so their effects cancel out. But if the material is placed in a strong magnetic field, the atomic current loops experience a torque and align themselves with the magnetic field (see Figure 19-14b). As a result, the material behaves like a much larger current loop. If the magnetic field is turned off, random thermal motion will cause the atomic current loops to return to their original, nonaligned orientations. Materials that display this behavior are called **paramagnetic**. Everyday paramagnetic materials include aluminum and sodium. The net magnetic effect in a paramagnetic material is generally quite small; while an empty can made of (paramagnetic) aluminum acts like a current loop when brought next to a magnet, the magnetic force on the aluminum can is so small that a magnet can't pick it up.

In a handful of materials the interactions between adjacent atomic current loops are very strong. As a result, once the material is placed in a magnetic field the atomic current loops not only align with the field but can *remain* aligned after the field is turned off, leaving the material permanently magnetized. Iron is the most common of these materials, which are called **ferromagnetic** ("ferro" derives from the Latin word for iron). Any permanent magnet, such as one you might use to attach notes to a refrigerator door, is made of a ferromagnetic material. A permanent magnet can pick up objects made of a ferromagnetic material, such as a steel paper clip. The field of the permanent magnet causes the atomic current loops in the paper clip to align, making the paper clip a magnet itself. The magnetized paper clip is then attracted to the permanent magnet.

WATCH OUT! Earth is not a permanent magnet.

Our planet's core is made primarily of iron and nickel, both of which are ferromagnetic materials. It's common to conclude from this that the core is magnetized like a permanent magnet and that this gives rise to our planet's magnetic field. However, this cannot be true. Any ferromagnetic material loses its magnetism if it is heated above a certain temperature specific to that material: This critical temperature is 773°C for iron and 354°C for nickel. The temperature in Earth's core is in excess of 4400°C, so the iron and nickel in the core do *not* act like ferromagnetic materials. Instead, Earth's magnetic field is caused by electric currents in the molten material that makes up the outer regions of the core.

Most materials are neither paramagnetic nor ferromagnetic because their atoms have zero net electron current. When placed in a magnetic field, a small amount of atomic current appears, but the current loops end up aligned in the direction *opposite* to what happens for paramagnetic or ferromagnetic materials. (This is a consequence of *electromagnetic induction*, which we'll discuss in Chapter 20.) As a result, these materials, called **diamagnetic**, are slightly repelled by magnets rather than being attracted. In most cases, however, the repulsion is very weak.

GOT THE CONCEPT? 19-7 Ampère's Law

The field lines in Figure 19-24 are closed curves. If the distance around one such curve is L and the current in the loop is i, what is the average value around the closed curve of the component of magnetic field $B_{\parallel}$ parallel to the curve? (a) $\mu_0 i$; (b) $\mu_0 i/L$; (c) $2\mu_0 i$; (d) $2\mu_0 i/L$; (e) not enough information given to decide.

TAKE-HOME MESSAGE FOR Section 19-7

✔ Ampère's law relates the current through a wire to the magnetic field it generates.

✔ Current through a long, straight wire produces a magnetic field with circular field lines centered on the wire.

✔ Current through a solenoid (a straight, helical coil of wire) produces a relatively uniform magnetic field along its axis that is proportional to the winding density as well as the current.

✔ A current loop produces a more complicated magnetic field. A magnetic material can be thought of as a collection of atomic current loops.

19-8 Two current-carrying wires exert magnetic forces on each other

We've seen that a current-carrying wire experiences a force when placed in a magnetic field, and also that a current-carrying wire generates a magnetic field. Let's put these ideas together and look at the magnetic interaction between *two* current-carrying wires. (Note that these two wires do not exert *electric* forces on each other. That's because each wire has as much positive charge as negative charge and so is electrically neutral.)

Experiment shows that two parallel, straight wires carrying current in the same direction attract each other, and that two parallel, straight wires carrying current in opposite directions repel. Let's see why this is the case.

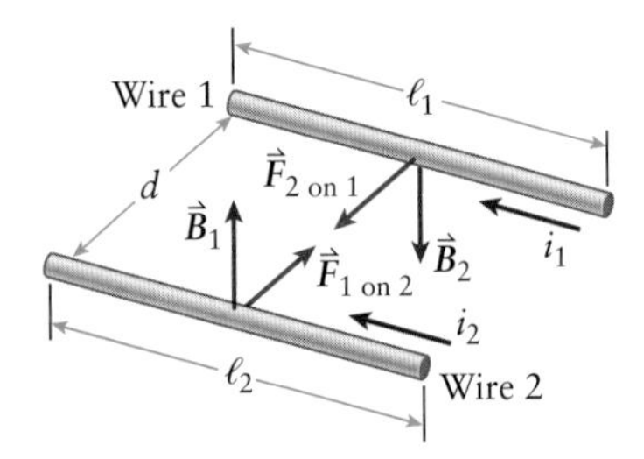

$\vec{B}_1$ = Magnetic field due to wire 1 at the position of wire 2

$\vec{B}_2$ = Magnetic field due to wire 2 at the position of wire 1

Figure 19-25 Magnetic forces between two current-carrying wires These two parallel wires carry current in the same direction and exert attractive magnetic forces on each other. If we reverse the direction of one of the currents, the forces become repulsive.

Figure 19-25 shows the situation. Wires 1 and 2 are long, straight, and parallel to each other and are separated by a distance d. The current i_1 in wire 1 sets up a magnetic field $\vec{B}_1$ at the position of wire 2, which carries current i_2. From Equation 19-9 the magnitude B_1 of this field is

$$B_1 = \frac{\mu_0 i_1}{2\pi d} \tag{19-20}$$

The right-hand rule for the field produced by a long, straight wire (Section 19-7) tells us that at the position of wire 2, $\vec{B}_1$ points upward and perpendicular to wire 2. To find the direction of the force $\vec{F}_{1\text{ on }2}$ that this field exerts on wire 2, use the right-hand rule for the direction of the magnetic force on a current-carrying wire (Section 19-5): This tells us that $\vec{F}_{1\text{ on }2}$ points toward wire 1, so the force attracts wire 2 to wire 1. The magnitude of the force on wire 2, of length ℓ_2, is given by Equation 19-5 with $\theta = 90°$ (since the direction of the current in wire 2 is perpendicular to the direction of $\vec{B}_1$):

$$F_{1\text{ on }2} = i_2\ell_2 B_1 \sin 90° = i_2\ell_2 B_1 \tag{19-21}$$

If we substitute B_1 from Equation 19-20 into Equation 19-21, we get

$$F_{1\text{ on }2} = i_2\ell_2\left(\frac{\mu_0 i_1}{2\pi d}\right) = \frac{\mu_0 i_1 i_2 \ell_2}{2\pi d} \tag{19-22}$$

The force per unit length on wire 2 is $F_{1\text{ on }2}$ (given by Equation 19-22) divided by the length ℓ_2 of wire 2:

$$\text{Magnetic force per unit length exerted by wire 1 on wire 2} = F_{1\text{ on }2}/\ell_2 = \frac{\mu_0 i_1 i_2}{2\pi d} \tag{19-23}$$

We can use the same procedure to find the magnetic force per unit length that wire 2 exerts on wire 1. The field $\vec{B}_2$ that the current i_2 in wire 2 produces at the position of wire 1 has magnitude $B_2 = \mu_0 i_2/(2\pi d)$ and points *downward* in Figure 19-25. From the right-hand rule for the force on a current-carrying wire, the force $\vec{F}_{2\text{ on }1}$ on wire 1 points toward wire 2 (the force is attractive); its magnitude is $F_{2\text{ on }1} = i_1\ell_1 B_2 = i_1\ell_1[\mu_0 i_2/(2\pi d)] = \mu_0 i_1 i_2 \ell_1/(2\pi d)$, where ℓ_1 is the length of wire 1. The force per unit length on wire 1 is then $F_{1\text{ on }2}$ divided by ℓ_1:

$$\text{Magnetic force per unit length exerted by wire 2 on wire 1} = F_{2\text{ on }1}/\ell_1 = \frac{\mu_0 i_1 i_2}{2\pi d} \tag{19-24}$$

The force magnitudes per unit length in Equations 19-23 and 19-24 are equal, and the forces $\vec{F}_{1\text{ on }2}$ and $\vec{F}_{2\text{ on }1}$ are opposite in direction. That's just what we would expect from Newton's third law.

What changes if we reverse the direction of the current i_1 in wire 1? The force *magnitudes* given by Equations 19-23 and 19-24 won't be affected, but the force *directions* will be. This will reverse the direction of the magnetic field $\vec{B}_1$ that wire 1 produces at the position of wire 2, and so will reverse the direction of the force $\vec{F}_{1\text{ on }2}$ on wire 2. So in this case wire 2 will be pushed away from wire 1 (it will be repelled). Reversing the direction of i_1 will also reverse the direction of the force $\vec{F}_{2\text{ on }1}$ that wire 2 exerts on wire 1, so this force will push wire 1 away from 2 (again, it will be repelled). So we conclude that

Two parallel current-carrying wires attract each other if they carry current in the same direction, and repel each other if they carry current in opposite directions.

EXAMPLE 19-6 Wires in a Computer

The two long, straight wires that run along the back of a computer case to power the cooling fan carry 0.110 A in opposite directions. The wires are separated by 5.00 mm. (a) Find the force per unit length (magnitude and direction) that these wires exert on each other. (b) The mass per unit length of the wire is 5.00×10^{-3} kg/m. What acceleration does one of the wires experience due to this force?

Set Up

We'll use Equation 19-23 to find the force per unit length that one wire exerts on the other. (As we saw with Equation 19-24, the force per unit length has the same magnitude for either wire.) If we assume that this force equals the net external force on the wire, we can use Newton's second law to calculate the acceleration of the wire.

Magnetic force per unit length exerted by wire 1 on wire 2:

$$F_{1\text{ on }2}/\ell_2 = \frac{\mu_0 i_1 i_2}{2\pi d} \qquad (19\text{-}23)$$

Newton's second law:

$$\sum \vec{F}_{\text{ext}} = m\vec{a} \qquad (4\text{-}2)$$

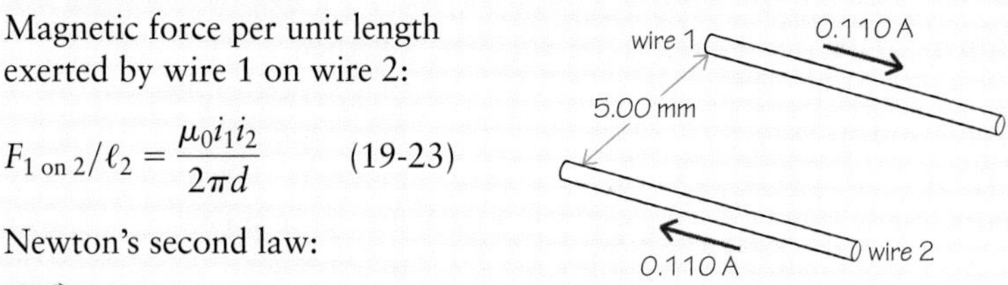

Solve

(a) The currents are in opposite directions, so the force is repulsive (it pushes the two wires apart). Use Equation 19-23 to find the magnitude of the force per unit length.

The two wires are separated by $d = 5.00\text{ mm} = 5.00 \times 10^{-3}\text{ m}$ and carry currents with the same magnitude: $i_1 = i_2 = 0.110\text{ A}$. The force per unit length on either wire is

$$= \frac{(4\pi \times 10^{-7}\text{ T}\cdot\text{m/A})(0.110\text{ A})(0.110\text{ A})}{2\pi(5.00 \times 10^{-3}\text{ m})}$$

$$= 4.84 \times 10^{-7}\text{ T}\cdot\text{A} = 4.84 \times 10^{-7}\text{ N/m}$$

(Recall that $1\text{ T} = 1\text{ N/(A}\cdot\text{m)}$.)

(b) Use Newton's second law to find the acceleration of the wire.

Our result for part (a) says that a 1-m length of wire would experience a force of magnitude 4.84×10^{-7} N. Since this wire has mass per unit length 5.00×10^{-3} kg/m, a 1-m length would have mass 5.00×10^{-3} kg. The acceleration is

$$a = |\vec{a}| = \frac{|\sum \vec{F}_{\text{ext}}|}{m} = \frac{4.84 \times 10^{-7}\text{ N}}{5.00 \times 10^{-3}\text{ kg}}$$

$$= 9.68 \times 10^{-5}\text{ m/s}^2$$

Reflect

The force and acceleration are both very gentle, so the effect on these wires will be almost imperceptible. In applications with very large currents, however, the magnetic forces between conductors can be substantial.

GOT THE CONCEPT? 19-8 Forces on a Current-Carrying Coil

A very flexible helical coil is suspended as shown in **Figure 19-26**. What will happen when a sizable current i is sent through the coil? (a) The coils will be pulled together. (b) The coils will be pushed apart. (c) Some of the coils will be pulled together, while others will be pulled apart. (d) There will be no net effect on the coils.

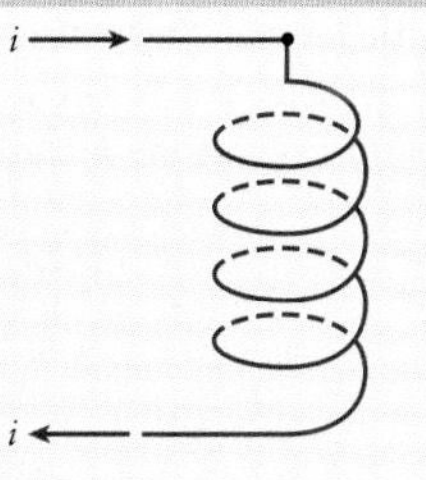

Figure 19-26 A hanging coil What happens to this coil when the current is turned on?

TAKE-HOME MESSAGE FOR Section 19-8

✔ Two wires attract each other when carrying current in the same direction and repel each other when carrying currents in opposite directions.

✔ The forces per unit length on the wires are equal in magnitude and opposite in direction, exactly as required by Newton's third law.

Key Terms

Ampère's law
Amperian loop
circulation
current loop
diamagnetic
electromagnetism
ferromagnetic
magnet
magnetic dipole
magnetic field
magnetic force
magnetic poles
magnetism
mass spectrometer
normal
paramagnetic
permeability of free space
right-hand rule
solenoid
tesla

Chapter Summary

Topic	Equation or Figure
Magnetism and magnetic forces: Magnetic forces are present whenever moving charged objects interact with each other. (By comparison, electric forces are present whenever charged objects interact with each other, whether moving or not.) A magnet is an object that contains charges in continuous motion. A magnet sets up a magnetic field in the space around it; a second magnet responds to that field and can be attracted or repelled, depending on its orientation.	(a) These magnets attract. 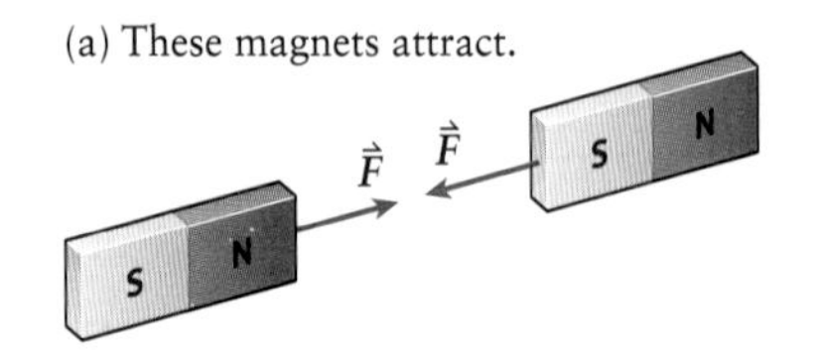 (b) These magnets repel. 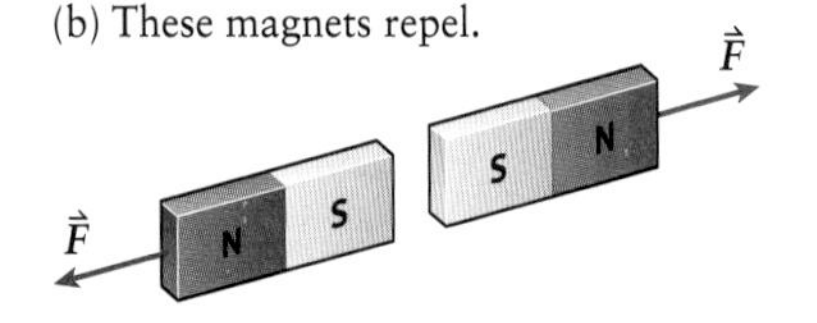 (Figure 19-3 a/b)
Magnetic force on a moving charged particle: A single charged particle can experience a magnetic force when moving in a magnetic field. The magnitude of the force depends on both the speed of the particle and the direction of the particle relative to the magnetic field. The direction of the force is perpendicular to both the velocity $\vec{v}$ and the magnetic field $\vec{B}$, and is given by a right-hand rule.	$F = \|q\|vB\sin\theta$ (19-1) Magnitude of the magnetic force on a moving charged particle Magnitude of the magnetic field Angle between the direction of the particle's velocity $\vec{v}$ and the direction of the magnetic field $\vec{B}$ **Magnitude** of the particle's **charge** **Speed** of the particle
	(a) 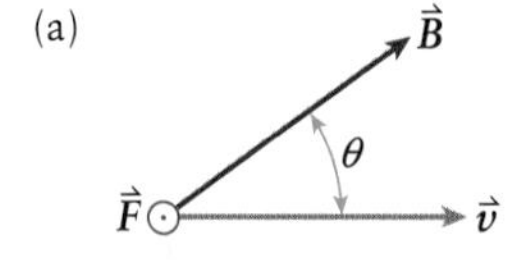 If $q > 0$, the force vector in this situation points out of the page (as indicated by the circle with a dot). (b) $\vec{B}$ θ $\vec{F}$ $\vec{v}$ If $q < 0$, the force vector in this situation points into the page (as indicated by the circle with an X). (Figure 19-6)

Particle trajectories in a magnetic field: A charged particle moving in a magnetic field and subject to no other forces can move in a circular trajectory whose radius depends on its speed, mass, and charge as well as the magnetic field magnitude. This is the principle of the mass spectrometer.

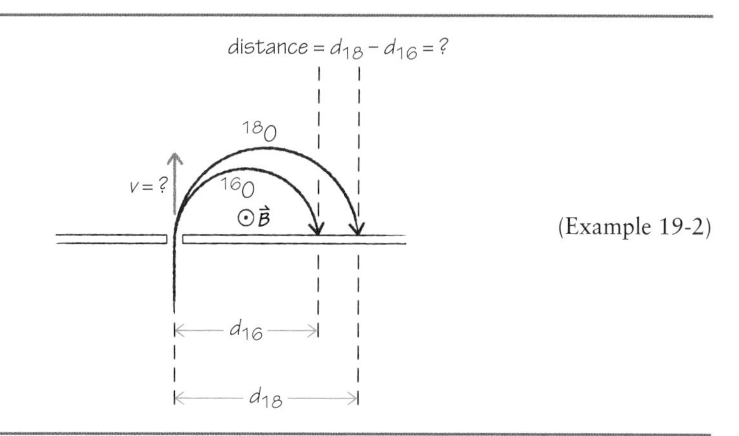

(Example 19-2)

Magnetic forces on current-carrying wires: If a wire carries a current and is placed in a magnetic field, it experiences a magnetic force. This force is the sum of the magnetic forces acting on the individual moving charges within the wire. The force magnitude depends on the amount of current and the orientation of the wire relative to the magnetic field; its direction is given by a right-hand rule.

$$F = i\ell B \sin\theta \quad (19\text{-}5)$$

- **Magnitude of the magnetic force** on a current-carrying wire (F)
- **Current** in the wire (i)
- **Length** of the wire (ℓ)
- **Magnitude of the magnetic field** (assumed uniform over the length of the wire) (B)
- **Angle** between the direction of the current and the direction of the magnetic field $\vec{B}$ (θ)

The force $\vec{F}$ on a current-carrying wire in a magnetic field $\vec{B}$ is perpendicular to both $\vec{B}$ and the length of the wire. The direction is given by a right-hand rule.

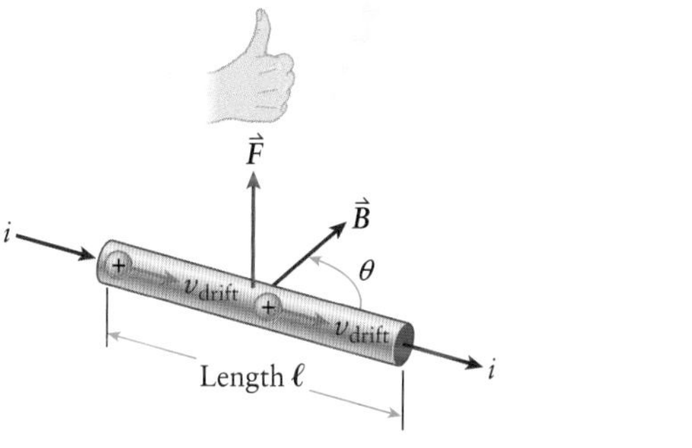

(Figure 19-10)

Magnetic torque on a current loop: A current-carrying loop of wire can experience a torque when placed in a magnetic field. The magnitude and direction of the torque depend on how the current loop is oriented relative to the direction of the magnetic field. Electric motors make use of this principle.

$$\tau = iAB \sin\phi \quad (19\text{-}8)$$

- **Magnitude of the magnetic torque** on a current loop (τ)
- **Current** in the loop (i)
- **Area** of the loop (A)
- **Magnitude** of the magnetic field (B)
- **Angle** between the normal to the plane of the loop and the direction of the magnetic field $\vec{B}$ (ϕ)

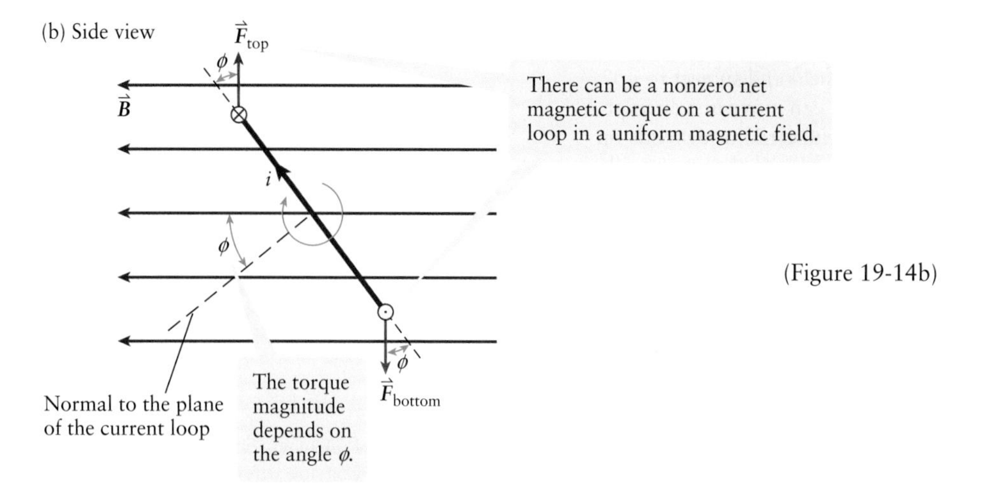

(Figure 19-14b)

Ampère's law and the field produced by moving electric currents: A long, straight, current-carrying wire produces a relatively simple magnetic field in the space around it. Ampère's law—which relates the circulation of magnetic field around a closed loop to the amount of current through that loop—can be used to find the magnetic field produced by currents with other simple geometries.

Magnitude of the magnetic field due to a long, straight wire

Permeability of free space

$$B = \frac{\mu_0 i}{2\pi r} \tag{19-9}$$

Current in the wire

Distance from the wire to the location where the field is measured

Circulation of magnetic field around an Amperian loop

Permeability of free space

$$\sum B_{||}\Delta \ell = \mu_0 i_{through} \tag{19-13}$$

Current through the Amperian loop

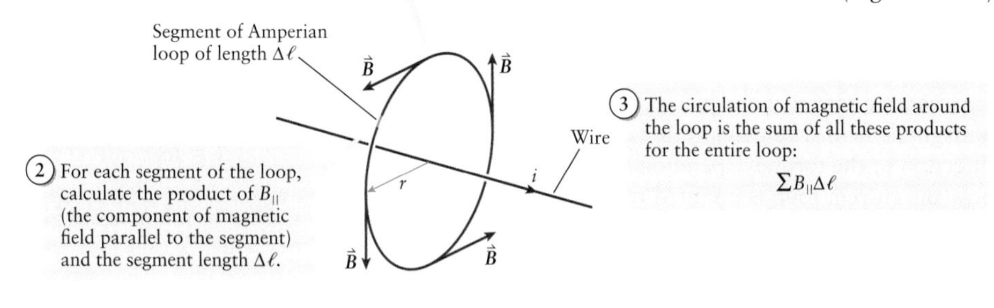

(Figure 19-20)

Current loops and magnetic materials: A current loop is called a magnetic dipole because the magnetic field that it produces is similar to the electric field produced by an electric dipole. A permanent magnet is a material in which the atoms behave like individual current loops, many of which are oriented in the same direction so that their individual magnetic fields add to make a strong field.

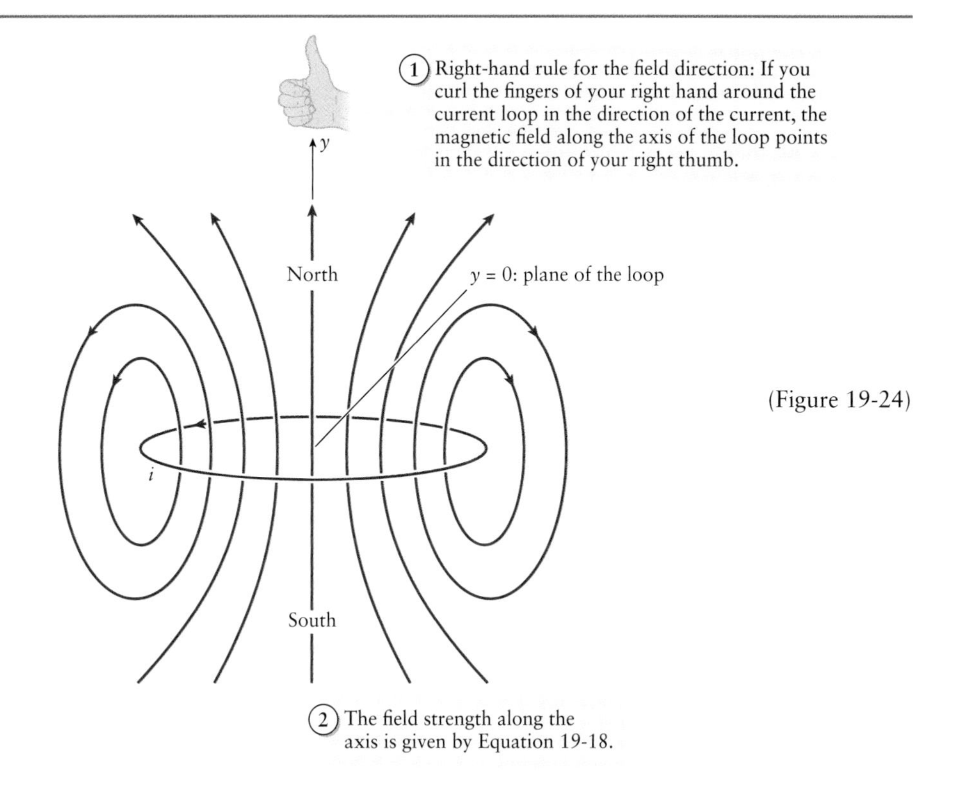

(Figure 19-24)

Force between current-carrying wires: Two parallel current-carrying wires exert magnetic forces on each other: The current in one wire produces a magnetic field, and the current in the other wire responds to that field. The two wires attract if the currents are in the same direction and repel if the currents are in opposite directions.

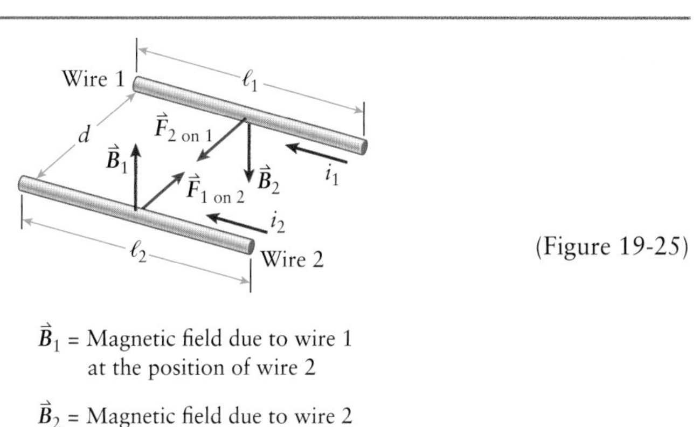

(Figure 19-25)

Answer to What do you think? Question

(c) Equation 19-1 (Section 19-3) gives the magnitude of the magnetic force on a charged particle moving in a magnetic field. The magnitude is proportional to sin θ, where θ is the angle between the magnetic field and the velocity of the particle. The sine function is greatest when $\theta = 90°$, in which case the particle is moving perpendicular to the field. This is in stark contrast to the *electric* force on a charged particle, which does not depend on the direction or magnitude of the particle's velocity.

Answers to Got the Concept? Questions

19-1 (b) Two charged objects exert *electric* forces on each other whether or not the objects are moving. But in order for these objects to exert *magnetic* forces on each other, *both* objects must be in motion. (One must be in motion to produce a magnetic field, and the other must be in motion to experience a force due to that field.) Since only one object is in motion, there is no magnetic force between the two objects.

19-2 (a) To apply the right-hand rule to the proton in Figure 19-7, start by pointing the fingers of your right hand in the direction of the proton's initial velocity. Orient your palm so that when you make a "slapping" motion with your hand, your palm points in the direction of the magnetic field (into the page). If you then stick your right thumb out straight, it points toward the top of the page. This is the direction of the force on the proton. This force makes the proton's trajectory bend upward in the figure.

19-3 (a), (c) Electrons have a negative charge, so the direction of the magnetic force each electron experiences is opposite to the direction your thumb points when applying the right-hand rule. For electron A, point the fingers of your right hand in the direction of the electron's initial velocity. Then orient your palm so that when you make a "slapping" motion with your hand, your palm points in the direction of the magnetic field (into the page). If you then stick your right thumb out straight, it points to the right, which means the electron feels a force to the left. So electron A feels a force toward the point at which the proton enters the field. For electron B, the right-hand rule has your right thumb pointing toward the bottom of the page, so the force on the electron is toward the top of the page. It does not bend toward the proton's entry point. The right-hand rule predicts the same direction for the magnetic force on electron C, but in this case a force toward the top of the page does bend the electron's trajectory toward the point where the proton enters the field. Finally, when you apply the right-hand rule to electron D, your right thumb sticks out to the left. For the negatively charged electron the magnetic force is therefore to the right, so the trajectory of electron D does not bend toward the proton's entry point into the field.

19-4 (c) Equation 19-3 shows that for a given speed v, the radius of the circular path followed by an ion in a mass spectrometer is $r = mv/(qB)$. This is directly proportional to the ratio of the ion mass m to its charge q. Compared to the value of the ratio m/q for a $^{16}O^{+}$ ion ($m = 16$ u, $q = e$), the value for a $^{16}O^{2+}$ ion is $1/2$ as great ($m = 16$ u, $q = 2e$), the value for a $^{32}S^{+}$ ion is twice as great ($m = 32$ u, $q = e$), and the value for a $^{32}S^{2+}$ ion is the same ($m = 32$ u, $q = 2e$).

19-5 (a) Force is zero; (b) $-z$ direction; (c) $+y$ direction; (d) $+z$ direction; (e) $-y$ direction; (f) force is zero. The force on a current-carrying wire is zero if the wire axis lies along the direction of the magnetic field, as in cases (a) and (f). In cases (b), (c), (d), and (e), we use the right-hand rule for the magnetic force on a current-carrying wire to find the direction of the force: Swing the extended fingers of your right hand, palm first, from the direction of the current to the direction of $\vec{B}$. Your extended right thumb then points in the direction of the force on the wire.

19-6 (e) Doubling the number of turns of wire will double the magnetic torque on the current loop. But this also doubles the mass and the moment of inertia of the current loop, so the angular acceleration—equal to the torque divided by the moment of inertia—will be unaffected. For a real electric motor, however, the moment of inertia is due partially to the mass of the coil and partially to the mass of what the motor is turning, such as the fan blades in Figure 19-13a. As a result, doubling the number of turns will increase the maximum torque by a factor of two but increase the moment of inertia by less than a factor of two, and the maximum angular acceleration will in fact increase.

19-7 (b) The net current through the closed curve is i, since the current loop passes once through the plane of the curve. From Ampère's law the circulation of the magnetic field around the closed curve is $\sum B_{\|}\Delta\ell = \mu_0 i_{\text{through}} = \mu_0 i$. The left-hand side of this equation is the average value of $B_{\|}$ multiplied by the total distance around the closed curve, so $(B_{\|})_{\text{average}}L = \mu_0 i$ and $(B_{\|})_{\text{average}} = \mu_0 i/L$.

19-8 (a) Each segment of the coil is a piece of wire and is attracted to the piece of wire in the coils directly above and below it (in each of which the current flows in the same direction). Each piece of wire is also repelled by the pieces of wire on the opposite side of the coil above it and the opposite side of the coil below it. But these pieces are at a greater distance d, so these repulsive forces are smaller than the attractive forces (see Equation 19-23). The net result is that the coils attract each other and so pull together.

Questions and Problems

In a few problems you are given more data than you actually need; in a few other problems you are required to supply data from your general knowledge, outside sources, or informed estimate.

Interpret as significant all digits in numerical values that have trailing zeros and no decimal points.

For all problems use $g = 9.80\ \text{m/s}^2$ for the free-fall acceleration due to gravity. Neglect friction and air resistance unless instructed to do otherwise.

•	Basic, single-concept problem
••	Intermediate-level problem; may require synthesis of concepts and multiple steps
•••	Challenging problem
SSM	*Solution is in Student Solutions Manual*
Example	*See worked example for a similar problem.*

Conceptual Questions

1. • You are given three iron rods. Two of them are magnets but the third one is not. How could you use the two magnets to find that the third rod is not magnetized? SSM

2. • In a lightning strike there is a negative charge moving rapidly from a cloud to the ground. In what direction is a lightning strike deflected by Earth's magnetic field?

3. • Physicists refer to crossed electric and magnetic fields as a *velocity selector*. In the same sense, the deflection of charged particles in a strong magnetic field perpendicular to their motion can be thought of as a *momentum selector*. Why?

4. • If a magnetic field exerts a force on moving charged particles, is it capable of doing work on the particles? Explain your answer.

5. •• A velocity selector consists of crossed electric and magnetic fields, with the magnetic field directed toward the top of the page. A beam of positively charged particles passing through the velocity selector from left to right is undeflected by the fields. (a) In what direction is the electric field? (left, right, toward the top of the page, toward the bottom of the page, into the page, out of the page) (b) The direction of the particle beam is reversed so that it travels from right to left. Is it deflected? If so, in what direction? (c) A beam of electrons (negatively charged) moving with the same speed is passed through from left to right. Is it deflected? If so, in what direction? SSM

6. • A current-carrying wire is in a region where there is a magnetic field, but there is no magnetic force acting on the wire. How can this be?

7. • A long, straight current-carrying wire is placed in a cubic region that has a uniform magnetic field as shown in **Figure 19-27**. Does the force on the wire depend on the width of the magnetic field? Explain your answer.

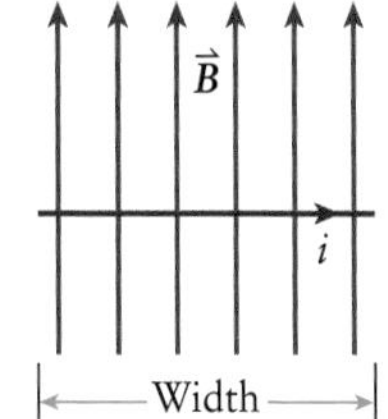

Figure 19-27 Problem 7

8. • How is it possible for an object that experiences no net magnetic force to experience a net magnetic torque?

9. • In telephone lines two wires carrying currents in opposite directions are twisted together. How does this reduce the magnetic fields surrounding the wires? SSM

10. • A power cord for an electronic device consists of two parallel straight wires carrying currents in opposite directions. Is there any force between them? Explain your answer.

11. • Parallel wires exert magnetic forces on each other. What about perpendicular wires? Explain your answer.

Multiple-Choice Questions

12. • The magnetic force on a charged moving particle
   A. depends on the sign of the charge on the particle.
   B. depends on the magnetic field at the particle's instantaneous position.
   C. is in the direction which is mutually perpendicular to the direction of motion of the charge and the direction of the magnetic field.
   D. is proportional both to the charge and to the magnitude of the magnetic field.
   E. is described by all of the above options, A through D.

13. • A proton traveling to the right enters a region of uniform magnetic field that points into the page. When the proton enters this region, it will be
   A. deflected out of the plane of page.
   B. deflected into the plane of page.
   C. deflected toward the top of the page.
   D. deflected toward the bottom of the page.
   E. unaffected in its direction of motion. SSM

14. • An electron is moving northward in a magnetic field. The magnetic force on the electron is toward the northeast. What is the direction of the magnetic field?
   A. up
   B. down
   C. west
   D. south
   E. This situation cannot exist because of the orientation of the velocity and force vectors.

15. • A proton with a velocity along the $+x$ axis enters a region where there is a uniform magnetic field $\vec{B}$ in the $+y$ direction. You want to balance the magnetic force with an electric field so that the proton will continue along a straight line. The electric field should be in the
   A. $+x$ direction.
   B. $-x$ direction.
   C. $+z$ direction.
   D. $-z$ direction.
   E. $-y$ direction.

16. • A circular flat coil that has N turns, encloses an area A, and carries a current i has its central axis parallel to a uniform magnetic field $\vec{B}$ in which it is immersed. The net force on the coil is
   A. zero.
   B. $NiAB$.
   C. NiB.
   D. iBA.
   E. NiA.

17. • A circular flat coil that has N turns, encloses an area A, and carries a current i has its central axis parallel to a uniform magnetic field $\vec{B}$ in which it is immersed. The net torque on the coil is
   A. zero.
   B. $NiAB$.
   C. NiB.
   D. iBA.
   E. NiA. SSM

18. • A very long, straight wire carries a constant current. The magnetic field a distance d from the wire and far from its ends varies with distance d according to
   A. d^{-3}.
   B. d^{-2}.
   C. d^{-1}.
   D. d.
   E. d^2.

19. • A solenoid carries a current. If the radius of the solenoid were doubled and all other quantities remained the same, the magnetic field inside the solenoid would
   A. remain the same.
   B. be twice as strong as initially.
   C. be half as strong as initially.
   D. be one-quarter as strong as initially.
   E. be four times as strong as initially.

20. • Two parallel wires carry currents in opposite directions, as shown in Figure 19-28. Which of the following statements is correct?

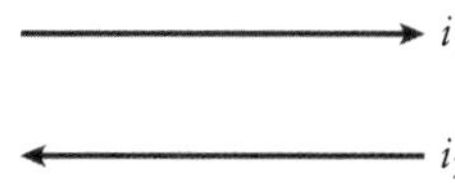

Figure 19-28 Problem 20

   A. The force on the i_2 wire is upward, and the force on the i_1 wire is upward.
   B. The force on the i_2 wire is downward, and the force on the i_1 wire is upward.
   C. The force on the i_2 wire is upward, and the force on the i_1 wire is downward.
   D. The force on the i_2 wire is downward, and the force on the i_1 wire is downward.
   E. Neither wire experiences a net force.

21. •• Two current-carrying wires are perpendicular to each other. One wire lies horizontally with the current directed toward the east. The other wire is vertical with the current directed upward. What is the direction of the net magnetic force on the horizontal wire due to the vertical wire?
   A. east
   B. west
   C. south
   D. north
   E. zero force SSM

Estimation/Numerical Analysis

22. •• Estimate the magnitude of the force per unit length acting on a pair of high-voltage transmission lines that carry current from a power plant to a distant substation for commercial use.

23. • Estimate the magnitude of the magnetic field at the ground level beneath standard power lines in your neighborhood.

24. • Estimate the magnitude of the magnetic field needed to separate two uranium ions (one has a mass of 235 u, one has a mass of 238 u) into circular paths with radii separated by 1 cm. Assume the two ions enter the magnetic field with the same velocity and the same charge.

25. • Honeybees can acquire a small net charge on the order of 1 pC as they fly through the air and interact with plants. Estimate

the magnetic force on a honeybee due to Earth's magnetic field as the bee flies near the ground from east to west.

26. •• Estimate the maximum torque on a coil in a typical electric drill motor.

27. • The magnetic field due to a current-carrying cylinder of radius 1 cm is measured at various points ($r < 1$ cm and $r > 1$ cm). Graph the magnitude of the magnetic field B as a function of r and use this graph to find the functional relationship between the magnetic field and the radial distance. *Hint:* The magnetic field will be described by two different functions—one for inside the cylinder and one for outside the cylinder. After making an initial graph, you may want to graph the data for the inside and outside separately.

r(m)	B(T)	r(m)	B(T)
0.001	0.00050	0.015	0.00353
0.002	0.00100	0.020	0.00250
0.003	0.00152	0.025	0.00200
0.004	0.00200	0.030	0.00180
0.005	0.00252	0.035	0.00143
0.006	0.00300	0.040	0.00125
0.007	0.00350	0.045	0.00110
0.008	0.00401	0.050	0.00103
0.009	0.00453	0.100	0.000502
0.010	0.00500		

Problems

19-1 Magnetic forces are interactions between two magnets
19-2 Magnetism is an interaction between moving charges
19-3 A moving point charge can experience a magnetic force

28. • Convert the units for the following expressions for magnetic fields as directed (recall 1 G = 10^{-4} T):

A. 5.00 T = ____________ G
B. 25,000 G = ____________ T
C. 7.43 mG = ____________ μT
D. 1.88 mT = ____________ G

29. • Determine the directions of the magnetic forces that act on positive charges moving in the magnetic fields as shown in **Figure 19-29.** SSM Example 19-1

Figure 19-29 Problem 29

30. • Determine the direction of the missing vector, $\vec{v}$, $\vec{B}$, or $\vec{F}$, in the scenarios shown in **Figure 19-30.** All moving charges are positive. Example 19-1

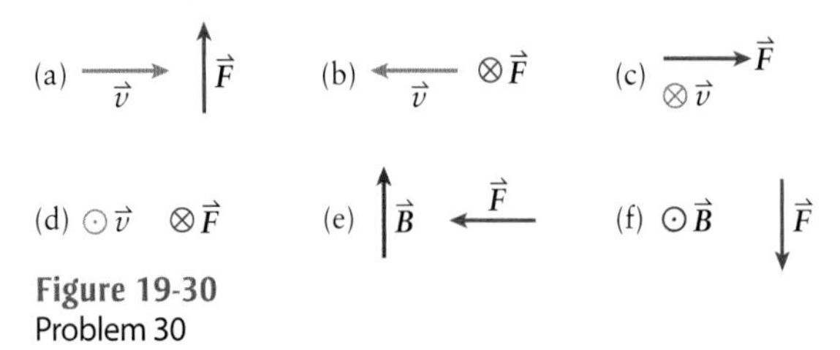

Figure 19-30 Problem 30

31. • A +1 C charge moving at 1 m/s makes an angle of 45° with a uniform, 1-T magnetic field. What is the magnitude of the magnetic force that the charge experiences? Example 19-1

32. • An electron is moving with a speed of 18 m/s in a direction parallel to a uniform magnetic field of 2.0 T. What are the magnitude and direction of the magnetic force on the electron? Example 19-1

33. • A proton P travels with a speed of 18 m/s toward the top of the page through a uniform magnetic field of 2.0 T directed into the page, as shown in **Figure 19-31.** What are the magnitude and direction of the magnetic force on the proton? SSM Example 19-1

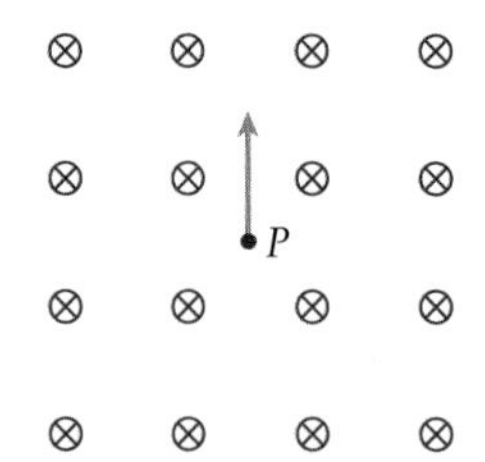

Figure 19-31 Problem 33

34. • A proton is propelled at 2×10^6 m/s perpendicular to a uniform magnetic field. If it experiences a magnetic force of 5.8×10^{-13} N, what is the magnitude of the magnetic field? Example 19-1

35. • An electron moves with a velocity of 1.0×10^7 m/s in the x–y plane at an angle of 45° to both the $+x$ and $+y$ axes. There is a magnetic field of 3.0 T in the $+y$ direction. Calculate the magnetic force (magnitude and direction) on the electron. SSM Example 19-1

36. •• There is a uniform magnetic field of magnitude 2.2 T in the $+z$ direction. Find the magnitude of the force on a particle of charge −1.2 nC if its velocity is (a) 1.0 km/s in the y–z plane in a direction that makes an angle of 40° with the z axis and (b) 1.0 km/s in the x–y plane in a direction that makes an angle of 40° with the x axis. Example 19-1

19-4 A mass spectrometer uses magnetic forces to differentiate atoms of different masses

37. •• A beam of protons is directed in a straight line along the $+z$ direction through a region of space in which there are crossed electric and magnetic fields. If the electric field is 500 V/m in the $-y$ direction and the protons move at a constant speed of 10^5 m/s, what must be the magnitude and direction of the magnetic field such that the beam of protons continues along its straight-line trajectory? Example 19-2

38. • A beam of ions (each ion has a charge $q = -2e$ and a kinetic energy of 4.00×10^{-13} J) is deflected by the magnetic field of a bending magnet as shown in **Figure 19-32.** The radius of curvature of the beam is 20.0 cm, and the strength of the magnetic field is 1.50 T. (a) What is the mass of one of the ions in the beam? (b) Sketch the path for the given ions in the given magnetic field as a reference path. Then sketch a path for a more massive doubly ionized negative ion and a less massive doubly ionized positive ion, both with the same speed as the given ions, for comparison. Example 19-2

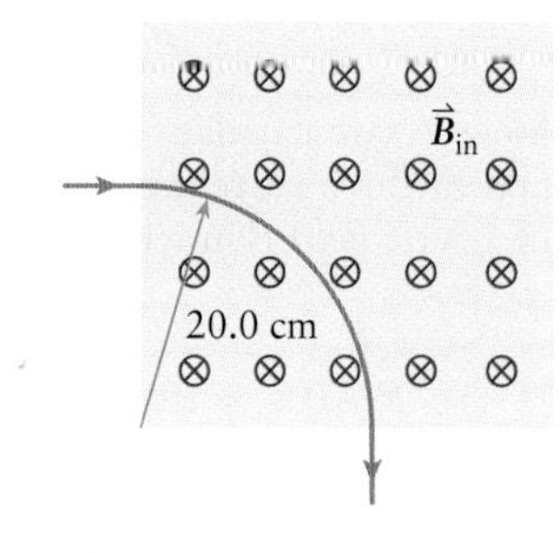

Figure 19-32 Problem 38

39. • A proton moves in a circle with a speed of 280 m/s through a uniform magnetic field of 2.0 T directed

perpendicular to the circular path. What is the radius of the circle? Example 19-2

19-5 Magnetic fields exert forces on current-carrying wires

40. • A 1.5-m length of straight wire experiences a maximum force of 2.0 N when in a uniform magnetic field that is 1.8 T. What current must be passing through it? Example 19-3

41. • A straight segment of wire 35.0 cm long carrying a current of 1.40 A is in a uniform magnetic field. The segment makes an angle of 53° with the direction of the magnetic field. If the force on the segment is 0.200 N, what is the magnitude of the magnetic field? SSM Example 19-3

42. • A straight wire of length 0.50 m is conducting a current of 2.0 A and makes an angle of 30° with a 3.0-T uniform magnetic field. What is the magnitude of the force exerted on the wire? Example 19-3

43. • A wire of length 0.50 m is conducting a current of 8.0 A in the $+x$ direction through a 4.0-T uniform magnetic field directed parallel to the wire. What are the magnitude and direction of the magnetic force on the wire? Example 19-3

44. • A wire of length 0.50 m is conducting a current of 8.0 A toward the top of the page and through a 4.0-T uniform magnetic field directed into the page as shown in **Figure 19-33**. What are the magnitude and direction of the magnetic force on the wire? Example 19-3

Figure 19-33 Problem 44

45. • A straight wire is positioned in a uniform magnetic field so that the maximum force on it is 4.0 N. If the wire is 80 cm long and carries a current that is 2 A, what is the magnitude of the magnetic field? Example 19-3

19-6 A magnetic field can exert a torque on a current loop

46. • A square loop 10.0 cm on a side with 100 turns of wire experiences a minimum torque of zero and a maximum torque of 0.0450 N · m in a uniform magnetic field. If the current in the loop is 2.82 A, calculate the magnetic field magnitude. Example 19-4

47. • What is the torque on a round loop of wire that carries a current of 1.00×10^2 A, has a radius of 10.0 cm, and whose plane makes an angle of 30.0° with a uniform magnetic field of 0.244 T? How does the answer change if the angle decreases to 10.0°? Increases to 50.0°? SSM Example 19-4

48. •• A wire loop with 50 turns is formed into a square with sides of length s. The loop is in the presence of a 1.5-T uniform magnetic field that points in the negative y direction. The plane of the loop is tilted off the x axis by 15° (**Figure 19-34**). If 2.0 A of current flows through the loop and the loop experiences a torque of magnitude 0.035 N · m, what are the lengths of the sides s of the square loop? Example 19-4

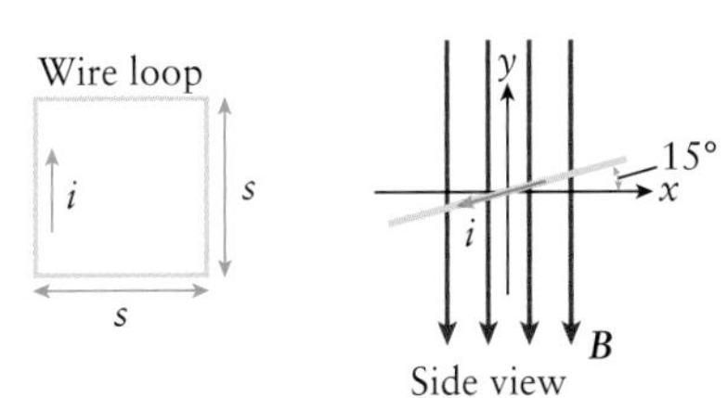

Figure 19-34 Problem 48

49. •• **Medical** MRI (Magnetic Resonance Imaging) scans are often prohibited for patients with an implanted device such as a pacemaker. There are multiple reasons for this, including electromagnetic interference that confuses the device. Here we consider another potential issue: the physical torque on the device due to the magnetic field. Suppose a patient's pacemaker implant requires 2.0 mA of current to function. The leads of the device travel close together until they reach the heart and then split off to touch the top and bottom of the heart. So we can approximate the size of the current loop as half the cross sectional area of the heart. Treat the heart as a box that is 12 cm long and 8.0 cm wide. Most MRI machines use a magnetic field of 1.5 T. Calculate the maximum torque on the current loop. Example 19-4

19-7 Ampère's law describes the magnetic field created by current-carrying wires

50. • A long, straight wire carries current in the $+z$ direction (out of the page). Determine the direction of the magnetic field due to the current at the points O, P, Q, and R (**Figure 19-35**). Example 19-5

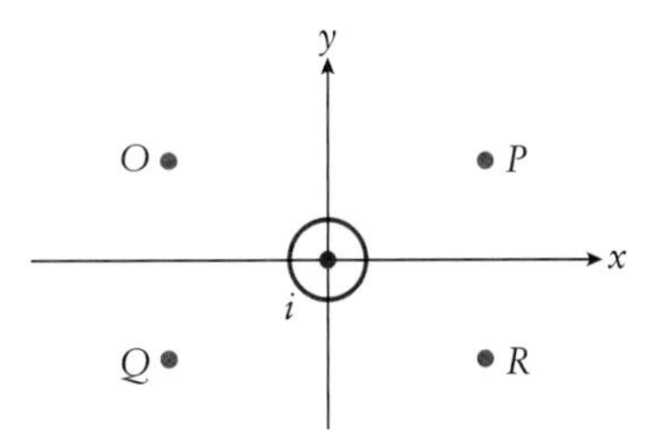

Figure 19-35 Problem 50

51. • A long, straight wire carries current in the $+x$ direction. Determine the direction of the magnetic field due to the current at the points O, P, Q, and R (**Figure 19-36**). Example 19-5

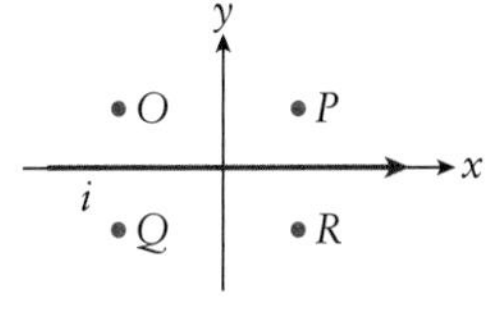

Figure 19-36 Problem 51

52. •• Using **Figure 19-37**, derive an expression for the magnetic field at the point C located at the center of the two circular, current-carrying arcs and the connecting radial lines. Assume the radii of the small and large arcs are r_1 and r_2, respectively.

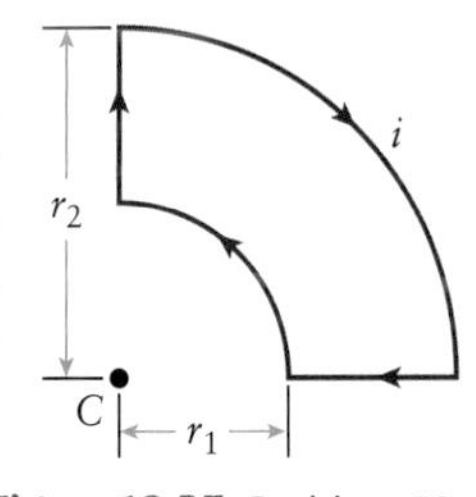

Figure 19-37 Problem 52

53. •• Derive an expression for the magnetic field at the point C at the center of the circular, current-carrying wire segments shown in **Figure 19-38**. SSM

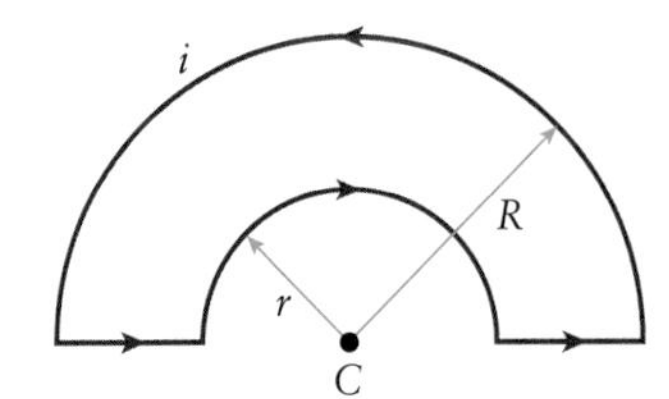

Figure 19-38 Problem 53

54. • Calculate the magnitude of the magnetic field at a perpendicular distance of 2.20 m from a long, copper pipe that has a diameter of 2.00 cm and carries a current of 20.0 A. Example 19-5

55. • Jerry wants to predict the magnetic field that a high-voltage line creates in his apartment. A current of 100 A passes through a wire that is 5 m from his window. Calculate the magnitude of the magnetic field. How does the field

compare to the magnitude of Earth's magnetic field of about 5×10^{-5} T in New York City? SSM Example 19-5

56. • A solenoid with 25 turns per centimeter carries a current of 25 mA. What is the magnitude of the magnetic field in the interior of the coils?

57. •• You want to wind a solenoid that is 3.5 cm in diameter, is 16 cm long, and will have a magnetic field of 0.0250 T when a current of 3.0 A is in it. What total length of wire do you need?

58. •• A coaxial cable consists of a solid inner conductor of radius R_i, surrounded by a concentric outer conducting shell of radius R_o. Insulating material fills the space between the conductors. The inner conductor carries current to the right, and the outer conductor carries the same current to the left down the outer surface of the cable (**Figure 19-39**). Using Ampère's law, derive an expression for the magnitude of the magnetic field in three separate regions of space: inside the inner conductor, between the two conductors, and outside of the outer conductor. Example 19-5

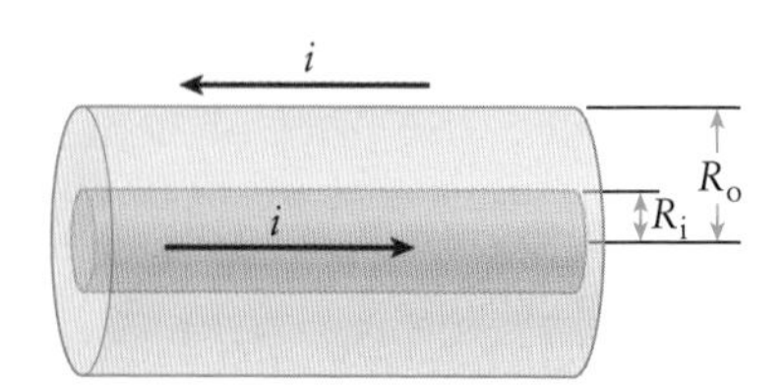

Figure 19-39 Problem 58

19-8 Two current-carrying wires exert magnetic forces on each other

59. • If wire 1 carries 2.00 A of current north, wire 2 carries 3.60 A of current south, and the two wires are separated by 1.40 m, calculate the force (magnitude and direction) acting on a 1.00-cm section of wire 1 due to wire 2. SSM Example 19-6

60. •• What is the net force (magnitude and direction) on the rectangular loop of wire that is 2.00 cm wide, 6.00 cm long, and is located 2.00 cm from a long, straight wire that carries $i = 40.0$ A of current as shown in **Figure 19-40**? Assume a current of 20.0 A in the loop. Example 19-6

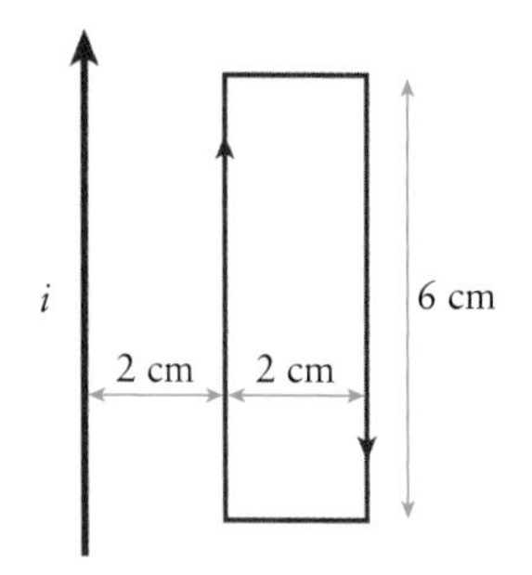

Figure 19-40 Problem 60

61. • The fasteners on overhead power lines are 50.0 cm long. What force must they be able to withstand if two high-voltage lines are 2.00 m apart, each carrying 2500 A in the same direction? Example 19-6

General Problems

62. • Horizontal electric power lines supported by vertical poles can carry large currents. Assume that Earth's magnetic field runs parallel to the surface of the ground from south to north with a magnitude of 0.50×10^{-4} T and that the supporting poles are 32 m apart. Find the magnitude and direction of the force that Earth's magnetic field exerts on a 32-m segment of power line (a wire) carrying 95 A if the current runs (a) from north to south, (b) from east to west, or (c) toward the northeast making an angle of 30.0° north of east. (d) Are any of the above forces large enough to have an appreciable effect on the power lines? Example 19-3

63. • A levitating train is three cars long (180 m) and has a mass of 100 metric tons (1 metric ton = 1000 kg). The current in the superconducting wires is about 500 kA, and even though the traditional design calls for many small coils of wire, assume for this problem that there is a 180-m-long wire carrying the current. Find the magnitude of the magnetic field needed to levitate the train. Example 19-3

64. •• An electron and a proton have the same kinetic energy upon entering a region of constant magnetic field, and their velocity vectors are perpendicular to the magnetic field. Suppose the magnetic field is strong enough to allow the particles to circle in the field. What is the ratio $r_{proton}/r_{electron}$ of the radii of their circular paths? Example 19-2

65. ••• In the mass spectrometer shown in **Figure 19-41** a particle with charge $-e = -1.60 \times 10^{-19}$ C enters a region of magnetic field that has a strength of 0.00242 T and points into the page. The velocity of the particle is confirmed with a velocity selector. The electric field is 9.00×10^4 V/m (down) and the magnetic field is 0.00530 T (into page) in the velocity selector. If the radius of curvature of the particle is 4.00 cm, calculate the mass of the particle. SSM Example 19-2

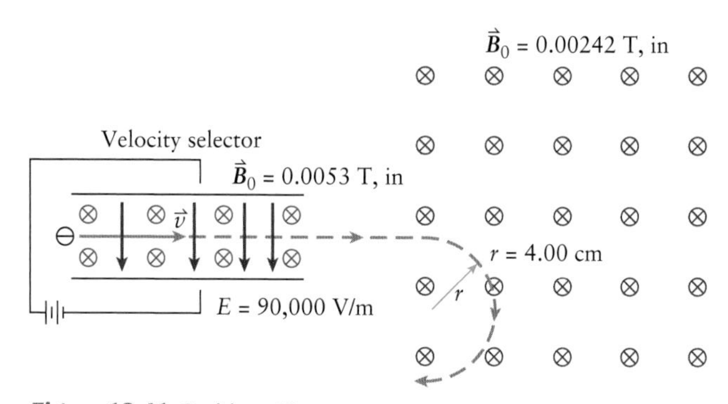

Figure 19-41 Problem 65

66. •• **Biology** The National High Magnetic Field Laboratory holds the world record for creating the largest magnetic field—100 T. To see if such a strong magnetic field could pose health risks for nearby workers, calculate the maximum acceleration the field could produce on Na^+ ions (of mass 3.8×10^{-26} kg) in blood traveling through the aorta. The speed of blood is highly variable, but 50 cm/s is reasonable in the aorta. Does your result indicate that it would be dangerous to expose workers to such a large magnetic field? Example 19-1

67. •• During electrical storms, a bolt of lightning can transfer 10 C of charge in 2.0 μs (the amount and time can vary considerably). We can model such a bolt as a very long current-carrying wire. (a) What is the magnetic field 1.0 m from such a bolt? What is the field 1.0 km away? How do the fields compare with Earth's magnetic field? (b) Compare the fields in part (a) with the magnetic field produced by a typical household current of 10 A in a very long wire at the same distances from the wire as in (a). (c) How close would you have to get to the wire in part (b) for its magnetic field to be the same as the field produced by the lightning bolt at 1.0 km from the bolt? Example 19-5

68. • A straight wire carries a current of 8.00 A toward the top of the page. What are the magnitude and direction of the magnetic field at point P, which is 8.00 cm to the right of the wire, as shown in **Figure 19-42**? Example 19-5

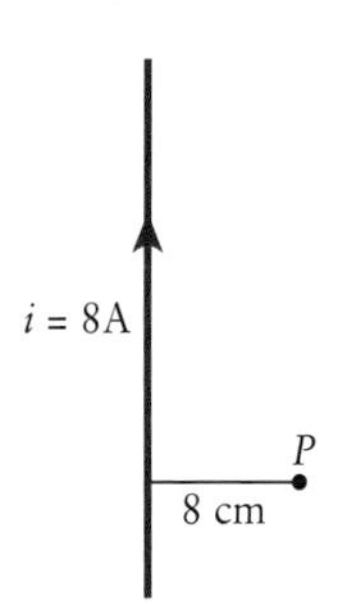

Figure 19-42 Problem 68

69. •• A long, straight wire carries a current as shown in **Figure 19-43**. A charged particle moving parallel to the wire experiences a force of 0.80 N at point P. Assuming the same charge and same velocity, what would be the magnitude of the magnetic force on the charge at point S? Example 19-5

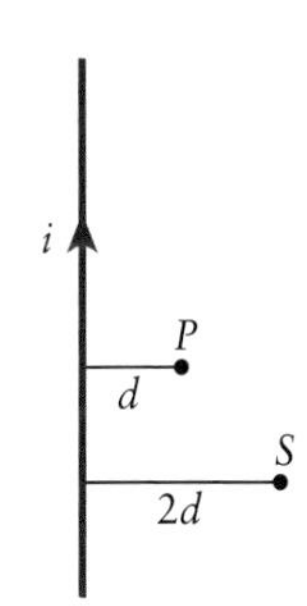

Figure 19-43 Problem 69

70. • Two long, straight wires parallel to the x axis are at $y = \pm 2.5$ cm (**Figure 19-44**). Each wire carries a current of 16 A in the $+x$ direction. Calculate the magnetic field on the y axis at (a) $y = 0$, (b) $y = 1.0$ cm, and (c) $y = 4.0$ cm. Example 19-5

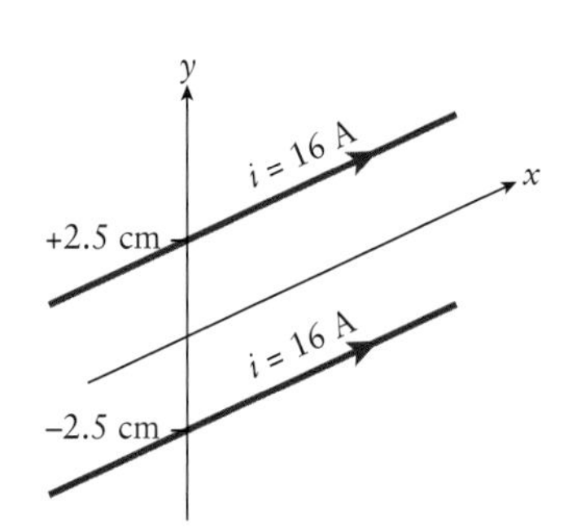

Figure 19-44 Problem 70

71. •• **Medical** Transcranial magnetic stimulation (TMS) is a noninvasive method to stimulate the brain using magnetic fields. It is used in treating strokes, Parkinson's disease, depression, and other physical conditions. In the procedure a circular coil is placed on the side of the forehead to generate a magnetic field inside the brain. Although values can vary, a typical coil would be about 15 cm in diameter and contain 250 thin circular windings. The magnetic field in the cortex (3.0 cm from the coil measured along a line perpendicular to the coil at its center) is typically 0.50 T. (a) What current in the coil is needed to produce the desired magnetic field inside the brain? (b) What is the magnetic field at the center of the coil at the forehead? (c) If the current needed in part (a) seems too large, how could you easily achieve the same magnetic field with a smaller current?

72. •• A wire of mass 40 g slides without friction on two horizontal conducting rails spaced 0.8 m apart. A steady current of 100 A is in the circuit formed by the wire and the rails. A uniform magnetic field of 1.2 T, directed into the plane of the drawing, acts on it. (a) In which direction in **Figure 19-45** will the wire accelerate? (b) What is the magnetic force on the wire? (c) How long must the rails be if the wire, starting from rest, is to reach a speed of 200 m/s? How would your answers differ if the magnetic field were (d) directed out of the page or (e) in the plane of the drawing, directed toward the top of the drawing? Example 19-3

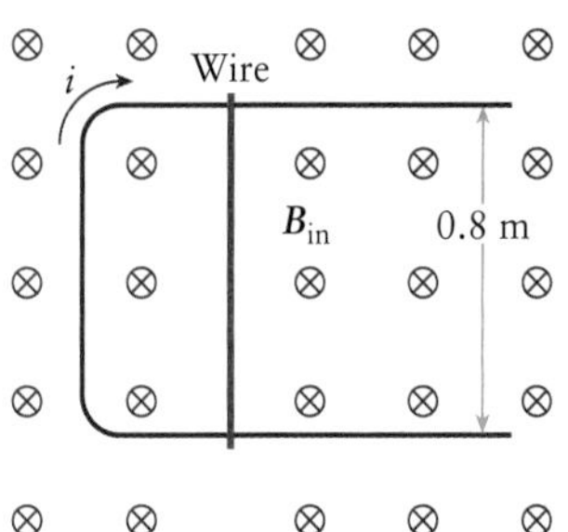

Figure 19-45 Problem 72

73. • A small 20-turn current loop with a 4.00-cm diameter is suspended in a region with a magnetic field of 1.00×10^3 G, with the plane of the loop parallel with the magnetic field direction. (a) What is the current in the loop if the torque exerted by the magnetic field on the loop is 4.00×10^{-5} N · m? (b) Describe the subsequent motion of the loop if it is allowed to rotate. Example 19-4

74. • A long, straight wire carries a current of 1.2 A toward the south. A second, parallel wire carries a current of 3.8 A toward the north and is 2.8 cm from the first wire. What is the magnetic force per unit length each wire exerts on the other? Example 19-6

75. ••• Two straight conducting rods, which are 1.0 m long, exactly parallel, and separated by 0.85 mm, are connected by an external voltage source and a 17-Ω resistance, as shown in **Figure 19-46**. The 0.5-Ω rod "floats" above the 2.5-Ω rod, in equilibrium. If the mass of each rod is 25 g, what must be the potential V of the voltage source? SSM Example 19-6

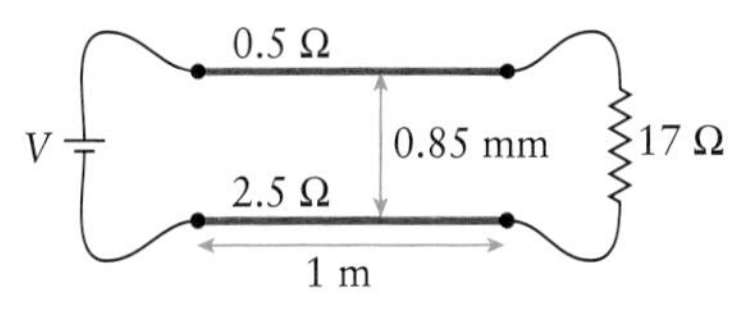

Figure 19-46 Problem 75

76. •• **Medical** When operated on a household 110-V line, typical hair dryers draw about 1650 W of power. We can model the current as a long, straight wire in the handle. During use, the current is about 3.0 cm from the user's head. (a) What is the current in the dryer? (b) What is the resistance of the dryer? (c) What magnetic field does the dryer produce at the user's head? Compare with Earth's magnetic field (5×10^{-5} T) to decide if we should have health concerns about the magnetic field created when using a hair dryer. Example 19-5

77. •• Three very long, straight wires lie at the corners of a square of side d, as shown in **Figure 19-47**. The magnitudes of the currents in the three wires are the same, but the two diagonally opposite currents are directed into the page while the other one is directed outward. Derive an expression for the magnetic field (magnitude and direction) at the fourth corner of the square. Example 19-5

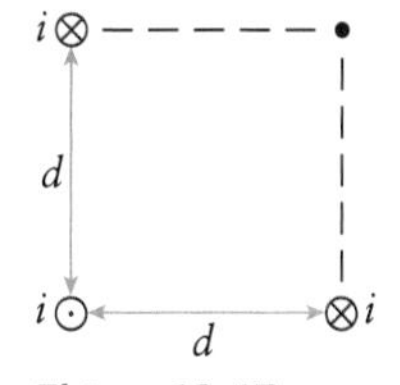

Figure 19-47 Problem 77

78. •• A 2.0-m lamp cord leads from the 110-V outlet to a lamp having a 75-W lightbulb. The cord consists of two insulated parallel wires 4.0 mm apart and held together by the insulation. One wire carries the current into the bulb, and the other carries it out. What is the magnitude of the magnetic field the cord produces (a) midway between the two wires and (b) 2.0 mm from one of the wires in the same plane in which

the two wires lie? (c) Compare each of the fields in parts (a) and (b) with Earth's magnetic field (5×10^{-5} T). (d) What magnetic force (magnitude and direction) do the two wires exert on one another? Is the force large enough to stress the insulation holding the wires together? Example 19-6

79. •• Some people have raised concerns about the magnetic fields produced by current-carrying high-voltage lines in residential neighborhoods. Currents in such lines can be up to 100 A. Suppose you have such a line near your house. If the wires are supported horizontally 5.0 m above the ground on vertical poles and your living room is 12 m from the base of the poles, what magnetic field strength does the wire produce in your living room if it carries 100 A? Express your answer in teslas and as a multiple of Earth's magnetic field (5×10^{-5} T). Does the magnetic field from such wires seem strong enough to cause health concerns? SSM Example 19-5

80. •• **Medical** Magnetoencephalography (MEG) is a technique for measuring changes in the magnetic field of the brain caused by external stimuli such as touching the body or viewing images of food. Such a change in the field occurs due to electrical activity (current) in the brain. During the process, magnetic sensors are placed on the skin to measure the magnetic field at that location. Typical field strengths are a few femtoteslas (1 femtotesla = 1 fT = 10^{-15} T). An adult brain is about 140 mm wide, divided into two sections (called hemispheres, although the brain is not truly spherical) each about 70 mm wide. We can model the current in one hemisphere as a circular loop, 65 mm in diameter, just inside the brain. The sensor is placed so that it is along the axis of the loop 2.0 cm from the center. A reasonable magnetic field is 5.0 fT at the sensor. According to this model, (a) what is the current in the brain and (b) what is the magnetic field at the center of the hemisphere of the brain?

81. •• Helmholtz coils are composed of two coils of wire that have their centers on the same axis, separated by a distance that is equal to the radius of the coils (**Figure 19-48**). The coils have N turns of wire that carry a current i in the same direction. If one coil is centered at the origin, and the other at $x = R$, derive expressions for the net magnetic field due to the coils at the points (a) $x = R/2$ and (b) $x = 2R$.

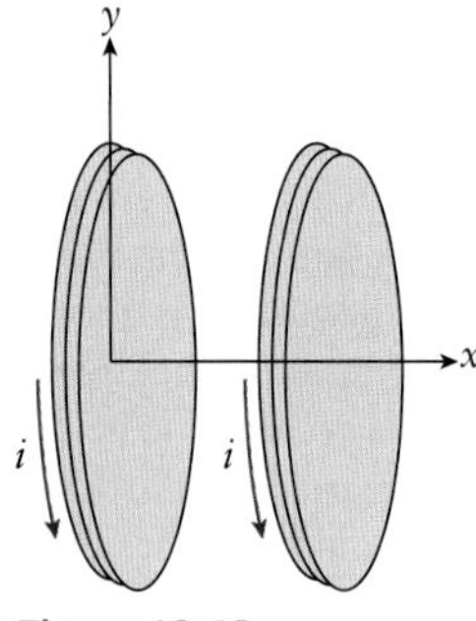

Figure 19-48 Problem 81

82. •• Geophysicists may use the gauss unit for magnetic field (10^4 G = 1 T). Earth's magnetic field at the equator can be taken as 0.7 G directed north. At the center of a flat circular coil that has 10 turns of wire and is 1.4 m in diameter, the coil's magnetic field exactly cancels Earth's field. (a) What must be the current in the coil? (b) How should the coil be oriented?

83. •• **Biology** Migratory birds use Earth's magnetic field to guide them. Some people are concerned that human-caused magnetic fields could interfere with bird navigation. Suppose that a pair of parallel high-voltage lines, each carrying 100 A, are 3.00 m apart and lie in the same horizontal plane. Find the magnitude and direction of the magnetic field the lines produce at a point 15.0 m above them equidistant from both lines in each of the following cases. (a) The lines run in the north–south direction, and both currents run from north to south. (b) Both lines run in the north–south direction, and the current in the eastern line runs northward while the current in the western line runs southward. (c) The lines run in the east–west direction, and both currents run from west to east. (d) Is it reasonable to think that the fields caused by the wires are likely to interfere with bird migration? Example 19-5

84. ••• A square loop of wire lies on a horizontal table, with one side of the loop constrained to stay on the table, as shown in **Figure 19-49**. The loop is in the presence of a magnetic field that is parallel to the surface of the table. When a current $i_1 = 0.350$ A flows through the loop, the loop lifts off the table to an angle $\theta_1 = 15.0°$ relative to the surface of the table. When a different current of $i_2 = 1.18$ A flows through the loop, the angle the loop makes relative to the table increases to $\theta_2 = 42.1°$. Calculate the ratio of the magnetic torque on the loop when current i_1 flows through it to the magnetic torque when current i_2 flows through the loop. Example 19-4

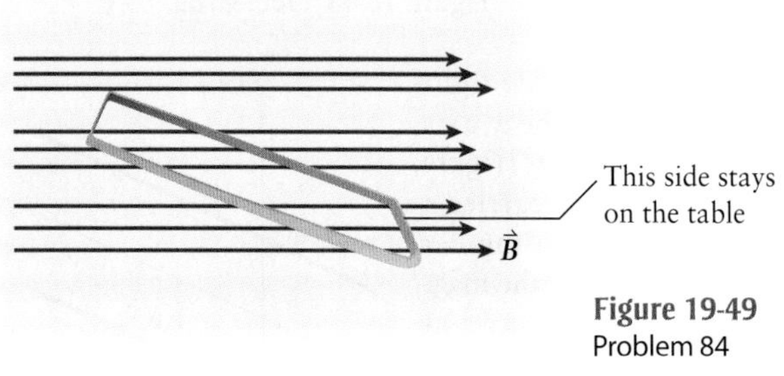

Figure 19-49 Problem 84

85. ••• A square, 35-turn, current-carrying loop is in a 0.75-T uniform magnetic field. The plane of the loop is parallel to the direction of the magnetic field, as shown in **Figure 19-50** (an edge-on view of the loop). The loop is kept from rotating by a stretched spring that is attached to one side of the loop. The spring constant is 450 N/m, and the spring is stretched 5.6 cm. (a) If the length of each side of the square loop is 42 cm, what is the magnitude of the current running through the loop? (b) When viewed from above, is the current in the loop clockwise or counterclockwise? Example 19-4

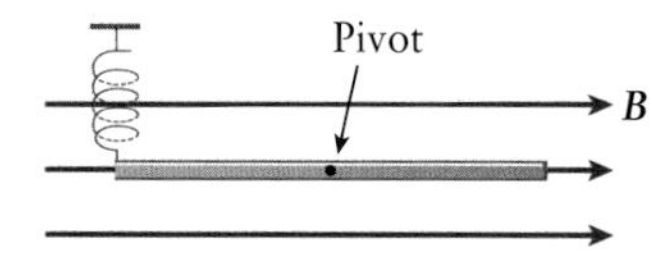

Figure 19-50 Problem 85

86. ••• A 300-turn wire loop with a 5.0 cm radius is connected to a 0.20-kg pulley by a lightweight rod. The pulley, which has a radius of 3.1 cm, is attached to a 4.2-kg block by a massless rope (**Figure 19-51**). The loop is in a 0.75-T magnetic field. When no current flows through the wire loop, the plane of the loop is parallel to the magnetic field, and the block attached to the pulley just touches the ground. However, when a 0.76-A current flows through the loop, magnetic forces cause the loop to rotate, which lifts the block off the ground. Calculate the angle the plane of the loop rotates relative to the magnetic field. How high off the ground is the block lifted? Example 19-4

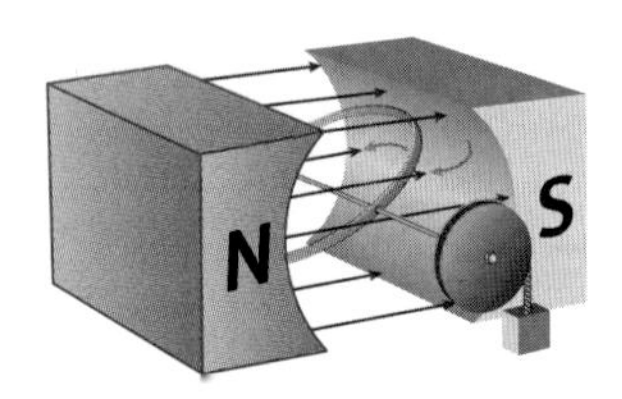

Figure 19-51 Problem 86

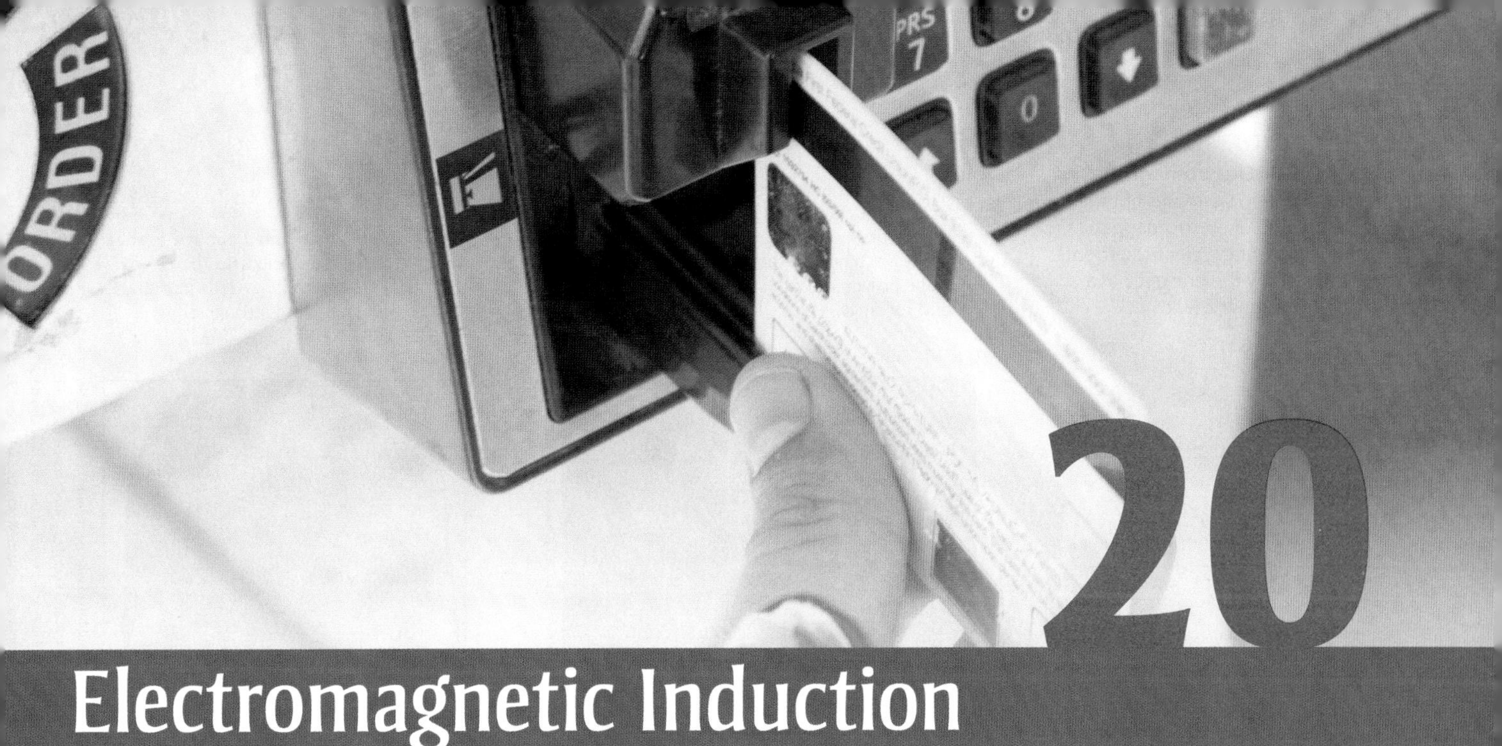

20 Electromagnetic Induction

pick-uppath/Getty Images

In this chapter, your goals are to:

- (20-1) Explain the importance of electromagnetic induction.
- (20-2) Describe what is meant by a motional emf and an induced emf.
- (20-3) Explain what determines the magnitude and direction of an emf in a circuit with a changing magnetic flux.
- (20-4) Define the key properties of an ac generator.

To master this chapter, you should review:

- (6-2) The work done by a constant force.
- (12-3) Simple harmonic motion.
- (16-6) The electric flux through an area.
- (18-6) The electric power for a resistor.
- (19-3, 19-5) The magnetic force on a moving charged particle and on a current-carrying wire.
- (19-7) The magnetic field produced by a current-carrying loop.

What do you think?

The stripe on the back of a credit card is magnetized in a pattern that encodes your account information. A credit card reader contains a loop of wire, and when you swipe the card through the reader, the magnetized card's motion generates an electric current in the wire that sends a signal to the credit card company. To make this current flow, what kind of force must act on electrons in the wires of the card reader? (a) A magnetic force; (b) an electric force; (c) a combination of electric and magnetic forces.

20-1 The world runs on electromagnetic induction

We've discussed electric circuits in which the current is caused by the emf provided by a battery. But many of the electric circuits around you are *alternating-current* circuits in which the current constantly changes direction. That includes the current in light fixtures, toasters, electric fans, and other devices plugged into wall sockets. (Alternating current also indirectly powers mobile devices like cell phones and laptop computers. These devices have batteries, but the batteries are recharged by plugging them into a wall socket.) What kind of emf produces an alternating current?

The answer to this question comes from a remarkable discovery made by physicists around 1830: *If the magnetic field in a region of space changes, the change gives rise to an electric field*. This electric field, called an *induced* field, is very different in character from the electric field produced by point charges that we described in Chapter 16: An induced electric field does not point away from positive charges and toward negative charges but instead has field lines that form closed loops like magnetic field lines. That's just what's needed for that induced electric field to push charges around a loop of wire and generate an electric current. As we'll see later in this chapter, it's easy to make this induced electric field flip its direction back and

Figure 20-1 Electromagnetic induction Two examples of the phenomenon of electromagnetic induction, in which electric currents are induced by the presence of a changing magnetic flux.

An electric generator produces current by electromagnetic induction: Coils of wire move relative to a magnetic field, which generates an emf in the coils. The motion can be powered by the wind, as in these wind turbines.

(a)

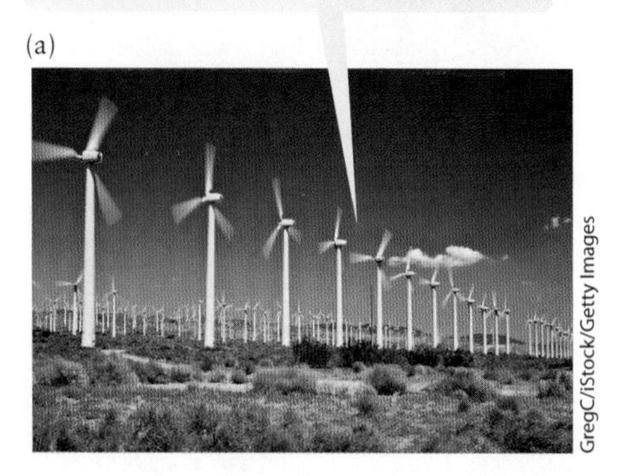

A changing magnetic field applied to the brain induces an electric field there, causing electric currents. Areas in red are where the currents are strongest.

(b)

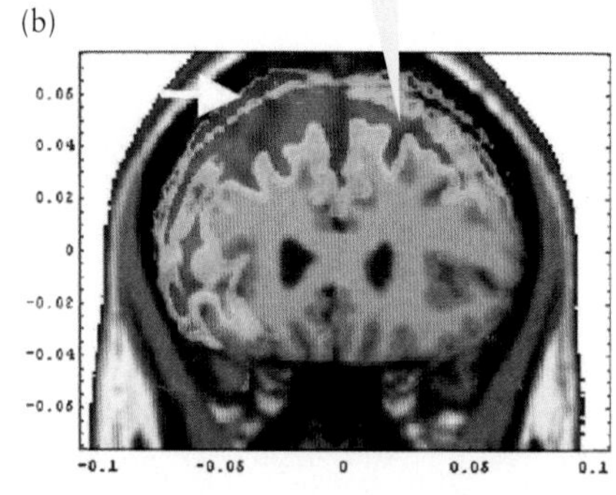

forth, which makes a current that flips back and forth—in other words, an alternating current. The vast amount of electric current used by our technological civilization is produced in this way (**Figure 20-1a**).

We use the term "**electromagnetic induction**" for the process whereby a changing magnetic field induces an electric field. (The word "electromagnetic" shows that this process involves both electric and magnetic fields.) Electromagnetic induction has many applications beyond producing an alternating current to be delivered to wall sockets. It's how a credit card reader decodes the information on the card's magnetized strip (see the photo that opens this chapter). It's also at the heart of a relatively new medical technique called *transcranial magnetic stimulation* (TMS), which allows physicians to stimulate electrical activity in the brain without sticking electrodes to the scalp or inserting them through the skull. In TMS, a time-varying magnetic field is produced inside the brain by current-carrying coils around the head. This causes an induced electric field, which in turn causes currents to flow within the brain (**Figure 20-1b**). TMS has been used with some success to treat cases of depression that have not responded to more conventional therapy.

BIO-Medical

In this chapter we'll begin by describing the relationship between a changing magnetic flux and the electric field that it induces. We'll introduce two important laws that describe electromagnetic induction. The first of these, Faraday's law, will tell us how the emf that appears in a closed loop (such as an electric circuit) due to an induced electric field is related to the rate of change of the magnetic flux through the loop. The second, Lenz's law, will tell us the direction of this induced emf. We'll see how induced emf makes possible the important device called a *generator*, which converts mechanical energy into electric energy and creates an alternating emf. (Each of the wind turbines shown in Figure 20-1 uses its spinning blades to run a generator.) In Chapter 21 we'll see how all of these ideas explain the behavior of alternating-current circuits.

TAKE-HOME MESSAGE FOR Section 20-1

✔ In electromagnetic induction, a time-varying magnetic flux in a certain region gives rise to an electric field in that same region.

✔ Electromagnetic induction is used to produce an alternating current.

20-2 A changing magnetic flux creates an electric field

Figure 20-2 shows an experiment that we can understand with the physics we already know. A loop of wire with an attached ammeter (a device for measuring the current in the loop) is placed near the south pole of a stationary magnet. No current flows if the loop is held stationary. That's not surprising, since there's no source of emf connected to the loop. But a current *does* flow in the loop when it is moved toward the magnet's south pole (**Figure 20-2a**) and flows in the opposite direction when the loop is moved away from the magnet's south pole (**Figure 20-2b**). What's happening is that the mobile charges within the loop are moving along with the loop through the magnetic field of the bar magnet and so experience a magnetic force that pushes the charges around the loop (**Figure 20-2c**). Reversing the direction in which the loop and its charges move also

(a) Moving the loop toward a stationary magnet
Ammeter
Motion of loop
Charges flow around the loop.
(b) Moving the loop away from a stationary magnet
Ammeter
Motion of loop
Charges flow in opposite direction around the loop.

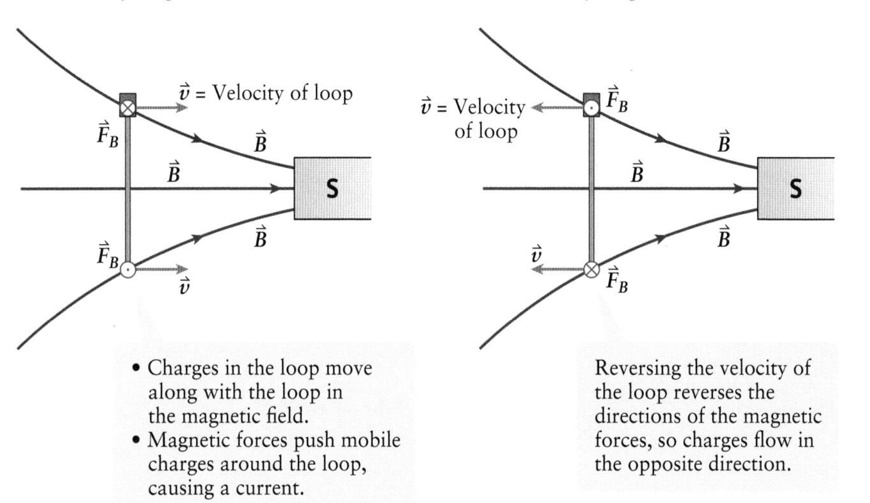
(c) Side view of loop moving toward a stationary magnet
$\vec{v}$ = Velocity of loop
$\vec{F}_B$
$\vec{B}$
S
• Charges in the loop move along with the loop in the magnetic field.
• Magnetic forces push mobile charges around the loop, causing a current.
(d) Side view of loop moving away from a stationary magnet
$\vec{v}$ = Velocity of loop
$\vec{F}_B$
$\vec{B}$
S
Reversing the velocity of the loop reverses the directions of the magnetic forces, so charges flow in the opposite direction.

Figure 20-2 A loop of wire moving with respect to a magnet If a loop of wire moves toward or away from a magnet, a current flows in the loop. The current is caused by magnetic forces.

reverses the direction of the magnetic force, so the charges are pushed in the opposite direction, and the current direction reverses (**Figure 20-2d**). In either case we say that the magnetic force on the mobile charges is equivalent to an emf that makes the current flow. There is no magnetic force, and hence no emf, if the loop and its charges are at rest. (Recall from Section 19-2 that magnetic forces act only on *moving* charges.) Because the loop must be in motion in order for the emf to appear, we call it a **motional emf.**

Figure 20-3 shows an experiment that looks similar but involves entirely different physics. Now we hold the loop stationary and move the south pole of the magnet either toward the loop (**Figure 20-3a**) or away from the loop (**Figure 20-3b**). In this case there can be no magnetic force on the mobile charges within the loop because those charges are at rest in the stationary loop. Nonetheless, there is an emf in the loop and a current flows around the loop in response, but only when the magnet is moving relative to the loop. Since there is no magnetic force in this situation, it must be that the emf is due to an *electric* force on the mobile charges (**Figures 20-3c** and **20-3d**).

What's happening is that when the magnet is moving, the magnetic field at the location of the loop is changing: Its magnitude increases when the magnet's south pole moves toward the loop (Figure 20-3a) and decreases when the magnet's south pole moves away from the loop (Figure 20-3b). So this experiment shows that an electric field is *induced* by the changing magnetic field. For this reason we call the emf in the experiment of Figure 20-3 an **induced emf.** We use the term "electromagnetic

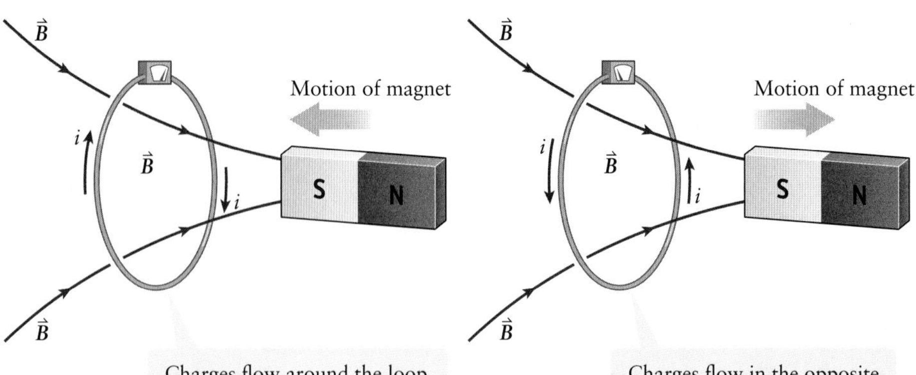

(a) Moving the magnet toward a stationary loop
(b) Moving the magnet away from a stationary loop
Motion of magnet
Motion of magnet
S
N
S
N
Charges flow around the loop.
Charges flow in the opposite direction around the loop.

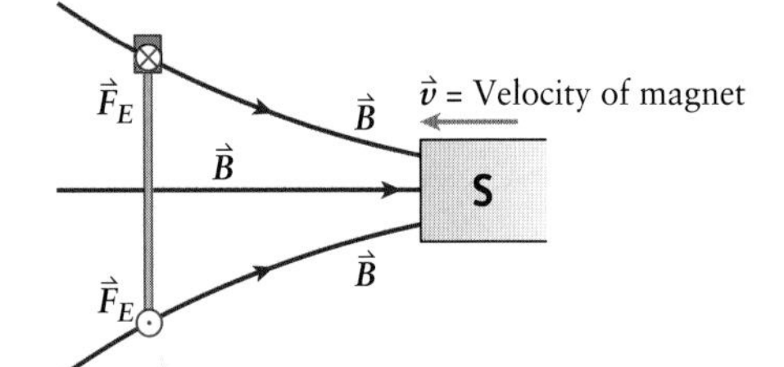

(c) Side view of magnet moving toward a stationary loop
v = Velocity of magnet
S
• Charges in the loop experience electric forces.
• These forces push mobile charges around the loop, causing a current.

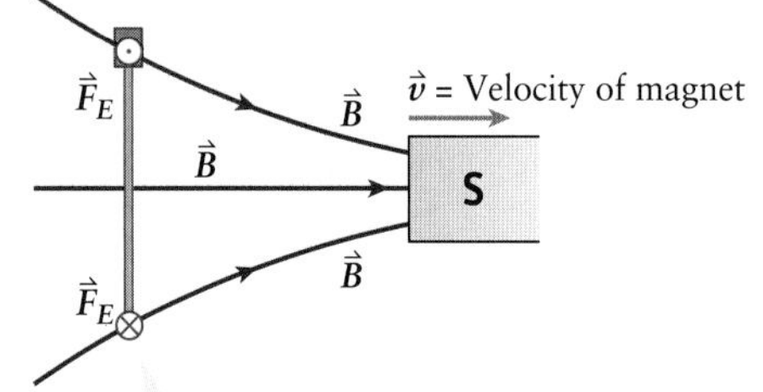

(d) Side view of magnet moving away from a stationary loop
v = Velocity of magnet
S
Reversing the velocity of the magnet reverses the directions of the electric forces, so charges flow in the opposite direction.

Figure 20-3 A magnet moving with respect to a loop of wire If a magnet moves toward or away from a loop of wire, a current flows in the loop. Magnetic forces cannot explain why this happens, so electric forces must be present to produce the current.

induction" for situations in which a changing magnetic field causes, or induces, an electric field.

Although the experiments in Figures 20-2 and 20-3 are different, they have the *same* result: Whether the loop moves toward the stationary magnet at 1 m/s, as in Figure 20-2a, or the magnet moves toward the stationary loop at 1 m/s, as in Figure 20-3a, the same emf appears in the loop. In fact, the same emf appears if the magnet and loop are both moving, as long as the magnet and loop approach each other at a relative speed of 1 m/s. Since the result is the same in each of these cases, we should be able to describe all of these effects in terms of a single equation. But what equation is that?

It turns out that we can describe the emf in any of these situations in terms of the change in *magnetic flux* through the loop in Figures 20-2 and 20-3. We define this in the same way that we defined *electric* flux in Section 16-6: It's the area A of the surface outlined by the loop, multiplied by $B \cos \theta$, the component of the magnetic field that's perpendicular to that surface (see part (a) of **Figure 20-4**). In equation form the **magnetic flux** Φ_B ("phi-sub-B") through the loop is

Magnetic flux (20-1)

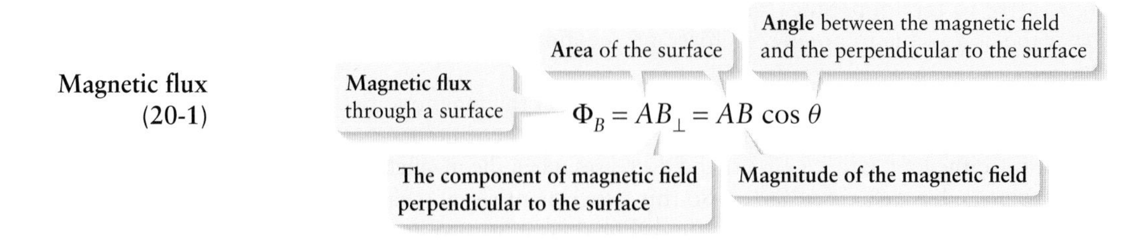

Area of the surface
Angle between the magnetic field and the perpendicular to the surface
Magnetic flux through a surface
The component of magnetic field perpendicular to the surface
Magnitude of the magnetic field

$$\Phi_B = AB_\perp = AB \cos \theta$$

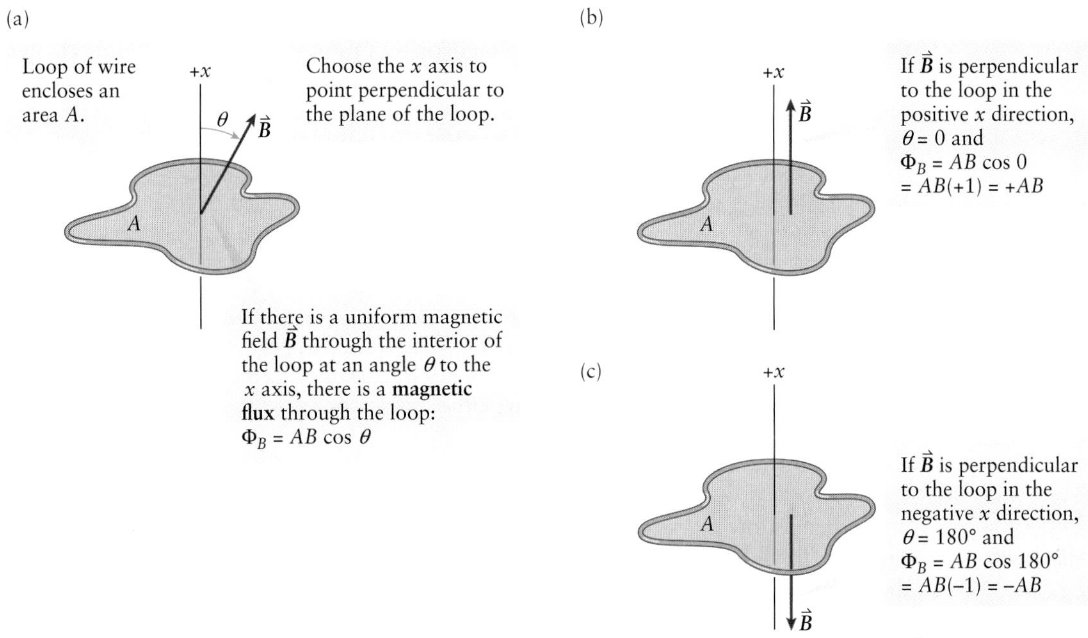

Figure 20-4 Magnetic flux (a) The magnetic flux through a loop. The flux is (b) most positive when $\vec{B}$ points along the perpendicular to the loop and (c) most negative when $\vec{B}$ points opposite to the perpendicular.

(The subscript "B" on the symbol Φ_B in Equation 20-1 reminds us that this is the flux of the magnetic field $\vec{B}$.) As parts (b) and (c) of Figure 20-4 show, the flux Φ_B can be positive or negative. Note that the choice of positive x direction is arbitrary; in Figure 20-4 we chose the positive x direction to be upward, so Φ_B is positive for the case shown in Figure 20-4b and negative for the case shown in Figure 20-4c. Had we chosen the positive x direction to be downward, we would have had $\Phi_B < 0$ in Figure 20-4b and $\Phi_B > 0$ in Figure 20-4c. It doesn't matter which one we choose, since the physics will turn out to be the same in either case. (In a similar way, in Chapter 2 we had to make a choice of positive x direction for analyzing motion in a straight line. The actual motion didn't depend on which direction we chose to be positive and which to be negative.)

See the Math Tutorial for more information on trigonometry.

In Figures 20-2 and 20-3 the magnetic field is not uniform over the area enclosed by the loop, so the perpendicular component $B_\perp = B \cos \theta$ has different values at different points on this area. In such a case $B_\perp$ in Equation 20-1 is the perpendicular component of the magnetic field *averaged* over the area A enclosed by the loop. Note that if the loop is actually a coil with N turns of wire, the net magnetic flux through the coil is N multiplied by the flux through one turn of the coil.

If the magnet and loop in Figures 20-2 and 20-3 are not moving with respect to each other, the magnetic flux through the loop remains the same. In this case there is no emf and no current in the loop. The flux changes, however, when either the loop moves relative to the magnet (Figure 20-2) or the magnet moves relative to the loop (Figure 20-3). In these cases there *is* an emf in the loop and the current. This suggests that *an emf appears in a loop when the magnetic flux through that loop changes*. This observation is known as **Faraday's law of induction**, named for the nineteenth-century English physicist Michael Faraday:

Magnitude of the induced or motional emf in a loop

Change in the magnetic flux through the surface outlined by the loop

$$|\varepsilon| = \left| \frac{\Delta \Phi_B}{\Delta t} \right|$$

Time interval over which the change in magnetic flux takes place

Faraday's law of induction (20-2)

This law states that the magnitude of the emf that appears in a loop is equal to the magnitude of the *rate of change* of the magnetic flux through the loop. If a large change in flux $\Delta\Phi_B$ happens in a short time interval Δt, the resulting emf has a large magnitude; if the change in flux is relatively small and happens over a long time interval, the resulting emf has a small magnitude.

Note that Equation 20-2 tells us only the *magnitude* of the emf, not its direction. In the following section we'll see how the direction is determined.

WATCH OUT! It's not the magnetic flux that causes an emf, but the rate at which the flux changes.

The mere presence of magnetic flux through a loop does not cause an emf to appear in the loop. If a flux is present but does not change, such as what happens when a magnet and loop are held stationary with respect to each other, there is *no* resulting emf. An emf appears only when the flux *changes*, such as when the magnet and loop in Figures 20-2 and 20-3 move either toward or away from each other.

Go to Interactive Exercise 20-1 for more practice dealing with motional emf.

In the following example we'll check Faraday's law. We'll do this by considering a situation in which we can use our knowledge of magnetic forces to calculate the emf, then compare this to the emf calculated using Equation 20-2.

EXAMPLE 20-1 Changing Magnetic Flux I: A Sliding Bar in a Magnetic Field

A copper bar of length L slides at a constant speed v along stationary, U-shaped copper rails (**Figure 20-5**). A uniform magnetic field of magnitude B is directed perpendicular to the plane of the bar and rails. The moving bar and stationary rails form a closed circuit, and an emf is produced in this circuit because the wire is moving in a magnetic field. Determine the emf in the circuit (a) by using the expression for the magnetic force on a charge in the moving wire and (b) by using Faraday's law of induction, Equation 20-2.

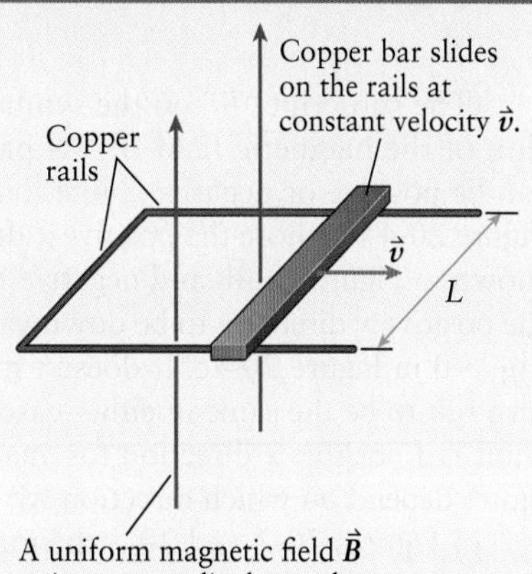

Figure 20-5 A sliding copper bar What emf is generated in the bar as it slides in the presence of a magnetic field $\vec{B}$?

Set Up

For a battery the magnitude of the emf equals the change in electric potential (potential energy per charge) for a charge that traverses the battery; that is, it's equal to the *work per charge* that the battery does on charges that travel from one terminal to the other. We'll use the same idea in part (a) to calculate the emf in terms of the work done by the magnetic force on a charged particle that travels the length of the moving bar. In part (b) we'll find the emf by instead using Equation 20-2. While the magnetic field doesn't change, the area of the loop outlined by the moving bar and the rails *does* change, and so the magnetic flux through this loop changes.

Magnetic force on a moving charged particle:

$$F = |q|vB \sin\theta \qquad (19\text{-}1)$$

Work done by a constant force that points in the same direction as the straight-line displacement:

$$W = Fd \qquad (6\text{-}1)$$

Magnetic flux:

$$\Phi_B = AB_\perp = AB\cos\theta \qquad (20\text{-}1)$$

Faraday's law of induction:

$$|\varepsilon| = \left|\frac{\Delta\Phi_B}{\Delta t}\right| \qquad (20\text{-}2)$$

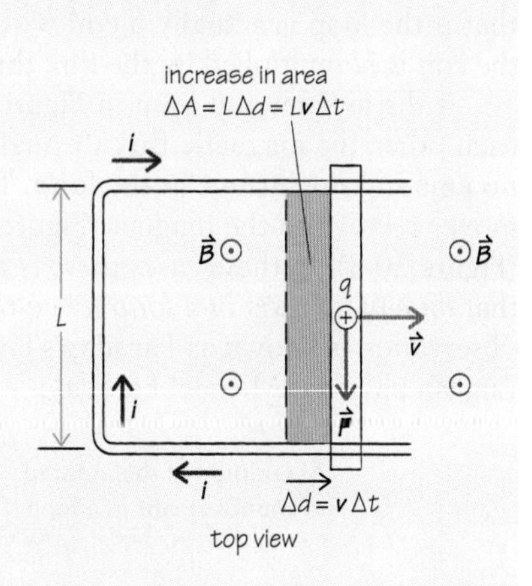

Solve

(a) Find the magnetic force on a charged particle moving along with the copper bar.

For a positive charge q moving with the bar, the velocity $\vec{v}$ is perpendicular to the magnetic field $\vec{B}$. So $\theta = 90°$ in Equation 19-1, and the magnetic force $\vec{F}$ on such a charge has magnitude

$$F = qvB \sin 90° = qvB\,(1) = qvB$$

The force $\vec{F}$ is perpendicular to both $\vec{v}$ and $\vec{B}$, so it is directed along the length of the moving bar.

Use the magnetic force on a charged particle to find the emf produced in the bar.

The magnetic force $\vec{F}$ on a charge q causes it to move along the length L of the bar. Since $\vec{F}$ is in the same direction as the displacement of the charge, the work done on the charge as it travels this length is

$$W = FL = qvBL$$

The magnitude of the emf in the bar equals the work done per charge:

$$|\varepsilon| = \frac{W}{q} = \frac{qvBL}{q} = vBL$$

(b) Find the emf using Faraday's law of induction.

The magnetic field $\vec{B}$ points perpendicular to the plane of the loop outlined by the moving copper bar and the copper rails. If the area of this loop is A and we take the positive x direction to point out of the plane of the above figure (in the same direction as $\vec{B}$), then $\theta = 0$ in Equation 20-1. The magnetic flux through the loop is then

$$\Phi_B = AB \cos 0 = AB(1) = AB$$

The magnetic field is constant, but the area A changes with time because the bar moves. The speed v of the bar is just the distance Δd that the bar moves divided by the time Δt that it takes to move that distance, so

$$v = \frac{\Delta d}{\Delta t} \quad \text{and} \quad \Delta d = v\,\Delta t$$

During time Δt the area A of the loop outlined by the moving bar and rails increases by an amount $\Delta A = L\,\Delta d = Lv\,\Delta t$. Therefore, the change in magnetic flux through the loop during this time is

$$\Delta\Phi_B = (\Delta A)B = (Lv\,\Delta t)B = vBL\,\Delta t$$

From Equation 20-2 the magnitude of the emf in the loop is

$$|\varepsilon| = \left|\frac{\Delta\Phi_B}{\Delta t}\right| = \left|\frac{vBL\,\Delta t}{\Delta t}\right| = vBL$$

Reflect

We find the same expression for the emf in both parts (a) and (b), as we must. This gives us added confidence that Equation 20-2 is valid, and a host of experiments backs up its validity.

EXAMPLE 20-2 Changing Magnetic Flux II: A Varying Magnetic Field

A uniform magnetic field of magnitude $B = 1.50$ T is directed at an angle of 60.0° to the plane of a circular loop of copper wire. The loop is 3.50 cm in diameter. (a) What is the magnetic flux through the loop? What is the induced emf in the loop if the magnetic field decreases to zero (b) in 10.0 s or (c) in 0.100 s?

Set Up

The magnetic flux is given by Equation 20-1. Note that θ in this equation is the angle between the direction of magnetic field $\vec{B}$ and the *perpendicular* to the loop, so $\theta = 90.0° - 60.0° = 30.0°$. The magnetic flux through the loop changes when the field magnitude changes, so an emf will be induced in the loop. We'll use Equation 20-2 to calculate the magnitude of this induced emf.

Magnetic flux:

$$\Phi_B = AB_\perp = AB\cos\theta \qquad (20\text{-}1)$$

Area of a circle of radius r:

$$A = \pi r^2$$

Faraday's law of induction:

$$|\varepsilon| = \left|\frac{\Delta\Phi_B}{\Delta t}\right| \qquad (20\text{-}2)$$

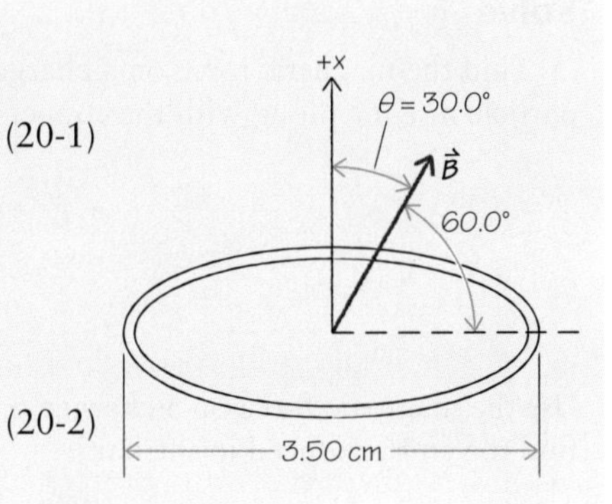

Solve

(a) Find the area of the loop, then use Equation 20-1 to calculate the magnetic flux through the loop.

The radius r of the loop is one-half of the diameter:

$$r = \frac{1}{2}(3.50 \text{ cm}) = 1.75 \text{ cm} = 1.75 \times 10^{-2} \text{ m}$$

The area of the loop is

$$A = \pi r^2 = \pi(1.75 \times 10^{-2} \text{ m})^2 = 9.62 \times 10^{-4} \text{ m}^2$$

From Equation 20-1 the magnetic flux through the loop is

$$\Phi_B = AB\cos\theta = (9.62 \times 10^{-4} \text{ m}^2)(1.50 \text{ T})\cos 30.0° = 1.25 \times 10^{-3} \text{ T}\cdot\text{m}^2$$

(b) The change in magnetic flux is the final value (zero) minus the initial value that we found in (a). Equation 20-2 tells us that to find the magnitude of the induced emf, we divide this change by the time $\Delta t = 10.0$ s over which the flux change takes place.

The change in magnetic flux is

$$\Delta\Phi_B = (\text{final flux}) - (\text{initial flux}) = 0 - 1.25 \times 10^{-3} \text{ T}\cdot\text{m}^2 = -1.25 \times 10^{-3} \text{ T}\cdot\text{m}^2$$

If the flux decreases to zero in $\Delta t = 10.0$ s, the magnitude of the induced emf is

$$|\varepsilon| = \left|\frac{\Delta\Phi_B}{\Delta t}\right| = \left|\frac{-1.25 \times 10^{-3} \text{ T}\cdot\text{m}^2}{10.0 \text{ s}}\right| = 1.25 \times 10^{-4} \text{ T}\cdot\text{m}^2/\text{s} = 1.25 \times 10^{-4} \text{ V}$$

(c) Repeat part (b) with $\Delta t = 0.100$ s.

If the flux decreases to zero in just $\Delta t = 0.100$ s, the magnitude of the induced emf is

$$|\varepsilon| = \left|\frac{\Delta\Phi_B}{\Delta t}\right| = \left|\frac{-1.25 \times 10^{-3} \text{ T}\cdot\text{m}^2}{0.100 \text{ s}}\right| = 1.25 \times 10^{-2} \text{ T}\cdot\text{m}^2/\text{s} = 1.25 \times 10^{-2} \text{ V}$$

Reflect

The induced emf is 100 times greater in part (c) than in part (b) because the same flux change takes place in 1/100 as much time. The faster the flux change, the greater the induced emf that results. Note that the emf is induced *only* during the time when the magnetic flux is changing. There is zero emf when the magnetic field is at its original value of 1.50 T, and there is zero emf when the magnetic field has stabilized at its final value of zero.

If we replace the loop by a coil of the same diameter with 500 turns of wire, the induced emf is 500 times greater: $500 \times 1.25 \times 10^{-4}$ V = 0.0625 V in part (b), $500 \times 1.25 \times 10^{-2}$ V = 6.25 V in part (c). The key to generating a large induced emf is to have many turns of wire and a rapid change in magnetic flux.

In this example we chose the positive x direction to be upward so that the angle θ between the magnetic field and the perpendicular to the loop was 30.0°. Can you show that we would have found the same results for emf had we chosen the positive x direction to be downward so that $\theta = 150.0°$?

GOT THE CONCEPT? 20-1 A Wooden Loop

? Suppose the loop in Example 20-2 were made out of wood rather than copper wire, but the magnetic field changes in the same manner as in part (b) of Example 20-2. Compared to the emf calculated in part (b) of Example 20-2, the emf induced in the wooden loop would be (a) zero; (b) much smaller but not zero; (c) slightly less; (d) the same; (e) greater.

TAKE-HOME MESSAGE FOR Section 20-2

✔ A motional emf appears in a conductor that moves in a magnetic field. The force that produces the emf is a magnetic one.

✔ An induced emf appears in any loop subjected to a changing magnetic field. The force that produces the emf is an electric one.

✔ Both motional emfs and induced emfs can be described by Faraday's law of induction: The magnitude of the emf in a loop is equal to the absolute value of the change in magnetic flux through the loop divided by the time over which the change takes place.

20-3 Lenz's law describes the direction of the induced emf

Equation 20-2 tells us the *magnitude* of the emf that appears in a loop when there is a change in magnetic flux through that loop: $|\varepsilon| = |\Delta\Phi_B/\Delta t|$. It does not, however, tell us the *direction* in which the emf tends to make current flow around that loop. As we'll see, there's a simple rule for determining this direction that works for both motional emfs (caused by a conductor moving in a magnetic field) and induced emfs (caused by a conductor being exposed to a changing magnetic field).

To learn about this rule let's think again about the loop of wire in Figure 20-2. In addition to the field $\vec{B}_{magnet}$, there's also a magnetic field produced by the current within the loop itself. As we learned in Section 19-7, a current-carrying loop produces a magnetic field of its own. The direction of the field $\vec{B}_{loop}$ due to the loop depends on the direction of the current around the loop and is given by a right-hand rule: Curl the fingers of your right hand around the loop in the direction of the current, and the extended thumb of your right hand will point in the direction of $\vec{B}_{loop}$ in the interior of the loop (see part (a) of **Figure 20-6**). So whenever an emf appears in a loop—either a motional emf as in Figure 20-2 or an induced emf as in Figure 20-3—the current produced by that emf generates a magnetic field $\vec{B}_{loop}$ whose direction depends on the direction of the current and emf. We call $\vec{B}_{loop}$ an **induced magnetic field.**

The field $\vec{B}_{loop}$ itself produces a magnetic flux through the loop, and it's the sense of this flux that will tell us the direction of the emf in the loop. Let's choose the positive x direction for the loop in Figure 20-6 to point to the right, perpendicular to the plane of the loop. If the loop is close to the south pole of a bar magnet, as in Figure 20-6b, the field of the magnet causes a positive magnetic flux through the loop (the field $\vec{B}_{magnet}$ points generally to the right, in the positive x direction). If the loop moves toward the magnet as in the left-hand side of Figure 20-6b, the field $\vec{B}_{magnet}$ inside the loop increases and the positive flux increases. Experiment shows that the current induced in the loop gives rise to an induced magnetic field $\vec{B}_{loop}$ within the loop, which points in the *opposite* direction to $\vec{B}_{magnet}$. So while the flux due to $\vec{B}_{magnet}$ becomes more positive, $\vec{B}_{loop}$ gives rise to a negative flux that opposes the change in the flux of $\vec{B}_{magnet}$.

If instead the loop moves away from the magnet as in the right-hand side of Figure 20-6b, the field $\vec{B}_{magnet}$ inside the loop decreases and the positive flux decreases. In this case the direction of the induced current in the loop is reversed, as is the direction of the induced magnetic field $\vec{B}_{loop}$: Now $\vec{B}_{loop}$ inside the loop points in the *same* direction as $\vec{B}_{magnet}$. The magnetic flux due to $\vec{B}_{loop}$ is now positive, which opposes the negative change (decrease) in the flux of $\vec{B}_{magnet}$.

In both cases shown in Figure 20-6b, the induced magnetic field is in a direction opposite to the *change* in flux of the external magnetic field (in this case the field due to the bar magnet). Many experiments show that this is always the case, no matter whether the induced magnetic field is due to a motional emf, an induced emf, or

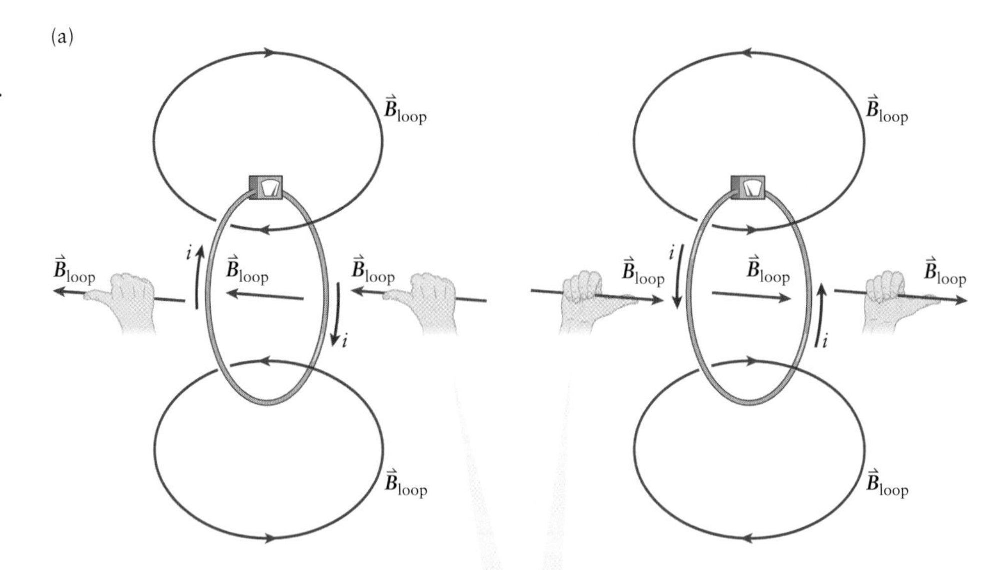
(a)
$\vec{B}_{loop}$
i
• A current-carrying loop generates a magnetic field $\vec{B}_{loop}$.
• To find the direction of $\vec{B}_{loop}$, curl the fingers of your right hand around the loop in the direction of the current i. Your extended right thumb points in the direction of $\vec{B}_{loop}$ in the interior of the loop.
• $\vec{B}_{loop}$ itself causes a magnetic flux through the loop.

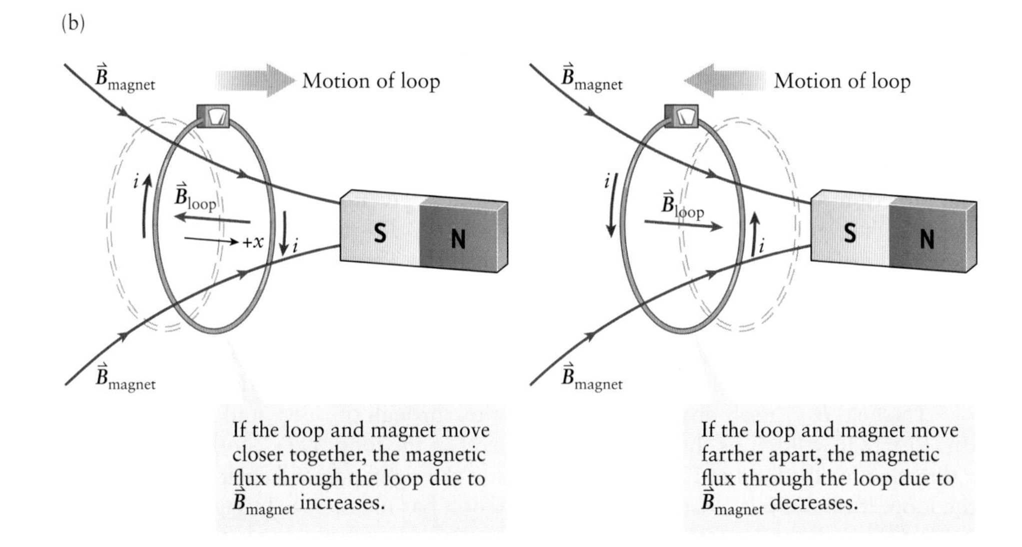
(b)
$\vec{B}_{magnet}$
Motion of loop
$\vec{B}_{loop}$
$+x$
i
S
N
If the loop and magnet move closer together, the magnetic flux through the loop due to $\vec{B}_{magnet}$ increases.
If the loop and magnet move farther apart, the magnetic flux through the loop due to $\vec{B}_{magnet}$ decreases.

Figure 20-6 Lenz's law The current induced in a loop by a change in flux always acts to oppose the flux change.

a combination of the two. The nineteenth-century Russian physicist Heinrich Lenz summarized these observations in a principle that we now call **Lenz's law:**

The direction of the magnetic field induced within a conducting loop opposes the change in magnetic flux that created it.

It's common to combine Faraday's law and Lenz's law into a single equation:

Faraday's law and Lenz's law for induction (20-3)

$$\varepsilon = -\frac{\Delta\Phi_B}{\Delta t}$$

Induced or motional emf in a loop

Change in the magnetic flux through the surface outlined by the loop

Time interval over which the change in magnetic flux takes place

The minus sign indicates that the current caused by the emf induces a magnetic field which opposes the change in flux.

WATCH OUT! Like Faraday's law, Lenz's law is about the *change* in flux.

Notice that Lenz's law refers to the direction of the *change* in magnetic flux ("Is the flux increasing or decreasing?"), not to the direction of the field that causes the flux. The field $\vec{B}_{\text{magnet}}$ points in the same direction in both of the situations shown in Figure 20-6b, but the flux change is different in the two situations, so the emf is in different directions as well.

We can check Lenz's law by revisiting the sliding copper bar from Example 20-1 (Section 20-2). The upward external magnetic field of magnitude B causes an upward magnetic flux through the loop formed by the sliding bar and the rails on which it slides (**Figure 20-7**). The area enclosed by this loop increases as the bar slides to the right, so the upward flux increases as well. By Lenz's law an induced current will flow in the loop in order to generate an induced magnetic field that opposes this change in flux. So this induced magnetic field must point downward, and to produce that induced field the current in the loop must be clockwise as seen from above the loop. That's just the direction of current flow that we depicted in the figure that accompanies Example 20-1. So this situation is consistent with Lenz's law.

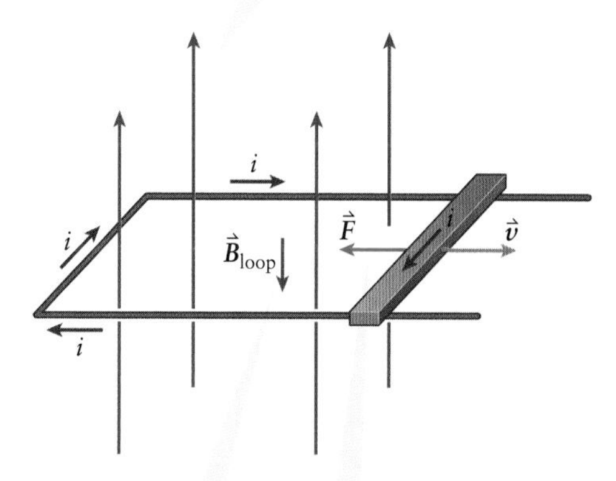

Figure 20-7 A sliding copper bar revisited Lenz's law helps explain the direction in which current flows in this situation.

The sliding copper bar in Figure 20-7 illustrates another aspect of Lenz's law. Once the induced current is flowing in the bar, the external magnetic field exerts a force on that current. Using the right-hand rule for this force (see Section 19-5), we see that this force points opposite to the direction in which the bar is moving. In other words, this force *opposes* the motion that gives rise to the change in flux through the loop made up of the bar and rails. That's always the case when a conductor moves through a magnetic field: A current is induced in the conductor, and the magnetic field exerts a force on the current that opposes the motion of the conductor. We can summarize this in an alternative statement of Lenz's law:

When the magnetic flux through a loop changes, current flows in a direction that opposes that change.

Since a magnetic force opposes the motion of the bar in Figure 20-7, we need to apply an external force to keep the bar in motion. If we make the bar move faster, the magnetic flux through the loop changes more rapidly, the emf and resulting current in the loop are greater, and the magnetic force opposing the motion of the bar is greater. (The magnetic force on the bar is proportional to the speed of the bar, just like the drag force on a microscopic object moving through a fluid; see Section 5-5.) So we must apply a greater force to make the bar slide at a faster speed.

This same effect explains the phenomenon of *magnetic braking*. If you try to make a magnet move past a conductor or a conductor move past a magnet, currents appear in the conductor. (These are called *eddy currents*, since their pattern resembles that of eddies in a body of water. The conductor does *not* need to be in the form of a loop for these currents to appear.) The magnetic force that the magnet exerts on the eddy currents opposes the motion of the conductor relative to the magnet, so by Newton's third law there is a force that opposes the motion of the magnet relative to the conductor. One application of this is to roller coasters. When a roller coaster car enters the part of the ride where it's supposed to slow down, a copper fin on the car passes through powerful permanent magnets mounted on the track. Eddy currents arise in the fin, and the interaction between the eddy currents and the field of the permanent magnets causes a force that smoothly brings the car to a slow speed. The car is then stopped by conventional mechanical braking.

Eddy currents are also used in an *electromagnetic flowmeter*, a device that can measure blood flow in an artery. Blood is an electrical conductor; eddy currents are

induced in the blood as it flows past magnets in the flowmeter. The device records the small but measurable magnetic fields due to these currents and uses them to determine the rate of flow. The advantage of an electromagnetic flowmeter is that it is noninvasive: No component of the device need be surgically introduced into the body.

Another application of eddy currents is *magnetic induction tomography*, a relatively new experimental technique for medical imaging. In this technique changing magnetic fields created by coils placed near a part of the body induce eddy currents. Observing the fields produced by these eddy currents is a way to monitor brain swelling. Eddy currents can also be used for the controlled, repeated delivery of medication. A capsule containing the drug is implanted in the body; the capsule is made from a gel that heats up slightly when there are eddy currents, opening pores through which the medication is released. As for an electromagnetic flowmeter, no implanted electronics are required.

GOT THE CONCEPT? 20-2 Induced Current I

In **Figure 20-8** a rectangular loop of wire moves to the right into a region of constant, uniform magnetic field. The field points into the plane of the figure, in a direction perpendicular to the plane of the loop. When the loop is entering the field region, as in Figure 20-8a, what is the direction of the current around the loop? (a) Clockwise; (b) counterclockwise; (c) the current is zero; (d) not enough information given to decide.

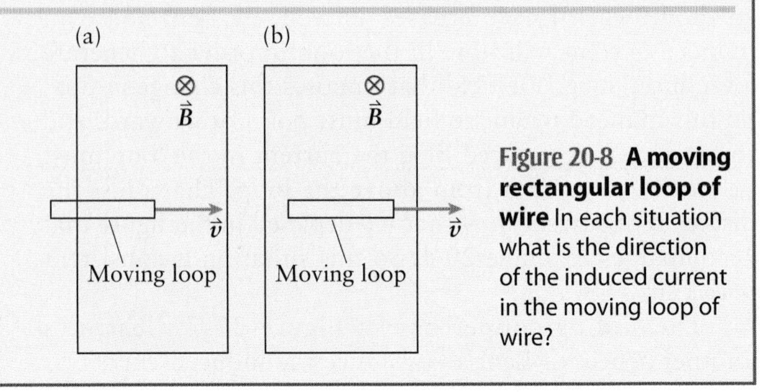

Figure 20-8 A moving rectangular loop of wire In each situation what is the direction of the induced current in the moving loop of wire?

GOT THE CONCEPT? 20-3 Induced Current II

In Figure 20-8 a rectangular loop of wire moves to the right into a region of constant, uniform magnetic field. The field points into the plane of the figure, in a direction perpendicular to the plane of the loop. When the loop is moving and completely inside the field region, as in Figure 20-8b, what is the direction of the current around the loop? (a) Clockwise; (b) counterclockwise; (c) the current is zero; (d) not enough information given to decide.

TAKE-HOME MESSAGE FOR Section 20-3

✔ An emf induced by a changing magnetic flux tends to cause a current to flow. This current generates a magnetic field of its own, called the induced magnetic field.

✔ The induced magnetic field is in a direction that opposes the change in flux that created the emf that gave rise to the induced field.

✔ Eddy currents arise whenever a conducting material, even a nonmagnetic one, moves relative to a magnetic field.

20-4 Faraday's law explains how alternating currents are generated

We learned in Section 18-2 about the importance of alternating current in technology. (If you're reading these words in a room lit by electric light, the light bulbs are powered by alternating current. If you're reading on a mobile device such as a tablet, the device's battery was charged by plugging it into a wall socket and using the alternating current delivered by that socket.) We now have the physics we need to understand how alternating current is produced.

Let's look at a coil of wire with N turns, each of which has area A. As **Figure 20-9** shows, this coil is free to rotate around an axis that lies along a diameter of the coil.

We place the coil in a region of uniform magnetic field $\vec{B}$, then rotate the coil at a constant angular speed ω. As the coil rotates the magnetic flux through each turn of the coil changes, so an emf is generated. As we will see, this is an alternating emf of just the sort required to generate an alternating current. That's why a rotating coil of the sort shown in Figure 20-9 is called an **ac generator**.

We begin by writing an equation for the magnetic flux through the rotating coil. The angle θ between the magnetic field $\vec{B}$ and the perpendicular to the coil changes as the coil rotates:

$$\theta = \omega t + \phi \qquad (20\text{-}4)$$

In Equation 20-4 ϕ is the value of the angle at $t = 0$. From Equation 20-1 the magnetic flux through the N turns of the coil is N times that through one turn:

$$\Phi_B = NAB \cos\theta = NAB \cos(\omega t + \phi) \qquad (20\text{-}5)$$

When $\theta = \omega t + \phi = 0$ so $\cos\theta = 1$, the perpendicular to the coil is in the same direction as $\vec{B}$ and the flux has its most positive value $\Phi_B = NAB$; when $\theta = \omega t + \phi = \pi/2$ so $\cos\theta = 0$, the coil is edge-on to the magnetic field and the flux is zero; when $\theta = \omega t + \phi = \pi$ and $\cos\theta = -1$, the perpendicular to the coil points opposite to $\vec{B}$ and the flux has its most negative value $\Phi_B = -NAB$; and so on. So the magnetic flux varies with time, and it follows that there will be an emf in the coil.

Faraday's law and Lenz's law (Equation 20-3) tell us that the emf is equal to the negative of the rate of change of Φ_B. We actually know how to find the rate of change of a cosine function like that in Equation 20-5. In Section 12-3 we saw that the position of an object undergoing simple harmonic motion with amplitude A is given by

$$x = A\cos(\omega t + \phi) \qquad (12\text{-}6)$$

The rate of change of position x is just the velocity v_x, which we found was equal to

$$v_x = -\omega A \sin(\omega t + \phi) \qquad (12\text{-}7)$$

You can see that Equation 20-5 for magnetic flux is identical to Equation 12-6 for position, with amplitude A replaced by NAB (note that A in Equation 20-5 denotes area, not amplitude). Making the same replacement in Equation 12-7 tells us that the rate of change of magnetic flux through the rotating coil is

$$\frac{\Delta\Phi_B}{\Delta t} = -\omega NAB \sin(\omega t + \phi) \qquad (20\text{-}6)$$

Substituting Equation 20-6 into Equation 20-3 then gives us the emf in the rotating coil:

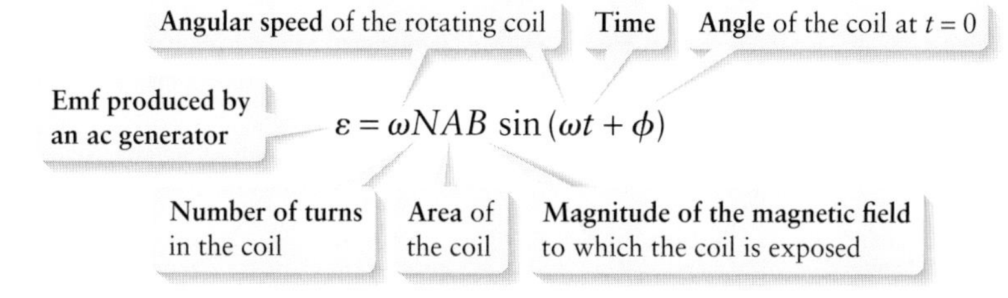

Emf in an ac generator (20-7)

The emf alternates with angular frequency ω, which is the same as the angular speed of the rotating coil. The maximum value of the emf is $\varepsilon_{max} = \omega NAB$, which shows that we can increase the maximum emf by increasing the angular speed ω, the number of turns N, the coil area A, the magnetic field magnitude B, or a combination of these.

While the emf produced by an ac generator changes from positive to negative, the power delivered by the generator does not. As an example, suppose an ac generator is connected to a circuit device (such as a light bulb or a toaster) that we can represent as a resistor with resistance R. If we ignore the internal resistance of the coil, the emf

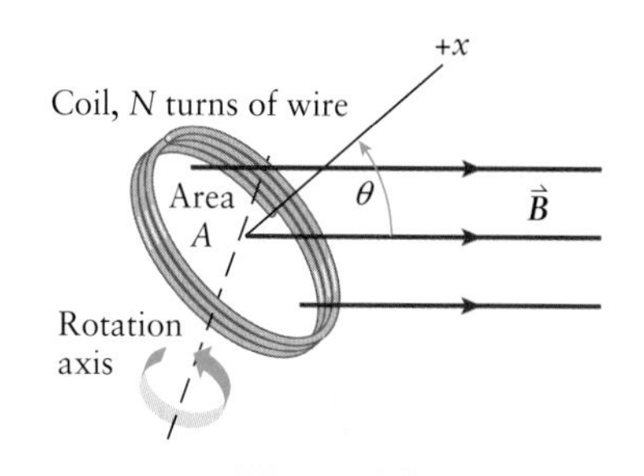

As the coil rotates with angular speed ω, the angle θ between the magnetic field $\vec{B}$ direction and the perpendicular to the coil changes: $\theta = \omega t + \phi$.

Figure 20-9 An ac generator As the coil rotates, an oscillating emf is generated in the turns of wire that make up the coil.

See the Math Tutorial for more information on trigonometry.

is equal to the voltage drop across the resistor: $\varepsilon = iR$. The current in the resistor is therefore

$$i = \frac{\varepsilon}{R} = \frac{\omega NAB}{R}\sin(\omega t + \phi) \quad (20\text{-}8)$$

The current in the resistor alternates with the same angular frequency ω as the emf. From Section 18-6 the power into such a resistor is

$$P = i^2 R \quad (18\text{-}24)$$

If we substitute Equation 20-8 into Equation 18-24, we get

$$P = \left(\frac{\omega NAB}{R}\sin(\omega t + \phi)\right)^2 R = \frac{\omega^2 N^2 A^2 B^2}{R}\sin^2(\omega t + \phi) \quad (20\text{-}9)$$

Figure 20-10 shows graphs of the emf ε (Equation 20-7) and resistor power P (Equation 20-9) as functions of time. The power P is *never* negative, which means that energy always flows from the ac generator into the resistor, never the other way. The average value of the function $\sin^2(\omega t + \phi)$ is $\frac{1}{2}$, so the average power into the resistor is

$$P_{\text{average}} = \frac{\omega^2 N^2 A^2 B^2}{2R} \quad (20\text{-}10)$$

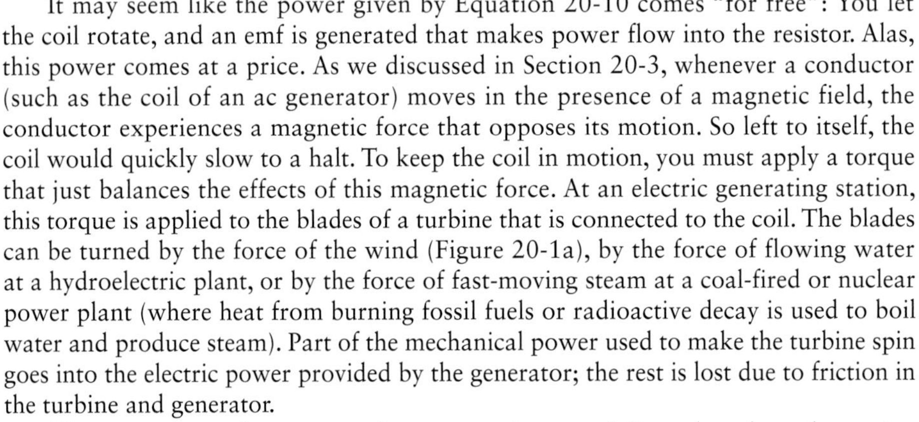

It may seem like the power given by Equation 20-10 comes "for free": You let the coil rotate, and an emf is generated that makes power flow into the resistor. Alas, this power comes at a price. As we discussed in Section 20-3, whenever a conductor (such as the coil of an ac generator) moves in the presence of a magnetic field, the conductor experiences a magnetic force that opposes its motion. So left to itself, the coil would quickly slow to a halt. To keep the coil in motion, you must apply a torque that just balances the effects of this magnetic force. At an electric generating station, this torque is applied to the blades of a turbine that is connected to the coil. The blades can be turned by the force of the wind (Figure 20-1a), by the force of flowing water at a hydroelectric plant, or by the force of fast-moving steam at a coal-fired or nuclear power plant (where heat from burning fossil fuels or radioactive decay is used to boil water and produce steam). Part of the mechanical power used to make the turbine spin goes into the electric power provided by the generator; the rest is lost due to friction in the turbine and generator.

Go to Interactive Exercise 20-2 for more practice dealing with coils.

The ac generator is one part of a system of power delivery based on alternating current. In Chapter 21 we'll explore in more detail the physics of alternating current in circuits.

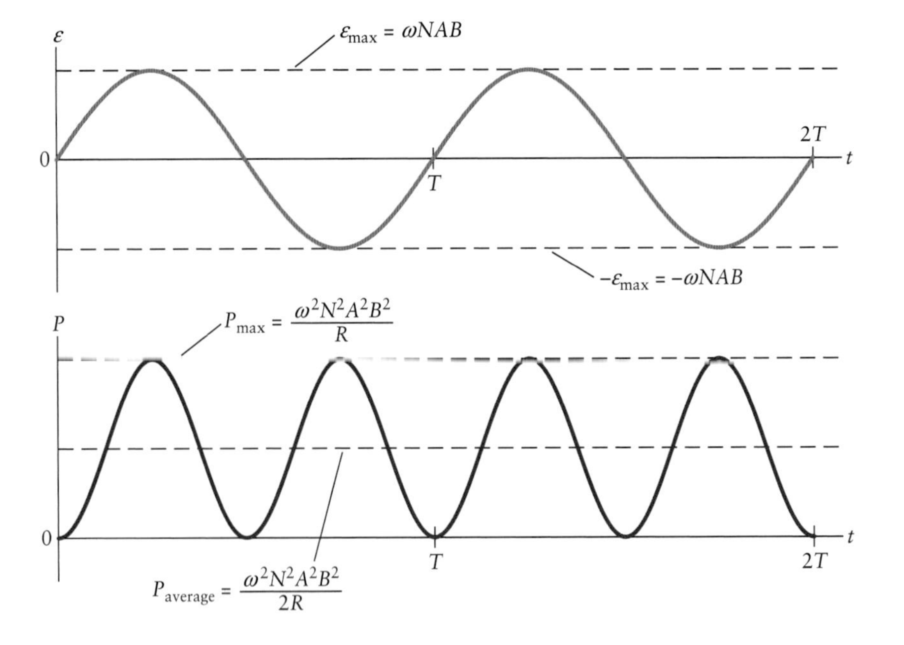

Figure 20-10 An ac generator: Emf and power These graphs show the emf ε generated in the coil shown in Figure 20-9 and the power P that this emf delivers to a resistor R connected to the coil. We assume $\phi = 0$ in Equations 20-7 and 20-9. Note that T is the time it takes for the coil to complete one rotation.

EXAMPLE 20-3 Lighting the Gym with a Bicycle

You attach the coil of an ac generator to an exercise bicycle so that as you work the pedals the coil turns, and an emf is generated. The generator is geared so that it makes 10 rotations for each rotation of the pedals. The coil has an area of 6.40×10^{-3} m^2, has 2000 turns of wire, and is in a magnetic field of magnitude 0.100 T. How many times a second must you turn the pedals in order to deliver an average power of 60.0 W to a light bulb with a resistance of 80.0 Ω?

Set Up

The situation is as shown in Figure 20-9. We'll use Equation 20-10 to solve for the angular speed ω of the generator coil. Since the coil makes 10 rotations for every rotation of the pedals, the angular speed of the pedals is equal to ω divided by 10.

Average power delivered by an ac generator to a resistor:

$$P_{\text{average}} = \frac{\omega^2 N^2 A^2 B^2}{2R} \qquad (20\text{-}10)$$

Solve

Rewrite Equation 20-10 to solve for ω.

Find the angular speed of the generator coil from Equation 20-10:

$$\omega^2 = \frac{2RP_{\text{average}}}{N^2 A^2 B^2}$$

$$\omega = \frac{\sqrt{2RP_{\text{average}}}}{NAB} = \frac{\sqrt{2(80.0\ \Omega)(60.0\ \text{W})}}{(2000)(6.40 \times 10^{-3}\ \text{m}^2)(0.100\ \text{T})}$$

$$= 76.5\ \text{rad/s}$$

Convert this from radians per second to revolutions per second:

$$\omega = \left(76.5\ \frac{\text{rad}}{\text{s}}\right)\left(\frac{1\ \text{rev}}{2\pi\ \text{rad}}\right) = 12.2\ \text{rev/s}$$

The generator turns 10 times faster than the pedals, so the pedals must turn at a rate of

$$(12.2\ \text{rev/s})/10 = 1.22\ \text{rev/s} = 73.1\ \text{rev/min}$$

Reflect

A cycling cadence of 73.1 rev/min isn't difficult for an amateur cyclist, so you can certainly power a light bulb in this way. Exercise bicycles with generators of this kind are commercially available and are used to return power to the electrical grid in the same manner as residential solar panels.

GOT THE CONCEPT? 20-4 A Flickering Fluorescent Lamp

Certain types of fluorescent lamps flicker rapidly. That's because these lamps emit a pulse of light every time a burst of electric power is provided to the lamp. If such a lamp is powered by a source of emf that oscillates at 60 Hz, what is the frequency at which the lamp will flicker? (a) 30 Hz; (b) 60 Hz; (c) 120 Hz; (d) 240 Hz; (e) 3600 Hz.

TAKE-HOME MESSAGE FOR Section 20-4

✔ An ac generator consists of a coil that rotates in a magnetic field.

✔ The emf produced by an ac generator oscillates at an angular frequency that equals the angular speed of the rotating coil.

Key Terms

ac generator
electromagnetic induction
Faraday's law of induction
induced emf
induced magnetic field
Lenz's law
magnetic flux
motional emf

Chapter Summary

Topic	Equation or Figure
Motional emf: An emf appears in a loop when that loop moves in the presence of a magnetic field. The emf is a result of magnetic forces on the mobile charges within the loop.	(c) Side view of loop moving toward a stationary magnet 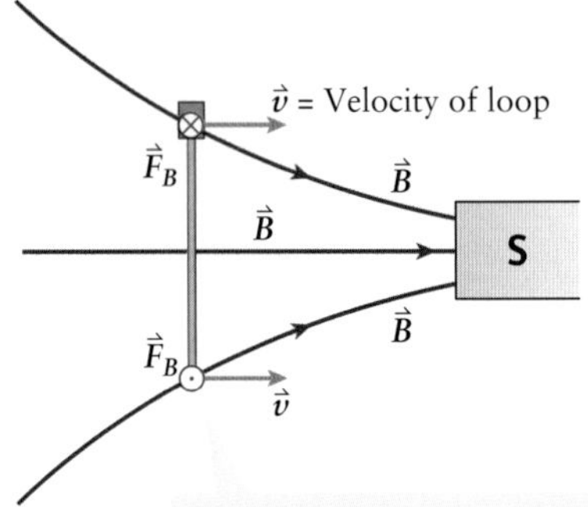 • Charges in the loop move along with the loop in the magnetic field. • Magnetic forces push mobile charges around the loop, causing a current. (Figure 20-2c)
Induced emf: An emf also appears in a loop when the magnetic field within the loop changes. Here the forces that create the emf are not magnetic, but electric; an electric field is produced by the changing magnetic field.	(c) Side view of magnet moving toward a stationary loop 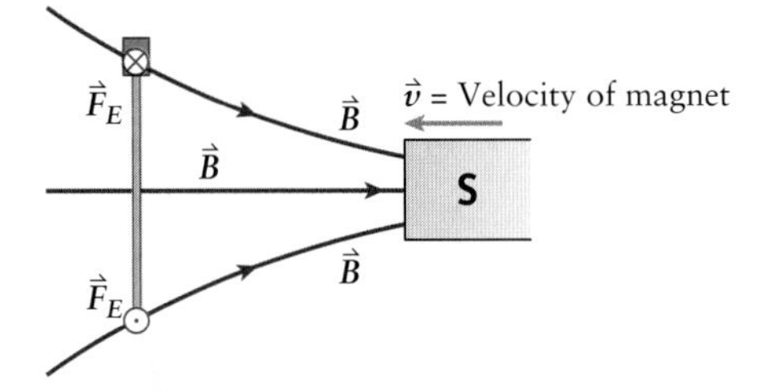 • Charges in the loop experience electric forces. • These forces push mobile charges around the loop, causing a current. (Figure 20-3c)

Faraday's law and Lenz's law: Faraday's law states that if the magnetic flux through a loop changes, an emf is induced around that loop. The magnitude of the induced emf equals the magnitude of the rate of change of the magnetic flux through the loop. Lenz's law states that the direction of the induced emf is such as to oppose the flux change: The induced current causes its own flux (which helps compensate for the change). A conductor and magnet that move relative to each other experience magnetic forces that oppose this relative motion.

$$\varepsilon = -\frac{\Delta \Phi_B}{\Delta t} \qquad (20\text{-}3)$$

Induced or motional emf in a loop

Change in the magnetic flux through the surface outlined by the loop

Time interval over which the change in magnetic flux takes place

The minus sign indicates that the current caused by the emf induces a magnetic field which opposes the change in flux.

Alternating current and an ac generator: To produce an alternating emf (the sort needed to cause an alternating current), rotate a coil in the presence of a constant magnetic field. The changing flux causes an emf that oscillates sinusoidally.

$$\varepsilon = \omega NAB \sin(\omega t + \phi) \qquad (20\text{-}7)$$

Angular speed of the rotating coil

Time

Angle of the coil at $t = 0$

Emf produced by an ac generator

Number of turns in the coil

Area of the coil

Magnitude of the magnetic field to which the coil is exposed

Answer to What do you think? Question

(b) The electrons in the wire are initially at rest and so do not experience a magnetic force. (Recall that a magnetic field exerts a force only on a charged object that is in motion.) So it must be an *electric* force that sets the electrons into motion. This is a consequence of electromagnetic induction: As you swipe the credit card through the reader, the wire loop is exposed to a varying magnetic field from the stripe on the back of the card. This causes a changing magnetic flux through the loop, which induces an electric field and an emf. Electrons move in the loop in response to this emf.

Answers to Got the Concept? Questions

20-1 (d) The induced emf does *not* depend on what the loop is made of. (Note that Equation 20-2 makes no reference to the properties of the loop material.) So the emf will be exactly the same whether the loop is made of copper, wood, silver, rubber, or even air. The difference is that because copper is a good conductor while wood is a very poor conductor, a current will be generated in the copper loop but not in the wooden loop.

20-2 (b) As the loop in Figure 20-8a moves into the field region, there is an increasing magnetic flux through the loop due to the magnetic field directed into the plane of the figure. According to Lenz's law, current will flow in the loop to induce a magnetic field $\vec{B}_{\text{loop}}$ that will oppose this increase, so $\vec{B}_{\text{loop}}$ in the interior of the loop must point out of the plane in Figure 20-8a. The right-hand rule depicted in Figure 20-6a tells us that to induce such a field, current must flow counterclockwise around the loop. (You can confirm this by using the right-hand rule for the magnetic force on a moving charge. A positive charge in the right-hand leg of the moving loop feels an upward magnetic force, and this force drives current in a counterclockwise direction around the loop. There are also magnetic forces on charges in the top and bottom legs of the loop, but these forces have zero component along the length of the wire and so do not induce a current.)

20-3 (c) Although the loop in Figure 20-8b is moving, the magnetic flux through the loop remains constant because it is moving through a region of constant, uniform magnetic field. Since there is no flux change, Faraday's law tells us that no emf is induced and so no current will be generated. (You can confirm this by using the right-hand rule for the magnetic force on a moving charge. A positive charge in the right-hand leg of the moving loop feels an upward magnetic force, and this force by itself would drive current in a counterclockwise direction around the loop. But a positive charge in the left-hand leg of the moving loop also feels an upward magnetic force, which by itself would drive current in a clockwise direction around the loop. There are also magnetic forces on charges in the top and bottom legs of the loop, but these forces have zero component along the length of the wire and so do not induce a current. The net effect is that there is *zero* current in the loop.)

20-4 (c) Figure 20-10 shows that the power delivered by an ac generator goes through two up-and-down cycles during the time T required for the emf to go through a single cycle. So if the emf varies at 60 Hz, the power delivered to the fluorescent lamp varies at $2 \times 60\ \text{Hz} = 120\ \text{Hz}$.

Questions and Problems

In a few problems you are given more data than you actually need; in a few other problems you are required to supply data from your general knowledge, outside sources, or informed estimate.

Interpret as significant all digits in numerical values that have trailing zeros and no decimal points.

For all problems use $g = 9.80 \text{ m/s}^2$ for the free-fall acceleration due to gravity. Neglect friction and air resistance unless instructed to do otherwise.

•	Basic, single-concept problem
••	Intermediate-level problem; may require synthesis of concepts and multiple steps
•••	Challenging problem
SSM	*Solution is in Student Solutions Manual*
Example	*See worked example for a similar problem*

Conceptual Questions

1. • Two conducting loops with a common axis are placed near each other, as shown in **Figure 20-11.** Initially the currents in both loops are zero. If a current is suddenly set up in loop a in the direction shown, is there also a current in loop b? If so, in which direction? What is the direction of the force, if any, that loop a exerts on loop b? Explain your answer.

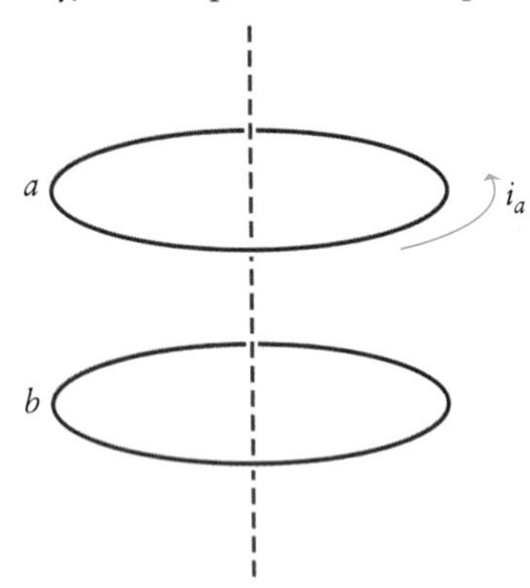

Figure 20-11 Problem 1

2. •• In a popular demonstration of electromagnetic induction, a metal plate is suspended in midair above a large electromagnetic coil, as shown in **Figure 20-12.** (a) How does this work? (b) If your professor does the demonstration, one thing you'll notice is that the plate gets quite hot. (In fact, you can end the demonstration by frying an egg on the plate!) Why does the plate become hot? (c) Would the trick work if the plate were made of an insulating material?

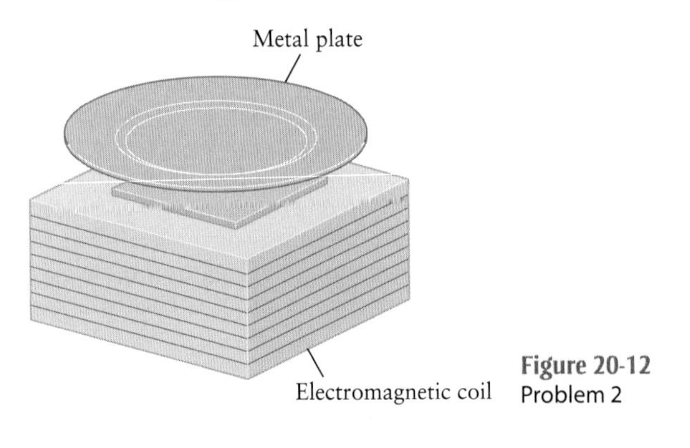

Figure 20-12 Problem 2

3. • A conducting rod slides without friction on conducting rails in a magnetic field, as shown in **Figure 20-13.** The rod is given an initial velocity $\vec{v}$ to the right. Describe its subsequent motion and justify your answer.

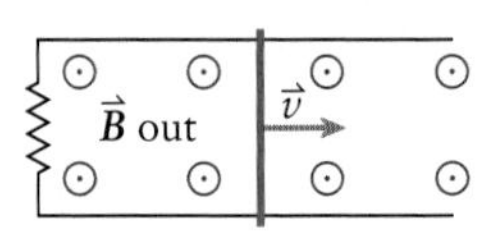

Figure 20-13 Problem 3

4. • A common physics demonstration is to drop a small magnet down a long, vertical aluminum pipe. Describe the motion of the magnet and the physical explanation for the motion. SSM

5. • **Medical** In hospitals with magnetic resonance imaging facilities and at other locations where large magnetic fields are present, there are usually signs warning people with pacemakers and other electronic medical devices not to enter. Why?

6. •• **Figure 20-14** depicts an electron in between the poles of an electromagnet. Explain how the electron is accelerated if the magnetic field is gradually being increased.

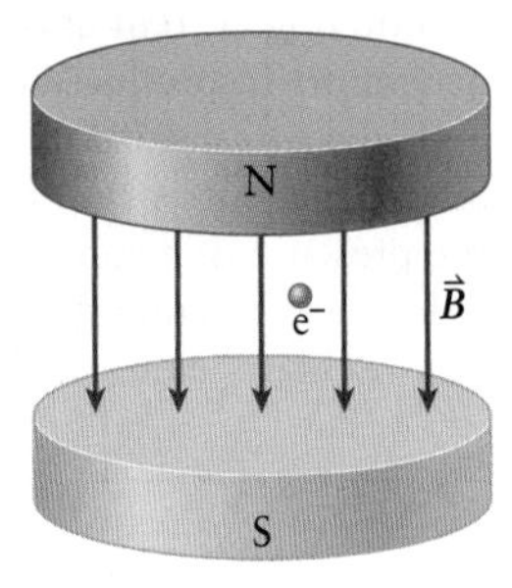

Figure 20-14 Problem 6

Multiple-Choice Questions

7. • **Figure 20-15** shows a sequence of sketches depicting a rectangular loop passing from left to right through a region of constant magnetic field. The field points out of the page and perpendicular to the plane of the loop. In which one of the sequences is the magnetic flux through the loop decreasing? SSM

   A. from left to right approaching the magnetic field
   B. entering the magnetic field
   C. inside the magnetic field
   D. leaving the magnetic field
   E. from left to right moving away from the magnetic field

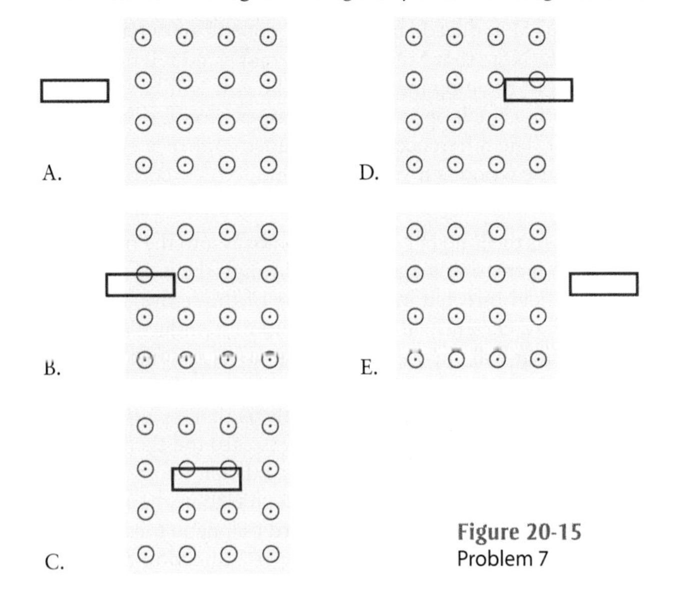

Figure 20-15 Problem 7

8. • On which variable does the magnetic flux depend?
   A. the magnetic field
   B. the area of a region through which the magnetic field passes
   C. the orientation of the field with respect to the region through which it passes
   D. all of the above
   E. none of the above

9. • Two metal rings with a common axis are placed near each other, as shown in **Figure 20-16**. If current i_a is suddenly set up and is increasing in ring a as shown, the current in ring b is
   A. zero.
   B. parallel to i_a.
   C. antiparallel to i_a.
   D. alternatively parallel and antiparallel to i_a.
   E. perpendicular to i_a. SSM

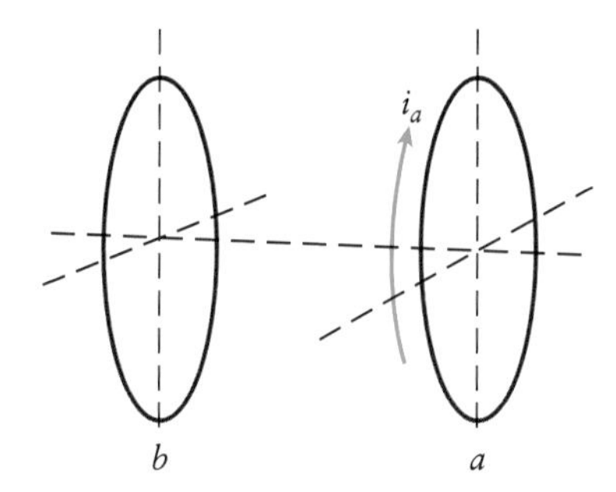

Figure 20-16 Problem 9

10. •• **Figure 20-17** shows two coils wound around an iron ring, which directs the magnetic field of each coil around the ring. Current appears in the coil on the right
   A. the moment the battery is connected by closing the switch.
   B. the entire time the battery is connected with the switch closed.
   C. the moment the battery is disconnected by opening the switch.
   D. the moment the battery is connected by closing the switch and the moment the battery is disconnected by opening the switch.
   E. the entire time the battery is disconnected with the switch open.

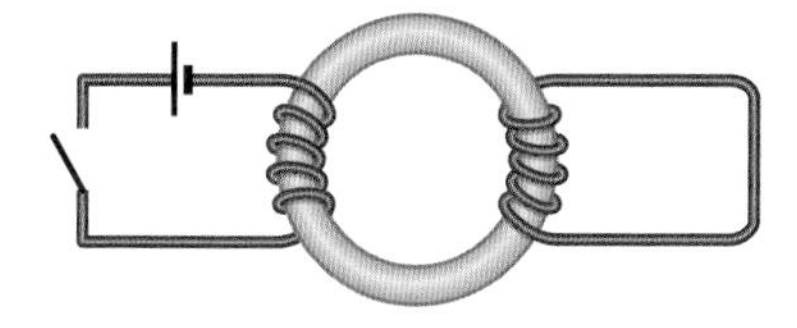

Figure 20-17 Problem 10

11. • The copper ring of radius R in **Figure 20-18** lies in a magnetic field pointed into the page. The field is uniformly decreasing in magnitude. The induced current in the ring is
   A. clockwise and constant.
   B. clockwise and changing.
   C. zero.
   D. counterclockwise and constant.
   E. counterclockwise and changing.

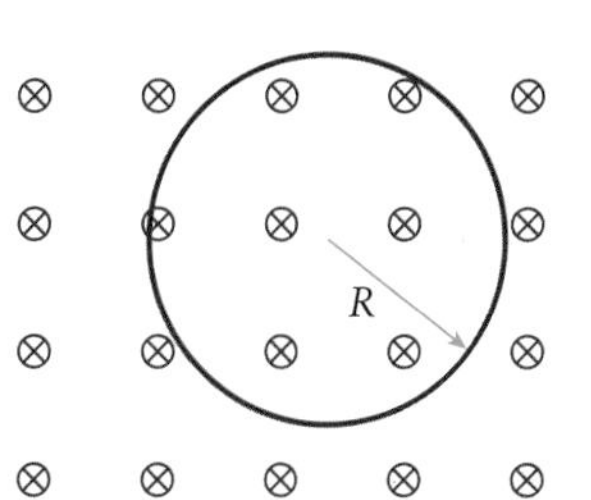

Figure 20-18 Problem 11

12. • A conducting loop moves at a constant speed parallel to a long, straight wire carrying a constant current, as shown in **Figure 20-19**.
   A. The induced current in the loop will be clockwise.
   B. The induced current in the loop will be only parallel to the current i.
   C. The induced current in the loop will be counterclockwise.
   D. The induced current in the loop will be alternately clockwise and then counterclockwise.
   E. There will be no induced current in the loop.

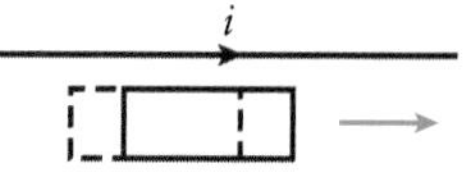

Figure 20-19 Problem 12

Estimation/Numerical Analysis

13. • Estimate the electric current in a generator in a large hydroelectric dam (such as Hoover Dam or the Oahe Dam).

14. • Estimate how many turbines (such as the ones at Hoover Dam) would be required to supply enough energy to power the United States.

15. • Give an estimate of the induced voltage created in transatlantic communication cables due to fluctuations in Earth's magnetic field.

16. • Estimate the magnitude of the fluctuations in Earth's geomagnetic field during times of solar flares.

17. •• The induced voltage versus time for a coil that has 100 circular turns of wire with radii 25 cm is given in the table. Plot $V(t)$ and use this graph to predict the graph of the magnetic field as a function of time $B(t)$ that is passing through the loop (assume the magnetic field is perpendicular to the plane of the loop).

t(s)	V(V)	t(s)	V(V)
0	0	9	2
1	2	10	4
2	4	11	2
3	2	12	0
4	0	13	−2
5	−2	14	−4
6	−4	15	−2
7	−2	16	0
8	0		

Problems

20-1 The world runs on electromagnetic induction

20-2 A changing magnetic flux creates an electric field

18. • A single-turn circular loop of wire that has a radius of 5.0 cm lies in the plane perpendicular to a spatially uniform magnetic field. During a 0.12-s time interval, the magnitude of the field increases uniformly from 0.20 to 0.40 T. Determine the magnitude of the emf induced in the loop during the time interval. Example 20-2

19. • A circular coil that has 100 turns and a radius of 10.0 cm lies in a magnetic field that has a magnitude of 0.0650 T directed perpendicular to the coil. (a) What is the magnetic flux through the coil? (b) The magnetic field through the coil is increased steadily to 0.100 T over a time interval of 0.500 s. What is the magnitude of the emf induced in the coil during the time interval? Example 20-2

20. • A 30-turn coil with a diameter of 6.00 cm is placed in a constant, uniform magnetic field of 1.00 T directed perpendicular to the plane of the coil. Beginning at time $t = 0$ s, the field is increased at a uniform rate until it reaches 1.30 T at $t = 10.0$ s. The field remains constant thereafter. What is the magnitude of the induced emf in the coil at (a) $t < 0$ s, (b) $t = 5.00$ s, and (c) $t > 10.0$ s? (d) Plot the

magnetic field and the induced emf as functions of time for the range $-5.00\text{ s} < t < 15.0\text{ s}$. Example 20-2

20-3 Lenz's law describes the direction of the induced emf

21. • Determine the direction of the induced current in the loop for each case shown in **Figure 20-20**.

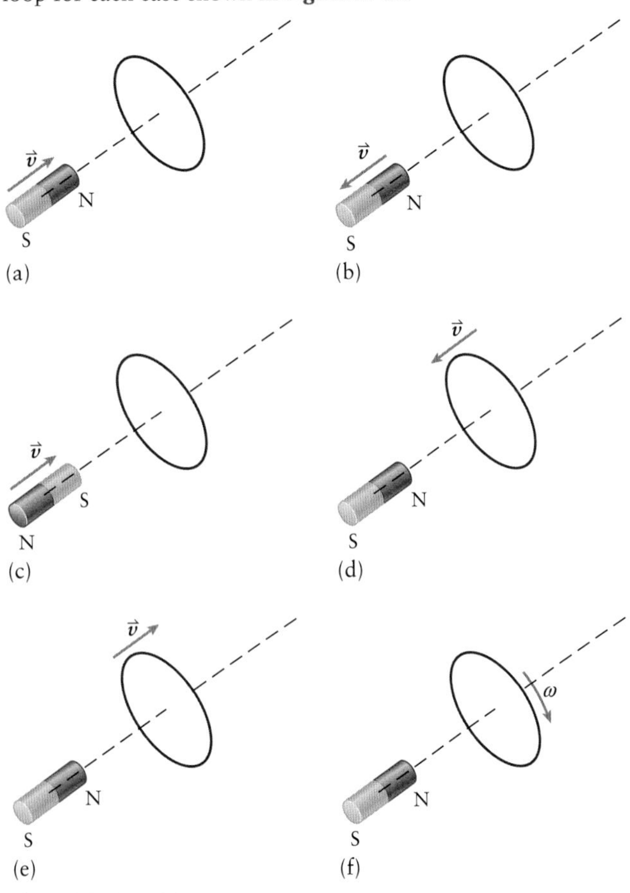

Figure 20-20 Problem 21

22. • A bar magnet is moved steadily through a wire loop, as shown in **Figure 20-21**. Make a qualitative sketch of the induced emf in the loop as a function of time (be sure to include the times t_1, t_2, and t_3). Consider the direction of positive emf to be as indicated in the figure.

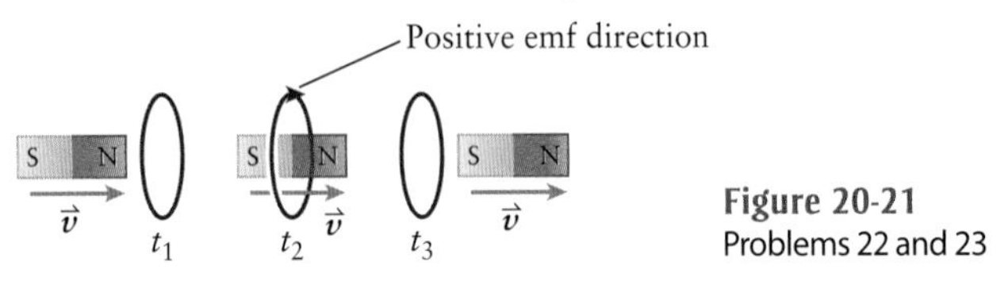

Figure 20-21 Problems 22 and 23

23. • A bar magnet is moved steadily through a wire loop, as shown in Figure 20-21, except that the leading edge of the magnet is the south pole instead of the north pole. Make a qualitative sketch of the induced voltage in the loop as a function of time (be sure to include the times t_1, t_2, and t_3). Consider the direction of positive emf to be as indicated in the figure. SSM

24. •• A square, 30-turn coil 10.0 cm on a side with a resistance of 0.820 Ω is placed between the poles of a large electromagnet. The electromagnet produces a constant, uniform magnetic field of 0.600 T directed into the page. As suggested by **Figure 20-22**, the field drops sharply to zero at the edges of the magnet. The coil moves to the right at a constant velocity of 2.00 cm/s. What is the current through the wire coil (a) before the coil reaches the edge of the field, (b) while the coil is leaving the field, and (c) after the coil leaves the field? (d) What is the total charge that flows past a given point in the coil as it leaves the field? (e) Plot the induced current in the loop as a function of the horizontal position of the right side of the current loop. Let the right-hand edge of the magnetic field region be $x = 0$. Your plot should be in the range of $-5.00\text{ cm} < x < 20.0\text{ cm}$.

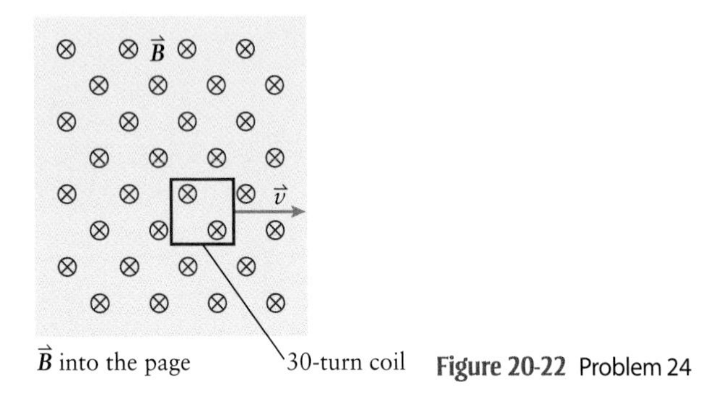

Figure 20-22 Problem 24

25. •• (a) Determine the magnitude and direction of the force on each side of the coil in Problem 24 for situations (a) through (c). (b) As the loop enters the field region from the left, what is the direction of the induced current and the resulting force on each segment of the coil?

20-4 Faraday's law explains how alternating currents are generated

26. • A rectangular coil with sides 0.10 m by 0.25 m has 500 turns of wire. It is rotated about its long axis in a magnetic field of 0.58 T directed perpendicular to the rotation axis. At what frequency must the coil be rotated for it to generate a maximum potential of 110 V?

27. • An electromagnetic generator consists of a coil that has 100 turns of wire, has an area of 400 cm^2, and rotates at 60 rev/s in a magnetic field of 0.25 T directed perpendicular to the rotation axis. What is the magnitude of the emf induced in the coil? SSM

28. •• You decide to build a small generator by rotating a coiled wire inside a static magnetic field of 0.30 T. You construct the apparatus by coiling wire into three loops of radius 0.16 m. If the coils rotate at 3.0 revolutions per second and are connected to a device with 1.0×10^2 Ω resistance, calculate the average power supplied to that device. Example 20-3

29. • Perhaps it has occurred to you that we could tap Earth's magnetic field to generate energy. One way to do this would be to spin a metal loop about an axis perpendicular to Earth's magnetic field. Suppose that the metal loop is a square that is 45.0 cm on each side and that we want to generate an electric potential

in the loop of amplitude 120 V at a place where Earth's magnetic field is 5×10^{-5} T. At what angular speed (in rev/s) would we have to spin the coil? Does this appear to be a feasible method to extract energy from Earth's magnetic field?

General Problems

30. ••• A long, rectangular loop of width w, mass m, and resistance R is being pushed into a magnetic field by a constant force $\vec{F}$ (**Figure 20-23**). Derive an expression for the speed of the loop while it is entering the magnetic field. Example 20-1

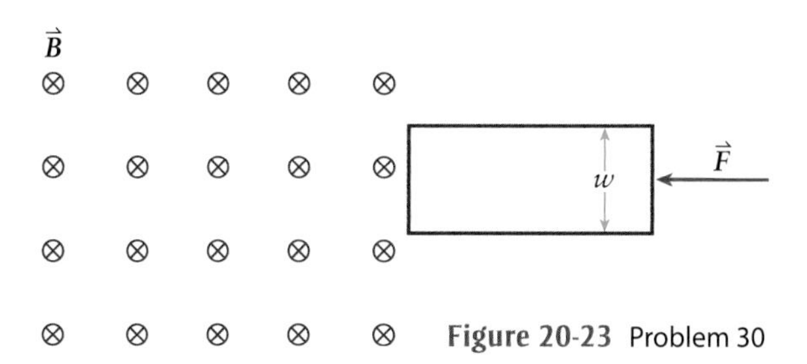

Figure 20-23 Problem 30

31. •• A magnetic field of 0.45 G is directed straight down, perpendicular to the plane of a circular coil of wire that is made up of 250 turns and has a radius of 20 cm. (a) If the coil is stretched, in a time of 15 ms, to a radius of 30 cm, calculate the emf induced in the coil during the process. (b) Assuming the resistance of the coil is a constant 25 Ω, what is the induced current in the coil during the process? (c) What is the direction of the induced current in the coil (clockwise or counterclockwise, as viewed from above)? SSM Example 20-1

32. • **Astronomy** Activity on the Sun, such as solar flares and coronal mass ejections, hurls large numbers of charged particles into space. When the particles reach Earth, they can interfere with communications and the power grid by causing electromagnetic induction. As one example, a current of millions of amps (known as the *auroral electrojet*) that runs about 100 km above Earth's surface can be perturbed. The change in the current causes a change in the magnetic field it produces at Earth's surface, which induces an emf along Earth's surface and in the power grid (which is grounded). Induced electric fields as high as 6.0 V/km have been measured. We can model the circuit at Earth's surface as a rectangular loop made up of the power lines completed by a path through the ground beneath. We can treat the magnetic field created by the electrojet as being uniform (but not constant). Consider a 1.0-km-long stretch of power line that is 5.0 m above the surface of Earth. If the induced emf in the Earth–power line loop is 6.0 V, at what rate must the magnetic field through the loop be changing? Example 20-2

33. • **Medical** During transcranial magnetic stimulation (TMS) treatment, a magnetic field typically of magnitude 0.50 T is produced in the brain using external coils. During the treatment the current in the coils (and hence the magnetic field in the brain) rises from zero to its peak in about 75 μs. Assume that the magnetic field is uniform over a circular area of diameter 2.0 cm inside the brain. What is the magnitude of the average induced emf around this area in the brain during the treatment? Example 20-2

34. •• A permanent bar magnet with the north pole pointing downward is dropped into a solenoid. (a) Determine the direction of the induced current that would be measured in the ammeter shown in **Figure 20-24**. (b) If the magnet is suddenly pulled upward through the solenoid, what is the direction of the induced current that would be measured in the ammeter?

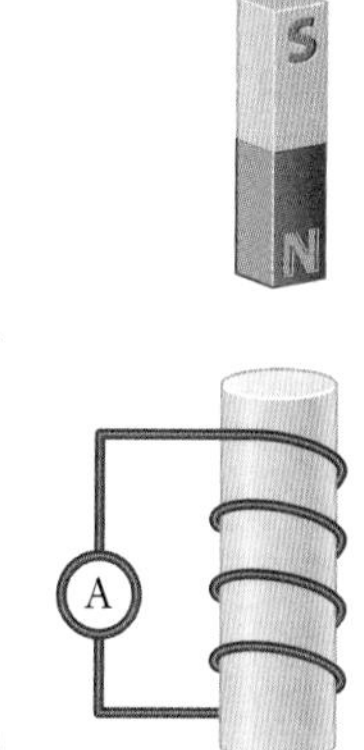

Figure 20-24 Problem 34

35. •• A pair of parallel conducting rails that are 12 cm apart lies at right angles to a uniform magnetic field of 0.8 T directed into the page, as shown in **Figure 20-25**. A 15-Ω resistor is connected across the rails. A conducting bar is moved to the right at 2 m/s across the rails. (a) What is the current in the resistor? (b) What direction is the current in the bar (up or down)? (c) What is the magnetic force on the bar? SSM Example 20-1

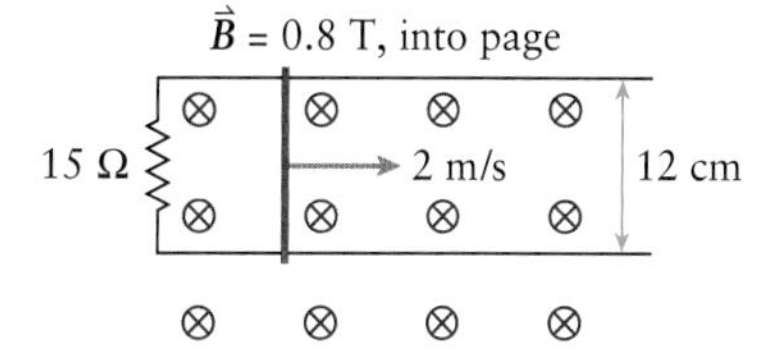

Figure 20-25 Problem 35

36. ••• A 50-turn square coil with a cross-sectional area of 5.00 cm^2 has a resistance of 20.0 Ω. The plane of the coil is perpendicular to a uniform magnetic field of 1.00 T. The coil is suddenly rotated about the axis shown in **Figure 20-26** through an angle of 60° over a period of 0.200 s. (a) What charge flows past a point in the coil during that time? (b) If the loop is rotated a full 360° around the axis, what is the net charge that passes the point in the loop? Explain your answer.

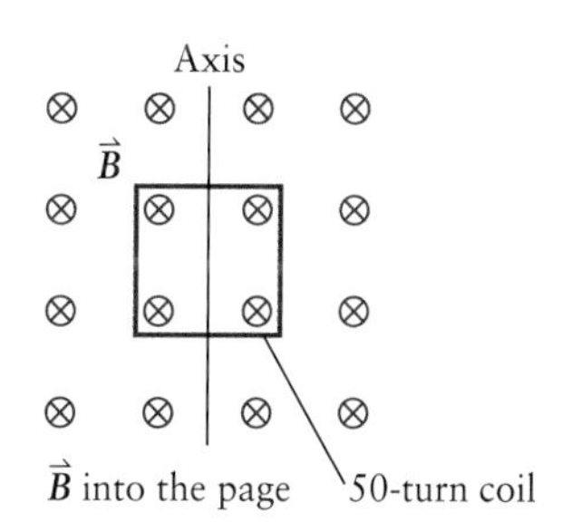

Figure 20-26 Problem 36

37. •• An ordinary gold wedding ring is tossed end over end into a running MRI machine with a 4.0 T field. The ring spends 1.3 s in the field, during which it completes 60 full rotations. If the ring has a resistance of 6.0 Ω and a diameter of 18 mm, what is the average power dissipated by the ring? Example 20-3

21 Alternating-Current Circuits

Deyan Georgiev/AGE Fotostock

What do you think?

These transformers use Faraday's law to raise and lower the voltage of alternating current. Can they also be used to raise or lower the voltage of *direct* current? (a) Yes; (b) yes, but only to raise the voltage; (c) yes, but only to lower the voltage; (d) no.

In this chapter, your goals are to:

- (21-1) Explain the importance of alternating current.
- (21-2) Describe what is meant by the root mean square value.
- (21-3) Calculate the voltage change produced by a transformer.
- (21-4) Describe why an inductor opposes changes in the current passing through it.
- (21-5) Explain the flow of energy in an *LC* circuit.
- (21-6) Describe what happens in a driven series *LRC* circuit when the driving frequency changes.
- (21-7) Discuss why current can flow in only one direction in a *pn* junction diode.

To master this chapter, you should review:

- (12-3, 12-4) Simple harmonic motion and the energy of a mass–spring system.
- (12-8) How a damped oscillator responds to different driving frequencies.
- (17-5, 17-6) The definition of capacitance and the electric energy stored by capacitors.
- (18-3, 18-6) The current through and voltage across a resistor, and the power into a resistor.
- (18-5) Using Kirchhoff's loop rule to analyze circuits.
- (19-7) The magnetic field created by a solenoid.
- (20-2, 20-3) How Faraday's law and Lenz's law describe the induced emf that opposes a change in magnetic flux through a loop.
- (20-4) How ac generators work and the power they generate.

21-1 Most circuits use alternating current

In Chapter 20 we learned the principles of electromagnetic induction and how they can be used to *generate* an alternating current. In this chapter we'll learn how to *manipulate* and *use* alternating current. Most of the electrical power on our planet is transmitted and used in the form of an alternating current, so by studying this chapter you'll learn an important aspect of how the world around you works.

We'll see how *mutual inductance*—in which a changing emf in one coil makes it possible to induce an emf in another coil—is key for understanding how electric power can be transmitted efficiently over long distances. This same principle explains how the relatively high voltage of 120 to 240 V available from a wall socket can be used to charge

Figure 21-1 Alternating current When you (a) use the transformer in your mobile phone charger or (b) tune your television to a different channel, you are using the physics of alternating current.

(a)

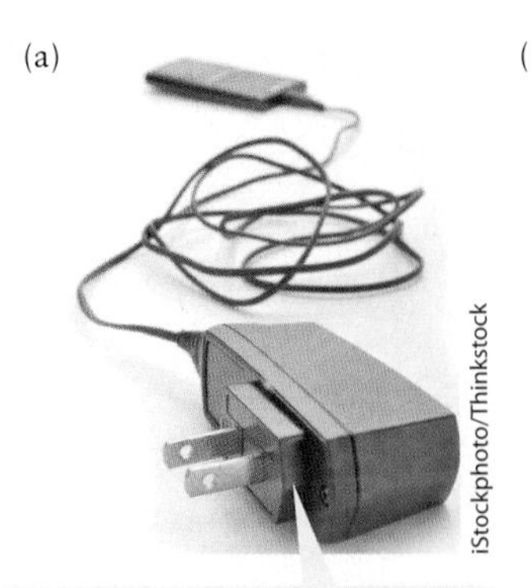

This transformer uses mutual inductance—in which a changing emf in one coil induces an emf in a second coil nearby—to reduce the voltage from a wall socket to a lower value suitable for charging a battery.

(b)

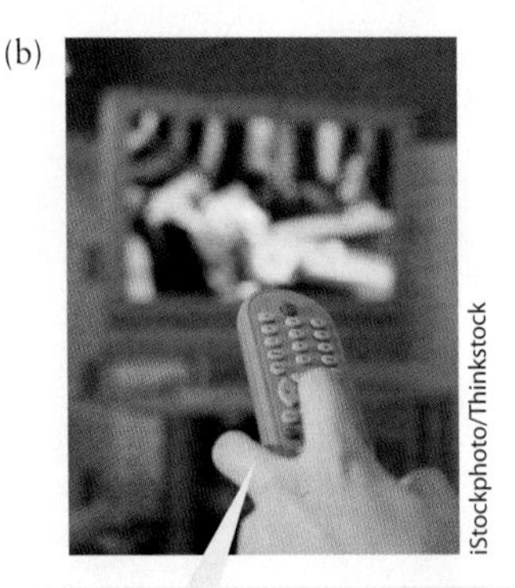

When you change the channel on a television, you're telling the TV tuning circuit to change its capacitance. This changes the natural frequency of the circuit so that it matches the carrier frequency of the channel you want to watch.

a cell phone that operates at low voltage, typically 5 V or less (**Figure 21-1a**). We'll also introduce *self-inductance*, an effect in which a changing emf in a coil produces an emf reaction in the coil itself. Self-inductance is at the heart of an important device called an *inductor* that has many applications in circuits.

Many circuits that are connected to an ac source (such as the ac voltage provided by a wall socket) can be modeled as a combination of resistors, capacitors, and inductors. We'll see what happens when any one of these circuit elements by itself is connected to an ac source. We'll then go on to examine an *LRC series circuit* in which all three of these circuit elements are present. Such circuits hold the key to understanding what happens when you tune a radio or television to receive a particular station (**Figure 21-1b**). Finally, we'll take a brief look at *semiconductors*, a class of material that is essential for the operation of two other important kinds of circuit elements, called *diodes* and *transistors*.

TAKE-HOME MESSAGE FOR Section 21-1

✔ The same principles of electromagnetic induction that explain how to produce an alternating current (ac) also tell us how to manipulate and use such currents.

✔ Many circuits that include an ac source can be modeled as a combination of resistors, capacitors, and inductors (a third type of circuit element).

21-2 We need to analyze ac circuits differently than dc circuits

Current arises in a circuit as a result of an applied voltage. The batteries that power your mobile phone or your flashlight are sources of (approximately) constant voltage. A 9-V battery, for example, introduces a roughly constant potential difference of 9 V between the two points at which it connects to a circuit. By convention, we refer to a circuit driven by a fixed voltage source, or one that does not change direction, as a **direct current**, or **dc** circuit. However, when you plug an electrical device into a wall socket, you are accessing an **alternating current**, or **ac** source.

The voltage from an ac source varies with time in a sinusoidal fashion, as we discussed in our description of ac generators in Section 20-4. In that section we wrote the emf delivered by an ac generator as

Emf in an ac generator (20-7)

$$\varepsilon = \omega NAB \sin(\omega t + \phi)$$

- **Emf produced by an ac generator**: ε
- **Angular speed** of the rotating coil: ω
- **Number of turns** in the coil: N
- **Area of** the coil: A
- **Magnitude of the magnetic field** to which the coil is exposed: B
- **Time**: t
- **Angle** of the coil at $t = 0$: ϕ

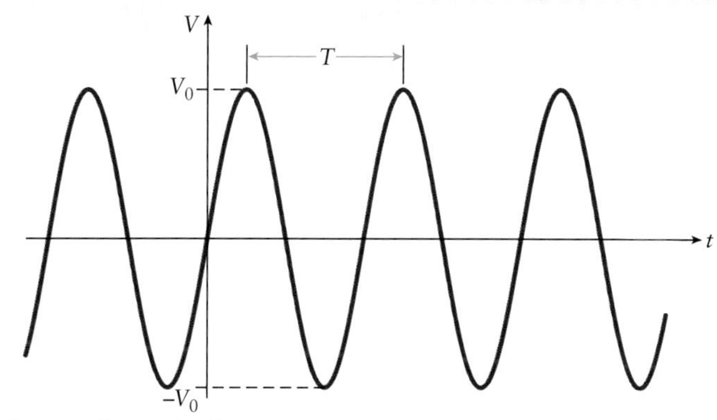

Figure 21-2 An ac voltage This figure graphs the ac voltage given by Equation 21-1. Note that at $t = 0$ the voltage is zero and increasing.

Let's choose the time $t = 0$ to be when the angle of the coil is $\phi = 0$, use the symbol V_0 for the combination of factors ωNAB, and use the symbol $V(t)$ for the time-varying emf. (We use V since emf is measured in volts.) If the generator is connected to two terminals, like the two terminals of a wall socket, then $V(t)$ represents the voltage between those terminals. We can then rewrite Equation 20-7 as a general equation that we'll use for ac sources of all kinds:

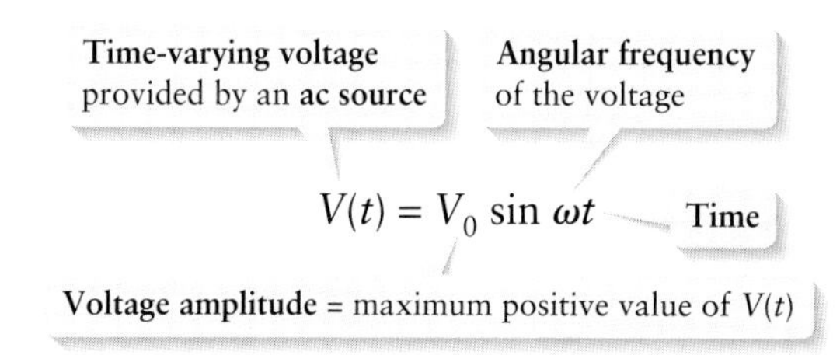

Voltage of an ac source (21-1)

Figure 21-2 graphs the voltage $V(t)$ as a function of time. The angular frequency ω of the voltage (in rad/s) is related to the frequency f of the voltage (in Hz) by the same relationship we used in Section 12-3 for simple harmonic motion:

$$\omega = 2\pi f \tag{21-2}$$

The period T of the oscillation is equal to $1/f$. For example, in the United States and Canada, ac voltage is applied at a frequency $f = 60$ Hz, so the period of oscillation is $T = 1/f = 1/(60 \text{ Hz}) = 0.017$ s, and the angular frequency is $\omega = 2\pi f = (2\pi \text{ rad})(60 \text{ Hz}) = 3.8 \times 10^2$ rad/s.

If we attach a source of ac voltage described by Equation 21-1 to a resistor of resistance R, we can still use Equation 18-24 to calculate the power that flows into the resistor:

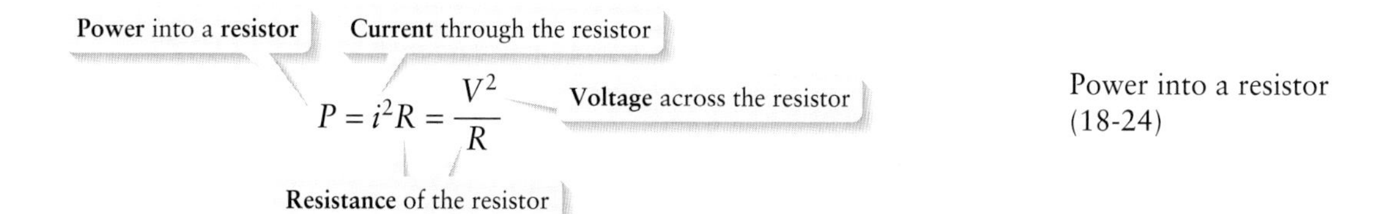

Power into a resistor (18-24)

The only difference between a resistor in a dc circuit and one in an ac circuit is that because the voltage varies with time, the power into the resistor also varies with time. The *instantaneous* power into the resistor at time t is

(21-3)
$$P(t) = \frac{V^2(t)}{R} = \frac{V_0^2 \sin^2(\omega t)}{R}$$

The value of $P(t)$ oscillates between zero (when $\sin \omega t = 0$) and V_0^2/R (when $\sin \omega t = 1$ or -1).

Often it's convenient to talk not about the instantaneous power $P(t)$ but the *average* power $P_{average}$. For example, if an appliance is designed to be powered by an ac voltage, its power rating in watts (such as a 1000-W microwave oven or a 1500-W hair dryer) is always stated in terms of the average power delivered to that appliance when in operation. The voltage amplitude V_0 and the resistance R are constants, so to find $P_{average}$ we have to figure out only the average value of $\sin^2(\omega t)$. As we discussed in Section 20-4, this average value is $\frac{1}{2}$ (see Figure 20-10). So from Equation 21-3 the average power into the resistor is

(21-4)
$$P_{average} = \frac{1}{2}\frac{V_0^2}{R}$$

The average power given by Equation 21-4 is one-half of the maximum value of the instantaneous power given by Equation 21-3.

Equation 21-4 suggests an alternative way to describe the voltage provided by an ac source. Comparing this equation to Equation 21-3, we can write

$$\left(\frac{V^2(t)}{R}\right)_{average} = \frac{1}{2}\frac{V_0^2}{R}$$

If we multiply both sides of this equation by the resistance R, we get

$$(V^2(t))_{average} = \frac{V_0^2}{2}$$

The left-hand side of this equation is called the *mean square value* of the ac voltage: We take the square of the voltage and then calculate the average (or *mean*) of that quantity. The square root of the mean square is called the **root mean square value**, or **rms value**, of the voltage:

The **root mean square (rms) value** of $V(t)$...

...is the square root of the average value of $V^2(t)$.

Root mean square value (21-5)
$$V_{rms} = \sqrt{(V^2(t))_{average}} = \frac{V_0}{\sqrt{2}}$$

If $V(t)$ is a sinusoidal function, its rms value equals the maximum value of $V(t)$ (the amplitude) divided by $\sqrt{2}$.

As an example, in the United States the standard value for the ac voltage supplied by a wall socket is 120 V; this is actually the rms voltage, so $V_{rms} = 120$ V. The voltage amplitude, or peak voltage, is larger by a factor of $\sqrt{2}$, so $V_0 = V_{rms}\sqrt{2} = (120 \text{ V})\sqrt{2} = 170$ V.

In terms of the rms voltage, we can write the average power from Equation 21-4 as

(21-6)
$$P_{average} = \frac{1}{2}\frac{V_0^2}{R} = \left(\frac{V_0^2}{2}\right)\frac{1}{R} = \left(\frac{V_0}{\sqrt{2}}\right)^2\frac{1}{R} = \frac{V_{rms}^2}{R}$$

This says that Equation 18-24, $P = V^2/R$, tells us the *average* power if we replace V by the rms value V_{rms} of the voltage.

WATCH OUT! The average voltage is zero, but the average power is not.

The average voltage of an ac source described by Equation 21-1 is zero. But the average power delivered by that source is *not* zero because power is proportional to the square of the voltage, which is always positive.

EXAMPLE 21-1 Traveling Abroad with Your Hair Dryer

Seasoned travelers know that an electric device that works in one country may not work in another. A certain hair dryer made in the United States, where the rms wall voltage is 120 V, is rated at 1.5 kW. Find the power consumed by the hair dryer when connected to a wall outlet in Australia, where the voltage amplitude is 325 V. Treat the resistance of the hair dryer as the same regardless of the wall voltage. (In reality, the resistance depends somewhat on the temperature of the dryer's heating elements, which depends on the operating voltage.)

Set Up

We'll use Equation 21-6 to find the resistance of the hair dryer from the given values of P_{average} and V_{rms} in the United States. We'll then use this same equation to find the average power in Australia. To do this we'll need to know the rms voltage in Australia; we'll find this from the given value of the voltage amplitude V_0 by using Equation 21-5.

Average power into a resistor:

$$P_{\text{average}} = \frac{V_{\text{rms}}^2}{R} \tag{21-6}$$

Root mean square (rms) voltage:

$$V_{\text{rms}} = \sqrt{(V^2(t))_{\text{average}}} = \frac{V_0}{\sqrt{2}} \tag{21-5}$$

Solve

Use Equation 21-6 to determine the resistance of the hair dryer. We assume this has the same value no matter what the operating voltage.

Solve Equation 21-6 for resistance R:

$$R = \frac{V_{\text{rms}}^2}{P_{\text{average}}}$$

Substitute the values in the United States, $V_{\text{rms}} = 120$ V and $P_{\text{average}} = 1.5$ kW $= 1.5 \times 10^3$ W:

$$R = \frac{(120\text{ V})^2}{1.5 \times 10^3\text{ W}} = 9.6\ \Omega$$

(Recall that $1\text{ W} = 1\text{ V}^2/\Omega$.)

Find the value of V_{rms} in Australia; then use this to determine the average power into the hair dryer when used in Australia.

In Australia the voltage amplitude is $V_0 = 325$ V, so the rms voltage is

$$V_{\text{rms}} = \frac{V_0}{\sqrt{2}} = \frac{325\text{ V}}{\sqrt{2}} = 230\text{ V}$$

With this value of V_{rms}, the average power into the hair dryer is

$$P_{\text{average,Australia}} = \frac{V_{\text{rms,Australia}}^2}{R} = \frac{(230\text{ V})^2}{9.6\ \Omega} = 5.5 \times 10^3\text{ W} = 5.5\text{ kW}$$

Reflect

Our result of 5.5 kW is *considerably* more power than the hair dryer is designed to take in. You may know how hot the air from a 1500-W hair dryer can be, or perhaps you've felt how hot a 100-W bulb can get after it's been on for a while. So you can probably imagine how hot 5500 W would make the heating element of the hair dryer. To radiate that much power, the required temperature of the heating element might well exceed the melting point of the material! We'll see in the following section how this hair dryer can be used safely even with the higher voltage provided in Australia.

GOT THE CONCEPT? 21-1 Average and Instantaneous Power

A resistor is connected to a source of ac voltage described by Equation 21-1. At which of the following times is the instantaneous power into the resistor equal to the average power into the resistor? (a) $t = 0$; (b) $t = T/8$; (c) $t = T/4$; (d) more than one of these; (e) none of these.

TAKE-HOME MESSAGE FOR Section 21-2

✔ The arithmetic mean of a sinusoidally varying voltage is zero because it is negative as often as it is positive. The average power is not zero because it is proportional to the square of the voltage, which cannot be negative.

✔ It is common to characterize ac voltage by its root mean square (rms) value. The peak voltage is $\sqrt{2}$ times the rms voltage.

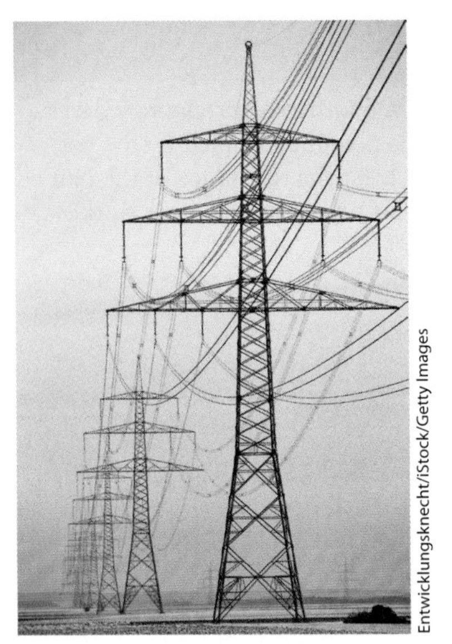

Figure 21-3 Power lines When electric energy is transmitted over long distances in the form of an alternating current, it is most efficient to use very high voltage (typically 110 kV or higher). The power lines shown here are elevated so that people and vehicles cannot touch them, which would present a safety hazard.

21-3 Transformers allow us to change the voltage of an ac power source

If you plug your mobile phone's charger into a wall socket, energy from an ac generator that may be hundreds of kilometers away flows through power lines (**Figure 21-3**) into your home and into your phone. The *voltage* associated with this energy has different values at different places in the circuit, however; the voltage in the power lines is typically hundreds of kilovolts, the voltage provided by a wall socket in the United States or Canada is 120 V, and the voltage supplied to your phone by the charger is 5 V. In this section we'll see the reason why different voltages are used in this way, and we'll learn about an important device called a *transformer* that makes it possible to raise or lower an alternating-current voltage.

To understand why high voltages are used in power lines, recall this relationship from Section 18-6:

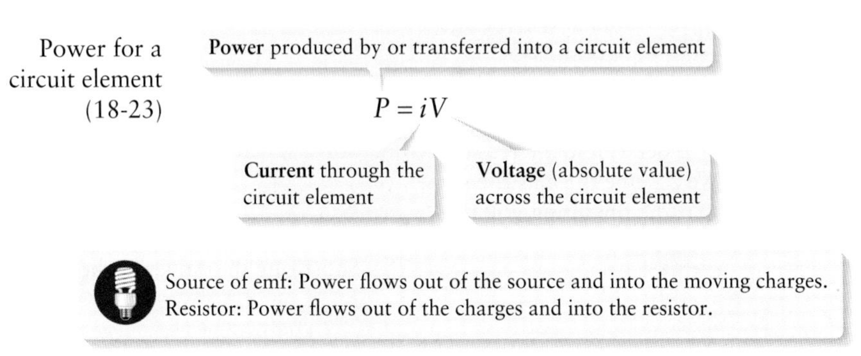

Power for a circuit element (18-23)
Power produced by or transferred into a circuit element
$P = iV$
Current through the circuit element
Voltage (absolute value) across the circuit element
Source of emf: Power flows out of the source and into the moving charges.
Resistor: Power flows out of the charges and into the resistor.

Equation 18-23 tells us that the same electric power P can be delivered at any voltage V, provided that the product of current and voltage remains the same. This is true for ac circuits as well as dc circuits. For a power line it's best to use high voltage and low current. The reason is that a long power line has a substantial resistance R, so some of the energy being carried by the current goes into heating the power line rather than being transmitted to the end user. The rate at which energy is dissipated in a resistor with resistance R is given by Equation 18-24 (see Section 21-2), which we can write as $P = i^2R$. To minimize this energy loss the current i should be as small as possible, so from Equation 18-23 the voltage V must be as large as possible.

WATCH OUT! In power lines, don't confuse the voltage between adjacent cables with the voltage between the ends of a single cable.

You might think "Equation 18-24 also says that $P = V^2/R$ for a resistor. So why isn't it preferable to have the voltage V be small to minimize energy loss?" The explanation is that in Equation 18-23, $P = iV$, V represents the voltage between the *two* cables that make up a power line. (Figure 21-3 shows that the cables come in pairs.) You need both cables to make a complete circuit, just as the power cord for a household appliance has two wires in it (one wire connects to one prong of the plug at the end of the cord and the other wire to the other prong). This is the voltage that must be large in order to make the current small. By contrast, in Equation 18-24 V is the potential difference between two ends of a *single* cable and is related to the current in the cable by the relationship $V = iR$ for resistors (Equation 18-9; see Section 18-3). This voltage is low if the current is small, so the power $P = i^2R = V^2/R$ that goes into heating the resistor is small.

In the home, however, the voltage must be kept small. That's because there is a risk that a person could be exposed to that voltage (for instance, by touching an appliance whose internal wiring had failed). If a voltage V is applied between two parts of your body, the resulting current through your body is proportional to V. A current of

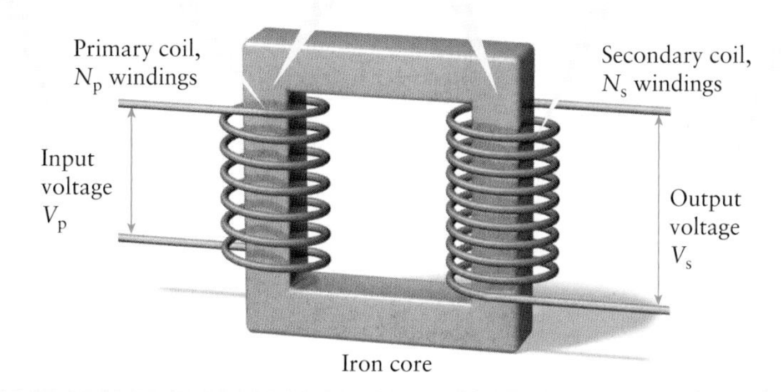

1 An input ac voltage with amplitude V_p and angular frequency ω causes an alternating current in the primary coil.
2 The alternating current in the primary coil creates an alternating magnetic field of the same angular frequency ω. The field lines are "trapped" by the iron core and all pass through the secondary coil.
3 The alternating magnetic flux through the windings of the secondary coil induces an alternating emf, or output voltage, of the same angular frequency ω but of amplitude V_s.
Primary coil, N_p windings
Input voltage V_p
Secondary coil, N_s windings
Output voltage V_s
Iron core

The amplitude V_s of the output voltage equals (N_s/N_p) times the amplitude V_p of the input voltage.

- If $N_s > N_p$, then $V_s > V_p$. This is a step-up transformer.
- If $N_s < N_p$, then $V_s < V_p$. This is a step-down transformer.

Figure 21-4 A transformer This device uses two coils and a piece of magnetic material such as iron to raise or lower the voltage from an ac source.

just 0.1 A can cause ventricular fibrillation, and a current of 0.2 A or more can cause severe burns and stop the heart altogether. To reduce the risk of accident, the voltage supplied to the home has to be kept at a much lower value (120 to 240 V) than that used in power lines (more than 110 kV = 110,000 V).

A device that can raise or lower an ac voltage to a desired value is called a **transformer.** The transformer shown in **Figure 21-4** consists of two coils formed by winding separate wires around a piece of iron. The **primary coil** on the left is connected to the input voltage. The **secondary coil** on the right is the source of the output voltage, even though there's no direct electric connection between the two coils. Let's see how this works.

When an ac voltage is applied to the primary coil in the transformer, the current in the windings gives rise to a magnetic field. This field oscillates with the same angular frequency ω as the ac voltage and current and so creates a time-varying magnetic flux through the windings of the primary coil. Because iron is a magnetic material, the iron core confines the magnetic field lines so that the same time-varying flux that passes through each winding of the primary coil also passes through each winding of the secondary coil. Faraday's law (see Sections 20-2 and 20-3) then tells us that a time-varying emf is induced in the secondary coil, which means this coil acts as a voltage source. The output voltage produced by the secondary coil oscillates at the same frequency as the input voltage that feeds into the primary coil. This effect, in which a change in the current in one coil induces an emf and current in a second coil, is called **mutual inductance.**

The *amplitude* of the output voltage in Figure 21-4 will be different from that of the input voltage—that is, the amplitude will be "transformed" by the transformer—if there are different numbers of windings in the primary and secondary coils (N_p and N_s, respectively, in Figure 21-4). To see how this can be, first notice that the time-varying magnetic flux Φ_B in the primary coil induces an emf there. From Equation 20-2 the magnitude of the emf induced in each of the windings of the primary coil is $|\Delta\Phi_B/\Delta t|$, the magnitude of the rate of change of the magnetic flux through that winding. There are N_p windings in the primary coil, so the total emf in that coil has magnitude $N_p|\Delta\Phi_B/\Delta t|$. By Kirchhoff's loop rule (Section 18-5), the sum of the voltage drops around the left-hand circuit that includes the primary coil must be zero, so the emf in the primary coil must have the same magnitude as the input voltage $V_p(t)$. (We're ignoring any voltage drops due to the resistance of the wires in this circuit.) So we can write

$$|\text{input voltage}| = |V_p(t)| = N_p\left|\frac{\Delta\Phi_B}{\Delta t}\right| \tag{21-7}$$

Since the iron core ensures that there is the same magnetic flux Φ_B through each winding of the secondary coil as through each winding of the primary coil, the rate of change of the flux is the same and so the emf per winding is the same as well. There are N_s windings in the secondary coil, so the total emf in the secondary coil has magnitude $N_s|\Delta\Phi_B/\Delta t|$. This emf is the output voltage $V_s(t)$, so the magnitude of the output voltage is

$$|\text{output voltage}| = |V_s(t)| = N_s\left|\frac{\Delta\Phi_B}{\Delta t}\right| \tag{21-8}$$

If we divide Equation 21-8 by Equation 21-7, the factors of $|\Delta\Phi_B/\Delta t|$ cancel:

$$\frac{|\text{output voltage}|}{|\text{input voltage}|} = \frac{|V_s(t)|}{|V_p(t)|} = \frac{N_s}{N_p} \tag{21-9}$$

Equation 21-9 is valid for the *instantaneous* values of the input and output voltages, so it must also be true for the *maximum* values. These are the voltage amplitudes, which we call V_p and V_s for the input (primary) voltage and output (secondary) voltage, respectively. Equation 21-9 is also true for the *rms* values of these voltages. So we are left with the result

Input and output voltages for a transformer

$$\frac{V_s}{V_p} = \frac{V_{s,rms}}{V_{p,rms}} = \frac{N_s}{N_p} \tag{21-10}$$

Amplitudes of the input (p) and output (s) voltages

Number of windings in the primary coil (p) and secondary coil (s)

Rms values of the input (p) and output (s) voltages

In other words, in a transformer the ratio of the number of windings in the secondary coil (N_s) to the number of windings in the primary coil (N_p) determines the output voltage relative to the input voltage. In a **step-up transformer** N_s is greater than N_p, so the output voltage is greater than the input voltage. In a **step-down transformer,** N_s is less than N_p, and the output voltage is less than the input voltage. A step-up transformer is used at an electric power generation station to raise the voltage provided by the generator (which is connected to the primary coil) to a much larger value in the transmission lines (which are connected to the secondary coil). Close to where the electric power is to be used in home or industry, step-down transformers are used to reduce the voltage to a safe value. The charger for your mobile phone or other electronic device is also a step-down transformer that lowers the voltage provided by the wall socket to a much lower value suitable for the battery to be recharged.

The *current* in the secondary coil is also different from that in the primary coil. If we neglect any energy losses in the transformer (for instance, due to heating of the iron core by eddy currents), conservation of energy requires that the rate at which energy is delivered to the primary coil by the input voltage must equal the rate at which energy is transferred from the primary to the secondary coil. The rate of energy delivery is power, and power equals current times voltage (Equation 18-23), so

$$\text{input power in primary coil} = \text{output power into secondary coil}$$

$$i_p(t)V_p(t) = i_s(t)V_s(t) \tag{21-11}$$

In Equation 21-11, $i_p(t)$ and $V_p(t)$ are the current and voltage in the primary coil at time t, and $i_s(t)$ and $V_s(t)$ are the corresponding quantities in the secondary coil. If we rearrange Equation 21-11 and compare to Equation 21-9, we see that

$$\frac{|\text{output current}|}{|\text{input current}|} = \frac{|i_s(t)|}{|i_p(t)|} = \frac{|V_p(t)|}{|V_s(t)|} = \frac{N_p}{N_s} \qquad (21\text{-}12)$$

Compare Equation 21-12 with Equation 21-10. For a step-up transformer, for which N_s is greater than N_p, the output voltage is greater than the input voltage, but the output current is smaller than the input current. The reverse is true for a step-down transformer.

Go to Interactive Exercise 21-1 for more practice dealing with coils and current.

EXAMPLE 21-2 A High-Voltage Transformer

Each of the 17 generators employed in the hydroelectric power plant at Hoover Dam in Colorado can generate up to 133 MW of average power. This power is delivered at 8.00 kV rms to a transformer that connects to a long-distance transmission line that operates at 5.00×10^2 kV rms. (a) What is the ratio of the number of windings in the secondary coil of the transformer to the number in the primary coil? (b) What is the rms current in the high-voltage transmission line?

Set Up

In part (a) we'll use Equation 21-10 to relate the ratio N_s/N_p (the number of windings in the secondary coil divided by the number in the primary coil) to the input and output rms voltages. In part (b) we'll assume that the transmission line delivers its power to a resistor, so we can relate the current, voltage, and power using Equations 18-9 and 18-23.

Input and output voltages for a transformer:

$$\frac{V_s}{V_p} = \frac{V_{s,rms}}{V_{p,rms}} = \frac{N_s}{N_p} \qquad (21\text{-}10)$$

Relationship among potential difference, current, and resistance:

$$V = iR \qquad (18\text{-}9)$$

Power for a circuit element:

$$P = iV \qquad (18\text{-}23)$$

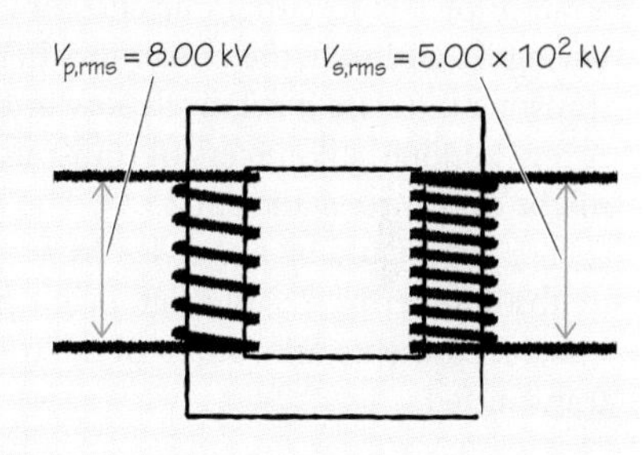

Solve

(a) We know both the primary and secondary rms voltages, so we can rearrange Equation 21-10 to solve for the ratio of the number of windings.

From Equation 21-10

$$\frac{N_s}{N_p} = \frac{V_{s,rms}}{V_{p,rms}} = \frac{5.00 \times 10^2 \text{ kV}}{8.00 \text{ kV}} = 62.5$$

There are 62.5 times as many windings in the secondary coil as in the primary coil (or, equivalently, 125 windings in the secondary for every two windings of the primary).

(b) Since the resistance R is a constant, the current and voltage both have the same $\sin \omega t$ time dependence. We use this to relate the average power $P_{average} = 133$ MW to the rms values of the current and voltage in the transmission line.

If the voltage is $V(t) = V_0 \sin \omega t$, from Equation 18-9 the current is

$$i(t) = \frac{V(t)}{R} = \frac{V_0 \sin \omega t}{R} = \frac{V_0}{R} \sin \omega t = i_0 \sin \omega t$$

In this expression i_0 is the amplitude of the current. From Equation 18-23 the instantaneous power is

$$P(t) = i(t)V(t) = (V_0 \sin \omega t)(i_0 \sin \omega t) = i_0 V_0 \sin^2 \omega t$$

To find the average power $P_{average}$, replace $\sin^2 \omega t$ by its average value $\frac{1}{2}$:

$$P_{average} = \frac{i_0 V_0}{2} = \left(\frac{i_0}{\sqrt{2}}\right)\left(\frac{V_0}{\sqrt{2}}\right) = i_{rms} V_{rms}$$

The average power is the product of the rms current and the rms voltage. Since $i(t)$ has the same time dependence as $V(t)$, the rms value of current is equal to its maximum value divided by $\sqrt{2}$, just as for the voltage. From this expression the rms current in the transmission line is

$$i_{s,rms} = \frac{P_{average}}{V_{s,rms}} = \frac{133\text{ MW}}{5.00 \times 10^2\text{ kV}} = \frac{133 \times 10^6\text{ W}}{5.00 \times 10^5\text{ V}} = 266\text{ A}$$

Reflect

Ratio of coil windings to step down 5.00×10^2 kV to 120 V:

$$\frac{N_p}{N_s} = \frac{V_p}{V_s} = \frac{5.00 \times 10^2\text{ kV}}{120\text{ V}} = 4170$$

In the process of delivering power to the end user, the voltage must ultimately be stepped down to 120 V. This process is not normally done with a single transformer. One reason is that if only one transformer were used for this stepping-down process, there would need to be more than 4000 windings in the primary coil for every one in the secondary.

The voltage between the two ends of the transmission line is $V_{ends} = iR$, where $R = 25\ \Omega$ is the resistance of the line. Use the formula we derived in part (b) to find the power lost in the transmission line:

$$P_{lost,average} = i_{rms}V_{ends,rms} = i_{rms}(i_{rms}R)$$
$$= i_{rms}^2 R = (266\text{ A})^2(25\ \Omega) = 1.8 \times 10^6\text{ W} = 1.8\text{ MW}$$

Express this as a percentage of the power that is transmitted:

$$\text{percentage} = \frac{\text{power lost}}{\text{power transmitted}} \times 100\%$$
$$= \frac{1.8\text{ MW}}{133\text{ MW}} \times 100\% = 1.4\%$$

We found a current of 266 A in the transmission line. Although that sounds large, it is only about a factor of 10 more than the current in household wiring. In addition, the power loss in the transmission wire is relatively low at this current. A few hundred kilometers of transmission wire might have a resistance of 25 Ω. The loss in such a line would be about 1.8 MW. Although in absolute terms this is a significant power loss, it represents only 1.4% of the total power transmitted.

GOT THE CONCEPT? 21-2 Transforming a Transformer

A certain step-up transformer has 100 windings in its primary coil and 250 windings in its secondary coil. You reverse the connections to the transformer so that the primary coil becomes the secondary and vice versa. If you now apply an ac input voltage of 120 V rms to the reversed transformer, what will be the rms output voltage? (a) 40 V; (b) 48 V; (c) 100 V; (d) 250 V; (e) 300 V.

TAKE-HOME MESSAGE FOR Section 21-3

✔ A transformer uses mutual inductance to raise or lower an ac voltage.

✔ In a step-up transformer, there are more windings in the secondary coil than in the primary coil. The output voltage is greater than the input voltage, and the output current is less than the input current.

✔ In a step-down transformer, there are fewer windings in the secondary coil than in the primary coil. The output voltage is less than the input voltage, and the output current is greater than the input current.

21-4 An inductor is a circuit element that opposes changes in current

We saw in Section 21-3 that a transformer works by the mutual inductance of two coils: A change in current in one coil causes a change in the magnetic flux through the other coil, which induces an emf. The same physics applies equally well to individual windings within a *single* coil. You can think of each turn of the coil as a separate loop; as a changing current passes through any loop, the changing magnetic field that arises induces an emf in the other loops. This induced emf, according to Lenz's law, opposes the change in flux that created it. So if the current is increasing in the coil, the induced emf will oppose an increase in current. If the current is decreasing, the induced emf will be directed so that the current is augmented. This **self-inductance** has the net effect of opposing a change in current in a coil. Let's see how to determine the emf induced in this way.

Inductance of a Coil

Consider a coil of N windings in which there is a current i. This current causes the coil to produce a magnetic field, and the flux of this field through each winding of the coil is Φ_B. The total flux through all N windings is then $N\Phi_B$. We define the **inductance** of the coil (symbol L) as the total flux divided by the current that produces it:

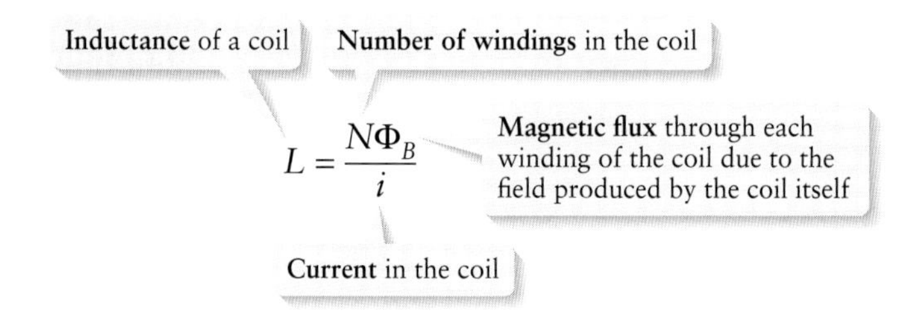

Definition of inductance (21-13)

The units of inductance are henrys, abbreviated as H: $1\ \text{H} = 1\ \text{T}\cdot\text{m}^2/\text{A}$.

The magnetic field is directly proportional to the current i that produces it, and the same is true of the flux of that magnetic field. So the numerator in Equation 21-13 is proportional to the current i. Since the denominator in Equation 21-13 is equal to i, the current will cancel out when we calculate the inductance L. The inductance therefore does not depend on i but only on geometrical factors such as the dimensions of the coil and physical constants. As an example, let's calculate the inductance of a long, straight coil or solenoid. We learned in Section 19-7 that the magnetic field of an ideal solenoid (one whose length is much greater than its diameter) is uniform and has a magnitude given by Equation 19-17:

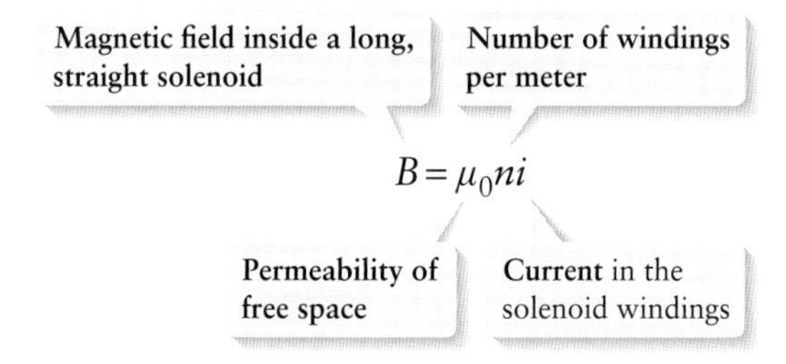

Magnetic field inside a long, straight solenoid (19-17)

If the solenoid has length ℓ, the quantity n in Equation 19-17 equals the total number of windings N divided by ℓ. We can then write the magnetic flux through the cross-sectional area A of each winding of the solenoid as

$$\Phi_B = BA = (\mu_0 n i)A = \left(\frac{\mu_0 N i}{\ell}\right)A$$

If we substitute this into Equation 21-13, we get the following expression for the inductance of the solenoid:

$$L = \frac{N\Phi_B}{i} = \frac{N}{i}\left(\frac{\mu_0 N i}{\ell}\right)A = \frac{\mu_0 N^2 A}{\ell} \quad (21\text{-}14)$$

(inductance of a solenoid)

For the solenoid the inductance depends only on its dimensions and a physical constant (μ_0, the permeability of free space); inductance does not depend on current.

The quantity $\mu_0 = 4\pi \times 10^{-7}\,\text{T}\cdot\text{m/A}$ appears in all expressions for the inductance of a coil. Since $1\,\text{H} = 1\,\text{T}\cdot\text{m}^2/\text{A}$, it follows that $1\,\text{T}\cdot\text{m/A} = 1\,\text{H/m}$. So an alternative way to express the permeability of free space is $\mu_0 = 4\pi \times 10^{-7}\,\text{H/m}$. This is useful for problems that involve inductance, like the following example.

EXAMPLE 21-3 How Large Is One Henry?

You decide to create a solenoid with inductance of 1.00 H by tightly wrapping wire around a plastic pipe. The wire has a diameter of 2.60 mm, and the pipe has an outside radius of 3.00 cm. How long a pipe will you need?

Set Up

Winding the wire tightly around the pipe will create a solenoid. The length of the solenoid may likely be large compared to its diameter, so we'll treat it as ideal. Then we can use Equation 21-14 to calculate the inductance of this solenoid. The pipe has a circular cross section, so to find A we'll use the expression for the area of a circle.

Inductance of a solenoid:

$$L = \frac{\mu_0 N^2 A}{\ell} \qquad (21\text{-}14)$$

Area of a circle of radius r:

$$A = \pi r^2$$

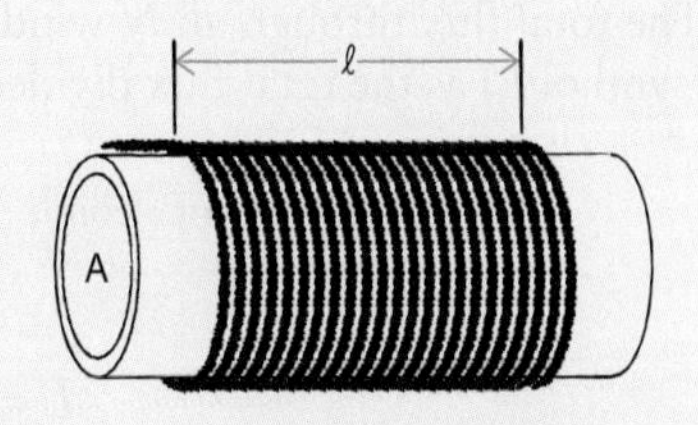

Solve

We don't know how many windings the solenoid will have. However, because the solenoid is tightly wrapped, we can write the number of windings as the total length of the solenoid divided by the diameter of the wire. We substitute this expression for N into the inductance equation and then solve for the length of the pipe.

The number of windings that will fit equals the length ℓ of the pipe divided by the diameter D_W of the wire:

$$N = \frac{\ell}{D_W}$$

Substitute this into Equation 21-14 for the inductance of the solenoid:

$$L = \frac{\mu_0(\ell/D_W)^2 A}{\ell} = \frac{\mu_0 \ell A}{D_W^2}$$

The length of the pipe is therefore

$$\ell = \frac{L D_W^2}{\mu_0 A}$$

The cross-sectional area of the pipe of radius 3.00 cm is

$$A = \pi r^2 = \pi(3.00\text{ cm})^2 = \pi(3.00 \times 10^{-2}\text{ m})^2 = 2.83 \times 10^{-3}\text{ m}^2$$

We are given $L = 1.00$ H, $D_W = 2.60 \times 10^{-3}$ m, and $\mu_0 = 4\pi \times 10^{-7}$ H/m, so

$$\ell = \frac{(1.00\text{ H})(2.60 \times 10^{-3}\text{ m})^2}{(4\pi \times 10^{-7}\text{ H/m})(2.83 \times 10^{-3}\text{ m}^2)}$$

$$= 1.90 \times 10^3\text{ m} = 1.90\text{ km}$$

Reflect

We would need a pipe nearly 2 km long to construct a 1.00-H inductor! One henry is a *very* large inductance; practical inductors have much smaller values of inductance. (Note that the length we determined is indeed much greater than the radius of the pipe, so we were justified in treating this as an ideal solenoid.)

The emf of an Inductor

An inductor in a circuit produces an emf if the current through the inductor changes. We can find an expression for this emf if we combine the definition of inductance, Equation 21-13, with Faraday's law and Lenz's law (Section 20-3). From Equation 21-13 the total flux through an inductor is

$$N\Phi_B = Li$$

Faraday's and Lenz's laws tell us that the emf produced by an inductor is equal to the negative of the rate of change of the total flux through the inductor (Equation 20-3). So we can write the induced emf in the inductor as

$$\varepsilon = -\frac{\Delta(N\Phi_B)}{\Delta t} = -\frac{\Delta(Li)}{\Delta t} \tag{21-15}$$

Let's ignore the resistance of the wire that makes up the inductor, so this is an *ideal* inductor. Then ε in Equation 21-15 is also equal to the voltage V across the inductor (there is no additional voltage drop due to resistance, since there is zero resistance). Since the inductance L does not change with time, we can write Equation 21-15 as

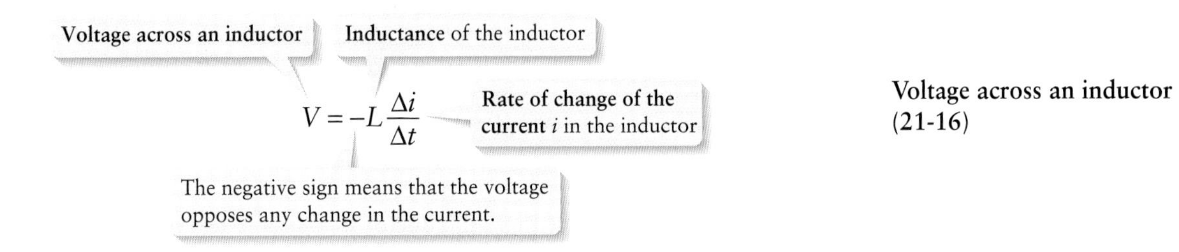

Voltage across an inductor (21-16)

Let's examine Equation 21-16 to see what it means. In **Figure 21-5a** an inductor (symbolized by a coil) carries a constant current i. The rate of change of this constant current is zero, so from Equation 21-16 there is zero voltage across the inductor: $V = 0$. (Remember, we are ignoring the resistance of the wires that make up the inductor.) In **Figure 21-5b** the current through the inductor increases, so in Equation 21-16 $\Delta i/\Delta t > 0$ and $V < 0$. The negative value of V means that there is a voltage *drop* from where the current enters the inductor to where it leaves the inductor. That's like the voltage drop for current that passes through a resistor, which means that the inductor is now resisting the current—or, more properly, resisting the increase in current. So in this case the voltage opposes the change in current. In **Figure 21-5c** the current through the inductor decreases, so in Equation 21-16 $\Delta i/\Delta t < 0$ and $V > 0$. The positive value of V means that there is a voltage *gain* from where the current enters the inductor to where it leaves. That's like the voltage gain for current directed through a battery (a source of emf) from the negative terminal to the positive terminal, so in this case the inductor is assisting the current and so

(a) An inductor with constant current

- The symbol for an inductor is a coil.
- An ideal inductor has zero resistance.

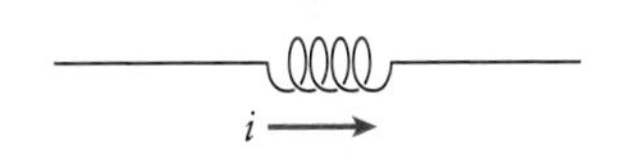

If the current in the inductor is constant, there is zero voltage across the inductor: $V = 0$. This includes the case where there is no current.

(b) An inductor with increasing current

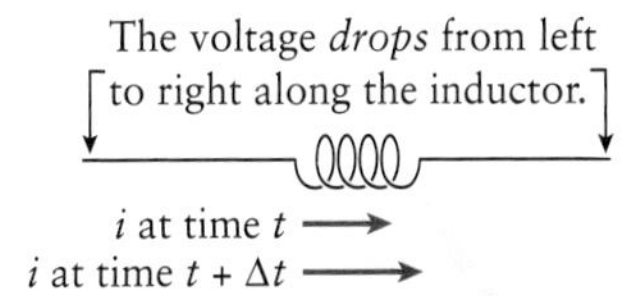

If current in the inductor is increasing so that $\Delta i/\Delta t > 0$, there is negative voltage across the inductor: $V < 0$. This voltage opposes the increase in the current.

(c) An inductor with decreasing current

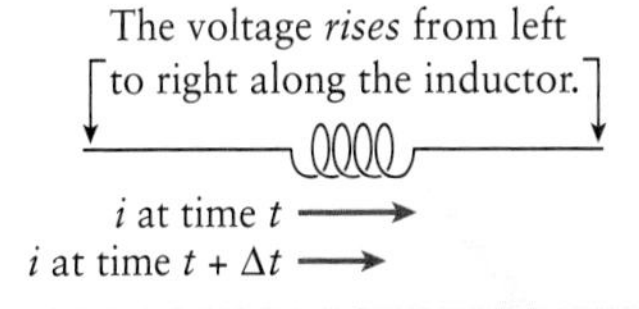

If current in the inductor is decreasing so $\Delta i/\Delta t < 0$, there is positive voltage across the inductor: $V > 0$. This voltage opposes the decrease in the current.

Figure 21-5 Voltage across an inductor The voltage across an ideal inductor (a coil with no resistance) depends on whether the current through the inductor is (a) constant, (b) increasing, or (c) decreasing.

is trying to prevent the current from decreasing. Again the voltage opposes the change in current. Let's summarize these observations:

The voltage across an inductor opposes any change in the current through the inductor. If the current is not changing, the voltage is zero.

This is just what we would expect from Lenz's law, which says that the induced emf in any coil (including an inductor) always acts to oppose any changes. A useful rule is that V in Equation 21-16 represents the voltage *gain* across the inductor when traveling in the direction of the current. If V is negative, there is a voltage drop.

WATCH OUT! Don't confuse inductors, capacitors, and resistors.

It's useful to contrast an inductor with two other devices used in circuits, a capacitor (Section 17-5) and a resistor (Section 18-3). The voltage across a capacitor is $V = q/C$, where q is the magnitude of the charge on either plate of the capacitor and C is the capacitance. The greater the amount of charge on each capacitor plate, the greater the capacitor voltage. The voltage across a resistor that carries current i is $V = iR$, where R is the resistance of the resistor. This voltage is proportional to the rate at which charge moves through the resistor, that is, the current i. The greater the current, the greater the rate at which charge flows through the resistor and the greater the resistor voltage. By contrast, for an inductor Equation 21-16 tells us that the voltage is proportional to the rate of change of current—that is, to the rate of change of the rate at which charge flows through the inductor.

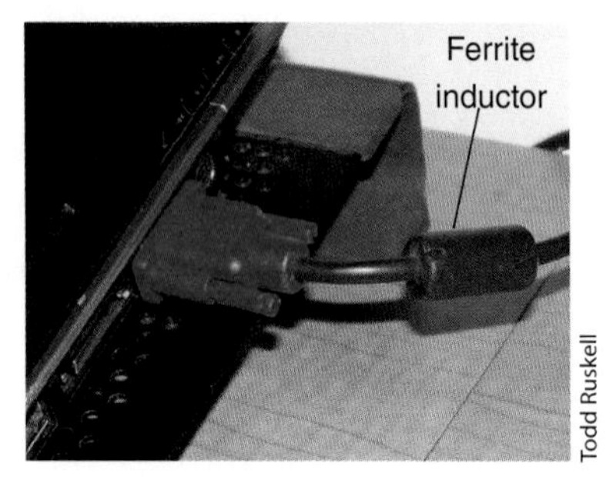

Figure 21-6 Inductor on a cable The ferrite inductor on this monitor cable reduces the interference from external high-frequency electromagnetic signals.

Our observations show that an inductor in a circuit acts to oppose changes in the current within the circuit. The more rapid the change in current—that is, the greater the value of the quantity $\Delta i/\Delta t$ in Equation 21-16—the greater the voltage that will oppose the change. Some computer cables have a built-in inductor (**Figure 21-6**) that is designed to deal with high-frequency interference from signals radiated from other devices. These high-frequency signals would cause rapidly changing currents in the cables that could compromise the operation of the computer, so it's important to suppress them. If such a signal appears, the inductor will set up a voltage that opposes the signal. The inductor is made of a special material called a *ferrite* that also has a large resistance to high-frequency currents, so the energy in the stray signal is absorbed and dissipated as heat.

As we'll see in the following sections, inductors aren't just for suppressing currents. In the right circumstances they can be used for *sustaining* an alternating current of just the right frequency. This is important in tuning a television or radio receiver.

EXAMPLE 21-4 Inductor Voltage

A 0.500-mH inductor is in a dc circuit that carries a current of 0.500 A. If the current increases to 0.900 A in 0.150 ms due to a fault in the circuit, how large is the voltage that appears across the inductor? From the perspective of a positive charge moving through the inductor, is there a voltage rise or drop across the inductor?

Set Up

We'll use Equation 21-16 to calculate the voltage induced across the inductor. To decide whether there's a voltage rise or drop, we'll use the idea that the voltage across the inductor acts to oppose the change in current.

Voltage across an inductor:

$$V = -L\frac{\Delta i}{\Delta t} \quad (21\text{-}16)$$

$i = 0.500$ A
increasing

Solve

Find the voltage that appears in the inductor.

The current increases from 0.500 to 0.900 A in 0.150 ms = 0.150×10^{-3} s, so

$$\frac{\Delta i}{\Delta t} = \frac{0.900\ \text{A} - 0.500\ \text{A}}{0.150 \times 10^{-3}\ \text{s}} = 2.67 \times 10^{3}\ \text{A/s}$$

From Equation 21-16 the voltage across the inductor is

$$V = -L\frac{\Delta i}{\Delta t} = -(0.500 \times 10^{-3}\ \text{H})(2.67 \times 10^{3}\ \text{A/s})$$
$$= -1.33\ \text{H} \cdot \text{A/s} = -1.33\ \text{V}$$

The magnitude of the voltage is 1.33 V. The value of V is negative, which means that there is a voltage *drop* of 1.33 V across the inductor. Thus, the voltage opposes the flow of current and so opposes the increase in current (compare Figure 21-5b).

Reflect

The inductor voltage is comparable to that produced by a AA or AAA battery. This voltage is present only during the time when the current is *changing*, however; when the current is constant, the inductor voltage is zero.

GOT THE CONCEPT? 21-3 Inductors Versus Capacitors Versus Resistors

For which of the following devices does the voltage across that device depend on the current that flows into that device? (a) An inductor; (b) a capacitor; (c) a resistor; (d) more than one of these; (e) none of these.

TAKE-HOME MESSAGE FOR Section 21-4

✔ The inductance of a coil equals the net magnetic flux through the coil due to the current in the coil, divided by the current.

✔ An inductor is a circuit device that opposes any change in the current through that device. The voltage across an inductor is proportional to its inductance and to the rate of change of current through the inductor.

21-5 In a circuit with an inductor and capacitor, charge and current oscillate

Figure 21-7 shows a circuit made up of an inductor of inductance L and a capacitor of capacitance C connected by ideal, zero-resistance wires. This is called an **LC circuit.** With the switch open, positive charge $+Q_0$ is placed on the upper plate of the capacitor, and negative charge $-Q_0$ is placed on the lower plate. This can be done by transferring a number of electrons from the upper plate to the lower plate. The switch is then closed, completing the circuit so that charge can flow. What happens? Remarkably, we'll see that the charge in this circuit behaves very much like the block attached to an ideal spring that we studied in Section 12-3: The charge *oscillates* back and forth in simple harmonic motion. Among other applications, this effect is used to generate the oscillating electric field that a microwave oven uses to cook food.

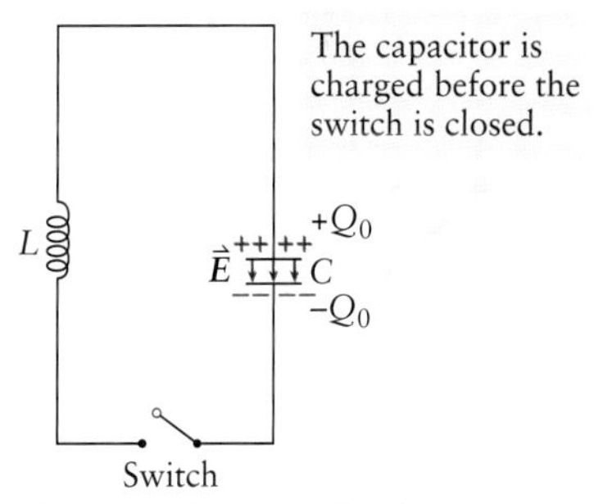

Figure 21-7 An LC circuit An inductor (inductance L) and a capacitor (capacitance C) are connected as shown. Figure 21-8 shows what happens when we close the switch.

An LC Circuit and Simple Harmonic Motion

You might think that the excess electrons would move along the wires from the lower plate to the upper plate until the two plates are electrically neutral, at which point the motion of electrons would stop. Because there's an inductor in the circuit, however, what happens is far more interesting, as **Figure 21-8** shows. Once the switch is closed (step 1 in Figure 21-8), a current is established around the circuit in the counterclockwise direction. (Remember that by convention the direction of current is that in which positive charges would flow.) The current can't suddenly increase, however, because a voltage appears across the inductor to oppose the increase in current. Instead the current increases gradually and reaches a maximum value when the capacitor is fully

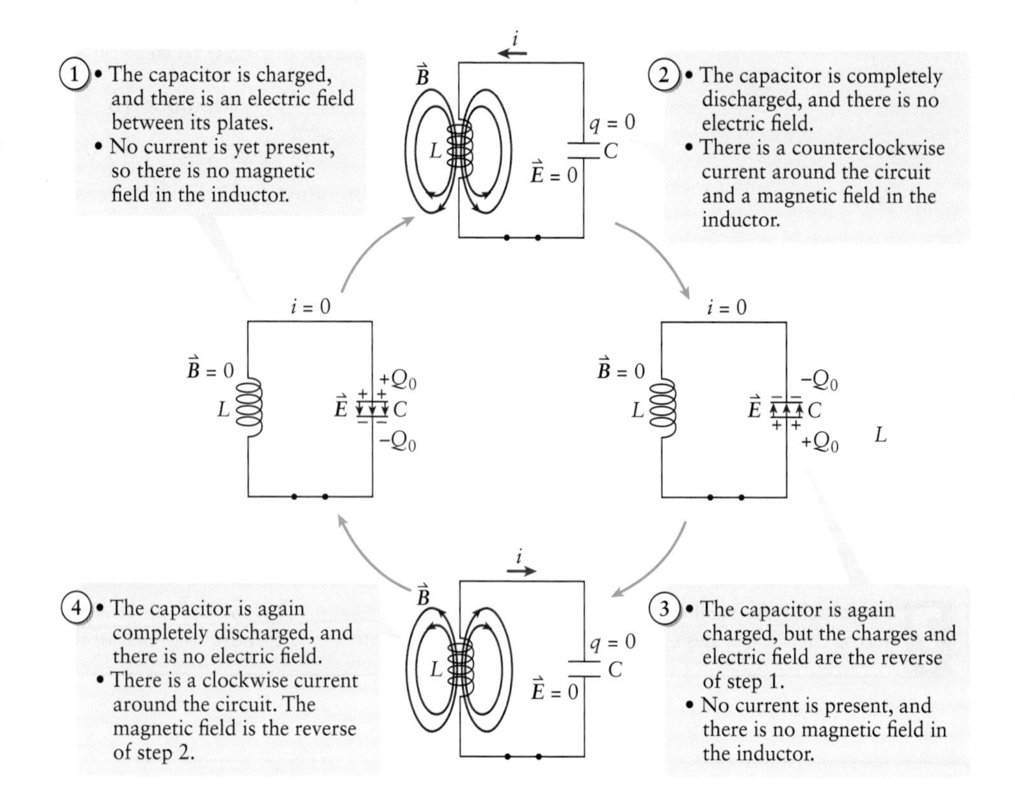

Figure 21-8 Oscillations of an *LC* circuit When the switch in Figure 21-7 is closed, the capacitor charge oscillates back and forth through the circuit. These electrical oscillations are directly analogous to the mechanical oscillations of a block attached to an ideal spring (Section 12-3).

discharged (step 2 in Figure 21-8). The current can't just stop, however; the inductor always opposes any change in the current, so once the current starts to decrease, a voltage will appear across the inductor to slow that decrease. As a result, the motion of charge continues. When the current finally decreases to zero, the capacitor is again charged but with the opposite polarity to its initial configuration (step 3 in Figure 21-8). Now a current is established just as it was between steps 1 and 2, but in the opposite direction: The current is clockwise, increasing gradually until the current is maximum and the capacitor is again fully discharged (step 4 in Figure 21-8). Again the inductor keeps the current from coming to a sudden stop, so the current decreases gradually until the capacitor is again fully charged and has returned to the configuration in step 1. The whole process then starts over!

What's happening here is an *oscillation* of charge and current. These oscillations are another example of *simple harmonic motion*, the special kind of oscillatory motion that we studied in Section 12-3. To see this let's apply Kirchhoff's loop rule to the *LC* circuit in Figure 21-8. This says that if we travel around the circuit and measure the voltage across each element of the circuit, the sum of those voltages must be zero.

We'll begin our trip around the circuit at the upper right-hand corner and move around the circuit clockwise so that we first pass downward through the capacitor. If we let q be the charge on the upper plate of the capacitor at a given instant, the charge on the lower plate at that same instant is $-q$. So if we start at the upper right-hand corner of the circuit and travel downward through the capacitor, we'll encounter a voltage drop

$$V_C = -\frac{q}{C} \quad (LC \text{ circuit}) \tag{21-17}$$

This voltage drop is negative if q is positive, since then we travel from the positive charge to the negative charge and so will encounter a decrease in electric potential (that is, a voltage drop). If q is negative, so that there's negative charge on the upper plate and positive charge on the lower plate, we'll encounter a voltage rise.

As we continue around the circuit, we'll encounter zero voltage across the wires (which have zero resistance). We next pass upward through the inductor on the left-hand side of the circuit. We'll take the current i to be positive if it also flows upward through the inductor (that is, clockwise around the circuit). If the current is positive, positive charge will flow onto the upper plate of the capacitor. The rate at which the charge q on the upper plate increases, measured in coulombs per second or amperes, is just equal to the current:

$$i = \frac{\Delta q}{\Delta t} \quad (LC \text{ circuit}) \tag{21-18}$$

From Equation 21-16 the voltage across the inductor is

$$V_L = -L\frac{\Delta i}{\Delta t} \quad (LC \text{ circuit}) \tag{21-19}$$

If the current through the inductor is positive (upward, or clockwise around the circuit) and increasing in magnitude (becoming more positive), then $\Delta i/\Delta t$ is positive and V_L is negative: We'll encounter a voltage drop as we move upward through the inductor. The same is true if the current through the inductor is negative (downward, or counterclockwise around the circuit) and decreasing in magnitude (becoming less negative). If instead the current through the inductor is either positive (clockwise) and decreasing or negative (counterclockwise) and becoming more negative, then $\Delta i/\Delta t$ is negative and V_L is positive: We'll encounter a voltage rise as we move upward through the inductor.

Kirchhoff's loop rule says that the sum of the voltages V_C from Equation 21-17 and V_L from Equation 21-19 must be zero:

$$V_C + V_L = -\frac{q}{C} - L\frac{\Delta i}{\Delta t} = 0$$

If we add $L(\Delta i/\Delta t)$ to both sides of this equation, we get

$$L\frac{\Delta i}{\Delta t} = -\frac{q}{C} \quad (LC \text{ circuit}) \tag{21-20}$$

Equations 21-18 and 21-20 look like very difficult equations to solve, since both the capacitor charge q and the current i depend on time. But in fact these are equations that we've solved before, although with different symbols. Recall from Section 12-3 that if a block of mass m is attached to an ideal spring of spring constant k and displaced from equilibrium by x, the force that the spring exerts on the block is $F_x = -kx$. (Positive x means that the spring is stretched, and negative x means that the spring is compressed.) If the block is then released, it oscillates. The velocity of the block is the rate of change of the displacement x:

$$v_x = \frac{\Delta x}{\Delta t} \quad (\text{block and ideal spring}) \tag{21-21}$$

Newton's second law says that the force $-kx$ on the block is equal to the mass of the block multiplied by its acceleration a_x. Acceleration is the rate of change of velocity, so we can write Newton's second law for the block as

$$m\frac{\Delta v_x}{\Delta t} = -kx \quad (\text{block and ideal spring}) \tag{21-22}$$

You can see that Equations 21-21 and 21-22 for a block and ideal spring are the *same* as Equations 21-18 and 21-20 for an LC circuit, with the names of the variables changed as listed in the first four rows of **Table 21-1**.

For a block attached to an ideal spring, the spring provides a restoring *force*: Whether the spring is compressed or stretched, this force tries to return the system to equilibrium. The block, however, has *inertia*: Once in motion, it tends to remain in motion. As a result, when the spring reaches the equilibrium position, it does not stop but instead overshoots. As a result, the block oscillates back and forth around its equilibrium position. In a similar way the capacitor in an LC circuit provides a restoring *voltage* $-q/C$; whether q is positive or negative, this voltage drives charge around the circuit in a direction that tries to neutralize both plates. The inductor opposes changes in current; once a current is in motion, the inductor tends to keep it in motion. As a result, when the capacitor is in equilibrium with both of its plates neutral, the current

TABLE 21-1 Comparing a Block on an Ideal Spring to an *LC* Circuit

Quantity for a block on an ideal spring	Corresponding quantity for an *LC* circuit
spring displacement, x	capacitor charge, q
block velocity, $v_x = \Delta x/\Delta t$	current, $i = \Delta q/\Delta t$
spring constant of the spring, k	reciprocal of the capacitance, $1/C$
mass of the block, m	inductance, L
oscillation amplitude, A	maximum capacitor charge, Q_0
potential energy of the spring, $U_{\text{spring}} = \frac{1}{2}kx^2$	electric energy in the capacitor, $U_E = \frac{q^2}{2C}$
kinetic energy of the block, $K = \frac{1}{2}mv_x^2$	magnetic energy in the inductor, $U_B = \frac{1}{2}Li^2$

"overshoots," and the capacitor ends up charged again with reversed polarity. The result is an oscillation that's *exactly* like the simple harmonic motion of a block on an ideal spring. It's like watching the same play but with different actors in the cast: The inductor plays the role of the block, and the capacitor plays the role of the spring.

We can use Table 21-1 to tell us the angular frequency, period, and frequency for the oscillations of an LC circuit. From Equation 12-11 these quantities for a block on a spring are

Angular frequency, period, and frequency for a block attached to an ideal spring (12-11)

Angular frequency $$\omega = \sqrt{\frac{k}{m}}$$

Period $$T = \frac{2\pi}{\omega} = 2\pi\sqrt{\frac{m}{k}}$$

Frequency $$f = \frac{1}{T} = \frac{1}{2\pi}\sqrt{\frac{k}{m}} = \frac{\omega}{2\pi}$$

k = **spring constant** of the spring

m = **mass** of the object connected to the spring

Table 21-1 tells us that k corresponds to $1/C$, and m corresponds to L. So the angular frequency, period, and frequency for the oscillations of an LC circuit are

Angular frequency, period, and frequency for an LC circuit (21-23)

Angular frequency $$\omega = \sqrt{\frac{1}{LC}}$$

Period $$T = \frac{2\pi}{\omega} = 2\pi\sqrt{LC}$$

Frequency $$f = \frac{1}{T} = \frac{1}{2\pi}\sqrt{\frac{1}{LC}} = \frac{\omega}{2\pi}$$

C = **capacitance**

L = **inductance**

The smaller the product LC of inductance and capacitance, the greater the angular frequency and frequency and the shorter the oscillation period.

As an example, imagine an LC circuit made with a 2.00-μH inductor and a 10.0-nF capacitor. Then $L = 2.00 \times 10^{-6}$ H and $C = 10.0 \times 10^{-9}$ F, and from the last of Equations 21-23 the oscillation frequency is

$$f = \frac{1}{2\pi}\sqrt{\frac{1}{LC}} = \frac{1}{2\pi}\sqrt{\frac{1}{(2.00 \times 10^{-6}\ \text{H})(10.0 \times 10^{-9}\ \text{F})}}$$
$$= 1.13 \times 10^{6}\ \text{Hz} = 1.13\ \text{MHz}$$

The charges in this circuit oscillate back and forth 1.13 million times per second and complete one oscillation in a time $T = 1/f = 8.89 \times 10^{-7}$ s $= 0.889$ μs. Such oscillations are much more rapid than those of a block attached to a spring.

The capacitor charge q is analogous to the displacement x for a block on a spring, so the maximum capacitor charge Q_0 is analogous to the amplitude A (the maximum displacement of the block on a spring). We can use this observation to write expressions for the capacitor charge q and current i as functions of time. In Section 12-3 we found that the displacement x and velocity v_x for a block on a spring are given by

$$x(t) = A\cos(\omega t + \phi) \quad \text{(block and ideal spring)} \tag{12-6}$$

$$v_x(t) = -\omega A \sin(\omega t + \phi) \quad \text{(block and ideal spring)} \tag{12-7}$$

Since A is analogous to Q_0, displacement x is analogous to q, and velocity v_x is analogous to i, we can write the following expressions for an LC circuit:

$$q(t) = Q_0 \cos(\omega t + \phi) \quad (LC \text{ circuit}) \tag{21-24}$$

$$i(t) = -\omega Q_0 \sin(\omega t + \phi) \quad (LC \text{ circuit}) \tag{21-25}$$

The angular frequency ω is given by the first of Equations 21-23: $\omega = \sqrt{1/LC}$. The cosine is maximum when the sine is zero and vice versa, so the capacitor charge is maximum when the current is zero and the capacitor charge is zero when the current is maximum. That's just what Figure 21-8 shows.

Since $\sin(\omega t + \phi)$ has values between $+1$ and -1, the maximum value of the current in Equation 21-25 is

$$i_{\max} = \omega Q_0 = \frac{Q_0}{\sqrt{LC}} \quad (LC \text{ circuit}) \tag{21-26}$$

As an example, consider the LC circuit we described above with $L = 2.00$ μH $= 2.00 \times 10^{-6}$ H and $C = 10.0$ nF $= 10.0 \times 10^{-9}$ F. If the maximum capacitor charge is $Q_0 = 0.400$ μC $= 4.00 \times 10^{-7}$ C, the maximum current in the LC circuit is

$$i_{\max} = \frac{Q_0}{\sqrt{LC}} = \frac{4.00 \times 10^{-7}\ \text{C}}{\sqrt{(2.00 \times 10^{-6}\ \text{H})(10.0 \times 10^{-9}\ \text{F})}} = 2.83\ \text{A}$$

Even though only a tiny amount of charge oscillates back and forth between the capacitor plates, the oscillation is so rapid that the maximum current is appreciable.

Energy in an *LC* Circuit

An electric circuit is first and foremost a means of transferring energy from one part of the circuit to another. The same is true for an LC circuit. To see what kinds of energy are involved, let's first look at the total mechanical energy E in a system made up of a block attached to an ideal spring. From Section 12-4 this is the sum of the kinetic energy K of the moving block and the elastic potential energy U_{spring} in the spring:

$$E = K + U_{\text{spring}} = \frac{1}{2}mv_x^2 + \frac{1}{2}kx^2$$

The kinetic and potential energies change, but the total energy E is constant and equal to $(1/2)kA^2$ (A is the amplitude, or maximum displacement).

Table 21-1 tells us how to find the analogous expression for the total energy of an LC circuit: Just replace m with L, k with $1/C$, x with q, and v_x with i. The result is

Energy in an LC circuit

Magnetic energy U_B in the inductor

Electric energy U_E in the capacitor

$$E = \frac{1}{2}Li^2 + \frac{q^2}{2C}$$

Charge on the capacitor

Inductance of the inductor

Current in the inductor

Capacitance of the capacitor

Magnetic and electric energies in an LC circuit (21-27)

We recognize the second term on the right-hand side of Equation 21-27, $q^2/2C$, from Section 17-6: It's just the electric potential energy stored in the capacitor. Since a capacitor plays the same role in an LC circuit as the spring does in a block–spring combination, this electric potential energy is analogous to the elastic potential energy of a spring. We use the symbol U_E for this electric energy. You can think of U_E as the energy required to take an uncharged capacitor and move charges $+q$ and $-q$ to the plates of the capacitor, thereby setting up an electric field $\vec{E}$ between the plates.

The first term on the right-hand side of Equation 21-27, $(1/2)Li^2$, is one that we haven't seen before. This term represents energy stored in the inductor as a result of the presence of current. To see how this energy arises, recall from Section 21-4 that an inductor sets up an emf that opposes any change in the current through the inductor. If we want to make current flow through an inductor where none was flowing before, we have to do work against that emf. The quantity $(1/2)Li^2$ is exactly equal to the amount of work we have to do. By building up the current from zero to i, we also create a magnetic field $\vec{B}$, so we can think of $(1/2)Li^2$ as the energy required to set up a magnetic field in and around the inductor. That's why we call this **magnetic energy** and denote it by the symbol U_B (the subscript B reminds us that a $\vec{B}$-field is involved).

WATCH OUT! Magnetic energy is not the same as kinetic energy.

Table 21-1 shows that the magnetic energy in an inductor, $U_B = (1/2)Li^2$, is analogous to the kinetic energy $K = (1/2)mv_x^2$ of the block in an oscillating block–spring system. This magnetic energy is present only if charges are in motion in the inductor, so i is nonzero. However, U_B is *not* the kinetic energy of the moving charges. That kinetic energy is very small because the moving electrons have very little mass. Instead, U_B is the energy stored in the magnetic field of the inductor that is caused by inductor current i.

In an LC circuit the energy oscillates between the electric and magnetic forms, just as the energy in a block–spring combination oscillates between the potential and kinetic forms (see Figure 12-10 in Section 12-4). In Figure 21-8 the energy is purely electric in steps 1 and 3 (where the capacitor charge is maximum and the current is zero) and purely magnetic in steps 2 and 4 (where the capacitor charge is zero and the current is maximum). If we substitute $q(t)$ from Equation 21-24 and $i(t)$ from Equation 21-25 into Equation 21-27, we get an expression for the energy of the LC circuit at any time:

$$\begin{aligned} E = U_B + U_E &= \frac{1}{2}L[-\omega Q_0 \sin(\omega t + \phi)]^2 + \frac{[Q_0 \cos(\omega t + \phi)]^2}{2C} \\ &= \frac{1}{2}L\omega^2 Q_0^2 \sin^2(\omega t + \phi) + \frac{Q_0^2}{2C}\cos^2(\omega t + \phi) \end{aligned} \tag{21-28}$$

The first of Equations 21-23 tells us that $\omega = 1/\sqrt{LC}$, so $\omega^2 = 1/LC$ and $L\omega^2 = 1/C$. If we substitute this into the first term on the right-hand side of Equation 21-28 and recall that $\sin^2\theta + \cos^2\theta = 1$, we get

$$\begin{aligned} E &= \frac{Q_0^2}{2C}\sin^2(\omega t + \phi) + \frac{Q_0^2}{2C}\cos^2(\omega t + \phi) \\ &= \frac{Q_0^2}{2C}[\sin^2(\omega t + \phi) + \cos^2(\omega t + \phi)] \\ &= \frac{Q_0^2}{2C} \quad (LC \text{ circuit}) \end{aligned} \tag{21-29}$$

The electric and magnetic energies in an oscillating LC circuit both vary with time, but the total energy $E = Q_0^2/2C$ remains constant.

EXAMPLE 21-5 Analyzing an *LC* Circuit

An oscillating LC circuit is made of a 2.00-μH inductor and a 10.0-nF capacitor. The maximum charge on the capacitor is 0.400 μC. (a) Find the total energy in the circuit. (b) At an instant when one-quarter of the total energy is electric energy in the capacitor, find the absolute values of the capacitor charge and the current in the inductor.

Set Up

Equation 21-29 tells us the total energy of the LC circuit. This equals the sum of the magnetic and electric energies, which depend on the current i and capacitor charge q as Equation 21-27 tells us. We'll use this to find the absolute values of i and q at the instant in question.

Energy in an LC circuit:

$$E = \frac{Q_0^2}{2C} \quad (21\text{-}29)$$

Magnetic and electric energies in an LC circuit:

$$E = U_B + U_E = \frac{1}{2}Li^2 + \frac{q^2}{2C} \quad (21\text{-}27)$$

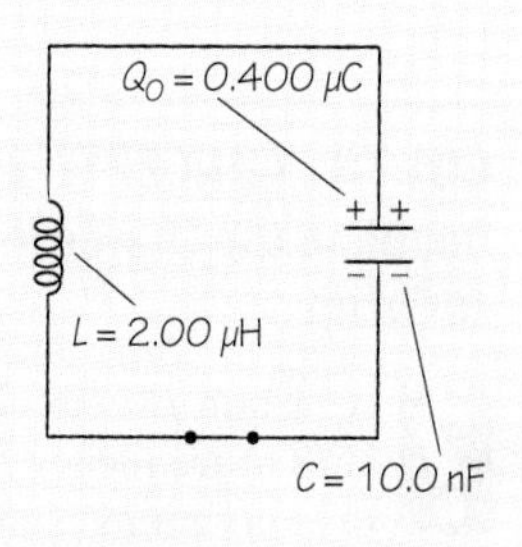

Solve

(a) Use Equation 21-29 to find the total energy.

The maximum capacitor charge is $Q_0 = 0.400\ \mu C = 0.400 \times 10^{-6}$ C, and the capacitance is $C = 10.0$ nF $= 10.0 \times 10^{-9}$ F. The total energy of the oscillating circuit is

$$E = \frac{Q_0^2}{2C} = \frac{(0.400 \times 10^{-6}\ C)^2}{2(10.0 \times 10^{-9}\ F)} = 8.00 \times 10^{-6}\ J = 8.00\ \mu J$$

(b) The total energy remains constant. So if one-quarter of the energy is in the electric form, the other three-quarters must be in the magnetic form. Use this to determine the absolute values of the current and capacitor charge.

The magnetic energy is $\frac{3}{4}$ of the total energy:

$$U_B = \frac{1}{2}Li^2 = \frac{3}{4}E = \frac{3}{4}(8.00 \times 10^{-6}\ J) = 6.00 \times 10^{-6}\ J$$

The inductance is 2.00 μH $= 2.00 \times 10^{-6}$ H. Solve for the absolute value of the current:

$$i^2 = \frac{2U_B}{L} = \frac{2(6.00 \times 10^{-6}\ J)}{2.00 \times 10^{-6}\ H} = 6.00\ A^2$$

$$|i| = \sqrt{i^2} = \sqrt{6.00\ A^2} = 2.45\ A$$

The electric energy is $\frac{1}{4}$ of the total energy:

$$U_E = \frac{q^2}{2C} = \frac{1}{4}E = \frac{1}{4}(8.00 \times 10^{-6}\ J) = 2.00 \times 10^{-6}\ J$$

Solve for the absolute value of the capacitor charge:

$$q^2 = 2CU_E = 2(10.0 \times 10^{-9}\ F)(2.00 \times 10^{-6}\ J) = 4.00 \times 10^{-14}\ C^2$$

$$|q| = \sqrt{q^2} = \sqrt{4.00 \times 10^{-14}\ C^2} = 2.00 \times 10^{-7}\ C = 0.200\ \mu C$$

Reflect

We can check our result for $|q|$ by noting that U_E is proportional to q^2. If the electric energy U_E has its maximum value when $q = Q_0 = 0.400\ \mu C$, then U_E will have $\frac{1}{4}$ of this value when $q = Q_0/2 = 0.200\ \mu C$.

Note that all that we can determine in this problem are the absolute values of i and q. That's because the magnetic and electric energies are proportional to the squares of i and q, respectively. If the current were negative (counterclockwise in Figure 21-8) so that $i = -2.45$ A, the magnetic energy would be the same as if $i = +2.45$ A (clockwise current in Figure 21-8); if the capacitor charge were negative (so the upper capacitor plate in Figure 21-8 were negatively charged) so $q = -0.400\ \mu C$, the electric energy would be the same as if $q = +0.400\ \mu C$ (so the upper capacitor plate were positively charged).

You use the physics of LC circuits whenever you use a microwave oven. At the heart of any microwave oven is a device called a *cavity resonator*, which is basically a hollow metal tube closed at both ends by metal caps. When equal amounts of positive and negative charges are placed on the caps, the charge flows back and forth between the caps along the inner surfaces of the tube. This gives rise to time-varying electric and magnetic fields inside the tube. The electric field is greatest when the magnetic field is zero and vice versa, just as in Figure 21-8. (The geometry of these fields is more complicated than in Figure 21-8, however.) The frequency at which the fields oscillate—typically 2.45 GHz, or 2.45×10^9 Hz, in a home microwave oven and 0.915 GHz in a large commercial oven—is determined by the size and shape of the cavity resonator. A small hole in the cavity resonator is connected to one end of a metal pipe, whose other end leads into the interior of the oven. The oscillating fields "flow" along the pipe into the oven, where the energy of the electric field is absorbed by water molecules in the food. This sets the molecules into vibration and raises the temperature of the food, cooking it.

There is resistance in a cavity resonator, just as there is in any real circuit. As a result, the oscillations will die away just as do the oscillations of a block on a spring when friction is present. To sustain the oscillations of a circuit, we need to *drive* the circuit with an alternating voltage, in much the same way you sustain the back-and-forth oscillations of a child on a playground swing by pushing on the child once per cycle. We'll analyze such *driven* ac circuits in the following section.

GOT THE CONCEPT? 21-4
Doubling the Charge in an *LC* Circuit

? A certain LC circuit oscillates at frequency f when the maximum capacitor charge is Q_0. If you double the value of Q_0, the new frequency of oscillation is (a) $2f$; (b) $f\sqrt{2}$; (c) $f/\sqrt{2}$; (d) $f/2$; (e) none of these.

TAKE-HOME MESSAGE FOR Section 21-5

✔ An LC circuit, made up of an inductor connected to a capacitor, is the electrical analog of a block attached to a spring. The inductor plays the role of the block, and the capacitor plays the role of the spring.

✔ In an LC circuit the capacitor charge and inductor current both oscillate with a frequency determined by the values of the inductance L and capacitance C.

✔ As the charge and current oscillate, the energy in the circuit oscillates between magnetic energy in the inductor and electric energy in the capacitor. If there is no resistance in the circuit, the energy remains constant.

21-6 When an ac voltage source is attached in series to an inductor, resistor, and capacitor, the circuit can display resonance

When you turn on the radio in a car, you expect to listen to only one station at a time. But the car antenna is being simultaneously bombarded by signals from *all* of the local radio stations. How does the radio "know" which signal to play through the car's speakers and which signals to ignore? The explanation is that the radio circuit is essentially a combination of an inductor, resistor, and capacitor in series, and this circuit is driven by an ac voltage coming from the radio signal. As we'll see, such a **series *LRC* circuit** can be designed to give a large current in response to a voltage at one particular frequency, while responding very little to voltages at other frequencies. When you tune a radio, you're adjusting the radio circuit so that the frequency at which it has the greatest response matches the frequency of the station you want to hear.

A driven series LRC circuit behaves in much the same way as a damped, driven mechanical oscillator like the one we studied in Section 12-8. (This would be a good time to review that section.) In order to understand the properties of a driven series LRC circuit, it's useful to first look at three simpler ac circuits: one with just an ac source and a resistor, one with just an ac source and a capacitor, and one with just an ac source and an inductor.

An ac Source and a Resistor

In **Figure 21-9a** an ac source is connected to a resistor of resistance R. The voltage provided by the source is

$$V(t) = V_0 \sin \omega t \tag{21-30}$$

What is the current $i(t)$ in the circuit, and how much power does the source deliver to the resistor? To find out, let's apply Kirchhoff's loop rule, which states that the sum of the voltage drops around the circuit is zero. The voltage across the resistor is $i(t)$ multiplied by R, so the loop rule says that

$$V(t) - i(t)R = 0$$

See the Math Tutorial for more information on trigonometry.

(a) An ac source connected to a resistor

- Current i is in phase with source voltage V.
- Since i and V always have the same sign, the power $P = iV$ delivered by the source is always positive.

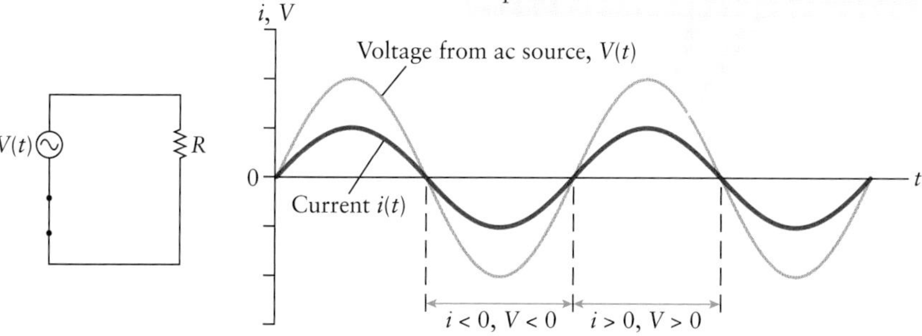

(b) An ac source connected to a capacitor

- Current i leads the source voltage V by 1/4 cycle.
- Half of the time i and V have the same sign, so the power $P = iV$ delivered by the source is positive; the other half of the time i and V have opposite signs, and P is negative.

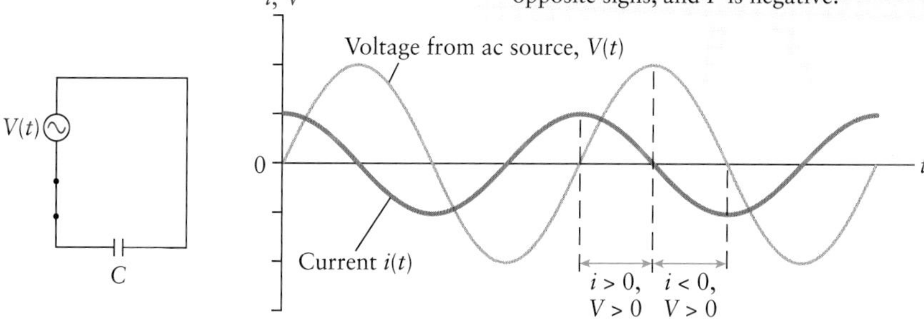

(c) An ac source connected to an inductor

- Current i lags the source voltage V by 1/4 cycle.
- Half of the time i and V have the same sign, so the power $P = iV$ delivered by the source is positive; the other half of the time i and V have opposite signs, and P is negative.

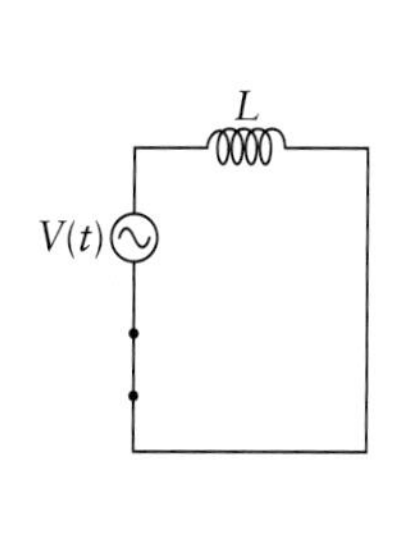

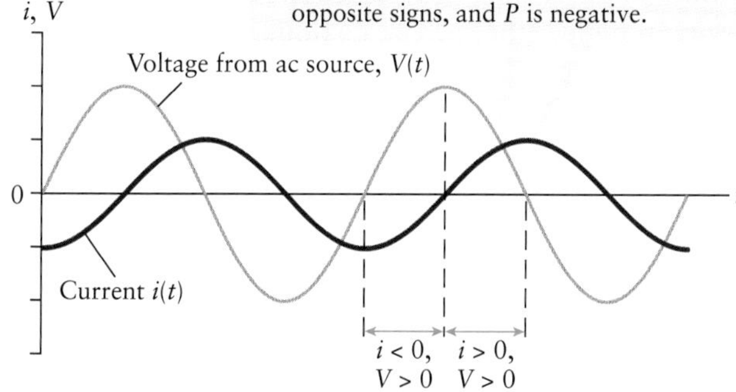

Figure 21-9 Three ac circuits The graphs show the source voltage and currents for three circuits: an ac source connected to (a) a resistor, (b) a capacitor, and (c) an inductor.

and so

$$i(t) = \frac{V(t)}{R} = \frac{V_0}{R} \sin \omega t \quad \text{(resistor and ac source)} \tag{21-31}$$

The graph of $V(t)$ and $i(t)$ versus time in Figure 21-9a shows that the current is in phase with the source voltage. From Equation 18-23, the power that the source delivers to the circuit (in this case to the resistor) is equal to the product of the current through the source multiplied by the voltage across the source: $P(t) = i(t)V(t)$. The graph in Figure 21-9a shows that $i(t)$ and $V(t)$ always have the same sign (both positive or both negative), so the product $i(t)V(t)$ is always positive. Thus power is always being transferred from the ac source into the resistor.

An ac Source and a Capacitor

The circuit shown in **Figure 21-9b** is the same as in Figure 21-9a, except that we've replaced the resistor by a capacitor. The voltage provided by the source is still given by Equation 21-30, but the voltage across the capacitor is $q(t)/C$, where $q(t)$ is the capacitor charge. The loop rule now says that

$$V(t) - \frac{q(t)}{C} = 0$$

The capacitor charge is therefore

$$q(t) = CV(t) = CV_0 \sin \omega t \quad \text{(capacitor and ac source)} \tag{21-32}$$

The capacitor charge oscillates between $+CV_0$ and $-CV_0$.

To find the current in the circuit, two trigonometric identities that will be of use are $\sin \theta = \cos(\theta - \pi/2)$ and $\cos \theta = -\sin(\theta - \pi/2)$. Using the first of these, we can rewrite Equation 21-32 as

$$q(t) = CV_0 \cos\left(\omega t - \frac{\pi}{2}\right) \quad \text{(capacitor and ac source)}$$

The current $i(t)$ is the rate of change of the capacitor charge $q(t)$, just as it was for the LC circuit that we discussed in Section 21-5. In that section we saw that if $q(t) = Q_0 \cos(\omega t + \phi)$, then $i(t) = -\omega Q_0 \sin(\omega t + \phi)$ (see Equations 21-24 and 21-25). That's the same situation we have here, with Q_0 replaced by CV_0 and ϕ equal to $-\pi/2$. So the current in the circuit of Figure 21-9b is

$$i(t) = -\omega CV_0 \sin\left(\omega t - \frac{\pi}{2}\right)$$

Using the second trigonometric identity that we stated above, we can write this as

$$i(t) = \omega CV_0 \cos \omega t \quad \text{(capacitor and ac source)} \tag{21-33}$$

From Equation 21-32, the maximum charge on the capacitor is CV_0 no matter what the frequency. If the frequency is low and the period long, there's plenty of time for the charge to build up from zero to CV_0, so the required current will be low. Equation 21-33 shows the same thing: If the frequency is low, the angular frequency ω is likewise low and the current amplitude ωCV_0 is small. So a capacitor connected to an ac source acts to reduce the current amplitude at low frequency.

The graph in Figure 21-9b shows the source voltage and the current versus time. We say that the current *leads* the voltage by 1/4 cycle: The graph of current has its first peak at $t = 0$, but the graph of voltage reaches its first peak 1/4 cycle later. As a result, for half of the time the current and voltage have the same sign, while for the other half of the time the signs of $i(t)$ and $V(t)$ are opposite. Therefore, for half of the time there is positive power $P(t) = i(t)V(t)$ out of the source and into the capacitor, while for the other half of the time the power is negative and energy is flowing out of the capacitor back into the source. What's happening is that the capacitor is alternately charging and discharging; when it's charging the electric energy U_E in the capacitor is increasing, and when it's discharging the electric energy is decreasing. Over one complete cycle, there's *zero* net flow of energy out of the source.

An ac Source and an Inductor

Figure 21-9c shows a third circuit, one that includes only an ac source and an inductor. Again the source voltage is given by Equation 21-30. The voltage drop across the inductor is $L(\Delta i/\Delta t)$, so the loop rule says that

$$V(t) - L\frac{\Delta i(t)}{\Delta t} = 0$$

We can solve this for $\Delta i(t)/\Delta t$, which is the rate of change of the current:

$$\frac{\Delta i(t)}{\Delta t} = \frac{V(t)}{L} = \frac{V_0}{L}\sin\omega t \quad \text{(inductor and ac source)} \tag{21-34}$$

Given Equation 21-34 for the rate of change of current, what is the current itself? As we did for the capacitor circuit shown in Figure 21-9b, let's think for a minute about Equation 21-24, $q(t) = Q_0 \cos(\omega t + \phi)$, and Equation 21-25, $i(t) = -\omega Q_0 \sin(\omega t + \phi)$. In these equations $i(t)$ is the rate of change of $q(t)$. So these equations tell us that if the rate of change of a function is given by a constant multiplied by $\sin(\omega t + \phi)$, the function itself is given by the same constant divided by $-\omega$ and multiplied by $\cos(\omega t + \phi)$. Using this with Equation 21-34, in which the constant is V_0/L and $\phi = 0$, we conclude that the current for the circuit in Figure 21-9c is

$$i(t) = -\frac{V_0}{\omega L}\cos\omega t \quad \text{(inductor and ac source)} \tag{21-35}$$

Equation 21-35 says that the amplitude of the current is $V_0/(\omega L)$. Since ω is in the denominator, the current amplitude is small when the frequency, and hence the angular frequency ω, is high. That agrees with the idea that an inductor opposes changes in current. High frequency corresponds to rapid changes in current, which the inductor suppresses by making the current small. So an inductor connected to an ac source acts to reduce the current amplitude at high frequency. That's in sharp contrast to a capacitor, which acts to reduce the current amplitude at low frequency.

The graph in Figure 21-9c shows $V(t)$ and $i(t)$ versus time for the circuit with an inductor and an ac source. In this circuit the current *lags* behind the voltage by 1/4 cycle; if you look at any peak of the voltage, you'll see that a peak of the current occurs 1/4 cycle later. Just as for the capacitor circuit in Figure 21-9b, for half of the time the current and voltage have the same sign, so power flows out of the source and adds energy to the inductor. In this case it's *magnetic* energy that's being added as the current through the inductor increases in magnitude. For the other half of the time, current and voltage have opposite signs, the source power is negative, and energy flows out of the inductor as the current decreases in magnitude. As for the capacitor, over one cycle there is zero net energy delivered to the inductor.

A Driven Series *LRC* Circuit

Let's now combine all of the devices shown in Figure 21-9 into a single circuit (**Figure 21-10**). Again the voltage of the ac source is given by Equation 21-30, $V(t) = V_0 \sin\omega t$. We expect that when the switch is closed there will be an oscillating current in the circuit with angular frequency ω. We can also expect that at low frequencies the presence of the capacitor will keep the current amplitude small, while at high frequencies the current amplitude will be small due to the presence of the inductor. At some frequency that's not too low and not too high, the current amplitude will be largest. That's the frequency where the voltage of the ac source produces the greatest response. One of our tasks is to find just what this frequency is.

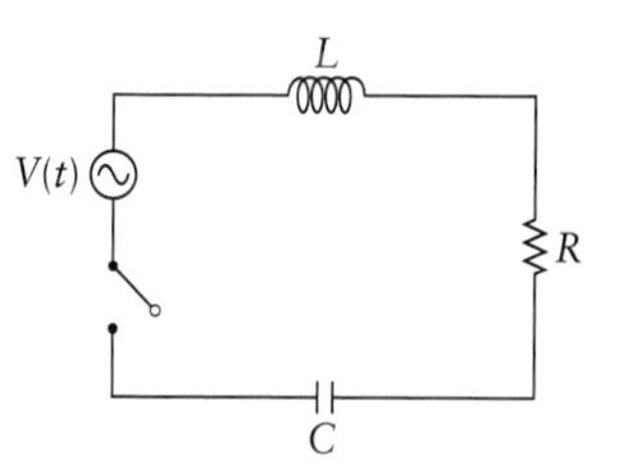

Figure 21-10 A driven series *LRC* circuit When the switch is closed, an oscillating current is set up in this circuit. The amplitude and phase of this current depend on the values of the inductance L, resistance R, and capacitance C, as well as the amplitude V_0 and angular frequency ω of the voltage provided by the ac source.

If we apply Kirchhoff's loop rule to the circuit shown in Figure 21-10, we get the following equation:

$$V(t) - L\frac{\Delta i(t)}{\Delta t} - i(t)R - \frac{q(t)}{C} = 0 \quad \text{(driven series } LRC \text{ circuit)} \tag{21-36}$$

This is a complicated equation, since it involves the capacitor charge $q(t)$; the current $i(t)$, which is the rate of change of the capacitor charge; and the rate of change of the current, $\Delta i(t)/\Delta t$. It is possible to solve Equation 21-36 using some tricks of trigonometry, but it's a rather strenuous exercise. Instead, let's just present the result for the current $i(t)$ and see what it tells us:

Current in a driven *LRC* circuit (21-37)

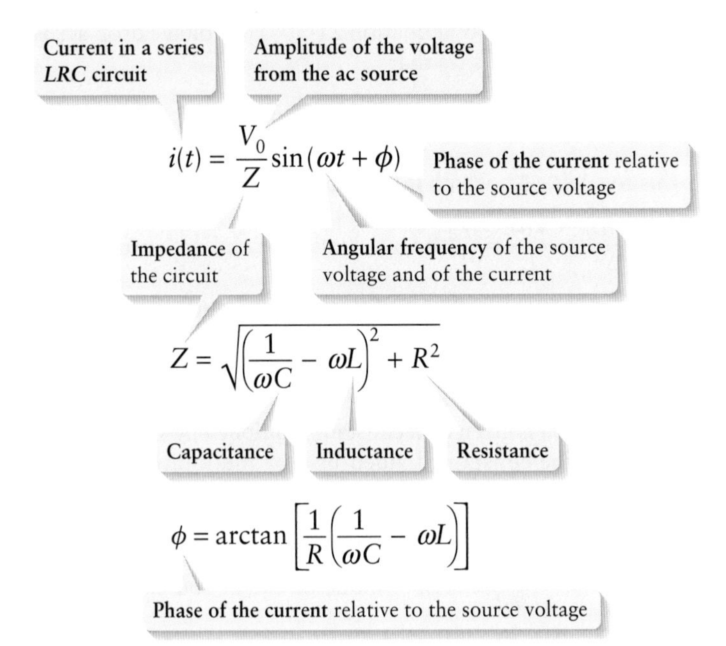

$$i(t) = \frac{V_0}{Z}\sin(\omega t + \phi)$$

$$Z = \sqrt{\left(\frac{1}{\omega C} - \omega L\right)^2 + R^2}$$

$$\phi = \arctan\left[\frac{1}{R}\left(\frac{1}{\omega C} - \omega L\right)\right]$$

The first of Equations 21-37 shows that the greater the value of the **impedance** Z, the smaller the amplitude V_0/Z of the current. Like resistance, impedance has units of ohms. So you can think of it as similar to an "effective resistance" of the circuit to an ac current. Unlike resistance, however, impedance depends on the angular frequency as given by the second of Equations 21-37. The quantity $[1/(\omega C) - \omega L]^2$ is large at both low and high angular frequencies; that's because the term $1/(\omega C)$ (due to the capacitor) gets very large for low values of ω, while the term ωL (due to the inductor) gets very large for high values of ω. So the impedance Z is large and the current amplitude V_0/Z is small for very low angular frequencies thanks to the capacitor, as well as for very high angular frequencies thanks to the inductor. That's just what we predicted above.

The impedance is smallest, and the current amplitude V_0/Z largest, when the quantity $[1/(\omega C) - \omega L]^2$ equals zero, so

$$\frac{1}{\omega C} - \omega L = 0$$

To solve for the angular frequency at which this happens, multiply this equation by ω/L:

$$\frac{\omega}{L}\left(\frac{1}{\omega C} - \omega L\right) = \frac{1}{LC} - \omega^2 = 0$$

So $\omega^2 = 1/(LC)$, or

$$\omega = \sqrt{\frac{1}{LC}} \quad (21\text{-}38)$$

We saw this same angular frequency in Equation 21-23: It's the angular frequency at which an LC circuit oscillates. We'll refer to $\sqrt{1/(LC)}$ as the **natural angular frequency** of the circuit and give it the symbol ω_0. It's the angular frequency at which the current would oscillate if the ac source and resistor weren't in the circuit at all!

In Section 12-8 we saw this same effect, called **resonance**, for an oscillating mechanical system. Equation 21-38 tells us that the current in a driven series LRC circuit is greatest, or in resonance, if the *driving* angular frequency ω equals the *natural* angular frequency $\omega_0 = \sqrt{1/(LC)}$. When $\omega = \omega_0 = \sqrt{1/(LC)}$ the impedance in Equation 21-37 is equal to R and the current amplitude is $V_0/Z = V_0/R$.

Why does resonance happen in a series LRC circuit? The explanation is that at low frequencies the capacitor suppresses the current, and at high frequencies the

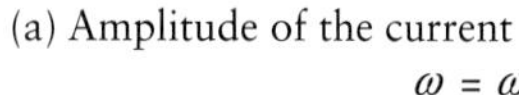

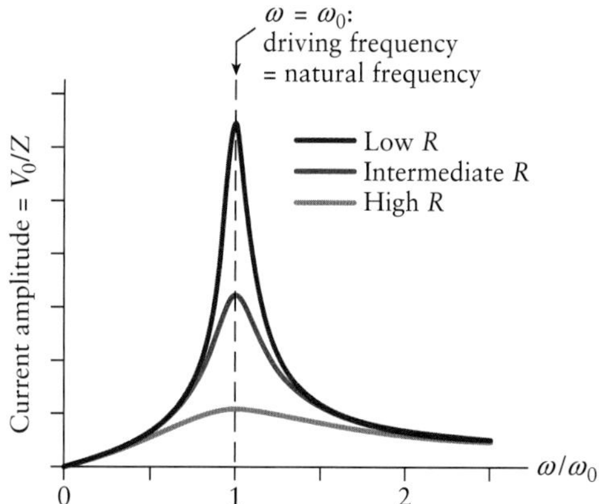

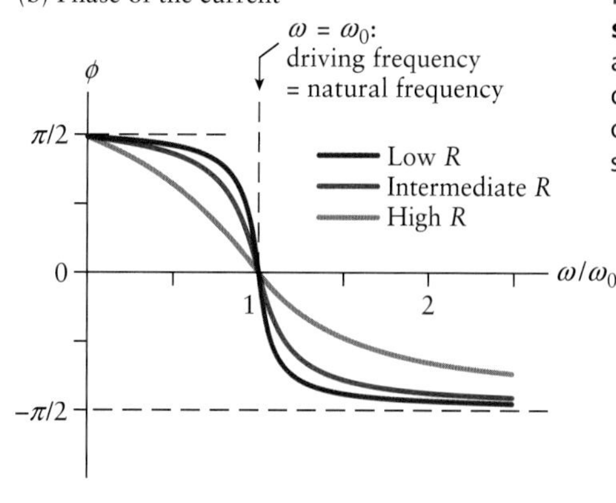

Figure 21-11 **Resonance in a driven series *LRC* circuit** (a) The amplitude and (b) the phase of the current in a driven series *LRC* circuit both depend on the angular frequency ω of the source voltage.

- The current amplitude is maximum at $\omega = \omega_0$ (resonance).
- The smaller the resistance, the greater the current amplitude at resonance.

- At $\omega = \omega_0$ (resonance), $\phi = 0$ and the current is in phase with the source voltage.
- If $\omega < \omega_0$, $\phi > 0$ and the current leads the source voltage. At very low frequencies ϕ approaches $\pi/2$ (current leads voltage by 1/4 cycle).
- If $\omega > \omega_0$, $\phi < 0$ and the current lags behind the source voltage. At very high frequencies ϕ approaches $-\pi/2$ (current lags voltage by 1/4 cycle).

inductor suppresses the current. But at one special frequency—the natural frequency of the circuit—the effects of the capacitor and inductor cancel exactly so that the current in the circuit has its maximum amplitude. **Figure 21-11a** shows the current amplitude as a function of the driving frequency ω for three different values of the resistance R. For any value of R the current amplitude is maximum when $\omega/\omega_0 = 1$ and $\omega = \omega_0 = \sqrt{1/(LC)}$. As R decreases, the peak becomes higher (the maximum current amplitude increases).

The third of Equations 21-37 describes the phase ϕ of the current (see **Figure 21-11b**). The value of ϕ tells us how much the current leads the source voltage. At resonance, where $\omega/\omega_0 = 1$ and $\omega = \omega_0 = \sqrt{1/(LC)}$, we have $(1/(\omega C)) - \omega L = 0$ and $\phi = \arctan 0 = 0$. So at resonance the current is in phase with the source voltage, just as happens in a circuit with only an ac source and a resistor (Figure 21-9a).

For very low angular frequencies ω, the quantity $1/(\omega C)$ is large while ωL is small, so $(1/(\omega C)) - \omega L$ is large and positive. In this case ϕ is near $\pi/2$ radians, so at low frequencies the current leads the voltage by about 1/4 of a cycle (recall that there are 2π radians in a cycle). That's just like the behavior of a circuit with only an ac source and a capacitor (Figure 21-9b). By contrast, at very high angular frequencies the quantity $1/(\omega C)$ is small while ωL is large. Then $(1/(\omega C)) - \omega L$ is large and negative, and ϕ is near $-\pi/2$ radians or $-1/4$ cycle. The negative value of ϕ means that at high frequencies the current *lags* the source voltage by about 1/4 cycle. The circuit with just an ac source and an inductor in Figure 21-9c behaves in the same way.

As Figure 21-11b shows, changing the resistance doesn't affect the angular frequency at which $\phi = 0$. This always happens when $\omega = \omega_0 = \sqrt{1/(LC)}$. Changing the value of R does change the shape of the graph of ϕ versus ω, however; the smaller the resistance, the closer the driving angular frequency ω must be to the natural angular frequency ω_0 in order for ϕ to be close to zero.

These ideas show what's involved in tuning a radio. The signal from a broadcast station is at a particular frequency. The radio antenna detects that signal and converts it to an ac voltage with that same frequency. In order for the circuit of the radio receiver to respond to that frequency, the natural frequency of the receiver circuit must match the driving frequency of the signal from the radio station—that is, the radio receiver must be *tuned* to be in resonance with the station to which you want to listen. Other radio signals from other stations will be present in the circuit as well, but their driving frequencies won't match the natural frequency of the circuit. Hence these undesired signals will produce very little response in the radio (see Figure 21-11a), and you'll hear only the sound of the station to which you've tuned the radio. The following example illustrates this idea.

Go to Interactive Exercise 21-2 for more practice dealing with LRC circuits.

EXAMPLE 21-6 Tuning an FM Radio

The tuner knob on an FM radio moves the plates of an adjustable capacitor. This capacitor is in series with a 0.130-μH inductor and a net resistance of 755 Ω. The peak current induced in this circuit by a radio wave becomes large when the natural frequency of the circuit matches the carrier frequency of the radio wave. (a) What is the frequency of the FM station that is tuned in when the capacitor in the radio is adjusted to 19.6 pF? (b) If the peak operating voltage in the tuning circuit is 9.00 V, what is the peak current?

Set Up

The tuning circuit resonates when driven by a radio wave with an angular frequency ω that equals the natural angular frequency ω_0 of the circuit. We'll use Equation 21-38 to determine the value of ω_0 for this circuit and convert it to an ordinary frequency in Hz. We'll find the peak current, or current amplitude, using Equations 21-37.

Natural angular frequency of the tuning circuit:

$$\omega_0 = \sqrt{\frac{1}{LC}} \qquad (21\text{-}38)$$

Relationship between frequency and angular frequency:

$$f = \frac{\omega}{2\pi}$$

Amplitude of the oscillating current:

$$i_{0,\max} = \frac{V_0}{Z}$$

$$Z = \sqrt{\left(\frac{1}{\omega C} - \omega L\right)^2 + R^2} \qquad (21\text{-}37)$$

R = 755 Ω
V(t) from radio station: f = ?
C = 19.6 pF
L = 0.130 μH

Solve

(a) Use the given values of capacitance and inductance to determine the natural frequency of the circuit.

We are given $L = 0.130\ \mu\text{H} = 0.130 \times 10^{-6}$ H and $C = 19.6$ pF $= 19.6 \times 10^{-12}$ F. From Equation 21-38 the natural angular frequency is $\omega_0 = \sqrt{1/(LC)}$, and the natural frequency is ω_0 divided by 2π:

$$f_0 = \frac{\omega_0}{2\pi} = \frac{1}{2\pi}\sqrt{\frac{1}{LC}} = \frac{1}{2\pi}\sqrt{\frac{1}{(0.130 \times 10^{-6}\ \text{H})(19.6 \times 10^{-12}\ \text{F})}}$$

$$= 99.7 \times 10^6\ \text{Hz} = 99.7\ \text{MHz}$$

(b) Find the amplitude of the oscillating current.

At resonance the term $(1/(\omega C)) - \omega L$ in the expression for impedance is equal to zero. Then Equation 21-37 tells us that $Z = R = 755\ \Omega$. If the voltage amplitude is 9.00 V, the current amplitude is

$$i_{0,\max} = \frac{V_0}{Z} = \frac{9.00\ \text{V}}{755\ \Omega} = 0.0119\ \text{A} = 11.9\ \text{mA}$$

Reflect

The carrier frequencies of FM radio stations lie in the megahertz (MHz) range. So the frequency we found, 99.7×10^6 Hz, is 99.7 MHz on your FM radio dial. The peak current, about 10 mA, is typical for a portable FM radio.

Note that you tune a television set in the same way. When you use the remote control to change the channel (Figure 21-1b), you're commanding a circuit in the television to change its capacitance. This adjusts the natural frequency of the circuit to match the carrier frequency of the channel you want to watch.

GOT THE CONCEPT? 21-5 Modifying a Series *LRC* Circuit

? We've discussed the behavior of a series *LRC* circuit with $R = 1.00\ \Omega$, $C = 5.00\ \mu$F, $L = 20.0\ \mu$H, and $V_0 = 1.00$ V. If we increased the resistance to 2.00 Ω but kept all the other values the same, which of the following would be affected? (a) The natural angular frequency of the circuit; (b) the current amplitude at resonance; (c) the current amplitude when the driving angular frequency is twice the natural angular frequency; (d) two of (a), (b), and (c); (e) all of (a), (b), and (c).

TAKE-HOME MESSAGE FOR Section 21-6

✔ A driven series *LRC* circuit is made up of an inductor, capacitor, and resistor connected in series to a source of ac voltage.

✔ The response of a driven series *LRC* circuit depends on the frequency of the ac source. At low frequencies, the capacitor dominates the circuit and the current is small. The current is also small at high frequencies, where the inductor dominates the circuit. When the frequency of the source equals the natural frequency of the circuit, the current is large and maximum power flows from the source into the resistor. This is resonance.

✔ As the charge and current oscillate, the energy in the circuit oscillates between magnetic energy in the inductor and electric energy in the capacitor. If there is no resistance in the circuit, the energy remains constant.

21-7 Diodes are important parts of many common circuits

In our discussion of electric circuits, we've seen two different types of voltage sources (batteries that provide a constant emf and ac sources that provide a time-varying emf) and three kinds of circuit elements (resistors, capacitors, and inductors). Before leaving the subject of circuits, let's take a look at another important circuit device called a *diode*. This device is possible because of *semiconductors*, a class of materials whose electrical properties are intermediate between those of conductors and insulators.

A common semiconductor is the element silicon (chemical symbol Si). Each silicon atom has four outer or *valence* electrons. In a conducting material such as copper or silver the valence electrons are free to move throughout the material, so the resistivity of the material is low. In silicon, however, the atoms are formed into a crystal structure in which all four of an atom's valence electrons are involved in chemical bonds with adjacent silicon atoms. As a result, the valence electrons are able to move only with great difficulty. That's why pure silicon has about 10^{10} times the resistivity of copper (though only about 10^{-13} the resistivity of an insulating material like rubber). That can change dramatically by **doping** the silicon—that is, by adding small amounts of a different kind of atom to solid silicon. Two common elements used for this purpose are phosphorous and boron.

Figure 21-12a depicts a silicon crystal doped with a small amount of phosphorous (chemical symbol P), so that the phosphorous atoms replace a few of the silicon atoms. An atom of phosphorous has five valence electrons, four of which form bonds to the adjacent silicon atoms. The fifth electron can move relatively easily through the material, and so the doped silicon has much lower resistivity than does pure silicon. Adding just one part per million of phosphorous can decrease the resistivity by several powers of ten. Since the mobile charges in phosphorous-doped silicon are negatively charged electrons, we call such a material an ***n*-type semiconductor** (*n* for negative).

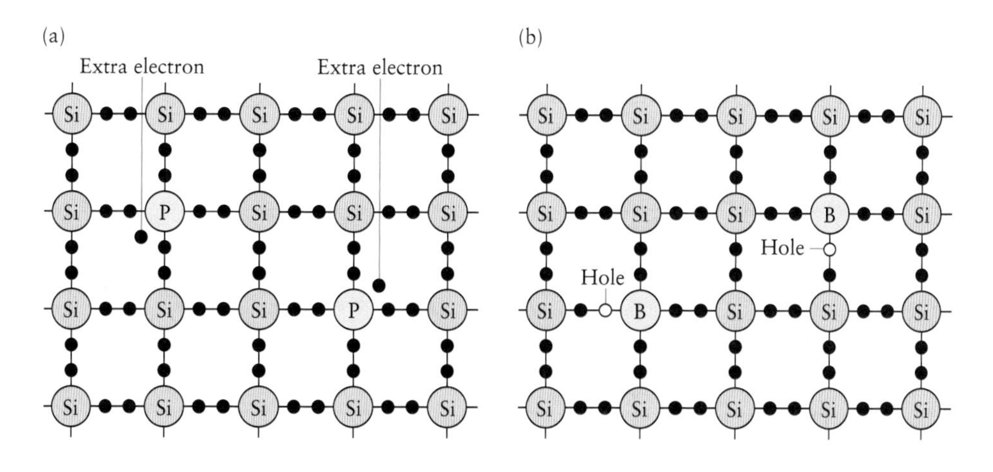

Figure 21-12 Doped semiconductors (a) Silicon (Si) doped with phosphorous (P) is an *n*-type semiconductor. Because phosphorous has five valence electrons, there is an extra, weakly bound electron that can contribute to electrical conduction. (b) Silicon doped with boron (B) is a *p*-type semiconductor. Because boron has only three valence electrons, there is a hole in one of its bonds. The hole can move through the semiconductor as though it were a positive charge, contributing to electrical conduction.

Figure 21-12b shows a different kind of doping in which some of the silicon atoms have been replaced with atoms of boron (chemical symbol B). Boron has only three valence electrons, so for each boron atom there is a "hole" in the electron population. These **holes** are actually free to travel through the boron-doped silicon. (A very rough analogy is to think of these holes as traveling like bubbles of carbon dioxide through a carbonated beverage.) Since a hole represents an absence of negative charge, it's equivalent to a *positive* charge. For this reason boron-doped silicon is called a ***p*-type semiconductor** (*p* for positive). Adding holes by doping the silicon with even a small amount of boron also causes a substantial decrease in resistivity.

Now let's take a crystal of silicon and dope one side of the crystal with boron and one side with phosphorous. The side with boron is the *p* side and the side with phosphorous is the *n* side (the two sides are *p*-type and *n*-type semiconductors, respectively). The region where the two sides meet is called a ***pn* junction**. Some of the mobile holes from the *p* side diffuse across the junction to the *n* side, and some of the mobile electrons from the *n* side diffuse to the *p* side. The semiconductor isn't a particularly good conductor, so neither the electrons nor the holes get very far from the junction. The net result is a double layer of charge around the junction, a positive layer of holes on the *n* side and a negative layer of electrons on the *p* side (**Figure 21-13**).

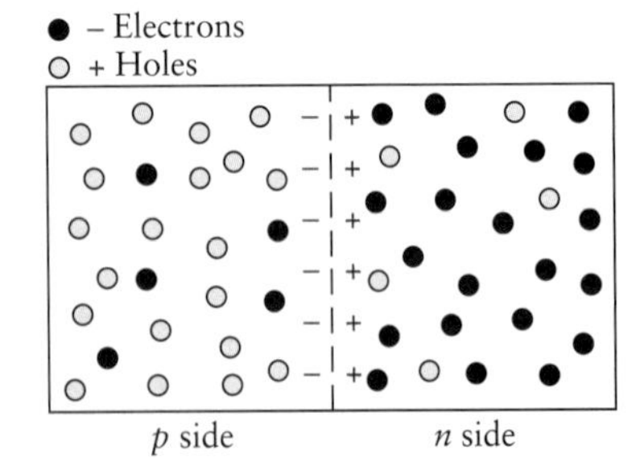

Figure 21-13 A *pn* junction The two sides of this semiconductor are doped with different atoms. Holes diffuse from the *p* side to the *n* side, and electrons diffuse from the *n* side to the *p* side. The result is a double layer of charge at the junction between the two sides.

To see what makes a *pn* junction useful, imagine placing it in a dc circuit with a source of emf ε and a resistor. If the polarity of the source is as shown in **Figure 21-14a**, a current flows through the junction and around the circuit. What happens is that the source of emf creates an electric field within the *pn* junction that points from left to right, which helps to drive positively charged holes to the right across the junction and negatively charged electrons to the left across the junction. This enhanced diffusion of electrons and holes gives rise to a net current from left to right in the semiconductor. In this case the *pn* junction is said to be *forward biased*. If we reverse the polarity of the source as shown in **Figure 21-14b**, however, the source creates an electric field that points from right to left within the *pn* junction. This tends to push the holes leftward (back into the *p* side from which they came) and the electrons rightward (back into the *n* side from which they came). In this case the *pn* junction is said to be *reverse biased*: The emf suppresses the diffusion of electrons and holes, and there is little or no current. Essentially the junction conducts in only one direction! A single-junction semiconductor device like this is called a **diode**.

Diodes have many uses, one of which is converting alternating current into direct current. If we replace the source of emf in Figure 21-14 with an ac source, the diode will conduct electricity when the alternating voltage is positive (forward bias) but not when the alternating voltage is negative (reverse bias). In this case the current through the resistor in Figure 21-14 won't be constant, but it will be in one direction only.

Another use for *pn* junctions is in **photovoltaic solar cells**. Light shining on the *p* side of the junction can excite electrons and produce new holes in the process. If the electrons happen to migrate to the junction, they will be accelerated into the *n* side by the electric field between the double layer of charge at the junction. The result is an excess negative charge on the *n* side and an excess positive charge on the *p* side. This charge imbalance produces a potential difference between the two regions. If one terminal of a resistor is connected to the *p* side and the other terminal to the *n* side, a current will flow through the resistor due to the potential difference and energy will be transferred to the resistor. The net effect is that the energy of sunlight is converted

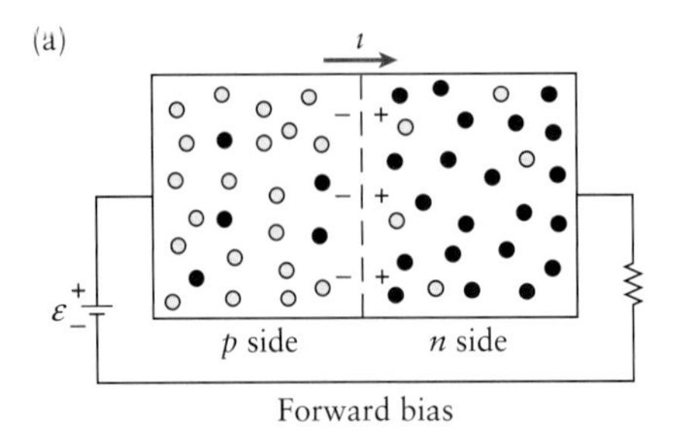

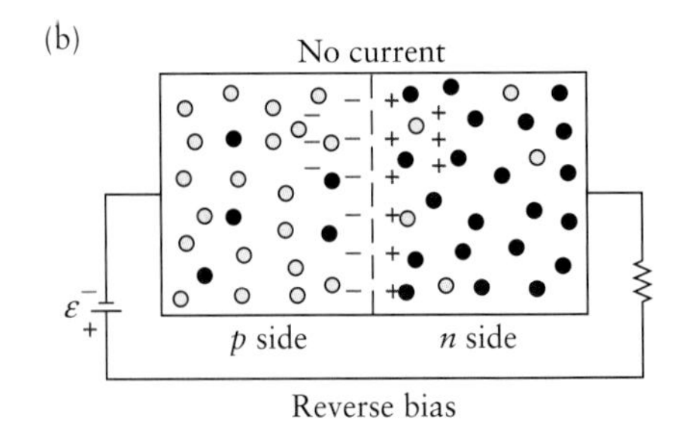

Figure 21-14 A *pn*-junction diode This device is like a one-way valve for electric current. (a) Current will flow through the *pn* junction if the emf is applied as shown here (forward bias). (b) Little or no current will flow through the *pn* junction if the emf is reversed (reverse bias).

to electric energy. All solar cells designed for home and industrial use operate on this basic principle.

The basic process used in solar cells can also be run in reverse. In certain kinds of *pn* junctions, a forward bias like that shown in Figure 21-14a produces large concentrations of electrons on the *p* side and holes on the *n* side. When the electrons and holes diffuse across the barrier and recombine, energy is released in the form of light. Such a device is called a **light-emitting diode** or **LED**. LEDs are extremely efficient light sources: They produce much more light for a given power input than do incandescent light bulbs or fluorescent lamps. A television remote control (Figure 21-1b) sends commands to the television in the form of infrared light emitted by an LED.

Many other important semiconductor devices are used in circuits. In a *transistor*, a weak electrical current is used to modulate a stronger current. For example, the alternating current from a music player carries all the musical information about the song you want to hear, but the current is too feeble to drive a speaker or even a pair of earbuds. By using transistors, however, this feeble current can be used to modulate a much stronger current from the music player's battery. This modulated current *is* large enough to drive a speaker or earbuds, and that's what produces the music that you hear.

TAKE-HOME MESSAGE FOR Section 21-7

✔ A semiconductor can be made a better conductor of electricity by doping it with a small amount of a different kind of atom. In an *n*-type semiconductor, there are extra electrons that are able to move through the material and carry a current. In a *p*-type semiconductor, the mobile charges are holes where there is an electron missing. Holes behave like positively charged objects.

✔ In a *pn* junction one side of the semiconductor is *p*-type and the other side is *n*-type. This has the property that it is much easier to establish a current through the junction in one direction compared to the opposite direction.

Key Terms

alternating current (ac)
diode
direct current (dc)
doping
hole
impedance
inductance
LC circuit
light-emitting diode (LED)
magnetic energy
mutual inductance
natural angular frequency
n-type semiconductor
photovoltaic solar cell
pn junction
primary coil
p-type semiconductor
resonance
root mean square value (rms value)
secondary coil
self-inductance
series *LRC* circuit
step-down transformer
step-up transformer
transformer

Chapter Summary

Topic	Equation or Figure	
Root mean square values for alternating current: To find the root mean square or rms value of a quantity, first find the average (mean) value of the square of that quantity, then take the square root of that average. Alternating voltages are often described in terms of their rms values rather than their amplitudes.	The **root mean square (rms) value** of $V(t)$... ...is the square root of the average value of $V^2(t)$. $V_{rms} = \sqrt{(V^2(t))_{average}} = \frac{V_0}{\sqrt{2}}$ If $V(t)$ is a sinusoidal function, its rms value equals the maximum value of $V(t)$ (the amplitude) divided by $\sqrt{2}$.	(21-5)

Transformers: A transformer uses electromagnetic induction to raise or lower the voltage of an alternating current. The ratio of the output voltage to the input voltage depends on the number of windings in the coils of the transformer. Raising the voltage lowers the current and vice versa. In an ideal transformer, no power is lost in this process.

$$\frac{V_s}{V_p} = \frac{V_{s,rms}}{V_{p,rms}} = \frac{N_s}{N_p} \tag{21-10}$$

Amplitudes of the input (p) and output (s) voltages

Rms values of the input (p) and output (s) voltages

Number of windings in the primary coil (p) and secondary coil (s)

Inductors: The simplest inductor is a coil of wire. If the current in the coil changes, there will be a change in the magnetic flux through the coil and an emf will be induced to oppose that change. So an inductor in a circuit always acts to oppose any change in the current in that circuit. The inductance of a coil depends on the coil's geometry.

$$L = \frac{N\Phi_B}{i} \tag{21-13}$$

Inductance of a coil

Number of windings in the coil

Magnetic flux through each winding of the coil due to the field produced by the coil itself

Current in the coil

$$V = -L\frac{\Delta i}{\Delta t} \tag{21-16}$$

Voltage across an inductor

Inductance of the inductor

Rate of change of the current i in the inductor

The negative sign means that the voltage opposes any change in the current.

***LC* circuits:** A circuit made up of an inductor and capacitor is the electric analog of a block oscillating at the end of an ideal spring. The angular frequency of the oscillation is $\omega = \sqrt{1/(LC)}$. The energy in the *LC* circuit oscillates between electric energy in the capacitor and magnetic energy in the inductor.

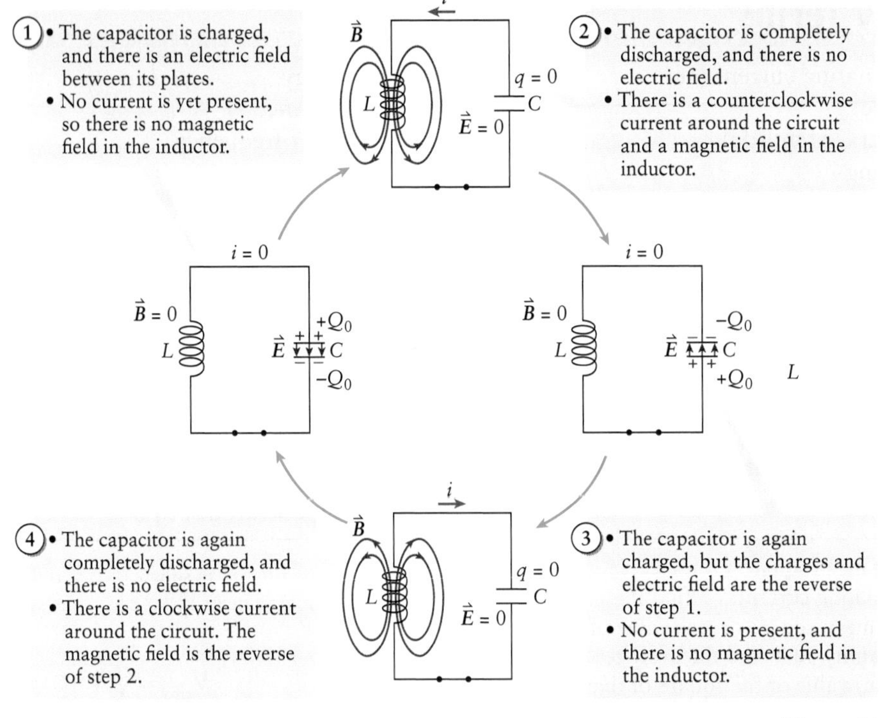

(Figure 21-8)

Driven series *LRC* circuits: If an ac source is connected in series to an inductor, resistor, and capacitor to make a circuit, the current in the circuit oscillates at the angular frequency of the source. The voltage across the inductor leads the current by 1/4 cycle, the voltage across the resistor is in phase with the current, and the voltage across the capacitor lags the current by 1/4 cycle.

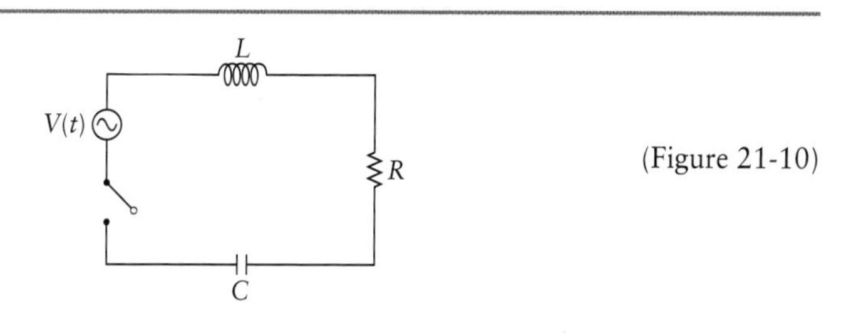

(Figure 21-10)

***LRC* circuit resonance:** The current in a driven series *LRC* circuit has its maximum amplitude when the source (driving) angular frequency equals the natural angular frequency $\omega_0 = \sqrt{1/(LC)}$ of the circuit. When this happens the circuit is in resonance. At resonance the current is in phase with the ac source voltage, and maximum power is delivered to the resistor. In addition, at resonance the voltages across the inductor and capacitor have equal amplitude but are 1/2 cycle out of phase with each other, so their voltages cancel.

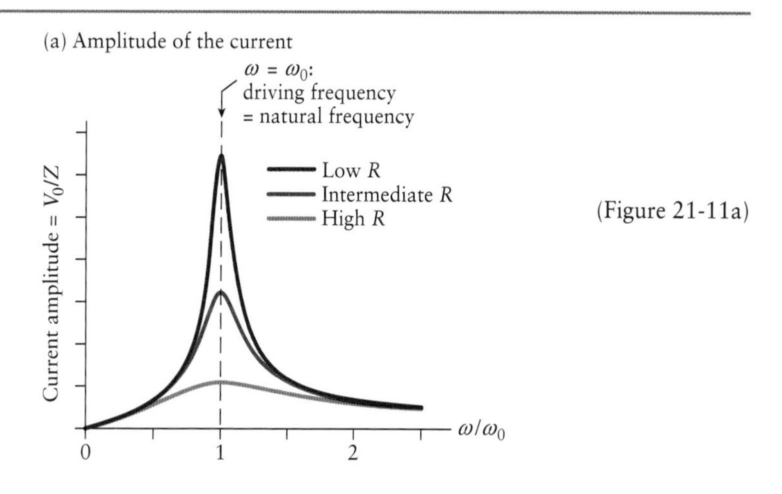

(Figure 21-11a)

Semiconductors and diodes: A semiconductor can be made a better conductor by adding impurities that contribute mobile electrons (an *n*-type semiconductor) or mobile holes (a *p*-type semiconductor). A semiconductor with one side that is *p*-type and one side that is *n*-type can act as a diode that conducts current easily from the *p* side to the *n* side, but conducts poorly in the opposite direction.

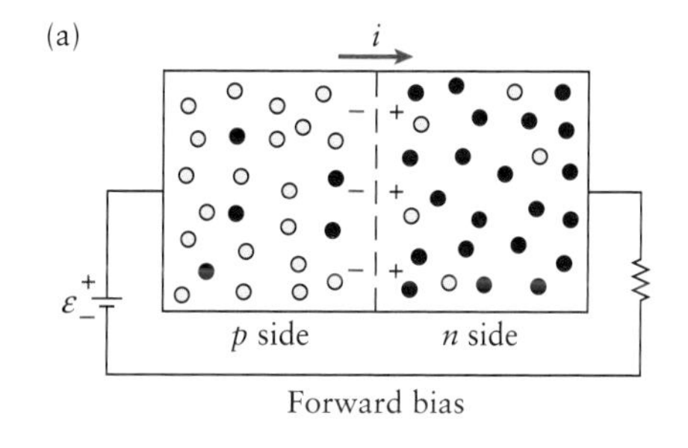

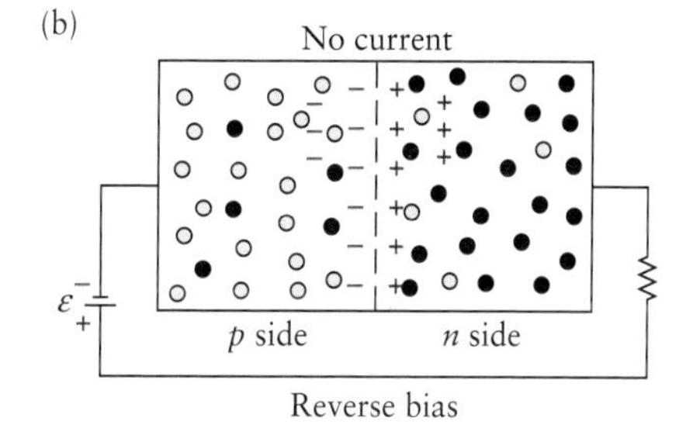

(Figure 21-14)

Answer to What do you think? Question

(d) As we learned in Chapter 20, Faraday's law comes into play only when currents are changing. If the current is constant, as in the case of direct current, Faraday's law has no effect. (See Section 21-3.)

Answers to Got the Concept? Questions

21-1 (b) The voltage provided by the ac source is $V(t) = V_0 \sin \omega t$, where the angular frequency ω is related to the period T by $\omega = 2\pi/T$. The instantaneous power into the resistor is given by Equation 21-3, $P(t) = V^2(t)/R$. At $t = 0$, $V(0) = V_0 \sin 0 = 0$, so $P(0) = 0$. At $t = T/8$, $\omega t = (2\pi/T)(T/8) = \pi/4$, so $V(T/8) = V_0 \sin(\pi/4) = V_0(1/\sqrt{2})$ and $P(T/8) = (V_0/\sqrt{2})^2/R = V_0^2/(2R)$. At $t = T/4$, $\omega t = (2\pi/T)(T/4) = \pi/2$, so $V(T/4) = V_0 \sin(\pi/2) = V_0(1) = V_0$ and $P(T/4) = V_0^2/R$. Equation 21-4 tells us that the average power is $P_{\text{average}} = V_0^2/(2R)$, which is equal to the instantaneous power at $t = T/8$ and not at the other times. Note also that at $t = T/8$, the instantaneous voltage V is equal to the rms value $V_{\text{rms}} = V_0/\sqrt{2}$.

21-2 (b) With the connections reversed, the number of windings in the primary coil is $N_p = 250$, and the number of windings in the secondary coil is $N_s = 100$. With $V_{p,rms} = 120$ V Equation 21-10 tells us that the output rms voltage is $V_{s,rms} = V_{p,rms}(N_s/N_p) = (120 \text{ V})(100/250) = 48$ V.

21-3 (c) The voltage across an inductor depends on the rate at which the current through the inductor is changing, not on the value of current itself. The voltage across a capacitor depends on the amount of charge on either plate, not on the current. (The current determines how rapidly the capacitor charge is changing and so how rapidly the voltage is changing, but does not affect the present value of the voltage.) The voltage $V = iR$ across a resistor is directly proportional to the current i, so this does depend on the value of current.

21-4 (e) As Equations 21-23 show, the oscillation frequency of an LC circuit depends on the values of inductance and capacitance alone. It does not depend on the amount of charge placed on the capacitor plates. This is the direct analog of the behavior of a block attached to an ideal spring, which we studied in Section 12-3: The oscillation frequency depends on the spring constant and the mass of the block, but does not depend on the amplitude of the oscillation.

21-5 (d) The natural angular frequency $\sqrt{1/(LC)}$ does not depend on the resistance. The other two quantities, however, do. You can see this from Figure 21-11a: Increasing the resistance decreases the current amplitude at *all* angular frequencies, not just at resonance.

Questions and Problems

In a few problems you are given more data than you actually need; in a few other problems you are required to supply data from your general knowledge, outside sources, or informed estimate.

Interpret as significant all digits in numerical values that have trailing zeros and no decimal points.

For all problems use $g = 9.80 \text{ m/s}^2$ for the free-fall acceleration due to gravity. Neglect friction and air resistance unless instructed to do otherwise.

•	Basic, single-concept problem
••	Intermediate-level problem; may require synthesis of concepts and multiple steps
•••	Challenging problem
SSM	*Solution is in Student Solutions Manual*
Example	*See worked example for a similar problem.*

Conceptual Questions

1. • Explain why ac voltage is often described using the root mean square value rather than the average voltage.

2. • Does Equation 18-9 apply to alternating currents? If not, is there any manner in which it can be amended so that it will be valid?

3. • Search the Internet to determine the root mean square value of voltage in a common wall receptacle for five countries that are not mentioned in this chapter.

4. • Give a simple explanation as to why an electric appliance, which is supposed to be operated with a certain root mean square voltage, operated with a slightly increased root mean square voltage will lead to catastrophic results.

5. • Examine the label on the power converter for a laptop computer, digital camera, or cell phone. If you took the converter to a country where the rms voltage is 240 V, would you need a power transformer? Explain your answer.

6. • Why does a transformer whose primary coil has 10 times as many turns as its secondary coil normally deliver about 1/10 of the voltage it receives?

7. • (a) What is a transformer? (b) How does it change the voltage input to some different voltage output? (c) Will a transformer work with a dc input?

8. • **Chemistry** Some of the most toxic substances that have been widely used in the United States (and many other countries in the world) are known as polychlorinated biphenyls or PCBs. (PCBs are a class of organic compounds that have two to 10 chlorine atoms attached to biphenyl, a molecule composed of two benzene rings.) Liquid PCBs were often used to fill the interior of transformers. Research PCBs and explain why.

9. • Why is electric power for domestic use in the United States, Canada, and most of the Western Hemisphere transmitted at very high voltages and stepped down to 120 V by a transformer near the point of consumption?

10. • A given length of wire is wound into a solenoid. How will its self-inductance be changed if it is rewound into another coil of (a) twice the length or (b) twice the diameter? SSM

11. • When the switch S is opened in the RL circuit shown in **Figure 21-15**, a spark jumps between the switch contacts. Why?

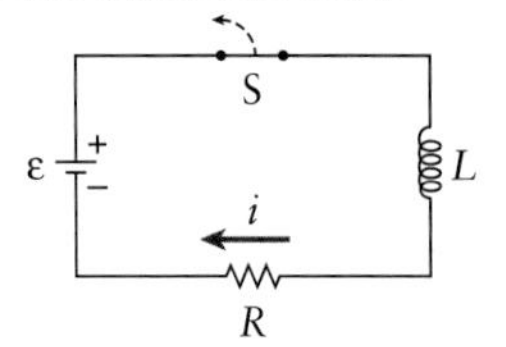

Figure 21-15 Problem 11

12. • What change must you make in the current in an inductor to double the energy stored in it?

13. • Discuss the storage of energy in an ideal LC circuit with no losses. SSM

14. • Compare the expressions for the energy stored in an inductor and the energy stored in a capacitor and explain the similarities.

15. • Is the current through a resistor in an ac circuit always in phase with the potential applied to the circuit? Why or why not?

16. • What does this statement mean: The voltage drop across an inductor leads the current by 1/4 cycle? SSM

17. • Define the concept of impedance for an LRC circuit. What are the units of impedance?

Multiple-Choice Questions

18. • The most important advantage of ac over dc electrical signals is that

A. electric power can't be delivered by a dc source.
B. dc could only be used in the early days of electrical power distribution.
C. dc results in more power loss in wire than ac.

D. it is relatively straightforward to change the voltage delivered by an ac source.
E. ac is safer.

19. • In a sinusoidal ac circuit with rms voltage V, the peak-to-peak voltage equals
A. $\sqrt{2}V$.
B. $2V$.
C. $V/\sqrt{2}$.
D. $2\sqrt{2}V$.
E. $V/2$. SSM

20. • The common electrical receptacle voltage in North America is often referred to as "120 volts ac." One hundred twenty volts is
A. the arithmetic mean of the voltage as it varies with time.
B. the root mean square (rms) average of the voltage as it varies with time.
C. the peak voltage from an ac wall receptacle.
D. the average voltage over many weeks of time.
E. one-half the peak voltage.

21. • If a power utility were able to replace an existing 500-kV transmission line with one operating at 1 MV, it would change the amount of heat produced in the transmission line to
A. 1/4 of the previous value.
B. 1/2 of the previous value.
C. 2 times the previous value.
D. 4 times the previous value.
E. 0.

22. • Two solenoids have the same cross-sectional area and length, but the first one has twice as many turns per unit length as the second one. The ratio of the inductance of the second to the first is
A. 1:1.
B. $1:\sqrt{2}$.
C. 1:1/2.
D. 1:4.
E. 2:1.

23. • For an LC circuit when the charge on the capacitor is 1/2 of the maximum charge, the energy stored in the capacitor is
A. the total energy.
B. 1/2 of the total energy.
C. 1/4 of the total energy.
D. 1/8 of the total energy.
E. twice the total energy. SSM

24. • If the current through an inductor were doubled, the energy stored in the inductor would be
A. halved.
B. the same.
C. doubled.
D. quadrupled.
E. 1/4 as much.

25. • The voltage leads the current
A. in circuits with an ac source and a resistor.
B. in circuits with an ac source and a capacitor.
C. in circuits with an ac source and an inductor.
D. in any ac circuit.
E. in circuits with an ac source and both a capacitor and an inductor.

Estimation/Numerical Analysis

26. • Estimate the power lost in a length of wire when the same power is transmitted by dc compared to ac.

27. • Estimate the ac current in an average household appliance. How much current does an average U.S. household draw? SSM

28. • Estimate the number of turns of the secondary compared to the number of turns of the primary for a transformer that is used to power your hair dryer that you purchased in the United States and used in Europe.

29. •• **Medical** Just as every circuit has a small amount of extra resistance in the wires, which we usually neglect, every circuit also has a small amount of stray capacitance and self-inductance because of the geometry of the wires. Consider a pacemaker implant in which the leads travel close together from the device to the heart and then separate and connect to the top and bottom of the heart. The circuit completes through the middle of the heart, so take the area of the current loop to be half the cross-sectional area of the heart, roughly a circle of radius 4 cm. Approximate the magnetic field as constant inside the loop and equal to the value at the center of the loop. Use this field to get the magnetic flux through the loop and hence estimate the stray self-inductance of the loop.

30. • A technician drives an LRC circuit at various frequencies with an ac voltage supply set to an rms voltage of $V_{rms} = 2.0$ V and records the resulting rms current in a table as shown. She knows that the capacitance in the circuit is $C = 20\ \mu$F and the resistance in the circuit is $R = 5\ \Omega$. (a) Use the current data to estimate the resonant angular frequency ω_0 and the inductance L in the circuit. (b) Use the rms current at the resonant frequency and the given rms voltage to verify that the resistance is, in fact, 5 Ω.

ω (krad/sec)	I_{rms} (A)	ω (krad/sec)	I_{rms} (A)
0.5	0.020	5.5	0.374
1.0	0.041	6.0	0.323
1.5	0.065	6.5	0.274
2.0	0.093	7.0	0.236
2.5	0.126	7.5	0.206
3.0	0.170	8.0	0.183
3.5	0.226	8.5	0.164
4.0	0.297	9.0	0.149
4.5	0.368	9.5	0.137
5.0	0.400	10.0	0.126

Problems

21-1 Most circuits use alternating current
21-2 We need to analyze ac circuits differently than dc circuits

31. • A sinusoidally varying voltage is represented by $V(t) = (75.0\text{ V})\sin(120\pi t)$. What are its frequency and peak voltage? Example 21-1

32. • Write an expression for the instantaneous voltage delivered by an ac generator supplying 120 V rms at 60 Hz. Example 21-1

33. • What is the rms current provided to a 60-W light bulb that is plugged into a 120-V rms wall receptacle? Example 21-1

34. • The maximum potential difference across the terminals of a 60.0-Hz sinusoidal ac source is +17.0 V at $t = 1/4$ cycle. If at $t = 0$ the potential difference is zero, calculate the potential difference at $t = 2.00$ ms. SSM Example 21-1

35. • An ac voltage is represented by $V(t) = (200 \text{ V}) \sin(120\pi t)$. What is its rms voltage? Example 21-1

36. • The rms current passing through a 50.0-Ω resistor in a sinusoidal ac circuit is 12.0 A. What is the maximum voltage drop across the resistor? Example 21-1

37. • The peak-to-peak current passing through a 150-Ω resistor is 24.0 A. Find the maximum voltage across the resistor and the rms current through the resistor. Example 21-1

38. •• Derive a relationship between the rms current through and the maximum voltage across a resistor R. Apply your result to the following situations: (a) $R = 100\ \Omega$, $V_{max} = 50.0$ V, $i_{rms} = ?$, (b) $R = 200\ \Omega$, $i_{rms} = 2.50$ A, $V_{max} = ?$, and (c) $V_{max} = 28.0$ V, $i_{rms} = 127$ mA, $R = ?$ Example 21-1

21-3 Transformers allow us to change the voltage of an ac power source

39. • The primary coil of a transformer makes 240 turns around its core, and the secondary coil of that transformer makes 80 turns. If the primary voltage is 120 V rms, what is the secondary voltage? Example 21-2

40. • The 400-turn primary coil of a step-down transformer is connected to an ac line that is 120 V rms. The secondary coil voltage is 6.50 V rms. Calculate the number of turns in the secondary coil. SSM Example 21-2

41. • The primary coil of a transformer makes 240 turns, and the secondary coil of that transformer makes 80 turns. If an alternating current of 3.00 A (rms) passes through the primary coil of the transformer, what current passes through the secondary coil of that transformer? Assume the transformer is ideal. Example 21-2

42. • An ideal transformer produces an output voltage that is 500% larger than the input voltage. If the input current is 10 A (rms), what is the output current? Example 21-2

43. • A neon sign in a shop operates at around 12 kV rms. A transformer is to step up the 120 V rms line voltage to that value. The secondary coil of the transformer has 20,000 turns. How many turns must be placed into the primary coil? Example 21-2

44. •• The 400-turn primary coil of a step-down transformer is connected to an ac line that is 120 V rms. The secondary coil is to supply 15.0 A at 6.30 V rms. Assuming no power loss in the transformer, calculate (a) the number of turns in the secondary coil and (b) the current in the primary coil. SSM Example 21-2

45. •• The ratio of the number of turns in the primary to the number of turns in the secondary of a step-down transformer is 25:1. If the input voltage across the primary coil is 750 V rms and the current in the output circuit is 25.0 A (rms), how much power is delivered to the secondary? Assume the transformer is 100% efficient. Example 21-2

46. •• **Chemistry** A high-voltage discharge tube is often used to study atomic spectra. The tubes require a large voltage across their terminals to operate. To get the large voltage, a step-up transformer is connected to a line voltage (120 V rms) and is designed to provide 5000 V rms to the discharge tube and to dissipate 75.0 W. (a) What is the ratio of the number of turns in the secondary to the number of turns in the primary? (b) What are the rms currents in the primary and secondary coils of the transformer? (c) What is the effective resistance that the 120-V source is subjected to? Example 21-2

21-4 An inductor is a circuit element that opposes changes in current

47. • How much voltage is produced by an inductor of value 25 μH if the time rate of change of the current is 58 mA/s? Example 21-4

48. • Uncle Leo tunes an old-fashioned radio that has an antenna made from a 3.0-cm-long solenoid with a cross-sectional area of 0.50 cm^2, composed of 300 turns of fine copper wire. Calculate the inductance of the coil assuming it is air-filled. SSM Example 21-3

49. • A tightly wound solenoid of 1600 turns, cross-sectional area of 6.00 cm^2, and length of 20.0 cm carries a current of 2.80 A. (a) What is its inductance? (b) If the cross-sectional area is doubled, does anything happen to the value of the inductance? Explain your answer. Example 21-3

21-5 In a circuit with an inductor and capacitor, charge and current oscillate

50. • What energy is stored in a 250-mH inductor with a current of 0.055 A? Example 21-5

51. • You have a 1.0-mH inductor. What size capacitor should you choose to make an oscillator with a natural frequency of 980 kHz?

52. • An LC circuit is formed with a 15-mH inductor and a 1000-μF capacitor. Calculate the frequency of oscillation f for the circuit.

53. • A 200-pF capacitor is charged to 120 V and then quickly connected to an inductor. Calculate the maximum energy stored in the magnetic field of the inductor as the circuit oscillates. Example 21-5

54. • LC circuits are used for filters in electronics. The selectivity of an LC circuit is defined as the ratio L/C. Calculate the selectivity of the bandwidth for an LC circuit composed of an inductor ($L = 0.250$ H) and a capacitor ($C = 875$ μF).

55. •• If the ratio of the energy stored in a capacitor compared to the total energy stored in an LC circuit is 0.5, calculate the ratio of the charge stored on the capacitor compared to the maximum charge stored on the capacitor in that circuit. Example 21-5

56. •• When the charge on the capacitor in an LC circuit is one-half of the maximum stored charge, calculate the ratio of the energy stored in the capacitor compared to the total energy in both the inductor and the capacitor. Example 21-5

21-6 When an ac voltage source is attached in series to an inductor, resistor, and capacitor, the circuit can display resonance

57. • An LRC circuit contains a 1.00-μF capacitor, a 5.00-mH inductor, and a 100-Ω resistor. What is its resonant frequency? Example 21-6

58. • An LRC circuit contains a 500-Ω resistor, a 5.00-H coil, and an unknown capacitor. The circuit resonates at 1000 Hz. What is the value of the capacitance? SSM Example 21-6

59. • The resonant frequency of an LRC circuit is 250 Hz. If the resistance is 200 Ω and the capacitance is 125 nF, what is the value of the inductance? Example 21-6

60. • A sinusoidal voltage of 120 V rms and a frequency of 60 Hz are applied to a 50.0-μF capacitor. Calculate the peak value of the current.

61. • A sinusoidal voltage that is 120 V rms and has a frequency of 60 Hz is applied to a 0.20-H inductor. Calculate the peak value of the current.

62. •• A sinusoidal voltage of 50.0 V (peak) at a frequency of 400 Hz is applied to a capacitor of unknown capacitance. The current in the circuit is 400 mA (rms). (a) What is the capacitance? (b) If the frequency of the voltage is increased, what, if anything, will happen to the rms value of the current in the circuit? Why?

63. • A 100-Ω resistor is connected across a 120-V rms, 60-Hz ac power line. Calculate (a) i_{rms}, (b) i_{max}, and (c) the average power dissipated in the resistor.

64. •• A sinusoidal voltage of 40.0 V rms and a frequency of 100 Hz is applied to (a) a 100-Ω resistor, (b) a 0.200-H inductor, and (c) a 50.0-μF capacitor. Calculate the peak value of the current and the average power delivered in each case. SSM

65. • A potential of 40.0 V rms and a frequency of 100 Hz is applied to (a) a 0.200-H inductor and (b) a 50.0-μF capacitor. In each case, find the peak value of the current.

66. •• A 35-mH inductor with 0.20-Ω resistance is connected in series to a 200-μF capacitor and a 60-Hz, ac, 45-V source. Calculate the (a) rms current and (b) phase angle for the circuit.

67. •• Assume that the circuit in **Figure 21-16** has L equal to 0.60 H, R equal to 250 Ω, and C equal to 3.5 μF. At a frequency of 60 Hz, what are the impedance and the phase angle between the current and voltage?

68. •• Assume that the circuit in **Figure 21-16** has L equal to 0.60 H, R equal to 280 Ω, and C equal to 3.5 μF. The amplitude of the driving voltage is 150 V rms. At a frequency of 60 Hz, what is the rms current in the circuit?

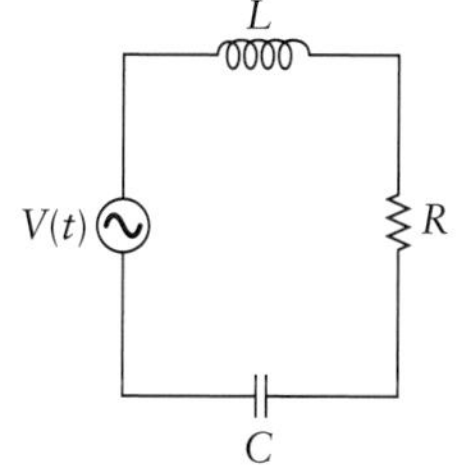

Figure 21-16
Problems 67 & 68

69. •• In an LRC circuit, the voltage amplitude and frequency of the source are 100 V and 500 Hz, respectively. The resistance has a value of 500 Ω, the inductance has a value of 0.20 H, and the capacitance has a value of 2.0 μF. (a) What is the impedance of the circuit? (b) What is the amplitude of the current from the source? (c) If the voltage of the source is given by $V(t) = (100 \text{ V}) \sin(1000\pi t)$, how does the current vary with time?

70. • Compare the rms current that is created in a circuit with only a 75-nF capacitor if the frequency of the 120 V rms source is 60 Hz versus 100 Hz.

71. •• The tuner knob on an FM radio moves the plates of an adjustable capacitor that is in series with a 0.400-μH inductor and a net resistance of 1000 Ω. The peak current induced in the circuit by a radio wave of a specific carrier frequency becomes large when the natural frequency of the circuit matches the carrier frequency. What is the frequency of the FM station that is tuned in when the capacitor in the radio is adjusted to 5.80 pF? If the peak operating voltage in the tuning circuit is 9.00 V, what is the peak current? Example 21-6

72. • A series LRC tuning circuit in a TV receiver resonates at 58 MHz. The circuit uses an 18-pF capacitor. What is the inductance of the circuit? Example 21-6

73. • An LRC series circuit contains a 500-Ω resistor, a 5.00-H inductor, and a capacitor. What value of capacitance will cause the circuit to resonate at 1000 Hz? SSM Example 21-6

74. • Hearing aids can be tuned to filter out or amplify either high- or low-frequency sounds depending on the frequency range in which a user has suffered hearing loss. If, for instance, a user needed to amplify low-frequency sounds and the hearing aid had a capacitance of 10 μF, what inductance should it have in order to produce peak signals at 1500 Hz? Example 21-6

General Problems

75. •• A 150-Ω resistor connected across a 60-Hz ac power supply of voltage amplitude 75 V produces heat in the resistor at a certain rate. If you want to replace the ac source with a dc power supply and still produce the same rate of heating, what should be the voltage of the dc source? Example 21-1

76. ••• You construct an LRC ac circuit using a 125-Ω resistor, a 12.5-mH inductor, and a parallel plate capacitor having a plate separation of 2.10 mm with a plastic material completely filling the region between the plates. The rectangular capacitor plates each measure 4.25 by 6.20 cm. (a) If you want the maximum rms current through the circuit at an ac frequency of 55.0 Hz, what should be the dielectric constant of the plastic in the capacitor? (b) Consult Table 17-1 to see if it appears feasible to achieve the desired results for the circuit and explain why or why not. Would the use of an ultracapacitor (which can have a capacitance of up to several farads) allow you to achieve the desired results? Example 21-6

77. •• In Europe the standard voltage is 240 V rms at 60 Hz. Suppose you take your 5.00-W electric razor to Rome and plug it into the receptacle (with an adapter to fit the receptacle but not to change the voltage). (a) What power will it draw in Rome? (b) What rms current will run through the razor in the United States and in Rome? Is the razor in danger of being damaged by using it in Rome without a voltage adapter? (c) If you want to use the razor in Rome without damaging it, what type of transformer would you need? Be as quantitative as you can. (d) What is the resistance of your razor? Example 21-1

78. •• **Medical** A dc current of 60 mA can cause paralysis of the body's respiratory muscles and hence interfere with breathing, but only 15 mA (rms) of ac current will do the same thing. Suppose a person is working with electrical power lines on a warm humid day and therefore has a low body resistance of 1000 Ω. What dc and what ac (amplitude and rms) potentials would it take to cause respiratory paralysis?

79. •• A power cord has a resistance of 0.080 Ω and is used to deliver 1500 W of power. (a) If the power is delivered at 12 V rms, how much power is dissipated in the power cord (assuming the current and voltage are in phase)? (b) If the power is delivered at 120 V rms, how much power is dissipated in the power cord (again assuming the current and voltage are in phase)? (c) Which voltage would you prefer to use to power your electrical device? Why? Example 21-1

80. ••• An ac electrical generator is made by turning a flat coil in a uniform constant magnetic field of 0.225 T. The coil consists of 33 square windings, and each winding is 15.0 cm on each side. It rotates at a steady rate of 745 rpm about an

axis perpendicular to the magnetic field passing through the middle of the coil and parallel to two of its opposite sides. An 8.50-Ω light bulb is connected across the generator. (a) Find the voltage and current amplitudes for the light bulb. (b) At what average rate is heat generated in the light bulb? (c) How much energy is consumed by the light bulb every hour? SSM Example 21-1

81. •• A small ac heating coil consisting of 750 windings is 8.50 cm long. It has a diameter of 1.25 cm and a resistance of 2.15 Ω. The coil is connected in series with a 2240-μF ultracapacitor, and the combination is plugged into a household outlet (60.0 Hz and 120 V rms). (a) What is the impedance of the circuit? (b) What is the rms current through the circuit? (c) What is the maximum (or peak) current through the circuit?

82. •• An *LRC* circuit consists of a 15.0-μF capacitor, a resistor, and an inductor connected in series across an ac power source of variable frequency having a voltage amplitude of 25.0 V. You observe that when the power source frequency is adjusted to 44.5 Hz, the rms current through the circuit has its maximum value of 65.0 mA. What will be the rms current if you change the frequency of the power source to 60.0 Hz? Example 21-6

83. ••• The circuit in **Figure 21-17** is known as a low-pass filter because it allows low-frequency signals to pass and it attenuates higher frequencies. Suppose its input is an ac signal that is composed of a broad range of frequencies. In this case the voltage across the capacitor is the output that is detected. Show that the ratio of the output voltage to the input voltage is $\frac{V_o}{V_i} = \frac{1}{\sqrt{1 + (\omega RC)^2}}$. SSM

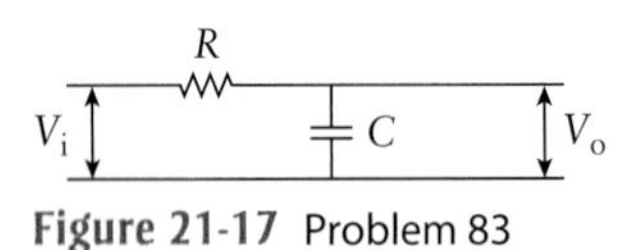

Figure 21-17 Problem 83

84. ••• The circuit in **Figure 21-18** is known as a high-pass filter because it allows high-frequency signals to pass and it attenuates lower frequencies. Suppose its input is an ac signal that is composed of a broad range of frequencies. The voltage across the resistor is the output that is detected. Show that the ratio of the output voltage to the input voltage is $\frac{V_o}{V_i} = \frac{1}{\sqrt{1 + 1/(\omega RC)^2}}$.

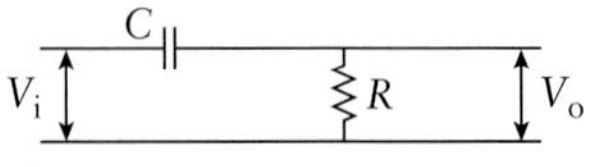

Figure 21-18 Problem 84

85. •• **Medical** Resonant wireless charging is an emerging technology where a powered transmitting circuit induces a current in a receiving circuit by the magnetic interaction between the circuits' inductors. Similar to how transformers work, what makes this process unique is that the receiving circuit can be several centimeters to a few meters away from the transmitting circuit, depending on the system. This induction with distance works only when both circuits, which can be treated like effective *LRC* circuits, operate at the same resonant frequency. One such medical benefit of this technology is the ability to remotely charge an internal battery in a medical implant. **Figure 21-19** shows a graph of the induced current in such a medical implant (the receiving *LRC* circuit) that is operating at its resonant frequency. (a) What is the rms value of the current? (b) What is the implant's resonant frequency ω_0?

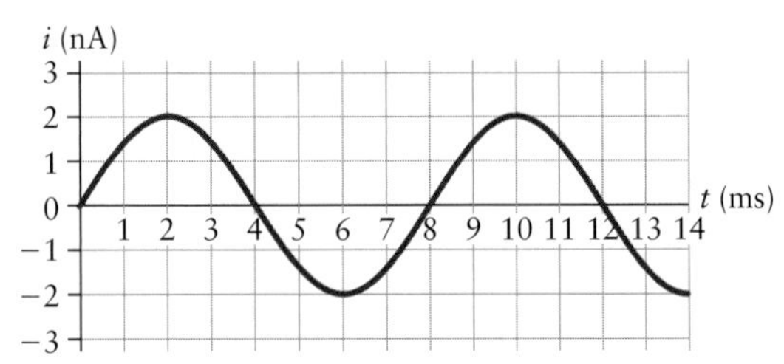

Figure 21-19 Problem 85

Matt Champlin/Moment Open/Getty Images

22 Electromagnetic Waves

What do you think?

Our eyes are sensitive to the light from the setting sun, while the receivers in a cell phone tower (at left) are sensitive to radio waves coming from mobile phones. Both visible light and radio waves are kinds of electromagnetic waves. Compared to radio waves, visible light has (a) much higher frequency and much faster speed; (b) much lower frequency and much slower speed; (c) about the same frequency and about the same speed; (d) much higher frequency and about the same speed; (e) much lower frequency and about the same speed.

In this chapter, your goals are to:

- (22-1) Define an electromagnetic wave.
- (22-2) Discuss how speed, frequency, and wavelength are related for electromagnetic waves, and describe the structure of an electromagnetic plane wave.
- (22-3) Explain what Maxwell's equations are and what they tell us about electromagnetic waves.
- (22-4) Calculate the energy density and intensity of an electromagnetic wave, and the energy of a photon.

To master this chapter, you should review:

- (13-2, 13-3, and 13-4) The properties of mechanical waves
- (16-6) Electric flux and Gauss's law for the electric field
- (17-5, 17-6) Capacitors
- (19-7) Ampère's law
- (20-2, 20-3) Magnetic flux and Faraday's law
- (21-2) Root-mean-square (rms) values
- (21-4, 21-5, and 21-6) Inductors

22-1 Light is just one example of an electromagnetic wave

What is light? The answer to this question was not discovered until the nineteenth century, when the Scottish physicist James Clerk Maxwell realized that light is an **electromagnetic wave**—a traveling disturbance that, unlike a sound wave, does not require a physical material through which to propagate. Instead, an electromagnetic wave involves oscillating electric and magnetic fields. These can exist even in the vacuum of space, which is why we can see the light from distant stars (**Figure 22-1a**).

In this chapter we'll study the properties of electromagnetic waves. We'll examine the broad variety of electromagnetic waves, which also includes x rays, microwaves, and radio waves. We'll see how wavelength, frequency, and speed are related for electromagnetic waves that propagate in a vacuum. We'll also look at the inner workings

We can see the light from this cluster of stars even though it is separated from us by 100,000 light years (about 10^{18} km) of nearly empty space. This is because electromagnetic waves, including visible light, can propagate in a vacuum.
(a)
S.Kafka and K.Honeycutt, IndianaUniversity/WIYN/NOAO/NSF

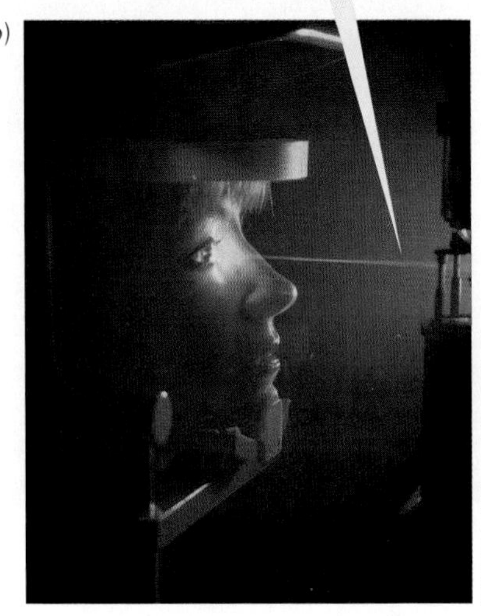

The intensity of an electromagnetic wave—including this laser beam used in ophthalmic surgery—depends on the amplitudes of the electric and magnetic fields that make up the wave.
(b)
Will & Deni McIntyre/Science Source

Figure 22-1 Electromagnetic waves Unlike sound waves or water waves, electromagnetic waves do not require the oscillation of any material substance. Instead, what oscillates are electric and magnetic fields. This explains (a) why these waves can propagate in a vacuum and (b) what determines their intensity.

of a particularly simple kind of electromagnetic wave called a *sinusoidal plane wave*. We'll discover how such waves are possible by examining the fundamental equations that govern electric and magnetic fields. We'll then use these equations to help us understand the amount of energy carried by an electromagnetic wave and the wave intensity (**Figure 22-1b**). We'll find that the energy of an electromagnetic wave comes in small packets called *photons*, whose properties help explain why certain kinds of electromagnetic waves can be harmful while others are not.

TAKE-HOME MESSAGE FOR Section 22-1

✔ Visible light is an example of an electromagnetic wave. The properties of electromagnetic waves are explained by the equations of electricity and magnetism.

22-2 In an electromagnetic plane wave, electric and magnetic fields both oscillate

Experiment shows that in a vacuum all electromagnetic waves—including radio waves, x rays, and others—propagate at the same speed. This is the speed of light, to which we give the symbol c:

$$c = 2.99792458 \times 10^8 \text{ m/s} \ (= 3.00 \times 10^8 \text{ m/s to 3 significant figures}) \tag{22-1}$$

Different kinds of electromagnetic waves have different frequencies and wavelengths. In Section 13-3 we learned that for a mechanical wave the frequency f and wavelength λ are related to the propagation speed of the wave v_p by $v_p = f\lambda$ (Equation 13-2). The same relationship holds for electromagnetic waves in a vacuum with v_p replaced by c:

Propagation speed, frequency, and wavelength of an electromagnetic wave (22-2)

$$c = f\lambda$$

Speed of light in a vacuum; **Frequency** of an electromagnetic wave; **Wavelength** of the wave in vacuum

Equation 22-2 tells us that the product of frequency f and wavelength λ has the same value, c, for *all* electromagnetic waves in a vacuum. The longer the wavelength, the lower the frequency; the shorter the wavelength, the higher the frequency.

Electromagnetic waves of any wavelength are possible. **Figure 22-2a** shows the names given to different wavelength ranges, which we refer to collectively as the **electromagnetic spectrum**. The human eye is sensitive to only a very narrow range

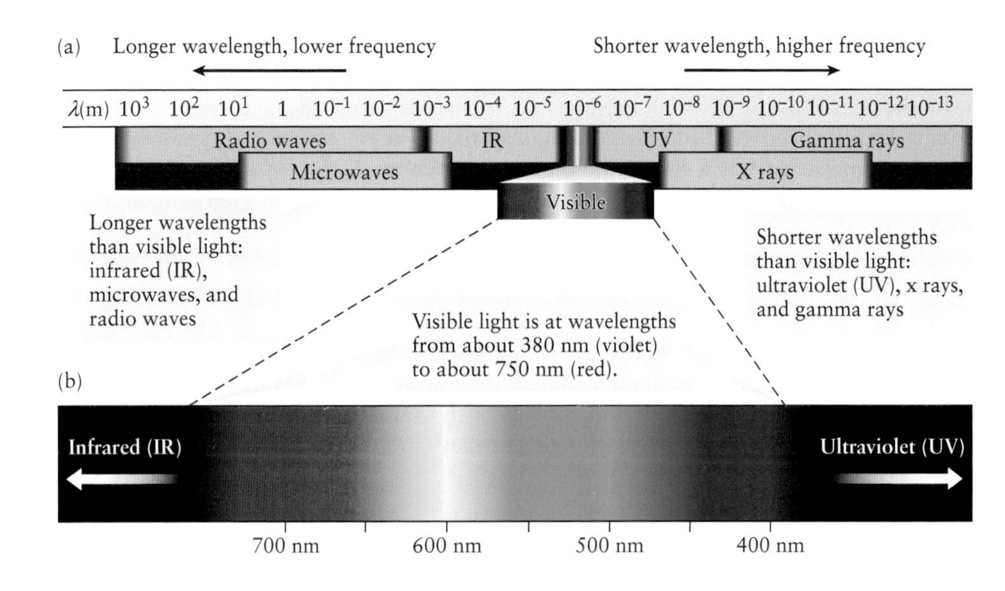

Figure 22-2 The electromagnetic spectrum (a) We classify electromagnetic waves according to their wavelength. (b) Visible light makes up a tiny portion of the entire electromagnetic spectrum.

of wavelengths known as **visible light** (**Figure 22-2b**). Visible light encompasses wavelengths from about $\lambda = 380$ nm to about $\lambda = 750$ nm (1 nm = 1 nanometer = 10^{-9} m). We perceive light of different wavelengths as having different colors; the shortest-wavelength light we can see is violet, and the longest-wavelength light we can see is red. At wavelengths shorter than visible light are ultraviolet light (UV), x rays, and gamma rays. At wavelengths longer than visible light are infrared light (IR), microwaves, and radio waves.

Other species can detect wavelengths longer or shorter than those visible to humans. Certain snakes (including pythons and rattlesnakes) have special pit organs on their heads that can sense infrared light. This enables these snakes to detect the radiation that both predators and prey emit due to their body temperature (see Section 14-7). Other species, such as damselfish, are able to sense ultraviolet light (**Figure 22-3**).

Note that in a medium other than a vacuum, electromagnetic waves propagate at speeds slower than c. For example, visible light travels at about 2.2×10^8 m/s $(0.73c)$ in water and at about $0.9998c$ in air. The propagation speed in a medium other than vacuum also depends on the wave frequency; for example, in ordinary glass blue light travels slightly slower than does red light. (We'll see in Chapter 23 that this explains why a prism is able to break white light into colors.) In a vacuum, however, electromagnetic waves of all frequencies propagate at c.

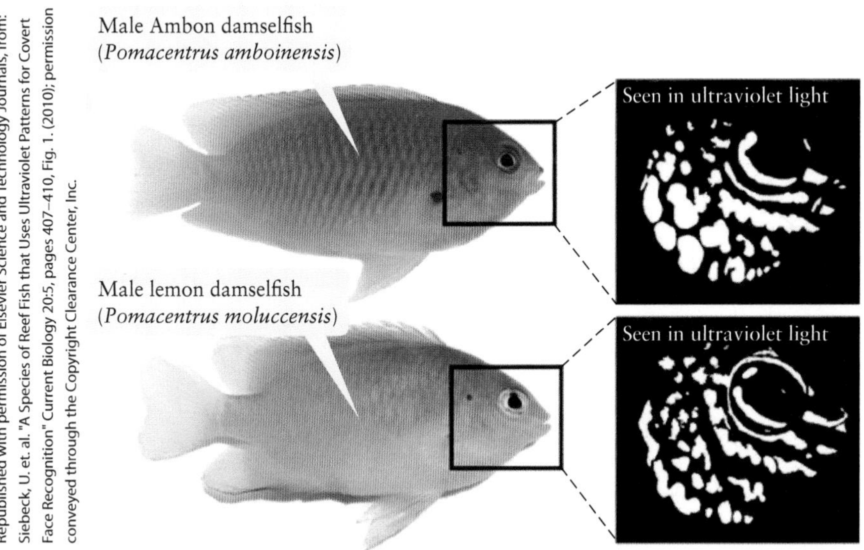

Figure 22-3 Using ultraviolet light for face recognition The Ambon damselfish (*Pomacentrus amboinensis*, a reef fish native to the western Pacific) has the ability to detect ultraviolet light. Where you see only dark and light bands on a fish's face, an Ambon sees an intricate pattern. This enables the territorial male Ambon damselfish (top) to identify and attack another Ambon in order to defend its territory but ignore a male lemon damselfish (*Pomacentrus moluccensis*, bottom) with its slightly different facial pattern.

The structure of electromagnetic waves can be quite complex. For example, the waves that make up the beam of laser light shown in Figure 22-1b are strong near the center of the beam, then taper off in strength toward the edge of the beam. A simpler, idealized kind of wave that has all of the key properties of a more general electromagnetic wave is a **sinusoidal plane wave** (**Figure 22-4a**). As the name suggests, the disturbance in such a wave oscillates in a sinusoidal fashion. There are actually two disturbances in this wave, an electric field $\vec{E}$ with amplitude E_0 and a magnetic field $\vec{B}$ with amplitude B_0. Both of these fields are *transverse*: They are perpendicular to the direction of propagation, just like the disturbance for waves propagating along a stretched rope or string (Sections 13-3 and 13-4). The fields are also perpendicular to each other. In a snapshot of the wave, as shown in Figure 22-4a, the $\vec{E}$ and $\vec{B}$ fields repeat over the same distance and so have the same wavelength λ. As the wave passes by a given point in space, the $\vec{E}$ and $\vec{B}$ fields both oscillate with the same frequency f. Equation 22-2 then tells us that both fields propagate at the same speed $c = f\lambda$.

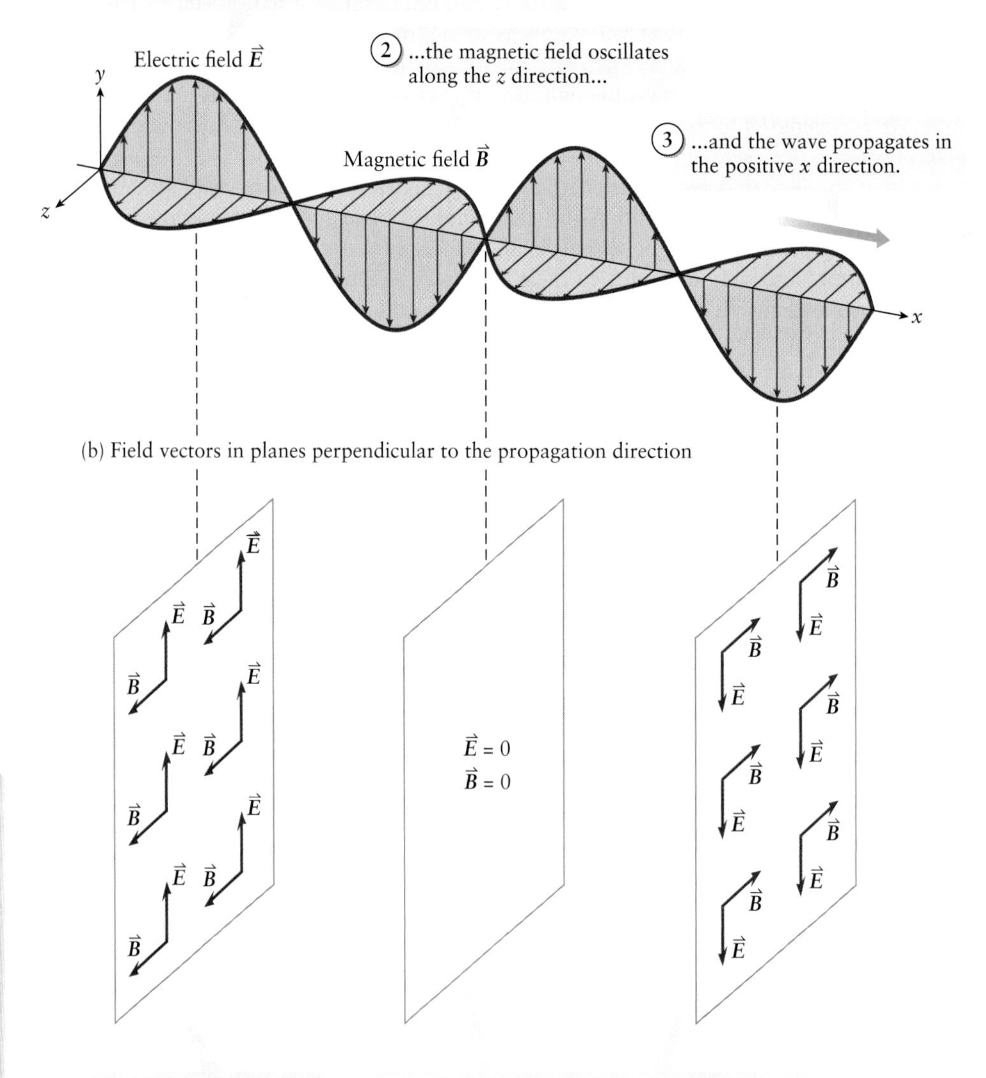

Figure 22-4 A sinusoidal electromagnetic plane wave (a) The characteristics of a simple electromagnetic wave. This illustration shows two complete wavelengths of the wave. (b) How the electric field $\vec{E}$ and magnetic field $\vec{B}$ extend through space.

- An electromagnetic wave is transverse: The electric and magnetic fields (the wave disturbances) are perpendicular to the propagation direction.
- In this sinusoidal plane wave, the electric and magnetic fields oscillate in phase.
- This is a plane wave because on any plane perpendicular to the direction in which the wave propagates, $\vec{E}$ has the same value at all points and $\vec{B}$ has the same value at all points.

WATCH OUT! The fields of an electromagnetic plane wave extend beyond the axis of propagation.

Figure 22-4a may give you the misleading impression that the $\vec{E}$ and $\vec{B}$ fields are present only along the x axis (the axis along which the wave propagates). Such a wave would be like an infinitely narrow laser beam. In fact, in a plane wave the fields have the same value at *all* points on a plane perpendicular to the direction of propagation (**Figure 22-4b**). That's the origin of the term *plane wave*. A true plane wave would extend infinitely far beyond the x axis in Figure 22-4a, so this is an idealization like a frictionless incline or a massless rope. But the fields shown in Figure 22-4 are a good approximation of the actual fields found near the center of the laser beam in Figure 22-1b.

In Section 13-3 we wrote a *wave function* that describes the disturbance associated with a wave on a rope. In the same fashion we can write wave functions for the electric and magnetic fields depicted in Figure 22-4a:

$$E_y(x,t) = E_0 \cos(kx - \omega t + \phi)$$
$$B_z(x,t) = B_0 \cos(kx - \omega t + \phi) \tag{22-3}$$

(sinusoidal electromagnetic plane wave)

In Equations 22-3 we use the same symbols that we used for waves on a rope in Section 13-3: $k = 2\pi/\lambda$ is the angular wave number, $\omega = 2\pi f$ is the angular frequency, and ϕ is the phase angle (which tells us what point in the oscillation cycle corresponds to $x = 0$, $t = 0$). Note that $E_y(x,t)$ and $B_z(x,t)$ both depend on position x and time t in the same way, which tells us that the electric and magnetic fields oscillate in phase with the same wavelength and frequency. In addition, the electric field amplitude E_0 and the magnetic field amplitude B_0 in a plane wave are directly proportional to each other:

$$B_0 = \frac{E_0}{c} \tag{22-4}$$

(sinusoidal electromagnetic plane wave)

EXAMPLE 22-1 A Radio Wave

A certain FM radio station broadcasts at a frequency of 98.7 MHz (1 MHz $= 10^6$ Hz). In the wave that reaches the radio in your car, the electric field amplitude is 6.00×10^{-2} V/m. Calculate the wavelength of the wave and the amplitude of the magnetic field.

Set Up

We are given the wave frequency f and the electric field amplitude E_0. We use Equation 22-2 to find the wavelength λ and Equation 22-4 to find the magnetic field amplitude B_0.

Propagation speed, frequency, and wavelength of an electromagnetic wave:

$$c = f\lambda \tag{22-2}$$

Relation between the electric and magnetic field amplitudes in an electromagnetic wave:

$$B_0 = \frac{E_0}{c} \tag{22-4}$$

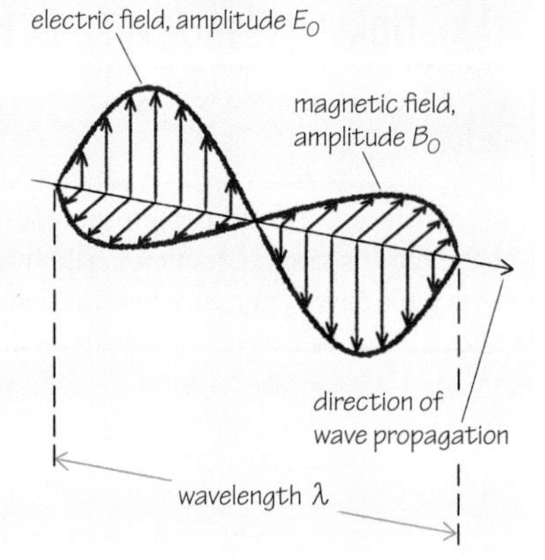

Solve

Use Equation 22-2 and the given frequency $f = 98.7$ MHz $= 98.7 \times 10^6$ Hz to solve for the wavelength.

From Equation 22-2

$$\lambda = \frac{c}{f} = \frac{3.00 \times 10^8 \text{ m/s}}{98.7 \times 10^6 \text{ Hz}} = 3.04 \frac{\text{m}}{\text{s} \cdot \text{Hz}}$$

Recall that 1 Hz $= 1 \text{ s}^{-1}$, so the units of s and Hz cancel:

$$\lambda = 3.04 \text{ m}$$

Use Equation 22-4 and the value $E_0 = 6.00 \times 10^{-2}$ V/m to solve for the magnetic field amplitude.

From Equation 22-4

$$B_0 = \frac{E_0}{c} = \frac{6.00 \times 10^{-2}\ \text{V/m}}{3.00 \times 10^8\ \text{m/s}} = 2.00 \times 10^{-10} \left(\frac{\text{V}}{\text{m}}\right)\left(\frac{\text{s}}{\text{m}}\right)$$

We learned in Section 17-3 that 1 V/m = 1 N/C, and in Section 19-3 we learned that 1 T = 1 (N·s)/(C·m). So

$$B_0 = 2.00 \times 10^{-10} \left(\frac{\text{N}}{\text{C}}\right)\left(\frac{\text{s}}{\text{m}}\right) = 2.00 \times 10^{-10} \frac{\text{N}\cdot\text{s}}{\text{C}\cdot\text{m}}$$

$$= 2.00 \times 10^{-10}\ \text{T}$$

Reflect

Because the speed of light c has such a large value in m/s, the magnetic field amplitude B_0 in tesla (T) is much smaller than the electric field amplitude E_0 in volts per meter (V/m). As we'll see later in the chapter, however, the electric and magnetic fields prove to be equally important in an electromagnetic wave in vacuum.

In this example we've seen how to relate the units of magnetic field (T) to those of electric field (V/m): $1\ \text{T} = 1\ (\text{V/m})\cdot(\text{s/m}) = 1\ \text{V}\cdot\text{s/m}^2$. We'll make use of this result in later examples.

Why must an electromagnetic wave propagating in vacuum be transverse? Would it be possible to have a longitudinal component of the wave, with an oscillating electric or magnetic field along the direction of propagation? For that matter, why is it necessary for an electromagnetic wave to include both electric and magnetic fields? Couldn't there be a wave that included an oscillating electric field but no magnetic field, or an oscillating magnetic field but no electric field? To answer these questions we need to look at the fundamental equations that govern electric and magnetic fields.

GOT THE CONCEPT? 22-1 Electromagnetic Wave Speeds

Four electromagnetic waves in vacuum have different frequencies. Which of these propagates at the fastest speed?

(a) $f = 3.95 \times 10^6$ Hz; (b) $f = 2.44 \times 10^7$ Hz; (c) $f = 1.26 \times 10^{11}$ Hz; (d) $f = 2.26 \times 10^8$ Hz; (e) all have the same speed.

TAKE-HOME MESSAGE FOR Section 22-2

✔ Electromagnetic waves in vacuum propagate at the speed of light.

✔ Different kinds of electromagnetic waves have different wavelengths and frequencies. The human eye can detect only a very narrow range of wavelengths.

✔ In a sinusoidal electromagnetic plane wave, the electric and magnetic fields both oscillate in phase with the same wavelength and frequency. The field directions are perpendicular to each other and perpendicular to the direction of propagation.

22-3 Maxwell's equations explain why electromagnetic waves are possible

By the middle of the nineteenth century, scientists understood a great deal about electricity and magnetism. The work of Gauss, Ampère, and Faraday had established fundamental relationships that describe electric and magnetic phenomena; for example, it was understood that electric charges give rise to electric fields and that electric currents give rise to magnetic fields. It was the Scottish physicist James Clerk Maxwell who added to these fundamental relationships and forged our understanding of electricity and magnetism into a unified theory. In this section we'll look at four basic equations, known as **Maxwell's equations**, which describe *all* electromagnetic phenomena. We'll then see how these equations help us understand the nature of electromagnetic waves like the sinusoidal plane wave that we discussed in Section 22-2.

Gauss's Laws for Electricity and Magnetism

The first of Maxwell's equations is **Gauss's law for the electric field**, which we first encountered in Section 16-6. It states that the net electric flux through a closed surface (called a *Gaussian surface*) is proportional to the total amount of electric charge enclosed within that surface:

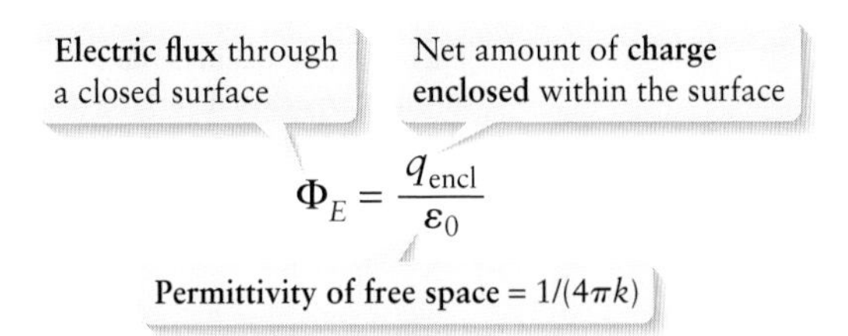
Electric flux through a closed surface
Net amount of charge enclosed within the surface
$\Phi_E = \frac{q_{encl}}{\varepsilon_0}$
Permittivity of free space = $1/(4\pi k)$

Gauss's law for the electric field (16-9)

In this equation we've added a subscript E to remind us that the quantity on the left-hand side is the flux of the electric field $\vec{E}$.

Figure 22-5a illustrates Gauss's law for the electric field. If a surface encloses a net positive charge, as for Gaussian surface 1 in Figure 22-5a, there is a net outward

(a) Gauss's law for the electric field: $\Phi_E = q_{encl}/\varepsilon_0$

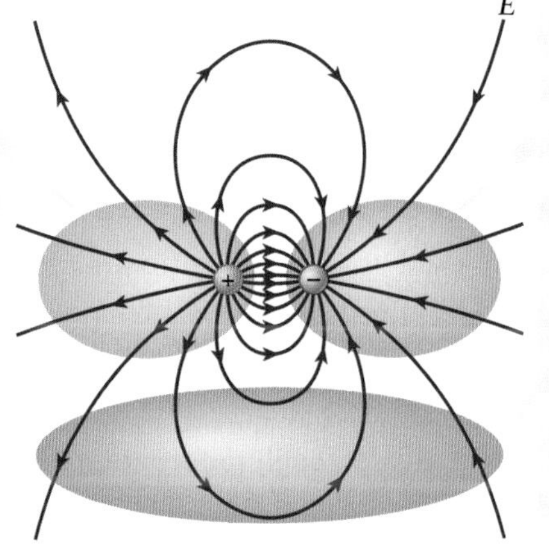
Gaussian surface 1 encloses positive charge. Field lines point out of this surface, and there is a net outward (positive) electric flux.
$\vec{E}$
Gaussian surface 2 encloses negative charge. Field lines point into this surface, and there is a net inward (negative) electric flux.
Gaussian surface 3 encloses zero charge. Each field line that points into this surface at one place points out at another place. There is zero net electric flux.

(b) Gauss's law for the magnetic field: $\Phi_B = 0$

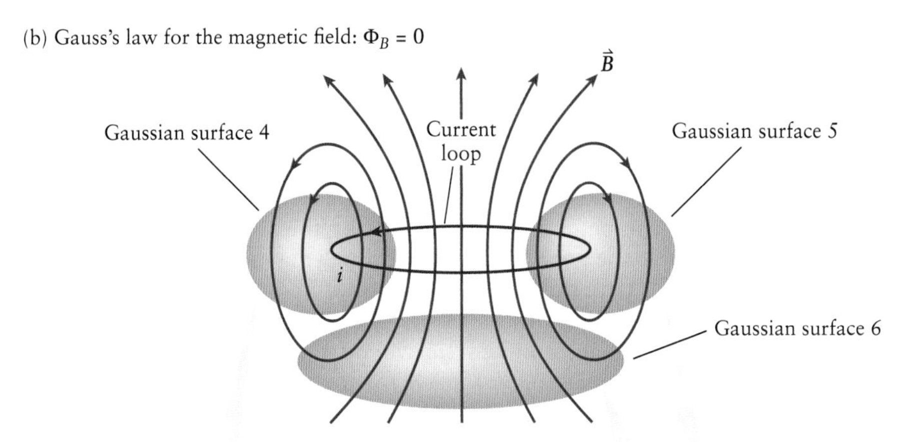
$\vec{B}$
Gaussian surface 4
Current loop
Gaussian surface 5
i
Gaussian surface 6

Gaussian surfaces 4 and 5 enclose part of the current loop. **Gaussian surface 6** encloses none of the current loop. For each of these surfaces each field line that points into the surface at one place points out at another place. There is zero net magnetic flux for each surface.

Figure 22-5 **Gauss's laws for the electric and magnetic fields** (a) The electric flux through a closed surface depends on the charge enclosed by that surface. (b) The magnetic flux through any closed surface is zero.

- In a region of space where there are no electric charges, there is zero net electric flux through any Gaussian surface in that region.
- In any region of space, there is zero net magnetic flux through any Gaussian surface.

(positive) electric flux and electric field lines point out of the surface. If instead there is a net negative charge inside the surface, as for Gaussian surface 2 in Figure 22-5a, the net electric flux is inward (negative) and electric field lines point into the surface. If there is no charge at all inside the surface, as for Gaussian surface 3 in Figure 22-5a, there is zero net electric flux: Each field line that enters the surface at one point exits it at another point.

In addition to Gauss's law for the electric field, there is **Gauss's law for the magnetic field.** This states that for *any* closed Gaussian surface, there is *zero* net flux of the magnetic field $\vec{B}$ through that surface:

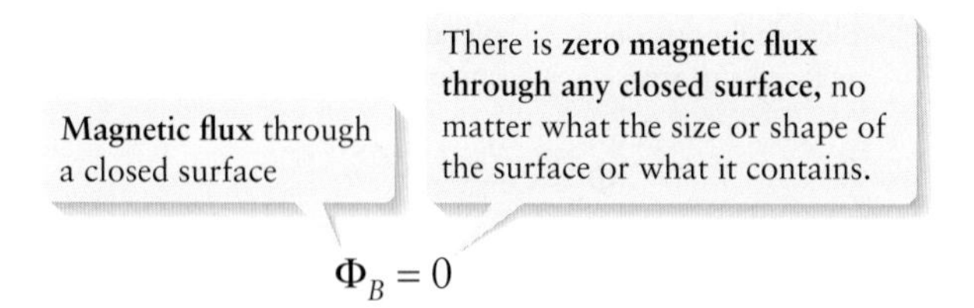

Gauss's law for the magnetic field (22-5)

$$\Phi_B = 0$$

As an example, **Figure 22-5b** shows three Gaussian surfaces in the magnetic field of a current loop. For all three surfaces, each field line that enters the surface at one point exits it at another point. This is true whether there is current enclosed within the Gaussian surface (as for surfaces 4 and 5) or if there is no enclosed current (as for surface 6). Just as for the electric flux for Gaussian surface 3 in Figure 22-5a, this implies that there is zero net flux of magnetic field for all three surfaces. (There *would* be a nonzero net magnetic flux through a Gaussian surface if the surface enclosed an isolated north magnetic pole that had no associated south pole, or an isolated south pole that had no associated north pole. These would be the magnetic analogs of an isolated positive or negative electric charge. Physicists have searched for decades for such isolated poles, called *magnetic monopoles*. No confirmed observations have yet been made.)

What Gauss's Laws Tell Us About Electromagnetic Waves

Equations 16-9 and 22-5 are valid not just for static situations in which the electric and magnetic fields are constant, but also in situations where the $\vec{E}$ and $\vec{B}$ fields are changing. So Gauss's laws for the electric and magnetic fields must also hold true for electromagnetic waves. In **Figure 22-6a** we've drawn a Gaussian surface that encloses part of the sinusoidal plane wave shown in Figure 22-4. Because the wave is transverse, there is no electric or magnetic flux through the front or back face of the Gaussian surface. Although the fields change, at any instant there is as much outward flux of $\vec{E}$ or $\vec{B}$ on one edge of the Gaussian surface as there is inward flux on the opposite edge. So the net electric flux and the net magnetic flux through this surface is zero, and both of Gauss's laws are obeyed.

Gauss's laws also explain why electromagnetic waves in vacuum *must* be transverse. **Figure 22-6b** shows part of an electromagnetic wave with an oscillating *longitudinal* electric field (one that points along or opposite to the direction in which the wave propagates). For the Gaussian surface that we've chosen and at the instant of time shown, there is a net nonzero outward flux of electric field. But the wave is in a vacuum, so there can be no charge enclosed by the surface. Equation 16-9 then tells us that there *cannot* be a net electric flux through the surface in Figure 22-6b. We're forced to conclude that the longitudinal wave shown in Figure 22-6b is **impossible** because it contradicts Gauss's law for the electric field. Likewise, it would be impossible for an electromagnetic wave to have a longitudinal magnetic field, since this would violate Gauss's law for the magnetic field as given by Equation 22-5.

Let's now examine the two remaining members of Maxwell's equations and see what they tell us about the nature of electromagnetic waves.

(a) Gauss's laws say that a transverse electromagnetic wave is permitted.

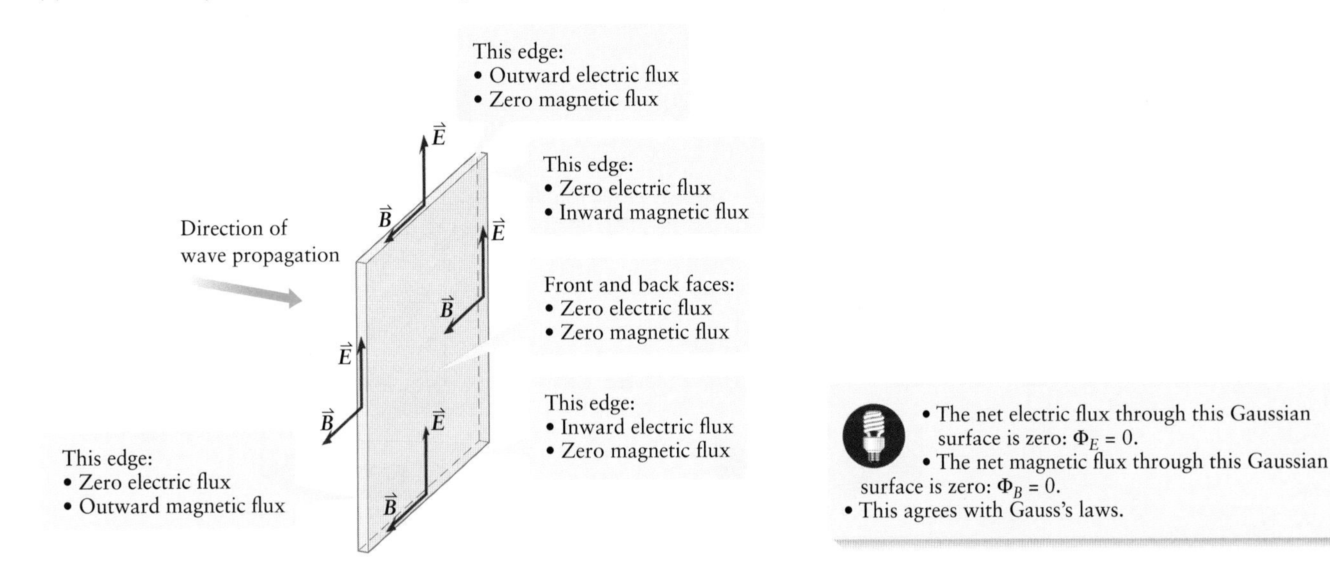

(b) Gauss's laws say that a longitudinal electromagnetic wave is not permitted.

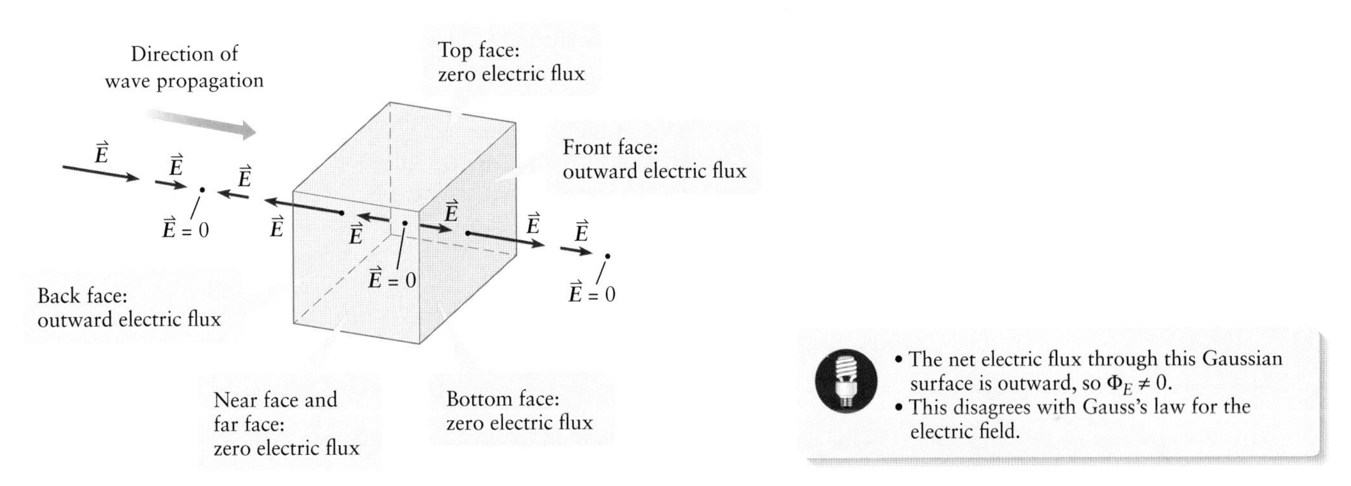

Figure 22-6 Gauss's laws applied to a sinusoidal electromagnetic plane wave (a) A transverse electromagnetic wave in vacuum is compatible with Gauss's laws, but (b) a longitudinal electromagnetic wave in vacuum is not.

Faraday's Law: A Changing Magnetic Field Generates an Electric Field

The third of Maxwell's equations is **Faraday's law**, which we introduced in Sections 20-2 and 20-3. It states that an emf is induced in a loop if the magnetic flux through that loop changes:

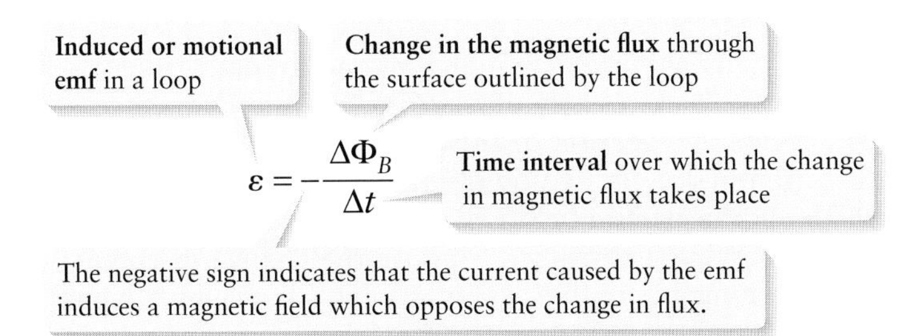

Faraday's law and Lenz's law for induction
(20-3)

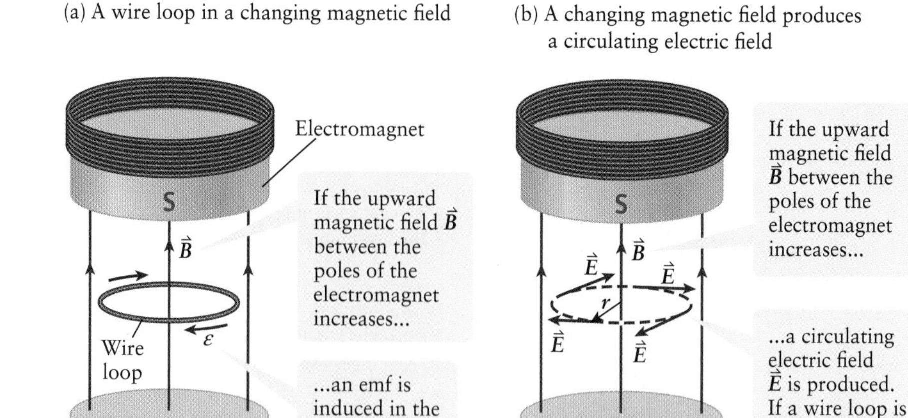

Figure 22-7 A changing $\vec{B}$ produces a circulating $\vec{E}$ (a) Faraday's law says that an emf is induced in the loop when the magnetic field changes. (b) Fundamentally what happens when the magnetic field changes is that a circulating electric field is produced.

As an example, **Figure 22-7a** shows a wire loop placed between the poles of an electromagnet. As the current in the electromagnet increases, the magnetic field increases and the increasing magnetic flux through the wire loop induces an emf in the loop. It's important to note that the emf is present whether the loop is made of a conductor like copper wire, a semiconductor like silicon, or an insulator like wood; the only difference is the amount of current that's established in response to the emf. The emf is even present if there is no material substance there at all!

Fundamentally, what happens is that the changing magnetic field generates a circulating electric field $\vec{E}$, as **Figure 22-7b** shows. The emf around a loop is just the *circulation* of the electric field around the loop. We define this in exactly the same way that we defined the circulation of magnetic field in Section 19-7. First imagine breaking the loop into a number of small segments of length $\Delta\ell$; then find the component $E_{\|}$ of the electric field that's tangent to each segment; then sum the products $E_{\|}\Delta\ell$:

$$\text{circulation of the electric field} = \sum E_{\|}\Delta\ell \tag{22-6}$$

(One simple case is the circulating electric field shown in Figure 22-7b. Here $\vec{E}$ is tangential to the circular loop of radius r, so the circulation is just the magnitude E multiplied by the total distance around the loop—that is, its circumference $2\pi r$.)

We can now rewrite Faraday's law by replacing the induced emf ε in Equation 20-3 with the circulation of the electric field, as given by Equation 22-6:

Faraday's law in terms of circulation (22-7)

$$\sum E_{\|}\Delta\ell = -\frac{\Delta\Phi_B}{\Delta t} \tag{22-7}$$

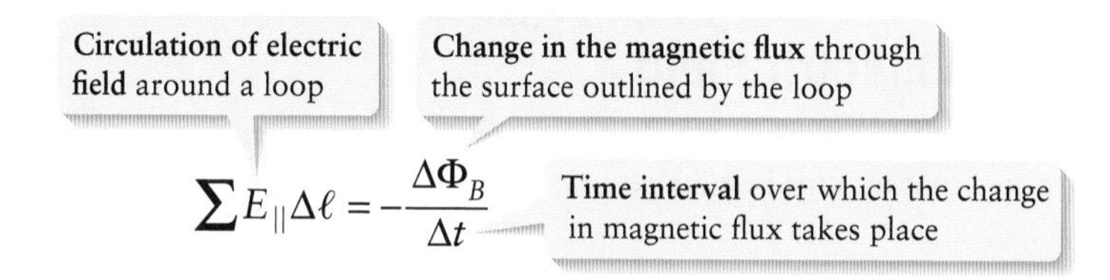

Equation 22-7 tells us that a changing magnetic flux through a loop causes an electric field that circulates around that loop. This is true even if there is no material substance at the location of the loop. The minus sign in Equation 22-7 tells us the direction of the circulating electric field: If a conducting wire is present around the loop, the electric field will cause charges to flow, and the magnetic field due to those moving charges will oppose the change in magnetic field that caused the change in magnetic flux. (This is Lenz's law, which we introduced in Section 20-3.)

EXAMPLE 22-2 An Electric Field Due to a Changing Magnetic Field

The uniform magnetic field shown in Figure 22-7b decreases in magnitude from 1.50 T to zero in a time t, inducing an electric field. What is the magnitude of this electric field around a loop 3.50 cm in diameter if (a) $t = 10.0$ s and (b) $t = 0.100$ s?

Set Up

Equation 20-1 tells us the magnetic flux through the loop of diameter 3.50 cm. The magnetic field $\vec{B}$ is perpendicular to the plane of the loop, so θ (the angle between the direction of $\vec{B}$ and the perpendicular to the loop) is $\theta = 0$.

The electric field in this case has the same magnitude E all the way around the circle and is tangent to the circle, so $E_{\|}$ in Equation 22-7 is equal to E.

Magnetic flux:

$$\Phi_B = AB_{\perp} = AB\cos\theta \qquad (20\text{-}1)$$

Faraday's law in terms of circulation:

$$\sum E_{\|}\Delta\ell = -\frac{\Delta\Phi_B}{\Delta t} \qquad (22\text{-}7)$$

Area of a circle of radius r:

$$A = \pi r^2$$

Circumference of a circle of radius r:

$$C = 2\pi r$$

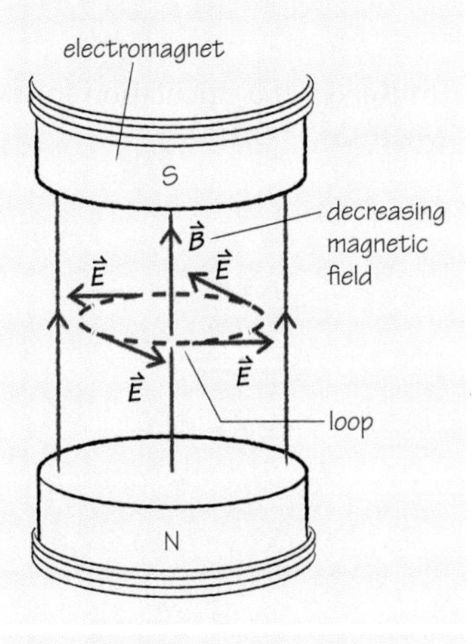

Solve

(a) First determine the initial and final values of magnetic flux and the change in magnetic flux for a circle of diameter 3.50 cm.

The radius of the loop is half the diameter:

$$r = \frac{1}{2}(3.50 \text{ cm}) = 1.75 \text{ cm} = 1.75 \times 10^{-2} \text{ m}$$

The area of the loop is

$$A = \pi r^2 = \pi\,(1.75 \times 10^{-2} \text{ m})^2 = 9.62 \times 10^{-4} \text{ m}^2$$

From Equation 20-1 the initial magnetic flux is

$$\Phi_B = AB\cos 0 = (9.62 \times 10^{-4} \text{ m}^2)(1.50 \text{ T})(1)$$
$$= 1.44 \times 10^{-3} \text{ T}\cdot\text{m}^2$$

The final magnetic field is zero, so the final magnetic flux is zero as well. The change in magnetic flux is

$$\Delta\Phi_B = (\text{final flux}) - (\text{initial flux}) = 0 - 1.44 \times 10^{-3} \text{ T}\cdot\text{m}^2$$
$$= -1.44 \times 10^{-3} \text{ T}\cdot\text{m}^2$$

Use Equation 22-7 to calculate the circulation of the electric field in the case where $\Delta t = 10.0$ s. Since the field magnitude E has the same value around the circle and $E_{\|} = E$, we can use this to calculate E.

The circulation of the electric field is equal to $-\Delta\Phi_B/\Delta t$:

$$\text{circulation of the electric field} = \sum E_{\|}\Delta\ell = -\frac{\Delta\Phi_B}{\Delta t}$$

$$= -\frac{-1.44 \times 10^{-3} \text{ T}\cdot\text{m}^2}{10.0 \text{ s}} = 1.44 \times 10^{-4} \frac{\text{T}\cdot\text{m}^2}{\text{s}}$$

In Example 22-1 (Section 22-2) we saw that

$1 \text{ T} = 1 \dfrac{\text{V}\cdot\text{s}}{\text{m}^2}$, so the magnitude of the circulation is

$$1.44 \times 10^{-4} \frac{\text{T}\cdot\text{m}^2}{\text{s}} = 1.44 \times 10^{-4}\left(\frac{\text{V}\cdot\text{s}}{\text{m}^2}\right)\left(\frac{\text{m}^2}{\text{s}}\right) = 1.44 \times 10^{-4} \text{ V}$$

Since $E_{\|} = E$ and E has the same value at all points around the circle, we can write the circulation as

$$\sum E_{\|}\Delta\ell = \sum E\Delta\ell = E\sum\Delta\ell$$

The sum $\sum \Delta \ell$ is the total distance around the loop, equal to the loop circumference $2\pi r$. So

$$E\,(2\pi r) = 1.44 \times 10^{-4}\ \text{V}$$

$$E = \frac{1.44 \times 10^{-4}\ \text{V}}{2\pi r} = \frac{1.44 \times 10^{-4}\ \text{V}}{2\pi(1.75 \times 10^{-2}\ \text{m})} = 1.31 \times 10^{-3}\ \text{V/m}$$

(b) Repeat the calculation for the case where $\Delta t = 0.100$ s.

Equation 22-7 tells us that the circulation of the electric field, and hence the electric field itself, is inversely proportional to the time Δt over which the magnetic flux changes. If the magnetic field drops to zero in $\Delta t =$ 0.100 s rather than $t = 10.0$ s, the elapsed time is smaller by a factor of

$$\frac{0.100\ \text{s}}{10.0\ \text{s}} = 1.00 \times 10^{-2}$$

and so the induced field is larger by a factor of

$$\frac{1}{1.00 \times 10^{-2}} = 1.00 \times 10^{2}$$

Therefore, the electric field in the case where $\Delta t = 0.100$ s is

$$E = (1.00 \times 10^{2})(1.31 \times 10^{-3}\ \text{V/m}) = 0.131\ \text{V/m}$$

Reflect

Our results show that the more rapid the change in magnetic field, the greater the magnitude of the electric field that is induced.

If a circular loop of conducting wire were placed along the circular path of diameter 3.50 cm, a current would be generated so as to produce a magnetic field that would oppose the change in magnetic flux. The upward magnetic field in the figure decreases, so the magnetic field produced in this way would have to be upward. The induced current that produces this magnetic field is in the same direction as the circulating electric field. So the electric field and current must both have the direction shown.

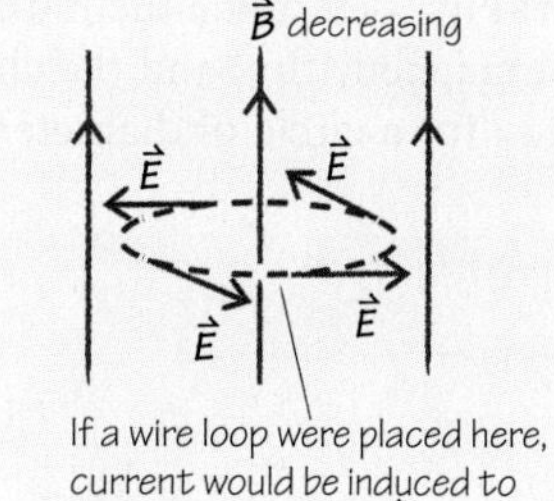
$\vec{B}$ decreasing
$\vec{E}$
$\vec{E}$
$\vec{E}$
$\vec{E}$
If a wire loop were placed here, current would be induced to produce an upward $\vec{B}$.

The Maxwell–Ampère Law: A Changing Electric Field Generates a Magnetic Field

Equation 22-7 tells us that a circulating electric field is produced by a magnetic field that changes over time. In Section 19-7 we learned that a circulating *magnetic* field is produced by electric charges in motion, that is, by a current. The mathematical expression of this statement is *Ampère's* law:

Ampère's law (19-13)

$$\sum B_{||}\Delta \ell = \mu_0 i_{\text{through}}$$

Circulation of magnetic field around an Amperian loop

Permeability of free space

Current through the Amperian loop

Figure 22-8a shows an application of Ampère's law that we introduced in Section 19-7: the magnetic field due to a long, straight, current-carrying wire. The current through each loop in the figure is equal to the current in the wire, so for each loop $i_{\text{through}} = i$. As a result, there is a magnetic field that circulates around each loop, and the circulation $\sum B_{||}\Delta \ell$ of the magnetic field is equal to $\mu_0 i$.

Now suppose we break the wire in Figure 22-8a and insert two metal disks to form a parallel-plate capacitor (**Figure 22-8b**). If the same steady current exists as

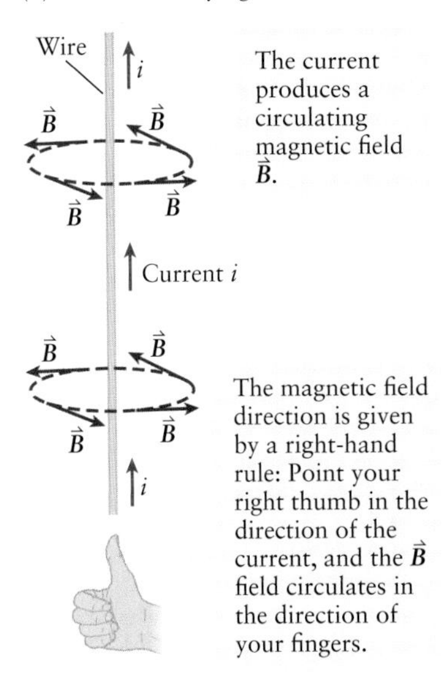

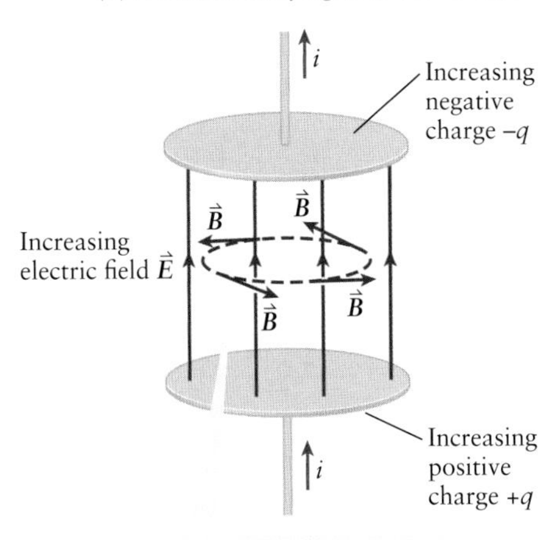

Figure 22-8 A changing $\vec{E}$ produces a circulating $\vec{B}$ (a) Ampère's law says that a magnetic field circulates around a current-carrying wire. (b) A circulating magnetic field can also be produced by a changing electric field.

in Figure 22-8a, positive charge will build up on the lower plate, negative charge will build up on the upper plate, and the electric field between the two plates will increase in magnitude. There is no current from the lower plate to the upper plate, so $i_{\text{through}} = 0$ for the loop between the plates in Figure 22-8b. So Equation 19-13 predicts that there should be no circulating magnetic field between the plates. Yet experiment shows that there *is* a circulating magnetic field between the plates! How can this be?

James Clerk Maxwell's great insight was to propose that if a changing magnetic field could produce a circulating electric field, then a changing *electric* field could produce a *magnetic* field. That's what's happening between the capacitor plates shown in Figure 22-8b: The electric field is increasing in magnitude as charge builds on the plates, and a circulating magnetic field results. To explain this effect, Maxwell expanded on Ampère's law as given by Equation 19-13 to include an additional term on the right-hand side. The result is called the **Maxwell–Ampère law:**

Circulation of magnetic field around an Amperian loop

Permittivity of free space

Change in the electric flux through the surface outlined by the loop

$$\sum B_{\parallel}\Delta\ell = \mu_0\left(i_{\text{through}} + \varepsilon_0\frac{\Delta\Phi_E}{\Delta t}\right) \quad (22\text{-}8)$$

Maxwell–Ampère law (22-8)

Permeability of free space

Current through the Amperian loop

Time interval over which the change in electric flux takes place

The electric flux Φ_E through a loop, calculated using Equation 16-6, is defined in precisely the same manner as the magnetic flux Φ_B through a loop. The quantity $\varepsilon_0(\Delta\Phi_E/\Delta t)$ has units of amperes and is called the **displacement current**. Experiment confirms Equation 22-8: A circulating magnetic field can be produced by an electric current, by a changing electric field, or a combination.

We saw in Section 17-5 that we can write the permittivity of free space as $\varepsilon_0 = 8.85 \times 10^{-12}$ F/m. Since 1 F = 1 C/V and 1 A = 1 C/s, we can write this as

$$\varepsilon_0 = 8.85 \times 10^{-12}\,\frac{\text{F}}{\text{m}} = 8.85 \times 10^{-12}\,\frac{\text{C}}{\text{V}\cdot\text{m}} = 8.85 \times 10^{-12}\,\frac{\text{A}\cdot\text{s}}{\text{V}\cdot\text{m}} \quad (22\text{-}9)$$

We'll make use of this equation in the following example.

EXAMPLE 22-3 A Magnetic Field Due to a Changing Electric Field

A parallel-plate capacitor like that shown in Figure 22-8b has circular plates 5.00 cm in diameter. The electric field between the plates increases by 8.00×10^5 V/m in 1.00 s, inducing a magnetic field. What is the magnitude of that magnetic field around a loop 3.50 cm in diameter in the space between the capacitor plates?

Set Up

No charge actually moves through the loop in question, so $i_{\text{through}} = 0$ in Equation 22-8. However, the electric flux through the loop changes as the electric field changes, so the term $\Delta\Phi_E/\Delta t$ in Equation 22-8 is not zero, and a circulating magnetic field will result. This example is very similar to Example 22-2, except that now we need to calculate the change in *electric* flux to determine the magnitude of the circulating *magnetic* field.

Equation 16-6 tells us the electric flux through the 3.50-cm diameter loop; the electric field $\vec{E}$ is perpendicular to the plane of the loop, so $\theta = 0$. The magnetic field has the same magnitude B all the way around the loop and is tangent to the loop, so $B_{\parallel} = B$ in Equation 22-8.

Electric flux:

$$\Phi_E = AE_{\perp} = AE\cos\theta \qquad (16\text{-}6)$$

Maxwell–Ampère law:

$$\sum B_{\parallel}\Delta\ell = \mu_0\left(i_{\text{through}} + \varepsilon_0\frac{\Delta\Phi_E}{\Delta t}\right) \qquad (22\text{-}8)$$

Permittivity of free space:

$$\varepsilon_0 = 8.85 \times 10^{-12}\ \frac{\text{A}\cdot\text{s}}{\text{V}\cdot\text{m}} \qquad (22\text{-}9)$$

Permeability of free space:

$$\mu_0 = 4\pi \times 10^{-7}\ \text{T}\cdot\text{m/A}$$

Area of a circle of radius r:

$$A = \pi r^2$$

Circumference of a circle of radius r:

$$C = 2\pi r$$

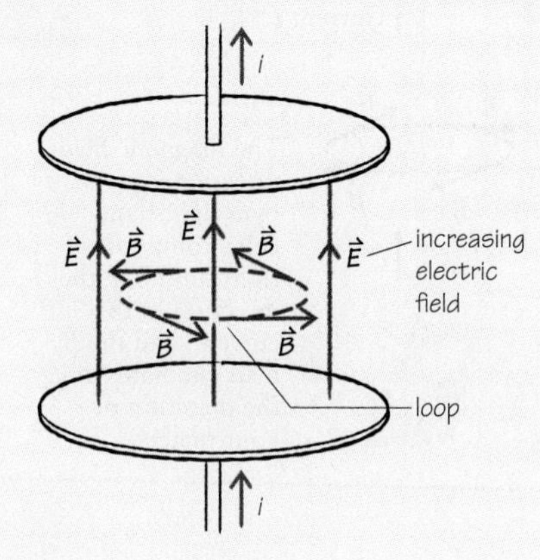

Solve

Find the displacement current $\varepsilon_0(\Delta\Phi_E/\Delta t)$ associated with the change in electric flux through the loop.

The change in electric flux in a time $\Delta t = 1.00$ s is equal to the area of the loop multiplied by the change in electric field in that time (recall that $\theta = 0$). The loop has radius $r = (1/2) \times (3.50\ \text{cm}) = 1.75\ \text{cm} = 1.75 \times 10^{-2}$ m, so

$$\begin{aligned}\Delta\Phi_E &= A(\Delta E)\cos 0 = A(\Delta E)(1)\\ &= (\pi)(1.75 \times 10^{-2}\ \text{m})^2(8.00 \times 10^5\ \text{V/m})(1)\\ &= 7.70 \times 10^2\ \text{V}\cdot\text{m}\end{aligned}$$

The displacement current through the loop is

$$\begin{aligned}\varepsilon_0\frac{\Delta\Phi_E}{\Delta t} &= \left(8.85 \times 10^{-12}\ \frac{\text{A}\cdot\text{s}}{\text{V}\cdot\text{m}}\right)\left(\frac{7.70 \times 10^2\ \text{V}\cdot\text{m}}{1.00\ \text{s}}\right)\\ &= 6.81 \times 10^{-9}\ \text{A}\end{aligned}$$

Use Equation 22-8 to determine the magnitude of the circulating magnetic field.

Since $i_{\text{through}} = 0$, the circulation of the magnetic field is equal to μ_0 times the displacement current:

$$\text{circulation of the magnetic field} = \sum B_{\parallel}\Delta\ell = \mu_0\left(\varepsilon_0\frac{\Delta\Phi_E}{\Delta t}\right)$$

$$= \left(4\pi \times 10^{-7}\ \frac{\text{T}\cdot\text{m}}{\text{A}}\right)(6.81 \times 10^{-9}\ \text{A}) = 8.56 \times 10^{-15}\ \text{T}\cdot\text{m}$$

Since $B_{\parallel} = B$ and B has the same value at all points around the circle, we can write the circulation as

$$\sum B_{\parallel}\Delta\ell = \sum B\ \Delta\ell = B\sum\Delta\ell$$

The sum $\sum \Delta\ell$ is the total distance around the loop, equal to the loop circumference $2\pi r$. So

$$B(2\pi r) = 8.56 \times 10^{-15}\ \mathrm{T \cdot m}$$

$$B = \frac{8.56 \times 10^{-15}\ \mathrm{T \cdot m}}{2\pi(1.75 \times 10^{-2}\ \mathrm{m})} = 7.78 \times 10^{-14}\ \mathrm{T}$$

Reflect

The induced magnetic field is very weak (7.78×10^{-14} T) because the displacement current that produces it is very small (6.81×10^{-9} A). If the electric field between the plates were to change more rapidly, the displacement current would be greater and the induced magnetic field stronger.

The displacement current is in the same direction as the current that brings positive charge to the lower plate of the capacitor. The right-hand rule for using Ampère's law (Section 19-7) tells us that the induced magnetic field is in the direction shown.

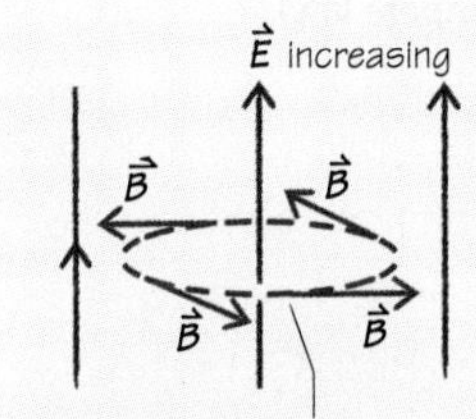

The magnetic field due to an increasing $\vec{E}$ circulates in the same direction as the electric field due to a decreasing $\vec{B}$ (see Example 22-2).

In a vacuum, where no electric currents are present, we can write Faraday's law and the Maxwell–Ampère law as

$$\sum E_{\parallel}\Delta\ell = -\frac{\Delta\Phi_B}{\Delta t}$$

$$\sum B_{\parallel}\Delta\ell = +\mu_0\varepsilon_0\frac{\Delta\Phi_E}{\Delta t} \quad (22\text{-}10)$$

Notice that the first of Equations 22-10 (Faraday's law) has a minus sign, while the second of these equations (the Maxwell–Ampère law) does not. This says that the circulating electric field produced by a *decreasing* magnetic flux (so $\Delta\Phi_B/\Delta t < 0$) is in the same direction as the circulating magnetic field produced by an *increasing* electric flux (so $\Delta\Phi_E/\Delta t > 0$). That's why the electric field that we found in Example 22-2 due to a decreasing magnetic field circulates in the same direction as the magnetic field that we found in Example 22-3 due to an increasing electric field.

What Faraday's Law and the Maxwell–Ampère Law Tell Us about Electromagnetic Waves

Faraday's law says that a varying magnetic field gives rise to a circulating electric field, and the Maxwell–Ampère law says that a varying electric field gives rise to a circulating magnetic field. Taken together, these two laws tell us how electromagnetic waves are possible and why they involve both electric and magnetic fields.

Imagine that you take an electric charge and oscillate it up and down. This will cause the electric field due to the charge to oscillate as well. By the Maxwell–Ampère law, it follows that a magnetic field will be produced which will circulate in one direction when the electric field is increasing and in the other direction when the electric field is decreasing. In other words, the magnetic field will itself oscillate. Faraday's law then tells us that this magnetic field will itself generate an electric field whose direction changes, circulating in one direction when the magnetic field is increasing and in the other direction when the magnetic field is decreasing. These two laws together tell us that a combination of an oscillating electric field and an oscillating magnetic field can sustain each other and will continue to sustain each other even after the original source of the oscillating fields—the oscillating charge—has stopped moving.

Let's look at this in a little more detail for the electromagnetic plane wave depicted in Figure 22-4a. **Figure 22-9a** shows five snapshots of the wave as it propagates from

Figure 22-9 Flux and circulation in a sinusoidal electromagnetic plane wave (a) As an electromagnetic plane wave propagates past the dashed rectangular loop, the electric field circulates around the loop when the magnetic flux through the loop changes (Faraday's law). (b) For this loop, the magnetic field circulates around the loop when the electric flux through the loop changes (the Maxwell–Ampère law).

(a) Viewing the plane of the $\vec{E}$ field

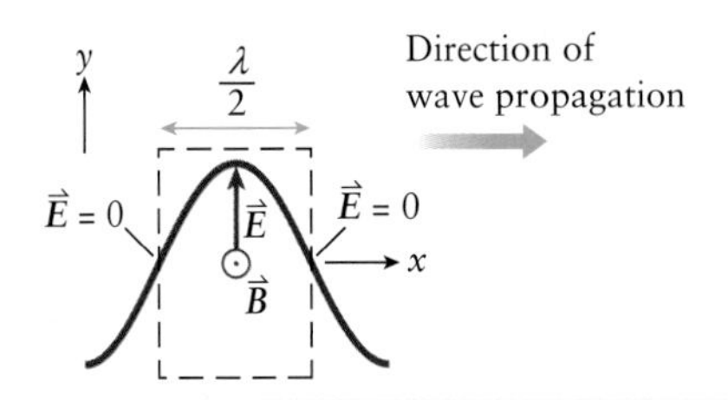

Magnetic flux through loop is maximum outward; zero circulation of $\vec{E}$.

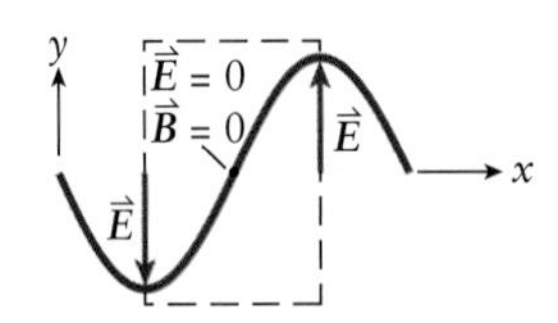

Magnetic flux through loop is changing from outward to inward; counterclockwise circulation of $\vec{E}$.

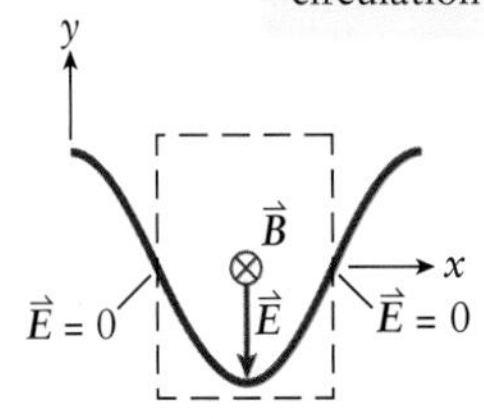

Magnetic flux through loop is maximum inward; zero circulation of $\vec{E}$.

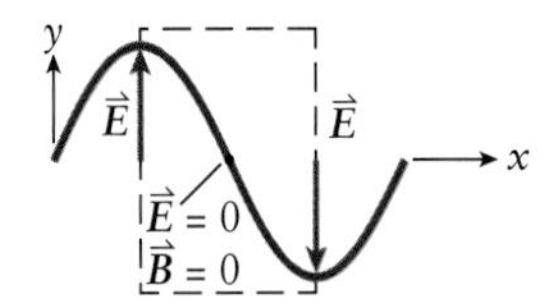

Magnetic flux through loop is changing from inward to outward; clockwise circulation of $\vec{E}$.

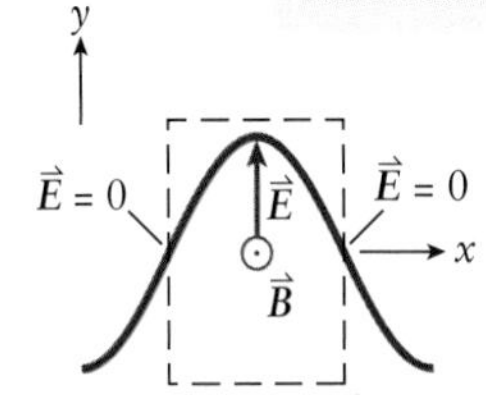

Again magnetic flux through loop is maximum outward; zero circulation of $\vec{E}$.

(b) Viewing the plane of the $\vec{B}$ field

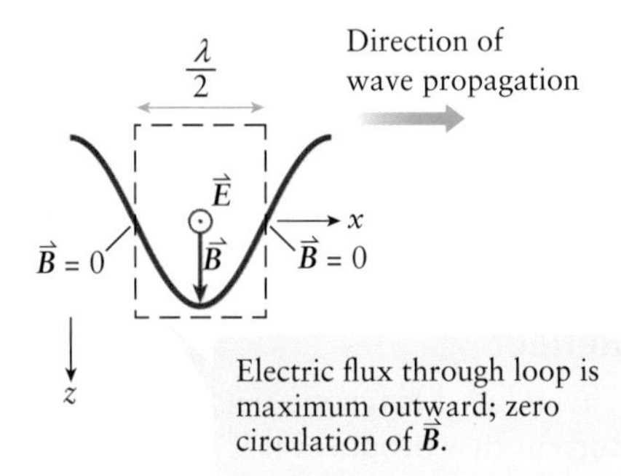

Electric flux through loop is maximum outward; zero circulation of $\vec{B}$.

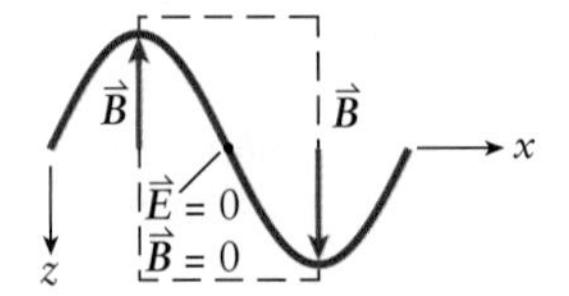

Electric flux through loop is changing from outward to inward; clockwise circulation of $\vec{B}$.

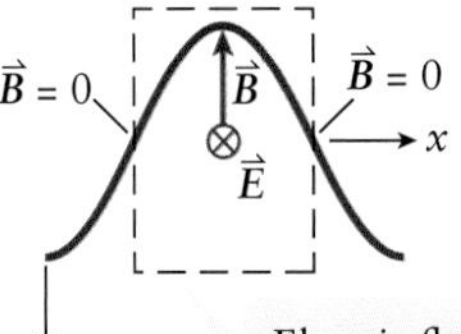

Electric flux through loop is maximum inward; zero circulation of $\vec{B}$.

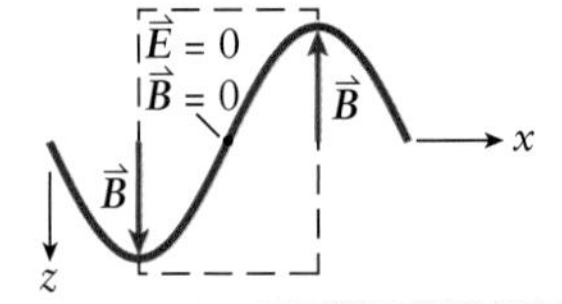

Electric flux through loop is changing from inward to outward; counterclockwise circulation of $\vec{B}$.

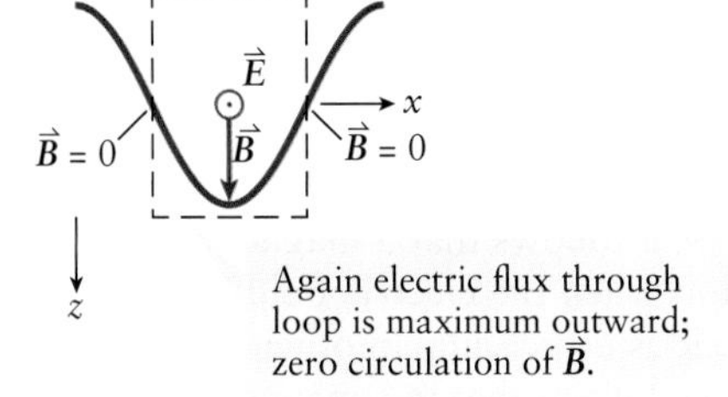

Again electric flux through loop is maximum outward; zero circulation of $\vec{B}$.

left to right in the positive x direction. In each snapshot we're looking at the xy plane, and in each snapshot we've drawn a stationary rectangular loop one-half of a wavelength in width. Because the magnetic field has only a z component, which is perpendicular to the plane of the loop, as the wave propagates there is a varying magnetic flux Φ_B through this loop. When the magnetic flux Φ_B is changing, there is a circulation of electric field around the loop; when the magnetic flux Φ_B has its maximum value either into or out of the xy plane, the flux is instantaneously not changing, and there is zero circulation of electric field. That's just what Faraday's law tells us must be true.

Figure 22-9b shows five snapshots of the wave at the same instants as in Figure 22-9a, but in these snapshots we're looking at the xz plane. The rectangular loop in this part of the figure is the same size as those in part (a), but there's a varying *electric* flux Φ_E though this loop because the electric field has only a y component and so is perpendicular to the plane of this loop. When the magnetic flux Φ_E is changing, there is a circulation of magnetic field around the loop; when the magnetic flux Φ_E has its maximum value either into or out of the xz plane, the flux is instantaneously not changing, and there is zero circulation of magnetic field. That's in agreement with the Maxwell–Ampère law. Notice also that the circulation of the electric field in Figure 22-9a is counterclockwise when the magnetic flux Φ_B through the loop is changing from outward to inward, while the circulation of the magnetic field in Figure 22-9b is counterclockwise when the electric flux Φ_E is changing from inward to outward. That's a consequence of there being a minus sign in Faraday's law but no minus sign in the Maxwell–Ampère law (see our discussion above of Equations 22-10).

Faraday's law and the Maxwell–Ampère law show why *both* the electric field and magnetic field are necessary for the propagation of a wave: The varying electric field sustains the magnetic field, and the varying magnetic field sustains the electric field. Once an electromagnetic wave is started, it will continue to propagate through a vacuum even across immense distances (Figure 22-1a). These laws also show why the electric and magnetic fields are naturally perpendicular to each other and why the two fields oscillate together with the same frequency.

Using Faraday's law and the Maxwell–Ampère law, it's possible to calculate the speed c at which such a wave should propagate through a vacuum. It turns out that the speed is determined by the values of the permittivity of free space ε_0 and the permeability of free space μ_0, both of which appear in Equations 22-10:

$$c = \frac{1}{\sqrt{\mu_0 \varepsilon_0}}$$

Substitute $\varepsilon_0 = 8.85 \times 10^{-12}\,(\mathrm{A \cdot s})/(\mathrm{V \cdot m})$ and $\mu_0 = 4\pi \times 10^{-7}\,\mathrm{T \cdot m/A}$:

$$c = \frac{1}{\sqrt{\left(4\pi \times 10^{-7}\,\frac{\mathrm{T \cdot m}}{\mathrm{A}}\right)\left(8.85 \times 10^{-12}\,\frac{\mathrm{A \cdot s}}{\mathrm{V \cdot m}}\right)}} = 3.00 \times 10^{8}\,\sqrt{\frac{\mathrm{V}}{\mathrm{T \cdot s}}}$$

These are very odd units! Note, however, that $1\ \mathrm{V} = 1\ \mathrm{N \cdot m/C}$ and $1\ \mathrm{T} = 1\ \mathrm{N \cdot s/(C \cdot m)}$, so

$$1\sqrt{\frac{\mathrm{V}}{\mathrm{T \cdot s}}} = 1\sqrt{\left(\frac{\mathrm{N \cdot m}}{\mathrm{C}}\right)\left(\frac{\mathrm{C \cdot m}}{\mathrm{N \cdot s}}\right)\left(\frac{1}{\mathrm{s}}\right)} = 1\sqrt{\frac{\mathrm{m}^2}{\mathrm{s}^2}} = 1\,\frac{\mathrm{m}}{\mathrm{s}}$$

The above expression for the speed of an electromagnetic wave then becomes

Speed of light in a vacuum

$$c = \frac{1}{\sqrt{\mu_0 \varepsilon_0}} = 3.00 \times 10^{8}\ \mathrm{m/s}$$

Permeability of free space | Permittivity of free space

Speed of light in a vacuum (22-11)

This agrees with Equation 22-1 for the speed of light in a vacuum. The same mathematical analysis that leads to Equation 22-11 also relates the amplitudes E_0 and B_0 of the electric and magnetic fields in the wave: The result is that $B_0 = E_0/c$, the same result that we stated in Section 22-2 (Equation 22-4).

In the following section we'll use our insight into the nature of electromagnetic waves to help us analyze the energy carried by such waves.

WATCH OUT! The speed of light in a medium other than a vacuum is less than *c*.

As we noted in Section 22-2, electromagnetic waves propagate at speed c only in a vacuum. We determined the value of c in Equation 22-11 using ε_0, the permittivity of free space, and μ_0, the permeability of free space; "free space" is equivalent to "in a vacuum." Different materials have different values of permittivity and permeability, which is why the speed of light is different in them.

GOT THE CONCEPT? 22-2 Maxwell's Equations

According to Maxwell's equations, which of the following situations is *possible*? (a) A closed surface that has a net outward magnetic flux through its surface; (b) a closed surface that has a net outward electric flux through its surface; (c) a loop that has zero electric flux through the interior of the loop and has a magnetic field that circulates around the loop; (d) more than one of (a), (b), and (c); (e) none of (a), (b), or (c).

TAKE-HOME MESSAGE FOR Section 22-3

✔ Maxwell's equations are the four basic equations of electromagnetism.

✔ Gauss's law for the electric field says that the net electric flux through a closed surface is proportional to the charge enclosed by that surface. Gauss's law for the magnetic field says that the net magnetic flux through any closed surface must be zero.

✔ Gauss's laws explain why an electromagnetic wave in a vacuum must be a transverse wave.

✔ Faraday's law says that a circulating electric field is produced around a loop by a time-varying magnetic flux through the loop. The Maxwell–Ampère law states that a circulating magnetic field is produced around a loop by a current through the loop or a time-varying electric flux through the loop.

✔ Faraday's law and the Maxwell–Ampère law explain how an electromagnetic wave sustains itself and why the electric and magnetic fields are mutually perpendicular.

22-4 Electromagnetic waves carry both electric and magnetic energy, and come in packets called photons

Electromagnetic waves carry energy. You can feel the energy delivered to your skin by sunlight (one kind of electromagnetic wave) on a sunny day. A microwave oven is useful for cooking because the water molecules found in food of all kinds absorb the energy in microwaves, which are at wavelengths between infrared and radio (Figure 22-2). And the energy in a laser beam is so tightly concentrated that it can be used as a surgical tool (Figure 22-1b).

The energy in an electromagnetic wave is actually contained within the electric and magnetic fields themselves. To see what this means, let's return to the physics of capacitors (Sections 17-5 and 17-6) and inductors (Sections 21-4 and 21-5). These devices will give us insight into the energy content of electromagnetic waves.

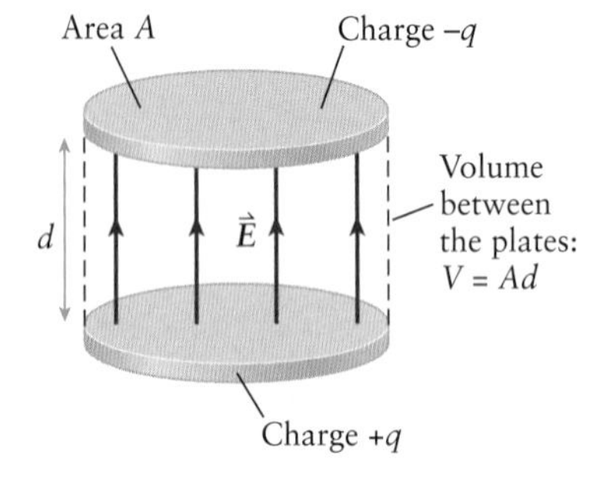

Figure 22-10 Calculating electric energy density The energy stored in this charged capacitor can be thought of as residing in the electric field $\vec{E}$ between the plates.

Energy in Electric and Magnetic Fields

Figure 22-10 shows a parallel-plate capacitor with plates of area A separated by a distance d. If the two plates are closely spaced, the electric field between the plates is approximately uniform and fills the volume between the plates (that is, there is very little field outside this volume). With a charge $+q$ on one plate and a charge $-q$ on the other plate, the magnitude of this field is

$$E = \frac{q}{\varepsilon_0 A} \tag{17-10}$$

It takes work to separate the charges $+q$ and $-q$, and this work goes into the electric potential energy U_E stored in the capacitor. The amount of stored energy is

$$U_E = \frac{q^2}{2C} \tag{17-17}$$

where C, the capacitance of the capacitor, is

$$C = \frac{\varepsilon_0 A}{d} \tag{17-13}$$

We can think of the electric energy in the capacitor as being stored in the electric field itself. To motivate this idea, let's substitute Equation 17-13 into Equation 17-17 and rearrange:

$$U_E = \frac{q^2}{2}\left(\frac{d}{\varepsilon_0 A}\right) = \frac{1}{2}\varepsilon_0\left(\frac{q^2}{\varepsilon_0^2 A^2}\right)Ad = \frac{1}{2}\varepsilon_0\left(\frac{q}{\varepsilon_0 A}\right)^2 Ad \tag{22-12}$$

The quantity $q/(\varepsilon_0 A)$ in parentheses on the far right-hand side of Equation 22-12 is just the electric field magnitude E from Equation 17-10. The quantity Ad is the volume V that the electric field occupies. So we can rewrite Equation 22-12 as

$$U_E = \left(\frac{1}{2}\varepsilon_0 E^2\right)V \tag{22-13}$$

Equation 22-13 tells us that the energy stored in the capacitor is equal to the volume V occupied by the electric field multiplied by a quantity $(1/2)\varepsilon_0 E^2$ that depends on the electric field magnitude E. This quantity has units of energy per volume (J/m^3) and is called the **electric energy density:**

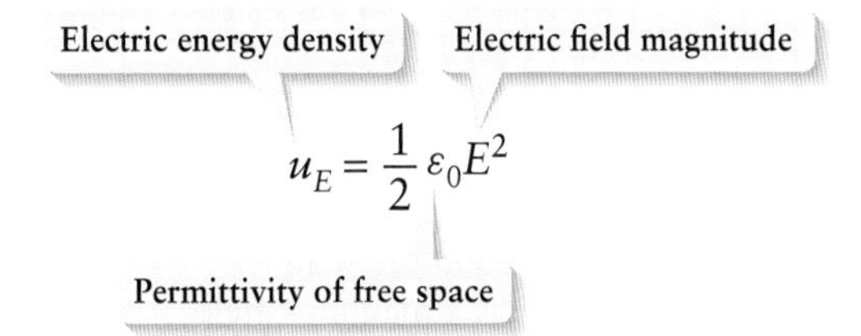

$$u_E = \frac{1}{2}\varepsilon_0 E^2$$

Electric energy density (22-14)

Equation 22-14 says that wherever there is an electric field, there is energy. This motivates the idea that the energy stored in a capacitor is stored in the field itself. We derived Equation 22-14 for the special case of a parallel-plate capacitor, but it turns out to be valid in *any* situation where an electric field is present.

We can come to a similar conclusion about the magnetic energy stored in an inductor like the one shown in **Figure 22-11**. Table 21-1 (Section 21-5) tells us that if an inductor has inductance L and carries a current i, the magnetic energy stored in the inductor is

$$U_B = \frac{1}{2}Li^2 \tag{22-15}$$

For an inductor like that in Figure 22-11, which is a long solenoid with N turns of wire, length ℓ, and cross-sectional area A, the inductance is

$$L = \frac{\mu_0 N^2 A}{\ell} \tag{21-14}$$

Like the electric field inside the parallel-plate capacitor in Figure 22-10, the magnetic field inside the solenoid in Figure 22-11 is nearly uniform and confined to the volume inside the solenoid. From Equation 19-17 the magnitude of this magnetic field is $B = \mu_0 ni$, where n is the number of turns of wire per meter. This is just the total

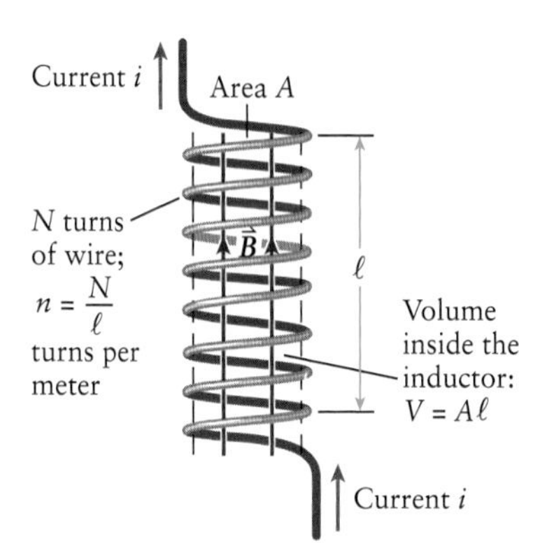

Figure 22-11 Calculating magnetic energy density The energy stored in this current-carrying inductor can be thought of as residing in the magnetic field $\vec{B}$ inside the inductor.

number of turns N divided by the length ℓ of the solenoid: $n = N/\ell$. So we can write the magnitude of the magnetic field inside the solenoid as

$$B = \frac{\mu_0 N i}{\ell} \tag{22-16}$$

Let's substitute Equation 21-14 for L into Equation 22-15 for the energy in the solenoid, then rearrange:

$$U_B = \frac{1}{2}\left(\frac{\mu_0 N^2 A}{\ell}\right)i^2 = \frac{1}{2\mu_0}\left(\frac{\mu_0^2 N^2 i^2}{\ell^2}\right)A\ell = \frac{1}{2\mu_0}\left(\frac{\mu_0 N i}{\ell}\right)^2 A\ell \tag{22-17}$$

The quantity $\mu_0 Ni/\ell$ in parentheses on the far right of Equation 22-17 is the magnitude B of the magnetic field inside the solenoid. The quantity $A\ell$ is the volume V inside the solenoid, which is also the volume that the magnetic field occupies. So Equation 22-17 becomes

$$U_B = \left(\frac{B^2}{2\mu_0}\right)V \tag{22-18}$$

Compare this to Equation 22-13 for the electric energy U_E. We see that according to Equation 22-18, the energy stored in the inductor equals the volume V occupied by the magnetic field multiplied by a quantity $B^2/(2\mu_0)$ that depends on the magnetic field magnitude B and has units J/m^3. We call $B^2/(2\mu_0)$ the **magnetic energy density**:

Magnetic energy density

Magnetic energy density | Magnetic field magnitude

$$u_B = \frac{B^2}{2\mu_0} \tag{22-19}$$

Permeability of free space

Just as Equation 22-14 tells us that there is electric energy wherever there is an electric field, Equation 22-19 tells us that there is magnetic energy wherever there is a magnetic field. This is true for an inductor, so we can think of the energy stored in a current-carrying inductor as being stored in the magnetic field within the inductor. But Equation 22-19 is valid *wherever* there is a magnetic field.

Energy in an Electromagnetic Plane Wave

Let's apply Equations 22-14 and 22-19 for the electric and magnetic energy densities to the sinusoidal electromagnetic plane wave that we introduced in Section 22-2. From Equation 22-3 the electric field of this plane wave has only a y component and the magnetic field has only a z component:

$$\begin{aligned} E_y(x,t) &= E_0 \cos(kx - \omega t + \phi) \\ B_z(x,t) &= B_0 \cos(kx - \omega t + \phi) \end{aligned} \tag{22-3}$$

Substituting these into Equations 22-14 and 22-19, we find that the electric and magnetic energy densities in the plane wave are

$$\begin{aligned} u_E &= \frac{1}{2}\varepsilon_0 E^2 = \frac{1}{2}\varepsilon_0 E_0^2 \cos^2(kx - \omega t + \phi) \\ u_B &= \frac{B^2}{2\mu_0} = \frac{B_0^2}{2\mu_0}\cos^2(kx - \omega t + \phi) \end{aligned} \tag{22-20}$$

Note that u_E and u_B both depend on position x and time t in the same way. It may appear from Equation 22-20 that there are different amounts of energy in the electric and magnetic forms, since the coefficients $(1/2)\varepsilon_0 E_0^2$ and $B_0^2/2\mu_0$ are different. But we know from Equations 22-4 and 22-11 that $B_0 = E_0/c$ and $c = 1/\sqrt{\mu_0\varepsilon_0}$, so

$$\frac{B_0^2}{2\mu_0} = \frac{(E_0/c)^2}{2\mu_0} = \frac{(E_0\sqrt{\mu_0\varepsilon_0})^2}{2\mu_0} = \frac{1}{2}\left(\frac{\mu_0\varepsilon_0}{\mu_0}\right)E_0^2 = \frac{1}{2}\varepsilon_0 E_0^2 \tag{22-21}$$

In other words, the coefficients of u_E and u_B in Equation 22-20 are equal. This means that at any position x and at any time t, an electromagnetic wave in vacuum has *equal* amounts of electric energy density and magnetic energy density:

$$u_E = u_B = \frac{1}{2}\varepsilon_0 E_0^2 \cos^2(kx - \omega t + \phi) = \frac{B_0^2}{2\mu_0}\cos^2(kx - \omega t + \phi) \tag{22-22}$$

The *total* energy density u in the wave is the sum of u_E and u_B. Equation 22-22 tells us that $u_E = u_B$, so u is equal to $2u_E$ or $2u_B$:

$$u = u_E + u_B = \varepsilon_0 E_0^2 \cos^2(kx - \omega t + \phi) = \frac{B_0^2}{\mu_0}\cos^2(kx - \omega t + \phi) \tag{22-23}$$

The value of u at any position varies with time, so it's often more useful to state its *average* value. The average value of the cosine function squared is $1/2$, so

Average energy density in an electromagnetic wave | Electric field magnitude | Electric field rms value | Magnetic field magnitude | Magnetic field rms value

$$u_{\text{average}} = \frac{1}{2}\varepsilon_0 E_0^2 = \varepsilon_0 E_{\text{rms}}^2 = \frac{B_0^2}{2\mu_0} = \frac{B_{\text{rms}}^2}{\mu_0}$$

Permittivity of free space | Permeability of free space

Average energy density in an electromagnetic wave (22-24)

In Equation 22-24 we've used the root-mean-square (rms) values of the oscillating electric and magnetic fields, $E_{\text{rms}} = E_0/\sqrt{2}$ and $B_{\text{rms}} = B_0/\sqrt{2}$ (see Section 21-2).

An even more useful way to express the energy carried by an electromagnetic wave is in terms of the wave **intensity**, or average power per unit area. **Figure 22-12** shows a portion of a wave that has cross-sectional area A and length ℓ. The energy in this portion of the wave is u_{average} from Equation 22-24 multiplied by the volume $A\ell$ of this portion. This entire portion of the wave moves at speed c through the cross-sectional area A in a time t. The power equals the energy in the wave portion divided by the time $t = \ell/c$ that it takes this portion of the wave to travel at speed c through the cross-sectional area A. The intensity, to which we give the symbol S_{average}, equals the power divided by the area A. So

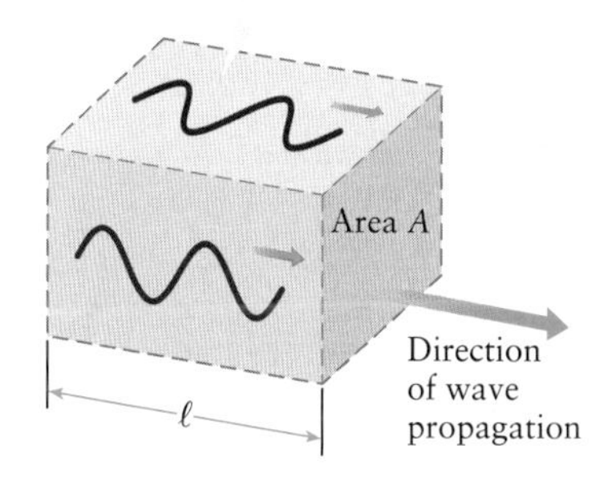

Figure 22-12 Calculating wave intensity The intensity of an electromagnetic wave equals the amount of wave energy that crosses an area A per unit time, divided by the area A.

$$S_{\text{average}} = \frac{\text{energy}}{\text{time}} \times \frac{1}{\text{area}} = \frac{(u_{\text{average}} A\ell)}{\ell/c} \times \frac{1}{A} = u_{\text{average}}\, c$$

Using Equation 22-24, we can rewrite this as

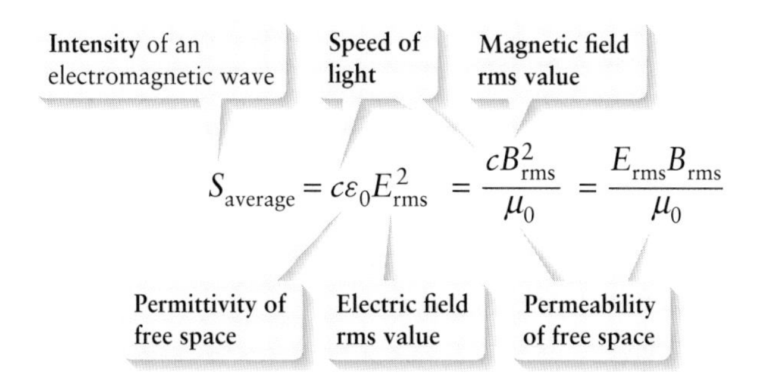

$$S_{\text{average}} = c\varepsilon_0 E_{\text{rms}}^2 = \frac{cB_{\text{rms}}^2}{\mu_0} = \frac{E_{\text{rms}} B_{\text{rms}}}{\mu_0}$$

Intensity of an electromagnetic wave (22-25)

(To write the last expression in Equation 22-25, we used the result that since $B_0 = E_0/c$, it follows that $B_{\text{rms}} = E_{\text{rms}}/c$.)

The units of intensity are watts per square meter, or W/m^2. The intensity of sunlight that reaches Earth is about $1.36 \times 10^3\ \text{W/m}^2$. As the following example shows, the intensities of other common electromagnetic waves can be much smaller.

EXAMPLE 22-4 Energy Density and Intensity in a Radio Wave

In Example 22-1 (Section 22-2) we considered an FM radio wave of frequency 98.7 MHz with an electric field amplitude 6.00×10^{-2} V/m. We found that the magnetic field has amplitude 2.00×10^{-10} T. Calculate the rms values of the electric and magnetic field, the average energy density in the wave, and the wave intensity.

Set Up

Each rms value is just equal to the amplitude divided by $\sqrt{2}$. Given the rms values, we'll use Equation 22-24 to calculate the energy density and Equation 22-25 to calculate the intensity.

Root-mean-square values:

$$E_{\text{rms}} = \frac{E_0}{\sqrt{2}}$$

$$B_{\text{rms}} = \frac{B_0}{\sqrt{2}}$$

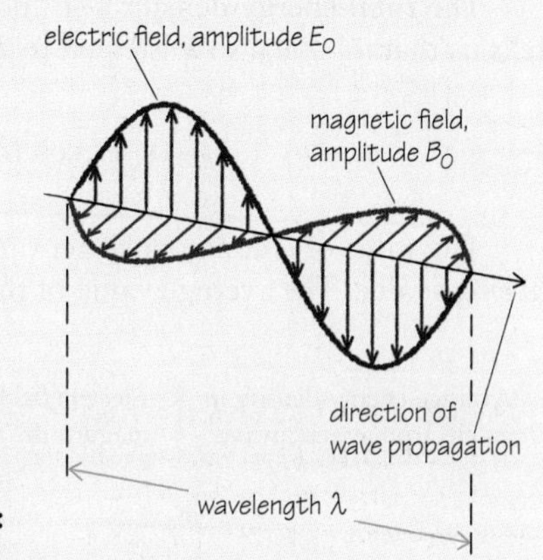

Average energy density in an electromagnetic wave:

$$u_{\text{average}} = \varepsilon_0 E_{\text{rms}}^2 = \frac{B_{\text{rms}}^2}{\mu_0} \quad \text{(22-24)}$$

Intensity of an electromagnetic wave:

$$S_{\text{average}} = c\varepsilon_0 E_{\text{rms}}^2 = \frac{cB_{\text{rms}}^2}{\mu_0} = \frac{E_{\text{rms}}B_{\text{rms}}}{\mu_0} \quad \text{(22-25)}$$

Permittivity of free space:

$$\varepsilon_0 = 8.85 \times 10^{-12} \frac{\text{C}}{\text{V} \cdot \text{m}} \quad \text{(22-9)}$$

Permeability of free space:

$$\mu_0 = 4\pi \times 10^{-7}\,\text{T} \cdot \text{m/A}$$

Solve

Calculate the rms values of the electric and magnetic fields.

We are given $E_0 = 6.00 \times 10^{-2}$ V/m and $B_0 = 2.00 \times 10^{-10}$ T. The corresponding rms values are

$$E_{\text{rms}} = \frac{E_0}{\sqrt{2}} = \frac{6.00 \times 10^{-2}\,\text{V/m}}{\sqrt{2}} = 4.24 \times 10^{-2}\,\text{V/m}$$

$$B_{\text{rms}} = \frac{B_0}{\sqrt{2}} = \frac{2.00 \times 10^{-10}\,\text{T}}{\sqrt{2}} = 1.41 \times 10^{-10}\,\text{T}$$

Use the value of E_{rms} to calculate the average energy density in the wave.

From Equation 22-24

$$u_{\text{average}} = \varepsilon_0 E_{\text{rms}}^2 = \left(8.85 \times 10^{-12} \frac{\text{C}}{\text{V} \cdot \text{m}}\right)\left(4.24 \times 10^{-2} \frac{\text{V}}{\text{m}}\right)^2$$

$$= 1.59 \times 10^{-14} \frac{\text{C} \cdot \text{V}}{\text{m}^3}$$

A coulomb times a volt is a joule: $1\,\text{C} \cdot \text{V} = 1$ J. So

$$u_{\text{average}} = 1.59 \times 10^{-14}\,\text{J/m}^3$$

Find the intensity of the wave.

Comparing Equations 22-24 and 22-25 shows that the wave intensity is c times the average energy density:

$$S_{\text{average}} = c\varepsilon_0 E_{\text{rms}}^2 = cu_{\text{average}}$$

$$= (3.00 \times 10^8\,\text{m/s})(1.59 \times 10^{-14}\,\text{J/m}^3)$$

$$= 4.78 \times 10^{-6} \frac{\text{J}}{\text{m}^2 \cdot \text{s}}$$

A joule per second is a watt: 1 J/s = 1 W. So

$$S_{\text{average}} = 4.78 \times 10^{-6}\,\text{W/m}^2$$

Reflect

The energy density and intensity are both very small quantities. It's a testament to the sensitivity of radio receivers that such a wave is quite easy to detect.

We can check our results by using the alternative expressions for u_{average} and S_{average} given in Equations 22-24 and 22-25. As an example, here's a check on the value of S_{average}.

From Equation 22-25,

$$S_{\text{average}} = \frac{E_{\text{rms}}B_{\text{rms}}}{\mu_0} = \frac{(4.24 \times 10^{-2}\ \text{V/m})(1.41 \times 10^{-10}\ \text{T})}{4\pi \times 10^{-7}\ \text{T}\cdot\text{m/A}}$$

$$= 4.78 \times 10^{-6}\ \frac{\text{V}\cdot\text{A}}{\text{m}^2}$$

One volt times one ampere is one watt ($1\ \text{V}\cdot\text{A} = 1\ \text{W}$), so

$$S_{\text{average}} = 4.78 \times 10^{-6}\ \text{W/m}^2$$

This agrees with our calculation above, as it must.

Photons

Equation 22-25 says that the intensity of an electromagnetic wave depends on the strength of the electric and magnetic fields that make up the wave but not on the wave frequency. (The frequency f doesn't appear anywhere in this equation.) But everyday experience suggests that wave frequency *does* play a role in the energy carried by an electromagnetic wave. As an example, ultraviolet light can trigger a chemical reaction in the skin that causes a suntan or sunburn, but visible light cannot. (That's why sunscreen contains a substance that allows visible light to pass but blocks ultraviolet light.) X rays, with even higher frequency than ultraviolet light, can pierce soft tissue but not bone; as a result, an x-ray image allows a physician to diagnose a broken bone and a dentist to see cavities in your teeth. Gamma rays, with higher frequency than x rays, can damage DNA, cause cancer, and even kill cells. How can we explain these differences?

The explanation is that the energy of an electromagnetic wave propagates as small packets called **photons**. The energy of an individual photon is proportional to the wave frequency, and the proportionality constant h is called **Planck's constant:**

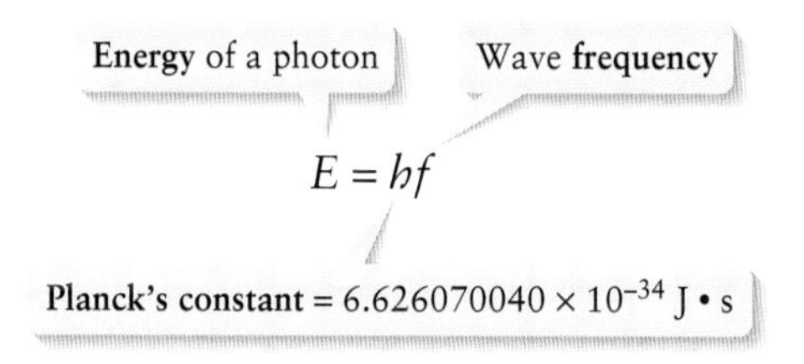

Energy of a photon (22-26)

To three significant figures $h = 6.63 \times 10^{-34}\ \text{J}\cdot\text{s}$.

Equation 22-26 explains why ultraviolet light causes a suntan or sunburn, but light in the visible spectrum does not. In order for tanning or burning to take place in your skin, individual molecules must absorb a certain minimum amount of energy from light to trigger a chemical change. A given molecule must absorb this energy in the form of a single photon, and a visible-light photon lacks sufficient energy to trigger this chemical change. An ultraviolet photon, by contrast, can trigger the change because it has a shorter wavelength, a higher frequency, and more energy per photon. An x-ray photon is more energetic still, which is why it is able to penetrate soft tissue.

Gamma-ray photons, x-ray photons, and short-wavelength, high-frequency ultraviolet photons have enough energy that they can dislodge an electron from an atom. Such **ionizing radiation** breaks apart molecules by pulling electrons from the chemical bonds that hold atoms together. Although ionizing radiation can directly break DNA molecules in living tissue, it's more likely to disrupt some other, more common molecule, such as water, to create highly reactive free radicals that then damage DNA. Depending on the severity of the damage, the cell may be able to recover. However, if the damage cannot be repaired, or if it is repaired incorrectly, the resulting mutations may be lethal. These effects generally go unnoticed until the next time the cell tries to divide. Because cancerous cells divide more frequently than most other cells in the body, they are more susceptible to radiation damage than most healthy cells.

Planck's constant is very small, so a single photon carries only a miniscule amount of energy. As an example, a photon of red light with wavelength $\lambda = 750$ nm $= 7.50 \times 10^{-7}$ m has frequency $f = c/\lambda = (3.00 \times 10^8 \text{ m/s})/(7.50 \times 10^{-7} \text{ m}) = 4.00 \times 10^{14}$ Hz and energy $E = hf = (6.63 \times 10^{-34} \text{ J} \cdot \text{s})(4.00 \times 10^{14} \text{ Hz}) = 2.65 \times 10^{-19}$ J. (Recall that 1 Hz $=$ 1 s^{-1}.) That's so small that you don't notice individual photons in the light from a lamp, just as you don't notice individual air molecules in a breeze against your face.

It's common to express photon energies in electron volts (eV). We introduced this unit in Section 17-6: 1 eV $= 1.60 \times 10^{-19}$ J. For a red photon of wavelength 750 nm, the energy is $E = (2.65 \times 10^{-19} \text{ J})/(1.60 \times 10^{-19} \text{ J/eV}) = 1.66$ eV. As the following example shows, radio photons have even less energy.

EXAMPLE 22-5 Photons in a Radio Wave

For the radio wave of Examples 22-1 and 22-4, calculate (a) the energy per photon, (b) the number of photons per cubic meter, and (c) the number of photons per second that strike a receiver antenna of area 10.0 cm^2.

Set Up

The radio wave has frequency 98.7 MHz $=$ 98.7×10^6 Hz. We'll use this and Equation 22-26 to determine the energy of a single radio photon. From Example 22-4 we know that the energy density (energy per unit volume) of the wave is $u_{\text{average}} = 1.59 \times 10^{-14} \text{ J/m}^3$ and the intensity (energy per area per time) is $S_{\text{average}} = 4.78 \times 10^{-6} \text{ W/m}^2$. We'll use these and our calculated value of the photon energy to determine the number of photons per unit volume and the number of photons striking the antenna per time.

Energy of a photon:

$$E = hf \qquad (22\text{-}26)$$

antenna, area 10.0 cm^2

wave

Solve

(a) Calculate the energy of an individual photon of frequency 98.7 MHz.

From Equation 22-26

$$E = hf = (6.63 \times 10^{-34} \text{ J} \cdot \text{s})(98.7 \times 10^6 \text{ Hz})$$
$$= 6.54 \times 10^{-26} \text{ J} \cdot \text{s} \cdot \text{Hz}$$

Since 1 Hz $= 1$ s^{-1}, 1 J $\cdot$ s $\cdot$ Hz $=$ 1 J and so

$$E = 6.54 \times 10^{-26} \text{ J or}$$

$$E = \frac{6.54 \times 10^{-26} \text{ J}}{1.60 \times 10^{-19} \text{ J/eV}} = 4.08 \times 10^{-7} \text{ eV}$$

This is much smaller than the energy of a visible-light photon (2.65×10^{-19} J $=$ 1.66 eV for red light) because the frequency of the radio wave is much less than the frequency of visible light (4.00×10^{14} Hz for red light).

(b) Calculate the photon density (number of photons per unit volume).

The energy density in the wave is $u_{\text{average}} = 1.59 \times 10^{-14} \text{ J/m}^3$ and the energy per photon is $E = 6.54 \times 10^{-26}$ J/photon. The number of photons per unit volume is

$$\frac{\text{photons}}{\text{volume}} = \frac{\text{energy}}{\text{volume}} \times \frac{\text{photon}}{\text{energy}} = \frac{\left(\dfrac{\text{energy}}{\text{volume}}\right)}{\left(\dfrac{\text{energy}}{\text{photon}}\right)}$$

$$= \frac{1.59 \times 10^{-14} \text{ J/m}^3}{6.54 \times 10^{-26} \text{ J/photon}} = 2.43 \times 10^{11} \text{ photons/m}^3$$

Each cubic meter of this wave contains 2.43×10^{11} (243 billion) photons.

(c) Calculate the number of photons that strike an area of 10.0 cm^2 in 1.00 s.

The intensity (energy per area per time) of the wave is $S_{average} = 4.78 \times 10^{-6}\ W/m^2$, so the rate at which energy arrives at the antenna of area $A = 10.0\ cm^2$ is

$$\frac{\text{energy}}{\text{area} \cdot \text{time}} \times \text{area} = S_{average}A$$

$$= (4.78 \times 10^{-6}\ W/m^2)(10.0\ cm^2)\left(\frac{1\ m}{100\ cm}\right)^2$$

$$= 4.78 \times 10^{-9}\ W = 4.78 \times 10^{-9}\ J/s$$

The rate at which photons arrive at the antenna is

$$\frac{\text{photons}}{\text{time}} = \frac{\text{energy}}{\text{time}} \times \frac{\text{photon}}{\text{energy}} = \frac{\left(\frac{\text{energy}}{\text{time}}\right)}{\left(\frac{\text{energy}}{\text{photon}}\right)}$$

$$= \frac{4.78 \times 10^{-9}\ J/s}{6.54 \times 10^{-26}\ J/\text{photon}} = 7.30 \times 10^{16}\ \text{photons/s}$$

In one second 7.30×10^{16} (73 quadrillion) photons arrive at the antenna.

Reflect

Even this relatively low-intensity wave contains a tremendous number of photons per cubic meter and delivers an astronomical number of photons per second to a receiver.

Our results give us insight into the concerns that some people have expressed about mobile phones causing cancer. The frequencies used by mobile phones are about 7 to 27 times higher (about 700 to 2700 MHz), but even those frequencies correspond to very low photon energies (about 2.9×10^{-6} to 1.1×10^{-5} eV). As we described above, the sort of ionizing radiation that can cause cancer has very high frequency and very high photon energy of several electron volts. The photon energies associated with mobile phones are millions of times smaller, which should be far too low to have any carcinogenic effect.

WATCH OUT! Photons are both particles and waves.

A common *incorrect* way to think about photons is to visualize them as small particles like miniature marbles and to imagine that a large number of photons acting together behave like a wave. The reality is far different! Each individual photon has aspects of *both* wave and particle, and those particles are very different in character from ordinary objects such as marbles. We'll learn more about the curious properties of photons in Chapter 26.

GOT THE CONCEPT? 22-3 Photons

Three lasers have equal power output. The first emits a pure violet light, the second emits a pure green light, and the third emits a pure red light. Which laser emits the greater number of photons per second? (a) The violet laser; (b) the green laser; (c) the red laser; (d) all emit the same number of photons per second; (e) answer depends on the value of the power output.

TAKE-HOME MESSAGE FOR Section 22-4

✔ Energy is associated with both electric and magnetic fields. An electromagnetic wave in a vacuum has equal amounts of electric energy and magnetic energy per volume.

✔ The intensity of an electromagnetic wave is the average power per unit area. In a vacuum, the intensity equals the average energy per volume in the wave multiplied by the speed of light *c*.

✔ The energy of an electromagnetic wave comes in packets called photons. The energy of a single photon is proportional to the wave frequency.

Key Terms

displacement current
electric energy density
electromagnetic spectrum
electromagnetic wave
Faraday's law
Gauss's law for the electric field
Gauss's law for the magnetic field
intensity
ionizing radiation
magnetic energy density
Maxwell–Ampère law
Maxwell's equations
photon
Planck's constant
sinusoidal plane wave
speed of light
visible light

Chapter Summary

Topic	Equation or Figure	
Speed of electromagnetic waves: In a vacuum all electromagnetic waves propagate at the same speed $c = 3.00 \times 10^8$ m/s. The shorter the wavelength, the higher the frequency of the wave. Our eyes are sensitive to only a narrow band of wavelengths known as the visible spectrum.	$c = f\lambda$ (**Speed of light in a vacuum**; **Frequency** of an electromagnetic wave; **Wavelength** of the wave in vacuum)	(22-2)
Electromagnetic plane waves: The simplest electromagnetic wave in vacuum is a sinusoidal plane wave. The electric and magnetic fields oscillate in phase, are perpendicular to each other, and are transverse (both are perpendicular to the direction of wave propagation). The amplitude B_0 of the magnetic field equals the amplitude E_0 of the electric field divided by the speed of light c.	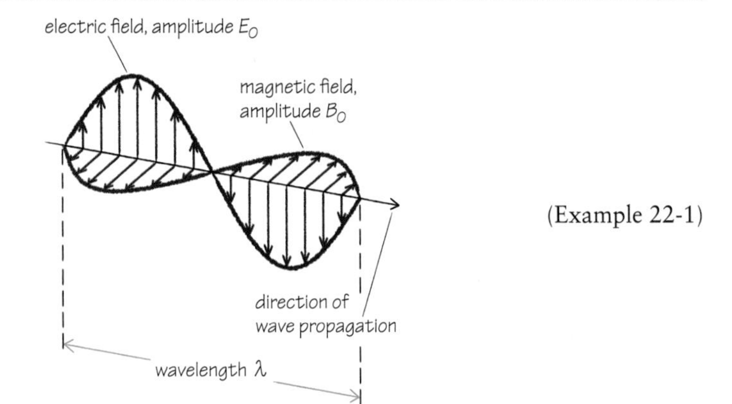	(Example 22-1)
Maxwell's equations: Four equations govern the behavior of electric and magnetic fields in all situations. The two Gauss's laws explain why electromagnetic waves in vacuum are transverse. Faraday's law and the Maxwell–Ampère law explain how the oscillations of the magnetic field produce the electric field and how the oscillations of the electric field produce the magnetic field. As a result, the wave must have both electric and magnetic aspects, and the electric and magnetic fields are naturally perpendicular to each other. Maxwell's equations also successfully predict the value of the speed of light in terms of the permittivity of free space ε_0 and the permeability of free space μ_0.	$\Phi_E = \frac{q_{encl}}{\varepsilon_0}$ (**Electric flux** through a closed surface; Net amount of **charge enclosed** within the surface; **Permittivity of free space** = $1/(4\pi k)$)	(16-9)
	$\Phi_B = 0$ (**Magnetic flux** through a closed surface; There is **zero magnetic flux through any closed surface,** no matter what the size or shape of the surface or what it contains.)	(22-5)
	$\sum E_{\parallel}\Delta\ell = -\frac{\Delta\Phi_B}{\Delta t}$ (**Circulation of electric field** around a loop; **Change in the magnetic flux** through the surface outlined by the loop; **Time interval** over which the change in magnetic flux takes place)	(22-7)

Circulation of magnetic field around an Amperian loop; Permittivity of free space; Change in the electric flux through the surface outlined by the loop

$$\sum B_{||}\Delta \ell = \mu_0\left(i_{\text{through}} + \varepsilon_0 \frac{\Delta \Phi_E}{\Delta t}\right) \quad (22\text{-}8)$$

Permeability of free space; Current through the Amperian loop; Time interval over which the change in electric flux takes place

Electric and magnetic field energy in electromagnetic waves: The energy density (energy per volume) associated with an electric field is $u_E = (1/2)\varepsilon_0 E^2$, and the energy density associated with a magnetic field is $u_B = B^2/2\mu_0$. In an electromagnetic wave in vacuum, there are equal amounts of electric energy and magnetic energy. The average energy density in an electromagnetic wave can be expressed in terms of either the field amplitudes or their rms values. The intensity of the wave is the average power per unit area of the wave.

Average energy density in an electromagnetic wave; Electric field magnitude; Electric field rms value; Magnetic field magnitude; Magnetic field rms value

$$u_{\text{average}} = \frac{1}{2}\varepsilon_0 E_0^2 = \varepsilon_0 E_{\text{rms}}^2 = \frac{B_0^2}{2\mu_0} = \frac{B_{\text{rms}}^2}{\mu_0} \quad (22\text{-}24)$$

Permittivity of free space; Permeability of free space

Intensity of an electromagnetic wave; Speed of light; Magnetic field rms value

$$S_{\text{average}} = c\varepsilon_0 E_{\text{rms}}^2 = \frac{cB_{\text{rms}}^2}{\mu_0} = \frac{E_{\text{rms}}B_{\text{rms}}}{\mu_0} \quad (22\text{-}25)$$

Permittivity of free space; Electric field rms value; Permeability of free space

Photons: The energy of an electromagnetic wave comes in packets called photons that have properties of both wave and particle. The higher the wave frequency, the more energy there is per photon.

Energy of a photon; Wave frequency

$$E = hf \quad (22\text{-}26)$$

Planck's constant $= 6.626070040 \times 10^{-34}$ J • s

Answer to What do you think? Question

(d) As we describe in Section 22-2, in a vacuum all varieties of electromagnetic wave propagate at the same speed. Visible light has a much shorter wavelength (λ = about 400 to 700 nm) than radio waves (λ = a few centimeters to several meters) and so has a much higher frequency.

Answers to Got the Concept? Questions

22-1 (e) No matter what the frequency of an electromagnetic wave, its propagation speed in a vacuum is the same: the speed of light $c = 3.00 \times 10^8$ m/s.

22-2 (d) Situation (a) is impossible: Gauss's law for the magnetic field states that the net magnetic flux through any closed surface must be zero. Situation (b) is possible; according to Gauss's law for the electric field, there will be a net outward electric flux through a closed surface that encloses positive charge. Situation (c) is also possible: One example is the loop around the current-carrying wire in Figure 22-8a. There is zero electric field and hence zero electric flux through the interior of the loop, but the current gives rise to a magnetic field that circulates around the loop.

22-3 (c) All three lasers emit the same amount of electromagnetic wave energy per second. However, the energy per photon is different because each laser emits light of a different wavelength and frequency. Figure 22-2 shows that violet has the shortest wavelength and so the highest frequency and highest photon energy; red has the longest wavelength and so the lowest frequency and lowest photon energy. To emit the same amount of energy per second, the red laser must emit more of its low-energy photons per second than the other lasers.

Questions and Problems

In a few problems you are given more data than you actually need; in a few other problems you are required to supply data from your general knowledge, outside sources, or informed estimate.

Interpret as significant all digits in numerical values that have trailing zeros and no decimal points.

For all problems use $g = 9.80 \text{ m/s}^2$ for the free-fall acceleration due to gravity. Neglect friction and air resistance unless instructed to do otherwise.

- • Basic, single-concept problem
- •• Intermediate-level problem; may require synthesis of concepts and multiple steps
- ••• Challenging problem
- SSM *Solution is in Student Solutions Manual*
- Example *See worked example for a similar problem.*

Conceptual Questions

1. • (a) Rank the following electromagnetic waves from the lowest to the highest wavelength: (a) microwaves, (b) red light, (c) ultraviolet light, (d) infrared light, and (e) gamma rays. (b) Which wavelength has the highest energy per photon? Which has the lowest?

2. • Changing electric fields create changing magnetic fields. These oscillating fields propagate at the speed of light. Describe the orientation (directions) of the fields and the velocity of the electromagnetic wave. SSM

3. • Describe how the frequency at which the changing electric and magnetic fields oscillate in electromagnetic waves is related to the speed of light.

4. • Does a wire connected to a dc source, such as a battery, emit an electromagnetic wave?

5. • James Clerk Maxwell is credited with compiling the three laws of electricity and magnetism (Gauss's law, Ampère's law, and Faraday's law), adding his own law (also called Gauss's law for magnetism), modifying Ampère's law, and understanding the connections between electricity, magnetism, and optics. Were Maxwell's efforts more important than the individual discoveries of Gauss, Ampère, and Faraday? Explain your answer.

6. • Match the equations that were conceived of by Carl Friedrich Gauss, André Ampère, Michael Faraday, and James Clerk Maxwell with the corresponding written statements:

i. $\sum B_{\parallel} \Delta \ell = \mu_0 \left(i_{\text{through}} + \varepsilon_0 \frac{\Delta \Phi_E}{\Delta t} \right)$

ii. $\sum E_{\parallel} \Delta \ell = -\frac{\Delta \Phi_B}{\Delta t}$

iii. $\Phi_B = 0$

iv. $\Phi_E = \frac{q_{\text{encl}}}{\varepsilon_0}$

A. One source of an electric field is an electric charge.
B. There are no magnetic monopoles.
C. Changing magnetic fields induce changing electric fields.
D. Changing electric fields induce changing magnetic fields.

7. • The energy of an ultraviolet light photon is unrelated to the speed of the fundamental electromagnetic waves that make up such radiation. Explain how this is possible. SSM

8. • Name three types of electromagnetic energy that you used today.

9. • Describe how the frequency of an electromagnetic wave is related to the energy of the photons of that wave.

Multiple-Choice Questions

10. • In comparison to x rays in vacuum, visible light in vacuum has
A. a speed that is faster.
B. wavelengths that are longer.
C. wavelengths that are equal.
D. wavelengths that are shorter.
E. frequencies that are equal.

11. • X rays and gamma rays in vacuum
A. have the same frequency.
B. have the same wavelength.
C. have the same speed.
D. have the same "color."
E. None of the above SSM

12. • In comparison to radio waves in vacuum, visible light in vacuum has
A. a speed that is faster.
B. wavelengths that are longer.
C. wavelengths that are equal.
D. wavelengths that are shorter.
E. frequencies that are equal.

13. • Which of the following requires a physical medium through which to travel?
A. radio waves
B. light
C. x rays
D. sound
E. gamma rays SSM

14. • In an *RC* circuit the capacitor begins to discharge. In the region of space between the plates of the capacitor,
A. there is an electric field but no magnetic field.
B. there is a magnetic field but no electric field.
C. there are both electric and magnetic fields.
D. there are no electric and magnetic fields.
E. there is an electric field whose strength is one-half that of the magnetic field.

15. • Maxwell's equations apply
A. to both electric fields and magnetic fields that are constant over time.
B. only to electric fields that are time-dependent.
C. only to magnetic fields that are constant over time.
D. to both electric fields and magnetic fields that are time-dependent.
E. to both time-independent and time-dependent electric and magnetic fields. SSM

16. • The phase difference between the electric and magnetic fields in an electromagnetic wave is
A. 90°.
B. 180°.
C. 0°.
D. alternately 90° and 180°.
E. alternately 0° and 90°.

Estimation/Numerical Analysis

17. • Estimate the number of photons per second emitted inside a household microwave oven.

18. • Estimate the average wavelength of radio waves that are received by (a) an AM radio and (b) an FM radio.

19. • Estimate the photon energy associated with the visible light (green) to which the human eye is most sensitive. SSM

20. • Suppose human colonists embark on a journey to the nearest star from our sun on a spaceship that travels at a speed of $0.01c$. Estimate how many generations will pass before the colonists arrive.

21. • Estimate the wavelength of electromagnetic radiation that would potentially be classified as ionizing radiation.

22. • Estimate the number of photons that are emitted in the 1000-h lifetime of a 100-W light bulb.

23. • A television station sends a live broadcast up to a satellite in geostationary orbit, which bounces the signal back down to a viewer's home three time zones away from the station. Estimate the shortest possible time delay in the broadcast, assuming the sender and receiver are both on the equator and the satellite is directly above the halfway point between them.

24. The wavelength of light roughly determines the level of detail you can see with it. Estimate the minimum energy (in eV) of a photon required to see a virus. Compare this energy to the ionization energy of a typical atom.

25. ••• Use a computer-based, graphical program to plot graphs of the electric field (E) versus time (t) and the magnetic field (B) versus time (t) on a three-dimensional graph (see the table). Let t be on the x axis, E be on the y axis, and B be on the z axis. Explain the shape of the graph.

E (N/C)	B ($\times 10^{-9}$ T)	$t(s)$
100	333	0
70.7	236	$\pi/4$
0	0	$\pi/2$
−70.7	−236	$3\pi/4$
−100	−333	π
−70.7	−236	$5\pi/4$
0	0	$3\pi/2$
70.7	236	$7\pi/4$
100	333	2π
70.7	236	$9\pi/4$
0	0	$5\pi/2$
−70.7	−236	$11\pi/4$
−100	−333	3π
−70.7	−236	$13\pi/4$
0	0	$7\pi/2$
70.7	236	$15\pi/4$
100	333	4π
70.7	236	$17\pi/4$
0	0	$9\pi/2$
−70.7	−236	$19\pi/4$
−100	−333	5π
−70.7	−236	$21\pi/4$
0	011	$\pi/2$
70.7	236	$23\pi/4$
100	333	6π

Problems

22-1 Light is just one example of an electromagnetic wave

22-2 In an electromagnetic plane wave electric and magnetic fields both oscillate

26. • Calculate the wavelengths of the electromagnetic waves with the following frequencies and classify the electromagnetic radiation of each (x ray, radio, etc.). Example 22-1

A. $f = 4.14 \times 10^{15}$ Hz
B. $f = 7.00 \times 10^{14}$ Hz
C. $f = 8.00 \times 10^{16}$ Hz
D. $f = 3.00 \times 10^{13}$ Hz
E. $f = 9.00 \times 10^{12}$ Hz
F. $f = 3.44 \times 10^{17}$ Hz
G. $f = 8.23 \times 10^{15}$ Hz
H. $f = 6.00 \times 10^{15}$ Hz

27. • Calculate the wavelengths of the electromagnetic waves with the following frequencies. Example 22-1

A. $f = 7.50 \times 10^{15}$ Hz
B. $f = 6.00 \times 10^{14}$ Hz
C. $f = 5.00 \times 10^{14}$ Hz
D. $f = 4.29 \times 10^{14}$ Hz
E. $f = 7.50 \times 10^{16}$ Hz
F. $f = 2.66 \times 10^{16}$ Hz
G. $f = 8.23 \times 10^{17}$ Hz
H. $f = 6.00 \times 10^{18}$ Hz

28. • Calculate the frequencies of the electromagnetic waves that have the following wavelengths: Example 22-1

A. $\lambda = 700$ nm
B. $\lambda = 600$ nm
C. $\lambda = 500$ nm
D. $\lambda = 400$ nm
E. $\lambda = 100$ nm
F. $\lambda = 0.0333$ nm
G. $\lambda = 500$ μm
H. $\lambda = 63.3$ pm SSM

29. • Calculate the frequencies of the electromagnetic waves that have the following wavelengths: Example 22-1

A. $\lambda = 800$ nm
B. $\lambda = 650$ nm
C. $\lambda = 550$ nm
D. $\lambda = 450$ nm
E. $\lambda = 2.22$ nm
F. $\lambda = 1.10 \times 10^{-8}$ m
G. $\lambda = 50.0$ μm
H. $\lambda = 33.4$ mm

30. • The FM radio band is from 88 to 108 MHz. Calculate the corresponding range of wavelengths.

31. • The antenna for an AM radio station is a 75-m-high tower whose height is equivalent to one-quarter wavelength. At what frequency does the station transmit? Example 22-1

32. • How far does light travel in a vacuum in 10 ns? SSM

33. • How long does it take light to travel 300 km in a vacuum?

34. • How long does it take a radio signal from Earth to reach the Moon, which has an orbital radius of approximately 3.84×10^8 m?

35. •• The frequency and wavelength of an electromagnetic wave are related to the speed of light by

$$c = f\lambda$$

Starting with this expression derive a similar relationship between the speed of light, the angular frequency ($\omega = 2\pi f$), and the wave number ($k = 2\pi/\lambda$).

36. •• The magnetic field of an electromagnetic wave is given by

$$B(x,t) = (0.7\,\mu\text{T})\sin[(8\pi \times 10^6\,\text{m}^{-1})x - (2.40\pi \times 10^{15}\,\text{s}^{-1})t]$$

Calculate (a) the amplitude of the electric field, (b) the speed, (c) the frequency, (d) the period, and (e) the wavelength. Example 22-1

37. •• Suppose the electric field associated with the radio transmissions of a medical helicopter is given by

$$E(x,t) = (400\,\mu\text{N/C})\cos[(40\pi\,\text{nm}^{-1})x - (12\pi\,\text{s}^{-1})t]$$

Determine (a) the wave number, (b) the angular frequency, (c) the wavelength, and (d) the frequency of the electromagnetic wave associated with the electric field. Example 22-1

22-3 Maxwell's equations explain why electromagnetic waves are possible

38. •• The electric field in a region of space increases from 0 to 3000 N/C in 5.00. What is the magnitude of the induced magnetic field around a circular area with a diameter of 1.00 m oriented perpendicular to the electric field? Example 22-3

39. ••• A parallel-plate capacitor has closely spaced plates. Charge is flowing onto the positive plate and off the negative plate at the rate $i = \Delta q/\Delta t = 2.8$ A. What is the displacement current through the capacitor between the plates? SSM Example 22-3

40. •• An 8-cm-diameter parallel-plate capacitor has a 1.0-mm gap. The electric field between the plates is increasing at the rate 1.0×10^6 V/(m·s). What is the magnetic field strength between the plates of the capacitor a distance of 5.0 cm from the axis of the capacitor? Example 22-3

41. •• Charge flows onto the positive plate of a 6.0-cm-diameter parallel-plate capacitor at the rate $i = \Delta q/\Delta t = 1.5$ A. What is the magnetic field between the plates at a distance of 3.0 cm from the axis of the plates? Example 22-3

42. •• A ring is in a region of space that contains a uniform magnetic field directed perpendicular to the plane of the ring. The diameter of the ring is 1.5 cm. **Figure 22-13** is a graph of the magnetic field as a function of time. What is the magnitude of the electric field along the perimeter of the ring for the time intervals $t = 0$ to 4.0 ms, 4.0 to 10.0 ms, and 10.0 to 14.0 ms? Example 22-2

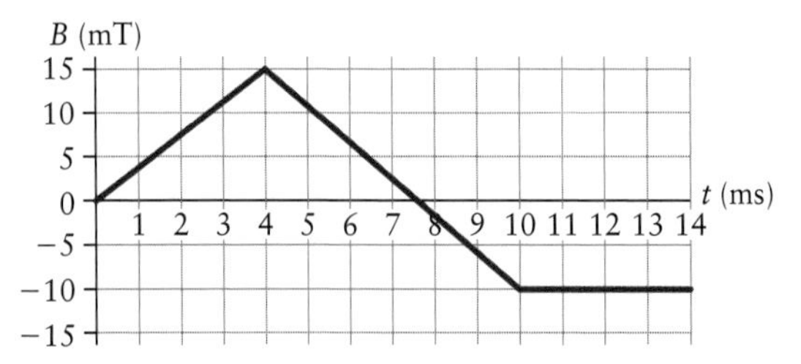

Figure 22-13 Problem 42

43. •• A 20.0-cm long solenoid, which has a radius of 1.00 cm and consists of 2500 turns of wire, has a current flowing through it that changes from 13.0 to 86.0 mA over a time of 1.20 s. What is the resulting electric field strength inside the solenoid at a distance of 0.500 cm from the axis of the solenoid? Example 22-2

22-4 Electromagnetic waves carry both electric and magnetic energy, and come in packets called photons

44. •• An electromagnetic plane wave has an intensity $S_{\text{average}} = 200$ W/m^2. (a) What are the rms values of the electric and magnetic field? (b) What are the amplitudes of the electric and magnetic fields? Example 22-4

45. •• The amplitude of an electromagnetic wave's electric field is 200 V/m. Calculate (a) the amplitude of the wave's magnetic field and (b) the intensity of the wave. Example 22-4

46. •• The rms value of an electromagnetic wave's magnetic field is 400 T. Calculate (a) the amplitude of the wave's electric field and (b) the intensity of the wave. Example 22-4

47. • Calculate the wavelengths and frequencies of photons that have the following energies. Example 22-5

A. $E_{\text{photon}} = 2.33 \times 10^{-19}$ J
B. $E_{\text{photon}} = 4.50 \times 10^{-19}$ J
C. $E_{\text{photon}} = 3.20 \times 10^{-19}$ J
D. $E_{\text{photon}} = 8.55 \times 10^{-19}$ J
E. $E_{\text{photon}} = 63.3$ eV
F. $E_{\text{photon}} = 8.77$ eV
G. $E_{\text{photon}} = 1.98$ eV
H. $E_{\text{photon}} = 4.55$ eV

48. • Calculate the wavelengths and frequencies of the photons that have the following energies. Example 22-5

A. $E_{\text{photon}} = 3.45 \times 10^{-19}$ J
B. $E_{\text{photon}} = 4.80 \times 10^{-19}$ J
C. $E_{\text{photon}} = 1.28 \times 10^{-18}$ J
D. $E_{\text{photon}} = 4.33 \times 10^{-20}$ J
E. $E_{\text{photon}} = 931$ MeV
F. $E_{\text{photon}} = 2.88$ keV
G. $E_{\text{photon}} = 7.88$ eV
H. $E_{\text{photon}} = 13.6$ eV

General Problems

49. • An important news announcement is transmitted by radio waves to people who are 300 km away and sitting next to their radios and also by sound waves to people sitting 3 m from the newscaster in a newsroom. Who receives the news first? Explain your answer. SSM

50. •• (a) Calculate the wave number of an electromagnetic wave that has an angular frequency of 6.28×10^{15} rad/s. (b) Calculate the angular frequency, the frequency, and the wavelength of a photon that has a wave number of $k = 4\pi \times 10^6$ rad/m. SSM Example 22-1

51. • **Biology** A recent study found that electrons having energies between 3.0 and 20 eV can cause breaks in a DNA molecule even though they do not ionize the molecule. If the energy were to come from light, (a) what range of wavelengths (in nanometers) could cause DNA breaks, and (b) in what part of the electromagnetic spectrum does the light lie? Example 22-5

52. •• **Medical** A dental x ray typically affects 200 g of tissue and delivers about 4.0 μJ of energy using x rays that have wavelengths of 0.025 nm. What is the energy (in electron volts) of such x-ray photons, and how many photons are absorbed during the dental x ray? Assume the body absorbs all of the incident x rays. Example 22-5

53. •• A HeNe laser produces a cylindrical beam of light with a diameter of 0.750 cm. The energy is pulsed, lasting for 1.50 ns, and each burst contains an energy of 2.00 J. (a) What is the length of each pulse of laser light? (b) What is the average energy per unit volume for each pulse? SSM Example 22-4

54. •• (a) What is the energy of a photon of green light that has a wavelength of 525 nm? Give your answer in joules and electron volts. (b) What is the wave number of the photon? Example 22-5

23 Wave Properties of Light

123dartist/Shutterstock

What do you think?

Diamonds are renowned for how they reflect light and produce a rainbow of colors. Diamonds have these properties because the speed of light in diamond (a) is faster than in air; (b) is slower than in air; (c) depends on the color of the light; (d) both (a) and (c); (e) both (b) and (c).

In this chapter, your goals are to:

- (23-1) Describe some key properties of light.
- (23-2) Explain Huygens' principle and what it tells us about the laws of reflection and refraction.
- (23-3) Recognize the special circumstances under which total internal reflection can take place.
- (23-4) Explain how a prism is able to break white light into its component colors.
- (23-5) Calculate how the intensity of light is affected by passing through a polarizing filter.
- (23-6) Use the idea of path length difference to calculate what happens in thin-film interference.
- (23-7) Explain two-slit interference in terms of the wave properties of light.
- (23-8) Explain why light spreads out when it passes through a narrow opening.
- (23-9) Calculate how the angular resolution of an optical device is limited by diffraction.

To master this chapter, you should review:

- (13-5) Constructive and destructive interference of waves
- (22-2) Plane waves and the electromagnetic nature of light
- (22-4) Photons

23-1 The wave nature of light explains much about how light behaves

In Chapter 22 we explored how light is an electromagnetic wave, with electric and magnetic fields that oscillate in phase. In this chapter we'll explore several of the consequences of the wave nature of light. For most of this exploration we won't need the details about electric and magnetic fields; what's important is simply that in many cases light can be treated as a wave. As a result, many of the properties of light waves that we'll encounter apply equally well to sound and other types of waves.

We'll begin by introducing *Huygens' principle*, a simplified model that describes how waves propagate through space. We'll use Huygens' principle to explain what happens when light is reflected (**Figure 23-1a**). Light waves travel at different speeds in different transparent materials, and we'll see how Huygens' principle explains

Figure 23-1 Light waves in nature (a) Reflection, (b) dispersion, and (c) interference are among the many phenomena that light waves exhibit in the natural world.

Cats can see even in very low light levels thanks to reflection by a layer called the *tapetum lucidum* at the back of each eye. Incoming light that isn't absorbed by the retina reflects straight back, and some of the reflected light is detected on the second pass.
(a)
David Tauck

The colors of the rainbow are caused by dispersion: The speed of light in water depends on the frequency of the light. As a result, each color of sunlight follows a different path as it enters a raindrop and undergoes refraction, reflects off the back of the raindrop, and undergoes refraction again as it exits the raindrop.
(b)
iStockphoto/Thinkstock

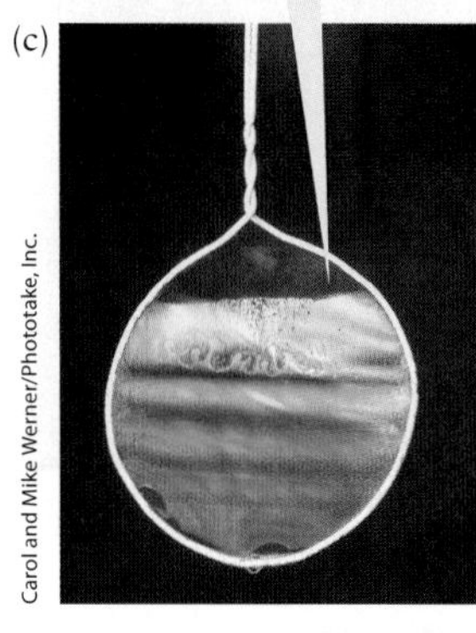

The colors of this soap film are caused by interference. Some light reflects from the front surface of the film, and some enters the film and reflects from the back surface. If the wavelength is just right, the two waves interfere constructively and produce a bright band.
(c)
Carol and Mike Werner/Phototake, Inc.

TAKE-HOME MESSAGE FOR Section 23-1

✔ Many of the key properties of light can be understood simply by using the idea that light can be treated as a wave.

refraction—the bending of light when it moves from one transparent material to another. We'll see that in certain circumstances light can be trapped inside a transparent material, just as if that material had mirrored surfaces. This effect, called *total internal reflection*, is essential for the medical technique of endoscopy. We'll see that the speed of light in a transparent material also depends on the wavelength of the light. This phenomenon, called *dispersion*, explains the vivid colors of a rainbow (**Figure 23-1b**).

One aspect of light that depends on its being a transverse wave is its *polarization*, which describes how the electric field vector of the wave is oriented. We'll see how light can become polarized by scattering or reflection, and we'll examine how polarizing filters work and why they're used in sunglasses.

We'll look at the phenomenon of *interference*, in which two light waves can add together constructively or destructively. Interference explains the colors seen in a thin film of soapy water (**Figure 23-1c**), as well as why cats and other animals have reflective eyes (see Figure 23-1a). We'll finish with a discussion of *diffraction*, an important effect in which light waves spread out when they pass through a small aperture.

George Resch/Fundamental Photographs

Figure 23-2 Reflection and refraction When a beam of incident light in air strikes the surface of water at an angle θ_1 from the normal, some light is reflected at the same angle ($\theta_1' = \theta_1$) and some light goes into the water at a different angle θ_2.

23-2 Huygens' principle explains the reflection and refraction of light

Light travels, or propagates, in a straight line if the material through which the light travels—called the *medium* for the light—is uniform in its properties. But the direction in which light propagates changes when the light strikes a *boundary* between two different media, such as that between air and water (**Figure 23-2**). In general, some of the light reflects off the boundary, while the remainder travels into the second material at a different angle. The **law of reflection** states that the angle of the reflected light is the same as the angle of the incoming, or **incident**, light: $\theta_1' = \theta_1$. **Refraction** is the change in direction of the light that travels into the second medium: the angle θ_2 for the refracted light is not equal to the angle θ_1 for the incident light.

WATCH OUT! In reflection and refraction, angles are always measured from the normal to the boundary between two media.

Note that in Figure 23-2 the angles θ_1, θ_1', and θ_2 are all measured relative to the **normal** to the boundary, which is a line perpendicular to the boundary at the point where the incident light hits the boundary. A common mistake is to measure these angles relative to the boundary itself. If you make that mistake, you'll end up getting the wrong answer when you use the formulas that we'll derive in this section!

Why is the law of reflection true? And what determines how the direction of the light changes when it travels from one medium into another? We'll answer both of these questions using a model of waves introduced by the seventeenth-century Dutch scientist Christiaan Huygens. (Huygens' model precedes by two centuries Maxwell's complete description of light as an electromagnetic wave, but is consistent with it.) We'll see that what determines the difference between the incident and refracted angles are the *speeds* at which light travels in the two media. Refraction isn't just for visible light but occurs for waves of all kinds: Radio waves, sound waves, and water waves may refract when crossing from one medium to another under the right circumstances.

Huygens' Principle and Reflection

Huygens considered waves that travel in two dimensions (like ripples on the surface of a pond) or in three dimensions (like light waves). He suggested that each point on a wave crest, or **front**, at time t can be treated as a source of tiny **wavelets** that themselves move at the speed of the wave (**Figure 23-3**). The wave front at a later time $t + \Delta t$ is then the superposition of all of these wavelets emitted at time t, and is tangent to the leading edges of the wavelets. This idea is called **Huygens' principle**. As Figure 23-3 shows, Huygens' principle helps explain how circular waves in water retain their shape as they propagate outward from a splash in a pond and how light waves spread out in spherical wave fronts from a light source.

To analyze what's happening to the light beams in Figure 23-2, let's apply Huygens' principle to a *plane* wave like the ones we introduced in Section 22-2. A plane wave propagates in a single direction, so it is a good description of the light beams shown in Figure 23-2. For each beam in Figure 23-2, the **ray** is an arrow that points in the direction of light propagation. Note that the ray is always perpendicular to the wave front.

Figure 23-4a shows how to apply Huygens' principle to understand the law of reflection. A plane wave with wave front ABC is directed at an angle toward the boundary between medium 1 (say, air) and medium 2 (say, water or glass). At time t point A on the wave front has just arrived at the boundary. A time Δt later, the wavelets from points A, B, and C have each spread outward by a distance $v_1 \Delta t$, where v_1 is the speed at which waves propagate in medium 1. The wavelets from points B and C propagate forward through medium 1, while the wavelet from point A is reflected at the boundary and so propagates *backward* from the boundary into medium 1. (We haven't drawn the wavelets that propagate into medium 2. We'll return to those a little later to help us understand refraction.)

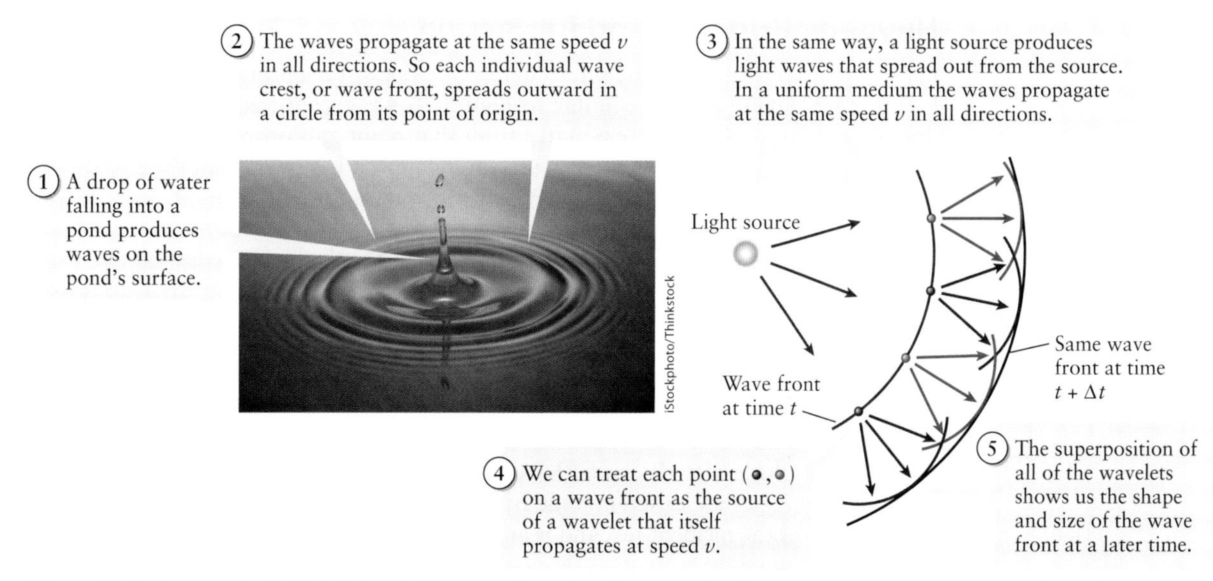

Figure 23-3 Huygens' principle This principle provides a simple way to visualize wave propagation in terms of wavelets.

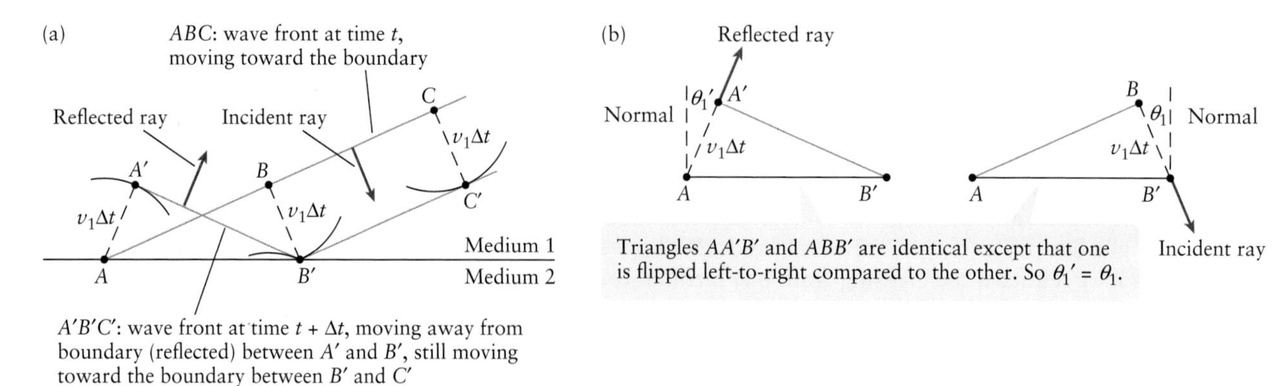

Figure 23-4 Huygens' principle and reflection (a) A wave front reflected at a boundary between two media. (b) Finding the law of reflection, $\theta_1' = \theta_1$.

See the Math Tutorial for more information on trigonometry.

If we draw a new wave front that's tangent to the leading edges of the wavelets that emanate from points A, B, and C, the result is $A'B'C'$ in Figure 23-4a. The wave front from B' to C' is still propagating toward the boundary; this represents the light still incident on the boundary. But the wave front from A' to B' is propagating away from the boundary and so represents the light reflected from the boundary.

The line BB' in Figure 23-4a is perpendicular to the incident wave front and so points in the direction of the incident ray. Likewise, the line AA' is perpendicular to the reflected wave front and so points in the direction of the reflected ray. To see how these directions are related to each other, notice that triangles ABB' and $AA'B'$ are both right triangles, both have the same hypotenuse of length AB', and both have one side of length $v_1\Delta t$ (**Figure 23-4b**). These two right triangles are identical, except that triangle $AA'B'$ has been flipped left-to-right compared to triangle ABB'. So the angle θ_1' of the line AA' measured from the vertical (that is, from the normal to the boundary) must be the same as the angle θ_1 of the line BB' measured from the vertical. We conclude that

The law of reflection for light waves at a boundary (23-1)

When light reflects at the boundary between two media, the **angle of the reflected ray** from the normal...

$$\theta_1' = \theta_1$$

...is equal to the **angle of the incident ray** from the normal.

This is just the law of reflection that we mentioned above. We'll use Equation 23-1 extensively in Chapter 24 when we study the properties of mirrors.

Huygens' Principle and Refraction

Let's now use Huygens' principle to determine the direction of the refracted ray in Figure 23-2. **Figure 23-5a** is similar to Figure 23-4a, except that for point A we've drawn only the wavelet that emanates from that point and propagates into medium 2. We've assumed that the wave speed v_2 in medium 2 is slower than the speed v_1 in medium 1. So in a time Δt the wavelet that propagates into medium 2 travels a short distance $v_2\Delta t$ while the wavelets in medium 1 travel a longer distance $v_1\Delta t$.

If we again draw a new wave front that's tangent to the leading edges of the wavelets from A, B, and C, the result is $A''B'C'$. As in Figure 23-4a, the wave front from B' to C' represents incident light in medium 1 that is still propagating toward the boundary at speed v_1. The wave front from A'' to B' represents *refracted* light that is propagating in medium 2 at speed v_2. The angle of the wave front, and hence the angle of the ray that's perpendicular to the wave front, has changed because the wave speed has changed.

In **Figure 23-5b** we've redrawn the right triangles ABB' and $AA''B'$ from Figure 23-5a. Both triangles have the same hypotenuse of length AB', but the angles of the two triangles are different. Side BB' of triangle ABB' has length $v_1\Delta t$, points in the direction of the incident ray, and is at an angle θ_1 from the normal to the boundary.

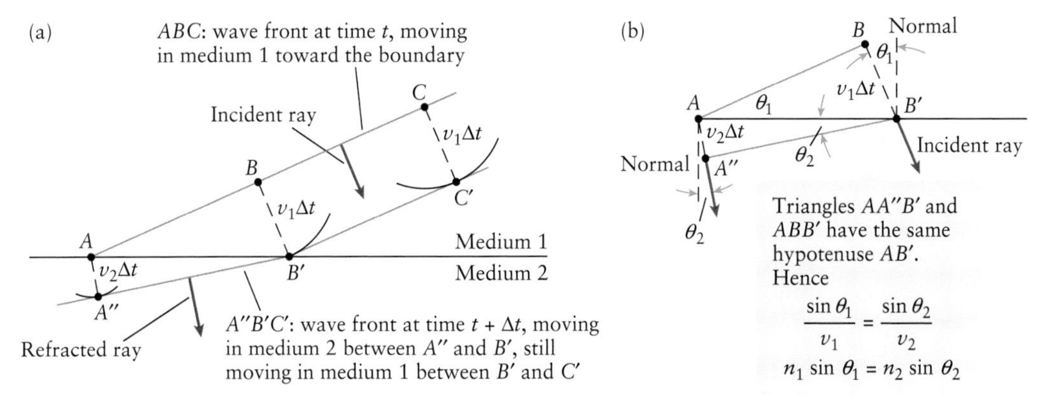

Figure 23-5 Huygens' principle and refraction (a) A wave front refracted at a boundary between two media. (b) Finding the law of refraction.

Because ABB' is a right triangle, you can see that the angle of side AB from the horizontal is also θ_1, the same as the angle of side BB' from the vertical (the normal to the boundary). If you look now at triangle $AA''B'$, you'll see that side AA'' has length $v_2\Delta t$, points in the direction of the refracted ray, and is at a different angle θ_2 from the normal to the boundary. And because $AA''B'$ is a right triangle, the angle of side $A''B'$ from the horizontal is θ_2, the same as the angle of side AA'' from the normal.

Recall that the sine of an angle in a right triangle equals the length of the side opposite to that angle divided by the length of the hypotenuse. So the sine of the angle θ_1 in triangle ABB' is

$$\sin\theta_1 = \frac{\text{length of side } BB'}{\text{length of side } AB'} = \frac{v_1\Delta t}{\text{length of side } AB'} \quad \text{so} \quad \frac{\sin\theta_1}{v_1} = \frac{\Delta t}{\text{length of side } AB'} \tag{23-2}$$

Similarly the sine of the angle θ_2 in triangle $AA''B'$ is

$$\sin\theta_2 = \frac{\text{length of side } AA''}{\text{length of side } AB'} = \frac{v_2\Delta t}{\text{length of side } AB'} \quad \text{so} \quad \frac{\sin\theta_2}{v_2} = \frac{\Delta t}{\text{length of side } AB'} \tag{23-3}$$

If you compare Equations 23-2 and 23-3, you can see that

$$\frac{\sin\theta_1}{v_1} = \frac{\sin\theta_2}{v_2} \tag{23-4}$$

Equation 23-4 tells us that the relationship between the angles θ_1 and θ_2 is determined by the speeds v_1 and v_2 of the wave in medium 1 and medium 2, respectively.

It's common to express the speed of light in a given medium in terms of a quantity called the **index of refraction**:

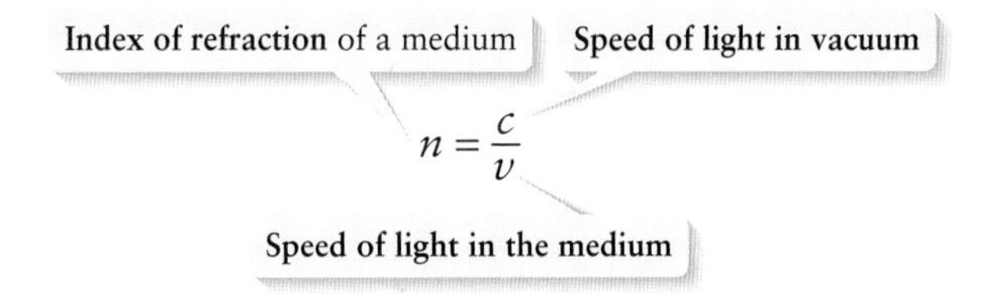

$$n = \frac{c}{v} \tag{23-5}$$

Index of refraction for light waves in a medium (23-5)

The index of refraction of vacuum is 1, since light travels at the speed of light c, so $v = c$ and $n = c/c = 1$. In any material medium light travels slower than c, so $v < c$ and $n > 1$. The greater the value of the index of refraction n in a given medium, the slower the speed v at which light propagates in that medium. **Table 23-1** lists the index of refraction of some common materials. Note that the index of refraction of air is equal to one to three significant digits, so we'll often take $n_{\text{air}} = 1$ in calculations.

Equation 23-5 tells us that $1/v = n/c$, so we can rewrite Equation 23-4 as

$$\left(\frac{n_1}{c}\right)\sin\theta_1 = \left(\frac{n_2}{c}\right)\sin\theta_2$$

If we cancel the factors of c on both sides of this equation, we get

Snell's law of refraction for light waves at a boundary (23-6)

$$n_1 \sin \theta_1 = n_2 \sin \theta_2$$

Angle of the incident ray from the normal

Angle of the refracted ray from the normal

Index of refraction for the medium with the incident light

Index of refraction for the medium with the refracted light

Go to Interactive Exercise 23-1 for more practice dealing with refraction.

Equation 23-6 is known as **Snell's law of refraction.** (This law is named for the Dutch scientist Willebrord Snellius but was in fact first discovered by the Persian scientist Ibn Sahl in 984, more than 600 years before Snellius.)

Snell's law tells us that when a ray of light crosses from one medium to another, the product of the index of refraction and the sine of the angle the ray makes to the normal remains constant. When light passes into a material of higher index of refraction—for example, from air into glass—so that the speed of light is slower in the second medium and $n_2 > n_1$, the sine of the refracted angle and the angle itself both decrease. In this case $\theta_2 < \theta_1$, and the light bends closer to the normal (**Figure 23-6a**). When light instead passes into a material of lower index of refraction—for example, from glass into air—so that the speed of light is faster in the second medium and $n_2 < n_1$, the sine of the refracted angle and the angle itself both increase. In this situation $\theta_2 > \theta_1$, and the light bends away from the normal (**Figure 23-6b**).

BIO-Medical The fraction of incident light that is reflected and the fraction that is refracted depend in part on the indices of refraction of the two media. (They also depend on the incident angle and on how the electric field vectors in the light wave are oriented relative to the boundary.) The index of refraction of a medium is often a function of the medium's density; one example is blood plasma, the density and index of refraction of which depend on the concentration of dissolved protein. Veterinarians can use this to estimate protein levels in livestock at the clinic or on the farm by measuring how much light refracts as it passes through a sample of an animal's plasma. Winemakers determine the amount of sugar in their grapes by using the same technique.

(a) Refraction from one medium into a slower one

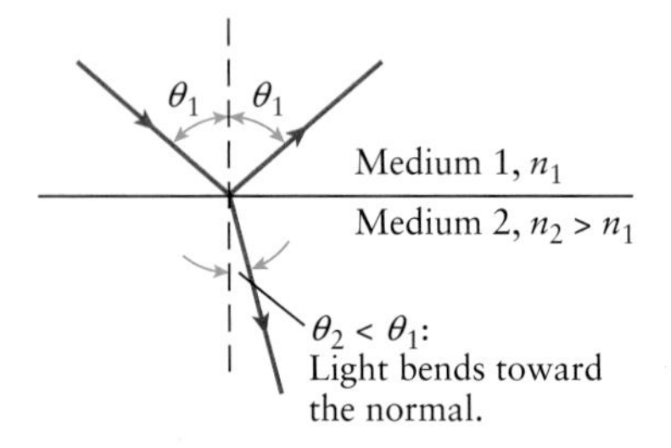

(b) Refraction from one medium into a faster one

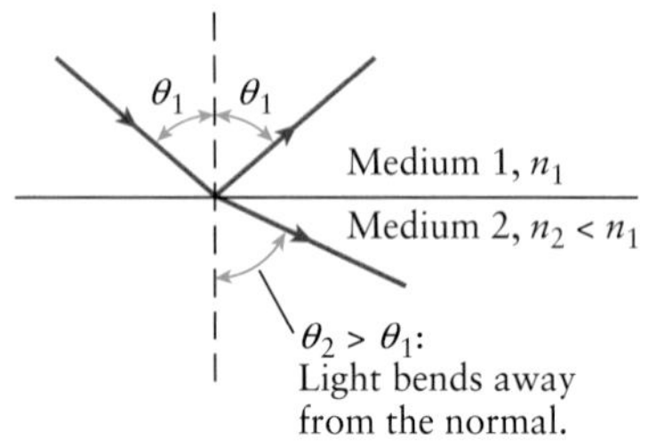

Figure 23-6 Refraction toward and away from the normal Which way the refracted ray bends depends on whether the speed of light in the second medium is (a) slower or (b) faster than in the first medium.

TABLE 23-1 Indices of Refraction

Material	Index of refraction
vacuum	1.00000
air at 20°C, 1 atm pressure	1.00029
ice	1.31
water at 20°C	1.33
acetone	1.36
ethyl alcohol	1.36
eye, cornea	1.38
eye, lens	1.41
sugar water (high concentration)	1.49
Plexiglas	1.49
typical crown glass	1.52
sodium chloride	1.54
sapphire	1.77
diamond	2.42

EXAMPLE 23-1 Seeing under Water

A surveyor (labeled S) looking at an aqueduct is just able to see the underwater edge at point F where the far wall meets the bottom (**Figure 23-7**). If the aqueduct is 4.2 m wide and her line of sight to the near, top edge at point N is 25° above the horizontal, find the depth of the aqueduct.

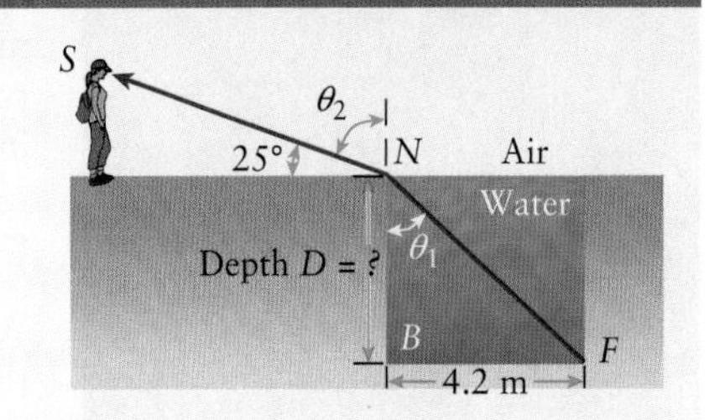

Figure 23-7 A refracted view If the surveyor just sees the bottom edge of the far wall of the aqueduct, how deep is the aqueduct?

Set Up

A light beam from point F refracts when it reaches the boundary between the water (medium 1) and the air (medium 2) at point N. Figure 23-7 shows that the angle of the refracted ray from the *normal* (shown as a vertical dashed line) is $\theta_2 = 90° - 25° = 65°$. We'll first use Snell's law to determine the angle θ_1 of the incident ray, and then use trigonometry and the given width of the aqueduct (4.2 m) to find the depth D.

Snell's law of refraction:

$$n_1 \sin \theta_1 = n_2 \sin \theta_2 \qquad (23\text{-}6)$$

Solve

Find the angle θ_1 of the incident ray using Snell's law.

Solve Equation 23-6 for the sine of the incident angle θ_1:

$$\sin \theta_1 = \frac{n_2}{n_1} \sin \theta_2$$

Figure 23-7 shows that $\theta_2 = 90° - 25° = 65°$, and from Table 23-1 we see that $n_1 = n_{\text{water}} = 1.33$ and $n_2 = n_{\text{air}} = 1.00$. So

$$\sin \theta_1 = \frac{1.00}{1.33} \sin 65° = 0.68$$

$$\theta_1 = \sin^{-1} 0.68 = 43°$$

Use trigonometry to determine the depth D of the aqueduct.

The incident ray that travels from point F to point N is the hypotenuse of a right triangle NFB with vertical dimension D and horizontal dimension 4.2 m. The side of length 4.2 m is the side opposite the angle θ_1, and the side of length D is the side adjacent to this angle. The tangent of θ_1 equals the opposite side divided by the adjacent side:

$$\tan \theta_1 = \frac{4.2 \text{ m}}{D}$$

Solve for the distance D:

$$D = \frac{4.2 \text{ m}}{\tan \theta_1} = \frac{4.2 \text{ m}}{\tan 43°} = 4.5 \text{ m}$$

The aqueduct is 4.5 m deep.

Reflect

The refracted angle $\theta_2 = 65°$ is larger than the incident angle $\theta_1 = 43°$, just as in Figure 23-6b. This makes sense, since $n_2 < n_1$ (the index of refraction for air is smaller than the index for water).

Our brains are used to the idea that light travels in straight lines. If we extend the refracted ray backwards, we see that to the surveyor's eye the light from the far edge of the bottom of the aqueduct appears to be coming from a shallower depth than $D = 4.5$ m. You can easily see this effect in a swimming pool.

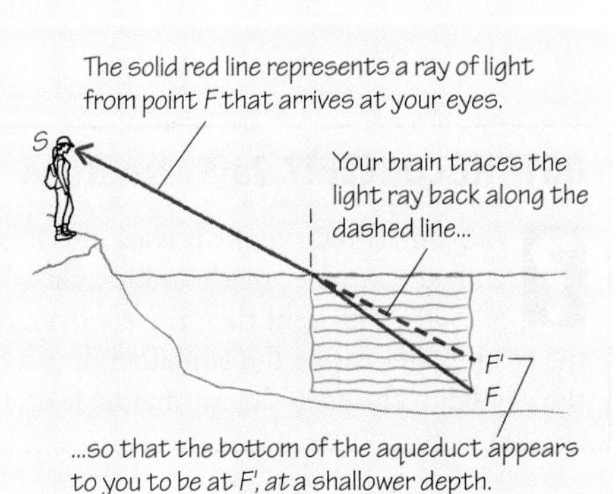

David Tauck

Figure 23-8 A refracted chopstick Due to refraction, the submerged part of the chopstick on the right appears displaced from its actual position.

The situation shown in **Figure 23-8** involves the same effect that makes the aqueduct in Example 23-1 appear shallower than it really is. When light from the submerged part of the chopstick passes from water to air at the boundary between the two, the light rays refract. As a result, it appears to our eyes that the submerged part is in a different position than its true location.

Frequency and Wavelength in Refraction

When a wave travels from one medium to another, the frequency of the wave remains the same. (In a given time interval, as many crests arrive at the boundary as leave the boundary. If this were not true, there would be a "traffic jam" of wave crests at the boundary.) However, since the wave speed is different in the two media, the wavelength must change. This follows from the relationship among the propagation speed v of the wave, the frequency f, and the wavelength λ. From Equation 13-2

$$v = f\lambda \quad \text{so} \quad \lambda = \frac{v}{f}$$

In a vacuum light waves travel at speed $v = c$, so the wavelength is

$$\lambda_{\text{vacuum}} = \frac{c}{f} \qquad (23\text{-}7)$$

In a medium with index of refraction n, the wave speed from Equation 23-5 is $v = c/n$. Then we can write the wavelength of light in a medium as

Wavelength of light in a medium (23-8)

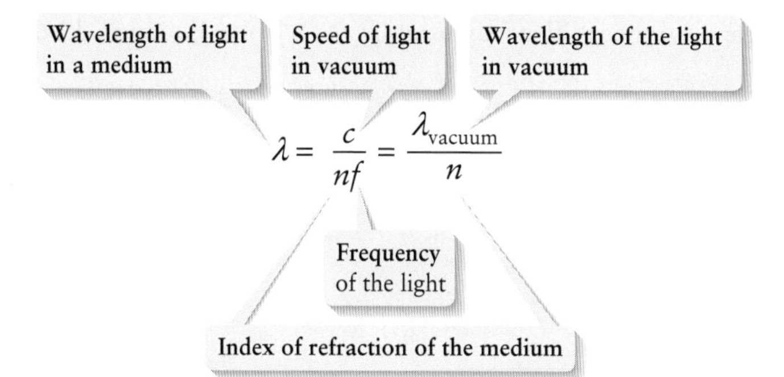

$$\lambda = \frac{c}{nf} = \frac{\lambda_{\text{vacuum}}}{n}$$

Equation 23-8 says that the wavelength is shorter in a medium with a higher index of refraction, where the propagation speed is slower. For example, red light that has a wavelength $\lambda_{\text{vacuum}} = 750$ nm in vacuum has a wavelength in water ($n = 1.33$) equal to $\lambda = \lambda_{\text{vacuum}}/n = (750 \text{ nm})/(1.33) = 564$ nm. The frequency of this light is the same in both media: $f = c/\lambda_{\text{vacuum}} = (3.00 \times 10^8 \text{ m/s})/(750 \times 10^{-9} \text{ m}) = 4.00 \times 10^{14}$ Hz in vacuum and $f = v/\lambda = c/(n\lambda) = (3.00 \times 10^8 \text{ m/s})/((1.33)(564 \times 10^{-9} \text{ m})) = 4.00 \times 10^{14}$ Hz in water.

We learned in Section 22-4 that the energy of an electromagnetic wave comes in packets called *photons*. A single photon has energy $E = hf$, where h is Planck's constant and f is the frequency (Equation 22-26). Since f is unchanged when a photon passes from one medium to another, the photon energy also remains unchanged. This is one important way that photons are very different from ordinary particles such as marbles, whose kinetic energy changes when their speed changes.

GOT THE CONCEPT? 23-1 Refraction

A beam of light in air travels into a transparent, flat-walled container made of Plexiglas. The incident angle is 5.00°. The light then travels from the Plexiglas into the fresh water inside the container. In each refraction, does the ray bend closer to the normal or farther away from it?

(a) Closer in both refractions; (b) farther away in both refractions; (c) closer going from air to Plexiglas, farther away going from Plexiglas to water; (d) farther away going from air to Plexiglas, closer going from Plexiglas to water; (e) in at least one of the refractions, the light does not bend at all.

TAKE-HOME MESSAGE FOR Section 23-2

✔ Huygens' principle says that each point on a wave front acts as a source of wavelets. The new wave front is the superposition of the individual wavelets.

✔ When waves encounter a boundary between two media in which the wave speed is different, the waves can bounce back into the first medium (reflect) or pass into the second medium at a different angle (refract).

✔ The angle of the reflected ray is the same as the angle of the incident ray (law of reflection).

✔ Snell's law describes the direction of the refracted ray. If the wave speed is lower in the second medium than in the first, the refracted ray bends toward the normal. If the wave speed is faster in the second medium, the refracted ray bends away from the normal.

✔ The index of refraction is a measure of the speed of light in a medium relative to the speed of light in vacuum. The slower light travels in a medium, the larger its index of refraction.

23-3 In some cases light undergoes total internal reflection at the boundary between media

In the photograph shown in Figure 23-2, light in air encounters a boundary with water on the other side. Some of the incident light is reflected at the boundary, and some of it is refracted from the first medium (air) into the second medium (water). There are situations, however, in which *none* of the light is refracted into the second medium. **Figure 23-9** shows a light beam in glass that encounters a boundary with air on the other side. As the photograph shows, 100% of the light is reflected back into the glass. This effect is called **total internal reflection.**

Figure 23-10 shows how total internal reflection arises. Light travels more slowly in glass than air, so the index of refraction of glass is higher than the index of refraction of air. As light crosses the boundary from glass to air, it is bent away from the normal as in **Figure 23-10a**. As the incident angle of the light increases (**Figure 23-10b**), the refracted light gets farther from the normal and decreases in intensity. When the incident angle equals the **critical angle** θ_c, the refracted light lies exactly in the plane of the surface (**Figure 23-10c**). At this angle the intensity of the refracted light is zero. If the incident angle is greater than the critical angle, as in **Figure 23-10d**, the light is completely reflected back into the glass. This is total internal reflection.

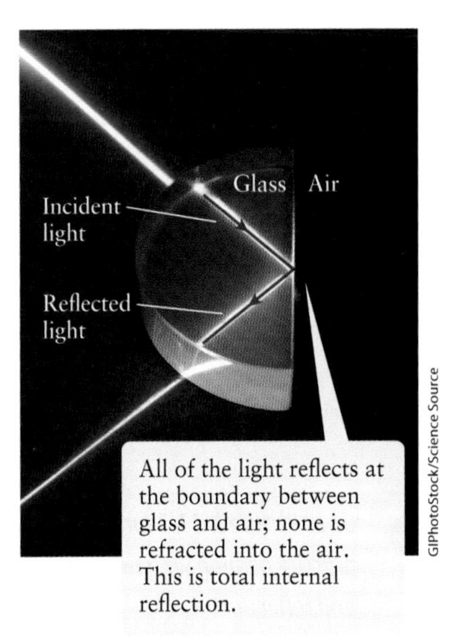

Figure 23-9 Total internal reflection There is nothing unusual about the place where the light beam strikes the glass, yet none of the light escapes into the air on the other side.

(a) Light refracts as it passes from glass into air. The index of refraction of glass is greater than the index of refraction of air, so $\theta_2 > \theta_1$.

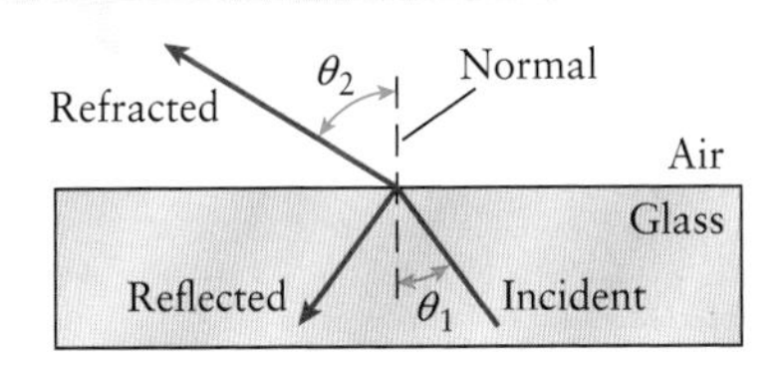

(b) As the angle θ_1 of the incident light increases, so does the angle θ_2 of the refracted light.

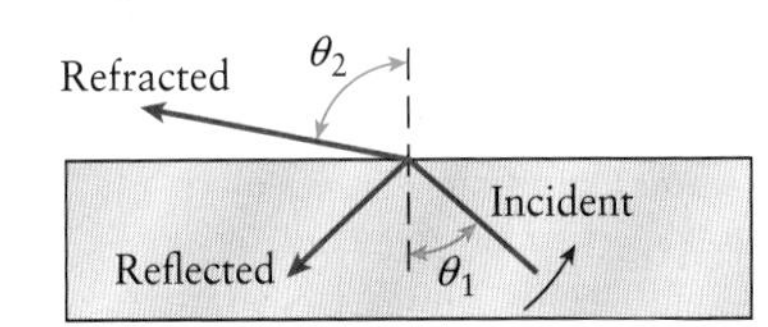

(c) When the angle θ_1 of the incident light equals the critical angle θ_c, the refracted angle is $\theta_2 = 90°$. The intensity of the refracted light becomes zero.

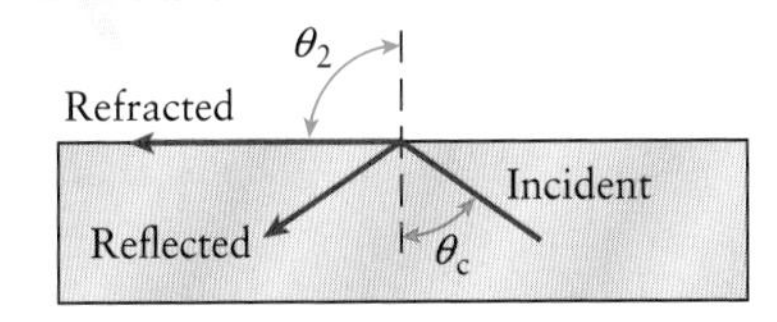

(d) When the angle θ_1 of the incident light is greater than the critical angle θ_c, the light is completely reflected back into the glass. This is total internal reflection.

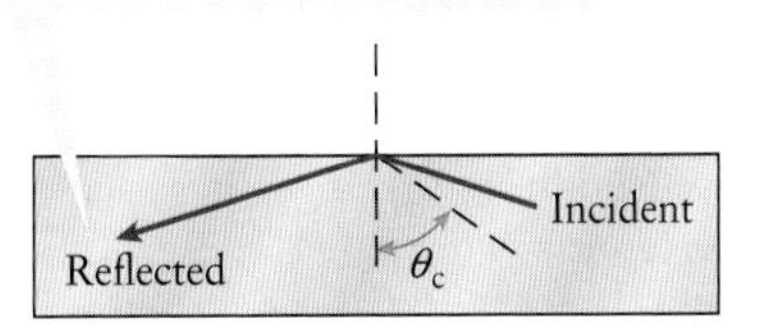

Figure 23-10 Approaching total internal reflection (a) and (b) Light approaches a boundary between glass and air at an incident angle θ_1 less than the critical angle θ_c. (c) When $\theta_1 = \theta_c$, the light ray is refracted along the surface. (d) Total internal reflection occurs for incident angles greater than the critical angle.

Total internal reflection is possible only when the first medium (in Figure 23-10, glass) has a higher index of refraction than the second medium (in Figure 23-10, air), so $n_1 > n_2$. Then the refracted angle θ_2 is greater than the incident angle θ_1, and the refracted angle can reach 90°, as in Figure 23-10c. If the first medium has a lower index of refraction than the second medium, so $n_1 < n_2$, the refracted angle θ_2 is less than the incident angle θ_1, and the refracted angle can never reach 90°.

We can calculate the critical angle θ_c using Snell's law of refraction, Equation 23-6. When the incident angle θ_1 equals the critical angle θ_c, the refracted angle θ_2 equals 90°. If we substitute these into Equation 23-6 and recall that sin 90° = 1, we get

$$n_1 \sin \theta_c = n_2 \sin 90° = n_2 \quad \text{so} \quad \sin \theta_c = \frac{n_2}{n_1} \qquad (23\text{-}9)$$

Note that the sine of an angle between 0 and 90° is between 0 and 1. If $n_1 > n_2$, the ratio n_2/n_1 is less than 1, and there will be some angle θ_c for which Equation 23-9 is satisfied. In this case total internal reflection is possible. But if $n_1 < n_2$, the ratio n_2/n_1 is greater than 1 and Equation 23-9 has no solution. This is another way of seeing that total internal reflection is possible only if $n_1 > n_2$.

If we solve Equation 23-9 for the critical angle θ_c, we get

Critical angle for total internal reflection of light waves at a boundary (23-10)

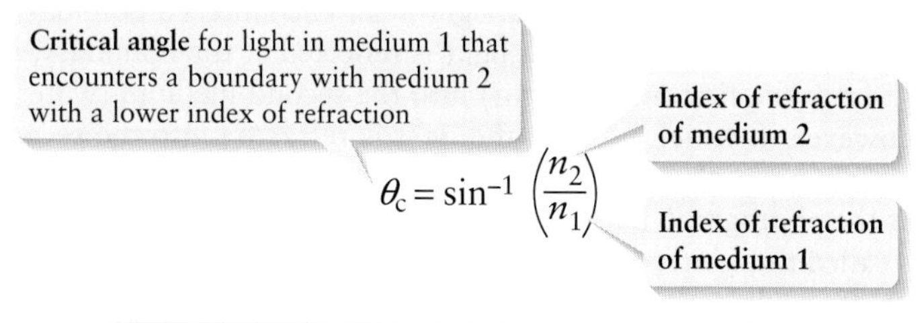

$$\theta_c = \sin^{-1}\left(\frac{n_2}{n_1}\right)$$

If the incident angle is greater than the critical angle, there is total internal reflection.

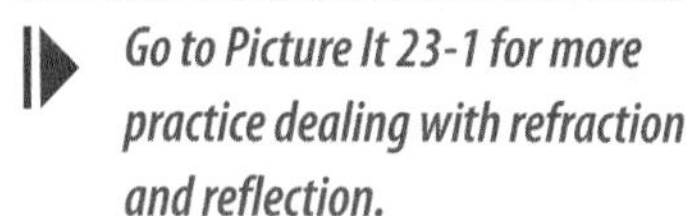

Go to Picture It 23-1 for more practice dealing with refraction and reflection.

Notice that the critical angle depends on the indices of refraction of the media on *both* sides of a boundary.

EXAMPLE 23-2 Critical Angles

(a) A laser is aimed from under the water toward the surface, as in **Figure 23-11**. Find the critical angle of the light incident in the water beyond which total internal reflection occurs.
(b) Find the critical angle if the liquid in the tank were replaced by water containing a high concentration of dissolved sugar.

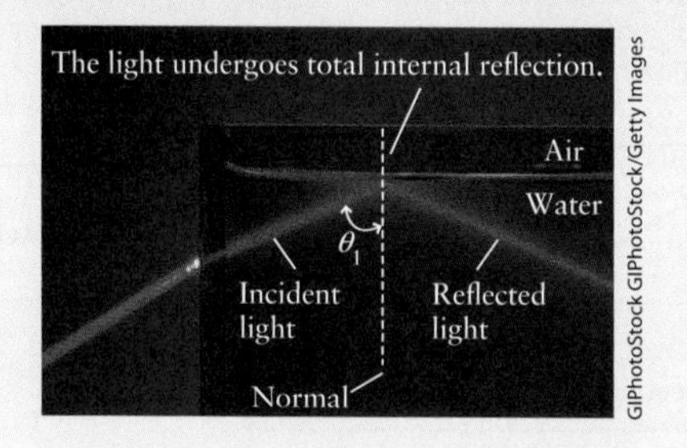

Figure 23-11 Total internal reflection in water Light from a laser aimed from under the water toward the surface is totally internally reflected. What is the minimum incident angle for which total internal reflection will occur?

Set Up

Before hitting the surface (the boundary between water and air), the light is propagating in water, so this is medium 1. In part (a) medium 1 is ordinary water with index of refraction $n_1 = 1.33$; in part (b) medium 1 is sugar water with $n_1 = 1.49$. In both parts, medium 2 on the other side of the surface is air with index of refraction $n_2 = 1.00$. We'll use Equation 23-10 to solve for the critical angle in each case.

Critical angle for total internal reflection:

$$\theta_c = \sin^{-1}\left(\frac{n_2}{n_1}\right) \qquad (23\text{-}10)$$

Solve

(a) Find the critical angle if medium 1 is water.

With $n_1 = 1.33$ and $n_2 = 1.00$, the critical angle is

$$\theta_c = \sin^{-1}\left(\frac{1.00}{1.33}\right) = \sin^{-1}0.752 = 48.8°$$

If the angle of incidence θ_1 is 48.8° or greater, the light will undergo total internal reflection and no light will go into the air. (Note that θ_1 in Figure 23-11 is approximately 60°, which is indeed greater than 48.8°.)

(b) Find the critical angle if medium 2 is sugar water.

With $n_1 = 1.49$ and $n_2 = 1.00$, the critical angle is

$$\theta_c = \sin^{-1}\left(\frac{1.00}{1.49}\right) = \sin^{-1}0.671 = 42.2°$$

Reflect

Adding sugar to water results in a lower speed of light, which increases the index of refraction. The minimum angle of incidence for which total internal reflection occurs is therefore smaller for sugar water than for pure water. The smaller the critical angle, the larger the range of angles at which light experiences total internal reflection.

Total internal reflection explains the brilliance of a cut diamond (see the image that opens this chapter). If light enters the front surface of a cut diamond, it will undergo total internal reflection at the back surface of the diamond (a boundary with air) if the incident angle is greater than the critical angle. Table 23-1 shows that diamond has a very high index of refraction of 2.42, so from Equation 23-10 the critical angle is very small: $\theta_c = \sin^{-1}(1.00/2.42) = 24.4°$. A talented jeweler cuts a diamond so that two things happen: First, light entering the front of the diamond over a broad range of angles will strike the back at an incident angle greater than $\theta_c = 24.4°$ so that total internal reflection occurs there; and second, this reflected light strikes the *front* surface of the diamond at an incident angle *less* than 24.4° so that this light can escape and be seen by you. The result is a gem that sparkles with brilliant reflections.

Total internal reflection also explains the *mirage* that happens when you look down the highway on a hot day and see what appear to be puddles of water on the road. You might even see the reflection of cars in these "puddles," as in **Figure 23-12**. The explanation is that air sits in layers above the road, each layer a bit warmer, less dense, and with a lower index of refraction than the one above it. Light from the sky is refracted as it encounters the boundary between one layer of air and the next (**Figure 23-13**). Because the index of refraction of the layer of air closer to the road is lower, the light is bent farther from the normal. The normal direction in these refractions is vertical, so the light is refracted closer to horizontal. Light that strikes the boundary between layers at a large angle with respect to the normal (a grazing angle

The Sydney Morning Herald/Getty Images

Figure 23-12 A mirage What causes these shimmering patches on the road that look like puddles?

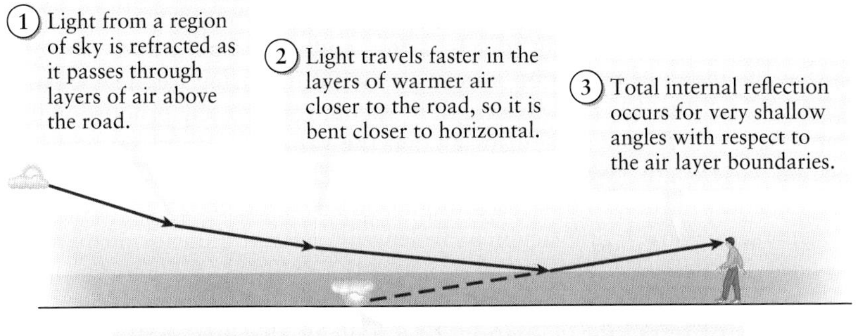

(1) Light from a region of sky is refracted as it passes through layers of air above the road.

(2) Light travels faster in the layers of warmer air closer to the road, so it is bent closer to horizontal.

(3) Total internal reflection occurs for very shallow angles with respect to the air layer boundaries.

(4) An image of the region of sky is formed close to the ground that gives the appearance of shimmery, blue puddles on the road.

Figure 23-13 Explaining a mirage The "puddles" in Figure 23-12 are an illusion caused by total internal reflection.

as measured from the air layer boundary) can experience total internal reflection and reflect back up into the higher layer. Your eyes trace the light rays back along straight lines, so the light appears to be coming from a point close to the road. What looks like a puddle is actually a refracted image of the blue sky.

Endoscopy is a medical procedure used to see inside the body. It relies on a light fiber, an optical device used to carry light and sometimes data encoded in pulses of light, from one place to another. A beam of light sent down a light fiber experiences total internal reflection at the surface of the fiber, which results in multiple reflections that keep the beam inside the fiber. The bent bar of Plexiglas in **Figure 23-14** carries light in a similar way. Notice in Figure 23-14 that no light leaks out of the bar at the points where the beam hits the surface. This is because the angle of incidence at the surface is greater than the critical angle, so total internal reflection occurs. To maximize total internal reflection, most light fibers are made by surrounding a central core with one or two layers of a material of lower index of refraction than the core. Many endoscopes actually use bundles of several fibers. Light is sent down some fibers to illuminate the subject. Other fibers carry the image back to a camera.

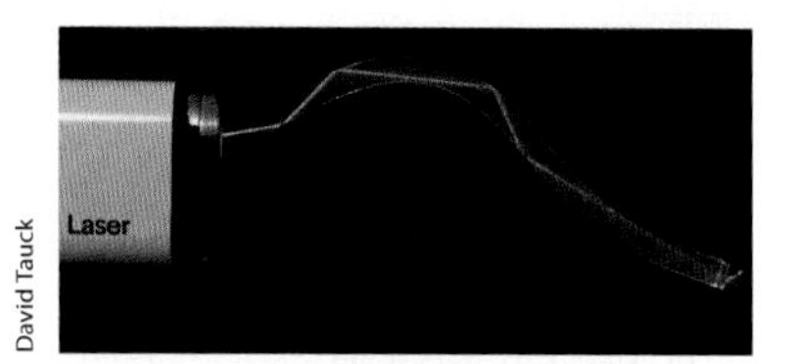

David Tauck

Figure 23-14 Light trapped by total internal reflection Light propagates through the interior of a curved bar of Plexiglas by a series of total internal reflections.

GOT THE CONCEPT? 23-2 Total Internal Reflection

It's possible for a beam of light in Plexiglas to undergo total internal reflection at a boundary if the material on the other side of the boundary is (a) ethyl alcohol; (b) sapphire; (c) diamond; (d) more than one of these; (e) none of these.

TAKE-HOME MESSAGE FOR Section 23-3

✔ When light traveling in a medium with index of refraction n_1 reaches a boundary with a second medium of index of refraction n_2, total internal reflection can happen if $n_1 > n_2$.

✔ Total internal reflection takes place only if the incident angle measured from the normal at the boundary is greater than the critical angle θ_c given by Equation 23-10. In this case none of the light is refracted into the second medium.

23-4 The dispersion of light explains the colors from a prism or a rainbow

White light is a mixture of all the colors of the visible spectrum. You can see these colors by allowing sunlight to pass through a glass prism, as in **Figure 23-15**: The different colors emerge in different directions. This happens because the speed of light in a medium other than vacuum, such as the glass in a prism, is different for different frequencies of light. (In vacuum the speed is equal to c for all frequencies.) This is a result of how a light wave interacts with the atoms of the medium. This variation of speed with frequency is called **dispersion**.

GIPhotoStock/Science Source

Figure 23-15 Dispersion Light of different frequencies, and hence different colors, propagates at different speeds through the glass of which this prism is made. As a result, different colors refract along slightly different paths.

Since the speed v of light waves in a medium depends on the frequency, the index of refraction $n = c/v$ (Equation 23-5) depends on frequency as well. In most transparent materials the speed decreases with increasing frequency, from red to yellow to violet (see Figure 22-2): Red light travels fastest, and violet light travels slowest. This means that the index of refraction n increases with increasing frequency. Note that Equation 23-7 tells us that the wavelength in vacuum is inversely proportional to frequency: $\lambda_{vacuum} = c/f$. So we can also say that the value of n decreases with increasing vacuum wavelength. (We specify *vacuum* wavelength, since the wavelength in the medium depends on the value of n; see Equation 23-8.) **Figure 23-16** shows how the index of refraction varies with vacuum wavelength for four different transparent materials. (The indices of refraction given in Table 23-1 are for yellow light, near the middle of the visible spectrum.)

Snell's law of refraction, Equation 23-6, tells us that the angle at which light refracts as it crosses the boundary between two transparent media depends on their indices of refraction. So it follows that different colors of light, with different vacuum wavelengths, refract at different angles. **Figure 23-17** shows this for light passing from vacuum into glass. The higher the index of refraction of the second medium, the more the refracted light is bent toward the normal. In common crown glass the index of refraction is about 1.51 for red light and about 1.53 for blue light. Hence the blue light bends more toward the normal than does the red light, and different colors of light are spread out or dispersed. (This is the origin of the term "dispersion.")

The same effect explains the appearance of a rainbow (Figure 23-1b). When raindrops in midair are illuminated by the Sun, sunlight enters each raindrop, is partly reflected off the back of the drop, and then exits out the front of the drop. The index of refraction of water is different for different wavelengths, so each color of light emerges in a slightly different direction to form a rainbow.

The amount of dispersion is different for different transparent materials. In crown glass, for example, the index of refraction varies by 0.02 from red ($n = 1.51$) to blue ($n = 1.53$), while for diamond the index of refraction varies by 0.04 from red ($n = 2.41$) to blue ($n = 2.45$). As a result, the colors of white light are spread out over a wider angle by a cut diamond than by a piece of glass cut to the same shape. This high value of dispersion contributes to the "sparkly" character of a cut diamond.

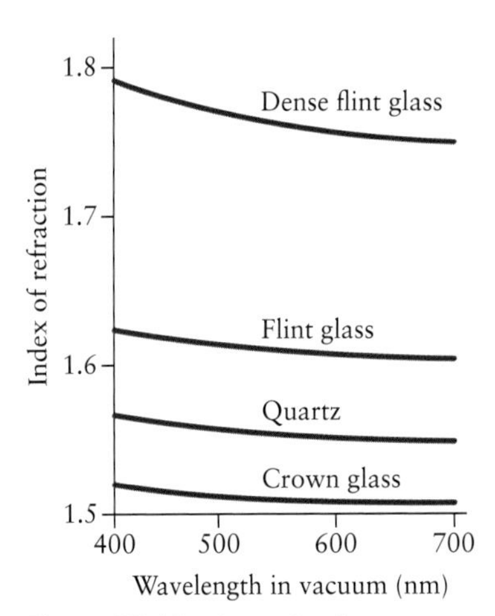

Figure 23-16 Dispersion in different materials The index of refraction in glass varies with the vacuum wavelength of light and with the type of glass.

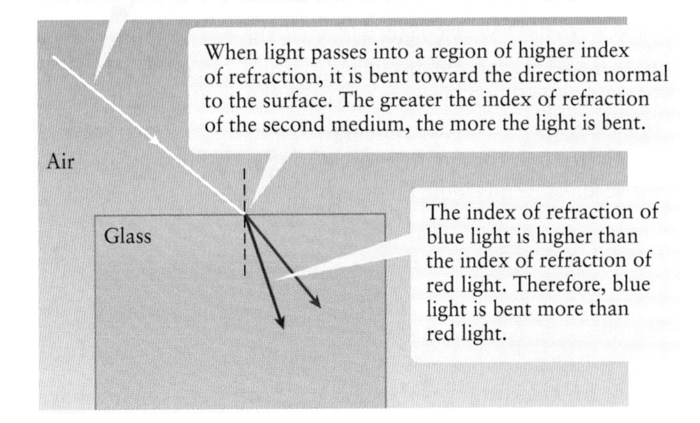

Figure 23-17 Analyzing dispersion The difference in refracted angle between the red and blue light is greatly exaggerated for clarity.

EXAMPLE 23-3 Dispersion in Dense Flint Glass

A narrow beam of white light enters a rectangular block of dense flint glass at 60.0° from the normal. The block is 0.500 m on a side. How far apart will the red and blue parts of the visible spectrum be when the light leaves the glass? The index of refraction of dense flint glass is 1.75 for red light and 1.79 for blue light.

Set Up

Both colors of light are incident on the block at the same angle: $\theta_1 = 60.0°$. However, the red and blue refract at different angles because the index of refraction is different for the two colors. We'll use Snell's law, Equation 23-6, to calculate the angles θ_{red} and θ_{blue}, and then use trigonometry to find the distances d_{red} and d_{blue} shown in the figure. The difference between these distances tells us how far apart the points are where the two colors emerge from the glass.

Snell's law of refraction:

$$n_1 \sin \theta_1 = n_2 \sin \theta_2 \quad (23\text{-}6)$$

Solve

Find the refracted angles θ_{red} and θ_{blue} for the two colors.

For both colors $n_1 = n_{air} = 1.00$ and $\theta_1 = 60.0°$. For red light $n_2 = 1.75$, so Equation 23-6 becomes

$$1.00 \sin 60.0° = 1.75 \sin \theta_{red}$$

$$\sin \theta_{red} = \frac{1.00 \sin 60.0°}{1.75} = 0.4948$$

$$\theta_{red} = \sin^{-1} 0.4948 = 29.66°$$

(We've kept an extra significant digit in our result; we'll round off at the end of the calculation.) Do the same calculation for blue light, for which $n_2 = 1.79$:

$$1.00 \sin 60.0° = 1.79 \sin \theta_{blue}$$

$$\sin \theta_{blue} = \frac{1.00 \sin 60.0°}{1.79} = 0.4838$$

$$\theta_{blue} = \sin^{-1} 0.484 = 28.93°$$

Find the distances d_{red} and d_{blue} for the two colors of light, and from these find the separation between the two colors as they exit the glass.

For each color of light, the ray that extends from where the light enters the block of glass to where it exits the block forms the hypotenuse of a right triangle. The other two sides are the vertical dimension of the block, $w = 0.500$ m, and the distance d_{red} or d_{blue} that the light is displaced horizontally. In each case the side of length w is adjacent to the angle θ_{red} or θ_{blue}, and the side of length d_{red} or d_{blue} is opposite to that angle. In a right triangle the length of the opposite side divided by the length of the adjacent side equals the tangent of the angle, so

$$\tan \theta_{red} = \frac{d_{red}}{w}$$

$$\tan \theta_{blue} = \frac{d_{blue}}{w}$$

Solve for the distances d_{red} and d_{blue} using the angles that we calculated above:

$$d_{red} = w \tan \theta_{red} = (0.500 \text{ m}) \tan 29.66° = 0.2847 \text{ m}$$

$$d_{blue} = w \tan \theta_{blue} = (0.500 \text{ m}) \tan 28.93° = 0.2764 \text{ m}$$

The distance between where the red light exits the glass and where the blue light exits the glass is

$$d_{red} - d_{blue} = 0.2847 \text{ m} - 0.2764 \text{ m} = 0.0083 \text{ m} = 8.3 \text{ mm}$$

Reflect

The separation between the two colors is fairly substantial, so this block of flint glass does a good job of spreading white light into its constituent colors. We invite you to repeat this calculation for crown glass, the kind of glass commonly used to make windows, for which $n_{red} = 1.51$ and $n_{blue} = 1.53$. Since crown glass has a smaller index of refraction than flint glass, and because the difference between the values of the two indices is smaller for crown glass than for flint glass, you'll find that the distance $d_{red} - d_{blue}$ is smaller for crown glass. Which type of glass would be a better choice for a prism intended to spread apart the different colors of light, as in Figure 23-16?

GOT THE CONCEPT? 23-3 Dispersion

A flash of white light (containing all of the colors of the visible spectrum) in air shines straight down on the surface of a pond of water. Which color of light from the flash reaches the bottom of the pond first? (a) The blue light; (b) the yellow light; (c) the red light; (d) all reach the bottom at the same time.

TAKE-HOME MESSAGE FOR Section 23-4

✔ The speed of light in a material medium depends on the frequency of the light. This is called dispersion. In most materials the index of refraction for visible light increases from the red end of the spectrum (low frequency, long wavelength) to the blue end of the spectrum (high frequency, short wavelength).

✔ When light crosses a boundary into a medium of different index of refraction, different colors refract by different angles. This causes the colors to spread apart by an amount that depends on how strongly the index of refraction varies with wavelength.

23-5 In a polarized light wave the electric field vector points in a specific direction

When a honeybee finds nectar, it communicates the location to other bees in the hive. In the 1940s Austrian ethologist Karl von Frisch established that if bees can see even a small patch of blue sky, they can use the position of the Sun to describe the path back to the food from the hive (**Figure 23-18**). How is it that bees can know the position of the Sun even if they can't see it directly?

The explanation is that light is a transverse electromagnetic wave, with an oscillating electric field $\vec{E}$ and magnetic field $\vec{B}$ that are perpendicular to each other and to the direction of wave propagation (see Figure 22-4). The orientation of the $\vec{E}$ field is called the **polarization** of the light wave. (We don't need to separately state the orientation of $\vec{B}$, since we know that it's perpendicular to both the direction of propagation and the orientation of $\vec{E}$.) Natural light such as that emitted by the Sun or an ordinary light bulb is **unpolarized**: The orientation of the electric field changes randomly from one moment to the next. For example, if the wave is propagating in the positive x direction, at one moment $\vec{E}$ may be oriented along the y axis, a short time later it may be oriented along the z axis, a short time after that it may be oriented at 23.7° to the y axis, and so on. The reason for this is that a source of natural light emits light in the form of a stream of photons (see Section 22-4), and the orientation of $\vec{E}$ varies randomly from one photon to another.

When sunlight scatters from molecules or small particles in the atmosphere, however, the scattered light that we see has its $\vec{E}$ field oriented predominantly in one direction (**Figure 23-19**). (Scattering is stronger for short-wavelength light than for long-wavelength light, which is why the color of the sky is dominated by short-wavelength

Medioimages/Photodisc/Getty Images

Figure 23-18 Navigating by the light of the sky Honeybees have the ability to detect the polarization of light entering their eyes—that is, the orientation of the electric field $\vec{E}$ in the light wave. The polarization helps them determine the position of the Sun in the sky, even if the Sun isn't directly visible. Karl von Frisch shared the 1973 Nobel Prize in Physiology or Medicine for this and other discoveries about bees and their behavior.

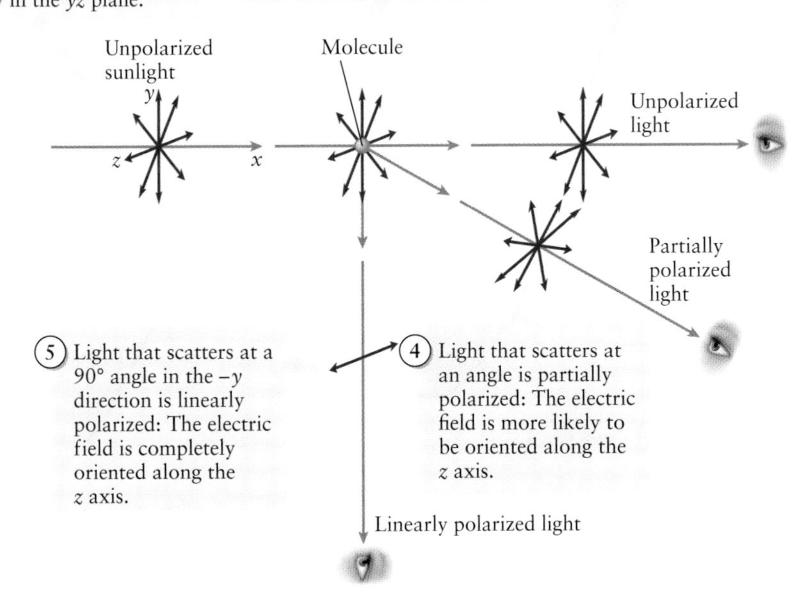

1 Sunlight propagates in the x direction. The light is unpolarized, so the orientation of the electric field (shown by the blue arrows) changes randomly in the yz plane.
2 When light strikes a molecule in the atmosphere, the polarization and direction of propagation can both change.
3 Light that scatters in the forward direction (that is, is not deflected) remains unpolarized.
Unpolarized sunlight
y
z
x
Molecule
Unpolarized light
Partially polarized light
5 Light that scatters at a 90° angle in the −y direction is linearly polarized: The electric field is completely oriented along the z axis.
4 Light that scatters at an angle is partially polarized: The electric field is more likely to be oriented along the z axis.
Linearly polarized light

Figure 23-19 Polarization by scattering The extent to which sunlight is polarized by scattering depends on the scattering angle.

blue light.) Light in which the orientation of the $\vec{E}$ field changes randomly, but is more likely to be in one orientation than in other orientations, is called **partially polarized.** Light for which the $\vec{E}$ field is oriented *completely* along one direction is called **linearly polarized.** For example, the $\vec{E}$ field for the light wave shown in Figure 22-4 has only a y component, so we say this light is linearly polarized along the y axis. Honeybees (Figure 23-18) have the ability to detect the polarization of light coming from the sky, and by using this they can infer the position of the Sun. (Human eyes, by contrast, are only weakly sensitive to polarization.)

Polarizing Light with a Polarizing Filter

A simple way to make polarized light from unpolarized light is by using a **polarizing filter.** This is a transparent sheet which contains long-chain molecules that are all oriented in the same direction. These molecules absorb light whose polarization direction is along the axis of the molecules, but have no effect on light that is polarized perpendicular to that direction. Equivalently, you can think of a polarizing filter as having slots that allow waves to pass if they are polarized along the direction of the slots, called the *polarization direction*. Waves that are polarized perpendicular to the slots are blocked (**Figure 23-20**). If we send unpolarized light into the filter, at any instant the electric field $\vec{E}$ has a component along the polarization direction of the filter and a component perpendicular to that direction. Only the component of $\vec{E}$ along the polarization direction of the filter will pass through, so the light that emerges from the filter will be linearly polarized.

Since a polarizing filter absorbs some of the incident light that falls on it, the light exiting the filter is in general less intense than the incident light. Suppose the incident light has an oscillating electric field with amplitude E_0 that is oriented at an angle θ to the polarization direction of the filter (**Figure 23-21**). Only the component of this field that is aligned with the polarization direction is allowed to pass through the filter:

$$E_{\text{transmitted}} = E_0 \cos \theta \tag{23-11}$$

The intensity of the light is proportional to the square of the electric field amplitude. Equation 23-11 says that the amplitude $E_{\text{transmitted}}$ of the transmitted light equals $\cos \theta$ times the amplitude E_0 of the incident light. Hence the intensity I of the transmitted light equals $\cos^2\theta$ times the intensity I_0 of the incident light:

$$I = E_0^2 \cos^2 \theta = I_0 \cos^2 \theta \tag{23-12}$$

If the incident light is unpolarized, the value of the angle θ will vary randomly between 0 and 360°. The average value of $\cos^2\theta$ over this range is 1/2, so if unpolarized light incident on a polarizing filter has intensity I_0, the intensity of the light that emerges from the filter will be $I = I_0/2$.

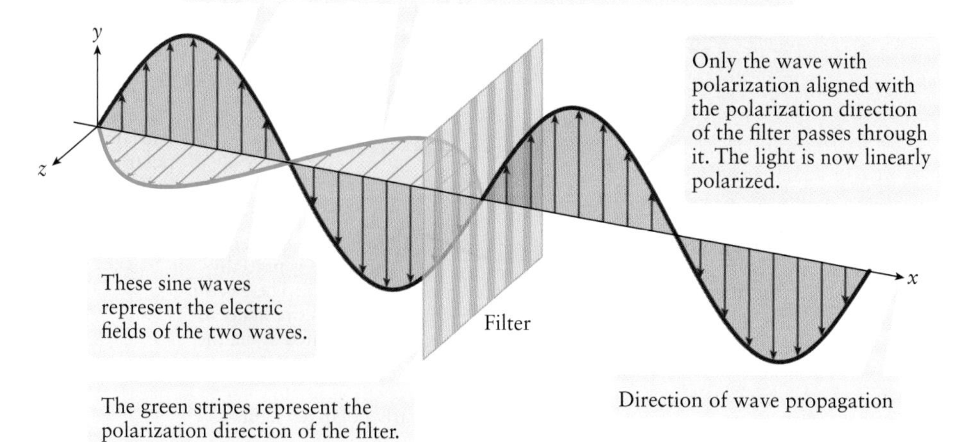

Figure 23-20 A linear polarizing filter | A filter of this kind allows only light with an electric field component aligned in a certain direction, called the polarization direction, to pass through it.

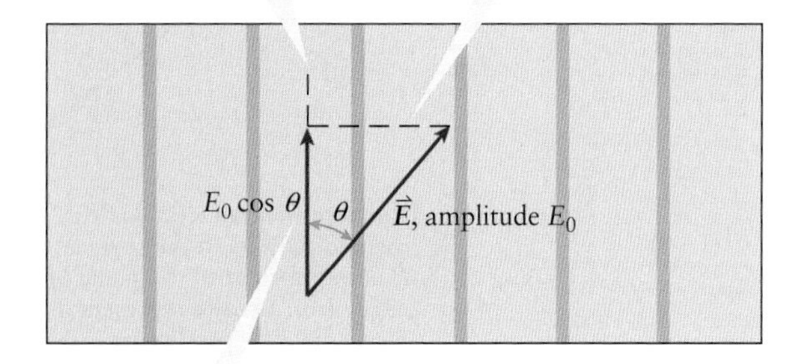
1 The polarizing filter allows light polarized along this direction to pass. It blocks light polarized in the perpendicular direction.
2 As a result, this component of the electric field of a light wave is blocked by the filter...
$E_0 \cos\theta$
θ
$\vec{E}$, amplitude E_0
3 ...and only this component is allowed to pass through the filter.

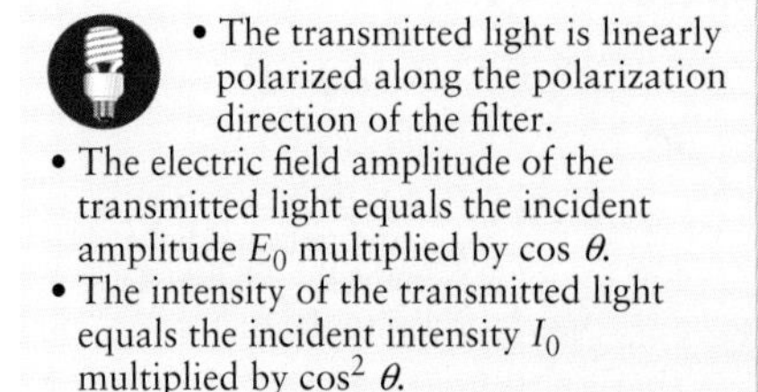
• The transmitted light is linearly polarized along the polarization direction of the filter.
• The electric field amplitude of the transmitted light equals the incident amplitude E_0 multiplied by cos θ.
• The intensity of the transmitted light equals the incident intensity I_0 multiplied by $\cos^2 \theta$.

Figure 23-21 A linear polarizing filter II What happens to light that enters a polarizing filter with its electric field $\vec{E}$ at an angle to the polarization direction?

Polarizing Light by Reflection

Another way to convert unpolarized light into light that is at least partially polarized is by *reflection*. When light strikes the boundary between two media, a fraction of the light is reflected and the remainder is refracted. The fraction that is reflected depends not only on the incident angle of the light but also on the polarization of the incident light. In general, light polarized parallel to the boundary surface is reflected more strongly than light polarized in the perpendicular direction. So if the incident light is unpolarized, the reflected light will be at least partially polarized in the orientation parallel to the boundary surface.

As an example, sunlight that reflects from the windshield and the hood of the car in **Figure 23-22a** is partially polarized parallel to the reflecting surface, so the electric field of the reflected light has a strong horizontal component and a weak vertical component. In **Figure 23-22b** we've placed a polarizing filter in front of the camera with its polarizing direction oriented vertically, so light with horizontal polarization is blocked. Hence the reflected light is greatly suppressed. Polarizing sunglasses use this same principle to minimize reflections from the road or the surface of a lake or ocean: Sunlight reflected from these horizontal surfaces is predominantly polarized in the horizontal direction, so to block this reflected light, the filters that make up the sunglass lenses have a vertical polarization direction. If you look at this reflected light through polarizing sunglasses and tilt your head to one side until one eye is directly above the other, you'll see the reflections reappear. That's because the polarization direction of the lenses is now horizontal, the same as the orientation of the electric field of the reflected light.

(a) Photographed without a filter

David Tauck

(b) Photographed with a polarizing filter oriented vertically, to block light with horizontal polarization

David Tauck

Figure 23-22 Reducing reflections with a polarizing filter Reflected light is linearly polarized. The polarizing lenses often used for sunglasses can dramatically reduce glare from reflected light.

When unpolarized light is incident on the boundary between two media, there is one particular angle of incidence for which the reflected light is *completely* polarized parallel to the boundary. The angle of incidence θ_1 that results in this special condition is called **Brewster's angle** θ_B. When $\theta_1 = \theta_B$, it turns out that the sum of the incident angle θ_1 and the refracted angle θ_2 equals 90° (**Figure 23-23**):

$$\theta_1 + \theta_2 = 90° \quad \text{when } \theta_1 = \theta_B \tag{23-13}$$

We can use this to solve for the value of Brewster's angle θ_B in terms of the indices of refraction n_1 and n_2 of the two media. Snell's law of refraction, Equation 23-6, gives us this relationship between θ_1 and θ_2:

$$n_1 \sin\theta_1 = n_2 \sin\theta_2 \tag{23-6}$$

If θ_1 equals Brewster's angle θ_B, Equation 23-13 tells us that

$$\theta_2 = 90° - \theta_1 = 90° - \theta_B$$

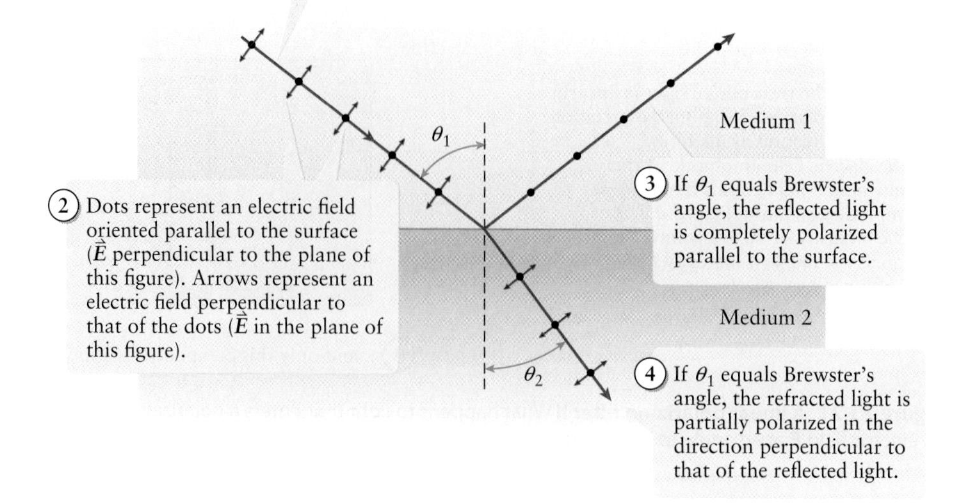

① Unpolarized light propagating in medium 1 strikes the boundary with medium 2 at an incident angle θ_1.
② Dots represent an electric field oriented parallel to the surface ($\vec{E}$ perpendicular to the plane of this figure). Arrows represent an electric field perpendicular to that of the dots ($\vec{E}$ in the plane of this figure).
θ_1
Medium 1
③ If θ_1 equals Brewster's angle, the reflected light is completely polarized parallel to the surface.
Medium 2
θ_2
④ If θ_1 equals Brewster's angle, the refracted light is partially polarized in the direction perpendicular to that of the reflected light.

Figure 23-23 Polarization by reflection Light reflected from a surface is partially polarized and becomes completely polarized when the incident angle equals Brewster's angle.

If we substitute this into Equation 23-6, we get

$$n_1 \sin \theta_B = n_2 \sin (90° - \theta_B)$$

We know from trigonometry that $\sin (90° - \theta_B) = \cos \theta_B$. So

$$n_1 \sin \theta_B = n_2 \cos \theta_B$$

Divide both sides of this equation by n_1, then divide both sides by $\cos \theta_B$. The result is

$$\frac{\sin \theta_B}{\cos \theta_B} = \frac{n_2}{n_1}$$

The sine of an angle divided by its cosine equals the tangent of the angle. So $\tan \theta_B = n_2/n_1$, or

Brewster's angle for polarization by reflection (23-14)

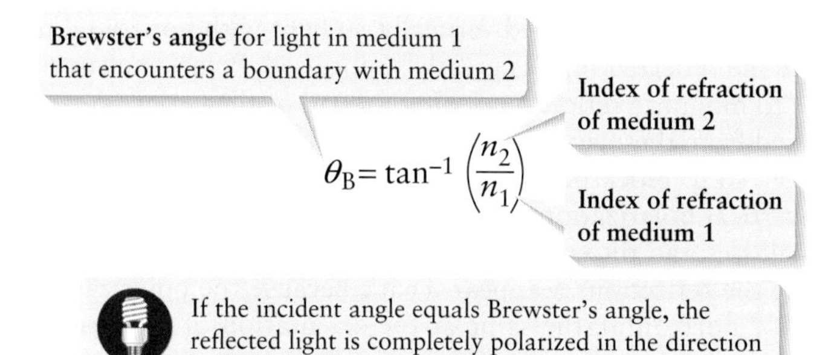

$$\theta_B = \tan^{-1}\left(\frac{n_2}{n_1}\right)$$

If the incident angle equals Brewster's angle, the reflected light is completely polarized in the direction parallel to the surface of the boundary.

Go to Interactive Exercise 23-2 for more practice dealing with polarization.

If the angle of incidence for unpolarized light is something other than θ_B, the reflected ray is partly polarized. If unpolarized light strikes the boundary perpendicular to it, so $\theta_1 = 0$, there is no change in polarization of the reflected light.

WATCH OUT! Brewster's angle versus the critical angle.

Be careful not to confuse Equation 23-14 for Brewster's angle with the similar-looking Equation 23-10 for the critical angle. For a boundary between two transparent media, Brewster's angle (which involves an inverse *tangent* function) is the *one and only* angle of incidence for which the reflected light is completely polarized. This can happen whether the index of refraction n_1 for the first medium is larger or smaller than the index of refraction n_2 for the second medium. The critical angle (which involves an inverse *sine* function) is the *minimum* angle of incidence for which there is total internal reflection (Section 23-3); there is also total internal reflection if the angle of incidence is greater than the critical angle. Total internal reflection is possible only if n_1 is greater than n_2.

EXAMPLE 23-4 Brewster's Angle for Air to Water

At what incident angle must light strike the surface of a pond so that the reflected light is completely polarized?

Set Up

This is the situation shown in Figure 23-23, with air as medium 1 and water as medium 2. The reflected light will be completely polarized if the incident angle θ_1 is equal to Brewster's angle θ_B given by Equation 23-14. Table 23-1 tells us the value of the indices of refraction: $n_1 = n_{air} = 1.00$ and $n_2 = n_{water} = 1.33$.

Brewster's angle for polarization by reflection:

$$\theta_B = \tan^{-1}\left(\frac{n_2}{n_1}\right) \qquad (23\text{-}14)$$

Solve

Use Equation 23-14 to find Brewster's angle for this situation.

With $n_1 = 1.00$ and $n_2 = 1.33$, Brewster's angle is

$$\theta_B = \tan^{-1}\left(\frac{1.33}{1.00}\right) = 53.1°$$

The reflected light will be completely polarized if the incident angle θ_1 is equal to $\theta_B = 53.1°$.

Reflect

The boundary between the two media (air and water) is the horizontal surface of the pond, so the normal to this surface is vertical. Our result tells us that if light from the sky strikes the surface at an angle of 53.1° from the vertical, the reflected light will be completely polarized. Light striking close to this angle will be strongly polarized but not 100% polarized.

You should repeat this calculation for light coming from below the water (say, from a diver's flashlight) and striking the surface of the pond from below. Can you show that in this case, the light that reflects back into the water will be completely polarized if the incident light is at an angle of 36.9° to the vertical?

GOT THE CONCEPT? 23-4 Polarizing Filters I

Two polarizing filters, *A* and *B*, are placed one behind the other with their polarization directions perpendicular to each other. A beam of unpolarized light is directed at filter *A*. What fraction of the original light intensity will remain after the light passes through both filters *A* and *B*? (a) 1/2; (b) 1/4; (c) 1/8; (d) zero; (e) none of these.

GOT THE CONCEPT? 23-5 Polarizing Filters II

Two polarizing filters, *A* and *B*, are placed one behind the other with their transmission directions perpendicular to each other. A third polarizing filter, *C*, is placed in between them. The transmission direction of *C* is halfway between those of filters *A* and *B* (that is, at 45° to that of filter *A* and at 45° to that of filter *B*). A beam of unpolarized light is directed at filter *A*. What fraction of the original light intensity will remain after the light passes through filters *A*, *B*, and *C*? (a) 1/2; (b) 1/4; (c) 1/8; (d) zero; (e) none of these.

TAKE-HOME MESSAGE FOR Section 23-5

✔ The orientation of the oscillating electric field in a light wave tells you the polarization of that wave.

✔ In natural light the direction of the electric field changes randomly and is equally likely to be in any direction perpendicular to the propagation direction of the wave. Such light is called unpolarized.

✔ Light that has its electric field oriented in one specific direction is called linearly polarized. Unpolarized light can be polarized by scattering, by passing it through a polarizing filter, or by reflection.

23-6 Light waves reflected from the surfaces of a thin film can interfere with each other, producing dazzling effects

Figure 23-24 Butterfly interference The iridescent colors on the wings of this *Morpho menelaus* butterfly result from the interference of light waves that reflect from the wing surfaces.

The wings of the butterfly *Morpho menelaus* (**Figure 23-24**) show brilliant colors. Yet the material of which the wings are made is colorless! The explanation for this seeming contradiction is that the colors are produced by the interference of light, a process similar to the interference of sound waves we investigated in Chapter 13. In Section 13-5 we saw that sound waves interfere constructively or destructively depending on how the peaks and troughs of the two waves align. The same is true for light waves.

To begin our investigation of light wave interference, let's consider what happens when light of a single wavelength, and hence a single color—called **monochromatic** light—strikes a thin layer of a transparent material. An example of such a transparent *thin film* is the *tapetum lucidum* ("shining carpet") that lines the back of some animals' eyes behind the retina (see Figure 23-1a). The material behind the *tapetum lucidum* is opaque (light cannot pass through it), so we can think of the *tapetum lucidum* as a **thin film** against an opaque backing.

As **Figure 23-25a** shows, some of the light waves that strike the front surface of the thin film are reflected. The remaining light enters the thin film (**Figure 23-25b**), strikes the opaque backing at the back surface of the film, and is reflected back up. (Some light is also absorbed by the backing.) As **Figure 23-25c** shows, the light waves reflected from the front surface and the light waves reflected from the back surface both end up above the thin film and traveling in the same direction.

The single light wave that was incident on the thin film has now been split into two reflected, outgoing waves that have traveled different paths. These two waves were in phase before they encountered the thin film because they were part of the same incident wave. But in general the two reflected waves are *not* in phase when they recombine above the front surface of the thin film. That's because the light that enters the film and reflects off the back surface travels farther than the light that reflects off the front surface.

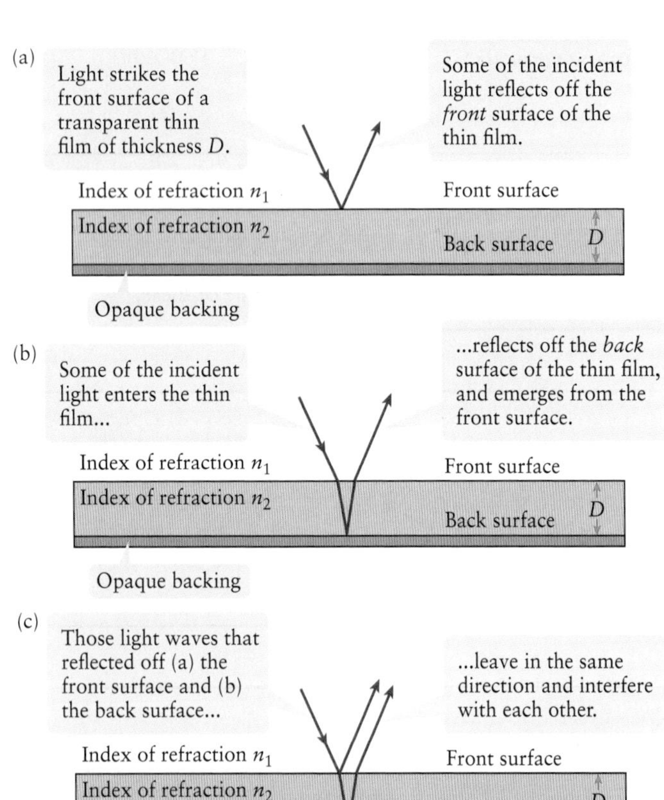

Figure 23-25 Interference from a thin film I For clarity in this figure we've drawn light rays that reach the surface of a thin film at an angle. In our calculations we'll assume that the light hits the surface face-on.

Whether the interference that occurs between the two reflected waves is constructive or destructive depends on the number of wave cycles that fit into the extra distance traveled by the light that enters the film. If an integer number of cycles (1, 2, 3, ... cycles) fit into that extra distance, then the two outgoing waves are in phase and constructively interfere. The surface of the film appears bright. If an odd number of half cycles (1/2, 3/2, 5/2, ... cycles) fit into the extra distance, however, the two outgoing waves are 180° out of phase and destructively interfere. (We specify an *odd* number of half cycles because an even number of half cycles is the same as an integer number of full cycles; for example, $4/2 = 2$. This case gives constructive, not destructive, interference.) When destructive interference occurs, the two outgoing waves cancel each other out and the surface of the film appears dark.

Let's consider the case in which the incident light is normal to the surface of the thin film (that is, the light strikes the film face-on). Then the path length difference Δ_{pl} for the two waves—one that reflects off the front of the film and the other that reflects off the back—is twice the thickness D of the film:

$$\Delta_{pl} = 2D \quad (23\text{-}15)$$

Constructive interference occurs when an integer number of wavelengths of light fit into this path length difference of $2D$. Destructive interference occurs when an odd number of half-wavelengths fit into the distance $2D$. However, the wavelength of light inside the film is not the same as the wavelength outside the film. Recall from Section 23-2 that the wavelength of light in a medium with index of refraction n is

$$\lambda = \frac{\lambda_{\text{vacuum}}}{n} \tag{23-8}$$

As in Figure 23-25, let's say that the material outside the film has index of refraction n_1, and the material of which the film is made has index of refraction n_2. Then Equation 23-8 tells us that the wavelengths in the two materials are

$$\lambda_1 = \frac{\lambda_{\text{vacuum}}}{n_1} \text{ outside the film}$$

$$\lambda_2 = \frac{\lambda_{\text{vacuum}}}{n_2} \text{ inside the film}$$

Comparing these two, we see that

$$\lambda_2 = \frac{n_1 \lambda_1}{n_2} \tag{23-16}$$

We now know everything we need to determine how the light reflected from the front surface of the thin film interferes with the light that enters the film and is reflected from the back surface. *Constructive* interference occurs when the path difference equals an integer number of wavelengths λ_2 (the wavelength of light inside the film):

$$2D = m\lambda_2 = m\frac{n_1 \lambda_1}{n_2}, \quad m = 1, 2, 3, \ldots \tag{23-17}$$

(constructive interference, thin film with an opaque backing)

If we set $m = 1$ in Equation 23-17 and solve for D, we get the *minimum* thickness that a thin film must have to give constructive interference when light of wavelength λ_1 strikes the front surface face-on:

$$D_{\text{min}} = \frac{n_1 \lambda_1}{2n_2} \tag{23-18}$$

(minimum thickness for constructive interference, thin film with an opaque backing)

Destructive interference occurs when the path difference equals an odd number of one-half the wavelength λ_2 inside the film:

$$2D = (2m - 1)\frac{\lambda_2}{2} = (2m - 1)\frac{n_1 \lambda_1}{2n_2}, \quad m = 1, 2, 3, \ldots \tag{23-19}$$

(destructive interference, thin film with an opaque backing)

If we set $m = 1$ in Equation 23-19 and solve for D, we get the minimum thickness that a thin film must have to give destructive interference when light of wavelength λ_1 strikes the front surface face-on:

$$D_{\text{min}} = \frac{n_1 \lambda_1}{4n_2} \tag{23-20}$$

(minimum thickness for destructive interference, thin film with an opaque backing)

The *tapetum lucidum* of an animal's eye appears shiny because it preferentially reflects certain wavelengths—to be specific, those that satisfy Equation 23-17, where n_2 is the index of refraction of the material of which the *tapetum lucidum* is made. This gives rise to the phenomenon of *eyeshine* that you can see in Figure 23-1a. The *tapetum lucidum* for the cat in that photo preferentially reflects green light because an integer number of wavelengths of that color fit into the path difference (twice the thickness of the *tapetum lucidum*). Other colors with other wavelengths do not satisfy that relationship perfectly, so they are not reflected as strongly. The eyes of different animals shine with different colors depending on the thickness of the *tapetum lucidum*.

Go to Interactive Exercise 23-3 for more practice dealing with films.

The same effect explains the colors of the *Morpho* butterfly shown in Figure 23-24. The wings of *Morpho* are covered with microscopic scales that act like a thin film. The thickness of these scales is such that there is constructive interference for reflected light at wavelengths in the blue-green part of the spectrum. As a result, *Morpho* appears to glow at those wavelengths.

EXAMPLE 23-5 A Soapy Film

Monochromatic light that has wavelength 560 nm in air strikes a layer of soapy water, which has an index of refraction of 1.40 and rests on a bathroom tile. (a) If the layer of soapy water is 700 nm thick, does constructive interference, destructive interference, or neither occur when the light strikes the surface close to the normal? (b) What is the minimum thickness of soapy water that would result in no (or minimum) reflection from the surface?

Set Up

The situation is the same as in Figure 23-25. Interference occurs between (i) light that is reflected from the top surface of the soapy layer and (ii) light that enters the soapy layer and is eventually reflected back out. In part (a) we'll compare the given values of wavelength and film thickness to Equations 23-17 and 23-19 to decide whether the interference is constructive, destructive, or something in between. In part (b) we'll use Equation 23-20 to find the minimum thickness for destructive interference. In both parts, Equation 23-16 will help us relate the wavelength of the light in air to its wavelength in the soapy water.

Constructive interference:

$$2D = m\lambda_2 = m\frac{n_1\lambda_1}{n_2}, \quad m = 1, 2, 3, \ldots \qquad (23\text{-}17)$$

Destructive interference:

$$2D = (2m-1)\frac{\lambda_2}{2} = (2m-1)\frac{n_1\lambda_1}{2n_2}, \quad m = 1, 2, 3, \ldots \qquad (23\text{-}19)$$

Minimum thickness for destructive interference:

$$D_{\text{min}} = \frac{n_1\lambda_1}{4n_2} \qquad (23\text{-}20)$$

Wavelength in two different media:

$$\lambda_2 = \frac{n_1\lambda_1}{n_2} \qquad (23\text{-}16)$$

Solve

(a) In this situation medium 1 is air ($n_1 = 1.00$) and medium 2, of which the film is made, is soapy water ($n_2 = 1.40$). The wavelength in the soapy water is therefore shorter than in air.

Wavelength in air: $\lambda_1 = 560$ nm

Wavelength in soapy water:

$$\lambda_2 = \frac{n_1\lambda_1}{n_2} = \frac{(1.00)(560\text{ nm})}{1.40} = 400\text{ nm}$$

Find how many wavelengths of the wavelength λ_2 in the soapy water fit into the path length difference $\Delta_{\text{pl}} = 2D$, where $D = 700$ nm is the film thickness.

The path length difference between the waves that reflect off the top and bottom surfaces of the film is

$$\Delta_{\text{pl}} = 2D = 2(700\text{ nm}) = 1400\text{ nm}$$

The number of wavelengths that fit into this path length difference equals Δ_{pl} divided by λ_2:

$$\frac{\Delta_{\text{pl}}}{\lambda_2} = \frac{1400\text{ nm}}{400\text{ nm}} = 3.5 = \frac{7}{2}$$

The path length difference is an odd number of half wavelengths. So there is destructive interference between the light that reflects from the top surface of the soapy water and the light that reflects from the bottom surface (where the soapy water touches the tile).

(b) The minimum thickness required for destructive interference is such that the path length difference $2D$ is one half-wavelength, so $2D = \lambda_2/2$ and $D = \lambda_2/4$.

From Equation 23-20, the minimum thickness for destructive interference is

$$D_{\text{min}} = \frac{n_1\lambda_1}{4n_2} = \frac{(1.00)(560\text{ nm})}{4(1.40)} = 100\text{ nm}$$

Alternatively, since $\lambda_2 = n_1\lambda_1/n_2$ from Equation 23-16,

$$D_{\text{min}} = \frac{\lambda_2}{4} = \frac{400\text{ nm}}{4} = 100\text{ nm}$$

Reflect

When light of wavelength 560 nm in air strikes the soapy layer close to the normal, destructive interference occurs both when the layer is 100 nm thick (so twice the thickness is 200 nm, or 1/2 the wavelength in the soapy water) and when the layer is 700 nm thick (so twice the thickness is 1400 nm, or 7/2 the wavelength in the soapy water). For these thicknesses, reflections from the surface would be minimized, and the surface would look dark.

Destructive interference always occurs when twice the thickness of the layer is an odd multiple of one-half of the wavelength. Can you see that this would also occur if the film of soapy water were either 300 nm or 500 nm thick?

Figure 23-1c shows another example of thin-film interference. This film was made by dipping an open ring into soapy water and then holding the ring vertically. Some of the light that strikes the soap film is reflected from the front surface, while some passes into the film before being reflected at the back surface. When these two light waves recombine, the wavelength of light (the color) that results in constructive interference appears bright. The thickness of the film increases from top to bottom, so the path difference for the two waves is different at different places on the film. That's why the brightest color you see (corresponding to the wavelength for which there is constructive interference) is different at different positions. The thickness of the film is relatively constant *across* the film, however, so the colors appear in bands.

Note that the very top of the soap film in Figure 23-1c appears dark. That may come as a surprise because the top of the film is very thin (far thinner than a wavelength of visible light), so the path difference between light that reflects from the front and back surfaces should be negligible. As a result, there should be constructive interference at the top of the film, and the top should appear bright rather than dark. Why is our prediction incorrect?

To see the explanation, we need to go back to our discussion in Chapter 13 of how waves are reflected from a boundary. In Section 13-6 we saw that a wave pulse on a rope is inverted as it reflects from a fixed boundary. This inversion is equivalent to a phase shift of one-half of a wavelength; the position along the wave that arrived at the boundary as a peak has been reflected as a trough. In general, a wave traveling in one medium is inverted when it reflects (either partially or completely) from a second medium in which the wave speed is lower. (For the rope, the wave speed is zero on the other side of the boundary, which is definitely lower than the speed along the rope.) An inversion (a one-half wavelength phase shift) happens to the light waves in **Figure 23-26a** when light waves reflect off the front surface of the thin film because light travels slower in the film of index of refraction $n_2 > n_1$. The same phase shift happens to the light waves that reflect off the opaque background at the back surface of the thin film. Since both waves undergo the same phase shift due to reflection, any phase difference between the two waves is due to the path difference, as we discussed above.

But as shown in **Figure 23-26b**, things are different for the soap film in Figure 23-1c. Again there is a one-half wavelength phase shift for waves that reflect off the front surface of the film. But there is *no* phase shift for waves that reflect off the back surface, since light waves travel *faster* in the medium on the other side of that boundary. So even if the film were extremely thin, so there were no path difference between the light waves that reflect off the front and back surfaces of the film, there would still be a one-half wavelength phase difference between these two waves. This would result in destructive interference. That's just what we see at the top of the soap film in Figure 23-1c,

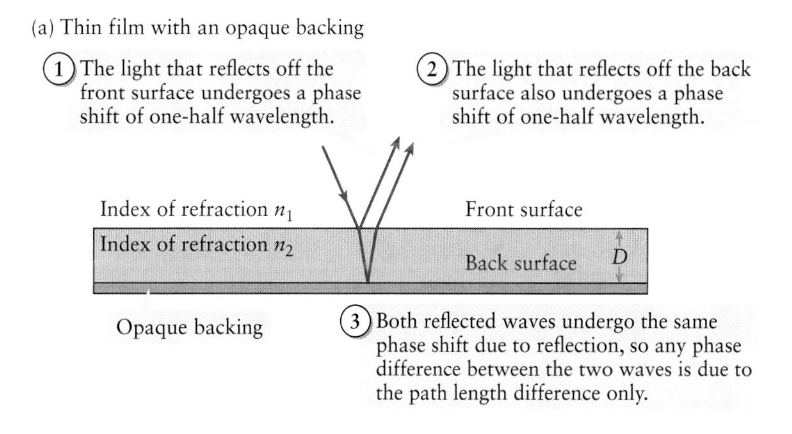

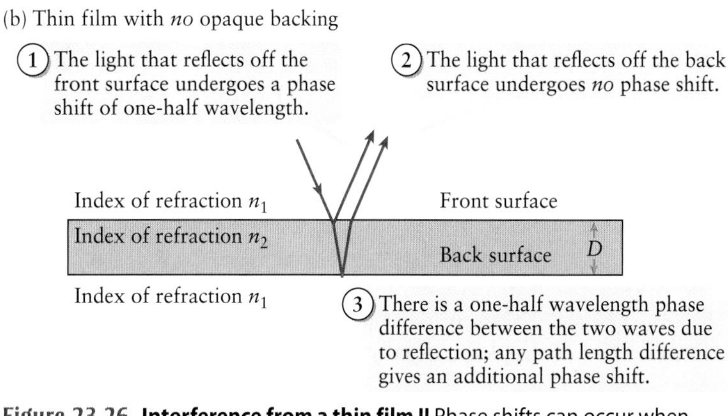

(b) Thin film with *no* opaque backing
1 The light that reflects off the front surface undergoes a phase shift of one-half wavelength.
2 The light that reflects off the back surface undergoes *no* phase shift.
Index of refraction n_1
Front surface
Index of refraction n_2
Back surface
D
Index of refraction n_1
3 There is a one-half wavelength phase difference between the two waves due to reflection; any path length difference gives an additional phase shift.

Figure 23-26 Interference from a thin film II Phase shifts can occur when light reflects from the surfaces of a thin film.

where the film is very thin compared to the wavelength of the light. This explains the dark band across the top of the film.

Thanks to the additional half-wavelength phase shift for a film like that in Figure 23-1c and Figure 23-26b, Equation 23-17 no longer tells us the condition for constructive interference, and Equation 23-19 no longer tells us the conditions for destructive interference. Instead the roles of these equations are reversed: Equation 23-17 is now the condition for *destructive* interference, and Equation 23-19 is now the condition for *constructive* interference.

EXAMPLE 23-6 Reducing the Reflection

The glass in an LCD display is sometimes coated with a transparent thin film to minimize glare. Interference from the front and back surfaces of such an *antireflective coating* minimizes the amount of light reflected from the display, particularly light that strikes normal to the display's surface. One material used in such coatings is zirconium acrylate, which has an index of refraction of 1.54. What is the minimum thickness of zirconium acrylate that will accomplish the desired reduction in reflection for light that has wavelength 560 nm in air? Note that the index of refraction of zirconium acrylate is higher than that of glass.

Set Up

We want the light reflected from the surface of the coating to interfere destructively with the light reflected from the zirconium acrylate-to-glass boundary. This requires that when these two light waves recombine, they are shifted by an odd number of half-wavelengths. One half-wavelength shift occurs because the light in air is inverted when it reflects off the surface of the zirconium acrylate, but the light in zirconium acrylate undergoes no shift when it reflects off the glass. (That's because light travels more slowly in zirconium acrylate than in air, but faster in glass than in zirconium acrylate.) Any additional shift arises from the path difference of $2D$ between the two waves. For the total shift to be equivalent to an odd number of half-wavelengths, the shift due to the path difference must be an *integer* number of wavelengths. The minimum thickness corresponds to $2D$ equal to one wavelength λ_2, the wavelength in the zirconium acrylate as given by Equation 23-16.

Wavelength in two different media:

$$\lambda_2 = \frac{n_1\lambda_1}{n_2} \qquad (23\text{-}16)$$

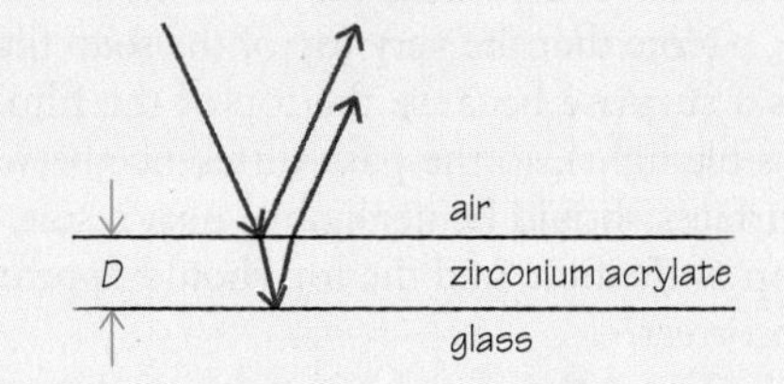

Solve

First calculate the wavelength of the light in zirconium acrylate.

We are given the wavelength in air ($n_1 = 1.00$): $\lambda_1 = 560$ nm. The wavelength in zirconium acrylate ($n_2 = 1.54$) is given by Equation 23-16:

$$\lambda_2 = \frac{n_1\lambda_1}{n_2} = \frac{(1.00)(560 \text{ nm})}{1.54} = 364 \text{ nm}$$

The minimum film thickness D_{min} for destructive interference is such that $2D_{min}$ equals λ_2. Use this to solve for D_{min}.

The condition for the minimum thickness that leads to destructive interference is

$$2D_{min} = \lambda_2 = 364 \text{ nm}$$

$$D_{min} = \frac{364 \text{ nm}}{2} = 182 \text{ nm}$$

Reflect

For most people light sensitivity peaks in the range of 555 to 565 nm, which we perceive as yellow. That's why antireflective coatings are usually optimized for yellow light (which includes the 560-nm wavelength we've used here).

Why have we been emphasizing interference due to light reflecting from a *thin* film? The explanation is that in this section we've assumed that a steady, continuous train of light waves is incident on the film. However, the light from ordinary sources such as light bulbs and the Sun is emitted in a sequence of short bursts, each of which is a segment of wave no more than a few micrometers to about a millimeter in length. This is called the *coherence length* of the light. The phase of the wave changes randomly from one burst to the next. If the thickness of the film is small compared to the coherence length, then the two waves that interfere—the light that reflects from the back of the film and the light that reflects from the front of the film—are part of the same burst. In this case the phase relationships we have developed in this section (which assumed a steady train of waves) are valid. But if the thickness of the film is large compared to the coherence length, the two waves are likely to be from different wave bursts and will differ in phase by a random and rapidly changing amount. As a result, the interference between the waves will be neither always constructive nor always destructive, and any interference effects will be wiped out. That's why you won't see interference effects like those we've described in this section from a thick film like an ordinary pane of glass, which is several millimeters deep. (However, you *can* see interference effects for a glass pane if the light source is a laser. A laser produces light in a very different way from an ordinary light bulb, and the coherence length can be several meters.)

GOT THE CONCEPT? 23-6 Inversion on Reflection

Figure 23-27 shows light shining on a thin layer of oil that has an index of refraction of 1.4. The oil layer is atop a piece of glass that has an index of refraction of 1.5. On the underside of the glass is air. At which boundary or boundaries does the reflected light undergo an inversion? (a) The air–oil boundary; (b) the oil–glass boundary; (c) the glass–air boundary; (d) more than one of these; (e) none of these.

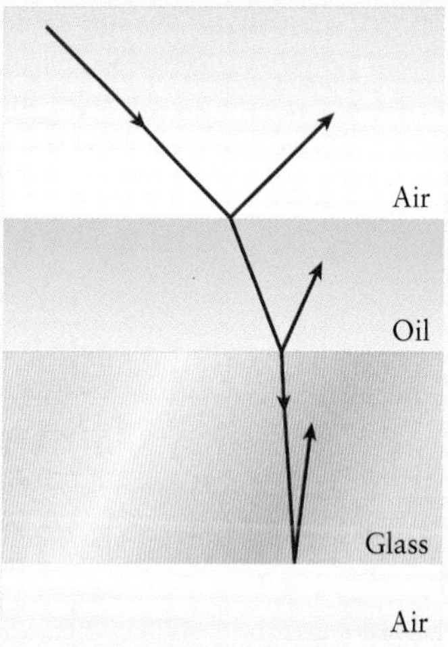

Air

Oil

Glass

Air

Figure 23-27 Layers of reflection At which boundary or boundaries does the reflected light undergo an inversion?

TAKE-HOME MESSAGE FOR Section 23-6

✔ Light waves can interfere constructively or destructively depending on how the peaks and troughs of two waves align when they combine.

✔ When a single beam of light strikes a thin film of a transparent material, some of the light is reflected from the front surface, while some enters the film and is reflected off its back surface. These two reflected light waves may not be in phase when they recombine, resulting in interference. Whether there is destructive interference, constructive interference, or neither depends on the number of wave cycles that fit into the extra distance traveled by the light that enters the thin layer. It also depends on what material is in contact with the back surface of the film.

23-7 Interference can occur when light passes through two narrow, parallel slits

Figure 23-28 shows a remarkable experiment into the properties of light. Light from a laser of a single wavelength λ (that is, monochromatic light) is directed at an opaque obstacle that has two narrow slits, S_1 and S_2, separated by a distance d. The light that emerges from the slits falls on a screen a distance L away.

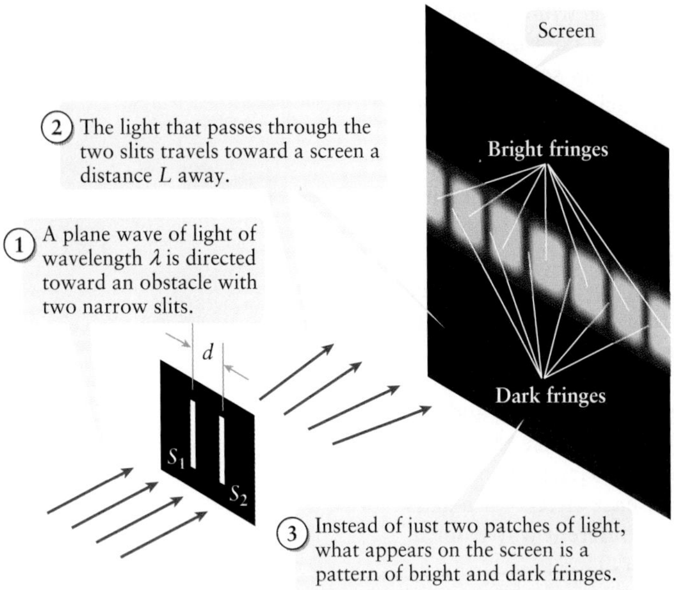

Screen

(2) The light that passes through the two slits travels toward a screen a distance L away.

Bright fringes

(1) A plane wave of light of wavelength λ is directed toward an obstacle with two narrow slits.

d

Dark fringes

S_1

S_2

(3) Instead of just two patches of light, what appears on the screen is a pattern of bright and dark fringes.

Figure 23-28 The double-slit experiment The light waves that pass through slits S_1 and S_2 produce a pattern of bright and dark fringes.

You might expect that there would be just *two* bright patches on the screen, one produced by the light that passed through slit S_1 and one produced by the light that passed through slit S_2. But in fact what we see is a *series* of bright patches with dark patches between them, collectively called **fringes**. We can explain this curious result using two key ideas about light waves: Huygens' principle and interference.

According to Huygens' principle (Section 23-2), each point on the incident plane wave that strikes the slits in Figure 23-28 acts as a source of wavelets. Since the two slits S_1 and S_2 are very narrow, we can treat each slit as a *single* source of this kind. Each crest of the incident plane wave strikes both slits simultaneously, so the wavelets of light that emanate from each slit start off in phase with each other (**Figure 23-29**). But because the wavelets travel different distances to reach various locations on the screen, they may or may not arrive at the screen in phase. There are three possibilities:

1. At certain locations on the screen the two wavelets arrive in phase with each other, so crests from S_1 and S_2 arrive at the same time. So at these locations there is constructive interference, the total wave has maximum amplitude, and the light reaching the screen is brightest. These **interference maxima** correspond to the centers of the bright fringes shown in Figures 23-28 and 23-29.

2. At certain other locations on the screen, the two wavelets arrive out of phase, so a crest from S_1 arrives at the same time as a trough from S_2 and vice versa. At these locations there is destructive interference, the total wave is zero, and no light reaches the screen. The centers of the dark fringes shown in Figure 23-28 and 23-29 correspond to these **interference minima.**

3. At all other locations on the screen, the interference between the wavelets from S_1 and S_2 is neither completely constructive nor completely destructive. At these locations a crest from S_1 does not arrive at the same time as either a crest or a trough from S_2, so the total wave has neither maximum amplitude nor zero amplitude. These are the locations in Figure 23-28 and 23-29 between the center of a bright fringe and the center of a dark fringe.

(2) Wavelets emanate from slit S_1 (shown in blue) and from slit S_2 (shown in red).

(3) Where wavelets from the two slits arrive at the screen *in phase*, there is an interference *maximum* (a *bright* fringe).

(1) Successive crests of a plane wave of light are incident on the two slits S_1 and S_2.

(4) Where wavelets from the two slits arrive at the screen *out of phase*, there is an interference *minimum* (a *dark* fringe).

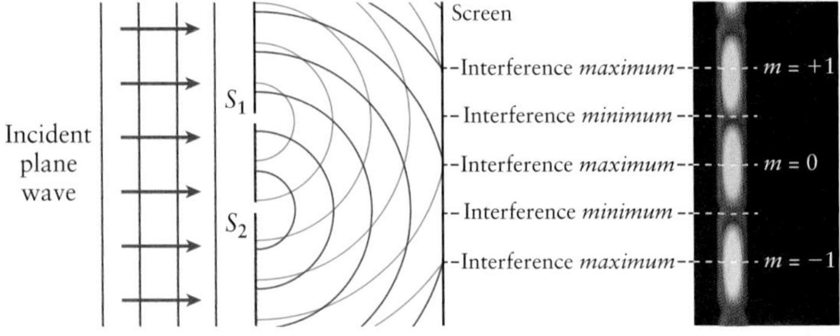

Figure 23-29 Huygens' principle and double-slit interference The pattern of bright and dark fringes arises from interference between wavelets emanating from slit S_1 and from slit S_2.

Because the wavelets that interfere with each other come from a pair of slits, the experiment shown in Figures 23-28 and 23-29 is called **double-slit interference** and the pattern of bright and dark fringes is called an **interference pattern.** Such interference patterns are conclusive evidence that light is a wave. (In Section 13-5 we discussed the same sort of interference that occurs when there are water waves or sound waves produced by two sources, analogous to the light waves emanating from the two slits in Figures 23-28 and 23-29.)

You may be wondering why it's important to use a laser to demonstrate double-slit interference. The reason is that the waves that enter slit S_1 and the waves that enter slit S_2 must have a constant phase relationship with each other. As we mentioned at the end of Section 23-6, laser light has a high degree of coherence, so the waves entering the two closely spaced slits are guaranteed to have such a constant relationship. (In Figure 23-29 the relationship is that the two waves are in phase.) If you were to use an ordinary light bulb with a color filter to make the light monochromatic before entering the slits, the waves entering slits S_1 and S_2 would be from different short bursts emitted by the light bulb. The phase relationship between these bursts would change from one moment to the next. Hence there would be no definite interference pattern, and on the screen in Figure 23-28 you would see two closely spaced bright spots, one caused by light from slit S_1 and one from slit S_2.

Locating the Interference Maxima and Minima

Let's see how to determine the locations of the double-slit interference maxima and minima on the screen in Figures 23-28 and 23-29. Just as for the thin-film interference that we studied in Section 23-6, what determines whether the interference is completely constructive, completely destructive, or intermediate between those two extremes is the *path difference* $\Delta_{pl} = D_2 - D_1$, the difference between the distance D_1 that a wave from slit S_1 travels to a point P on the screen and the distance D_2 that a wave from slit S_2 travels to that same point (**Figure 23-30a**). The conditions for constructive and destructive interference are

Constructive interference:

$$\Delta_{pl} = \text{a whole number of wavelengths} = 0, \pm\lambda, \pm 2\lambda, \pm 3\lambda, \ldots$$

Destructive interference:

$$\Delta_{pl} = \text{an odd number of half-wavelengths} = \pm\lambda/2, \pm 3\lambda/2, \pm 5\lambda/2, \ldots \qquad (23\text{-}21)$$

Note that the path difference Δ_{pl} is positive for locations in the upper half of the screen. These locations are farther from slit S_2 than from slit S_1, so D_2 is greater than D_1 (the case shown in Figure 23-29). Locations in the lower half of the screen are closer to slit S_2 than to slit S_1, so for these locations D_2 is less than D_1 and the path difference Δ_{pl} is negative.

In typical double-slit experiments the spacing d between the slits is about a millimeter and the distance L from the slits to the screen is a meter or more. Since L is so

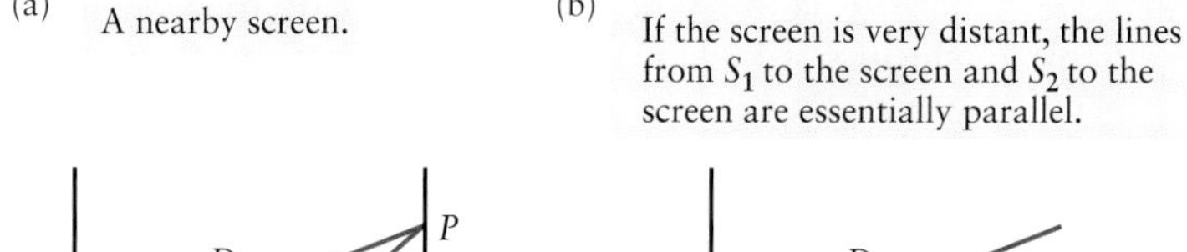

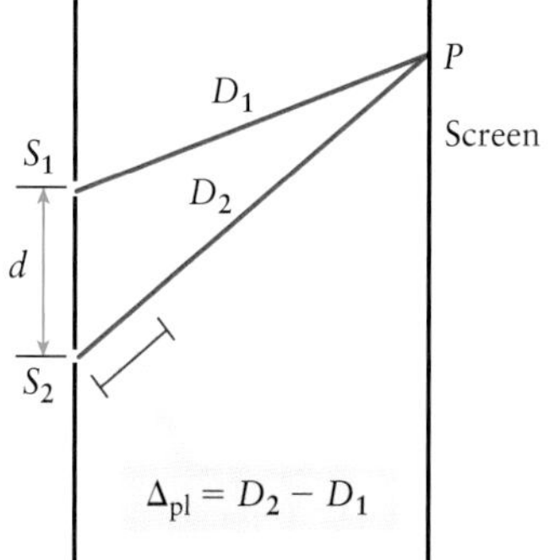

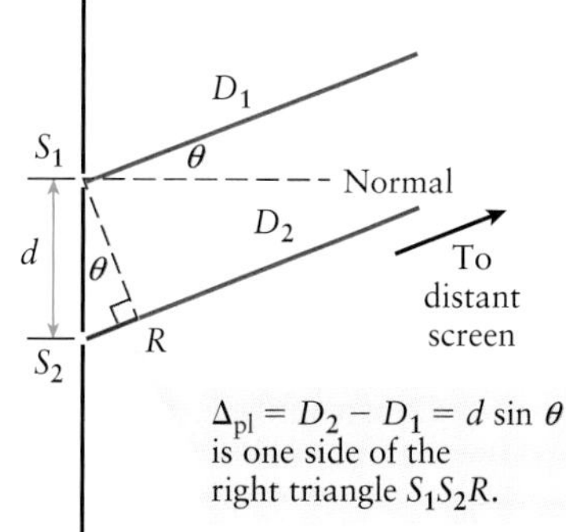

Figure 23-30 Calculating the path difference for double-slit interference (a) The path difference Δ_{pl} for a point P on the screen. (b) If the distance to the screen is large compared to the distance d between the slits, Δ_{pl} is related to d and the angle θ to the point P.

large compared to d, it's a good approximation to treat the screen as being infinitely far away. Then the straight lines from slit S_1 to the point P on the screen and from slit S_2 to point P are parallel to each other, and are both at an angle θ from the normal to the opaque obstacle with the two slits (**Figure 23-30b**). The path difference Δ_{pl} is then one side of a right triangle S_1S_2R with hypotenuse d. The angle opposite the side of length Δ_{pl} is θ, and the sine of θ equals the length of this opposite side divided by the length d of the hypotenuse:

$$\sin\theta = \frac{\Delta_{pl}}{d} \quad \text{so} \quad d\sin\theta = \Delta_{pl} \tag{23-22}$$

If we combine Equations 23-21 and 23-22, we get the following conditions for a bright fringe (constructive interference) and for a dark fringe (destructive interference):

Double-slit experiment: Condition for constructive interference (23-23)

Constructive interference (bright fringes): $d \sin\theta = m\lambda$

Distance between the two slits

Wavelength of the light illuminating the two slits

Angle between the normal to the two slits and the location of a **bright fringe** on the screen

Number of the bright fringe: $m = 0, \pm1, \pm2, \pm3, \ldots$

Double-slit experiment: Condition for destructive interference (23-24)

Destructive interference (dark fringes): $d \sin\theta = \left(m + \frac{1}{2}\right)\lambda$

Distance between the two slits

Wavelength of the light illuminating the two slits

Angle between the normal to the two slits and the location of a **dark fringe** on the screen

Number of the dark fringe: $m = 0, \pm1, \pm2, \pm3, \ldots$

Equation 23-23 gives the angles θ at which the bright fringes are found. Each of the bright fringes in Figure 23-29 is labeled with the corresponding values of m. For example, for the $m = 0$ bright fringe it follows that $d \sin\theta = 0$, so $\sin\theta = 0$ and $\theta = 0$. Hence the $m = 0$ bright fringe is at the center of the pattern. For the other bright fringes, $\sin\theta = m\lambda/d = +\lambda/d$ (for $m = +1$), $-\lambda/d$ (for $m = -1$), $+2\lambda/d$ (for $m = +2$), $-2\lambda/d$ (for $m = -2$), ...

WATCH OUT! Be careful with numbering the dark fringes.

! Equation 23-24 gives the angles θ at which the dark fringes are found. Note that $m = 0$ in this equation does *not* correspond to the center of the interference pattern (where there is a bright fringe). Instead this is a dark fringe given by $d \sin\theta = (1/2)\lambda$, so $\sin\theta = \lambda/2d$. This dark fringe lies between the $m = 0$ bright fringe and the $m = +1$ bright fringe shown in Figure 23-29. For the $m = +1$ dark fringe, $\sin\theta = [1 + ½]\lambda/d = +3\lambda/2d$; this dark fringe lies between the $m = +1$ and $m = +2$ bright fringes. For the $m = -1$ dark fringe, $\sin\theta = [-1 + ½]\lambda/d = -\lambda/2d$; this dark fringe lies between the $m = 0$ and $m = -1$ bright fringes.

If we divide both sides of Equation 23-23 and both sides of Equation 23-24 by d, we see that that the value of $\sin\theta$ for either a given bright fringe or a given dark fringe is proportional to λ/d, the ratio of the wavelength to the slit spacing:

$$\text{Constructive interference (bright fringe): } \sin\theta = m\frac{\lambda}{d}$$

$$\text{Destructive interference (dark fringe): } \sin\theta = \left[m + \frac{1}{2}\right]\frac{\lambda}{d} \tag{23-25}$$

Equations 23-25 show that if we increase the wavelength λ, the value of $\sin\theta$ and hence of θ for each fringe increases. This means that the entire interference pattern becomes broader. If instead we increase the slit spacing d, the value of $\sin\theta$ and hence of θ for each fringe decreases. In this case the entire interference pattern becomes narrower.

EXAMPLE 23-7 Measuring Wavelength Using Double-Slit Interference

You send light from a laser through two narrow slits spaced 0.100 mm apart, producing an interference pattern on a screen 5.00 m away. The angle from the central bright fringe to the next bright fringe is 0.300°. (a) What is the wavelength of the light? (b) What is the distance on the screen from the center of the interference pattern to the first dark fringe?

Set Up

We are given the slit spacing $d = 0.100$ mm and the angle $\theta = 0.300°$ for the first bright fringe, which corresponds to $m = 1$ in Equation 23-23. In part (a) we'll use this equation to solve for the wavelength λ. Given the wavelength, in part (b) we'll find the angle of the first dark fringe using Equation 23-24 with $m = 0$. We'll then use trigonometry to find the desired distance.

Bright fringes: $d\sin\theta = m\lambda$ (23-23)

Dark fringes: $d\sin\theta = [m + \frac{1}{2}]\lambda$ (23-24)

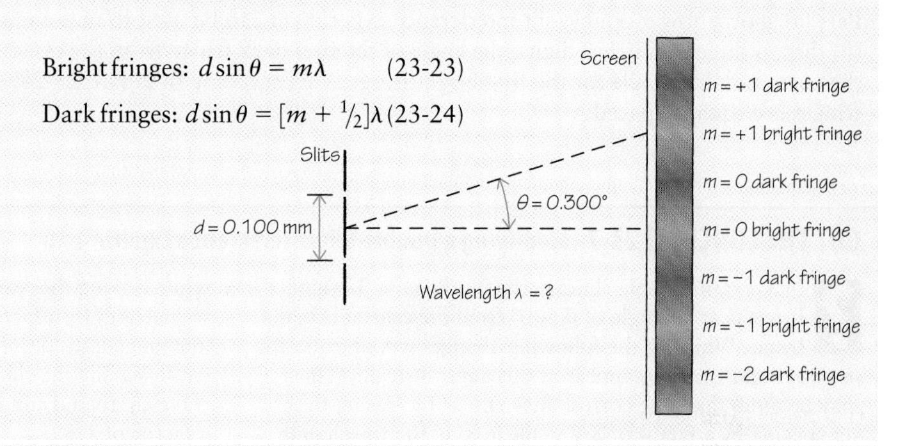

Solve

(a) Solve Equation 23-23 for the wavelength λ.

We have $d = 0.100$ m $= 1.00 \times 10^{-4}$ m for the slit spacing, and $\theta = 0.300°$ for the angle of the first bright fringe away from the central bright fringe. From Equation 23-23 with $m = 1$,

$$\lambda = d\sin\theta = (1.00 \times 10^{-4}\ \text{m})\sin 0.300°$$
$$= (1.00 \times 10^{-4}\ \text{m})(5.24 \times 10^{-3})$$
$$= 5.24 \times 10^{-7}\ \text{m} = 524\ \text{nm}$$

This is in the green part of the visible spectrum.

(b) Use Equation 23-24 to find the angle of the first dark fringe.

The first dark fringe corresponds to $m = 0$ in Equation 23-24. Solve this for θ:

$$d\sin\theta = (0 + \tfrac{1}{2})\lambda = \lambda/2$$

$$\sin\theta = \frac{\lambda}{2d} = \frac{5.24 \times 10^{-7}\ \text{m}}{2(1.00 \times 10^{-4}\ \text{m})}$$
$$= 2.62 \times 10^{-3}$$
$$\theta = \sin^{-1}(2.62 \times 10^{-3}) = 0.150°$$

The point P on the screen where the first dark fringe lies, the center of the interference pattern C, and the point M midway between the two slits define a right triangle. The side of this triangle adjacent to the angle $\theta = 0.150°$ is the distance $L = 5.00$ m from the slits to the screen, and the side opposite this angle is the distance D that we want.

From trigonometry,

$$\tan\theta = \frac{\text{opposite side}}{\text{adjacent side}} = \frac{D}{L}$$

Solve for the distance D from the center of the interference pattern to the first dark fringe:

$$D = L \tan\theta = (5.00 \text{ m}) \tan 0.150°$$
$$= (5.00 \text{ m})(2.62 \times 10^{-3}) = 0.0131 \text{ m} = 1.31 \text{ cm}$$

Reflect

Part (a) shows how a double-slit interference experiment can be used to measure the wavelength of light. Note from part (b) that to three significant digits, the angle of the first dark fringe from the center of the interference pattern is one-half the corresponding angle for the first bright fringe. This agrees with Figure 23-28, which shows that the bright and dark fringes are equally spaced.

GOT THE CONCEPT? 23-7 Modifying a Double-Slit Interference Experiment

In a certain double-slit experiment the $m = 3$ bright fringe is at an angle of 0.500° from the central bright fringe. Which of the following changes would cause the $m = 2$ bright fringe to appear at this angle instead? (There may be more than one correct answer.) (a) Increasing the slit spacing by a factor of 3/2 while leaving the wavelength unchanged; (b) decreasing the slit spacing by a factor of 2/3 while leaving the wavelength unchanged; (c) increasing the wavelength by a factor of 3/2 while leaving the slit spacing unchanged; (d) decreasing the wavelength by a factor of 2/3 while leaving the slit spacing unchanged; (e) doubling the wavelength and simultaneously decreasing the slit spacing by a factor of 1/3.

TAKE-HOME MESSAGE FOR Section 23-7

✔ When light waves strike a pair of narrow slits, the waves that emerge from the two slits give rise to an interference pattern with bright and dark fringes.

✔ The positions of the fringes are determined by the ratio of the slit spacing to the wavelength.

23-8 Diffraction is the spreading of light when it passes through a narrow opening

In Section 23-7 we used Huygens' principle to understand the bright and dark interference fringes formed when light passes through two narrow slits. Huygens' principle will also allow us to understand **diffraction**, in which waves tend to spread out when they pass through a narrow opening or near the sharp edge of an object. We will see that diffraction can also give rise to a pattern of bright and dark fringes.

Consider a wave front of a plane wave that passes through an opening in an obstacle, as in **Figure 23-31a**. The opening is wide compared to the wavelength of the wave. Many of the wavelets pass through the opening, as in **Figure 23-31b**. The resulting wave on the other side of the obstacle is mostly a new plane wave, though the ends of the wave front are curved. This means that most of the wave energy continues straight through the opening, with only a small fraction "leaking" to the sides.

Something rather different happens when the opening is comparable in size to or smaller than the wavelength. That's the situation in **Figure 23-32a**, in which the opening is much narrower than in Figure 23-31a. Because the opening is narrow, only a very few of the wavelets that comprise each wave front pass through it. And because so few wavelets get through, they cannot reproduce the straight wave front that was incident

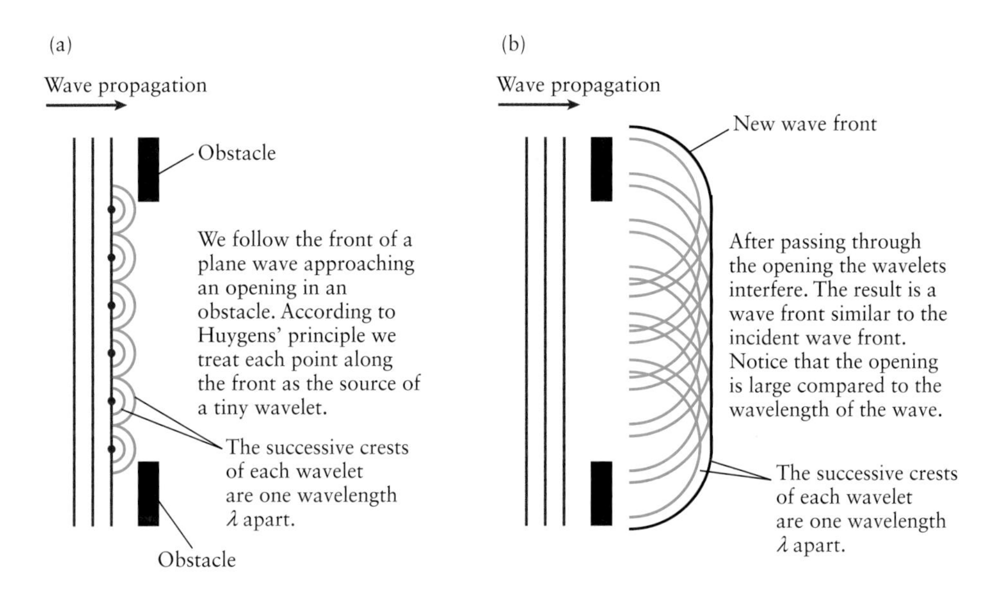

Figure 23-31 Diffraction I: A wide opening Huygens' principle predicts that if the width of the opening is large compared to the wavelength, most of the wave continues straight ahead through the opening. A slight amount diffracts to the sides.

on the opening. After passing through the narrow opening, the wave spreads out; it has undergone diffraction (**Figure 23-32b**). Diffraction is also present in Figure 23-31b, as shown by the waves that "leak" to the side of the much wider opening, but to a much smaller greatly reduced extent.

Figure 23-33 shows these effects for water waves passing through an opening in an obstacle. If the width of the opening is large compared to the wavelength, as in Figure 23-33a, there are almost no effects of diffraction. But if the opening is comparable in size to the wavelength as in Figure 23-33b, the wave spreads out. This explains why you hear someone talking on the other side of an open door, even if you're not directly in front of it. The wavelengths of sound used in human speech are in the range of a few meters, comparable in size to the width of a typical door (about one meter). As a result sound waves emerging from a door spread out like the water waves in Figure 23-33b, making it easy to eavesdrop on conversations. However, you can't *see* through an open door if you stand to one side because the wavelengths of visible light (around 550 nm, or 5.5×10^{-7} m) are very small compared to the width of the door. So light waves passing through the door behave like the water waves in Figure 23-33a.

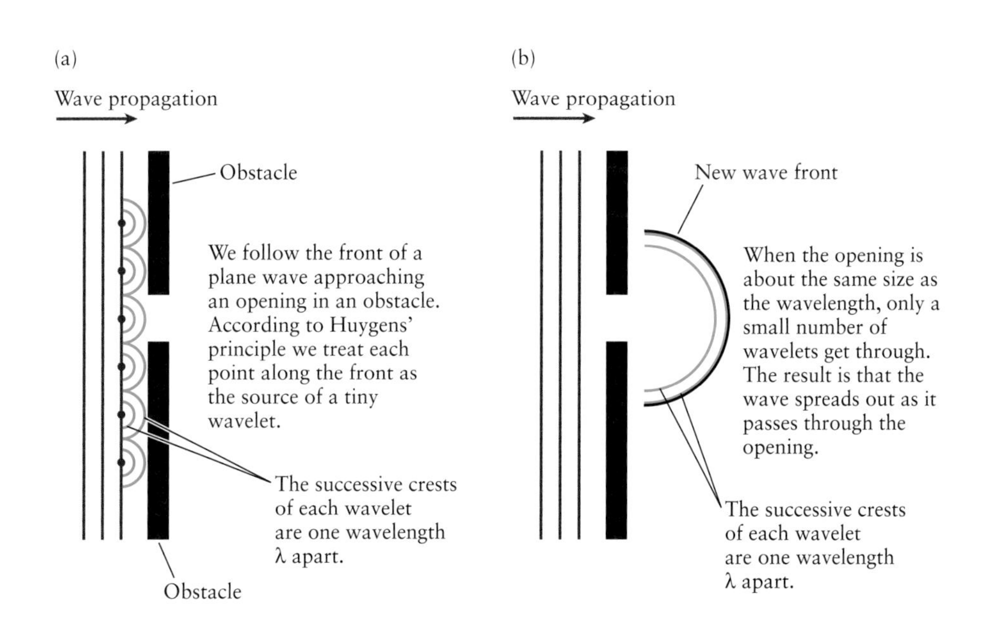

Figure 23-32 Diffraction II: A narrow opening If the width of the opening is comparable to the wavelength, the wave spreads out substantially after exiting the opening. (Compare Figure 23-31.)

Figure 23-33 Water wave diffraction Compare these photographs to Figures 23-31 and 23-32.

(a) Water waves pass through a wide opening.

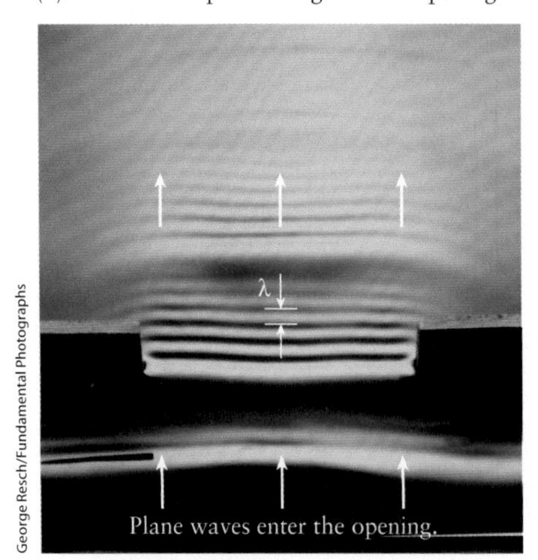
λ
Plane waves enter the opening.

George Resch/Fundamental Photographs

- The width of the opening is large compared to the wavelength λ.
- Most of the wave energy continues straight ahead, so there is very little diffraction.

(b) Water waves pass through a narrow opening.

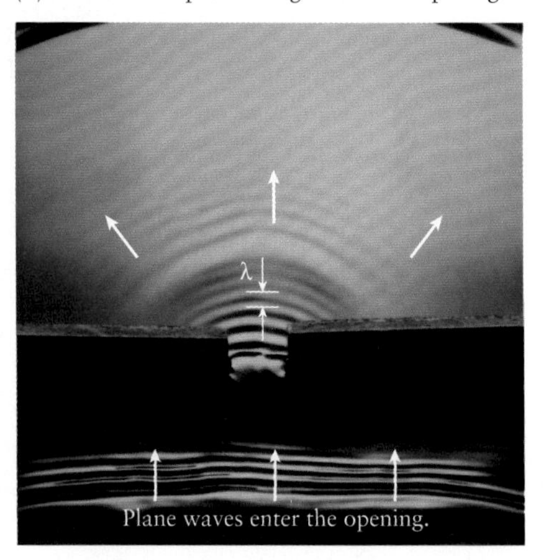
λ
Plane waves enter the opening.

- The width of the opening is comparable in size to the wavelength λ.
- The wave spreads out after passing through the opening; diffraction is important.

To consider diffraction in more detail, imagine that we send a beam of light of a single wavelength λ through a long, narrow slit of width w. After passing through the slit the light falls on a screen a distance L away (**Figure 23-34**). You might expect that the pattern on the screen would be a single blob of light. But as the photo in Figure 23-34 shows, what actually appears is a series of bright and dark fringes. In Section 23-7 we saw how such fringes arise in the *double*-slit experiment due to interference between Huygens wavelets emanating from the first slit and wavelets emanating from the second slit. In this **single-slit diffraction** experiment, the fringes arise due to interference between wavelets emanating from different parts of the *same* slit. Huygens wavelets emanate from each part of the slit. At some locations, these wavelets interfere destructively; these are the locations of the dark fringes, also called **diffraction minima**. At locations between the dark fringes, the wavelets interfere more or less constructively and we see bright fringes, also called **diffraction maxima**.

Let's see how to determine the positions of the dark fringes. The first dark fringe is found where light waves from the upper half of the slit, of width $w/2$, and the lower half, also of width $w/2$, interfere destructively. This means that the waves from one half of the slit must travel one half-wavelength farther from the slit to the screen than do the waves from the other half, so the path length difference Δ_{pl} equals $\lambda/2$. We can use this to easily determine the position of the first dark fringe if we assume that the distance L to the screen is much greater than the slit width w. With this assumption,

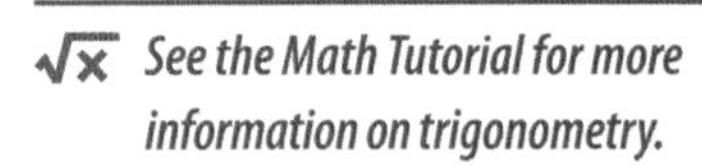
√x See the Math Tutorial for more information on trigonometry.

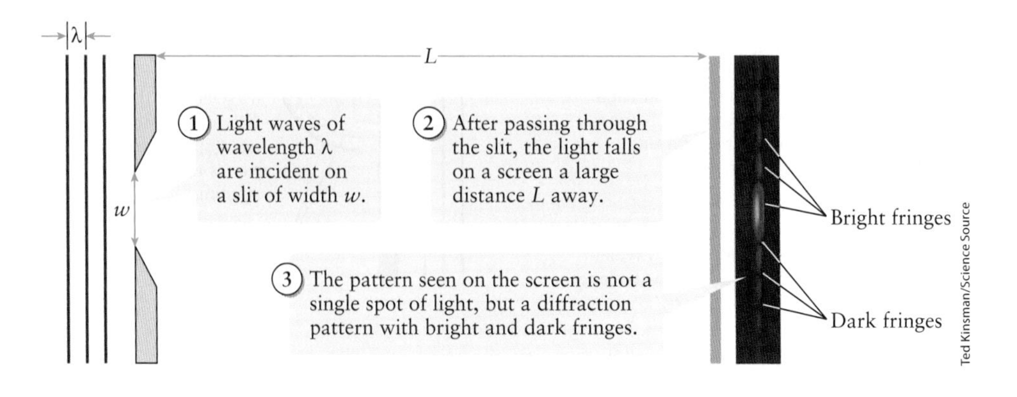
λ
L
w
1 Light waves of wavelength λ are incident on a slit of width w.
2 After passing through the slit, the light falls on a screen a large distance L away.
3 The pattern seen on the screen is not a single spot of light, but a diffraction pattern with bright and dark fringes.
Bright fringes
Dark fringes

Ted Kinsman/Science Source

Figure 23-34 Diffraction through a single slit A diffraction pattern arises when monochromatic light passes through a narrow slit.

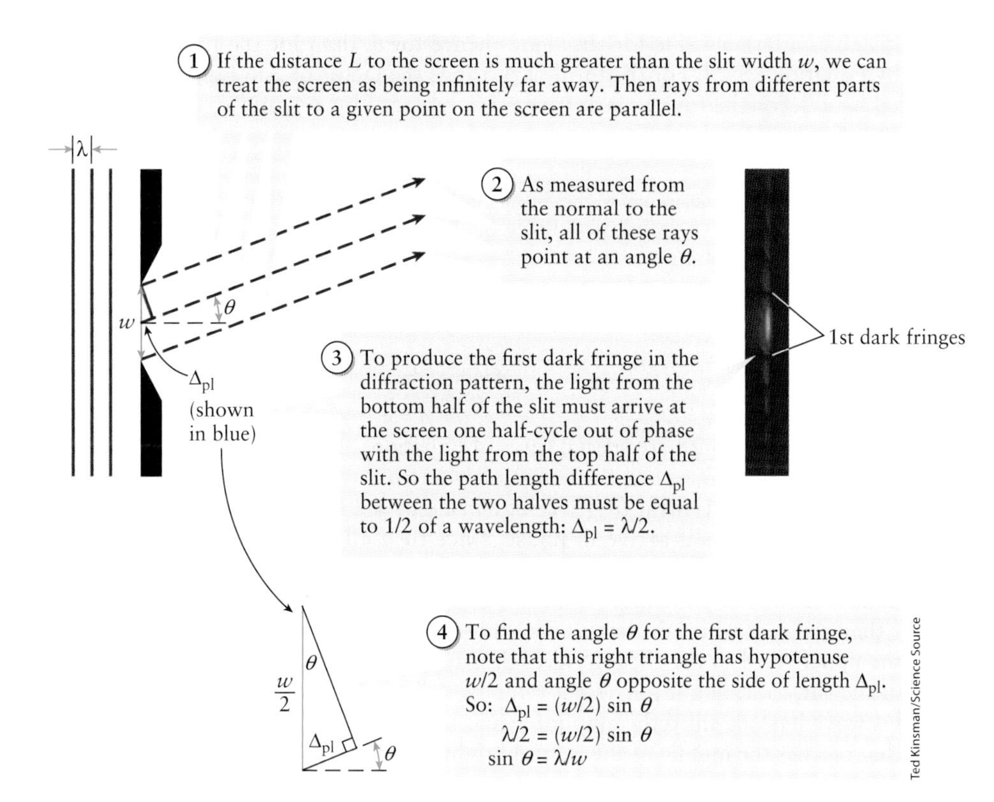

Figure 23-35 Calculating the single-slit diffraction pattern Dark fringes are found where light from each part of the slit interferes destructively with light from another part.

Figure 23-35 shows that for a given position on the screen, Δ_{pl} is related to the angle θ between the normal to the slit and a line to that position:

$$\Delta_{pl} = \left(\frac{w}{2}\right)\sin\theta \tag{23-26}$$

To find the angle for the first dark fringe, substitute $\Delta_{pl} = \lambda/2$ (the condition for destructive interference) into Equation 23-26:

$$\frac{\lambda}{2} = \left(\frac{w}{2}\right)\sin\theta$$

so

$$\sin\theta = \frac{\lambda}{w} \tag{23-27}$$

(first dark fringe in single-slit diffraction)

There are two of these first dark fringes, one on either side of the central bright fringe (see Figure 23-35).

To see how the second dark fringe arises, imagine breaking the slit into quarters, each of width $w/4$. Destructive interference occurs when light from the first quarter arrives at the screen out of phase with light from the second quarter, and light from the third quarter arrives at the screen out of phase with light from the fourth quarter. The path length difference Δ_{pl} between light from adjacent quarters is given by Equation 23-26 with $w/2$ replaced by $w/4$. The condition that $\Delta_{pl} = \lambda/2$ then tells us that

$$\frac{\lambda}{2} = \left(\frac{w}{4}\right)\sin\theta$$

If we solve this for $\sin\theta$, we get

$$\sin\theta = \frac{2\lambda}{w} \tag{23-28}$$

(second dark fringe in single-slit diffraction)

This gives a larger value for sin θ, and hence for θ, than for the first dark fringe given by Equation 23-27. For the third dark fringe the factor of 2 in Equation 23-28 is replaced by a 3, for the fourth dark fringe it is replaced by a 4, and so on. In general, the angle of the mth dark fringe is given by

Dark fringes in single-slit diffraction (23-29)

$$\sin\theta = \frac{m\lambda}{w}$$

Angle between the normal to the slit and the location of the mth **dark fringe**

Number of the dark fringe: $m = 1, 2, 3, \ldots$

Wavelength of the light

Width of the slit

Equations 23-27 and 23-28 are special cases of Equation 23-29, with $m = 1$ and $m = 2$ respectively.

The middle of the central *bright* fringe in Figures 23-34 and 23-35 is where all of the waves from the slit arrive in phase, since they all travel essentially the same distance to this point. This point, which corresponds to $\theta = 0$, is where the intensity is maximum. The next bright fringe, located between the first and second dark fringes, is fainter than the central bright fringe. Roughly speaking, at this bright fringe the light from the upper third of the slit interferes destructively with the light from the middle third of the slit, canceling it out. What remains is the light from the lower third of the slit. This wave has one-third the amplitude and hence $(1/3)^2 = 1/9$ the intensity of the wave that reaches the middle of the central bright fringe. (Recall from Section 22-4 that the intensity of an electromagnetic wave is proportional to the square of the amplitude of the electric field.) Successive bright fringes are even fainter. Note that the point of greatest intensity in the central bright fringe is at its center; for the other bright fringes the point of greatest intensity is close to (but slightly displaced from) a point halfway between the adjacent dark fringes.

Go to Interactive Exercise 23-4 for more practice dealing with diffraction.

EXAMPLE 23-8 Diffraction Through a Slit

A green laser pointer emits light at a wavelength of 532 nm. You aim the beam from this laser at a slit 1.50 μm wide. Find the angles of the first, second, and third dark fringes in the diffraction pattern.

Set Up

We are given the wavelength λ and the slit width w, and we want to find the value of the angle θ for the $m = 1, 2,$ and 3 dark fringes. We'll use Equation 23-29 for this purpose.

Dark fringes in single-slit diffraction:

$$\sin\theta = \frac{m\lambda}{w} \quad (23\text{-}29)$$

Solve

Find the angle of the first dark fringe ($m = 1$).

We have $\lambda = 532 \text{ nm} = 5.32 \times 10^{-7}$ m and $w = 1.50\ \mu\text{m} = 1.50 \times 10^{-6}$ m. From Equation 23-29, with $m = 1$,

$$\sin\theta_1 = \frac{\lambda}{w} = \frac{5.32 \times 10^{-7}\text{ m}}{1.50 \times 10^{-6}\text{ m}} = 0.355$$

$$\theta_1 = \sin^{-1} 0.355 = 20.8°$$

Find the angle of the second dark fringe ($m = 2$).

From Equation 23-29, with $m = 2$,

$$\sin\theta_2 = \frac{2\lambda}{w} = \frac{2(5.32 \times 10^{-7}\text{ m})}{1.50 \times 10^{-6}\text{ m}} = 0.709$$

$$\theta_2 = \sin^{-1} 0.709 = 45.2°$$

Find the angle of the third dark fringe ($m = 3$).

From Equation 23-29, with $m = 3$,

$$\sin \theta_3 = \frac{3\lambda}{w} = \frac{3(5.32 \times 10^{-7}\text{ m})}{1.50 \times 10^{-6}\text{ m}} = 1.06$$

This equation has *no* solution! The sine of an angle cannot be greater than 1, so there is no value of θ_3 that satisfies this equation. We are forced to conclude that the diffraction pattern in this situation has a first dark fringe at $\theta_1 = 20.8°$ and a second dark fringe at $\theta_2 = 45.2°$, but there is no third dark fringe before the end of the pattern at $\theta = 90°$.

Reflect

The number of dark fringes present in the diffraction pattern of a slit depends on the relative sizes of the wavelength λ and the slit width w. We explore this further below.

Figure 23-36 shows the intensity in the diffraction pattern of a slit as a function of the angle θ. As the width of the slit is increased from $w = \lambda$ (Figure 23-36a) to $w = 4\lambda$ (Figure 23-36b) to $w = 8\lambda$ (Figure 23-36c), the pattern becomes narrower. Note that if $w = \lambda$, as in Figure 23-36a, the first dark fringe corresponds to $\sin \theta = \lambda/w = 1$, so $\theta = \sin^{-1} 1 = 90°$; there are no dark fringes at smaller angles. The light intensity is spread out very broadly over all angles, much like what happens to the water waves in Figure 23-33b. If the slit width is large compared to the wavelength, as in Figure 23-36c, the vast majority of the light emerging from the slit goes straight ahead (toward $\theta = 0$) or very nearly so. That's just like what happens to the water waves in Figure 23-33a, for which there is very little diffraction.

The intensity pattern of single-slit diffraction also plays a role in *double-slit* interference, as **Figure 23-37** shows. The closely spaced bright fringes in this image are due to interference between light coming from the two slits. However, not all of these fringes are equally bright. The explanation is that light also diffracts as it emanates from each of the slits. Each of the slits has a width w that is larger than the wavelength λ, so the effect of diffraction is that the intensity of light emanating from each slit varies with angle θ as shown in Figures 23-36b and 23-36c. Hence the brightness of the bright double-slit interference fringes in Figure 23-37 rises and falls with angle θ just as the intensity curves in Figures 23-36b and 23-36c rise and fall with θ.

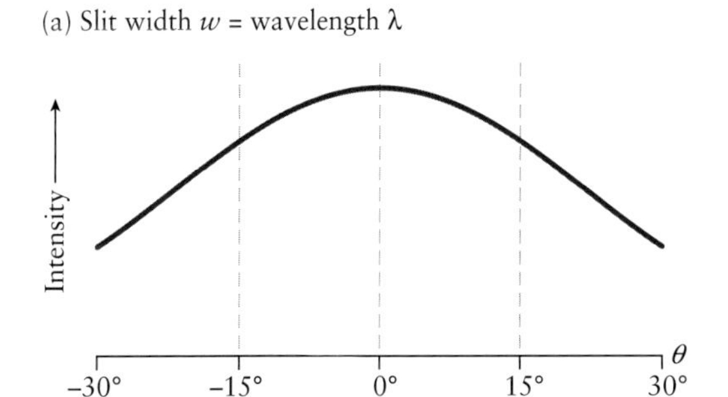

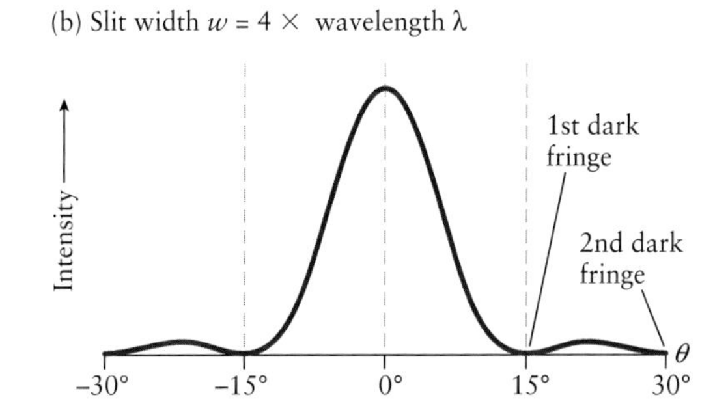

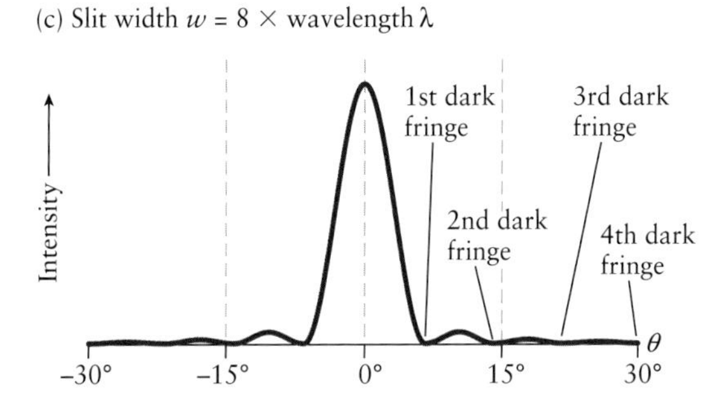

Figure 23-36 Intensity in single-slit diffraction The intensity in the diffraction pattern from a narrow slit depends on the relative size of the slit width w and the wavelength λ.

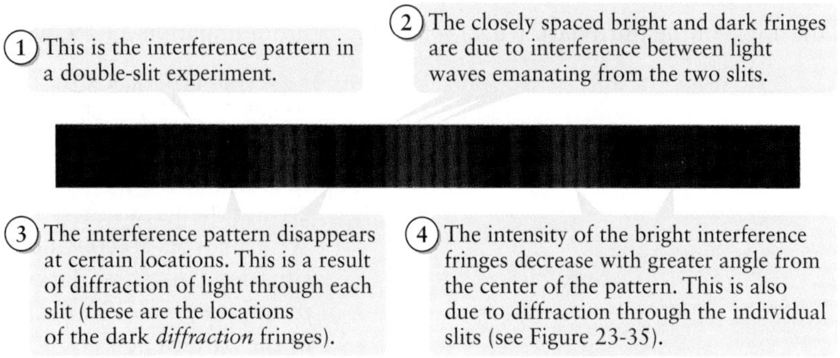

Figure 23-37 Single-slit diffraction affects double-slit interference In a double-slit experiment there is both interference of the light waves emanating from the two slits and diffraction of light waves through each individual slit. The diffraction affects the brightness of the bright interference fringes.

(1) This is the interference pattern in a double-slit experiment.

(2) The closely spaced bright and dark fringes are due to interference between light waves emanating from the two slits.

(3) The interference pattern disappears at certain locations. This is a result of diffraction of light through each slit (these are the locations of the dark *diffraction* fringes).

(4) The intensity of the bright interference fringes decrease with greater angle from the center of the pattern. This is also due to diffraction through the individual slits (see Figure 23-35).

In this section we've concentrated on the diffraction that takes place when waves pass through a narrow opening. But diffraction can also happen when waves encounter an obstacle. One example is a sound wave coming from one side of your head. As we mentioned above, sound waves used in speech have wavelengths of a meter or more, which is large compared to the diameter of a human head. As a result, these sound waves are able to diffract around your head, so you can hear the sound with both ears. Because the sound wave must travel a greater distance to one ear than to the other, there will be a phase difference between the waves that the two ears detect. Your brain detects and processes this information about phase and uses this to help determine the direction from which the sound is coming.

GOT THE CONCEPT? 23-8 Comparing Three Slits

The photographs in **Figure 23-38** show the diffraction pattern that is created when red laser light passes through a narrow slit. The wavelength of the light is the same in all three photographs, but the width of the slit is different. Order the photographs from the widest slit to the narrowest one.

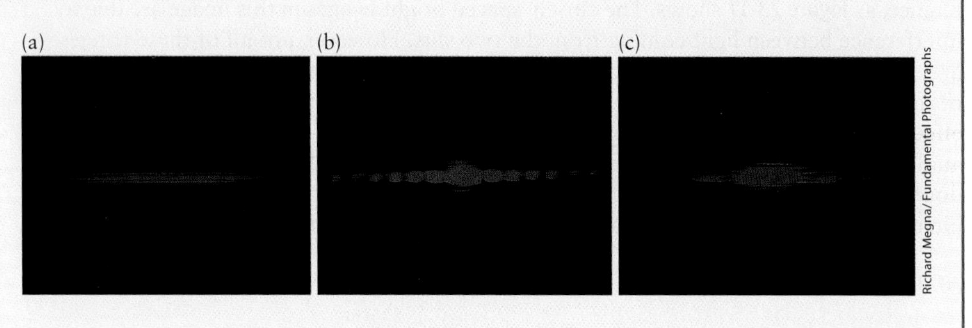

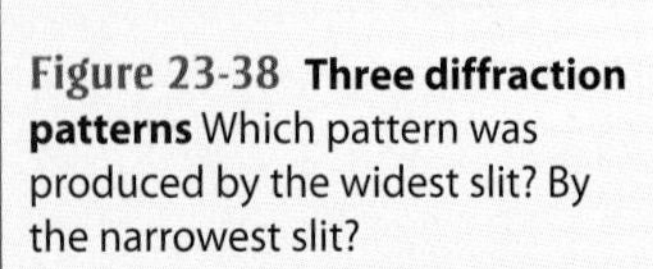

Figure 23-38 Three diffraction patterns Which pattern was produced by the widest slit? By the narrowest slit?

TAKE-HOME MESSAGE FOR Section 23-8

✔ Waves passing through a narrow opening tend to spread out, or diffract. The diffraction is more important the smaller the size of the opening.

✔ The diffraction pattern caused by waves passing through a narrow slit has bright and dark fringes. The positions of these depend on the relative size of the wavelength and the slit width.

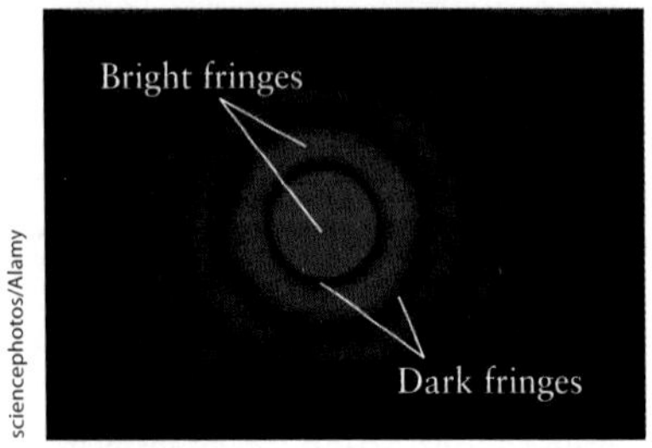

Bright fringes

Dark fringes

sciencephotos/Alamy

Figure 23-39 Diffraction by a circular aperture I Compare the diffraction pattern from a circular aperture with that for a narrow slit (Figure 23-34).

23-9 The diffraction of light through a circular aperture is important in optics

An important real-life application of diffraction is the case of light passing through a *circular* aperture. That's what happens whenever light enters the lens of a microscope, the circular mirror of an astronomical telescope, or the pupil of a human eye. **Figure 23-39** shows the diffraction pattern produced by red laser light passing through a circular aperture. Like the diffraction pattern of a slit (Section 23-8), there are bright and dark fringes. But because the aperture is circular, the fringes are circles.

Figure 23-40 shows how we define the angle θ of a point in the diffraction pattern of a circular aperture of diameter D through which passes light of wavelength λ. For

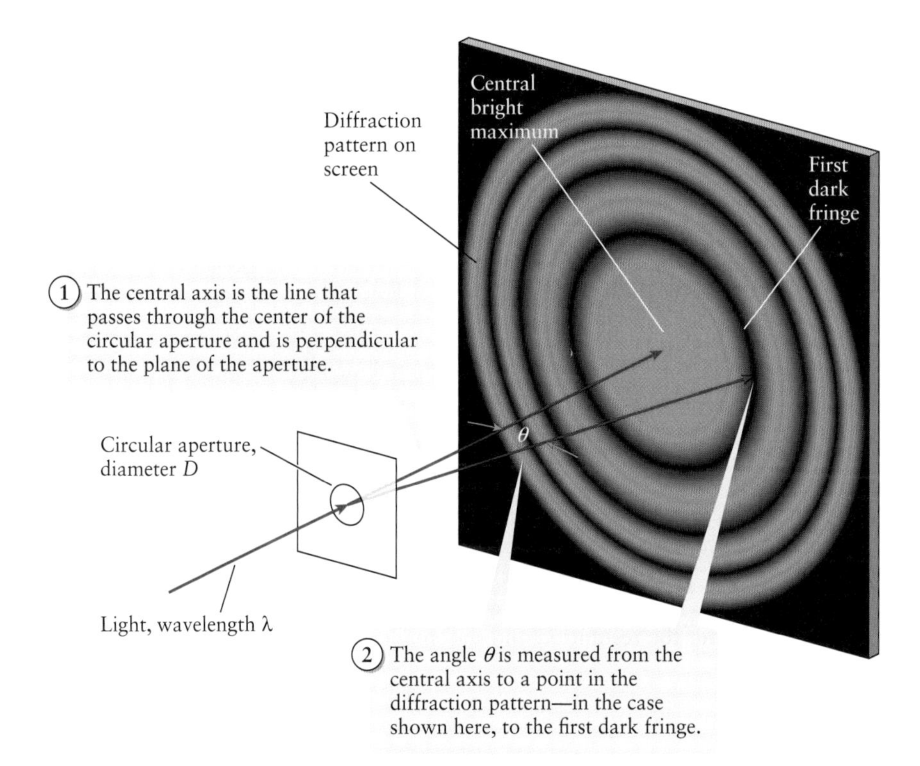

Figure 23-40 Diffraction by a circular aperture II The size of the first dark fringe—and hence the size of the central bright maximum that it surrounds—depends on the ratio of wavelength λ to aperture diameter D.

a narrow slit we found that the diffraction pattern depends on the relative sizes of the wavelength λ and the slit width w; in the same way, the diffraction pattern of a circular aperture depends on the relative sizes of λ and D. In particular, the location of the center of the first dark fringe is given by

$$\sin\theta = 1.22\frac{\lambda}{D} \tag{23-30}$$

(circular aperture, first dark fringe)

This is similar to Equation 23-27 for the first dark fringe formed by light passing through a narrow slit, with the slit w replaced by the diameter D. The factor of 1.22 results from the different geometry of a circular opening versus a rectangular one.

Figure 23-41 shows the diffraction pattern made by two pointlike objects as the objects are moved closer and closer together. In Figure 23-41a the two objects are so far apart we see only one object and its associated diffraction pattern in the field of view. In Figure 23-41b a second object has been brought close to the first; the diffraction patterns of the two objects overlap but are still distinct. In Figure 23-41c, however, the two objects are very close together. Their diffraction patterns overlap so much that it is barely possible to tell the two objects apart. We have run into the limit on our ability to **resolve**, or optically distinguish, the two objects.

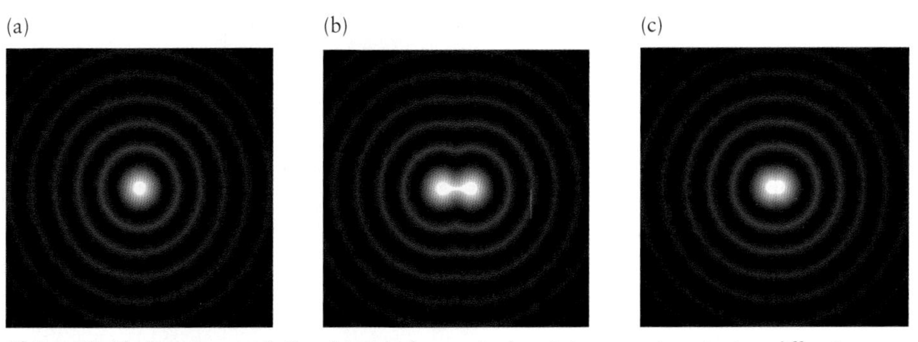

Figure 23-41 Angular resolution (a) Light from a single point source gives rise to a diffraction pattern when it passes through a circular aperture. (b) These diffraction patterns from two distant point sources partially overlap but are distinct. (c) At the limit of our ability to resolve two distant point sources, the diffraction patterns formed as their light passes through a circular aperture overlap and are barely distinguishable.

The nineteenth-century English physicist John William Strutt, 3rd Baron Rayleigh, proposed that two pointlike objects observed through a circular aperture can be resolved when the central maximum of one coincides with the center of the first dark fringe of the other. The angle θ_R that separates two point objects that are just barely resolved through a circular aperture, known as the **angular resolution** of the aperture, is then just the angle given by Equation 23-30:

Rayleigh's criterion for resolvability (23-31)

$$\sin\theta_R = 1.22\frac{\lambda}{D}$$

Angle between two pointlike objects that can **barely be resolved** through an optical device

Wavelength of the light

Diameter of the circular aperture of the device

A physician's eye chart is a device for measuring the value of θ_R for each of your eyes. An unaided eye with normal vision can distinguish objects (such as the lines that make up the letter "E" on an eye chart) that are separated by an angle of as small as 1/60 of a degree. So $\theta_R = (1/60)°$ for a person with normal vision. The smaller the value of θ_R, the better the resolution and the smaller the details that can be resolved.

EXAMPLE 23-9 The Hubble Space Telescope

The Hubble Space Telescope (HST) has a circular aperture 2.4 m in diameter. What is the theoretical angular resolution of the HST for light of wavelength 550 nm?

Set Up

Equation 23-31 gives the angular resolution, the smallest angular separation of two objects that can be resolved as a result of diffraction effects.

Rayleigh's criterion for resolvability:

$$\sin\theta_R = 1.22\frac{\lambda}{D} \quad (23\text{-}31)$$

Solve

We are given $\lambda = 550$ nm and $D = 2.4$ m. We calculate θ_R from these using Equation 23-31.

From Equation 23-31,

$$\sin\theta_R = 1.22\frac{(550\times10^{-9}\text{ m})}{(2.4\text{ m})} = 2.8\times10^{-7}$$

$$\theta_R = \sin^{-1}(2.8\times10^{-7}) = 1.6\times10^{-5}\text{ degree}$$

Reflect

The photograph shown here is a Hubble Space Telescope (HST) image of the minor planet Pluto and its largest moon, Charon. Pluto and Charon are separated by about 25×10^{-5} degrees in this image. They are easily resolved by HST, for which $\theta_R = 1.6\times10^{-5}$ degrees. By contrast, a telescope of the same diameter on Earth would have a much poorer resolution of about 0.03 degrees due to the blurring effects of the atmosphere. Such a telescope would not be able to resolve Pluto and Charon. This illustrates one important rationale for orbiting telescopes, which operate high above the atmosphere.

NASA, ESA, and STScI

GOT THE CONCEPT? 23-9 Angular Resolution

Rank the following telescopes from best to worst angular resolution. Assume that diffraction is the only limiting factor. (a) A radio telescope with an effective diameter of 27 km that observes waves at wavelength 21 cm. (b) A telescope in orbit that has a mirror of 0.85 m diameter and that observes infrared light of wavelength 4.5 μm. (c) An inexpensive telescope for amateur astronomers with a lens of 5.0 cm diameter that observes visible light of wavelength 550 nm.

TAKE-HOME MESSAGE FOR Section 23-9

✔ The angular resolution of an optical device is limited by diffraction, which blurs the images of even pointlike objects.

✔ Rayleigh's criterion states that two pointlike objects can just be resolved if the center of the bright maximum for one object coincides with the first dark fringe for the second object.

Key Terms

angular resolution
Brewster's angle
critical angle
diffraction
diffraction maxima
diffraction minima
diffraction pattern
dispersion
double-slit interference
fringe
front
Huygens' principle
incident
index of refraction
interference maxima
interference minima
interference pattern
law of reflection
linearly polarized
monochromatic
normal
partially polarized
polarization
polarizing filter
ray
refraction
resolve
single-slit diffraction
Snell's law of refraction
thin film
total internal reflection
unpolarized
wavelet

Chapter Summary

Topic	Equation or Figure
Speed of light in a medium: In a medium (transparent material) other than vacuum, the speed of light is less than c. This is described in terms of the index of refraction n of the material. The value of n is different for different materials. The wavelength also changes when light enters a medium.	Index of refraction of a medium; Speed of light in vacuum; Speed of light in the medium: $n = \frac{c}{v}$ (23-5)
	Wavelength of light in a medium; Speed of light in vacuum; Wavelength of the light in vacuum; Frequency of the light; Index of refraction of the medium: $\lambda = \frac{c}{nf} = \frac{\lambda_{\text{vacuum}}}{n}$ (23-8)

Huygens' principle, reflection, and refraction: Each point on a wave front (or wave crest) acts as a source of spherical waves called wavelets. This principle helps us explain the laws of reflection and refraction, which describe what happens when light encounters the boundary between two media. The angle of the reflected light is always equal to the angle θ_1 of the incident light, and the angle θ_2 of the refracted light is given by Snell's law.

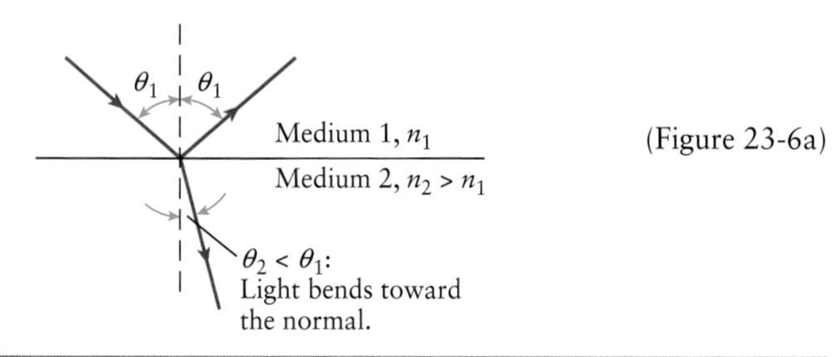

(Figure 23-6a)

Angle of the incident ray from the normal

Angle of the refracted ray from the normal

$$n_1 \sin \theta_1 = n_2 \sin \theta_2 \quad (23\text{-}6)$$

Index of refraction for the medium with the incident light

Index of refraction for the medium with the refracted light

Total internal refraction: If light in one medium reaches a boundary with a second medium of lower index of refraction ($n_1 > n_2$), total internal reflection is possible. It will take place only if the incident angle is greater than the critical angle for the two media.

Critical angle for light in medium 1 that encounters a boundary with medium 2 with a lower index of refraction

$$\theta_c = \sin^{-1}\left(\frac{n_2}{n_1}\right) \quad (23\text{-}10)$$

Index of refraction of medium 2

Index of refraction of medium 1

If the incident angle is greater than the critical angle, there is total internal reflection.

Dispersion: For a given material the index of refraction depends on the frequency of the light (or, equivalently, the wavelength in vacuum). This is the reason why a prism or water droplets can break white light into its constituent colors.

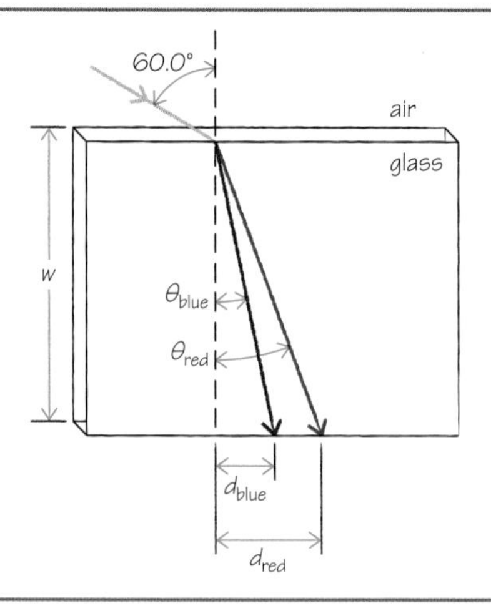

(Example 23-3)

Polarization: The polarization of a light wave is a description of the orientation of its electric field vector. In unpolarized light this orientation changes randomly. In linearly polarized light the orientation is in a fixed direction. Unpolarized light can become polarized by scattering from the atmosphere, by passing through a polarizing filter, or by reflection at Brewster's angle.

Brewster's angle for light in medium 1 that encounters a boundary with medium 2

$$\theta_B = \tan^{-1}\left(\frac{n_2}{n_1}\right) \quad (23\text{-}14)$$

Index of refraction of medium 2

Index of refraction of medium 1

If the incident angle equals Brewster's angle, the reflected light is completely polarized in the direction parallel to the surface of the boundary.

Thin-film interference: If a thin, transparent film is illuminated with monochromatic light, the light that reflects from the back surface of the film interferes with light that reflects from the front surface. If the two waves emerge in phase, the interference is constructive; if they are out of phase, it is destructive. The details of the interference depend on whether there is an inversion of the wave on each reflection.

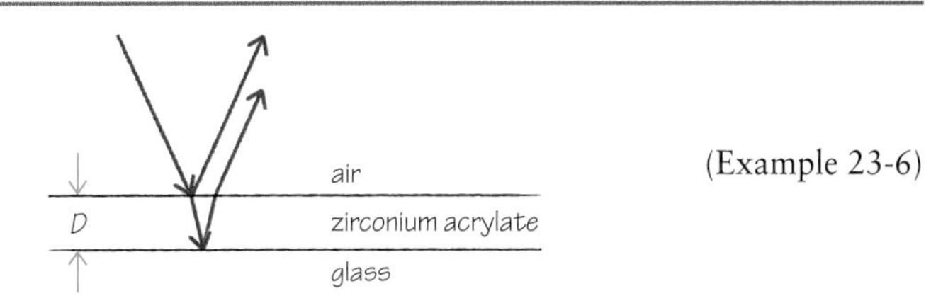

(Example 23-6)

Double-slit interference: When monochromatic light illuminates a pair of closely spaced narrow slits, a pattern of bright and dark fringes results. The bright fringes appear where light from the two slits interfere constructively; the dark fringes appear where light from the two slits interfere destructively. The closer together the two slits, the broader the interference pattern.

Constructive interference (bright fringes): $$d \sin\theta = m\lambda \qquad (23\text{-}23)$$

- **Distance between the two slits** (d)
- **Wavelength** of the light illuminating the two slits (λ)
- **Angle** between the normal to the two slits and the location of a **bright fringe** on the screen (θ)
- **Number of the bright fringe:** $m = 0, \pm1, \pm2, \pm3, \ldots$

Destructive interference (dark fringes): $$d \sin\theta = \left(m + \frac{1}{2}\right)\lambda \qquad (23\text{-}24)$$

- **Distance between the two slits** (d)
- **Wavelength** of the light illuminating the two slits (λ)
- **Angle** between the normal to the two slits and the location of a **dark fringe** on the screen (θ)
- **Number of the dark fringe:** $m = 0, \pm1, \pm2, \pm3, \ldots$

Single-slit diffraction: A pattern of bright and dark fringes also results when monochromatic light illuminates a single narrow slit. When monochromatic light illuminates a narrow slit, a pattern of bright and dark fringes results. The dark fringes appear where light from any one part of the slit interferes destructively with light from some other part of the slit, so that zero net electromagnetic wave reaches the position of the dark fringe. The narrower the width of the slit, the broader the diffraction pattern.

$$\sin\theta = \frac{m\lambda}{w} \qquad (23\text{-}29)$$

- **Angle** between the normal to the slit and the location of the mth **dark fringe** (θ)
- **Number of the dark fringe:** $m = 1, 2, 3, \ldots$
- **Wavelength** of the light (λ)
- **Width of the slit** (w)

Diffraction by a circular aperture: Light also diffracts when it passes through a circular aperture. This sets a fundamental limit on the angular resolution of an optical device: If two objects are closer together than an angle θ_R, it will be impossible to tell with the device whether they are two objects or a single object.

$$\sin\theta_R = 1.22\,\frac{\lambda}{D} \qquad (23\text{-}31)$$

- **Angle** between two pointlike objects that can **barely be resolved** through an optical device (θ_R)
- **Wavelength** of the light (λ)
- **Diameter of the circular aperture** of the device (D)

Answer to What do you think? Question

(e) Light travels more slowly in diamond than in air. So once light enters the front of a diamond, there can be total internal reflection of this light from the cut back surfaces (Section 23-3). This strongly reflected light then emerges from the front of the diamond. The speed of light in the diamond also depends on its color (frequency), so light waves of different colors follow slightly different paths on entering and emerging from the diamond (Section 23-4). Hence you see different colors of light reflected from a cut diamond as you view it at different angles.

Answers to Got the Concept? Questions

23-1 (c) For the first refraction, medium 1 is air and medium 2 is Plexiglas. Table 23-1 shows that the index of refraction of Plexiglas (1.49) is greater than that of air (1.00), so $n_2 > n_1$. Snell's law of refraction (Equation 23-6) tells us that in this situation $\theta_2 < \theta_1$, so the angle θ_2 that the light makes to the normal in Plexiglas is less than the angle θ_1 the light makes in air. So the light bends toward the normal as it crosses the boundary from air into Plexiglas. For the second refraction medium 1 is Plexiglas and medium 2 is water. Since water has a lower index of refraction (1.33) than Plexiglas (1.49), in this case $n_2 < n_1$, and Snell's law tells us that in this situation $\theta_2 > \theta_1$. The angle θ_2 that the light makes to the normal in water is greater than the angle θ_1 the light makes in Plexiglas, so the light bends away from the normal as it crosses the boundary from Plexiglas into water. Can you show that the angle is 3.35° in Plexiglas and 3.76° in water?

23-2 (a) Total internal reflection at a boundary between media is possible only if the second medium has a lower index of refraction than the first medium. Here the first medium is Plexiglas, so from Table 23-1, $n_1 = 1.49$. Total internal reflection is possible for ethyl alcohol, for which $n_2 = 1.36$ is less than n_1. It is not possible for sapphire ($n_2 = 1.77$) or diamond ($n_2 = 2.42$), since both of these have an index of refraction greater than $n_1 = 1.49$.

23-3 (c) For water, like most materials, the index of refraction n is greater, and so the wave speed $v = c/n$ is slower for light of greater frequency and shorter wavelength. This means that the blue light is slowest, and the red light is fastest, so the red light reaches the bottom of the pond slightly before the other colors.

23-4 (d) Once the light has passed through filter A, it will have one-half the intensity of the incident unpolarized light and will be completely polarized in the transmission direction of that filter. Filter B is set so that its transmission direction is perpendicular to that of filter A. So the angle between the polarization direction of the light that reaches filter B and the transmission direction of filter B is $\theta = 90°$. So from Equation 23-12 the intensity of the light that emerges from filter B will be the intensity of light reaching it from filter A multiplied by $\cos^2 90° = 0$. In other words, *no* light emerges from filter B.

23-5 (c) Once the light has passed through filter A, it will have one-half the intensity of the incident unpolarized light and will be completely polarized in the transmission direction of that filter. The transmission direction of filter C is at 45° to that of filter A, so from Equation 23-12 the light that emerges from filter C is reduced in intensity by an additional factor of $\cos^2 45° = (1/\sqrt{2})^2 = 1/2$. This light is polarized in the transmission direction of filter C. This light now enters filter B, which has a transmission direction at 45° to the direction of filter C. So the light that emerges from filter B is polarized in the transmission direction of filter B and has $\cos^2 45° = (1/\sqrt{2})^2 = 1/2$ the intensity of the light that reached it from filter C. So compared to the unpolarized light that was incident on filter A, the polarized light that emerges from filter B has $(1/2) \times (1/2) \times (1/2) = 1/8$ the intensity.

23-6 (d) When light is reflected from a boundary going from a material of higher wave speed to one of lower wave speed, the wave is inverted. No inversion occurs when light is reflected from a boundary where the transition is from a medium of lower speed to one of higher speed. The higher a material's index of refraction, the lower the speed of light is in that material. Another way to state the rule is that a light wave is inverted when it is reflected from a boundary going from a material that has a lower index of refraction to one that has a higher index of refraction. So here the speed of light is highest in air, lower in the oil, and lowest in glass. Light striking the air-to-oil boundary is therefore inverted upon reflection, as is light striking the oil-to-glass boundary. In both cases the boundary separates a material of higher light speed from one of lower light speed. No inversion occurs at the glass-to-air boundary, however, because the speed of light in glass is lower than the speed in air.

23-7 (b), (c) From the first of Equations 23-25, the angle of the $m = 3$ bright fringe is given by $\sin\theta = 3\lambda/d$, and the angle of the $m = 2$ bright fringe is given by $\sin\theta = 2\lambda/d$. To move the $m = 2$ bright fringe to the position of the $m = 3$ bright fringe therefore requires that you increase the ratio λ/d by a factor of 3/2. You can do this by decreasing d by a factor of 2/3 while leaving λ alone [choice (b)] or by increasing λ by a factor of 3/2 while leaving d unchanged [choice (c)]. You could also increase λ by a factor of 3 and decrease d by a factor of ½; note that this is not what choice (e) says, however.

23-8 (b), (c), (a) The narrower the slit, the more the diffracted light is spread out (see Figure 23-36). A good way to compare the three photographs is by looking at the width of the central bright maximum. The central bright maximum is widest in (a), so this is the narrowest slit. Of the other two photos, (c) has the wider central maximum, so the slit width is narrower in photo (c) than in photo (b). Hence the order from the widest to the narrowest slit is (b), (c), (a).

23-9 (b), (a), (c) The smaller the value of θ_R as given by Equation 23-31, the better the diffraction-limited angular resolution. For (a) $\lambda = 21$ cm $= 0.21$ m and $D = 27$ km $= 2.7 \times 10^4$ m, so $\sin\theta_R = 1.22\lambda/D = 9.5 \times 10^{-6}$ and $\theta_R = 5.4 \times 10^{-4}$ degrees. (Note that 27 km is not the diameter of a single telescope. To improve the angular resolution, multiple radio telescopes many kilometers apart are linked together so that they act like a single, gigantic telescope.) For (b) $\lambda = 4.5\ \mu$m $= 4.5 \times 10^{-6}$ m and $D = 0.85$ m, so $\sin\theta_R = 6.5 \times 10^{-6}$ and $\theta_R = 3.7 \times 10^{-4}$ degrees. For (c) $\lambda = 550$ nm $= 550 \times 10^{-9}$ m and $D = 5.0$ cm $= 0.050$ m, so $\sin\theta_R = 1.3 \times 10^{-5}$ and $\theta_R = 7.7 \times 10^{-4}$ degrees. (The actual resolution would be worse due to atmospheric turbulence.) So (b) is the best, (a) is second, and (c) is worst.

Questions and Problems

In a few problems you are given more data than you actually need; in a few other problems you are required to supply data from your general knowledge, outside sources, or informed estimate.

Interpret as significant all digits in numerical values that have trailing zeros and no decimal points.

For all problems use $g = 9.80 \text{ m/s}^2$ for the free-fall acceleration due to gravity. Neglect friction and air resistance unless instructed to do otherwise.

•	Basic, single-concept problem
••	Intermediate-level problem; may require synthesis of concepts and multiple steps
•••	Challenging problem
SSM	*Solution is in Student Solutions Manual*
Example	*See worked example for a similar problem.*

Conceptual Questions

1. • What is Huygens' principle, and why is it necessary to understand Snell's law of refraction?

2. • Does the refraction of light make a swimming pool seem deeper or shallower? Explain your answer.

3. • Does the depth of a pool determine the critical angle that a light ray will have as it travels from the bottom of the pool and heads toward the air above the water? Explain your answer.

4. • Give two common uses of total internal reflection.

5. • In your own words explain why the phenomenon of total internal reflection occurs only when light moves from a medium with a larger index of refraction toward a medium with a smaller index of refraction.

6. • Sunlight striking a diamond throws rainbows of color in every direction. From where do the colors come? SSM

7. • Recently researchers have created materials with a negative index of refraction. Explain what happens to the angle of refraction if light enters such a material from air. SSM

8. • Why do you expect the last color of the sunset to be on the red end of the visible spectrum?

9. • Explain why the Moon appears to change colors during a total lunar eclipse (when Earth's shadow completely blocks the light coming from the Sun). (See problem 8.) SSM

10. • Describe the physical interactions that take place when unpolarized light is passed through a polarizing filter. Be sure to describe the electric field of the light before and after the filter as well as the incident and transmitted intensities of the light source.

11. • Describe how polarized sunglasses work. Why do such sunglasses have *vertically* polarized lenses (as opposed to *horizontally* polarized lenses)?

12. • Linearly polarized light is incident at Brewster's angle on the surface of an optical medium. What can be said about the refracted and reflected beams if the incident beam is polarized (a) parallel to the plane of the surface and (b) perpendicular to the plane of the surface?

13. • Give two or three examples of thin-film interference.

14. • A thin layer of gasoline floating on water appears brightly colored in sunlight. What is the origin of these colors?

15. In a two-slit interference experiment, why must the incident light striking one slit have the same wavelength and phase as that striking the other slit?

16. • Does the phenomenon of diffraction apply to wave sources other than light? Give an example if it does.

17. • The sound waves used in speech have wavelengths of a few meters. Explain why some of the sound waves emerging from a person's mouth go to the sides so that they can be heard even if the listener is not in front of the person.

Multiple-Choice Questions

18. A red laser shines on a pair of slits separated by a distance d, producing an interference pattern on a screen. A green laser of the same intensity is then used to illuminate the same slit pair. Compared to the red laser, the green laser will produce an interference pattern with maxima that are

A. in the same locations.
B. brighter.
C. farther apart.
D. dimmer.
E. closer together.

19. • Which kind of wave can refract when crossing from one medium to another with the wave having a different speed in the two media?

A. electromagnetic waves
B. sound waves
C. water waves
D. electromagnetic, sound, and water waves
E. only electromagnetic and sound waves

20. • When light enters a piece of glass from air with an angle of θ with respect to the normal to the boundary surface,

A. it bends with an angle larger than θ with respect to the normal to the boundary surface.
B. it bends with an angle smaller than θ with respect to the normal to the boundary surface.
C. it does not bend.
D. it bends with an angle equal to two times θ with respect to the normal to the boundary surface.
E. it bends with an angle equal to one-half θ with respect to the normal to the boundary surface.

21. • Which phenomenon would cause monochromatic light to enter the prism and follow along the path as shown in **Figure 23-42**?

A. reflection
B. refraction
C. interference
D. diffraction
E. polarization

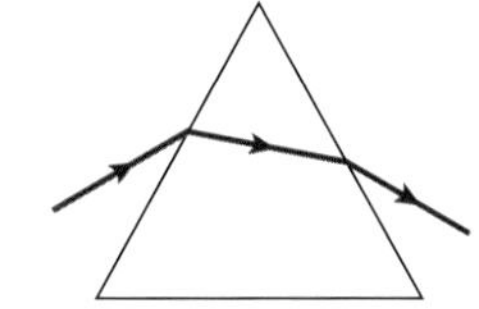

Figure 23-42 Problem 21

22. • Which color of light, red or blue, travels faster in crown glass?

A. red
B. blue
C. their speeds are the same
D. depends on the material surrounding the glass
E. blue if the glass is thin

23. • Two linear polarizing filters are placed one behind the other so that their transmission directions are parallel to one another. A beam of unpolarized light of intensity I_0 is directed at the two filters. What fraction of the light will pass through both filters?

A. 0
B. $(1/2)I_0$
C. I_0
D. $(1/4)I_0$
E. $2I_0$ SSM

24. • Two linear polarizing filters are placed one behind the other so that their transmission directions form an angle of 45°. A beam of unpolarized light of intensity I_0 is directed at the two filters. What fraction of the light will pass through both?

A. 0
B. $(1/2)I_0$
C. I_0
D. $(1/4)I_0$
E. $2I_0$

25. • A monochromatic light passes through a narrow slit and forms a diffraction pattern on a screen behind the slit. As the wavelength of the light decreases, the diffraction pattern

A. shrinks with all the fringes getting narrower.
B. spreads out with all the fringes getting wider.
C. remains unchanged.
D. spreads out with all the fringes getting alternately wider and then narrower.
E. becomes dimmer.

26. • A monochromatic light passes through a narrow slit and forms a diffraction pattern on a screen behind the slit. As the slit width increases, the diffraction pattern

A. shrinks with all the fringes getting narrower.
B. spreads out with all the fringes getting wider.
C. remains unchanged.
D. spreads out with all the fringes getting alternately wider and then narrower.
E. becomes dimmer.

27. • **Figure 23-43** shows two single-slit diffraction patterns created with the same source. The distance between the slit and the viewing screen is the same in both cases. Which of the following is true about the widths, w_a and w_b, of the two slits?

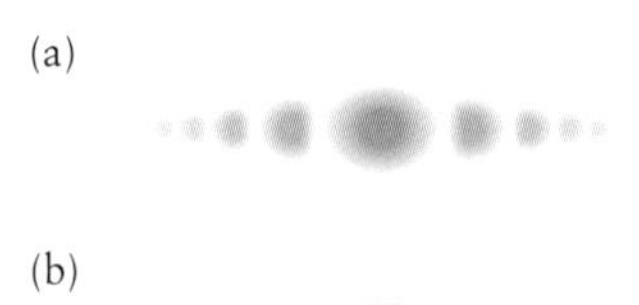

Figure 23-43 Problem 27

A. $w_a > w_b$
B. $w_a < w_b$
C. $w_a = w_b$
D. $w_a = (1/2)w_b$
E. $w_a = (1/4)w_b$ SSM

Estimation/Numerical Analysis

28. • The closest star to our sun (Proxima Centauri) is 4.0×10^{16} m away. It has a planet (Proxima b) in orbit around it at about 1/20 of the radius of Earth's orbit around our sun. Estimate the minimum diameter of the aperture of a space telescope that would be required to resolve Proxima b from its star, for light in the visible part of the spectrum.

29. • Estimate the range of values for the index of refraction that you will use in this class.

30. • Estimate the time required for light to travel from (a) the Moon to Earth, (b) the Sun to Earth, and (c) Earth to Proxima Centauri (the nearest star to our solar system). (See problem 28.)

31. •• A swimmer lies at the bottom of a 3-m-deep Olympic-size swimming pool, near the center and looking straight up. Estimate what percentage of the pool's surface does not look like a mirror to the swimmer, that is, what percentage shows the outside world.

32. •• Suppose you are measuring double-slit interference patterns using an optics kit that contains the following options that you can mix and match: a red laser or a green laser; a slit width of 0.04 or 0.08 mm; a slit separation of 0.25 or 0.50 mm. Estimate how far you should place a screen from the double slit to give you an interference pattern on the screen that you can accurately measure using an ordinary ruler.

33. •• A beam of light travels from medium 1 to medium 2 in the x–y plane. Medium 1 is found in quadrants 2 and 3; medium 2 is in quadrants 1 and 4. The beam touches each of the points in the x–y plane given in the table below. Calculate the ratio of the index of refraction of medium 2 to medium 1.

x (cm)	y (cm)	x (cm)	y (cm)
−4.00	−2.00	+1.00	+0.296
−3.00	−1.52	+2.00	+0.595
−2.00	−1.02	+3.00	+0.901
−1.00	−0.514	+4.00	+1.20
0	0		

Problems

23-1 The wave nature of light explains much about how light behaves

23-2 Huygens' principle explains the reflection and refraction of light

34. • The speed of light in a newly developed plastic is 1.97×10^8 m/s. Calculate the index of refraction.

35. • The index of refraction for a vacuum is 1.00000. The index of refraction for air is 1.00029. Determine the ratio of time required for light to travel through 1000 m of air to the time required for light to travel through 1000 m of vacuum.

36. • Calculate the speed of light for each of the following materials:

A. ice
B. acetone
C. Plexiglas
D. sodium chloride
E. sapphire
F. diamond
G. water
H. crown glass

37. •• The speed of light in methylene iodide is 1.72×10^8 m/s. The index of refraction of water is 1.33. Through what distance of methylene iodide must light travel such that the time

to travel through the methylene iodide is the same as the time required for light to travel through 1000 km of water? SSM

38. •• Light travels from air toward water. If the angle that is formed by the light beam in air with respect to the normal line between the two media is 27°, calculate the angle of refraction of the light in the water. Example 23-1

39. • Determine the unknown angle in each of the situations in **Figure 23-44.** Example 23-1

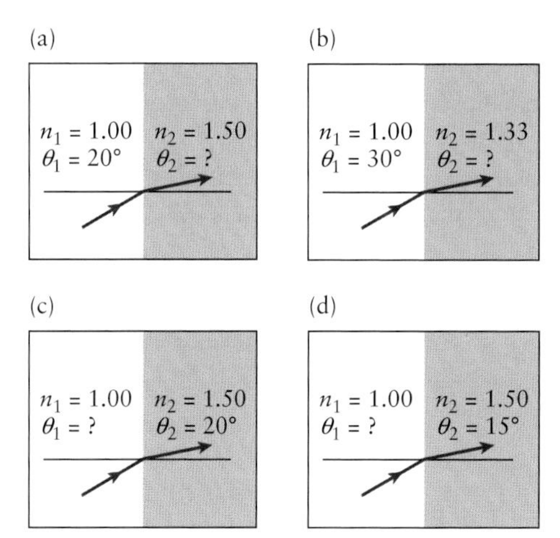

Figure 23-44 Problem 39

40. • Determine the unknown index of refraction in each of the situations in **Figure 23-45.** Example 23-1

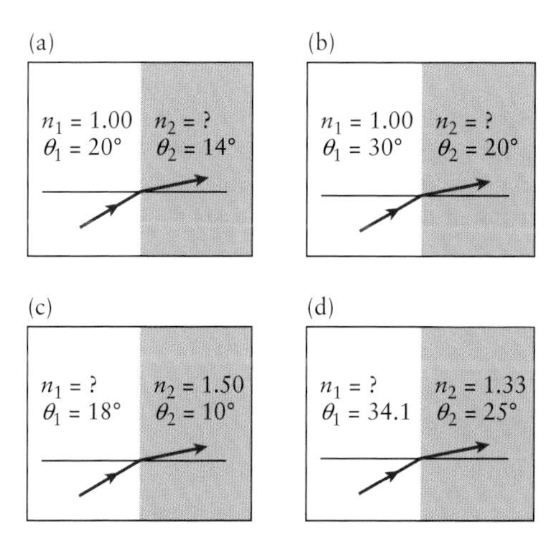

Figure 23-45 Problem 40

41. •What is the wavelength of red light (700 nm in air) when it is inside a glass slab with $n = 1.55$?

23-3 In some cases light undergoes total internal reflection at the boundary between media

42. • Calculate the critical angle for the following:
   A. Light travels from plastic ($n = 1.50$) to air ($n = 1.00$).
   B. Light travels from water ($n = 1.33$) to air ($n = 1.00$).
   C. Light travels from glass ($n = 1.56$) to water ($n = 1.33$).
   D. Light travels from air ($n = 1.00$) to glass ($n = 1.55$).
   SSM Example 23-2

43. • For each of the critical angles given, calculate the index of refraction for the optical materials that light travels out of toward air ($n = 1.00$): Example 23-2
   A. $\theta_c = 48.5°$
   B. $\theta_c = 47.0°$
   C. $\theta_c = 42.6°$
   D. $\theta_c = 35.0°$
   E. $\theta_c = 55.7°$
   F. $\theta_c = 38.5°$
   G. $\theta_c = 22.2°$
   H. $\theta_c = 75.0°$

44. • What is the critical angle for light traveling from sapphire to air? Example 23-2

45. •• What is the largest angle θ_1 that will ensure that light is totally internally reflected in the fiber-optic pipe made of acrylic ($n = 1.50$) shown in **Figure 23-46?** Example 23-2

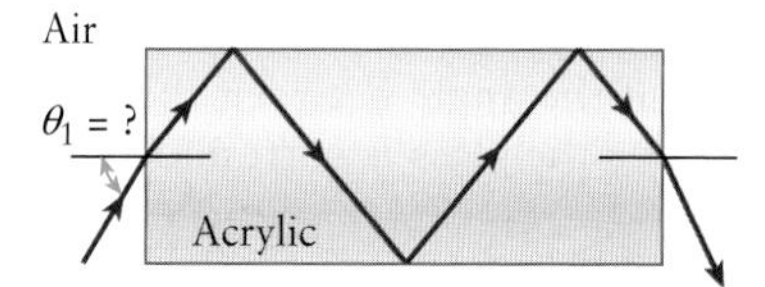

Figure 23-46 Problem 45

46. • At what angle with respect to the vertical must a scuba diver look in order to see her friend standing on the very distant shore? Take the index of refraction of the water to be $n = 1.33$. SSM Example 23-2

47. •• A point source of light is 2.50 m below the surface of a pool. What is the diameter of the circle of light that a person above the water will see? Assume the water has an index of refraction of $n = 1.33$. Example 23-2

48. • A block of glass that has an index of refraction of 1.55 is completely immersed in water ($n = 1.33$). What is the critical angle for light traveling from the glass to the water? Example 23-2

23-4 The dispersion of light explains the colors from a prism or a rainbow

49. •• For a certain optical medium the speed of light varies from a low value of 1.90×10^8 m/s for violet light to a high value of 2.00×10^8 m/s for red light. (a) Calculate the range of the index of refraction of the material for visible light. (b) A white light is incident on the medium from air, making an angle of 30.0° with the normal. Compare the angles of refraction for violet light and red light. (c) Repeat the previous part when the incident angle is 60.0°. Example 23-3

50. ••• A beam of light shines on an equilateral glass prism at an angle of 45° to one face (**Figure 23-47a**). (a) What is the angle at which the light emerges from the opposite face given that $n_{glass} = 1.57$? (b) Now consider what happens when dispersion is involved (**Figure 23-47b**). Assume the incident ray of light spans the spectrum of visible light between 400 and 700 nm (violet to red, respectively). The index of refraction for violet light in the glass prism is 1.572, and it is 1.568 for red light in the glass prism. Find the distance along the right face of the prism between the points where the red light and violet light emerge back into air. Assume the prism is 10.0 cm on a side and the incident ray hits the midpoint of the left face. Example 23-3

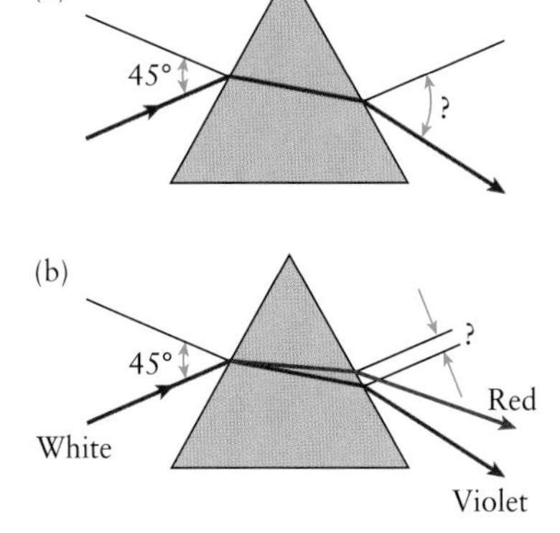

Figure 23-47 Problem 50

51. • A light beam strikes a piece of glass with an incident angle of 45.0°. The beam contains two colors: 450 nm and an unknown wavelength. The index of refraction for the 450-nm light is 1.482. Determine the index of refraction for the unknown wavelength if the angle between the two refracted rays is 0.275°. Assume the glass is surrounded by air. Example 23-3

52. • Blue light (500 nm) and yellow light (600 nm) are incident on a 12-cm-thick slab of glass as shown in **Figure 23-48**. In the glass the index of refraction for the blue light is 1.545, and for the yellow light it is 1.523. What distance along the glass slab (side AB) separates the points at which the two rays emerge back into air? SSM Example 23-3

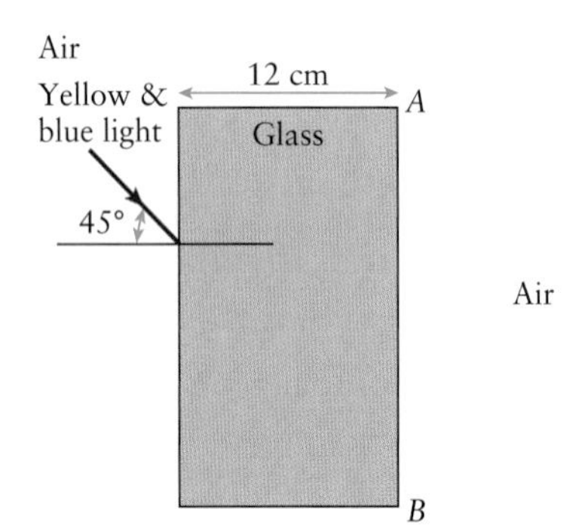

Figure 23-48 Problem 52

23-5 In a polarized light wave, the electric field vector points in a specific direction

53. • Unpolarized light is passed through an optical filter that is oriented in the vertical direction. If the incident intensity of the light is 78 W/m^2, what are the polarization and intensity of the light that emerges from the filter?

54. • What angle(s) does vertically polarized light make relative to a polarizing filter that diminishes the intensity of the light by 25%? SSM

55. •• Light that passes through a series of three polarizing filters emerges from the third filter horizontally polarized with an intensity of 250 W/m^2. If the polarization angle between the filters increases by 25° from one filter to the next, find the intensity of the incident beam of light, assuming it is initially unpolarized.

56. •• Vertically polarized light that has an intensity of 400 W/m^2 is incident on two polarizing filters. The first filter is oriented 30.0° from the vertical, while the second filter is oriented 75.0° from the vertical. Predict the intensity and polarization of the light that emerges from the second filter.

57. •What is Brewster's angle when light in water is reflected off a glass surface? Assume $n_{water} = 1.33$ and $n_{glass} = 1.55$. Example 23-4

58. •• The critical angle between two optical media is 60.0°. What is Brewster's angle at the same interface between the two media? Example 23-4

59. • (a) What would be Brewster's angle for reflections off the surface of water when the light source is beneath the surface? (b) Compare that answer to the angle for total internal reflection when light starts in water and reflects off air. Example 23-4

60. •• At what angle θ above the horizontal is the Sun when a person observing its rays reflected off water finds them linearly polarized along the horizontal (**Figure 23-49**)? SSM Example 23-4

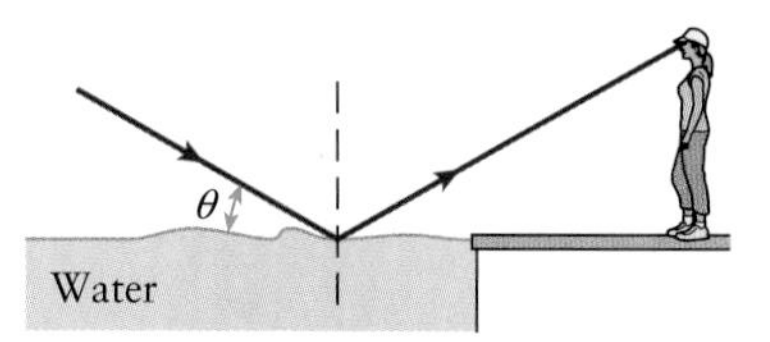

Figure 23-49 Problem 60

23-6 Light waves reflected from the layers of a thin film can interfere with each other, producing dazzling effects

61. • A ray of normal-incidence light is reflected from a thin film back into air. If the film is actually a coating on a slab of glass ($n_{film} < n_{glass}$), describe the phase changes that the reflected ray undergoes (a) as it reflects off the front surface of the film and (b) as it reflects off the back surface of the film (the film–glass interface). Example 23-5

62. •• When white light illuminates a thin film with normal incidence, it strongly reflects both indigo light (450 nm in air) and yellow light (600 nm in air) (**Figure 23-50**). Calculate the minimum thickness of the film if it has an index of refraction of 1.28 and it sits atop a slab of glass that has $n = 1.50$. Example 23-5

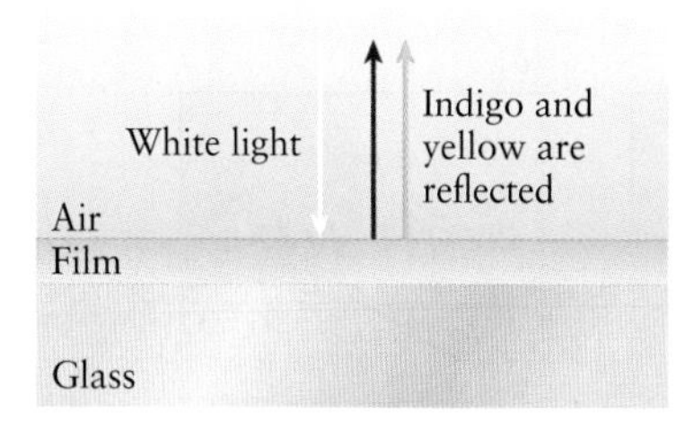

Figure 23-50 Problem 62

63. ••When white light illuminates a thin film normal to the surface, it strongly reflects both blue light (500 nm in air) and red light (700 nm in air) (**Figure 23-51**). Calculate the minimum thickness of the film if it has an index of refraction of 1.35 and it "floats" on water with $n = 1.33$. SSM Example 23-6

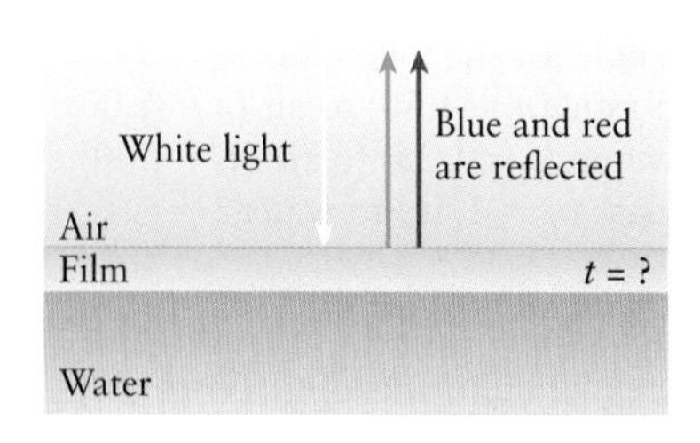

Figure 23-51 Problem 63

64. •• A soap bubble is suspended in air. If the thickness of the soap is 625 nm and both blue light (500 nm in air) and red light (700 nm in air) are *not* observed to reflect from the soap film when viewed normal to the surface of the film, what is the index of refraction of the thin film? Example 23-6

65. •• A thin film of cooking oil ($n = 1.38$) is spread on a puddle of water ($n = 1.33$). What are the minimum and the next three thicknesses of the oil that will strongly reflect blue light having a wavelength in air of 518 nm at normal incidence? Example 23-6

66. •• What is the minimum thickness of a nonreflective coating of magnesium chloride ($n = 1.39$) so that no normal-incident light centered around 550 nm in air will reflect back off a glass lens ($n = 1.56$)? Example 23-5

67. •• Water ($n = 1.33$) in a shallow pan is covered with a thin film of oil that is 450 nm thick and has an index of refraction of 1.45. What visible wavelengths will *not* be present in the reflected light when the pan is illuminated with white light and viewed from straight above? SSM Example 23-6

23-7 Interference can occur when light passes through two narrow, parallel slits

68. • Light that has a 650-nm wavelength is incident upon two narrow slits that are separated by 0.500 mm. An interference pattern from the slits is projected onto a screen that is 3.00 m away. (a) What is the separation distance on the screen of the first bright fringe from the central bright fringe? (b) What is the separation distance on the screen of the second dark fringe from the central bright fringe? Example 23-7

69. •• Conducting an experiment with a 532-nm wavelength green laser, a researcher notices a slight shift in the image generated and suspects the laser is unstable and switching between two closely spaced wavelengths, a phenomenon known as mode-hopping. To determine if this is true, she decides to shine the laser on a double-slit apparatus and look for changes in the pattern. Measuring to the first bright fringe on a screen 0.5 m away, with a slit separation of 80 μm, she measures a distance of 3.325 mm. When the laser shifts, so does the pattern and she then measures the same fringe spacing to be 3.375 mm. What wavelength is the laser "hopping" to? Example 23-7

70. • An interference pattern from a double-slit experiment displays 12 bright fringes per centimeter on a screen that is 8.5 m away. The wavelength of light incident on the slits is 550 nm. What is the distance between the two slits? Example 23-7

71. •• An argon laser that has a wavelength of 455 nm shines on a double-slit apparatus, which produces an interference pattern on a screen that is 10.0 m away from the slits. The slit separation distance is 70.0 μm. (a) How many bright fringes are there on the screen within an angle of $\pm 1°$ relative to the central axis? (b) How many dark fringes are there on the screen within an angle of $\pm 2°$ relative to the central axis? Be careful to count *all* the fringes. Example 23-7

23-8 Diffraction is the spreading of light when it passes through a narrow opening

72. •Light that has a wavelength of 550 nm is incident on a single slit that is 10.0 μm wide. Determine the angular location of the first three dark fringes that are formed on a screen behind the slit. Example 23-8

73. • Light that has a wavelength of 475 nm is incident on a single slit that is 800 nm wide. Calculate the angular location of the first three dark fringes that are formed on a screen behind the slit. Example 23-8

74. • What is the highest order dark fringe that is found in the diffraction pattern for light that has a wavelength of 633 nm and is incident on a single slit that is 1500 nm wide? Example 23-8

75. • The highest order dark fringe found in a diffraction pattern is 6. Determine the wavelength of light that is used with the single slit that has a width of 3500 nm. Example 23-8

76. •• When blue light ($\lambda = 500$ nm) is incident on a single slit, the central bright spot has a width of 8.75 cm. If the screen is 3.55 m distant from the slit, calculate the slit width. Example 23-8

77. • A helium–neon laser illuminates a narrow, single slit that is 1850 nm wide. The first dark fringe is found at an angle of 20.0° from the central peak. Determine the wavelength of the light from the laser. SSM Example 23-8

78. •• Yellow light that has a wavelength of 625 nm produces a central maximum peak that is 24.0 cm wide on a screen that is 1.58 m from a single slit. Calculate the width of the slit. Example 23-8

23-9 The diffraction of light through a circular aperture is important in optics

79. •• Light from a helium–neon laser with a wavelength of 633 nm passes through a 0.180-mm-diameter hole and forms a diffraction pattern on a screen 2.0 m behind the hole. Calculate the diameter of the central maximum. Example 23-9

80. • **Biology** The average pupil is 5.0 mm in diameter, and the average normal-sighted human eye is most sensitive at a wavelength of 555 nm. What is the eye's angular resolution in radians? Example 23-9

81. • **Astronomy** The telescope at Mount Palomar has an objective mirror that has a diameter of 508 cm. What is the angular limit of resolution due to diffraction for 560-nm light in degrees and radians? Example 23-9

82. • **Astronomy** The Hubble Space Telescope has a diameter of 2.4 m. What is the angular limit of resolution due to diffraction when a wavelength of 540 nm is viewed? Example 23-9

83. •• The distance from the center of a circular diffraction pattern to the first dark ring is 15,000 wavelengths on a screen that is 0.85 m away. What is the size of the aperture? SSM Example 23-9

84. •• **Biology** Assume your eye has an aperture diameter of 3.00 mm at night when bright headlights are pointed at it. At what distance can you see two headlights separated by 1.50 m as distinct? Assume a wavelength of 550 nm, near the middle of the visible spectrum. Example 23-9

General Problems

85. •• One way of describing the speed of light in an optical material is to specify the ratio of the time that is required for light to travel through a length of vacuum to the time required for light to travel through the same length of the optical material. For example, if light travels through a material in 150% of the time for light to travel through a vacuum, the speed of light in the material would be $2/3 = 1/1.5$ that in a vacuum. Complete the table by giving the speed of light and the index of refraction for each of the following optical materials, listed with the corresponding percentage.

Optical material with percentage of time required for light to pass through compared to an equal length of vacuum	Speed of light	Index of refraction
100%		
125%		
150%		
200%		
500%		
1000%		

86. •• (a) Determine the index of refraction for medium 2 if the distance between points B and C in **Figure 23-52** is 0.75 cm. Assume the index of refraction in medium 1 is 1.00. (b) Suppose $n_2 = 1.55$. Calculate the distance between points B and C. Example 23-1

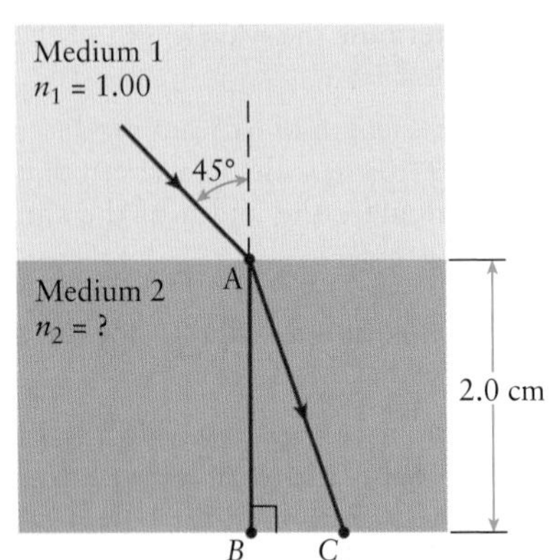

Figure 23-52 Problem 86

87. ••• Prove that in the case where there are more than two optically different media sandwiched together, with air on the left and air on the right, the angle at which light returns to air is independent of the indices of refraction of the interior media (**Figure 23-53**). In other words, the refraction angle, θ_n, depends *only* on n_1, θ_1, and n_i (not $n_2, n_3, n_4, \ldots$). Example 23-1

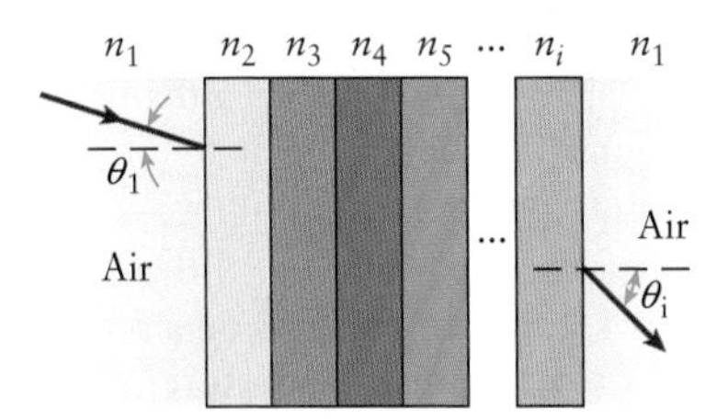

Figure 23-53 Problem 87

88. • One of the world's largest aquaria is the Monterey Bay Aquarium. The viewing wall is made of acrylic and is 0.33 m thick, and the tank holds 1.2 million gallons of water (**Figure 23-54**). If a ray of light is directed into the plastic from air at an angle of 40.0°, calculate the angle that the ray will make (a) when it enters the plastic and (b) when it enters the seawater. The indices of refraction for air, acrylic, and seawater are listed on the figure. Example 23-1

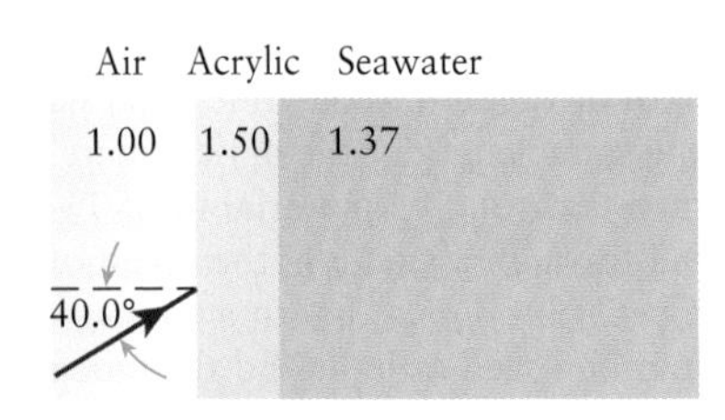

Figure 23-54 Problem 88

89. •• A flat glass surface ($n = 1.54$) has a layer of water ($n = 1.33$) of uniform thickness directly above the glass. At what minimum angle of incidence must light in the glass strike the glass–water interface for the light to be totally internally reflected at the water–air interface? SSM Example 23-2

90. ••• Light rays in air fall normally on the vertical surface of a glass prism ($n = 1.55$), as shown in **Figure 23-55**. (a) What is the largest value of θ such that the ray is totally internally reflected at the slanted face? (b) Repeat the calculation if the prism is immersed in water with $n = 1.33$. Example 23-2

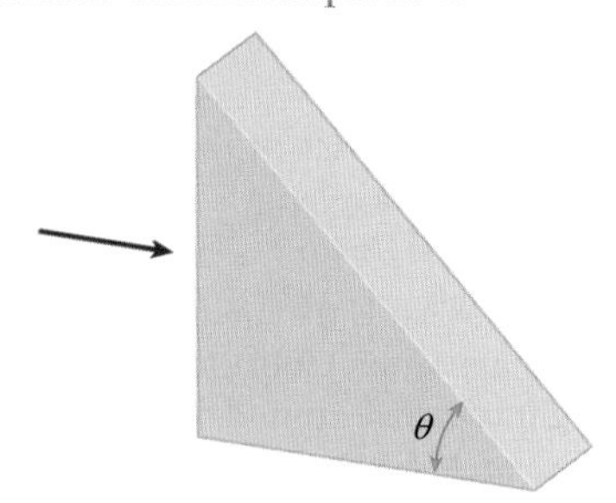

Figure 23-55 Problem 90

91. •• The object in **Figure 23-56** is a depth $d = 0.85$ m below the surface of clear water. (a) How far from the end of the dock, distance D in the figure, must the object be if it cannot be seen from any point on the end of the dock? The index of refraction of water is 1.33. (b) If you could change the index of refraction of the water, how would you change it so that the object could be seen at any distance from the dock? Example 23-2

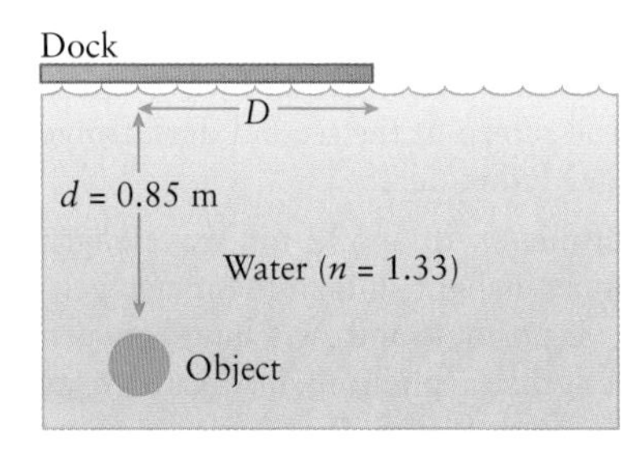

Figure 23-56 Problem 91

92. •• The polarizing angle for light that passes from water ($n = 1.33$) into a certain plastic is 61.4°. What is the critical angle for total internal reflection of the light passing from the plastic into air? Example 23-4

93. •• A baseball is hit into a round pool of water that is 4.00 m deep and 17.0 m across (**Figure 23-57**). It lands right in the center of the pool. A large round raft shaped like a lily pad is floating in the pool, concentrically on top of the location of the ball. What minimum diameter must the raft have in order to completely obscure the ball from sight? Assume that water has an index of refraction of 1.33. Example 23-2

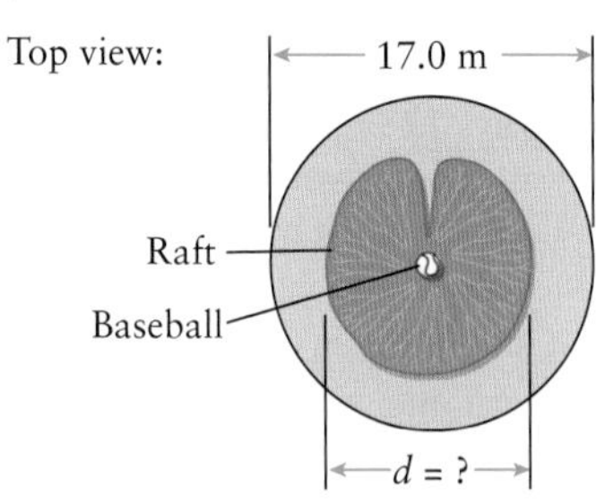

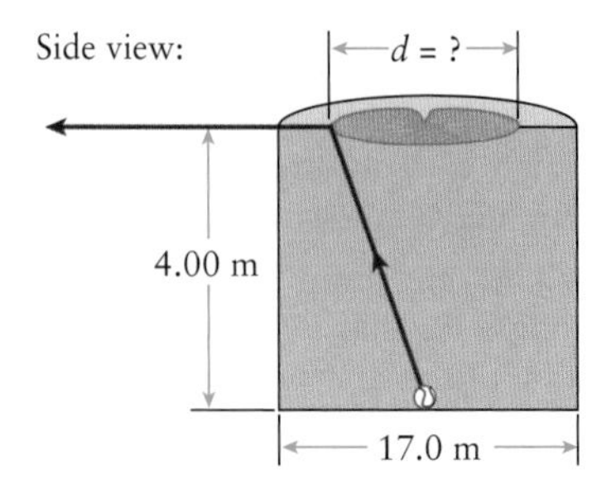

Figure 23-57 Problem 93

94. •• A glass lens that has an index of refraction equal to 1.57 is coated with a thin layer of transparent material that has an index of refraction equal to 2.10. If white light strikes the lens at near-normal incidence, light of wavelengths 495 nm in air and 660 nm in air are absent from the reflected light. What is the thinnest possible layer of material for which this can be accomplished? Example 23-6

95. •• Unpolarized light of intensity 100 W/m^2 is incident on two ideal polarizing sheets that are placed with their transmission axes perpendicular to each other. An additional polarizing sheet is then placed between the two, with its transmission axis oriented at 30° to that of the first. (a) What is the intensity of the light passing through the stack of polarizing sheets? (b) What orientation of the middle sheet enables the three-sheet combination to transmit the greatest amount of light? SSM

96. •• Unpolarized light that has an intensity of 850 W/m^2 is incident on a series of polarizing filters as shown in **Figure 23-58**. If the intensity of the light after the final filter is 75 W/m^2, what is the orientation of the second filter relative to the x axis? *Hint:* $\cos(90 - \theta) = \sin\theta$ and $2\sin\theta\cos\theta = \sin 2\theta$.

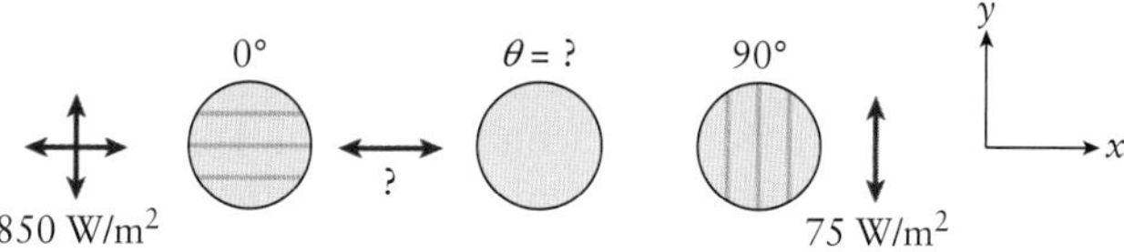

Figure 23-58 Problem 96

97. •• A thin film of soap solution ($n = 1.33$) has air on either side and is illuminated normally with white light. Interference minima are visible in the reflected light only at wavelengths of 400, 480, and 600 nm in air. What is the minimum thickness of the film? Example 23-6

98. ••• A brass sheet has a thin slit scratched in it. At room temperature (22.0°C) a laser beam illuminates the slit, and you observe that the first dark diffraction spot occurs at ±25.0° on either side of the central maximum. The brass sheet is then immersed in liquid nitrogen at 77.0 K until it reaches the same temperature as the nitrogen. It is removed and the same laser is shined on the slit. At what angle will the first dark spot now occur? Consult Table 14-1 for the thermal properties of brass. Example 23-8

99. •• A wedge-shaped air film is made by placing a small slip of paper between the edges of two thin plates of glass 12.5 cm long. Light of wavelength 600 nm in air is incident normally on the glass plates. If interference fringes with a spacing of 0.200 mm are observed along the plate, how thick is the paper? This form of interferometry is a very practical way of measuring small thicknesses. SSM Example 23-6

100. ••A thin layer of SiO, having an index of refraction of 1.45, is used as a coating on certain solar cells. The refractive index of the cell itself is 3.5. (a) What is the minimum thickness of the coating needed to cancel visible light of wavelength 400 nm in the light reflected from the top of the coating in air? Are any other visible wavelengths also canceled? (b) Suppose that technological limitations require you to make the coating 3.0 times as thick as in part (a). Which, if any, visible wavelengths in the reflected light will be canceled in air and which, if any, will be reinforced? Example 23-6

101. •• You want to coat a pane of glass that has an index of refraction of 1.54 with a 155-nm-thick layer of material that has an index of refraction greater than that of the glass. The purpose of the coating is to cancel light reflected off the top of the film having a wavelength (in air) of 550 nm. (a) What should be the index of refraction of the coating? (b) If, due to technological difficulties, you cannot achieve a uniform coating at the desired thickness, what are the next three thicknesses of the coating you could use? Example 23-6

102. •• Two point sources of light each of which has a wavelength of 500 nm are photographed from a distance of 100 m using a camera with a 50.0-mm focal length lens. The camera aperture is 1.05 cm in diameter. What is the minimum separation of the two sources if they are to be resolved in the photograph, assuming the resolution is limited only by diffraction? Example 23-9

103. •• In 2009 researchers reported on evidence that a giant tsunami had hit the eastern coast of the Mediterranean Sea (present-day Lebanon and Israel) around 1600 BCE, causing huge damage to the civilizations located there. It is believed that the tsunami was caused by the eruption of the Thera volcano near the island of Crete. The waves would have passed through the 100-mile-wide opening between Crete and Rhodes, which would cause them to diffract and spread out. Satellite observations of tsunamis show that the waves measure about 250 mi from a crest to the adjacent trough, and the time between successive crests is typically 60 min. (a) How fast do tsunami waves travel? (b) How long after the eruption of Thera would the tsunami reach the eastern shore of the Mediterranean Sea, 600 mi from Thera? (c) For these waves, could we apply the formula $w\sin\theta = m\lambda$ to find the angles at which the waves cancel after passing through the 100-mile "slit" between Crete and Rhodes? Explain why or why not. SSM Example 23-8

104. ••**Biology** The pupil (the opening through which light enters the lens) of a house cat's eye is round under low light, but ciliary muscles narrow it to a thin vertical slit in very bright light. Assume that bright light of wavelength 550 nm in air is entering the eye perpendicular to the lens and that the pupil has narrowed to a slit that is 0.500 mm wide. What are the three smallest angles on either side of the central maximum at which no light will reach the cat's retina (a) if we imagine the eye is filled with air and (b) if we take into consideration that in reality the eye is filled with a fluid having index of refraction of approximately 1.4? (c) Would the cat be aware of the pattern of alternating dark and light fringes? Why or why not? Example 23-8

105. •• If you peek through a 0.75-mm-diameter hole at an eye chart, you will notice a decrease in visual acuity. Calculate the angular limit of resolution if the wavelength is taken as 575 nm. Compare the result to a 4-mm pupil of the eye that has an angular resolution of 1.75×10^{-4} rad. Example 23-9

106. •• **Biology** The pupil of the eye is the circular opening through which light enters. Its diameter can vary from about 2.0 to about 8.0 mm to control the intensity of the light reaching the retina. (a) Calculate the angular resolution, θ_R, of the eye for light that has a wavelength of 550 nm in both bright light and dim light. In which light can you see more sharply, dim or bright? (b) You probably have noticed that when you squint, objects that were a bit blurry suddenly become somewhat

clearer. In light of your results in part (a), explain why squinting helps you see an object more clearly. Example 23-9

107. •• **Biology** Under bright light the pupil of the eye (the circular opening through which light enters) is typically 2.0 mm in diameter. The diameter of the eye is about 25 mm. Suppose you are viewing something with light of wavelength 500 nm. Ignore the effect of the lens and the vitreous humor in the eye. (a) At what angles (in radians and degrees) will the first three diffraction dark rings occur on either side of the central bright spot on the retina at the back of the eye? (b) Approximately how far (in millimeters) from the central bright spot would the dark rings in part (a) occur? (c) Explain why we do not actually observe such diffraction effects in our vision. Example 23-9

108. •• **Astronomy** Sometime around 2024, astronomers at the European Southern Observatory hope to begin using the E-ELT (European Extremely Large Telescope), which is planned to have a primary mirror 42 m in diameter. Let us assume that the light it focuses has a wavelength of 550 nm. (a) What is the most distant Jupiter-sized planet the telescope could resolve, assuming its resolution is limited only by diffraction? Express your answer in meters and light years. (b) The nearest known exoplanets (planets beyond the solar system) are around 20 light years away. What would have to be the minimum diameter of an optical telescope to resolve a Jupiter-sized planet at that distance using light of wavelength 550 nm? (1 light year = 9.461×10^{15} m) Example 23-9

109. •• **Astronomy** Under the best atmospheric conditions at the premium site for land-based observing (Mauna Kea, Hawaii, elevation ~4.27 km), an optical telescope can resolve celestial objects that are separated by one-fourth of a second of arc (1 second of arc = 1 arcsec = 1/3600 of a degree). The viewing never gets any better than this because of atmospheric turbulence, which makes the images jitter. (a) What minimum diameter aperture is necessary to provide 1/4-arcsec resolution, assuming the resolution is limited only by diffraction? (b) Is there ever any point in building a telescope much bigger than this? Explain your answer. SSM Example 23-9

110. ••**Astronomy** The Herschel infrared telescope, launched into space in 2009, made observations from 2010 to 2013. Its primary mirror is 3.5 m in diameter, and the telescope focuses infrared light in the range of 55 to 672 μm. Because this telescope operated above Earth's atmosphere, its resolution was limited only by diffraction. (a) What wavelength in its observing range will give the maximum angular resolution? What is that maximum resolution (in radians and seconds, $1° = 60'$ and $1' = 60''$)? (b) To achieve the same resolution as in part (a) using visible light of wavelength 550 nm, what should be the mirror diameter of an optical telescope? (c) What is the smallest infrared source that the Herschel infrared telescope can resolve at a distance of 150 light years? (A light year is the distance that light travels in one year—about 9.461×10^{15} m.) Example 23-9

111. •• **Astronomy** The world's largest refracting telescope is at Yerkes Observatory in Williams Bay, Wisconsin. Its objective is 1.02 m in diameter. Suppose you could mount the telescope on a spy satellite 200 km above the ground. (a) Assuming that the resolution is limited only by diffraction, what minimum separation of two objects on the ground could it resolve? Take 550 nm as a representative wavelength for visible light. (b) Because of atmospheric turbulence, objects on the surface of Earth can be distinguished only if their angular separation is at least 1.00 arcsec. (1 arcsec = 1/3600 of a degree.) How far apart would two objects on Earth's surface be if they subtended an angle of 1.00 arcsec as measured from the satellite? Compare this with your answer to part (a). Example 23-9

112. ••• Light from a helium–neon laser that has a wavelength of 633 nm shines on a planar surface with two small slits separated by a distance of 4.00 μm. What is the maximum number of bright fringes that can be seen on a screen 10 m away? Example 23-7

113. ••• In a realistic two-slit experiment, you see two interference effects described in this chapter. There is the large-scale interference pattern from diffraction that depends on the width of each slit, and within that pattern there is the small-scale interference pattern that depends on the separation distance between the slits. For a slit width of 0.04 mm and a slit separation of 0.50 mm, how many of the bright two-slit interference fringes fit inside the central bright fringe of the diffraction pattern (from the first dark fringe on one side of the center to the first dark fringe on the other side) if the laser has a wavelength of (a) 650 nm or (b) 530 nm? Example 23-7

114. •• An equilateral triangular glass prism rests on a table (**Figure 23-59**). A beam of white light is incident on one face of the prism, at an angle of 60° below the normal to the surface. If the index of refraction of the glass is 1.51 for red light and 1.53 for blue light, what is the angular separation of the red and blue portions of the spectrum that emerge from the prism? Example 23-3

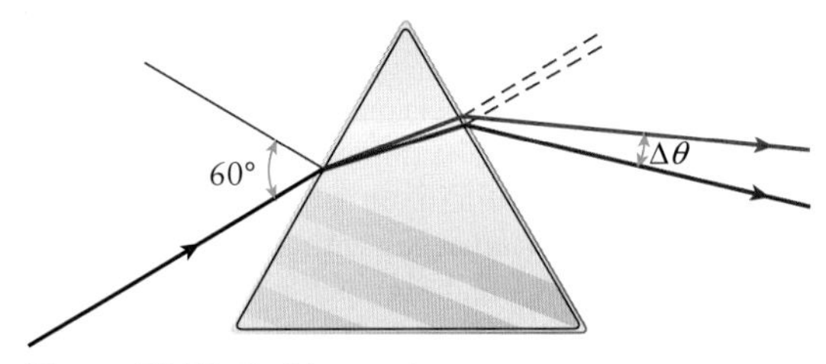

Figure 23-59 Problem 114

Geometrical Optics

moodboard/Alamy

What do you think?

Many older adults can see distant objects clearly but must wear corrective eyeglasses to see nearby objects (for example, to read a book). Are the lenses of these eyeglasses (a) thicker at the middle, (b) thicker at the edges, or (c) of uniform thickness?

In this chapter, your goals are to:

- (24-1) Explain the importance of optical devices.
- (24-2) Describe how a plane mirror forms an image.
- (24-3) Use ray diagrams to explain how the image formed by a concave mirror depends on the position of the object.
- (24-4) Calculate the position and height of an image made by a concave mirror.
- (24-5) Explain the differences between the images made by a convex mirror and a concave mirror.
- (24-6) Calculate the position and height of an image made by a convex mirror.
- (24-7) Describe how the curved surfaces of a lens make light rays converge or diverge.
- (24-8) Calculate the focal length of a lens based on its composition and shape.
- (24-9) Explain how the eye forms images on the retina, and how corrective lenses are used to compensate for deficiencies in vision.

To master this chapter, you should review:

- (3-7) Calculating the length of a circular arc.
- (23-2) The law of reflection and Snell's law of refraction.
- (23-4) Dispersion.

24-1 Mirrors or lenses can be used to form images

We saw in Chapter 23 that the direction of a light ray changes when it either reflects from a surface or refracts as it passes from one transparent medium to another. **Geometrical optics** is the branch of physics that uses the laws of reflection and refraction to understand **optical devices**—instruments that change the direction of light rays in a regular way. The simplest optical devices are reflective objects, or *mirrors*, and pieces of transparent material with carefully shaped surfaces, or *lenses*. Mirrors or lenses whose surfaces are curved in just the right way can be used to change the apparent sizes of objects (**Figure 24-1a**). The lens of your eye is an essential part of vision and needs to be replaced if it becomes clouded with age (**Figure 24-1b**).

As the name *geometrical optics* suggests, to study optical devices all we need besides the laws of reflection and refraction is a little bit of geometry. We won't need to refer at all to the wave properties of light. (Indeed, much of the basic physics of mirrors and lenses was deduced before it was understood that light is a wave.)

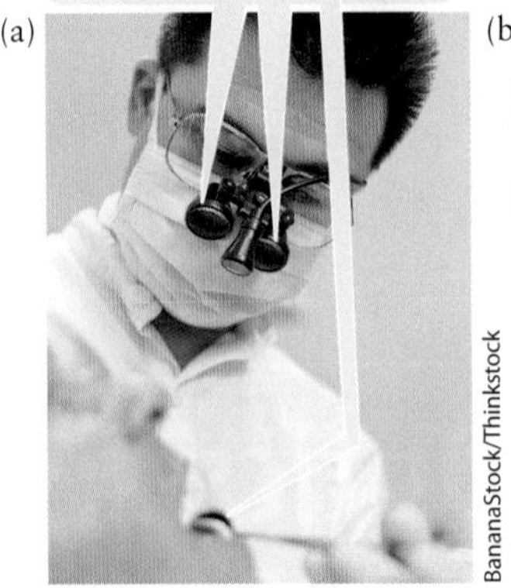

BananaStock/Thinkstock

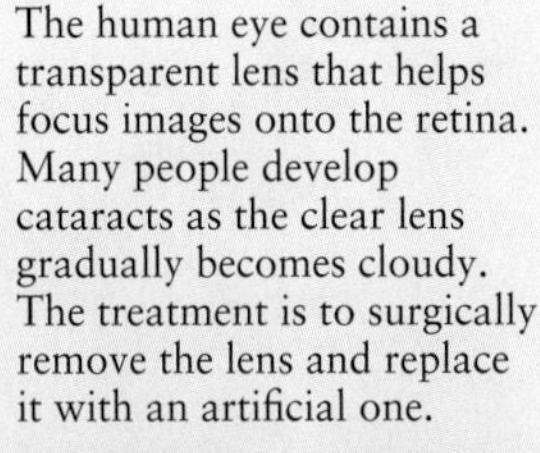

The human eye contains a transparent lens that helps focus images onto the retina. Many people develop cataracts as the clear lens gradually becomes cloudy. The treatment is to surgically remove the lens and replace it with an artificial one.

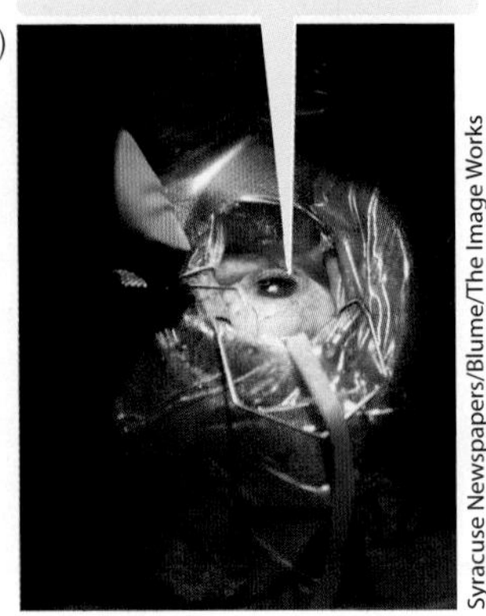

Syracuse Newspapers/Blume/The Image Works

Figure 24-1 Mirrors and lenses Mirrors and lenses are two different devices used to form images.

TAKE-HOME MESSAGE FOR Section 24-1

✔ A mirror forms images by the reflection of light from the mirror's surface.

✔ A transparent lens forms images by the refraction of light as it enters and exits the lens.

We'll begin by considering a simple plane mirror. We'll see how the law of reflection explains the kind of image that it forms. We'll then use similar ideas to understand the kind of images formed by a mirror with a surface that's curved either inward or outward. In contrast to mirrors, lenses use the refraction of light to form an image: Light rays can change direction when they enter a lens and again when they exit the lens. Happily, we'll find that many of the same ideas that we'll develop by considering curved mirrors apply equally well to lenses. We'll conclude with a discussion of the human eye, how it forms images, and how deficiencies in vision can be corrected.

24-2 A plane mirror produces an image that is reversed back to front

Our visual system can detect only objects that emit or reflect light. (We can't see in the dark!) Although it's easy to find examples of luminous things, such as the Sun or the screen of a mobile device, most of what we see only reflects light. As we learned in Section 23-2, a ray of light that strikes a surface always reflects in such a way that the angle of incidence equals the angle of reflection. (This is the law of reflection.) However, because most objects have uneven surfaces, the light that they reflect goes off in many seemingly random directions. This is called **diffuse** reflection (**Figure 24-2a**).

If an object has a flat surface, however, light rays that strike that surface are all reflected in the same general direction (**Figure 24-2b**). Such a flat, reflecting surface is called a **plane mirror**, and reflections from such a surface are called *specular* (from the Latin word for mirror, *speculum*). Your reflection in a bathroom mirror is an example of **specular reflection**. Let's look more closely at how a plane mirror creates an image.

BIO-Medical **Figure 24-3** shows how your eye and brain interpret light coming from an **object** (a term that refers to anything that acts as a source of light rays for an optical device). Some of the light rays coming from the object go to your eyes. Your brain traces those rays backward to a common origin and interprets that origin as the location of the object. **Figure 24-4** shows a similar situation, except that we have added a plane mirror. Some of the light from the object strikes the mirror and is reflected toward your eye. These rays appear to be coming from a point behind the mirror, so your brain interprets the light as coming from that point. We say that the mirror has formed an **image** of the object, and this image lies behind the mirror.

As you can see in Figure 24-4, no light rays actually pass through the location of the image. For this reason, the image formed by a plane mirror is said to be a **virtual image**. We will shortly encounter some optical devices that cause light rays to bend

(a)

Light that reflects from an object's surface in many random directions is called diffuse. Uneven surfaces produce diffuse light even when the incident rays come from the same direction.

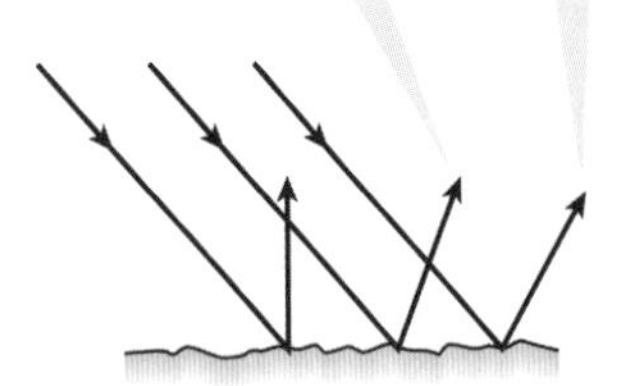

(b)

Light that reflects from a smooth surface is called specular. Light rays coming from the same general direction all reflect in the same general direction.

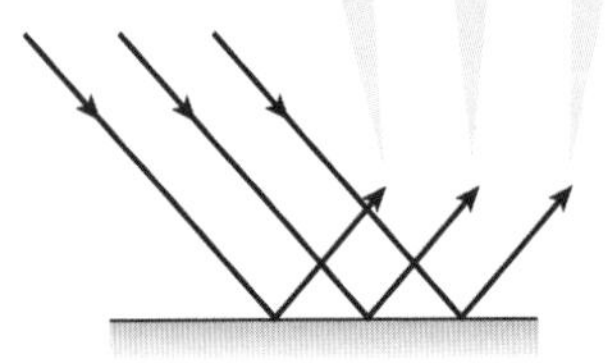

Figure 24-2 Diffuse and specular reflection Light reflecting from (a) an uneven surface and (b) a smooth surface.

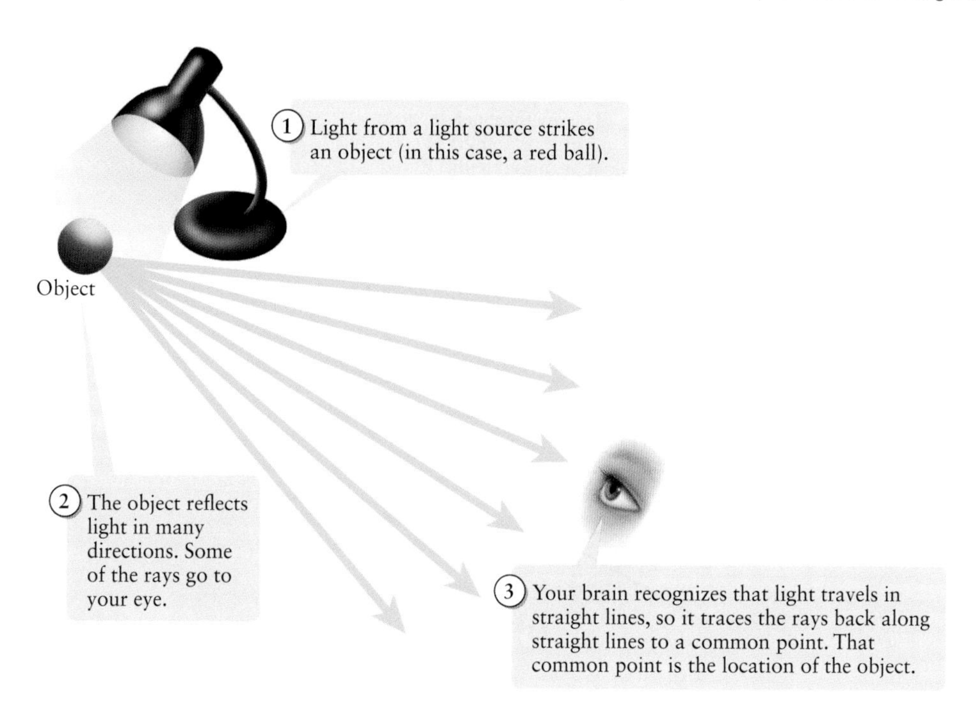

Figure 24-3 Locating an object When light strikes our eyes, we trace the rays of light back along straight lines to an apparent common source.

toward each other so that the image forms where light rays do actually meet. An image formed by light rays coming together is called a **real image.**

We can determine the position of the image made by a plane mirror by using a **ray diagram (Figure 24-5)**. In such a diagram, we draw a few light rays coming from the object and show how they reflect from the mirror. We've drawn the object as a red arrow of height h_O (the **object height**) located a distance d_O (the **object distance**) from the mirror. To determine the image position we must draw at least two rays that emanate from the tip of the object arrow. (We've drawn three in Figure 24-5.) When each ray strikes the mirror, it obeys the law of reflection: The angle of the reflected ray equals the angle of the incident ray. Note that the horizontal ray strikes the mirror face-on, so it is reflected back horizontally toward the tip of the object arrow.

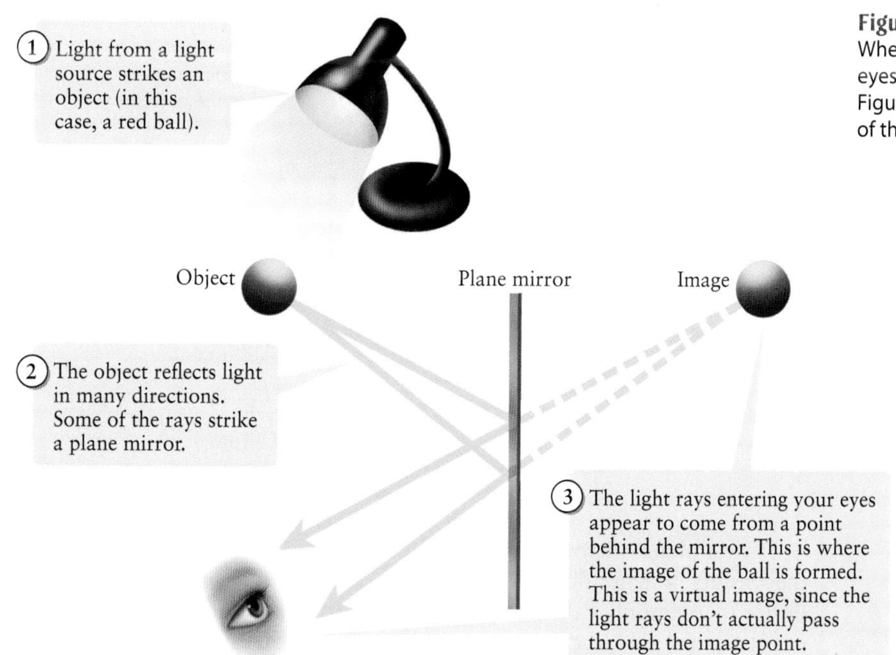

Figure 24-4 Locating an image When light from a mirror strikes our eyes, we use the same technique as in Figure 24-3 to determine the position of the image made by the mirror.

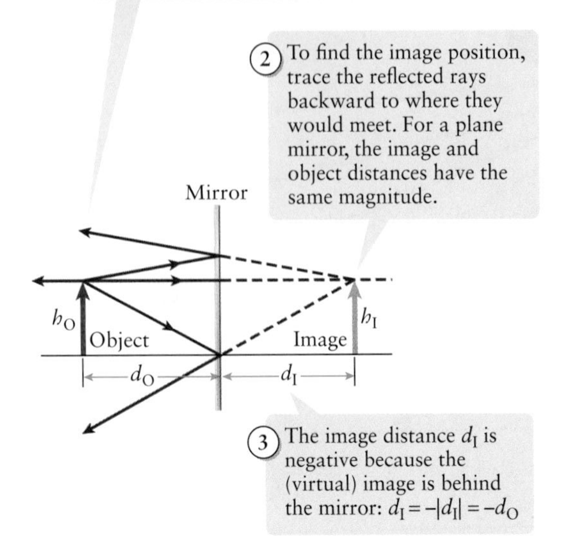

1 Light rays from the object reflect from the plane mirror. (The ray that hits the mirror face-on reflects back along its initial path.)
2 To find the image position, trace the reflected rays backward to where they would meet. For a plane mirror, the image and object distances have the same magnitude.
Mirror
h_O
Object
h_I
Image
d_O
d_I
3 The image distance d_I is negative because the (virtual) image is behind the mirror: $d_I = -|d_I| = -d_O$

Figure 24-5 A ray diagram for a plane mirror This diagram helps us determine the position, orientation, and size of the image.

The reflected rays diverge from each other and never actually meet. But if we trace these rays back to where they would meet, as shown by the dashed lines in Figure 24-5, we find the position of the tip of the image arrow. The geometry of the light rays requires that the image be as far behind the front of the mirror as the object is in front of the mirror. In other words, the **image distance** d_I has the same magnitude as the object distance d_O. We'll use the convention that a point on the reflective side of the mirror (to the left of the mirror in Figure 24-5) is at a positive distance, while a point on the back side of the mirror (to the right of the mirror in Figure 24-5) is at a negative distance. So in Figure 24-5 the object distance d_O is positive, but the image distance d_I is negative. We can write the relationship between d_I and d_O for a plane mirror as

Image distance for a plane mirror (24-1)

The **image distance** is negative. The **object distance** is positive.

$$d_I = -|d_I| = -d_O$$

The negative value of d_I indicates that the image is on the opposite side of the mirror from the object.

For example, if you stand a distance $d_O = 1.0\,\text{m}$ in front of a plane mirror, your image is at $d_I = -1.0\,\text{m}$—that is, 1.0 m behind the mirror.

Figure 24-5 also shows that the **image height** h_I is the same as the object height h_O. We define the **lateral magnification** m as the ratio of the image height to the object height. (We will often refer to m simply as the "magnification.")

Lateral magnification (24-2)

Lateral magnification Image height

$$m = \frac{h_I}{h_O}$$

Object height

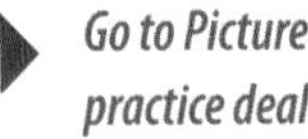

Go to Picture It 24-1 for more practice dealing with mirrors.

For a plane mirror $h_I = h_O$, so the magnification is $m = 1$. For an optical device that results in magnification greater than 1, the image is larger than the object. Such a device is commonly called a *magnifier*; it forms a magnified image of the object. When m is less than one, the image formed by the optical device is smaller than the object. As we'll see in later sections, a *curved* mirror can form an image that is either larger or smaller than an object placed in front of it.

WATCH OUT! The image formed by a plane mirror appears *behind* the mirror.

The image does not form "on" a plane mirror but rather behind it. This is evident from the ray diagrams in Figure 24-5. But if this diagram doesn't convince you, you can prove it to yourself by taping a bit of paper to a plane mirror, then looking at your reflection in the mirror from about 1 m away. You'll find it difficult, likely impossible, to focus your eyes on both your image in the mirror and the piece of paper at the same time. That's because your image in the mirror is twice as far from your eyes as the paper on the surface of the mirror. The image is behind the mirror, not on it.

WATCH OUT! The image formed by a plane mirror is reversed back to front, *not* left to right.

It's a common misconception that your image in a plane mirror is reversed left to right. A better description is that your image is reversed *back to front*. As an example, in **Figure 24-6** a rectangular box *ABCDEFGH* sits in front of a plane mirror, making an image *A′B′C′D′E′F′G′H′*. Note that the face *A′B′C′D′* of the image has the same orientation as the face *ABCD* of the object and is the same distance from the mirror; the same is true of the face *E′F′G′H′* of the image and the corresponding face *EFGH* of the object. You can see that the net result is that the image is identical to the object but flipped from back to front.

Each point on the image behind the mirror is directly opposite the corresponding point on the object in front of the mirror: *A′* opposite *A*, *B′* opposite *B*, and so on.

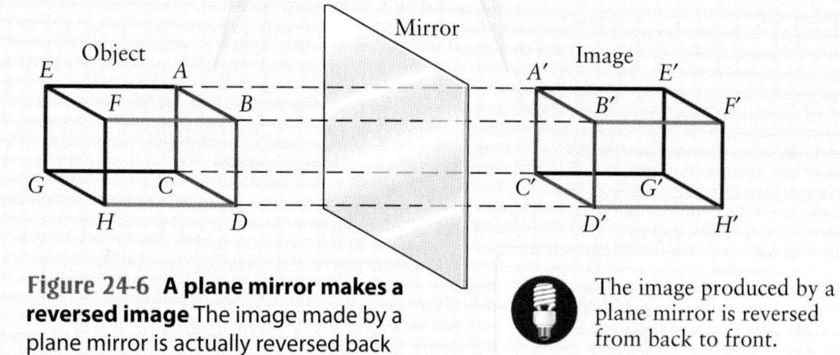

Figure 24-6 A plane mirror makes a reversed image The image made by a plane mirror is actually reversed back to front, not left to right.

The image produced by a plane mirror is reversed from back to front.

GOT THE CONCEPT? 24-1 See You, See Me I

You look into a mirror hanging near the corner of a hallway and see the eyes of someone standing around the corner. Is she able to see you? (a) Yes; (b) no; (c) not enough information given to decide.

GOT THE CONCEPT? 24-2 See You, See Me II

You look into a mirror hanging near the corner of a hallway. You see the right hand of someone standing around the corner, but not his eyes. Is he able to see you? (a) Yes; (b) no; (c) not enough information given to decide.

TAKE-HOME MESSAGE FOR Section 24-2

✔ When light rays coming from the same general direction hit a plane mirror, they tend to be reflected in the same general direction.

✔ If an object is placed in front of a plane mirror, the image is as far behind the mirror as the object is in front. The image is the same size as the object, but reversed from back to front.

✔ A plane mirror makes a virtual image; the light rays coming from the image do not actually pass through the image position.

24-3 A concave mirror can produce an image of a different size than the object

Figure 24-7 shows a jalapeño pepper placed in front of two curved mirrors. The mirror in Figure 24-7a is **convex** (its reflective surface is curved outward) and produces an image that is smaller than the object. The mirror in Figure 24-7b is **concave** (its

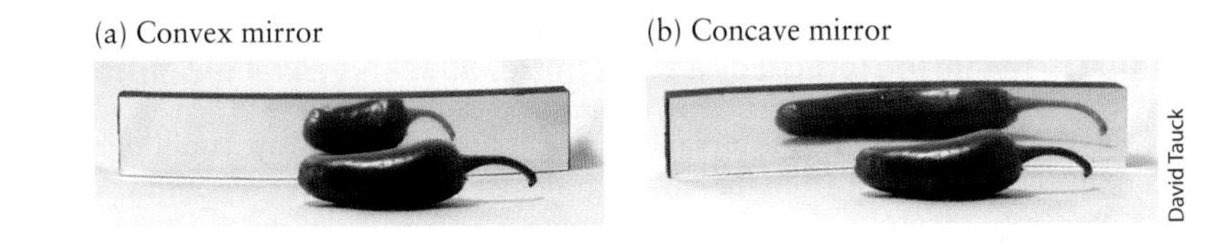

Figure 24-7 Images from curved mirrors A curved mirror can produce an image that is a different size than the object.

reflective surface is curved inward, or "caved in") and produces an image that is larger than the object. By contrast, the plane mirror that we studied in Section 24-2 always produces an image of the same size as the object. Let's analyze what happens with curved mirrors of this kind.

Figure 24-8 shows ray diagrams for parallel light rays striking two *spherical* mirrors—that is, mirrors that are shaped like a section cut from a complete sphere. **Figure 24-8a** shows that parallel light rays converge after they strike a concave mirror, while **Figure 24-8b** shows that parallel light rays diverge after they strike a convex mirror. By comparison, parallel light rays that strike a plane mirror remain parallel after striking it (see Figure 24-2b). It's important to consider such light rays because light rays that come from distant objects are either parallel or nearly so. The Sun's rays, for example, are essentially parallel when they strike Earth. You can see this from the crisp shadows cast by an object placed in the Sun, such as the hanging frame in **Figure 24-9a**. In contrast the shadow cast by a light bulb in **Figure 24-9b** is fuzzy because the light rays from the light bulb are not parallel.

The concave mirror has many practical uses. For example, a bathroom mirror used for applying makeup or for shaving is concave so that it gives an enlarged image of your face (like the enlarged image of the pepper in Figure 24-7b). Telescopes used by professional and amateur astronomers have a large concave mirror that brings the light from distant objects to a focus, forming an image. Automobile headlights use the same principle in reverse: A light bulb is placed in front of a concave mirror, and

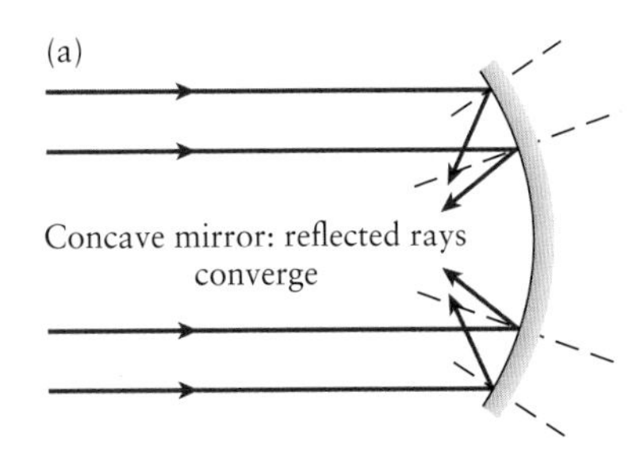

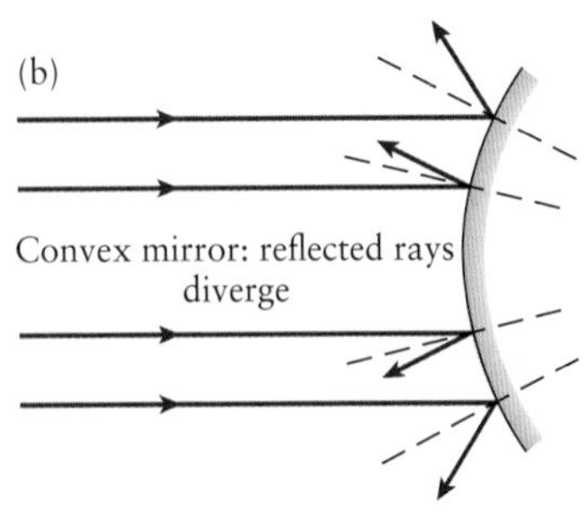

Figure 24-8 Concave and convex mirrors (a) Parallel light rays converge after reflecting from a concave mirror but (b) diverge after reflecting from a convex mirror.

(a)

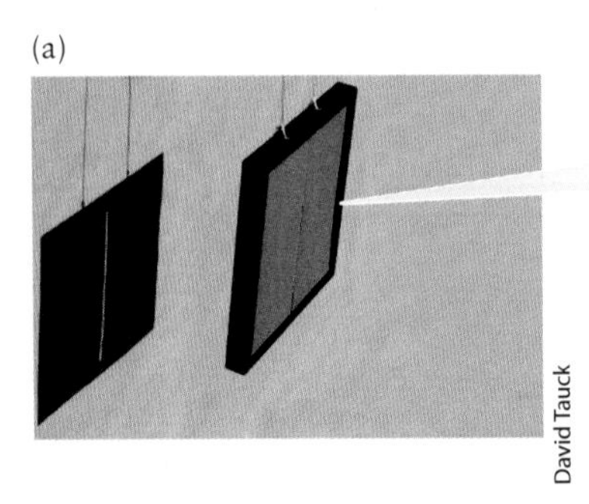

Rays of light from the distant Sun strike the hanging frame. Notice the sharp edges on all of the shadows, including those of the strings holding the frame.

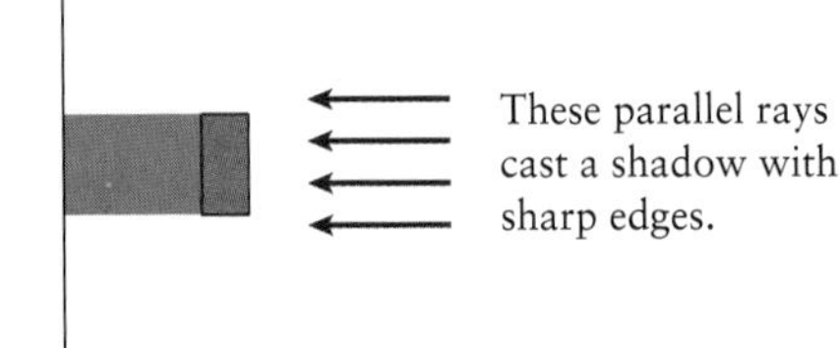

(b)

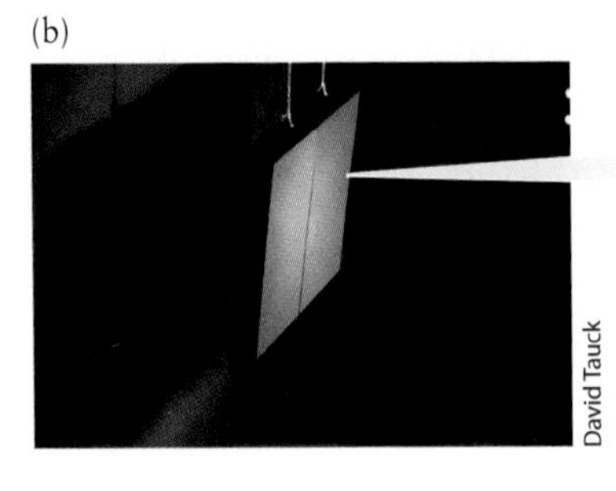

Light from a lamp strikes the frame. Because the light bulb is close to the frame, the central region of the shadow is darker than the outer regions.

These nonparallel rays cast a shadow with fuzzy edges.

Figure 24-9 Parallel and nonparallel light rays These photographs show the difference between light rays that are (a) parallel and (b) nonparallel.

the mirror reflects the light forward to illuminate the road ahead. In this section and the next, we'll look at the concave mirror in detail; we'll return to the convex mirror in Sections 24-5 and 24-6.

Figure 24-10 shows what happens when a series of parallel light rays strike a concave mirror formed from a fairly large section of a sphere. The incoming rays are parallel to each other and also parallel to the **principal axis** of the mirror, the axis that runs through the center of the sphere and also the center of the mirror. Notice that while the reflected rays generally converge along the principal axis of the mirror, they do not converge to the exact same point. This unfortunate characteristic of spherical mirrors is referred to as *spherical aberration*.

- A concave, spherical mirror causes parallel light rays to nearly converge along the principal axis of the mirror.
- If we consider only rays close to the principal axis, or if the mirror is only a small arc of the complete sphere, the light rays all converge to essentially a single point.

To avoid spherical aberration, we limit the reflective surface of the mirrors we consider to a relatively small section of a sphere. Here "relatively small" means that the size of the reflective surface is small compared to the radius of the sphere. There is no specific cut-off value; rather, the smaller the mirror compared to the radius, the more tightly focused the reflected rays will be. For the rest of this chapter, we will deal with spherical mirrors small enough that all rays parallel to the principal axis are focused to essentially a single point.

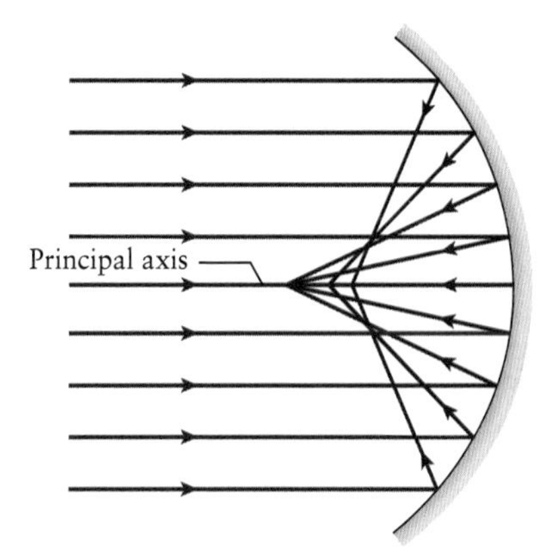

Figure 24-10 Reflection from a spherical mirror How parallel light rays behave when they strike a concave, spherical mirror.

Figure 24-11 defines many of the variables we use to describe a concave, spherical mirror. The **focal point** of the mirror is the point F along the principal axis at which incident rays parallel to the principal axis converge to a common focus when they reflect off the mirror. The distance from the focal point to the center of the mirror is f, the **focal length**. The mirror is a small section of a full sphere, and the point C—the **center of curvature** of the mirror—is the center of that sphere. The distance from C to any point on the mirror is the radius r of the sphere, also called the **radius of curvature** of the mirror.

For a concave mirror small enough that all rays parallel to the principal axis are focused at the focal point, the focal length is *exactly* half of the radius:

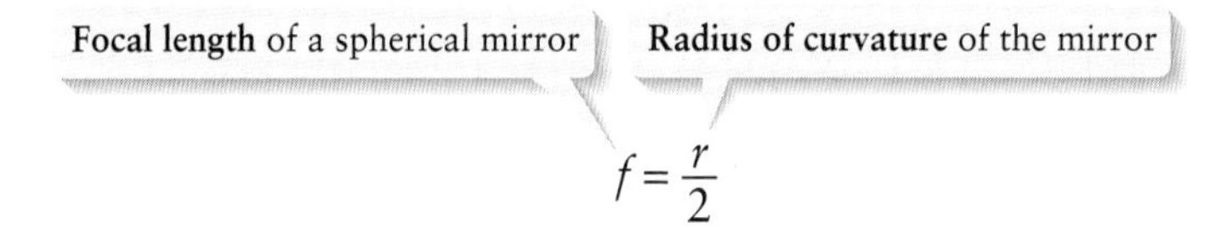

$$f = \frac{r}{2}$$

Focal length of a spherical mirror (24-3)

(Figure 24-11 shows this relationship.) Equation 24-3 says that the tighter the curve of a spherical mirror and hence the smaller the radius of curvature r, the shorter the focal length f. (We'll prove the relationship $f = r/2$ in Section 24-4.)

To see how a concave mirror forms an image, let's put our standard arrow at a point far from the mirror, as in **Figure 24-12a**. "Far" in this case means that the object distance d_O is greater than the radius of the mirror r; in other words, the base of the arrow is farther from the mirror, along the principal axis, than the center of curvature C. Now let's trace two rays of light from the tip of the arrow as they strike and then are reflected from the mirror. The image of the arrow's tip forms where those two rays intersect. Although any two light rays that originate at the tip of the arrow and that strike the mirror will work, we choose two that are particularly convenient. In **Figure 24-12b**, we trace a ray that starts parallel to the principal axis because all such rays are reflected through the focal point. In **Figure 24-12c** we add the trace of a ray that strikes the center of the mirror. The normal to the surface at this point lies along the principal axis, so it's easy to apply the law of reflection; the incident and reflected rays are symmetric around the principal axis.

Where does the image of the *base* of the arrow form? The base is on the principal axis, which is normal to the mirror's surface at the center of the mirror. So a light ray coming from the base of the arrow is reflected straight back from the center of the mirror, which means the image of the base of the arrow forms along the principal axis. **Figure 24-12d** shows the final image of the arrow. This is an **inverted image**: It's flipped upside down compared to the object. The image is also smaller than the object. Finally, the image is real; that is, it forms in front of the mirror where light rays reflected from different parts of the mirror's surface meet.

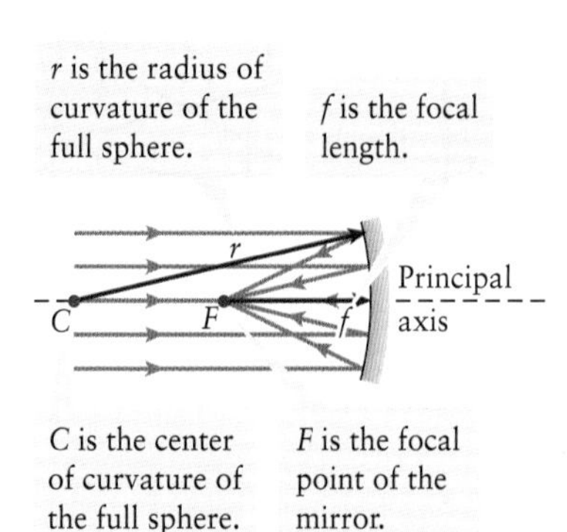

Figure 24-11 Mirror nomenclature This drawing defines many of the variables we use to describe a concave, spherical mirror.

(a)
An arrow is placed in front of a spherical, concave mirror. We will trace two light rays from the tip of the arrow to the mirror and as they reflect from it.

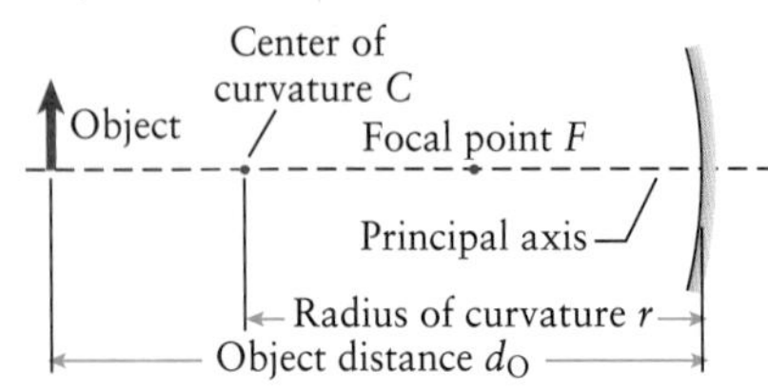

(b)
First we draw a light ray from the tip of the arrow to the mirror, parallel to the principal axis. It is reflected so that it passes through the focal point.

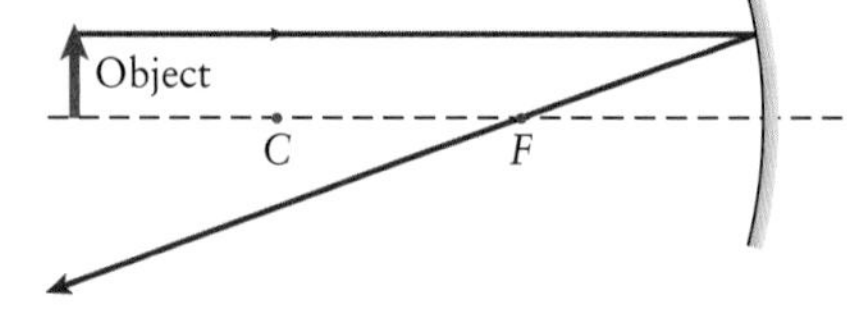

(c)
The angle of incidence of a ray that strikes the center of the mirror always equals the angle of reflection. This is easy to draw for this light ray.

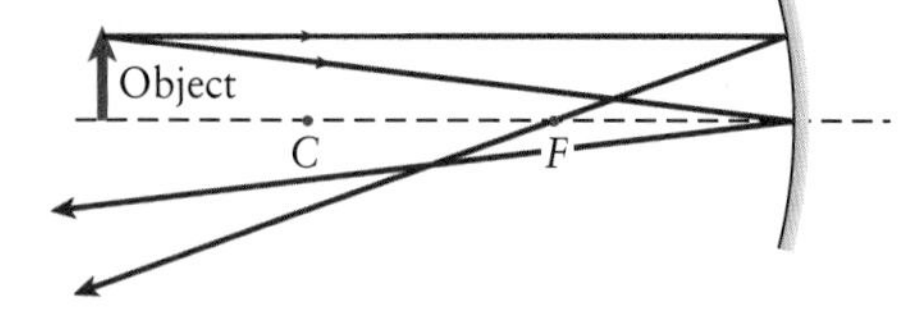

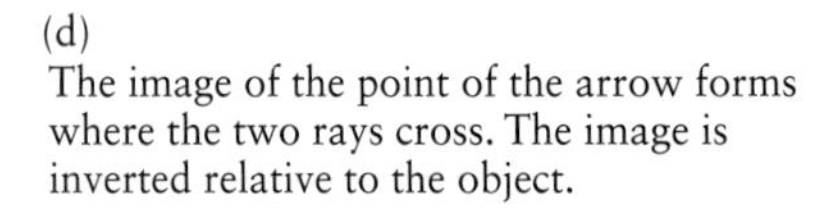
(d)
The image of the point of the arrow forms where the two rays cross. The image is inverted relative to the object.

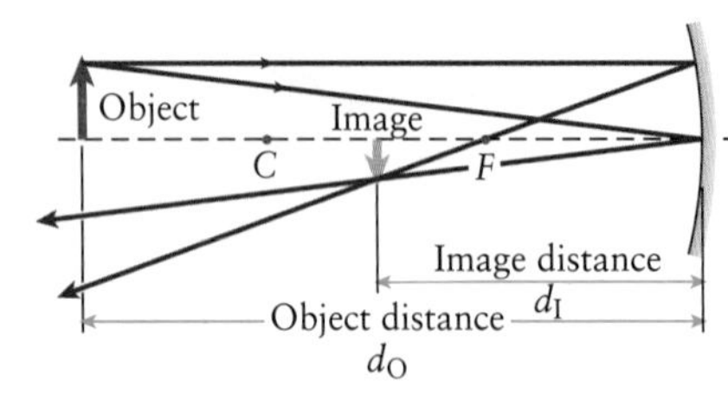

(e)

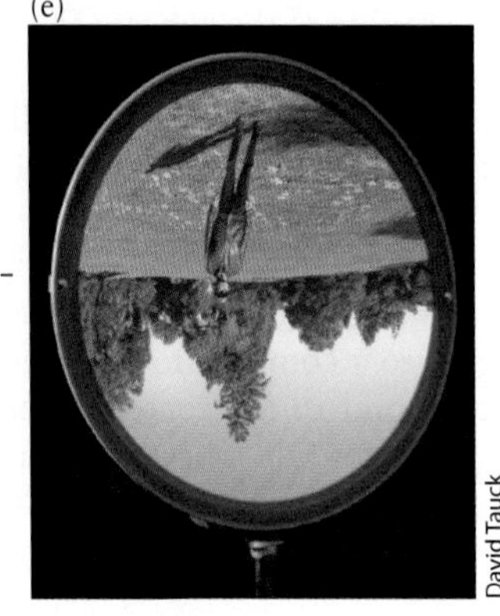

When the the object distance is greater than the radius of curvature of the mirror, the resulting real image is inverted, smaller than the object, and located between the center of curvature and the focal point.

Figure 24-12 **Ray diagram for a concave mirror I** (a) A distant object in front of a concave mirror. (b), (c), (d) Locating the image of this object. (e) The image of a person standing far from a concave mirror.

Figure 24-12d shows both the object distance d_O and the image distance d_I; both distances are positive, since the object and image are both on the reflective side of the mirror. It also shows that if d_O is large, the image is closer to the mirror than the object is, so $d_I < d_O$. Note that the image distance is greater than the focal length (the distance from the mirror to the focal point F). This is how a concave mirror is used in a telescope: The object is very far away, while the image is formed very close to the mirror and is much smaller than the object. (For example, the Moon is 3.84×10^5 km distant and 1738 km in radius, but the Moon's image made by an amateur astronomer's telescope is formed only a meter or so from the mirror and is less than a centimeter in radius.) **Figure 24-12e** shows the inverted image of one of the authors of this book standing far away from a concave mirror.

Let's see what happens if we move the object closer to the mirror. **Figure 24-13a** shows the object arrow placed at the center of curvature C, so the object distance equals the radius of curvature r. From Equation 24-3 the focal length $f = r/2$, so for the case shown in Figure 24-13, $d_O = r = 2f$. We trace the same two light rays as in Figure 24-12; where the two rays meet and the image forms is now farther from the mirror. This image is the same size as the object and, as in the previous case, is both real (the light rays pass through the image) and inverted. With the object placed at the center of curvature, the image forms at the center of curvature, so $d_I = d_O$. In **Figure 24-13b** one of the authors is standing at the center of curvature of a concave mirror; note that his inverted image is larger than that shown in Figure 24-12e.

In **Figure 24-14a** we've moved the arrow closer still to the mirror so that it sits between the center of curvature C and the focal point F. In this case the object

Figure 24-13 **Ray diagram for a concave mirror II** (a) We move the object from Figure 24-12 to the center of curvature of the concave mirror. (b) The image of a person standing at the center of curvature of a concave mirror.

(a)

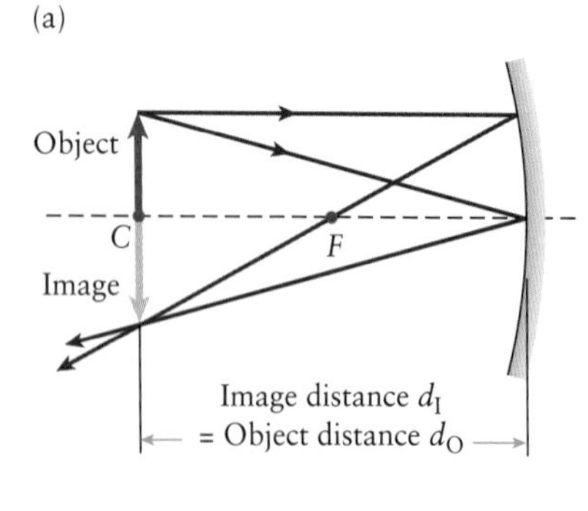

(b)

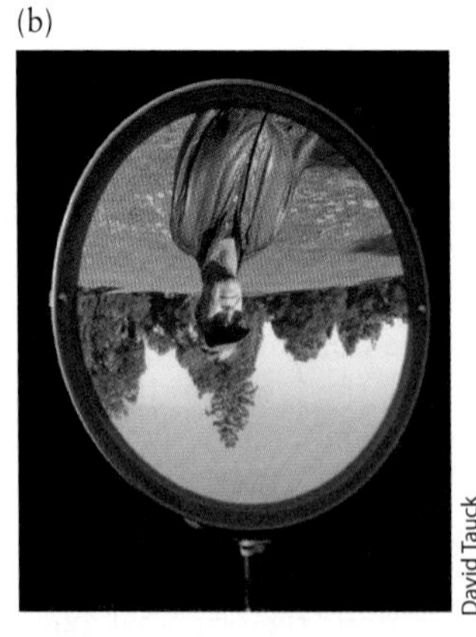

When the object is located at the radius of curvature of the mirror, the resulting real image is inverted, the same size as the object, and located at the radius of curvature.

distance is between $2f$ and f: $2f > d_O > f$. Again we've traced the same two light rays that we considered in the previous two cases. The image is still both real and inverted but has moved farther still from the mirror, so it is now farther from the mirror than the object (so $d_I > d_O$) and larger than the object. **Figure 24-14b** shows such an image of one of the authors; compare Figures 24-12e and 24-13b.

Comparing Figures 24-12, 24-13, and 24-14 shows that the image gets larger and moves farther from the mirror as we move a distant object closer to the focal point. In **Figure 24-15a** we place the object *at* the focal point so that the object distance equals the focal length, or $d_O = f$. After reflection, light rays that emanate from the tip of the arrow are parallel. (This is how a concave mirror is used in an automobile headlight: The lamp itself is placed close to the focal point of the curved mirror behind it, and the reflected light forms a beam of nearly parallel light rays.) With the object at the focal point, no image is formed because the parallel reflected rays never meet. The author in **Figure 24-15b** is standing close to the focal point of the concave mirror; there is no sharp image. If the object is slightly outside the focal point (d_O is slightly larger than f), the image is formed very far from the mirror and is much larger than the object.

What happens when the object is placed inside the focal point, so the object distance is less than the focal length ($d_O < f$)? In this case the reflected rays never actually meet but rather appear to meet at a point behind the mirror (**Figure 24-16a**). Hence the image is virtual, like the image formed by the plane mirror of Section 24-2, and the image distance d_I is negative: $d_I = -|d_I| < 0$. The image is larger than the object and is an **upright image** (it has the same orientation as the object). That's the kind of enlarged image you see when you look in a curved bathroom mirror for shaving or

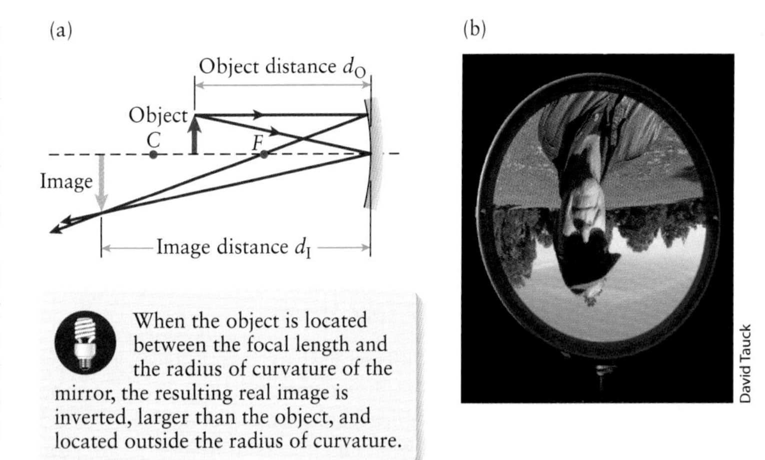
(a)
(b)
Object distance d_O
Object
C
F
Image
Image distance d_I
When the object is located between the focal length and the radius of curvature of the mirror, the resulting real image is inverted, larger than the object, and located outside the radius of curvature.
David Tauck

Figure 24-14 Ray diagram for a concave mirror III (a) We move the object from Figures 24-12 and 24-13 to a point between the center of curvature and the focal point of the concave mirror. (b) The image of a person standing at such a point.

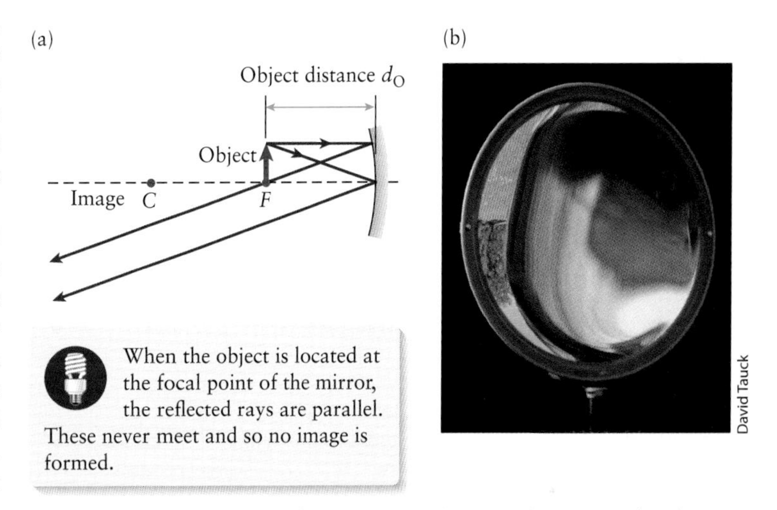
(a)
(b)
Object distance d_O
Object
Image
C
F
When the object is located at the focal point of the mirror, the reflected rays are parallel. These never meet and so no image is formed.
David Tauck

Figure 24-15 Ray diagram for a concave mirror IV (a) We move the object from Figures 24-12, 24-13, and 24-14 to the focal point of the concave mirror. (b) The image of a person standing near the focal point.

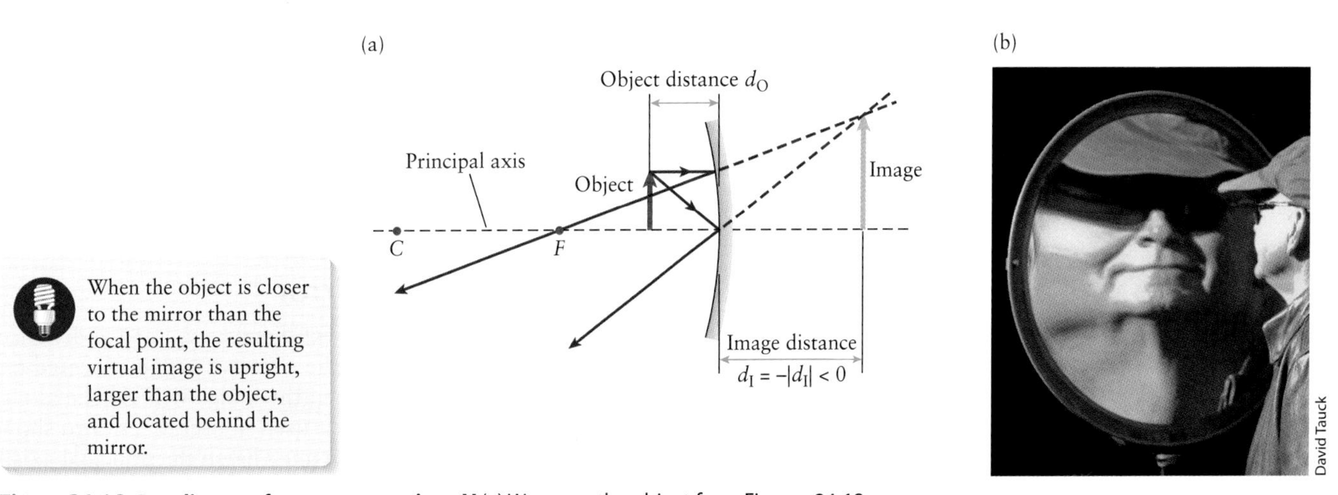
(a)
(b)
Object distance d_O
Principal axis
Object
Image
C
F
Image distance
$d_I = -|d_I| < 0$
When the object is closer to the mirror than the focal point, the resulting virtual image is upright, larger than the object, and located behind the mirror.
David Tauck

Figure 24-16 Ray diagram for a concave mirror V (a) We move the object from Figures 24-12, 24-13, 24-14, and 24-15 to a point inside the focal point of the concave mirror. (b) The image of a person standing at such a point.

applying makeup. **Figure 24-16b** shows such an enlarged image of one of the authors (compare Figures 24-12e, 24-13b, 24-14b, and 24-15b). The photograph in Figure 24-7b shows another such enlarged image.

So far our results for image formation by concave mirrors have been qualitative only. In the following section we'll take a more *quantitative* look at how the position of an object placed in front of a concave mirror determines the position and size of the resulting image.

GOT THE CONCEPT? 24-3 A Soup Spoon

A shiny spoon is not so different in shape than a spherical mirror: It's concave on one side and convex on the other. Your reflection from the concave side of the spoon, when held at arm's length, is upside down and appears to float in front of the spoon. When you hold the spoon about 6 cm from your eye, you see only a blur reflected in the spoon, but when you hold it about 4 cm from your eye your reflection is right side up and appears to be behind the spoon. What is the approximate focal length of the spoon? (a) More than 6 cm; (b) 6 cm; (c) between 4 and 6 cm; (d) 4 cm; (e) less than 4 cm.

GOT THE CONCEPT? 24-4 Covering a Concave Mirror

You place a light bulb oriented vertically (with its base on the bottom) just outside the focal point of a concave mirror. The resulting real image of the light bulb falls on a wall far from the mirror. This image is inverted and larger than the light bulb. If you were to paint the bottom half of the mirror black so that light rays that strike this half of the mirror would not be reflected, the image would show (a) only the top of the light bulb; (b) only the bottom of the light bulb; (c) only the left side of the light bulb; (d) only the right side of the light bulb; (e) the entire light bulb.

TAKE-HOME MESSAGE FOR Section 24-3

✔ A concave spherical mirror is in the shape of the inner surface of a section of a sphere.

✔ When an object is outside the focal point of a concave mirror, the image is outside the focal point, real and inverted.

✔ As the object is moved closer to the focal point, the image becomes larger and forms farther from the mirror.

✔ When the object is at the focal point, the reflected rays are parallel. Effectively, the image is at infinity and infinitely large.

✔ If the object is placed inside the focal point, the image is virtual and upright.

24-4 Simple equations give the position and magnification of the image made by a concave mirror

Let's return to the concave spherical mirror that we considered in the previous section. We'll see that given the radius of the mirror and the object distance, we can determine exactly how far from the mirror the image forms and how much magnification the mirror provides. All we need are a little geometry and the law of reflection.

The Mirror Equation for a Concave Mirror

To find the mathematical relationship among the object distance d_O, image distance d_I, and the radius of curvature r, we'll imagine that the object is a point located on the principal axis of the mirror as in **Figure 24-17**. Then the image will also lie along the principal axis. (The explanation is the same one we used in Section 24-3 to show why the image of the base of the arrow in Figure 24-12 must lie on the principal axis.) Figure 24-17 shows an arbitrarily chosen light ray coming from the object at point O and reflecting off the mirror at point P. The image forms where this reflected ray intercepts the principal axis at point I.

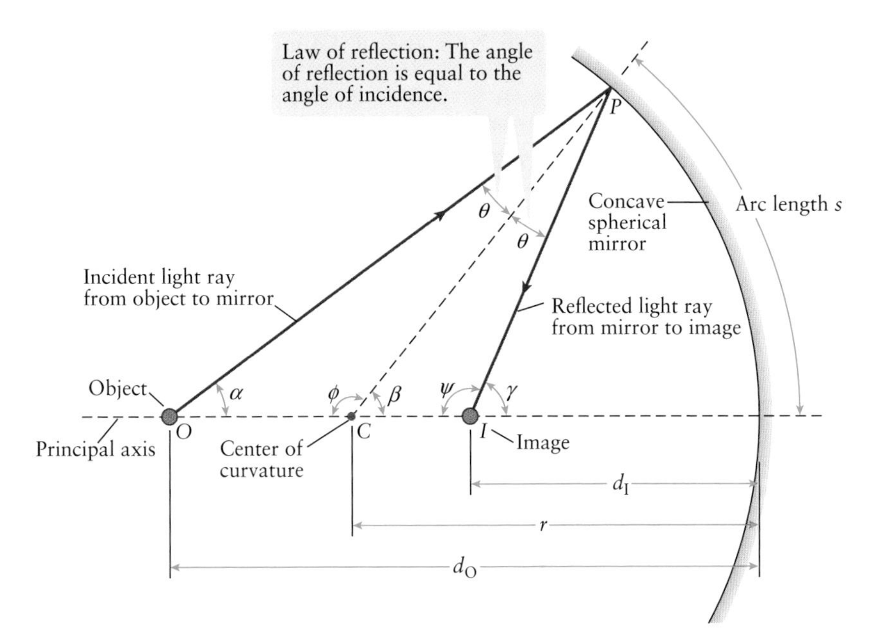

Figure 24-17 Analyzing a concave spherical mirror A ray diagram for a point object on the principal axis of a concave spherical mirror.

WATCH OUT! The path from object to mirror to image isn't just for one special light ray.

There's nothing special about the light ray we've drawn in Figure 24-17. *Any* ray that emanates from the object point O will reflect off the mirror and pass through the *same* image point I, provided the ray is at a shallow enough angle α to the principal axis of the mirror. So every part of the mirror contributes to forming the image. (Note that in Figure 24-17 we've drawn the angle α as fairly large, even though this angle is actually quite small. We've done this just to make the angles in the figure easier to see.)

We can find a relationship among d_O, d_I, and r by first finding a relationship among the angles in Figure 24-17 labeled α (the angle between the principal axis and the light ray from O to P), β (the angle between the principal axis and a line from the center of curvature C to P), and γ (the angle between the principal axis and the light ray from P to I). Since we're considering α to be a small angle, then necessarily β and γ are small angles as well. For that reason we can treat each of the three regions formed by one of these angles and the arc length s of the mirror as a sector (a pie-like slice) of a circle. The arc length is common to all three sectors. The length of the arc of a circle of radius r subtended by an angle θ equals $r\theta$ (see Section 3-7), provided the angle θ is in radians. So in Figure 24-17

√x *See the Math Tutorial for more information on geometry.*

$$s = d_O\alpha; \quad s = r\beta; \quad s = d_I\gamma \tag{24-4}$$

Note also that the sum of the angles in any triangle equals 180° or π radians. For the triangle OPC we have

$$\alpha + \theta + \phi = \pi \tag{24-5}$$

However, the angles ϕ and β in Figure 24-17 must add to 180°, or π radians (together they make up half a circle). So $\phi + \beta = \pi$ and $\phi = \pi - \beta$, and Equation 24-5 becomes

$$\alpha + \theta + \pi - \beta = \pi \quad \text{or} \quad \theta = \beta - \alpha \tag{24-6}$$

Similarly, for the triangle OPI the sum of the angles is π:

$$\alpha + 2\theta + \psi = \pi \tag{24-7}$$

Figure 24-17 shows that the angles ψ and γ also make up half a circle, so $\psi + \gamma = \pi$ and $\psi = \pi - \gamma$. Triangle CPI shows that $\beta + \theta + \psi = \pi$. Comparing with Equation 24-7 it follows that $\alpha + 2\theta = \beta + \theta$. Equation 24-7 can then be written as

$$\beta + \theta + \pi - \gamma = \pi \quad \text{or} \quad \theta = \gamma - \beta \tag{24-8}$$

Equations 24-6 and 24-8 are two different expressions for the angle θ in Figure 24-17. If we set these equal to each other, we get

$$\beta - \alpha = \gamma - \beta \quad \text{or} \quad \alpha + \gamma = 2\beta \tag{24-9}$$

From Equations 24-4 we have $\alpha = s/d_O$, $\beta = s/r$, and $\gamma = s/d_I$. Substituting these into Equation 24-9 gives

$$\frac{s}{d_O} + \frac{s}{d_I} = 2\frac{s}{r}$$

To simplify we divide through by the arc length s, giving us the relationship we've been seeking among the object distance, image distance, and radius of curvature of the mirror:

$$\frac{1}{d_O} + \frac{1}{d_I} = \frac{2}{r} \tag{24-10}$$

To help interpret Equation 24-10, recall from Figure 24-11 that if the incident light rays are parallel, they come to a focus at the focal point a distance f from the mirror. So in this case $d_I = f$. The incident rays will be parallel if the object is infinitely far away, so $d_O \to \infty$ and $1/d_O \to 0$. Then Equation 24-10 becomes

$$0 + \frac{1}{f} = \frac{2}{r} \quad \text{or} \quad f = \frac{r}{2}$$

This justifies Equation 24-3: The focal length of a concave spherical mirror is one-half of the radius of curvature.

If we replace $2/r$ with $1/f$ in Equation 24-10, we get the final form of the **mirror equation:**

Mirror equation and lens equation relating object distance, image distance, and focal length (24-11)

$$\frac{1}{d_O} + \frac{1}{d_I} = \frac{1}{f} \tag{24-11}$$

Focal length

Object distance

Image distance

We also call Equation 24-11 the *lens equation* because, as we shall see in Section 24-8, it's also applicable to the image formed by a lens.

We take the focal length f of a concave mirror to be positive because the center of curvature C of the mirror is in front of the mirror, that is, on its reflective side (to the left in Figure 24-17). Likewise, since the object point O is in front of the mirror, the object distance d_O is positive. The image distance d_I, however, can be positive if the image is real (in front of the mirror) or negative if the image is virtual (behind the mirror). To see when the image distance is positive and when it is negative, let's rewrite Equation 24-11 to solve for d_I:

$$\frac{1}{d_I} = \frac{1}{f} - \frac{1}{d_O} = \frac{d_O}{d_O f} - \frac{f}{d_O f} = \frac{d_O - f}{d_O f} \quad \text{or} \quad d_I = \frac{d_O f}{d_O - f} \tag{24-12}$$

In Equation 24-12, d_O and f are both positive for a concave mirror, so the numerator $d_O f$ is positive. However, the denominator $d_O - f$ can be positive or negative depending on whether d_O is larger or smaller than f.

If the object is outside the focal point so that $d_O > f$, then the denominator $d_O - f$ in Equation 24-12 is positive and the image distance is positive. Therefore, in this case the image made by the mirror is real. As the object moves closer to the focal point, the difference between d_O and f gets smaller and so d_I given by Equation 24-12 gets larger. Hence the image moves farther away from the mirror. If the object is inside the focal

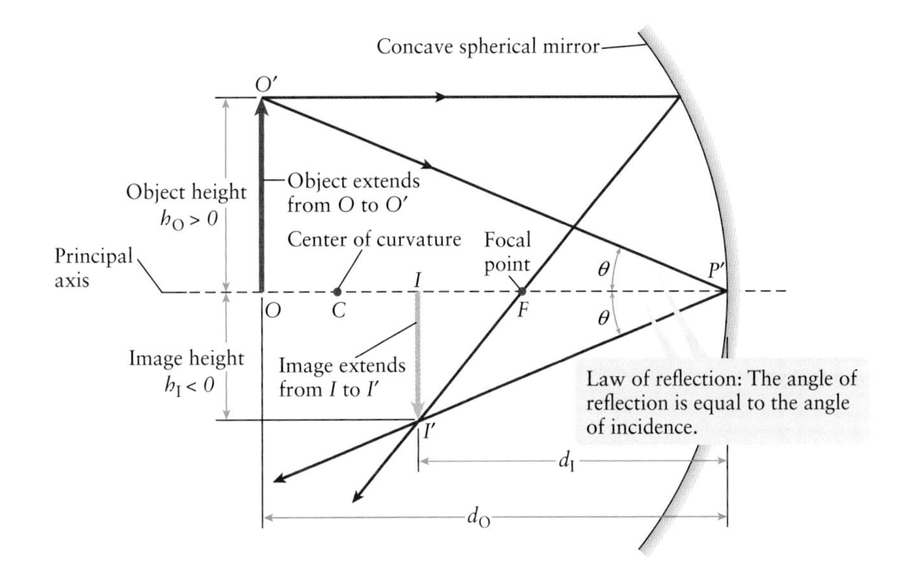

Figure 24-18 Magnification for a concave spherical mirror We replace the point object in Figure 24-17 with an object arrow.

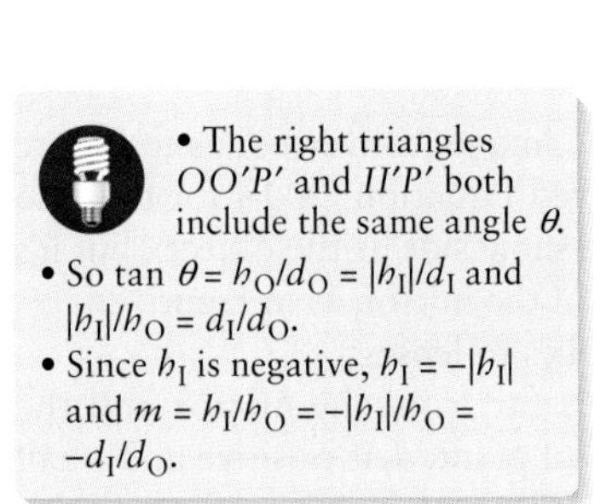

point, however, then $d_O < f$ and the denominator $d_O - f$ is negative. Then d_I is negative as well, and the image is virtual. That's exactly the behavior that we deduced in Section 24-3 by analyzing ray diagrams.

• The right triangles OO′P′ and II′P′ both include the same angle θ.
• So $\tan\theta = h_O/d_O = |h_I|/d_I$ and $|h_I|/h_O = d_I/d_O$.
• Since h_I is negative, $h_I = -|h_I|$ and $m = h_I/h_O = -|h_I|/h_O = -d_I/d_O$.

As a further check on Equations 24-11 and 24-12, note that if the object is exactly two focal lengths from the mirror so $d_O = 2f$, the image distance is

$$d_I = \frac{d_O f}{d_O - f} = \frac{(2f)f}{2f - f} = \frac{2f^2}{f} = 2f$$

So when the object is a distance $2f$ from the mirror—which, because $r = 2f$, is at the center of curvature—the image is at the same position. That's the same conclusion we came to by using the ray diagram in Figure 24-13.

Magnification for a Concave Mirror

We can also use simple geometry to find an expression for the lateral magnification of the image produced by a concave spherical mirror. In **Figure 24-18** we've replaced the point object at O with an upright arrow of height h_O that extends from O (on the principal axis) to O′. Just as in Figure 24-12 we draw two rays coming from the tip of the object arrow at O′ to determine the position I' of the tip of the image arrow.

Since the object in Figure 24-18 is outside the focal point F, the image is real and inverted and so the height of the image is negative: $h_I = -|h_I|$. By the law of reflection the ray from O′ to the point P' at the center of the mirror makes the same angle θ with the mirror's principal axis (shown as a dashed line in Figure 24-18) as does the reflected ray from P' to I'. If you look at the right triangle $OO'P'$, you'll see that the tangent of θ (the opposite side divided by the adjacent side) is $\tan\theta = h_O/d_O$; if you do the same for the right triangle $II'P'$, you'll see that $\tan\theta = |h_I|/d_I$. Setting these two expressions equal to each other, we see that

$$\frac{h_O}{d_O} = \frac{|h_I|}{d_I} \quad \text{or} \quad \frac{|h_I|}{h_O} = \frac{d_I}{d_O}$$

Since $h_I = -|h_I|$, we can rewrite this as

$$-\frac{h_I}{h_O} = \frac{d_I}{d_O} \quad \text{or} \quad \frac{h_I}{h_O} = -\frac{d_I}{d_O} \qquad (24\text{-}13)$$

From Equation 24-2 the lateral magnification is $m = h_I/h_O$. So Equation 24-13 tells us that

Lateral magnification for a mirror or lens (24-14)

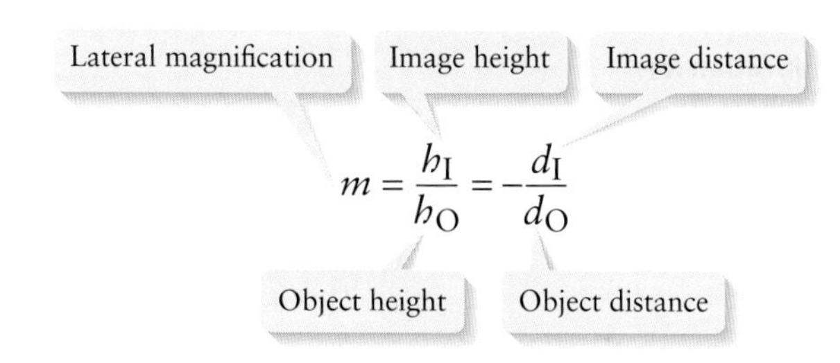

$$m = \frac{h_I}{h_O} = -\frac{d_I}{d_O}$$

Go to Interactive Exercise 24-1 for more practice dealing with concave mirrors.

A negative value of the magnification m means that the image is inverted, as in Figure 24-18. If m is positive, the image is upright. We've derived Equation 24-14 for the case in which the image is real and in front of the mirror (the same side as the object), but it's also true when the image is virtual and on the back side of the mirror, as in Figure 24-16. (As we'll see in Section 24-8, the same equation also applies to lenses.)

When the object is far from the mirror and the image is close to the focal point, d_I is positive and small compared to d_O, so from Equation 24-14 m is small and negative; the mirror produces a reduced, inverted image (see Figure 24-12). As the object distance decreases, the image distance increases and the ratio $m = -d_I/d_O$ increases in absolute value. As we saw above, when $d_O = 2f$ the image distance d_I is also equal to $2f$; then $m = -1$ and the inverted image is as large as the object (see Figure 24-13). If we move the object even closer to the focal point, but still outside it, the image distance is greater than the object distance and the absolute value of the magnification is greater than 1. Hence the inverted image is larger than the object (see Figure 24-14). If we move the object inside the focal point so that $d_O < f$, the image distance is negative (the image is behind the mirror) and the image is virtual. In this case Equation 24-14 tells us that m is positive, so the virtual image is upright (see Figure 24-16). We see that Equation 24-14 gives us the same results as we deduced from the ray diagrams in Section 24-3.

Table 24-1 summarizes when the radius of curvature r, focal length f, image distance d_I, image height h_I, and magnification m are positive and when they are negative.

TABLE 24-1 Sign Conventions for Mirrorss

Quantity	Sign convention
mirror radius of curvature r	• positive for a concave mirror • negative for a convex mirror
focal length f	• positive for a concave mirror • negative for a convex mirror
image distance d_I	• positive if on the reflective side of the mirror (the same side as the object); the image is then a real image • negative if on the nonreflective side of the mirror (the opposite side from the object); the image is then a virtual image
image height h_I	• positive if the image is upright (the same orientation as the object) • negative if the image is inverted (flipped upside down compared to the object)
lateral magnification m	• positive if the image is upright (the same orientation as the object) • negative if the image is inverted (flipped upside down compared to the object)

EXAMPLE 24-1 Images Made by a Concave Mirror

An object is 1.50 cm tall and is placed in front of a spherical concave mirror that has a radius of curvature equal to 20.0 cm. Find the image distance and height if the object is (a) 14.0 cm from the mirror; (b) 6.00 cm from the mirror. In each case, draw a ray diagram as part of the solution.

Set Up

Our mathematical tools are the expression for the focal length of the mirror, the mirror equation, and the equation for lateral magnification. We're given the radius of curvature $r = 20.0$ cm and the object height $h_O = 1.50$ cm; our goal is to find the values of the image distance d_I and image height h_I for the cases $d_O = 14.0$ cm and $d_O = 6.00$ cm.

Focal length of a spherical mirror:

$$f = \frac{r}{2} \quad (24\text{-}3)$$

Mirror equation:

$$\frac{1}{d_O} + \frac{1}{d_I} = \frac{1}{f} \quad (24\text{-}11)$$

Lateral magnification of a mirror

$$m = \frac{h_I}{h_O} = -\frac{d_I}{d_O} \quad (24\text{-}14)$$

Solve

First calculate the focal length of the mirror.

From Equation 24-3,

$$f = \frac{r}{2} = \frac{20.0 \text{ cm}}{2} = 10.0 \text{ cm}$$

(a) The object distance $d_O = 14.0$ cm is greater than the focal length $f = 10.0$ cm but less than $2f = 20.0$ cm. That is, the object is inside the center of curvature but outside the focal point. We've drawn a ray diagram that's similar to Figure 24-14. This tells us to expect that the image will be real, inverted, farther from the mirror than the object is, and larger than the object.

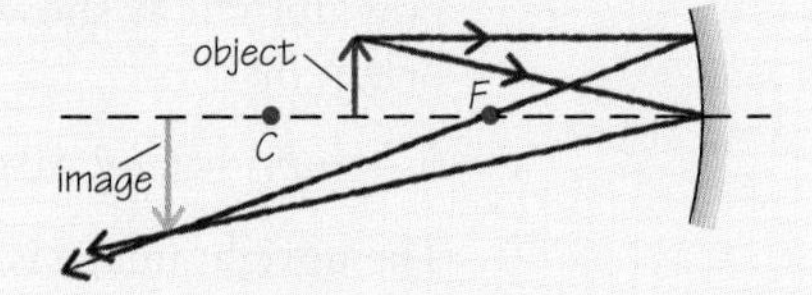

Calculate the image distance using the mirror equation.

From Equation 24-11

$$\frac{1}{d_I} = \frac{1}{f} - \frac{1}{d_O} = \frac{1}{10.0 \text{ cm}} - \frac{1}{14.0 \text{ cm}}$$

$$= 0.1000 \text{ cm}^{-1} - 0.0714 \text{ cm}^{-1} = 0.0286 \text{ cm}^{-1} \text{ so}$$

$$d_I = \frac{1}{0.0286 \text{ cm}^{-1}} = 35.0 \text{ cm}$$

The image distance is positive, so the image is 35.0 cm in front of the mirror. An image that forms in front of the mirror is a real image.

Calculate the image height using the magnification equation.

From Equation 24-14

$$m = \frac{h_I}{h_O} = -\frac{d_I}{d_O} = -\frac{35.0 \text{ cm}}{14.0 \text{ cm}} = -2.50$$

The image is 2.50 times larger than the object and, as the minus sign shows, inverted. Solve for the image height h_I:

$$h_I = mh_O = (-2.50)(1.50 \text{ cm}) = -3.75 \text{ cm}$$

The inverted image is 3.75 cm high.

(b) Now the object distance $d_O = 6.00$ cm is less than the focal length $f = 10.0$ cm, so the object is inside the focal point. We've drawn a ray diagram that's similar to Figure 24-16. This tells us to expect that the image will be virtual, upright, farther from the mirror, and larger than the object.

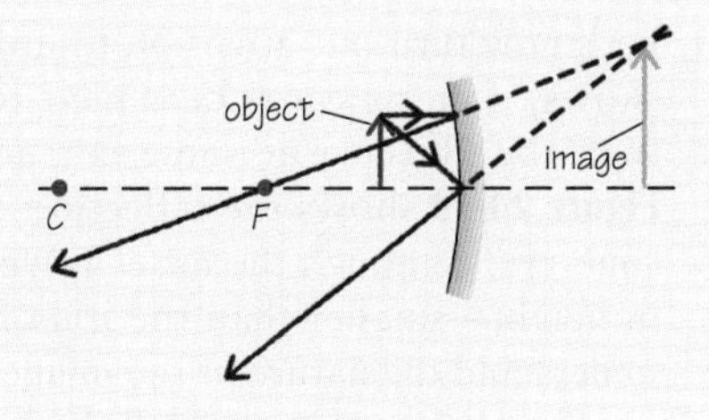

Calculate the image distance using the mirror equation.

From Equation 24-11

$$\frac{1}{d_I} = \frac{1}{f} - \frac{1}{d_O} = \frac{1}{10.0 \text{ cm}} - \frac{1}{6.00 \text{ cm}}$$

$$= 0.1000 \text{ cm}^{-1} - 0.1667 \text{ cm}^{-1} = -0.0667 \text{ cm}^{-1} \text{ so}$$

$$d_I = \frac{1}{-0.0667 \text{ cm}^{-1}} = -15.0 \text{ cm}$$

The image distance is negative, so the image is 15.0 cm behind the mirror. An image that forms behind the mirror is a virtual image.

Calculate the image height using the magnification equation.

From Equation 24-14

$$m = \frac{h_I}{h_O} = -\frac{d_I}{d_O} = -\frac{(-15.0 \text{ cm})}{6.00 \text{ cm}} = +2.50$$

Again the image is 2.50 times larger than the object but is now upright as shown by the plus sign. Solve for the image height h_I:

$$h_I = mh_O = (+2.50)(1.50 \text{ cm}) = +3.75 \text{ cm}$$

The upright image is 3.75 cm high.

Reflect

In both parts the position and size of the image are consistent with our ray diagrams. It's always a good idea to draw such diagrams as a check on your calculations using the mirror equation and magnification equation.

GOT THE CONCEPT? 24-5 A Concave Mirror

If an object is placed 12.0 cm from a concave mirror, the resulting real image is 2.00 times as large as the object. What is the focal length of the mirror? (a) 6.00 cm; (b) 8.00 cm; (c) 16.0 cm; (d) 20.0 cm; (e) 24.0 cm.

TAKE-HOME MESSAGE FOR Section 24-4

✔ The mirror equation relates the focal length of a concave mirror and the positions of the object and image.

✔ A positive value of the image distance d_I indicates that the image is in front of the mirror and is real. A negative value of d_I indicates that the image is behind the mirror and virtual (the light rays never actually go there).

✔ The magnification of the image is positive if the image is upright and negative if the image is inverted.

24-5 A convex mirror always produces an image that is smaller than the object

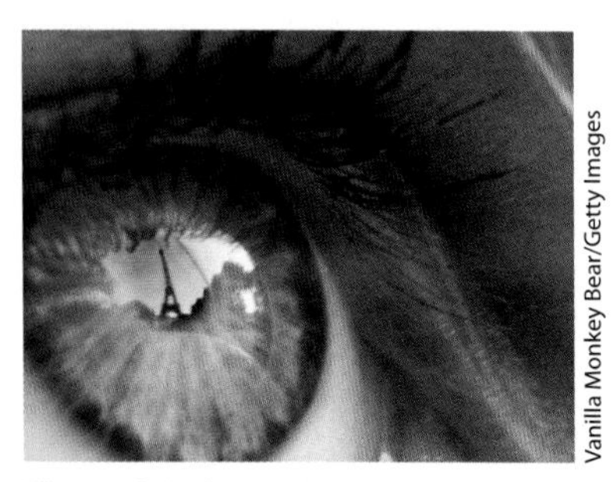

Vanilla Monkey Bear/Getty Images

Figure 24-19 Parisian reflections The Eiffel Tower can be seen reflected from the surface of this person's eye. The eye is relatively spherical, so its outer surface acts like a convex mirror.

Let's now turn our attention to images produced by a *convex* mirror. As Figure 24-7a shows, if an object is held next to a convex mirror, the resulting image is smaller than the object. The same is true if an object is far away from a convex mirror. **Figure 24-19** shows the reflection of the Eiffel Tower in the convex surface of a person's eye. Although the Eiffel Tower is 324 m tall, its image is only about a centimeter in height—smaller than the iris of the eye. In this section we'll use ray diagrams to understand the nature of the images formed by a convex mirror.

Recall from Figure 24-8 the key difference between concave and convex mirrors: Parallel light rays that reflect from a concave mirror converge toward a point in front of the mirror, while parallel light rays that reflect from a convex mirror diverge from a point on the back side of the mirror. (If the mirrors are spherical, the rays don't truly converge on or diverge from a single point. But this is a good description of what happens if the rays are all close to the principal axis of the mirror. We'll make this assumption—the same that we made in Sections 24-3 and 24-4 for concave mirrors—throughout this section.)

Because the focal point of a convex mirror is behind the mirror, we say that the focal length f is negative. For a convex mirror, we call the focal point *virtual* because parallel light rays that reflect from the mirror seem to emanate from that point but don't actually pass through it.

Figure 24-20 Ray diagram for a convex mirror How the image is formed for an object in front of a convex mirror.

- A convex mirror has a negative focal length: $f = -|f|$.
- For any object distance d_O a convex mirror produces a virtual image behind the mirror.
- The image distance d_I is negative: $d_I = -|d_I|$.
- The virtual image is upright and smaller than the object.

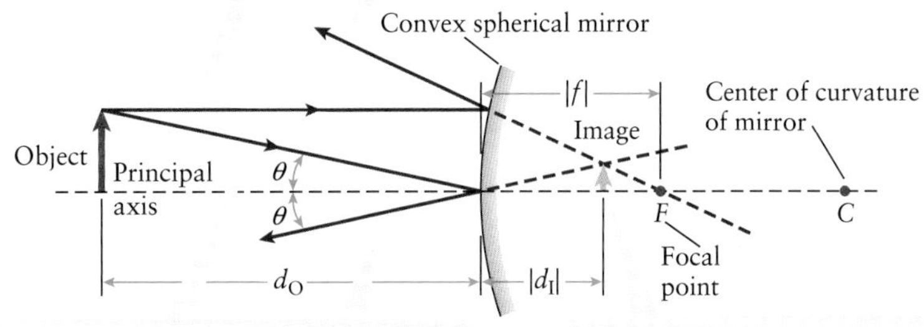

Figure 24-20 shows an object placed a distance d_O in front of a convex mirror. The focal point F lies a distance $|f|$ behind the mirror. (Since the focal length f is negative, the distance is the absolute value of f.) The center of curvature C is also behind the mirror, so the radius of curvature r is also negative. It turns out that the focal length f is equal to one-half of the radius of curvature r, just as for a concave mirror:

$$f = \frac{r}{2} \qquad (24\text{-}3)$$

(We'll justify this statement in Section 24-6.) For a concave mirror f and r are both positive; for a convex mirror f and r are both negative.

To find the location of the image in Figure 24-20, we draw two rays that emanate from the tip of the object arrow, just as we did for the ray diagrams in Section 24-3 for a concave mirror. The image of the arrow tip forms where the two reflected rays appear to meet. The image of the base of the arrow must lie along the principal axis. (A light ray coming from the base of the arrow and traveling along the principal axis strikes the mirror normal to the surface and so is reflected straight back. The image of the base of the arrow must therefore form on the principal axis.) As Figure 24-20 shows, the image is virtual because it forms behind the mirror, just like the image formed by a plane mirror (Section 24-2): The reflected rays don't actually cross there. Hence the image distance d_I is negative. We saw in Section 24-3 that a concave mirror produces a virtual image only if the object is inside the focal point; a convex mirror produces a virtual image for *any* position of the object.

Figure 24-20 shows that the image formed by a convex mirror is smaller than the object, no matter what the object distance is. Figure 24-20 also shows that the image formed by the convex mirror will always be upright and will always be closer to the mirror than the virtual focal point. **Figure 24-21** shows that as the object distance d_O decreases, the image becomes larger and closer to the convex mirror, but remains virtual and upright. If the object were moved all the way to the surface of the mirror, the image would be exactly the same size as the object.

Because a convex mirror makes objects appear smaller, they provide a wide-angle view. Rear view mirrors on automobiles and trucks often include a convex mirror to allow the driver to see as much of the area behind the vehicle as possible. (Any mirror with the label "Objects in mirror are closer than they appear" is a convex mirror.)

In the following section we'll look at these ideas more quantitatively and see how to calculate the image size and position for a convex mirror.

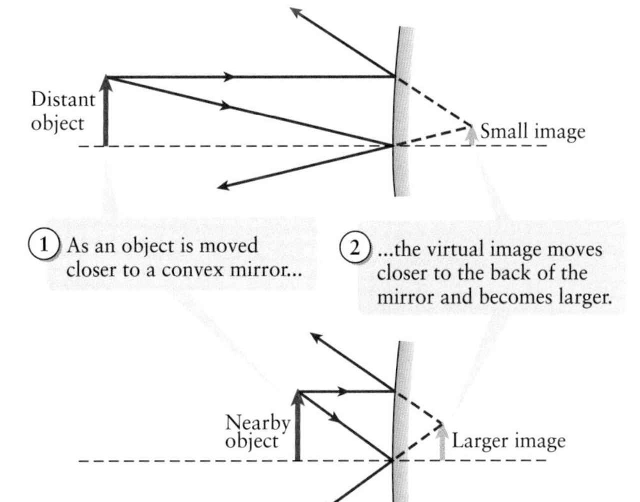

Figure 24-21 Moving closer to a convex mirror As the object far from a convex mirror is moved closer to the mirror, the image increases in size.

GOT THE CONCEPT? 24-6 Solar Cooking

The Sun is so far from Earth that rays of sunlight are effectively parallel. Sunlight reflected from a mirror can be focused onto a pot, raising its temperature enough to pasteurize water in the pot or cook food. What kind of spherical mirror could be used for this purpose? (a) A convex mirror; (b) a concave mirror; (c) either a convex or concave mirror.

TAKE-HOME MESSAGE FOR Section 24-5

✔ Parallel light rays that strike a convex mirror diverge and appear to emanate from a virtual focal point behind the mirror.

✔ If an object is placed anywhere in front of a convex mirror, the resulting image is virtual, upright, closer to the mirror than the object is, and smaller than the object.

24-6 The same equations used for concave mirrors also work for convex mirrors

Just as we did for the concave mirror in Section 24-4, we can use geometry and the law of reflection to find an equation that relates the object and image distances and the radius of curvature for a convex mirror. We'll also find an equation that relates the sizes of the image and object for a convex mirror. Remarkably we'll see that these equations are exactly the same as those for a concave mirror, provided we're careful with the signs of the radius of curvature, focal length, and image distance.

The Mirror Equation for a Convex Mirror

In **Figure 24-22** we've placed a point object on the principal axis of a convex mirror with (negative) radius of curvature r. This is analogous to Figure 24-17, in which we placed a point object on the principal axis of a concave mirror. The object is at position O, a distance d_O from the mirror. We've drawn an arbitrarily chosen light ray that travels away from the object at an angle α from the principal axis and reflects off the mirror at P; if we extend the reflected ray backward, the extension (shown as a dashed blue line) crosses the principal axis at I. A second light ray (not shown) that travels away from the object along the principal axis will be reflected straight back along that axis. If we extend this reflected ray backward, it will meet the extension of

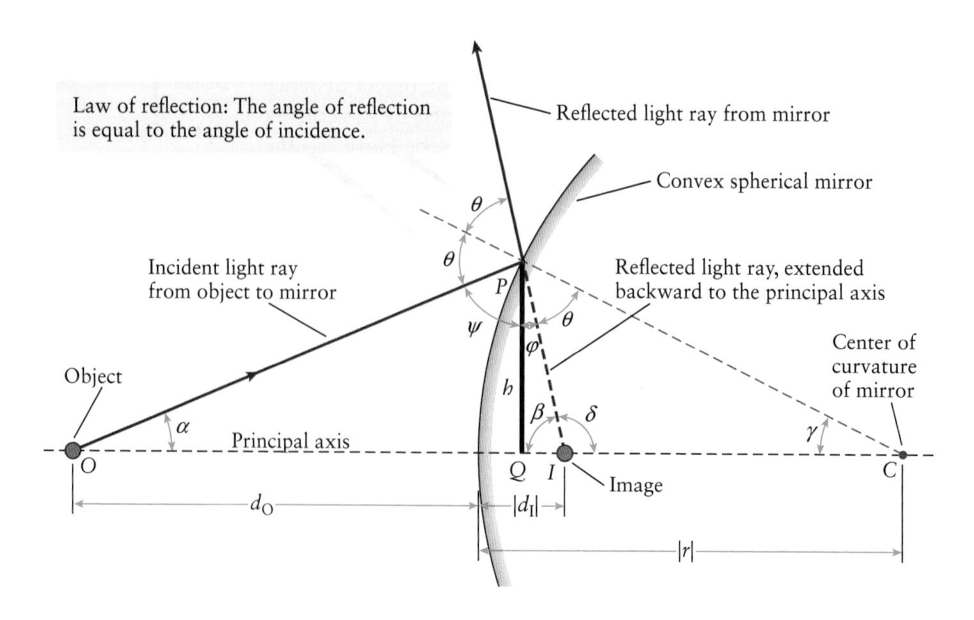

Figure 24-22 Analyzing a convex spherical mirror A ray diagram for a point object on the principal axis of a convex spherical mirror.

the first reflected ray at I. Hence I is the position of the image made by the mirror. The image is behind the mirror, so the image distance is negative. The center of curvature C of the mirror also lies behind the mirror, so the radius of curvature is also negative. Mathematically, $d_I = -|d_I| < 0$ and $r = -|r| < 0$.

Note that the red dashed line CP in Figure 24-22 is normal to the mirror at P, so by the law of reflection the incident ray and the reflected ray are both at the same angle θ relative to this normal. It follows that θ is also the angle between the line CP and the backward extension PI of the reflected ray.

To relate the object distance d_O, image distance d_I, and radius of curvature r, we'll first find a relationship among α, the angle between the principal axis and the light ray incident on the mirror; β, the angle between the principal axis and the reflected ray; and γ, the angle between the principal axis and the red dashed line CP that defines the normal at the point P where the light ray is reflected. Notice that each of these three angles is part of a right triangle for which one side is the thick black line PQ in Figure 24-22: triangle OPQ for angle α, triangle IPQ for angle β, and triangle CPQ for angle γ. In each case the line PQ, of length h, is the side opposite the angle. Since we're considering only light rays that are very close to the principal axis, the reflection point P is also very close to the principal axis. Hence point Q (directly below point P) is essentially at the point where the principal axis touches the mirror surface. As a result, we can treat d_O as the base of triangle OPQ, $|d_I|$ as the base of triangle IPQ, and $|r|$ as the base of triangle CPQ. For each triangle the tangent of the angle equals the length h of the side opposite the angle, divided by the length of the triangle's base. So

See the Math Tutorial for more information on trigonometry.

$$\tan\alpha = \frac{h}{d_O}, \quad \tan\beta = \frac{h}{|d_I|}, \quad \tan\gamma = \frac{h}{|r|} \tag{24-15}$$

If the incident ray OP is at a small angle to the principal axis, then α, β, and γ are all small angles. (We've drawn the angle α fairly large in Figure 24-22 to make the geometry easier to visualize. In reality, it must be quite small to conform to the approximation that the rays are nearly parallel to the principal axis.) The tangent of a small angle is approximately equal to the angle in radians, so Equation 24-15 becomes

$$\alpha = \frac{h}{d_O}, \quad \beta = \frac{h}{|d_I|}, \quad \gamma = \frac{h}{|r|} \tag{24-16}$$

To relate the angles α, β, and γ, we'll do as in Section 24-4 and use the result that the sum of the angles of any triangle must be 180°, or π radians. For the triangle IPC that connects the image point I, reflection point P, and center of curvature C, the three angles are γ, θ, and δ. So

$$\gamma + \theta + \delta = \pi \tag{24-17}$$

Figure 24-22 shows that the angles β and δ must add to π radians (together they make up half a circle around the point I). So $\beta + \delta = \pi$, $\delta = \pi - \beta$, and Equation 24-17 becomes

$$\gamma + \theta + \pi - \beta = \pi \quad \text{or} \quad \theta = \beta - \gamma \tag{24-18}$$

For the triangle OPC that connects the object point O, reflection point P, and center of curvature C, the angles are α, $(\psi + \theta + \phi)$, and γ, so we have

$$\alpha + (\psi + \theta + \phi) + \gamma = \pi \tag{24-19}$$

We can simplify Equation 24-19 by noting from Figure 24-22 that the angles θ, ψ, ϕ, and θ form a half-circle around the point P, so their sum is π radians: $\theta + \psi + \phi + \theta = \pi$, or $\psi + \theta + \phi = \pi - \theta$. If we substitute this expression for $(\psi + \theta + \phi)$ into Equation 24-19, we get

$$\alpha + \pi - \theta + \gamma = \pi \quad \text{or} \quad \theta = \alpha + \gamma \tag{24-20}$$

Equations 24-18 and 24-20 are both expressions for the angle θ. If we set these equal to each other, we get

$$\beta - \gamma = \alpha + \gamma \quad \text{or} \quad \alpha - \beta = -2\gamma \tag{24-21}$$

Equation 24-21 is the relationship among the angles α, β, and γ we've been looking for. We can now get a relationship among the object distance d_O, image distance d_I, and radius of curvature r by substituting the expressions for α, β, and γ from Equation 24-16 into Equation 24-21:

$$\frac{h}{d_O} - \frac{h}{|d_I|} = -\frac{2h}{|r|}$$

If we divide through by the height h of the black line in Figure 24-22 and recall that the image distance and radius of curvature are both negative, so that $d_I = -|d_I|$ and $r = -|r|$, this becomes

$$\frac{1}{d_O} + \frac{1}{d_I} = \frac{2}{r} \tag{24-22}$$

Equation 24-22 is *exactly* the same as Equation 24-10, which we derived for a *concave* mirror. It's reassuring that we get the same expression for both kinds of mirrors.

Note that if the object is infinitely far away, so that $d_O \to \infty$ and $1/d_O \to 0$, the rays from the object to the mirror will all be parallel to the axis and the virtual image will be formed at the focal point F, a distance $|f|$ behind the mirror. Then $d_I = -|f| = f$ (recall that the focal length f is negative). For this situation, Equation 24-22 becomes

$$\frac{1}{f} = \frac{2}{r} \quad \text{or} \quad f = \frac{r}{2}$$

This result is exactly the same as that for a concave spherical mirror: The focal length f is equal to one-half of the radius of curvature r (Equation 24-3). The only difference is that for a convex mirror f and r are both negative.

If we substitute $1/f = 2/r$ into Equation 24-22, we get the mirror equation for a convex mirror:

$$\frac{1}{d_O} + \frac{1}{d_I} = \frac{1}{f} \tag{24-11}$$

This is Equation 24-11, the *same* mirror equation that we derived for a *concave* mirror in Section 24-4. This equation works equally well whether the focal length is positive (for a concave mirror) or negative (for a convex mirror).

We can use Equation 24-11 to explore the properties of the image made by a convex mirror. We saw in Section 24-4 that this equation can be rewritten as

$$d_I = \frac{d_O f}{d_O - f} \tag{24-12}$$

Using $f = -|f|$ for the negative focal length of a convex mirror, Equation 24-12 becomes

$$d_I = \frac{d_O(-|f|)}{d_O - (-|f|)} = -\left(\frac{d_O}{d_O + |f|}\right)|f| \tag{24-23}$$

The right-hand side of Equation 24-23 is always negative, so the image distance d_I for a convex mirror will always be negative (the image will always be behind the mirror). Note that the fraction in parentheses is always less than or equal to 1, so the image distance is always between 0 and $-f$. That is, the image always forms somewhere between the mirror and the focal point.

Magnification for a Convex Mirror

Figure 24-23 again shows the image made by a convex spherical mirror, but now the object is an upright arrow of height h_O a distance d_O in front of the mirror. Just as for a concave mirror (see Figure 24-18 in Section 24-4), we draw two rays coming from the tip of the object arrow at O'. One ray is parallel to the principal axis of the mirror;

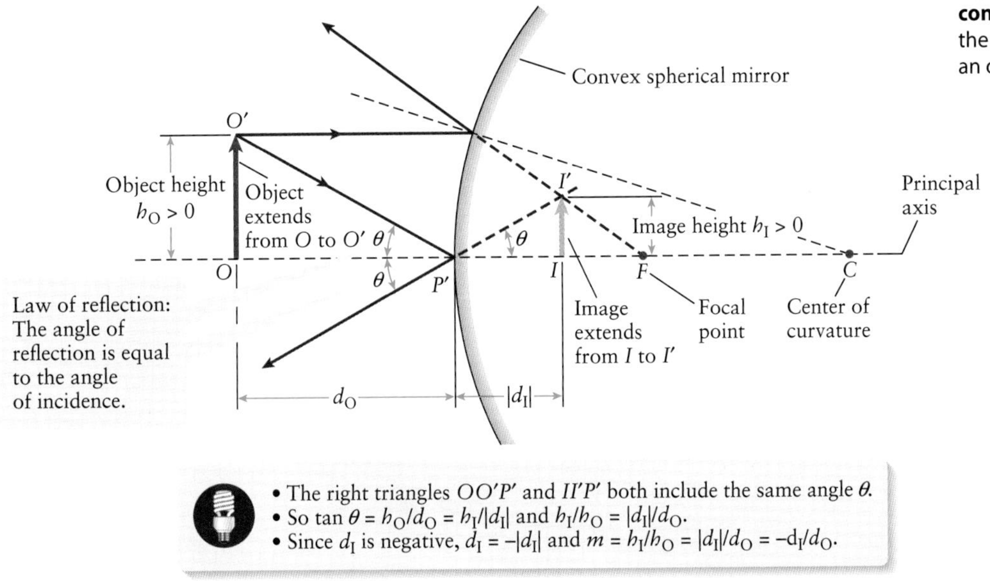

Convex spherical mirror

O′

Object height $h_O > 0$

Object extends from O to O′

Principal axis

I′

Image height $h_I > 0$

θ

O

P′

I

F

C

Law of reflection: The angle of reflection is equal to the angle of incidence.

Image extends from I to I'

Focal point

Center of curvature

d_O

$|d_I|$

- The right triangles $OO'P'$ and $II'P'$ both include the same angle θ.
- So $\tan\theta = h_O/d_O = h_I/|d_I|$ and $h_I/h_O = |d_I|/d_O$.
- Since d_I is negative, $d_I = -|d_I|$ and $m = h_I/h_O = |d_I|/d_O = -d_I/d_O$.

Figure 24-23 Magnification for a convex spherical mirror We replace the point object in Figure 24-22 with an object arrow.

after reflection this travels away from the mirror as though it had been emitted from the focal point F. The other ray strikes the center of the mirror at P', and the reflected ray makes the same angle θ with the principal axis as the incident ray from O' to P'. If we extend this reflected ray backward, the extension is at the same angle θ to the principal axis. The extensions of the two reflected rays meet at I', the position of the tip of the image arrow. The base of the image arrow is at I, directly underneath I' on the principal axis. The image is behind the mirror, so the image distance d_I is negative. This image is also upright, so the image height h_I is positive.

To relate the image and object heights, note that the right triangles $OO'P'$ and $II'P'$ both include the same angle θ. For $OO'P'$ the tangent of θ (the opposite side divided by the adjacent side) equals h_O/d_O; for $II'P'$ the tangent of θ equals $h_I/|d_I|$. These two expressions for $\tan\theta$ must be equal, so

$$\frac{h_O}{d_O} = \frac{h_I}{|d_I|} \quad \text{or} \quad \frac{h_I}{h_O} = \frac{|d_I|}{d_O} = -\frac{d_I}{d_O} \quad (\text{since } d_I = -|d_I|)$$

The lateral magnification m equals h_I/h_O, so it follows that for a convex mirror

$$m = \frac{h_I}{h_O} = -\frac{d_I}{d_O} \tag{24-14}$$

This is the *same* expression for magnification as for a concave mirror. Since the image distance d_I is negative for a convex mirror, Equation 24-14 tells us that the magnification is positive and so the image is upright. In addition, for a convex mirror the image is closer to the mirror than is the object, so $|d_I| < d_O$, and the value of m is less than 1. That is, the image made by a convex mirror is always smaller than the object, just as we saw in Section 24-5.

Go to Interactive Exercise 24-2 for more practice dealing with concave and convex mirrors.

WATCH OUT! Positive magnification indicates that an image is upright.

Remember that the *sign* of the magnification of a mirror doesn't indicate whether the image is larger or smaller than the object, only whether the image is upright or inverted. An upright image, for example, is always associated with a positive magnification, regardless of whether the image is larger than or smaller than the object.

EXAMPLE 24-2 An Image Made by a Convex Mirror

An object is 1.50 cm high and is placed in front of a spherical convex mirror with a radius of curvature of magnitude 48.0 cm. Find the image distance and height if the object is (a) 68.0 cm from the mirror; (b) 3.00 cm from the mirror.

Set Up

This is similar to Example 24-1 in Section 24-4. The key difference is that the mirror is now convex, so the radius of curvature is negative: $r = -48.0$ cm. We're given the object height $h_O = 1.50$ cm, and want to find the values of the image distance d_I and image height h_I for the cases $d_O = 68.0$ cm and $d_O = 3.00$ cm. Our tools are Equations 24-3, 24-11, and 24-14, which apply to convex mirrors as well as concave ones.

Focal length of a spherical mirror:

$$f = \frac{r}{2} \quad (24\text{-}3)$$

Mirror equation:

$$\frac{1}{d_O} + \frac{1}{d_I} = \frac{1}{f} \quad (24\text{-}11)$$

Lateral magnification of a mirror

$$m = \frac{h_I}{h_O} = -\frac{d_I}{d_O} \quad (24\text{-}14)$$

Solve

Use Equation 24-3 to calculate the focal length of the mirror.

From Equation 24-3

$$f = \frac{r}{2} = \frac{-48.0 \text{ cm}}{2} = -24.0 \text{ cm}$$

The negative value of focal length means that the focal point is behind the mirror.

In both cases we expect that the image will be virtual (behind the mirror), smaller than the object, and closer to the mirror than the object is. The specific position and size of the image will be different in the two cases, however.

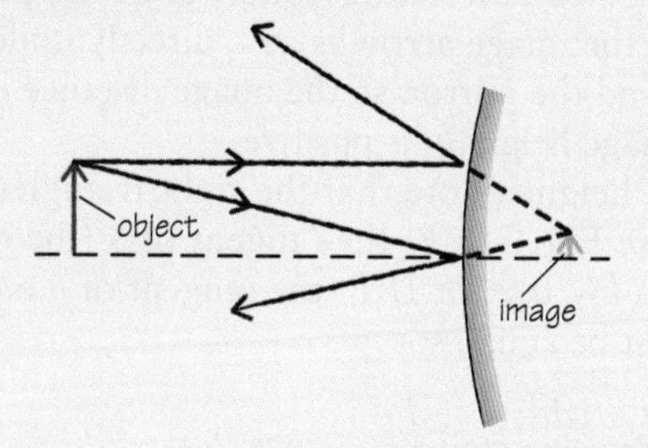

(a) With an object distance $d_O = 68.0$ cm, we use Equation 24-11 to find the image distance and Equation 24-14 to find the image size.

From the mirror equation (Equation 24-11)

$$\frac{1}{d_I} = \frac{1}{f} - \frac{1}{d_O} = \frac{1}{(-24.0 \text{ cm})} - \frac{1}{68.0 \text{ cm}}$$

$$= -0.0417 \text{ cm}^{-1} - 0.0147 \text{ cm}^{-1} = -0.0564 \text{ cm}^{-1}, \text{ so}$$

$$d_I = \frac{1}{(-0.0564 \text{ cm}^{-1})} = -17.7 \text{ cm}$$

The image distance is negative, so the image is 17.7 cm behind the mirror. An image that forms behind the mirror is a virtual image. The magnification of the image is, from Equation 24-14,

$$m = \frac{h_I}{h_O} = -\frac{d_I}{d_O} = -\frac{(-17.7 \text{ cm})}{68.0 \text{ cm}} = +0.261$$

Since $m > 0$, the image is upright. The image is 0.261 as tall as the object, so its height is

$$h_I = mh_O = (+0.261)(1.50 \text{ cm}) = 0.391 \text{ cm}$$

(b) Repeat the calculations of part (a) with $d_O = 3.00$ cm.

Calculate the image distance:

$$\frac{1}{d_I} = \frac{1}{f} - \frac{1}{d_O} = \frac{1}{(-24.0 \text{ cm})} - \frac{1}{3.00 \text{ cm}}$$

$$= -0.0417 \text{ cm}^{-1} - 0.333 \text{ cm}^{-1} = -0.375 \text{ cm}^{-1}, \text{ so}$$

$$d_I = \frac{1}{(-0.375 \text{ cm}^{-1})} = -2.67 \text{ cm}$$

Again the image distance is negative. Note that with a smaller object distance than in (a), the image distance is also smaller.

The magnification of the image is

$$m = \frac{h_I}{h_O} = -\frac{d_I}{d_O} = -\frac{(-2.67 \text{ cm})}{3.00 \text{ cm}} = +0.889$$

Again $m > 0$, and the image is upright. The image is 0.889 as tall as the object; its height is

$$h_I = m h_O = (+0.889)(1.50 \text{ cm}) = +1.33 \text{ cm}$$

Reflect

As the object is moved closer to the convex mirror, the image moves closer to the mirror and increases in size.

GOT THE CONCEPT? 24-7 A Convex Mirror

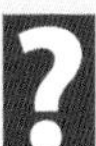

If the image made by a convex mirror of focal length f is 1/2 the height of the object, the distance from the object to the mirror must be equal to (a) $4|f|$; (b) $2|f|$; (c) $|f|$; (d) $|f|/2$; (e) $|f|/4$.

TAKE-HOME MESSAGE FOR Section 24-6

✔ The mirror equation that relates the image distance, object distance, and focal length of a convex mirror is the same as for a concave mirror. The same is true for the equation that relates the heights of the image and the object to the image and object distances, and the equation that relates the focal length and radius of curvature.

✔ The fundamental difference between concave and convex mirrors is that a convex mirror has a negative radius of curvature and so a negative focal length.

24-7 Convex lenses form images like concave mirrors and vice versa

A curved mirror forms images by reflection. A **lens**—a piece of glass or other transparent material with a curved surface on its front side, back side, or both sides—forms images by the *refraction* of light as the light enters and leaves the lens. Just as for a curved mirror, the images made by a lens can be larger or smaller than the object (**Figure 24-24**). In this section we'll explore how lenses form images.

The key idea that we need to understand lenses is Snell's law of refraction. As we learned in Section 23-2, a light ray changes direction when it moves from one transparent medium to a second medium in which light travels at a different speed. The ray bends toward the normal if the speed of light is slower in the second medium (**Figure 24-25a**) and bends away from the normal if the speed of light is faster in the second medium (**Figure 24-25b**).

Figure 24-24 Lenses can magnify or shrink Different types of lenses can make (a) large or (b) small images.

(a) This rodent appears larger when viewed through a magnifying glass, a lens that is convex on both sides.

Monika Graff/The Image Works

(b) The ruler appears smaller than actual size when viewed through this lens, which is concave on both sides.

JEROME WEXLER/Science Source/Getty Images

Figure 24-25 Snell's law of refraction A light ray crossing the boundary between two transparent media can refract either (a) toward or (b) away from the normal, depending on how the speed of light compares in the two media.

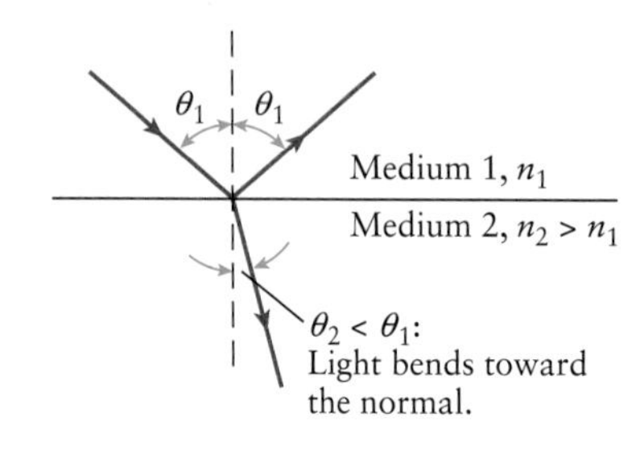

(a) Refraction from one medium into a slower one

Medium 1, n_1
Medium 2, $n_2 > n_1$
$\theta_2 < \theta_1$: Light bends toward the normal.

(b) Refraction from one medium into a faster one

Medium 1, n_1
Medium 2, $n_2 < n_1$
$\theta_2 > \theta_1$: Light bends away from the normal.

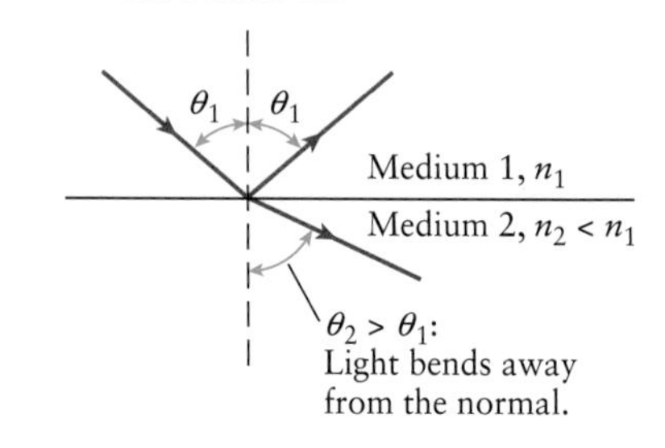

Figure 24-26 shows how parallel light rays refract when they enter or exit a glass sphere (in which the speed of light is relatively slow) surrounded by air (in which the speed of light is almost as fast as in vacuum). Both the front and back surfaces of the sphere are convex (they bulge outward). Parallel light rays converge when they enter the glass through the left-hand convex surface, and also converge when they exit the glass through the right-hand convex surface. **Figure 24-27** shows what happens when parallel light rays enter or exit a piece of glass with a concave surface (one that bulges inward). In this case parallel rays diverge as they cross the concave surface into the glass, and diverge as they exit the glass through the concave surface.

Figure 24-26 Refraction by a glass sphere If a glass sphere is surrounded by air, parallel light rays converge whether they (a) enter the sphere or (b) exit the sphere.

(a) Parallel light rays entering a glass sphere

Each dashed line represents the normal to the surface at the point where a light ray enters the sphere.

The incident angle θ_1 and the refracted angle θ_2 obey the law of refraction.

θ_2 is less than θ_1 for light crossing from air to glass, so the rays are bent toward the principal axis of the sphere. Hence the light rays converge as they enter the sphere.

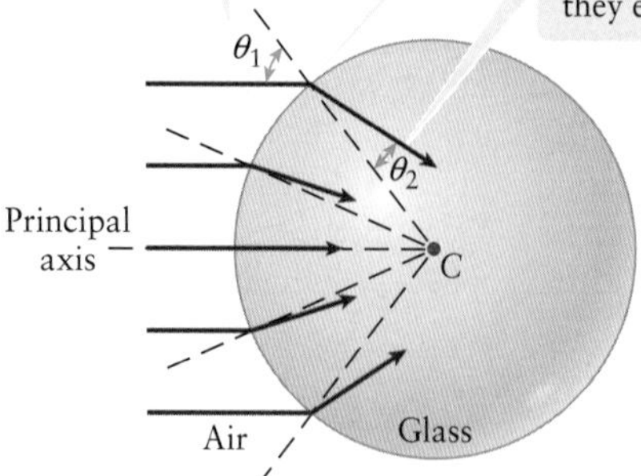

(b) Parallel light rays exiting a glass sphere

θ_2 is greater than θ_1 for light crossing from glass to air, so the rays are bent toward the principal axis of the sphere. Hence the light rays converge as they exit the sphere.

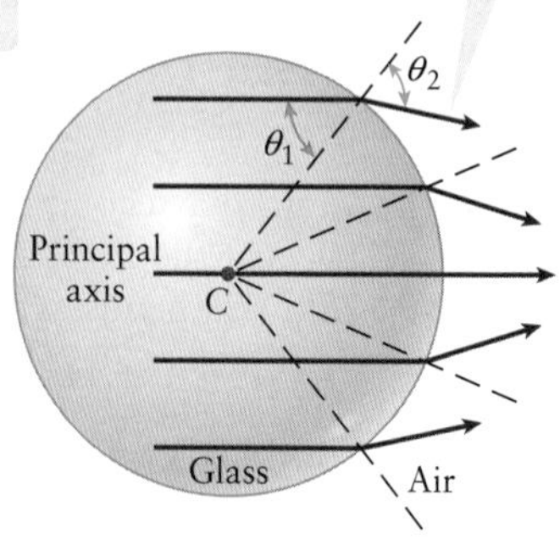

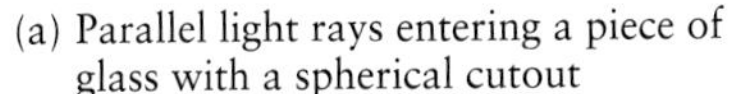

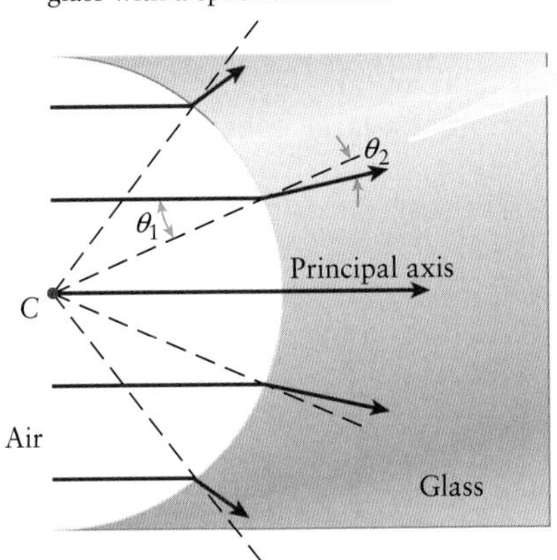

(b) Parallel light rays exiting a piece of glass with a spherical cutout

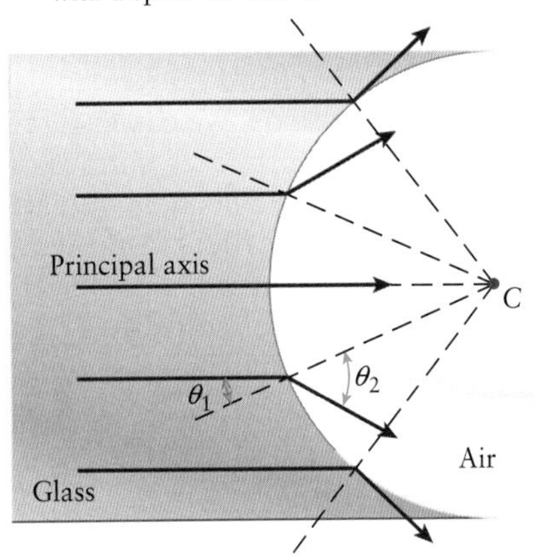

Figure 24-27 Refraction by glass with a spherical cutout If a piece of glass with a spherical cutout is surrounded by air, parallel light rays diverge whether they (a) enter the glass through the cutout or (b) exit the glass through the cutout.

The light rays in Figures 24-26 and 24-27 undergo only a single refraction when they enter or exit a glass object with a curved surface. In a lens there are *two* refractions: once when the light enters the front surface of the lens, and once when it exits the back surface. Figure 24-26 shows that the refractions will make the rays converge if each surface is convex, and Figure 24-27 shows that the refractions will make the rays diverge if each surface is concave. So a lens with two convex surfaces—called a *convex* lens—will be a **converging lens** that takes incoming parallel light rays and makes them converge toward the principal axis. The same is true for a lens with one convex surface and one flat surface, called a *plano-convex* lens. The refraction at the flat surface by itself causes neither convergence nor divergence by itself. Similarly, a lens with two concave surfaces—called a *concave* lens—will be a **diverging lens** that takes incoming parallel light rays and makes them diverge away from the principal axis. A lens with one concave surface and one flat surface, called a *plano-concave* lens, will also be a diverging lens (**Figure 24-28**).

These are converging lenses: Parallel light rays that enter either of these lenses will exit the lens converging toward each other.

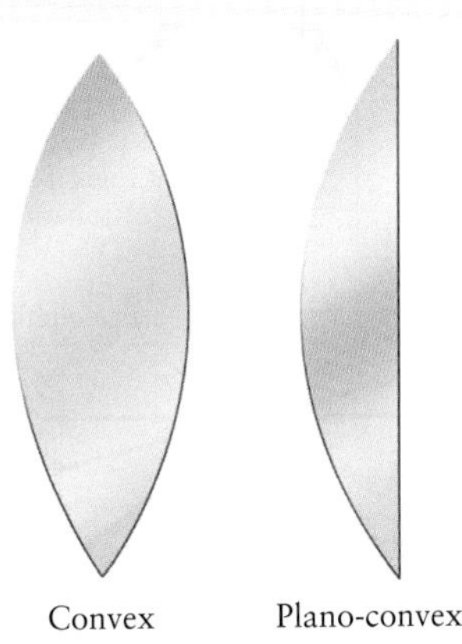

These are diverging lenses: Parallel light rays that enter either of these lenses will exit the lens diverging away from each other.

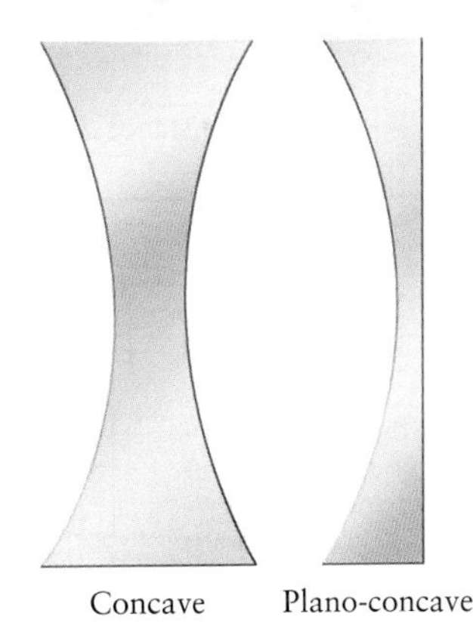

Figure 24-28 Converging and diverging lenses Two examples of converging lenses and two examples of diverging lenses.

WATCH OUT! The curvature of a lens has the opposite effect to the same curvature in a mirror.

We saw in Sections 24-3 and 24-5 that a concave mirror causes light rays to converge on reflection, while a convex mirror causes light rays to diverge on reflection. By contrast, a concave lens causes light rays to *diverge* as they pass through, and a convex lens causes light rays to *converge* as they pass through. Mirrors and lenses are different!

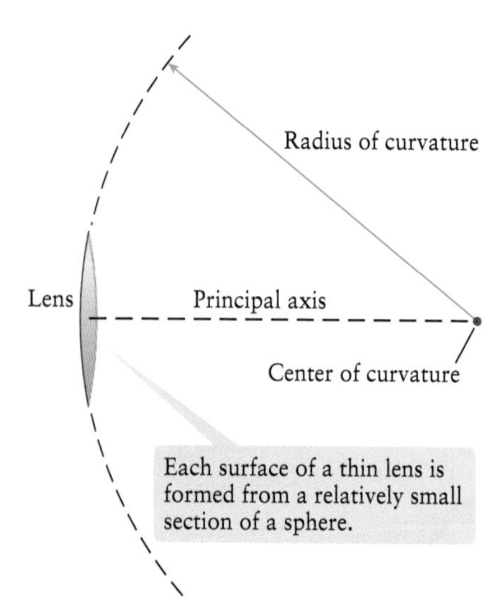

Figure 24-29 A thin lens Analyzing image formation by a lens is much easier if we assume that the lens is thin and has spherical surfaces.

A lens with spherical surfaces suffers from the same spherical aberration as a spherical mirror (Section 24-3): Parallel light rays do not all focus to the same point for a converging lens or appear to originate from a single point for a diverging lens. However, we can neglect this effect if the rays of light are all close to the principal axis. We can ensure this by using only a small section of a large spherical surface to form each surface of the lens (**Figure 24-29**). We'll also assume that there is very little thickness of material between the front and back surfaces of the lens. The result is called a **thin lens.** We'll consider only thin lenses for the rest of this chapter so that we can neglect spherical aberration and treat each lens as if parallel rays are focused to a single point. Eyeglasses and contact lenses are everyday examples of thin lenses.

Go to Picture It 24-2 for more practice dealing with converging lenses.

Ray Diagrams for Converging Lenses

We saw earlier that ray diagrams are powerful tools for visualizing how a curved mirror produces an image. Let's see how to draw a ray diagram to help us locate the position of the image made by a converging lens.

Figure 24-30 shows a thin convex lens. Notice that we have marked *two* focal points F. If parallel light rays enter the left-hand side of the lens, they converge at the focal point on the right-hand side; if parallel rays enter the lens on its right-hand side, they converge at the focal point on the left-hand side. For a thin lens the focal length, the distance from the center of the lens to the focal point, is the same on both sides of the lens even if the two surfaces have a different radius of curvature. That's why we've drawn the two focal points F in Figure 24-30 the same distance from the center of the lens.

In Figure 24-30 we've drawn a red arrow as our object on the left-hand side of the lens and placed this arrow outside the left-hand focal point. We've also drawn three representative light rays emanating from the tip of the object arrow. While each light ray actually refracts twice, once on entering the lens and once on exiting, for simplicity we've drawn the rays as if they refract only once, along the centerline of the lens (the vertical line that runs through the center of the lens). The three rays are:

(1) A ray that arrives at the lens traveling parallel to the principal axis is refracted so that it passes through the far (right-hand) focal point F.

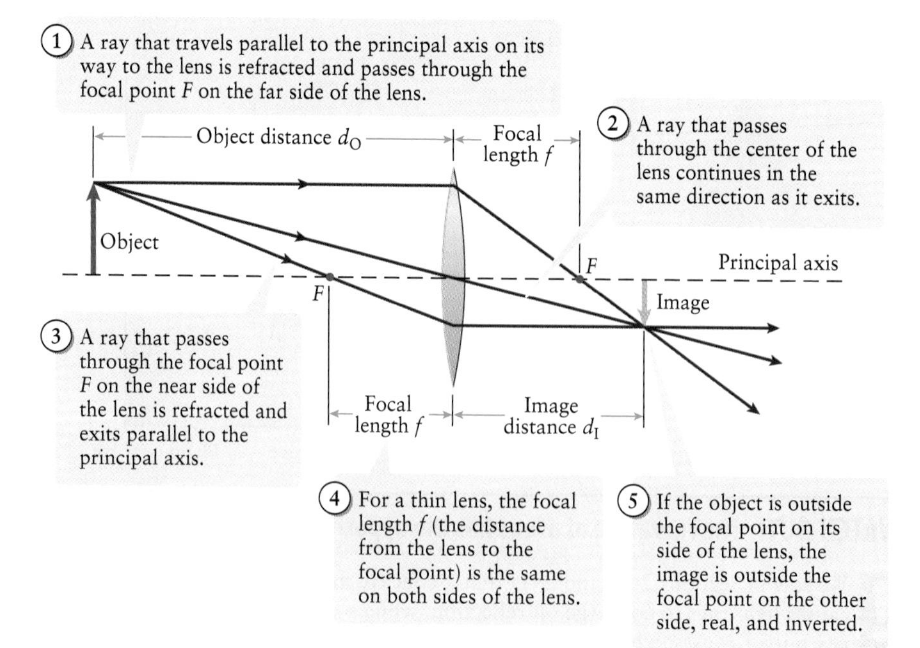

Figure 24-30 Ray diagram for a converging lens I How the image is formed for an object placed outside the focal point of a converging lens.

(2) A ray that enters the lens directly at its center continues in a straight line as it exits. This is only an approximation because it assumes that the ray undergoes no refraction on entering or exiting the lens. But this is a good approximation for a thin lens, which is nearly flat at its very center.

(3) A ray that passes through the near (left-hand) focal point F on its way to the lens exits the lens traveling parallel to the principal axis. To see why this is the case, imagine that we could record a video of light traveling along this path, then run the video backward. Then we would see a light ray coming from the far right in Figure 24-30 and traveling parallel to the principal axis. After this ray passes through the lens from right to left, it would naturally pass through the left-hand focal point F. If we now run that same video forward, we see light from the object following the path shown in Figure 24-30.

The image of the tip of the arrow in Figure 24-30 forms where the rays coming from the tip cross. The image of the base of the arrow must form along the principal axis because a ray that comes from the base and travels along the axis strikes the center of the lens and therefore continues in the same direction. You can see that the image of the arrow is real (the light rays actually cross at the image point) and is inverted.

Note that in Figure 24-30 the object is relatively far from the lens (the object distance d_O is greater than twice the focal length f), and the image is real, inverted, *and* smaller than the object. That's the same result we found for a concave mirror in Section 24-3 (see Figure 24-12). This reinforces the idea that a convex lens behaves similarly to a concave mirror. **Figure 24-31** shows the small, inverted image formed when light from a distant object passes through a convex lens. You can also see this in Figure 11-39 (Section 11-12), in which a sphere of water acts as a convex lens.

Marka/Getty Images

Figure 24-31 A real, inverted image made by a converging lens The lens in the person's hand makes a real, inverted, and small image of these pyramids in Sudan.

WATCH OUT! A real image formed by a lens is on the side of the lens opposite to the object.

We've defined a real image as one that forms where reflected or refracted light rays converge. A concave mirror can cause light rays to converge only on the reflective side of the mirror, which is where we would put an object, so this is the only side on which a real image can form. A lens, however, can only cause light rays to converge on the side of the lens *opposite* to the object, as in Figure 24-30. As we'll see below, only a converging lens can produce a real image, and only if the object is outside the focal point as in Figure 24-30.

Recall that for mirrors our convention was that the image distance is positive if the image is real and so is on the same side of the mirror as the object. For a lens we'll also say that the image distance is positive if the image is real. This means that for a lens, a positive image distance d_I implies that the image is on the *opposite* side of the lens from the object. For example, the image distance is positive in Figure 24-30.

In **Figure 24-32** we've moved the object to a point between the lens and the focal point, so the object distance is less than the focal length: $d_O < f$. To locate the image, we've drawn just two light rays from the tip of the arrow. One ray travels parallel to the principal axis of the lens and therefore passes through the focal point on the far side after being refracted by the lens. The other ray we have drawn enters the center of the lens and so continues straight. (In Figure 24-30 we drew a third ray through the focal point on the same side of the lens as the object. We don't draw that ray here, since it leads away from the lens rather than toward it.) Notice that these two rays do not meet. But if we extend the rays backward, as shown by the dashed lines, they appear to meet on the same side of the lens as the object. Where they meet is the location of the image. The rays do not actually cross there, so this is a *virtual* image. Because the image is on the same side of the lens as the object, the image distance is negative: $d_I < 0$.

As in Figure 24-30, the base of the image arrow must lie on the principal axis of the lens. It follows that the image is upright. Note that the image is also larger than the object. This is the kind of image shown in Figure 24-24a; when we hold a convex lens so that the rodent is inside the focal point of the lens, we see an enlarged, upright image of the rodent. This is very similar to the image made by a concave mirror when an object is placed inside the focal point of the mirror (see Figure 24-16).

Figure 24-32 Ray diagram for a converging lens II How the image is formed for an object placed inside the focal point of a converging lens.

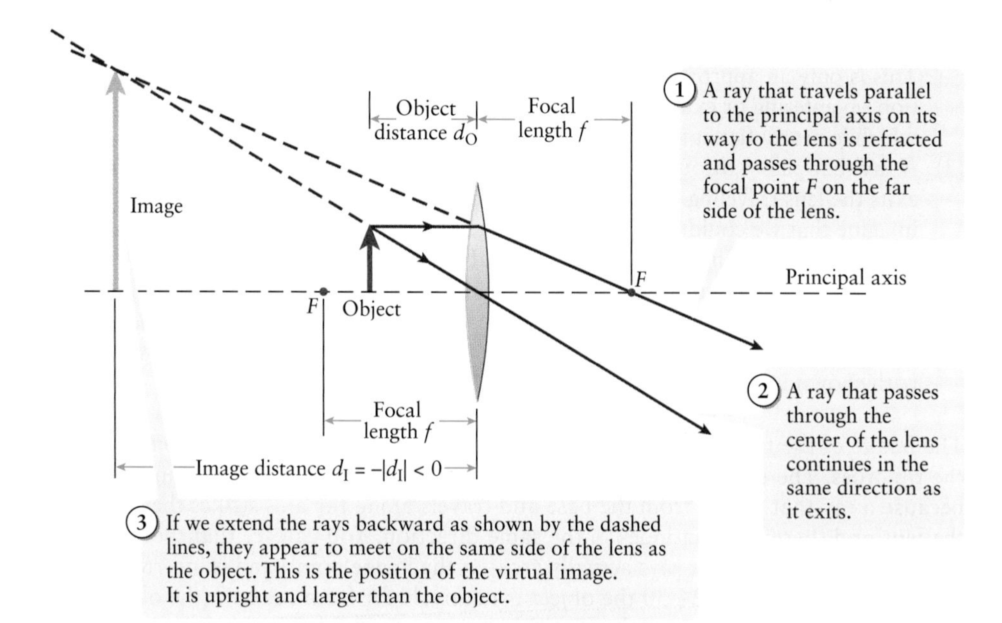

Ray Diagrams for Diverging Lenses

We can also use ray diagrams to learn about the image made by a diverging lens. In **Figure 24-33** we've placed an object arrow in front of a thin lens with two concave surfaces. Since this lens causes parallel light rays to diverge, we say that the focal length is negative: $f = -|f| < 0$. (The focal length of a convex mirror, which also causes parallel light rays to diverge, is also negative.)

We've drawn two light rays to determine where the image forms, and as we did for the thin, convex lens, we've drawn the rays as if they refract only once along the centerline of the lens. The refracted rays don't actually meet, but if we extend these rays backward, we find the location where the extensions meet. The image of the tip of the arrow forms at this point; the image is virtual (since the rays don't actually meet there), upright, and smaller than the object. This is the same behavior we saw in Section 24-5 for convex mirrors. As for a convex mirror, the image is virtual, upright, and smaller no matter what the object distance.

Figure 24-24b shows an image of a ruler made by a concave lens. Just like the image in Figure 24-33, this image is upright and smaller than the object.

Figure 24-33 Ray diagram for a diverging lens How the image is formed for an object placed in front of a diverging lens.

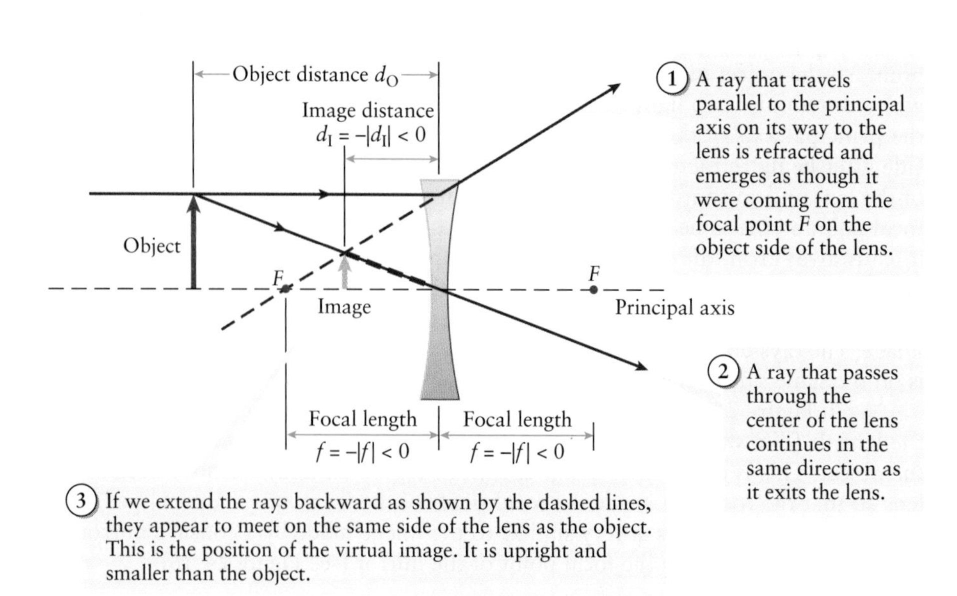

GOT THE CONCEPT? 24-8 A Lens in Sugar Water

? You make a lens with two convex surfaces out of ice ($n = 1.31$). You then submerge the lens in a large tank of concentrated sugar water ($n = 1.49$). If you place an object in the tank and in front of the lens, an image is formed at a position inside the tank. What kind of image is this? (a) A real image; (b) a virtual image that is larger than the object; (c) a virtual image that is smaller than the object; (d) either (a) or (b), depending on the distance from the object to the lens; (e) any of (a), (b), or (c), depending on the distance from the object to the lens.

TAKE-HOME MESSAGE FOR Section 24-7

✔ When parallel light rays enter one side of a convex lens, they exit the lens converging toward the focal point on the other side of the lens.

✔ When parallel light rays enter a concave lens, they exit the lens diverging away from the focal point on the same side of the lens that the rays entered.

✔ A convex lens produces a real image if the object is outside the focal point and produces a virtual image if the object is inside the focal point.

✔ A concave lens produces a virtual image no matter what the position of the object.

24-8 The focal length of a lens is determined by its index of refraction and the curvature of its surfaces

As we did for concave mirrors and convex mirrors, we'd like to find equations for the focal length of a lens; for the relationship among the focal length of a lens, the object distance, and the image distance; and for the magnification of an image produced by a lens. As we'll see, the latter two equations turn out to be identical to those for curved mirrors. The expression for the focal length, however, is a little more complicated.

Focal Length of a Thin Lens

The focal length f of a thin lens depends on the index of refraction n of the material of which it is made. The greater the value of n, the more sharply a light ray is bent as it passes either from the surrounding air (for which the index of refraction is essentially 1) into the lens or from the lens into the air. The value of f also depends on how the front and back surfaces of the lens are curved. The mathematical expression of these relationships for a lens in air is called the **lensmaker's equation:**

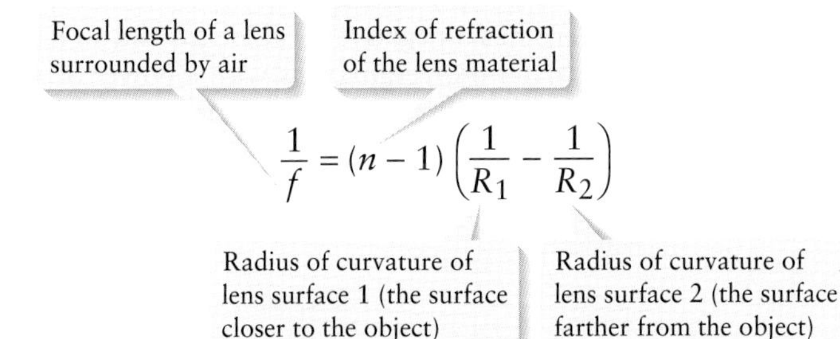

$$\frac{1}{f} = (n-1)\left(\frac{1}{R_1} - \frac{1}{R_2}\right)$$

Lensmaker's equation for the focal length of a thin lens (24-24)

Equation 24-24 can be derived by applying Snell's law of refraction (Equation 23-6) to a ray of light refracted by both surfaces of a lens. The derivation is beyond our scope, however.

The values of the radii of curvature R_1 and R_2 depend on how sharply and in what direction the two surfaces of the lens are curved. As **Figure 24-34** shows, we take a radius to be positive if the center of curvature is on the other side of the lens from the object but negative if the center of curvature is on the same side as the object. For example, in Figure 24-34 the radius R_1 of surface 1 is positive, but the radius R_2 of surface 2 is negative.

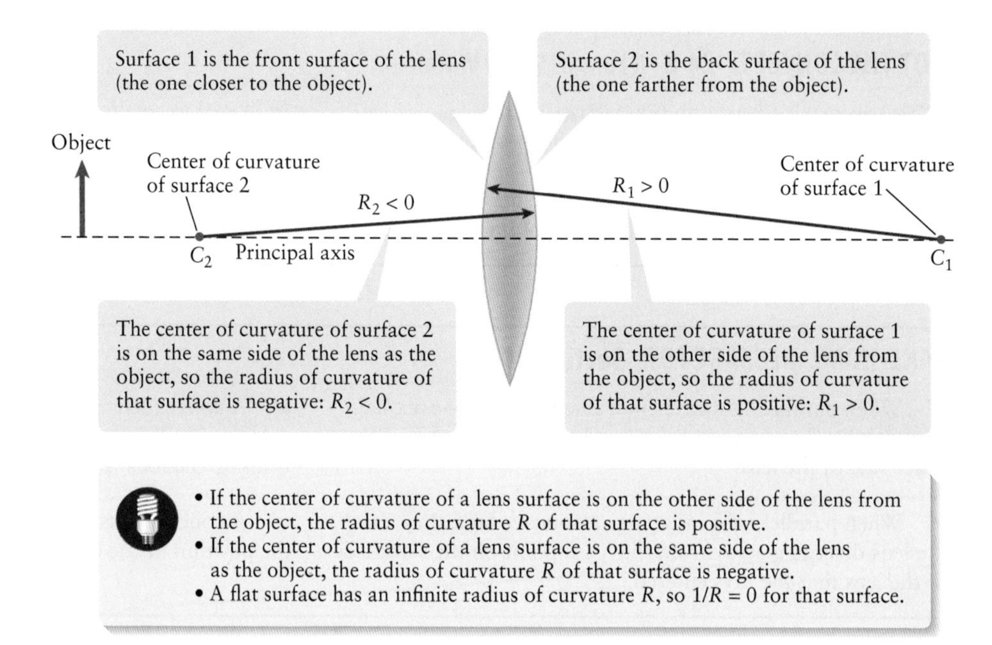

Surface 1 is the front surface of the lens (the one closer to the object).
Surface 2 is the back surface of the lens (the one farther from the object).
Object
Center of curvature of surface 2
$R_2 < 0$
$R_1 > 0$
Center of curvature of surface 1
C_2
Principal axis
C_1
The center of curvature of surface 2 is on the same side of the lens as the object, so the radius of curvature of that surface is negative: $R_2 < 0$.
The center of curvature of surface 1 is on the other side of the lens from the object, so the radius of curvature of that surface is positive: $R_1 > 0$.
• If the center of curvature of a lens surface is on the other side of the lens from the object, the radius of curvature R of that surface is positive.
• If the center of curvature of a lens surface is on the same side of the lens as the object, the radius of curvature R of that surface is negative.
• A flat surface has an infinite radius of curvature R, so $1/R = 0$ for that surface.

Figure 24-34 Interpreting the lensmaker's equation For each surface of a thin lens, the sign of the radius of curvature depends on where the center of curvature is located.

Equation 24-24 tells us that the more tightly curved the surfaces of a lens are, the *smaller* the magnitudes of the radii R_1 and R_2 and the *smaller* the magnitude of the focal length f. (We saw a similar result for the focal length of a spherical mirror in Section 24-3.) The smaller the magnitude of f, the more sharply light rays are refracted by passing through the lens.

The lensmaker's equation is expressed in terms of the reciprocal of the focal length f. For that reason it's common, especially among optometrists, to characterize lenses (and also mirrors) in terms of the reciprocal of the focal length:

$$P = \frac{1}{f} \qquad (24\text{-}25)$$

The quantity P is called the **power** of the lens or mirror. The units of P are m^{-1}; $1\ \text{m}^{-1}$ is known as 1 **diopter**. The larger the power of a lens, the greater the amount of refraction it causes (that is, the more "powerfully" it causes light rays to bend). If parallel light rays enter a thin, convex lens of power 2 diopters, they come to a focus 1/2 m (0.5 m) behind the lens; if instead these parallel light rays enter a 5-diopter converging lens, they are bent more sharply and come to a focus just 1/5 m (0.2 m) behind the lens. A thin, concave lens has a negative focal length and so has a negative power P: When parallel light rays enter a diverging lens with a power of −5 diopters, they emerge from the lens as though they were emanating from a point 1/5 m (0.2 m) in front of the lens (see Figure 24-33). In general, the smaller the magnitude of the focal length of a lens, the greater the magnitude of its power.

EXAMPLE 24-3 Calculating Focal Length for a Convex Lens

In Figure 24-34 the front surface (surface 1) of the lens has a radius of curvature of magnitude 15.0 cm, and the back surface (surface 2) has a radius of curvature of magnitude 25.0 cm. The lens is made of crown glass with an index of refraction 1.520. (a) Calculate the focal length of the lens. (b) Now flip the orientation of the lens so that the front surface is now the one with a radius of curvature of magnitude 25.0 cm and the back surface is now the one with a radius of curvature of magnitude 15.0 cm. Calculate the focal length of the lens in this case.

Set Up

For each situation we'll draw the lens and determine on which side of the lens the centers of curvature lie; that will tell us whether R_1 and R_2 are positive or negative. We'll then use Equation 24-24 to calculate the focal length.

Lensmaker's equation for the focal length of a thin lens:

$$\frac{1}{f} = (n-1)\left(\frac{1}{R_1} - \frac{1}{R_2}\right) \qquad (24\text{-}24)$$

Solve

(a) Both surfaces are convex. Hence the center of curvature of each surface is on the other side of the lens from that surface. The center of curvature C_1 of the front surface (on the left in the figure) is on the far side of the lens, so R_1 is positive: $R_1 = +15.0$ cm. The center of curvature C_2 of the back surface (on the right in the figure) is on the near side of the lens, so R_2 is negative: $R_2 = -25.0$ cm.

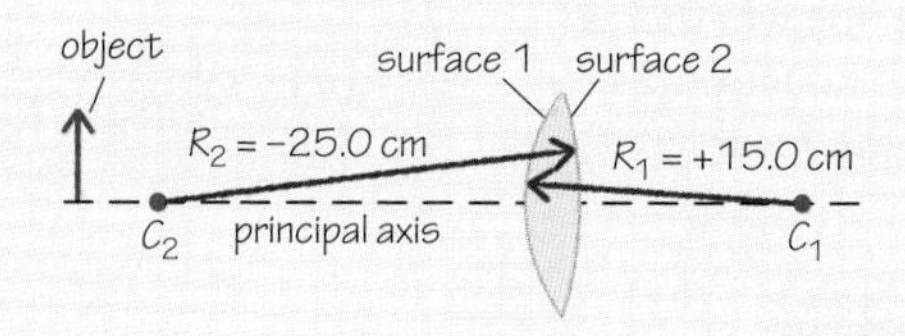

Calculate the focal length.

From Equation 24-24

$$\frac{1}{f} = (1.520 - 1)\left[\frac{1}{(+15.0 \text{ cm})} - \frac{1}{(-25.0 \text{ cm})}\right]$$
$$= 0.520[(0.0667 \text{ cm}^{-1}) - (-0.0400 \text{ cm}^{-1})]$$
$$= 0.0555 \text{ cm}^{-1}$$
$$f = \frac{1}{0.0555 \text{ cm}^{-1}} = 18.0 \text{ cm} = 0.180 \text{ m}$$

Although R_2 is negative, we subtract rather than add $1/R_2$ in Equation 24-24, so both the $1/R_1$ term and the $1/R_2$ term—that is, both surface 1 and surface 2—contribute to giving $1/f$ a positive value. As a result, the focal length is positive, as we expect for a convex lens.

(b) Because we have flipped the lens around, we have interchanged surfaces 1 and 2. The object is still to the left of the lens, however. As in part (a), R_1 is positive and R_2 is negative, but now $R_1 = +25.0$ cm and $R_2 = -15.0$ cm.

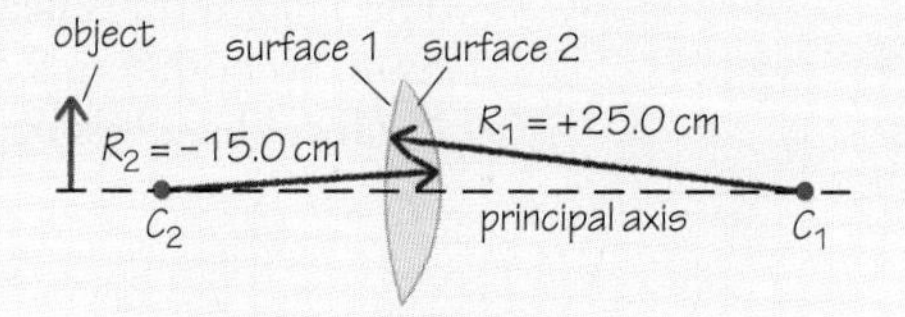

Calculate the focal length.

From Equation 24-24

$$\frac{1}{f} = (1.520 - 1)\left[\frac{1}{(+25.0 \text{ cm})} - \frac{1}{(-15.0 \text{ cm})}\right]$$
$$= 0.520[(0.0400 \text{ cm}^{-1}) - (-0.0667 \text{ cm}^{-1})]$$
$$= 0.0555 \text{ cm}^{-1}$$
$$f = \frac{1}{0.0555 \text{ cm}^{-1}} = 18.0 \text{ cm} = 0.180 \text{ m}$$

Again both the $1/R_1$ term and the $1/R_2$ term contribute to giving $1/f$ a positive value. The focal length has the same positive value as in part (a).

Reflect

The focal length of this thin lens stays the same after we flip it back to front. As we stated in the previous section, the focal points on either side of the lens are both the same distance (that is, the same focal length) from the lens. Note that the power of this lens is $P = 1/f = 1/(0.180 \text{ m}) = 5.55 \text{ m}^{-1}$, or 5.55 diopters.

EXAMPLE 24-4 Calculating Focal Length for a Concave Lens

A certain lens made of crown glass ($n = 1.520$) has concave front and back surfaces. The front surface has a radius of curvature of magnitude 15.0 cm, and the back surface has a radius of curvature of magnitude 25.0 cm. Calculate the focal length of the lens.

Set Up

As in the previous example, we'll first draw the lens and use our drawing to decide whether R_1 and R_2 are positive or negative. Equation 24-24 will then allow us to calculate the focal length.

Lensmaker's equation for the focal length of a thin lens:

$$\frac{1}{f} = (n - 1)\left(\frac{1}{R_1} - \frac{1}{R_2}\right) \tag{24-24}$$

Solve

Both surfaces are concave. Hence the center of curvature of each surface is on the same side of the lens as that surface. The center of curvature C_1 of the front surface (on the left in the figure) is on the near side of the lens, so R_1 is negative: $R_1 = -15.0$ cm. The center of curvature C_2 of the back surface (on the right in the figure) is on the far side of the lens, so R_2 is positive: $R_2 = +25.0$ cm.

Calculate the focal length.

From Equation 24-24,

$$\frac{1}{f} = (1.520 - 1)\left[\frac{1}{(-15.0 \text{ cm})} - \frac{1}{(+25.0 \text{ cm})}\right]$$

$$= 0.520[(-0.0667 \text{ cm}^{-1}) - (0.0400 \text{ cm}^{-1})]$$

$$= -0.0555 \text{ cm}^{-1}$$

$$f = \frac{1}{(-0.0555 \text{ cm}^{-1})} = -18.0 \text{ cm} = -0.180 \text{ m}$$

The $1/R_1$ term is negative, and we subtract from it the positive $1/R_2$ term. So both terms contribute to giving $1/f$ a negative value so that the focal length is negative.

Reflect

The focal length is negative, just as we expect for a concave lens. The power of such a lens is negative: $P = 1/f = 1/(-0.180 \text{ m}) = -5.55 \text{ m}^{-1}$, or -5.55 diopters. You should repeat the calculation with the lens reversed back to front, as in part (b) of Example 24-3; you should get the same result for the focal length. Do you?

Image Position and Magnification for a Thin Lens

To see how the image distance is related to the object distance and the focal length for a thin, convex lens, let's look again at a ray diagram like Figure 24-30. We've drawn such a diagram in **Figure 24-35**. Let's see how to use trigonometry to find relationships between the distances of interest.

The shaded regions in Figure 24-35a are similar triangles. That's because the ray from the tip of the object at O' to the tip of the image at I' passes through the center of the lens at C without deflection, so the angle θ is the same in both triangle $O'C'C$ and triangle ICI'. The ratio of the heights of the two triangles is therefore equal to the ratio of their bases; that is

$$\frac{|h_I|}{h_O} = \frac{d_I}{d_O} \tag{24-26}$$

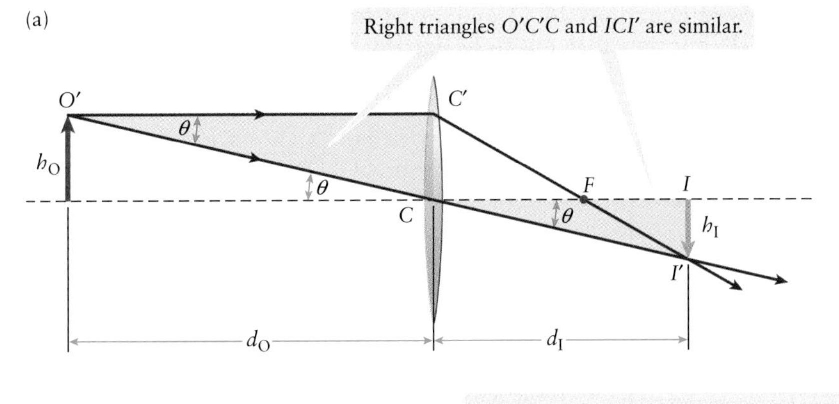

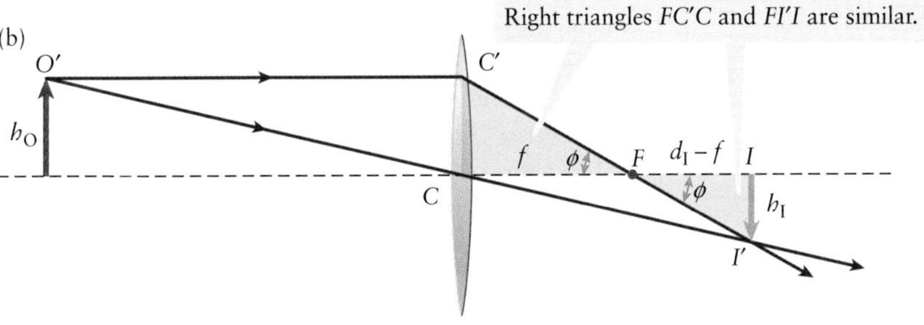

Figure 24-35 Analyzing a converging lens Ray diagrams for an object arrow on the principal axis of a converging lens. The similar triangles in (a) and (b) help us determine the position and magnification of the image.

(The height h_I of the image in Figure 24-35a is negative because the image is inverted. That's why we've used $|h_I|$ in Equation 24-26 for the distance from I to I'.) The two shaded regions in Figure 24-35b—the right triangles $FC'C$ and $FI'I$—are also similar triangles. That's because the straight ray from C' to I' passes through the focal point, so the angle ϕ is the same on either side of F. So for these two triangles as well, the ratio of their heights is equal to the ratio of their bases:

$$\frac{|h_I|}{h_O} = \frac{d_I - f}{f} \tag{24-27}$$

If we combine Equation 24-26 and Equation 24-27, we find a relationship among the image distance d_I, the object distance d_O, and the focal length f:

$$\frac{d_I}{d_O} = \frac{d_I - f}{f} \quad \text{or} \quad \frac{d_I}{d_O} = \frac{d_I}{f} - 1 \quad \text{or} \quad \frac{d_I}{d_O} + 1 = \frac{d_I}{f}$$

Divide both sides by d_I:

$$\frac{1}{d_O} + \frac{1}{d_I} = \frac{1}{f} \quad \text{(lens equation)} \tag{24-11}$$

This is the same equation that we deduced for spherical mirrors in Section 24-4. It applies just as well to thin, spherical convex lenses. Now we see why we were justified in calling Equation 24-11 the mirror and lens equation—it works for both.

Equation 24-26 above also tells us the magnification of the image. Since the image height h_I is negative in Figure 24-35a (the image is inverted), h_I is equal to $-|h_I|$ and $|h_I| = -h_I$. If we substitute this into Equation 24-26, we get

$$\frac{(-h_I)}{h_O} = \frac{d_I}{d_O} \quad \text{or} \quad \frac{h_I}{h_O} = -\frac{d_I}{d_O}$$

Lateral magnification is equal to the ratio of image height to object height: $m = h_I/h_O$ (Equation 24-2). So the magnification of the image produced by a thin lens is

$$m = \frac{h_I}{h_O} = -\frac{d_I}{d_O} \quad \text{(lateral magnification)} \tag{24-14}$$

That's the same Equation 24-14 that we derived in Section 24-4 for a curved mirror.

Go to Interactive Exercise 24-3 for more practice dealing with lenses.

We've derived Equations 24-11 and 24-14 for a thin convex lens with a positive focal length, but they turn out to be equally valid for a thin concave lens with a negative focal length. As for mirrors, it's important to keep track of the signs of quantities in these equations. **Table 24-2** summarizes when the quantities in the lensmaker's equation (Equation 24-24), the lens equation (Equation 24-11), and the magnification equation (Equation 24-14) are positive and when they are negative.

TABLE 24-2 Sign Conventions for Lenses

lens surface radius of curvature R_1 or R_2	• positive if the center of curvature is on the side of the lens opposite from the object • negative if the center of curvature is on the same side of the lens as the object
focal length f	• positive for a converging lens • negative for a diverging lens
image distance d_I	• positive if on the side of the lens opposite from the object; the image is then a real image • negative if on the same side of the lens as the object; the image is then a virtual image
image height h_I	• positive if the image is upright (the same orientation as the object) • negative if the image is inverted (flipped upside down compared to the object)
lateral magnification m	• positive if the image is upright (the same orientation as the object) • negative if the image is inverted (flipped upside down compared to the object)

Because the equations for mirrors and thin lenses are effectively identical, many of the same conclusions that we came to for images made by a curved mirror also apply to images made by a thin lens:

- If the lens has a positive focal length (a converging lens) and an object is placed outside the focal point, the image is real and inverted. Depending on the object distance, the image can be smaller, larger, or the same size as the object.
- If the lens has a positive focal length (a converging lens) and an object is placed inside the focal point of a converging lens, the image is virtual, upright, and larger than the object.
- If the lens has a negative focal length (a diverging lens), the image is virtual, upright, and smaller than the object. This is true for any object distance.

One key difference between mirrors and lenses is in the position of the image. For a mirror a real image is on the same side of the mirror as the object, and a virtual image is on the other side of the mirror. For a lens a real image is on the opposite side of the lens from the object, and a virtual image is on the same side of the lens as the object.

EXAMPLE 24-5 An Image Made by a Convex Lens

A thin convex lens has a focal length of 15.0 cm. How far from the lens does the image form when an object is placed 9.00 cm from the center of the lens? Is the image virtual or real? Is the image inverted or upright? By what factor is the image magnified relative to the object?

Set Up

We'll begin by drawing a ray diagram to help us visualize the kind of image that will be produced. We'll use Equation 24-11 to calculate the position of this image, and Equation 24-14 to calculate the image height compared to the object height.

Lens equation:

$$\frac{1}{d_O} + \frac{1}{d_I} = \frac{1}{f} \quad (24\text{-}11)$$

Magnification:

$$m = \frac{h_I}{h_O} = -\frac{d_I}{d_O} \quad (24\text{-}14)$$

Solve

We begin by drawing a ray diagram. We've drawn one ray from the object that enters the lens parallel to the principal axis, exits the lens, and passes through the focal point F on the other side of the lens. We've also drawn a ray that passes through the center of the lens. The rays never meet on the other side of the lens, but their extensions do meet on the same side of the lens as the object. So the image will be virtual. As the diagram shows, the image is also upright and larger than the object.

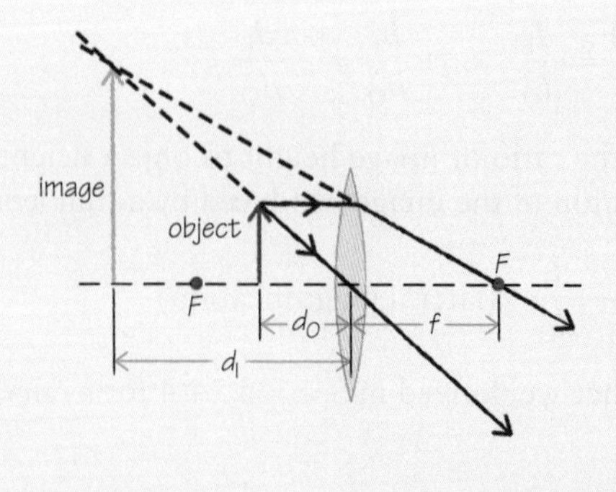

Find the image distance using the lens equation.

We are given $d_O = 9.00$ cm and $f = 15.0$ cm. From Equation 24-11

$$\frac{1}{d_I} = \frac{1}{f} - \frac{1}{d_O} = \frac{1}{15.0\text{ cm}} - \frac{1}{9.00\text{ cm}}$$
$$= 0.06667\text{ cm}^{-1} - 0.111\text{ cm}^{-1}$$
$$= -0.0444\text{ cm}^{-1}$$
$$d_I = \frac{1}{(-0.0444\text{ cm}^{-1})} = -22.5\text{ cm}$$

The negative value of d_I means that the image is on the same side of the lens of the object, just as the ray diagram shows. It must therefore be a virtual image.

Find the image height using the magnification equation.

From Equation 24-14

$$m = \frac{h_I}{h_O} = -\frac{d_I}{d_O} = -\frac{(-22.5\text{ cm})}{9.00\text{ cm}} = +2.50$$

The plus sign means the image is upright, in agreement with the ray diagram. The image is 2.50 times as large as the object.

Reflect

Although we didn't try to draw the ray diagram to exact scale, by eye the ratio of the focal length to the object distance looks to be about 2 to 1, which is certainly consistent with the actual values $f = 15.0$ cm and $d_O = 9.00$ cm. So we expect that the image distance in the diagram is about one and half times the focal length (22.5 cm compared to 15.0 cm)—and it is. Can you verify that the height of the image in the diagram is about 2.5 times the height of the object?

GOT THE CONCEPT? 24-9 A Magnifying Glass

? By placing an object closer to your eyes, it looks larger and you can see it in more detail. The object has an angular size θ_0 (**Figure 24-36a**). If you're typical, however, you can focus on an object only if it's about 25 cm or farther from your eyes, which may not allow you to see small details clearly. A **magnifying glass**, a convex lens with a relatively short focal length, is the solution (see Figure 24-24a). When a magnifying glass is placed in between the object and your eye and the object is closer to the lens than the focal point, an upright virtual image forms with angular size θ_M, which is larger than θ_0. In addition, the image forms farther from your eye than the 25 cm limit so that you can focus clearly on it. Draw a ray diagram that shows the principle of a magnifying glass, starting from the sketch in **Figure 24-36b**.

(a) When an object is placed close to the eye it spans an angular size θ_0.

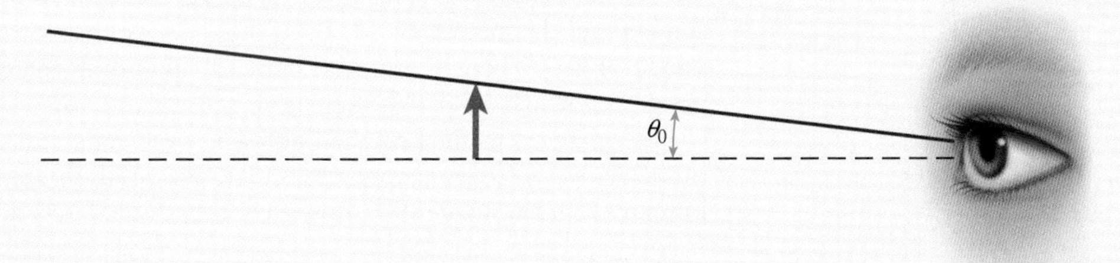

(b) A magnifying glass is placed between the object and the eye, in order to see more detail without moving the object closer to the eye. Where does the image form?

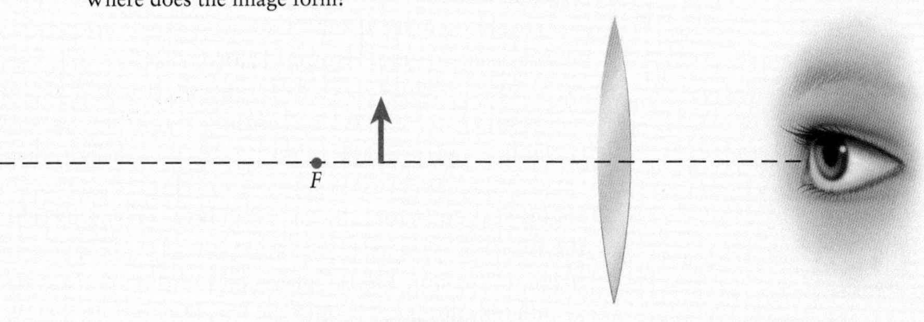

Figure 24-36 A magnifying glass (a) The angular size of an object. (b) Use this setup to draw a ray diagram that demonstrates the principle of a magnifying glass.

GOT THE CONCEPT? 24-10 Microscopes

You've used microscopes many times, but do you know how they work? By applying what you've learned about lenses, you can figure it out for yourself! A simple compound microscope has two lenses, the objective and the eyepiece. An object is placed just beyond the focal point of the objective in order to form a magnified image. Using the sketch in **Figure 24-37**, determine the location and size of the image by tracing rays from the tip of the arrow to the eye. Is the image real or virtual?

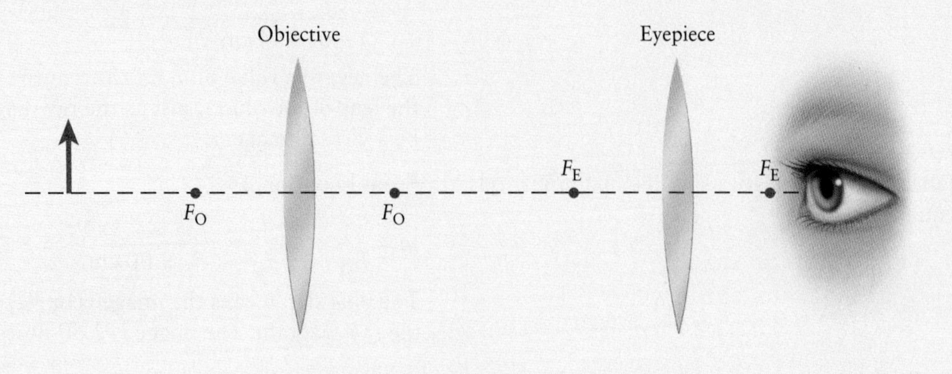

Figure 24-37 A compound microscope Use this setup to draw a ray diagram that demonstrates the principle of a compound microscope.

TAKE-HOME MESSAGE FOR Section 24-8

✔ The focal length of a thin lens is determined by the index of refraction of the lens material and the radii of curvature of the front and back surfaces of the lens. These radii can be positive or negative.

✔ The more sharply curved the surfaces of a lens, the smaller the magnitudes of their radii of curvature and the smaller the magnitude of the focal length of the lens.

✔ The same equation that relates object distance, image distance, and focal length for a spherical mirror also applies to thin lenses. The magnification equation is also the same for spherical mirrors and thin lenses.

✔ Image distance d_I is positive when the image is real and negative when the image is virtual. The focal length f is positive for a thin, convex lens and negative for a thin, concave lens.

24-9 A camera and the human eye use different methods to focus on objects at various distances

We can now apply the ideas of the last two sections to two important optical devices: a camera and the human eye. Both of these devices use refraction to form a real, inverted image of an object. In a digital camera (including the camera in a smartphone) the image is formed on a light-sensitive sensor and stored electronically; in the human eye the image is formed on the light-sensitive retina and sent via the optic nerve to the brain. As we will see, however, there are essential differences in how a camera and the human eye form images and how they focus on objects at different distances.

The Camera

Figure 24-38a shows a simplified cross section of a typical camera. Unlike the thin lenses we discussed in Sections 24-7 and 24-8, the lens of a camera can be relatively thick and made up of several individual pieces or *elements*. (Figure 24-38a shows a two-element lens; many smartphone cameras have five or more elements, and lenses used by professional photographers may have more than 15 elements.) The elements typically have different shapes and are made of different kinds of glass. The shapes are chosen to minimize spherical aberration (see Section 24-7) and to provide a sharp image at all points on the light-sensitive sensor.

Unfortunately, glass suffers from dispersion (Section 23-4): The index of refraction is slightly different for light of different wavelengths. The focal length of a lens depends on the index of refraction (Equation 24-24), so if all of the elements of a camera lens were made of the same kind of glass, light of different wavelengths would be brought to a focus at different distances behind the lens. As a result, if the yellow colors of an object formed a sharp image on the sensor, the blue and red colors would be slightly blurred (**Figure 24-38b**). To avoid such *chromatic aberration*, lens designers use different kinds of glass for different lens elements. These kinds of glass are chosen so that the different dispersion of the individual elements largely cancel out. The result is that light of all colors comes to the same focus (**Figure 24-38c**).

The lens equation, Equation 24-11, tells us that for a camera lens of a given focal length f, changing the distance d_O from lens to object will change the distance d_I from lens to image:

$$\frac{1}{d_O} + \frac{1}{d_I} = \frac{1}{f} \tag{24-11}$$

If the object is very distant, d_O is very large and $1/d_O$ is nearly zero. Then Equation 24-11 becomes $1/d_I = 1/f$, which tells us that the image distance d_I equals the focal length f. Hence the camera's light-sensitive sensor is placed a distance f behind the lens so that distant objects will be in sharp focus (**Figure 24-39a**).

As the object approaches the lens, the object distance d_O decreases, so the image distance d_I increases. Hence the sensor must be farther from the lens to record a sharp image. In most professional cameras *focusing* the camera on a nearby object is accomplished by moving the lens farther away from the sensor (**Figure 24-39b**). On some digital cameras you can actually see the lens move outward from the camera body to focus on a nearby object.

In the camera on a smartphone, a different method is used to focus on nearby objects. The front element of the lens remains fixed, but one or more of the rear elements move to decrease the effective focal length of the elements acting in combination.

(a) A schematic camera

Parallel light rays from a distant object

Camera body

Electronic sensor at the focal point of the lens

Lens elements

(b) A single-element lens has chromatic aberration

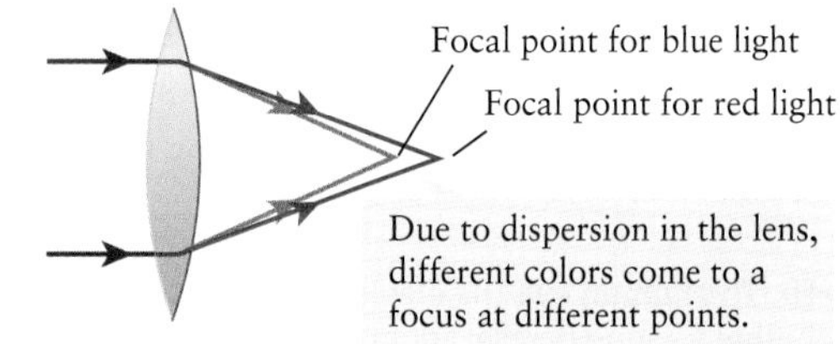

(c) Correcting chromatic aberration

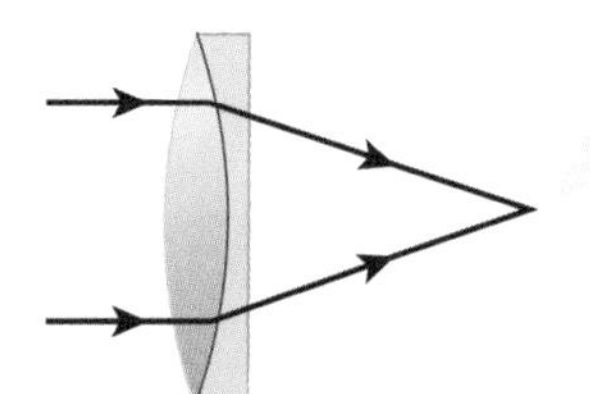

By using lens elements with different dispersion, all colors are brought to focus at the same focal point.

Figure 24-38 A camera and its lens (a) The lens of a camera is composed of two or more elements. (b) A single-element lens produces chromatic aberration. (c) A lens with more than one element of different indices of refraction can largely eliminate chromatic aberration.

The Human Eye

Like a camera lens, the human eye is composed of different elements with different indices of refraction (**Figure 24-40**). The *cornea* is the front surface of the eye and has index of refraction $n = 1.376$, which is substantially higher than that of the air outside the eye ($n = 1.000$). Behind the cornea is the *aqueous humor*, a fluid that is 98% water and whose index of refraction ($n = 1.336$) is consequently very close to that of pure water ($n = 1.330$). Since the indices of refraction of the cornea and aqueous humor have fairly similar values, most of the refraction caused by the cornea takes place at its front surface rather than its rear surface.

At the rear of the aqueous humor is the *iris*. This is a diaphragm that opens and closes to regulate the amount of light falling on the sensitive retina. After passing through the aperture of the iris, light rays undergo further refraction when they enter and leave the crystalline *lens*. This is a converging lens about 9 mm in diameter and about 4 mm thick. It is made up of transparent cells called lens fibers arranged in concentric layers rather like the layers of an onion.

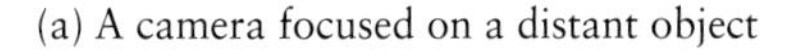

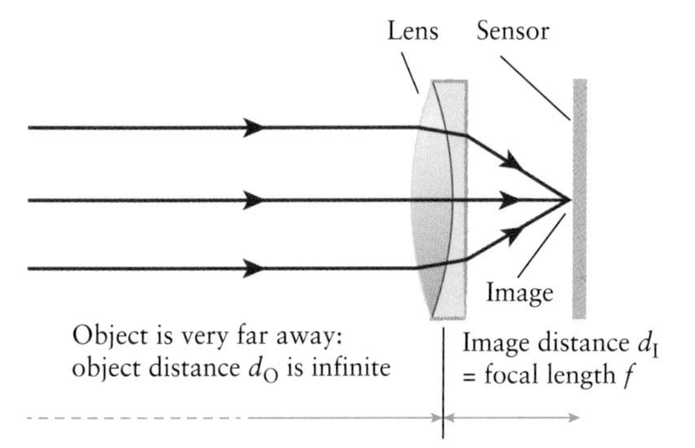

(b) The same camera focused on a nearby object

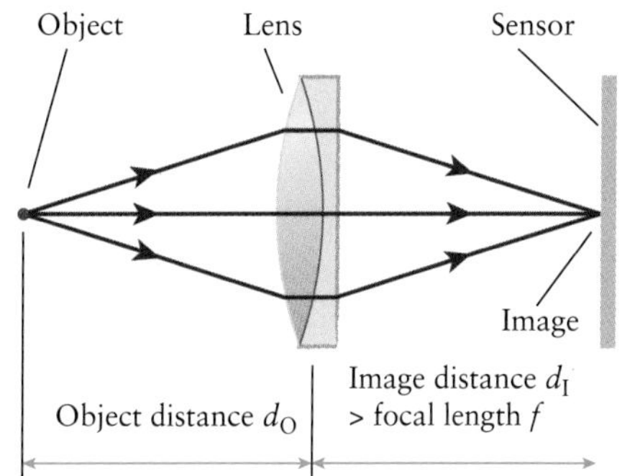

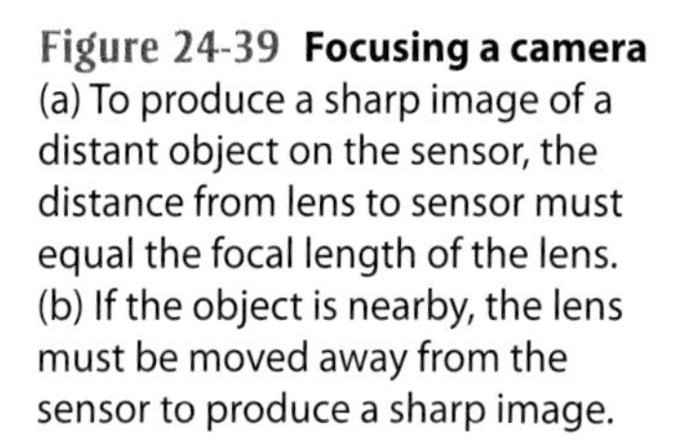

Figure 24-39 Focusing a camera (a) To produce a sharp image of a distant object on the sensor, the distance from lens to sensor must equal the focal length of the lens. (b) If the object is nearby, the lens must be moved away from the sensor to produce a sharp image.

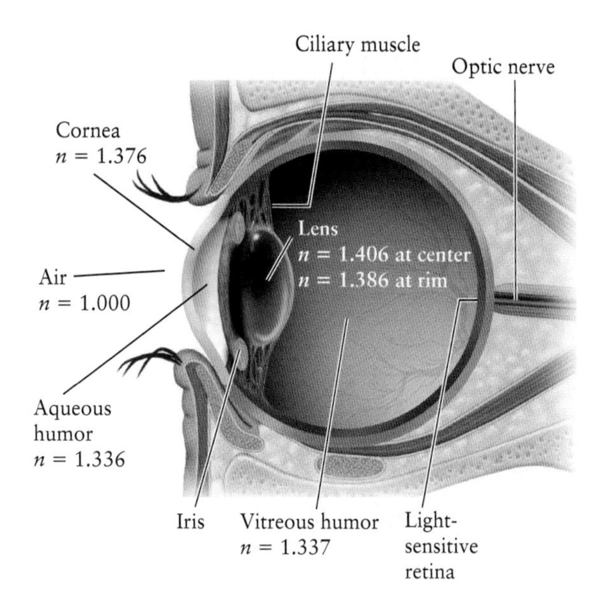

Figure 24-40 The human eye The elements of the human eye work together to bring parallel light rays from a distant object to a focus on the light-sensitive retina.

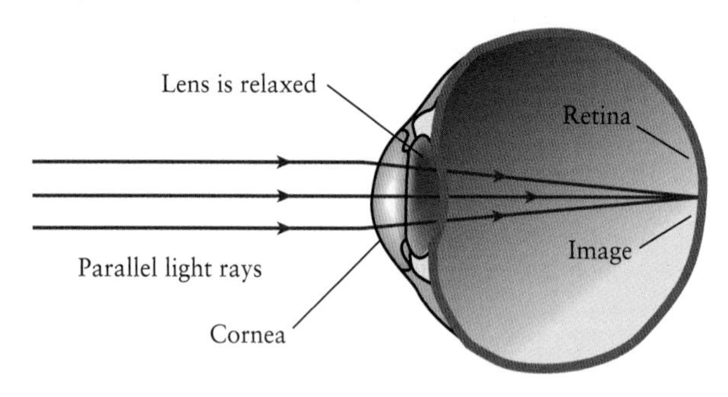

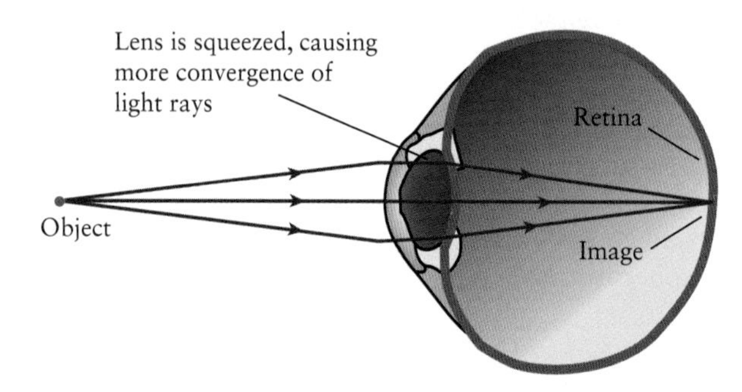

Figure 24-41 Focusing the human eye (a) When the lens is relaxed, parallel light rays from a distant object come to a focus on the light-sensitive retina. (b) In order to make a sharp image on the retina of a nearby object, the lens must change shape.

Unlike the lenses we studied in Sections 24-7 and 24-8, the index of refraction of the lens is *not* the same throughout its volume: Its value varies from about $n = 1.406$ at the center of the lens to about $n = 1.386$ near its outer rim. This actually enhances the ability of the lens to make light rays converge, and helps to minimize spherical aberration.

The index of refraction of the lens is only slightly greater than that of the aqueous humor in front of the lens ($n = 1.336$) or that of the *vitreous humor* ($n = 1.337$), the transparent gel that lies behind the lens and fills most of the volume of the eye. Hence the lens causes less convergence of light rays than does the cornea. Roughly 70% of the refraction needed to bend parallel light rays coming from a distant object and focus them on the retina comes from the cornea; the remaining 30% comes from the refraction provided by the lens (**Figure 24-41a**).

Unlike in a camera lens, the human eye does not focus on nearby objects by moving its elements forward or back: The distance from the cornea to the center of the lens remains the same, as does the distance from the center of the lens to the retina. Instead, the lens (which is made of flexible biological material) changes shape! To focus on a close object, the *ciliary muscles* that hold the lens in place squeeze the lens around its rim. This causes the lens to deform, as shown in **Figure 24-41b**. The front and back surfaces of the lens become more sharply curved, so their radii of curvature are reduced in magnitude. Hence the focal length f of the lens decreases, the power $P = 1/f$ of the lens increases (see Equation 24-25), and light rays are bent more sharply as they traverse the lens. Hence an image of the nearby object can be brought to a focus on the retina.

In many people the focusing mechanism shown in Figure 24-41 does not work perfectly. One common example is the condition known as *hyperopia* or *farsightedness*, in which the cornea and lens provide too little bending of light rays. As a result, parallel light rays that enter the eye have not yet come to a focus when they reach the retina, so objects appear blurred (**Figure 24-42a**). The lenses of a person with hyperopia must be squeezed, as in Figure 24-41b, to focus both on distant objects and on nearby ones, which causes eye strain and headaches. Even with the lenses squeezed as much as possible, the nearest object that can be seen in sharp focus can still be quite far away. Hyperopia can occur if the eye is too short from front to back, if the cornea has the wrong shape, or if the ciliary muscles are too weak. To compensate for the inadequate convergence of light rays provided by a hyperopic eye, the treatment is to wear converging (convex) eyeglasses or contact lenses.

More common than hyperopia is *myopia* or *nearsightedness*. In this condition the cornea and lens actually do too good of a job of bending light rays and making them converge. As a result, parallel light rays from a distant object come to a focus at a point in front of the retina (**Figure 24-42b**). Nearby objects do appear in focus, however. Myopia can occur if the eye is too long from front to back, or if the cornea or lens has an incorrect shape. The treatment is to wear diverging (concave) eyeglasses or contact lenses. Then parallel light rays coming from a distant object will be diverging when they enter the cornea, and this divergence compensates for the excess convergence caused by the cornea and lens.

Even if you have eyes without hyperopia or myopia, you may need eyeglasses by the time you reach middle age. The reason is that the lenses in our eyes lose some of their flexibility as we age. As a result, the lens is unable to squeeze as in Figure 24-41b, so light from nearby objects cannot be focused on the retina and these objects appear blurred. This condition, called *presbyopia*, is treated by wearing reading glasses (see the photograph that opens this chapter) with convex lenses. Just as in hyperopia,

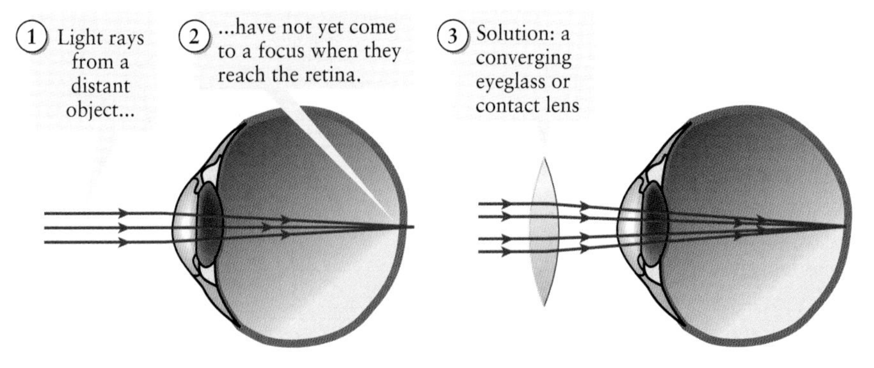
(a) Hyperopia (farsightedness): light rays converge too little
1 Light rays from a distant object...
2 ...have not yet come to a focus when they reach the retina.
3 Solution: a converging eyeglass or contact lens

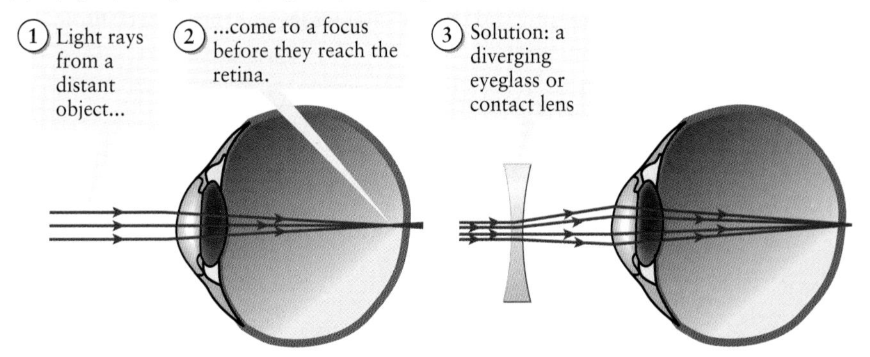
(b) Myopia (nearsightedness): light rays converge too much
1 Light rays from a distant object...
2 ...come to a focus before they reach the retina.
3 Solution: a diverging eyeglass or contact lens

Figure 24-42 Correcting deficiencies of vision (a) Hyperopia is corrected with a converging lens; (b) myopia is corrected with a diverging lens.

these eyeglasses compensate for the inadequate converging power of the aging, less flexible lens. Unlike hyperopia, however, a person with presbyopia may not need to use eyeglasses to see distant objects clearly. Many older people have both myopia (which requires concave, diverging eyeglasses to see distant objects) *and* presbyopia (which requires convex, converging eyeglasses to see nearby objects). The solution to this problem is *bifocal* eyeglasses, which have a concave shape in their upper part for distant vision and a convex shape in their lower part for reading.

With age the lenses of the eyes can also become clouded by cataracts. In this case the lenses are surgically removed and replaced by artificial lenses (see Figure 24-1b).

EXAMPLE 24-6 Correcting for Farsightedness and Nearsightedness

(a) Even with the lenses of her eyes squeezed as much as possible, a certain farsighted person has sharp vision only for objects that are 3.00 m or farther away from her corneas. What should be the focal length and power of a contact lens that will allow her to clearly see an object 0.250 m away from her corneas? (b) A certain nearsighted person can only see objects sharply if they are no more than 1.25 m in front of his corneas. What should be the focal length and power of a contact lens that will allow him to see very distant objects clearly?

Set Up

Since a contact lens sits on the cornea and is very thin, the distance d_O from the contact lens to the object in each case is essentially the same as from the cornea to the object [0.250 m in (a), infinity in (b)]. In part (a) the contact lens must make the eye "think" that an object 0.250 m away is actually 3.00 m away. So this contact lens must produce a virtual image of the object that's 3.00 m in front of the contact lens. Similarly, in part (b) the contact lens must make the eye "think" that a distant object is only 1.25 m away. To do this, the contact lens must produce a virtual image of the object 1.25 m in front of the contact lens. In each case we'll use Equation 24-11 to find the required focal length f from the specified object and image distances. Equation 24-25 then tells us the power of the contact lens.

Lens equation:

$$\frac{1}{d_O} + \frac{1}{d_I} = \frac{1}{f} \qquad (24\text{-}11)$$

Lens power:

$$P = \frac{1}{f} \qquad (24\text{-}25)$$

Solve

(a) We draw a ray diagram with a point object a distance $d_O = 0.250$ m from the contact lens. (We draw a point object rather than an object arrow, since we're not concerned with the size of the image.) The light rays emerging from this contact lens must appear to emanate from a point 3.00 m in front of the lens, so the contact lens must make the light rays converge. Hence it must be a converging contact lens with a positive focal length.

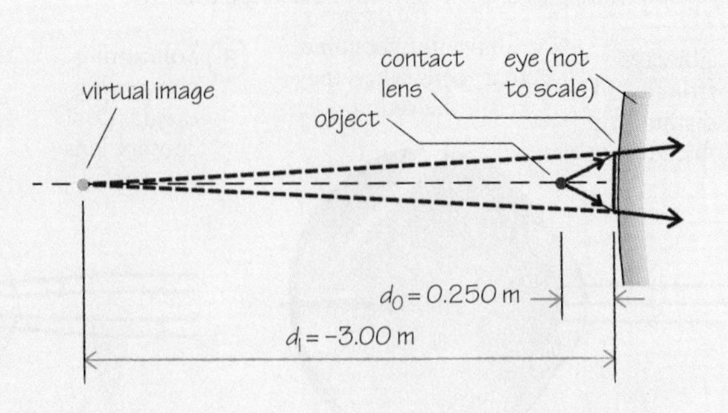

Solve for the focal length of the contact lens using the lens equation.

The distance to the object is $d_O = 0.250$ m, and the distance to the image formed by the contact lens is $d_I = -3.00$ m (negative because the image is on the same side of the contact lens as the object). From Equation 24-11

$$\frac{1}{f} = \frac{1}{d_O} + \frac{1}{d_I} = \frac{1}{0.250\text{ m}} + \frac{1}{(-3.00\text{ m})}$$

$$= 4.00\text{ m}^{-1} - 0.333\text{ m}^{-1}$$

$$= 3.67\text{ m}^{-1}$$

$$f = \frac{1}{3.67\text{ m}^{-1}} = 0.273\text{ m}$$

The positive focal length means that as predicted, this is a converging contact lens. The power of this contact lens is

$$P = \frac{1}{f} = 3.67\text{ m}^{-1} = 3.67\text{ diopters}$$

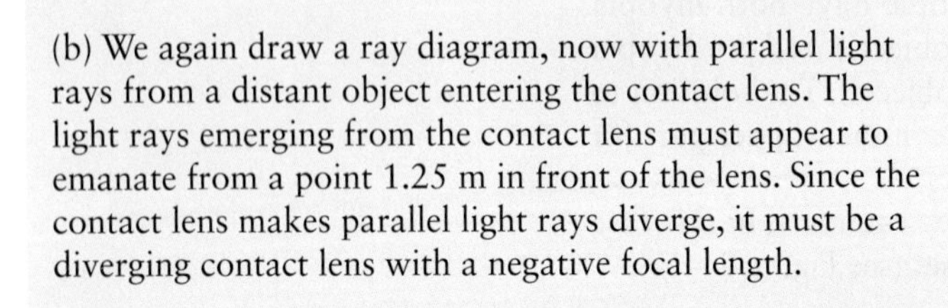

(b) We again draw a ray diagram, now with parallel light rays from a distant object entering the contact lens. The light rays emerging from the contact lens must appear to emanate from a point 1.25 m in front of the lens. Since the contact lens makes parallel light rays diverge, it must be a diverging contact lens with a negative focal length.

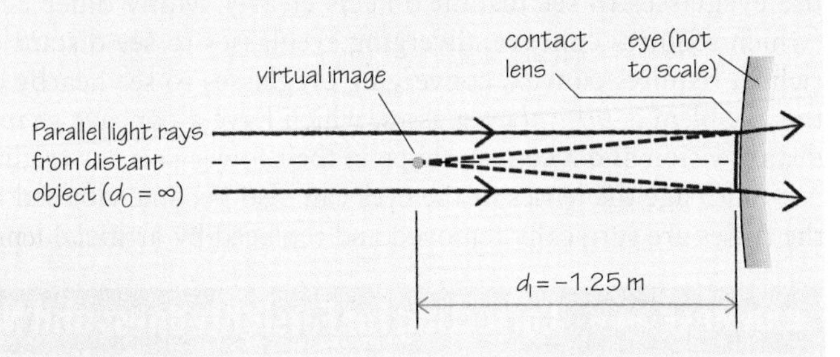

Again use the lens equation to solve for the focal length of the contact lens.

The distance d_O to the object is infinite (so $1/d_O = 0$), and the image formed by the contact lens is at $d_I = -1.25$ m. (This is negative because the image is on the same side of the contact lens as the object.) From Equation 24-11

$$\frac{1}{f} = \frac{1}{d_O} + \frac{1}{d_I} = 0 + \frac{1}{(-1.25\text{ m})}$$

$$\frac{1}{f} = \frac{1}{(-1.25\text{ m})}$$

Take the reciprocal of both sides:

$f = -1.25$ m

The negative value means that this is a diverging contact lens, as we expect. The power of this contact lens is

$$P = \frac{1}{f} = \frac{1}{(-1.25\text{ m})} = -0.800\text{ m}^{-1} = -0.800\text{ diopter}$$

Reflect

This example reinforces the idea that you should draw a ray diagram for *any* problem that involves image formation by lenses or mirrors. Note that in part (a) the converging contact lens produces a virtual image. This can happen only if the object is placed inside the focal point of the contact lens, so that the object distance is less than the focal length. This is indeed the case in part (a): The object distance was $d_O = 0.250$ m, and we found that the focal length was $f = 0.273$ m.

GOT THE CONCEPT? 24-11 A Corrective Intraocular Lens

? A patient develops cataracts in her myopic eye. As an ophthalmic surgeon, your task is to choose a replacement artificial lens that will also correct for her myopia. To accomplish this, how should the artificial lens compare to the patient's natural lens? (There may be more than one correct answer.) (a) Same shape but a higher index of refraction; (b) same shape but a lower index of refraction; (c) same index of refraction but with more sharply curved surfaces; (d) same index of refraction but with more gently curved surfaces.

TAKE-HOME MESSAGE FOR Section 24-9

✔ A camera lens refracts parallel light rays from a distant object so that they come to a focus on a light-sensitive sensor. To focus on nearby objects, either the lens is moved away from the sensor, or the focal length of the lens is decreased.

✔ In the human eye, refraction takes place in both the cornea and the lens to focus light rays on the retina. Adjusting the focus from distant to nearby objects is done by reshaping the lens.

✔ Corrective lenses (eyeglasses or contact lenses) are used to either increase or decrease the amount by which light rays are forced to converge within the eye on their way to the retina.

Key Terms

center of curvature
concave
converging lens
convex
diffuse light
diopter
diverging lens
focal length
focal point
geometrical optics
image
image distance
image height
inverted image
lateral magnification
lens equation
lensmaker's equation
magnifying glass
mirror equation
object
object distance
object height
optical device
plane mirror
power (of a lens or mirror)
principal axis
radius of curvature
ray diagram
real image
specular reflection
thin lens
upright image
virtual image

Chapter Summary

Topic	Equation or Figure
Plane mirrors: A plane mirror makes an upright, virtual image of any object. The image is the same size as the object, so the lateral magnification is $m = 1$. The image is reversed back to front, not side to side.	The **image distance** is negative. The **object distance** is positive. $$d_I = -\lvert d_I\rvert = -d_O \quad (24\text{-}1)$$ The negative value of d_I indicates that the image is on the opposite side of the mirror from the object.

Lateral magnification; Image height; Object height

$$m = \frac{h_I}{h_O} \tag{24-2}$$

Spherical mirrors: If we consider only parallel light rays that are close to the principal axis of a spherical concave mirror, the reflected rays all converge at the focal point. The distance from the center of the mirror to the focal point is the focal length. If the mirror is convex rather than concave, parallel light rays diverge rather than converge after reflection. The radius of curvature and the focal length are both negative for a convex mirror.

Focal length of a spherical mirror; **Radius of curvature** of the mirror

$$f = \frac{r}{2} \tag{24-3}$$

Image formation by a concave mirror: We can locate the image made by a concave mirror by drawing a ray diagram. If the object is outside the focal point, the image is real and inverted; its size depends on how far the object is from the mirror. If the object is inside the focal point, the image is virtual, upright, and larger than the object.

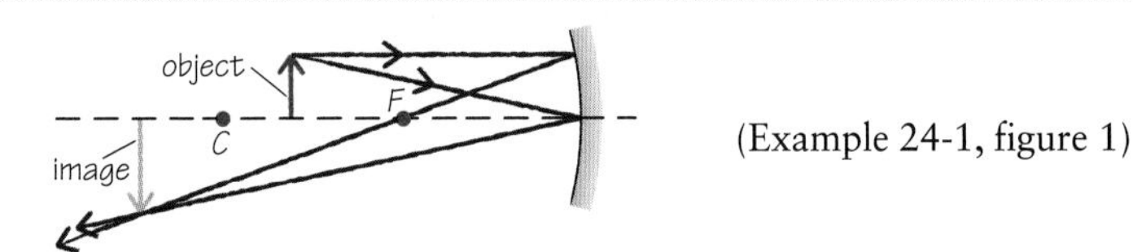

(Example 24-1, figure 1)

(Example 24-1, figure 2)

Image formation by a convex mirror: A ray diagram also helps us locate the image made by a convex mirror. No matter where the object is placed, the image is virtual, upright, and smaller than the object.

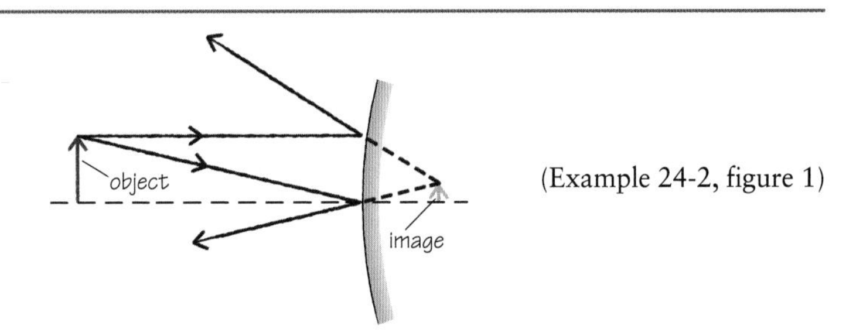

(Example 24-2, figure 1)

The mirror equation and lens equation: The mirror equation and lens equation relate the object distance, image distance, and focal length for either a concave or convex mirror or a converging or diverging lens. The lateral magnification depends on the image and object distances and can be positive (if the image is upright) or negative (if the image is inverted).

$$\frac{1}{d_O} + \frac{1}{d_I} = \frac{1}{f} \tag{24-11}$$

Focal length; Object distance; Image distance

Lateral magnification; Image height; Image distance

$$m = \frac{h_I}{h_O} = -\frac{d_I}{d_O} \tag{24-14}$$

Object height; Object distance

Lenses: Due to refraction, light rays converge as they enter or exit a converging glass lens and diverge as they enter or exit a diverging glass lens. A converging lens behaves similarly to a concave mirror, and a diverging lens behaves similarly to a convex mirror. The focal length of a lens is given by the lensmaker's equation, which involves the index of refraction of the lens material and the radii of curvature of the lens surfaces. This equation is valid if the lens is thin.

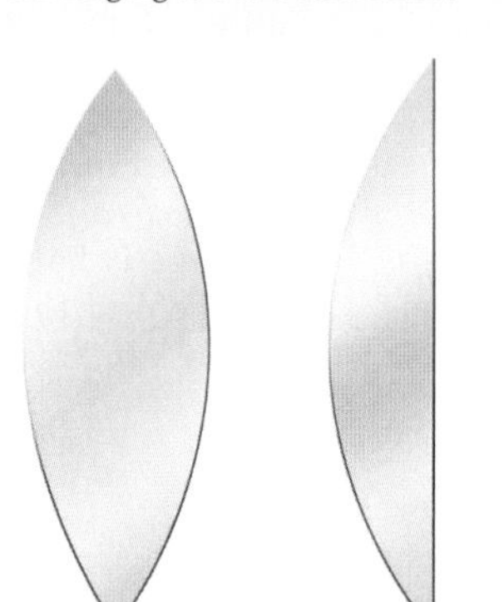

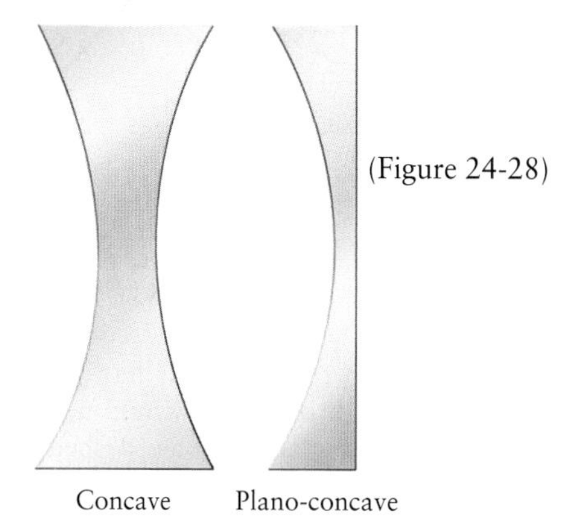

(Figure 24-28)

Focal length of a lens surrounded by air

Index of refraction of the lens material

$$\frac{1}{f} = (n-1)\left(\frac{1}{R_1} - \frac{1}{R_2}\right) \qquad (24\text{-}24)$$

Radius of curvature of lens surface 1 (the surface closer to the object)

Radius of curvature of lens surface 2 (the surface farther from the object)

Image formation by lenses: A converging lens produces a real, inverted image if the object is outside the focal point and a virtual, upright, enlarged image if the object is inside the focal point. A diverging lens always produces a virtual, upright, reduced image. The same equation that relates object distance d_O, image distance d_I, and focal length f for a spherical mirror also applies to thin lenses, as does the equation for lateral magnification.

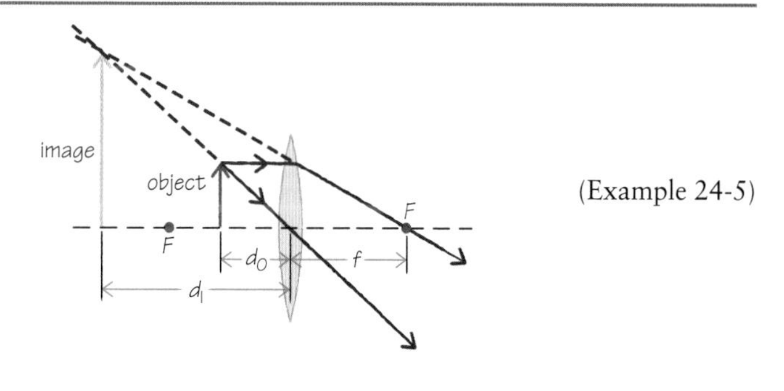

(Example 24-5)

Answer to What do you think? Question

(a) These eyeglasses are used by people with presbyopia. Their eyes are able to take essentially parallel light rays from distant objects and make them converge onto the retina to form an image but aren't able to do the same with the diverging light rays coming from a nearby object. The corrective eyeglasses take these diverging light rays and make them nearly parallel, so a converging lens is needed. A converging lens is thicker at the middle than it is at the edges. We discuss the physics of the eye and of corrective lenses in Section 24-9.

Answers to Got the Concept? Questions

24-1 (a) If you see someone's eyes in a mirror, she can see you as well. This must be true because if light rays can reflect off the mirror from her eyes to yours, then light can follow that path in the opposite direction and reflect from your eyes to hers, too.

24-2 (c) From the information given it's not possible to know whether the person around the corner can see you. We've sketched this situation in **Figure 24-43**. Light rays that reflect off the mirror from his right hand to your eyes, shown

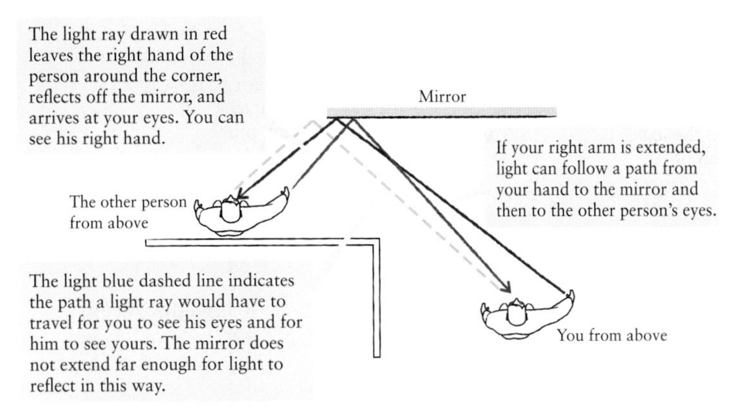

The light ray drawn in red leaves the right hand of the person around the corner, reflects off the mirror, and arrives at your eyes. You can see his right hand.
Mirror
If your right arm is extended, light can follow a path from your hand to the mirror and then to the other person's eyes.
The other person from above
The light blue dashed line indicates the path a light ray would have to travel for you to see his eyes and for him to see yours. The mirror does not extend far enough for light to reflect in this way.
You from above

Figure 24-43

in red, enable you to see his hand. You can't see his eyes because the path that light would need to follow for this to happen, shown by the light blue, dashed line, does not intercept the mirror. Whether the other person can see some part of you, however, depends on how far out you are holding your right hand. If it is far enough away from your body, light rays can follow the path drawn in dark blue, from your hand to his eyes.

24-3 (b) When an object is placed farther from a concave mirror than the focal point, the image is real and inverted. Note that a real image forms in front of a mirror; that is, it appears to be in front of the mirror. So when you hold the spoon at arm's length, the object (you!) must therefore be farther from the reflective surface than the focal point. When you hold the spoon 4 cm from your eye, however, you (your eye) must be closer to the spoon than its focal point. An object placed closer to a concave reflective surface than the focal point forms a virtual, inverted image. Because a virtual image is formed at the point where light rays from any point on an object appear to meet (but don't actually), the image forms behind the mirror and is upright. The focal point of the spoon must be about 6 cm from the spoon. At the focal point no image forms because light rays from any point on your face are reflected parallel to each other.

24-4 (e) A concave mirror makes an image because light rays from a point on the object that strike anywhere on the mirror are all reflected to the same point on the image. If you black out part of the mirror, there are fewer positions on the mirror that contribute to the image, but they still form an image at the same place. This is true for any point on the object, so the image will still show the entire light bulb. The only difference is that the image will be dimmer because only half as much light energy is reflected by the half-painted mirror.

24-5 (b) A real image made by a concave mirror is also an inverted image, so the magnification is $m = -2.00$. Equation 24-14 then tells us the image distance: $m = -d_I/d_O$, so $d_I = -md_O = -(-2.00)(12.0 \text{ cm}) = +24.0 \text{ cm}$. (The positive value of the image distance tells us that the image is in front of the mirror, as must be true for a real image.) The mirror equation, Equation 24-11, then tells us the focal length: $1/f = 1/d_O + 1/d_I = 1/(12.0 \text{ cm}) + 1/(24.0 \text{ cm}) = 0.0833 \text{ cm}^{-1} + 0.0417 \text{ cm}^{-1} = 0.125 \text{ cm}^{-1}$, so $f = 1/(0.125 \text{ cm}^{-1}) = 8.00 \text{ cm}$. Note that the object distance of 12.0 cm is greater than the focal length, as must be the case if the image is to be real.

24-6 (b) A concave mirror can focus parallel light rays; a convex mirror cannot, so it cannot concentrate solar energy onto a pot for cooking. Another way to see this is to think of the mirror as making an image of the Sun that lies at the position of the pot so as to concentrate the Sun's light there. This is possible with a concave mirror, which makes a real image of a distant object. A convex mirror, by contrast, can make only a virtual image whose rays never actually cross. Hence a convex mirror is no good for concentrating solar energy.

24-7 (c) We can solve this problem using the mirror equation, Equation 24-11, and the expression for magnification, Equation 24-14. The image is one-half the height of the object, so the lateral magnification is $m = h_I/h_O = 1/2$. From Equation 24-14 the magnification m is also equal to $-d_I/d_O$, so $-d_I/d_O = 1/2$ and $d_I = -d_O/2$; that is, the image is half as far from the mirror as the object is and is behind the mirror (because d_I is negative). Substitute $d_I = -d_O/2$ into the mirror equation and solve for d_O:

$$\frac{1}{d_O} + \frac{1}{d_I} = \frac{1}{f} \quad \text{so} \quad \frac{1}{d_O} + \left(-\frac{2}{d_O}\right) = -\frac{1}{d_O} = \frac{1}{f}$$

$$\text{and} \quad d_O = -f = |f|$$

(The focal length of a convex mirror is negative, so $f = -|f|$.) The distance from the object to the mirror equals the distance from the mirror to the focal point.

24-8 (c) The lenses we described in Section 24-7 are made of a material in which light travels more slowly than in the surrounding air—that is, a material with a higher index of refraction than the surroundings. In this situation, however, the lens is made of a material with a *lower* index of refraction than the surroundings. Hence light rays refract in the opposite sense as they enter and exit the lens, which means that this convex lens is actually a diverging lens. Like the diverging lens in Figure 24-33, the image made by this lens is virtual and smaller than the object.

24-9 **Figure 24-44** shows the ray diagram.

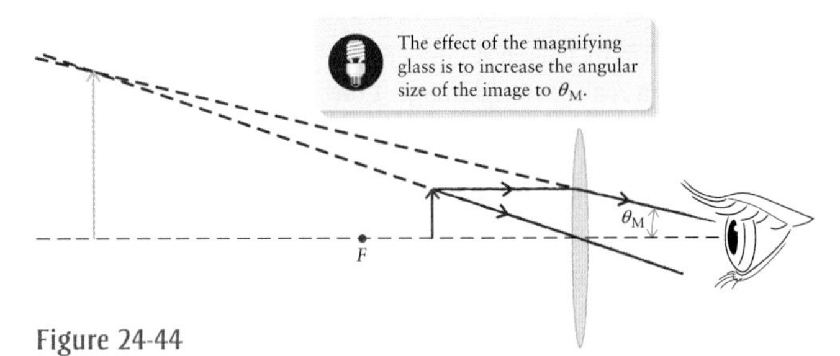

Figure 24-44

24-10 **Figure 24-45** shows the ray diagram.

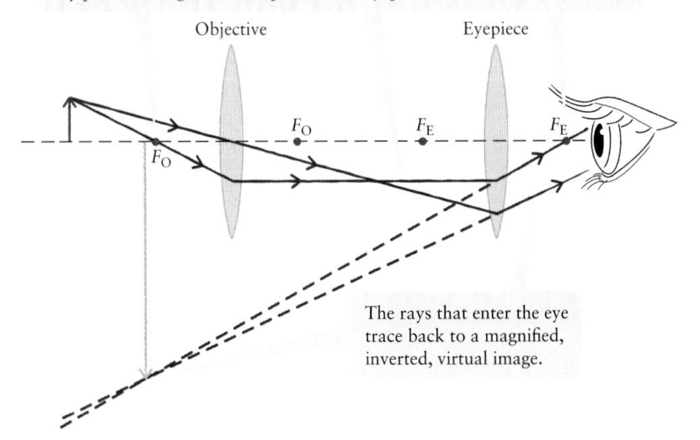

Figure 24-45

24-11 (b), (d) In myopia (nearsightedness) the cornea and lens make light rays converge too sharply, so parallel light rays from a distant object come to a focus in front of the retina, as in Figure 24-42b. The replacement lens should therefore cause less convergence of light and have a longer focal length (that is, a lower power). This can be accomplished by reducing the index of refraction of the lens to be closer in value to that of the aqueous and vitreous humors, so that less refraction takes place. It can also be accomplished by giving the lens surfaces a more gentle curve, which means radii of curvature of greater magnitude and hence a longer focal length.

Questions and Problems

In a few problems you are given more data than you actually need; in a few other problems you are required to supply data from your general knowledge, outside sources, or informed estimate.

Interpret as significant all digits in numerical values that have trailing zeros and no decimal points.

For all problems use $g = 9.80 \text{ m/s}^2$ for the free-fall acceleration due to gravity. Neglect friction and air resistance unless instructed to do otherwise.

•	Basic, single-concept problem
••	Intermediate-level problem; may require synthesis of concepts and multiple steps
•••	Challenging problem
SSM	*Solution is in Student Solutions Manual*
Example	*See worked example for a similar problem.*

Conceptual Questions

1. • A plane mirror seems to invert your image left and right but not up and down. Is this what it really does? SSM

2. • Explain why reflected light is usually diffuse.

3. • What is the difference between a real image and a virtual image?

4. • When you view a car's side mirror, you see a smaller image than you would if the mirror were flat. Is the mirror concave or convex? Explain your answer.

5. • Explain the meaning, in terms of physics, of the phrase etched on the right side mirror of most cars: "Objects in mirror are closer than they appear."

6. • What is the radius of curvature of a plane mirror? Explain your answer.

7. • For a certain glass lens in air, both radii of curvature are positive. Is it a converging lens or a diverging lens, or do you need additional information to tell? Explain. SSM

8. • A convex lens made of clear ice ($n = 1.31$) acts as a *converging* lens when it is in air, but as a *diverging* lens when it is surrounded by acetone ($n = 1.36$). Explain why.

9. • A laptop computer is connected to a video projector that projects an image on a screen. If the lens of the projector is half covered, what happens to the image? Explain your answer.

10. • **Biology** The image focused on your retina is actually inverted (sketch a simple ray diagram showing this observation). What does this fact say about our definitions of "right side up" and "upside down"?

11. • **Biology** Experimental subjects who wear inverting lenses (glasses that invert all images) for several days adapt to their new perception of the world so well that they can even ride a bicycle. Several days after the glasses are removed, their perceptions return to normal. Discuss this phenomenon and comment.

12. • **Medical** Explain why converging lenses are used to correct farsightedness (hyperopia) while diverging lenses are used for nearsightedness (myopia).

13. • **Biology** Discuss why nearsightedness is not found in all people, but virtually everyone eventually has difficulty focusing on nearby objects as they age.

14. •• Explain why looking through a small opening often provides visual acuity even to an extremely nearsighted person.

15. • Explain why the lens in a digital camera must move away from the light sensor in order to focus on a nearby object.

Multiple-Choice Questions

16. • Which is true when an object is moved farther from a plane mirror?
   A. The height of the image decreases, and the image moves farther from the mirror.
   B. The height of the image stays the same, and the image moves farther from the mirror.
   C. The height of the image increases, and the image moves farther from the mirror.
   D. The height of the image stays the same, and the image moves closer to the mirror.
   E. The height of the image decreases, and the image moves closer to the mirror. SSM

17. • A real image can form in front of
   A. a plane mirror.
   B. a concave mirror.
   C. a convex mirror.
   D. any type of mirror.
   E. no mirror of any type.

18. • When an object is placed a little farther from a concave mirror than the focal length, the image is
   A. magnified and real.
   B. magnified and virtual.
   C. smaller and real.
   D. smaller and virtual.
   E. smaller and reversed.

19. • If you want to start a fire using sunlight, which kind of mirror would be most efficient?
   A. a plane mirror
   B. a concave mirror
   C. a convex mirror
   D. any type of plane, concave, or convex mirror
   E. It is not possible to start a fire using sunlight and a mirror; you must use a concave lens.

20. • An object is placed at the center of curvature of a concave mirror. The image is
   A. real and upright.
   B. real and inverted.
   C. virtual and upright.
   D. virtual and inverted.
   E. nonexistent; no image is formed. SSM

21. • When an object is placed farther from a convex mirror than the absolute value of the focal length, the image is

A. larger and real.
B. larger and virtual.
C. smaller and real.
D. smaller and virtual.
E. smaller and reversed.

22. • **Medical** When a dentist needs a mirror to see an enlarged, upright image of a patient's tooth, what kind of mirror should she use?

A. a plane mirror
B. a concave mirror
C. a convex mirror
D. either a plane mirror or a concave mirror
E. either a plane mirror or a convex mirror

23. • A magnifying lens allows one to look at a very near object by forming an image of it farther away. The object appears larger. To create a magnifying lens, one would use a

A. short focal length ($f < 1$ m) converging lens.
B. short focal length ($|f| < 1$ m) diverging lens.
C. long focal length ($f > 1$ m) converging lens.
D. long focal length ($|f| > 1$ m) diverging lens.
E. either a converging or a diverging lens.

24. •• A compound microscope is a two-lens system used to look at very small objects. Which of the following statements is correct?

A. The objective lens and the eyepiece both have the same focal length, and both serve as magnifying lenses.
B. The objective lens is a short focal length, converging lens and the eyepiece functions as a magnifying lens.
C. The objective lens is a long focal length, converging lens and the eyepiece functions as a magnifying lens.
D. The objective is a short focal length, diverging lens and the eyepiece functions as a magnifying lens.
E. The objective lens is a long focal length, diverging lens and the eyepiece functions as a magnifying lens. SSM

25. An engineer would like to modify a consumer camera to better capture images in the near infrared, just beyond the visible reds. To save money, the decision is made to keep the same lens for producing the image. The index of refraction of the lens is smaller for infrared radiation than for visible light. Which of the following changes could be made to produce a sharp image in the near infrared?

A. Relocate the detector closer to the lens.
B. Relocate the detector farther from the lens.
C. A sharp image cannot be produced with the same lens system.
D. There is no need to change anything.

Estimation/Numerical Analysis

26. • Estimate the focal length of a convex blind spot mirror that is often added to a vehicle's outside mirror.

27. • Estimate the radius of curvature of a spherical mirror that is typically found in the corridors of busy hospitals, for example, to help prevent collisions when going around a corner.

28. • Give three examples of spherical concave mirrors in your daily life and estimate their approximate focal lengths. SSM

29. • Give three examples of spherical convex mirrors in your daily life and estimate their approximate focal length.

30. • Estimate the focal length of the lens in a mobile phone's camera when it is focused on a distant object.

31. • Determine the focal length for an unknown lens with the following object and image distances:

Object Distance (cm)	Image Distance (cm)	Object Distance (cm)	Image Distance (cm)
30	98	60	37
35	67	65	35
40	53	70	34
45	47	75	33
50	42	80	32
55	38		

Problems

24-1 Mirrors or lenses can be used to form images
24-2 A plane mirror produces an image that is reversed back to front

32. • If the angle of incidence on a flat mirror is 0°, what is the angle of reflection?

33. • Two flat mirrors are perpendicular to each other. An incoming beam of light makes an angle of $\theta = 30°$ with the first mirror, as shown in **Figure 24-46**. What angle will the outgoing beam make with respect to the normal of the second mirror? SSM

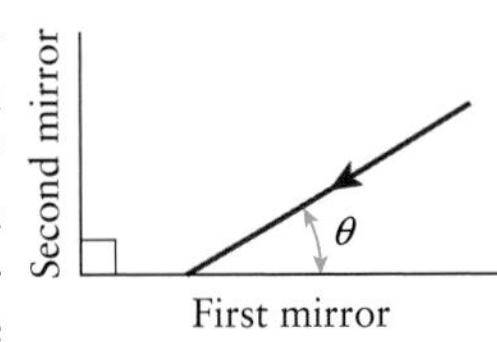

Figure 24-46 Problem 33

34. • A 1.8-m-tall man stands 2.0 m in front of a vertical plane mirror. How tall will the image of the man be?

35. • What must be the minimum height of a plane mirror in order for a 1.8-m-tall person to see a full image of himself?

36. •• A plane mirror is 10 m away from and parallel to a second plane mirror (**Figure 24-47**). Find the locations of the first five images formed by each mirror when an object is positioned exactly in the middle between the two mirrors.

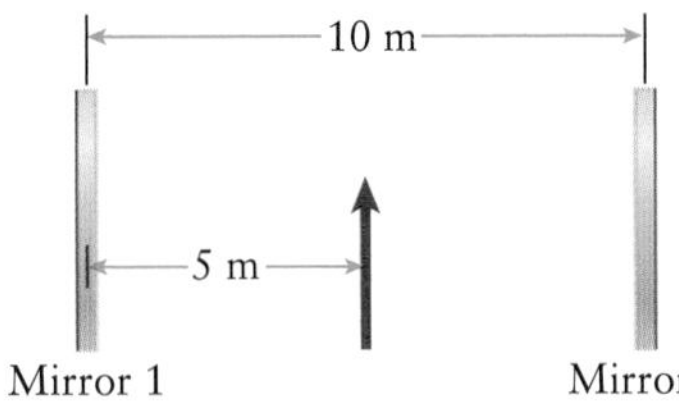

Figure 24-47 Problem 36

37. •• A plane mirror is 10 m away from and parallel to a second plane mirror (**Figure 24-48**). Find the locations of the first five images formed by each mirror when an object is positioned 3 m from one of the mirrors. SSM

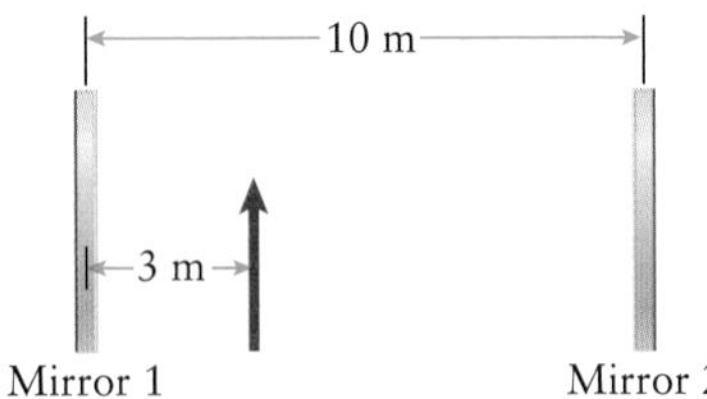

Figure 24-48 Problem 37

38. • At which of the points A through E will the image of the face be visible in the plane mirror of length L (**Figure 24-49**)? The distance x equals $L/2$. Points A through E are collinear and separated by a distance of $3L/4$, and point C lies on a line bisecting the mirror. Assume the face lies directly on point C.

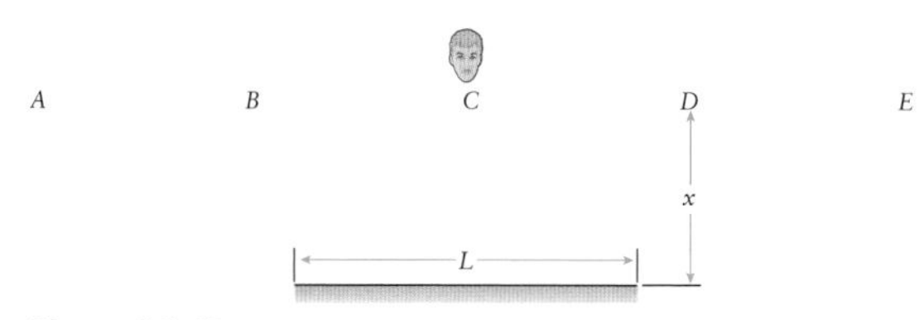

Figure 24-49 Problem 38

39. • Using a ruler, a protractor, and the law of reflection, show the location of the image of your face when you stand a short distance in front of a plane mirror, as shown in **Figure 24-50**.

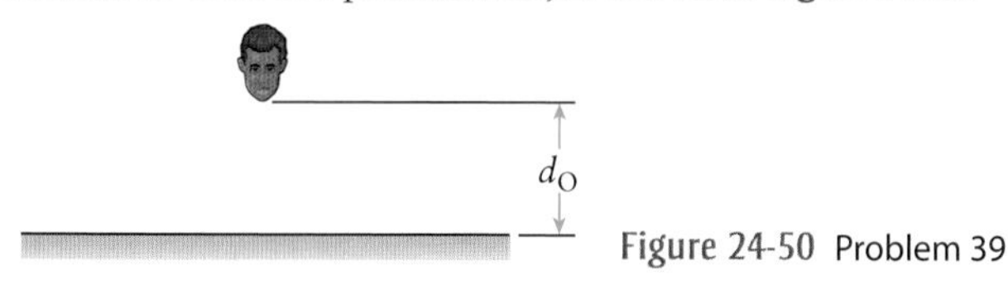

Figure 24-50 Problem 39

40. • One person is looking in a plane mirror at the image of a second person (**Figure 24-51**). Using a ruler, a protractor, and the law of reflection, show the location of the image as seen by the first person.

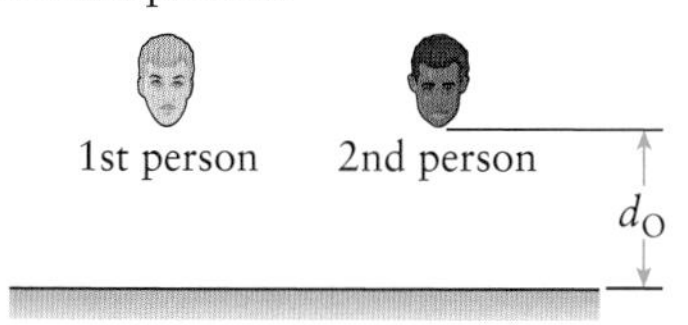

Figure 24-51 Problem 40

41. • You hold an autofocusing camera 2 m in front of a plane mirror. Describe the self-portrait that is photographed when you snap a picture of yourself with this camera. An autofocus camera sends infrared waves from a transmitter and receives the reflected waves that bounce off of the objects in front of the camera to determine the distance at which to focus the lens.

24-3 A concave mirror can produce an image of a different size than the object

42. • Describe the difference between the images seen in a spherical, concave mirror when the object is "up close" (closer to the image than the focal length) compared to "far away" (outside the focal length).

43. • Are there any situations where a real image is formed in a spherical, concave mirror? Describe the location of the object relative to the focal point for such situations. SSM

44. • Describe the image you see of yourself if your head is at the center of curvature of a spherical concave mirror.

24-4 Simple equations give the position and magnification of the image made by a concave mirror

45. • An object is placed 8.0 cm in front of a concave mirror with a 10.0-cm radius of curvature. Calculate the image distance and the magnification of the image. Determine whether the image is real or virtual and whether it is inverted or upright by using (a) a ray diagram and (b) the mirror equation. SSM Example 24-1

46. • An object 1.0 cm tall is placed 3.0 cm in front of a spherical concave mirror with a radius of curvature equal to 10.0 cm. Calculate the image distance and height by using (a) a ray diagram and (b) the mirror equation. Example 24-1

47. • An object 1.0 cm tall is placed 6.0 cm in front of a spherical concave mirror with a radius of curvature equal to 10.0 cm. Calculate the image distance and height by using (a) a ray diagram and (b) the mirror equation. Example 24-1

48. • The radius of curvature of a spherical concave mirror is 20.0 cm. Describe the image formed when a 10.0-cm-tall object is (a) 5.0 cm from the mirror, (b) 20.0 cm from the mirror, (c) 50.0 cm from the mirror, and (d) 100.0 cm from the mirror. For each case give the image distance, the image height, the type of image (real or virtual), and the orientation of the image (upright or inverted). Example 24-1

49. • The radius of curvature of a spherical concave mirror is 15.0 cm. Describe the image formed when a 20.0-cm-tall object is (a) 10.0 cm from the mirror, (b) 20.0 cm from the mirror, and (c) 100.0 cm from the mirror. For each case give the image distance, the image height, the type of image (real or virtual), and the orientation of the image (upright or inverted). SSM Example 24-1

50. • An object is 24.0 cm from a spherical concave mirror of unknown focal length. The image that is formed is 30.0 cm from the mirror. (a) Calculate the focal length. (b) Is the image real or virtual? (c) If the object is 10.0 cm tall, determine the height of the image(s). Example 24-1

51. • Construct ray diagrams to locate the images in the following cases. (a) A 10.0-cm-tall object is located 5.0 cm in front of a spherical concave mirror with a radius of curvature of 20.0 cm. (b) A 10.0-cm-tall object is located 10.0 cm in front of a spherical concave mirror with a radius of curvature of 20.0 cm. (c) A 10.0-cm-tall object is located 20.0 cm in front of a spherical concave mirror with a radius of curvature of 20.0 cm. Example 24-1

52. •• Derive a relationship between the radius of curvature of a spherical, concave mirror and the object distance that gives an upright image that is four times as tall as the object. Example 24-1

24-5 A convex mirror always produces an image that is smaller than the object

53. • Describe the difference between the images seen in a spherical, convex mirror when the object is "up close" (a shorter distance from the mirror than the focal distance of the mirror) compared to "far away" (a longer distance from the mirror than the focal distance of the mirror).

54. • Are there any situations where a real image is formed in a spherical, convex mirror? Why or why not?

55. • How can you remember the difference between the shapes of a spherical *concave* mirror and a spherical *convex* mirror?

56. • Construct ray diagrams to locate the images and estimate the image height in each of the following cases. (a) A 10-cm-tall object located 5 cm in front of a spherical convex mirror with a radius of curvature of 20 cm. (b) A 10-cm-tall object located 10 cm in front of a spherical convex mirror with a radius of curvature of 20 cm. (c) A 10-cm-tall object located 20 cm in front of a spherical convex mirror with a radius of curvature of 20 cm.

24-6 The same equations used for concave mirrors also work for convex mirrors

57. • The radius of curvature of a spherical convex mirror is 20.0 cm. Describe the image formed when a 10.0-cm-tall object is positioned (a) 20.0 cm from the mirror, (b) 50.0 cm from the mirror, and (c) 100.0 cm from the mirror. For each case, provide the image distance, the image height, the type of image (real or virtual), and the orientation of the image (upright or inverted). SSM Example 24-2

58. • The radius of curvature of a spherical convex mirror is 15.0 cm. Describe the image formed when a 20.0-cm-tall object is positioned (a) 5.0 cm from the mirror, (b) 20.0 cm from the mirror, and (c) 100.0 cm from the mirror. For each case, provide the image distance, the image height, the type of image (real or virtual), and the orientation of the image (upright or inverted). Example 24-2

59. • A car's convex rearview mirror has a radius of curvature equal to 15.0 m. What are the magnification, type, and location of the image that is formed by an object that is 10.0 m from the mirror? Example 24-2

60. • An 18.0-cm-long pencil is placed beside a convex spherical mirror, and its image is 10.5 cm in length. If the radius of curvature of the mirror is 88.4 cm, find the image distance, the object distance, and the magnification of the pencil. Example 24-2

61. • A 1-cm-long horse fly hovers 1.0 cm from a shiny sphere with a radius of 25.0 cm. Calculate the location of the image of the fly, its type (real or virtual), and its length. Example 24-2

62. • A spherical convex mirror is placed at the end of a driveway on a corner with a limited view of oncoming traffic. The mirror has a radius of curvature of 1.85 m. Where will the image of a car that is 12.6 m from the mirror appear? Example 24-2

63. •• Using the mirror equation, prove that all images in spherical convex mirrors are virtual. SSM Example 24-2

64. • A girl sees her image in a shiny glass sphere tree ornament that has a diameter of 10.0 cm. The image is upright and is located 1.5 cm behind the surface of the ornament. How far from the ornament is the child located? Example 24-2

65. • A shiny sphere, 30.0 cm in diameter, is placed in a garden for aesthetic purposes. Determine the type, location, and height of the image of a 6.00-cm-tall squirrel located 40.0 cm in front of the sphere. Example 24-2

24-7 Convex lenses form images like concave mirrors and vice versa

66. • Under what circumstances will the images formed by converging or diverging lenses be designated as "real"? Indicate the type or types of lenses and the required position of the object.

67. • A real image that is created due to reflection in a spherical mirror appears in front of the mirrored surface. Is this also the case for a real image that is created due to refraction in a lens? SSM

68. • Where does the bending of light physically take place in a typical concave or convex lens? Is this how we draw ray diagrams? Why or why not?

69. • What *minimum* number of rays are required to locate the image that is formed by a lens in a ray diagram? Explain.

24-8 The focal length of a lens is determined by its index of refraction and the curvature of its surfaces

70. • A 10.0-cm-tall object is located in front of a converging lens with a power of 5.00 diopters. Describe the image created (type, location, height) and draw the ray diagrams if the object is located (a) 5.0 cm from the lens, (b) 10.0 cm from the lens, (c) 20.0 cm from the lens, and (d) 50.0 cm from the lens. SSM Example 24-5

71. • A 10.0-cm-tall object is located in front of a diverging lens with a power of −5.00 diopters. Describe the type, location, and height of the image created, and draw the ray diagrams if the object is located (a) 5.0 cm from the lens, (b) 10.0 cm from the lens, (c) 20.0 cm from the lens, and (d) 50.0 cm from the lens. Example 24-5

72. • A 2.00-cm-tall object is located 18.0 cm in front of a converging lens with a focal length of 30.0 cm. (a) Use the lens equation and (b) a ray diagram to describe the type, location, and height of the image that is formed. Example 24-5

73. • A lens is formed from a plastic material that has an index of refraction of 1.55. If the radius of curvature of one surface is 1.25 m and the radius of curvature of the other surface is 1.75 m, use the lens maker's equation to calculate the focal length and the power of the lens. Example 24-3

74. •• A glass lens ($n = 1.60$) has a focal length of −31.8 cm and a plano-concave shape. (a) Calculate the radius of curvature of the concave surface. (b) If a lens is constructed from the same glass to form a plano-convex shape with the same radius of curvature, what will the focal length be? SSM Example 24-4

75. • A 2.00-cm-tall object is 30.0 cm in front of a converging lens that has a focal length of 18.0 cm. (a) Use the lens equation and (b) a ray diagram to describe the type, location, and height of the image that is formed. Example 24-5

76. ••• **Biology** Calculate the overall magnification of a compound microscope that uses an objective lens with a focal length of 0.50 cm, an eyepiece with a focal length of 2.50 cm, and a distance of 18 cm between the two.

24-9 A camera and the human eye use different methods to focus on objects at various distances

77. • Gbenga needs to get glasses to correct his farsightedness. His eyes currently cannot focus on objects that are within 2 ft (or 61 cm) of his eyes. This is in contrast to people with normal vision who can focus on objects as close as 25 cm in front of them. If the glasses that Gbenga will get will sit 1.6 cm in front of his eyes, what lens focal length and power would correct his vision? That is, what lens focal length and power would allow Gbenga to focus on objects that are 25 cm in front of his eyes? Will they be converging or diverging lenses? Example 24-6

78. • Suppose a given cell phone camera has a single lens and a light sensor that can move to change its distance from the lens. If the camera can focus only on an object 6.5 cm or farther away and the lens has a focal length of 4.3 mm, what are the maximum and minimum distances of the light sensor from the lens? Example 24-5

79. •• At a distance of 7.0 cm, a 51 mm by 89 mm business card fills the screen of a cell phone in camera mode. (a) If the focal length of the camera lens is 4.3 mm, what are the length and height of the camera's light sensor? (b) The light sensor is made up of 16×10^6 individual picture elements, or pixels (so this is a 16-megapixel sensor). What is the area of each pixel? If the pixels are square, what is the length of each side of a pixel? Example 24-5

80. • Andrea, who is nearsighted, wears glasses with lenses that have a power of –1.60 diopters. If the glasses Andrea wears sit 1.40 cm in front of her eyes, what is the farthest distance that objects can be in order for Andrea to see them clearly without her glasses? Example 24-6

General Problems

Note: In these problems "infinity" means a very large distance compared to the focal length of a lens.

81. •• The opposite walls of a barber shop are covered by plane mirrors, so that multiple images arise from multiple reflections, and you see many reflected images of yourself receding to infinity. The width of the shop is 6.50 m, and you are standing 2.00 m from the north wall. (a) How far apart are the first two images of you behind the north wall? (b) What is the separation of the first two images of you behind the south wall? Explain your answer.

82. • An object is 40.0 cm from a concave spherical mirror whose radius of curvature is 32.0 cm. Locate and describe the type and magnification of the image formed by the mirror (a) by calculating the image distance and lateral magnification and (b) by drawing a ray diagram. On the ray diagram draw an eye in a position from which it can view the image. Example 24-1

83. •• **Biology** A typical human eye is nearly spherical and usually about 2.5 cm in diameter. Suppose a person first looks at a coin that is 2.3 cm in diameter, located 30.0 cm from her eye, and then looks up at her friend who is 1.8 m tall and 3.25 m away. (a) Find the approximate size of each image (coin and friend) on her retina. (*Hint:* Just consider rays from the top and bottom of the object that pass through the center of the lens.) (b) Are the images in part (a) upright or inverted, and are they real or virtual?

84. •• **Biology** A typical human lens has an index of refraction of 1.41. The lens has a double convex shape, but its curvature can be varied by the ciliary muscles acting around its rim. At minimum power, the radius of the front of the lens is 10.0 cm, while that of the back is 6.00 mm. At maximum power the radii are 6.00 mm and 5.50 mm, respectively. (The numbers can vary somewhat.) If the lens were in air, (a) what would be the ranges of its focal length and its power (in diopters)? (b) At maximum power, where would the lens form an image of an object 25 cm in front of the front surface of the lens? (c) Would the image fall on the retina of a human eye? The retina is located approximately 2.5 cm from the lens. Example 24-3

85. •• **Biology** A typical person's eye is 2.5 cm in diameter and has a near point (the closest an object can be and still be seen in focus) of 25 cm, and a far point (the farthest an object can be and still be in focus) of infinity. (a) What is the range of the effective focal lengths of the focusing mechanism (lens plus cornea) of the typical eye? (b) Is the equivalent focusing mechanism of the eye a diverging or a converging lens? Justify your answer without using any mathematics, and then see if your answer is consistent with your result in part (a). Example 24-5

86. ••• A geneticist looks through a microscope to determine the phenotype of a fruit fly. The microscope is set to an overall magnification of 400× with an objective lens that has a focal length of 0.60 cm. The distance between the eyepiece and objective lenses is 16 cm. Find the focal length of the eyepiece lens assuming a near point of 25 cm (the closest an object can be to the eye and still be seen in focus). SSM

87. •• A thin lens made of glass that has a refractive index equal to 1.60 has surfaces with radii of curvature that have magnitudes equal to 12.0 and 18.0 mm. These surfaces could be either convex or concave. What are the possible values for its focal length? Sketch a cross-sectional view of the lens for each possible combination, making sure to label the radii of curvature of each surface of the lens and the associated focal length of the entire lens. Example 24-3

88. • You are designing lenses that consist of small double convex pieces of plastic having surfaces with radii of curvature of magnitudes 3.50 cm on one side and 4.25 cm on the other side. You want the lenses to have a focal length of 1.65 cm in air. What should be the index of refraction of the plastic to achieve the desired focal length? Example 24-3

89. ••• A lens of focal length +15.0 cm is 10.0 cm to the left of a second lens of focal length −15.0 cm. (a) Where is the final image of an object that is 30.0 cm to the left of the positive lens? (b) Is the image real or virtual? (c) How do the image's size and orientation compare to those of the original object? Explain your answer. (d) Where should your eye be located to see the image?

90. ••• A thin, diverging lens having a focal length of magnitude 45.0 cm has the same principal axis as a concave mirror with a radius of 60.0 cm. The center of the mirror is 20.0 cm from the lens, with the lens in front of the mirror. An object is placed 15.0 cm in front of the lens. (a) Where is the final image due to the lens–mirror combination? (b) Is the final image real or virtual? Upright or inverted? (c) Suppose now that the concave mirror is replaced by a convex mirror of the same radius. Repeat parts (a) and (b) for the new lens–mirror combination. SSM

91. ••• When you place a bright light source 36.0 cm to the left of a lens, you obtain an upright image 14.0 cm from the lens and also a faint inverted image 13.8 cm to the left of the lens that is due to reflection from the front surface of the lens. When the lens is turned around, a faint inverted image is 25.7 cm to the left of the lens. What is the index of refraction of the material? Example 24-4

92. ••• A thin, converging lens having a focal length of magnitude 25.00 cm is placed 1.000 m from a plane mirror that is oriented perpendicular to the principal axis of the lens. A flower, 8.400 cm tall, is 1.450 m from the mirror on the principal axis of the lens. (a) Where is the final image of the flower produced by the lens–mirror combination? Is it real or virtual? Upright or inverted? How tall is the image? (b) If the converging lens is replaced by a diverging lens having a focal length of the same magnitude as the original lens, what will be the answers to part (a)?

93. ••• **Biology** A compound microscope has a tube length of 20.0 cm and an objective lens of focal length 8.0 cm. (a) If it is to have a magnifying power of 200×, what should be the focal length of the eyepiece? (b) If the final image is viewed at infinity, how far from the objective should the object be placed?

94. •• **Biology** You may have noticed that the eyes of cats appear to glow green in low light. This effect is due to the reflection of light by the *tapetum lucidum*, a highly reflective membrane just behind the retina of the eye (see Figure 23-1a). Light that has passed through the retina without hitting photoreceptors is reflected back to the retina, thus enabling the animal to see much better than humans in low light. The eye of a typical cat is about 1.25 cm in diameter. Assume that the light enters the eye traveling parallel to the principal axis of the lens. (a) If some of the light reflected off the *tapetum lucidum* escapes being absorbed by the retina, where will it be focused? (b) The refractive index of the liquid in the eye is about 1.4. How does this affect the location of the image in part (a)? SSM

95. •• **Medical** A nearsighted eye is corrected by placing a diverging lens in front of the eye. The lens will create a virtual image of a distant object at the far point (the farthest an object can be and still be in focus) of the myopic viewer where it will be clearly seen. In the traditional treatment of myopia, an object at infinity is focused to the far point of the eye. If an individual has a far point of 70 cm, prescribe the correct power of the lens that is needed. Example 24-6

96. •• **Medical** A farsighted eye is corrected by placing a converging lens in front of the eye. The lens will create a virtual image that is located at the near point (the closest an object can be and still be in focus) of the viewer when the object is held at a comfortable distance (usually taken to be 25 cm). If a person has a near point of 75 cm, what power reading glasses should be prescribed to treat this hyperopia? Example 24-6

97. ••• **Medical** (a) Prove that when two thin lenses of focal lengths f_1 and f_2 are pressed next to one another, the effective focal length $f_{combined}$ of the two lenses acting together is given by

$$\frac{1}{f_{combined}} = \frac{1}{f_1} + \frac{1}{f_2}$$

(b) Describe how this relates to a prescription for a contact lens which is placed directly on the eye? (Assume there is no significant separation between the contact lens and the lens of the eye.) (c) Why would an eyeglass prescription that is identical to a contact lens prescription give a very subtle difference in the image seen? Example 24-6

98. •• **Medical** Without glasses, a certain person needs to have his eyes 15.0 cm from a book to read comfortably and can focus clearly only on distant objects up to 2.75 m away, but no farther. A typical normal eye should be able to focus on objects that are between 25.0 cm (the near point) and infinity (the far point) from the eye. (a) What type of correcting lenses does the person need: single focal length or bifocals? Why? (b) What should an optometrist specify as the focal length(s) of the correcting contact lens or lenses? (c) What is the power (in diopters) of the correcting lens or lenses? SSM Example 24-6

99. ••• **Medical** One of the inevitable consequences of aging is a decrease in the flexibility of the lens. This leads to the farsighted condition called *presbyopia* (elder eye). Almost every aging human will experience it to some extent. However, for the myopic person, at some point, it is possible that far vision will be limited by a subpar far point *and* near vision will be hampered by an expanding near point. One solution is to wear bifocal lenses that are diverging in the upper half to correct the nearsightedness and converging in the lower half to correct the farsightedness.

Suppose one such individual asks for your help. The patient complains that she can't see far enough to safely drive (her far point is 112 cm) and she can't read the font of her smart phone without holding it beyond arm's length (her near point is 83 cm). Prescribe the bifocals that will correct the visual issues for your patient. Example 24-6

100. •• A common zoom lens for a digital camera covers a focal length range of 18 mm to 200 mm. For the purposes of this problem, treat the lens as a thin lens. If the lens is zoomed out to 200 mm and is focused on a petroglyph that is 15.0 m away and 38 cm wide, (a) how far is the lens from the light sensor of the camera and (b) how wide is the image of the petroglyph on the sensors? (c) If the closest that the lens can get to the sensor at its 18 mm focal length is 5.2 cm, what is the closest object it can focus on at that focal length? SSM

101. •• A macro lens is designed to take very close-range photographs of small objects such as insects and flowers. At its closest focusing distance, a certain macro lens has a focal length of 35.0 mm and forms an image on the light sensor of the camera that is 1.09 times the size of the object. (a) How close must the object be to the lens to achieve this maximum image size? (b) What is the magnification if the object is twice as far from the lens as in part (a)? For this problem, treat the lens as a thin lens.

102. • **Astronomy** A refracting astronomical telescope, or refractor, consists of an eyepiece lens at one end of a cylindrical tube and an objective lens at the other end. The objective lens gathers light from a distant object (such as a planet) and focuses it at the focal point of the eyepiece lens. The eyepiece basically acts as a magnifying lens to create a virtual image of the objective's image. The overall magnification, M, is found to be $M = -f_o/f_e$, where f_o is the focal length of the objective and f_e is the focal length of the eyepiece. (a) Calculate the magnification of the 36-in refractor at the University of California's Lick Observatory on Mount Hamilton near San Jose, California. The focal length of the objective lens is 17.37 m, and the focal length of the eyepiece is 22 mm. (b) What is the significance of the negative sign in the magnification equation?

103. • A certain lens element inside a camera is a converging lens that has symmetric convex sides ($R_1 = 4.300$ cm and $R_2 = -4.300$ cm) and can be modeled as a thin lens. The lens's index of refraction for blue light is $n_{blue} = 1.588$ and for red light is $n_{red} = 1.582$. Calculate the magnitude of the separation distance of the lens's focal lengths for blue and red light, $|f_{blue} - f_{red}|$. Example 24-3

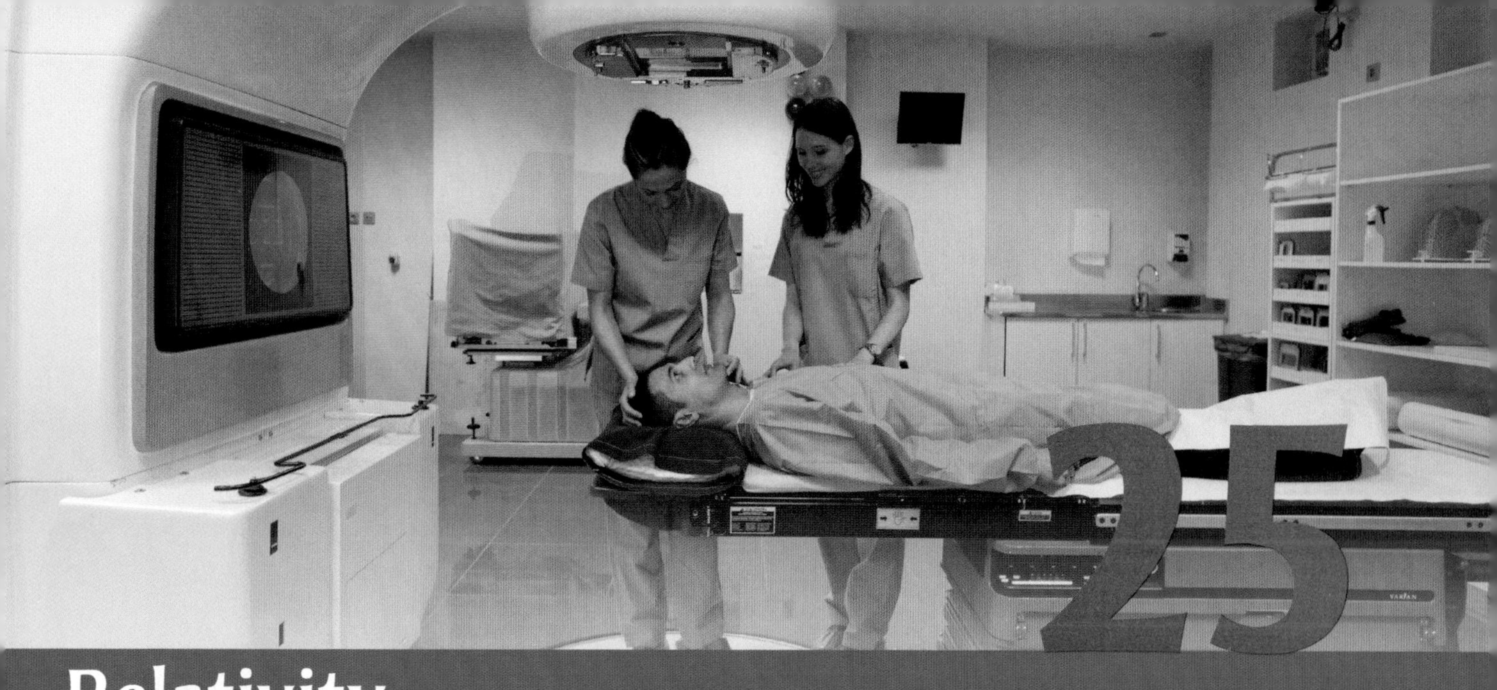

Relativity

age fotostock/Alamy Stock Photo

What do you think?

In cancer radiotherapy, a beam of electrons is accelerated to nearly the speed of light. The kinetic energy of the electrons is used to create intense x ray beams that can be accurately targeted on the location of the cancerous tissue. Which requires more energy: (a) accelerating an electron from rest to 90% of the speed of light, or (b) further accelerating that electron from 90% to 99% of the speed of light?

In this chapter, your goals are to:

- (25-1) Identify some circumstances under which the physics you know breaks down.
- (25-2) Describe how different observers view the same motion in Newtonian physics.
- (25-3) Explain how the Michelson–Morley experiment helped rule out the ether model of the propagation of light.
- (25-4) Describe how the time interval between two events can have different values in different frames of reference.
- (25-5) Calculate how the dimensions of an object change when it is in motion.
- (25-6) Explain why the speed of light in a vacuum is an ultimate speed limit.
- (25-7) Calculate the rest energy of an object with mass.
- (25-8) Explain what the principle of equivalence tells us about the nature of gravity.

To master this chapter, you should review:

- (2-6) The motion of objects in free fall.
- (3-6) Projectile motion.
- (4-2, 4-3) Newton's first and second laws.
- (6-3) Kinetic energy and the work-energy theorem.
- (7-4) Elastic collisions.
- (7-6) The relationship between external forces and momentum change.
- (22-3) The electromagnetic nature of light.

25-1 The concepts of relativity may seem exotic, but they're part of everyday life

Most of our everyday experiences involve objects that move at speeds that are slow compared to the speed of light in a vacuum. In the late nineteenth century, scientists began to realize that the laws of physics that we've developed so far in this book don't properly describe light or objects moving at speeds near the speed of light. Albert Einstein first understood the way to extend physics into this regime of extremely high speed.

Figure 25-1 Relativity in your world (a) A GPS-equipped mobile phone and (b) the light from the Sun illustrate applications of the theory of relativity.

(a) In order for a GPS receiver to accurately determine its position, it must use relativity to adjust for time flowing at a different rate aboard a GPS satellite than on Earth.

iStockphoto/Thinkstock

(b) Sunlight is a result of reactions deep inside the Sun that convert mass into energy—a process predicted by the special theory of relativity.

Comstock/Thinkstock

TAKE-HOME MESSAGE FOR Section 25-1

✔ The physics we have learned so far must be modified for objects moving at speeds comparable to the speed of light.

✔ These modifications lead us to new ideas about space, time, and energy.

In this chapter we'll focus on Einstein's *special theory of relativity*. We'll see how discoveries concerning the nature of light helped motivate the central ideas of this theory. We'll also see how a simple postulate—that the speed of light does not depend on the motion of either the emitter or the observer of the light—leads to a radical transformation of our understanding of space and time. We'll discover that the speed of light is an ultimate speed limit, and it's impossible to accelerate an object beyond that speed. We'll also see that mass is simply another form of energy. We'll conclude with a look at Einstein's *general theory of relativity*, which provides new insights into the nature of gravity.

Although the effects of the special theory of relativity are most pronounced for objects moving at very high speeds, they can be seen in the world around you. Your mobile phone probably has the ability to determine its location using the Global Positioning System, or GPS (**Figure 25-1a**). A GPS receiver detects signals from a collection of satellites in Earth's orbit and calculates its position by timing those signals. However, the satellites move at about 28,000 km/h (18,000 mi/h) relative to Earth, and special relativity tells us that a moving clock or timekeeper runs at a different rate than a stationary one (see Section 25-4). Your mobile phone has to be able to correct for this in order to give you accurate positioning information. Ordinary sunlight is also a consequence of relativity: The Sun shines by converting a fraction of its mass into electromagnetic energy, a direct application of the idea that objects have energy simply as a consequence of having mass (**Figure 25-1b**).

Eitan Simanor / Alamy

Figure 25-2 Relative motion This is the view of one train as seen from the window of another. Which train is moving?

25-2 Newton's mechanics includes some ideas of relativity

You're on a train, looking out the window at a second train right next to yours. One of the trains is moving (**Figure 25-2**). Is your train moving and the other one stationary, or vice versa? Perhaps both are moving. How can you tell?

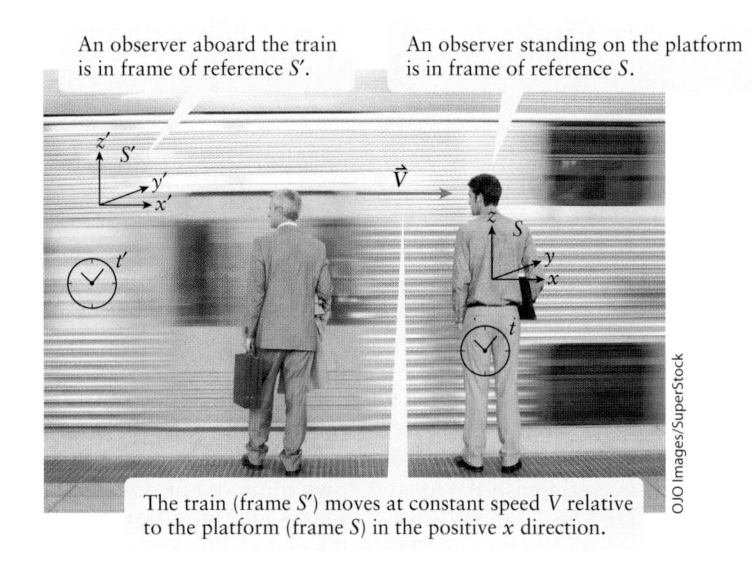

Figure 25-3 Two observers Observers on the platform in a train station use coordinates x, y, and z and measure time t. An observer aboard the train uses coordinates x', y', and z' and measures t'.

To address this question let's return to Newton's first law, which we introduced in Section 4-3:

If the net external force on an object is zero... ...the object does not accelerate...

$$\text{If } \sum \vec{F}_{\text{ext}} = 0, \text{ then } \vec{a} = 0 \text{ and } \vec{v} = \text{constant}$$

...and the velocity of the object remains constant. If the object is at rest, it remains at rest; if it is in motion, it continues in motion in a straight line at a constant speed.

Newton's first law of motion (4-6)

The first law states that an object that experiences no net force could *either* be at rest *or* in motion at a constant velocity. So although we might make a distinction between an object at rest and one in (uniform) motion with respect to us, the laws of physics do not. In the case of the two trains, this tells us something profound: If you have no reference to the ground or the tracks on which the trains move, there is *no* experiment or measurement that would tell you whether the other train is moving at a constant velocity with respect to yours or your train is moving a constant velocity with respect to the other one.

A useful way to think about this idea is to introduce the concept of a **frame of reference** (also called a *reference frame* or simply *frame*). This is a coordinate system with respect to which we can make observations or measurements. **Figure 25-3** shows two different frames of reference. A person standing on the train platform is in frame S and measures the positions of objects using the coordinates x, y, and z. This person has a watch, and the time of a certain event as measured on this watch is t. A person riding on the train is in frame S' and measures the positions of objects using the coordinates x', y', and z'. The time of an event as measured by this person is t'. If neither frame of reference is accelerating, the message of Newton's first law is that *both* frames are equally good for making measurements.

As an example, let's consider a child riding on the train (**Figure 25-4**). The train is on a straight track and is moving relative to the platform at a constant speed V. As measured by a person on the platform (frame of reference S), the child is moving at a constant velocity with zero acceleration, so the net force on the child must be zero: The upward normal force exerted by the seat exactly balances the downward gravitational force that Earth exerts on the child. As measured by another passenger seated on board the train (frame of reference S'), the child is at rest (the child isn't moving relative to that passenger). So as measured in S', the net force on the child must again be zero. The two observers—one in frame of reference S and one in frame of reference S'—disagree about *how* the child moves. But they agree that the child's motion is in accordance with Newton's first law, Equation 4-6.

(a)
As measured in S

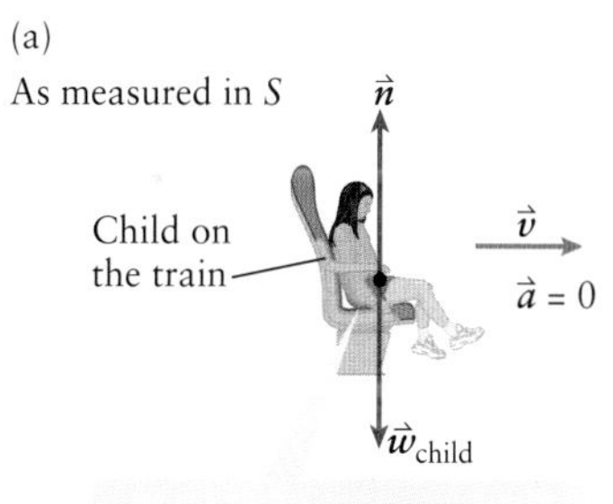

net force on the child = 0
acceleration of the child = 0

(b)
As measured in S'

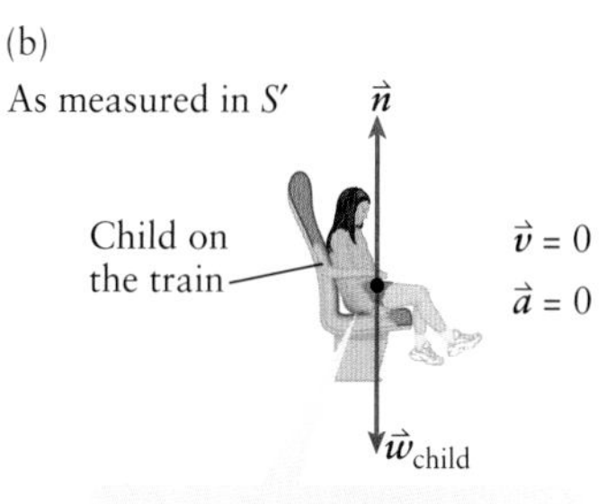

net force on the child = 0
acceleration of the child = 0

Figure 25-4 Newton's first law in two frames of reference The free-body diagram for a child on the train as measured in the two frames of reference shown in Figure 25-3.

It's not just Newton's first law that applies in both frames of reference depicted in Figure 25-3. Newton's *second* law also applies in both frames:

Newton's second law of motion (4-2)

If a **net external force** acts on an object...

...the object accelerates. The acceleration is in the same direction as the net force.

$$\sum \vec{F}_{ext} = m\vec{a}$$

The magnitude of acceleration that the net external force causes depends on the mass m of the object (the quantity of material in the object). The greater the mass, the smaller the acceleration.

(a)
As measured in S'

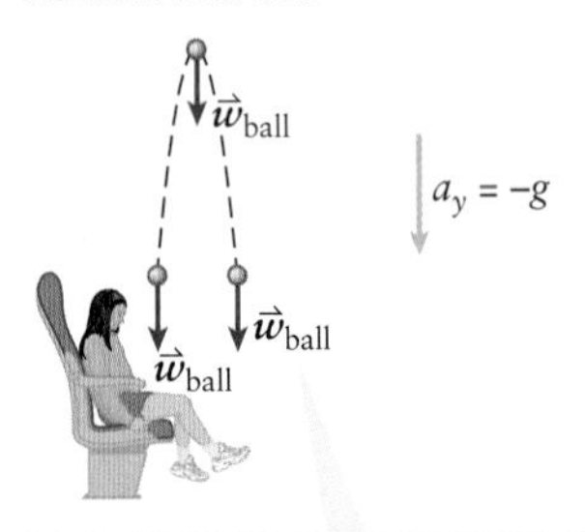

net force on ball = gravitational force
Ball experiences free fall.

(b)
As measured in S

net force on ball = gravitational force
Ball experiences projectile motion.

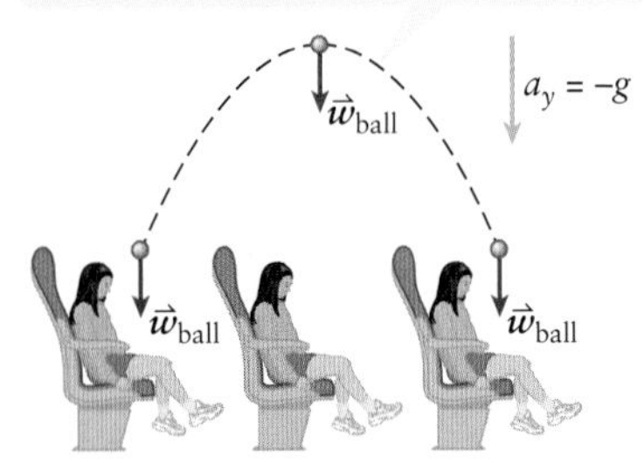

Figure 25-5 A tossed ball in two frames of reference The child on the train from Figure 25-4 tosses a ball straight up and down relative to her.

As an illustration, suppose the child riding on the train is tossing a ball up and down (**Figure 25-5**). A passenger seated in the train (frame of reference S') sees the motion of the ball as purely vertical (we draw the ball's motion with a slight horizontal displacement to distinguish the up and down motion, but this motion is really only along a vertical line). By contrast, a person on the platform (frame of reference S) sees the ball following a parabolic path: The ball has the same horizontal component of velocity V as the train, and it maintains that horizontal velocity during its flight. As for the child in Figure 25-4, the observers in the two frames of reference disagree about how the ball moves. Both observers, however, agree that the ball obeys Newton's second law: When the ball is in flight, only the gravitational force acts on it, so the acceleration is downward and has magnitude g. In frame S' the straight up-and-down motion is free fall, as we described in Section 2-6; in frame S the ball is in projectile motion, as we described in Section 3-6. Each description is correct for the frame of reference in which the motion is observed.

We refer to a frame of reference attached to an object that does not accelerate as an **inertial frame.** An inertial frame of reference is one in which Newton's first law is valid: If the net force on an object is zero, it either remains at rest or moves with a constant velocity relative to an observer in that frame of reference. If one frame of reference S is inertial, a second frame of reference S' is also inertial if it moves at a constant velocity relative to S. That's the case for the two frames of reference depicted in Figure 25-3.

By contrast, a frame of reference attached to an accelerated object is a **noninertial frame.** To an observer in a noninertial frame, Newton's first law does *not* hold true. An example is a frame of reference attached to a car that is accelerating forward. A ball sitting on the floor of this car has zero net force on it (the upward normal force exerted by the floor balances the downward gravitational force), yet the ball accelerates toward the back of the car. A rotating frame of reference, such as a carnival merry-go-round, is also noninertial because an object that follows a circular path is accelerating. Just like a ball in a car that accelerates forward, a ball placed on the merry-go-round floor has zero net force acting on it, yet this ball will tend to roll to the outside of the merry-go-round. In the frame of reference of a person riding on the merry-go-round, Newton's first law does not hold true.

Strictly speaking, an observer at rest on Earth's surface, such as the person in frame S standing on the platform in Figure 25-3, is in a noninertial frame. That's because Earth rotates on its axis like a merry-go-round and also moves along a roughly circular orbit around the Sun. However, the accelerations involved with those motions are so small (each is a small fraction of g) that for many purposes we can ignore them. As a result, we can safely regard the frame of reference S in Figure 25-3 as an effectively inertial one, and likewise for the frame S' attached to the train.

We've seen that Newton's first and second laws of motion work equally well in both inertial frame S and inertial frame S'. The same should be true for *any* inertial frame of reference. This statement is called the **principle of Newtonian relativity:**

The laws of motion are the same in all inertial frames of reference.

The word "relativity" means that measurements made relative to one inertial frame of reference are just as valid as those made relative to another inertial frame. Since the laws of motion don't distinguish between two inertial frames S and S', it's meaningless to ask which frame is "really" moving and which frame is "really" at rest.

(You may think that frame S is the one that's really at rest because it's stationary with respect to the platform. But remember that the platform is on Earth and that our entire planet is in motion through the solar system.) So another way to express the principle of Newtonian relativity is:

There is no way to detect absolute motion. Only motion relative to a selected frame of reference can be detected.

Let's apply the principle of Newtonian relativity to the ball shown in Figure 25-5. Suppose you are standing on the platform as the train goes by, so you are in reference frame S. You see the ball following a parabolic path, and you make observations of the x, y, and z coordinates of the ball as functions of time t. The child riding in the train is in reference frame S' and sees the ball moving straight up and down relative to her. As the ball moves she measures the x', y', and z' coordinates of the ball as functions of time t'. The two sets of coordinates have the same orientation, as **Figure 25-6** shows. To calibrate the two clocks—one in frame S, the other in frame S'—you and the child both set your clocks to read zero at the instant that you pass each other, when the origins of your frames of reference coincide. So at this instant $t = t' = 0$.

Suppose you and the child both measure the same **event**—that is, the ball being at a certain point in its motion, such as the high point in its path. In your frame S the coordinates of this event in space and time are x, y, z, and t, and the coordinates of this same event as measured in frame S' are x', y', z', and t'. (Note that we are thinking of time as a fourth coordinate of an event.) As Figure 25-6 shows, a simple set of equations relates the coordinates of the same event in the two frames:

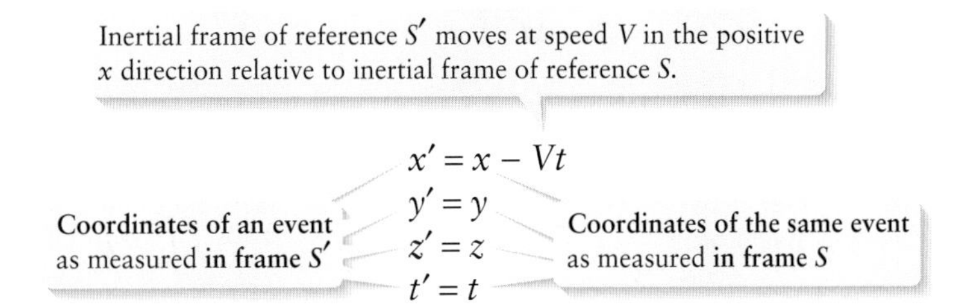
Inertial frame of reference S' moves at speed V in the positive x direction relative to inertial frame of reference S.
Coordinates of an event as measured in frame S'
Coordinates of the same event as measured in frame S

$$x' = x - Vt$$
$$y' = y$$
$$z' = z$$
$$t' = t$$

Galilean coordinate transformation (25-1)

This set of equations is known as the **Galilean transformation**. Note that the relative motion of the two frames of reference along the positive x axis does not affect the y and z coordinates, which are the same in both reference frames.

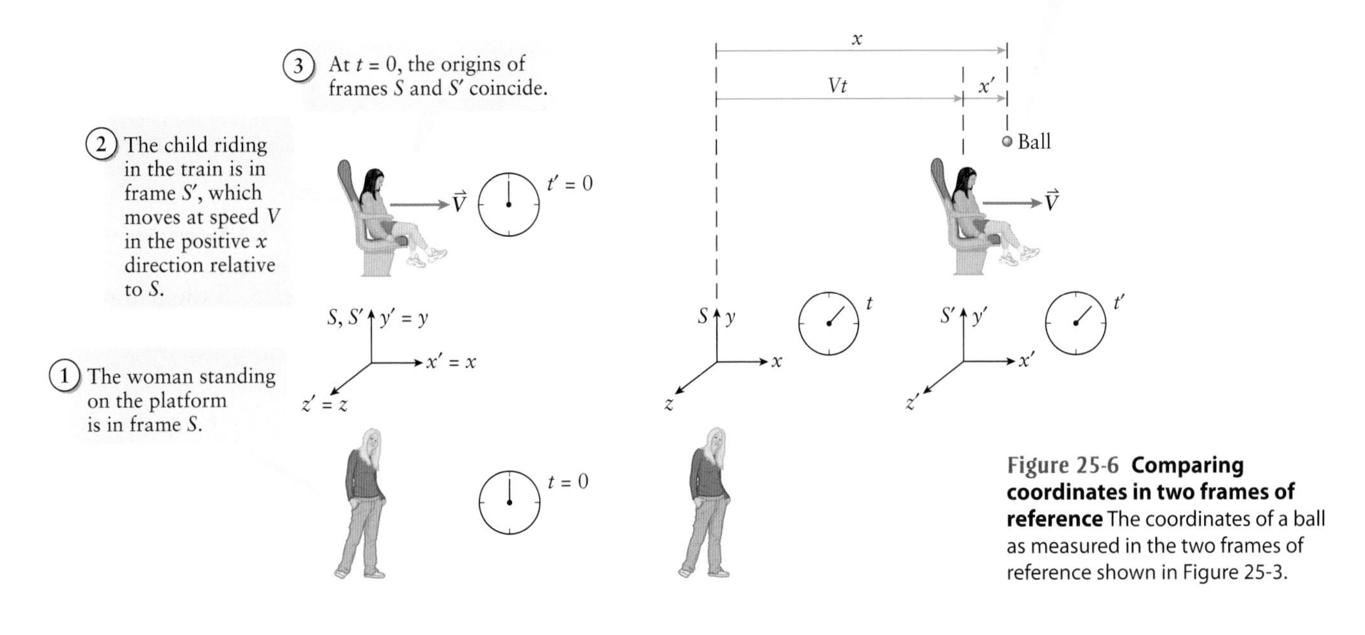
(1) The woman standing on the platform is in frame S.
(2) The child riding in the train is in frame S′, which moves at speed V in the positive x direction relative to S.
(3) At t = 0, the origins of frames S and S′ coincide.
(4) The child tosses a ball. At time t, the coordinate of the ball in S is x, and the coordinate of the ball in S′ is x′ = x − Vt.
t′ = 0
t = 0
S, S′ y′ = y
x′ = x
z′ = z
x
Vt
x′
Ball
S y
x
z
t
S′ y′
x′
z′
t′

Figure 25-6 Comparing coordinates in two frames of reference The coordinates of a ball as measured in the two frames of reference shown in Figure 25-3.

We can use the Galilean transformation to compare the velocity of the ball as measured in frame S' to its velocity as measured in frame S. To do this we'll look at *two* events in the motion of the ball separated by a short time interval from t_1 to t_2 as measured in frame S. During this time interval the coordinates of the ball as measured in frame S change from x_1, y_1, and z_1 to x_2, y_2, and z_2. The components of the ball's velocity $\vec{v}$ in frame S are

$$v_x = \frac{\Delta x}{\Delta t} = \frac{x_2 - x_1}{t_2 - t_1}$$

(25-2) $$v_y = \frac{\Delta y}{\Delta t} = \frac{y_2 - y_1}{t_2 - t_1}$$

$$v_z = \frac{\Delta z}{\Delta t} = \frac{z_2 - z_1}{t_2 - t_1}$$

For the same two events as measured in frame S', the time interval is from t'_1 to t'_2, and the coordinates change from x'_1, y'_1, and z'_1 to x'_2, y'_2, and z'_2. So in frame S' the components of the object's velocity $\vec{v}'$ are

$$v'_x = \frac{\Delta x'}{\Delta t'} = \frac{x'_2 - x'_1}{t'_2 - t'_1}$$

(25-3) $$v'_y = \frac{\Delta y'}{\Delta t'} = \frac{y'_2 - y'_1}{t'_2 - t'_1}$$

$$v'_z = \frac{\Delta z'}{\Delta t'} = \frac{z'_2 - z'_1}{t'_2 - t'_1}$$

Now substitute the expressions for x', y', z', and t' from Equations 25-1 into Equations 25-3. We get

$$v'_x = \frac{(x_2 - Vt_2) - (x_1 - Vt_1)}{t_2 - t_1} = \frac{(x_2 - x_1) - V(t_2 - t_1)}{t_2 - t_1} = \frac{(x_2 - x_1)}{t_2 - t_1} - V$$

(25-4) $$v'_y = \frac{y_2 - y_1}{t_2 - t_1}$$

$$v'_z = \frac{z_2 - z_1}{t_2 - t_1}$$

If we now compare Equations 25-4 for the velocity components in frame S' to Equations 25-2 for the velocity components in frame S, we see that

Galilean velocity transformation (25-5)

Inertial frame of reference S' moves at speed V in the positive x direction relative to inertial frame of reference S.

$$v_x' = v_x - V$$
$$v_y' = v_y$$
$$v_z' = v_z$$

Velocity components of an object as measured in frame S'

Velocity components of the same object as measured in frame S

Equations 25-5, which relate the velocity of an object in frame S' to the velocity of the same object in frame S, are called the **Galilean velocity transformation.**

As an example, think again of the ball shown in Figure 25-5. If the ball moves straight up and down as measured in frame S', then in that frame the ball is in free fall with only a y component of velocity. The other two components are zero: $v'_x = v'_z = 0$. As measured in frame S, the velocity of the ball has components

$$v_x = v'_x + V = V$$
$$v_y = v'_y$$
$$v_z = v'_z = 0$$

As measured in frame S, the ball moves up and down along the y direction with the same velocity as measured in frame S': $v_y = v'_y$. At the same time, as measured in frame S, the ball maintains a constant velocity V in the x direction. That's just the behavior we expect for a projectile: Its motion is a combination of up-and-down free fall and constant-velocity horizontal motion (see Section 3-6).

WATCH OUT! There is nothing special about either the S frame or the S' frame.

In Figures 25-5 and 25-6 we've chosen to think of frame S as at rest and frame S' as moving. This selection is purely our choice because the principle of Newtonian relativity says that *all* inertial frames are equivalent. We could just as well say that frame S' is stationary and frame S is moving with speed V in the negative x direction.

You may wonder why we've spent so much time and effort explaining motion as seen from two different inertial frames of reference. As we'll discover in the next few sections, the reason is that something remarkable happens when the relative speed V of the two frames is comparable to c, the speed of light in a vacuum. In that case we'll find that the Galilean transformations do *not* hold true: As measured in the two different frames, the time interval between events can be different and objects can have different dimensions. These remarkable observations will radically transform our notions of the nature of time and space themselves.

EXAMPLE 25-1 Two Cars

You observe two race cars approaching you. A red car is in one lane moving at 24 m/s relative to you. In a second lane a blue car is moving at 36 m/s relative to you. (a) What is the velocity of the red car as measured by the driver of the blue car? (b) What is the velocity of the blue car as measured by the driver of the red car?

Set Up

We'll use Equations 25-5 to transform the velocity of a car as measured in one frame of reference to the velocity of the same car as measured in a different frame of reference. Note that these equations assume that frame S' is moving relative to frame S at speed V in the positive x direction. We'll use this to decide which frame of reference corresponds to S and which to S'.

Galilean velocity transformation:

$$v'_x = v_x - V$$
$$v'_y = v_y$$
$$v'_z = v_z \qquad (25\text{-}5)$$

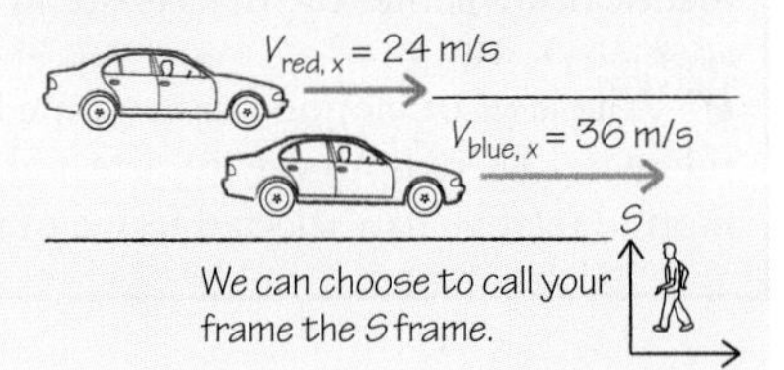
The given speeds of both cars are measured relative to you, that is, in your frame.
$V_{\text{red},x} = 24$ m/s
$V_{\text{blue},x} = 36$ m/s
S
We can choose to call your frame the S frame.

Solve

(a) Let's take the positive x direction to be in the direction both cars are moving relative to you. Then we'll take S to be your frame of reference and S' to be the frame of reference of the driver of the blue car. The relative speed of these two frames is $V = 36$ m/s (the speed of the blue car relative to you). All of the motions in this example are along the x axis, so we don't need the y or z members of Equations 25-5.

Use the x equation from Equations 25-5 to relate the velocity of the red car as measured by you ($v_{\text{red},x} = +24$ m/s) to its velocity as measured by the driver of the blue car ($v'_{\text{red},x}$):

$$v'_{\text{red},x} = v_{\text{red},x} - V$$
$$= +24 \text{ m/s} - 36 \text{ m/s}$$
$$= -12 \text{ m/s}$$

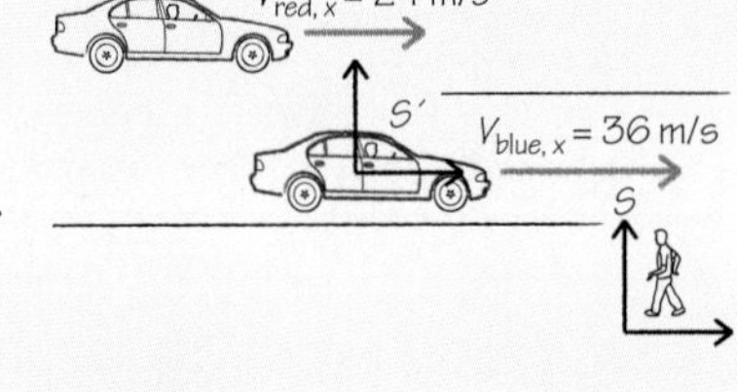
$V_{\text{red},x} = 24$ m/s
S'
$V_{\text{blue},x} = 36$ m/s
S

As measured by the driver of the blue car, the red car is moving in the negative x direction—that is, backward—at 12 m/s.

(b) Again we take S to be your frame of reference, but now S' is the frame of reference of the driver of the red car. The relative speed of these two frames is $V = 24$ m/s (the speed of the red car relative to you).

Use the x equation from Equations 25-5 to relate the velocity of the blue car as measured by you ($v_{\text{blue},x} = +36$ m/s) to its velocity as measured by the driver of the red car ($v'_{\text{blue},x}$):

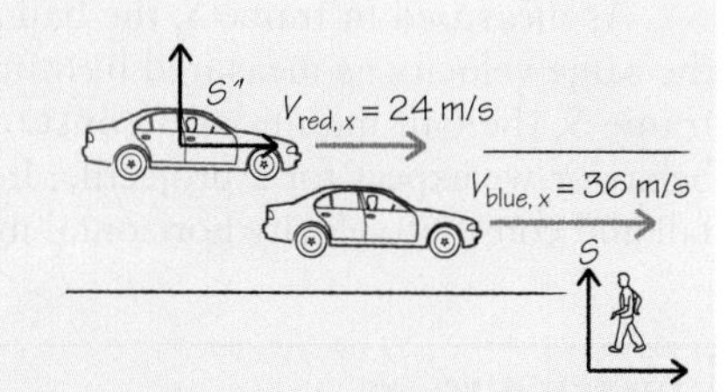

$$\begin{aligned} v'_{\text{blue},x} &= v_{\text{blue},x} - V \\ &= +36 \text{ m/s} - 24 \text{ m/s} \\ &= +12 \text{ m/s} \end{aligned}$$

As measured by the driver of the red car, the blue car is moving in the positive x direction—that is, forward—at 12 m/s.

Reflect

The driver of the blue car sees the red car falling farther and farther behind her at a rate of 12 m/s, while the driver of the red car sees the blue car moving farther and farther in front of him at a rate of 12 m/s. Note that the *magnitude* of the two answers is the same: Both drivers agree that the other driver is moving at a relative speed of 12 m/s.

GOT THE CONCEPT? 25-1 Groundspeed versus Airspeed

A typical jet airliner has a cruise airspeed—that is, its speed relative to the air through which it is flying—of 900 km/h. If the wind at the airliner's cruise altitude is blowing at 100 km/h from west to east, what is the speed of the airliner relative to the ground if the airplane is flying from west to east? From east to west? (a) 800 km/h west to east, 1000 km/h east to west; (b) 1000 km/h west to east, 800 km/h east to west; (c) 800 km/h in both directions; (d) 900 km/h in both directions; (e) 1000 km/h in both directions.

TAKE-HOME MESSAGE FOR Section 25-2

✔ Newton's laws of motion treat all nonaccelerating objects identically, whether the objects are in motion or at rest.

✔ The laws of motion are the same in all inertial frames of reference. There is no way to detect absolute motion; only motion relative to a selected frame of reference can be detected.

✔ The Galilean transformations allow you to convert the coordinates and velocity of an object observed in one inertial frame of reference to the coordinates and velocity of the same object observed in a different inertial frame.

25-3 The Michelson–Morley experiment shows that light does not obey Newtonian relativity

We saw in the preceding section that Newton's laws of motion are the same in all inertial frames. Is the same true for the other laws of physics? Physicists asked this very question during the second half of the nineteenth century, specifically about the laws of electromagnetism.

We learned in Section 22-3 that the laws of electromagnetism explain how electromagnetic waves, including visible light, are possible. These laws also predict that the speed of electromagnetic waves in a vacuum is

Speed of light in a vacuum (22-11)

Speed of light in a vacuum

$$c = \frac{1}{\sqrt{\mu_0 \varepsilon_0}} = 3.00 \times 10^8 \text{ m/s}$$

Permeability of free space — Permittivity of free space

Our discussion in Section 25-2 tells us that we can measure relative motion but not absolute motion. So we are forced to ask this question: Relative to what inertial frame of reference is the speed of light in a vacuum equal to c?

In the nineteenth century the most common answer to this question was to imagine a substance that fills all space. This substance, which was called the *luminiferous ether*, was thought to be the medium for electromagnetic waves, just as air is the medium for sound waves in our atmosphere. This substance must be of extraordinarily low density so that its presence is almost undetectable. (The word "luminiferous" comes from the Latin for "light-bearing." "Ether" in this phrase has nothing to do with the organic compounds of the same name.) In this model c is the speed of electromagnetic waves relative to the frame in which the luminiferous ether is at rest.

How can we test whether this model is correct? To see the answer, note that the speed of sound waves in dry air is 343 m/s, but you will measure a different speed if a wind is blowing (that is, the air is moving relative to you). If a sound wave is traveling from west to east and a wind is blowing past you at 10 m/s from west to east, the sound will move relative to you at 343 m/s + 10 m/s = 353 m/s. If the sound wave is traveling from west to east and a 10-m/s wind is blowing from east to west, the speed of the sound wave relative to you will be 343 m/s − 10 m/s = 333 m/s. The same should be true for light waves if the luminiferous ether is moving past you so that there is an "ether wind." If a light wave is traveling from west to east and the ether is moving from west to east relative to you at 10 m/s, you would measure the speed of the wave to be c + 10 m/s; if the ether is instead moving from east to west relative to you at 10 m/s, you would measure the speed of the wave to be c − 10 m/s. If we can detect these small changes in the speed of light, that would be evidence that the luminiferous ether really exists.

Nineteenth-century scientists looked to Earth's motion around the Sun as a source of "ether wind." Our planet moves around its orbit at an average speed of 29.8 km/s = 2.98×10^4 m/s, or about $10^{-4}\ c$. If the luminiferous ether is at rest relative to the solar system as a whole, we should experience an "ether wind" that blows past our moving planet at $10^{-4}\ c$. Depending on the direction of that "ether wind" relative to the direction of light propagation, we would expect the speed of light to vary between $(1 + 10^{-4})c$ and $(1 - 10^{-4})c$. The challenge is to design an experiment that can detect such small changes in the speed of light.

In 1887 the American scientists Albert Michelson and Edward Morley carried out the first definitive experiment of this kind. **Figure 25-7** shows a simplified version of the

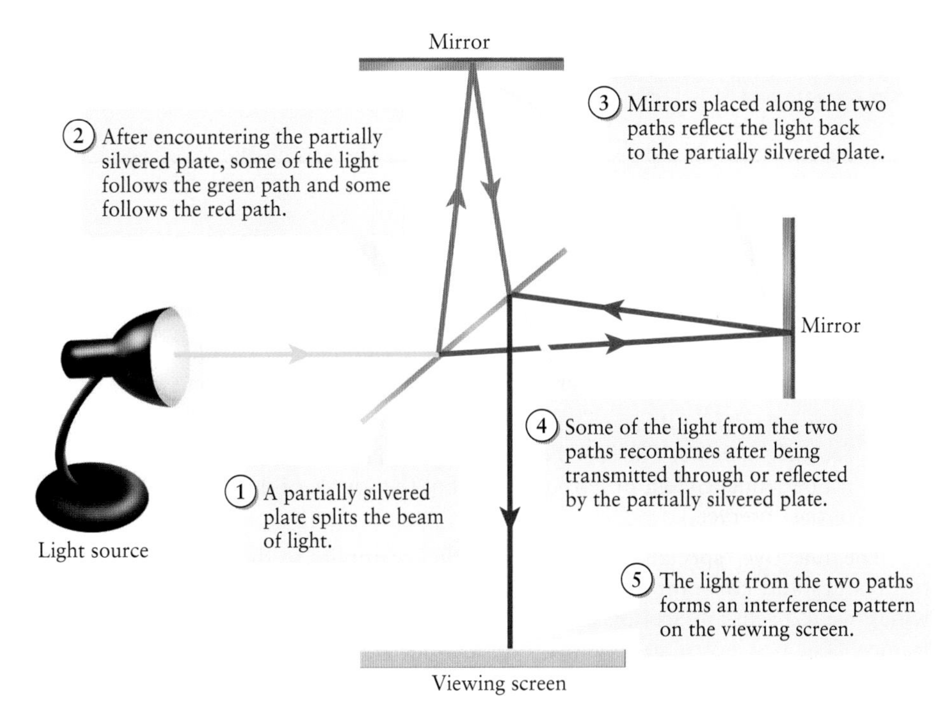
Mirror
(2) After encountering the partially silvered plate, some of the light follows the green path and some follows the red path.
(3) Mirrors placed along the two paths reflect the light back to the partially silvered plate.
Mirror
(4) Some of the light from the two paths recombines after being transmitted through or reflected by the partially silvered plate.
(1) A partially silvered plate splits the beam of light.
Light source
(5) The light from the two paths forms an interference pattern on the viewing screen.
Viewing screen

Figure 25-7 **The Michelson–Morley experiment simplified** If the luminiferous ether exists and is moving relative to this apparatus, its presence will be apparent in the interference pattern on the viewing screen.

Michelson–Morley experiment. Their apparatus split a beam of light into two, sent the two beams along perpendicular paths, and then allowed them to recombine at a viewing screen. What is seen on the viewing screen is an interference pattern between the waves in the two beams. The nature of this pattern depends on the difference in length between the two paths and also on whether the speeds at which light travels along each path are the same or different. If there is an "ether wind" that is more nearly aligned with one leg of the interferometer than the other, the speed of light should indeed be different along the two legs. Their apparatus was sensitive enough that Michelson and Morley should have been able to measure the effects of an "ether wind" due to Earth's motion around the Sun.

Michelson and Morley found *no* effect due to an "ether wind." They and other scientists refined and repeated the experiment many times, but the results were always the same: There is no evidence for the existence of the luminiferous ether. The conclusion from this and other experiments is that electromagnetic waves do not require the presence of a material medium but can propagate in a complete vacuum.

Under Newtonian relativity, to explain a constant value of the speed of light in a vacuum required that the equations of electromagnetism hold true only in a specific reference frame, one that is at rest relative to the medium for electromagnetic waves. But if there is no luminiferous ether, there is no such medium and hence no such special frame of reference. So physicists had no choice but to conclude that Newtonian relativity does *not* apply to electromagnetic waves. It was left to Albert Einstein to modify the ideas of Newtonian relativity and find a new way to look at the relationships between measurements made in different inertial frames of reference. We'll explore Einstein's simple yet radical ideas in the following section.

BIO-Medical **GOT THE CONCEPT? 25-2 Speed of Sound in Water**

A bottlenose dolphin (*Tursiops truncates*) can swim at speeds up to about 10 m/s. These dolphins produce sounds used for social communication and for echolocation (using sound to detect objects around them while swimming in dark or murky waters). If a swimming dolphin travels at top speed and produces a sound wave that propagates forward, how fast does that wave travel relative to the dolphin? The speed of sound in water is 1500 m/s. (a) 10 m/s; (b) 1490 m/s; (c) 1500 m/s; (d) 1510/m/s; (e) answer depends on the frequency of the sound wave.

TAKE-HOME MESSAGE FOR Section 25-3

✔ Experiment shows that electromagnetic waves do not require a material medium; they can easily propagate in a perfect vacuum.

✔ As a result, we conclude that electromagnetic waves do not obey Newtonian relativity.

25-4 Einstein's relativity predicts that the time between events depends on the observer

In 1905 German-born theoretical physicist Albert Einstein published his **special theory of relativity**, or *special relativity* for short. Einstein based his theory on two postulates:

- *First postulate:* All laws of physics are the same in all inertial frames.
- *Second postulate:* The speed of light in a vacuum is the same in all inertial frames, independent of both the speed of the source of the light and the speed of the observer.

The adjective "special" means that the theory applies to the special case of inertial frames and constant-velocity motion. (Einstein's *general* theory of relativity, which we'll discuss in Section 25-8, extends these ideas to accelerating, noninertial frames.) The first postulate extends the principle of relativity beyond Newton's laws of motion to include all aspects of physics, including electromagnetic waves. Einstein's second postulate derives from our conclusion in Section 25-3 that there is no special frame of reference in which the speed of light in a vacuum is equal to $c = 3.00 \times 10^8$ m/s.

Perhaps the most astounding consequence of the second postulate of special relativity is that the time interval between two events is *not* an absolute. Rather, this time interval depends on the motion of the frame of reference from which time is measured. To see how this comes about, we'll do a thought experiment.

Imagine a *light clock*, a special clock that uses light to measure intervals of time (**Figure 25-8**). A laser fires an extremely brief burst, or pulse, of light straight downward. This is reflected by a mirror and arrives at a light-sensitive detector next to the laser. For simplicity we can treat the laser and detector as being at the same position. One tick of the clock is the time interval between the pulse leaving the laser (event 1) and the pulse arriving at the detector (event 2). In the frame of reference of the clock, these two events occur at the same point in space. We use the term "**proper time**" for the time interval between two events that occur at the same place, and we denote it by the symbol Δt_{proper}. Another way to think of proper time is that it is the time interval as measured in a frame of reference in which the clock is at rest. You can think of the rest frame of the clock as being "attached" to the clock.

Figure 25-8 A light clock The duration of each "tick" of this clock is the time between the start of the pulse and the time when the reflected light arrives at the detector.

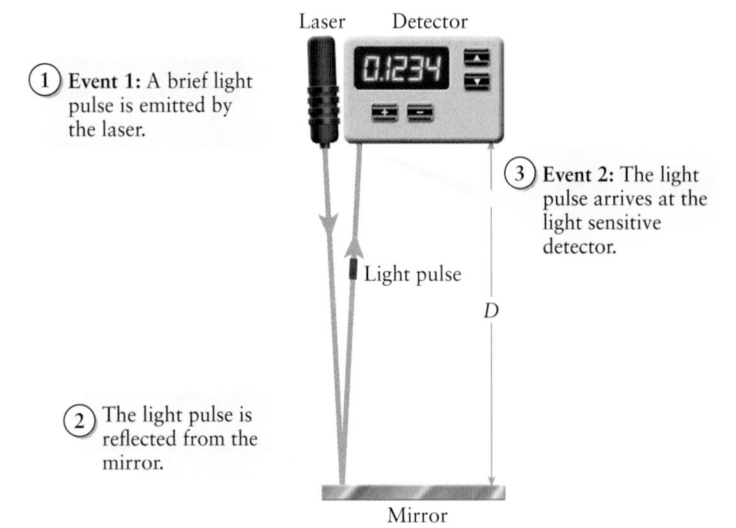
Laser
Detector
0.1234
(1) Event 1: A brief light pulse is emitted by the laser.
(3) Event 2: The light pulse arrives at the light sensitive detector.
Light pulse
D
(2) The light pulse is reflected from the mirror.
Mirror

> **WATCH OUT! Every object—including a moving object—has a rest frame.**
>
> An object is always at rest with respect to itself, which means that an object is not moving in a reference frame that is attached to it. Notice, however, that the statement is a relative one—the object is not moving *relative* to its own rest frame. We can still define any number of other frames with respect to which the object *is* in motion. A person standing on the sidewalk is at rest in her own rest frame but is moving as measured from the frame of reference of a car driving past.

During one tick of our light clock, a light pulse travels the distance D to the mirror and then the same distance back to the detector, for a total distance of $2D$. We imagine that the clock is placed in a vacuum so that the speed of the light pulse is c. The time interval for the tick is then

$$\Delta t_{\text{proper}} = \frac{2D}{c} \tag{25-6}$$

We now let the light clock move at speed V relative to us. In keeping with the way we named frames in discussing Newtonian relativity, we attach a frame S' to the clock and consider ourselves in the S frame. The S' frame therefore moves at speed V relative to the S frame. **Figure 25-9** shows the process of a single tick of the clock as observed from our S frame. The entire clock moves during the time that it takes for the light pulse to move from the laser to the mirror and from the mirror to the detector. The total distance L that the clock moves during this time equals the product of the speed V and the time interval of one clock tick. Be careful, however: We *cannot* assume that the time interval for one tick as measured in S is Δt_{proper} because Δt_{proper} is measured in frame S' at rest with respect to the clock. Instead we use the symbol Δt for the time interval of one tick as observed from the S frame. So

$$L = V\Delta t \tag{25-7}$$

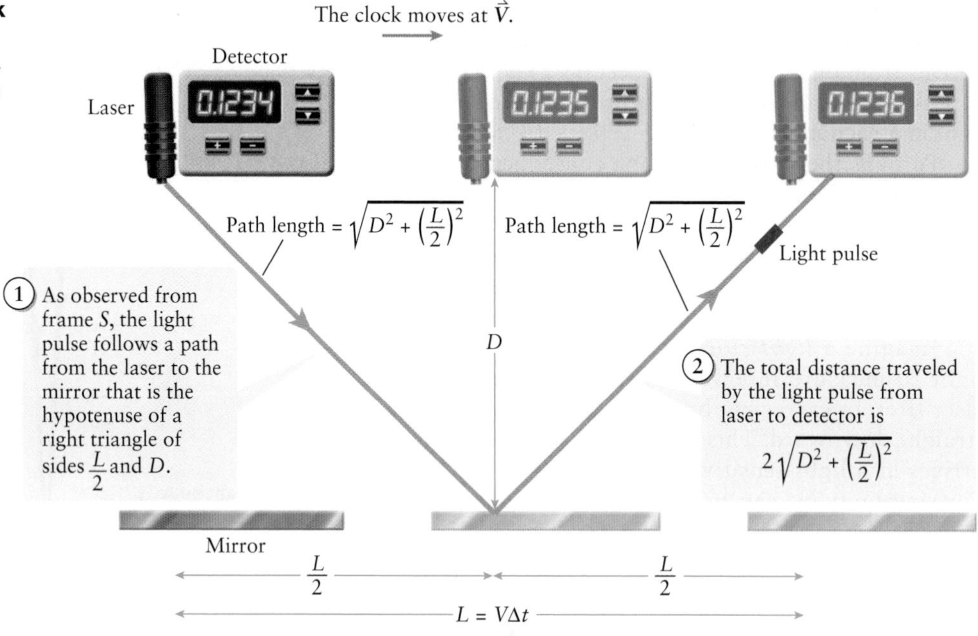

Figure 25-9 A moving light clock and time dilation When the light clock depicted in Figure 25-8 moves, the distance the light pulse travels is longer than twice the distance from the laser to the mirror.

See the Math Tutorial for more information on trigonometry.

From our vantage point in the S frame, the path followed by the light pulse during time Δt traces out two sides of a triangle (Figure 25-9). The time interval measured in S is therefore equal to the sum of the lengths of these two sides—the distance the light pulse travels—divided by the speed at which the light pulse travels in frame S. The second postulate of special relativity assures us that the speed of the light pulse is c in *all* inertial frames, and so is the same in frame S as in frame S'. Thus

$$\Delta t = \frac{2\sqrt{D^2 + \left(\frac{L}{2}\right)^2}}{c} \tag{25-8}$$

We now have two expressions for the time interval between the light pulse leaving the laser and the same pulse arriving at the detector. As measured in the clock rest frame S', this time interval is the proper time Δt_{proper} as given by Equation 25-6; as measured in our rest frame S, the time interval is Δt as given by Equation 25-8. To compare these two time intervals, first note that the distance L does not appear in the expression for Δt_{proper}. To eliminate L from the expression for Δt, substitute Equation 25-7, $L = V\Delta t$, into Equation 25-8:

$$\Delta t = \frac{2\sqrt{D^2 + \left(\frac{V\Delta t}{2}\right)^2}}{c}$$

Square both sides of this equation and rearrange to bring the two Δt terms together:

$$c^2\Delta t^2 = 4\left(D^2 + \left(\frac{V\Delta t}{2}\right)^2\right) = 4D^2 + V^2\Delta t^2$$

$$c^2\Delta t^2 - V^2\Delta t^2 = 4D^2 \quad \text{or} \quad (c^2 - V^2)\Delta t^2 = 4D^2$$

Solve for Δt:

$$\Delta t^2 = \frac{4D^2}{c^2 - V^2} = \frac{4D^2}{c^2\left(1 - \frac{V^2}{c^2}\right)} \quad \text{or} \quad \Delta t = \frac{2D}{c\sqrt{1 - \frac{V^2}{c^2}}} \tag{25-9}$$

If we compare Δt (Equation 25-9) to Δt_{proper} (Equation 25-6), we get

Time interval between two events as measured in frame S', in which those events occur at the same place

$$\Delta t = \frac{\Delta t_{\text{proper}}}{\sqrt{1 - \dfrac{V^2}{c^2}}} \qquad (25\text{-}10)$$

Time interval between the same two events as measured in frame S

Speed of S' relative to S

Speed of light in a vacuum

Time dilation (25-10)

An observer in S moving relative to the clock measures a time interval Δt for one tick of the light clock. An observer in S' who is at rest with respect to the clock measures the time interval for one tick to be Δt_{proper}. Equation 25-10 shows that Δt and Δt_{proper} are *not* equal: An observer in the S frame sees time running at a *different* rate than an observer in the S' frame!

For any nonzero value of the speed V of one inertial frame relative to another, the denominator in Equation 25-10 is less than 1. It follows that Δt is greater than Δt_{proper}. That is, the time interval as measured in S is longer than as measured in S'. We say that the time interval as measured in S has been expanded or *dilated* (the same term used to refer to an increase in size of the pupil of the eye). That's why this effect of special relativity is known as **time dilation.** If $\Delta t_{\text{proper}} = 1$ s, the time interval Δt for one tick as measured in S will be greater than 1 s. Equivalently if Δt as measured in S equals 1 s, Δt_{proper} as measured in S' will be less than 1 s. So an observer in S says that the light clock, which is moving past her at speed V, is ticking off time slowly: After 1 s has elapsed according to the observer in S, the moving clock has ticked off less than 1 s. So Equation 25-10 says that *a moving clock runs slowly.*

It may seem that Equation 25-10 is valid only for time intervals measured using a light clock. But the first postulate of relativity says that *all* laws of physics are the same in every inertial frame of reference. This implies that time dilation is also valid for time intervals measured using a mechanical clock, such as a grandfather clock that keeps time with an oscillating pendulum or a wristwatch that uses an oscillating piece of quartz (Figure 12-1b). Imagine a pendulum clock on your desk, with its pendulum swinging back and forth once every second. If the clock is moving, then the swing of the pendulum is slower, so the clock ticks more slowly. As we will see, however, the time dilation effect is very small unless V is very large.

BIO-Medical

WATCH OUT! Time dilation occurs whether or not a clock is present.

You don't need to have a clock per se for the effects of time dilation to occur. Imagine, say, that you can place a *Caenorhabditis elegans* worm (**Figure 25-10**) on a spacecraft that will fly past Earth at high speed. *C. elegans* is popular among geneticists because it grows to adulthood in a series of easily identifiable developmental stages that all together take less than 3 days. Each stage lasts 8 to 12 hours. Were you to observe a *C. elegans* worm as it moved past you at high speed, each of those stages might take days or even years, measured on a clock at rest with respect to you.

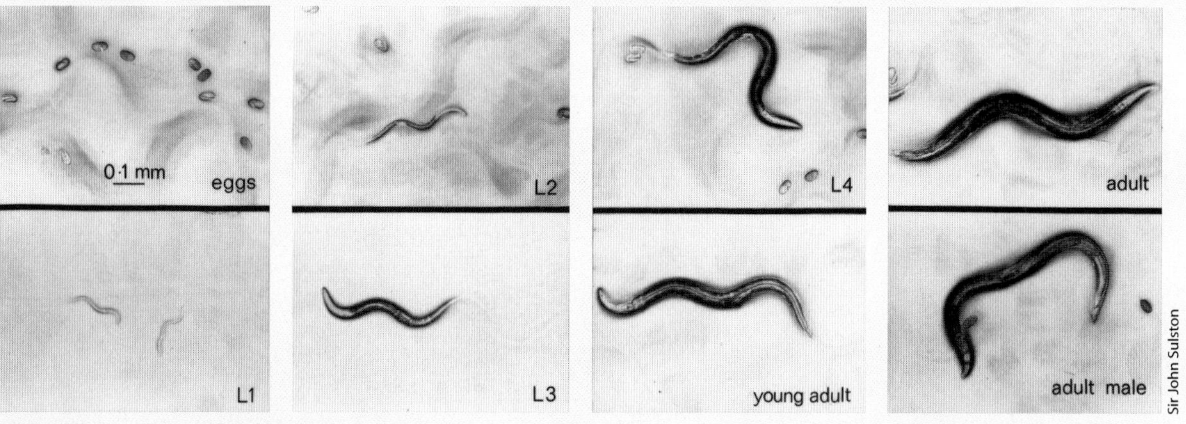

Figure 25-10 A worm "clock" The worm *Caenorhabditis elegans* grows to adulthood in a series of easily identifiable developmental stages, each of which lasts a well-defined period of time.

WATCH OUT! The effect of time dilation arises only when events in one frame are viewed from a second frame in motion relative to the first.

We imagined, above, a pendulum clock that ticks once every time the pendulum makes one full oscillation. If the clock sits on your desk, you see it swinging back and forth once every second. If the clock moves relative to you, you see the pendulum swing more slowly. But if you and your clock are moving *together* relative to some other specific object or frame, you will see *no* time dilation effect on your clock, regardless of how fast you and the clock are moving relative to that other object or frame. You will still see the pendulum swinging back and forth once every second. There is no time dilation because you and the clock are not moving relative to each other.

We do not observe time dilation in everyday life because most objects travel very slowly compared to the speed of light. As an example, the greatest launch speed ever given to a spacecraft is 16,260 m/s = 58,536 km/h = 36,373 mi/h relative to Earth. Even this tremendous speed is only 5.42×10^{-5} of the speed of light c, and the factor $1/\sqrt{1 - V^2/c^2}$ in Equation 25-10 is larger than 1 by only 1.47×10^{-9}. Due to time dilation, a clock on board the spacecraft does indeed run slowly as measured from Earth, but by only one second every 21.5 years! As the following two examples show, however, time dilation can be substantial if the speeds involved are **relativistic speeds**—that is, an appreciable fraction of the speed of light.

EXAMPLE 25-2 A Moving Clock

A clock moves past you. What must be the speed of the clock relative to you so that you see it as running at one-half (0.500) the rate of the clock on your cell phone in your hand?

Set Up

We want a moving clock to be observed as running at 0.500 the rate of a clock at rest with respect to you. This means making Δt (the time interval measured by you) equal to twice Δt_{proper} (the time interval measured in the rest frame of the moving clock). That implies the factor $1/\sqrt{1 - V^2/c^2}$ in Equation 25-10 must be equal to $1/0.500 = 2.00$. We'll use this to solve for the speed V of the moving clock relative to you.

Time dilation:

$$\Delta t = \frac{\Delta t_{\text{proper}}}{\sqrt{1 - \dfrac{V^2}{c^2}}} \qquad (25\text{-}10)$$

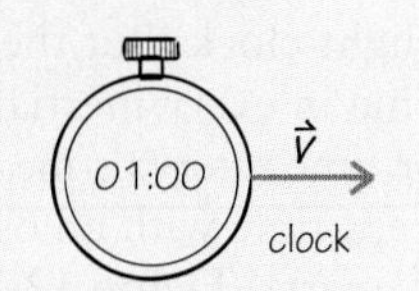

Solve

Determine the value of V that makes $\Delta t = 2.00\Delta t_{\text{proper}}$.

From Equation 25-10 we want

$$\Delta t = \frac{\Delta t_{\text{proper}}}{\sqrt{1 - \dfrac{V^2}{c^2}}} = 2.00\Delta t_{\text{proper}} \quad \text{so} \quad \frac{1}{\sqrt{1 - \dfrac{V^2}{c^2}}} = 2.00$$

Solve for the value of the speed V:

$$\sqrt{1 - \frac{V^2}{c^2}} = \frac{1}{2.00} = 0.500$$

$$1 - \frac{V^2}{c^2} = (0.500)^2 = 0.250$$

$$\frac{V^2}{c^2} = 1 - 0.250 = 0.750$$

$$\frac{V}{c} = \sqrt{0.750} = 0.866$$

$$V = 0.866c$$
$$= 0.866(3.00 \times 10^8 \text{ m/s}) = 2.60 \times 10^8 \text{ m/s}$$

Reflect

For you to observe the moving clock running a factor of 2.00 slower than the clock at rest with respect to you, the relative speed between you and the clock would need to be 86.6% of the speed of light in a vacuum. This is more than 10^4 times faster than the top speed of any craft built by humans.

EXAMPLE 25-3 Muon Decay

Our planet is continually bombarded by fast-moving subatomic particles from space (mostly protons). When these particles collide with the atoms and molecules that make up Earth's upper atmosphere, they can produce a new subatomic particle called the *muon*. These muons travel at high speed, around $0.994c$. Muons naturally decay, however; if you create a number of muons in the lab, on average half of them will decay after 1.56 μs (1 μs $= 10^{-6}$ s). This time is known as the *half-life* of the muon. Of the remaining muons, another half will decay after another 1.56 μs, and so on. If 1.00×10^6 muons are created at an altitude of 15.0 km, (a) how many would you expect to strike Earth if time-dilation effects were ignored? (b) How many would you expect to strike Earth when time dilation is taken into account?

Set Up

The production and decay of the muon occur at the same place in the rest frame of the muon, so the half-life of 1.56 μs is a proper time interval in the muon frame. We'll call this frame S'. This frame moves at speed $V = 0.994c$ relative to our frame on Earth, so we measure a different half-life Δt as given by Equation 25-10. We'll use the idea that the muon population decreases by 1/2 in each half-life to determine how many reach our planet's surface.

Time dilation:

$$\Delta t = \frac{\Delta t_{\text{proper}}}{\sqrt{1 - \frac{V^2}{c^2}}} \quad (25\text{-}10)$$

muon produced

15.0 km

$V = 0.994c$

ground

Solve

(a) If there were no time dilation, the half-life of the muon in the Earth frame S would be $\Delta t = \Delta t_{\text{proper}} = 1.56$ μs, the same as in the muon frame S'. Use this to calculate the number of muons that successfully reach Earth's surface.

Time for a muon to travel a distance $d = 15.0$ km at $V = 0.994c$:

$$T = \frac{d}{V} = \frac{15.0 \text{ km}}{0.994(3.00 \times 10^8 \text{ m/s})}\left(\frac{10^3 \text{ m}}{1 \text{ km}}\right)$$

$$= 5.03 \times 10^{-5} \text{ s}\left(\frac{1\ \mu\text{s}}{10^{-6} \text{ s}}\right) = 50.3\ \mu\text{s}$$

Express this as a multiple of the half-life:

$$\frac{T}{\Delta t_{\text{proper}}} = \frac{50.3\ \mu\text{s}}{1.56\ \mu\text{s}} = 32.2 \text{ half-lives}$$

After one half-life, the number of muons has decreased to 1/2 of its initial value; after two-half-lives, to $(1/2) \times (1/2) = 1/2^2$ of its initial value; after three half-lives, to $(1/2) \times (1/2) \times (1/2) = 1/2^3$ of its initial value; and so on. So after 32.2 half-lives, the number of muons remaining would be the original number of 1.00×10^6 multiplied by $1/2^{32.2}$:

$$\text{muons remaining} = (1.00 \times 10^6)\left(\frac{1}{2^{32.2}}\right)$$

$$= (1.00 \times 10^6)(1.97 \times 10^{-10})$$

$$= 1.97 \times 10^{-4}$$

Much less than 1 muon, on average, survives the trip to Earth's surface.

(b) With time dilation, we first calculate the half-life in the Earth frame using Equation 25-10. We then use the same method as in part (a) to calculate the number of muons that reach Earth's surface.

Accounting for time dilation, the half-life as measured in the Earth frame S is the half-life measured in the muon frame S' divided by $\sqrt{1 - V^2/c^2}$:

$$\Delta t = \frac{\Delta t_{\text{proper}}}{\sqrt{1 - \frac{V^2}{c^2}}} = \frac{1.56\ \mu\text{s}}{\sqrt{1 - \frac{(0.994c)^2}{c^2}}} = \frac{1.56\ \mu\text{s}}{\sqrt{1 - (0.994)^2}}$$

$$= \frac{1.56\ \mu\text{s}}{\sqrt{0.0120}} = \frac{1.56\ \mu\text{s}}{0.109}$$

$$= 14.3\ \mu\text{s}$$

Due to time dilation the half-life as measured in the Earth frame is about 9 times longer than the half-life as measured in the muon frame.

The effect of time dilation causes time to run more slowly for the muon than we measure in our own frame, so fewer half-lives elapse for a given distance traveled. Because fewer half-lives have elapsed, fewer muons have decayed.

From part (a), the time for a muon to travel to the surface is 50.3 μs. Expressed as a multiple of the time-dilated half-life, this is

$$\frac{T}{\Delta t} = \frac{50.3\ \mu\text{s}}{14.3\ \mu\text{s}} = 3.53 \text{ half-lives}$$

The number of muons remaining at the surface is the original number of 1.00×10^6 multiplied by $1/2^{3.53}$:

$$\text{muons remaining} = (1.00 \times 10^6)\left(\frac{1}{2^{3.53}}\right)$$
$$= (1.00 \times 10^6)\ (0.0868)$$
$$= 8.68 \times 10^4$$

About 1 in 11 of the muons produced in the upper atmosphere reaches Earth's surface, far more than would be the case were there no time dilation.

Reflect

One of the earliest confirmations of Einstein's special theory of relativity was a 1941 experiment that compared the number of muons observed in an hour at two elevations in Colorado, 3240 and 1616 m above sea level. Without time dilation, the number of muons observed at the lower elevation would have been about 9% of the number at the higher elevation. The measured number at the lower elevation was over 80%, in accordance with Einstein's prediction. This experiment provided dramatic evidence that the bizarre phenomenon of time dilation is very real.

The Twin Paradox

Imagine two twins named Bertha and Eartha. Bertha takes a round trip from Earth to another star and back on a fast spaceship that travels at $v = 0.866c$, while her twin Eartha stays at home on Earth. Eartha's clock ticks off 10 years during the time that Bertha is gone. As we saw in Example 25-2, if a clock moves past you at this speed, you will observe it as running at one-half the rate of a clock in your hand. So Eartha expects that when her twin returns, Bertha's clock will have ticked off only 5 years, not 10, and that Bertha will have aged only 5 years. So Eartha predicts that *Bertha* will actually be younger than Eartha after the round trip is done.

But from *Bertha's* frame of reference, *Eartha's* clock is the one that's moving at $0.866c$, so Bertha observes Eartha's clock to be running slow. Hence Bertha expects that when the she returns to Earth, she will find that Eartha's clock has ticked off half as much time as Bertha's clock on the spaceship. Therefore, Bertha predicts that Eartha will have aged only half as much as Bertha has and that *Eartha* will be the younger twin after the round trip is done.

It seems like the experiences of the two twins are exactly symmetrical: Each twin sees the other one moving, and hence aging more slowly, and predicts that the other one will have aged less. But clearly both twins can't be correct. So how can we resolve this *twin paradox*?

The explanation is that the two observers, Eartha and Bertha, are *not* perfectly symmetrical. Eartha remained in the same inertial reference frame (the Earth) at all times during Bertha's round trip, but Bertha did not: On the outbound leg Bertha was in a frame moving at $0.866c$ in the direction from Earth toward the other star, while on the return leg Bertha was moving at $0.866c$ in the *opposite* direction from the star back toward Earth. It turns out that because Bertha changed her reference frame and her velocity—that is, because she *accelerated*—she is the twin who ages less during the round trip. So Eartha is correct, and after the round trip Eartha will have aged 10 years, while Bertha will have aged only 5 years.

This effect has been experimentally verified many times. If an unstable particle like a muon (Example 25-3) is produced in the laboratory and sent at relativistic speed along a circular round-trip path, its half-life is longer than if the particle is produced at rest; in other words, the particle that takes the round trip ages more slowly. The effect has also been tested using very precise clocks, one that remained in the laboratory and one that was flown on a round trip in a fast-moving airplane. The clock that traveled out and back ticked off a fraction of a second less time than did the clock that stayed at rest, just as predicted by special relativity.

GOT THE CONCEPT? 25-3 Proper Time

? A clock is placed aboard Starship *Alpha*, which Albert flies past Earth at half the speed of light relative to Earth. Barbara flies Starship *Beta* alongside *Alpha* at the same velocity. George pilots Starship *Gamma* past Earth at half the speed of light relative to Earth, but in the opposite direction. Elena observes from Earth. For which of these observers is the time interval between ticks of the clock equal to the proper time? (a) Albert; (b) Barbara; (c) George; (d) Elena; (e) more than one of these.

GOT THE CONCEPT? 25-4 Comparing Clocks

? A clock is placed aboard Starship *Alpha*, which Albert flies past Earth at half the speed of light relative to Earth. Elena observes from Earth, where she has an identical clock. Which pair of words correctly fills in the blanks in this statement: "Elena measures Albert's clock as running _____, and Albert measures Elena's clock as running _____." (a) slow, fast; (b) slow, slow; (c) fast, fast; (d) fast, slow.

TAKE-HOME MESSAGE FOR Section 25-4

✔ Einstein based his special theory of relativity on two postulates: first that all laws of physics are the same in all inertial frames and second that the speed of light in a vacuum is the same in all frames and independent of both the speed of the source of the light and the speed of the observer.

✔ If two events happen at the same place in one frame, the time interval between these events as measured in that frame is called the proper time. As measured from a second frame moving relative to the first one, the time interval between those two events is longer than the proper time. This is called time dilation.

25-5 Einstein's relativity also predicts that the length of an object depends on the observer

We learned in the preceding section that the time interval between two events is not an absolute but depends on the motion of the observer. As we will see in this section, the *length* of an object is also not an absolute: Different observers will measure the same object as having different dimensions. So the nature of space and time is very different from what had been thought previous to the work of Einstein.

We'll conclude this section by looking at the *Lorentz transformation*, a set of equations that allows us to convert the space and time coordinates of an event in one inertial frame of reference to the coordinates in a second inertial frame. This transformation is an extension of the Galilean transformation that we explored in Section 25-2. We'll see that the Galilean transformation is actually just a special case of the Lorentz transformation valid for two frames that are moving with respect to each other at a speed far less than c.

Length Contraction

Figure 25-11 shows a thought experiment in which we look at the same motion in two different frames of reference, much as we did in Figures 25-8 and 25-9. In **Figure 25-11a** a rod is moving to the right at speed V relative to frame S, which you can think of as our frame of reference. The right-hand end of the rod is at $x = 0$ at time $t = 0$, and the left-hand end of the rod is at $x = 0$ at a later time $t = \Delta t_S$. The length L of the rod in our frame is therefore the speed V of the rod multiplied by the time Δt_S needed to travel its own length:

$$L = V\Delta t_S \tag{25-11}$$

(length of the rod in frame S)

Figure 25-11b shows the same process as observed in the frame of reference S' in which the rod is at rest. In this frame the rod has length L_{rest}, and the frame S moves to the left at speed V. The point $x = 0$ on frame S travels from one end of the rod to the other in a time $\Delta t_{S'}$. The length of the rod equals the speed V of frame S multiplied by the time $\Delta t_{S'}$ that frame S needs to travel this length:

$$L_{rest} = V\Delta t_{S'} \tag{25-12}$$

(length of the rod in its own rest frame S')

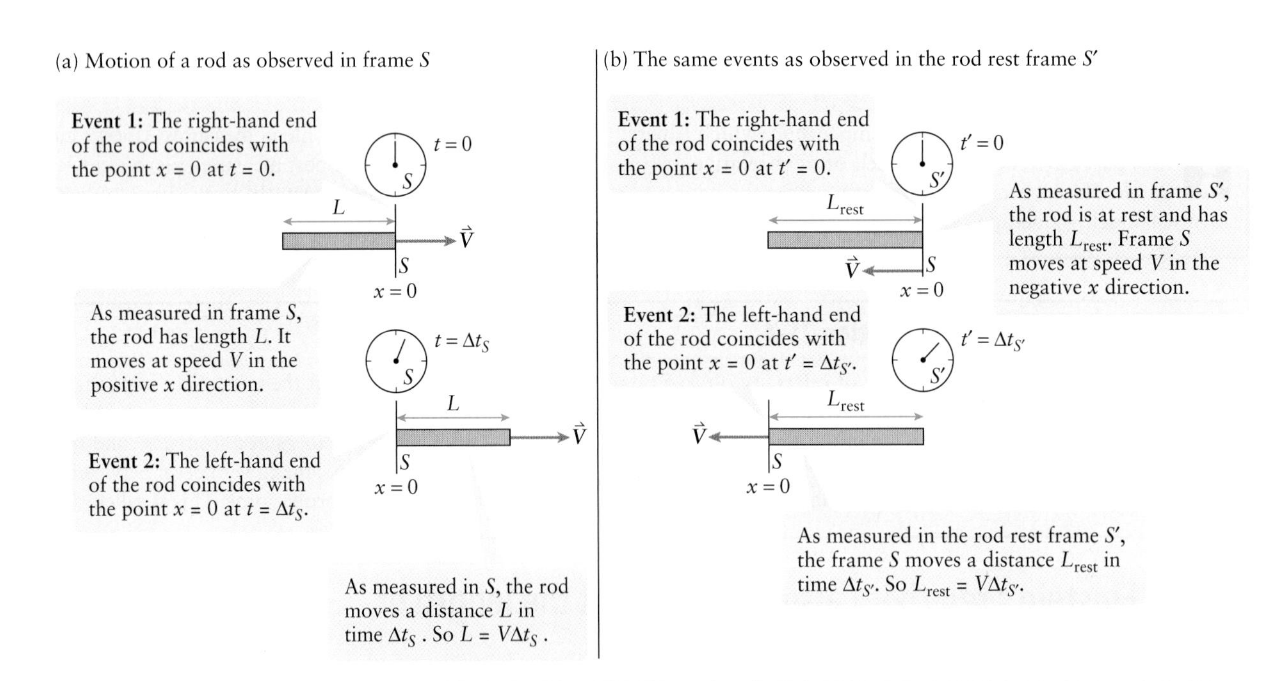

(a) Motion of a rod as observed in frame S
Event 1: The right-hand end of the rod coincides with the point x = 0 at t = 0.
t = 0
S
L
V
S
x = 0
As measured in frame S, the rod has length L. It moves at speed V in the positive x direction.
t = Δt_S
L
Event 2: The left-hand end of the rod coincides with the point x = 0 at t = Δt_S.
As measured in S, the rod moves a distance L in time Δt_S . So L = VΔt_S .
(b) The same events as observed in the rod rest frame S′
Event 1: The right-hand end of the rod coincides with the point x = 0 at t′ = 0.
t′ = 0
S′
L_rest
As measured in frame S′, the rod is at rest and has length L_rest. Frame S moves at speed V in the negative x direction.
Event 2: The left-hand end of the rod coincides with the point x = 0 at t′ = Δt_S′.
t′ = Δt_S′
As measured in the rod rest frame S′, the frame S moves a distance L_rest in time Δt_S′. So L_rest = VΔt_S′.

- The two events are (1) the right-hand end of the rod coinciding with $x = 0$ and (2) the left-hand end of the rod coinciding with $x = 0$.
- These two events happen at the same place in frame S, so Δt_S is the proper time interval between these events. The time interval $\Delta t_{S'}$ must therefore be greater than Δt_S (time dilation).
- The distance L_{rest} must therefore be greater than the distance L. So the length L of the moving rod is less than the length L_{rest} of the rod in a frame where it is at rest. This is length contraction.

Figure 25-11 A moving rod and length contraction A rod as observed in (a) a frame in which the rod is moving along its length and (b) a frame in which the rod is at rest.

To see how the lengths as measured in the two frames compare, note that the length measurements in S and S' both involve the same pair of events: the right-hand end of the rod coinciding with the point $x = 0$ in frame S, and the left-hand end of the rod coinciding with that same point (events 1 and 2 in Figure 25-11). These two events happen at the same location in frame S but at different locations in frame S'. So Δt_S is the proper time interval between these events. The time interval $\Delta t_{S'}$ measured in frame S' is therefore longer ("dilated") compared to the time interval Δt_S measured in frame S, and the two time intervals are related by Equation 25-10:

$$\Delta t_{S'} = \frac{\Delta t_S}{\sqrt{1 - \frac{V^2}{c^2}}} \tag{25-13}$$

Multiply both sides of Equation 25-13 by V then replace $V\Delta t_S$ by L in accordance with Equation 25-11 and replace $V\Delta t_{S'}$ by L_{rest} according to Equation 25-12:

$$V\Delta t_{S'} = \frac{V\Delta t_S}{\sqrt{1 - \frac{V^2}{c^2}}} \quad \text{or} \quad L_{\text{rest}} = \frac{L}{\sqrt{1 - \frac{V^2}{c^2}}}$$

We can rewrite this as

Length of an object in frame S', in which it is at rest

$$L = L_{\text{rest}}\sqrt{1 - \frac{V^2}{c^2}}$$

Speed of S' relative to S

Speed of light in a vacuum

Length of the same object in frame S, in which the object is moving along its length

Length contraction (25-14)

If the rod is moving, V is greater than zero and the factor $\sqrt{1 - V^2/c^2}$ is less than 1. So the length L of the moving rod is less than the length L_{rest} of the rod at rest. This is called **length contraction:** *A moving object is shortened along the direction in which it is moving.*

WATCH OUT! Length contraction occurs only along the direction of motion.

There is *no* change in length of a moving object in any direction other than the direction of motion. For example, if the rod in Figure 25-11 is moving in the x direction relative to frame S, an observer in S will measure the rod as having a shorter length in the x direction than will an observer in S' at rest with respect to the rod. But both observers will agree about the height and width of the rod (its dimensions in the y and z directions).

Length contraction gives us an alternative way to understand the results of Example 25-3 in the preceding section, in which we used time dilation to explain how short-lived muons produced at an altitude of 15.0 km are able to survive their trip to Earth's surface. Imagine a vertical rod that extends 15.0 km upward from the ground to where the muons are produced. This rod is stationary relative to Earth, so its length as measured by you on the ground is its rest length: $L_{\text{rest}} = 15.0 \text{ km} = 1.50 \times 10^4 \text{ m}$. But as seen from the frame of a descending muon, this rod is moving upward along its length at $V = 0.994c$. As measured by a muon, the length of this rod is contracted in accordance with Equation 25-14:

$$\begin{aligned} L &= L_{\text{rest}}\sqrt{1 - \frac{V^2}{c^2}} = (15.0 \text{ km})\sqrt{1 - \frac{(0.994c)^2}{c^2}} \\ &= (15.0 \text{ km})\sqrt{1 - (0.994)^2} \\ &= (15.0 \text{ km})\sqrt{0.0120} = (15.0 \text{ km})(0.109) \\ &= 1.64 \text{ km} = 1.64 \times 10^3 \text{ m} \end{aligned}$$

Moving at $V = 0.994c$, this contracted rod travels past the muon in a time

$$\frac{1.64 \times 10^3 \text{ m}}{0.994c} = \frac{1.64 \times 10^3 \text{ m}}{(0.994)(3.00 \times 10^8 \text{ m/s})} = 5.50 \times 10^{-6} \text{ s} = 5.50 \ \mu\text{s}$$

The half-life of the muon in its own rest frame is 1.56 μs, so from the muon's perspective it takes just $(5.50 \ \mu\text{s})/(1.56 \ \mu\text{s}) = 3.53$ half-lives for the contracted rod to move past it—that is, for the muon to move from where it is produced to Earth's surface. That's exactly the result that we found in Example 25-3 using the ideas of time dilation. The ideas of length contraction and time dilation are mutually consistent!

EXAMPLE 25-4 A Flying Meter Stick I

A meter stick (length 1.00 m) hurtles through space at a speed of $0.800c$ relative to you, with its length aligned with the direction of motion. What do you measure as the length of the meter stick?

Set Up

The meter stick's length along its direction of motion is 1.00 m as measured in its own rest frame S', so $L_{rest} = 1.00$ m. Your frame is frame S, and the two frames are moving relative to each other at $V = 0.800c$. We'll use Equation 25-14 to find the length L as measured in your frame.

Length contraction:

$$L = L_{rest}\sqrt{1 - \frac{V^2}{c^2}} \quad (25\text{-}14)$$

Solve

Substitute $L_{rest} = 1.00$ m and $V = 0.800c$ into Equation 25-14 and calculate L.

From Equation 25-14,

$$L = L_{rest}\sqrt{1 - \frac{V^2}{c^2}} = (1.00 \text{ m})\sqrt{1 - \frac{(0.800c)^2}{c^2}}$$
$$= (1.00 \text{ m})\sqrt{1 - (0.800)^2} = (1.00 \text{ m})\sqrt{1 - 0.640}$$
$$= (1.00 \text{ m})\sqrt{0.360} = (1.00 \text{ m})(0.600)$$
$$= 0.600 \text{ m} = 60.0 \text{ cm}$$

Reflect

As measured in your frame of reference, the meter stick is only 60.0 cm in length. This is not an optical illusion; the meter stick really is only 60.0% as long in your frame of reference as in the rest frame of the meter stick.

EXAMPLE 25-5 A Flying Meter Stick II

The meter stick from the preceding example again hurtles through space at a speed of $0.800c$ relative to you, but now it is tilted at an angle of 30.0° as measured in its rest frame with respect to the direction of motion. Now what do you measure as the length of the meter stick?

Set Up

As measured in the meter stick's rest frame S', the meter stick is inclined at an angle $\theta' = 30.0°$ to its direction of motion relative to frame S. The x dimension of the stick undergoes length contraction given by Equation 25-14, but the y dimension does not. We'll calculate the x and y dimensions of the stick in your frame S, then use the Pythagorean theorem to find the length of the stick in frame S.

Length contraction:

$$L = L_{rest}\sqrt{1 - \frac{V^2}{c^2}} \quad (25\text{-}14)$$

Solve

First calculate the x and y dimensions of the meter stick in its rest frame.

In the rest frame S' of the stick, its dimensions are

$$L_{rest,x} = L_{rest}\cos\theta' = (1.00 \text{ m})\cos 30.0° = 0.866 \text{ m}$$
$$L_{rest,y} = L_{rest}\sin\theta' = (1.00 \text{ m})\sin 30.0° = 0.500 \text{ m}$$

Calculate the x and y dimensions of the meter stick in your frame, in which the stick is moving at $V = 0.800c$ along the x direction.

In your frame S, the x dimension of the stick is contracted:

$$L_x = L_{\text{rest},x}\sqrt{1-\frac{V^2}{c^2}} = (0.866\text{ m})\sqrt{1-\frac{(0.800c)^2}{c^2}}$$
$$= (0.866\text{ m})\sqrt{1-0.640} = (0.866\text{ m})\sqrt{0.360}$$
$$= 0.520\text{ m}$$

Length contraction occurs only along the direction of motion, so the y dimension of the stick is *not* contracted:

$$L_y = L_{\text{rest},y} = 0.500\text{ m}$$

Use the Pythagorean theorem to calculate the length of the meter stick in your frame.

From the Pythagorean theorem, the length of the meter stick in your frame S is

$$L = \sqrt{L_x^2 + L_y^2}$$
$$= \sqrt{(0.520\text{ m})^2 + (0.500\text{ m})^2}$$
$$= \sqrt{(0.520\text{ m})^2 + (0.500\text{ m})^2}$$
$$= \sqrt{0.520\text{ m}^2} = 0.721\text{ m}$$

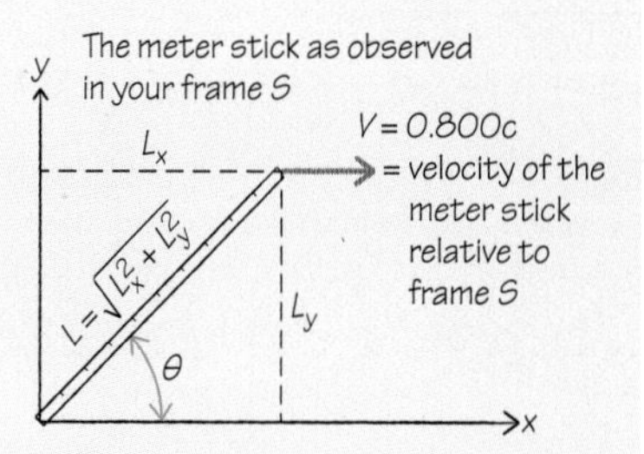

Reflect

Because the meter stick is not aligned with the direction of motion, the amount of contraction is not as great as in Example 25-4, in which the meter stick was completely aligned with the direction of motion. Notice also that the angle θ of the meter stick in your frame is different from the angle $\theta' = 30.0°$ in the stick's rest frame. Can you show that $\theta = 43.9°$?

The Lorentz Transformation

The Galilean transformation that we presented in Section 25-2 is not consistent with the postulates of special relativity. A set of transformation equations that *is* consistent with relativity is the **Lorentz transformation.** As in Figure 25-3, we take frame S' to be moving at speed V in the positive x direction relative to frame S. The origins of the two frames coincide at $t = 0$ in frame S and $t' = 0$ in frame S'. If an event takes place at coordinates x, y, z, t in frame S, the coordinates of that same event in frame S' are

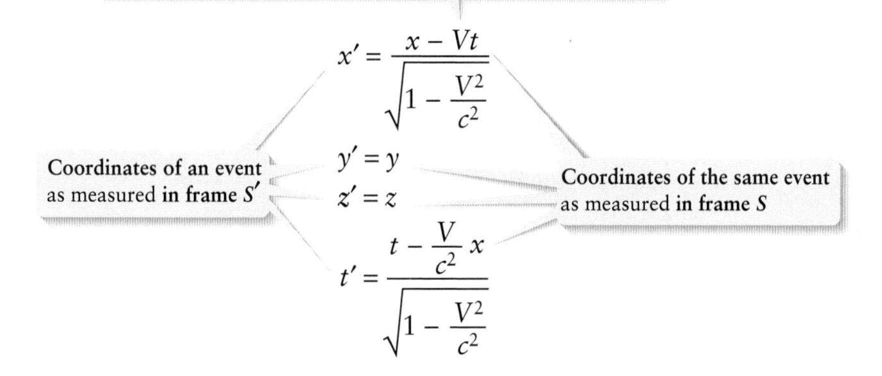

$$x' = \frac{x - Vt}{\sqrt{1-\frac{V^2}{c^2}}}$$
$$y' = y$$
$$z' = z$$
$$t' = \frac{t - \frac{V}{c^2}x}{\sqrt{1-\frac{V^2}{c^2}}}$$

Lorentz transformation (25-15)

Note that at speeds that are far less than the speed of light, the ratio V/c is very small and can be treated as essentially zero. Then $\sqrt{1 - V^2/c^2}$ is essentially equal to 1, and Equations 25-15 become

$$x' = x - Vt, \quad y' = y, \quad z' = z, \quad t' = t$$

These are just the Galilean transformation equations that we presented in Section 25-2. So at speeds far slower than the speed of light, the equations of Einstein's special theory of relativity reduce to the equations of Newtonian relativity.

Equations 25-15 are useful for finding the coordinates of an event in frame S' if we know the event's coordinates in frame S. If we instead want to determine the coordinates of an event in frame S from the coordinates in frame S', it's most convenient to use the *inverse* Lorentz transformation:

Inverse Lorentz transformation (25-16)

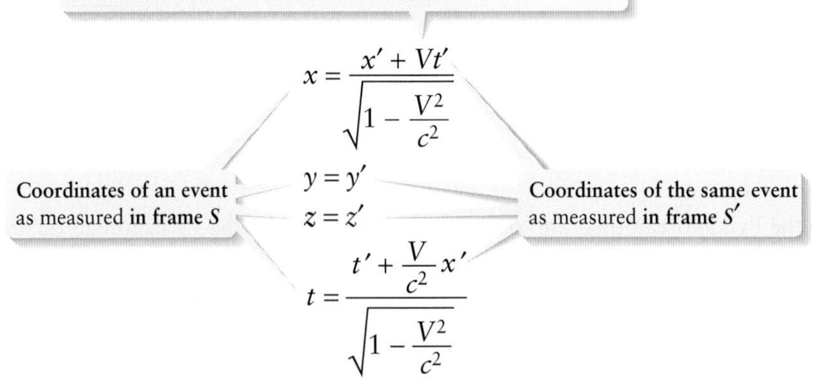

Inertial frame of reference S′ moves at speed V in the positive x direction relative to inertial frame of reference S.
Coordinates of an event as measured in frame S
Coordinates of the same event as measured in frame S′

$$x = \frac{x' + Vt'}{\sqrt{1 - \frac{V^2}{c^2}}}$$

$$y = y'$$

$$z = z'$$

$$t = \frac{t' + \frac{V}{c^2}x'}{\sqrt{1 - \frac{V^2}{c^2}}}$$

It's possible to derive the equations for time dilation (Equation 25-10) and length contraction (Equation 25-14) from the Lorentz transformation equations. Instead let's look at another remarkable result of the special theory of relativity that we can deduce from the Lorentz transformation equations.

EXAMPLE 25-6 Simultaneity Is Relative

In your reference frame you have a meter stick that is oriented along the x axis, with one end at $x = 0$ and the other end at $x = 1.00$ m. There is a light bulb at each end of the stick, and you make the two bulbs flash simultaneously (as measured by you) at $t = 0$. A spacecraft flies past you at $V = 0.800c$ in the positive x direction. (a) According to an observer in the spacecraft, are the two light flashes simultaneous? If not, which flash happens first as measured by her? (b) On board the spacecraft is an identical meter stick with a light bulb at each end. The observer on board the spacecraft places the two ends of the stick at $x' = 0$ and $x' = 1.00$ m, and she makes the two bulbs flash simultaneously (as measured by her) at $t' = 0$. According to you are the two light flashes simultaneous? If not, which flash happens first as measured by you?

Set Up

In part (a) we're given the coordinates in frame S of two events, the flash of the left-hand bulb at $x = 0$ and $t = 0$ and the flash of the right-hand bulb at $x = 1.00$ m and $t = 0$. We'll use the t' equation from the Lorentz transformation, Equations 25-15, to determine the times of these two events as measured in the spaceship frame S'. Similarly in part (b) we're given the coordinates in frame S' of two other events, the flash of the left-hand bulb at $x' = 0$ and $t' = 0$ and the flash of the right-hand bulb at $x' = 1.00$ m and $t' = 0$. We'll find the times of these events as measured in your S frame using the t equation from the inverse Lorentz tranfsormation, Equations 25-16.

Time equation from the Lorentz transformation:

$$t' = \frac{t - \frac{V}{c^2}x}{\sqrt{1 - \frac{V^2}{c^2}}} \qquad (25\text{-}15)$$

Time equation from the inverse Lorentz transformation:

$$t = \frac{t' + \frac{V}{c^2}x'}{\sqrt{1 - \frac{V^2}{c^2}}} \qquad (25\text{-}16)$$

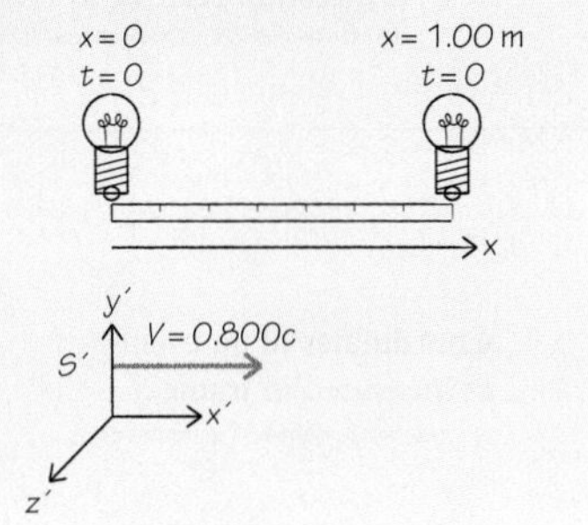

(a) Meter stick at rest in frame S:
What does an observer in frame S′ measure?
x = 0
t = 0
x = 1.00 m
t = 0
x
y′
V = 0.800c
S′
x′
z′

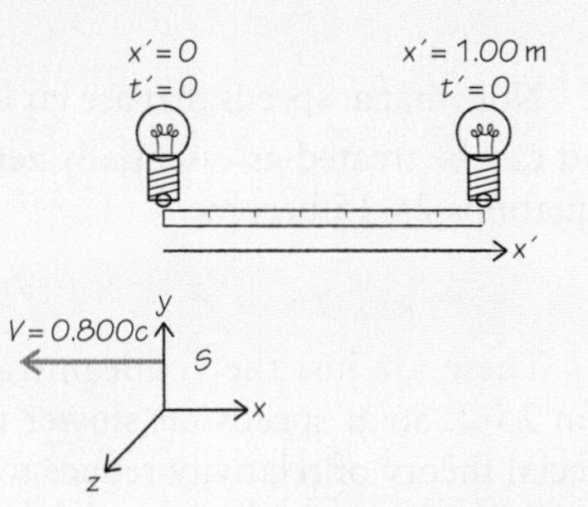

(b) Meter stick at rest in frame S′:
What does an observer in frame S measure?
x′ = 0
t′ = 0
x′ = 1.00 m
t′ = 0
x′
y
V = 0.800c
S
x
z

Solve

(a) Use the Lorentz transformation to calculate the times of the simultaneous flashes in S as measured in S'.

The left-hand bulb on the meter stick at rest in frame S flashes at $x = 0, t = 0$. In frame S' this bulb flashes at

$$t'_{\text{left}} = \frac{(0) - \frac{V}{c^2}(0)}{\sqrt{1 - \frac{V^2}{c^2}}} = 0$$

The right-hand bulb on the meter stick at rest in frame S flashes at $x = 1.00$ m, $t = 0$. In frame S' this bulb flashes at

$$t'_{\text{right}} = \frac{(0) - \frac{V}{c^2}(1.00\text{ m})}{\sqrt{1 - \frac{V^2}{c^2}}} = \frac{-\frac{(0.800c)}{c^2}(1.00\text{ m})}{\sqrt{1 - \frac{(0.800c)^2}{c^2}}}$$

$$= \frac{-0.800\text{ m}}{c\sqrt{1 - (0.800)^2}} = \frac{-0.800\text{ m}}{(3.00 \times 10^8\text{ m/s})(0.600)}$$

$$= -4.44 \times 10^{-9}\text{ s}$$

As observed from the spaceship frame S', the two events are *not* simultaneous: The right-hand bulb flashes 4.44×10^{-9} s *before* the left-hand bulb.

(b) Use the inverse Lorentz transformation to calculate the times of the simultaneous flashes in S' as measured in S.

The left-hand bulb on the meter stick at rest in frame S' flashes at $x' = 0, t' = 0$. In frame S this bulb flashes at

$$t_{\text{left}} = \frac{(0) + \frac{V}{c^2}(0)}{\sqrt{1 - \frac{V^2}{c^2}}} = 0$$

The right-hand bulb on the meter stick at rest in frame S' flashes at $x' = 1.00$ m, $t' = 0$. In frame S this bulb flashes at

$$t_{\text{right}} = \frac{(0) + \frac{V}{c^2}(1.00\text{ m})}{\sqrt{1 - \frac{V^2}{c^2}}} = \frac{+\frac{(0.800c)}{c^2}(1.00\text{ m})}{\sqrt{1 - \frac{(0.800c)^2}{c^2}}}$$

$$= \frac{+0.800\text{ m}}{c\sqrt{1 - (0.800)^2}} = \frac{+0.800\text{ m}}{(3.00 \times 10^8\text{ m/s})(0.600)}$$

$$= +4.44 \times 10^{-9}\text{ s}$$

As observed from your frame S, the two events are *not* simultaneous: The right-hand bulb flashes 4.44×10^{-9} s *after* the left-hand bulb.

Reflect

This example illustrates yet another counterintuitive consequence of Einstein's special theory of relativity: Two events that are simultaneous to one observer need not be simultaneous to another observer. So even the simple statement "Two things happen at the same time" has to be qualified by stating in which frame of reference it holds true. Time in relativity is not an absolute!

GOT THE CONCEPT? 25-5 No Contraction?

A rod with a rest length of 1.00 m whizzes past you at $0.995c$. Is it possible that you could measure its length to be equal to its rest length? (a) Yes; (b) no.

TAKE-HOME MESSAGE FOR Section 25-5

✔ When an object moves relative to an observer, its length in the direction of motion is contracted compared to the length measured in the object's rest frame. This is called length contraction.

✔ The Lorentz transformation allows you to calculate the coordinates of an event measured in one inertial frame based on the coordinates of that event measured in another inertial frame. Unlike the Galilean transformation, the Lorentz transformation is consistent with the postulates of relativity.

25-6 The speed of light is the ultimate speed limit

Suppose you are an outfielder running to catch a batted baseball (**Figure 25-12a**). The baseball is traveling at 30.0 m/s relative to the ground, and you are running toward the ball at 10.0 m/s relative to the ground. The Galilean velocity transformation that we learned in Section 25-2 says that, relative to you, the baseball travels at 30.0 m/s plus 10.0 m/s, or 40.0 m/s; the velocities simply add. But suppose instead that you are an astronaut flying in your spaceship at 1.00×10^8 m/s, as in **Figure 25-12b**, and you are moving toward a light beam aimed at you by a stationary astronaut. The same idea that we applied to the baseball predicts that relative to you the light beam travels at $c = 3.00 \times 10^8$ m/s (the speed of the light beam relative to the other astronaut) plus 1.00×10^8 m/s (the speed of your spaceship relative to the other astronaut), or 4.00×10^8 m/s. But that *cannot* be correct: The second postulate of the special theory of relativity says that the speed of light in vacuum is the same to all inertial observers.

(a)

As seen by the outfielder the ball is approaching her at (30.0 m/s) + (10.0 m/s) = 40.0 m/s.

(b)

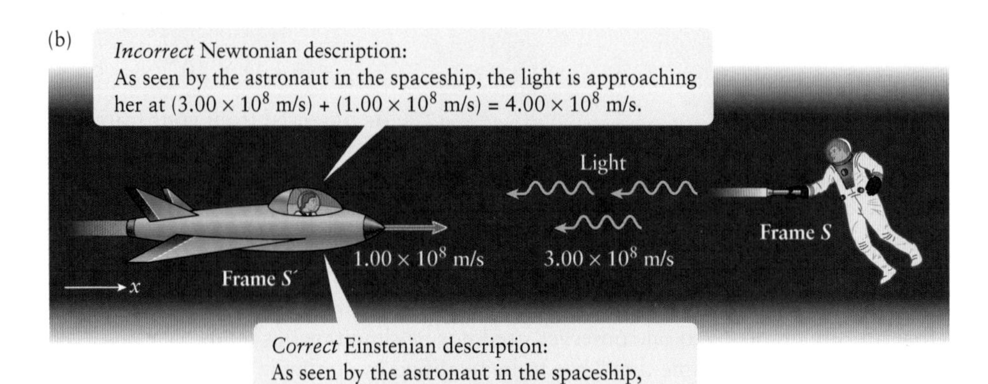

Figure 25-12 Velocity addition—Newtonian and Einsteinian (a) An outfielder running toward a batted baseball. (b) An astronaut in her spaceship flying toward a light beam.

So the light beam must also travel at speed $c = 3.00 \times 10^8$ m/s relative to you. Clearly the Galilean transformation for velocities is inadequate. In this section we'll explore the *Lorentz velocity transformation*, which allows us to combine velocities in a way that is consistent with Einsteinian relativity.

It's possible to derive the transformation of velocities from the Lorentz transformation of coordinates that we introduced in Section 25-5 (see Equations 25-15 and 25-16). We'll skip over the derivation and just present the result for the special case in which all motions are along the same line, as in Figure 25-12. Suppose that inertial frame S' is moving at speed V in the positive x direction relative to inertial frame S. An object is moving relative to frame S along the x direction, with an x component of velocity v_x. In frame S' the same object has an x component of velocity given by

Inertial frame of reference S' moves at speed V in the positive x direction relative to inertial frame of reference S.

$$v_x' = \frac{v_x - V}{1 - \dfrac{V}{c^2} v_x}$$

x component of velocity of an object moving along the x axis as measured in frame S'

x component of velocity of the same object as measured in frame S

Lorentz velocity transformation (25-17)

This is called the **Lorentz velocity transformation.** Notice that if the speed V of one frame relative to the other is small compared to c, the ratio V/c is much less than 1 and the denominator is essentially equal to 1. The same thing happens if the velocity v_x of the object relative to frame S is small compared to c. So if either of the speeds involved is a small fraction of the speed of light, Equation 25-17 reduces to $v_x' = v_x - V$, which is just the Galilean velocity transformation (the first of Equations 25-5). This justifies our use of the Galilean velocity transformation for slow-moving objects.

Equation 25-17 allows us to find the object's velocity relative to frame S' if we know its velocity relative to frame S. If instead we know the object's velocity relative to frame S' and want to calculate its velocity relative to frame S, we use the **inverse Lorentz velocity transformation:**

Inertial frame of reference S' moves at speed V in the positive x direction relative to inertial frame of reference S.

$$v_x = \frac{v_x' + V}{1 + \dfrac{V}{c^2} v_x'}$$

x component of velocity of an object moving along the x axis as measured in frame S

x component of velocity of the same object as measured in frame S'

Inverse Lorentz velocity transformation (25-18)

Let's see what the Lorentz velocity transformation tells us about the situation shown in Figure 25-12b. The astronaut with the flashlight is in frame S, and you and your spaceship are in frame S'. You are moving to the right (in the positive x direction), which is just how frame S' must move relative to frame S in order to use Equation 25-17 or 25-18. The relative speed of the two frames is V. We know that the light travels relative to the astronaut with the flashlight at the speed of light in the negative x direction, so its velocity relative to S is $v_x = -c$. From Equation 25-17, the velocity v_x' of the light relative to you in frame S' is

$$v_x' = \frac{v_x - V}{1 - \dfrac{V}{c^2} v_x} = \frac{-c - V}{1 - \dfrac{V}{c^2}(-c)} = \frac{-(c + V)}{1 + \dfrac{V}{c}}$$

We can simplify this by factoring c out of the numerator:

$$v_x' = \frac{-c\left(1 + \dfrac{V}{c}\right)}{1 + \dfrac{V}{c}} = -c$$

This says that as measured by you in frame S', the light travels at the speed of light c in the negative x direction just as in frame S. Note that while $V = 1.00 \times 10^8$ m/s in Figure 25-12b, our result doesn't depend on the value of V: No matter what the relative speed of the two frames, each observer will see light propagating in a vacuum at the same speed c.

A direct consequence of this calculation is that *no object can move faster than c in any inertial frame of reference.* If an object (light) is traveling at speed c in one inertial frame, it is traveling at c in all inertial frames; if an object is traveling slower than c in one inertial frame, it is traveling slower than c in all inertial frames. Thus the speed of light in a vacuum represents an ultimate speed limit.

EXAMPLE 25-7 Baseball for Superheroes

Suppose the baseball game in Figure 25-12 is being played by superheroes. The outfielder has super speed, and can run at $0.300c$ relative to the ground. The batter has super strength and can bat the ball with such force that the ball ends up traveling horizontally at $0.900c$ relative to the ground. (The bat and ball are made of super materials that can withstand the tremendous forces required.) How fast is the ball moving relative to the outfielder?

Set Up

This is nearly the same situation that we discussed above with the two astronauts and the beam of light, except that the astronauts have been replaced by (super) baseball players and the beam of light has become a baseball. We use Equation 25-17 to calculate the velocity of the ball relative to the outfielder in frame S'.

Lorentz velocity transformation:

$$v'_x = \frac{v_x - V}{1 - \frac{V}{c^2}v_x} \quad (25\text{-}17)$$

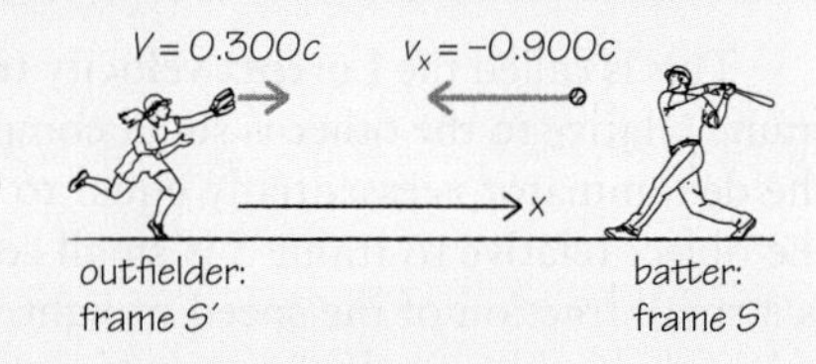

Solve

Use the velocity of the ball relative to the batter (v_x) and the velocity of the outfielder relative to the batter (V) to find the velocity of the ball relative to the outfielder (v'_x).

The outfielder (frame S') is moving in the positive x direction relative to the batter (frame S) at $V = 0.300c$. The velocity of the ball relative to the batter (frame S) is $v_x = -0.900c$ (negative because the ball is moving in the negative x direction). From Equation 25-17 the velocity of the ball relative to the outfielder is

$$v'_x = \frac{v_x - V}{1 - \frac{V}{c^2}v_x} = \frac{-0.900c - 0.300c}{1 - \frac{0.300c}{c^2}(-0.900c)}$$

$$= \frac{-1.200c}{1 + (0.300)(0.900)} = \frac{-1.200c}{1.27}$$

$$= -0.945c$$

Relative to the outfielder the ball is moving at $0.945c$ in the negative x direction (to the left).

Reflect

In the Galilean velocity transformation the velocity of the ball relative to the outfielder would have been $v'_x = v_x - V = -0.900c - 0.300c = -1.200c$, which is faster than c. Thanks to the $1 - (V/c^2)v_x$ term in the denominator of Equation 25-17, the actual speed of the ball relative to the outfielder is slower than c. The batter may be super-powered, but the ball cannot exceed the speed of light in a vacuum as measured by any observer.

GOT THE CONCEPT? 25-6 Returning a Super Baseball

Suppose the super outfielder in Example 25-7 catches the ball and throws it back toward the super batter, who is still standing at home plate. If the outfielder is still running at $0.300c$ and she throws the ball at $0.700c$ relative to her, what is the speed of the ball relative to the batter? (a) $0.331c$; (b) $0.506c$; (c) $0.826c$; (d) $1.00c$; (e) $1.26c$.

TAKE-HOME MESSAGE FOR Section 25-6

✔ At speeds that are an appreciable fraction of the speed of light, we must use the Lorentz velocity transformation to calculate relative velocities. This transformation respects the rule that the speed of light in a vacuum is the same in all inertial frames of reference.

✔ No object can move faster than c, the speed of light in a vacuum, in any inertial frame of reference.

25-7 The equations for kinetic energy and momentum must be modified at very high speeds

We have seen that the speed of light in a vacuum c is an ultimate speed limit: No object can travel faster than c in any inertial frame of reference. As we'll see in this section, this tells us that the expressions we learned earlier in this book for kinetic energy K (Chapter 6) and momentum $\vec{p}$ (Chapter 7) *cannot* be entirely correct. They fail at speeds that are a reasonable fraction of c. We'll see why this must be so and encounter a new kind of energy called *rest energy* that is intrinsic to any object with mass.

In Section 6-3 we introduced kinetic energy through the work-energy theorem:

Work done on an object by the net force on that object

$$W_{\text{net}} = K_{\text{f}} - K_{\text{i}}$$

Kinetic energy of the object after the work is done on it

Kinetic energy of the object before the work is done on it

The work-energy theorem (6-9)

This says that if an object starts at rest so that its initial kinetic energy K_{i} is zero, the more work the net force on an object does, the greater the final kinetic energy K_{f} of the object. In principle there is no limit to how much kinetic energy an object can acquire. There is a problem, however: Using Newtonian physics we found that the expression for the kinetic energy K of an object of mass m moving at speed v is

$$K = \frac{1}{2}mv^2$$

Since the speed of light in a vacuum c is the maximum speed an object can acquire, this expression says that the maximum kinetic energy that an object can acquire is $(1/2)mc^2$. This contradicts the idea that there should be no limit on an object's kinetic energy. Clearly we need an improved equation for kinetic energy.

There is a similar problem with the Newtonian expression for momentum. We saw in Section 7-6 that the change in momentum of an object is determined by the external forces on the object and the duration of the time interval over which the forces act:

The sum of all external forces acting on an object

Duration of a time interval over which the external forces act

$$\left(\sum \vec{F}_{\text{external on object}}\right)\Delta t = \vec{p}_{\text{f}} - \vec{p}_{\text{i}} = \Delta\vec{p}$$

Change in the momentum $\vec{p}$ of the object during that time interval

External force and momentum change for an object (7-23)

Suppose an object starts at rest so that its initial momentum $\vec{p}_{\text{i}}$ is zero and a constant net force acts on it. Then Equation 7-23 says that the longer the time interval Δt that the force acts, the greater the final momentum $\vec{p}_{\text{f}}$ of the object. In principle there is no limit on how long the force can act, so there should be no upper limit on an object's momentum. However, this can't be reconciled with the Newtonian expression for the momentum of an object of mass m with velocity $\vec{v}$:

$$\vec{p} = m\vec{v}$$

According to this expression, the maximum magnitude of momentum that an object of mass m can have is $p = mc$, which would be attained only when an object is moving at the speed of light. This directly contradicts the notion that there should be no upper limit on momentum. Just as for kinetic energy, we need a new expression for momentum that's consistent with the special theory of relativity.

It's possible to derive the correct expressions for K and $\vec{p}$ by analyzing an elastic collision in which both mechanical energy and momentum are conserved (see Section 7-5). The derivation is beyond our scope, so we'll just look at the results. For a particle moving with velocity $\vec{v}$, both the expression for the kinetic energy and the expression for the momentum involve a dimensionless quantity called **relativistic gamma:**

Relativistic gamma (25-19)

Relativistic gamma for a particle moving at speed v

$$\gamma = \frac{1}{\sqrt{1 - \dfrac{v^2}{c^2}}}$$

Speed of the particle

Speed of light in a vacuum

Relativistic gamma is equal to 1 when $v = 0$ and becomes infinitely large as v approaches c (**Figure 25-13**). In terms of relativistic gamma, we can write the correct expressions for kinetic energy and momentum as

Einsteinian expressions for kinetic energy and momentum (25-20)

Kinetic energy of a particle of mass m and velocity $\vec{v}$

$$K = (\gamma - 1)mc^2$$

$$\vec{p} = \gamma m\vec{v}$$

Relativistic gamma for speed v

Momentum of a particle of mass m and velocity $\vec{v}$

At speeds that are a small fraction of the speed of light c, these are approximately equal to the Newtonian expressions

$$K = \frac{1}{2}mv^2 \text{ and } \vec{p} = m\vec{v}$$

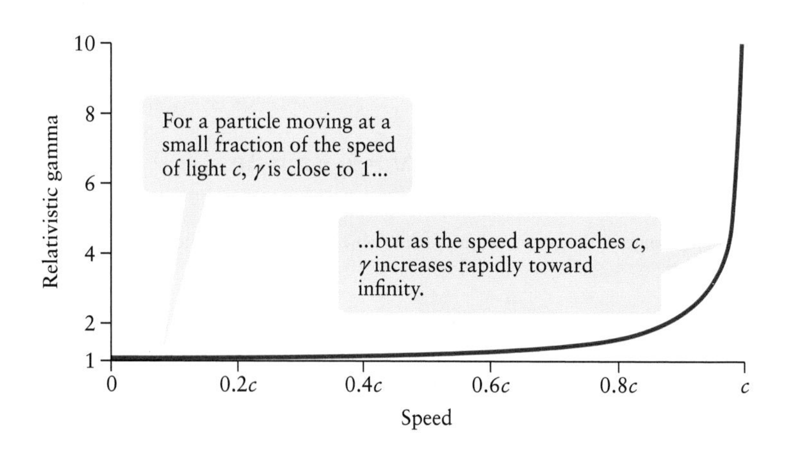

Figure 25-13 Relativistic gamma The quantity $\gamma = 1/\sqrt{1 - (v^2/c^2)}$ increases dramatically as speed v approaches the speed of light.

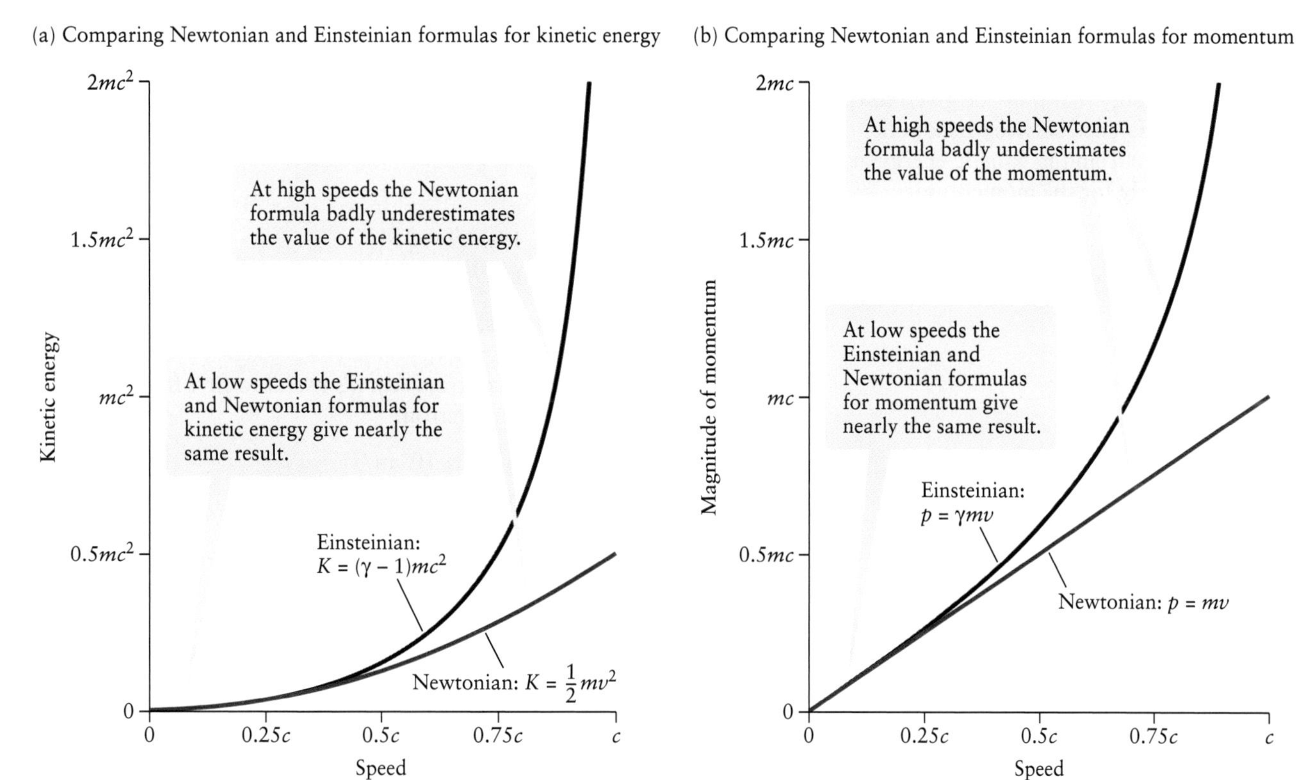

Figure 25-14 Kinetic energy and momentum in relativity These graphs show how the Newtonian and Einsteinian expressions for (a) the kinetic energy of a particle and (b) the momentum of a particle depend on the speed of the particle.

Figure 25-14 compares the kinetic energy K and the magnitude of momentum p from Equations 25-20 to the Newtonian expressions for these quantities. If v is a small fraction of the speed of light, the Einsteinian and Newtonian expressions give essentially identical results. This justifies our use of these expressions in earlier chapters, in which we considered only relatively slow-moving objects. But as the speed v approaches c, the Einsteinian expressions from Equations 25-20 approach infinity. So even though the speed of light is an upper limit to the speed of an object, there is *no* upper limit on the kinetic energy or momentum of an object.

EXAMPLE 25-8 The Energy and Momentum Costs of High Speed

Electrons can be accelerated to speeds very close to the speed of light. The mass of an electron is 9.11×10^{-31} kg. (a) How much kinetic energy must be given to an electron to accelerate it from rest to $0.900c$? How much momentum? (b) How much additional kinetic energy must be given to the electron to accelerate it from $0.900c$ to $0.990c$? How much additional momentum?

Set Up

We use Equations 25-20 to find the kinetic energy K and magnitude of momentum p for each speed. Equation 25-19 tells us the value of relativistic gamma for each speed.

Relativistic gamma:

$$\gamma = \frac{1}{\sqrt{1 - \frac{v^2}{c^2}}} \quad (25\text{-}19)$$

Einsteinian kinetic energy and momentum:

$$K = (\gamma - 1)\,mc^2$$

$$\vec{p} = \gamma m\vec{v} \quad (25\text{-}20)$$

Solve

(a) The kinetic energy that must be given to the electron is the difference between its kinetic energy at $v = 0.900c$ and its kinetic energy at $v = 0$, and similarly for the momentum.

At $v = 0$

$$\gamma = \frac{1}{\sqrt{1 - \frac{v^2}{c^2}}} = \frac{1}{\sqrt{1-0}} = \frac{1}{1} = 1$$

$$K = (\gamma - 1)\,mc^2 = (1 - 1)\,mc^2 = 0$$

$$p = \gamma mv = (1)\,m\,(0) = 0$$

Just as in Newtonian physics, a particle at rest has zero kinetic energy and zero momentum.

To calculate K and p at nonzero speeds, it's useful to first find the values of mc and mc^2 for an electron:

$$mc = (9.11 \times 10^{-31}\ \text{kg})(3.00 \times 10^{8}\ \text{m/s}) = 2.73 \times 10^{-22}\ \text{kg} \cdot \text{m/s}$$

$$mc^2 = (9.11 \times 10^{-31}\ \text{kg})(3.00 \times 10^{8}\ \text{m/s})^2 = 8.20 \times 10^{-14}\ \text{kg} \cdot \text{m}^2/\text{s}^2 = 8.20 \times 10^{-14}\ \text{J}$$

At $v = 0.900c$,

$$\gamma = \frac{1}{\sqrt{1 - \frac{v^2}{c^2}}} = \frac{1}{\sqrt{1 - \frac{(0.900c)^2}{c^2}}} = \frac{1}{\sqrt{1 - (0.900)^2}} = \frac{1}{\sqrt{0.190}} = 2.29$$

$$K = (\gamma - 1)\,mc^2 = (2.29 - 1)(8.20 \times 10^{-14}\ \text{J}) = 1.06 \times 10^{-13}\ \text{J}$$

$$p = \gamma mv = (2.29)\,m\,(0.900c) = (2.29)(0.900)\,mc = (2.29)(0.900)(2.73 \times 10^{-22}\ \text{kg} \cdot \text{m/s}) = 5.64 \times 10^{-22}\ \text{kg} \cdot \text{m/s}$$

The electron begins with $K = 0$ and $p = 0$, so it must be given 1.06×10^{-13} J of kinetic energy and 5.64×10^{-22} kg·m/s of momentum to accelerate it from rest to $0.900c$.

(b) Repeat the calculation in part (b) for the additional kinetic energy and momentum that must be given to the electron to accelerate it from $0.900c$ to $0.990c$.

At $v = 0.990c$

$$\gamma = \frac{1}{\sqrt{1 - \frac{v^2}{c^2}}} = \frac{1}{\sqrt{1 - \frac{(0.990c)^2}{c^2}}} = \frac{1}{\sqrt{1 - (0.990)^2}} = \frac{1}{\sqrt{0.0199}} = 7.09$$

$$K = (\gamma - 1)\,mc^2 = (7.09 - 1)\,(8.20 \times 10^{-14}\ \text{J}) = 4.99 \times 10^{-13}\ \text{J}$$

$$p = \gamma mv = (7.09)\,m\,(0.990c) = (7.09)(0.990)\,mc = (7.09)(0.990)(2.73 \times 10^{-22}\ \text{kg} \cdot \text{m/s}) = 1.92 \times 10^{-21}\ \text{kg} \cdot \text{m/s}$$

The difference between these values and the values of K and p at $v = 0.900c$ from part (a) tells us the additional kinetic energy and momentum that must be given to the electron:

$$\Delta K = 4.99 \times 10^{-13}\ \text{J} - 1.06 \times 10^{-13}\ \text{J} = 3.93 \times 10^{-13}\ \text{J}$$

$$\Delta p = 1.92 \times 10^{-21}\ \text{kg} \cdot \text{m/s} - 5.64 \times 10^{-22}\ \text{kg} \cdot \text{m/s} = 1.35 \times 10^{-21}\ \text{kg} \cdot \text{m/s}$$

Reflect

Compared to accelerating an electron from rest to $0.900c$, accelerating that same electron from $0.900c$ to $0.990c$ requires 3.70 times as much additional kinetic energy and 2.40 times as much additional momentum. As the speed gets closer and closer to c, it requires ever-greater amounts of kinetic energy and momentum to cause an ever-smaller speed increase. You can see that an object can never be accelerated from rest to the speed of light. This would require adding *infinite* amounts of kinetic energy and momentum to the object.

We mentioned above that the Einsteinian expressions for kinetic energy and momentum, Equations 25-20, come from an analysis of elastic collisions. In particular we must demand that if energy and momentum are conserved in one inertial frame of reference, they must be conserved in *all* inertial frames. That's required if energy and momentum are to be consistent with the first of the postulates of special relativity. It turns out that in order for this to be the case, we must also include a term mc^2 in the total energy of a particle of mass m. This quantity is called the **rest energy** of a particle, since it is present even when the particle is not in motion.

Rest energy of a particle — Mass of the particle

$$E_0 = mc^2$$

Speed of light in a vacuum

Rest energy (25-21)

Equation 25-21 is one of the most famous in science, and one of the most misunderstood. Rest energy is not potential energy; potential energy is associated with a force an object experiences. (For example, gravitational potential energy is associated with the gravitational force.) It is not kinetic energy; kinetic energy is associated with motion. Rest energy is energy that is intrinsic to an object because of its mass.

Equation 25-21 tells us that mass is simply one possible manifestation of energy. Stated another way, the mass of an object is a measure of its rest energy content. **Figure 25-15** is visual evidence of the equivalence of mass and energy. This image shows the result of a head-on collision between two protons, each of which was moving at just under the speed of light. Dozens of new particles appear after the collision. These are not fragments of the colliding protons but rather new particles that were created in the collision. This is possible because some of the kinetic energy of the colliding protons was converted into the rest energy of these new particles; another portion of this kinetic energy went into the kinetic energies of the new particles, which fly away from the collision site at high speeds.

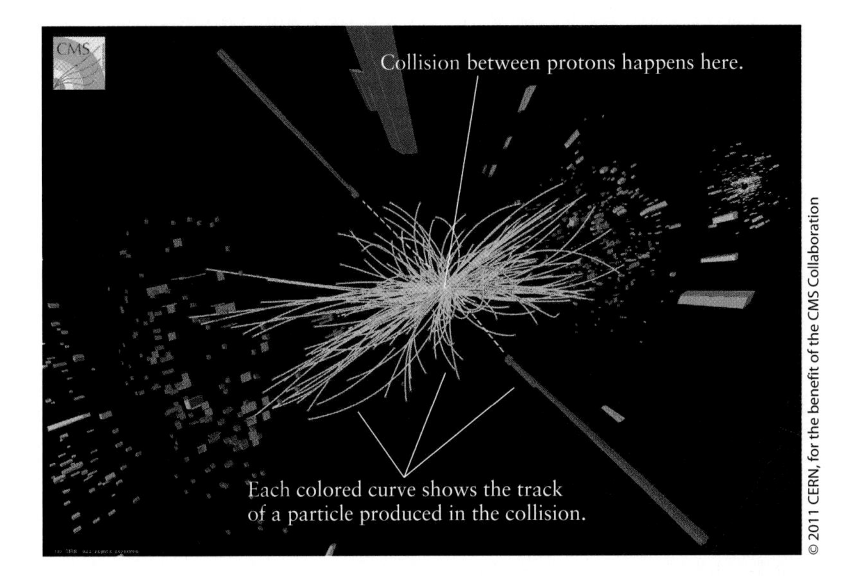

Figure 25-15 Converting kinetic energy to new particles This image from the Large Hadron Collider at CERN in Geneva, Switzerland, shows dozens of new particles produced by a collision between energetic protons.

You can observe the equivalence of mass and energy in action whenever you go outside on a sunny day or look up at the stars on a clear night. The Sun and stars shine thanks to a process occurring in their interiors in which hydrogen nuclei are fused together to form nuclei of helium. The mass of the products of such a reaction is slightly less than the mass of the hydrogen nuclei present before the reaction. The "lost" mass is converted to energy in the form of electromagnetic radiation, and it is that radiation that we see in the form of sunlight and starlight.

EXAMPLE 25-9 Converting Mass to Energy in the Sun

The Sun emits 3.84×10^{26} J of energy every second. (a) At what rate (in kg/s) is the Sun's mass decreasing? (b) How much mass has the Sun lost since it was formed 4.56×10^9 years ago? Assume that it has emitted energy at the same rate over its entire history. Compare to the present-day mass of the Sun, 1.99×10^{30} kg.

Set Up

The emitted energy comes from the rest energy of mass that is "lost" in nuclear reactions in the Sun's interior. We use Equation 25-21 to find the amount of mass equivalent to the energy emitted in 1 second. The total mass lost over the Sun's lifetime equals this amount of mass multiplied by the number of seconds that have elapsed since the Sun formed.

Rest energy:

$$E_0 = mc^2 \qquad (25\text{-}21)$$

Solve

(a) Calculate the amount of mass lost by the Sun per second.

Mass equivalent of 3.84×10^{26} J:

$$m = \frac{E_0}{c^2} = \frac{3.84 \times 10^{26}\ \text{J}}{(3.00 \times 10^8\ \text{m/s})^2} = 4.27 \times 10^9\ \text{J} \cdot \text{s}^2/\text{m}^2$$

Since $1\ \text{J} = 1\ \text{kg} \cdot \text{m}^2/\text{s}^2$, we can write this as

$$m = 4.27 \times 10^9\ \text{kg}$$

The mass of the Sun decreases at a rate of 4.27×10^9 kg/s.

(b) Calculate the total amount of mass lost by the Sun in its history.

The age of the Sun in seconds is

$$(4.56 \times 10^9\ \text{y})\left(\frac{365.25\ \text{d}}{1\ \text{y}}\right)\left(\frac{24\ \text{h}}{1\ \text{d}}\right)\left(\frac{60\ \text{min}}{1\ \text{h}}\right)\left(\frac{60\ \text{s}}{1\ \text{min}}\right) = 1.44 \times 10^{17}\ \text{s}$$

In this number of seconds the total mass lost by the Sun is

$$\left(4.27 \times 10^9\ \frac{\text{kg}}{\text{s}}\right)(1.44 \times 10^{17}\ \text{s}) = 6.14 \times 10^{26}\ \text{kg}$$

As a fraction of the Sun's present-day mass, the amount of mass lost is $(6.14 \times 10^{26}\ \text{kg})/(1.99 \times 10^{30}\ \text{kg}) = 3.09 \times 10^{-4} = 0.0309\%$ of the total mass.

Reflect

In more than 4 billion years of producing energy at a prodigious rate, the Sun has lost only a tiny fraction of its total mass. This is a testament to how much energy can be released by converting even a small amount of mass.

GOT THE CONCEPT? 25-7 Can We Build a Starship?

? The total amount of electric energy produced per year in the United States from all sources is about 1.5×10^{19} J. You propose to use all of this energy to accelerate a spacecraft to $0.990c$ in order to travel to other stars. About how massive could your proposed starship be? (a) About 3×10^6 kg (the mass of an ocean liner); (b) about 3×10^5 kg (the mass of a large airliner); (c) about 3×10^3 kg (the mass of a sport utility vehicle); (d) about 30 kg (the mass of a kayak or canoe); (e) about 3 kg (the mass of a skateboard).

TAKE-HOME MESSAGE FOR Section 25-7

✔ The mathematical expressions for kinetic energy and momentum have to be modified to be consistent with special relativity. Both the kinetic energy and momentum of an object increase without limit as the object's speed approaches *c*.

✔ Any object with mass has a kind of energy called rest energy.

25-8 Einstein's general theory of relativity describes the fundamental nature of gravity

Sitting in a chair in the patent office in Bern, Switzerland, in 1907, a young Albert Einstein had what he would call "the happiest thought of my life." He imagined a man falling freely from the roof of a house and realized that "at least in his immediate surroundings—there exists no gravitational field." If the man released an object, for example, it would accelerate at the same rate as the man accelerated, and because the man would "not feel his own weight," it would appear to him that neither he nor the object was experiencing a gravitational force.

Einstein's thought experiment led him to postulate a new principle called the **principle of equivalence.** This principle states that

> *A gravitational field is equivalent to an accelerated frame of reference in the absence of gravity.*

The principle of equivalence dictates that it is not possible to distinguish experimentally between a system in an accelerating frame and a system under the influence of gravity. In other words if you were to drop a ball in a windowless elevator car that makes no noise and doesn't shake, you could not tell from the motion of the ball whether the elevator car were sitting stationary on the surface of a planet and experiencing its gravity or accelerating in empty space, far from sources of gravity.

Let's explore physics in this imaginary elevator car further. In **Figure 25-16a** a ball is thrown horizontally while the elevator car is stationary near Earth's surface. Due to the force of gravity, the ball accelerates downward, following a familiar parabolic arc. The figure shows the positions of the ball at five instants, spanning four equal time intervals. What if the stationary elevator car were far from Earth and from any other massive object that could exert a noticeable gravitational force on the ball? In that case the ball would travel along a straight line, as shown in **Figure 25-16b.**

Now consider the situation shown in **Figure 25-17.** The elevator car is again far from any object that could exert a noticeable gravitational force on the ball, but now the car is accelerating "upward" (toward the top of the page). Because there is no discernible gravitational force, the ball travels in a straight line, as in Figure 25-16b. However, because the elevator car is accelerating, an observer *in the car* sees the ball trace out a parabolic arc with respect to the walls and floor of the car. Remember that there are no windows in the elevator car, and it makes no noise and does not vibrate as it moves. An observer in the car cannot, therefore, detect its motion. The observer observes the effect of the principle of equivalence: It appears that the ball is falling under the influence of gravity.

Einstein's theory of gravitation is therefore a theory of accelerating frames of reference. Because this is more general than the case of inertial frames, the type that was at the heart of the special theory of relativity, this expanded theory is called the **general theory of relativity.**

Predictions of General Relativity

Einstein's general theory of relativity does more than give us a way to think about accelerating frames of reference. It also makes remarkable predictions about the behavior of light and the nature of time and space, predictions that have been verified by careful experiment and observation. Let's look at a few of these.

Gravity bends light. Imagine that in Figure 25-17 we replace the thrown ball with a beam of light fired horizontally. As seen by an outside observer in an inertial frame of reference, the light beam moves in a horizontal straight line toward the far wall. But because

(a)

(b)

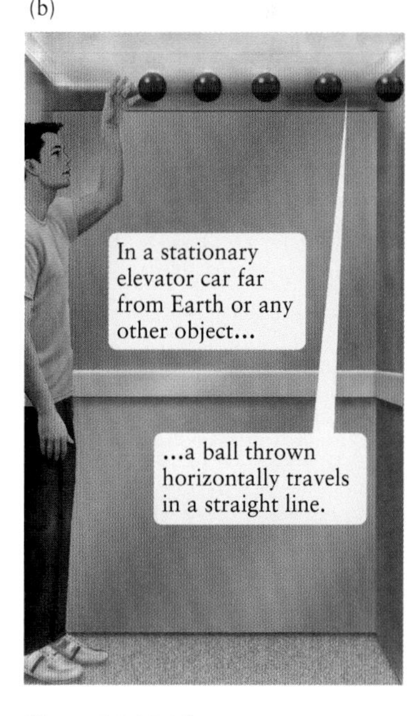

Figure 25-16 Elevator cars on Earth and in space A ball is thrown horizontally in (a) an elevator car on Earth's surface and (b) in an elevator car far from any massive object.

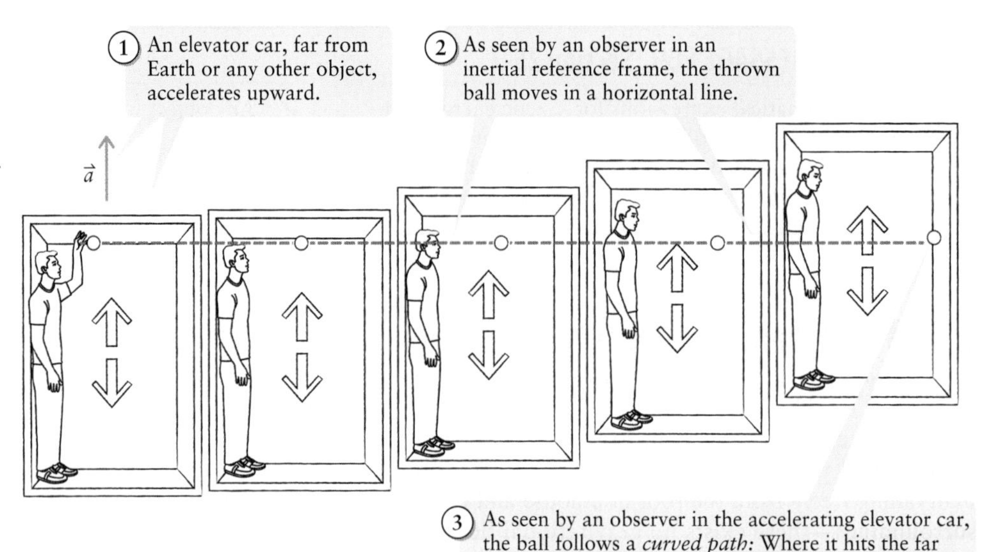

Figure 25-17 An accelerating elevator car in space When a ball is thrown horizontally in an elevator car, which is both far from any massive object and also accelerating, it follows a straight path. An observer in the elevator, however, would see the ball follow a parabolic path with respect to the floor and walls of the elevator car.

The trajectory of the ball in the accelerated frame of reference of the elevator car is identical to that of a ball in a uniform gravitational field. This agrees with Einstein's principle of equivalence.

the elevator car is accelerating upward, an observer in the car will see the light beam follow a curved path and hit the far wall below the height from which it started. (The curvature of the beam's path will be very much than less than that of the trajectory of the ball in Figure 25-17 because the light travels so much faster. Nevertheless, it *will* curve.)

According to the principle of equivalence, the effects of acceleration are indistinguishable from the effects of gravitation. So we conclude that a light beam fired horizontally on Earth will curve downward, just as the path of a ball thrown horizontally curves. This effect is called the **gravitational bending of light.**

Because the speed of light is so great, the gravitational bending of light predicted by Einstein is too small to measure on Earth. For example, a light beam traversing the width of a typical elevator would bend downward by less than 10^{-15} m, about the diameter of a proton. But the bending *is* measurable if a light beam is acted on by a much stronger gravitational field that acts over a much greater distance. The first measurement of gravitational bending of light was made in 1919 during a total solar eclipse. During totality, when the Moon blocked out the Sun's disk, astronomers photographed the stars around the Sun. These stars were shifted from their usual positions by 4.86×10^{-4} of a degree, consistent with Einstein's predictions.

A more stunning phenomenon associated with gravitational bending of light occurs when multiple images of a distant star form as light is bent by a closer, massive celestial object. **Figure 25-18** depicts such *gravitational lensing*, in which a massive object is positioned directly between Earth and a distant star. Light from the star cannot reach Earth directly, but the lensing effect results in light that was initially not propagating toward Earth to be bent back toward us. In this way light from the star can approach Earth from many directions—for example, the two directions shown in the figure. **Figure 25-19a** presents an

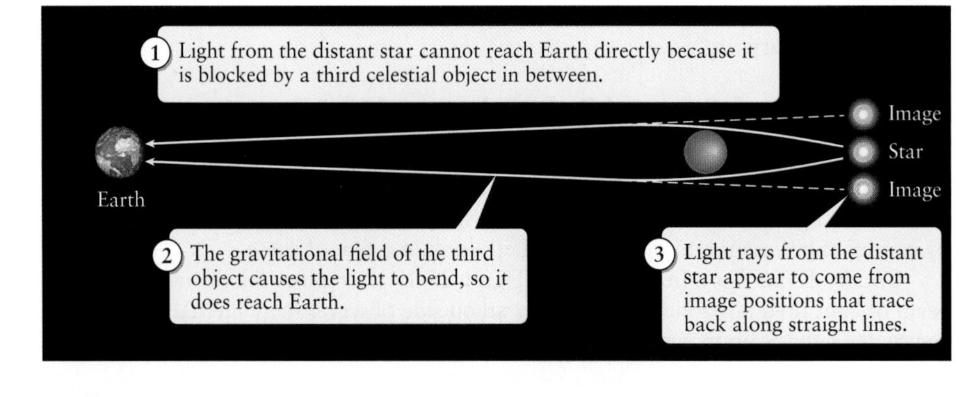

Figure 25-18 Gravity deflects light A massive object positioned between Earth and a distant star bends light from the star so that it can be seen on Earth. This is called gravitational lensing.

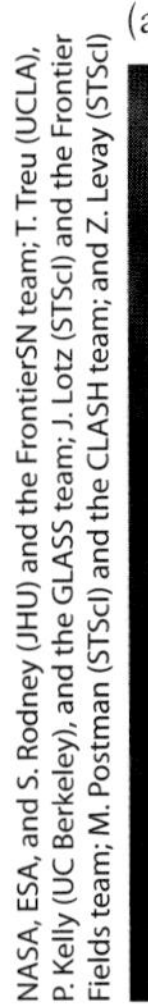

Figure 25-19 Gravitational lenses Two examples of a massive, distant galaxy acting as a gravitational lens. (a) This gravitational lens makes four images of an even more distant supernova (an exploding star). (b) A more distant galaxy is located directly along our light of sight to the gravitational lens. The resulting image of the more distant galaxy is a nearly perfect circular ring.

example of gravitational lensing showing four distinct images. When light is bent around the intervening object and reaches Earth from a full circle around it, the star's light is spread out into a ring around the lensing object, as in **Figure 25-19b**.

Space is curved, and gravitational waves propagate through space. A way to interpret the gravitational bending of light is that light in fact travels in a straight line in empty space but that *space itself is curved* by the presence of a massive object. Indeed, Einstein envisioned gravity as being caused by a curvature of space. In this picture, a massive object like Earth curves the space around it, and a light beam bends because it follows that curvature. Furthermore, an object like a ball falls toward Earth because it responds to the curvature of space that Earth produces. (A 0.1-kg ball and a 10-kg ball sense the same curvature of space due to Earth, which explains why gravity produces the same acceleration on objects of different mass.) Einstein's full mathematical formulation of the general theory of relativity is a set of equations that describe how the curvature of space *and* time, collectively referred to as *spacetime*, are affected by the presence of mass and energy.

How can we test this idea? The general theory of relativity predicts not only that spacetime curves in response to the presence of massive objects, but also that ripples in spacetime called **gravitational waves**—that is, small variations in the curvature of spacetime—should spread away from massive objects that are oscillating in a certain way (**Figure 25-20a**). (Newton's theory of gravitation makes no such prediction.) Such gravitational waves were detected for the first time in 2015 by the Laser Interferometer Gravitational-Wave Observatory (LIGO), a set of two detectors located 3000 km apart in the U.S. states of Washington and Louisiana (**Figure 25-20b**).

The operating principle of each LIGO detector is the same as that of the Michelson–Morley apparatus shown schematically in Figure 25-7. A light beam is split in two, sent along two perpendicular arms of the detector, reflected from mirrors at the end of each arm (in the case of the LIGO detectors, 4 km away), and then recombined. If a gravitational wave from space passes through a LIGO detector, the change in the curvature of space-time will slightly change the length of each arm of the detector and hence the distance that each light beam travels, and this changes the interference pattern formed when the beams recombine. Local disturbances such as an earthquake could also trigger a change in these distances, which would produce a false signal. That's why LIGO has two detectors located 3000 km apart: Only a disturbance coming from space will produce the same signal in both detectors.

As of this writing, LIGO has made five confirmed detections of gravitational wave events, most of them caused by a pair of objects many times more massive than the Sun colliding with each other. (These objects are thought to be *black holes*. In a black hole the material has been so compressed that the gravitational field near the black hole is so strong that nothing, not even light, can escape. Black holes are another prediction of general relativity.) Although such a cataclysm emits an immense amount of gravitational wave energy (equivalent to converting several times the mass of the Sun completely into energy), the signal detected by LIGO is miniscule: In each detection each 4-km arm of the detector changed in length by only about 10^{-18} m, about 10^{-3} as large

(a)

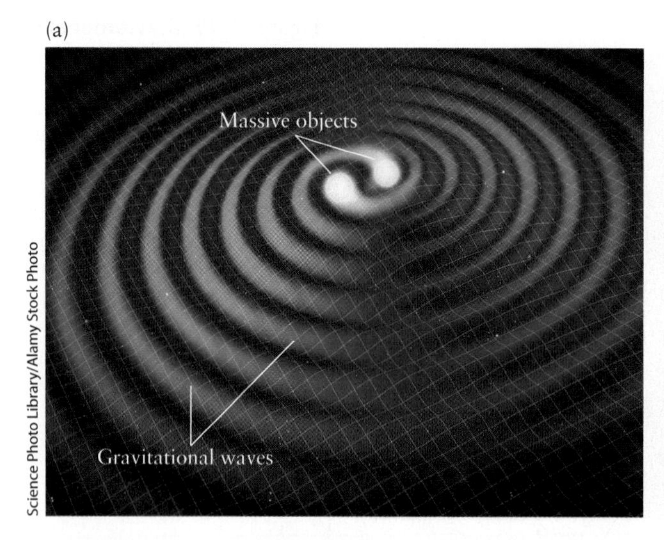

Science Photo Library/Alamy Stock Photo

(b)

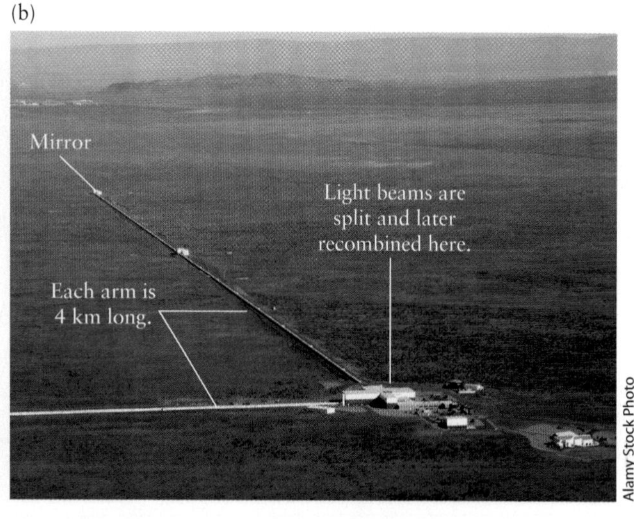

Alamy Stock Photo

Figure 25-20 Gravitational waves (a) As two massive celestial objects orbit each other, they produce ripples in space-time called gravitational waves. These gravitational waves carry away energy, causing the objects to spiral inward and emit a strong burst of gravitational waves when they finally collide and merge. (b) This LIGO gravitational wave detector is near Hanford, Washington. (The other is in Livingston, Louisiana). A laser beam is split into two, with each individual beam then sent along one of the 4-km arms and reflected back by a mirror at the end of the arm. The passage of a gravitational wave through the detector changes the lengths of each of the arms, which causes a change in the interference pattern formed when the two light beams are recombined.

as the radius of a proton. The 2017 Nobel Prize in Physics was awarded to Rainer Weiss, Barry Barish, and Kip Thorne for their work on LIGO.

Gravity affects time. We saw in Sections 25-4 and 25-5 that in the special theory of relativity, both time intervals and distances are affected by motion. Similarly, in the general theory of relativity, a massive object such as Earth affects time as well as curving space. Einstein predicted that clocks on the ground floor of a building should tick slightly more slowly than clocks on the top floor, which are farther from Earth. This **gravitational slowing of time** has been measured even for differences in height as small as 33 cm (1 ft): an experiment in 2010 using extremely precise clocks showed that the lower clock would fall behind by about 9×10^{-8} s over a 79-year human lifetime, in agreement with Einstein's prediction.

An important application of the gravitational slowing of time is to the Global Positioning System, or GPS (Figure 25-1a). A GPS receiver uses signals from orbiting satellites, each of which carries extremely accurate clocks, in order to triangulate its position on Earth. The effects of gravity slow down time on Earth's surface compared to the satellites' clocks, so the general theory of relativity *must* be taken into account for an accurate GPS result. If the gravitational slowing of time were not taken into account, a GPS receiver would accumulate errors of more than 10 km per day!

The general theory of relativity has never made an incorrect prediction. It now stands as our most accurate and complete description of gravity.

GOT THE CONCEPT? 25-8 Time Passages

? You use your spaceship to take a clock from Earth's surface to the Moon's. Which of these will affect how much time elapses on this clock compared to how much time elapses on an identical clock that you left on Earth? (a) Earth's gravitation; (b) the Moon's gravitation; (c) the speed of your spaceship; (d) both (a) and (b), but not (c); (e) all of (a), (b), and (c).

TAKE-HOME MESSAGE FOR Section 25-8

✔ The principle of equivalence, a central postulate of Einstein's general theory of relativity, states that a gravitational field is equivalent to an accelerated frame of reference in the absence of gravity.

✔ It is not possible to distinguish between a system in an accelerating frame and a system under the influence of gravity.

✔ General relativity predicts the gravitational bending of light, the existence of gravitational waves, and the gravitational slowing of time. All of these have been observed.

Key Terms

event
frame of reference (reference frame)
Galilean transformation
Galilean velocity transformation
general theory of relativity
gravitational bending of light
gravitational slowing of time
gravitational waves
inertial frame
inverse Lorentz velocity transformation
length contraction
Lorentz transformation
Lorentz velocity transformation
Michelson–Morley experiment
noninertial frame
principle of equivalence
principle of Newtonian relativity
proper time
reference frame (frame of reference)
relativistic gamma
relativistic speed
rest energy
special theory of relativity
time dilation

Chapter Summary

Topic — **Equation or Figure**

Newtonian relativity: The principle of Newtonian relativity states that the laws of motion are the same in all inertial frames of reference. The Galilean transformation relates the coordinates of an event in one frame to the coordinates of the same event in another frame, and the Galilean velocity transformation relates the velocity of an object relative to one frame to its velocity relative to another frame. These transformations are valid only for speeds that are small compared to the speed of light c.

Inertial frame of reference S' moves at speed V in the positive x direction relative to inertial frame of reference S.

$$x' = x - Vt \quad y' = y \quad z' = z \quad t' = t \qquad (25\text{-}1)$$

Coordinates of an event as measured **in frame S'** (left side); **Coordinates of the same event** as measured **in frame S** (right side).

Inertial frame of reference S' moves at speed V in the positive x direction relative to inertial frame of reference S.

$$v_x' = v_x - V \quad v_y' = v_y \quad v_z' = v_z \qquad (25\text{-}5)$$

Velocity components of an object as measured in frame S' (left side); **Velocity components of the same object as measured in frame S** (right side).

The Michelson–Morley experiment: In the nineteenth century it was hypothesized that space was filled with a material medium, called the luminiferous ether, that was required in order for light to propagate through space. The Michelson–Morley experiment provided evidence that the ether does not exist and motivated Einstein's special theory of relativity.

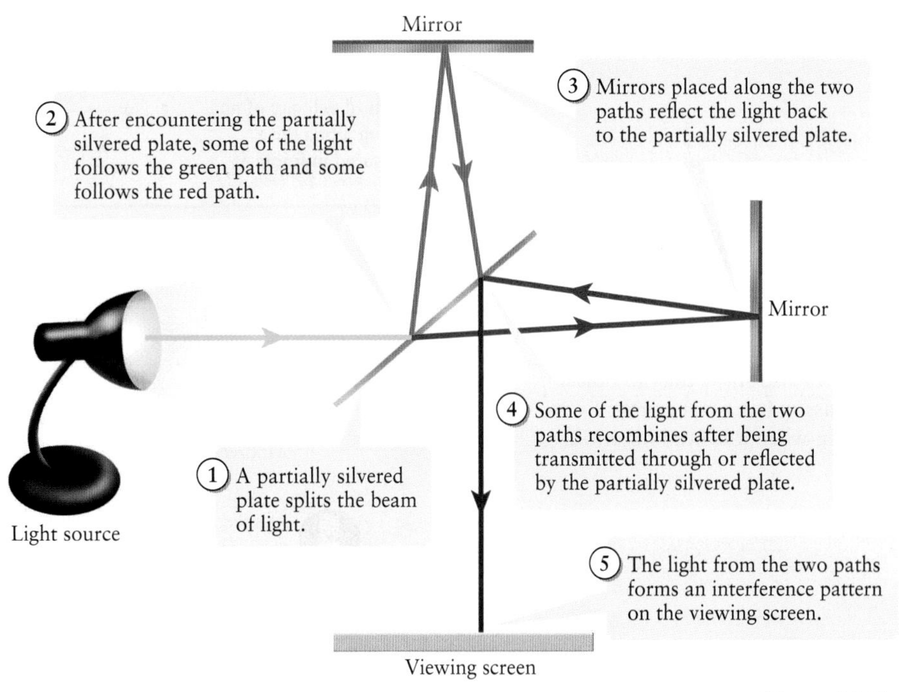
Mirror
2 After encountering the partially silvered plate, some of the light follows the green path and some follows the red path.
3 Mirrors placed along the two paths reflect the light back to the partially silvered plate.
Mirror
4 Some of the light from the two paths recombines after being transmitted through or reflected by the partially silvered plate.
Light source
1 A partially silvered plate splits the beam of light.
5 The light from the two paths forms an interference pattern on the viewing screen.
Viewing screen

(Figure 25-7)

The special theory of relativity: Einstein's special theory of relativity is based on the postulates that all laws of physics are the same in all inertial frames, and the speed of light in a vacuum is the same in all inertial frames. Two consequences are that moving clocks run slow, and moving objects are shortened along their direction of motion.

$$\Delta t = \frac{\Delta t_{proper}}{\sqrt{1 - \frac{V^2}{c^2}}} \quad (25\text{-}10)$$

- Δt_{proper}: Time interval between two events as measured in frame S', in which those events occur at the same place
- Δt: Time interval between the same two events as measured in frame S
- V: Speed of S' relative to S
- c: Speed of light in a vacuum

$$L = L_{rest}\sqrt{1 - \frac{V^2}{c^2}} \quad (25\text{-}14)$$

- L_{rest}: Length of an object in frame S', in which it is at rest
- L: Length of the same object in frame S, in which the object is moving along its length
- V: Speed of S' relative to S
- c: Speed of light in a vacuum

The Lorentz transformation: The Lorentz transformation is a generalization of the Galilean transformation that is valid for all speeds. The Lorentz velocity transformation is a generalization of the Galilean velocity transformation: It shows that the speed of light c is an ultimate speed limit and that an object at rest can never be accelerated to c.

Inertial frame of reference S' moves at speed V in the positive x direction relative to inertial frame of reference S.

$$x' = \frac{x - Vt}{\sqrt{1 - \frac{V^2}{c^2}}} \qquad y' = y \qquad z' = z \qquad t' = \frac{t - \frac{V}{c^2}x}{\sqrt{1 - \frac{V^2}{c^2}}} \quad (25\text{-}15)$$

- x', y', z', t': Coordinates of an event as measured in frame S'
- x, y, z, t: Coordinates of the same event as measured in frame S

Inertial frame of reference S' moves at speed V in the positive x direction relative to inertial frame of reference S.

$$v_x' = \frac{v_x - V}{1 - \frac{V}{c^2}v_x} \quad (25\text{-}17)$$

- v_x': x component of velocity of an object moving along the x axis as measured in frame S'
- v_x: x component of velocity of the same object as measured in frame S

Kinetic energy, momentum, and rest energy: The formulas for kinetic energy and momentum must be modified to account for c being the ultimate speed limit. The relativity postulates also show that an object with mass m has a rest energy $E_0 = mc^2$ even when it is not in motion.

$$K = (\gamma - 1)mc^2 \qquad \vec{p} = \gamma m\vec{v} \quad (25\text{-}20)$$

- K: Kinetic energy of a particle of mass m and velocity $\vec{v}$
- $\vec{p}$: Momentum of a particle of mass m and velocity $\vec{v}$
- γ: Relativistic gamma for speed v

At speeds that are a small fraction of the speed of light c, these are approximately equal to the Newtonian expressions

$$K = \frac{1}{2}mv^2 \text{ and } \vec{p} = m\vec{v}$$

The general theory of relativity: The principle of equivalence states that a gravitational field is equivalent to an accelerated frame of reference in the absence of gravity. This principle shows that light is affected by gravity just as objects with mass are.

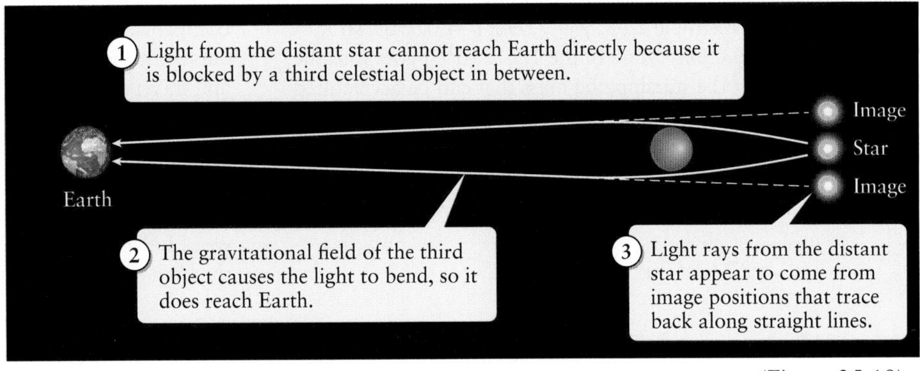

(Figure 25-18)

Answer to What do you think? Question

(b) As an object approaches the speed of light, it becomes increasingly difficult to further increase its speed. See Example 25-8 in Section 25-7.

Answers to Got the Concept? Questions

25-1 (b) Let S be the frame of reference of an observer on the ground and let S' be the frame of reference of an observer moving with the air (for example, an observer floating in a balloon). Then S' is moving due east at $V = 100$ km/h, so we take the positive x direction to be to the east. First suppose the airliner is flying west to east, so its velocity relative to the air is $v'_x = +900$ km/h. From the first of Equations 25-5, the velocity of the airplane relative to the ground in this situation is $v_x = v'_x + V = +900 \text{ km/h} + 100 \text{ km/h} = +1000 \text{ km/h}$; relative to the ground the airliner is moving east at 1000 km/h. If instead the airliner is flying east to west, its velocity relative to the air is $v'_x = -900$ km/h and its velocity relative to the ground is $v_x = v'_x + V = -900 \text{ km/h} + 100 \text{ km/h} = -800 \text{ km/h}$; the airliner is moving west at 800 km/h relative to the ground. In North America high-altitude winds typically blow from west to east, as in this example, so a west-to-east airliner trip is typically faster than an east-to-west trip between the same two airports.

25-2 (b) This is similar to Got the Concept? Question 25-1. Let S be the frame of reference of the water and let S' be the frame of reference of the dolphin. The dolphin swims in the positive x direction at speed $V = 10$ m/s. As measured in frame S, the sound wave moves in the positive x direction at velocity $v_x = 1500$ m/s. From the first of Equations 25-5, the velocity of the sound wave relative to S', the frame of reference of the dolphin, is $v'_x = v_x - V = 1500 \text{ m/s} - 10 \text{ m/s} = 1490 \text{ m/s}$. Because the dolphin is "chasing" the sound wave, the speed of the wave relative to the dolphin is less than its speed relative to the water. As we will see in Section 25-6, light waves do not behave like this: If you shoot a laser beam forward from a fast-moving vehicle, the speed of the light wave relative to the vehicle is $c = 3.00 \times 10^8$ m/s no matter how fast the vehicle is moving.

25-3 (e) The proper time is measured in any frame that is at rest with respect to the clock. Albert aboard Starship *Alpha* is at rest with respect to the clock, so he measures the clock's proper time. Although Barbara is not on the same craft as the clock, she is also not moving relative to it, so she also measures the clock's proper time. George's speed is the same as the clock's but not in the same direction. So George is moving with respect to the clock and therefore does not observe proper time intervals. The same is true for Elena on Earth, who is moving relative to the clock. We conclude that the time interval between clock ticks is the proper time for both Albert and Barbara, but it is not the proper time for either George or Elena.

25-4 (b) The statement of time dilation is that moving clocks run slow. Albert's clock is moving relative to Elena, and Elena's clock is moving at the same speed relative to Albert. So *both* observers will see the other person's clock running slow. There is no contradiction, since the two observers are looking at different pairs of events: Elena is looking at the time interval between two clicks of Albert's clock, while Albert is looking at the time interval between two clicks of Elena's clock.

25-5 (a) The length you measure will be the same as the rest length if the rod is oriented so that its length is perpendicular to the direction of motion. There is no length contraction along a perpendicular direction. (The *width* of the rod will be contracted along the direction of motion, but that's not what the question is about.)

25-6 (c) The outfielder's frame is S', so the velocity of the ball relative to her is $v'_x = +0.700c$ (positive, since the ball is now moving in the positive x direction). She is still moving at $V = 0.300c$ relative to the batter. The velocity of the ball relative to the batter is v_x, given by the inverse Lorentz velocity transformation (Equation 25-18):

$$v_x = \frac{v'_x + V}{1 + \dfrac{V}{c^2}v'_x} = \frac{0.700c + 0.300c}{1 + \dfrac{(0.300c)}{c^2}(0.700c)}$$

$$= \frac{1.00c}{1 + 0.210} = 0.826c$$

25-7 (d) The kinetic energy of the spacecraft will be $K = (\gamma - 1)mc^2$. From Example 25-8, $\gamma = 7.09$ if $v = 0.990c$, so $K = (7.09 - 1)mc^2 = 6.09mc^2$. If all of the 1.5×10^{19} J goes into the kinetic energy K of the starship, the mass that can be accelerated to $0.990c$ will be

$$m = \frac{K}{6.09c^2} = \frac{1.5 \times 10^{19}\ \text{J}}{(6.09)(3.00 \times 10^8\ \text{m/s})^2}$$

$$= 27\ \text{J} \cdot \text{s}^2/\text{m}^2 = 27\ \text{kg}$$

That's about the mass of a kayak or canoe, not including any occupants. This calculation shows that it's simply not feasible with current technology to build a spacecraft that can carry humans and travel at speeds approaching the speed of light.

25-8 (e) To account for everything that could cause a difference in time recorded by the two clocks, you would have to include the different gravitational slowing of time on Earth and the Moon (which has weaker gravity than Earth). You would also have to include the effects of time dilation due to the high-speed travel from Earth to Moon.

Questions and Problems

In a few problems you are given more data than you actually need; in a few other problems you are required to supply data from your general knowledge, outside sources, or informed estimate.

Interpret as significant all digits in numerical values that have trailing zeros and no decimal points.

For all problems use $g = 9.80\ \text{m/s}^2$ for the free-fall acceleration due to gravity. Neglect friction and air resistance unless instructed to do otherwise.

•	Basic, single-concept problem
••	Intermediate-level problem; may require synthesis of concepts and multiple steps
•••	Challenging problem
SSM	*Solution is in Student Solutions Manual*
Example	*See worked example for a similar problem*

Conceptual Questions

1. • How would you change the following Galilean transformation equations in the case where a frame was moving both in the x and the y direction?

$$x' = x - Vt$$
$$y' = y$$
$$z' = z$$
$$t' = t$$

2. • What is a frame of reference? What is an inertial frame of reference? SSM

3. • What kind of reference frame is Earth's surface? Explain your answer.

4. • What does the phrase "all motion is relative, there is no absolute motion" mean?

5. • Describe the Michelson and Morley experiment. Why do you think it was repeated so many times?

6. • What are the two postulates of the special theory of relativity?

7. • Explain how the measurement of time enters into the determination of the length of an object. SSM

8. • Is it possible for one observer to find that event A happens after event B and another observer to find that event A happens before event B? Explain your answer.

9. • Is it possible to accelerate an object to the speed of light in a real situation? Explain your answer.

10. • Gene Roddenberry created the popular TV and film series *Star Trek*. Once at a public lecture he was asked if he thought we would ever be able to travel faster than light. He responded, "No, and that's a good thing because it means that we can always do it in science fiction." How have science fiction writers gotten around the cosmic speed limit?

11. • What is the fundamental postulate of the general theory of relativity?

12. • Describe one of the predictions of general relativity.

13. • Why do you think there was so much resistance from the established physics community when Einstein proposed his new theory of relativistic motion in 1905?

Multiple-Choice Questions

14. • Which of the following statements are true?
   A. The laws of motion are the same in all inertial frames of reference.
   B. The laws of motion are the same in all reference frames.
   C. There is no way to detect absolute motion.
   D. All of the above statements are true.
   E. Only two of the above statements are true.

15. • If we used radio waves to communicate with an alien spaceship approaching Earth at 10% of the speed of light, we would receive their signals at a speed of
   A. $0.10c$.
   B. $0.90c$.
   C. $0.99c$.
   D. $1.00c$.
   E. $1.10c$. SSM

16. • Time dilation means that
   A. the slowing of time in a moving frame of reference is only an illusion resulting from motion.
   B. time really does pass more slowly in a frame of reference moving relative to a frame of reference at relative rest.
   C. time really does pass more slowly in a frame of reference at rest relative to a frame of reference that is moving.
   D. time is unchanging regardless of the frame of reference.
   E. no two clocks at rest can ever read the same time.

17. • A meter stick hurtles through space at a speed of $0.95c$ with its length perpendicular to the direction of motion. You measure its length to be equal to
   A. 0 m.
   B. 0.05 m.
   C. 0.95 m.
   D. 1.00 m.
   E. 1.05 m.

18. • A particle has an Einsteinian momentum of p. If its speed doubles, the Einsteinian momentum will be
   A. greater than $2p$.
   B. equal to $2p$.
   C. less than $2p$, but not equal to $2p/c$.
   D. equal to $2p/c$.
   E. equal to p.

19. • Consider two atomic clocks, one at the GPS ground control station near Colorado Springs (elevation 1830 m) and the other one in orbit in a GPS satellite (altitude 20,200 km). According to the general theory of relativity, which atomic clock runs slow?

A. The clock in Colorado runs slow.
B. The clock in orbit runs slow.
C. The clocks keep identical time.
D. The orbiting clock is 95% slower than the clock in Colorado.
E. The orbiting clock alternately runs slow and then fast depending on where the Sun is. SSM

20. Aamir, Bob, and Cesar are all the same age when they start an interstellar shipping business to a colony several light years from Earth. Aamir takes the first round trip alone while the others stay home on Earth. On the next trip he takes Cesar with him while Bob stays on Earth. Bob and Cesar take the third trip together while Aamir takes some well-deserved rest on Earth. Assuming each trip is identical, what is the order of their ages after the third trip?

A. Aamir > Bob > Cesar
B. Aamir > Bob = Cesar
C. Cesar = Aamir > Bob
D. Bob > Aamir = Cesar
E. Bob = Cesar > Aamir

21. A massive particle speeds up and slows down as it travels along at an appreciable fraction of the speed of light. The relativistic kinetic energy

A. will always be less than the classical kinetic energy.
B. will always be the same as the classical kinetic energy.
C. will always be greater than the classical kinetic energy.
D. cannot be compared with the classical kinetic energy without more information.
E. will always be equal to the rest energy.

Estimation/Numerical Analysis

22. • Estimate the difference in a 10,000-s time interval as measured by a proper observer and a relative observer traveling on a commercial jetliner.

23. •• Estimate the relativistic correction factor gamma for an electron in a scanning electron microscope, which accelerates the electrons through a potential difference of several tens of kV.

24. • Estimate the percent error between the Newtonian momentum and the Einsteinian momentum at (a) typical macroscopic speeds and (b) speeds that are a substantial fraction of c.

25. • Are there any numerical values of v/c that lend themselves to relatively simple calculations of γ (without the use of a calculator)?

26. • If all effects due to general relativity are neglected, and the effects of time dilation are cumulative, how many trips into outer space would an astronaut have to make to experience a 10% change in the length of his life (as measured by an observer at rest on Earth)? Is the time longer or shorter by 10% according to the observer on Earth?

27. • Estimate the percentage of the world's population that has a basic understanding of Einstein's special theory of relativity (assume your understanding of the concepts in this chapter constitutes a basic understanding). SSM

28. A car is driving along the freeway. Estimate the ratio of its kinetic energy to its rest energy.

29. • This problem is intended to help you to understand the behavior of relativistic variables such as time, length, and momentum. To begin, construct a table of relativistic gamma γ vs v/c for the following values of v/c:

v/c	v/c
0	0.7
0.1	0.8
0.2	0.9
0.3	0.99
0.4	0.999
0.5	0.9999
0.6	

Now plot a graph of γ versus v/c. As long as γ is close to unity (1.0), the effects of Einstein's special theory of relativity go unnoticed. (a) From your graph (or your table) determine the maximum value of v/c for which γ is less than 1% larger than unity (1.0). (b) What numerical value does the slope of your graph take at $v/c = 0.1$, $v/c = 0.5$, $v/c = 0.9$, and $v/c = 0.999$?

Problems

25-1 The concepts of relativity may seem exotic, but they're part of everyday life

25-2 Newton's mechanics includes some ideas of relativity

30. • A bicyclist rides at 8.00 m/s toward the north. A car is moving at 25.0 m/s, also toward the north, and is initially behind the rider. A truck is moving at 15.0 m/s toward the south, approaching the bicycle and the car. (a) Make a sketch of the three vehicles and label their respective velocity vectors, relative to the ground. (b) Calculate the relative velocities of each vehicle compared to the other two. Example 25-1

31. • A frame of reference, S, is fixed on the surface of Earth with the x axis pointing toward the east, the y axis pointing toward the north, and the z axis pointing up. A second frame of reference, S', is moving at a constant 4.00 m/s toward the east. At time $t = t' = 0$ the origins of both frames of reference coincide. (a) Describe mathematically the relationships between x' and x, y' and y, and z' and z. (b) A picture is taken in the S frame of reference at $t = 4.00$ s at the point (2 m, 1 m, 0 m). Calculate the corresponding values of x', y', and z' for the same event in the S' frame. SSM Example 25-1

32. • A boat sails from the pier at Fisherman's Wharf in San Francisco at 4.00 m/s, directly toward Alcatraz. A kite rider heads directly away from Alcatraz toward the boat at a relative speed of 6.00 m/s, according to the skipper of the sailboat. Calculate the velocity of the kite rider relative to the pier. Example 25-1

33. •• A radio-controlled model car travels at 15.0 m/s, to the right, relative to the parking lot that it is driving on. A girl on her scooter chases after the model car at 4.00 m/s. The parking lot represents reference frame S, the model car is in frame S', and the girl is described by frame S''. Assume that the car and the girl are both at the origin of the parking lot at $t = 0$ s. Write down the Galilean transformation between (a) S and S', (b) S and S'', and (c) S' and S''. Example 25-1

34. • Assume that the origins of S and S' coincide at $t = t' = 0$ s. An observer in inertial frame S measures the space and time coordinates of an event to be $x = 750$ m, $y = 250$ m, $z = 250$ m, and $t = 2.0$ μs. What are the space coordinates of the event in inertial frame S', which is moving in the $+x$ direction at a speed of $0.01c$ relative to S? Example 25-1

35. • At time $t' = 4.00 \times 10^{-3}$ s, as measured in S', a particle is at the point $x' = 10$ m, $y' = 4$ m, and $z' = 6$ m. Compute the corresponding values of x, y, and z, as measured in S, for (a) a relative velocity between S' and S of $+500$ m/s and (b) a relative velocity between S' and S of -500 m/s. Assume that the origins of S and S' coincide at $t = 0$ and that the motion lies along the x and x' axes. Example 25-1

36. • Suppose that at $t = 6.00 \times 10^{-4}$ s, the space coordinates of a particle are $x = 100$ m, $y = 10$ m, and $z = 30$ m according to coordinate system S. Compute the corresponding values as measured in the frame S' if the relative velocity between S' and S is 150,000 m/s along the x and x' axes. The origins of the reference frames coincide at $t = 0$. Example 25-1

37. ••• A floatplane lands on a river. The velocity of the airplane relative to the air is 30 m/s, due east; the velocity of the wind is 20 m/s, due north; and the current in the river is 5 m/s, due south. Calculate the velocity of the plane relative to the water just before it lands. SSM Example 25-1

38. • A spaceship moves by Earth at 2.4×10^8 m/s. A satellite moves by Earth in the opposite direction at 1.6×10^8 m/s. Use the Galilean transformation to calculate the speed of the satellite relative to the spaceship and comment on your answer. Example 25-1

25-3 The Michelson–Morley experiment shows that light does not obey Newtonian relativity

39. • The Michelson–Morley experiment was performed hundreds of times in a futile attempt to find the luminiferous ether. Why was the experiment performed on an enormous slab of marble that was floated in a pool of mercury?

40. •• Consider an airplane traveling at 50.0 m/s airspeed between two points that are 200 km apart. What is the round-trip time for the plane (a) if there is no wind? (b) if there is a wind blowing at 10.0 m/s along the line joining the two points? Example 25-1

41. •• Suppose that the distance in the Michelson–Morley experiment (Figure 25-7) from the partially silvered plate to either mirror is 20,000 m. (a) How much time would light, moving at speed c, need to travel from the partially silvered plate to one mirror and back again? How much *additional* time would be required *if the ether existed*, if the Galilean velocity transformation were valid, and if Earth moved relative to the ether along the direction from the partially silvered plate to this mirror at the following speeds: (b) at $0.01c$, (c) at $0.1c$, (d) at $0.5c$, and (e) at $0.9c$?

25-4 Einstein's relativity predicts that the time between events depends on the observer

42. • An observer in reference frame S observes that a lightning bolt strikes the origin, and 10^{-4} s later a second lightning bolt strikes the same location. What is the time separation between the two lightning bolts determined by a second observer in reference frame S' moving at a speed of $0.8c$ along the collinear x–x' axis? Example 25-2

43. • A radioactive particle travels at $0.80c$ relative to the laboratory observers who are performing research. Calculate the half-life of the particle as measured in the laboratory frame compared to the half-life according to the proper frame of the particle. Example 25-3

44. • Muons at rest in the laboratory have a half-life of 1.56 μs. What is the average lifetime, measured in the laboratory, of muons traveling at $0.99c$ with respect to the laboratory? Example 25-3

45. • When certain unstable subatomic particles traveled through the laboratory at a speed of $0.98c$, scientists measured an average half-life of 11 μs for their decay. If these particles were at rest in the laboratory, what would be their half-life? SSM Example 25-3

46. •• The time dilation effect between two frames of reference is measured in the lab to have a 0.01% difference between the relative time and the proper time $[\Delta t = (t - t_{\text{proper}})/t_{\text{proper}} \times 100\% = 0.01\%]$. Calculate the relative speed of the two reference frames. Example 25-2

47. •• **Astronomy** The Andromeda galaxy is a spiral galaxy that is a distance of 2.54 million light years from Earth. Is there any possible speed that a spaceship can achieve to deliver a human being to this galaxy? Consider the lifetime of a human to be 80 years. Example 25-2

48. • **Astronomy** The nearest star to our own sun is Proxima Centauri, 4.25 light years away. If a spaceship travels at $0.75c$, how much time is required for a one-way trip (a) according to an Earth observer and (b) according to the captain of the ship? Example 25-2

49. • A subatomic particle is traveling with a relativistic gamma, $\gamma = 1/\sqrt{1-(v/c)^2}$, of 20 as observed by a radiation monitor in a nuclear power plant. The particle is observed to decay 30 ns after it is observed by the plant. What is the proper lifetime of this particle? Example 25-3

50. • **Astronomy** A spaceship travels at $0.95c$ toward Alpha Centauri. According to Earthlings, the distance is 4.37 light years. (a) From the perspective of the space travelers, how long does it take to reach this star if the ship starts at Earth? (b) How long do Earthlings measure for the trip? Example 25-3

25-5 Einstein's relativity also predicts that the length of an object depends on the observer

51. • A car travels in the positive x direction in the reference frame S. The reference frame S' moves at a speed of $0.80c$, along the x axis. The proper length of the car is 3.20 m. Calculate the length of the car according to observers in the S' frame. SSM Example 25-4

52. • A moving spaceship is measured to have a length that is two-thirds of its proper length. What is the speed of the spaceship relative to the observer? Example 25-4

53. • A stick moves past an observer at a speed of $0.44c$. According to the observer the stick is oriented parallel to the direction of motion and is 0.88 m long. Determine the proper length of the stick. Example 25-4

54. •• A standard tournament domino is 1.5 in wide and 2.5 in long. Describe how you might orient a domino so that it will measure 1.5 in by 1.5 in as it moves by. What relative speed is required? Example 25-5

55. •• How fast must a positive pion (an unstable particle) be moving to travel 100 m (according to the laboratory frame) before it has a 50% chance of decaying? The half-life, at rest, of a positive pion is 1.80×10^{-8} s. Give your answer in units of meters per second (m/s) and as a fraction of the speed of light (in other words, $v = ?$ and $v/c = ?$). SSM Example 25-5

56. •• You stand on the ground halfway between two mountains, one on your left side and one on your right side. Atop both mountain peaks are bright lamps, which are separated by a distance of 2.00 km. You observe them both turn on, but at slightly different times. As you observe this, a pilot, who is flying a super-advanced rocket plane at a speed of $0.750c$, flies parallel to a line that connects both mountain peaks. According to the pilot, who is flying right-to-left in your reference frame, both lamps turn on at the same time. Which lamp do you observe turning on first, and by how much time does it precede the other lamp? Example 25-6.

57. •• A spacecraft passes you traveling at $0.600c$. Your alien friend Gaar on the spaceship measures the length of the ship as 60 m from front to back. He stands in the middle of the ship and fires one photon backward and one photon forward, and each photon strikes a detector at each end of the ship. (a) According to Gaar, what is the travel time of each photon? Verify that he reports the detection events as simultaneous. (b) According to you, which photon registers first? (c) Use the Lorentz transformation, and the fact that the detection events are simultaneous according to Gaar, to find the time difference according to you. Example 25-6.

25-6 The relative velocity of two objects is constrained by the speed of light, the ultimate speed limit

58. •• Spaceship A moves at $0.8c$ toward the right, while spaceship B moves in the opposite direction at $0.7c$ (both speeds are measured relative to Earth). (a) Calculate the velocity of Earth relative to spaceship A. (b) Calculate the velocity of Earth relative to spaceship B. (c) Calculate the velocity of spaceship A relative to spaceship B. Example 25-7

59. •• A (very fast) car traveling with a velocity of $+0.35c$ to the right passes an observer sitting on the side of the road. A truck traveling with a velocity of $+0.25c$ passes the same observer. Determine the relative velocity between the car and the truck. Give your answer in terms of "the velocity of the truck relative to the car is..." and "the velocity of the car relative to the truck is...." Example 25-7

60. •• A spaceship flies by Earth at $0.92c$. It fires a rocket at $0.75c$ in the forward direction, relative to the spaceship. What is the velocity of the rocket relative to Earth? Example 25-7

61. •• Suppose the spaceship in problem 61 continues to fly by Earth at $0.92c$. This time, however, it fires a rocket at $0.75c$ in the backward direction relative to the spaceship. What is the velocity of the rocket relative to Earth? SSM Example 25-7

62. •• Prove that the relative velocity of a laser fired from a spaceship that is moving at $0.92c$ past Earth will be c from the perspective of Earth. Example 25-7

63. •• A proton travels at 99.999954% of the speed of light. An antiproton travels in the opposite direction at the same speed. What is the relative velocity of the two particles? Example 25-7

25-7 The equations for kinetic energy and momentum must be modified at very high speeds

64. • A 2.00-kg object moves at 400,000 m/s. (a) Calculate the Newtonian momentum of the object. (b) Calculate the Einsteinian momentum of the object. (c) Which of the answers is correct? What is the percent difference? Example 25-8

65. •• An electron travels at $0.444c$. Calculate (a) its Einsteinian momentum, (b) its Einsteinian kinetic energy, (c) its rest energy, and (d) the total energy of the electron. SSM Example 25-8

66. • A proton ($m = 1.673 \times 10^{-27}$ kg) is traveling at $0.5c$. Calculate its Einsteinian momentum and its Einsteinian kinetic energy. Example 25-8

67. • A particle is traveling with respect to an observer such that its Einsteinian energy is twice its rest energy. How fast is it moving with respect to the observer? Example 25-8

68. •• A particle has a rest energy of 5.33×10^{-13} J and a total energy of 9.61×10^{-13} J. Calculate the momentum of the particle. Example 25-8

69. •• A proton has a rest energy of 1.50×10^{-10} J and a momentum of 1.07×10^{-19} kg·m/s. Calculate its speed. SSM Example 25-8

70. • Recent home energy bills indicate that a household used 411 kWh of electrical energy and 201 therms for gas heating and cooking in a period of one month. Given that 1.0 therm is equal to 29.3 kWh, how many milligrams of mass would need to be converted directly to energy each month to meet the energy needs for the home? SSM Example 25-9

25-8 Einstein's general theory of relativity describes the fundamental nature of gravity

71. •• What would an observer inside an elevator measure for the free-fall acceleration near the surface of Earth if the elevator accelerates downward at 8.00 m/s^2?

72. •• What would an observer inside an elevator measure for the free-fall acceleration near the surface of Earth if the elevator accelerates downward at 18.0 m/s^2? SSM

General Problems

73. •• A super rocket car traverses a straight track 2.40×10^5 m long in 10^{-3} s as measured by an observer next to the track. (a) How much time elapses on a clock in the rocket car during the run? (b) What is the distance traveled in traversing the track as determined by the driver of the rocket car? Example 25-3

74. •• Observers in reference frame S see an explosion located at $x_1 = 580$ m. A second explosion occurs 4.5 μs later at $x_2 = 1500$ m. In reference frame S', which is moving along the $+x$ axis at speed v, the explosions occur at the same point in space. What is the separation in time between the two explosions as measured in S'? SSM Example 25-6

75. •• A radioactive nucleus traveling at a speed of $0.8c$ in a laboratory decays and emits an electron in the same direction as the nucleus is moving. The electron travels at a speed of $0.6c$ relative to the nucleus. (a) How fast is the electron moving according to an observer in the laboratory? (b) Rocket A travels away from Earth at $0.6c$, and rocket B travels away from Earth in exactly the opposite direction at $0.8c$. What is the speed of rocket B as measured by the pilot of rocket A? (c) Why did you get the same answer that you did for part (a)? Example 25-7

76. •• A beam of positive pions (unstable particles) has a speed of $0.88c$. Their half-life, as measured in the reference frame of the laboratory, is 1.80×10^{-8} s. What is the distance traveled by the laboratory, as measured by the pion, during one half-life? Example 25-3

77. ••• Muons have a proper half-life of 1.56×10^{-6} s. Suppose a muon is formed at an altitude of 3000 m and travels at a speed of $0.950c$ straight toward Earth. (a) Does the muon reach Earth's surface within one half-life? Complete the problem from both perspectives: the muon's point of view and Earth's reference frame. (b) Calculate the minimum speed of the muon so that it *just barely* reaches Earth's surface after one half-life. Example 25-3

78. •• Two students, Nora and Allison, are both the same age when Allison hops aboard a flying saucer and blasts off to achieve a cruising speed of $0.800c$ for 20 years (according to Allison). Neglecting the acceleration of the ship during blastoff, landing, and turnarounds, find the difference in age between Nora and Allison when they are reunited. Who is younger? Example 25-2

79. •• Suppose a jet plane flies at 300 m/s relative to an observer on the ground. Using only special relativity, determine how far the plane must travel, as measured by an observer on the ground, before the clocks aboard the plane are 10 s behind clocks on the ground. Assume the two clocks were originally synchronized to start. SSM Example 25-2

80. •• At the end of the linear accelerator at the Stanford Linear Accelerator Center (SLAC), electrons have a speed of $0.99999999995c$. (a) Calculate the value of γ for an electron at SLAC. (b) What time interval would an observer at rest relative to the accelerator measure for a time interval of 1.66 μs measured from the electron's perspective? Example 25-3

81. •• The half-life of a certain subatomic particle at rest is about 1×10^{-8} s. How fast is a beam of these particles moving if one-half of them decay in 6×10^{-8} s as measured in the laboratory? Example 25-3

82. •• Based on experiments in your lab, you know that a certain radioactive isotope has a half-life of 2.25 μs. As a high-speed spaceship passes your lab, you measure that the same isotope at rest inside the spaceship takes 3.15 s for one-half of it to decay. (a) What is the half-life of the isotope as measured by an astronaut working inside the spaceship? (b) How fast is the spaceship traveling relative to Earth? Example 25-3

83. • We know 1 kg of trinitrotoluene (TNT) yields an energy of 4.2 MJ. The energy released comes from the chemical bonds in the material. How much mass would be required to create an explosion equivalent to 1.8×10^9 kg TNT? Assume all the energy released comes simply from the rest energy of the material. Example 25-9

84. •• **Astronomy** High-speed cosmic rays strike atoms in Earth's upper atmosphere and create secondary showers of particles. Suppose a particle in one of the showers is created 25.0 km above the surface traveling downward at 90.0% the speed of light. Consider the following two events: "a particle is created in the upper atmosphere" and "a particle strikes the ground." We can view the events from two reference frames, one fixed on Earth and one traveling with the created particle. (a) In which of the reference frames are the proper time and the rest length between the two events measured? Explain your reasoning. (b) In the particle's reference frame how long after its creation does it take it to reach the ground? (c) In Earth's reference frame how long after creation does it take the particle to reach the ground? (d) Show that the times in parts (b) and (c) are consistent with time dilation. Example 25-4

85. ••• A rocket 642 m long is traveling parallel to Earth's surface at $0.5c$ from left to right. At time $t = 0$ a light flashes for an instant at the center of the rocket. Detectors at opposite ends of the rocket record the arrival of the light signal. Call event A the light striking the left detector and event B the light striking the right detector. Observers at rest in the rocket and on Earth record the events. (a) What is the speed of the light signal as measured by the observer (i) at rest in the rocket and (ii) at rest on Earth? (b) At what time after the flash do events A and B occur as measured by (i) the observer in the rocket and (ii) the observer at rest on Earth? Which event occurs first in each case? (c) Show that the results in part (b) are consistent with time dilation. Example 25-6

86. •• **Biology** Twin astronauts, Harry and Larry, have identical pulse rates of 70 beats/min on Earth. Harry remains on Earth, but Larry is assigned to a space voyage during which he travels at $0.75c$ relative to Earth. What will be Larry's pulse rate as measured by (a) Harry on Earth and (b) the physician in Larry's spacecraft? Example 25-2

87. •• A rocket is traveling at speed v relative to Earth. Inside a lab in the rocket, two laser beams are turned on, one pointing in the forward direction and the other pointing in the backward direction relative to the rocket's velocity. (a) What is the speed of each laser beam relative to the laboratory in the rocket? (b) Use the Lorentz velocity transformation to find the velocity of each laser beam as measured by an observer at rest on Earth. (c) As observed from Earth, how fast are the two laser beams separating from *each other*? Example 25-7

88. ••• A 1000-kg rocket is flying at $0.90c$ relative to your lab. Calculate the kinetic energy of the rocket using the Einsteinian formula and the ordinary Newtonian formula. What is the percent error if we use the Newtonian formula? Does the Newtonian formula overestimate or underestimate the kinetic energy? Example 25-8

89. ••• A spaceship is traveling at $0.50c$ relative to Earth. Inside the ship, a cylindrical piston that is 50.0 cm long and 4.50 cm in diameter contains 1.25 mol of ideal gas at 25.0°C under 2.20 atm of pressure. The cylinder is oriented with its axis parallel to the direction in which the spaceship is flying. What is the particle density (in molecules per cubic meter) of the gas in the cylinder as measured by (a) an astronaut in the rocket ship's lab and (b) a scientist in an Earth lab? Example 25-5

90. ••• **Astronomy** In December 2009 the discovery was announced of a planet that may have a large amount of water and hence would be a good candidate for possible life. The planet, GJ 1214b, orbits a small star that is 42 light years from Earth. In the future we might decide to send some astronauts to explore the planet. When they arrive there we want them to be young enough to perform tests. Suppose that the captain is 25 years old at launch time. (a) What is the minimum speed the spaceship will need for the captain to be no more than 60 years old at arrival? (b) As soon as the spaceship arrives at the planet, the captain has orders to send a radio signal to Earth to notify Mission Control that the trip was successful. How many years after launch from Earth will it be when the signal arrives at Earth? You can ignore acceleration times and any motion of Earth and GJ 1214b. SSM Example 25-4

26 Quantum Physics and Atomic Structure

ESO

What do you think?

Some of the stars in this photo are blue, while others are red. Based on the color alone, you can conclude that the blue stars are (a) hotter than the red stars; (b) cooler than the red stars; (c) made of different materials than the red stars; (d) both (a) and (c); (e) both (b) and (c).

In this chapter, your goals are to:

- (26-1) Recognize the limitations of classical physics for explaining the properties of light and matter.
- (26-2) Describe how the photoelectric effect and blackbody radiation provide evidence for the photon picture of light.
- (26-3) Explain why the wavelength of a photon increases if it scatters from an electron.
- (26-4) Calculate the wavelength of a particle such as an electron.
- (26-5) Describe why atoms absorb light at only certain wavelengths, and why they emit and absorb light at the same wavelengths.
- (26-6) Explain how the Bohr model of the atom explains the spectrum of hydrogen.

To master this chapter, you should review:

- (3-7) Uniform circular motion.
- (4-2) Newton's second law.
- (8-8) Angular momentum of a moving particle.
- (14-3) The energy of molecules in an ideal gas.
- (14-7) Heat transfer by radiation.
- (17-3) Electric potential.
- (22-2, 22-3, and 22-4) The electromagnetic nature of light.
- (25-7) Relativistic kinetic energy and momentum.

26-1 Experiments that probe the nature of light and matter reveal the limits of classical physics

In Section 22-4 we introduced the idea that light and other electromagnetic waves come in particle-like packets of energy called photons. Is there solid evidence for this? Can a photon strike a particle such as an electron and "bounce" or scatter the way a cue ball does when it strikes a billiard ball? And if waves have a particle-like nature, do particles ever exhibit properties we associate with waves?

TAKE-HOME MESSAGE FOR Section 26-1

✔ Light waves have some of the characteristics of particles, and particles such as electrons have some of the characteristics of waves.

✔ The energies of atoms are quantized (restricted to certain specific values).

Night vision goggles "amplify light" using the photoelectric effect. Even faint light causes a surface to emit electrons, and the current of these electrons can be amplified in a circuit to generate a brighter image that the wearer of the goggles can see.
(a)
Archive Image/Alamy

The characteristic color of neon lights is due to the structure of the neon atom. When excited by an electric current, neon atoms in the light tube jump to a higher energy level. When they jump down to a lower energy level, they emit photons of a specific frequency, wavelength, and color.
(b)
VACANCY
iStockphoto/Thinkstock

Figure 26-1 Photons, electrons, and atoms Two examples of the interaction between light and matter. These can only be understood if we use the ideas that light has particle aspects and matter has wave aspects.

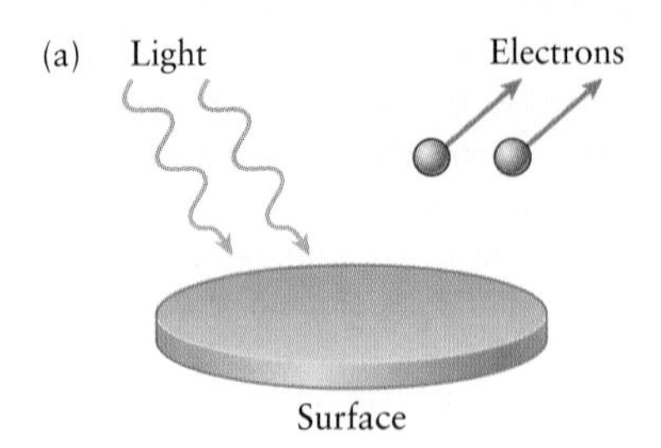

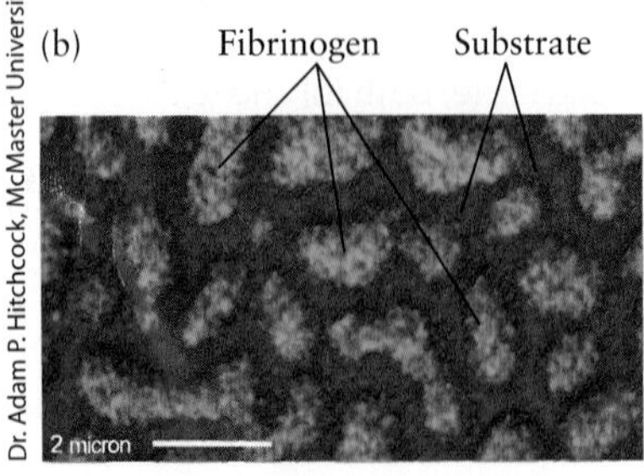

Figure 26-2 The photoelectric effect (a) In the photoelectric effect electrons escape from a surface when the surface is illuminated with light. (b) An example of photoemission electron microscopy. Protein interactions with polymers are critical to understanding the compatibility of synthetic materials with blood (such as medical implants). Here the spatial distribution of fibrinogen, a blood protein, is mapped relative to a PS/PMMA polymer substrate, using photoemission electron microscopy (PEEM). Data measured at the Advanced Light Source, Lawrence Berkeley National Laboratory. This image could be made because fibrinogen and the substrate emit different numbers of electrons when illuminated with ultraviolet light.

In this chapter we'll see that the answer to each of these questions is "yes." The photoelectric effect (**Figure 26-1a**), in which light striking a surface causes the surface to eject electrons, can be understood only if light comes in the form of photons. The same is true for blackbody radiation, the light emitted by an object as a consequence of its temperature. We'll see direct evidence that photons really do behave like tiny "cue balls" when they scatter from electrons. And we'll learn that particles like electrons also have a wavelike aspect. This wave nature of matter will help us understand the structure of atoms and the manner in which atoms absorb and emit light (**Figure 26-1b**). We'll find that the energies of atoms are *quantized*—that is, they can have only certain very definite values. The lesson of this chapter is that the microscopic world of atoms and light is very different in character from the macroscopic world of ordinary-sized objects that we see around us.

26-2 The photoelectric effect and blackbody radiation show that light is absorbed and emitted in the form of photons

By the end of the nineteenth century, it was recognized that light was an electromagnetic wave. Maxwell's equations (Section 22-3) describe the properties of these waves in great detail, and so physicists were confident that they had a deep understanding of the nature of light. But experimental studies of two very different phenomena showed that the true nature of light is actually more complex. In one of these phenomena, the *photoelectric effect*, a material absorbs light and the absorbed energy is used to eject electrons; in the other, called *blackbody radiation*, a material emits electromagnetic radiation when it is heated. The discoveries made by studying these two phenomena radically altered how physicists answered the question, "What is light?"

The Photoelectric Effect

BIO-Medical When light strikes certain materials, electrons can be ejected from the surface of those materials (**Figure 26-2a**). This **photoelectric effect**, discovered in 1886, plays an important role in biological research through a technique called *photoemission electron microscopy* or PEEM. In this technique a biological sample is illuminated with ultraviolet light or x rays. Different materials in the surface layer of the sample will emit fewer or greater numbers of electrons in response to this light. By recording the differences in the number of emitted electrons, called **photoelectrons**, it's possible to construct an image of the sample surface that shows where the different materials are located. **Figure 26-2b** shows an example of an image made in this way.

What makes the photoelectric effect so remarkable is that it does not behave in accordance with the idea that light is an electromagnetic wave. It takes energy to liberate an electron from the surface of a material, and in the photoelectric effect this energy is provided by the electric and magnetic energy in the light absorbed by the surface. We learned in Section 22-4 that according to Maxwell's equations, the intensity of such a wave depends on the rms values of the electric and magnetic fields in that wave, but not on the frequency of the wave:

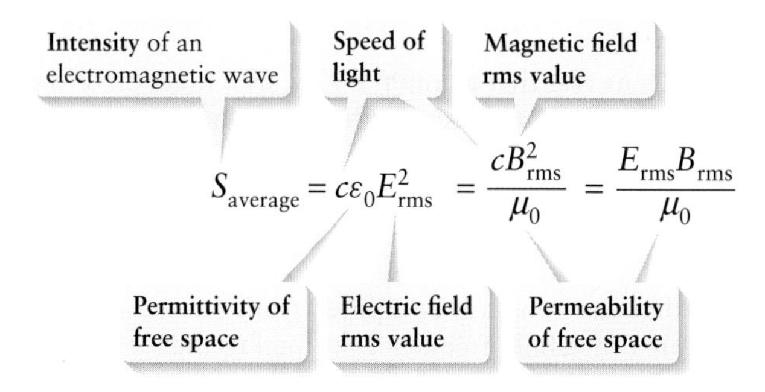

Intensity of an electromagnetic wave (22-25)

Equation 22-25 suggests that light of *any* frequency should be able to liberate an electron from the surface of a material, provided the light wave is sufficiently intense. Experiment shows that this is not the case. For example, if the biological sample shown in Figure 26-2 is illuminated with red light, no electrons are ejected no matter how intense the light. But if instead we illuminate the sample with x rays, which have a higher frequency than red light, electrons *are* ejected. (Figure 26-2b was made by using x rays.) This is impossible to understand on the basis of Maxwell's equations.

In 1905, the same year that he published his special theory of relativity, Albert Einstein proposed a simple but radical explanation for the strange behavior of the photoelectric effect. He suggested that light of frequency f comes in small packets, each with an energy E that is directly proportional to the frequency. Today these packets are called **photons**. We first encountered this idea in Section 22-4:

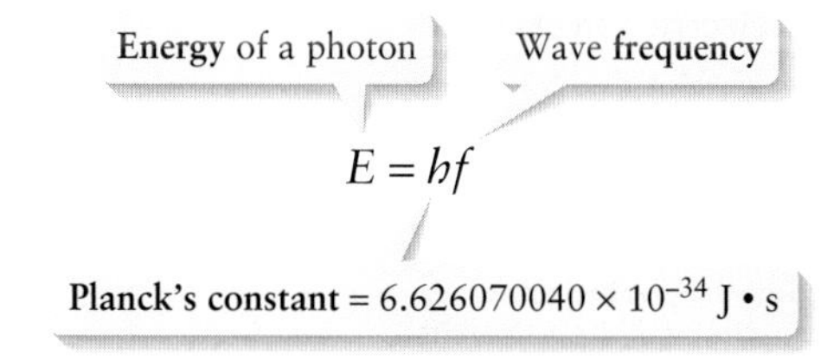

Energy of a photon (22-26)

We'll see shortly how the value of Planck's constant h is determined from the photoelectric effect.

Let's see how Einstein's idea explains the properties of the photoelectric effect. The minimum amount of energy required to remove a single electron from a material is called the **work function** Φ_0 of the material (Φ is the uppercase Greek letter "phi"). The value of Φ_0 varies from one material to another; it is small if electrons are easy to remove and large if electrons are hard to remove. In a given material, some electrons will be more difficult to remove, but Φ_0 represents the energy required to remove the most easily dislodged electron. Einstein proposed that an electron can absorb only a single photon at a time. So for even the most easily dislodged electron to be ejected from the material, it must absorb a photon with an energy equal to or greater than Φ_0. If the energy of the absorbed photon is greater than Φ_0, the energy that remains after the electron is ejected goes into the kinetic energy of the electron as it flies away from the material. Electrons that require more energy to be ejected will emerge from the material with less kinetic energy, but those with *maximum* kinetic energy will be those that were the easiest to dislodge. So in Einstein's picture the most energetic electrons ejected from the material will emerge with kinetic energy K_{max} given by

Maximum kinetic energy of an electron in the photoelectric effect (26-1)

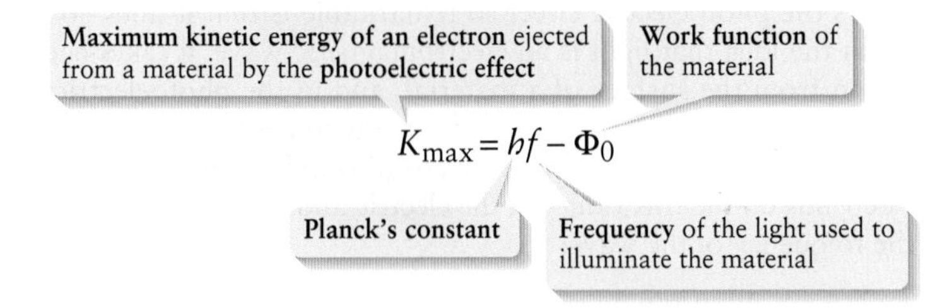

Equation 26-1 tells us that that a graph of K_{max} as a function of the light frequency f should be a straight line of slope h (see **Figure 26-3**). Since kinetic energy can never be negative, Equation 26-1 also tells us that electrons will be emitted only if $hf - \Phi_0 > 0$, or

$$f > f_0 = \frac{\Phi_0}{h} \tag{26-2}$$

Equation 26-2 says that electrons will be emitted from the surface only if the light frequency is greater than a threshold frequency f_0 equal to the work function Φ_0 divided by Planck's constant h. This agrees with the observation that no electrons can be ejected from a surface by light of too low a frequency, no matter how intense the light.

The graph shown in Figure 26-3 turns out to be an excellent match to experimental measurements of the maximum kinetic energy of ejected electrons for different light frequencies f. The slope of the graph tells us the value of Planck's constant h. Equation 26-1 also shows that if we extend the graph of K_{max} as a function of f to (unphysical) values of K_{max} less than zero, the graph intercepts the vertical axis at $-\Phi_0$. If we repeat the experimental measurements for a second material with a different work function Φ_0, the straight line intercepts the vertical axis at a different point but has the same slope h as for the first material. This reinforces the idea that Planck's constant is a universal constant; its value does not depend on the properties of the material that absorbs the photons.

Go to Interactive Exercise 26-1 for more practice dealing with the photoelectric effect.

The remarkable fit of Einstein's theory to experiment is powerful evidence that light is indeed absorbed in the form of photons with energy $E = hf$ as given by Equation 22-26. Einstein was awarded the Nobel Prize in Physics in 1921 for his explanation of the photoelectric effect.

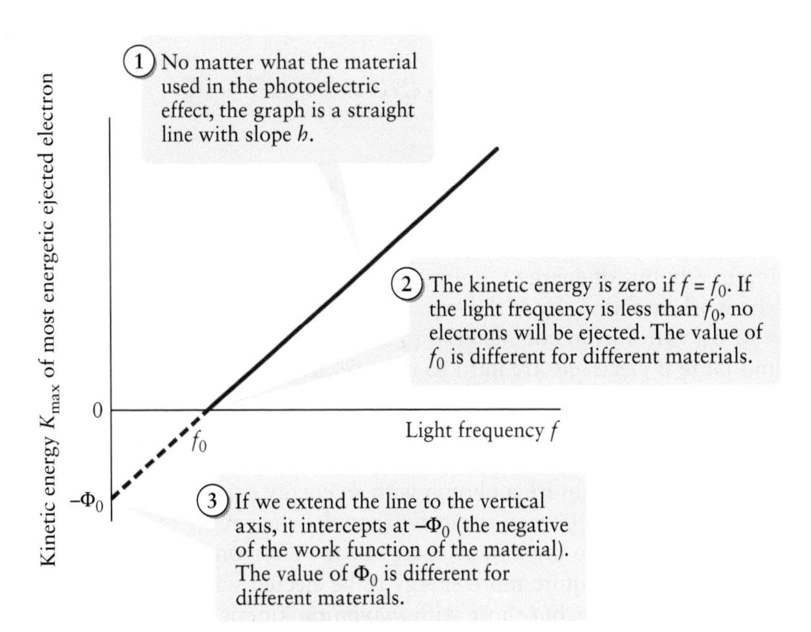

Figure 26-3 **Electron kinetic energy versus frequency in the photoelectric effect** The kinetic energy of the most energetic electron ejected in the photoelectric effect depends on the frequency of the light. This cannot be explained using the wave model of light but can be explained by the photon concept.

WATCH OUT! Some electrons require more energy than Φ_0 to be ejected.

The work function Φ_0 is the smallest amount of energy required to eject an electron from a given material under the most favorable conditions. Other electrons in the material require more energy to eject and will emerge from the material with less kinetic energy than the value K_{max} given by Equation 26-1.

EXAMPLE 26-1 The Photoelectric Effect with Cesium

The work function for a sample of cesium is 3.43×10^{-19} J. (a) What is the minimum frequency of light that will result in electrons being ejected from this sample by the photoelectric effect? (b) What is the maximum wavelength of light that will result in electrons being ejected from this sample by the photoelectric effect?

Set Up

Equation 26-1 tells us that the minimum energy required to eject an electron corresponds to having $K_{max} = 0$, so the electrons just barely make it out of the cesium. We'll use this to find the threshold frequency f_0 that just barely allows an electron to be ejected. We'll find the corresponding wavelength using Equation 22-2.

Maximum kinetic energy of an electron in the photoelectric effect:

$$K_{max} = hf - \Phi_0 \quad (26\text{-}1)$$

Propagation speed, frequency, and wavelength of an electromagnetic wave:

$$c = f\lambda \quad (22\text{-}2)$$

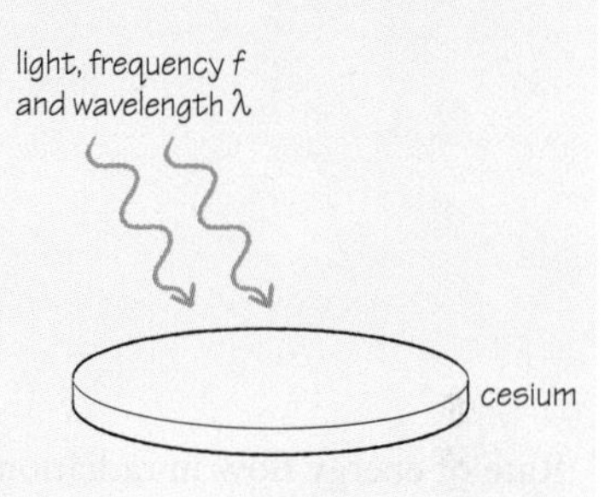

Solve

(a) Use Equation 26-1 to calculate the frequency f that corresponds to $K_{max} = 0$.

From Equation 26-1 with $K_{max} = 0$,

$$0 = hf - \Phi_0$$
$$hf = \Phi_0$$

This says that the photon energy hf is just enough to remove the most easily dislodged electron from the material (which requires energy Φ_0), with nothing left over to give the electron any kinetic energy. So $hf = \Phi_0$ is the minimum photon energy that will eject an electron, and the frequency f of this photon is the minimum (threshold) frequency that will do the job:

$$f_0 = f = \frac{\Phi_0}{h}$$
$$= \frac{3.43 \times 10^{-19}\ \text{J}}{6.63 \times 10^{-34}\ \text{J}\cdot\text{s}} = 5.17 \times 10^{14}\ \text{s}^{-1} = 5.17 \times 10^{14}\ \text{Hz}$$

(b) Find the wavelength that corresponds to the frequency that we calculated in part (a).

We can rewrite Equation 22-2 as

$$\lambda = \frac{c}{f}$$

In words, this says that wavelength is inversely proportional to frequency. So the *minimum* frequency of light that will eject an electron corresponds to the *maximum* wavelength that will eject an electron:

$$\lambda_{max} = \frac{c}{f_0} = \frac{3.00 \times 10^8\ \text{m/s}}{5.17 \times 10^{14}\ \text{Hz}}$$
$$= 5.80 \times 10^{-7}\text{m} = 580\ \text{nm}$$

(Recall that 1 nm = 10^{-9} m.)

Reflect

Figure 22-2 shows that a wavelength of 580 nm is in the yellow-green part of the visible spectrum. If we illuminate cesium with light of higher frequency and shorter wavelength than this (for example, blue or violet light), the photons will have more energy than the minimum and electrons will be ejected from the cesium. If instead we illuminate cesium with light of lower frequency and longer wavelength (for example, orange or red light), the photons will have less energy than the required minimum and no electrons will be ejected.

Blackbody Radiation

The photoelectric effect shows that light is *absorbed* in the form of photons. If we are to fully believe the photon concept, however, it must also be true that light is *emitted* in the form of photons. We learned in Section 14-7 that ordinary objects emit electromagnetic radiation as a result of their temperature. If we can find evidence that this emission is in the form of photons, we will have further evidence that the photon description of light is the correct one. Let's take a closer look at radiation of this kind.

Experiment shows that the rate at which an object emits radiation is proportional to its surface area A and to the fourth power of its Kelvin temperature T:

Rate of energy flow in radiation (14-22)

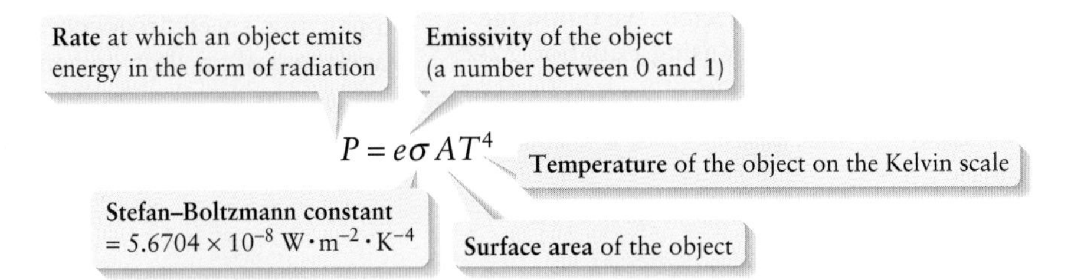

The higher the temperature of an object of a given size, the greater the radiated power P and so the more brightly it glows.

Experiment also shows that the *color* of the radiation emitted by an object depends on its temperature T (**Figure 26-4**). A heated object emits light at all wavelengths, but emits most strongly at a particular frequency called the *frequency of maximum emission*. As the temperature increases, the frequency of maximum emission increases.

Equation 14-22 shows that the radiated power also depends on a quantity e called the *emissivity*, which depends on the properties of the object's surface. This has its greatest value ($e = 1$) for an idealized type of dense object called a **blackbody.** An ideal blackbody does not reflect any light at all but absorbs all radiation falling on it. If a

- A hot, dense object emits electromagnetic radiation. The idealized case is called **blackbody radiation.**
- The frequency of maximum emission is directly proportional to the Kelvin temperature T of the object: The higher the temperature T, the greater the frequency of maximum emission.

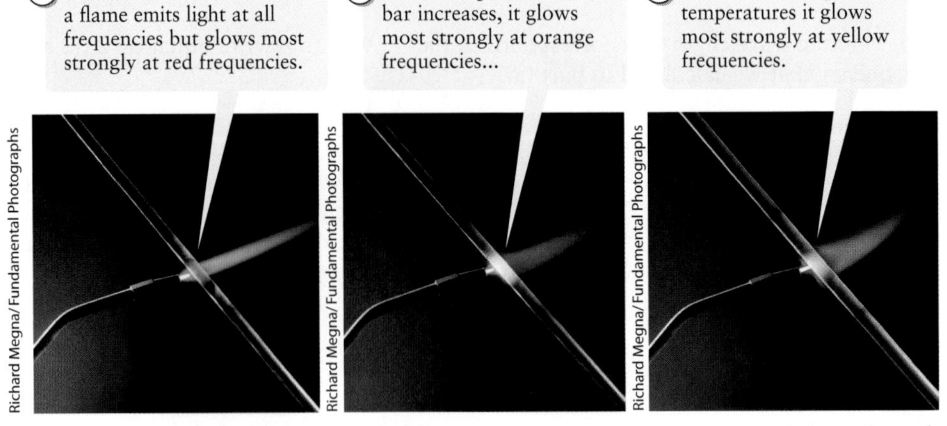

(1) This metal bar heated with a flame emits light at all frequencies but glows most strongly at red frequencies.
(2) As the temperature of the bar increases, it glows most strongly at orange frequencies...
(3) ...and at even higher temperatures it glows most strongly at yellow frequencies.
Richard Megna/Fundamental Photographs
Richard Megna/Fundamental Photographs
Richard Megna/Fundamental Photographs

Figure 26-4 Radiation from heated objects The color of the light from a heated object depends on its temperature.

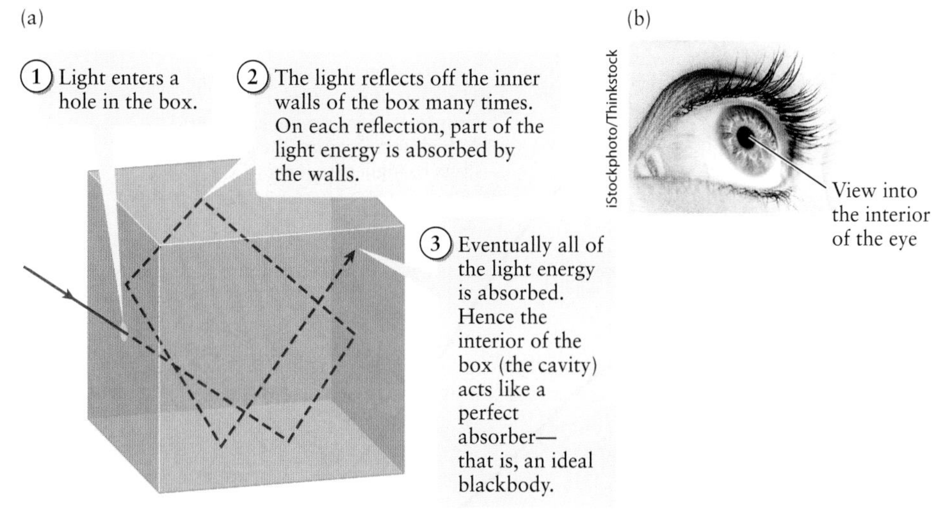

(a)
(b)
1 Light enters a hole in the box.
2 The light reflects off the inner walls of the box many times. On each reflection, part of the light energy is absorbed by the walls.
3 Eventually all of the light energy is absorbed. Hence the interior of the box (the cavity) acts like a perfect absorber—that is, an ideal blackbody.
iStockphoto/Thinkstock
View into the interior of the eye

Figure 26-5 A blackbody cavity (a) A box with a small hole in one side is a good approximation to an ideal blackbody. (b) The interior of the eye has very similar properties to the box in (a).

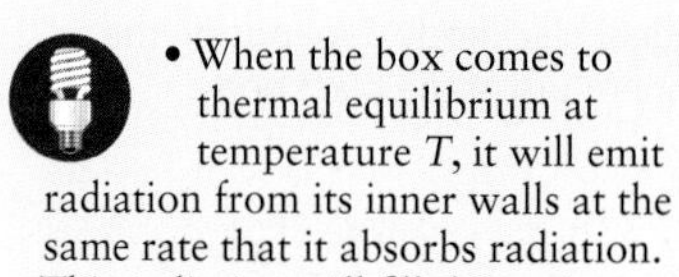

- When the box comes to thermal equilibrium at temperature T, it will emit radiation from its inner walls at the same rate that it absorbs radiation.
- This radiation will fill the cavity inside the box.
- Some of the radiation will leak out of the hole. Because the box acts as a perfect blackbody, the radiation that leaks out will be blackbody radiation.

blackbody is in thermal equilibrium with its surroundings, it must emit energy at the same rate that it absorbs it in order for its temperature T to remain constant. So in addition to being a perfect absorber of energy, an ideal blackbody in thermal equilibrium with its surroundings is also a perfect emitter of energy because it emits as much energy as it absorbs.

Ordinary objects, such as tables, textbooks, and people, are not ideal blackbodies; they reflect light, which is why they are visible. (Even a piece of wood darkened with soot or painted a dull black reflects *some* light.) But it is possible to make a nearly ideal blackbody simply by building a box and drilling a small hole in one side (**Figure 26-5a**). Light that enters the hole will reflect around inside the box, with part of the light energy being absorbed by the walls on each reflection. Eventually all of the light energy will be absorbed, so the interior of the box acts like a perfect absorber and is effectively an ideal blackbody. You can see this effect if you look into another person's eye (**Figure 26-5b**). The pupil at the center of the iris appears black, even though the tissues that line the interior of the eye are pinkish in color. That's because after multiple reflections, those tissues almost completely absorb light that enters the eye through the pupil.

If the box in Figure 26-5a is in thermal equilibrium at temperature T, the rate at which the walls absorb energy in the form of radiation must be equal to the rate at which the walls emit energy. The cavity in the interior of the box will be filled with this radiation, which will itself be in thermal equilibrium with the walls of the box. Since the walls act as an ideal blackbody, the light that fills the cavity is effectively **blackbody radiation**—the kind of light that would be emitted by a perfect blackbody of emissivity $e = 1$ at temperature T. We can study this light by examining the small fraction of light that emerges from the hole in Figure 26-5a.

Figure 26-6 shows the experimentally observed *spectrum* of blackbody radiation—that is, the relative amount of light energy present at different frequencies—for two different temperatures. The high-temperature curve lies above the low-temperature curve, which tells us that the higher the temperature of a blackbody, the greater the amount of radiation at all frequencies. Furthermore, as the blackbody temperature increases, the peak of the curve shifts to a higher frequency. This agrees with our observation about how the frequency of maximum emission varies with an object's temperature (Figure 26-4). Figure 26-6 explains why we can't see the radiation from objects at room temperature, about $T = 300$ K. At this relatively low temperature, the frequency of maximum emission is in the infrared, which our eyes cannot see. There is some emission at visible frequencies (in Figure 26-6, to the right of the peak of the curve), but the amount of emission at $T = 300$ K is so low that our eyes can't detect it.

In the late nineteenth century physicists tried to understand the shape of the blackbody spectrum shown in Figure 26-6 using their knowledge of thermodynamics

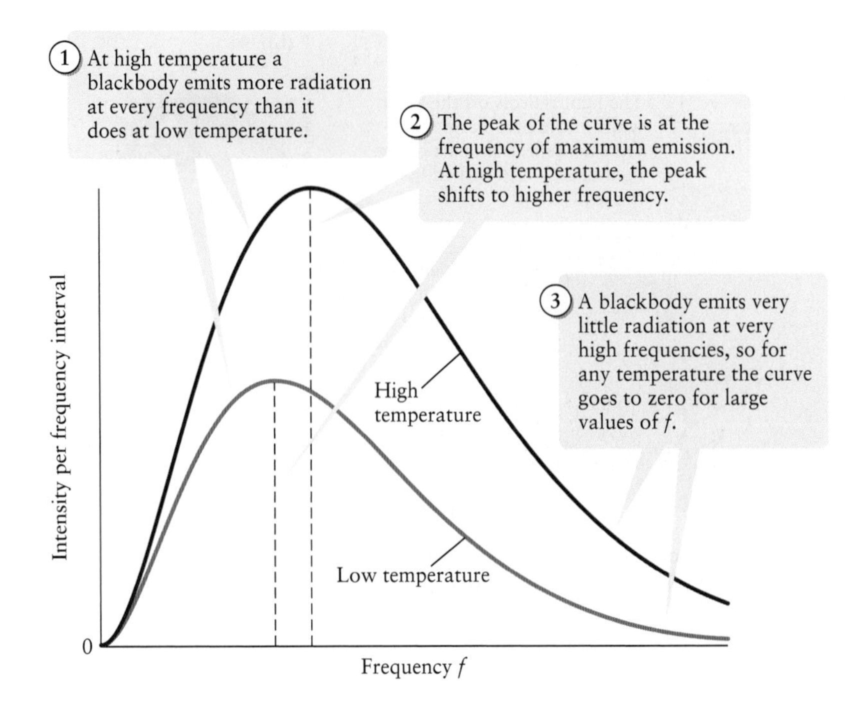

1 At high temperature a blackbody emits more radiation at every frequency than it does at low temperature.
2 The peak of the curve is at the frequency of maximum emission. At high temperature, the peak shifts to higher frequency.
3 A blackbody emits very little radiation at very high frequencies, so for any temperature the curve goes to zero for large values of f.
Intensity per frequency interval
High temperature
Low temperature
0
Frequency f

Figure 26-6 Blackbody spectra The spectrum of light emitted by an ideal blackbody depends on the temperature of the blackbody.

and electromagnetic waves. Their efforts ended in failure. To understand how they failed, we begin by noting that the electromagnetic waves inside the cavity in Figure 26-5a should be in the form of *standing* waves. That's because the waves will bounce back and forth between the walls of the cavity. We learned in Section 13-6 that when waves bounce back and forth between the ends of a string that's tied down at both ends, steady wave patterns arise for waves of certain wavelengths and frequencies. The same is true for electromagnetic waves in a cavity. The standing wave patterns for electromagnetic waves in a cavity are more complex than for waves on a string: Electromagnetic waves are three-dimensional, not just one-dimensional like those along the length of a string, and involve two varying quantities, the electric field and the magnetic field. For our purposes, however, all we need to know is that these standing waves exist.

The shape of the blackbody spectrum in Figure 26-6 indicates how much energy is present in the standing waves in each frequency range: There is relatively little energy at very low frequencies, more energy at frequencies near the frequency of maximum emission, and again relatively little energy at high frequencies. However, nineteenth-century physics suggested that *every* possible standing wave in the cavity should contain on average the same amount of energy. This conclusion came from the equipartition theorem, which we first encountered in Section 14-3. This theorem states that a molecule in a gas at a Kelvin temperature T has, on average, an amount of energy $(1/2)kT$ for each degree of freedom of the molecule, where $k = 1.381 \times 10^{-23}$ J/K is the Boltzmann constant. Arguments from thermodynamics suggest that for the same reason a standing wave inside a cavity in thermal equilibrium at temperature T should also possess an average amount of energy equal to $(1/2)kT$ per degree of freedom. There are two degrees of freedom per standing wave, one for the electric field and one for the magnetic field, so the total average energy per standing wave should be kT. It turns out that the number of standing waves in a given frequency interval increases with increasing frequency. So according to nineteenth-century physics, the total energy per frequency interval (equal to the energy kT per standing wave multiplied by the number of standing waves per frequency interval) should increase with increasing frequency and *never* decrease (**Figure 26-7**). This is in profound disagreement with the experimentally observed shape of the spectrum.

If we now introduce the photon concept, however, we *can* match the experimentally observed spectrum. We still use the idea that the average energy available for a standing wave of frequency f is kT, but now this energy goes into photons of that frequency which each have energy $E = hf$. This energy fundamentally comes from the walls of the box in Figure 26-5a, since these walls emit the photons. At low frequencies the photon energy hf is small compared to the available energy kT, so there will be many photons present for a low-frequency standing wave. But at high frequencies hf is much larger than kT, which means that the energy required to produce a photon is larger than the average energy available to produce one. Hence the average number of photons present for that standing wave will be very small. (It need not be zero, since kT is only the *average* energy available. From time to time the available energy will be greater than kT, and some photons can be produced.) So even though energy is *available* for a high-frequency standing wave, that energy can't be used to create photons, so the amount of energy *present* is quite small. That's just the effect we need to make the theoretical curve in Figure 26-7 decline at large frequencies.

Using a somewhat different version of this photon argument, in 1900 the German physicist Max Planck was able to make a theoretical prediction for the blackbody spectrum that was in excellent agreement with the experimental spectrum shown in Figure 26-7. Planck's theoretical formula was the first to involve the new quantity h that now bears his name, and the value of h given in Equation 22-26 is the one that gives the best match between this formula and the experimental data. Planck was awarded the 1918 Nobel Prize in Physics for his achievement.

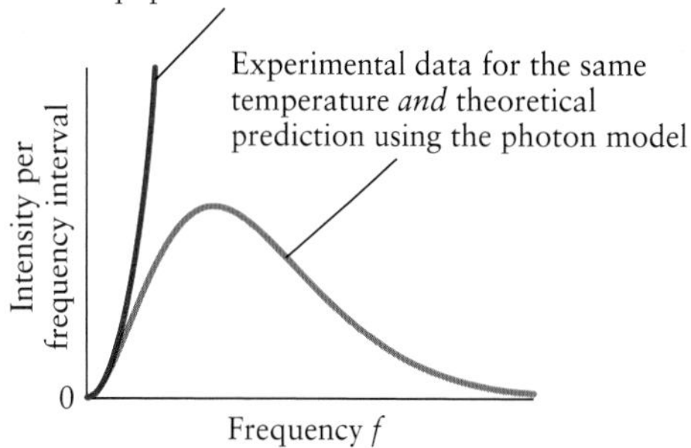

- A photon model accurately describes the spectrum of blackbody radiation.
- The model that does not use the photon concept fails to describe the spectrum.

Figure 26-7 The photon model explains blackbody radiation A model for blackbody radiation that does not use the photon concept predicts (incorrectly) that the intensity should increase without limit as the frequency increases.

The explanation of blackbody spectra in terms of photons is the evidence we were seeking that light is emitted in the form of photons. In the following section we'll see even more compelling evidence for the photon picture of light.

GOT THE CONCEPT? 26-1 Blackbody Radiation

? Two objects of the same size are both perfect blackbodies. One is at a temperature of 3000 K, so its frequency of maximum emission is in the infrared part of the electromagnetic part of the spectrum; the other is at a temperature of 12,000 K, so its frequency of maximum emission is in the ultraviolet part of the spectrum. Compared to the object at 3000 K, the object at 12,000 K (a) emits more infrared light; (b) emits more visible light; (c) emits more ultraviolet light; (d) two of (a), (b), and (c); (e) all of (a), (b), and (c).

TAKE-HOME MESSAGE FOR Section 26-2

✔ In the photoelectric effect, an electron in a material can absorb light energy that strikes the surface and as a result be ejected from the surface.

✔ For any material, the light must have a certain minimum frequency in order for electrons to be ejected. This is evidence that light comes in the form of photons, with an energy proportional to their frequency.

✔ A perfect blackbody is an ideal absorber of light and also an ideal emitter of light. The spectrum of light emitted by a blackbody depends on its temperature.

✔ The spectrum of light emitted by a blackbody can be understood only if we use the idea that light is emitted in the form of photons.

26-3 As a result of its photon character, light changes wavelength when it is scattered

Blackbody radiation and the photoelectric effect suggest that photons can be treated like tiny bundles of energy, and so have a particle-like nature. In the photoelectric effect, a photon strikes an electron and is *absorbed*. But if a photon is like a particle, is it possible that, as in the collisions we studied in Chapter 7, a photon could strike an electron and *bounce off*? If so, based on our experience with collisions, linear momentum would be conserved in such an interaction and the momentum of the photon should change as a result. This effect, called *Compton scattering*, is further evidence that light does indeed come in the form of photons.

Let's first see how to express the momentum of a photon. We saw in Section 25-7 that for a particle of mass m, we can write the kinetic energy and momentum as

Einsteinian expressions for kinetic energy and momentum (25-20)

Kinetic energy of a particle of mass m and velocity $\vec{v}$

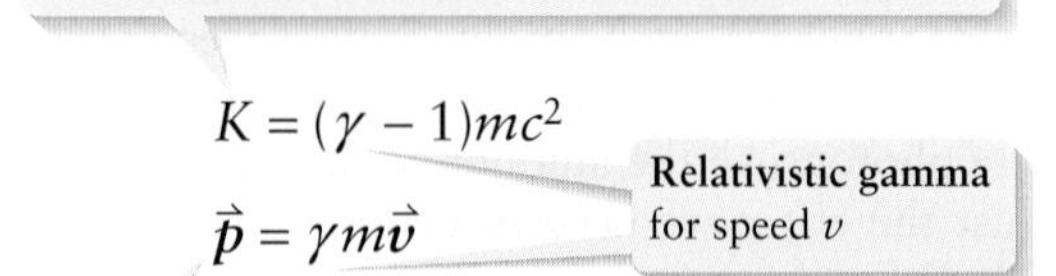

$$K = (\gamma - 1)mc^2$$

$$\vec{p} = \gamma m\vec{v}$$

Relativistic gamma for speed v

Momentum of a particle of mass m and velocity $\vec{v}$

At speeds that are a small fraction of the speed of light c, these are approximately equal to the Newtonian expressions:

$$K = \frac{1}{2}mv^2 \text{ and } \vec{p} = m\vec{v}$$

In these expressions the quantity γ (relativistic gamma) is

Relativistic gamma (25-19)

Relativistic gamma for a particle moving at speed v

$$\gamma = \frac{1}{\sqrt{1 - \frac{v^2}{c^2}}}$$

Speed of the particle

Speed of light in a vacuum

We also saw that an object of mass m has a rest energy E_0 that is present even when it is not moving:

Rest energy (25-21)

Rest energy of a particle

Mass of the particle

$$E_0 = mc^2$$

Speed of light in a vacuum

If we combine the first of Equations 25-20 with Equations 25-19 and 25-21, we find that the total energy of a particle (kinetic energy plus rest energy) is

(26-3)
$$E = K + E_0 = (\gamma - 1)mc^2 + mc^2 = \gamma mc^2 = \frac{mc^2}{\sqrt{1 - \frac{v^2}{c^2}}}$$

From the second of Equations 25-20, the magnitude of the momentum of a particle is

(26-4)
$$p = \gamma mv = \frac{mv}{\sqrt{1 - \frac{v^2}{c^2}}}$$

Comparing Equations 26-3 and 26-4, we see that

$$p = \frac{Ev}{c^2} \quad \text{(momentum of a particle of total energy } E\text{)} \qquad (26\text{-}5)$$

A photon has zero mass, which is why it can travel at the speed of light. (Any object with nonzero mass would require an infinite amount of kinetic energy to reach $v = c$, as we described in Section 25-7.) As such, we can't apply Equation 26-3 or 26-4 directly to photons. But we can use the combination in Equation 26-5, in which mass does not appear explicitly. Setting $v = c$ in Equation 26-5, we get the following relationship for the momentum of a photon:

$$p = \frac{Ec}{c^2} = \frac{E}{c} \quad \text{(momentum of a photon)} \qquad (26\text{-}6)$$

From Equation 22-26 we can write $E = hf$, and we know that for light waves $c = f\lambda$ (Equation 22-2). If we substitute these into Equation 26-6, we get an alternative expression for the momentum of a photon:

$$p = \frac{hf}{f\lambda}$$

or, simplifying,

$$p = \frac{E}{c} = \frac{h}{\lambda} \qquad (26\text{-}7)$$

Magnitude of the **momentum** of a photon; **Energy** of the photon; Planck's constant; Speed of light in a vacuum; **Wavelength** of the photon

Momentum and energy of a photon (26-7)

A photon's momentum is directly proportional to the energy that it carries and inversely proportional to its wavelength. Thus a violet photon of wavelength 400 nm has twice the momentum, as well as twice the energy, of an infrared photon of wavelength 800 nm.

In the early 1920s American physicist Arthur Compton showed conclusively that photons have momentum and that the momentum is inversely proportional to the wavelength, as stated in Equation 26-7. In his experiments, an x-ray photon collided with an electron in a carbon atom, a process now called **Compton scattering**. Compton detected both the electron, which is knocked out of the atom, and the scattered photon. Compton could account for the directions and energies of the electron and the scattered photon by requiring that the total momentum of the electron and the photon be conserved in the collision. In other words, he showed that the photon description of light applies not just to the absorption and emission of light but also to what happens to light when it is scattered. For revealing this fundamental aspect of light, Compton was awarded the Nobel Prize in Physics in 1927.

Figure 26-8 shows the situation that Compton studied, the collision of a photon (symbol γ_i) and a stationary electron (symbol e^-). The electron is scattered at angle ϕ relative to the initial direction of the photon, and the photon scatters at angle θ relative to its initial direction. Although we have labeled the photon γ_f after the collision, the scattered photon is the same photon that collided with the electron. The subscripts "i" for "initial" and "f" for "final" instead imply that the energy, wavelength, and other quantities associated with the photon have changed as a result of the collision. For example, because energy

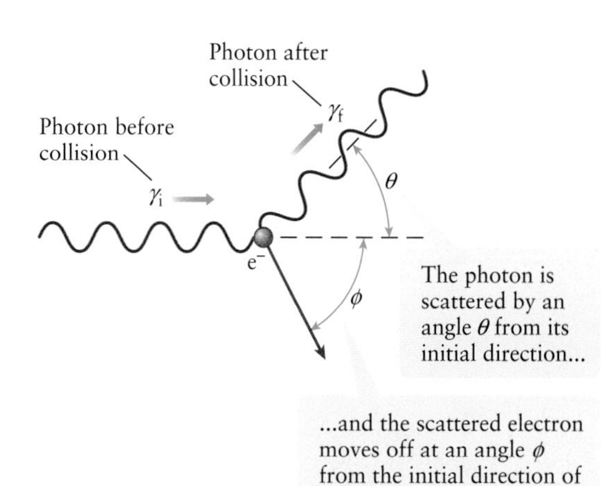

Figure 26-8 Compton scattering A photon undergoes a wavelength shift when it scatters from an electron that is initially at rest.

is conserved during the collision and because some of the photon's energy is almost always transferred to the electron, the outgoing photon carries less energy than it had initially. From Equation 26-7 the energy of a photon is

$$E = pc = \frac{hc}{\lambda}$$

Because the photon has less energy after the collision than it had initially, E_f is less than E_i, and the final wavelength λ_f is greater than the initial wavelength λ_i. In other words, as the energy of the photon decreases, its wavelength increases. Compton found that the increase in wavelength $\Delta\lambda$ from λ_i to λ_f is a function of the angle θ at which the photon scatters:

Compton scattering equation (26-8)

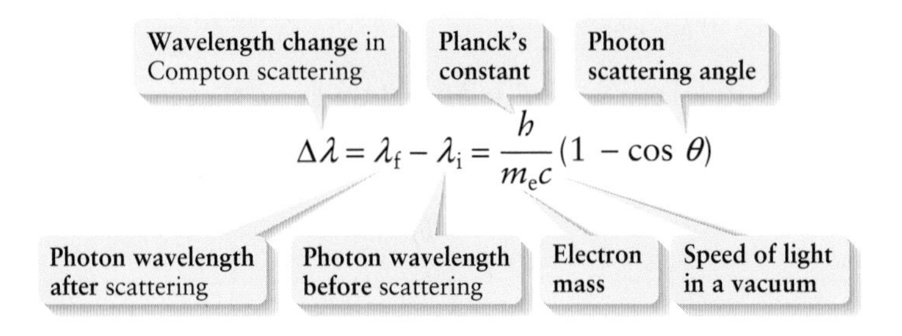

The proportionality constant in Equation 26-8 is known as the Compton wavelength λ_C:

$$\lambda_C = \frac{h}{m_e c} \qquad (26\text{-}9)$$

See the Math Tutorial for more information on trigonometry.

Go to Picture It 26-1 for more practice dealing with Compton scattering.

The scattering angle θ of the photon ranges from 0° (straight forward) to 180° (straight back). When the photon continues in the same direction after the collision, so that θ equals 0° and $\cos\theta$ equals 1, $\Delta\lambda$ equals zero. In other words, if $\theta = 0°$ there is no change in the photon's wavelength and no change in the photon's energy.

The maximum change in a photon's wavelength and energy when it undergoes Compton scattering occurs when it scatters straight back, in the direction opposite to the one in which it approached the electron. In this case $\theta = 180°$, so $\cos\theta = -1$ and the term in parentheses in Equation 26-8 equals 2. The maximum possible change in the photon's wavelength is therefore $2\lambda_C$, or twice the Compton wavelength. Using the known values of h, m_e, and c, we find

$$\lambda_C = \frac{h}{m_e c} = \frac{6.63 \times 10^{-34}\ \text{J}\cdot\text{s}}{(9.11 \times 10^{-31}\ \text{kg})(3.00 \times 10^{8}\ \text{m/s})}$$
$$= 2.43 \times 10^{-12}\ \text{m} = 2.43 \times 10^{-3}\ \text{nm} = 0.00243\ \text{nm}$$

(Recall that 1 nm $= 10^{-9}$ m.) The small value of λ_C means that Compton scattering has a negligible effect on the wavelength of visible-light photons. The wavelength of visible light ranges between about 380 and 750 nm; the Compton wavelength is considerably smaller. For example, if a photon of wavelength 400 nm is scattered by an electron, its wavelength changes by a value on the order of 0.00243 nm. Such a tiny change is very difficult to measure, so we can ignore the wavelength change of a visible-light photon due to Compton scattering. But for an x-ray photon with a wavelength on the order of 10^{-11} m $= 0.01$ nm, a wavelength shift of 0.00243 nm corresponds to a large percentage of the initial wavelength. That's why Compton first noticed this effect in the scattering of x rays.

If light did not have a photon aspect, we would expect *no* wavelength change on scattering. A light wave of frequency f and wavelength λ encountering an electron would make the electron oscillate at the same frequency f, and the electron would emit radiation with the same frequency f and hence the same wavelength λ as the initial light wave. The change in wavelength that Compton observed is unambiguous evidence that light does indeed have a particle character.

EXAMPLE 26-2 Compton Scattering

A photon carries 2.00×10^{-14} J of energy. It undergoes Compton scattering in a block of carbon. What is the largest fractional change in energy the photon can undergo as a result?

Set Up

Given the initial photon energy $E_i = 2.00 \times 10^{-14}$ J, we can calculate its wavelength λ_i using Equation 26-7. We use Equation 26-8 to calculate the change in wavelength due to scattering; this will be maximum if $\theta = 180°$, so $\cos\theta = -1$. Once we know the final wavelength λ_f, we can use Equation 26-7 again to find the final photon energy. Comparing this to the initial photon energy tells us the fractional change in energy.

Momentum and energy of a photon:

$$p = \frac{E}{c} = \frac{h}{\lambda} \qquad (26\text{-}7)$$

Compton scattering equation:

$$\Delta\lambda = \lambda_f - \lambda_i = \frac{h}{m_e c}(1 - \cos\theta) \qquad (26\text{-}8)$$

Before:

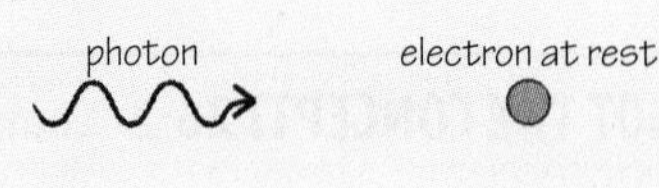

After:

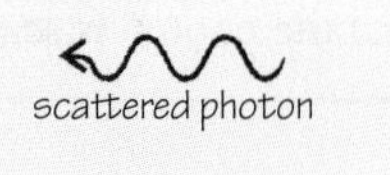

Solve

First calculate the wavelength of the initial photon using Equation 26-7.

From Equation 26-7 the wavelength of the initial photon is

$$\lambda_i = \frac{hc}{E_i} = \frac{(6.63 \times 10^{-34}\ \text{J}\cdot\text{s})(3.00 \times 10^{8}\ \text{m/s})}{2.00 \times 10^{-14}\ \text{J}}$$

$$= (9.95 \times 10^{-12}\ \text{m})\left(\frac{1\ \text{nm}}{10^{-9}\ \text{m}}\right) = 9.95 \times 10^{-3}\ \text{nm}$$

Calculate the wavelength shift using Equation 26-8.

The maximum wavelength shift is with $\theta = 180°$ and $\cos\theta = -1$:

$$\Delta\lambda_{\text{max}} = \lambda_f - \lambda_i = \frac{h}{m_e c}(1 - \cos 180°)$$

$$= (2.43 \times 10^{-3}\ \text{nm})[1 - (-1)] = 4.86 \times 10^{-3}\ \text{nm}$$

The wavelength of the final photon equals the wavelength of the initial photon plus the shift $\Delta\lambda$. Use this to find the energy of the final photon.

The final wavelength is

$$\lambda_f = \lambda_i + \Delta\lambda = 9.95 \times 10^{-3}\ \text{nm} + 4.86 \times 10^{-3}\ \text{nm}$$

$$= 1.481 \times 10^{-2}\ \text{nm} = 1.481 \times 10^{-11}\ \text{m}$$

The energy of the final photon is

$$E_f = \frac{hc}{\lambda_f} = \frac{(6.63 \times 10^{-34}\ \text{J}\cdot\text{s})(3.00 \times 10^{8}\ \text{m/s})}{1.481 \times 10^{-11}\ \text{m}}$$

$$= 1.34 \times 10^{-14}\ \text{J}$$

This is less than the energy of the initial photon. The lost energy has gone into the kinetic energy of the scattered electron.

Express the energy change as a fraction of the initial photon energy.

The fractional energy change is the energy change $E_f - E_i$ divided by the initial energy E_i:

$$\text{fractional energy change} = \frac{E_f - E_i}{E_i}$$

$$= \frac{1.34 \times 10^{-14}\ \text{J} - 2.00 \times 10^{-14}\ \text{J}}{2.00 \times 10^{-14}\ \text{J}}$$

$$= -0.328 = -32.8\%$$

Reflect

An initial photon of this high energy and short wavelength can lose as much as 32.8% (nearly one-third) of its initial energy when it undergoes Compton scattering.

Example 26-2 suggests why x rays are useful in cancer radiation therapy. If an x-ray photon strikes an electron in a water molecule within a cancerous cell, the photon can scatter and transfer a substantial amount of energy to the electron. This transferred energy is great enough that the electron escapes from the molecule, leaving the water molecule in an ionized state. These ionized water molecules damage the DNA of the cancerous cell and cause cell death.

GOT THE CONCEPT? 26-2 Compton Scattering

Suppose a photon has a wavelength equal to the Compton wavelength λ_C. If this photon collides with an electron, and the photon is scattered through an angle of 90°, what will be the wavelength of the photon after the collision? (a) Zero; (b) $\lambda_C/2$; (c) λ_C; (d) $2\lambda_C$; (e) $3\lambda_C$.

TAKE-HOME MESSAGE FOR Section 26-3

✔ A photon has zero mass but does have momentum. The magnitude of the momentum is proportional to the photon energy and inversely proportional to the wavelength.

✔ In Compton scattering, a photon scatters from an electron. The photon loses energy and momentum, and these are transferred to the electron. The change in wavelength of the photon depends on the angle through which it is scattered.

26-4 Matter, like light, has aspects of both waves and particles

We have seen that light has a dual nature, with attributes of both waves and particles. Light comes in the form of particles (photons), but these particles have a wave aspect: Associated with them is a frequency f, which determines the photon energy $E = hf$, and a wavelength λ, which determines the photon momentum $p = h/\lambda$. Is it possible that ordinary matter, which we know is made of particles such as electrons, protons, and neutrons, also has a dual nature? Could these particles also have a wave aspect?

In 1924 French graduate student Louis de Broglie (pronounced "de broy") proposed precisely that idea. In particular, he suggested that the relationship $p = h/\lambda$ between momentum p and wavelength λ that applies to photons should also apply to particles such as electrons (**Figure 26-9**). The wavelength of a particle is called its **de Broglie wavelength:**

de Broglie wavelength (26-10)

$$\lambda = \frac{h}{p}$$

A particle has a **de Broglie wavelength...** ...equal to **Planck's constant...** ...divided by the **momentum of the particle.** The greater the momentum, the shorter the de Broglie wavelength.

How large should we expect the wavelength of an electron to be? Let's examine the case of an electron of charge $q = -e$ that gains its momentum by moving through a potential difference of V, so the electron starts at position a where the potential is zero and moves to a position b where the potential has a positive value V. (You may want to review the discussion of electric potential in Section 17-3.) The electron potential energy then changes from $U_a = qV_a = (-e)(0)$ to $U_b = qV_b = (-e)V = -eV$. The change in electric potential energy is

$$\Delta U = U_b - U_a = (-eV) - 0 = -eV < 0$$

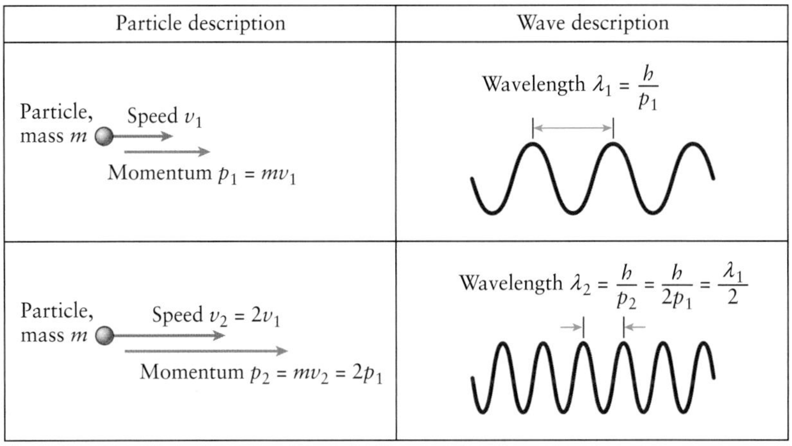

Particle description	Wave description
Particle, mass m; Speed v_1; Momentum $p_1 = mv_1$	Wavelength $\lambda_1 = \frac{h}{p_1}$
Particle, mass m; Speed $v_2 = 2v_1$; Momentum $p_2 = mv_2 = 2p_1$	Wavelength $\lambda_2 = \frac{h}{p_2} = \frac{h}{2p_1} = \frac{\lambda_1}{2}$

Figure 26-9 Wave-particle duality Matter has both particle aspects (speed and momentum) and wave aspects (wavelength).

The de Broglie wavelength is inversely proportional to the momentum. If the momentum is doubled, the wavelength decreases by 1/2.

The electric potential energy decreases by an amount eV. Mechanical energy is conserved if the only force acting on the electron is the (conservative) electric force, so the decrease in electric potential energy equals the gain in kinetic energy of the electron:

$$\Delta K = -\Delta U = -(-eV) = +eV > 0 \qquad (26\text{-}11)$$

If the electron starts at rest, its initial kinetic energy is zero and its final kinetic energy is $K = (1/2)mv^2$, so $\Delta K = (1/2)mv^2 - 0 = (1/2)mv^2$. (We're assuming that the electron is moving at a speed much slower than the speed of light c, so we don't have to use the Einsteinian expression for kinetic energy.) Then, from Equation 26-11, the final kinetic energy of the electron is

$$K = \frac{1}{2}mv^2 = eV$$

Solve for the final speed of the electron:

$$v^2 = \frac{2K}{m} = \frac{2eV}{m} \quad \text{so} \quad v = \sqrt{v^2} = \sqrt{\frac{2eV}{m}} \qquad (26\text{-}12)$$

The final momentum of the electron is $p = mv$, and its final wavelength is $\lambda = h/p = h/mv$ from Equation 26-10. If we substitute v from Equation 26-12 into the formula for the wavelength of the electron, we get

$$\lambda = \frac{h}{mv} = \frac{h}{m}\sqrt{\frac{m}{2eV}} = \frac{h}{\sqrt{2meV}} \qquad (26\text{-}13)$$

Suppose that the electron is accelerated through a potential difference $V = 50.0$ V. Substituting this into Equation 26-13 along with the values of Planck's constant h, the electron mass m, and the magnitude e of the electron charge, we get

$$\lambda = \frac{6.63 \times 10^{-34}\ \text{J}\cdot\text{s}}{\sqrt{2(9.11 \times 10^{-31}\ \text{kg})(1.60 \times 10^{-19}\ \text{C})(50.0\ \text{V})}}$$
$$= 1.74 \times 10^{-10}\ \text{m} = 0.174\ \text{nm}$$

To see how to measure such a short electron wavelength, note that a photon with this wavelength is in the x-ray region of the electromagnetic spectrum (see Figure 22-2). It was known in the 1920s that x rays show interference effects when they reflect from adjacent atoms in a crystal: At certain angles waves that reflect from one atom will interfere constructively (so the reflected intensity is high) with those that reflect from neighboring atoms, while at other angles they interfere destructively (so the reflected

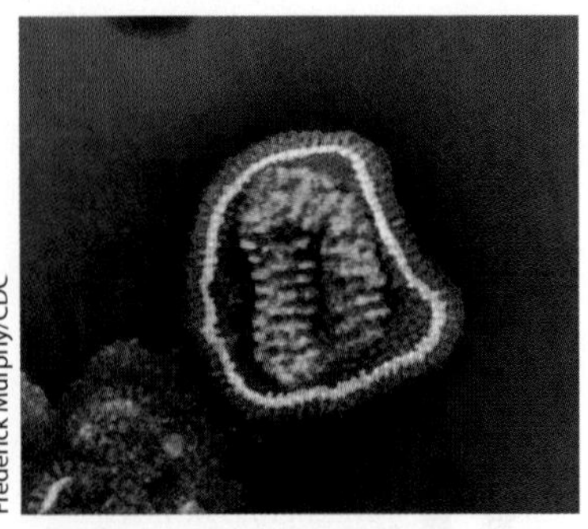
Frederick Murphy/CDC

Figure 26-10 An electron micrograph When a beam of low-energy electrons is shot through a thin slice of a specimen in a transmission electron microscope (TEM), the pattern formed by the diffracted electrons forms an image. A TEM captured this (false-color) image of an influenza virus particle, which is only about 100 nm in diameter. Because the wavelengths of low-energy electrons are so much shorter than those of light, a TEM is capable of significantly better resolution than light microscopes (better than 0.005 nm for a TEM, compared to about 0.2 μm with the most powerful optical microscopes).

intensity is near zero). This can happen because the spacing between adjacent atoms in a crystal is around 0.1 nm, comparable to the wavelength of the x rays. So if electrons have a wave aspect, we expect that a beam of electrons that have been accelerated from rest through 50.0 V should display the same kind of interference effects as a beam of x rays.

In 1927 the American physicists Clinton Davisson and Lester Germer performed precisely this kind of experiment using a beam of electrons directed at a target of crystalline nickel. They found that the intensity of reflected electrons was greater for certain angles, just as for x rays. What's more, the angles at which this maximum intensity occurred were precisely those expected if the wavelength of electrons was given by the de Broglie relation, Equation 26-10. This groundbreaking result was quickly confirmed in experiments carried out by the British physicist G. P. Thomson. These results resoundingly confirmed de Broglie's remarkable hypothesis and showed that matter does indeed have a wave aspect. (The 1927 Nobel Prize in Physics went to de Broglie; the 1937 prize was shared by Davisson and Thomson.) The dual character of *both* light and matter, which have both wave and particle characteristics, is called **wave-particle duality**.

Wave-particle duality is both surprising and counterintuitive. It is also of tremendous practical use. One important application that has revolutionized biology is the *electron microscope*. A major limitation of ordinary microscopes is that the smallest detail that can be resolved is about the size of a wavelength of visible light (about 380 to 750 nm). This makes microscopes useless for studying the structure of viruses, for example, which range in size from 5 to 300 nm. But as our above example of the 50.0-V electron shows, the wavelength of electron waves can be a fraction of a nanometer. So images made with an *electron microscope* can reveal details that are forever hidden from an ordinary visible-light microscope. **Figure 26-10** is an electron microscope image of an influenza virus, in which details smaller than a nanometer across can be seen.

If particles have wave aspects, why don't we notice these for objects that are large enough to see with the naked eye? The answer is that an object large enough to see has a relatively large momentum, so its de Broglie wavelength (which is inversely proportional to momentum) is infinitesimal. As an example, a dust mote floating in the air (such as you might see when a shaft of sunlight comes through the window) has a mass of about 8×10^{-10} kg. If it drifts at a speed of 1 mm/s = 10^{-3} m/s, its momentum is

$$p = mv = (8 \times 10^{-10}\ \text{kg})(10^{-3}\ \text{m/s}) = 8 \times 10^{-13}\ \text{kg} \cdot \text{m/s}$$

and its de Broglie wavelength is

$$\lambda = \frac{h}{p} = \frac{6.63 \times 10^{-34}\ \text{J} \cdot \text{s}}{8 \times 10^{-13}\ \text{kg} \cdot \text{m/s}} = 8 \times 10^{-22}\ \text{m}$$

That's about 10^{-6} of the diameter of a proton! It's impossible to see wave effects from a wave with such a tiny wavelength. To see diffraction of such a wave, we would have to create a slit whose width is much smaller than the width of a single proton. Wave effects are even smaller for larger objects (greater mass m) moving faster (greater speed v). So the wave aspect of matter is generally noticeable only on the atomic or subatomic scale. Particles such as electrons exhibit noticeable wave properties; objects such as dust motes, baseballs, and humans do not.

EXAMPLE 26-3 Finding the Wavelength of a Room-Temperature Neutron

A nuclear reactor emits *thermal neutrons*. These are neutrons that behave as though they were particles of an ideal gas at Kelvin temperature T. We learned in Section 14-3 that the average kinetic energy of a particle in an ideal gas at temperature T is $(3/2)kT$, where $k = 1.381 \times 10^{-23}$ J/K is the Boltzmann constant. Calculate the de Broglie wavelength of an average neutron at 293 K (room temperature). The mass of a neutron is 1.67×10^{-27} kg.

Set Up

We'll first calculate the kinetic energy of an average neutron using Equation 14-13, which we learned in Section 14-3. From this we can find the speed and magnitude of momentum of an average neutron. Equation 26-10 will then allow us to calculate the de Broglie wavelength of such a neutron.

de Broglie wavelength:

$$\lambda = \frac{h}{p} \quad (26\text{-}10)$$

neutron m v = $\lambda = ?$

Temperature and average translational kinetic energy of an ideal gas particle:

$$K_{\text{translational, average}} = \frac{1}{2}m(v^2)_{\text{average}} = \frac{3}{2}kT \quad (14\text{-}13)$$

Solve

Calculate the translational kinetic energy, speed, and momentum of an average neutron at $T = 293$ K.

From Equation 14-13

$$K_{\text{translational, average}} = \frac{3}{2}kT = \frac{3}{2}(1.381 \times 10^{-23}\ \text{J/K})(293\ \text{K})$$
$$= 6.07 \times 10^{-21}\ \text{J}$$

Calculate the speed of a neutron with this kinetic energy:

$$K_{\text{translational, average}} = \frac{1}{2}mv^2 \text{ so}$$

$$v = \sqrt{\frac{2K_{\text{translational, average}}}{m}} = \sqrt{\frac{2(6.07 \times 10^{-21}\ \text{J})}{1.67 \times 10^{-27}\ \text{kg}}}$$
$$= 2.70 \times 10^3\ \text{m/s}$$

The magnitude of momentum of a neutron with this speed is

$$p = mv = (1.67 \times 10^{-27}\ \text{kg})(2.70 \times 10^3\ \text{m/s})$$
$$= 4.50 \times 10^{-24}\ \text{kg}\cdot\text{m/s}$$

Calculate the de Broglie wavelength of such a neutron.

From Equation 26-10

$$\lambda = \frac{h}{p} = \frac{6.63 \times 10^{-34}\ \text{J}\cdot\text{s}}{4.50 \times 10^{-24}\ \text{kg}\cdot\text{m/s}} = 1.47 \times 10^{-10}\ \text{m} = 0.147\ \text{nm}$$

(Recall that $1\ \text{J} = 1\ \text{kg}\cdot\text{m}^2/\text{s}^2$ and $1\ \text{nm} = 10^{-9}$ m.)

Reflect

The de Broglie wavelength of a thermal neutron is about 0.147 nm, a distance that is typical of the size of atoms and of the spacing between atoms within a molecule. For this reason thermal neutrons are useful for studying the structure of complex molecules such as proteins. When the neutrons scatter from a protein molecule, they diffract and produce a diffraction pattern that is characteristic of the particular arrangement of atoms in the molecule. X rays can have the same wavelength, but they interact with the charges within atoms and so scatter only weakly from the relatively small atoms (with a small amount of internal charge) such as hydrogen, carbon, nitrogen, and oxygen found in proteins. Neutrons, by contrast, are electrically neutral and actually scatter more strongly from smaller atoms than from larger ones. This makes neutrons superior to x rays for studies of protein structure.

GOT THE CONCEPT? 26-3 Ranking de Broglie Wavelengths

Rank the following objects in order of their de Broglie wavelength, from longest to shortest. (a) A proton moving at 2.00×10^3 m/s; (b) a proton moving at 4.00×10^3 m/s; (c) an electron moving at 2.00×10^3 m/s; (d) an electron moving at 4.00×10^3 m/s.

TAKE-HOME MESSAGE FOR Section 26-4

✔ Particles can exhibit wave properties such as diffraction.

✔ The wavelength of a particle is inversely proportional to its momentum. Hence wave effects are noticeable only for very small particles such as electrons, for which the momentum is very small.

26-5 The spectra of light emitted and absorbed by atoms show that atomic energies are quantized

We have seen that the late nineteenth and early twentieth centuries were years of tremendous change in the study of physics. Studying the photoelectric effect and blackbody radiation led to the revolutionary concept that light has particle aspects, and de Broglie introduced the no less revolutionary idea that matter has wave aspects. During this same time a key set of experiments radically transformed our understanding of the nature of atoms.

The Nuclear Atom

The early Greeks introduced the idea of the atom, a unit of matter so small that it could not be subdivided. (The word "atom" is derived from the Greek term for indivisible.) In 1897 the British physicist J. J. Thomson discovered that atoms are not in fact indivisible but have an internal structure: All atoms contain negatively charged particles (electrons) that can be removed from the atom. It was known that atoms are electrically neutral, so there must also be positively charged material inside an atom. But what form does this positive charge take?

Thomson proposed that most of the mass of the atom is in the form of electrons and that the positively charged material is a low-density sort of jelly in which the electrons are embedded. This model is sometimes called the plum pudding model, since Thomson envisioned electrons scattered throughout the positive charge much like raisins in the traditional English dessert. (If you're not familiar with plum pudding, think of electrons as pieces of fruit embedded in a gelatin dessert or salad.)

A crucial experiment that tested this model was carried out in 1909 at the University of Manchester in England by the New Zealand–born British chemist and physicist Ernest Rutherford with his colleagues Hans Geiger and Ernest Marsden. They fired subatomic particles called *alpha particles* (which were known to be positive helium ions, thousands of times more massive than an electron) at a very thin gold foil. Rutherford expected that if the gold atoms had the structure described in Thomson's model, the alpha particles would be only slightly deflected from their initial direction as a result of passing through the diffuse, positive "pudding" of the atoms. Instead, Rutherford was startled to discover that alpha particles were sometimes scattered at large angles with respect to the initial direction, occasionally leaving the gold foil directly *backward*. (In reflecting on this experiment Rutherford later said, "It was quite the most incredible event that ever happened to me in my life. It was as incredible as if you fired a 15-in. shell [a large projectile fired from a military weapon] at a piece of tissue paper and it came back and hit you. On consideration, I realized that this scattering backwards must be the result of a single collision, and when I made calculations I saw that it was impossible to get anything of that order of magnitude unless you took a system in which the mass of the atom was concentrated in a minute nucleus."

To account for this, Rutherford proposed a model of the atom in which negatively charged electrons orbit a small, positively charged *nucleus* that contains nearly all of the atom's mass. In Rutherford's model most of the volume of each atom is empty, so most alpha particles fired at the gold foil would experience only slight deflections as they passed through. But once in a while, about one time out of every 10,000, an alpha particle would approach a gold nucleus almost head on and be scattered at a large angle, sometimes directly backward.

Unlike planets orbiting the Sun, electrons orbiting an atomic nucleus fit into a well-defined organizational structure. We'll explore this structure from three perspectives and in historical order: first the early clues that hinted at the structure, then the

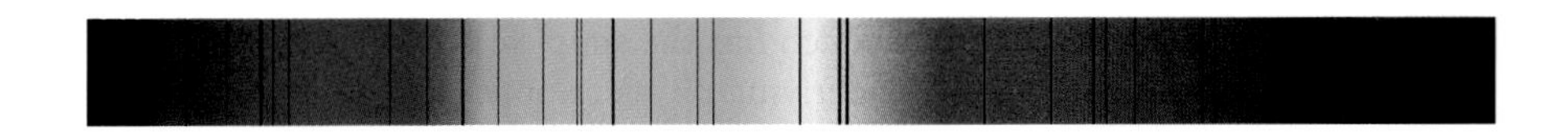

Figure 26-11 The absorption spectrum of the Sun The dark lines in the spectrum of sunlight indicate that certain wavelengths are absorbed when light from the solar interior passes through the Sun's atmosphere.

development of mathematical models that describe the structure, and finally (in the following section) the theory that explains it.

The Discovery of Atomic Spectra

In the early part of the nineteenth century, the English scientist William Hyde Wollaston and the German physicist Joseph von Fraunhofer independently discovered dark lines in the spectrum of visible light coming from the Sun (**Figure 26-11**). These lines are always in the same locations within the spectrum. Some years later Gustav Kirchhoff, the same physicist we encountered in our study of electric circuits, was able to reproduce these same dark lines in the laboratory. He passed light from a lamp (made to simulate sunlight) through vapors created by heating sodium. The light from the lamp itself had a continuous spectrum like that of a blackbody, but the spectrum of light that had passed through the sodium vapor had two dark lines. Kirchhoff concluded that the dark lines result from certain specific colors of light being absorbed by the sodium vapor. What is more, these lines were at the same position as the closely spaced pair of lines in the yellow-orange region of the Sun's spectrum. (You can easily find these lines in Figure 26-11.) The same mechanism must therefore be happening with sunlight: The light coming from the solar interior has a continuous, blackbody-like spectrum, but certain wavelengths of that light are absorbed by atoms in the Sun's atmosphere. The Sun's atmosphere must contain sodium atoms identical to those in Kirchhoff's laboratory. In light of Kirchhoff's discovery, a spectrum like that shown in Figure 26-11 is called an **absorption spectrum**, and the dark lines are called **absorption lines**.

Scientists soon discovered that each element produces its own characteristic absorption lines when light passes through a vapor containing atoms of that element. Thus an absorption spectrum acts as a "fingerprint" of the chemical composition of the vapor that produced the absorption lines. (This is how we know the chemical composition of the Sun's atmosphere. It's also how we determine the chemical composition of the atmospheres of distant stars like those in the photograph that opens this chapter, and how we know that all stars have basically the same chemical makeup.) What was not understood was *why* atoms should selectively absorb only light of certain wavelengths and why the absorbed wavelengths should be different for atoms of different elements.

An important clue about the mystery of absorption spectra came from studying the light *emitted* by atoms. Physicists of the nineteenth century discovered that light created by heating a vapor gives rise to an **emission spectrum**, a spectrum that consists only of specific emitted wavelengths. What is more, if the vapor contains atoms of a certain element, the wavelengths in the emission spectrum from those atoms (the **emission lines**) are precisely the same as the wavelengths in the absorption spectrum of that same element (**Figure 26-12**). To explore this, we'll concentrate on the absorption

(a) The absorption spectrum produced by passing white light through a gas of hydrogen atoms

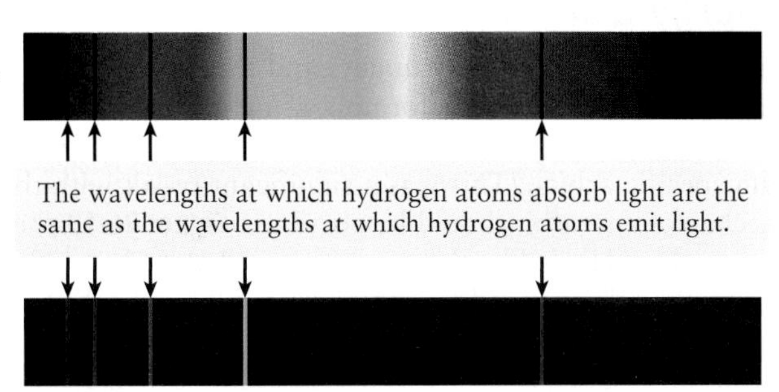

The wavelengths at which hydrogen atoms absorb light are the same as the wavelengths at which hydrogen atoms emit light.

(b) The emission spectrum produced by a heated gas of hydrogen atoms

- Atoms of each element absorb and emit light at wavelengths that are characteristic of that element.
- The characteristic wavelengths differ from one element to another.
- Hydrogen has the simplest arrangement of characteristic wavelengths of any element.

Figure 26-12 Absorption and emission spectra of atomic hydrogen (a) When light passes through a gas, light of specific wavelengths is absorbed, forming dark lines. (b) When a gas is made to glow by passing an electric current through it, it emits only specific wavelengths of light.

and emission spectra of hydrogen, which has the simplest set of absorption and emission lines of any element. But the underlying physics applies to all elements.

Johann Balmer, a Swiss mathematician, made an analysis of the lines in the absorption and emission spectra of hydrogen. He devised a formula that both reproduced the wavelengths of lines that had been reported and correctly predicted the wavelengths of spectral lines that had not yet been observed. This was later extended by the Swedish physicist Johannes Rydberg. The Rydberg formula for the hydrogen spectral lines is

Rydberg formula for the spectral lines of hydrogen (26-14)

Wavelength of an absorption or emission line in the spectrum of atomic hydrogen

$$\frac{1}{\lambda} = R_{\text{H}} \left(\frac{1}{n^2} - \frac{1}{m^2} \right)$$

Rydberg constant = $1.09737 \times 10^7\ \text{m}^{-1}$

n and m are integers: n can be 1, 2, 3, 4, ... and m can be any integer greater than n.

The value of the constant R_{H} in Equation 26-14, called the *Rydberg constant*, is chosen to match the experimental data. To four significant figures $R_{\text{H}} = 1.097 \times 10^7\ \text{m}^{-1}$. As an example, the hydrogen absorption and emission lines shown in Figure 26-12 all correspond to $n = 2$ in Equation 26-14. The series of wavelengths for which $n = 2$ are called the *Balmer series*. For example, to get the wavelength of the red spectral line in Figure 26-12, set $n = 2$ and $m = 3$ in Equation 26-14 and then take the reciprocal:

$$\frac{1}{\lambda} = R_{\text{H}}\left(\frac{1}{2^2} - \frac{1}{3^2}\right) = (1.097 \times 10^7\ \text{m}^{-1})\left(\frac{1}{4} - \frac{1}{9}\right) = 1.524 \times 10^6\ \text{m}^{-1}$$

$$\lambda = \frac{1}{1.524 \times 10^6\ \text{m}^{-1}} = 6.563 \times 10^{-7}\ \text{m} = 656.3\ \text{nm}$$

The spectral line to the left of this one in Figure 26-12 (in the blue-green part of the spectrum) corresponds to $n = 2$ and $m = 4$; you can show that for this spectral line, $\lambda = 486.2$ nm. The wavelengths with $n = 1$ are all in the ultraviolet and are called the *Lyman series*; the wavelengths with $n = 3$ are all in the infrared and are called the *Paschen series*.

What Balmer and Rydberg did not know was *why* the spectral hydrogen lines were given by this relatively simple formula. It was left to the Danish physicist Niels Bohr to provide the explanation.

Energy Quantization

Bohr realized that to fully understand the structure of the hydrogen atom, he had to be able to derive Balmer's formula using the laws of physics. He first made the rather wild assumption that the electron in a hydrogen atom can orbit the nucleus only in certain specific orbits. (This was a significant break with the ideas of Newton, in whose mechanics any orbit should be possible.) **Figure 26-13** shows the four smallest of these **Bohr orbits**, labeled by the numbers $n = 1$, $n = 2$, $n = 3$, and so on.

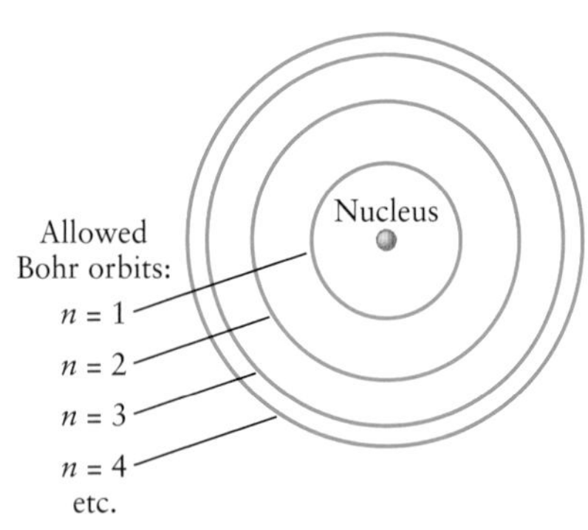

Figure 26-13 Bohr orbits in the hydrogen atom In the model devised by Niels Bohr, electrons in the hydrogen atom are allowed to be in certain orbits only. (The radii of the orbits are not shown to scale.)

Although confined to one of these allowed orbits while circling the nucleus, an electron can jump from one Bohr orbit to another. For an electron to do this, the hydrogen atom must gain or lose a specific amount of energy. The atom must absorb energy for the electron to go from an inner to an outer orbit; the atom must release energy for the electron to go from an outer to an inner orbit. As an example, **Figure 26-14** shows an electron jumping between the $n = 2$ and $n = 3$ orbits of a hydrogen atom as the atom absorbs or emits a photon.

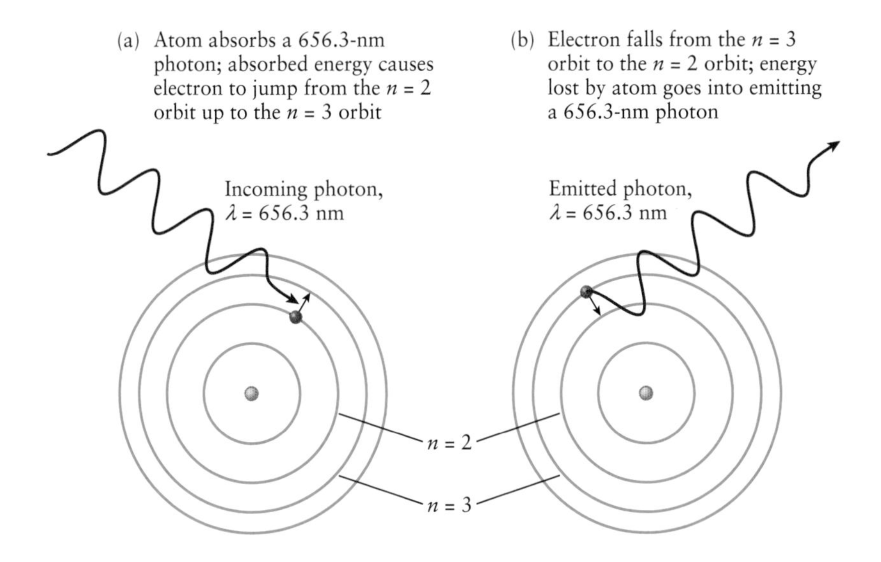

Figure 26-14 The Bohr model explains absorption and emission spectra When a photon is (a) absorbed or (b) emitted by a hydrogen atom, the electron makes a transition or jump between two allowed orbits. The photon energy equals the difference in energy between the upper and lower electron orbits.

When the electron jumps from one orbit to another, the energy of the photon that is emitted or absorbed equals the difference in energy between these two orbits. This energy difference, and hence the photon energy, is the same whether the jump is from a low orbit to a high orbit (Figure 26-14a) or from the high orbit back to the low one (Figure 26-14b). According to Einstein, if two photons have the same energy E, the relationship $E = hf$ (Equation 22-26) tells us that they must also have the same frequency f and hence the same wavelength $\lambda = c/f$. It follows that if an atom can emit photons of a given energy and wavelength, it can also absorb photons of precisely the same energy and wavelength. Thus Bohr's picture explains Kirchhoff's observation that atoms emit and absorb the same wavelengths of light.

The Bohr picture also helps us visualize what happens to produce an emission spectrum. When a gas is heated its atoms move around rapidly and can collide forcefully with each other. These energetic collisions excite the atoms' electrons into high orbits. The electrons then cascade back down to the innermost possible orbit, emitting photons whose energies are equal to the energy differences between different Bohr orbits. In this fashion a hot gas produces an emission line spectrum with a variety of different wavelengths.

To produce an absorption spectrum, begin with a relatively cool gas so that the electrons in most of the atoms are in inner, low-energy orbits. If a beam of light with a continuous spectrum is shone through the gas, most wavelengths will pass through undisturbed. Only those photons will be absorbed whose energies are just right to excite an electron to an allowed outer orbit. Hence only certain wavelengths will be absorbed, and dark lines will appear in the spectrum at those wavelengths.

As in Figure 26-14, the energy of the photon that is absorbed or emitted in a jump between orbits must be equal to the *difference* between the energy of the atom with the electron in the larger-radius, higher-energy orbit and the energy of the atom with the electron in the smaller-radius, lower-energy orbit. Bohr concluded that the numbers n and m in the Rydberg formula correspond to the numbers of the orbits between which an electron jumps as it absorbs or emits a photon. The value of n is the number of the lower orbit, and the value of m is the number of the upper orbit. For example, the jump shown in Figure 26-14 corresponds to $n = 2$ and $m = 3$.

We can better understand Bohr's idea by combining the Rydberg formula, Equation 26-14, with the expressions $E = hf$ and $f = c/\lambda$ for a photon. Together these latter two expressions say that the energy of a photon of wavelength λ is $E = hc/\lambda$. So if we multiply Equation 26-14 by hc, we get an expression for the energy of a photon absorbed or emitted by a hydrogen atom:

$$E_{\text{photon}} = \frac{hc}{\lambda} = hcR_{\text{H}}\left(\frac{1}{n^2} - \frac{1}{m^2}\right) = \frac{hcR_{\text{H}}}{n^2} - \frac{hcR_{\text{H}}}{m^2} = \left(-\frac{hcR_{\text{H}}}{m^2}\right) - \left(-\frac{hcR_{\text{H}}}{n^2}\right) \tag{26-15}$$

If we say that a hydrogen atom has energy $E_{\text{atom},n} = -hcR_{\text{H}}/n^2$ when the electron is in the nth orbit and has energy $E_{\text{atom},m} = -hcR_{\text{H}}/m^2$ when the electron is in the mth orbit, where m is greater than n, then we can rewrite Equation 26-15 as

$$E_{\text{photon}} = E_{\text{atom},m} - E_{\text{atom},n} \tag{26-16}$$

Equation 26-16 uses the idea that the energy of a hydrogen atom is **quantized:** That is, the energy can only have certain values. These energies are given by

$$E_{\text{atom},n} = -\frac{hcR_{\text{H}}}{n^2} \quad \text{where } n = 1, 2, 3, 4, \ldots \tag{26-17}$$

Note that the energy of the atom is negative, and greater values of n correspond to energies that are less negative (that is, closer to zero). This agrees with the idea that the larger the orbit and the greater the value of n for that orbit, the higher the energy.

Each quantized value of the energy is called an **energy level. Figure 26-15** shows several of the energy levels of the hydrogen atom, along with vertical arrows that show the energy of the photon that must be absorbed or emitted in a jump or transition between levels. The transitions that correspond to the Lyman series involve a photon with a very large amount of energy, so these photons have a high frequency and short wavelength: They are all in the ultraviolet part of the spectrum. By contrast, the transitions that correspond to the Paschen series involve a photon with a very small amount of energy, which is why these photons have a low frequency and long wavelength and are in the infrared part of the spectrum. The Balmer series is intermediate between these two; the wavelengths are either in the visible range (380 to 750 nm) or the ultraviolet range.

Elements other than hydrogen also absorb and emit light at specific wavelengths, although those wavelengths do not follow a simple mathematical pattern like the characteristic wavelengths of hydrogen given by Equation 26-14. The conclusion is that there are quantized energy levels for atoms of other elements, but the arrangement of energy levels is more complex than for hydrogen. In the following section we'll see how Niels Bohr justified the quantization of energy for the relatively simple case of the hydrogen atom, and how he was able to reproduce Equation 26-17 for the energy levels. We'll then use these ideas to gain insight into the structure of other atoms.

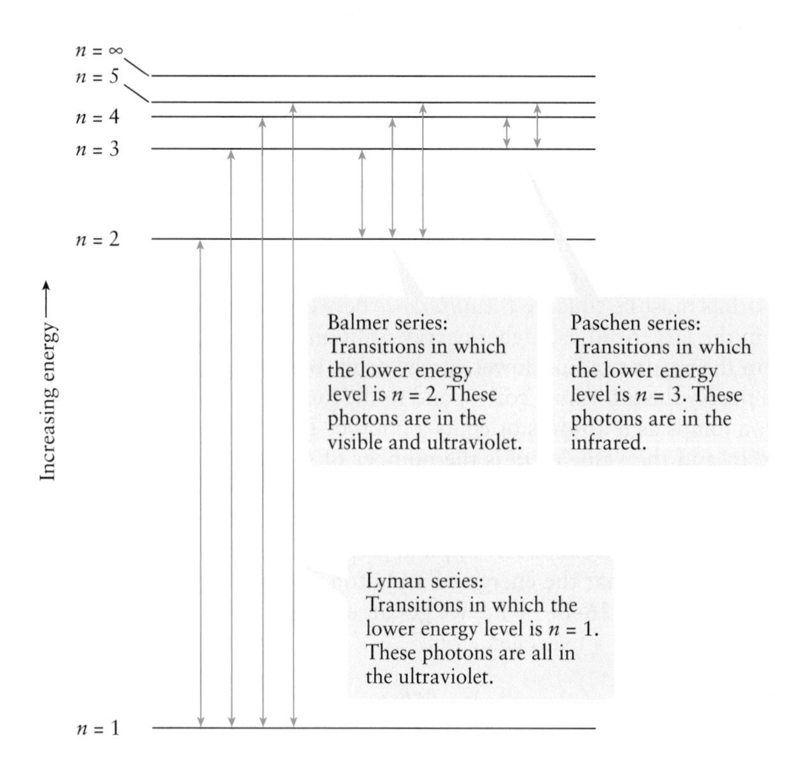

Figure 26-15 Hydrogen energy levels This figure shows some of the lower-lying energy levels of the hydrogen atom and possible transitions between those levels. Note that the energy difference is greatest between the $n = 1$ and $n = 2$ levels, less between the $n = 2$ and $n = 3$ levels, and even less between the $n = 3$ and $n = 4$ levels.

WATCH OUT! Energy quantization is not just an obscure effect in atomic physics.

Discrete energy levels play an important role in modern technology. One example is a *laser*, a device that emits an intense beam of light of a very specific wavelength. This is possible because the laser contains a material with two distinct energy levels. Light of the laser's characteristic wavelength is emitted when transitions take place from the upper level to the lower one. What makes the laser unique is that these transitions occur coherently rather than at random. Energy is added to the material to pump its molecules into excited states. The excited molecules naturally want to transition to a lower state, but if left on their own, would do so at random times. However, when a photon of energy equal to the difference between two states is sent into the material, it can *stimulate* this transition and cause a second photon of that same energy to be released. This photon can stimulate further emission, so the number of emitted photons increases. The emitted photons are all in phase and all travel in the same general direction. The net result is an intense beam. If materials did not have quantized energy levels, the lasers found in Blu-ray and DVD players (which scan the video information encoded on the disc) could not exist.

Fluorescent light bulbs also depend on energy quantization. The material inside the bulb emits ultraviolet photons when an electric current passes through it. These photons are absorbed by the white coating on the inner surface of the bulb. Since ultraviolet photons are very energetic, this excites the material of the coating to very high energy levels. The coating then drops down to its initial energy level in a series of small steps. (It's like taking a big leap to the top of a staircase, then coming carefully down the staircase one step at a time.) Each small step between closely spaced energy levels emits a low-energy, visible-light photon. There are so many such energy levels, with a variety of spacing between them, that the net result is that a mixture of almost all visible colors—that is, white light—is emitted from the bulb.

EXAMPLE 26-4 Photon Possibilities

A collection of hydrogen atoms is excited to the $n = 3$ energy level. What are the possible wavelengths that these atoms could emit as they return to the lowest-energy ($n = 1$) level?

Set Up

There are two routes that an atom can take from the $n = 3$ level to the $n = 1$ level. One, it could drop down to the $n = 1$ level in a single step by emitting a single photon whose energy is equal to the difference between the energies of the $n = 3$ and $n = 1$ levels. The wavelength of this photon is given by Equation 26-14 with $n = 1$ and $m = 3$. Two, the atom could first drop to the $n = 2$ level by emitting a photon (with a wavelength given by Equation 26-14 with $n = 2$ and $m = 3$), then emit a second photon as it drops from the $n = 2$ level to the $n = 1$ level (with a wavelength given by Equation 26-14 with $n = 1$ and $m = 2$). So three different wavelengths can be emitted by the excited atoms. We'll calculate each of these in turn.

Rydberg formula for the spectral lines of hydrogen:

$$\frac{1}{\lambda} = R_H\left(\frac{1}{n^2} - \frac{1}{m^2}\right) \quad (26\text{-}14)$$

Solve

Use Equation 26-14 to calculate each of the three possible wavelengths.

For the $n = 3$ to $n = 1$ transition,

$$\frac{1}{\lambda} = R_H\left(\frac{1}{1^2} - \frac{1}{3^2}\right) = (1.097 \times 10^7\ \text{m}^{-1})\left(1 - \frac{1}{9}\right)$$

$$= 9.751 \times 10^6\ \text{m}^{-1}$$

$$\lambda = \frac{1}{9.751 \times 10^6\ \text{m}^{-1}} = 1.026 \times 10^{-7}\ \text{m} = 102.6\ \text{nm}$$

This is an ultraviolet wavelength.

For the $n = 3$ to $n = 2$ transition,

$$\frac{1}{\lambda} = R_H\left(\frac{1}{2^2} - \frac{1}{3^2}\right) = (1.097 \times 10^7\ \text{m}^{-1})\left(\frac{1}{4} - \frac{1}{9}\right)$$

$$= 1.524 \times 10^6\ \text{m}^{-1}$$

$$\lambda = \frac{1}{1.524 \times 10^6\ \text{m}^{-1}} = 6.563 \times 10^{-7}\ \text{m} = 656.3\ \text{nm}$$

This is a visible wavelength (in the red part of the spectrum).

For the $n = 2$ to $n = 1$ transition,

$$\frac{1}{\lambda} = R_H\left(\frac{1}{1^2} - \frac{1}{2^2}\right) = (1.097 \times 10^7\ \text{m}^{-1})\left(1 - \frac{1}{4}\right)$$

$$= 8.228 \times 10^6\ \text{m}^{-1}$$

$$\lambda = \frac{1}{8.228 \times 10^6\ \text{m}^{-1}} = 1.215 \times 10^{-7}\ \text{m} = 121.5\ \text{nm}$$

This is another ultraviolet wavelength.

Reflect

Comparing with Figure 26-15 shows that the 656.3-nm wavelength represents an emission line of the Balmer series, while the 102.6-nm and 121.5-nm wavelengths represent emission lines of the Lyman series.

GOT THE CONCEPT? 26-4 Ranking Hydrogen Transitions

Rank the following transitions between hydrogen energy levels in terms of the energy of the photon involved, from highest to lowest energy. (a) An atom drops from the $n = 4$ level to the $n = 2$ level. (b) An atom rises from the $n = 3$ level to the $n = 5$ level. (c) An atom drops from the $n = 3$ level to the $n = 1$ level. (d) An atom drops from the $n = 4$ to the $n = 3$ level.

TAKE-HOME MESSAGE FOR Section 26-5

✔ Atoms are composed of electrons orbiting a positively charged nucleus that has most of the mass of the atom.

✔ In the Bohr model an electron's orbits around the nucleus can have only certain well-defined energies. Thus the energy of the atom is quantized and can have only certain values. The atom cannot have energies intermediate between those values.

✔ Electrons that make a transition from one allowed orbit to another either radiate or absorb a photon of a well-defined energy. This gives rise to the lines in emission and absorption spectra.

26-6 Models by Bohr and Schrödinger give insight into the intriguing structure of the atom

The force that pulls the Moon toward Earth is directed toward Earth's center, yet the Moon does not fall into Earth. In the same way, the negatively charged electron in an atom experiences a Coulomb force that points directly toward the positively charged protons at the center of the atom. Why does the electron orbit rather than fall in to the proton? You likely have an intuitive answer that both the Moon and the electron *orbit* rather than fall in. (In a real sense each *is* falling in, but it is always missing the target!) Niels Bohr based his model of the hydrogen atom, for which he received the Nobel Prize in Physics in 1922, on the physics of atomic orbits. The **Bohr model** provides a theoretical foundation for the physics of atomic spectra that we encountered in the previous section.

The Bohr Model

Bohr made two fundamental assumptions in describing the orbit of an electron around a positive atomic nucleus. First he modeled the orbit as uniform circular motion, that is, as an electron moving at constant speed in a circular path. This is not exactly correct but more than satisfactory to provide a broad understanding of the atom. Bohr's second assumption was that only specific values are allowed for the angular momentum of an orbiting electron. We will return to this second assumption shortly.

Bohr considered a single electron orbiting a nucleus of charge $+Ze$, where Z, the *atomic number* of an atom, is the number of protons in the nucleus. The orbiting electron experiences only one force, the Coulomb attraction between it and the protons in the atomic nucleus. The magnitude of the Coulomb force on an electron orbiting in a circle of radius r is, from Equation 16-1,

$$F = \frac{k(Ze)(e)}{r^2} = \frac{kZe^2}{r^2}$$

Here $k = 8.99 \times 10^9\ \text{N} \cdot \text{m}^2/\text{C}^2$ is the Coulomb constant. Newton's second law (Equation 4-2) requires that this force equal the mass of the electron m_e multiplied by the acceleration it experiences. For an object in uniform circular motion at speed v, the magnitude of the acceleration is, from Equation 3-17,

$$a = \frac{v^2}{r}$$

Combining the above two equations into Newton's second law gives

$$\frac{kZe^2}{r^2} = \frac{m_e v^2}{r}$$

If we multiply both sides of this equation by r, we get

$$\frac{kZe^2}{r} = m_e v^2 \tag{26-18}$$

Let's set Equation 26-18 aside for a moment and examine Bohr's second assumption—a requirement that only specific values are allowed for the angular momentum associated with the electron. Bohr recognized that the dimensions of Planck's constant h are those of angular momentum, so he constrained the electron's angular momentum to be only multiples of h. Specifically, this requirement is

$$L = n\left(\frac{h}{2\pi}\right) = n\hbar \tag{26-19}$$

(orbital angular momentum in the Bohr model)

where L is the electron's orbital angular momentum and n is any integer starting from 1. The constant $\hbar$ (pronounced "h bar") is defined to be h divided by 2π.

To relate angular momentum to Equation 26-18, we express L in terms of the mass of the electron m_e, its speed v, and its distance from the center of the atom r using Equation 8-23:

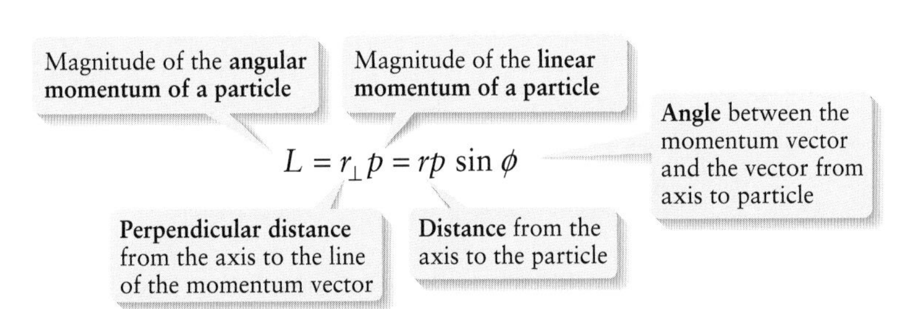

Magnitude of the angular momentum of a particle (8-23)

The magnitude of the electron's linear momentum is $p = m_e v$. Since the electron moves in a circle, the vector from the rotation axis to the particle always has the same radius r and is always perpendicular to the momentum vector. So $\phi = 90°$, $\sin\phi = 1$, and

$$L = rm_e v \tag{26-20}$$

We can now rewrite Equation 26-18 in terms of the angular momentum L by multiplying the right-hand side by 1 in the form of $(r^2/r^2)(m_e/m_e)$:

$$\frac{kZe^2}{r} = \frac{r^2 m_e^2 v^2}{m_e r^2} = \frac{L^2}{m_e r^2}$$

Bohr's requirement that the electron's angular momentum is an integer multiple of $\hbar$ then gives

$$\frac{kZe^2}{r} = \frac{(n\hbar)^2}{m_e r^2}$$

or

$$r_n = \frac{n^2\hbar^2}{m_e kZe^2} \quad \text{where } n = 1, 2, 3,... \tag{26-21}$$

(orbital radii in the Bohr model)

We add the subscript "n" to the variable r to indicate that the radius can take on only specific values and that the allowed values of radius depend on n. Because the values of r_n are proportional to the square of an integer, the orbital radii of electrons in an atom are quantized.

Notice that both n and Z in Equation 26-21 are dimensionless. Because r_n is a distance, all of the other terms on the right-hand side, taken as they appear in the equation, must have dimensions of distance as well. This distance, usually written as a_0 and called the Bohr radius, is

Bohr radius (26-22)

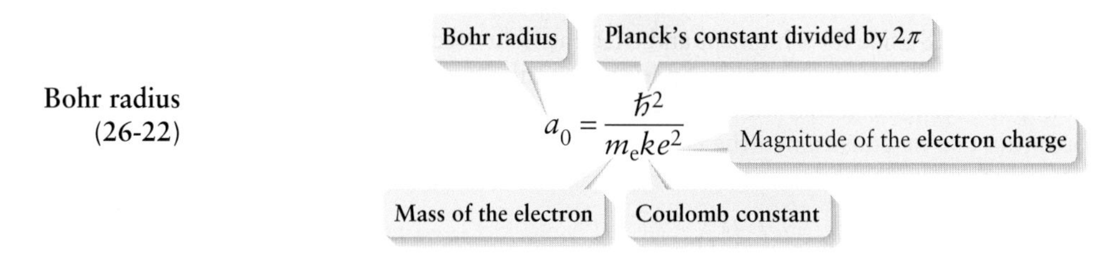

$$a_0 = \frac{\hbar^2}{m_e ke^2}$$

Using the best measured values for $\hbar$, m_e, k, and e, we find that the value of the Bohr radius a_0 is approximately

$$a_0 = 0.529 \times 10^{-10} \text{ m} = 0.0529 \text{ nm}$$

In terms of a_0 the quantized radii of the electron orbits (Equation 26-21) are

Orbital radii in the Bohr model (26-23)

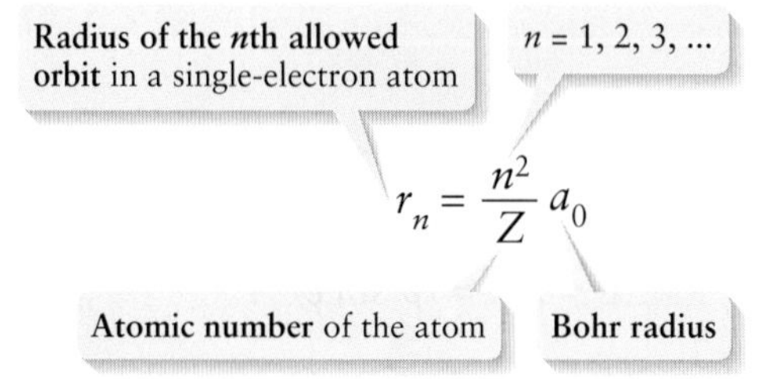

$$r_n = \frac{n^2}{Z} a_0$$

The integer n identifies the orbit, where the $n = 1$ orbit is the closest to the nucleus. Because every element is distinguished by the number of protons it carries, the atomic number Z specifies a particular element. So, for example, setting Z equal to 1 gives the radii of the electron orbits in a hydrogen atom, and setting Z equal to 1 and n equal to 1 gives the radius of the first electron orbit in hydrogen. This is the normal state, usually referred to as the ground state, of hydrogen. Moreover, notice that the radius of the ground state orbit of hydrogen is the Bohr radius. In other words, the radius of a typical hydrogen atom is about 0.05 nm. We can also conclude from Equation 26-23 together with the value of the Bohr radius that, in general, atoms are no more than a few nanometers in radius. The Bohr model sets the scale for atomic sizes.

What is the energy of an electron orbiting an atomic nucleus according to the Bohr model? This is the sum of its kinetic energy and its electric potential energy. For two point charges q_1 and q_2, $U_{\text{electric}} = kq_1q_2/r$ (Equation 17-3), so

$$E = \frac{1}{2}m_e v^2 + \frac{k(-e)(Ze)}{r_n} \tag{26-24}$$

We can write the kinetic energy term in terms of angular momentum in a way that is similar to our approach in developing the relationship for r_n. Using Equation 26-20 shows that the kinetic energy term becomes

$$\frac{1}{2}m_e v^2 = \frac{1}{2}\frac{L^2}{m_e r_n^2} = \frac{n^2\hbar^2}{2m_e r_n^2}$$

so the total electron energy from Equation 26-24 is

$$E = \frac{n^2\hbar^2}{2m_e r_n^2} - \frac{kZe^2}{r_n}$$

Substituting Equation 26-23 for r_n and Equation 26-22 for a_0 gives the energy in terms of only the physical constants and the counting integer n:

$$E = \frac{n^2\hbar^2}{2m_e}\left(\frac{m_e kZe^2}{n^2\hbar^2}\right)^2 - kZe^2\frac{m_e kZe^2}{n^2\hbar^2}$$

Simplifying this is straightforward when you notice that the numerator of both terms is $m_e(kZe^2)^2$ and that both terms have $n^2\hbar^2$ in the denominator:

$$E = \frac{m_e(kZe^2)^2}{2n^2\hbar^2} - \frac{m_e(kZe^2)^2}{n^2\hbar^2}$$

or

$$E_n = -\frac{m_e(kZe^2)^2}{2n^2\hbar^2}, \quad n = 1, 2, 3, \ldots \tag{26-25}$$

(electron energies in the Bohr model)

We add the subscript "n" to the variable E to indicate that the orbital energy of the electron can take on only specific values that depend on n. Because the values of E_n are proportional to $1/n^2$, the orbital energy of electrons in an atom is quantized. This is in agreement with our conclusion in the previous section that for atomic spectral lines to occur only at specific wavelengths, electrons must orbit the hydrogen nucleus with specific, well-defined energies. Note also that the energy is equal to a negative constant divided by n^2, exactly in accordance with Equation 26-17. (We'll see below that the numerical value of the constant is the same in Equation 26-25 as in Equation 26-17.)

The value of energy of an electron in orbit around an atomic nucleus is negative but closer and closer to zero for increasing values of n. In other words, the lowest energy orbit is the one closest to the nucleus, as we would expect. That the energy is negative emphasizes that the electron is bound to the nucleus, and that energy must be supplied in order to either move the electron to a higher orbit or to break the electron free from the nucleus altogether.

Because E_n is an energy, and because both n and Z are dimensionless, the combination of other quantities on the right-hand side of Equation 26-25 must have dimensions of energy as well. This energy, usually written as E_0, is called the Rydberg energy:

Rydberg energy (26-26)

Coulomb constant
Magnitude of the electron charge
Mass of the electron
Rydberg energy
Planck's constant divided by 2π

$$E_0 = \frac{m_e(ke^2)^2}{2\hbar^2} = 2.18 \times 10^{-18}\ \text{J} = 13.6\ \text{eV}$$

In Equation 26-26 we've given the value of the Rydberg energy in joules and in electron volts (eV). The energy that an electron acquires when it moves through a potential difference of 1 V is 1 eV, or approximately 1.602×10^{-19} J. To give you an idea of the amount of energy 1 eV represents, a photon of visible light carries between about 1.5 and 3 eV.

Using Equation 26-26 we can write Equation 26-25 for the quantized electron orbital energy in terms of the Rydberg energy E_0:

Electron energies in the Bohr model (26-27)

Electron energy for the nth allowed orbit in a single-electron atom
Atomic number of the atom
$n = 1, 2, 3, \ldots$
Rydberg energy = 13.6 eV

$$E_n = -\frac{Z^2}{n^2} E_0$$

Again the integer n identifies the orbit, and the atomic number Z specifies a particular element. Setting Z equal to 1 and n equal to 1 therefore tells us that the energy of the ground state of hydrogen is -13.6 eV. We can also conclude from Equation 26-27 that, in general, the energy of electrons in orbit around an atomic nucleus is between about -10 eV and, for the largest elements (for which Z is about 100), -10^5 eV. The Bohr model sets the scale for atomic electron energies.

What does the Bohr model predict for the atomic spectral lines of hydrogen? Every line results from a transition between two electron orbits; that is, the energy of the emitted photon is the energy difference ΔE between two allowed orbits (see Equation 26-16). Since the energy of a photon is hf and its frequency $f = c/\lambda$, the wavelength λ of the photon is then

$$\lambda = \frac{hc}{\Delta E}$$

and the reciprocal of the wavelength (which is what appears in the Rydberg formula) is

(26-28)
$$\frac{1}{\lambda} = \frac{\Delta E}{hc}$$

Let's consider the transition of an electron from a higher orbit m down to a lower orbit n. We can determine the energy difference between these two orbits by applying Equation 26-27. Atomic number Z equals 1 for hydrogen, so

$$\Delta E = -\frac{1}{m^2}E_0 - \left(-\frac{1}{n^2}\right)E_0$$

or

$$\Delta E = \left(\frac{1}{n^2} - \frac{1}{m^2}\right)E_0$$

Substituting this into Equation 26-28 yields

$$\frac{1}{\lambda} = \frac{E_0}{hc}\left(\frac{1}{n^2} - \frac{1}{m^2}\right)$$

Compare this to the Rydberg formula, Equation 26-14. It has exactly the same form!

How does the value of E_0/hc compare to the value of R_H, which equals approximately $1.10 \times 10^7\ m^{-1}$? In SI units

$$\frac{E_0}{hc} = \frac{2.18 \times 10^{-18}\ J}{(6.63 \times 10^{-34}\ J \cdot s)(3.00 \times 10^8\ m/s)} = 1.10 \times 10^7\ m^{-1}$$

The ratio E_0/hc *equals* R_H, so $E_0 = hcR_H$. That's just what we expect if we compare Equation 26-17 (in which we deduced the energies from the Rydberg equation) and Equation 26-27 (in which we derived the energies from the Bohr assumptions). We conclude that the Bohr model is entirely consistent with the observed spectral lines of hydrogen as described by the Rydberg formula.

EXAMPLE 26-5 Lowest Energy Level of a Lithium Ion

Find the (a) radius, (b) energy, and (c) speed of an electron in the lowest energy level of the doubly ionized lithium ion. An atom of lithium ($Z = 3$) has three electrons, so this ion has just one electron.

Set Up

For this ion, $Z = 3$; for the lowest energy level $n = 1$. We'll use Equation 26-23 to calculate the radius of this orbit, Equation 26-27 to calculate the energy, and Equation 26-19 for angular momentum to determine the electron speed.

Orbital radii in the Bohr model:

$$r_n = \frac{n^2 a_0}{Z} \quad \text{where } n = 1, 2, 3, \ldots \tag{26-23}$$

Electron energies in the Bohr model:

$$E_n = -\frac{Z^2}{n^2}E_0 \quad \text{where } n = 1, 2, 3, \ldots \tag{26-27}$$

Orbital angular momentum in the Bohr model:

$$L = n\left(\frac{h}{2\pi}\right) = n\hbar \tag{26-19}$$

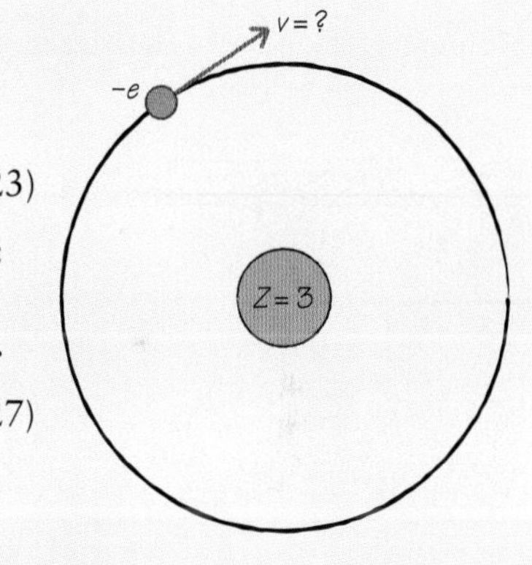

Solve

(a) Calculate the radius of the orbit.

From Equation 26-23 with $Z = 3$ and $n = 1$,

$$r_1 = \frac{1^2 a_0}{3} = \frac{a_0}{3} = \frac{0.0529\ nm}{3} = 0.0176\ nm$$

(b) Calculate the energy.

From Equation 26-27 with $Z = 3$ and $n = 1$,

$$E_n = -\frac{3^2}{1^2}E_0 = -9E_0 = -9(13.6\ eV) = -122\ eV$$

(c) To find an expression for the electron speed, combine Equation 26-19 with the equation for the angular momentum of a particle moving in a circular orbit.

Equation 26-20 tells us that the angular momentum of an electron of mass m_e moving in an orbit of radius r at speed v is $L = rm_e v$. Set this equal to the expression for L in Equation 26-19:

$$rm_e v = n\hbar \quad \text{so}$$

$$v = \frac{n\hbar}{rm_e}$$

We are interested in the case of $n = 1$ and calculated r_1 in part (a). Substituting the values for $\hbar = h/(2\pi)$ and the mass of the electron gives:

$$v = \frac{1(6.63 \times 10^{-34}\ \text{J} \cdot \text{s})}{2\pi\,(1.76 \times 10^{-11}\ \text{m})(9.11 \times 10^{-31}\ \text{kg})}$$

$$= 6.58 \times 10^{6}\ \text{m/s}$$

Reflect

The radius of an electron in the lowest level of the doubly ionized lithium ion ($Z = 3$) is one-third the radius of an electron in the ground state of hydrogen. Because there are three protons in the lithium nucleus, compared to one for hydrogen, the Coulomb force on the electron is greater, so it is reasonable that the orbital radius is smaller. Also, notice that the speed of the electron is about 2% of the speed of light in a vacuum.

Why did we insist that the problem be about a doubly ionized lithium ion, rather than a lithium atom? The difference is that when more than one electron is present, each electron is affected not only by the electric force from the nucleus but also by the electric force from the other electrons. We haven't taken the interaction between electrons into account in any of our calculations: The Bohr model is for *single-electron* atoms and ions only.

Although the Bohr model successfully predicts atomic spectra and other phenomena associated with hydrogen atoms, it nevertheless leaves us with an outstanding question. Why should the electron orbits be quantized in multiples of $\hbar$? For the answer we turn back to Louis de Broglie. Recall that de Broglie postulated that particles, such as electrons, have a wavelike nature. For a nonrelativistic electron of mass m_e moving at speed v, the momentum has magnitude $p = m_e v$. From the de Broglie relation, Equation 26-10, the wavelength of such an electron is

$$\lambda = \frac{h}{m_e v}$$

For an electron orbiting an atomic nucleus we express this in terms of angular momentum by making use of Equation 26-20:

$$\lambda = \frac{hr}{m_e vr} = \frac{hr}{L} = \frac{2\pi\hbar r}{L}$$

(We used $\hbar = h/2\pi$, so $h = 2\pi\hbar$.) Now let L be an integer multiple of $\hbar$, as Bohr required. Then

$$\lambda = \frac{2\pi\hbar r}{n\hbar}$$

or

$$n\lambda = 2\pi r$$

We recognize $2\pi r$ as the circumference of the orbital path of the electron. So Bohr's requirement that L be an integer multiple of $\hbar$ is equivalent to demanding that an integer number of electron wavelengths fit into the circumference of the orbit so that the wave joins smoothly onto itself. In a real sense, it is the wavelike nature of particles that results in the quantization of the energy of atomic electrons.

Confirming Energy Quantization

Bohr's model of the atom works relatively well for hydrogen, for a singly ionized helium ion (an atom with Z equal to 2 but only one electron), and for a doubly ionized lithium ion (an atom with Z equal to 3 but only one electron, as in Example 26-5). For atoms with more than one electron, it isn't possible to make calculations of energy levels using Bohr's physics. However, the general picture of the atom it provides, with electrons in quantized energy states, applies to all atoms. This was confirmed experimentally in 1914 by the German physicists James Franck and Gustav Hertz. (Franck

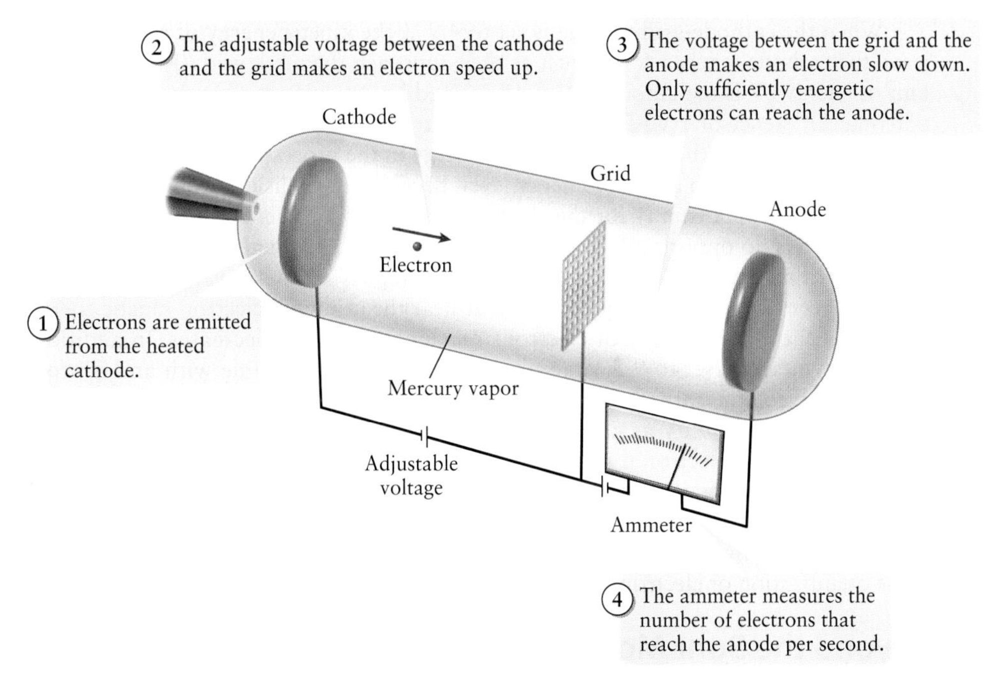

Figure 26-16 The Franck–Hertz experiment In this experiment, electrons are accelerated from a cathode toward a mesh grid in a tube filled with mercury vapor. Some electrons pass through the grid, and the current at the anode is measured by the ammeter.

and Hertz were awarded the 1925 Nobel Prize in Physics for their work. Gustav Hertz was a nephew of Heinrich Hertz, whom we encountered earlier.)

Franck and Hertz used a device similar to the one shown schematically in **Figure 26-16** to measure the effect of bombarding atoms in a gas with electrons. The cathode is heated in order to give off electrons, which are accelerated by a variable voltage toward the mesh grid. Some electrons pass through the grid and arrive at the anode, where the electron current is detected by the ammeter. Notice that a voltage is also applied between the grid and the anode, which acts against the electrons; only electrons that carry sufficient energy as they pass the grid will make it to the anode. These electrical components sit inside a tube filled with low-pressure mercury vapor, so collisions between an electron and a mercury (Hg) atom can occur. Franck and Hertz used mercury vapor because the spectral lines of low-pressure mercury gas were well studied; one prominent ultraviolet spectral line has a wavelength $\lambda = 253.7$ nm. The energy of this photon is

$$E = hf = \frac{hc}{\lambda} = \frac{(6.63 \times 10^{-34}\ \text{J}\cdot\text{s})(3.00 \times 10^{8}\ \text{m/s})}{253.7 \times 10^{-9}\ \text{m}}$$

$$= (7.84 \times 10^{-19}\ \text{J})\left(\frac{1\ \text{eV}}{1.60 \times 10^{-19}\ \text{J}}\right) = 4.90\ \text{eV}$$

This is the energy emitted when an atomic electron falls from an excited energy level down to a lower energy level. Now consider what happens in the collision of an electron and a mercury atom in the Franck–Hertz tube. In general, the more kinetic energy an electron carries as it approaches the grid, the more likely it will make it through. So anode current should grow as the voltage between the cathode and the grid is increased. However, if the energy levels of electrons in Hg (and all) atoms are quantized, when the kinetic energy of the incident electron equals the energy difference between two Hg energy levels, the Hg atom can absorb the electron's energy. When this happens it is less likely that the electron will reach the anode, so the anode current should decrease. **Figure 26-17** shows a typical curve of current versus voltage from the Franck–Hertz experiment. The general trend, as we would expect, is that the anode current increases as the voltage is increased. But the current drops dramatically at certain values of accelerating voltage, in this case, at 4.9 V, and at 9.8 V and 14.7 V, which are multiples of 4.9 V—the energy of the mercury spectral line.

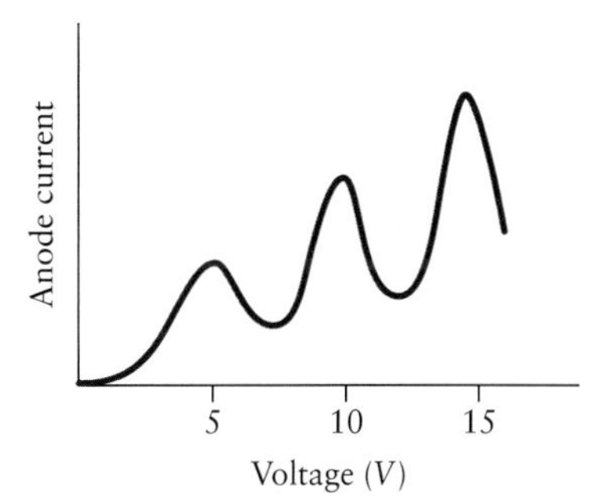

Figure 26-17 Evidence of energy quantization As the voltage between the cathode and the mesh grid in the Franck–Hertz experiment increases, the anode current increases. When the kinetic energy of the accelerated electrons equals the energy required to excite an electron in a mercury atom to a higher energy level, accelerated electrons lose energy. Not as many electrons reach the anode at the corresponding cathode-grid voltages, so the anode current decreases.

When the cathode-grid voltage is set to 4.9 V, the kinetic energy of electrons that leave the cathode reaches 4.9 eV just as they approach the mesh grid. (Recall that the unit of energy eV is the amount of energy gained by an electron when it experiences a potential difference of 1 V.) For this reason electrons that collide with a mercury atom in close proximity to the grid give up their energy in the process of causing an electron in the atom to jump to a higher energy level. Because the collision occurs close to the grid, there is no opportunity for the electron to undergo another acceleration; in other words, there is no opportunity for the electron to acquire enough energy to reach the anode. For this reason when the cathode-grid voltage is set to 4.9 V, so that the collisions between electrons and Hg atoms occur near the grid with electron kinetic energy equal to 4.9 eV, the number of electrons reaching the anode decreases. In addition, when the voltage is set to 9.8 V, accelerated electrons that collide with an Hg atom halfway between the cathode and the grid lose their energy and then accelerate up to a kinetic energy of 4.9 eV again by the time they reach the grid. A collision there once again results in the electron transferring its energy to an Hg atom and being unable to reach the anode. A similar phenomenon occurs when the voltage is set to any integer multiple of 4.9 V. This rise and fall in anode current versus cathode-grid voltage is shown dramatically in Figure 26-17. The underlying explanation of this phenomenon is the quantization of electron energy levels in the atom, as predicted by Bohr.

Beyond the Bohr Model: Quantum Mechanics

As powerful as the Bohr model is at providing an understanding of the atom and atomic spectra, it does not tell the whole story. The Bohr atom treats the electrons that orbit atomic nuclei as particles. But no theoretical description of the atom can be complete unless it accounts for the wave nature of the electrons. The theoretical underpinning of our understanding of atoms that includes these wave properties is found in an equation developed by Austrian physicist Erwin Schrödinger. The *Schrödinger equation* relies on matter waves to describe the state of a system as a function of time, much like Newton's laws do while treating physical systems as particles. The Schrödinger equation is thus the fundamental equation of **quantum mechanics**, in which matter is treated as intrinsically wavelike in nature. Schrödinger was awarded the 1933 Nobel Prize in Physics for his work. (The Schrödinger equation itself is too mathematically ornate for the purposes of this book.)

Perhaps the most notable difference between Schrödinger's quantum-mechanical description and Newton's classical description of physics is the ability to specify the position and velocity of objects. In Newton's description at any instant of time we can identify a specific position in space and a specific velocity vector for any particle in a system, for example, an electron orbiting an atomic nucleus. A wave, however, is not localized in space. The result is that while Newton's laws predict the position and velocity of an object, the Schrödinger equation predicts the *probability* of finding a certain value of position or velocity. For this reason, electrons in the quantum model of the atom are described not as tiny marbles orbiting the nucleus at fixed radii but rather as a charge distribution. This distribution, sometimes called a *probability cloud*, gives the probability of finding the electron at any given position; the denser the cloud in some region, the more likely it is that the electron will be found there.

Figure 26-18a shows the probability cloud associated with the ground state of hydrogen. The more dense the color in any region in this figure, the more likely it is that the electron will be found in that region. This probability distribution is spherically symmetric; that is, it varies only as a function of radius from the center of the nucleus. For that reason we can also express the same information in a curve of probability versus radius, as in **Figure 26-18b**. Notice that the most probable radius of an electron in the ground state of a hydrogen atom is a_0, the Bohr radius.

The Bohr model employs a single integer, or **quantum number**, to describe electron states. For the Bohr atom this integer is n, which determines the energy level of the electron. In the fully quantum-mechanical view of the atom, *four* quantum numbers are required. These are n, the principal quantum number; ℓ, the angular momentum (or orbital) quantum number; m_ℓ, the magnetic quantum number; and m_s, the electron spin quantum number. The specific values of each of these four quantum numbers completely describes the state of an electron in an atom.

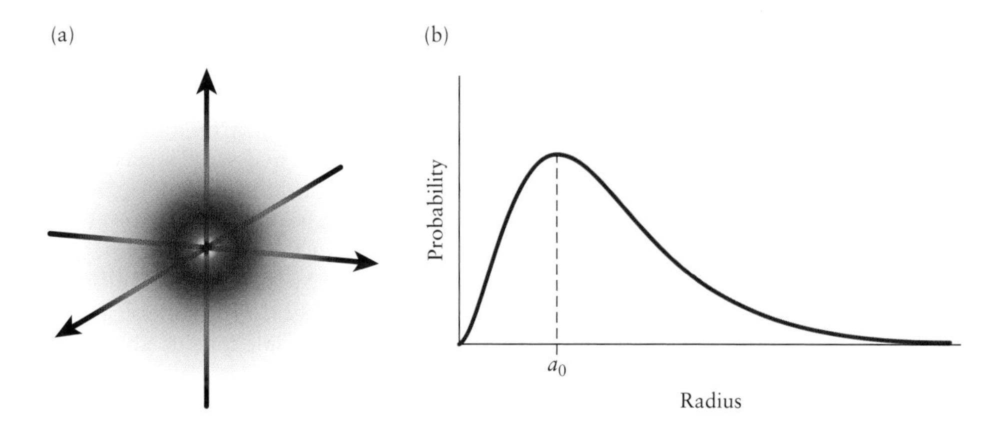

Figure 26-18 The Schrödinger picture of the hydrogen atom (a) In the state of lowest energy, an electron in a hydrogen atom can be found anywhere relative to the nucleus. The darker (denser) the color in a region of the probability cloud, the more likely it is that the electron will be found in that region. (b) The probability cloud in part (a) is spherically symmetric, so we can represent the probability of finding the electron as a function of radius from the center of the nucleus. The probability peaks at a distance from the nucleus equal to a_0, the Bohr radius.

The principal quantum number n plays a role in the quantum atom similar to that which n plays in the classical Bohr atom; in particular, it specifies the energy level or electron shell. The lowest energy state, or ground state, corresponds to n equal to 1.

The angular momentum quantum number ℓ is a measure of the angular momentum the electron carries. For any value of n, ℓ varies in integer steps from 0 to $n - 1$; each value of ℓ specifies a subshell, or electron orbital, within the energy level specified by n. The shape of each orbital is different. By convention we refer to the orbitals with letters rather than integer numbers; the values of ℓ equal to 0, 1, 2, 3, 4, and 5 correspond to the orbitals *s, p, d, f, g*, and *h*. It is also standard to refer to an electron subshell by giving both n and this orbital letter code together. For example, an electron in the p subshell of the $n = 2$ energy level is said to be in the $2p$ subshell. The electron's energy is slightly dependent on the value of ℓ, an effect called *fine structure* not found in the Bohr model.

The magnetic quantum number m_ℓ specifies an orientation of an electron's subshell. It is so called because its value determines the (very small) energy associated with the interaction of a moving charge—the electron—with the magnetic field of the nucleus. The larger the value of ℓ, the more orientations are allowed; m_ℓ can take integer values between $-\ell$ and $+\ell$, including 0. **Figure 26-19** shows the shapes of a number of electron orbitals.

The fourth quantum number m_s involves a new feature of the electron that was not discovered until the 1920s. Electrons have an intrinsic characteristic called **spin**, which is akin to the angular momentum of a rotating sphere. Even electrons that do not orbit an atomic nucleus possess spin, which can take on one of two values, often

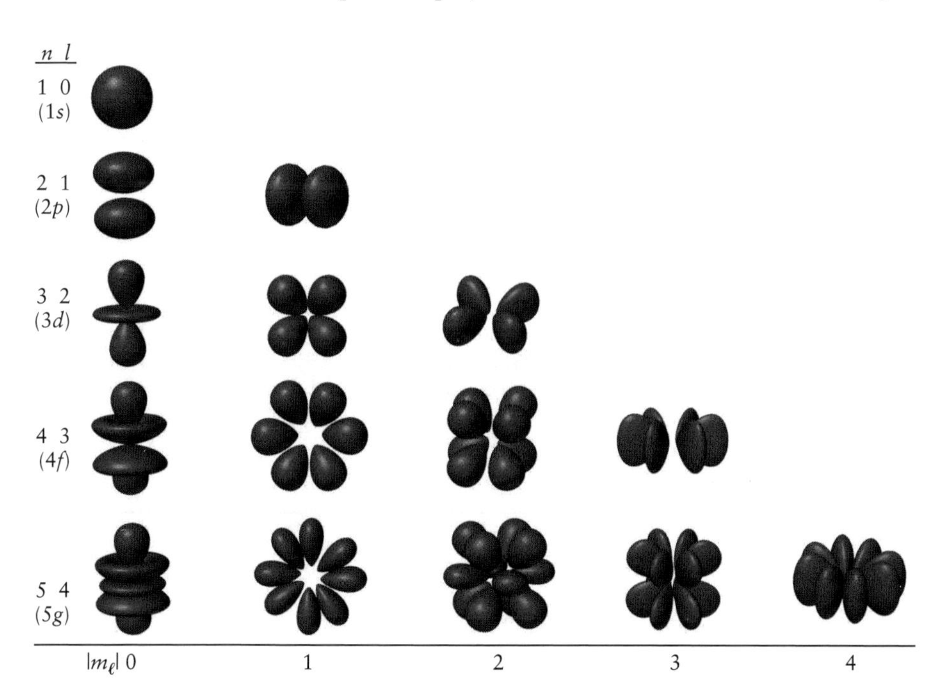

Figure 26-19 Probability clouds for different quantum numbers The probability cloud for each atomic orbital, specified by a specific value of the quantum numbers n, ℓ, and m_ℓ, has a different shape.

called spin "up" and spin "down." To fully describe an atomic electron, then, we must also specify its spin state. This is described by the electron spin quantum number m_s, which for an electron can be equal to either $+1/2$ or $-1/2$.

These quantum numbers play an important role in multi-electron atoms. Each electron in a multi-electron atom has a specific value of n, ℓ, m_ℓ, and m_s. (The details of the orbitals are affected by the presence of other electrons, but the same four quantum numbers still apply.) What is more, it turns out that there can be only *one* electron with a specific combination of these four quantum numbers. This fundamental restriction on electrons is called the **Pauli exclusion principle**, after the Swiss physicist Wolfgang Pauli, who deduced this principle in 1925. (Pauli received the 1945 Nobel Prize in Physics for his work.) Let's see what the Pauli exclusion principle tells us about the structure of multi-electron atoms.

For the $n = 1$ shell, only one value of ℓ (equal to 0, which is the s orbital) and therefore only one value of m_ℓ, is allowed. Two values of m_s are always possible, so the maximum number of electrons that can occupy the $n = 1$ shell is $1 \times 2 = 2$, the product of the number of possible m_ℓ values and the number of possible m_s values. That is, two electrons can occupy the 1*s* orbital. For $n = 2$, ℓ is allowed to be either 0 or 1. When $\ell = 0$, the only possible value of the magnetic quantum number is $m_\ell = 0$. So including the factor of 2 for the electron spin quantum number, one possible value of m_ℓ and two possible values of m_s means two electrons can occupy the 2*s* orbital. However, when ℓ equals 1 (the *p* orbital), m_ℓ can be -1, 0, or $+1$. Including the factor of 2 for the electron spin quantum number, that means that the number of electrons that can occupy the 2*p* orbital is 3×2, or 6. Because two electrons can occupy the 2*s* orbital and six electrons can occupy the 2*p* orbital, an atom can have at most eight electrons in the $n = 2$ energy level.

The pattern above repeats for $n = 3$, up through ℓ equals 2, the *d* orbital. For this orbital, five values (-2, -1, 0, $+1$, $+2$) are possible for m_ℓ, so the maximum number of electrons in this orbital is $5 \times 2 = 10$. Thus two electrons can occupy 3*s*, six can occupy 3*p*, and ten can occupy 3*d*. The maximum number of electrons in the $n = 3$ energy shell of an atom is therefore 18.

The arrangement of electrons, according to how many occupy which orbitals, is directly correlated to the chemical properties of the elements. Consider the 18 lightest elements and their electron configurations, listed in **Table 26-1**. (In the table the number of electrons that occupy a particular orbital is given as a superscript, for example, $2p^4$ indicates that four electrons occupy the 2*p* orbital.) In hydrogen, lithium, and sodium (as well as potassium, rubidium, cesium, and francium), the outermost, or valence, electron is a single electron in an *s* orbital. These *alkali metals* share common chemical properties and occupy a single column in the periodic table.

Three of the *noble gases* are listed in Table 26-1. These are helium, neon, and argon; in each the outermost subshell is completely full. As a result, all electrons are relatively tightly bound, so these elements do not easily gain, lose, or share electrons. For that reason, the noble gases are relatively inert. Helium, neon, and argon, as well as the other noble gases, occupy a single column in the periodic table.

WATCH OUT! As atomic number increases, electron orbitals do not fill in a continuous fashion.

The configuration of the 18 electrons of argon, the heaviest element listed in Table 26-1, is $1s^2 2s^2 2p^6 3s^2 3p^6$. The next heaviest element is potassium, for which the configuration of the 19 electrons is $1s^2 2s^2 2p^6 3s^2 3p^6 4s^1$. Notice that the additional electron does not occupy the 3*d* orbital, although *d* follows *p* in our ordering (*s*, *p*, *d*, *f*, and so on). Orbitals fill according to the increase in energy required, and for that reason do not fill according to counting up linearly in ℓ (orbital letters) and m_ℓ. Instead, orbitals fill in this order: 1*s*, 2*s*, 2*p*, 3*s*, 3*p*, 4*s*, 3*d*, 4*p*, 5*s*, 4*d*, 5*p*, 6*s*, 4*f*, 5*d*, 6*p*, 7*s*, 5*f*, 6*d*, 7*p*. Notice that 4*s* and not 3*d* follows the 3*p* orbital in this sequence, which is why the valence electron in potassium is in a 4*s* orbital.

Halogens are elements that are highly reactive; that is, they easily form bonds with certain other elements, especially the alkali metals, to form molecules. Two halogens, fluorine and chlorine, are listed in Table 26-1. The outermost shell in both is one electron short of being full, which means that an atom of one of these elements can readily share an electron with another atom. This is particularly true for an atom of an alkali metal, which has one valence electron that is easily shared. So bring a sodium atom near a chlorine atom, and they will readily bond to form NaCl (table salt).

The ideas of quantum mechanics find applications on scales even smaller than that of the atom. In the final two chapters of this book we will see how quantum-mechanical ideas help us understand the nature of the atomic nucleus and of the fundamental particles that are the essential building blocks of all ordinary matter.

Go to Interactive Exercise 26-2 for more practice dealing with atomic energy.

TABLE 26-1 Electron Configurations of Light Elements

Atomic number	Element	Electron configuration
1	hydrogen (H)	$1s^1$
2	helium (He)	$1s^2$
3	lithium (Li)	$1s^2 2s^1$
4	beryllium (Be)	$1s^2 2s^2$
5	boron (B)	$1s^2 2s^2 2p^1$
6	carbon (C)	$1s^2 2s^2 2p^2$
7	nitrogen (N)	$1s^2 2s^2 2p^3$
8	oxygen (O)	$1s^2 2s^2 2p^4$
9	fluorine (F)	$1s^2 2s^2 2p^5$
10	neon (Ne)	$1s^2 2s^2 2p^6$
11	sodium (Na)	$1s^2 2s^2 2p^6 3s^1$
12	magnesium (Mg)	$1s^2 2s^2 2p^6 3s^2$
13	aluminum (Al)	$1s^2 2s^2 2p^6 3s^2 3p^1$
14	silicon (Si)	$1s^2 2s^2 2p^6 3s^2 3p^2$
15	phosphorus (P)	$1s^2 2s^2 2p^6 3s^2 3p^3$
16	sulfur (S)	$1s^2 2s^2 2p^6 3s^2 3p^4$
17	chlorine (Cl)	$1s^2 2s^2 2p^6 3s^2 3p^5$
18	argon (Ar)	$1s^2 2s^2 2p^6 3s^2 3p^6$

GOT THE CONCEPT? 26-5 Ionization Energy

The *ionization energy* of an atom is the energy required to remove an electron from the atom. In terms of the Bohr model, it is equal to the energy difference between the $n = 1$ energy level (the level of lowest energy, in which the electron is closest to the nucleus) and the $n = \infty$ energy level (which has an infinite radius, so the electron has moved infinitely far away). Compared to the ionization energy of hydrogen, the energy required to remove the last electron from doubly ionized lithium is (a) 3 times greater; (b) 9 times greater; (c) 27 times greater; (d) 81 times greater; (e) 243 times greater.

TAKE-HOME MESSAGE FOR Section 26-6

✔ In the Bohr model of the atom, electrons follow Newtonian orbits around the nucleus but with quantized values of orbital angular momentum. As a result, only certain orbital radii and orbital energies are allowed.

✔ The Franck–Hertz experiment confirmed that atomic energies are quantized.

✔ The Schrödinger equation explains the hydrogen atom by describing the electron as a wave. It predicts the probability of finding the electron at a particular location within the atom.

✔ The Pauli exclusion principle allows us to understand the structure of multi-electron atoms.

Key Terms

absorption lines
absorption spectrum
blackbody
blackbody radiation
Bohr model
Bohr orbit
Compton scattering
de Broglie wavelength
emission lines
emission spectrum
energy level
Pauli exclusion principle
photoelectric effect
photoelectrons
photon
quantized
quantum mechanics
quantum number
spin
wave-particle duality
work function

Chapter Summary

Topic	Equation or Figure
The photoelectric effect: When light shines on a surface, the surface can emit electrons. However, no electrons are emitted if the frequency of the light is below a certain critical value. Einstein showed that this could be explained if light is absorbed in the form of photons whose energy is proportional to their frequency: $E = hf$, where h is Planck's constant.	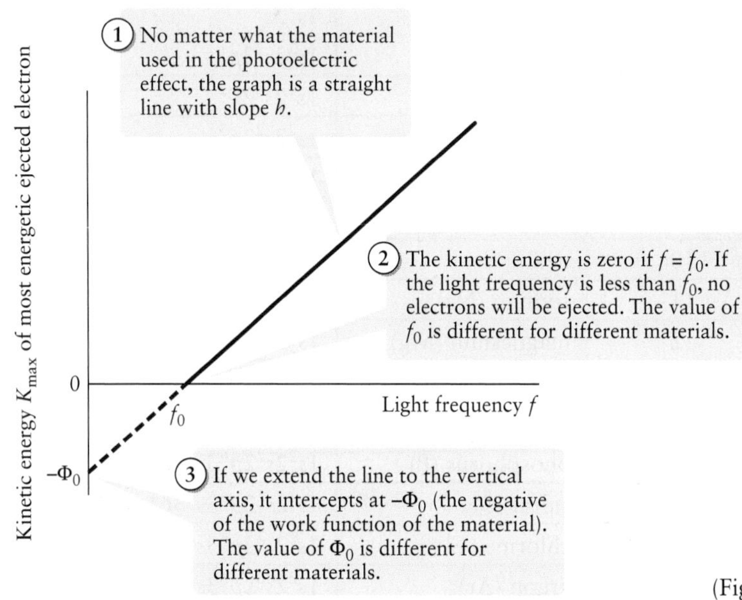 (Figure 26-3)
Blackbody radiation: Objects emit light due to their temperature. An ideal blackbody (one that does a perfect job of absorbing light) is also a perfect emitter of light. The details of the spectrum of a blackbody can be understood only if a blackbody emits light in the form of photons.	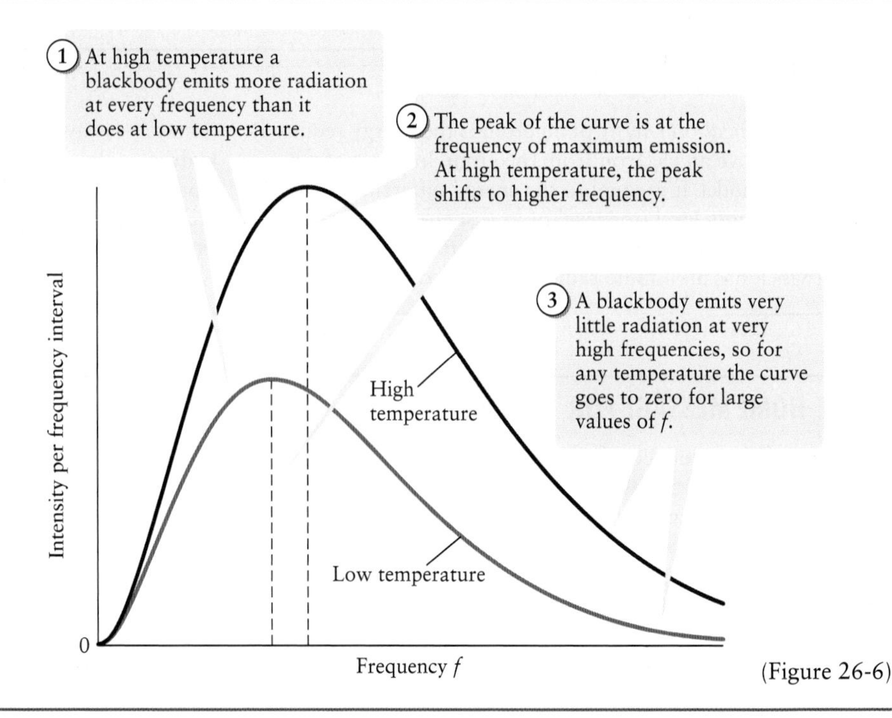 (Figure 26-6)

Compton scattering: Photons have momentum in inverse proportion to their wavelength. This is demonstrated by Compton scattering, in which a photon undergoes an increase in wavelength when it scatters from an electron.

Magnitude of the **momentum** of a photon

Energy of the photon

$$p = \frac{E}{c} = \frac{h}{\lambda} \quad (26\text{-}7)$$

Planck's constant

Speed of light in a vacuum

Wavelength of the photon

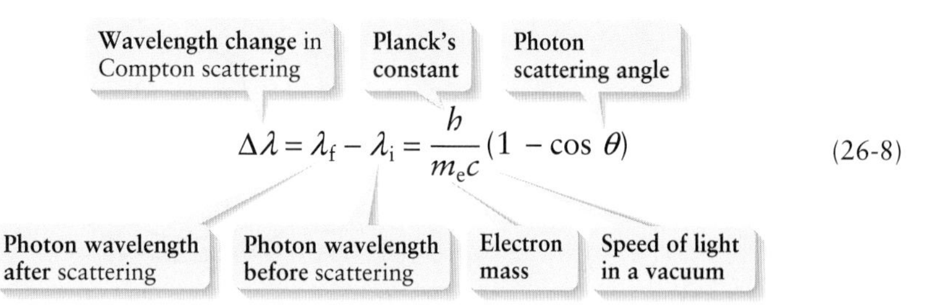

$$\Delta\lambda = \lambda_f - \lambda_i = \frac{h}{m_e c}(1 - \cos\theta) \quad (26\text{-}8)$$

Wave-particle duality: Just as photons have particle aspects, matter has wave aspects. The de Broglie wavelength of a particle is inversely proportional to the momentum (the same relationship between wavelength and momentum as for a photon).

A particle has a **de Broglie wavelength...**

...equal to **Planck's constant...**

$$\lambda = \frac{h}{p} \quad (26\text{-}10)$$

...divided by the **momentum of the particle.** The greater the momentum, the shorter the de Broglie wavelength.

Atomic structure and atomic spectra: Alpha particle scattering shows that the atom is made up of a small positive nucleus surrounded by electrons. The energy of an atom is quantized. It can have only certain definite values. The evidence for this comes from the spectra of atoms, which show that atoms absorb only specific wavelengths of light and that they emit the same wavelengths that they absorb.

(a) The absorption spectrum produced by passing white light through a gas of hydrogen atoms

The wavelengths at which hydrogen atoms absorb light are the same as the wavelengths at which hydrogen atoms emit light.

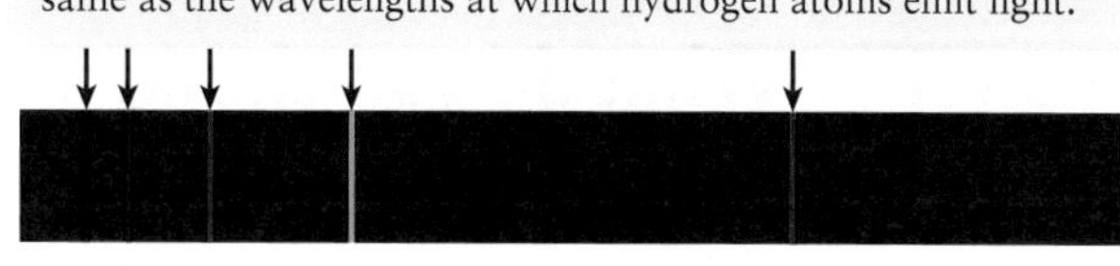

(b) The emission spectrum produced by a heated gas of hydrogen atoms

- Atoms of each element absorb and emit light at wavelengths that are characteristic of that element.
- The characteristic wavelengths differ from one element to another.
- Hydrogen has the simplest arrangement of characteristic wavelengths of any element.

(Figure 26-12)

The Bohr model of the atom: In the Bohr model of hydrogen, a single electron orbits the nucleus much like a satellite orbiting Earth. The difference is that the orbit can have only certain values of angular momentum, radius, and energy. Transitions between the allowed orbits are the cause of absorption and emission spectra.

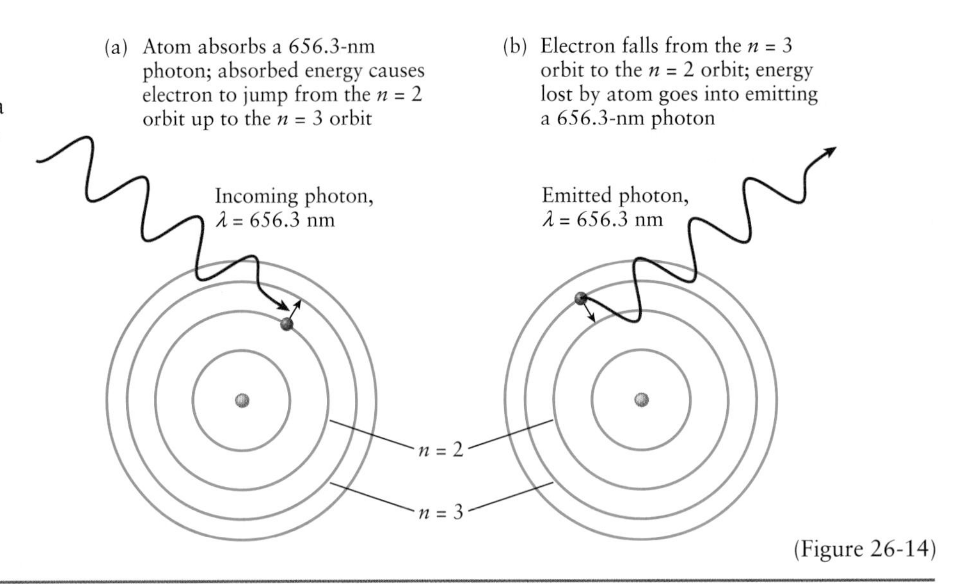

(Figure 26-14)

The Schrödinger equation and the Pauli exclusion principle: In the more complete Schrödinger description of the atom, the electron is described by a wave. Four quantum numbers describe the state of the electron, and a probability cloud describes the probability of finding an electron at different positions within the atom. The Pauli exclusion principle states that there can be only one electron per state; this helps explain the properties of multi-electron atoms.

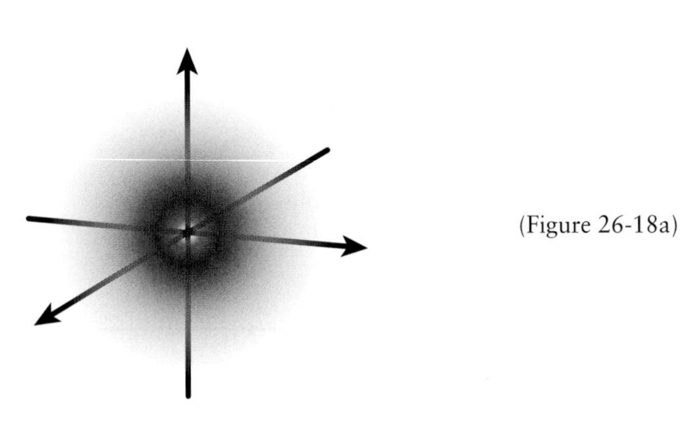

(Figure 26-18a)

Answer to What do you think? Question

(a) The spectrum of light from a star is very similar to the spectrum of an ideal blackbody (Section 26-2). Blue light has a higher frequency than red light, so the blue stars have a frequency of maximum emission (at which they emit most strongly) that is greater than that of the red stars. The frequency of maximum emission of a blackbody increases with increasing temperature, so the blue stars must be hotter than the red stars. To determine the chemical composition of the stars, the color of stars isn't enough information: We need to look at the absorption spectrum of the stars (Section 26-5). This can't be inferred from a photograph like the one that opens this chapter. In fact, it turns out that all of the stars in this image have almost the same chemical composition: predominantly hydrogen and helium, with trace amounts of other elements.

Answers to Got the Concept? Questions

26-1 (e) Figure 26-6 shows that the higher the temperature, the more light a blackbody of a given size emits at *all* frequencies. Note that both the 3000-K blackbody and the 12,000-K blackbody emit light at visible frequencies, not just the invisible frequencies at which they emit most strongly. That's why you can see radiation from these objects. The 3000-K blackbody emits more red light than any other visible color, so it will appear red; the 12,000-K blackbody emits more blue light than any other visible color, so it will appear blue.

26-2 (d) If $\theta = 90°$, then $\cos\theta = \cos 90° = 0$ and the wavelength shift in Compton scattering is $\Delta\lambda = (h/m_e c)(1 - \cos\theta) = (h/m_e c)(1 - 0) = h/m_e c = \lambda_C$. The initial wavelength is λ_C, so the final wavelength is $\lambda_f = \lambda_i + \Delta\lambda = \lambda_C + \lambda_C = 2\lambda_C$. That is, the final photon has twice the wavelength and so half the energy of the initial photon. (The lost energy is transferred to the electron from which the photon scattered.)

26-3 (c), (d), (a), (b) Equation 26-10 tells us that the de Broglie wavelength λ is inversely proportional to the momentum

p: $\lambda = h/p$. So a ranking from longest to shortest de Broglie wavelength is the same as a ranking from smallest to largest momentum. A proton has mass 1.67×10^{-27} kg and an electron has mass 9.11×10^{-31} kg, so the momentum in each of the four cases is (a) $(1.67 \times 10^{-27}\text{ kg})(2.00 \times 10^3\text{ m/s}) = 3.34 \times 10^{-24}\text{ kg}\cdot\text{m/s}$; (b) $(1.67 \times 10^{-27}\text{ kg})(4.00 \times 10^3\text{ m/s}) = 6.68 \times 10^{-24}\text{ kg}\cdot\text{m/s}$; (c) $(9.11 \times 10^{-31}\text{ kg})(2.00 \times 10^3\text{ m/s}) = 1.82 \times 10^{-27}\text{ kg}\cdot\text{m/s}$; (d) $(9.11 \times 10^{-31}\text{ kg})(4.00 \times 10^3\text{ m/s}) = 3.64 \times 10^{-27}\text{ kg}\cdot\text{m/s}$. So the ranking is (c) slow-moving electron, (d) fast-moving electron, (a) slow-moving proton, (b) fast-moving proton.

26-4 (c), (a), (b), (d) The photon energy is equal to the difference between the energies of the upper energy level of the atom and the lower energy level of the atom. To determine the ranking, just look at Figure 26-15 to see the spacing between the energy levels of the atom. The spacing is (c) greatest between the $n = 3$ and $n = 1$ levels, (a) second greatest between the $n = 4$ and $n = 2$ levels, (b) third greatest between the $n = 5$ and $n = 3$ levels, and (d) least between the $n = 4$ and $n = 3$ levels.

26-5 (b) The ionization energy is equal to $E_{\text{ionization}} = E_\infty - E_1$. From Equation 26-27

$$E_{\text{ionization}} = E_\infty - E_1 = \left(-\frac{Z^2}{\infty^2}E_0\right) - \left(-\frac{Z^2}{1^2}E_0\right)$$
$$= 0 - (-Z^2E_0) = Z^2E_0$$

(The reciprocal of infinity is zero.) A doubly ionized lithium ion is like a hydrogen atom—both have a single electron—but with $Z = 3$ instead of $Z = 1$ for hydrogen. Since the ionization energy is proportional to Z^2, the value for doubly ionized lithium is $3^2 = 9$ times greater than for hydrogen. You can see that for hydrogen $E_{\text{ionization}} = E_0 = 13.6$ eV, while for doubly ionized lithium $E_{\text{ionization}} = 9E_0 = 9(13.6\text{ eV}) = 122$ eV.

Questions and Problems

In a few problems you are given more data than you actually need; in a few other problems, you are required to supply data from your general knowledge, outside sources, or informed estimate. Interpret as significant all digits in numerical values that have trailing zeros and no decimal points.

For all problems use $g = 9.80\text{ m/s}^2$ for the free-fall acceleration due to gravity. Neglect friction and air resistance unless instructed to do otherwise.

•	Basic, single-concept problem
••	Intermediate-level problem; may require synthesis of concepts and multiple steps
•••	Challenging problem
SSM	*Solution is in Student Solutions Manual*
Example	*See worked example for a similar problem.*

Conceptual Questions

1. • Prior to Einstein's description of the photoelectric effect, light was thought to act like a wave. Explain why the existence of a frequency below which photoelectrons are not emitted favors a description of light as a particle instead.

2. • Describe how the number of photoelectrons emitted from a metal plate in the photoelectric effect would change if (a) the intensity of the incident radiation were increased, (b) the wavelength of the incident radiation were increased, and (c) the work function of the metal were increased.

3. • Is it possible to observe photoelectrons emitted from a metal plate with relativistic speeds?

4. • Consider the photoelectric emission of electrons induced by incident light of a single wavelength. The incoming photons all have the same energy, but the emitted electrons have a range of kinetic energies. Why?

5. • A markedly nonclassical feature of the photoelectric effect is that the energy of the emitted electrons doesn't increase as you increase the intensity of the light striking the metal surface. What change does occur as the intensity is increased?

6. • How does the intensity of light from a blackbody change when its temperature is increased? What changes occur in the body's radiation spectrum?

7. • The Compton effect is practically unobservable for visible light. Why?

8. • Which of the two Compton scattering experiments more clearly demonstrates the particle nature of electromagnetic radiation: a collision of the photon with an electron or a collision with a proton? Explain your answer.

9. • An electron and a proton have the same kinetic energy. Which has the longer wavelength? SSM

10. • Is the wavelength of an electron the same as the wavelength of a photon if both particles have the same total energy?

11. • Does the de Broglie wavelength of a particle increase or decrease as its kinetic energy increases?

12. • Why do we never observe the wave nature of particles for everyday objects such as birds or bumblebees, for example? SSM

13. • According to classical electromagnetic theory an accelerated charge emits electromagnetic radiation. What would this mean for the electron in the Bohr atom? What would happen to its orbit?

14. • What is the shortest wavelength of electromagnetic radiation that can be emitted by a hydrogen atom? SSM

15. • Why do you think Bohr's model was originally developed for the element hydrogen?

16. • Why do you think that Bohr's model of the hydrogen atom is still taught in undergraduate physics classes?

17. • Are there quantities in classical physics that are quantized?

Multiple-Choice Questions

18. • Atoms of an element emit a spectrum that
   A. is the same as all other elements.
   B. is evenly spaced.
   C. is unique to that element.
   D. is evenly spaced and unique to that element.
   E. is indistinguishable from most other elements.

19. • An ideal blackbody is an object that
   A. absorbs most of the energy that strikes it and emits a little of the energy it absorbs.

B. absorbs a little of the energy that strikes it and emits most of the energy it absorbs.
C. absorbs half of the energy that strikes it and emits half of the energy it absorbs.
D. absorbs all the energy that strikes it and emits all the energy it absorbs.
E. neither absorbs nor emits energy except at ultraviolet ("black light") wavelengths.

20. • The color of light emitted by a hot object depends on
A. the size of the object.
B. the shape of the object.
C. the material from which the object is made.
D. the temperature of the object.
E. the color of the object when it is cold.

21. • Which photon has more energy?
A. a photon of ultraviolet radiation.
B. a photon of green light.
C. a photon of yellow light.
D. a photon of red light.
E. a photon of infrared radiation. SSM

22. • Light that has a wavelength of 600 nm strikes a metal surface, and a stream of electrons is ejected from the surface. If light of wavelength 500 nm strikes the surface, the maximum kinetic energy of the electrons emitted from the surface will
A. be greater.
B. be smaller.
C. be the same.
D. be 5/6 smaller.
E. be unmeasurable.

23. • In the Compton effect experiment, the change in a photon's wavelength depends on
A. the scattering angle.
B. the initial wavelength.
C. the final wavelength.
D. the density of the scattering material.
E. the atomic number of the scattering material.

24. • The maximum change in a photon's energy when it undergoes Compton scattering occurs when its scattering angle is
A. 0°.
B. 45°.
C. 90°.
D. 135°.
E. 180°.

25. • As the scattering angle in the Compton effect increases, the energy of the scattered photon
A. increases.
B. stays the same.
C. decreases.
D. decreases by sin θ.
E. increases by sin θ. SSM

26. • The de Broglie wavelength depends only on
A. the particle's mass.
B. the particle's speed.
C. the particle's energy.
D. the particle's momentum.
E. the particle's charge.

Estimation/Numerical Analysis

27. • Estimate the order of magnitude of atomic ionization energies. Compare this with the order of magnitude for the nuclear binding energy, which is in the megaelectron volt (MeV) range.

28. • What is the approximate size of an atom?

29. • Estimate the amount of blackbody energy radiated from your body per second. SSM

30. • On average an electron will exist in any given state in the hydrogen atom for about 10^{-8} s before jumping to a lower level. Estimate the number of orbits around the nucleus that an electron makes in 10^{-8} s.

31. • Estimate the de Broglie wavelength of (a) a baseball that has been thrown as a fastball, and (b) a bullet fired from a rifle.

32. • Estimate the ionization energy of the outermost electron of lithium using (a) the Bohr model. This estimate is poor because the $n = 1$ electrons effectively "screen" the attractive force of the three protons in the nucleus. (b) Model this screening effect by subtracting the $n = 1$ electrons from the atomic number Z in your estimate of the ionization energy. The actual ionization energy is about 6 eV. How does it compare to your two estimates?

33. • A muon is a particle that is identical to an electron except its mass is about 200 times larger. Estimate the radius of a muonic hydrogen atom (a muon bound to a proton, the same thing as a hydrogen atom with the electron replaced by a muon).

34. •• In the early twentieth century, Max Planck used an empirical mathematical equation to describe the data that was collected in measuring blackbody radiation. Here's his equation for the intensity of light emitted per wavelength range:

$$\frac{\Delta I}{\Delta \lambda} = \frac{2\pi hc^2}{\lambda^5}\left(\frac{1}{e^{hc/(\lambda k_B T)} - 1}\right)$$

where λ is the wavelength of light emitted, h is Planck's constant (6.63×10^{-34} J·s), c is the speed of light (3.00×10^8 m/s), k is the Boltzmann constant (1.38×10^{-23} J/K), and T is the Kelvin temperature of the radiating cavity. Create a spreadsheet that generates at least at least 15 values of $\Delta I/\Delta\lambda$ (choose a range of at least 15 values of wavelength and calculate the value of the above function for $\Delta I/\Delta\lambda$ for some constant temperature). Once you have the set of data, plot it in a graph and apply an appropriate curve fit to depict the unique shape of the curve. Now change the temperature several times to see how the curve changes.

35. •• Using Einstein's explanation of the photoelectric effect ($K = hf - \Phi_0$), derive a numerical value for Planck's constant (h) by plotting a graph of the data in this table.

Wavelength of light (nm)	Maximum kinetic energy
400	1.60×10^{-19} J
450	1.06×10^{-19} J
500	6.09×10^{-20} J
550	2.00×10^{-20} J

Problems

26-1 Experiments that probe the nature of light and matter reveal the limits of classical physics

26-2 The photoelectric effect and blackbody radiation show that light is absorbed and emitted in the form of photons

36. • Calculate the range of photon frequencies and energies in the visible spectrum of light (approximately 380–750 nm). SSM Example 26-1

37. • What is the energy of a low-frequency 2000-Hz radio photon? Example 26-1

38. • What is the energy of a 0.200-nm x-ray photon? Example 26-1

39. • What are the wavelength and frequency of a 3.97×10^{-19}-J photon? Example 26-1

40. • **Biology** Under most conditions, the human eye will respond to a flash of light if 100 photons hit photoreceptors at the back of the eye. Determine the total energy of such a flash if the wavelength is 550 nm (green light). Example 26-1

41. •• The threshold wavelength for the photoelectric effect for silver is 262 nm. (a) Determine the work function for silver. (b) What is the maximum kinetic energy of an electron emitted from silver if the incident light has a wavelength of 222 nm? Example 26-1

42. •• Light that has a 195-nm wavelength strikes a metal surface, and photoelectrons are produced moving as fast as $0.004c$. (a) What is the work function of the metal? (b) What is the threshold wavelength for the metal above which no photoelectrons will be emitted? SSM Example 26-1

43. • What is the minimum frequency of light required to eject electrons from a metal with a work function of 6.53×10^{-19} J? Example 26-1

44. •• The work functions of aluminum, calcium, potassium, and cesium are 6.54×10^{-19} J, 4.65×10^{-19} J, 3.57×10^{-19} J, and 3.36×10^{-19} J, respectively. For which of the metals will photoelectrons be emitted when irradiated with visible light (wavelengths from 380 to 750 nm)? Example 26-1

45. • Wien's displacement law states that the wavelength of maximum emission for a blackbody is given by the following formula:

$$\lambda_{max} = \frac{0.290\ \text{K} \cdot \text{cm}}{T}$$

What is the wavelength (in nanometers) of maximum emission for a blackbody at a temperature of 400°C?

46. • MIG (metal inert gas) welders can be approximated as blackbodies. What is the wavelength of maximum emission from such a welder if it operates at a temperature of 4000 K? Compare your answer to the wavelength of maximum emission from a TIG (tungsten inert gas) welder that operates at 6000 K. The wavelength of maximum emission is given by Wien's displacement law, shown in problem 45.

47. • The Sun has a wavelength of maximum emission of 475 nm. Using Wien's displacement law (see problem 45), determine the corresponding temperature of the outer layer of the Sun, assuming it is a blackbody.

48. • (a) If the human body acts like a blackbody, use Wien's displacement law (see problem 45) to calculate the body's wavelength of maximum emission. Assume an average skin temperature of 34°C. (b) In what part of the electromagnetic spectrum is the light?

49. •• Suppose a blackbody at 400 K radiates just enough heat in 15 min to boil water for a cup of tea. How long will it take to boil the same water if the temperature of the radiator is 500 K?

26-3 As a result of its photon character, light changes wavelength when it is scattered

50. • The quantity $\lambda_C = h/(mc)$ is called the Compton wavelength. Calculate the numerical value of this quantity for (a) an electron, (b) a proton, and (c) a pi meson (which has a mass of 2.50×10^{-28} kg).

51. • What is the momentum of a photon if its wavelength is (a) 550 nm and (b) 0.0711 nm? SSM Example 26-2

52. •• X rays that have wavelengths of 0.125 nm are scattered off stationary electrons at an angle of 30.0°. (a) Calculate the wavelength of the scattered electromagnetic radiation. (b) Calculate the *fractional wavelength change* ($\Delta\lambda/\lambda_i = (\lambda_f - \lambda_i)/\lambda_i$) for the scattered x rays. Example 26-2

53. •• If a photon undergoes Compton scattering from a stationary electron and experiences a fractional wavelength change (see problem 52) of +7.25%, calculate the angle at which the scattered photons are directed if the original photons have a wavelength of 0.00335 nm. Example 26-2

54. • Photons that have a wavelength of 0.00225 nm are Compton scattered off stationary electrons at 45.0°. What is the energy of the scattered photons? Example 26-2

55. • X-ray photons that have a wavelength of 0.140 nm are scattered off carbon atoms (which possess essentially stationary electrons in their valence shells). What are the wavelengths of the Compton-scattered photons and the kinetic energies of the scattered electrons if the photons are scattered at angles of (a) 0.00°, (b) 30.0°, (c) 45.0°, (d) 60.0°, (e) 90.0°, and (f) 180°? SSM Example 26-2

56. • A 0.0750-nm photon Compton scatters off a stationary electron. Determine the maximum speed of the scattered electron. Example 26-2

57. •• A photon Compton scatters off a stationary electron at an angle of 60.0°. The electron moves away with 1.28×10^{-17} J of kinetic energy. Determine the initial wavelength of the photon. Example 26-2

58. •• An x ray source is incident on a collection of stationary electrons. The electrons are scattered with a speed of 4.50×10^5 m/s, and the photon scatters at an angle of 60.0° from the incident direction of the photons. Determine the wavelength of the x ray source. Example 26-2

59. • Arthur Compton scattered photons that had a wavelength of 0.0711 nm off a block of carbon during his famous experiment of 1923 at Washington University in St. Louis, Missouri. (a) Calculate the frequency and energy of the photons. (b) What is the wavelength of the photons that are scattered at 90.0°? (c) What is the energy of the photons that are scattered at 90.0°? (d) What is the energy of the electrons that recoil from the Compton scattering with $\theta = 90.0°$? Example 26-2

26-4 Matter, like light, has aspects of both waves and particles

60. • Calculate the de Broglie wavelength of a 0.150-kg ball moving at 40.0 m/s. Comment on the significance of the result. Example 26-3

61. • Calculate the de Broglie wavelength of an electron that has a speed of $0.00730c$. SSM Example 26-3

62. • What is the de Broglie wavelength of a proton ($m = 1.67 \times 10^{-27}$ kg) moving at 4.00×10^5 m/s? Example 26-3

63. •• What is the de Broglie wavelength of an electron that has a kinetic energy of (a) 1.60×10^{-19} J, (b) 1.60×10^{-18} J, (c) 1.60×10^{-17} J, and (d) 1.60×10^{-16}? SSM Example 26-3

64. •• Calculate the de Broglie wavelength of an alpha particle ($m_\alpha = 6.64 \times 10^{-27}$ kg) that has a kinetic energy of (a) 1.60×10^{-13} J and (b) 8.00×10^{-13} J. Example 26-3

65. •• Calculate the de Broglie wavelength of a thermal neutron that has a kinetic energy of about 6.41×10^{-21} J. Example 26-3

66. •• Write an expression that relates the Newtonian kinetic energy ($K = (1/2)mv^2$) and mass of a nonrelativistic particle to its de Broglie wavelength. (That is, complete the expression $\lambda(K, m) = ?$) Example 26-3

26-5 The spectra of light emitted and absorbed by atoms show that atomic energies are quantized

67. • The Balmer formula can be written as follows:

$$\lambda = (364.56 \text{ nm})\left(\frac{m^2}{m^2 - 4}\right)$$

where m is equal to any integer larger than 2. This represents the wavelengths of visible colors that are emitted from the hydrogen atom. Calculate the first four colors (wavelengths) that are observed in the spectrum of hydrogen due to the Balmer series. Example 26-4

68. •• **Astronomy** Hydrogen atoms in interstellar gas clouds emit electromagnetic radiation at a wavelength of 21 cm when an electron in the ground state of hydrogen switches spin states. Determine the energy difference between the two spin states in this transition. SSM Example 26-4

69. •• Prove that the Balmer formula is a special case of the Rydberg formula with n set equal to 2. Example 26-4

$$\text{Rydberg formula: } \frac{1}{\lambda} = R_H\left(\frac{1}{n^2} - \frac{1}{m^2}\right)$$

$$R_H = 1.09737 \times 10^7 \text{ m}^{-1}$$

$$\text{Balmer formula: } \lambda = b\left(\frac{m^2}{m^2 - 4}\right) \qquad b = 364.56 \text{ nm}$$

70. • A hypothetical atom has four unequally spaced energy levels in which a single electron can be found. Suppose a collection of the atoms is excited to the highest of the four levels. (a) What is the maximum number of unique spectral lines that could be measured as the atoms relax and return to the lowest, ground state? (b) Suppose the previous hypothetical atom has 10 energy levels. Now what is the maximum number of unique spectral lines that could be measured in the emission spectrum of the atom? Example 26-4

71. • The Lyman series results from transitions of the electron in hydrogen in which the electron ends at the $n = 1$ energy level. Using the Rydberg formula for the Lyman series, calculate the wavelengths of the photons emitted in the transitions that end in the $n = 1$ level and start in the energy levels that correspond to n equal to 2 through 6, and indicate the initial and final levels of the transition corresponding to each wavelength. State whether each wavelength is visible (380 to 750 nm), ultraviolet (shorter than 380 nm), or infrared (longer than 750 nm). Example 26-4

72. • The Balmer series results from transitions of the electron in hydrogen in which the electron ends at the $n = 2$ energy level. Using the Rydberg formula for the Balmer series, calculate the wavelengths of the photons emitted in the transitions that end in the $n = 2$ level and start in the energy levels that correspond to n equal to 3 through 6 and indicate the initial and final levels of the transition corresponding to each wavelength. State whether each wavelength is visible (380 to 750 nm), ultraviolet (shorter than 380 nm), or infrared (longer than 750 nm). SSM Example 26-4

73. • The Paschen series results from transitions of the electron in hydrogen in which the electron ends at the $n = 3$ energy level. Using the Rydberg formula for the Paschen series, calculate the wavelengths of the photons emitted in the transitions that end in the $n = 3$ level and start in the energy levels that correspond to n equal to 4 through 6 and indicate the initial and final levels of the transition corresponding to each wavelength. State whether each wavelength is visible (380 to 750 nm), ultraviolet (shorter than 380 nm), or infrared (longer than 750 nm). Example 26-4

74. •• Calculate the shortest wavelength (and the highest energy) associated with emitted photons in the (a) Lyman, (b) Balmer, and (c) Paschen series (see problems 71, 72, and 73). Example 26-4

75. •• Express the Balmer formula (see problem 67) in terms of the *frequency* of the photons that are emitted (rather than the wavelength). Extend all numerical values out to five significant figures.

76. •• Express the Rydberg formula in terms of the *frequency* of the photons that are emitted (rather than the wavelength). Extend all numerical values out to five significant figures. SSM

26-6 Models by Bohr and Schrödinger give insight into the intriguing structure of the atom

77. • For an electron in the nth state of the hydrogen atom, write expressions for (a) the angular momentum of the electron, (b) the radius of the electron's orbit, (c) the kinetic energy of the electron, (d) the total energy of the electron, and (e) the speed of the electron. Example 26-5

78. • Set up a chart for the five quantities listed in problem 77 and calculate the values for $n = 1, 2, 3, 4,$ and 5 (4 significant figures, SI units). See the following table. Example 26-5

n	L_n	r_n	K_n	E_n	v_n
1					
2					
3					
4					
5					

79. •• Devise a straightforward method that allows you to calculate (a) the speed and (b) the angular momentum of an electron in the nth Bohr orbit. Your expressions should contain the number of the orbit, n, and constants.

80. •• Using the formula that you devised in problem 79, calculate the speed of an electron and the angular momentum of an electron in the tenth Bohr orbit. SSM

81. •• **Astronomy** For carbon ($Z = 6$) the frequencies of spectral lines resulting from a single electron transition are increased over those for hydrogen by the ratio of the Rydberg constants:

$$\frac{R_C}{R_H} = \frac{1 - m_e/m_C}{1 - m_e/m_H}$$

The mass of a hydrogen atom (m_H) is 1837 times greater than the mass of the electron (m_e), and the mass of the carbon atom (m_C) is 12 times greater than the mass of the hydrogen atom. Find the difference in frequency between the carbon 272α transition and the hydrogen 272α transition. This spectral line refers to the transition from $n_i = 273$ to $n_f = 272$. Express your answer in megahertz (MHz).

82. • A hydrogen atom that has an electron in the $n = 2$ state absorbs a photon. (a) What wavelength must the photon possess to send the electron to the $n = 4$ state? (b) What possible wavelengths would be detected in the spectral lines that result from the deexcitation of the atom as it returns from $n = 4$ to the ground state? SSM Example 26-5

83. • How much energy is needed to ionize a hydrogen atom that starts in the Bohr orbit represented by $n = 3$? If an atom is ionized, its outer electron is no longer bound to the atom. Example 26-5

84. • Find the energies of the first 10 energy levels (sketch and label an energy-level diagram) for singly-ionized helium, He^+. Example 26-5

85. • In the Bohr model of hydrogen, there is just one possible state of the electron in the lowest energy level ($n = 1$) and just one possible state of the electron in the next energy level ($n = 2$). In the more accurate Schrödinger picture of the hydrogen atom, in which the electron is described by the four quantum numbers n, ℓ, m_ℓ, and m_s, there are two possible states in the $n = 1$ level and eight possible states in the $n = 2$ level. (a) Give the values of n, ℓ, m_ℓ, and m_s for each of the two possible states in the $n = 1$ energy level. (b) Give the values of n, ℓ, m_ℓ, and m_s for each of the eight possible states in the $n = 2$ energy level.

86. • Which of the following electron configurations for an excited atom of beryllium ($Z = 4$) are possible? Which are impossible? Explain your reasoning.

(i) $1s^2 2s^1 2p^1$
(ii) $1s^1 2s^3$
(iii) $1s^1 2p^3$
(iv) $1s^3 2s^1$

General Problems

87. • **Biology** Vitamin D is produced in the skin when 7-dehydrocholesterol reacts with UVB rays (ultraviolet B) having wavelengths between 270 and 300 nm. What is the energy range of the UVB photons? Example 26-1

88. •• A helium-neon laser produces light of wavelength 632.8 nm. The laser beam carries a power of 0.50 mW and strikes a target perpendicular to the beam. (a) How many photons per second strike the target? (b) At what rate does the laser beam deliver linear momentum to the target if the photons are all absorbed by the target? Example 26-1

89. • **Astronomy** In 2009 astronomers detected gamma ray photons having energy ranging from 700 GeV to around 5 TeV coming from supernovae (exploding giant stars) in the galaxy M82. (1 GeV = 10^9 eV, 1 TeV = 10^{12} eV.) (a) What is the range of wavelengths of the gamma ray photons detected from M82? (b) Calculate the ratio of the energy of a 5 TeV photon to the energy of a visible light photon having a wavelength of 500 nm. Example 26-1

90. ••• Derive an expression for the change in photon *energy* in Compton scattering (that is, an expression for the difference between the final and initial photon energies E_f and E_i) as a function of the photon scattering angle.

91. •• A photon of frequency 4.81×10^{19} Hz scatters off a free stationary electron. Careful measurements reveal that the photon goes off at an angle of 125° with respect to its original direction. (a) How much energy does the electron gain during the collision? (b) What percent of its original energy does the photon lose during the collision? Example 26-2

92. •• **Chemistry** A laboratory oven that contains hydrogen molecules H_2 and oxygen molecules O_2 is maintained at a constant temperature T. Each oxygen molecule is 16 times as massive as a hydrogen molecule. Find the ratio of the de Broglie wavelength of the hydrogen molecule to that of the oxygen molecule, assuming that each molecule has kinetic energy $(3/2)kT$. Example 26-3

93. •• A hydrogen atom makes a transition from the $n = 5$ state to the ground state and emits a single photon of light in the process. The photon then strikes a piece of silicon, which has a photoelectric work function of 4.8 eV. Is it possible that a photoelectron will be emitted from the silicon? If not, why not? If so, find the maximum possible kinetic energy of the photoelectron. Example 26-4

94. •• Suppose the electron in the hydrogen atom were bound to the proton by gravitational forces (rather than electrostatic forces). Find (a) the radius and (b) the energy of the first orbit.

95. •• **Biology** The *E. coli* bacterium is about 2.0 μm long. Suppose you want to study it using photons of that wavelength or electrons having that de Broglie wavelength. (a) What is the energy of the photon and the energy of the electron? (b) Which one would be better to use, the photon or the electron? Explain why. Example 26-3

96. ••• Use the de Broglie wave concept to fit circular waves into the orbits of the Bohr model of the hydrogen atom to prove Bohr's hypothesis of the quantization of angular momentum. Assume that in the first Bohr orbit, exactly one de Broglie wavelength matches up with the circumference, in the second orbit, two waves match up, in the third orbit, three waves match up, and so on (**Figure 26-20**).

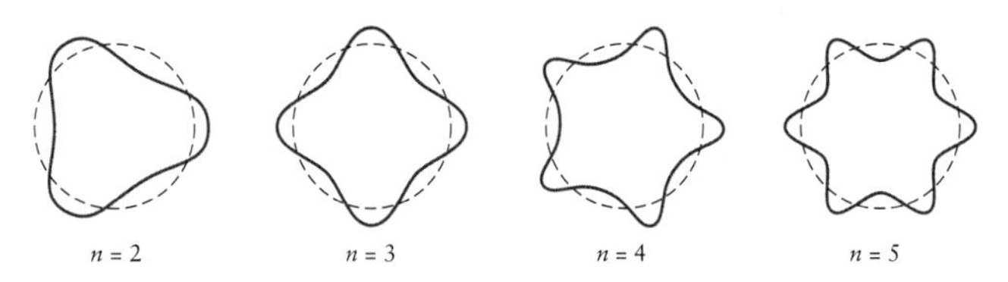

Figure 26-20 Problem 96

97. •• Restate de Broglie's formula for particle waves in the case that the speeds are relativistic. (You will need to use the relativistic relationship between speed and momentum.)

98. •• A relativistic electron has a de Broglie wavelength of 346 fm (1 fm = 10^{-15} m). Determine its speed. (You will

need to use the relativistic relationship between speed and momentum. See problem 97.)

99. • **Astronomy** Radio astronomers use radio frequency waves to identify the elements in distant stars. One of the standard lines that is often studied is designated the 272α line. This spectral line refers to the transition in hydrogen from $n_i = 273$ to $n_f = 272$. Calculate the wavelength and frequency of the electromagnetic radiation that is emitted for the 272α transition.

100. • The Lyman series ($n_f = 1$), the Balmer series ($n_f = 2$), and the Paschen series ($n_f = 3$) are commonly studied in basic chemistry and physics classes (see problems 71, 72, and 73). The Brackett series ($n_f = 4$) and the Pfund series ($n_f = 5$) are not so well known. (a) Calculate the shortest and longest wavelengths for the spectral lines that are part of the Brackett series. State whether each wavelength is visible (380 to 750 nm), ultraviolet (shorter than 380 nm), or infrared (longer than 750 nm). (b) Calculate the shortest and longest wavelengths for the spectral lines that are part of the Pfund series. State whether each wavelength is visible (380 to 750 nm), ultraviolet (shorter than 380 nm), or infrared (longer than 750 nm). Example 26-4

101. •• When x-ray photons are aimed at a carbon target, a photon can undergo Compton scattering from one of the electrons in a carbon atom. But a photon can undergo Compton scattering from any charged particle, and so can be scattered by a carbon nucleus. (a) Calculate the Compton wavelength $\lambda_c = h/mc$ for a carbon nucleus, which has a mass of 1.99×10^{-26} kg. Express your answer in nanometers (nm). (b) Find the wavelength shift $\Delta\lambda$ for a photon that undergoes a 180° scattering from a carbon nucleus. Express your answer in nm. Compare to the wavelength shift of 0.00486 nm for a photon that undergoes a 180° scattering from an electron. (c) The x-ray photons that Arthur Compton used in his 1923 experiment had a wavelength of 0.0711 nm. What fractional change in wavelength $\frac{\Delta\lambda}{\lambda}$ would such a photon undergo due to Compton scattering from an electron? From a carbon nucleus? Explain why Compton was only able to measure the Compton scattering of photons from electrons, not from nuclei. (d) Suppose you want a photon to undergo a 1.00% increase in wavelength (which is more easily measured) as a result of a 180° Compton scattering from a carbon nucleus. What would be the wavelength and energy of such a photon before scattering? (This is a gamma-ray photon, and such photons were not available to Compton when he did his experiments.)

102. •• **Biology** You want to use a microscope to study the structure of a mitochondrion about 1 μm in size. To be able to observe small details within the mitochondrion, you want to use a wavelength of 0.0500 nm. (a) If your microscope uses light of this wavelength, what is the momentum p of a photon? What is the energy E of a photon? (b) If instead your microscope uses electrons of this de Broglie wavelength, what is the momentum p of an electron? What is the speed v of an electron? (Recall $p = mv$.) What is the kinetic energy $K = (1/2)mv^2$ of an electron? (c) Comparing your results for the photon momentum and energy in (a) to your results for the electron momentum and kinetic energy in (b), what advantages can you see to using electrons rather than photons?

103. •• **Astronomy** Stars form within giant clouds of gas composed mostly of hydrogen. When the star begins to shine, it remains surrounded by a thin cloud or *nebula* of the gas. If the star has a sufficiently high surface temperature, the emitted light will cause hydrogen atoms in the nebula to become ionized. When an electron is recaptured by a hydrogen ion (a proton), it jumps down through the energy levels back to $n = 1$ and emits photons in the process. Hence the nebula emits light with an emission spectrum like that shown in Figure 26-12. (a) Suppose the star has a surface temperature of 25,000 K. Stars emit light like a blackbody, so use Wien's law (see problem 45) to find the wavelength of maximum emission λ_{max} (in nm) of this star and the energy (in eV) of a photon with this wavelength. (b) Because a blackbody emits light over a continuous spectrum of wavelengths, a star emits copious amounts of photons with energies several times greater than that of a photon with wavelength λ_{max}. Based on this, explain why you would expect that many hydrogen atoms in the nebula surrounding a 25,000-K star would become ionized, and so such a nebula would emit light with an emission spectrum. (c) Relatively few stars have surface temperatures of 25,000 K or higher. Repeat part (a) for a more common type of star with a surface temperature of 3000 K. Would you expect many hydrogen atoms in the nebula surrounding a 3000-K star to be ionized? Would you expect many hydrogen atoms in this nebula to be excited from the $n = 1$ level to the $n = 2$ level? Would you expect to see much emission from this nebula? (d) If you see a star that has a surrounding nebula that glows with an emission spectrum, what can you conclude about that star's surface temperature? Explain your answer.

104. •• (a) When a helium atom jumps down from its first excited energy level (electron configuration $1s^1 2p^1$) to its lowest energy level (electron configuration $1s^2$), it emits an ultraviolet photon of wavelength 58.4 nm. What is the energy of such a photon in eV? (b) Figure 26-17 shows that in a Franck-Hertz experiment using mercury vapor, the anode current drops off for cathode-grid voltages of 4.9 V, 2×4.9 V = 9.8 V, and 3×4.9 V = 14.7 V. If you repeat the Franck-Hertz experiment using helium gas rather than mercury vapor, what are the first three cathode-grid voltages at which you would expect to see a drop in the anode current?

Nuclear Physics

HIP/Art Resource, NY

What do you think?

This remarkable painting on a wall in the Chauvet-Pont-d'Arc Cave in southern France is known to be 30,000 to 33,000 years old. This age is determined from the radioactive decay of carbon-14 (a type of carbon with 6 protons and 8 neutrons in its nucleus), which has a half-life of 5730 years. Of the carbon-14 that was present when this painting was made, the fraction that remains today is somewhat less than (a) half; (b) 1/4; (c) 1/8; (d) 1/16; (e) 1/32.

In this chapter, your goals are to:

- (27-1) Explain why quantum ideas play an important role in nuclear physics.
- (27-2) Describe how we know that the force that holds the nucleus together is both strong and of short range.
- (27-3) Explain how and why the binding energy per nucleon in a nucleus depends on the size of the nucleus.
- (27-4) Calculate the energy released in nuclear fission.
- (27-5) Describe why very high temperatures are needed for nuclear fusion reactions.
- (27-6) Calculate what happens in the decay of a radioactive substance.

To master this chapter, you should review:

- (3-7) Uniform circular motion.
- (11-2) How to calculate the density of an object.
- (19-6, 19-7) Current loops and magnetic fields.
- (22-4) The energy of a photon.
- (25-7) The equivalence of mass and energy.
- (26-5) How an atom in an excited energy level emits a photon when the atom decays to a lower energy level.
- (26-6) The spin of an electron.

27-1 The quantum concepts that help explain atoms are essential for understanding the nucleus

We learned in Chapter 26 that a handful of radical ideas—the notion that energy is quantized, that electromagnetic waves come in packets called photons, and that the laws of quantum mechanics can specify only the probabilities that particles will behave in certain ways—are essential for understanding the nature and behavior of the atom. These same ideas apply on the even smaller scale of the atomic nucleus.

We'll see that unlike atoms, in which negatively charged electrons are bound to the positively charged nucleus by the attractive electric force, nuclei are bound by a *strong nuclear force* that keeps the protons and neutrons together. We'll also see that like electrons, nuclei can have an intrinsic angular momentum or *spin*; magnetic resonance imaging, or MRI, makes use of how nuclear spins respond to an external magnetic

(a)

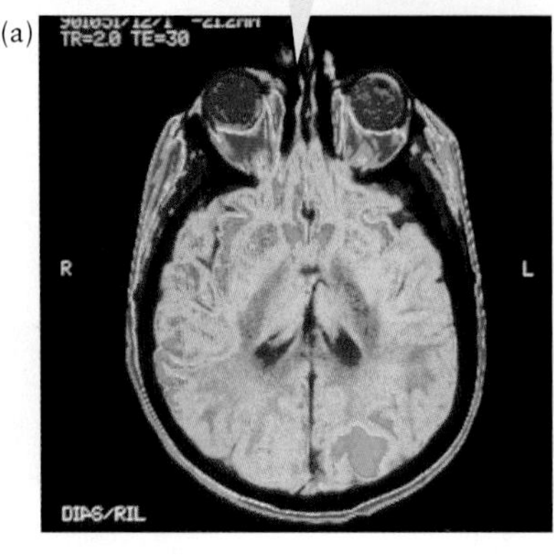

(b)

(c)

More than 50% of the energy that powers our planet's geological activity, including volcanic eruptions, comes from the radioactive decay of unstable nuclei in Earth's interior.
iStockphoto/Thinkstock

Figure 27-1 Applications of nuclear physics The properties of atomic nuclei explain (a) how magnetic resonance imaging (MRI) works, (b) how the Sun provides energy for life on Earth, and (c) how Earth sustains its internal energy.

field (**Figure 27-1a**). The interplay between the strong nuclear force and the electric force (which makes all of the protons in a nucleus try to repel each other) means that nuclei of different sizes are more tightly or loosely bound. We'll learn that as a result, the largest nuclei are prone to break into smaller fragments through a process called *fission*. We'll also learn that the smallest nuclei can release energy by *fusion*, in which two nuclei join together to form a larger one. Fusion reactions make the Sun shine, and the sunlight that they produce makes life on Earth possible (**Figure 27-1b**).

Understanding the binding energies of nuclei will also help us understand the three main types of radioactive decay: alpha, beta, and gamma. The energy released by alpha and beta decays in our planet's interior helps power geologic activity such as volcanic eruptions (**Figure 27-1c**). We'll find that the concept of *half-life* will help us understand the nature of radioactive decays of all kinds.

TAKE-HOME MESSAGE FOR Section 27-1

✔ Many of the same concepts that help us understand atomic physics are also important in nuclear physics.

✔ The balance between the strong nuclear force and the electric force determines the stability of nuclei.

27-2 The strong nuclear force holds nuclei together

As we learned in Chapter 26, an atom has a small, positively charged nucleus at its center. There are two ways you can see that the nucleus must be positively charged. First, because atoms are neutral, positive charge is required to balance the negative charge of the electrons. Second, the Coulomb attraction between the nucleus and the negatively charged electrons provides the force that holds the atom together.

The *repulsive* Coulomb forces between the protons, however, must be large because they are close together inside the nucleus. To see just how large these forces are, note that a typical radius of a nucleus is about 5 fm (1 fm = 1 femtometer = 10^{-15} m). The electric force between two protons (each of charge $+e = 1.60 \times 10^{-19}$ C) separated by a distance $r = 5.00$ fm is

$$F = \frac{k(+e)(+e)}{r^2} = \frac{(8.99 \times 10^9\ \mathrm{N\cdot m^2/C^2})(1.60 \times 10^{-19}\ \mathrm{C})^2}{(5.00 \times 10^{-15}\ \mathrm{m})^2}$$
$$= 9.21\ \mathrm{N}$$

This may not seem like a lot of force, but remember that this force is applied to a proton, which has a mass of only 1.67×10^{-27} kg. If the electric force were the only force acting on the protons in a nucleus, the nucleus would simply fly apart.

What's more, nuclei also have neutrons, the neutrally charged particle we mentioned briefly in Chapter 16. Neutrons and protons share many similar properties, including similar masses: The masses of the neutron and proton are 1.6749×10^{-27} kg and 1.6726×10^{-27} kg, respectively. Collectively, we refer to protons and neutrons as **nucleons**. Neutrons do not feel the electric force at all, so this force can't be responsible for keeping neutrons within the nucleus.

We conclude that there must be an additional *attractive* force that acts on all nucleons (both protons and neutrons) and that binds them together in the nucleus. This attractive force must be stronger than the repulsive electric force between protons, so we call it the **strong nuclear force**. Over short distances the strong nuclear force is hundreds of times stronger than the electrostatic force.

If the nuclear force is so strong compared to the electrostatic force, and if protons attract other protons by this force, why are neutrons necessary to help overcome the Coulomb repulsion between protons? The answer lies in the *range* of the strong nuclear force, which is the distance beyond which one nucleon no longer experiences a force due to another. Experiments show that the strong nuclear force between two nucleons has a range of only about 2.0 fm. The radius of a proton or neutron is about 0.85 fm, so two nucleons must almost be touching to experience the strong nuclear force. As a very rough analogy, you can think of protons and neutrons as tiny spheres coated with very strong Velcro, which makes the nucleons stick together if they are brought close enough to each other.

By contrast, the electrostatic repulsion between protons separated by a distance r is proportional to $1/r^2$, so is present even if the distance r is very large. We say that the strong nuclear force is a *short-range* force, whereas the electric force is a *long-range* force.

Because the range of the nuclear force is smaller than the diameter of most nuclei (a few to perhaps 15 fm), each nucleon exerts an attractive nuclear force only on its nearest neighbors. Each proton, however, exerts a repulsive force on *every other* proton in the nucleus. As a result, the nuclear force between neighboring protons cannot overcome the Coulomb repulsion between all of the protons. To prevent a nucleus from spontaneously breaking apart (that is, for it to be *stable*), the nucleus must also contain neutrons. You can think of the neutrons as "spacers" that increase the average distance between protons and so decrease their mutual Coulomb repulsion.

In the smallest stable nuclei, the number of neutrons is about the same as the number of protons; in nuclei with more than about 20 protons, there must be more neutrons than protons for the nucleus to be stable. An unstable nucleus will eventually undergo a spontaneous transformation, termed a *decay*, in which it either splits apart or gives off energy in some other way. **Figure 27-2** shows the number of neutrons versus the number of protons in known atomic nuclei. Stable nuclei are shown in black. Notice that as the number of protons increases, more additional neutrons are required for stability.

Nuclides, Isotopes, and Nuclear Sizes

Atoms of each element have a unique number of protons and the same number of electrons: Hydrogen has one, helium two, lithium three, and so on. Many properties of atoms are related to this number, usually designated as the **atomic number** Z. Many properties of nuclei arise from *both* the value of Z and the value of the **neutron number** N, which is the number of neutrons in the nucleus. Although the value of Z of an elemental species is fixed (changing the number of protons changes the element), the value of N is not fixed for each element. Each combination of N and Z specifies a **nuclide**. For each element, the most common configuration of N and Z corresponds to the most stable nuclide; nuclei of that element with a different number of neutrons are termed **isotopes**. For example, potassium has 19 protons (and 19 electrons). The most stable and most common nuclide of potassium has 20 neutrons, $N = 20$. We denote this nuclide with the symbol ^{39}K, commonly referred to as "potassium-39." The number of protons ($Z = 19$) is understood from the symbol K for potassium, and the total number of protons and neutrons, termed the **mass number** A, is given in the pre-superscript, 39. Note that the mass number of any nucleus equals the atomic number plus the neutron number: $A = Z + N$.

More than 93% of all potassium atoms have a ^{39}K nucleus. About 7% of potassium atoms have a ^{41}K nucleus, however, with 19 protons and 22 neutrons. We say that ^{39}K and ^{41}K are two isotopes of potassium.

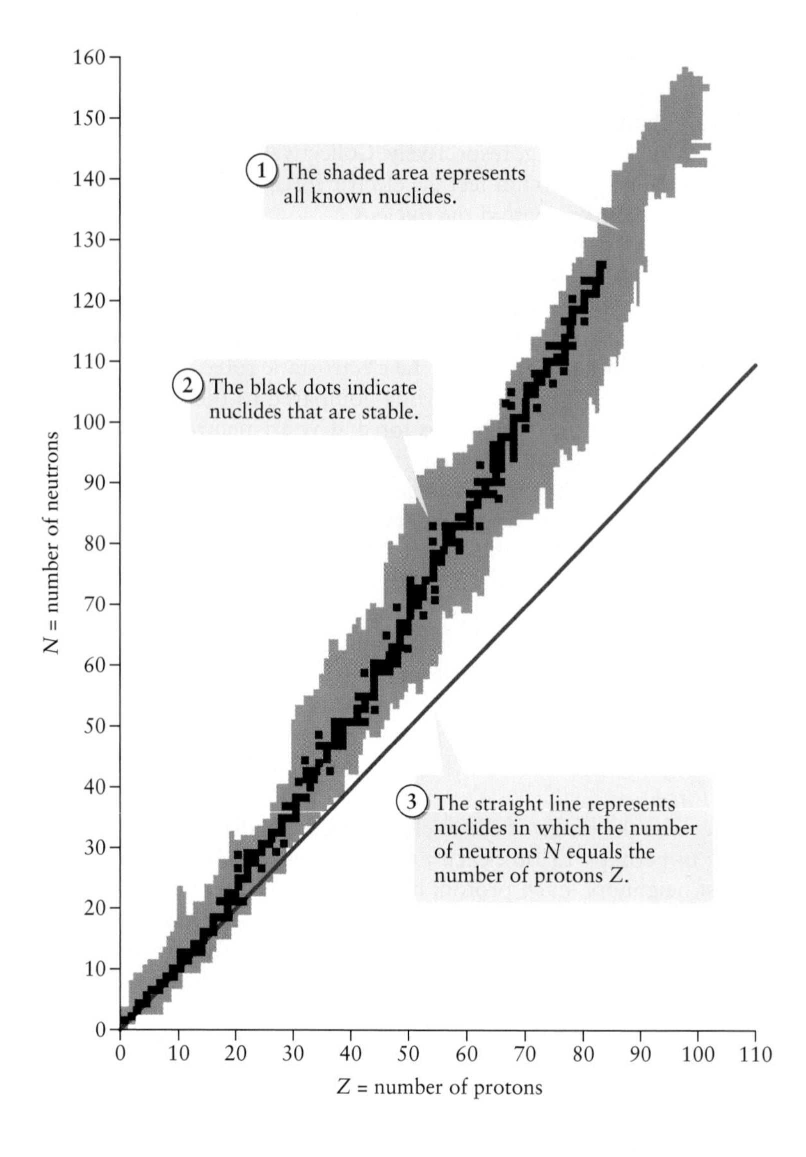

Figure 27-2 Neutrons versus protons in nuclei The interplay between the strong nuclear force and the electric force explains the relationship between the numbers of protons and neutrons in nuclei.

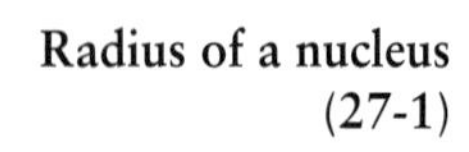

For most nuclides, more neutrons than protons must be present (N must be greater than Z) to prevent the protons from flying apart due to electrostatic repulsion.

The *size* of a nucleus (its radius and volume) is related to the mass number A. Experiments show that all nuclei are approximately spherical and have radii proportional to the cube root of A, or $A^{1/3}$:

Radius of a nucleus (27-1)

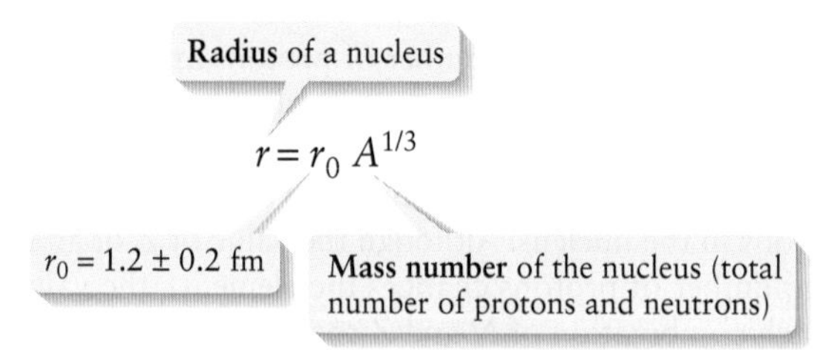

This equation states that if we increase the number of nucleons in a nucleus by a factor of 10, say from $A = 20$ to $A = 200$, the radius of the nucleus will increase by a factor of $10^{1/3} = 2.2$ (**Figure 27-3**).

The volume of a sphere is proportional to r^3. Hence the volume of a nucleus is directly proportional to $(A^{1/3})^3 = A$, which is the mass number of the nucleus and the number of nucleons within that nucleus. This is consistent with our model of nucleons as spheres covered with Velcro. In this model, increasing the number of nucleons by a factor of 10 would result in a ball of nucleons (that is, a nucleus) with 10 times the volume. Because the attractive force is short range, there is no tendency for the nucleons to be compressed together as the size of the nucleus increases.

It's useful to contrast nuclei with planets, which are held together by the *long* range gravitational force of attraction between all of the parts of the planet. As an example, compare the planets Jupiter and Saturn (which have the same chemical composition): Jupiter has 3.3 times more mass than Saturn, but Jupiter has only 1.7 times greater volume because it is more highly compressed and has a greater density. Nuclei of different sizes do *not* behave like planets of different sizes: All nuclei have basically the same density (see Example 27-2 below). This is further evidence that the strong nuclear force is short-range rather than long-range.

Can we simply add more protons and more neutrons to make larger and larger nuclei? The answer is "no," and for the same reason that neutrons are required for nuclear stability. Figure 27-2 shows that as we work our way up the periodic table to atoms that have more and more protons, more *additional* neutrons are required for stability. Each additional proton exerts a repulsive force on all the others, but neutrons and protons can only attract their nearest neighbors. Those additional neutrons cause the size of the nucleus to grow so that eventually too many neutrons are near the surface of the nucleus and therefore not completely surrounded by neighbors. At that point, the nuclear forces holding the nucleus together are not large enough to overcome the Coulomb repulsion between the protons, and the nucleus cannot be stable. The largest stable nuclide is lead-208 (^{208}Pb), which has 82 protons and 126 neutrons.

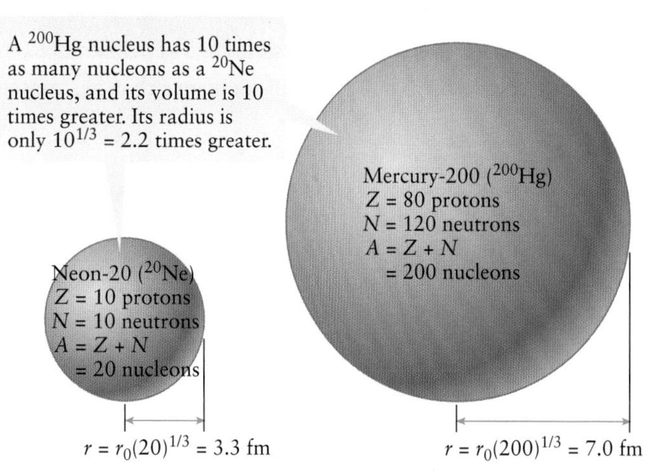

Figure 27-3 Nuclear sizes The volume of an atomic nucleus is proportional to the number of nucleons it contains (the mass number *A*). The nuclear radius is proportional to the cube root of *A*.

EXAMPLE 27-1 Nuclear Radii

Estimate the radius of the nucleus of ^{12}C, a relatively small nucleus; ^{118}Sn, a nucleus of medium size; and ^{236}U, a relatively large nucleus. The nuclides ^{12}C and ^{118}Sn are stable; ^{236}U, like all other isotopes of uranium, is unstable.

Set Up

In each case we use Equation 27-1 to calculate the radius of the nucleus. The value of *A* for each nucleus is given by the pre-superscript: $A = 12$ for carbon (C), $A = 118$ for tin (Sn), and $A = 236$ for uranium (U).

Radius of a nucleus:

$$r = r_0 A^{1/3} \quad (27\text{-}1)$$

Solve

Apply Equation 27-1 to each nuclide.

We'll use $r_0 = 1.2$ fm in our calculations.
For ^{12}C, which has 6 protons and 6 neutrons,

$$r(^{12}\text{C}) = (1.2 \text{ fm})(12)^{1/3} = 2.7 \text{ fm}$$

For ^{118}Sn, which has 50 protons and 68 neutrons,

$$r(^{118}\text{Sn}) = (1.2 \text{ fm})(118)^{1/3} = 5.9 \text{ fm}$$

For ^{236}U, which has 92 protons and 144 neutrons,

$$r(^{236}\text{U}) = (1.2 \text{ fm})(236)^{1/3} = 7.4 \text{ fm}$$

Reflect

This calculation shows that typical nuclei have radii of just a few femtometers. Although ^{236}U has $236/12 = 19.7$ times as many nucleons as ^{12}C, it is only larger in radius by a factor of $(7.4 \text{ fm})/(2.7 \text{ fm}) = 2.7$. That's a consequence of the $A^{1/3}$ factor in Equation 27-1: Note that $(19.7)^{1/3} = 2.7$.

Notice that in carbon, the smallest of the three nuclei, the number of neutrons equals the number of protons. In tin, which has almost 10 times the mass number of carbon, the number of neutrons required for nuclear stability is 36% higher than the number of protons. And even with nearly 60% more neutrons than protons, the relatively large ^{236}U nucleus is not stable. So we see direct evidence that in larger nuclei more and more additional neutrons, compared to the number of protons, are required for nuclear stability, and that at some size a nucleus is too large for the strong nuclear force to overcome the Coulomb repulsion between the protons.

EXAMPLE 27-2 Nuclear Density

Estimate the density (in kg/m^3) of a nucleus that has mass number A.

Set Up

The density of an object is its mass m divided by its volume V. We'll find the volume of a nucleus from Equation 27-1 and the formula for the volume of a sphere. To estimate the mass of a nucleus, we'll multiply the mass number A (the number of nucleons) by the average mass of a nucleon.

Radius of a nucleus:

$$r = r_0 A^{1/3} \qquad (27\text{-}1)$$

Definition of density:

$$\rho = \frac{m}{V} \qquad (11\text{-}1)$$

Volume of a sphere of radius r:

$$V = \frac{4}{3}\pi r^3$$

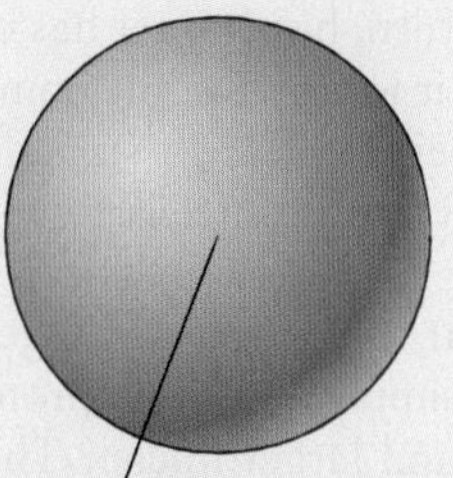

Solve

Find the volume of a nucleus of mass number A.

Substitute Equation 27-1 into the expression for the volume of a sphere of radius r:

$$V = \frac{4}{3}\pi r^3 = \frac{4}{3}\pi(r_0 A^{1/3})^3 = \frac{4}{3}\pi r_0^3 (A^{1/3})^3$$
$$= \frac{4}{3}\pi r_0^3 A$$

Use $r_0 = 1.2$ fm $= 1.2 \times 10^{-15}$ m, as in Example 27-1:

$$V = \frac{4}{3}\pi(1.2 \times 10^{-15}\,\text{m})^3 A$$
$$= (7.2 \times 10^{-45}\,\text{m}^3)A$$

Take the average mass of a nucleon to be the average of the proton mass m_p and the neutron mass m_n. Use this to write an expression for the mass of a nucleus of mass number A.

The average mass of a nucleon is

$$m_{avg} = \frac{m_p + m_n}{2} = \frac{(1.6726 \times 10^{-27}\,\text{kg}) + (1.6749 \times 10^{-27}\,\text{kg})}{2}$$
$$= 1.6738 \times 10^{-27}\,\text{kg}$$

A nucleus with A nucleons then has mass

$$m = m_{avg}A = (1.6738 \times 10^{-27}\,\text{kg})A$$

Calculate the density of the nucleus.

The density of the nucleus is

$$\rho = \frac{m}{V} = \frac{(1.6738 \times 10^{-27}\,\text{kg})A}{(7.2 \times 10^{-45}\,\text{m}^3)A} = \frac{1.6738 \times 10^{-27}\,\text{kg}}{7.2 \times 10^{-45}\,\text{m}^3}$$
$$= 2.3 \times 10^{17}\,\text{kg/m}^3$$

Reflect

Our final expression for the density ρ does not depend on A, the mass number of the nucleus. So the density of *all* nuclei is about the same. This agrees with our statements about the short-range character of the strong nuclear force. Note also that a block of solid iridium, the densest of all stable elements, has a density of 22,650 kg/m^3 (22.65 times the density of water). Our calculation shows that nuclei are 10^{13} times denser than iridium. Nuclei are *extremely* dense! This makes sense: Most of the mass of an atom is concentrated in its nucleus, which has a far smaller volume than the atom as a whole. So nuclear density (mass divided by volume) must be far greater than what we think of as the "ordinary" density of matter such as water or iridium.

Nuclear Spin and Magnetic Resonance Imaging

We learned in Section 26-6 that electrons have a type of intrinsic angular momentum called *spin*. (This is something of a misnomer because this angular momentum does not correspond directly to a spinning motion of the electrons.) Protons, and neutrons, too, have spin, and like an electron the spin of a proton or neutron can take on one of two values, often referred to as "spin up" and "spin down." In nuclei with an even number of nucleons, typically there are as many "spin up" nucleons as there are "spin down," and most such nuclei have zero net spin. (The orbital angular momentum of nucleons moving inside the nucleus can also contribute to the net spin of the nucleus.) But a nucleus with an odd number of nucleons must have a nonzero net spin. In particular, the nucleus of hydrogen—a single proton—has a net spin.

Measurements that make use of the spin of hydrogen nuclei enable us to localize hydrogen in an object or body, as well as to get information about the material in which those hydrogen atoms are embedded. That's the principle of **magnetic resonance imaging** (MRI). In an MRI scanner, spin information is used to form a three-dimensional map of the density of hydrogen atoms in a body. Living organisms are composed largely of water—our bodies, for example, are 60% to 70% water—and each water molecule contains two hydrogen atoms. As such, MRI is an ideal way to probe the internal structures in the body. The MRI scan in **Figure 27-4** shows the leg bones rubbing against each other in the knee of a person with osteoarthritis. Figure 27-1a shows an MRI scan of a patient's head.

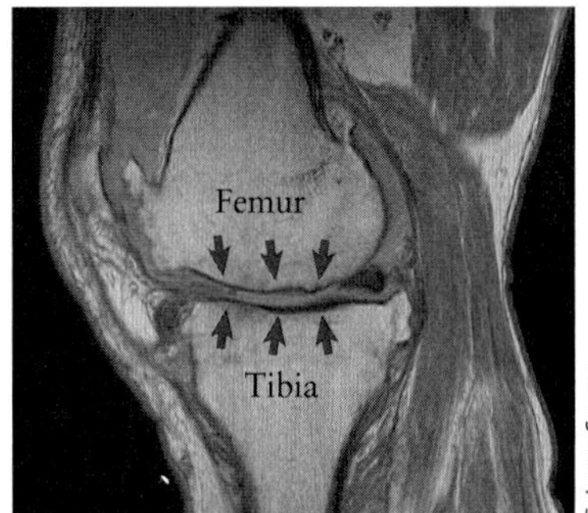

Figure 27-4 A magnetic resonance image The red arrows in this MRI image indicate where bones rub together in the knee of a patient suffering from osteoarthritis.

Protons and neutrons also have associated magnetic fields. This field is a dipole field like that of a current-carrying loop of wire (see Section 19-7). The line connecting the north and south poles of a proton or neutron is along the direction of its spin angular momentum vector. Now, we saw in Section 19-6, a current loop placed in an external magnetic field tends to align with its normal along the direction of the external field. In the same way, when a single proton or neutron is placed in a uniform external magnetic field, the magnetic force tends to align the spin direction either generally parallel to or antiparallel to the external field. The same is true for a nucleus with an odd number A of nucleons. To be precise, the spin direction aligns so that its component along the direction of the external field has a fixed positive or negative value. As such, the spin direction can rotate, or *precess*, around an axis defined by the field direction. **Figure 27-5a** shows this precession for a nucleus with spin aligned with the external field, and **Figure 27-5b** shows this precession for a nucleus with spin anti-aligned with (that is, opposed to) the external field.

Consider a large number of hydrogen atoms placed in a uniform magnetic field. About half of the protons will end up with spin aligned with the field and half with spin anti-aligned. Now, the energy of the spin-aligned orientation of a nucleus in an external magnetic field (Figure 27-5a) is higher than the energy of the anti-aligned orientation (Figure 27-5b) by an amount ΔE that depends on the magnetic field strength.

(a) A nucleus with its spin aligned with an external magnetic field

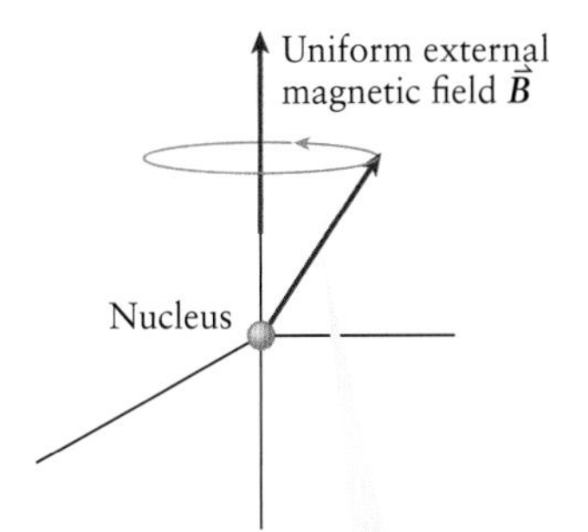

The spin angular momentum vector of the nucleus traces out a cone, but is generally aligned in the same direction as $\vec{B}$.

(b) A nucleus with its spin opposed to an external magnetic field

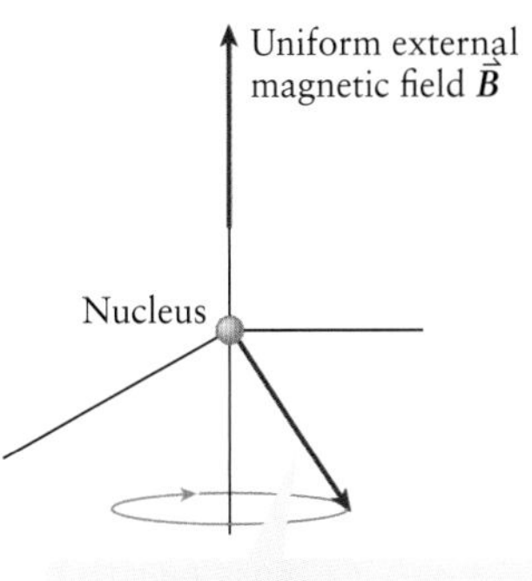

The spin angular momentum vector of the nucleus traces out a cone, but is generally aligned in the direction opposite to $\vec{B}$.

Figure 27-5 Nuclear spin When a nucleus is placed in an external magnetic field, the component of the spin along the direction of the external field is either (a) aligned with the external field or (b) anti-aligned with the external field.

Suppose we now bathe the atoms in an additional alternating magnetic field with frequency f. The alternating field is made up of photons of frequency f and energy hf, where h is Planck's constant (see Section 26-2). If we choose the frequency f so that hf is equal to ΔE, the photon energy is just equal to the energy difference between the two spin states. As a result, protons that are initially in the lower energy state, with their spin anti-aligned with the field, can absorb a photon of energy $\Delta E = hf$ and flip their spin to align with the external field. What is more, protons that are initially in the higher energy state, with their spin aligned with the field, can be stimulated by the alternating field to emit a photon of energy $\Delta E = hf$ and so flip their spin to the lower energy, anti-aligned state. (A similar sort of *stimulated* emission also occurs in a laser; see Section 26-5.)

The difference in energy between the two spin states of a hydrogen atom depends on the strength of the external field. With the high field strength required to be able to clearly observe the spin-flipping phenomenon, typically around 3 T, the energy difference ΔE is about 5×10^{-7} eV. From Equation 22-26 the frequency of a photon of this energy is

$$f = \frac{\Delta E}{h} = \frac{5 \times 10^{-7} \text{ eV}}{4.14 \times 10^{-15} \text{ eV} \cdot \text{s}} = 1.2 \times 10^{8} \text{ s}^{-1} = 120 \text{ MHz}$$

(We used the value of h in eV · s.) So the frequency of the oscillating field used to induce the hydrogen atoms to flip their spins in a strong magnetic field is around 100 MHz, which is in the radio-frequency part of the electromagnetic spectrum.

How do we "see" this spin flipping and so locate the hydrogen atoms? Initially there will be more atoms in the lower energy state than the higher energy state (lower energy is more likely for a system than higher energy), so when the hydrogen atoms in an external magnetic field are exposed to radio-frequency waves, there is a net absorption of the electromagnetic energy. When the radio-frequency signal is turned off, the spins begin to return to their initial (lower-energy) state, and in so doing they generate a radio-frequency field. The MRI device detects that radio-frequency field and uses it to map the density of hydrogen atoms in the body.

As we have mentioned, the difference in energy ΔE between the two hydrogen spin states depends on the strength of the external magnetic field. In an MRI device the magnetic field is made to vary over a body's volume, so the energy absorbed and then re-emitted in the spin–flip process also varies in different parts of the body. The exact frequencies of the radio energy detected by the MRI device therefore provide the information necessary to create images of high spatial resolution. In addition, the time it takes for the spins of the hydrogen nuclei to return to their equilibrium state depends on the particular molecules in the tissue. So timing information in an MRI device provides the means to differentiate one type of tissue from another.

GOT THE CONCEPT? 27-1 Nuclear Radius and Density

? A ^{20}Ne nucleus has 10 protons and 10 neutrons for a total of 20 nucleons, and a ^{160}Dy nucleus has 66 protons and 94 neutrons for a total of 160 nucleons. Compared to a ^{20}Ne nucleus, a ^{160}Dy nucleus has (a) double the radius and a greater density; (b) 8 times the radius and a greater density; (c) double the radius and the same density; (d) 8 times the radius and the same density; (e) double the radius and a lower density.

TAKE-HOME MESSAGE FOR Section 27-2

✔ The strong nuclear force binds nucleons (protons and neutrons) together in the nucleus of an atom.

✔ The volume of a nucleus is proportional to its mass number (the total number of nucleons in the nucleus).

✔ To be stable a light nucleus must have about as many neutrons as protons. More massive nuclei with more than about 20 protons require more neutrons than protons to be stable. Very large nuclei with mass number greater than 208 are always unstable.

✔ Protons and neutrons have spin. Magnetic resonance imaging uses the difference in energy between a state in which a nuclear spin is aligned with an external magnetic field and the state in which the spin is anti-aligned with the field.

27-3 Some nuclei are more tightly bound and more stable than others

Release a ball at the top of a hill, and it rolls down. Pull an object attached to the free end of a spring away from its equilibrium position, and it tends to return to that position. In both cases the systems are finding their way to a more stable configuration. All physical systems do the same; if a more stable configuration exists for a system, it will eventually find itself in that configuration as long as nature provides a mechanism for the transition to take place.

In this context, consider the nucleus of a helium atom, which consists of two protons and two neutrons. This configuration of these four nucleons must be more stable than when they are separate; otherwise the helium nucleus would end up broken apart. This stability results because the attraction of the strong nuclear force between the four nucleons overwhelms the electrostatic repulsive force between the two protons.

Let's be quantitative about *how* stable a given nucleus is. The total mass M_{tot} of the two protons and two neutrons in the nucleus of ^4He equals the sum of two proton masses (m_p) and two neutron masses (m_n):

$$\begin{aligned} M_{tot} &= 2m_p + 2m_n \\ &= 2(1.6726 \times 10^{-27}\ \text{kg}) + 2(1.6749 \times 10^{-27}\ \text{kg}) = 6.695 \times 10^{-27}\ \text{kg} \end{aligned}$$

But the actual mass of a helium nucleus is 6.645×10^{-27} kg, which is *less* than the total mass of the two protons and two neutrons. How is this possible? The answer is the key to nuclear stability; the energy equivalent of the difference in mass is tied up in binding the nucleons together. Recall from Section 25-7 that energy and mass are equivalent, and that an object of mass m has a rest energy $E_0 = mc^2$ (Equation 25-21). Since a ^4He nucleus has a smaller mass than its constituent nucleons, it also has a smaller rest energy. The difference between the rest energy of the ^4He nucleus and the rest energy of its constituent nucleons is the **binding energy** E_B. You can think of this as the energy that is released when the four nucleons come together to form a ^4He nucleus. Alternatively, you can think of the binding energy as the energy that would be required to separate a ^4He nucleus into two protons and two neutrons. The greater the binding energy *per nucleon* in a nucleus, the more tightly the nucleus is bound and therefore the more stable it is.

Figure 27-6 is a graph of the binding energy per nucleon (E_B/A) for different nuclides as a function of their mass number A. The energy values on the vertical axis are given in MeV, where 1 MeV = 10^6 eV. As we have learned before, one eV, or one electron volt, is the amount of energy acquired by an electron when it experiences a potential difference of one volt: 1 eV = 1.60×10^{-19} J. Figure 27-6 shows that the value of E_B/A for all nuclides is in the range from 1 to 9 MeV, which is why these units are convenient. (By comparison, the binding energy of a hydrogen *atom*—the energy required to separate the single electron in a hydrogen atom from the proton—is 13.6 eV, about 10^{-5} as great as the binding energy of even the most weakly bound nucleus. This indicates how small the forces are on electrons within atoms compared to the forces within the nucleus.)

Figure 27-6 shows that the nuclide with the smallest binding energy per nucleon is ^2H, an isotope of hydrogen with one proton and one neutron. (Just 0.0115% of hydrogen atoms are of this isotope.) As A increases, E_B/A increases rapidly because more nucleons are surrounded by other nucleons to which they are attracted and so are more tightly bound. The value of E_B/A peaks at about 8.8 MeV for A in the range of 56 to 62 and then decreases slowly for higher and higher values of A. This decrease happens because the number of protons Z also increases as A increases, and the electric repulsion between protons destabilizes the nucleus. The most stable nuclei are ^{56}Fe and ^{58}Fe (two isotopes of iron) and ^{62}Ni; the binding energy per nucleon of these nuclei places them near the peak of the E_B/A curve in Figure 27-6.

Whenever possible nuclei will rearrange themselves to maximize their stability and so maximize their binding energy per nucleon. Figure 27-6 shows that nuclei with relatively *large* values of mass number A (at the far right of the graph) can become more stable by *decreasing* the number of nucleons. One process of this kind is **nuclear**

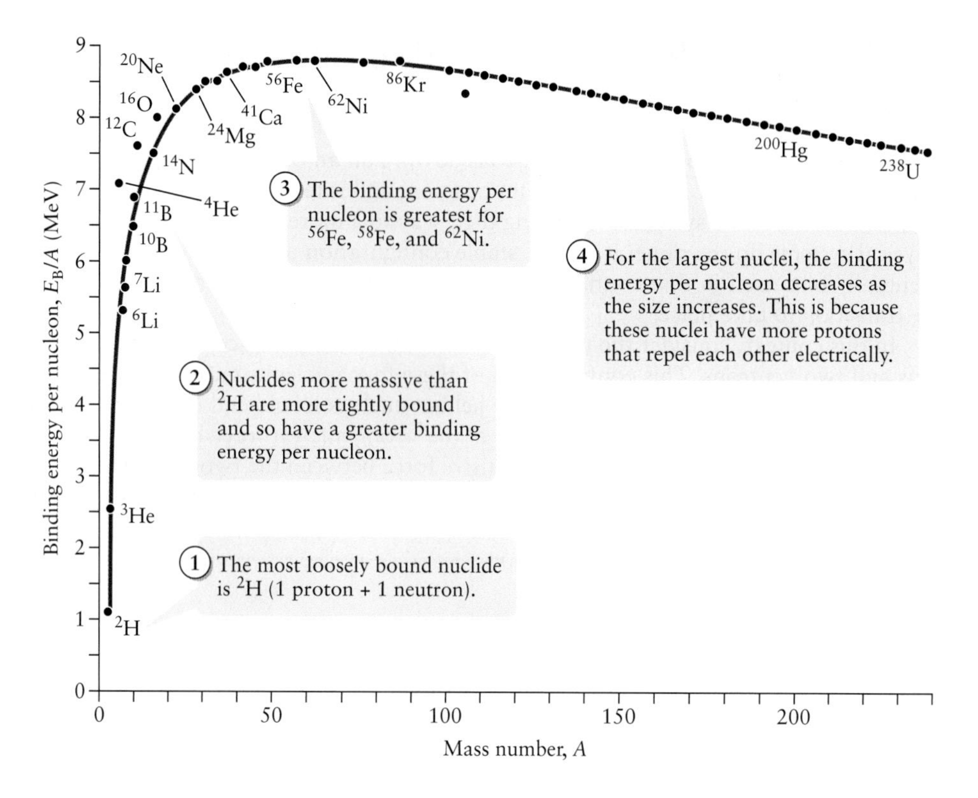

Figure 27-6 The curve of binding energy The binding energy per nucleon in a nucleus depends on the mass number *A*.

fission, in which nuclei split into smaller pieces. For example, atoms of curium-244 can spontaneously fission into xenon-135 and molybdenum-109. Curium has 96 protons, so ^{244}Cm has 148 neutrons. There are 54 protons in xenon and 42 in molybdenum or 96 total. The total number of nucleons in ^{135}Xe and ^{109}Mo, 244, equals the number of nucleons in ^{244}Cm. In other words, ^{135}Xe and ^{109}Mo contain the same 96 protons and 148 neutrons as in the original ^{244}Cm; the curium atom has split into two fragments. We'll discuss fission in more detail in Section 27-4.

Figure 27-6 also shows that nuclei with relatively *small* values of *A* (at the far left of the graph) can become more stable by *increasing* the number of nucleons. One way to do this is by **nuclear fusion**, in which two small nuclei join together to make a larger one. For example, the 12 protons and 12 neutrons in two carbon-12 nuclei can fuse to form a magnesium-24 nucleus. More than one nucleus can also be formed in a fusion process. For example, when a helium-3 nucleus fuses with a lithium-6 nucleus, the reaction forms two helium-4 nuclei and one hydrogen-1 nucleus. (Count the nucleons: ^{3}He has two protons and one neutron, and ^{6}Li has three protons and three neutrons, for a total of five protons and four neutrons. The two ^{4}He nuclei have two protons and two neutrons each, leaving one more proton as a ^{1}H nucleus.) In Section 27-5 we'll discuss nuclear fusion more carefully.

Let's see how to calculate the binding energy per nucleon of a nucleus such as ^{4}He. We'll do this by finding the total mass of the protons and neutrons that make up the nucleus and comparing it to the mass of the actual nucleus. We then convert that difference to the equivalent energy.

Instead of measuring mass in kilograms, it's most convenient to use units that take advantage of the equivalence between mass and energy. Since rest energy equals mass multiplied by c^2, we'll measure masses in units of MeV/c^2. Note that 1 MeV/c^2 = 1.7827×10^{-30} kg. In these units, the mass of a proton is 938.27 MeV/c^2, the mass of a neutron is 939.57 MeV/c^2, and the mass of a ^{4}He nucleus is 3727.4 MeV/c^2.

The difference Δ between (i) the mass of the two protons and two neutrons separately and (ii) the mass of the helium nucleus is

$$\begin{aligned}\Delta &= 2m_p + 2m_n - m_{He} \\ &= 2(938.27\ \text{MeV}/c^2) + 2(939.57\ \text{MeV}/c^2) - 3727.4\ \text{MeV}/c^2 \\ &= 28.3\ \text{MeV}/c^2\end{aligned}$$

The energy equivalent of any mass is obtained by multiplying it by c^2, so the energy equivalent of this difference is 28.3 MeV. (Notice how straightforward it is to find the energy equivalent of mass when we write mass in units of MeV/c^2.) In other words, the binding energy of ^4He is 28.3 MeV. There are four nucleons in the helium nucleus, so E_B/A is about 7.1 MeV. You can verify this result from the curve in Figure 27-6.

The binding energy per nucleon is much higher for ^4He than for other light nuclei. (Figure 27-6 shows that E_B/A is about 2.5 MeV for ^3He and about 5.3 MeV for ^6Li.) Thanks to its relatively high binding energy per nucleon, ^4He is far more stable than other light nuclei. For this reason, when large nuclei break apart to transform to a more stable, more energetically favorable state, in many cases they do so by emitting two protons and two neutrons bound together. In such processes the four bound protons and neutrons are referred to as an *alpha particle* (α particle). **Nuclear radiation** is the emission by a nucleus of either energy (in the form of a photon) or particles, such as an α particle. We'll discuss nuclear radiation in more detail in Section 27-6.

GOT THE CONCEPT? 27-2
Stability

? Rank these nuclides in order from most stable to least stable: (a) ^{11}B, (b) ^{20}Ne, (c) ^{86}Kr, (d) ^{200}Hg.

To find the binding energy of ^4He, we subtracted the mass of the nucleus from the mass of the two protons and two neutrons separately and then multiplied by c^2 to find the equivalent energy. In general, for a nucleus consisting of N neutrons and Z protons, E_B is

$$E_B = (Nm_n + Zm_p - m_{\text{nucleus}})c^2$$

where m_n is the mass of a neutron, m_p is the mass of a proton, and m_{nucleus} is the mass of the nucleus. In practice it's easier to measure the masses of neutral *atoms* (including their electrons) than the masses of isolated atomic nuclei. In terms of these masses, we can write the binding energy of a nucleus as

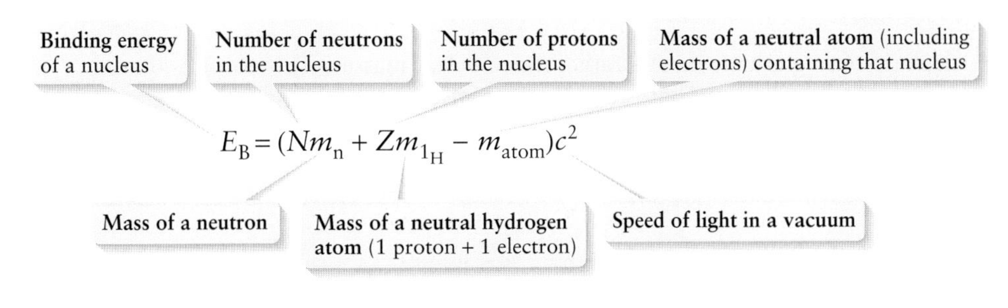

Binding energy of a nucleus (27-2)

The terms $Zm_{^1\text{H}}$ and m_{atom} each include the mass of Z electrons, so the electron masses cancel. The masses of neutral atoms are given in Appendix C. Note that the values in that appendix are given in units of atomic mass units (u or amu), where

$$1\ \text{u} = 931.494\ \text{MeV}/c^2$$

EXAMPLE 27-3 The Binding Energy of ^4He

Previously, we determined the binding energy per nucleon in the ^4He nucleus using the mass of the nucleus. Use the values from Appendix C to determine the binding energy per nucleon (in MeV/c^2) in the ^4He nucleus using the atomic mass of ^4He.

Set Up

We'll use Equation 27-2 and the values given in Appendix C for the neutron mass, the atomic mass of ^1H, and the atomic mass of ^4He.

Binding energy of a nucleus:

$$E_B = (Nm_n + Zm_{^1\text{H}} - m_{\text{atom}})c^2 \quad (27\text{-}2)$$

binding energy of ^4He = rest energy of two neutrons + rest energy of two ^1H atoms − rest energy of a ^4He atom

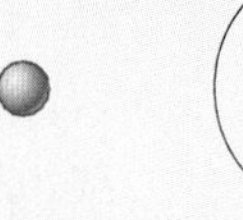

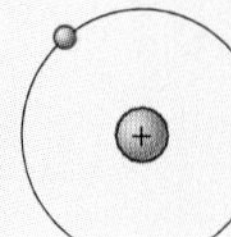

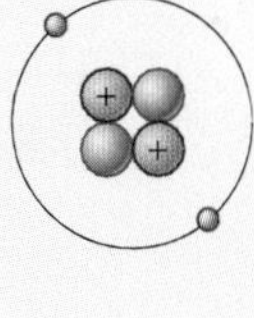

Solve

Calculate the binding energy of the ^{4}He nucleus, which has two neutrons ($N = 2$) and two protons ($Z = 2$).

From Appendix C:
neutron mass = $m_n = 1.008665$ u
atomic mass of ^{1}H = $m_{^1\text{H}} = 1.007825$ u
atomic mass of ^{4}He = $m_{^4\text{He}} = 4.002602$ u
Substitute these into Equation 27-2, with $m_{\text{atom}} = m_{^4\text{He}}$:

$$E_B = (2(1.008665\ \text{u}) + 2(1.007825\ \text{u}) - 4.002602\ \text{u})c^2$$
$$= 0.030378\ \text{u}c^2$$

Since 1 u = 931.494 MeV/c^2,

$$E_B = 0.030378\ \text{u}c^2\left(\frac{931.494\ \text{MeV}/c^2}{1\ \text{u}}\right)$$
$$= 28.297\ \text{MeV}$$

The binding energy per nucleon equals the binding energy of the nucleus divided by the number of nucleons in the nucleus.

The ^{4}He nucleus has four nucleons (two neutrons and two protons), so the binding energy per nucleon is

$$\frac{E_B}{A} = \frac{28.297\ \text{MeV}}{4} = 7.0742\ \text{MeV}$$

Reflect

Previously, we calculated $E_B/A = 7.1$ MeV to two significant figures using the mass of the ^{4}He nucleus; our new calculation is consistent with this.

Why would we do this kind of calculation using atomic masses rather than nuclear masses? The reason is that in general the masses of neutral atoms have been well measured, but precise measurements of the masses of atomic nuclei in isolation are difficult to obtain.

TAKE-HOME MESSAGE FOR Section 27-3

✔ The binding energy of a nucleus is the energy that would be required to separate it into its individual nucleons.

✔ The greater the binding energy per nucleon in a nucleus, the more tightly the nucleus is bound and therefore the more stable it is.

✔ The most stable nuclides are ^{56}Fe, ^{58}Fe, and ^{62}Ni. Smaller and larger nuclei have a lower binding energy per nucleon.

27-4 The largest nuclei can release energy by undergoing fission and splitting apart

As we saw in the previous section, nuclei with higher values of binding energy per nucleon (E_B/A) are more stable than those with lower values. Figure 27-6, a plot of the binding energy E_B per nucleon in nuclei versus mass number A, shows that E_B/A decreases as A increases beyond 60 or so. In other words, large nuclei are less stable than smaller ones for A greater than about 60. As a consequence of this instability, these large nuclei can undergo fragmentation or *fission* into smaller nuclei. Fission of a large nucleus can happen spontaneously, or it can be induced by imparting energy to the nucleus through a collision. In either case the smaller fragments have a higher value of E_B/A and are more stable.

Let's take a look at one of the most important processes of this kind, called **neutron-induced fission.** As an example, the collision of a neutron with a ^{235}U nucleus will cause it to fission. **Figure 27-7** shows one possible result. For a brief time the neutron and ^{235}U nucleus remain stuck together as ^{236}U*. (The asterisk indicates that this is an excited and short-lived state of ^{236}U.) This excited nucleus quickly fissions into fragments. In this particular reaction the fragments are an isotope of tellurium (^{134}Te), an

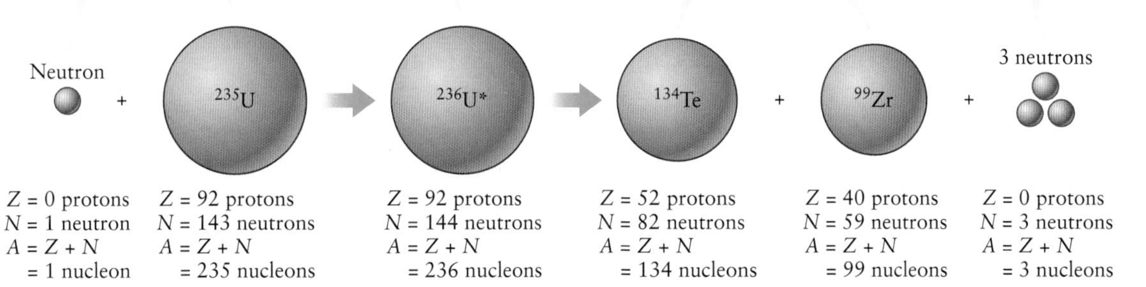

Energy is released in this fission reaction: The total kinetic energy of the fission fragments is much greater than the total kinetic energy of the initial neutron and ^{235}U nucleus.

Figure 27-7 Neutron-induced fission When one of the largest nuclei absorbs a slow-moving neutron, it can fission into smaller, more stable fragments.

isotope of zirconium (^{99}Zr), and three neutrons. Because these fragments are all more stable than the original nucleus, energy is released. The process described by Figure 27-7 occurs even when the colliding neutron is moving very slowly and so has essentially zero kinetic energy. So the released energy is almost entirely due to the change in binding energy between the initial ^{235}U nucleus and the fission products.

We can estimate the energy released in the process shown in Figure 27-7 by comparing the binding energy of the ^{235}U nucleus to the binding energies of the ^{134}Te and ^{99}Zr fragments. (There is no binding energy associated with the initial or final neutrons; they are not bound to any other particle.) If you make measurements on the graph in Figure 27-6, you'll see that the binding energy per nucleon E_B/A is about 7.6 MeV for $A = 235$, about 8.4 MeV for $A = 134$, and about 8.7 MeV for $A = 99$. The binding energy of each nucleus (E_B) equals the binding energy per nucleon (E_B/A) multiplied by the number of nucleons (A). The energy released in the fission reaction equals the difference between the total binding energy of the fragments and the binding energy of the initial ^{235}U nucleus:

$$\begin{aligned}
&(\text{energy released})\\
&\quad = (\text{binding energy of } ^{134}\text{Te}) + (\text{binding energy of } ^{99}\text{Zr})\\
&\qquad - (\text{binding energy of } ^{235}\text{U})\\
&\quad = (134)(8.4 \text{ MeV}) + (99)(8.7 \text{ MeV}) - (235)(7.6 \text{ MeV})\\
&\quad = 200 \text{ MeV}
\end{aligned}$$

We've given our result to just one significant figure because the values of E_B/A that we measured from Figure 27-6 are just estimates. The actual amount of energy released during this process is about 185 MeV, which is quite close to our estimate. This illustrates the tremendous amount of energy released in fission. By contrast, combustion (a chemical process that involves the electrons in molecules, not the nuclei of atoms) yields only a few eV for every molecule of fuel consumed. The energy release in fission is greater by a factor of several million!

When a heavy nucleus like ^{235}U undergoes fission, a wide variety of fragments can result. Figure 27-7 shows one possible result. Two others are

$$\text{n} + {}^{235}\text{U} \rightarrow {}^{236}\text{U}^* \rightarrow {}^{143}\text{Ba} + {}^{90}\text{Kr} + 3\text{n}$$
$$\text{n} + {}^{235}\text{U} \rightarrow {}^{236}\text{U}^* \rightarrow {}^{140}\text{Xe} + {}^{92}\text{Sr} + 4\text{n}$$

In each case the total number of protons and the total number of neutrons both remain the same. The ^{235}U nucleus has 92 protons and $235 - 92 = 143$ neutrons. In the first of these two processes, ^{143}Ba has 56 protons and 87 neutrons, and ^{90}Kr has 36 protons and 54 neutrons. The total number of protons after the fission has occurred is then 56 + 36 = 92. The total number of neutrons before the fission is 143 + 1 = 144 (including the neutron that starts the process). After the fission the number of neutrons is 87 + 54 + 3 = 144 (including the three neutrons released in the process). You can easily verify that the number of protons and the number of neutrons likewise remain the same in the second process above.

Uranium-235 can also undergo *spontaneous* fission, in which the nucleus fragments without undergoing a collision with a neutron. The following example shows how to calculate the energy released in this process.

EXAMPLE 27-4 Spontaneous Uranium Fission

Determine the energy released when a ^{235}U nucleus spontaneously undergoes fission to ^{140}Xe, ^{92}Sr, and three neutrons. The binding energy per nucleon in the nuclei of ^{235}U, ^{140}Xe, and ^{92}Sr are 7.59 MeV, 8.29 MeV, and 8.65 MeV, respectively.

Set Up

The energy released during the fission process is the difference between the binding energy of ^{140}Xe and ^{92}Sr nuclei and the binding energy of the ^{235}U nucleus. (There is no binding energy associated with the three neutrons.) The binding energy for each nucleus equals the binding energy per nucleon for that nucleus multiplied by the number of nucleons A.

(energy released)
= (total binding energy of fragments)
− (binding energy of original ^{235}U nucleus)

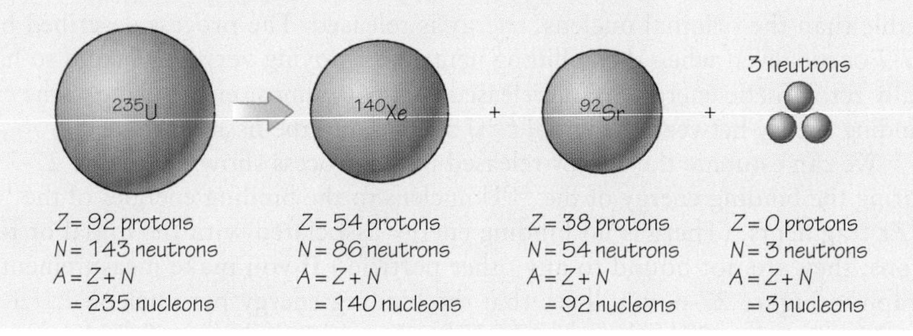

Solve

Calculate the binding energy for each of the nuclei, and then use these values to calculate the energy released.

The binding energies of the individual nuclei are

For ^{235}U: (235)(7.59 MeV) = 1784 MeV
For ^{140}Xe: (140)(8.29 MeV) = 1161 MeV
For ^{92}Sr: (92)(8.65 MeV) = 796 MeV

The energy released in the spontaneous fission is then

(energy released) = (binding energy of ^{140}Xe) + (binding energy of ^{92}Sr)
− (binding energy of ^{235}U)
= 1161 MeV + 796 MeV − 1784 MeV
= 173 MeV

Reflect

This result is consistent with our earlier claim that a typical amount of energy released in the fission of ^{235}U is around 200 MeV.

Spontaneous fission of ^{235}U is a *very* unlikely process: A given ^{235}U nucleus has a 50% chance of decaying in a period of 7.04×10^8 years, and the probability that it will decay by spontaneous fission is only 7.0×10^{-11}. (Here 7.04×10^8 y is the *half-life* of ^{235}U. We'll discuss this concept more carefully in Section 27-6. In that section we'll see that ^{235}U almost always decays by a different process called *alpha emission*.) By contrast, once a ^{235}U nucleus merges with a neutron to form an excited ^{236}U* nucleus, as in Figure 27.7, it typically undergoes fission within a fraction of a second.

All of the examples of fission that we've described result in the release of neutrons. Imagine what can happen if a large number of ^{235}U atoms are close to each other. Should one nucleus be struck by a neutron and fission, as shown in Figure 27-7, there would then be three neutrons moving among the atoms. Should each of these neutrons strike a ^{235}U nucleus and start a fission process, there would be nine neutrons, so possibly nine more fissions. With a sufficient number of ^{235}U atoms present, this *chain reaction* quickly grows, with an accompanying rapid increase in energy released.

Isotopes that are capable of sustaining a fission chain reaction are used as nuclear fuels. Such isotopes are termed *fissile*; the most common fissile nuclear fuels are ^{233}U, ^{235}U, ^{239}Pu, and ^{241}Pu. In the fission reactions they undergo, typically two or three neutrons are released in addition to larger fragments. These released neutrons don't have to be moving rapidly to trigger additional fission reactions: a chain reaction in a fissile material can be induced by a neutron carrying essentially zero kinetic energy. Indeed, in fuels such as ^{235}U slower, less energetic neutrons are more efficiently absorbed by the fissile nuclei.

Most nuclear reactors use the energy released in a fission chain reaction to heat water and produce steam to drive an electric generator (see Section 20-4). There are several challenges to producing energy in this way. First, a minimum amount, or *critical mass*, of fissile material must be present to sustain a chain reaction. As it happens, only about 0.7% of the naturally occurring uranium in the world is ^{235}U, and the other three commonly used fissile isotopes do not occur naturally. Most naturally occurring uranium is ^{238}U, which is not fissile. To use uranium as a nuclear fuel, then, it is necessary to separate the ^{235}U atoms from the ^{238}U, a costly and difficult process known as *enrichment*. In addition, the ^{238}U atoms that inevitably remain tend to absorb free neutrons and thereby inhibit a chain reaction. Once the critical mass of ^{235}U has been assembled, controlling the chain reaction is another challenge. If too many of the neutrons produced in the fissions result in a second fission, the energy released increases so rapidly that the fuel and whatever vessel is used to contain it can be damaged or even melt. To control a fission chain reaction in a nuclear reactor, control rods made of a substance that is a good absorber of neutrons are inserted between pieces of fuel.

Operating a fission nuclear reactor safely is a significant challenge. First, the fission fragments are radioactive. Many of these fragments, or the fragments produced when they decay, are long lived and tend to produce dangerous radiation for years, centuries, or even longer. In addition, reactors commonly use water as a *moderator*, a material that tends to slow the free neutrons (in order to make them more easily absorbed by a ^{235}U nucleus). Should the containment vessel rupture, this hot water can be released into the atmosphere in the form of steam carrying radioactive particles. Perhaps the most significant nuclear accident occurred at the Chernobyl nuclear power plant in Ukraine in 1986, in which an uncontrolled chain reaction caused a catastrophic power increase, leading to a series of explosions and the release of large quantities of radioactive steam, fuel, and smoke into the environment.

GOT THE CONCEPT? 27-3 Fission

? Consider the spontaneous fission process $^{20}Ne \rightarrow {}^{10}B + {}^{10}B$. This process does not occur in nature. Why not? (a) The number of protons does not remain constant. (b) The number of neutrons does not remain constant. (c) Both (a) and (b). (d) This process would absorb energy, not release it. (e) This process would neither release nor absorb energy.

TAKE-HOME MESSAGE FOR Section 27-4

✔ The binding energy per nucleon of nuclides with more than about 60 neutrons and protons decreases with increasing values of A. For this reason the fission process, in which a nucleus breaks up into smaller fragments, leads to more stable configurations of the nucleons in bigger nuclei.

✔ Fission can be triggered by allowing a slow-moving neutron to merge with a fissile nucleus.

27-5 The smallest nuclei can release energy if they are forced to fuse together

As we saw in the previous section, Figure 27-6 explains why the largest nuclei can undergo fission: The binding energy per nucleon E_B/A is maximum for mass numbers A around 60. By breaking into smaller fragments, nuclei with values of A much

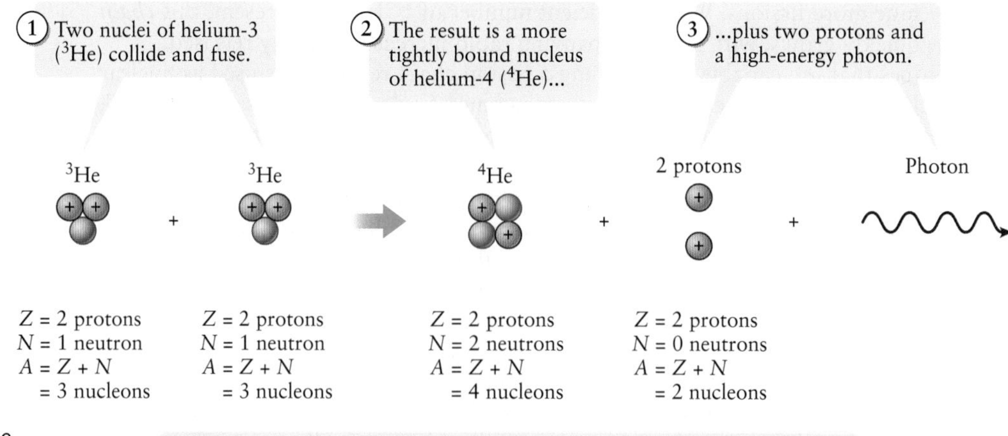

Figure 27-8 Nuclear fusion The fusion of two ^{3}He nuclei to make a ^{4}He nucleus is one of the energy-releasing reactions that take place in the core of the Sun.

Energy is released in this fusion reaction: The total kinetic energy of the ^{4}He nucleus and protons, plus the energy of the photon, is much greater than the total kinetic energy of the initial ^{3}He nuclei.

greater than 60 can therefore increase their binding energy per nucleon E_B/A and so become more stable. The opposite is true for the lightest nuclei, with A much less than 60. These nuclei can increase the value of E_B/A and become more stable by becoming *larger*. Processes in which two small nuclei combine to form a larger one are called *fusion* processes.

Figure 27-8 shows how two ^{3}He nuclei (each with two protons and one neutron) fuse together to form a ^{4}He nucleus (with two protons and two neutrons). Two protons are left over, and without more neutrons there is no way for the protons to be bound together in a single nucleus. As a result, these protons fly off separately. Energy is released during this process because the final configuration of the protons and neutrons is more stable than the initial configuration. A photon carries away the energy released.

How much energy is released in the process shown in Figure 27-8? As in fission, the energy released in fusion is the difference between the total binding energy of the final nuclei and the total binding energy of the original nuclei. From Figure 27-6, the binding energy per nucleon for ^{3}He is approximately 2.5 MeV. Each ^{3}He has three nucleons, so the total binding energy of ^{3}He is 3(2.5 MeV) = 7.5 MeV, and the two ^{3}He nuclei together have a combined binding energy equal to 7.5 MeV + 7.5 MeV = 15.0 MeV. In Section 27-3 we found that a ^{4}He nucleus has a binding energy of 28.3 MeV. The two protons are single particles, so they make no contribution to the binding energy. The energy released in the fusion process is therefore

$$\begin{aligned}(\text{energy released}) &= (\text{binding energy of } ^4\text{He nucleus}) \\ &\quad - (\text{binding energy of two } ^3\text{He nuclei}) \\ &= (28.3 \text{ MeV}) - (15.0 \text{ MeV}) = 13.3 \text{ MeV}\end{aligned}$$

The actual value, found using more accurate values of the binding energies, is closer to 12.86 MeV.

The process shown in Figure 27-8 is the final step in the *proton–proton cycle*, the fusion process that takes place near the center of the Sun and is the source of the Sun's energy. The cycle begins with the fusing of two protons (the nuclei of ^{1}H) to form ^{2}H. This nucleus fuses with another proton, forming ^{3}He, and finally, two ^{3}He nuclei fuse to form ^{4}He. We can summarize these three steps as

$$\begin{aligned}&\text{Step 1: } ^1\text{H} + {}^1\text{H} \rightarrow {}^2\text{H} + \text{e}^+ + \nu_\text{e} \\ &\text{Step 2: } ^2\text{H} + {}^1\text{H} \rightarrow {}^3\text{He} + \gamma \\ &\text{Step 3: } ^3\text{He} + {}^3\text{He} \rightarrow {}^4\text{He} + {}^1\text{H} + {}^1\text{H} + \gamma\end{aligned}$$

In step 1 the e^+ particle is a **positron**, a particle with the same mass as an electron but with positive charge $+e$. The particle named ν_e (the Greek letter "nu" with a subscript "e") is a **neutrino**, a nearly massless, neutral particle. Also note that this step involves a proton being converted into a neutron. We'll discuss this conversion, called beta-plus decay, in Section 27-6. Step 1 and the subsequent interaction of the positron with an electron in the Sun (in which the two particles annihilate each other and convert into photons) release 1.44 MeV of energy. The energy released in step 2 is 5.49 MeV. Both of these steps must occur twice before step 3 can occur (because step 3 requires two ^{3}He nuclei). So six protons are used in steps 1 and 2, and in step 3 two of those protons are returned along with a ^{4}He nucleus. The net result is therefore that four ^{1}H nuclei disappear and are replaced by one ^{4}He nucleus. The net energy release is 2(1.44 MeV) from step 1 happening twice, plus 2(5.49 MeV) from step 2 happening twice, plus 12.86 MeV from the fusion of two ^{3}He nuclei in step 3. The sum of these is a net energy release of 26.7 MeV as four ^{1}H nuclei are transformed into a ^{4}He nucleus.

Fission processes release more energy than the proton–proton fusion cycle, around 200 MeV compared to 26.7 MeV. Therefore, it might seem that fission is a more effective way to convert fuel to energy. Consider, however, that while 235 nucleons in ^{235}U are spent in order to release 200 MeV, in the fusion process described above, the 26.7 MeV released come at the expense of only four nucleons. Comparing energy per nucleon (think miles per gallon), fission provides less than 1 MeV per nucleon, while proton–proton fusion gives 26.7 MeV divided by 4, or nearly 7 MeV per nucleon. Fusion processes are *much* more efficient at releasing nuclear binding energy.

Hydrogen makes up about 75% of the Sun's mass of approximately 2×10^{30} kg. If the proton–proton cycle leads to more stability, why doesn't all of that hydrogen quickly fuse to form ^{4}He, leaving the Sun a gigantic (and cool) ball of helium gas? The answer lies in the same forces at play within nuclei: the Coulomb force that repels protons from each other and the short-ranged strong nuclear force that draws them together. In order for two protons to fuse to form ^{2}H, they must come within a few femtometers of each other, at which point the strong attraction is able to overcome the Coulomb repulsion. This requires the protons to have considerable kinetic energy, which can come from being at high temperature. A temperature of more than 4×10^{6} K is required for the proton–proton cycle to start. The temperature at the core of the Sun is around 15×10^{6} K, so the proton–proton cycle can and does occur there. Even at that temperature, however, the probability that two nearby protons will fuse is small. This means that only a small fraction—about 4×10^{-19}—of the hydrogen in the Sun is undergoing fusion at any one time. The Sun won't burn out for a long time.

As a star ages, its core temperature increases and additional fusion reactions become possible. For example, three ^{4}He nuclei can fuse to form a ^{12}C nucleus. This requires a temperature of about 10^{8} K because the ^{4}He nuclei are more massive than protons and repel each other more strongly due to their greater charge. At even higher temperatures a ^{4}He nucleus can fuse with a ^{12}C nucleus to form a ^{16}O nucleus, a ^{4}He nucleus can fuse with a ^{16}O nucleus to form a ^{20}Ne nucleus, and so on. So as stars age they manufacture heavier and heavier chemical elements. In the most massive stars, which have the highest core temperatures, so much kinetic energy is available at the very end of the star's evolution that fusion processes can produce even the heaviest nuclei up to uranium. Making these massive nuclei by fusion absorbs rather than releases energy, which is why it can happen only in very special circumstances.

BIO-Medical These fusion reactions in stars make life on Earth possible. Here's why: When the universe first originated some 13.8 billion years ago, almost all ordinary matter was in the form of hydrogen or helium. (We'll discuss the origin of the universe in Chapter 28.) All heavier elements had to be manufactured by fusion within stars. After an aging star produces elements heavier than helium and goes through its final stages of evolution, it disperses much of its material into interstellar space (**Figure 27-9**). This material, which is enriched in heavy elements, can then be incorporated into a later generation of stars. Our Sun is such a "second-generation" star, with an elevated abundance of elements heavier than helium. As part of the process by which the Sun formed, some of these elements went into forming Earth and the Sun's other planets. This means that all of the nuclei of carbon in the organic compounds that make up

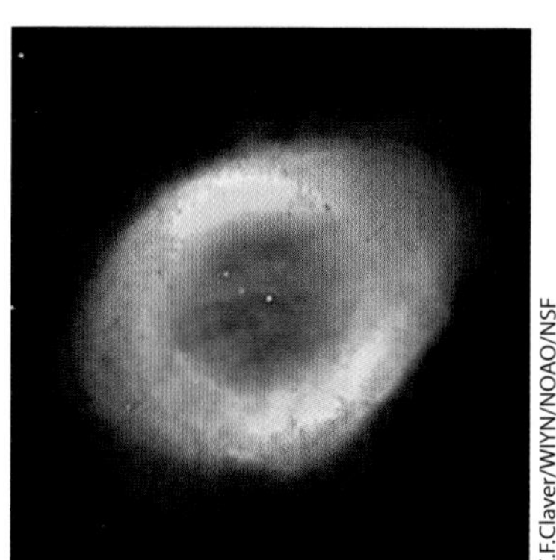

Figure 27-9 Seeding space with fusion products The Ring Nebula is a cloud of gas emitted by an aging star. The cloud includes nitrogen (shown in red) and oxygen (shown in green) produced by fusion reactions within the star.

your body, all of the nuclei in the oxygen that you breathe, and all of the nuclei of iron in the hemoglobin in your blood were manufactured in stars that died billions of years ago. This is one of the great lessons of nuclear physics: You are made of star-stuff.

GOT THE CONCEPT? 27-4 Fusion

Ordinary hydrogen gas is in the form of diatomic hydrogen (H_2), and more than 99.9% of the atoms in hydrogen gas have a nucleus that is a single proton (^1H). Why don't the two hydrogen nuclei in an H_2 molecule spontaneously undergo fusion as in Step 1 of the proton–proton cycle: $^1H + {}^1H \rightarrow {}^2H + e^+ + \nu_e$? (a) The nuclei are too far apart; (b) the nuclei are moving too slowly relative to each other; (c) the nuclei in an atom are different from those outside an atom; (d) both (a) and (b); (e) all of (a), (b), and (c).

TAKE-HOME MESSAGE FOR Section 27-5

✔ The binding energy per nucleon of nuclides that have fewer than about 60 neutrons and protons is larger for increasing values of mass number A. For this reason, fusion of two small nuclei to form a larger one leads to more stable configuration of the nucleons.

✔ Fusion can only take place if the fusing nuclei come very close to each other. Therefore very high temperatures are required so the nuclei can overcome their mutual electric repulsion.

27-6 Unstable nuclei may emit alpha, beta, or gamma radiation

For many people phrases such as "radioactivity" and "nuclear radiation" are synonymous with danger. Not all nuclear radiation is dangerous, however, and indeed several types of nuclear radiation have important practical applications. In this section we'll explore a number of aspects of nuclear radiation.

All naturally occurring nuclear processes take place because the final state is more energetically favorable than the initial state. This is true of fission and fusion, and it is true of radiation processes, too. A relatively few nuclides—266 out of over 3000—are stable. All the rest are **radioactive**; that is, they decay into another nuclide by radiating away one or more particles. It's also possible for a nucleus in an excited state to radiate energy in the form of a photon as the nucleus transitions to a less excited state.

The three most common kinds of radiation are *alpha*, *beta*, and *gamma* radiation. The terms were coined by Ernest Rutherford, who in his research between 1899 and 1903 classified radiation according to the depth that a radiation particle was able to penetrate other objects. Alpha particles penetrated the least, beta particles more, and gamma particles the most. (Rutherford received the 1908 Nobel Prize in Chemistry for this research, which was the first to show that one element can change into another through radioactive processes.) We now know that alpha particles are ^4He nuclei; beta particles are electrons; and gamma particles are photons with energies that can be millions of times greater than the energies of visible-light photons.

Before we consider the properties of these specific types of radiation, let's look at a concept that is common to all of them—the idea of *radioactive half-life*.

Radioactive Decay and Half-Life

Nuclear radiation of all kinds involves physics on the very small scale of the nucleus, so is governed by quantum mechanics. As we learned in Section 26-6, quantum mechanics cannot tell us the position or velocity of a particular object at any time. It can, however, tell us the *probability* that an object will be at a particular place or have a particular velocity at any given instant. In the same way, quantum mechanics cannot tell us when a given radioactive nucleus will decay, but it can predict the probability that this nucleus will decay within a given time interval.

As an example, consider the emission of a beta particle in the decay of ^{137}Cs (cesium), in which a neutron is converted into a proton:

$$^{137}\text{Cs} \rightarrow {}^{137}\text{Ba} + \text{e}^- + \bar{\nu}_e$$

The beta particle is the electron (e^-), and $\bar{\nu}_e$ is an **antineutrino**, related to the light, neutral neutrino particle we discussed in the context of fusion in the previous section. Every ^{137}Cs nucleus can, and eventually will, decay radioactively to a barium nucleus (^{137}Ba). Experiment shows that if a sample of ^{137}Cs contains, say, 10,000 atoms, after 30.17 years only about 5000 will be left. However, if we select any individual ^{137}Cs atom of those 10,000, we have no idea when its nucleus will decay. Perhaps it will decay in the next second or perhaps not for thousands of years.

While we can't make a definitive claim about when any particular atomic nucleus will decay, we can quantify the *rate* at which a group of radioactive atoms decays, that is, the number of decays per second. The probability λ that any one nucleus of a given type will decay in the next second is the same for all such nuclei. The quantity λ is called the **decay constant.** It has units of s^{-1}, since it refers to a probability per second. The value of λ is different for different radioactive nuclides: It is greater for nuclides that decay rapidly and smaller for nuclides that decay slowly.

If we have a sample of N such nuclei, the total number of decays that take place in the next second—that is, the *decay rate*—will be equal to the product of the decay constant λ (the probability that any one nucleus decays in the next second) and N (the number of radioactive nuclei present). So the decay rate of a radioactive sample is greater if the sample is larger (the number of nuclei N in the sample is greater) or the nuclei have a higher decay constant λ. The SI unit of decay rate is the **becquerel** (Bq), after the nineteenth-century French physicist Antoine Henri Becquerel, who along with Marie Skłodowska-Curie and Pierre Curie won the 1903 Nobel Prize in Physics for their discovery of radioactivity. The becquerel is equivalent to one radioactive decay per second. Physicists also commonly use units of curies (Ci) for decay rate; $1 \text{ Ci} = 3.7 \times 10^{10} \text{ Bq} = 3.7 \times 10^{10} \text{ decays/s}$.

After an elapsed time of Δt seconds, the total number of decays will be equal to $\lambda N \Delta t$. This means that in a time Δt the number of radioactive nuclei decreases by $\lambda N \Delta t$. So the *change* in the number of radioactive nuclei present is

$$\Delta N = -\lambda N \Delta t \qquad (27\text{-}3)$$

The minus sign in Equation 27-3 indicates that the number of nuclei decreases as a result of the decays. If we divide both sides of Equation 27-3 by the elapsed time Δt, we get an expression for the rate of change $\Delta N/\Delta t$ of the number of radioactive nuclei:

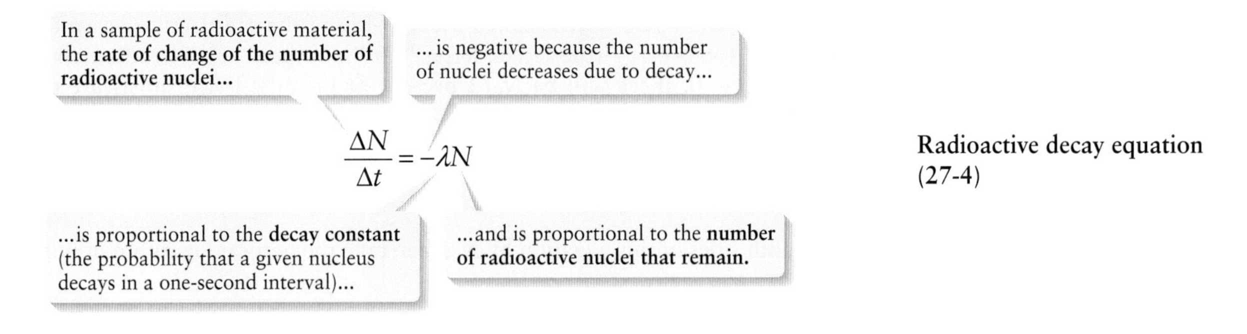

Radioactive decay equation (27-4)

Equation 27-4 tells us that as time goes by the decay rate will decrease because the number of radioactive nuclei will decrease. Using the tools of calculus we can solve Equation 27-4 to find the number of nuclei present as a function of time $N(t)$. The result is

$$N(t) = N_0 e^{-\lambda t} \qquad (27\text{-}5)$$

In Equation 27-5 N_0 is the number of nuclei present at a specific time that we choose to call $t = 0$. This equation tells us that the number of nuclei present decreases

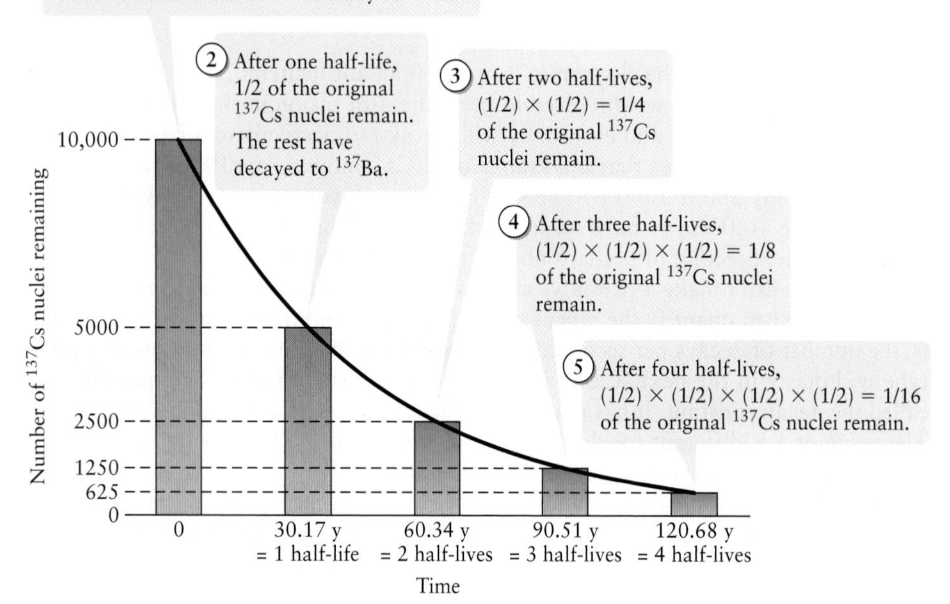

Figure 27-10 Nuclear decay and half-life This exponential curve shows the evolution of a sample that originally contains 10,000 cesium-137 (^{137}Cs) nuclei, which decay to barium-137 (^{137}Ba).

- The half-life of a nuclear decay is the time required for one-half of the nuclei present initially to decay.
- Radioactive decay is a statistical process: It's impossible to predict when any one individual unstable nucleus will decay.

exponentially, as **Figure 27-10** shows. The number of decays per second at time t is equal to the decay constant λ (the decay probability per second per nucleus), multiplied by $N(t)$, the number of nuclei remaining at time t. We can write this as

$$R(t) = \lambda N(t) = \lambda N_0 e^{-\lambda t}$$

In this equation λN_0 is equal to the decay rate at $t = 0$, which we call R_0. So the decay rate as a function of time is

$$R(t) = R_0 e^{-\lambda t} \quad (27\text{-}6)$$

Equation 27-6 tells us that the decay rate, too, will decrease exponentially as time goes by. With the passage of time, a radioactive sample will undergo fewer and fewer decays per second.

Figure 27-10 shows that the number of ^{137}Cs nuclei remaining decreases by one-half every 30.17 years due to beta decay. Because the decay rate is proportional to the number of ^{137}Cs nuclei remaining, the decay rate also decreases by one-half every 30.17 years. This time is called the **half-life** of the radioactive decay, to which we give the symbol $\tau_{1/2}$ (the Greek letter "tau"). If there are N_0 radioactive nuclei present at $t = 0$, there will be $N_0/2$ present at $t = \tau_{1/2}$. If we substitute this into Equation 27-5, we get

$$\frac{N_0}{2} = N_0 e^{-\lambda \tau_{1/2}}$$

Divide both sides of this equation by N_0, then take the natural logarithm of both sides:

$$\ln\left(\frac{1}{2}\right) = \ln\left(e^{-\lambda \tau_{1/2}}\right) \quad (27\text{-}7)$$

Why do we do this? The reason is that the natural logarithm "undoes" the exponential function: For any x, $\ln(e^x) = x$. Furthermore, $\ln(1/x) = -\ln x$ for any x. If we apply these to Equation 27-7, we get

$$-\ln 2 = -\lambda \tau_{1/2}$$

or

$$\tau_{1/2} = \frac{\ln 2}{\lambda} \quad (27\text{-}8)$$

Equation 27-8 says that the half-life is inversely proportional to the decay constant λ, the probability that a given nucleus of a certain type will decay in a 1-s interval. The greater the decay constant, the shorter the half-life and the more rapidly a sample of that nucleus will decay.

The half-life of radioactive sources varies widely, from far less than 1 s to billions of years or more. Common radioactive sources include ^{32}P, a beta emitter used in DNA research that has a half-life of 14.3 days; ^{241}Am, an alpha emitter often found in household smoke detectors that has a half-life of 432.2 y; and ^{238}U, an alpha emitter with a half-life of 4.47×10^9 y. Radioactive isotopes with half-lives on the order of a second or less aren't terribly useful because they don't stay around long enough.

EXAMPLE 27-5 Technetium

A form of an isotope of technetium, ^{99m}Tc, undergoes gamma decay with a half-life of 6.01 h. (The "m" stands for "metastable." Technetium-99 is widely used as a radioactive tracer for medical purposes because its gamma radiation is easily detected and because technetium doesn't stay in the body for long. Hence the total radiation delivered to the patient is low.) (a) What is the decay constant λ, the probability that a nucleus of ^{99m}Tc will decay per second, in this sample? Does λ change as time goes on? (b) What fraction of the initial number of ^{99m}Tc nuclei will be left after 1.00 day? (c) What fraction will be left after 4.00 days have elapsed?

Set Up

We'll use Equation 27-8 to relate the decay constant λ to the half-life. Equation 27-5 will tell us the number of ^{99m}Tc nuclei remaining after a time t in terms of the initial number of nuclei N_0.

Half-life of a radioactive substance:

$$\tau_{1/2} = \frac{\ln 2}{\lambda} \qquad (27\text{-}8)$$

Number of radioactive nuclei present at time t:

$$N(t) = N_0 e^{-\lambda t} \qquad (27\text{-}5)$$

Solve

(a) Determine the decay constant for ^{99m}Tc.

We can rewrite Equation 27-8 as

$$\lambda = \frac{\ln 2}{\tau_{1/2}}$$

Substitute $\tau_{1/2} = 6.01$ h:

$$\lambda = \frac{\ln 2}{(6.01 \text{ h})}\left(\frac{1 \text{ h}}{60 \text{ min}}\right)\left(\frac{1 \text{ min}}{60 \text{ s}}\right) = 3.20 \times 10^{-5} \text{ s}^{-1}$$

The probability that a given ^{99m}Tc nucleus will decay in a 1-s interval is 3.20×10^{-5}, corresponding to odds of 1 in $(1/3.20 \times 10^{-5}) = 31{,}200$. This value depends only on the half-life, so it does not vary with time.

(b) To find the fraction of nuclei remaining after $t = 1.00$ d, substitute this value of t into Equation 27-5.

The fraction of ^{99m}Tc nuclei remaining after a time t equals the number remaining $N(t)$ divided by the number N_0 present initially:

$$\frac{N(t)}{N_0} = \frac{N_0 e^{-\lambda t}}{N_0} = e^{-\lambda t}$$

From part (a) we know the value of λ in s^{-1}, so we need to express t in seconds:

$$t = (1.00 \text{ d})\left(\frac{24 \text{ h}}{1 \text{ d}}\right)\left(\frac{60 \text{ min}}{1 \text{ h}}\right)\left(\frac{60 \text{ s}}{1 \text{ min}}\right) = 8.64 \times 10^4 \text{ s}$$

The fraction of nuclei remaining is then

$$\frac{N(1.00\ \text{d})}{N_0} = e^{-(3.20\times10^{-5}\ \text{s}^{-1})(8.64\times10^{4}\ \text{s})}$$

$$= e^{-2.77} = 0.0628$$

(c) Repeat part (b) with $t = 4.00$ d.

The elapsed time is now

$$t = (4.00\ \text{d})\left(\frac{8.64\times10^{4}\ \text{s}}{1\ \text{d}}\right) = 3.46\times10^{5}\ \text{s}$$

and the fraction of nuclei remaining is

$$\frac{N(4.00\ \text{d})}{N_0} = e^{-(3.20\times10^{-5}\ \text{s}^{-1})(3.46\times10^{5}\ \text{s})}$$

$$= e^{-11.1} = 1.55\times10^{-5}$$

Reflect

We can check our results by noting that $t = 1.00$ d is almost exactly 4 times the 6.01-h half-life of $^{99\text{m}}$Tc, and $t = 4.00$ d is almost exactly 16 times the half-life. This check agrees with our calculations, as it should.

After each half-life, the number of $^{99\text{m}}$Tc nuclei remaining decreases by one-half. So after four half-lives, the fraction of $^{99\text{m}}$Tc nuclei remaining will be

$$\frac{1}{2}\times\frac{1}{2}\times\frac{1}{2}\times\frac{1}{2} = \frac{1}{2^4} = \frac{1}{16} = 0.0625$$

The fraction actually remaining at $t = 1.00$ d (slightly less than four 6.01-h half-lives) is 0.0628, very close to our estimate.

After 16 half-lives, the fraction of $^{99\text{m}}$Tc remaining will be

$$\frac{1}{2^{16}} = \frac{1}{65{,}536} = 1.53\times10^{-5}$$

The fraction actually remaining at $t = 4.00$ d (slightly less than 16 times the 6.01-h half-life) is 1.55×10^{-5}, which again is very close to our estimate.

Alpha Radiation

We saw in Section 27-4 that a large nucleus can increase the binding energy per nucleon E_B/A by breaking into smaller fragments. The most likely decay products are those that are more stable, that is, those with larger binding energy per nucleon. As Figure 27-6 shows, ^{4}He has a greater binding energy per nucleon than any other nucleus with a small value of A. So the nucleus of the ^{4}He atom is far more stable than other small nuclei and therefore a far more probable decay product of large nuclei. For this reason, the radioactive emission of a ^{4}He nucleus is the most likely decay process for a large nucleus. Another name for a ^{4}He nucleus is an **alpha particle** (symbol "α"), and emission of an alpha particle is called **alpha decay.**

The alpha radiation process reduces the number of protons Z of the initial, or parent, nucleus by two and reduces the number of neutrons of the parent by two. The result is a daughter nucleus that has an atomic number Z–2 and a mass number A–4, accompanied by an alpha particle with two protons and two neutrons (**Figure 27-11**).

The daughter nucleus and the alpha particle both carry away the energy released in alpha decay. However, the kinetic energy of the alpha particle is far greater than the kinetic energy of the daughter. We can confirm this by looking at the ratio of the kinetic energy K_α of the alpha particle to the kinetic energy K_D of the daughter. The alpha particle has mass m_α and is emitted with speed v_α, and

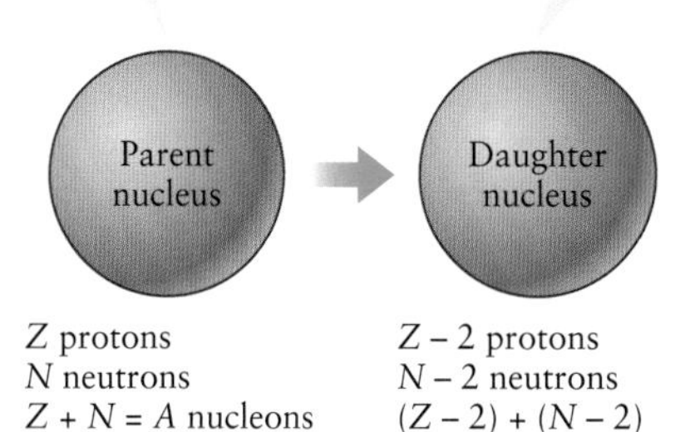
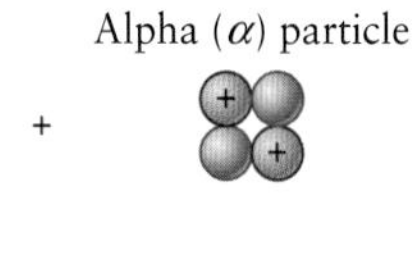

1 This parent nucleus is large and unstable.
2 The parent decays into a daughter nucleus with two fewer protons and two fewer neutrons...
3 ...plus an alpha particle (a ^{4}He nucleus).
Parent nucleus
Daughter nucleus
+
Alpha (α) particle
Z protons
N neutrons
$Z + N = A$ nucleons
$Z - 2$ protons
$N - 2$ neutrons
$(Z - 2) + (N - 2)$
$= A - 4$ nucleons
$Z = 2$ protons
$N = 2$ neutrons
$Z + N = 4$ nucleons

Figure 27-11 **Alpha decay** Large nuclei can increase their stability by emitting an alpha particle and becoming a smaller, more stable daughter nucleus.

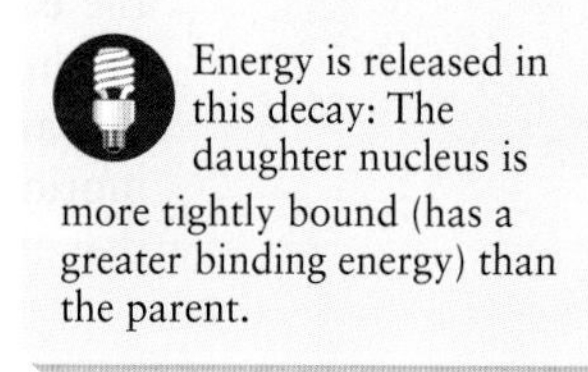

Energy is released in this decay: The daughter nucleus is more tightly bound (has a greater binding energy) than the parent.

the daughter nucleus has mass m_D and is emitted with speed v_D. The kinetic energies of the alpha particle and the daughter are then

$$K_\alpha = \frac{1}{2} m_\alpha v_\alpha^2$$

$$K_D = \frac{1}{2} m_D v_D^2$$

The ratio of these kinetic energies is

$$\frac{K_\alpha}{K_D} = \frac{(1/2) m_\alpha v_\alpha^2}{(1/2) m_D v_D^2} = \frac{m_\alpha}{m_D}\left(\frac{v_\alpha}{v_D}\right)^2 \tag{27-9}$$

We can relate the speeds v_α and v_D by noting that momentum must be conserved in the alpha decay (because no external forces act on the parent nucleus as it decays). If the parent nucleus is at rest, the total momentum is zero before the decay and so must be zero after the decay. The alpha particle and daughter nucleus must therefore fly off in opposite directions, and each must have the same magnitude of momentum:

$$m_\alpha v_\alpha = m_D v_D \quad \text{so} \quad \frac{v_\alpha}{v_D} = \frac{m_D}{m_\alpha}$$

If we substitute this into Equation 27-9, we find that the ratio of the alpha particle's kinetic energy to that of the daughter nucleus is

$$\frac{K_\alpha}{K_D} = \frac{m_\alpha}{m_D}\left(\frac{v_\alpha}{v_D}\right)^2 = \frac{m_\alpha}{m_D}\left(\frac{m_D}{m_\alpha}\right)^2 = \frac{m_D}{m_\alpha}$$

The mass of the daughter nucleus is much larger than the mass of the alpha particle, so the fraction m_D/m_α is much greater than one and K_α is large compared to K_D.

All nuclei with more than 82 protons are unstable and have some probability of alpha decay. As an example, the element thorium ($Z = 90$ protons) undergoes alpha decay to radium, which contains two fewer protons ($Z = 88$). The α decay of ^{228}Th, for example, is

$$^{228}\text{Th} \rightarrow {}^{224}\text{Ra} + \alpha$$

Just as ^{228}Th decays to ^{224}Ra, ^{224}Ra undergoes alpha decay to ^{220}Rn. This process continues until the daughter nucleus has $Z = 82$ or less. In this case the final alpha decay is to lead, with $Z = 82$:

$$^{228}\text{Th} \rightarrow {}^{224}\text{Ra} + \alpha \quad (\tau_{1/2\ \text{Th}} = 1.91\ \text{y})$$
$$\hookrightarrow {}^{220}\text{Rn} + \alpha \quad (\tau_{1/2\ \text{Ra}} = 3.63\ \text{d})$$
$$\hookrightarrow {}^{216}\text{Po} + \alpha \quad (\tau_{1/2\ \text{Rn}} = 55.6\ \text{s})$$
$$\hookrightarrow {}^{212}\text{Pb} + \alpha \quad (\tau_{1/2\ \text{Po}} = 0.145\ \text{s})$$

Note that in each alpha decay the daughter nucleus has two fewer protons and two fewer neutrons than its parent.

Another nucleus that decays by alpha radiation is ^{235}U, which has a half-life $\tau_{1/2} = 7.04 \times 10^8$ y. (In Example 27-4 in Section 27-4, we looked at the spontaneous fission of ^{235}U. This is a very rare decay mode; ^{235}U undergoes alpha decay rather than spontaneous fission almost 100% of the time.) Substantial amounts of ^{235}U, ^{238}U ($\tau_{1/2} = 4.47 \times 10^9$ y), and ^{232}Th ($\tau_{1/2} = 1.40 \times 10^{10}$ y) are present in Earth's core, and the energy released by the alpha decay of these isotopes helps to sustain our planet's high internal temperatures. All of Earth's geologic activity, including earthquakes, volcanic eruptions (Figure 27-1c), and the drifting of continents, is powered by the motions of our planet's interior. So alpha decay plays an important role in Earth's dynamic geology.

EXAMPLE 27-6 Alpha Decay of ^{238}U

The uranium isotope ^{238}U undergoes alpha decay to ^{234}Th. The binding energy per nucleon is 7.570 MeV in ^{238}U and 7.597 MeV in ^{234}Th. Find the energy released in the process $^{238}U \rightarrow {}^{234}Th + \alpha$.

Set Up

We'll use the same principle that we used in Example 27-4 (Section 27-4) to find the energy released in fission: The released energy equals the total binding energy of the nuclei present after the decay, minus the binding energy of the original (parent) nucleus. As in that example, we'll find the binding energy for each nucleus by multiplying the binding energy per nucleon times the number of nucleons A.

(energy released)

= (total binding energy of daughter plus alpha particle)
− (binding energy of original ^{238}U nucleus)

Solve

From Example 27-3 (Section 27-3) the binding energy of a ^4He nucleus (an alpha particle) is

$$E_B(^4\text{He}) = 28.297 \text{ MeV}$$

The binding energy of a ^{234}Th nucleus ($A = 234$) is

$$E_B(^{234}\text{Th}) = (234)(7.597 \text{ MeV}) = 1777.7 \text{ MeV}$$

and the binding energy of a ^{238}U nucleus ($A = 238$) is

$$E_B(^{238}\text{U}) = (238)(7.570 \text{ MeV}) = 1801.7 \text{ MeV}$$

The released energy is then

$$\begin{aligned} E_{\text{released}} &= E_B(^{234}\text{Th}) + E_B(^4\text{He}) - E_B(^{238}\text{U}) \\ &= 1777.7 \text{ MeV} + 28.297 \text{ MeV} - 1801.7 \text{ MeV} \\ &= 4.3 \text{ MeV} \end{aligned}$$

Reflect

The alpha particles emitted by radioactive isotopes with long half-lives, such as ^{238}U, tend to have kinetic energies in the 4 to 5 MeV range.

Beta Radiation

For most possible mass numbers A, there exist a number of nuclides with that same value of A. For example, molybdenum, technetium, ruthenium, and rhodium each have an isotope with 99 nucleons: ^{99}Mo has 42 protons and 57 neutrons, ^{99}Tc has 43 protons and 56 neutrons, ^{99}Ru has 44 protons and 55 neutrons, and ^{99}Rh has 45 protons and 54 neutrons. (It is also possible to create isotopes of other elements with A equal to 99.) The binding energy per nucleon in each is slightly different, however, so only one is the most stable: For $A = 99$, the most stable isotope is ^{99}Ru. If there were a process whereby ^{99}Tc could convert one of its neutrons into a proton, or ^{99}Rh could convert one of its protons into a neutron, either of these nuclei could transform into the more stable ^{99}Ru.

The process that makes this possible is called **beta decay**. There are actually two varieties of beta decay. In **beta-minus decay** a neutron (charge zero) changes into a proton (charge $+e$). The net charge cannot change, so an electron (charge $-e$), also known as a beta-minus (β^-) particle, is also produced and escapes from the nucleus. To account for other conservation requirements a third particle, the neutral and nearly massless antineutrino ($\bar{\nu}_e$), is also created in this process. The full process, then, is

$$\mathrm{n} \rightarrow \mathrm{p} + \mathrm{e}^- + \bar{\nu}_e \quad \text{(beta-minus decay)}$$

In **beta-plus decay** a proton (charge $+e$) changes into a neutron (charge zero). To conserve charge a positively charged electron or *positron*, also called a beta-plus (β^+) particle, is also produced and escapes from the nucleus, along with a neutral and nearly massless neutrino (which, for our purposes, is essentially the same particle as an antineutrino). This process is

$$\mathrm{p} \rightarrow \mathrm{n} + \mathrm{e}^+ + \nu_e \quad \text{(beta-plus decay)}$$

Figure 27-12a depicts the beta-minus decay of a nucleus with too many neutrons such as ^{99}Tc. We can write this process as

$$^{99}\mathrm{Tc} \rightarrow {}^{99}\mathrm{Ru} + \mathrm{e}^- + \bar{\nu}_e$$

The number of nucleons ($A = 99$) is the same before and after the decay, but the number of protons Z has increased by 1 (from 43 to 44) and the number of neutrons

(a) Nuclei with too many neutrons can undergo beta-minus (β^-) decay.

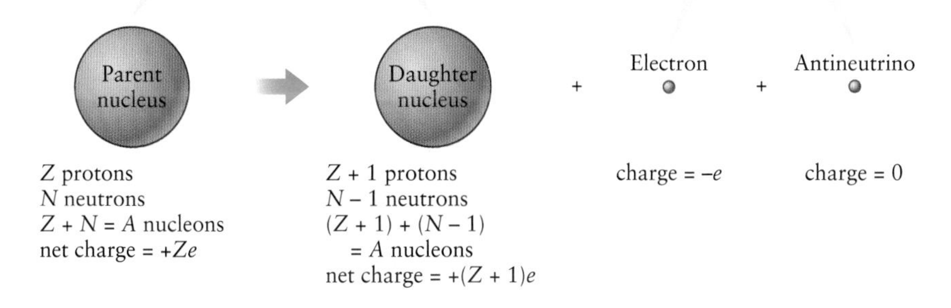

(b) Nuclei with too many protons can undergo beta-plus (β^+) decay.

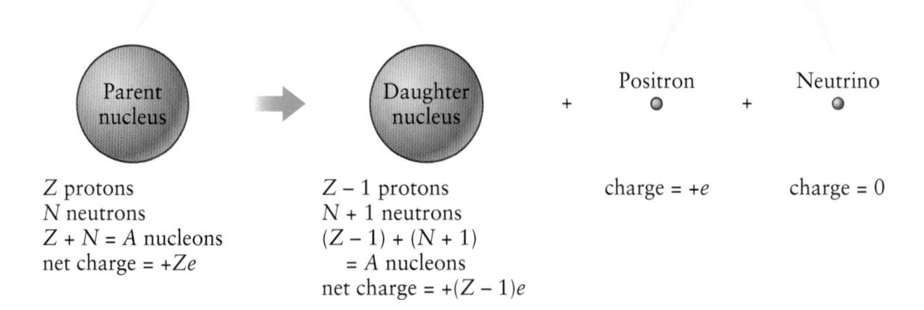

Figure 27-12 Beta decay Nuclei can increase their stability by (a) converting one neutron into a proton or (b) converting one proton into a neutron.

N has decreased by 1 (from 56 to 55). The decay of ^{137}Cs to ^{137}Ba, which we discussed at the beginning of this section, is another example of beta-minus decay. In this case Z increases from 55 to 56 and N decreases from 82 to 81.

One particularly important example of beta-plus decay is the decay of potassium-40 (^{40}K) to argon-40 (^{40}Ar). This process has a very long half-life of 1.25×10^9 y. Since potassium is abundant in Earth's interior, the energy released by this decay makes a substantial contribution to keeping our planet's interior in a fluid state and powering its geologic activity.

Figure 27-12b depicts the beta-plus decay of a nucleus with too many protons such as ^{99}Rh. We can write this process as

$$^{99}\text{Rh} \rightarrow {}^{99}\text{Ru} + \text{e}^{+} + \nu_e$$

As for beta-minus decay, the number of nucleons remains the same (in this case $A = 99$). But now the number of protons Z decreases by 1 (from 45 to 44), and the number of neutrons increases by 1 (from 54 to 55).

An important application of beta decay is in carbon-14 dating, a technique used to measure the age of objects that are composed, or partially composed, of organic matter. Almost all carbon atoms in Earth's atmosphere—for example, the carbon in carbon dioxide, CO_2—has a nucleus with a stable isotope of carbon, either ^{12}C (98.9%) or ^{13}C (1.1%). But about 1 in 10^{12} of those carbon atoms has a ^{14}C nucleus. This radioactive isotope of carbon is a β^- emitter with a half-life of 5730 y:

$$^{14}\text{C} \rightarrow {}^{14}\text{N} + \text{e}^{-} + \bar{\nu}_e$$

Carbon-14 is constantly produced in the atmosphere by cosmic rays slamming into ^{14}N nuclei. As a result, even though ^{14}C radioactively decays, the ratio of ^{14}C to ^{12}C in the atmosphere has remained relatively constant for at least tens of thousands of years. The ^{14}C/^{12}C ratio is the same in living organisms—for example, plants that breathe in CO_2—as it is in the atmosphere. However, once an organism dies, it no longer replenishes its supply of carbon, so the ^{14}C/^{12}C ratio decreases as the ^{14}C decays. As an example, a measurement of the ^{14}C/^{12}C ratio in, say, the smoke stains in the Chauvet-Pont-d'Arc Cave in southern France (see the image that opens this chapter), which contains the earliest known cave paintings, allows a determination of time since the firewood that created the smoke was part of a living tree. Carbon-14 dating tells us that the paintings were made between 30,000 and 33,000 years ago.

EXAMPLE 27-7 Ötzi the Iceman

In 1991, two German hikers discovered a human corpse in the Ötztal Alps on the border between Austria and Italy. The remains proved to be a well-preserved natural mummy of a man who lived during the last Ice Age. The rate of radioactive decay of ^{14}C in the mummy of "Ötzi the Iceman" was measured to be 0.121 Bq per gram (Bq/g). In a living organism, the rate of radioactive decay of ^{14}C is 0.231 Bq/g. How long ago did Ötzi the Iceman live?

Set Up

Once Ötzi died, his body stopped taking in ^{14}C. After that time the number $N(t)$ of ^{14}C nuclei in his body decreased due to beta-minus decay. The ^{14}C decay rate $R(t)$, which is proportional to $N(t)$, decreased in the same manner. We are given $R(t) = 0.121$ Bq/g for the present-day decay rate and $R_0 = 0.231$ Bq/g (the decay rate for a living organism and hence the decay rate at time $t = 0$, the last date on which Ötzi was still alive). We'll solve Equation 27-6 for the present time t (the elapsed time since Ötzi died). We'll also have to use Equation 27-8 to find the decay constant λ from the known half-life $\tau_{1/2} = 5730$ y of ^{14}C.

Decay rate as a function of time:

$$R(t) = R_0 e^{-\lambda t} \qquad (27\text{-}6)$$

Half-life of a radioactive substance:

$$\tau_{1/2} = \frac{\ln 2}{\lambda} \qquad (27\text{-}8)$$

Solve

Rearrange Equations 27-6 and 27-8 to find an expression for the time t since Ötzi died.

We know the present-day decay rate $R(t)$ and the initial decay rate in a living organism R_0. We want to find the time t since Ötzi died, so we rearrange Equation 27-6. Divide both sides by R_0:

$$\frac{R(t)}{R_0} = e^{-\lambda t}$$

Take the natural logarithm of both sides and recall that $\ln e^x = x$:

$$\ln\left(\frac{R(t)}{R_0}\right) = \ln e^{-\lambda t} = -\lambda t$$

Divide both sides by $-\lambda$:

$$t = -\frac{1}{\lambda}\ln\left(\frac{R(t)}{R_0}\right)$$

To get an expression for $1/\lambda$, divide both sides of Equation 27-8 by ln 2:

$$\frac{1}{\lambda} = \frac{\tau_{1/2}}{\ln 2}$$

Putting everything together, the time t since Ötzi died is

$$t = -\frac{\tau_{1/2}}{\ln 2}\ln\left(\frac{R(t)}{R_0}\right)$$

Substitute the given values into the expression for t.

We are given $\tau_{1/2} = 5730$ y for ^{14}C, $R(t) = 0.121$ Bq/g, and $R_0 = 0.231$ Bq/g:

$$t = -\frac{(5730\text{ y})}{\ln 2}\ln\left(\frac{0.121\text{ Bq/g}}{0.231\text{ Bq/g}}\right) = -\frac{(5730\text{ y})}{0.693}\ln 0.524$$

$$= -\frac{(5730\text{ y})}{0.693}(-0.647) = 5350\text{ y}$$

Reflect

Ötzi died 5350 y ago. His mummy thus gives us a unique look into life in prehistoric Europe.

Carbon-14 dating can be used only on objects less than about 50,000 years old, or about 8 to 10 half-lives of ^{14}C. For older objects the decay rate of ^{14}C has decreased to such a small value that it is hard to measure accurately, so any determination of age with this technique becomes difficult. For much older objects such as rocks, a similar approach is used but with isotopes with much longer half-lives. For example, the age of meteorites that fall to Earth is determined by looking at the ratio of uranium to lead (the endpoint of a series of alpha decays that starts with uranium); the oldest meteorites are more than 4.5×10^9 years old.

Gamma Radiation

A nucleus in an excited state radiates energy in a way analogous to the emission of a photon when an electron in an excited atomic state falls to a state of lower energy (see Section 26-5). Just like atoms, nuclei have excited states of definite energy, and when they transition from an excited state to a less excited one, they will emit a photon of energy equal to the difference in energy between the initial and final states.

The most common way that a nucleus can become excited is following an alpha decay or a beta decay. Although these decay processes result in a more stable configuration of the nucleons, the nucleons that remain in the daughter nucleus may not be, initially, in the most stable arrangement for that particular nuclide. This

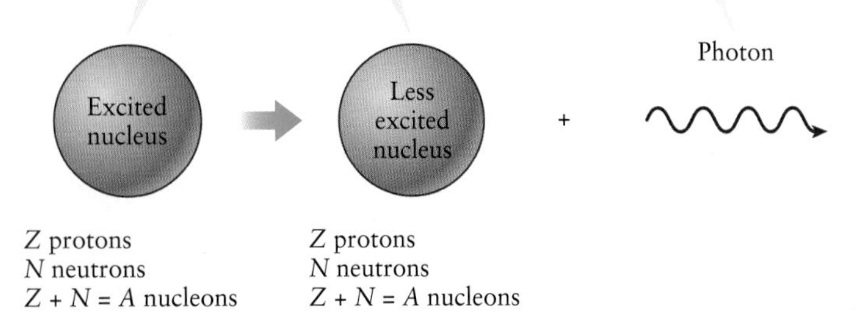

In gamma decay there is no change in the number of protons or the number of neutrons in the nucleus.

Figure 27-13 Gamma decay An excited nucleus can lower its energy by emitting a photon.

excited daughter nucleus decays to a more stable configuration, giving off energy in the form of a gamma (γ) ray (**Figure 27-13**). This process is called **gamma decay.**

The energy carried away when a nucleus in an excited state decays is on the order of 1 MeV. From Equation 22-26 in Section 22-4, $E = hf$, the frequency and wavelength of a photon of energy $E = 1.00$ MeV $= 1.00 \times 10^6$ eV are

$$f = \frac{E}{h} = \frac{1.00 \times 10^6 \text{ eV}}{4.14 \times 10^{-15} \text{ eV} \cdot \text{s}} = 2.42 \times 10^{20} \text{ Hz}$$

$$\lambda = \frac{c}{f} = \frac{3.00 \times 10^8 \text{ m/s}}{2.42 \times 10^{20} \text{ Hz}}$$

$$= 1.24 \times 10^{-12} \text{ m} = 1.24 \times 10^{-3} \text{ nm}$$

Recall that the wavelength of visible light photons is in the range from 380 to 750 nm; a photon emitted by an excited nucleus is in the gamma radiation range, far from the visible part of the spectrum.

Gamma radiation does not change the atomic number Z or the neutron number N of a nucleus; that is, the number of protons and the number of neutrons remain the same after a γ ray is emitted. As an example, earlier we considered the beta-minus decay of ^{137}Cs to ^{137}Ba. In 95% of those decays the ^{137}Ba nucleus is formed in an excited state that we denote as ^{137}Ba*:

$$^{137}\text{Cs} \rightarrow {}^{137}\text{Ba}^* + \text{e}^- + \bar{\nu}_e$$

The excited barium nucleus then decays to its ground state by emission of a photon of energy 0.662 MeV:

$$^{137}\text{Ba}^* \rightarrow {}^{137}\text{Ba} + \gamma$$

The values of $Z = 56$ and $N = 81$ for the barium nucleus do not change in this second step of the radiation process.

GOT THE CONCEPT? 27-5 Half-Lives

? A certain radioactive isotope has a half-life of 5 days. You are given a sample containing a number of nuclei of this isotope. About how long would you have to wait before about 1/1000 of the initial number of nuclei remained? (a) 20 days; (b) 40 days; (c) 50 days; (d) 100 days; (e) 1000 days.

TAKE-HOME MESSAGE FOR Section 27-6

✔ Radioactive decay is a statistical process. The number of nuclei and the decay rate both decrease exponentially with time, and both decrease by one-half in a time equal to one half-life.

✔ The three most common modes by which radiation occurs are alpha, beta, and gamma radiation.

✔ Alpha particles are ^{4}He nuclei and are emitted by large nuclei with $Z > 82$. In alpha emission the proton number and neutron number each decrease by 2.

✔ Beta particles are either negatively charged electrons or positively charged positrons. In beta-minus emission a neutron in the nucleus changes into a proton; in beta-plus emission a proton in the nucleus changes into a neutron.

✔ Gamma particles are high-energy photons, typically with energies of about 1 MeV. They are emitted when a nucleus decays from an excited state to a less excited one.

Key Terms

alpha decay
alpha particle
antineutrino
atomic number
becquerel
beta decay
beta-minus decay
beta-plus decay
binding energy
decay constant
gamma decay
half-life
isotope
magnetic resonance imaging
mass number
neutrino
neutron-induced fission
neutron number
nuclear fission
nuclear fusion
nuclear radiation
nucleon
nuclide
positron
radioactive
strong nuclear force

Chapter Summary

Topic	Equation or Figure
The strong nuclear force and nuclear sizes: The strong nuclear force is a short-range attractive force that acts between nucleons (protons or neutrons). Because this force has a short range, the volume of a nucleus is proportional to the mass number (number of nucleons). In larger nuclei, the number of neutrons must exceed the number of protons in order to counterbalance the electric repulsion between protons.	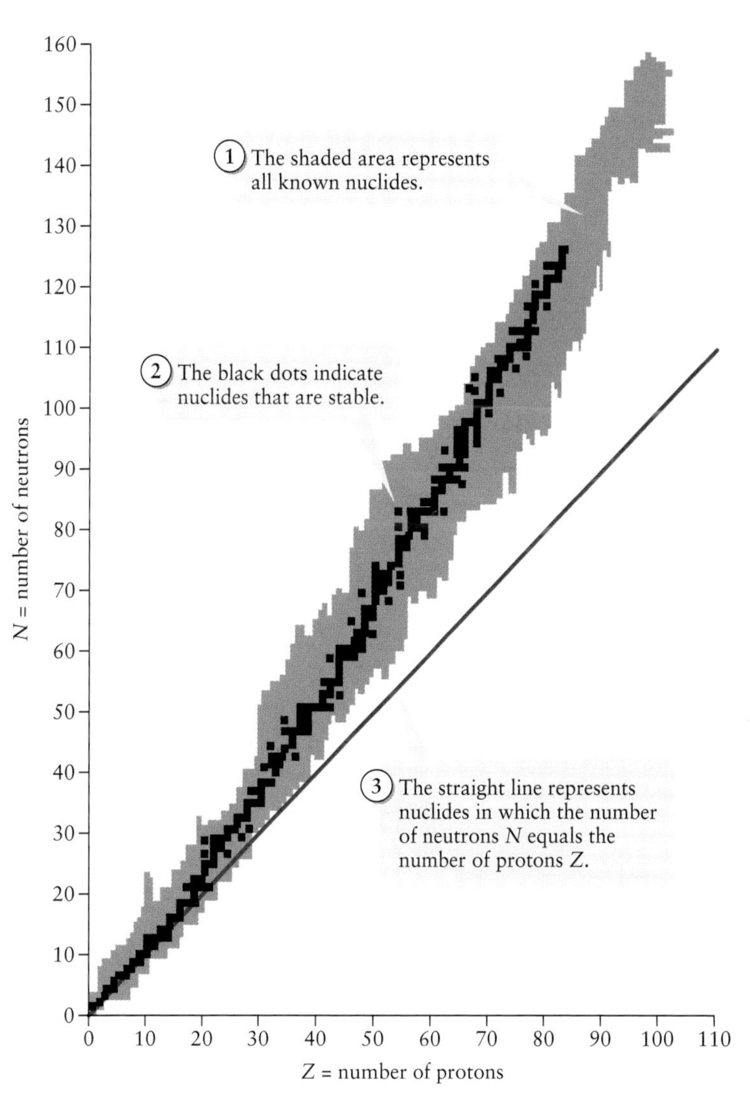 (Figure 27-2)

For most nuclides, more neutrons than protons must be present (N must be greater than Z) to prevent the protons from flying apart due to electrostatic repulsion.

Nuclear binding energy: The binding energy of a nucleus is the energy required to separate it into its constituent nucleons. The binding energy per nucleon is greatest for nuclei with around 60 nucleons; for larger nuclei the electric repulsion between protons makes nuclei less stable. The binding energy of a particular isotope can be calculated from the mass of a neutral atom containing that isotope.

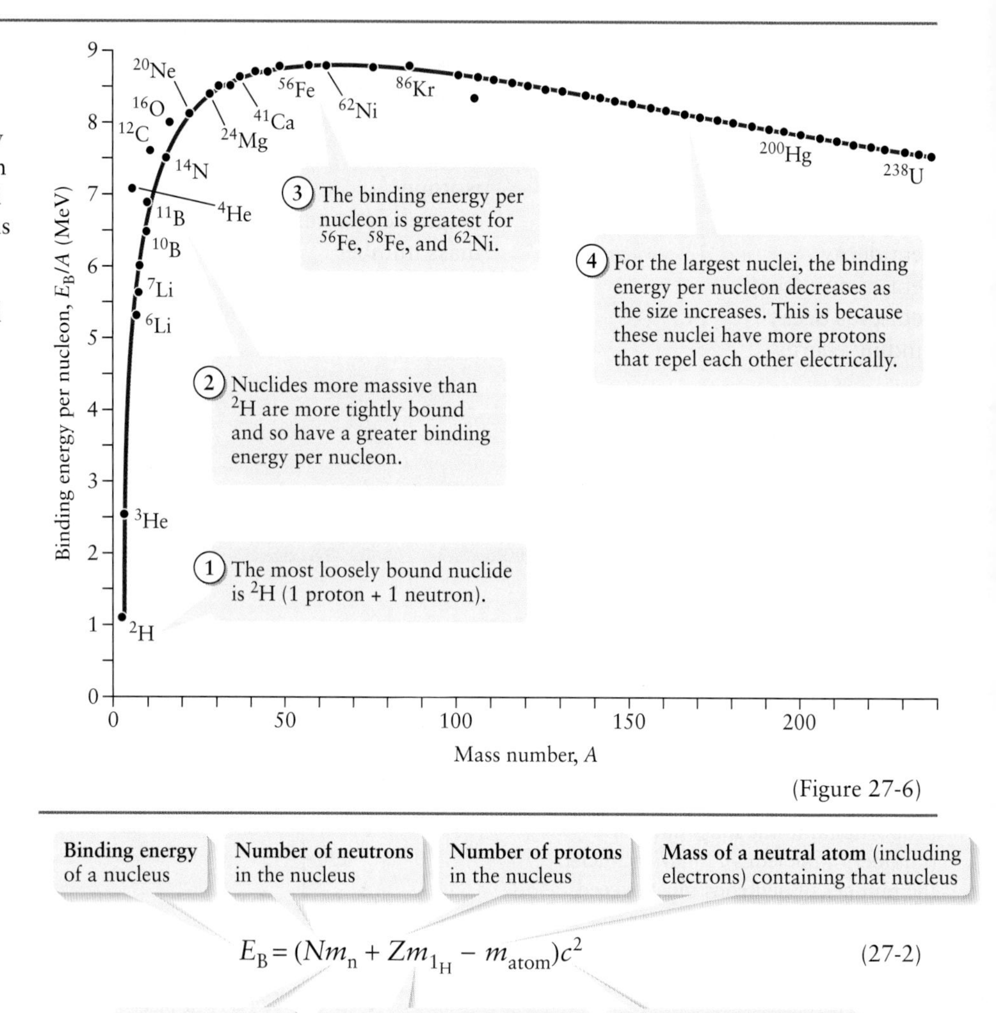

Binding energy per nucleon, E_B/A (MeV)
Mass number, A
20Ne
16O
12C
14N
11B
10B
7Li
6Li
4He
3He
2H
56Fe
62Ni
41Ca
24Mg
86Kr
200Hg
238U
(3) The binding energy per nucleon is greatest for 56Fe, 58Fe, and 62Ni.
(4) For the largest nuclei, the binding energy per nucleon decreases as the size increases. This is because these nuclei have more protons that repel each other electrically.
(2) Nuclides more massive than 2H are more tightly bound and so have a greater binding energy per nucleon.
(1) The most loosely bound nuclide is 2H (1 proton + 1 neutron).

(Figure 27-6)

Binding energy of a nucleus | **Number of neutrons** in the nucleus | **Number of protons** in the nucleus | **Mass of a neutral atom** (including electrons) containing that nucleus

$$E_B = (Nm_n + Zm_{^1H} - m_{atom})c^2 \quad (27\text{-}2)$$

Mass of a neutron | **Mass of a neutral hydrogen atom** (1 proton + 1 electron) | **Speed of light in a vacuum**

Nuclear fission: When the largest nuclei absorb a neutron, they fragment (fission) into two smaller nuclei plus a few neutrons. The fragments are more tightly bound than the original nucleus, so energy is released in this process. In a sustained fission reaction the released neutrons trigger other, nearby nuclei to also undergo fission.

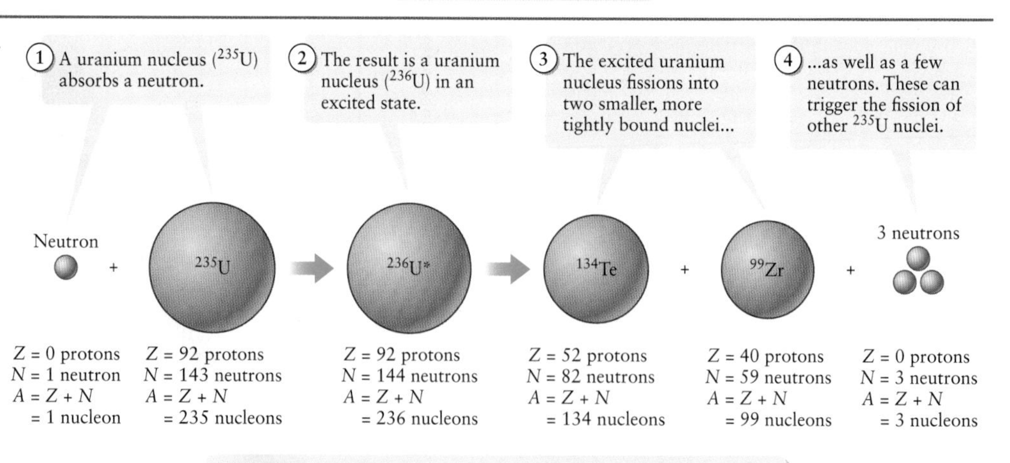

(1) A uranium nucleus (235U) absorbs a neutron.
(2) The result is a uranium nucleus (236U) in an excited state.
(3) The excited uranium nucleus fissions into two smaller, more tightly bound nuclei...
(4) ...as well as a few neutrons. These can trigger the fission of other 235U nuclei.
Neutron
235U
236U*
134Te
99Zr
3 neutrons
Z = 0 protons N = 1 neutron A = Z + N = 1 nucleon
Z = 92 protons N = 143 neutrons A = Z + N = 235 nucleons
Z = 92 protons N = 144 neutrons A = Z + N = 236 nucleons
Z = 52 protons N = 82 neutrons A = Z + N = 134 nucleons
Z = 40 protons N = 59 neutrons A = Z + N = 99 nucleons
Z = 0 protons N = 3 neutrons A = Z + N = 3 nucleons

Energy is released in this fission reaction: The total kinetic energy of the fission fragments is much greater than the total kinetic energy of the initial neutron and ^{235}U nucleus.

(Figure 27-7)

Nuclear fusion: The smallest nuclei can merge together to form a larger nucleus, releasing energy in the process. These fusion processes require very high temperatures so that the fusing nuclei have enough kinetic energy to overcome their mutual electric repulsion.

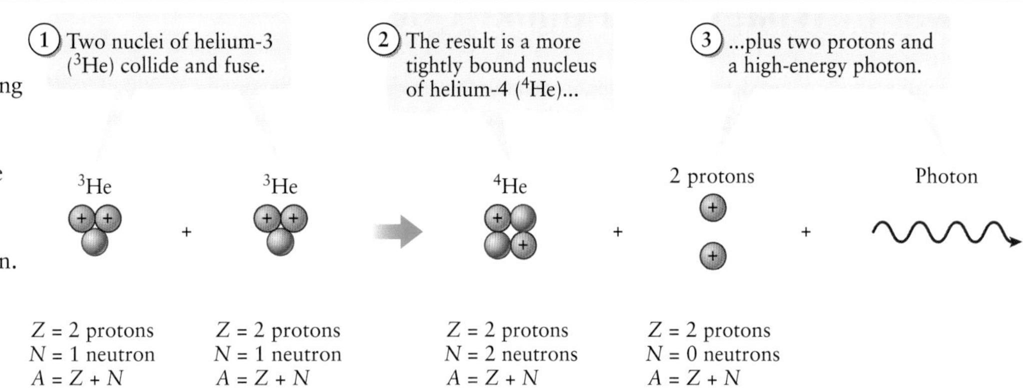

Energy is released in this fusion reaction: The total kinetic energy of the ^{4}He nucleus and protons, plus the energy of the photon, is much greater than the total kinetic energy of the initial ^{3}He nuclei.

(Figure 27-8)

Nuclear decay and half-life: The decay of unstable nuclei is a statistical process. This means that the rate of decay is proportional to the number of unstable nuclei present. As a result the number of nuclei and the decay rate both decline in an exponential manner. The time for the number of nuclei and the decay rate to decrease by one-half is called the half-life.

- The half-life of a nuclear decay is the time required for one-half of the nuclei present initially to decay.
- Radioactive decay is a statistical process: It's impossible to predict when any one individual unstable nucleus will decay.

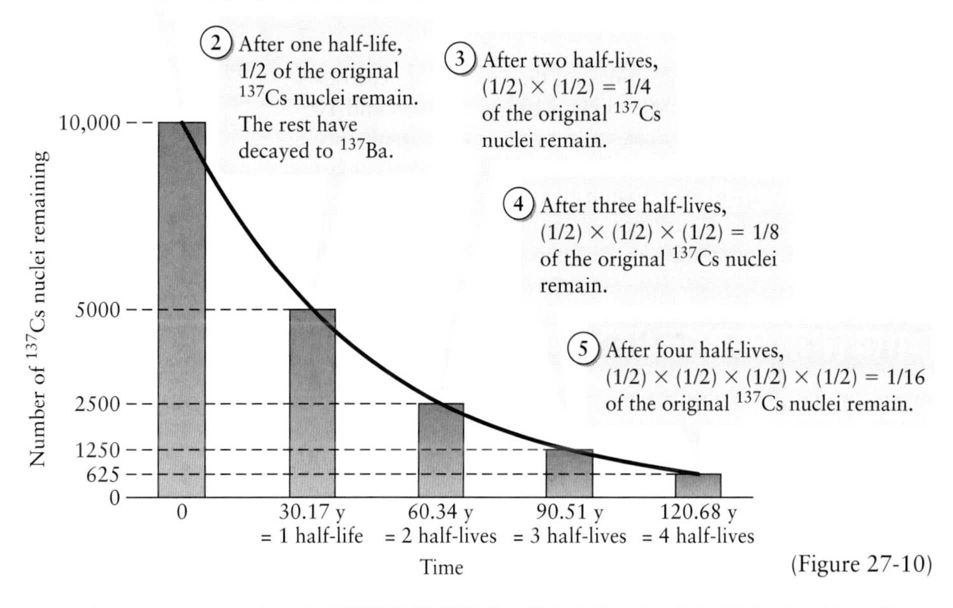

(Figure 27-10)

Alpha, beta, and gamma decays: Large nuclei release energy and become more stable by emitting an alpha particle (a ^{4}He nucleus). Other nuclei with too many neutrons undergo beta-minus decay, in which one neutron changes into a proton; those with too many protons undergo beta-plus decay, in which one proton changes into a neutron. In gamma decay an excited state of a nucleus (indicated by an asterisk) transitions to a less excited state and emits a gamma-ray photon (γ).

An alpha decay:

$$^{228}\text{Th} \rightarrow {}^{224}\text{Ra} + \alpha$$

A beta-minus decay:

$$^{99}\text{Tc} \rightarrow {}^{99}\text{Ru} + e^- + \bar{\nu}_e$$

A beta-plus decay:

$$^{99}\text{Rh} \rightarrow {}^{99}\text{Ru} + e^+ + \nu_e$$

A gamma decay:

$$^{137}\text{Ba}^* \rightarrow {}^{137}\text{Ba} + \gamma$$

Answer to What do you think? Question

(e) In a radioactive material, the amount that remains decreases by 1/2 after one half-life. After two half-lives, the amount that remains is $(1/2) \times (1/2) = (1/2)^2 = 1/4$; after three half-lives, $(1/2) \times (1/2) \times (1/2) = (1/2)^3 = 1/8$; and so on. The age of the cave painting is between (30,000 y)/(5730 y) = 5.2 half-lives and (33,000 y)/(5730 y) = 5.8 half-lives, so more than 5 half-lives have elapsed since the painting was made. Therefore the amount of carbon-14 remaining is less than $(1/2) \times (1/2) \times (1/2) \times (1/2) \times (1/2) = (1/2)^5 = 1/32$ of the original.

Answers to Got the Concept? Questions

27-1 (c) Equation 27-1 tells us that the radius of a nucleus is proportional to $A^{1/3}$, the cube root of the number of nucleons (the mass number). Compared to a ^{20}Ne nucleus, a ^{160}Dy nucleus has 160/20 = 8 times as many nucleons, so its radius is larger by a factor of $8^{1/3} = 2$. As we discussed in Example 27-2, nuclei with any value of A have the same density.

27-2 (c), (b), (d), (a) A ranking in order of stability is a ranking in order of binding energy per nucleon. Figure 27-6 shows that this is greatest for ^{86}Kr (more than 8.5 MeV), second greatest for ^{20}Ne (slightly more than 8 MeV), third greatest for ^{200}Hg (slightly less than 8 MeV), and least for ^{11}B (about 6.9 MeV).

27-3 (d) The number of protons and the number of neutrons both remain the same: A ^{20}Ne nucleus has 10 protons and 10 neutrons, and each ^{10}B nucleus has 5 protons and 5 neutrons. But the binding energy per nucleon for ^{10}B is *less* than for ^{20}Ne (see Figure 27-6). For fission to occur and energy to be released, the fragments must have a greater binding energy per nucleon than the initial nucleus. The process $^{20}\text{Ne} \rightarrow {}^{10}\text{B} + {}^{10}\text{B}$ would require a substantial amount of energy to be spontaneously added to the ^{20}Ne nucleus, and there's no way that can happen.

27-4 (d) In order for two ^{1}H nuclei to fuse, they must come to within a few femtometers (1 fm = 10^{-15} m) of each other. But in a typical molecule, including H_2, the separation between nuclei in adjacent atoms is on the order of 10^{-10} m. So the nuclei are normally too far apart to fuse. If they were moving rapidly with respect to each other and so had a large kinetic energy, the nuclei could overcome their mutual Coulomb repulsion and come close enough to fuse. But atomic nuclei move hardly at all within molecules, so fusion cannot happen this way. Option (c) is incorrect: The properties of a nucleus are affected not at all by the presence or absence of electrons orbiting the nucleus.

27-5 (c) After each 5-day half-life, the number of nuclei decreases by a factor of 1/2. For the number to decrease by a factor of 1/1000 requires about 10 half-lives, since $(1/2)^{10} = 1/2^{10} = 1/1024$.

Questions and Problems

In a few problems you are given more data than you actually need; in a few other problems you are required to supply data from your general knowledge, outside sources, or informed estimate.

Interpret as significant all digits in numerical values that have trailing zeros and no decimal points.

For all problems use $g = 9.80$ m/s^2 for the free-fall acceleration due to gravity. Neglect friction and air resistance unless instructed to do otherwise.

•	Basic, single-concept problem
••	Intermediate-level problem; may require synthesis of concepts and multiple steps
•••	Challenging problem
SSM	*Solution is in Student Solutions Manual*
Example	*See worked example for a similar problem.*

Conceptual Questions

1. • What is an isotope?

2. • What is the difference between atomic number and mass number?

3. • (a) Describe what is meant by the phrase "larger nuclei are neutron rich." (b) Why do most nuclei contain at least as many neutrons as protons?

4. • Some historians would claim that without Einstein's special theory of relativity, nuclear physics would never have developed. Explain why. SSM

5. • A simple idea of nuclear physics can be stated as follows: "The whole nucleus weighs less than the sum of its parts." Explain why.

6. • Describe two characteristics of the binding energy that are comparable to the work function (from the photoelectric effect) and two characteristics that are dissimilar to the concept of the work function.

7. • Describe the basic characteristics of the nuclear force that exists between nucleons. What other competing force between nucleons is present in the nucleus?

8. • What is the difference between fission and fusion?

9. • (a) Which elements in the periodic table are more likely to undergo nuclear fission? (b) Which are more likely to undergo nuclear fusion?

10. • **Astronomy** (a) Describe the nuclear reactions that occur in our Sun. (b) Discuss how the equilibrium state of the Sun is not permanent and discuss the eventual future of our solar system.

11. • Explain how conservation of energy and momentum would be violated if a neutrino were not emitted in beta decay. SSM

12. • The decay constant of a radioactive nucleus is just that, *constant*. It does not depend on the size of the nuclear sample, the temperature, or any external fields (such as gravity,

electricity, or magnetism). Define the decay constant and comment on how nuclear radioactivity would change if the quantity were dependent on temperature.

13. • At any given instant a sample of radioactive uranium contains many, many different isotopes of atoms that are *not* uranium. Explain why.

14. • (a) Explain how radioactive ^{14}C is used to determine the age of ancient artifacts. (b) Which types of artifacts can have their age determined in this way and which types cannot?

15. • If atomic masses are used, explain why the mass of a beta particle is *not* accounted for in the basic beta decay

$$n \rightarrow p + e^- + \bar{\nu}_e$$

Assume that the mass of the antineutrino ($\bar{\nu}_e$) is very small and can be neglected. SSM

16. • **Medical** Describe, in broad terms, the health risks associated with the three major forms of radioactivity: alpha, beta, and gamma. Focus on the dangers due to inherent health risks and the ability of each to penetrate shielding material.

Multiple-Choice Questions

17. • In an atomic nucleus, the nuclear force binds ______ together.
   A. electrons
   B. neutrons
   C. protons
   D. neutrons and protons
   E. neutrons, protons, and electrons

18. • The mass of a nucleus is ____________ the sum of the masses of its nucleons.
   A. always less than
   B. sometimes less than
   C. always more than
   D. always equal to
   E. sometimes equal to

19. • Which of the following statements is true?
   A. Fusion absorbs energy and fission releases energy.
   B. Fusion releases energy and fission absorbs energy.
   C. Both fusion and fission absorb energy.
   D. Both fusion and fission release energy.
   E. Both fusion and fission can release or absorb energy. SSM

20. • In fission processes, which of the following statements is true?
   A. Only the total number of nuclei remains the same.
   B. Only the total number of protons remains the same.
   C. Only the total number of neutrons remains the same.
   D. The total number of protons and the total number of nuclei both remain the same.
   E. The total number of protons and the total number of neutrons both remain the same.

21. • In a spontaneous fission reaction the total mass of the products is __________ the mass of the original element.
   A. greater than
   B. less than
   C. the same as
   D. double
   E. one-half

22. • What is the source of the Sun's energy?
   A. chemical reactions
   B. fission reactions
   C. fusion reactions
   D. gravitational collapse
   E. both fusion reactions and fission reactions

23. • In a fusion reaction the total mass of the products is __________ the mass of the original elements.
   A. greater than
   B. less than
   C. the same as
   D. double
   E. one-half SSM

24. • The decay constant λ depends only on
   A. the number of atoms at the initial time.
   B. the initial decay rate.
   C. the half-life.
   D. the binding energy per nucleon.
   E. whether the decay is alpha, beta, or gamma.

25. • The number of radioactive atoms in a radioactive sample
   A. decreases linearly with time.
   B. increases linearly with time.
   C. decreases exponentially with time.
   D. increases exponentially with time.
   E. remains constant.

26. • The decay rate for a sample of any isotope
   A. decreases linearly with time.
   B. increases linearly with time.
   C. decreases exponentially with time.
   D. increases exponentially with time.
   E. remains constant.

Estimation/Numerical Analysis

27. • Estimate the relative size of the nuclear force compared to the electrostatic force between two adjacent protons in the nucleus.

28. • Estimate the density of an atomic nucleus compared to the density of an atom. SSM

29. • Estimate the energy released in a typical nuclear reaction compared to that in a typical chemical reaction.

30. • Estimate the ratio of the nuclear force between two nucleons when they are separated by a distance of 1.0 fm compared to a separation distance of 2.0 fm.

31. • Estimate the number of half-lives that must go by before 10% of an isotope remains. What about 1%?

32. • Estimate the mass of an object the size of a basketball that has the density of an atomic nucleus.

33. • Estimate the number of nuclei that are present in a 50-kg human body.

34. • If the half-life of a radioactive isotope is 1 day, (a) estimate how long it takes before the sample is reduced to 62.5% of its original amount and (b) how long before it is reduced to 6.25% of its original amount. SSM

35. • (a) Using a spreadsheet or programmable calculator, calculate the binding energy per nucleon for the following isotopes of the five least massive elements. Masses given are atomic masses.

hydrogen-1: 1.007825 u	lithium-8: 8.022486 u	boron-8: 8.024605 u
hydrogen-2: 2.014102 u	lithium-9: 9.026789 u	boron-10: 10.012936 u
hydrogen-3: 3.016049 u	lithium-11: 11.043897 u	boron-11: 11.009305 u
helium-3: 3.016029 u	beryllium-7: 7.016928 u	boron-12: 12.014352 u
helium-4: 4.002602 u	beryllium-9: 9.012174 u	boron-13: 13.017780 u
helium-6: 6.018886 u	beryllium-10: 10.013534	boron-14: 14.025404 u
helium-8: 8.033922 u	beryllium-11: 11.021657 u	boron-15: 15.031100 u
lithium-6: 6.015121 u	beryllium-12: 12.026921 u	
lithium-7: 7.016003 u	beryllium-14: 14.024866 u	

(b) Now calculate the binding energy per nucleon for the following isotopes of the five most massive naturally occurring elements. The masses provided are atomic masses.

radium-221: 221.01391 u	thorium-228: 228.028716 u	uranium-231: 231.036264 u
radium-223: 223.018499 u	thorium-229: 229.031757 u	uranium-232: 232.037131 u
radium-224: 224.020187 u	thorium-230: 230.033127 u	uranium-233: 233.039630 u
radium-226: 226.025402 u	thorium-231: 231.036299 u	uranium-234: 234.040946 u
radium-228: 228.031064 u	thorium-232: 232.038051 u	uranium-235: 235.043924 u
actinium-227: 227.027749 u	thorium-234: 234.043593 u	uranium-236: 236.045562 u
actinium-228: 228.031015 u	protactinium-231: 231.035880 u	uranium-238: 238.050784 u
thorium-227: 227.027701 u	protactinium-234: 234.043300 u	uranium-239: 239.054290 u

(c) Compare your results and comment on any patterns or trends that are obvious.

Problems

Note: In all cases, unless otherwise stated, binding energy, energy released, and the like are to be expressed in MeV. Also, all atomic masses should be calculated to the nearest 10^{-6}.

27-1 The quantum concepts that help explain atoms are essential for understanding the nucleus

27-2 The strong nuclear force holds nuclei together

36. • Provide the elemental abbreviation (e.g., ^{16}O for oxygen-16) and give the number of protons, the number of neutrons, and the mass number for each of the following isotopes:

A. hydrogen-3 C. aluminum-26
B. beryllium-8 D. gold-197
E. technetium-100 G. osmium-190
F. tungsten-184 H. plutonium-239

37. • Calculate the radius of each of the nuclei in problem 36. Example 27-1

38. • Name the element and give the number of protons, the number of neutrons, and the mass number for each of the following nuclei:

A. ^{2}H D. ^{12}C G. ^{131}I
B. ^{4}He E. ^{56}Fe H. ^{235}U
C. ^{6}Li F. ^{90}Sr

39. • If our Sun (mass = 1.99×10^{30} kg, radius = 6.96×10^{8} m) were to collapse into a neutron star (an object composed of tightly packed neutrons with roughly the same density as a nucleus), what would the new radius of our "neutron-sun" be? Example 27-2

40. •• Given that a nucleus is approximately spherical and has a radius $r = r_0 A^{1/3}$ (where r_0 is about 1.2 fm), determine its approximate mass density. Express your answer in SI units and convert to tons per cubic inch, units that might be used in a news report. SSM Example 27-2

27-3 Some nuclei are more tightly bound and more stable than others

41. • Calculate the atomic mass of each of the isotopes listed below. Give your answer in atomic mass units (u) and in grams (g). The values will include the mass of Z electrons. Example 27-3

A. ^{1}H D. ^{12}C G. ^{131}I
B. ^{4}He E. ^{56}Fe H. ^{238}U
C. ^{9}Be F. ^{90}Sr

42. • What is the binding energy of carbon-12? Give your answer in MeV. SSM Example 27-3

43. • What is the binding energy per nucleon for the following isotopes? Example 27-3

A. ^{2}H D. ^{12}C G. ^{129}I
B. ^{4}He E. ^{56}Fe H. ^{235}U
C. ^{6}Li F. ^{90}Sr

44. • What minimum energy is needed to remove a neutron from ^{40}Ca and so convert it to ^{39}Ca? The atomic masses of the two isotopes are 39.96259098 and 38.97071972 u, respectively. Example 27-3

45. • What is the binding energy of the last neutron of carbon-13? The atomic mass of carbon-13 is 13.003355 u. Example 27-3

46. • **Medical** Iodine-131 is a radioactive isotope that is used in the treatment of cancer of the thyroid. The natural tendency of the thyroid to take up iodine creates a pathway for which radiation (β^- and γ) that is emitted from this unstable nucleus can be directed onto the cancerous tumor with very little collateral damage to surrounding healthy tissue. Another advantage of the isotope is its relatively short half-life (8 days). Calculate the binding energy of iodine-131 and the binding energy per nucleon. The mass of iodine-131 is 130.906124 u. Example 27-3

27-4 The largest nuclei can release energy by undergoing fission and splitting apart

47. • Calculate the energy released in the following nuclear fission reaction:

$$^{239}\text{Pu} + \text{n} \rightarrow {}^{98}\text{Tc} + {}^{138}\text{Sb} + 4\text{n}$$

The atomic masses are ^{239}Pu = 239.052157 u, ^{98}Tc = 97.907215 u, and ^{138}Sb = 137.940793 u. Example 27-4

48. • Complete the following nuclear fission reaction of thorium-232 and calculate the energy released in the reaction: $^{232}Th + n \rightarrow {}^{99}Kr + {}^{124}Xe +$ _? The atomic masses are $^{232}Th = 232.038051$ u, $^{99}Kr = 98.957606$ u, and $^{124}Xe = 123.905894$ u. SSM Example 27-4

49. • Complete the following fission reactions: Example 27-4

A. $^{235}U + n \rightarrow {}^{128}Sb + {}^{101}Nb +$ _?_

B. $^{235}U + n \rightarrow$ _?_ $+ {}^{116}Pd + 4n$

C. $^{238}U + n \rightarrow {}^{99}Kr +$ _?_ $+11n$

D. _?_ $+ n \rightarrow {}^{101}Rb + {}^{130}Cs + 8n$

50. • Complete the following fission reactions: Example 27-4

A. $^{242}Am +$ _?_ $\rightarrow {}^{90}Sr + {}^{149}La + 4n$

B. $^{244}Pa + n \rightarrow$ _?_ $+ {}^{131}Sb + 12n$

C. _?_ $+ n \rightarrow {}^{92}Se + {}^{153}Sm + 6n$

D. $^{262}Fm + n \rightarrow {}^{112}Rh +$ _?_ $+ 9n$

51. • Calculate the energy (in MeV) released in the following nuclear fission reaction:

$$^{242}Am + \_?\_ \rightarrow {}^{90}Sr + {}^{149}La + 4n$$

Start by completing the reaction and use the following nuclear masses: $^{242}Am = 242.059549$ u, $^{90}Sr = 89.9077387$ u, and $^{149}La = 148.934733$ u. Example 27-4

52. •• Assuming that in a fission reactor a neutron loses half its energy in each collision with an atom of the moderator, determine how many collisions are required to slow a 200-MeV neutron to an energy of 0.04 eV. Example 27-4

53. • Knowing that the binding energy per nucleon for uranium-235 is about 7.6 MeV/nucleon and the binding energy per nucleon for typical fission fragments is about 8.5 MeV/nucleon, find an average energy release per uranium-235 fission reaction in MeV. Example 27-4

54. •• How many kilograms of uranium-235 must completely fission to produce 1000 MW of power continuously for one year? SSM Example 27-4

55. • Repeat problem 54 in the more realistic case where the fission reactions are about 30% efficient in producing 1000 MW of power over 1 year of continuous operation. Example 27-4

56. • Calculate the number of fission reactions per second that take place in a 1000-MW reactor. Assume that 200 MeV of energy is released in each reaction. Example 27-4

27-5 The smallest nuclei can release energy if they are forced to fuse together

57. • Complete the following fusion reactions: Example 27-4

A. $^{2}H + {}^{3}H \rightarrow {}^{4}He +$ _?_

B. $^{4}He + {}^{4}He \rightarrow {}^{7}Be +$ _?_

C. $^{2}H + {}^{2}H \rightarrow {}^{3}He +$ _?_

D. $^{2}H + {}^{1}H \rightarrow \gamma +$ _?_

E. $^{2}H + {}^{2}H \rightarrow {}^{3}H +$ _?_

58. •• Calculate the energy released in each of the fusion reactions in problem 58. Give your answers in MeV. Example 27-4

59. •• **Astronomy** Consider the proton–proton cycle that occurs in most stars (including our own Sun):

Step 1: $^{1}H + {}^{1}H \rightarrow {}^{2}H + e^{+} + \nu_{e}$

Step 2: $^{2}H + {}^{1}H \rightarrow {}^{3}He + \gamma$

Step 3: $^{3}He + {}^{3}He \rightarrow {}^{4}He + 2\,{}^{1}H + \gamma$

Calculate the net energy released from the three steps. Do *not* ignore the mass of the positron in step 1. (You may ignore the mass of the neutrino.) Example 27-4

60. •• Each fusion reaction of deuterium (^{2}H) and tritium (^{3}H) releases about 20 MeV. What mass of tritium is needed to create 10^{14} J of energy, the same as that released by exploding 25,000 tons of TNT? Assume that an endless supply of deuterium is available. SSM Example 27-4

61. •• How many fusion reactions per second must be sustained to operate a deuterium–tritium fusion power plant that outputs 1000 MW, operating at 33% efficiency? Example 27-4

27-6 Unstable nuclei may emit alpha, beta, or gamma radiation

62. • Complete the following conversions:

A. 100 μCi = __________ Bq

B. 1500 decays/min = __________ Bq

C. 16,500 Bq = __________ Ci

D. 7.55×10^{10} Bq = __________ decays/min

63. • The curie unit is defined as 1 Ci = 3.7×10^{10} Bq, which is about the rate at which radiation is emitted by 1.00 g of radium. (a) Calculate the half-life of radium from the definition. (b) What does your calculation tell you about the radiation emission rate of radium? Example 27-5

64. • A certain radioactive isotope has a decay constant of 0.00334 s^{-1}. Find the half-life in seconds and days. SSM Example 27-5

65. •• A radioactive sample is monitored with a radiation detector which registers 5640 counts per minute. Twelve hours later, the detector reads 1410 counts per minute. Calculate the decay constant and the half-life of the sample. Example 27-5

66. • What fraction of a sample of ^{32}P will be left after 4 months? Its half-life is 14.3 days. Example 27-5

67. • What fraction of a radioactive sample will be left after 6 half-lives? What about 7.5 half-lives? Example 27-5

68. •• **Medical** A patient is injected with 7.88 μCi of radioactive iodine-131 that has a half-life of 8.02 days. Assuming that 90% of the iodine ultimately finds its way to the thyroid, what decay rate do you expect to find in the thyroid after 30 days? SSM Example 27-5

69. •• The ratio of carbon-14 to carbon-12 in living wood is 1.3×10^{-12}. How many decays per second are there in 550 g of wood? Example 27-7

70. •• You take a course in archaeology that includes field work. An ancient wooden totem pole is excavated from your archaeological dig. The beta decay rate is measured at 150 decays/min. If the totem pole contains 225 g of carbon and the ratio of carbon-14 to carbon-12 in living trees is 1.3×10^{-12}, what is the age of the pole? Example 27-7

71. •• Determine the decay rate for 500 g of carbon from a tree limb twelve centuries after it is cut off. Example 27-7

72. •• How many nuclei of radon-222 are present in a sample for which you measure 485 decays/min? SSM Example 27-5

73. •• The ages of rocks that contain fossils can be determined using the isotope ^{87}Rb. This isotope of rubidium undergoes beta decay with a half-life of 4.75×10^{10} y. Ancient samples contain a ratio of ^{87}Sr to ^{87}Rb of 0.0225. Given that ^{87}Sr is a stable product of the beta decay of ^{87}Rb, and there was

originally no ^{87}Sr present in the rocks, calculate the age of the rock sample. Assume that the decay rate is constant over the relatively short lifetime of the rock compared to the half-life of ^{87}Rb. Example 27-7

74. • Complete the following alpha decays: Example 27-6
A. $^{238}U \rightarrow \alpha + \_?\_$
B. $^{234}Th \rightarrow \_?\_ + {}^{?}_{-}Ra$
C. $\_?\_ \rightarrow \alpha + {}^{236}U$
D. $^{214}Bi \rightarrow \alpha + \_?\_$

75. • Complete the following beta decays:
A. $^{14}C \rightarrow e^- + \bar{\nu}_e + \_?\_$
B. $^{239}Np \rightarrow e^- + \bar{\nu}_e + \_?\_$
C. $\_?\_ \rightarrow e^- + \bar{\nu}_e + {}^{60}Ni$
D. $^{3}H \rightarrow e^- + \bar{\nu}_e + \_?\_$
E. $^{13}N \rightarrow e^+ + \_?\_ + \_?\_$

76. • Complete the following gamma decays:
A. $^{131}I^* \rightarrow \gamma + \_?\_$
B. $^{145}Pm^* \rightarrow {}^{145}Pm + \_?\_$
C. $\_?\_ \rightarrow \gamma + {}^{24}Na$

77. •• Nickel-64 has an excited state 1.34 MeV above the ground state. The atomic mass of the ground state of this isotope of nickel is 63.927967 u. (a) What is the mass of the atom when the nucleus is in this excited state? (b) What is the wavelength of the gamma ray that is emitted when the nucleus decays to the ground state?

General Problems

78. •• (a) What is the approximate radius of the ^{238}U nucleus? (b) What electric force do two protons on opposite ends of the ^{238}U nucleus exert on each other? (c) If the electric force in part (b) were the only force acting on the protons, what would be their acceleration just as they left the nucleus? (d) Why do the protons in part (b) not accelerate apart? SSM Example 27-1

79. • The semi-empirical binding energy formula is given as follows:

$$E_B = (15.8\text{ MeV})A - (17.8\text{ MeV})A^{2/3} - (0.71\text{ MeV})\frac{Z(Z-1)}{A^{1/3}} - (23.7\text{ MeV})\frac{(N-Z)^2}{A}$$

where A is the mass number, N is the number of neutrons, and Z is the number of protons. Using the formula, calculate the binding energy per nucleon for fermium-252. Compare your answer with the standard common expression for the binding energy: $E_B = (Nm_n + Zm_p - m_{whole})c^2$. Example 27-3

80. • The *fissionability parameter* is defined as the atomic number squared divided by the mass number for any given nucleus (Z^2/A). It can be shown that when this parameter is less than 44, a nucleus will be stable against small deformation; essentially, the nucleus will be stable against spontaneous fission. Calculate the value of this parameter for (a) ^{235}U, (b) ^{238}U, (c) ^{239}Pu, (d) ^{240}Pu, (e) ^{246}Cf, and (f) ^{254}Cf.

81. • The stable isotope of sodium is ^{23}Na. What kind of radioactivity would be expected from (a) ^{22}Na and (b) ^{24}Na?

82. •• In 2010 physicists first created element number 117 (tennessine, Ts) by colliding ^{48}Ca and ^{249}Bk nuclei. The result was two isotopes of the new element, one of which had a half-life of 14 ms and contained 176 neutrons. (a) What is the radius of the nucleus of the new element 117? (b) What percent of the newly created isotope was left 1.0 s after its creation? SSM Example 27-5

83. •• Natural uranium is made up of two isotopes: ^{235}U and ^{238}U. The half-life of ^{235}U is 7.04×10^8 y, and the half-life of ^{238}U is 4.47×10^9 y. Assuming that all uranium isotopes were created simultaneously and in equal amounts at the same time that Earth was formed, estimate the age of Earth. The current percent abundance of ^{235}U is 0.72% and for ^{238}U it is 99.28%. Example 27-7

84. •• **Astronomy** The atom technetium (Tc) has no stable isotopes, yet its spectral lines have been detected in red giant stars (stars at the end of their lifetimes). Tc can be produced artificially on Earth. Its longest-lived isotope, ^{98}Tc, has a half-life of 4.2 million years. (a) If any ^{98}Tc was present when Earth formed 4.5×10^9 y ago, what percentage of it is still present? (Careful! You cannot do this calculation with your calculator. You must use logarithms to express the answer in scientific notation.) (b) What percent of the original ^{98}Tc would be present in a red giant that is 10 billion years old? (Careful again! You'll need to use logarithms.) (c) Explain why the detection of technetium in old stars is strong evidence that stars manufacture the atoms in the universe. Example 27-5

85. • An old wooden bowl unearthed in an archeological dig is found to have one-fourth of the amount of carbon-14 present in a similar sample of fresh wood. Determine the age of the bowl. Example 27-7

86. • In an attempt to determine the age of the cave paintings in Chauvet-Pont-d'Arc Cave in France, scientists used carbon-14 dating to measure the age of bones of bears found in the cave. The bears are depicted in the paintings, so presumably the bones are approximately the same age as the paintings. The results showed that the level of ^{14}C was reduced to 2.35% of its present-day level. How old were the bones (and presumably the paintings)? SSM Example 27-7

87. • In one common type of household smoke detector, the radioactive isotope americium-241 decays by alpha emission. The alpha particles produce a small electrical current because they are charged. If smoke enters the detector, it blocks the alpha particles, which reduces the current and causes the alarm to go off. The half-life of ^{241}Am is 433 y, and its atomic weight is 241 g/mol. Typical decay rates in smoke detectors are 690 Bq. (a) Write the alpha decay reaction of ^{241}Am and identify the daughter nucleus. (b) By how much does the alpha particle current decrease in 1.0 y due to the decay of the americium? How much in 50 y? (c) How many grams of ^{241}Am are there in a typical smoke detector? Example 27-5

88. • In March 2011 a giant tsunami struck the Fukushima nuclear reactor in Japan, resulting in very large radiation leaks, including cesium-137. The isotope has a 30-y half-life and is a beta-minus emitter. (a) What daughter nucleus is left after cesium-137 decays? (b) How long after the release will it take for the decay rate of the cesium-137 to be reduced by 99%? Example 27-5

89. •• Three isotopes of aluminum are given in the following table:

Isotope	Atomic mass (u)	E_B/nucleon	Decay process
^{26}Al	25.986892		
^{27}Al	26.981538		stable
^{28}Al	27.981910		

Calculate the binding energy per nucleon for each isotope and make a prediction of the decay processes for the unstable isotopes aluminum-26 and aluminum-28. Example 27-3

90. •• **Biology** In February 2010 it was announced that water containing the carcinogen tritium (^{3}H) was leaking from aging pipes at 27 U.S. nuclear reactors. In one well in Vermont, contaminated water registered 70,500 pCi/L; the federal safety limit was 20,000 pCi/L. Tritium is a β^- emitter with a half-life of 12.3 y. (a) How many protons and neutrons does the tritium nucleus contain? (b) Write out the decay equation for tritium and identify the daughter nucleus. (c) If the leak at the Vermont site is stopped, how long will it take for the water in the contaminated well to reach the federal safety level? Example 27-5

91. • **Medical** Iodine-125 is used to treat, among other things, brain tumors and prostate cancer. It decays by gamma decay with a half-life of 59.4 days. Patients who fly soon after receiving ^{125}I implants are given medical statements from the hospital verifying such treatment because their radiation could set off radiation detectors at airports. If the initial decay rate was 525 μCi, (a) what will the rate be at the end of the first year, and (b) how many months after the treatment will the decay rate be reduced by 90%? Example 27-5

92. •• **Medical** Ruthenium-106 is used to treat melanoma in the eye. This isotope decays by β^- emission with a half-life of 373.59 days. One source of the isotope is reprocessed nuclear reactor fuel. (a) How many protons and neutrons does the ^{106}Ru nucleus contain? (b) Could we expect to find significant amounts of ^{106}Ru in ore mined from the ground? Why or why not? (c) Write the decay equation for ^{106}Ru and identify the daughter nucleus. (d) How many years after ^{106}Ru is implanted in the eye does it take for its decay rate to be reduced by 75%? Example 27-5

93. •• **Medical** You are asked to prepare a sample of ruthenium-106 for a radiation treatment. Its half-life is 373.59 days, it is a beta emitter, its atomic weight is 106/g/mol, and its density at room temperature is 12.45 g/cm^3. (a) How many grams will you need to prepare a sample having an activity rate of 125 μCi? (b) If the sample in part (a) is a spherical droplet, what will be its radius? Example 27-5

94. •• Electron capture by a proton is *not* allowed in nature. Explain why. Specifically, describe why the following nuclear reaction does *not* occur (and for good reason!): SSM Example 27-3

$$e^- + p \not\rightarrow n + \nu_e$$

95. •• Taking into account the recoil (kinetic energy) of the daughter nucleus, calculate the kinetic energy of the alpha particle in the following decay of a ^{235}U nucleus at rest: Example 27-6

$$^{235}\text{U} \rightarrow \alpha + {}^{231}\text{Th}$$

96. •• Several radioactive decay series are observed in nature. Four of the most well known are as shown here. The neptunium decay series actually is extinct. The first three all end in different, stable isotopes of lead.

A. Thorium decay series: $A = 4n$
Starting isotope: ^{232}Th Ending isotope: ^{208}Pb

B. Radium or uranium decay series: $A = 4n + 2$
Starting isotope: ^{238}U Ending isotope: ^{206}Pb

C. Actinium decay series: $A = 4n + 3$
Starting isotope: ^{235}U Ending isotope: ^{207}Pb

D. Neptunium decay series: $A = 4n + 1$
Starting isotope: ^{237}Np Ending isotope: ^{209}Bi

Trace out the pathway (keeping count of the total number of alpha and beta emissions) that terminates with a stable nucleus for each of the radioactive decay series above.

97. •• A friend suggests that the world's energy problems could be solved if only physicists were to pursue the fusion of *heavy* nuclei rather than the fusion of *light* nuclei. To prove his point he suggests that the following fusion reaction should be considered: $^{157}\text{Nd} + {}^{80}\text{Ge} \rightarrow {}^{235}\text{U} + 2\text{n}$.

Using the insights that you have acquired in this chapter, show that his argument is flawed. The atomic mass of neodymium-157 is 156.939032 u, and the mass of germanium-80 is 79.925373 u.

Philippe Plailly/Science Source

What do you think?

The collision of an electron and its antimatter equivalent, a positron, produces two sprays of particles at the center of an experiment at CERN, the European particle physics laboratory. Among the particles produced by such a collision, you are likely to find ones made of (a) matter; (b) antimatter; (c) a mixture of matter and antimatter; (d) both (a) and (b); (e) all of (a), (b), and (c).

In this chapter, your goals are to:

- (28-1) Explain why physicists examine both the smallest and largest objects in the universe.
- (28-2) Describe the difference between hadrons and leptons, and explain what hadrons are made of.
- (28-3) Explain how fundamental particles interact with each other, and the differences among the four fundamental forces.
- (28-4) Calculate the distance to a remote galaxy from its recessional velocity.

To master this chapter, you should review:

- (10-4) How the speed of an object in a gravitational orbit depends on the strength of the gravitational force (distance from the massive central object).
- (13-10) How the Doppler effect describes the shift in frequency of a sound wave coming from a moving object.
- (15-3) How an ideal gas cools when it undergoes an adiabatic expansion.
- (22-4) The energy of a photon.
- (25-7) The equivalence of mass and energy.
- (26-2) How the frequency spectrum of blackbody radiation depends on temperature.
- (26-5) Atomic spectra and Rutherford's discovery of the nucleus.
- (27-6) The process of beta decay.

28-1 Studying the ultimate constituents of matter helps reveal the nature of the universe

"What is the world made of?" "Where did I come from?" You probably asked questions like these when you were a small child. Physicists and astronomers ask these questions throughout their professional lives, in search of ever more sophisticated answers about the nature and origin of the physical universe. In this final chapter we'll take a brief look at our present understanding of **particle physics**, the branch of physics that concerns the fundamental constituents of matter and how they interact. We'll also see evidence that our universe began in a state of tremendously high temperature and immense density some 13.8 billion years ago.

We'll begin by looking at the fundamental particles that make up all of the ordinary matter that you see around you, including the matter that makes up your body.

Potatoes are a good source of potassium. A small fraction of the potassium is radioactive ^{40}K, which decays by emitting a positron—a bit of antimatter. This decay also involves a quark, a neutrino, and an exchange particle called a W^+.
(a)
John Scott/Getty Images

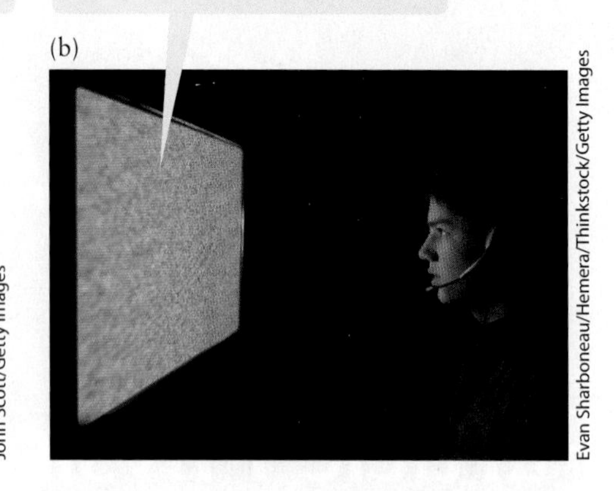

When a television is disconnected from the source, about 1% of the "static" on the screen is from cosmic background radiation—the afterglow of the Big Bang at the beginning of time.
(b)
Evan Sharboneau/Hemera/Thinkstock/Getty Images

Figure 28-1 Particle physics and cosmology Ordinary objects and common technology connect us to (a) particle physics, the study of fundamental particles and their interactions, and (b) cosmology, the study of the nature and evolution of the universe.

These include leptons, of which the electron is the best-known example, and quarks, which have the curious property that their charges are a fraction of the fundamental charge e. We'll see that these fundamental particles interact with each other by exchanging particles back and forth, a process that actually involves violating the law of conservation of energy (but in a way that's nonetheless compatible with the laws of physics). These interactions help explain the strong forces that hold the nucleus together, the electric forces that keep electrons in the atom, and the weak forces that cause radioactive beta decay (**Figure 28-1a**).

We'll conclude the chapter by redirecting our attention from the smallest particles to some of the largest structures in the universe, the galaxies. We'll learn that our universe is expanding and is filled with electromagnetic radiation that is left over from the first few hundred thousand years of the history of the universe (**Figure 28-1b**). We'll also find that most of the matter in the universe is in a form whose nature is almost a complete mystery and that most of the energy in the universe is even more mysterious.

TAKE-HOME MESSAGE FOR Section 28-1

✔ Ordinary matter such as atoms is composed of a small variety of different fundamental particles.

✔ Studying distant galaxies tells us about the nature and evolution of our universe.

28-2 Most forms of matter can be explained by just a handful of fundamental particles

By the late nineteenth century scientists had come to the conclusion that all matter was composed of atoms and that atoms could not be subdivided into more elementary particles. The notion was that a hydrogen atom was fundamentally different from a helium atom, which in turn was fundamentally different from a carbon atom, and so on. As we learned in Section 26-5, this conclusion was incorrect. In 1897 J. J. Thomson discovered the electron, a particle of charge $-e = -1.602 \times 10^{-19}$ C, which turned out to be a constituent of the atoms of every element. In 1909 Ernest Rutherford discovered the atomic nucleus, and in 1917 he found evidence that all nuclei contain a positively charged particle of charge $+e$ that is identical with a hydrogen nucleus—that is, what we now call a proton. In 1932 the English physicist James Chadwick discovered the neutron, which has zero charge. As we learned in Chapter 27, all nuclei are composed of protons and neutrons (referred to collectively as nucleons), and all atoms are made of nuclei plus electrons. So by 1932 it seemed that these three particles—electron, proton, and neutron—were the truly fundamental building blocks of which all matter is made. That conclusion, too, turned out to be wildly incorrect.

Since 1932 literally hundreds of other subatomic particles have been discovered. All of these are unstable and decay to other particles with a radioactive half-life of a fraction of a second. (In this aspect they resemble the neutron: A free neutron that is not incorporated into a nucleus undergoes beta decay with a half-life of about 15 minutes.)

But none of these additional particles can be regarded as simple combinations of protons, neutrons, and electrons. They include the neutrino, which has no charge, interacts hardly at all with other particles, and has a mass so close to zero that it has yet to be accurately measured; the muon, which resembles an electron in almost every way except that it is 207 times more massive; the pion, which like the proton and neutron experiences the strong nuclear force but has only about one-seventh the mass of a proton; and the delta, which resembles a proton or neutron but is about 30% more massive and comes in four varieties, with charges $+2e$, $+e$, 0, and $-e$. The discovery of these additional particles forced physicists to once again ask the question: What *are* the fundamental building blocks of matter?

Hadrons and Quarks

To answer this question, it's useful to distinguish between particles that experience the strong nuclear force, including the proton and neutron, and those that do not, such as the electron. Since protons and neutrons are relatively heavy and electrons are relatively light, we use the term "**hadrons**" (from the Greek word for stout or thick) for particles that experience the strong force and the term "**leptons**" (from the Greek word for small or delicate) for those that do not. (The photon is considered to be in a special category of its own, to which we will return later.) The vast majority of new particles discovered since 1932 are hadrons, so we'll look at these first.

In 1964 the American physicists Murray Gell-Mann and George Zweig independently proposed that all hadrons are made of more fundamental entities that Gell-Mann whimsically named **quarks**. The first evidence that quarks really exist came from experiments carried out at the Stanford Linear Accelerator Center, or SLAC, in 1967 (**Figure 28-2a**). These experiments were the same in principle as Rutherford's 1909 experiment that led to the discovery of the atomic nucleus. As we saw in Section 26-5, Rutherford aimed a beam of alpha particles at a target of gold. Had the charge inside the atom been distributed more or less uniformly, the alpha particles would have undergone only gentle deflections as they passed through the gold atoms. Instead, Rutherford found that some alpha particles were scattered by very large angles. This was evidence that charge was highly concentrated into a very small object within the atom—namely, the nucleus. The experiment at SLAC used electrons instead of alpha particles and aimed these electrons at a target of protons (the nuclei of hydrogen atoms inside a tank of liquid hydrogen). The 3.2-km-long accelerator gave each electron a tremendous kinetic energy—some 20,000 MeV, compared to the 7 MeV of Rutherford's alpha particles—and a correspondingly large momentum. This was done so that the electron would have a de Broglie wavelength much smaller than the size of the proton and so would be sensitive to fine details of the proton's internal structure as it passed through the proton.

Much as in the Rutherford experiment six decades before, many physicists expected that the electrons would undergo only small-angle deflections because they thought the charge inside the proton was distributed uniformly over its volume. And just as in the Rutherford experiment, what they found was that a substantial number of electrons were scattered by very large angles (**Figure 28-2b**). The conclusion was that the proton's charge is carried by smaller entities inside the proton, which are the quarks.

These experiments and a host of others confirm that Gell-Mann and Zweig were correct and that quarks are the fundamental building blocks of all hadrons. To explain all of the hundreds of hadron varieties currently known, we need six varieties or *flavors* of quarks. These are known as the up (u), down (d), charm (c), strange (s), top (t), and bottom (b) quarks. **Table 28-1** lists the six quarks, along with their masses and charges. The quarks are divided into three groups, or *generations*, of two quarks: u and d in the first generation, c and s in the second generation, and t and b in the third generation. As seen in the table, the quarks in each generation are more massive than those in the previous generation. Physicists usually write the quark generations as

$$\begin{pmatrix} \text{u} \\ \text{d} \end{pmatrix} \begin{pmatrix} \text{c} \\ \text{s} \end{pmatrix} \begin{pmatrix} \text{t} \\ \text{b} \end{pmatrix}$$

The quarks along the top row (u, c, and t) all have the same charge, as do the quarks along the bottom row (d, s, and b). However, all six quarks differ not only in

(a)

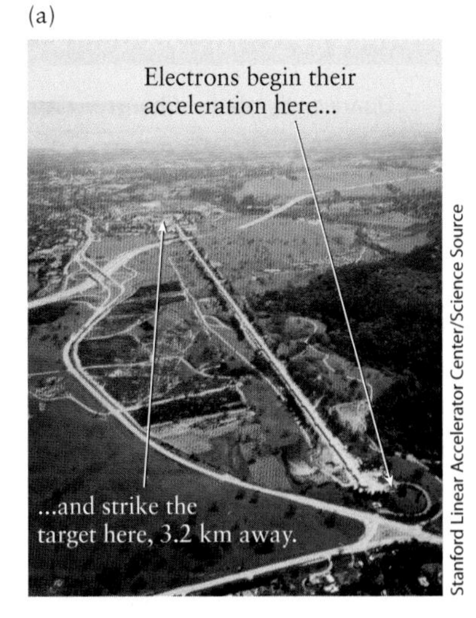
Electrons begin their acceleration here...
...and strike the target here, 3.2 km away.

Stanford Linear Accelerator Center/Science Source

(b)

When high-energy electrons scatter from protons, a substantial number scatter backward.

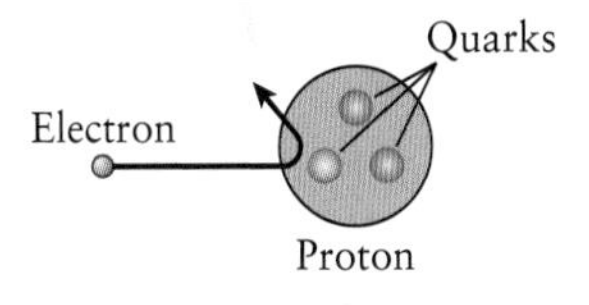
Quarks
Electron
Proton

This is evidence that there are small charged objects (quarks) inside the proton.

Figure 28-2 Discovering quarks (a) In a seminal experiment at the Stanford Linear Accelerator Center, electrons were accelerated to a kinetic energy of 20 GeV = 2×10^4 MeV and fired into a target containing protons. (b) Measuring how the electrons scattered from the protons provided evidence of the existence of quarks.

TABLE 28-1 The Six Quarks

Quark	Symbol	Charge	Approximate mass
up	u	$+\frac{2}{3}e$	$2\ \text{MeV}/c^2$
down	d	$-\frac{1}{3}e$	$5\ \text{MeV}/c^2$
charm	c	$+\frac{2}{3}e$	$1.3\ \text{GeV}/c^2$
strange	s	$-\frac{1}{3}e$	$0.1\ \text{GeV}/c^2$
top	t	$+\frac{2}{3}e$	$173\ \text{GeV}/c^2$
bottom	b	$-\frac{1}{3}e$	$5\ \text{GeV}/c^2$

Note: $1\ \text{GeV}/c^2 = 10^3\ \text{MeV}/c^2 = 10^9\ \text{eV}/c^2$.

(a) Proton (p)

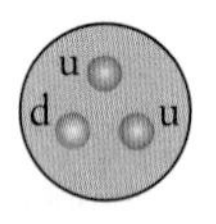

Net charge $= q_u + q_u + q_d$

$= \left(+\frac{2}{3}e\right) + \left(+\frac{2}{3}e\right) + \left(-\frac{1}{3}e\right) = +e$

(b) Neutron (n)

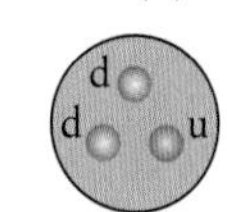

Net charge $= q_d + q_d + q_u$

$= \left(-\frac{1}{3}e\right) + \left(-\frac{1}{3}e\right) + \left(+\frac{2}{3}e\right) = 0$

(c) Delta-plus-plus (Δ^{++})

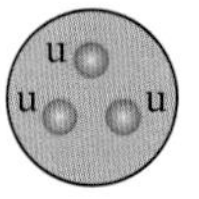

Net charge $= q_u + q_u + q_u$

$= \left(+\frac{2}{3}e\right) + \left(+\frac{2}{3}e\right) + \left(+\frac{2}{3}e\right) = +2e$

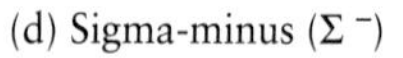

(d) Sigma-minus (Σ^-)

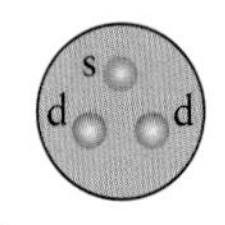

Net charge $= q_s + q_d + q_d$

$= \left(-\frac{1}{3}e\right) + \left(-\frac{1}{3}e\right) + \left(-\frac{1}{3}e\right) = -e$

Figure 28-3 Baryons All baryons, including (a) the proton and (b) the neutron, are made of combinations of three quarks. The proton is the only stable baryon; the neutron and all others, like (c) and (d), decay into simpler particles.

mass but also in other subtle properties. (These properties are beyond our scope in this brief introduction.) All attempts to find evidence of a fourth-generation quark have failed; as best we know, there are only three generations of quarks.

Note that each flavor of quark has a charge that is a *fraction* of e, either $+2e/3$ or $-e/3$. The explanation is that the proton, neutron, and other related particles are actually combinations of three quarks. **Figure 28-3** shows four examples of such combinations. Any hadron that can be made up of three quarks is called a **baryon**. Baryons always have a net charge that is an integer multiple of e. Two examples are the least massive baryons, the proton (**Figure 28-3a**) of net charge $+e$ and the neutron (**Figure 28-3b**) with net charge zero. This picture explains why the neutron, which has zero net charge, nonetheless produces a magnetic field of its own: The u quark and two d quarks within the neutron are each charged, and the motions of these charged particles within the neutron generate a magnetic field.

Note that in Table 28-1 we list only an *approximate* mass for each quark. We know quite precise values of the masses of other particles; for example, the mass of the proton is known to seven significant figures. By contrast, we know only rough values for the masses of the quarks. That's because, unlike protons, neutrons, or electrons, quarks are never found as solitary particles. Instead, they are found only in combinations like those shown in Figure 28-3. This situation is quite unlike that in atoms, in which electrons can be removed by ionization, or in nuclei, in which protons or neutrons can be removed from a nucleus in a sufficiently energetic collision. By contrast, it seems to be impossible to remove an individual quark from a baryon. This is called **quark confinement:** Quarks within baryons are confined there and cannot be removed and isolated.

Note also that the mass of a baryon is generally much greater than the sum of the masses of its constituent quarks. For example, a proton is composed of two u quarks and a d quark, which from Table 28-1 have a combined mass of approximately $2(2\ \text{MeV}/c^2) + 5\ \text{MeV}/c^2 = 9\ \text{MeV}/c^2$. Yet the mass of the proton is $938.1\ \text{MeV}/c^2$. The difference is associated with the forces that bind the quarks together and keep them confined within the proton. We'll explore these forces in the following section.

Other hadrons are made up of a quark and an **antiquark**, the **antimatter** version of a quark. For every type of matter particle there is an antimatter partner that is identical to it in every way except that it is oppositely charged. We've already encountered one example of antimatter, the positron (e^+) produced in β^+ decay (Section 27-5); the positron and electron have the same mass and are identical except for the sign of their charges ($-e$ for the electron, $+e$ for the positron). Each of the six quarks has an antiquark associated with it. Antiquarks are signified by adding "bar" to the name or placing a bar over the quark's symbol. For example, the antiquark associated with the up quark (u) is the up-bar or $\bar{u}$. The $\bar{u}$ antiquark has charge $-2e/3$, the opposite of the $+2e/3$ charge of the u quark. In the same way, the $\bar{d}$ antiquark carries charge $+e/3$, the opposite of the $-e/3$ charge of the d quark.

(a) Positive pion (π^+)

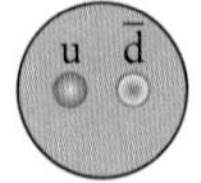
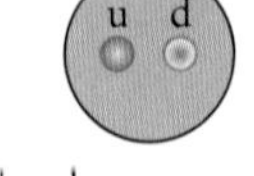

Net charge $= q_u + q_{\bar{d}}$

$= \left(+\frac{2}{3}e\right) + \left(+\frac{1}{3}e\right) = +e$

(b) Negative pion (π^-)

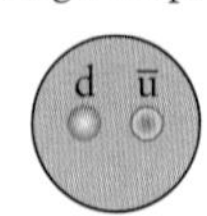

Net charge $= q_d + q_{\bar{u}}$

$= \left(-\frac{1}{3}e\right) + \left(-\frac{2}{3}e\right) = -e$

(c) Strange D-plus (D_s^+)

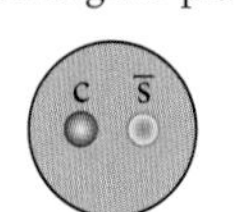

Net charge $= q_c + q_{\bar{s}}$

$= \left(+\frac{2}{3}e\right) + \left(+\frac{1}{3}e\right) = +e$

Figure 28-4 Mesons All mesons are made of a quark and an antiquark.

Figure 28-4 shows three examples of hadrons made up of a quark and antiquark. Such quark-antiquark combinations are called **mesons**. The charge of a meson is always an integer multiple of e. Pions are the least massive mesons; the π^+ (**Figure 28-4a**) and the π^- (**Figure 28-4b**) both have a mass of $139.6\ \text{MeV}/c^2$. There is also a neutral pion (π^0), which is made up of a combination of $u\bar{u}$ and $d\bar{d}$ and has a slightly lower mass of $135.0\ \text{MeV}/c^2$. All mesons, including pions, are unstable; they decay with a half-life that is a small fraction of a second.

When a particle and its antimatter partner meet, their total mass can be converted to its equivalent energy. For example, the collision of an electron and a positron results in two or more photons that carry the total energy of the two particles, a process known as *annihilation*. A quark and antiquark can also annihilate each other; we'll see examples of these processes in the next section.

Just as three quarks make up a baryon, three antiquarks make up an *antibaryon*. For each variety of baryon, there is a corresponding variety of antibaryon. For example, the proton has quark content uud and charge $2e/3 + 2e/3 + (-e/3) = e$; the corresponding antibaryon is the antiproton, which has quark content $\overline{\mathrm{u}}\overline{\mathrm{u}}\overline{\mathrm{d}}$ and charge $(-2e/3) + (-2e/3) + e/3 = -e$.

TABLE 28-2 The Six Leptons

Lepton	Symbol	Charge	Mass
electron	e^-	$-e$	$0.5110\ \mathrm{MeV}/c^2$
electron neutrino	ν_e	0	$<2\ \mathrm{eV}/c^2$
muon	μ^-	$-e$	$105.7\ \mathrm{MeV}/c^2$
muon neutrino	ν_μ	0	$<0.18\ \mathrm{MeV}/c^2$
tau	τ^-	$-e$	$1777\ \mathrm{MeV}/c^2$
tau neutrino	ν_τ	0	$<18.2\ \mathrm{MeV}/c^2$

Leptons

While all hadrons are composed of quarks or antiquarks, there are other particles that are not composed of quarks at all. The *leptons* are another category of particles that are as fundamental as the quarks; to the best of our knowledge, leptons are not made up of smaller constituents. The electron is a lepton, as is the muon that we encountered in Section 25-4 (see Example 25-3). The electron neutrino that is created in hydrogen fusion (Section 27-5) and in beta decay (Section 27-6) is also a lepton. There are six leptons, listed in **Table 28-2**, each of which has an antimatter partner.

Notice that in Table 28-2 we have listed the leptons in three groups of two. As is the case for the quarks, leptons form three generations, which we usually write as

$$\begin{pmatrix}\mathrm{e}^-\\ \nu_\mathrm{e}\end{pmatrix}\begin{pmatrix}\mu^-\\ \nu_\mu\end{pmatrix}\begin{pmatrix}\tau^-\\ \nu_\tau\end{pmatrix}$$

As for the quarks listed in Table 28-1, the leptons in each successive generation are more massive than their counterparts in the preceding generation. In many ways, the muon and tau are more massive versions of the electron, so they share many properties and interact with other particles in similar ways. One way in which electrons, muons, and tau particles *are* significantly different (in addition to the differences in mass) is that while the electron is stable and does not decay, the other two are unstable: The muon has a half-life of 1.56 μs (see Example 25-3), and the tau has a half-life of about 2.0×10^{-13} s.

Each lepton also has an antimatter particle associated with it. We have already encountered the electron and positron. In the same way, the antimuon (μ^+) and antitau (τ^+) have the same masses as the muon and tau, respectively, but the μ^+ and τ^+ have positive charge $+e$ rather than negative charge $-e$.

Unlike quarks, which appear in groups of three to form baryons or in quark-antiquark combinations to form mesons, leptons do not group together to form other particles. In the following section we'll see the reason for this. We'll also discover an essential third class of particles in addition to hadrons and leptons, the *exchange particles*.

Conservation Laws for Hadrons and Leptons

In earlier chapters we encountered a number of important conservation laws, including the conservation of energy and the conservation of momentum. There are several additional conservation laws that govern the behavior of hadrons and leptons.

- *Baryon number is conserved*. Every baryon (composed of three quarks) is assigned *baryon number* B equal to $+1$, and every antibaryon is assigned B equal to -1. Mesons are assigned $B = 0$, as are leptons. Experiments show that in every process that involves baryons, the sum of the values of B for all particles present before the process equals the sum of the values of B for all particles present after the process. Consider, for example, β^- decay of a neutron:

$$\begin{array}{rcccc} & \mathrm{n} \rightarrow & \mathrm{p} + & \mathrm{e}^- + & \overline{\nu}_\mathrm{e} \\ B = & 1 & 1 & 0 & 0 \end{array} \qquad (28\text{-}1)$$

The neutron and proton are both baryons, so each has $B = +1$. The electron and the antineutrino are leptons, each with $B = 0$. The total baryon number equals 1 before the decay and equals $1 + 0 + 0 = 1$ after the decay, so baryon number is conserved. Quarks have a fractional baryon number: Every quark has $B = +1/3$, and every antiquark has

$B = -1/3$. That's consistent with a baryon with three quarks having $B = 3(+1/3) = +1$ and a meson with a quark and antiquark having $B = (+1/3) + (-1/3) = 0$.

- *Lepton number is conserved.* Every electron (e^-) and electron neutrino (ν_e) is assigned *electron-lepton number* $L_e = +1$, and every positron (e^+) and electron antineutrino ($\bar{\nu}_e$) is assigned L_e equal to -1. Muons, muon neutrinos, tau particles, and tau neutrinos each have a similarly defined *muon-lepton number* L_μ or a *tau-lepton number* L_τ. A particle that is not a lepton is assigned L_e, L_μ, and L_τ equal to zero. Experiment shows that each of these lepton numbers is separately conserved.

As an example, consider the β^- decay of a neutron from Equation 28-1:

$$\begin{array}{ccccccc} \mathrm{n} & \rightarrow & \mathrm{p} & + & \mathrm{e}^- & + & \bar{\nu}_e \\ L_e = 0 & & 0 & & 1 & & -1 \end{array} \tag{28-2}$$

Both the neutron and the proton are baryons, so each has $L_e = 0$; the electron has $L_e = +1$, and the electron antineutrino has $L_e = -1$. The total value of L_e before the decay is zero, and afterward it is $0 + 1 + (-1) = 0$. The electron-lepton number is conserved in β^- decay. A second example is the decay of a tau into an electron, an electron antineutrino, and a tau neutrino:

$$\begin{array}{ccccccc} \tau^- & \rightarrow & \mathrm{e}^- & + & \bar{\nu}_e & + & \nu_\tau \\ L_e = 0 & & 1 & & -1 & & 0 \\ L_\tau = 1 & & 0 & & 0 & & 1 \end{array} \tag{28-3}$$

The tau (τ^-) and tau neutrino (ν_τ) each have $L_e = 0$ and $L_\tau = 1$, the electron (e^-) has $L_e = 1$ and $L_\tau = 0$, and the electron antineutrino ($\bar{\nu}_e$) has $L_e = -1$ and $L_\tau = 0$. You can see that the total electron-lepton number equals 0 before and after the decay, and the tau-lepton number equals 1 before and after the decay. Both of these lepton numbers are separately conserved in the β^- decay.

Physicists have searched for processes in which either baryon number or one of the lepton numbers is not conserved. No such process has ever been observed. It appears that these four laws—conservation of baryon number, electron-lepton number, muon-lepton number, and tau-lepton number—are as universal and fundamental as the law of conservation of electric charge (which says that the net electric charge has the same value before and after any process).

GOT THE CONCEPT? 28-1 Particles That May or May Not Exist

Which of the following particles could possibly exist? (a) A meson with charge $-2e$; (b) a baryon with charge $-2e$; (c) an antibaryon with charge $-2e$; (d) more than one of these; (e) none of these.

GOT THE CONCEPT? 28-2 Processes That May or May Not Happen

Electric charge is conserved in each of these processes, but some are never observed to occur. Which *can* occur? (a) $\mathrm{n} + \mathrm{p} \rightarrow \mathrm{n} + \mathrm{p} + \mathrm{p} + \bar{\mathrm{p}}$ ($\bar{\mathrm{p}}$ is an antiproton); (b) $\mathrm{n} + \mathrm{p} \rightarrow \mathrm{n} + \mathrm{n} + \mathrm{p}$; (c) $\pi^- \rightarrow \mu^- + \bar{\nu}_e$; (d) more than one of these; (e) none of these.

TAKE-HOME MESSAGE FOR Section 28-2

✔ Quarks are the constituents of hadrons, the particles that experience the strong nuclear force. There are six varieties of quarks, each of which has a charge that is a fraction of e.

✔ Baryons such as the proton are made up of three quarks. Mesons are made up of a quark and an antiquark.

✔ Leptons, which include electrons and neutrinos, have no constituent particles. They do not experience the strong force.

✔ In the interactions of subatomic particles, baryon number and the three lepton numbers are conserved.

28-3 Four fundamental forces describe all interactions between material objects

Since early in our study of physics we have seen the importance of *forces*, the pushes and pulls that one object exerts on another. We've encountered three fundamental kinds of forces: gravity, the electromagnetic force, and the strong nuclear force. Gravity is an attractive force that draws objects closer together. The gravitational force attracts you to Earth and keeps Earth in orbit around the Sun. Electric and magnetic forces, two manifestations of the electromagnetic force, cause charged objects to accelerate. As a result, electrons can be bound to atomic nuclei, and atoms can bond together. Any contact force, such as the normal force between two objects that touch or the force of friction, are electromagnetic in nature; they arise from the electromagnetic interactions between the atoms of the two surfaces in contact. As we explored in Chapter 27, the strong force binds protons and neutrons together to form atomic nuclei. To this list we will add a force that we have not yet named but the effects of which we encountered in Section 27-6; this **weak force** is at the heart of the interaction that governs beta decay.

Now that we are exploring the fundamental constituents of matter, we are in a position to ask a central question about force: How do objects exert forces on each other? And what is fundamentally different about the four different kinds of forces: gravitational, electromagnetic, strong, and weak? If we can understand how fundamental particles such as quarks and leptons exert forces of different kinds on each other, we will be closer to answering these questions.

As we will see, the manner in which fundamental particles exert forces on each other—that is, how they *interact*—comes from a remarkable aspect of nature: It is possible to violate the law of conservation of energy, provided we do it for a sufficiently short time.

The Heisenberg Uncertainty Principle

To see how it's possible to violate energy conservation, let's consider the photon. We learned in Section 22-4 that the energy of a photon is proportional to its frequency:

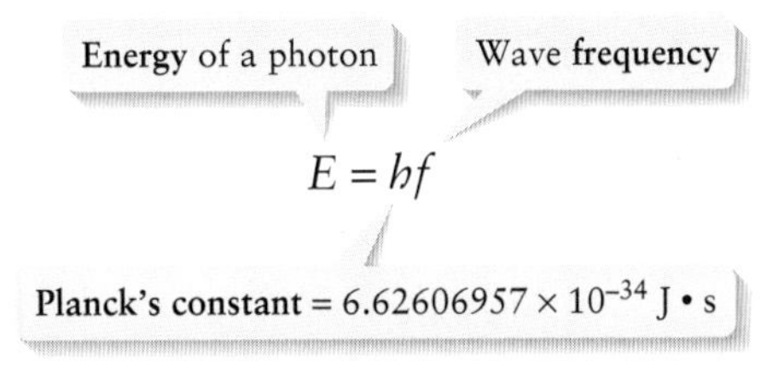

Energy of a photon (22-26)

For a photon to have a definite energy, it must therefore have a definite frequency. However, if a wave described by a function ψ has a definite frequency, it also has an infinite duration (**Figure 28-5a**). This means that the wave has always been present and will always be present. A more realistic description of a wave is one that has a finite duration; for example, the wave produced when you turn a source of waves (like a laser pointer) on and then off again. Mathematically, a wave of a finite duration Δt can be expressed as a sum of waves of infinite duration like the one shown in Figure 28-5a but with a range of frequencies of breadth Δf (**Figure 28-5b**). To make a shorter-duration wave, we have to add together infinite-duration waves from a broader range of frequencies (**Figure 28-5c**). We can express the relationship between Δt (the duration of a wave) and Δf (the breadth of frequencies that go into that wave) as

$$\Delta f \Delta t \geq \frac{1}{4\pi} \qquad (28\text{-}4)$$

Equation 28-4 says that the product of the wave duration Δt and the frequency breadth Δf cannot be less than $1/4\pi$. (This number arises from the specific way in which Δt and Δf are defined mathematically.) It says that to minimize the duration of a wave necessarily means increasing the range of frequencies that make up the wave. So the shorter the duration of a wave, the less precisely we can answer the question "What is the frequency of the wave?" In other words, a wave of finite duration does not have a single definite frequency.

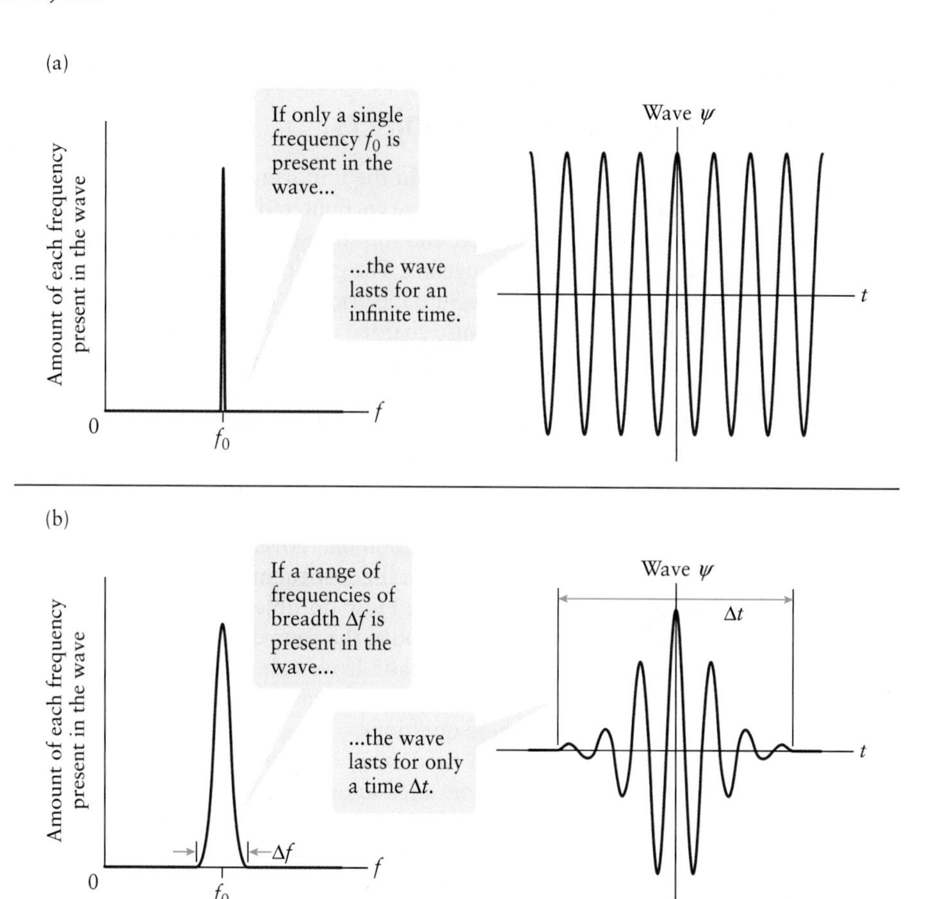

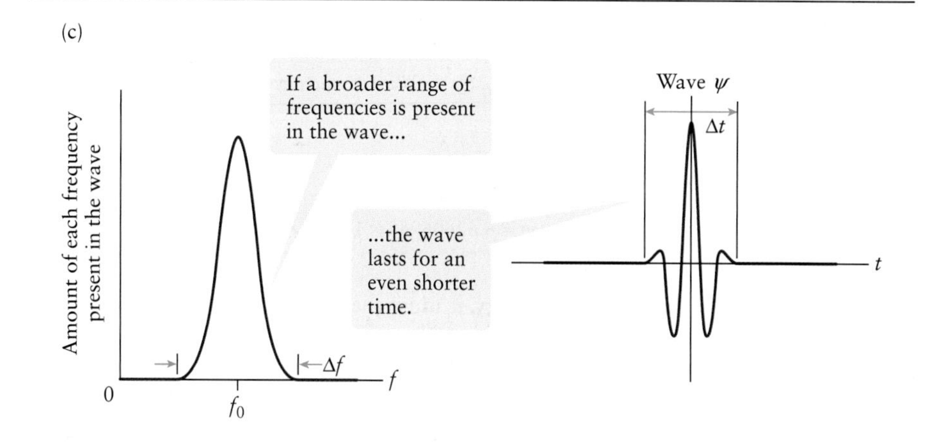

Figure 28-5 Wave frequency and wave duration (a) A wave of a single, discrete frequency has an infinite duration. (b), (c) Combining waves over a breadth of frequencies yields a total wave that has a finite duration. The shorter the wave duration, the greater the range of frequencies present in the wave.

Equation 28-4 is true for waves of all kinds, from ocean waves to sound waves to seismic waves. If we apply it to electromagnetic waves and multiply both sides of the equation by Planck's constant h, we get

$$h\Delta f \Delta t \geq \frac{h}{4\pi} \quad \text{or} \quad \Delta(hf)\Delta t \geq \frac{\hbar}{2} \tag{28-5}$$

Here $\Delta(hf)$ is the breadth of values of the quantity hf that must be included in the wave, and $\hbar = h/2\pi$. (We introduced $\hbar = h/2\pi$, or "h bar," in Section 26-6.) But Equation 22-26 tells us that $E = hf$ is the energy of a photon associated with the wave. So $\Delta(hf) = \Delta E$, the uncertainty in photon energy, and Equation 28-5 becomes

$$\Delta E \Delta t \geq \frac{\hbar}{2} \text{ for a photon} \tag{28-6}$$

This means that just as a wave of finite duration does not have a single definite frequency, a photon of finite duration does not have a single definite energy: It includes energy values that extend over a range of breadth ΔE. We can think of ΔE as the *uncertainty* in the energy of the photon. Equation 28-6 says that if a photon has a duration Δt, the product of Δt and uncertainty in energy ΔE of the photon cannot be less than $\hbar/2$, so ΔE cannot be less than $\hbar/(2\Delta t)$. This energy uncertainty is *not* a result of the limitations of an experimental apparatus that we might use to measure the energy of a photon. Rather, it is intrinsic to photons because of their wave nature.

We have used the equation $E = hf$ to apply to photons only. But this relationship applies to particles of *all* kinds: We can think of anything with an energy E as having an associated frequency f given by $E = hf$. Then Equation 28-6 applies to phenomena in general, not just to photons. This implies that any physical phenomenon that has a finite duration Δt will necessarily have an uncertainty ΔE in energy given by Equation 28-6. The minimum value of this uncertainty is found by replacing the $\geq$ sign (greater than or equal to) in Equation 28-6 with an equals sign. The shorter the duration Δt of the phenomenon, the greater the minimum value of the energy uncertainty ΔE and the more uncertain the energy of the phenomenon. The German physicist Werner Heisenberg first expressed this idea in 1927, which is why it is known as the **Heisenberg uncertainty principle:**

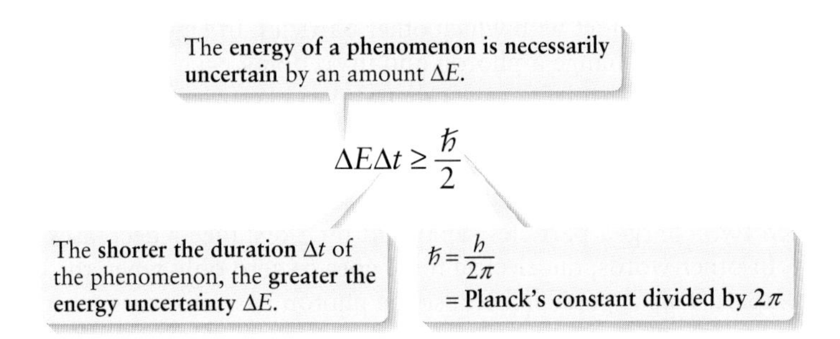

$$\Delta E \Delta t \geq \frac{\hbar}{2}$$

Heisenberg uncertainty principle
(28-7)

Another way to interpret Equation 28-7 is to say that it places limits on how precisely the law of conservation of energy must be obeyed. Suppose a system undergoes some kind of process that lasts for a time Δt. Equation 28-7 says that the energy of the system is necessarily uncertain during that process, and the minimum energy uncertainty is given by $\Delta E \Delta t = \hbar/2$ or $\Delta E = \hbar/(2\Delta t)$. So it's fundamentally impossible to measure the energy of the system with an uncertainty less than $\hbar/(2\Delta t)$. This means that during the process the energy of the system could actually vary by as much as $\hbar/(2\Delta t)$, and there would be no way that we could tell that the energy had changed value—that is, that the energy was not conserved. The shorter the duration Δt of the process, the greater the amount $\hbar/(2\Delta t)$ by which energy conservation can be (temporarily) violated during that time Δt. Stated another way, it's acceptable to violate the law of conservation of energy by an amount ΔE, provided the duration of time during which the law is violated is no more than $\Delta t = \hbar/(2\Delta E)$.

Exchange Particles: The Electromagnetic Force

The Heisenberg uncertainty principle helps us understand the following bold statement: *All forces result from the exchange of particles.* To see what we mean by this statement, first consider the electromagnetic force. Up to this point we've used the idea that charged particles exert electromagnetic forces on each other even at a distance, with no physical contact required. But a more sophisticated way to look at this force is to envision that when two charged particles exert an electric or magnetic force on each other, they do so by exchanging a photon: One of the charged particles emits the photon, and the other absorbs it. The exchanged photon has energy, and it violates the law of conservation of energy for this photon to spontaneously appear and be emitted by one of the charged particles. But as we have seen, it's perfectly acceptable to violate the law of conservation of energy by an amount ΔE, provided we do so for a time no longer than $\Delta t = \hbar/(2\Delta E)$. So the uncertainty principle proves that what we've described can take place provided the exchanged photon is absorbed by the second particle (and so disappears) within a time $\hbar/(2\Delta E)$ after the photon was emitted by the first particle.

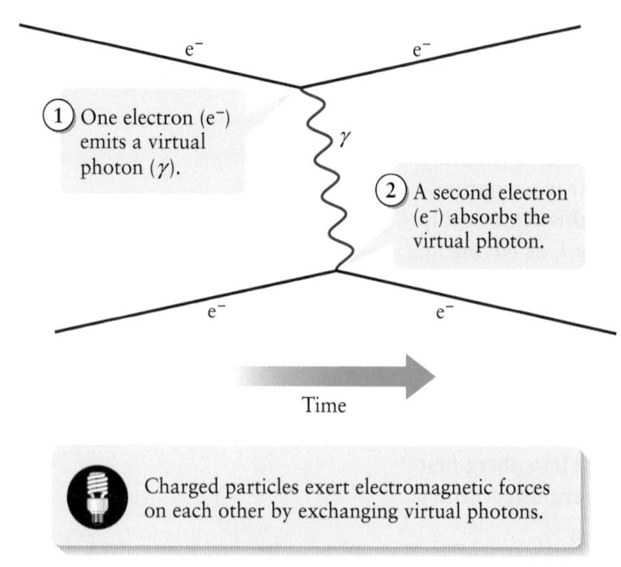

Figure 28-6 Photon exchange and the electromagnetic interaction All electric and magnetic interactions between charged particles involve the exchange of virtual photons.

In this picture we say that the electromagnetic force is *mediated* by the exchange of a photon and that the exchanged photon is the *mediator* of the force. The particles that mediate forces are called **exchange particles.** You should try to envision a continuous stream of photons going back and forth between charged particles so that the two particles continuously exert an electromagnetic force on each other. Any charged particle can and does emit and absorb photons exchanged in this way. These photons are not the same as the photons emitted by a light bulb or a laser, however; they can exist for only a finite time before they must disappear. We call them **virtual particles.**

We can represent the exchange process by a diagram such as the one shown in **Figure 28-6.** In this slightly simplified version of a *Feynman diagram*, invented by American physicist Richard Feynman to visualize and analyze processes that involve fundamental particles, time runs from left to right. The lines associated with each particle do not represent actual paths that the particles take through space but only indicate which particles interact with which other particles. In Figure 28-6 two electrons exchange a photon and in so doing each exerts an electromagnetic force on the other. A particle such as an electron is drawn as a solid, straight line in Feynman diagrams. An exchange particle is drawn as either a wavy line (as for the photon) or a spiral line.

This picture helps us understand Coulomb's law, which tells us that the electric force between two charged particles separated by a distance r decreases in proportion to $1/r^2$. In other words, the electric force goes to zero only when the charges are infinitely far apart. This is possible because the photon has zero mass, so its minimum energy $E = hf$ is zero. (If the photon did have a mass m, its minimum energy would be its rest energy mc^2.) If one particle violates conservation of energy by creating and emitting a photon of energy ΔE, the uncertainty principle says that this photon can exist no longer than $\Delta t = \hbar/(2\Delta E)$. Even traveling at the speed of light c, this photon can travel no farther than $c\Delta t = c\hbar/(2\Delta E)$, so that distance is the maximum separation at which two charged particles can interact by exchanging photons of energy ΔE. But because the lower limit on the energy of a photon is zero, the amount ΔE by which the particle violates conservation of energy can be as small as we like. So the distance $r = c\Delta t = c\hbar/(2\Delta E)$ can be arbitrarily large, and two charged particles can interact via the electromagnetic force at any distance out to infinity. But because only low-energy photons (with small ΔE) can be exchanged between charged particles separated by great distances r, the force is quite weak if r is large—just as in Coulomb's law.

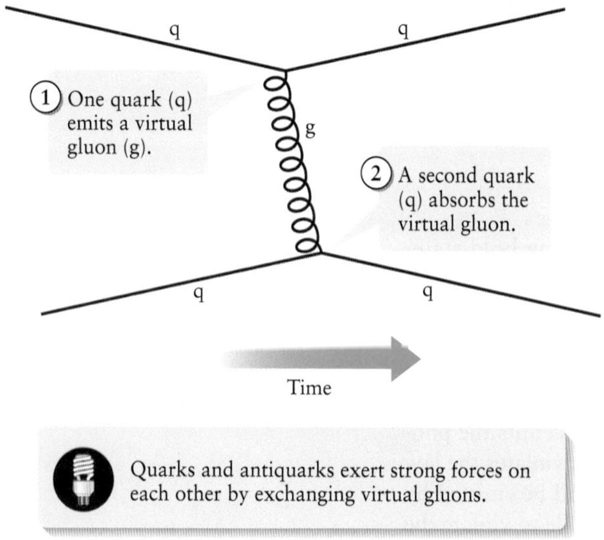

q
q
1 One quark (q) emits a virtual gluon (g).
g
2 A second quark (q) absorbs the virtual gluon.
q
q
Time

Quarks and antiquarks exert strong forces on each other by exchanging virtual gluons.

Figure 28-7 Gluon exchange and the strong interaction between quarks The forces that bind quarks together inside baryons and mesons are the result of the exchange of virtual gluons.

Exchange Particles: The Strong Force

We use a similar picture to explain the strong interaction between quarks. The Feynman diagram in **Figure 28-7** shows two quarks that exert forces on each other by exchanging a particle called the **gluon.** This rather whimsical name expresses the idea that the exchange of gluons provides the force that confines, or glues, quarks inside baryons. Gluon exchange is also how the quark and antiquark inside a meson interact with each other and how the antiquarks inside an antibaryon interact.

Like the photon the gluon has zero mass, so the lower limit on the energy of a gluon is zero. Using the same argument we made above for photons, it follows that quarks or antiquarks separated by any distance, no matter how large, can interact by exchanging gluons. But unlike photons, the kinds of particles that emit and absorb gluons are not simply those with electric charge: They are particles that have a different attribute called *color*. (This is yet another of the light-hearted names associated with quark physics. It has nothing to do with the wavelength of light or the perception of color by the human eye.) Quarks and antiquarks carry color, as do gluons themselves. So unlike photons, gluons can interact with each other by exchanging other gluons. (By contrast, photons have no electric charge and cannot interact directly with other photons.) As a consequence of this curious aspect of gluons, the force that gluons mediate between quarks gets stronger, not weaker, as the quarks move farther apart. This helps to explain why this force makes it impossible to remove an isolated quark from a hadron.

As we mentioned in Section 28-2, the mass of a nucleon (a proton or neutron) is much greater than the sum of the masses of the three quarks that constitute the nucleon. The gluon picture helps us understand why this is: Part of the additional mass is associated (through the relationship $E = mc^2$) with the energy of the virtual gluons that are continuously exchanged between the constituent quarks. The remaining mass is associated with the kinetic energies of the constituent quarks, which are in high-speed motion inside the nucleon.

Since quarks are charged, they also interact electromagnetically by exchanging photons. But these interactions have a relatively small effect compared to the dominant strong interaction mediated by gluons.

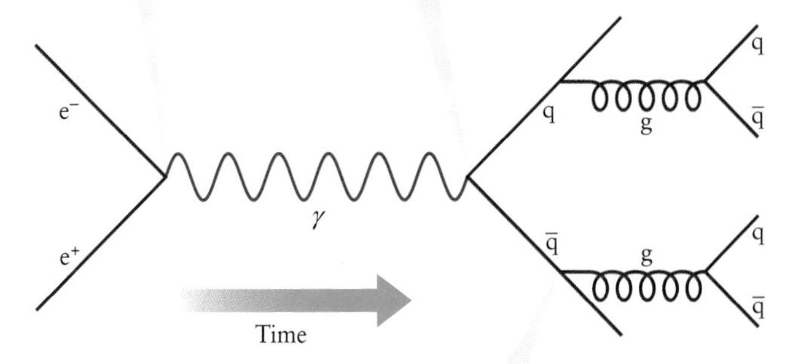
① In electron–positron annihilation, an electron (e⁻) and a positron (e⁺) collide and are transformed into a virtual photon.
② If the e⁻ and e⁺ have sufficient energy, the photon can transform into a quark (q) and antiquark (q̄).
e⁻
e⁺
γ
q
q̄
g
Time
③ The quark and antiquark can emit gluons, which transform into quark–antiquark pairs. The shower of quarks and antiquarks coalesces into hadrons, including baryons (three quarks), antibaryons (three antiquarks), and/or mesons (quark–antiquark pairs).

Figure 28-8 Electron–positron annihilation and hadron production When an electron and positron collide, the result can be a shower of hadrons produced via virtual photons and gluons.

As we mentioned in the previous section, when an electron and a positron (matter and antimatter) collide, they annihilate. The Feynman diagram in **Figure 28-8** shows one possible outcome of such an annihilation. The electron of charge $-e$ and the positron of charge $+e$ disappear and are replaced by a single photon. This *virtual* photon does not have the proper relationship between energy and momentum that a real photon must have, so it violates the conservation laws that govern the electron–positron collision. Like an exchange photon, this virtual photon can exist for only a finite time before it must disappear. In some cases the virtual photon will transform back into an electron–positron pair. But as Figure 28-8 shows, it's also possible for the virtual photon to become a quark and an antiquark of the same flavor (for instance, a u quark and a $\bar{\text{u}}$ antiquark, or an s quark and an $\bar{\text{s}}$ antiquark). The quark and antiquark can themselves emit virtual gluons, which can in turn transform into quark–antiquark pairs. Depending on how much energy is available in the original collision between electron and positron, many such quark–antiquark pairs can be produced. These will sort themselves into combinations of quarks (baryons), combinations of antiquarks (antibaryons), and quark–antiquark pairs (mesons). The result will be a shower of hadrons emanating from the site of the electron–positron collision. The image that opens this chapter shows "jets" of hadrons emerging from the site of just such a collision. Many varieties of hadrons were first identified in electron–positron collision experiments of this sort.

The notion of gluon exchange also helps us understand the strong force in the context in which we first encountered it in Section 27-2: a force that attracts nucleons (protons or neutrons) to each other. **Figure 28-9** shows how this force between nucleons arises. The meson that is produced in this way acts as an exchange particle between the nucleons, and this exchange gives rise to the attractive force between nucleons. The meson exchange particle has a substantial mass m_{exchange}, so the minimum energy that an exchanged meson can have is its rest energy $m_{\text{exchange}}c^2$. Producing such a meson means violating the conservation of energy by an amount of at least $\Delta E = m_{\text{exchange}}c^2$, which means that the meson can exist for a time no longer than $\Delta t = \hbar/(2\Delta E) = \hbar/2m_{\text{exchange}}c^2$. Even moving at the speed of light, the maximum distance that this exchange meson can travel during a time Δt is

$$c\Delta t = \frac{c\hbar}{2m_{\text{exchange}}c^2} = \frac{\hbar}{2m_{\text{exchange}}c} \tag{28-8}$$

(range of a force mediated by an exchange particle of mass m_{exchange})

Figure 28-9 The strong force between nucleons The force that binds protons and neutrons together has its origin in the interaction between quarks and gluons.

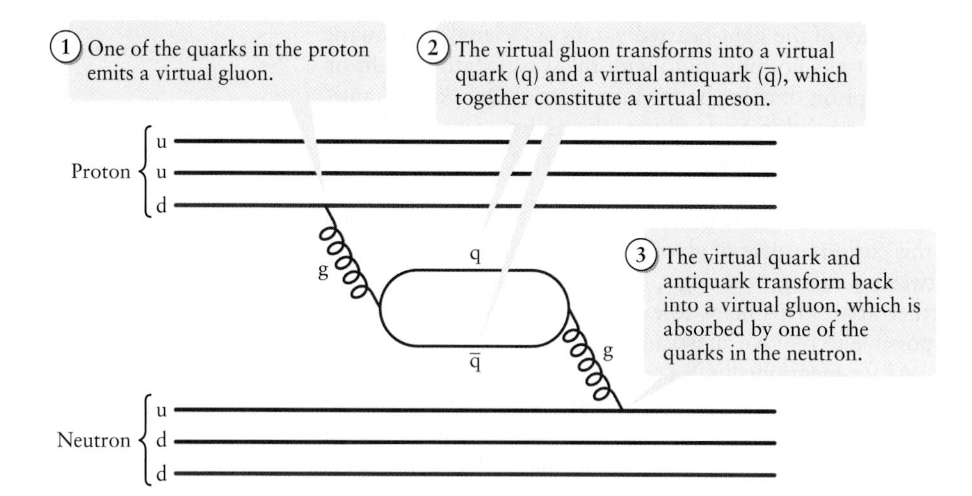

- Nucleons (protons and neutrons) exert the strong force on each other by exchanging virtual mesons.
- This exchange is actually due to interactions between quarks and gluons.

Equation 28-8 tells us the *range* of the force mediated by the exchange meson. There are many types of mesons, but the type whose exchange will have the longest range is the one with the smallest mass. (That's because the range in Equation 28-8 is inversely proportional to the mass of the exchange particle.) The mass of the lightest meson, the neutral pion (π^0), is $135.0 \text{ MeV}/c^2 = 135.0 \times 10^6 \text{ eV}/c^2$, so the maximum range of the strong force between nucleons should be

$$\frac{\hbar}{2m_{\text{neutral pion}}c} = \frac{\hbar c}{2m_{\text{neutral pion}}c^2} = \frac{hc}{4\pi m_{\text{neutral pion}}c^2}$$

$$= \frac{(4.136 \times 10^{-15} \text{ eV}\cdot\text{s})(3.00 \times 10^8 \text{ m/s})}{4\pi(135.0 \times 10^6 \text{ eV}/c^2)c^2}$$

$$= 7.31 \times 10^{-16} \text{ m} = 0.731 \text{ fm}$$

Because this range is so small, the strong force between nucleons is a short-range force, just as we discussed in Section 27-2. Indeed the rough value of 0.731 fm that we calculated here is of the same order of magnitude as the value we gave in Section 27-2 for the range of the strong force between nucleons.

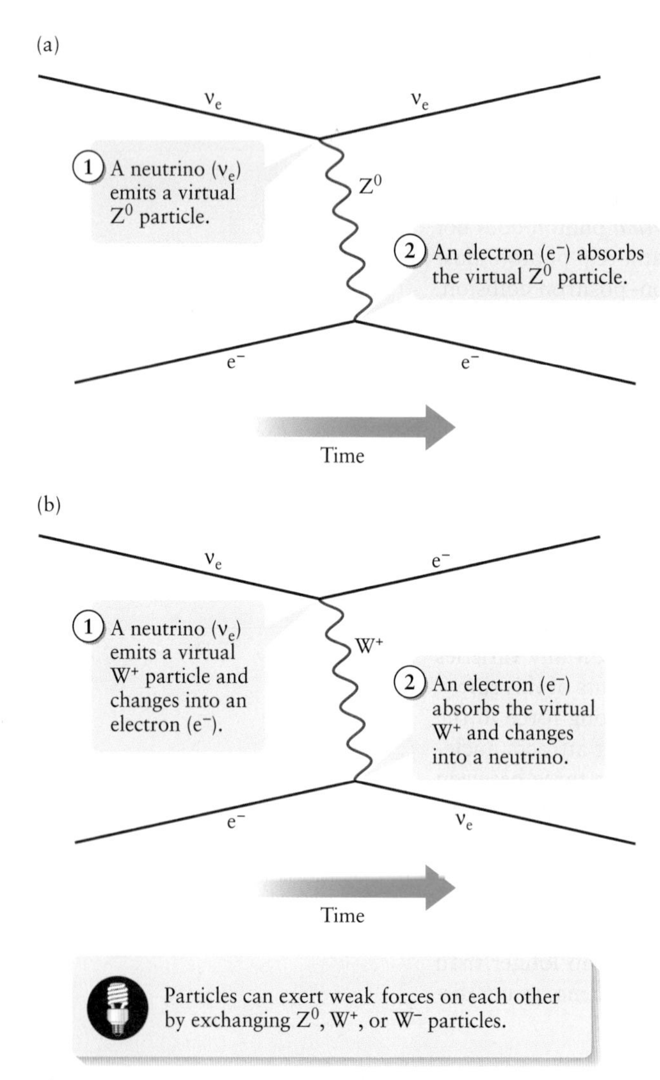

Particles can exert weak forces on each other by exchanging Z^0, W^+, or W^- particles.

Figure 28-10 The weak interaction The weak force between particles involves the exchange of massive virtual particles. These come in both (a) neutral and (b) charged varieties.

Exchange Particles: The Weak Force and the Gravitational Force

The weak force is mediated by *three* different exchange particles. These are the neutral Z^0 (charge zero) and the positively and negatively charged W particles, W^+ (charge $+e$) and W^- (charge $-e$). These particles are not massless: The Z^0 has mass $91.2 \text{ GeV}/c^2$, about 100 times the mass of a proton, and the W^+ and W^- both have mass $80.4 \text{ GeV}/c^2$. The calculation above shows that the range of the force mediated by an exchange particle of mass m_{exchange} is inversely proportional to the mass. So since the Z^0, W^+, and W^- have roughly 600 times the mass of the neutral pion ($135.0 \text{ MeV}/c^2$), the range of the weak force is roughly 1/600 that of the strong force between nucleons, or about $10^{-18} \text{ m} = 10^{-3} \text{ fm}$. The weak force is not only weak, it is of *extremely* short range!

It turns out that any two particles can exert a weak force on each other. For instance, **Figure 28-10** shows two

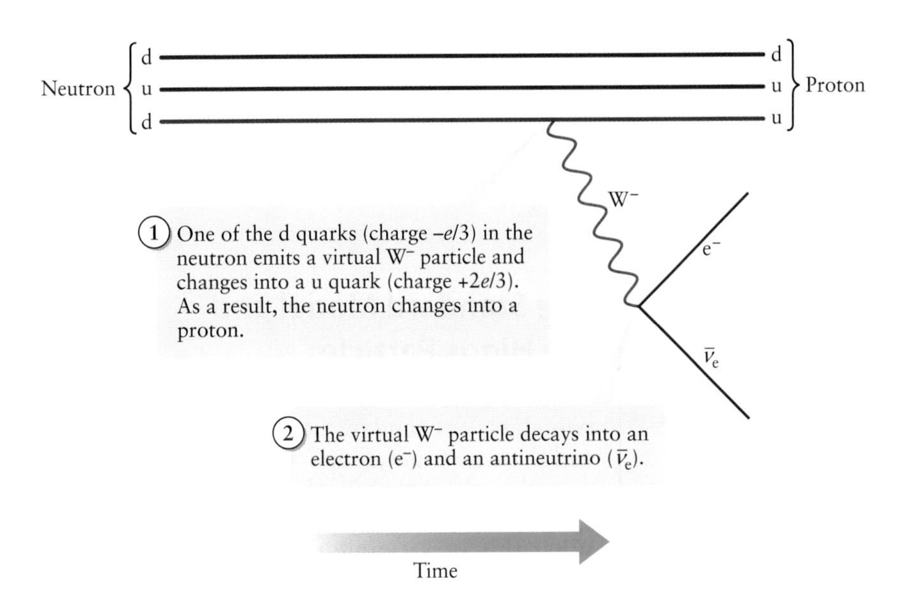

Figure 28-11 Beta decay of a neutron The weak interaction describes how a free neutron decays into a proton, an electron, and an antineutrino.

examples of the weak interaction between a neutrino and an electron. Note that even though the weak interaction does not directly involve electric charge, charge is still conserved. In **Figure 28-10a** the neutrino remains neutral, and the electron retains its charge of $-e$. In **Figure 28-10b** the neutrino emits a W^+ of charge $+e$ and becomes an electron of charge $-e$, so charge is conserved. The W^+ combines with an electron of charge $-e$ and becomes a neutrino of charge zero, and again charge is conserved.

Figure 28-11 shows how the weak force gives rise to the β^- decay of a neutron (quark content udd) to a proton (quark content uud):

$$\mathrm{n} \to \mathrm{p} + \mathrm{e}^- + \bar{\nu}_e$$

Compared to a u quark, a d quark is substantially more massive (see Table 28-1) and so has a substantially greater rest energy. So the d quark can lower its energy by becoming a u quark. It does this by emitting a W^- particle, as shown. This is a very short-lived virtual particle: The rest energy of the W^- ($80.4\ \mathrm{GeV}/c^2$) is far greater than the rest energy of the original d quark (about $5\ \mathrm{MeV}/c^2 = 5 \times 10^{-3}\ \mathrm{GeV}/c^2$), so emitting a W^- means violating energy conservation by quite a bit. The W^- of charge $-e$ quickly decays into an electron of charge $-e$ and a neutral antineutrino. (It has to be an *anti*neutrino to conserve electron–lepton number L_e: The W^- is not a lepton and therefore has $L_e = 0$, so the W^- must decay into particles with a net value of L_e equal to zero. The electron, with $L_e = +1$, must therefore be accompanied by a neutral particle with $L_e = -1$, which means an antineutrino rather than a neutrino.)

As we discussed in Section 27-6, in some nuclei it's energetically favorable for β^+ decay to take place, in which a proton changes to a neutron. This is similar to the process shown in Figure 28-11, except that one of the u quarks in a proton changes to a d quark by emitting a W^+ rather than a W^-. The W^+ then decays into a positron (e^+) and a neutrino (ν_e). One nucleus in which this happens is the potassium isotope ^{40}K, which is found in foods such as potatoes (see Figure 28-1a). The antimatter produced in this way—the positron—quickly encounters an atomic electron and annihilates. Happily, this happens very infrequently in your food (and releases very little energy when it does happen). So you needn't worry about biting into a bit of antimatter when you eat a potato!

A particle named the *graviton* is thought to mediate the gravitational force. So far no experimental evidence has confirmed the existence of this particle, although no experimental evidence has excluded it either. Because the gravitational force between two masses separated by a distance r is proportional to $1/r^2$, just like the electric force

TABLE 28-3 The Four Fundamental Forces

Force	Range	Mediator(s)	Strength relative to the strong force
gravity	infinite	graviton	10^{-40}
weak	$\sim 10^{-3}$ fm	Z^0, W^+, W^-	10^{-6}
electromagnetic	infinite	photon (γ)	10^{-2}
strong*	~ 1 fm	gluon (g)	1

* Gluons mediate the strong force between quarks; this force has infinite range. As described in the text, the strong force between *nucleons* (protons and neutrons) has a range of only about 1 fm because it involves not just gluons but also virtual meson exchange.

between two charged particles, the graviton is thought to have the same mass as the photon: zero. The graviton also carries no charge.

Table 28-3 summarizes the four fundamental forces and the exchange particles that mediate these forces.

The Standard Model and the Higgs Particle

The picture that Table 28-3 summarizes—in which two classes of fundamental particles, the six quarks and the six leptons, interact by means of six exchange particles, the graviton, the Z^0, the W^+, the W^-, the photon, and the gluon—is called the **Standard Model.** What this very simple table does not reflect is the decades of experimental and theoretical effort (and several Nobel Prizes in Physics) that have gone into constructing this model. One long-standing question about the Standard Model that has recently been answered is this: Of the exchange particles listed in Table 28-3, why do the Z^0, W^+, and W^- all have substantial masses (91.2, 80.4, and 80.4 GeV/c^2, respectively), while the graviton, photon, and gluon each have zero mass? The answer proposed in the 1960s was that there is a field that fills all of space, now called the **Higgs field,** and the interactions of the Z^0, W^+, and W^- particles with this field give these particles their masses. Unlike electric and magnetic fields, which are vectors with a magnitude and direction, the Higgs field is a scalar that has no direction. Because the Higgs field has the same nonzero value everywhere in space, the masses of the Z^0, W^+, and W^- have the same values no matter where in space these particles are found.

GOT THE CONCEPT? 28-3
A Quark Decay

A c quark can decay into an s quark via the weak force. In this process, what does the c quark emit?
(a) A W^+, which decays into a positron and a neutrino;
(b) a W^+, which decays into a positron and an antineutrino;
(c) a W^-, which decays into an electron and a neutrino;
(d) a W^-, which decays into an electron and an antineutrino.

This proposal may remind you of a debunked idea that we encountered in Section 25-3. In the nineteenth century, it was thought that there was a substance called the luminiferous ether that filled all space and that acted as a medium for the propagation of light waves. The Michelson–Morley experiment, which we also discussed in Section 25-3, demonstrated that no such luminiferous ether exists. To determine whether the Higgs field exists, twenty-first-century physicists used the idea that for each kind of field there is a corresponding particle. For example, the particle that corresponds to the electric and magnetic fields is the photon. The electric and magnetic forces are mediated by virtual photon exchange, and an electromagnetic wave is made up of photons. To explain the observed masses of the Z^0, W^+, and W^-, the **Higgs particle** that corresponds to the Higgs field would have to have zero charge, have a mass about 50% greater than that of the W^+ or W^-, and be unstable. If such a particle could be produced in a high-energy collision of other particles, it would suggest that the Higgs field does indeed exist.

In 2012, after decades of experimental effort, the Higgs particle was discovered in collisions between high-energy protons at the Large Hadron Collider in Switzerland. The observed Higgs particle mass of 125 GeV/c^2 is within the expected range of values, and as predicted the Higgs particle has zero charge and a half-life of about 10^{-22} s. This discovery provided strong experimental support that unlike the luminiferous ether, the Higgs field actually does exist throughout space. This field is now regarded as an essential part of the Standard Model.

The Higgs field is thought to explain not just the masses of the Z^0, W^+, and W^-, but also the masses of the leptons and the quarks. In this picture, the reason the muon has a greater mass than the electron is that muons interact more strongly with the Higgs field than electrons do. In future experiments, physicists will investigate how particles of various kinds interact with the Higgs *field* by measuring how these particles interact with Higgs *particles*. These and other aspects of the Standard Model are the subject of active research by physicists around the globe.

TAKE-HOME MESSAGE FOR Section 28-3

✓ Particles exert forces on each other through the exchange of virtual particles that are emitted by one particle and absorbed by another.

✓ The photon is the exchange particle that mediates the electromagnetic force; the gluon mediates the strong force; and the Z^0, W^+, and W^- particles mediate the weak force. The graviton, which has not yet been detected, is thought to mediate the gravitational force.

✓ The Standard Model is our picture of six types of quarks and six types of leptons that interact by means of these various exchange particles. A key part of the Standard Model is the Higgs field, a field that exists throughout all space and explains the large masses of the Z^0, W^+, and W^-.

28-4 We live in an expanding universe, and the nature of most of its contents is a mystery

The Standard Model describes all the normal matter in the world around us. All of the atoms in our bodies and our surroundings are made of leptons and hadrons, the hadrons are all made of quarks, and all of these particles interact by means of exchange particles. But why do we specify that this description applies to "normal" matter only? What other sort of matter could there be, and how much of the contents of the physical universe is so-called normal matter? To answer these questions, let's turn our attention from the smallest fundamental particles to galaxies, some of the largest structures in the universe, and to the nature of the universe itself—the subject of the science called **cosmology**.

WATCH OUT! A galaxy is not a solar system, nor is it the entire universe.

In everyday language, many people use the terms "galaxy," "solar system," and "universe" interchangeably. In fact, a galaxy is a large collection of stars and other matter, while a solar system is a single star and its retinue of planets. Our galaxy contains some 200 billion stars, many of which have planets, and the observable universe contains literally billions of galaxies.

The Hubble Law and the Expansion of the Universe

Figure 28-12 is an image made with the Hubble Space Telescope of a small portion of the night sky. At first glance it may appear that the bright dots in this image are individual stars. But closer inspection shows that each is actually a **galaxy**, a collection of a tremendous number of stars (typically 10^{10} to 10^{12}). Our Sun is one of approximately 2×10^{11} stars in the Milky Way, the local galaxy of which we are part.

In the 1920s the American astronomer Edwin Hubble showed that galaxies are very distant. (He did this by identifying within other galaxies certain types of stars that are found in our own galaxy and whose light output is known. These stars appear very dim when seen in other galaxies, so they must be very far away.) In collaboration with the astronomer Milton Humason, he also measured the spectra of many galaxies. These are really the spectra of the combined light from all of the stars that make up each galaxy, so they have absorption lines (see Section 26-5). Hubble and Humason found that the absorption lines in the spectra of almost all galaxies are shifted to longer wavelengths and lower frequencies than in the spectra of nearby stars in our own galaxy. We call this a **redshift** because in the visible part of the electromagnetic spectrum, red light has the longest wavelength and lowest frequency. What's more, the greater the distance to a galaxy, the greater the wavelength shift. What did this discovery mean?

NASA, ESA, and J. Lotz, M. Mountain, A. Koekemoer, and the HFF Team (STScI)

Figure 28-12 A universe of galaxies Each blob of light in this image from the Hubble Space Telescope is a galaxy, a grouping of billions of stars and other matter. The area shown is a very small patch of sky, only about 1/15 the apparent width of the full Moon.

To answer this question recall what we learned in Section 13-10 about the *Doppler effect*: If a source of waves is moving away from us, the wave that we receive from that source is shifted to a lower frequency and a longer wavelength. In Section 13-10 we introduced this idea in the context of sound waves, but the same effect also applies to electromagnetic waves. Hubble and Humason concluded that other galaxies are moving away from us, and so the light that we receive from those galaxies is at a longer wavelength. They found that the speed v at which a distant galaxy is receding from us is directly proportional to the distance d to that galaxy (**Figure 28-13**):

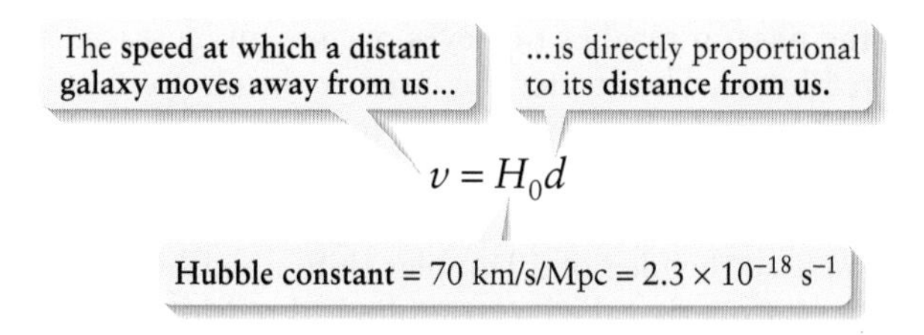

$$v = H_0 d$$

The Hubble law (28-9)

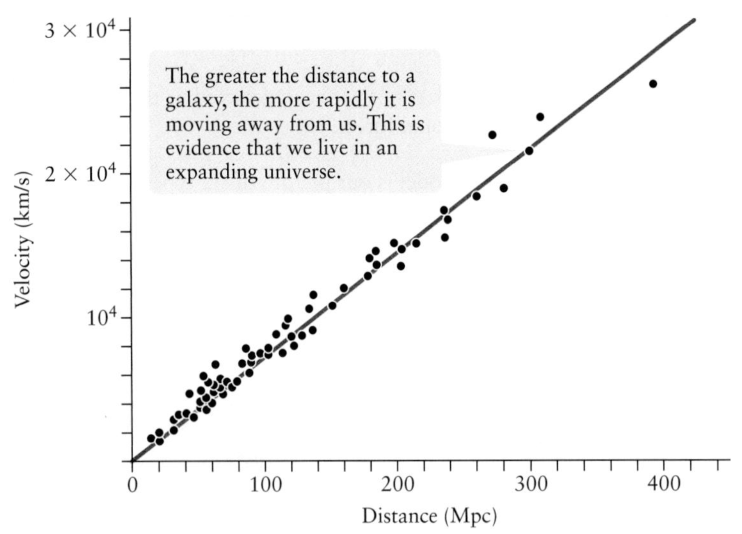

The greater the distance to a galaxy, the more rapidly it is moving away from us. This is evidence that we live in an expanding universe.
3 × 10⁴
2 × 10⁴
10⁴
Velocity (km/s)
0
100
200
300
400
Distance (Mpc)

Figure 28-13 The Hubble law Each black dot in this graph represents the speed at which an individual galaxy is moving away from us, as well as the distance to that galaxy. The straight line is the best fit to that data, corresponding to a value of $H_0 = 70$ km/s/Mpc in Equation 28-9.

This direct proportionality is called the **Hubble law**, and the constant H_0 in Equation 28-9 is called the **Hubble constant.** Astronomers typically measure distances in the cosmos in parsecs (pc), a unit equal to 3.26 light-years. One light-year (ly) is the distance that light travels in one year, equal to 3.09×10^{16} m. The current best value of H_0 is about 70 km/s/Mpc, where 1 Mpc = 1 megaparsec = 10^6 pc = 3.26×10^6 ly.

The interpretation of the Hubble law is that because remote galaxies are getting farther and farther apart as time goes on, the universe is expanding. A good analogy is a loaf of raisin bread being baked in an oven. As the loaf expands during baking, the amount of space between the raisins gets larger and larger. In the same way, as the universe expands, the amount of space between widely separated galaxies increases. The expansion of the universe is actually the expansion of space.

WATCH OUT! The universe is expanding, but the galaxies (and you) are not.

It's important to realize that the expansion of the universe occurs primarily in the vast spaces that separate clusters of galaxies. Just as the raisins in a loaf of raisin bread don't expand as the loaf expands in the oven, galaxies themselves do not expand. Einstein and others have established that an object that is held together by its own gravity, such as a galaxy or a cluster of galaxies, is always contained within a patch of nonexpanding space. A galaxy's gravitational field produces this nonexpanding region. Thus Earth and your body, for example, are not getting any bigger. Only the distance between widely separated galaxies increases with time.

Although distant galaxies are moving away from us, that does *not* mean that we are at the center of the universe. In an expanding universe, every galaxy moves away from every other galaxy, so an alien astronomer in a distant galaxy would see the same relationship between the speeds and distances of galaxies—that is, the same Hubble law—as does an Earth astronomer. Since every point in the universe appears to be at the center of the expansion, it follows that our universe has no center at all.

WATCH OUT! "If the universe is expanding, what is it expanding into?"

This commonly asked question arises only if we take too literally our raisin bread analogy, in which the loaf (representing the universe) expands in three-dimensional space into the surrounding air. But the actual universe includes *all* space; there is nothing "beyond" it because there is no "beyond." Asking "What lies beyond the universe?" is as meaningless as asking "Where on Earth is north of the North Pole?"

The ongoing expansion of space explains why the light from remote galaxies is redshifted. Imagine a photon coming toward us from a distant galaxy. As the photon travels through space, the space is expanding, so the photon's wavelength becomes stretched. When the photon reaches our eyes, we see an increased wavelength: The photon has been redshifted. The greater the distance the photon has had to travel and so the longer the amount of time it has had to travel to reach us, the more its wavelength will have been stretched. Thus photons from distant galaxies have larger redshifts than those of photons from nearby galaxies, as expressed by the Hubble law.

Although we've used the idea of a Doppler shift, a redshift caused by the expansion of the universe is properly called a **cosmological redshift.** It is *not* the same as a Doppler shift. Doppler shifts are caused by an object's motion through space, whereas a cosmological redshift is caused by the expansion of space.

EXAMPLE 28-1 Measuring the Distance to a Galaxy from Its Redshift

Find the distances to the following galaxies: (a) NGC 4889, whose redshifted spectrum shows that it is moving away from us at 6410 km/s (2.14% of the speed of light); (b) 1255-0, which has a larger redshift that shows it to be moving away from us at $0.822c$. Express the distances in megaparsecs and in light years.

Set Up

For each galaxy we know the speed v at which it is moving away from us due to the expansion of the universe. So we can use the Hubble law to determine the distance d to that galaxy.

The Hubble law:

$$v = H_0 d \qquad (28\text{-}9)$$

Solve

(a) Use Equation 28-9 and the value $H_0 = 70$ km/s/Mpc to calculate the distance to NGC 4889.

Rewrite Equation 28-9 to solve for the distance d:

$$d = \frac{v}{H_0}$$

We are given $v = 6410$ km/s, so

$$d = \frac{6410 \text{ km/s}}{70 \text{ km/s/Mpc}} = 92 \text{ Mpc}$$

$$= (92 \text{ Mpc})\left(\frac{3.26 \times 10^6 \text{ ly}}{1 \text{ Mpc}}\right) = 3.0 \times 10^8 \text{ ly}$$

(b) Repeat the calculations for the galaxy 1255-0, which is moving away at $0.822c$.

For this galaxy

$$v = 0.822c = 0.822(3.00 \times 10^5 \text{ km/s})$$

$$= 2.47 \times 10^5 \text{ km/s}$$

Use this to calculate the distance to the galaxy as in part (a):

$$d = \frac{2.47 \times 10^5 \text{ km/s}}{70/\text{km/s/Mpc}} = 3.5 \times 10^3 \text{ Mpc}$$

$$= (3.5 \times 10^3 \text{ Mpc})\left(\frac{3.26 \times 10^6 \text{ ly}}{1 \text{ Mpc}}\right) = 1.1 \times 10^{10} \text{ ly}$$

Reflect

NGC 4889 is some 300 million light years away, so the light we see from NGC 4889 left that galaxy 300 million years ago, before even the first dinosaurs appeared on Earth. Galaxy 1255-0 is far more distant, some 11 *billion* light years away. When the light we receive from 1255-0 left that galaxy some 11 billion years ago, our Earth—which is a mere 4.56×10^9 years old—had not yet formed. When we use telescopes to look at distant astronomical objects, we are not only looking out into space; we are also looking back in time.

The Big Bang, Cosmic Background Radiation, and the Origin of Matter

If the universe is expanding, it must be that in the past the matter in the universe must have been closer together and therefore denser than it is today. If we look far enough into the very distant past, there must have been a time when the density of matter was almost inconceivably high. This leads us to conclude that some sort of tremendous event caused ultradense matter to begin the expansion that continues to the present day. This event, called the **Big Bang**, marks the beginning of the universe. If we use the observed expansion rate of the universe and work backward, we find that the age of the universe is approximately 13.8 billion (13.8×10^9) years.

We learned in Section 15-3 that the temperature of an ideal gas decreases when it undergoes an adiabatic expansion (one in which there is no heat transfer to its surroundings). The expansion of the universe is much like an adiabatic expansion, so it follows that the average temperature of the universe has decreased over the past 13.8 billion years. (There are places in the universe that are at very high temperature, such as the interiors of stars. But the average temperature of space is very cold.) If we work backward

WATCH OUT! It's not correct to think of the Big Bang as an explosion.

When a bomb explodes pieces of debris fly off into space from a central location. If you could trace all the pieces back to their origin, you could find out exactly where the bomb had been. This process is not possible with the universe, however, because the universe itself always has and always will consist of all space. The present-day universe is infinite and was infinite when it first originated in the Big Bang.

in time, it follows that the early universe must have been at very high temperatures indeed. The hot early universe must therefore have been filled with many high-energy photons of high frequency and short wavelength. The properties of this radiation field depended on its temperature, as described by Planck's blackbody law (Section 26-2).

The universe has expanded so much since those ancient times that all those short-wavelength photons have had their wavelengths stretched by a tremendous factor. As a result, they have become low-energy, long-wavelength photons. The temperature of this cosmic radiation field is now only a few degrees above absolute zero, and the blackbody spectrum of this radiation has its peak intensity at low frequencies in the microwave part of the electromagnetic spectrum (wavelengths of approximately 1 mm). Hence this radiation field, which fills all of space, is called the **cosmic microwave background** or **cosmic background radiation**. It represents the "afterglow" of the very high-temperature conditions that prevailed when the universe was young.

It's actually possible to detect the cosmic background radiation using an ordinary television (Figure 28-1b). Using much more sensitive detectors, scientists have found that the cosmic background radiation does indeed have a blackbody spectrum with a temperature of 2.725 K, which we can regard as the average temperature of the present-day universe. However, we observe slightly different cosmic background radiation coming from different parts of the sky. Even when the effects of Earth's motion are accounted for, there remain variations in the temperature of the radiation field of about 300 μK (300 microkelvins, or 3×10^{-4} K) above or below the average 2.725 K temperature (**Figure 28-14**). These tiny temperature variations indicate that matter and radiation were not distributed in a totally uniform way in the early universe. Regions that were slightly denser than average were also slightly cooler than average; less dense regions were slightly warmer. Over time the denser regions evolved to form the first galaxies, and within them the first stars. By studying these nonuniformities we are really studying our origins.

The map of the sky shown in Figure 28-14 shows the universe as it was some 380,000 years after the Big Bang, when the universe was less than 0.003% of its present age. Prior to that time the universe was so dense as to be opaque. (Think of a hot, dense, luminous fog.) Events prior to that date are forever hidden from our direct view. But by using the laws of particle physics that we described in Sections 28-2 and 28-3, we can infer a great deal about what conditions must have been like in the very early universe.

Our best understanding is that the universe began at extremely high temperatures and that the immense energy gave rise to a sea of quarks, antiquarks, leptons, and antileptons. As the universe expanded and cooled, by 10^{-6} s after the Big Bang quarks and antiquarks had coalesced into baryons, antibaryons, and mesons. But now we have a dilemma. If there had been perfect symmetry between particles and antiparticles, then for every proton there should have been an antiproton. For every electron, there should likewise have been a positron. By the time the universe was 1 second old, every particle would have been annihilated by an antiparticle, leaving no matter at all in the universe! Obviously, this did not happen. But why didn't it?

The resolution to this dilemma is that the laws of physics are very slightly asymmetrical between matter and antimatter: For every 10^9 antimatter particles that were created out of the energy of the Big Bang, 10^9 plus one matter particles were created. When those 10^9 antimatter particles encountered 10^9 matter particles, annihilations resulted in no more particles and lots of energy in the form of photons. (These are the photons that gave rise to the cosmic background radiation that we see today.) That one extra matter particle avoided annihilation. All the normal matter that we see today is a result of the very slight imbalance between matter and antimatter in the early universe. Experiments on particle interactions are consistent with this very slight asymmetry between matter and antimatter.

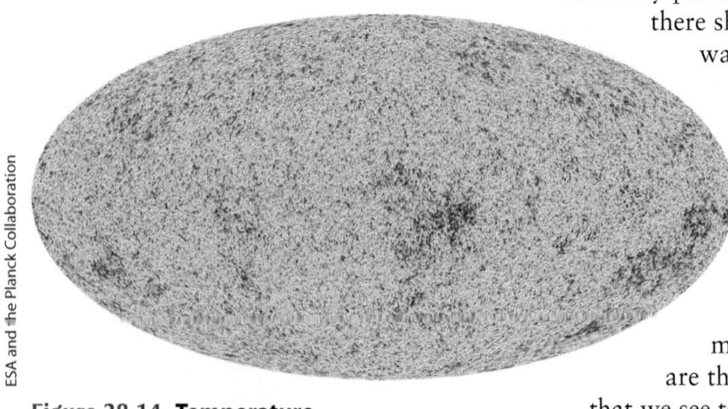

ESA and the Planck Collaboration

Figure 28-14 Temperature variations in the cosmic microwave background This map shows small variations in the temperature of the cosmic background radiation across the entire sky. Lower-temperature regions (shown in blue) show where the early universe was slightly denser than average.

By 15 minutes after the Big Bang, temperatures had dropped enough that protons and neutrons could coalesce into the first atomic nuclei. Only the four lightest elements (hydrogen, helium, lithium, and beryllium) were present in appreciable numbers. The heavier elements would be formed only much later, once stars had formed and nuclear reactions within those stars could manufacture carbon, nitrogen, oxygen, and all the other elements.

Keep in mind that only *nuclei*, not atoms, formed in the first 15 minutes of the history of the universe. It would be another 380,000 years before temperatures became low enough for these nuclei to combine with electrons to form atoms. When this happened the universe became transparent for the first time. The map of the cosmic background radiation in Figure 28-14 shows the universe at that moment in cosmic history.

Dark Matter and Dark Energy

At this point we might feel that we have understood, at least in broad outline, the nature and origin of matter. But have we in fact accounted for all of the matter in the universe? To answer this question we again look at galaxies, in particular *spiral galaxies* like the ones shown in **Figure 28-15**. These galaxies have a flattened disk shape, and all of the material in the galaxy rotates around the center. Each part of the galaxy is held in its orbit around the center by the gravitational attraction of the other parts of the galaxy; the greater that gravitational attraction on a given part, the faster that part must move to remain in a circular orbit (see Section 10-4). So if we can measure the speed at which parts of the galaxy orbit around the center, we can calculate the total amount of mass in the galaxy. Most of the atoms in the galaxy are in its stars, and we can estimate the combined mass of all the stars from the brightness of the galaxy. If the total amount of mass in the galaxy is a close match to the combined amount of mass in its stars, then we can conclude that ordinary matter accounts for most or all of the mass of the galaxy.

We can measure orbital speeds most directly for spiral galaxies that happen to be oriented edge-on to us, like the one shown in **Figure 28-15b**. We do this by looking at the spectrum of light from the two sides of the galaxy: The spectral lines will be Doppler shifted toward shorter wavelengths on the side where material is rotating toward us, and toward longer wavelengths on the side where material is rotating away from us. (These shifts are in addition to the cosmological redshift in the galaxy's spectrum as a whole, caused by the expansion of the universe.) The amount of Doppler shift tells us the speeds at which material in various parts of the galaxy are moving around the center.

These measurements have been made for a wide variety of spiral galaxies, and the results are always the same: The total mass of the galaxy deduced from the orbital speeds is about *five times greater* than the combined mass of all of the stars in the galaxy. So ordinary matter, which is concentrated in the galaxy's stars, is only a small fraction of the mass of the galaxy. The remaining mass is in a form that does not emit electromagnetic radiation of any kind and may not even be composed of the fundamental particles that we described in Sections 28-2 and 28-3. This mysterious matter, which is the dominant form of matter in spiral galaxies and apparently in the universe as a whole, is called **dark matter**. As of this writing, its nature remains a mystery to science.

There is yet another complication to our understanding of the universe. As we look at increasingly distant galaxies, and so look farther back in the history of the universe, we find that as expected these distant galaxies are moving away from us due to the expansion of the universe. But the expansion rate has *not* been constant; observations show that for the past 5 billion years, the expansion of the universe has been *speeding up*! To date the only viable explanation for this increased speed is that, in addition to ordinary matter and dark matter, the universe is suffused with a curious form of energy that causes an accelerated expansion of the universe. We cannot detect this energy from its gravitational effects (the technique astronomers use to detect dark matter), and it does not emit detectable radiation of any kind. We refer to this curious energy as **dark energy**. The nature of dark energy is even more mysterious than that of dark matter.

Since matter and energy are equivalent through the Einstein relation $E = mc^2$, we can compare the relative importance of ordinary matter, dark matter, and dark energy by asking what fraction of the total energy in the universe each one represents. The graph in **Figure 28-16** shows the results: Dark energy represents 69.4% of the total, dark matter another 25.8%, and "ordinary" normal matter just 4.8%.

(a) We see spiral galaxy NGC 3982 face-on.

All of the matter in the galaxy rotates around the center.
NASA, ESA, and the Hubble Heritage Team (STScI/AURA)

(b) We see spiral galaxy NGC 4013 edge-on.

Figure 28-15 Measuring the mass of a spiral galaxy (a) The material in a spiral galaxy is held in orbit around the center by the gravitational attraction of all the other material in the galaxy. (b) For a galaxy that is edge-on to us, we can measure the orbital speeds, and hence the mass of the galaxy, using the Doppler effect.

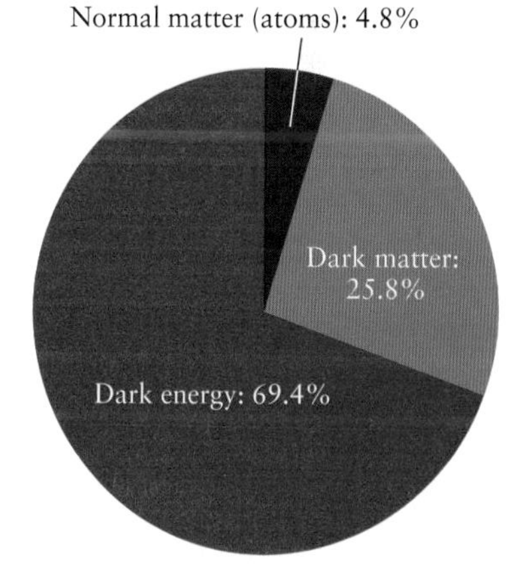

What we think of as "normal" matter makes up less than 5% of the contents of the universe.

Figure 28-16 Recipe for a universe This pie chart shows the three constituents of the universe as a whole. The numbers show what percentage of the total energy content of the universe each constituent represents.

This is a sobering graph, for it tells us that more than 95% of the energy in the universe is in forms that we do not understand! The quest to understand both dark energy and dark matter is one of the most challenging pursuits in contemporary physics.

GOT THE CONCEPT? 28-4 Dark Matter

In spiral galaxies like those shown in Figure 28-15, the ratio of dark matter to visible matter is about five to one. The percentage of the light from a spiral galaxy that is emitted by dark matter is closest to (a) 83%; (b) 80%; (c) 20%; (d) 17%; (e) zero.

TAKE-HOME MESSAGE FOR Section 28-4

✔ The redshifts of galaxies reveal that we live in an expanding universe.

✔ The universe began with a Big Bang, with tremendously high temperatures present in the early universe. The cosmic background radiation is the "afterglow" of that early epoch in the history of the universe.

✔ The dominance of matter over antimatter in the universe is the result of a small but important asymmetry in the laws of particle physics.

✔ Most of the matter in the universe is in the form of dark matter, and most of the energy in the universe is in the form of dark energy. The nature of dark matter and the nature of dark energy are unsolved problems in physics.

Key Terms

antimatter
antiquark
baryon
Big Bang
cosmic background radiation
cosmic microwave background
cosmological redshift
cosmology
dark energy
dark matter
exchange particle
galaxy
gluon
hadron
Heisenberg uncertainty principle
Higgs field
Higgs particle
Hubble constant
Hubble law
lepton
meson
particle physics
quark
quark confinement
redshift
Standard Model
virtual particle
weak force

Chapter Summary

Topic	Equation or Figure
Hadrons and leptons: Hadrons, including the proton and neutron, are particles that experience the strong force; leptons, including the electron and neutrino, are particles that do not. Hadrons are composed of more fundamental particles called quarks; these are confined to the interior of hadrons and cannot exist in isolation. Hadrons include baryons, which are made of three quarks, and mesons, which are made of a quark and antiquark. Leptons are themselves fundamental particles; they are not made of anything simpler.	Baryons (a) Proton (p) (b) Neutron (n) 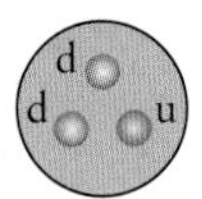 Net charge = $q_u + q_u + q_d = \left(+\frac{2}{3}e\right)+\left(+\frac{2}{3}e\right)+\left(-\frac{1}{3}e\right) = +e$ Net charge = $q_d + q_d + q_u = \left(-\frac{1}{3}e\right)+\left(-\frac{1}{3}e\right)+\left(+\frac{2}{3}e\right) = 0$ (Figure 28-3) Mesons (a) Positive pion (π^+) (b) Negative pion (π^-) 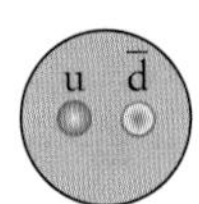 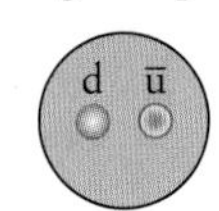 Net charge = $q_u + q_{\bar{d}} = \left(+\frac{2}{3}e\right)+\left(+\frac{1}{3}e\right) = +e$ Net charge = $q_d + q_{\bar{u}} = \left(-\frac{1}{3}e\right)+\left(-\frac{2}{3}e\right) = -e$ (Figure 28-4)

The four forces and exchange particles: All interactions between particles can be understood in terms of four basic forces. Each of these forces involves the exchange of virtual particles, which is permitted by the Heisenberg uncertainty principle. The strong force between quarks involves the exchange of gluons, and the electromagnetic force involves the exchange of photons; both gluons and photons have zero mass, so these are long-range forces. The weak force responsible for beta decay involves the exchange of massive particles, so this force has a very short range. The gravitational force involves the exchange of gravitons, which have not yet been detected experimentally.

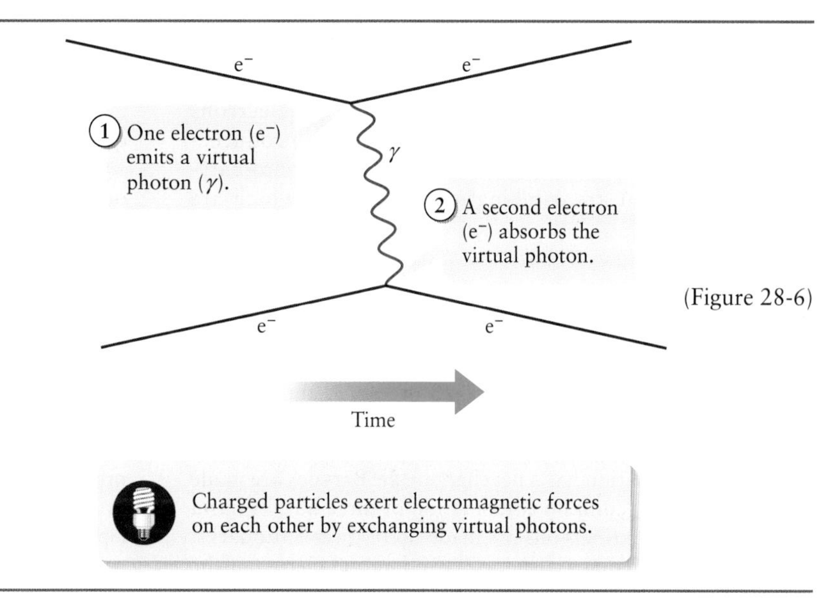
e⁻
e⁻
(1) One electron (e⁻) emits a virtual photon (γ).
γ
(2) A second electron (e⁻) absorbs the virtual photon.
e⁻
e⁻
Time
Charged particles exert electromagnetic forces on each other by exchanging virtual photons.

(Figure 28-6)

The **energy of a phenomenon is necessarily uncertain** by an amount ΔE.

$$\Delta E \Delta t \geq \frac{\hbar}{2} \quad (28\text{-}7)$$

The **shorter the duration** Δt of the phenomenon, the **greater the energy uncertainty** ΔE.

$\hbar = \dfrac{h}{2\pi}$ = **Planck's constant divided by** 2π

Cosmology, dark matter, and dark energy: The Hubble law—the more distant a galaxy is from us, the faster it moves away from us—tells us that the universe is expanding and was once very highly compressed and at very high temperature. The cosmic background radiation is a relic of this ancient epoch. Processes in the early universe had a slight preference for matter over antimatter, which is why the present-day universe contains matter but almost no antimatter. Normal matter is actually just a small component of the universe: Mysterious dark matter is about five times as prevalent. Even more important and even more mysterious is dark energy, which causes the expansion of the universe to speed up.

The **speed at which a distant galaxy moves away from us...** ...is directly proportional to its **distance from us.**

$$v = H_0 d \quad (28\text{-}9)$$

Hubble constant = 70 km/s/Mpc = 2.3×10^{-18} s^{-1}

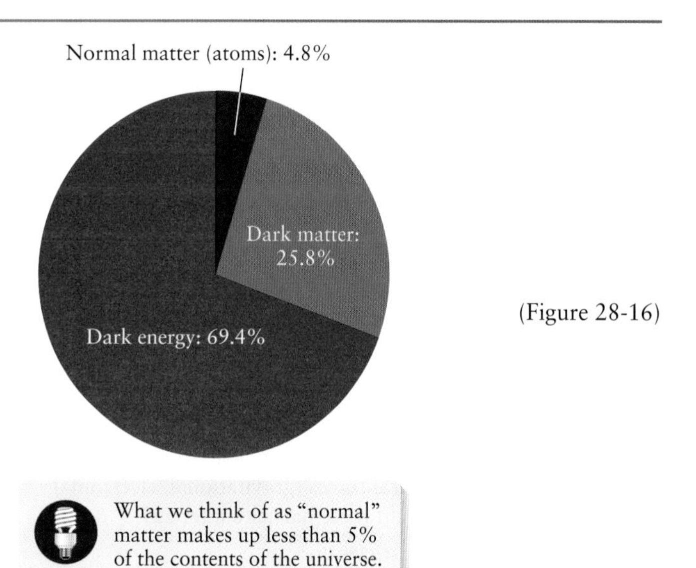
Normal matter (atoms): 4.8%
Dark matter: 25.8%
Dark energy: 69.4%
What we think of as "normal" matter makes up less than 5% of the contents of the universe.

(Figure 28-16)

Answer to What do you think? Question

(e) A collision such as this can produce other electrons (matter) and positrons (antimatter) but can also produce a shower of quarks and antiquarks. These can group into baryons (particles that are combinations of quarks, which are matter), antibaryons (particles that are combinations of antiquarks, which are antimatter), and mesons (particles that are a combination of a quark and an antiquark and hence a mixture of matter and antimatter).

Answers to Got the Concept? Questions

28-1 (c) A quark has charge $+2e/3$ or $-e/3$, and an antiquark has charge $-2e/3$ or $+e/3$ (see Table 28-1). Mesons are made up of one quark and one antiquark, and no quark–antiquark combination has a net charge $-2e$. Baryons are made up of three quarks, and no combination of three quarks has a net charge $-2e$. Antibaryons are made up of three antiquarks; if each of the three has charge $-2e/3$, the total charge of the antibaryon will be $(-2e/3) + (-2e/3) + (-2e/3) = -2e$.

28-2 (a) The neutron (n) and proton (p) each have baryon number $B = 1$, and the antiproton ($\overline{\text{p}}$) has $B = -1$. Baryon number is conserved in process (a), since total B equals $1 + 1 = 2$ before the process and equals $1 + 1 + 1 + (-1) = 2$ after the process. This process can occur if the initial neutron and proton have sufficient kinetic energy so that, when they collide, this energy can be converted into the rest energy of the additional proton and antiproton. Baryon number is *not* conserved in process (b), since total B equals $1 + 1 = 2$ before the process and $1 + 1 + 1 = 3$ after the process. So (b) cannot occur. In process (c) the π^- is a meson with $L_e = 0$ and $L_\mu = 0$, the muon μ^- has $L_e = 0$ and $L_\mu = 1$, and the electron antineutrino $\bar{\nu}_e$ has $L_e = -1$ and $L_\mu = 0$. The total electron–lepton number is 0 before the process and $0 + -1 = -1$ after the process, and the total muon–lepton number is zero before the process and $1 + 0 = 1$ after the process. So neither electron–lepton number L_e nor muon–lepton number L_μ is conserved in process (c), and process (c) cannot occur.

28-3 (a) Table 28-1 shows that the c quark has charge $+2e/3$ and the s quark has a charge $-e/3$. The quark must therefore emit a W^+ with charge $+e$ so that the total charge of the W^+ and s quark equals the initial charge of the c quark: $(+e) + (-e/3) = +2e/3$. The W^+ must decay into a positron (charge $+e$) to conserve charge. Since the W^+ has electron–lepton number $L_e = 0$ (it's not a lepton) and the positron has $L_e = -1$, the positron must be accompanied by a neutral particle with $L_e = +1$. That must be a neutrino, not an antineutrino (which has $L_e = -1$).

28-4 (e) Dark matter is so called because it emits *no* electromagnetic radiation of any kind.

Questions and Problems

In a few problems you are given more data than you actually need; in a few other problems you are required to supply data from your general knowledge, outside sources, or informed estimate.

Interpret as significant all digits in numerical values that have trailing zeros and no decimal points.

- • Basic, single-concept problem
- •• Intermediate-level problem; may require synthesis of concepts and multiple steps
- ••• Challenging problem
- SSM *Solution is in Student Solutions Manual*
- Example *See worked example for a similar problem*

Conceptual Questions

1. • Define these terms: (a) baryon, (b) meson, (c) quark, (d) lepton, and (e) antiparticle.

2. • A meson and a baryon come very close to one another. By which of the fundamental forces (gravitational, electromagnetic, weak, or strong) could these particles interact?

3. • What do exchange particles do?

4. • Discuss the similarities and differences between a photon and a gluon.

5. • Write the quark content for the antiparticle of each of the following: (a) n, (b) p, and (c) π^+.

6. • List all of the unique combinations of quarks resulting in a baryon that can be produced from just the up (u), down (d), and strange (s) quarks. (Neglect the antimatter baryons.) SSM

7. • A positron is stable; that is, it does not decay. Why, then, does a positron have only a short existence?

8. • When a positron and an electron annihilate at rest, why must more than one photon be created?

9. • Explain how we know the universe is expanding.

10. • Explain the evidence in favor of the existence of (a) dark matter and (b) dark energy.

Multiple-Choice Questions

11. • A particle composed of a quark and an antiquark is classified as a

A. baryon.
B. meson.
C. photon.
D. lepton.
E. antiparticle. SSM

12. • A particle composed of three quarks is classified as a

A. baryon.
B. meson.
C. photon.
D. lepton.
E. antiparticle.

13. • The quark composition of an antiproton is
   A. uud.
   B. $\text{uu}\overline{\text{d}}$.
   C. $\overline{\text{uu}}\text{d}$.
   D. $\overline{\text{uud}}$.
   E. uuu.

14. • The beta decay process is mediated by the
   A. gravitational force.
   B. electromagnetic force.
   C. strong force.
   D. weak force.
   E. more than one of A, B, C, and D.

15. • Through which force does the photon primarily interact with other particles?
   A. strong
   B. electromagnetic
   C. weak
   D. gravitational
   E. More than one of the above are equally important.

16. • Which of these particles interact using the strong force?
   A. neutrinos.
   B. leptons.
   C. photons.
   D. quarks.
   E. gravitons.

17. • Recent observations indicate that most of the energy in the universe is in the form of
   A. protons, neutrons, and electrons.
   B. dark matter.
   C. dark energy.
   D. photons.
   E. none of the above.

18. • What causes the expansion of the universe to accelerate?
   A. gravitational potential energy
   B. electric potential energy
   C. thermal energy
   D. dark energy
   E. kinetic energy

Estimation/Numerical Analysis

19. • One type of meson that can be exchanged between nucleons is the ρ (rho), with a mass of 775 MeV/c^2. Estimate the range of the interaction due to the exchange of ρ mesons.

20. • Estimate the distance to a galaxy that is moving away from us at one-tenth of the speed of light due to the expansion of the universe.

21. •• Suppose that a new fundamental force were discovered having a range about equal to the radius of a hydrogen atom. Estimate what you would expect to be the mass (in MeV/c^2) of the mediating particle of the new force. How does the mass compare with the mass of the electron?

Problems

28-1 Studying the ultimate constituents of matter helps reveal the nature of the universe
28-2 Most forms of matter can be explained by just a handful of fundamental particles

22. • Is the reaction $n \rightarrow \pi^+ + \pi^- + \mu^+ + \mu^-$ possible? Explain why or why not.

23. • Is the reaction $p \rightarrow e^+ + \gamma$ possible? Explain why or why not. SSM

24. • Is the reaction $e^- + p \rightarrow n + \bar{\nu}_e$ possible? Explain why or why not.

25. • What is the charge of the particle that is composed of (a) the quark combination uds? (b) The quark combination uss? (c) Which particle is likely to be more massive? Explain your answers.

26. • Two protons collide in a particle accelerator and generate new particles from their kinetic energy. Which of the following reactions are possible? Which are not possible? If a reaction is not possible, state which conservation law(s) is (are) violated.
   A. $p + p \rightarrow p + p + p + \bar{p}$
   B. $p + p \rightarrow p + p + n + \bar{n}$
   C. $p + p \rightarrow p + K^+$

27. • Which of the following two possibilities for the weak decay of a sigma particle (a baryon) are possible? Why?
   A. $\Sigma^- \rightarrow \pi^- + p$
   B. $\Sigma^- \rightarrow \pi^- + n$

28. • Which of the following reactions are possible? If a reaction is not possible, tell which conservation law(s) is (are) violated.
   A. $n \rightarrow p + e^- + \bar{\nu}_e$
   B. $\mu^- \rightarrow e^- + \bar{\nu}_e + \nu_\mu$
   C. $\pi^- \rightarrow \mu^- + \bar{\nu}_\mu$ SSM

29. •• A neutral η (eta) meson at rest decays into two photons:

$$\eta \rightarrow \gamma + \gamma$$

Calculate the energy, momentum, and wavelength of each of the photons. The mass of the eta particle is 547 MeV/c^2.

30. •• Two protons with the same speed collide head-on in a particle accelerator, causing the reaction:

$$p + p \rightarrow p + p + \pi^0$$

Calculate the minimum kinetic energy of each of the incident protons.

31. •• A high-energy photon in the vicinity of a nucleus can create an electron-positron pair:

$$\gamma \rightarrow e^- + e^+$$

(a) What minimum energy must the photon have? (b) Why is the nucleus needed? SSM

32. •• How much energy, in the form of gamma rays, would result due to the annihilation of a positron with kinetic energy 34 MeV and an electron with kinetic energy 16 MeV?

33. •• A proton–antiproton annihilation takes place, leaving two photons with a combined energy of 2.5 GeV. Find the kinetic energy of the proton and antiproton if the proton has (a) the same kinetic energy as the antiproton and (b) 1.25 times as much kinetic energy as the antiproton. SSM

34. •• The kinetic energy of a neutral pion (π^0) is 860 MeV. This pion decays to two photons, one of which has energy 640 MeV. Calculate the energy of the other photon.

28-3 Four fundamental forces describe all interactions between material objects

35. • Draw the Feynman diagram for beta-plus decay wherein a proton changes into a neutron, a positron, and a neutrino.

36. • Draw the Feynman diagram for the process described in Problem 31.

37. • Draw the Feynman diagram for annihilation of a proton and an antiproton into two photons. SSM

38. • Draw the Feynman diagram for the beta decay of the antineutron.

39. • Two protons with enough energy to possibly produce the Higgs particle are smashed together in a series of experiments. Each of these experiments produces a unique collection of signals corresponding to the presence of particles produced directly or indirectly as a result of the proton-proton collision. One experiment produces signals indicating the presence of an electron, two muons, two high-energy photons, and a Z boson. Could this collection of particles have been produced by the decay of a Higgs particle? Explain your answer.

28-4 We live in an expanding universe, and the nature of most of its contents is a mystery

40. • A galaxy is observed to recede from Earth with an approximate speed of $0.8c$. (a) Approximately how far from Earth is this galaxy? (b) How long ago was the light that we see emitted by the galaxy?

41. • Galaxy NGC 3982 (Figure 28-15a) is 6.8×10^7 ly from Earth. Approximately how fast is it receding from Earth?

42. • One proposal about dark matter is that it is made up of particles substantially more massive than a proton that have zero electric charge, so they respond to gravitational forces but not electromagnetic forces. Assume that a dark matter particle has 100 times the mass of proton, which has mass 1.67×10^{-27} kg. (a) The average density of dark matter in our Milky Way galaxy is thought to be about 5×10^{-23} kg/m^3. How many dark matter particles are there in one cubic meter? (b) There are about 2.7×10^{25} molecules per cubic meter in the air around you, and a total of about 10^{19} grains of sand on all the beaches on Earth. Relatively speaking, which is more likely: finding a dark matter particle in the air around you, or finding one particular grain of sand somewhere on one of Earth's beaches?

General Problems

43. • Suppose that a fifth fundamental force were discovered and that it was mediated by electrons and positrons. Estimate the range of the force.

44. •• The United States used approximately 4.12×10^{12} kWh of electrical energy during 2010. (a) If we could generate all the energy using a matter–antimatter reactor, how many kilograms of fuel would the reactor need, assuming 100% efficiency in the annihilation? (b) Suppose the fuel consisted of iron and anti-iron, each of density 7800 kg/m^3. If the iron and anti-iron were each stored as a cube, what would be the dimensions of each cube?

45. •• Suppose that an electron neutrino and an electron antineutrino, both of which are just barely moving, encounter each other in space and completely annihilate to form two photons of equal energy. In view of the uncertainty about the mass of the electron neutrino (see Table 28-2), what is the shortest wavelength of light that could be emitted by the annihilation? Would the light be visible to the human eye?

46. •• **Astronomy** To determine what took place during the first 10^{-43} s after the Big Bang, a time interval called the *Planck time* t_P, a theory that describes gravity in terms of quantum mechanics is required. (No such theory yet exists.) Calculate t_P by first using dimensional analysis and finding the proper combination of the fundamental constants: Newton's gravitational constant $G = 6.6738 \times 10^{-11}\ \text{N} \cdot \text{m}^2/\text{kg}^2$, the speed of light $c = 2.9979 \times 10^8$ m/s, and Planck's constant divided by 2π, $\hbar = 1.0546 \times 10^{-34}$ J · s. Start your calculation by asking what power would each of the three constants need to have in order to yield units of seconds: $[t_P] = [G]^x[c]^y[\hbar]^z$. Solve for x, y, and z to derive the formula for t_P. Then use the values of the three constants to calculate the precise value of t_P.

47. •• The very massive Higgs particle (mass 125 GeV/c^2) is created when two protons traveling at equally high speeds but in opposite directions collide head-on. The mass of a proton is 938.27 MeV/c^2. In order to make a Higgs particle when they collide, each proton must have a minimum kinetic energy of 62.5 GeV. (a) What is the minimum total energy of each proton? (b) In terms of the speed of light c, what speed is each proton traveling? Calculate the total energy and speed to five significant figures.

48. • A Higgs particle (mass 125 GeV/c^2) can decay into two photons. (a) If a Higgs particle produced in a high-energy collision is at rest, what will be the energy of each photon? (*Hint*: Since the Higgs particle has zero momentum, the two photons must have the same magnitude of momentum but must travel in opposite directions.) Give your answer in GeV and in joules. (b) What will be the wavelength of each photon? (c) In what part of the electromagnetic spectrum does this wavelength lie?

49. •• The photons that make up the cosmic microwave background (Figure 28-14) were emitted about 380,000 years after the Big Bang. Today, 13.8 billion years after the Big Bang, the wavelengths of these photons have been stretched by a factor of about 1100 since they were emitted because lengths in the expanding universe have increased by that same factor of about 1100. (a) Consider a cubical region of empty space in today's universe 1.00 m on a side, with a volume of 1.00 m^3. What was the length of each side of this same cubical region 380,000 years after the Big Bang? What was its volume? (b) Today the average density of ordinary matter in the universe is about 2.4×10^{-27} kg/m^3. What was the average density of ordinary matter at the time that the photons in the cosmic background radiation were emitted? Example 28-1

APPENDIX A
SI Units and Conversion Factors

Base Units*

Length	The *meter* (m) is the distance traveled by light in a vacuum in 1/299,792,458 s.
Time	The *second* (s) is the duration of 9,192,631,770 periods of the radiation corresponding to the transition between the two hyperfine levels of the ground state of the ^{133}Cs atom.
Mass	The *kilogram* (kg) is the mass of the international standard body preserved at Sèvres, France.
Mole	The *mole* (mol) is the amount of substance of a system which contains as many elementary entities as there are atoms in 0.012 kg of carbon-12.
Current	The *ampere* (A) is that constant current which, if maintained in two straight parallel conductors of infinite length, of negligible circular cross section, and placed 1 m apart in vacuum, would produce between the conductors a force equal to 2×10^{-7} N/m of length.
Temperature	The *kelvin* (K) is 1/273.16 of the thermodynamic temperature of the triple point of water.
Luminous intensity	The *candela* (cd) is the luminous intensity in a given direction, of a source that emits monochromatic radiation of frequency 540×10^{12} Hz and that has a radiant intensity, in that direction of 1/683 W/steradian.

*These definitions are found on the Internet at http://physics.nist.gov/cuu/Units/current.html.

Derived Units

Force	newton (N)	$1\ N = 1\ kg \cdot m/s^2$
Work, energy	joule (J)	$1\ J = 1\ N \cdot m$
Power	watt (W)	$1\ W = 1\ J/s$
Frequency	hertz (Hz)	$1\ Hz = cy/s$
Charge	coulomb (C)	$1\ C = 1\ A \cdot s$
Potential	volt (V)	$1\ V = 1\ J/C$
Resistance	ohm (Ω)	$1\ \Omega = 1\ V/A$
Capacitance	farad (F)	$1\ F = 1\ C/V$
Magnetic field	tesla (T)	$1\ T = 1\ N/(A \cdot m)$
Magnetic flux	weber (Wb)	$1\ Wb = 1\ T \cdot m^2$
Inductance	henry (H)	$1\ H = 1\ J/A^2$

Conversion Factors

Conversion factors are written as equations for simplicity; relations marked with an asterisk are exact.

Length

1 km = 0.6214 mi

1 mi = 1.609 km

1 m = 1.0936 yard = 3.281 ft = 39.37 in.

*1 in. = 2.54 cm

*1 ft = 12 in. = 30.48 cm

*1 yard = 3 ft = 91.44 cm

1 light-year = 1 $c \cdot$ y = 9.461×10^{15} m

*1 Å = 0.1 nm

Area

*1 m² = 10^4 cm²

1 km² = 0.3861 mi² = 247.1 acres

*1 in.² = 6.4516 cm²

1 ft² = 9.29×10^{-2} m²

1 m² = 10.76 ft²

*1 acre = 43 560 ft²

1 mi² = 640 acres = 2.590 km²

Volume

*1 m³ = 10^6 cm³

*1 L = 1000 cm³ = 10^{-3} m³

1 gal = 3.785 L

1 gal = 4 qt = 8 pt = 128 oz = 231 in.³

1 in.³ = 16.39 cm³

1 ft³ = 1728 in.³ = 28.32 L
= 2.832×10^4 cm³

Time

*1 h = 60 min = 3.6 ks

*1 d = 24 h = 1440 min = 86.4 ks

1 y = 365.25 day = 3.156×10^7 s

Speed

*1 m/s = 3.6 km/h

1 km/h = 0.2778 m/s = 0.6214 mi/h

1 mi/h = 0.4470 m/s = 1.609 km/h

1 mi/h = 1.467 ft/s

Angle and Angular Speed

*π rad = 180°

1 rad = 57.30°

1° = 1.745×10^{-2} rad

1 rev/min = 0.1047 rad/s

1 rad/s = 9.549 rev/min

Mass

*1 kg = 1000 g

*1 tonne = 1000 kg = 1 Mg

1 u = 1.6605×10^{-27} kg
931.49 MeV/c^2

1 kg = 6.022×10^{26} u

1 slug = 14.59 kg

1 kg = 6.852×10^{-2} slug

Density

*1 g/cm³ = 1000 kg/m³ = 1 kg/L

(1 g/cm³)g = 62.4 lb/ft³

Force

1 N = 0.2248 lb = 10^5 dyn

*1 lb = 4.448222 N

(1 kg)g = 2.2046 lb

Pressure

*1 Pa = 1 N/m²

*1 atm = 101.325 kPa = 1.01325 bar

1 atm = 14.7 lb/in.² = 760 mmHg
= 29.9 in.Hg = 33.9 ftH_2O

1 lb/in.² = 6.895 kPa

1 torr = 1 mmHg = 133.32 Pa

1 bar = 100 kPa

Energy

*1 kW $\cdot$ h = 3.6 MJ

*1 cal = 4.186 J

1 ft $\cdot$ lb = 1.356 J = 1.286×10^{-3} BTU

*1 L $\cdot$ atm = 101.325 J

1 L $\cdot$ atm = 24.217 cal

1 BTU = 778 ft $\cdot$ lb = 252 cal = 1054.35 J

1 eV = 1.602×10^{-19} J

1 u $\cdot$ c^2 = 931.49 MeV

*1 erg = 10^{-7} J

Power

1 horsepower = 550 ft $\cdot$ lb/s = 745.7 W

1 BTU/h = 2.931×10^{-4} kW

1 W = 1.341×10^{-3} horsepower
= 0.7376 ft $\cdot$ lb/s

Magnetic Field

*1 T = 10^4 G

Thermal Conductivity

1 W/(m $\cdot$ K) = 6.938 BTU $\cdot$ in./(h $\cdot$ ft² $\cdot$ °F)

1 BTU $\cdot$ in./(h $\cdot$ ft² $\cdot$ °F) = 0.1441 W/(m $\cdot$ K)

APPENDIX B
Numerical Data

Terrestrial Data	
Free-fall acceleration g	
Standard value (at sea level at 45° latitude)*	$9.806\ 65\ m/s^2$; $32.1740\ ft/s^2$
At equator*	$9.7804\ m/s^2$
At poles*	$9.8322\ m/s^2$
Mass of Earth M_E	5.98×10^{24} kg
Radius of Earth R_E, mean	6.38×10^6 m; 3960 mi
Escape speed	1.12×10^4 m/s; 6.96 mi/s
Solar constant†	$1.37\ kW/m^2$
Standard temperature and pressure (STP):	
Temperature	273.15 K
Pressure	101.3 kPa (1.00 atm)
Molar mass of air	28.97 g/mol
Density of air (273.15 K, 101.3 kPa), ρ_{air}	$1.29\ kg/m^3$
Speed of sound (273.15 K, 101.3 kPa)	331 m/s
Latent heat of fusion of H_2O (0°C, 1 atm)	334 kJ/kg
Latent heat of vaporization of H_2O (100°C, 1 atm)	2.26 MJ/kg

* Measured relative to Earth's surface.
† Average power incident normally on 1 m^2 outside Earth's atmosphere at the mean distance from Earth to the Sun.

Astronomical Data*	
Earth	
Distance to the Moon, mean†	3.844×10^8 m; 2.389×10^5 mi
Distance to the Sun, mean†	1.496×10^{11} m; 9.32×10^7 mi; 1.00 AU
Orbital speed, mean	2.98×10^4 m/s
Moon	
Mass	7.35×10^{22} kg
Radius	1.737×10^6 m
Period	27.32 day
Acceleration of gravity at surface	$1.62\ m/s^2$
Sun	
Mass	1.99×10^{30} kg
Radius	6.96×10^8 m

* Additional solar system data are available from NASA at http://nssdc.gsfc.nasa.gov/planetary/planetfact.html.
† Center to center.

Physical Constants*

Universal constant of gravitation	G	$6.674\,08(31) \times 10^{-11}\ \mathrm{N \cdot m^2/kg^2}$
Speed of light	c	$2.997\,924\,58 \times 10^{8}\ \mathrm{m/s}$
Fundamental charge	e	$1.602\,176\,6208(98) \times 10^{-19}\ \mathrm{C}$
Avogadro's constant	N_A	$6.022\,140\,857(74) \times 10^{23}$ particles/mol
Gas constant	R	$8.314\,4598(48)\ \mathrm{J/(mol \cdot K)}$
		$1.987\,2036(11)\ \mathrm{cal/(mol \cdot K)}$
		$8.205\,7338(47) \times 10^{-2}\ \mathrm{L \cdot atm/(mol \cdot K)}$
Boltzmann constant	$k = R/N_A$	$1.380\,648\,52(79) \times 10^{-23}\ \mathrm{J/K}$
		$8.617\,3303(50) \times 10^{-5}\ \mathrm{eV/K}$
Stefan-Boltzmann constant	$\sigma = (\pi^2/60)k^4/(\hbar^3 c^2)$	$5.670\,367(13) \times 10^{-8}\ \mathrm{W/(m^2 \cdot K^4)}$
Atomic mass constant	$m_u = (1/12)m(^{12}\mathrm{C})$	$1.660\,539\,040(20) \times 10^{-27}\ \mathrm{kg} = 1\ \mathrm{u}$
Permeability of free space	μ_0	$4\pi \times 10^{-7}\ \mathrm{N/A^2}$ $1.256\,637\ldots \times 10^{-6}\ \mathrm{N/A^2}$
Permittivity of free space	$\epsilon_0 = 1/(\mu_0 c^2)$	$8.854\,187\,817\ldots \times 10^{-12}\ \mathrm{C^2/(N \cdot m^2)}$
Coulomb constant	$k = 1/(4\pi\epsilon_0)$	$8.987\,551\,787\ldots \times 10^{9}\ \mathrm{N \cdot m^2/C^2}$
Planck's constant	h	$6.626\,070\,040(81) \times 10^{-34}\ \mathrm{J \cdot s}$
		$4.135\,667\,662(25) \times 10^{-15}\ \mathrm{eV \cdot s}$
	$\hbar = h/(2\pi)$	$1.054\,571\,800(13) \times 10^{-34}\ \mathrm{J \cdot s}$
		$6.582\,119\,514(40) \times 10^{-16}\ \mathrm{eV \cdot s}$
Mass of electron	m_e	$9.109\,383\,56(11) \times 10^{-31}\ \mathrm{kg}$
		$0.510\,998\,9461(31)\ \mathrm{MeV}/c^2$
Mass of proton	m_p	$1.672\,621\,898(21) \times 10^{-27}\ \mathrm{kg}$
		$938.272\,0813(58)\ \mathrm{MeV}/c^2$
Mass of neutron	m_n	$1.674\,927\,471(21) \times 10^{-27}\ \mathrm{kg}$
		$939.565\,4133(58)\ \mathrm{MeV}/c^2$
Bohr magneton	$m_B = e\hbar/(2m_e)$	$9.274\,009\,994(57) \times 10^{-24}\ \mathrm{J/T}$
		$5.788\,381\,8012(26) \times 10^{-5}\ \mathrm{eV/T}$
Nuclear magneton	$m_n = e\hbar/(2m_p)$	$5.050\,783\,699(31) \times 10^{-27}\ \mathrm{J/T}$
		$3.152\,451\,2550(15) \times 10^{-8}\ \mathrm{eV/T}$
Magnetic flux quantum	$\phi_0 = h/(2e)$	$2.067\,833\,831(13) \times 10^{-15}\ \mathrm{T \cdot m^2}$
Quantized Hall resistance	$R_K = h/e^2$	$2.581\,280\,745\,55(59) \times 10^{4}\ \Omega$
Rydberg constant	R_H	$1.097\,373\,156\,8508(65) \times 10^{7}\ \mathrm{m^{-1}}$
Josephson frequency–voltage quotient	$K_J = 2e/h$	$4.835\,978\,525(30) \times 10^{14}\ \mathrm{Hz/V}$
Compton wavelength	$\lambda_C = h/(m_e c)$	$2.426\,310\,2367(11) \times 10^{-12}\ \mathrm{m}$

* Updated values for these and other constants may be found on the Internet at http://physics.nist.gov/cuu/Constants/index.html. The numbers in parentheses represent the uncertainties in the last two digits. (For example, 2.044 43(13) stands for 2.044 43 ± 0.000 13.) Values without uncertainties are exact, including those values with ellipses (such as the value of π, which is exactly 3.1415. . .).

APPENDIX C

Periodic Table of Elements*

1	2	3	4	5	6	7	8	9	10	11	12	13	14	15	16	17	18
1 H																	2 He
3 Li	4 Be											5 B	6 C	7 N	8 O	9 F	10 Ne
11 Na	12 Mg											13 Al	14 Si	15 P	16 S	17 Cl	18 Ar
19 K	20 Ca	21 Sc	22 Ti	23 V	24 Cr	25 Mn	26 Fe	27 Co	28 Ni	29 Cu	30 Zn	31 Ga	32 Ge	33 As	34 Se	35 Br	36 Kr
37 Rb	38 Sr	39 Y	40 Zr	41 Nb	42 Mo	43 Tc	44 Ru	45 Rh	46 Pd	47 Ag	48 Cd	49 In	50 Sn	51 Sb	52 Te	53 I	54 Xe
55 Cs	56 Ba	57–71 Lanthanoids	72 Hf	73 Ta	74 W	75 Re	76 Os	77 Ir	78 Pt	79 Au	80 Hg	81 Tl	82 Pb	83 Bi	84 Po	85 At	86 Rn
87 Fr	88 Ra	89–103 Actinoids	104 Rf	105 Db	106 Sg	107 Bh	108 Hs	109 Mt	110 Ds	111 Rg	112 Cn	113 Nh	114 Fl	115 Mc	116 Lv	117 Ts	118 Og

Lanthanoids	57 La	58 Ce	59 Pr	60 Nd	61 Pm	62 Sm	63 Eu	64 Gd	65 Tb	66 Dy	67 Ho	68 Er	69 Tm	70 Yb	71 Lu
Actinoids	89 Ac	90 Th	91 Pa	92 U	93 Np	94 Pu	95 Am	96 Cm	97 Bk	98 Cf	99 Es	100 Fm	101 Md	102 No	103 Lr

* From https://iupac.org/what-we-do/periodic-table-of-elements/.

Atomic Numbers and Atomic Weights*

Atomic Number	Name	Symbol	Weight
1	Hydrogen	H	[1.007 84; 1.008 11]
2	Helium	He	4.002602(2)
3	Lithium	Li	[6.938; 6.997]
4	Beryllium	Be	9.0121831(5)
5	Boron	B	[10.806; 10.821]
6	Carbon	C	[12.009 6; 12.011 6]
7	Nitrogen	N	[14.006 43; 14.007 28]
8	Oxygen	O	[15.999 03; 15.999 77]
9	Fluorine	F	18.998403163(6)
10	Neon	Ne	20.1797(6)
11	Sodium	Na	22.98976928(2)
12	Magnesium	Mg	[24.304, 24.307]
13	Aluminum	Al	26.9815385(7)
14	Silicon	Si	[28.084; 28.086]
15	Phosphorus	P	30.973761998(5)
16	Sulfur	S	[32.059; 32.076]
17	Chlorine	Cl	[35.446; 35.457]
18	Argon	Ar	39.948(1)
19	Potassium	K	39.0983(1)
20	Calcium	Ca	40.078(4)
21	Scandium	Sc	44.955908(5)
22	Titanium	Ti	47.867(1)
23	Vanadium	V	50.9415(1)
24	Chromium	Cr	51.9961(6)
25	Manganese	Mn	54.938044(3)
26	Iron	Fe	55.845(2)
27	Cobalt	Co	58.933194(4)
28	Nickel	Ni	58.6934(4)
29	Copper	Cu	63.546(3)
30	Zinc	Zn	65.38 (2)
31	Gallium	Ga	69.723(1)
32	Germanium	Ge	72.630(8)
33	Arsenic	As	74.921595(6)
34	Selenium	Se	78.971(8)
35	Bromine	Br	[79.901, 79.907]
36	Krypton	Kr	83.798(2)
37	Rubidium	Rb	85.4678(3)
38	Strontium	Sr	87.62(1)
39	Yttrium	Y	88.90584(2)
40	Zirconium	Zr	91.224(2)
41	Niobium	Nb	92.90637(2)
42	Molybdenum	Mo	95.95(1)
43	Technetium	Tc	
44	Ruthenium	Ru	101.07(2)
45	Rhodium	Rh	102.90550(2)
46	Palladium	Pd	106.42(1)
47	Silver	Ag	107.8682(2)
48	Cadmium	Cd	112.414(4)
49	Indium	In	114.818(1)
50	Tin	Sn	118.710(7)
51	Antimony	Sb	121.760(1)
52	Tellurium	Te	127.60(3)
53	Iodine	I	126.90447(3)
54	Xenon	Xe	131.293(6)
55	Cesium	Cs	132.90545196(6)
56	Barium	Ba	137.327(7)
57	Lanthanum	La	138.90547(7)
58	Cerium	Ce	140.116(1)
59	Praseodymium	Pr	140.90766(2)
60	Neodymium	Nd	144.242(3)
61	Promethium	Pm	
62	Samarium	Sm	150.36(2)
63	Europium	Eu	151.964(1)
64	Gadolinium	Gd	157.25(3)
65	Terbium	Tb	158.92535(2)
66	Dysprosium	Dy	162.500(1)
67	Holmium	Ho	164.93033(2)
68	Erbium	Er	167.259(3)
69	Thulium	Tm	168.93422(2)
70	Ytterbium	Yb	173.045(10)
71	Lutetium	Lu	174.9668(1)
72	Hafnium	Hf	178.49(2)
73	Tantalum	Ta	180.94788(2)
74	Tungsten	W	183.84(1)
75	Rhenium	Re	186.207(1)
76	Osmium	Os	190.23(3)
77	Iridium	Ir	192.217(3)
78	Platinum	Pt	195.084(9)
79	Gold	Au	196.966569(5)
80	Mercury	Hg	200.592(3)
81	Thallium	Tl	[204.382; 204.385]
82	Lead	Pb	207.2(1)
83	Bismuth	Bi	208.98040(1)
84	Polonium	Po	
85	Astatine	At	
86	Radon	Rn	
87	Francium	Fr	
88	Radium	Ra	
89	Actinium	Ac	
90	Thorium	Th	232.0377(4)
91	Protactinium	Pa	231.03588(2)
92	Uranium	U	238.02891(3)
93	Neptunium	Np	
94	Plutonium	Pu	
95	Americium	Am	
96	Curium	Cm	
97	Berkelium	Bk	
98	Californium	Cf	
99	Einsteinium	Es	
100	Fermiun	Fm	
101	Mendelevium	Md	
102	Nobelium	No	
103	Lawrencium	Lr	
104	Rutherfordium	Rf	
105	Dubnium	Db	
106	Seaborgium	Sg	
107	Bohrium	Bh	
108	Hassium	Hs	
109	Meitnerium	Mt	
110	Darmstadtium	Ds	
111	Roentgenium	Rg	
112	Copernicium	Cn	
113	Nihonium	Nh	
114	Flerovium	Fl	
115	Moscovium	Mc	
116	Livermorium	Lv	
117	Tennessine	Ts	
118	Oganesson	Og	

*Some weights are listed as intervals ([*a*; *b*]; $a \le$ atomic weight $\le b$) because these weights are not constant but depend on the physical, chemical, and nuclear histories of the samples used. Atomic weights are not listed for some elements because these elements do not have stable isotopes. Exceptions are thorium, protactinium, and uranium. From http://www.ciaaw.org/atomic-weights.htm.

Table of Atomic Masses

Element	Symbol	Mass number (*indicates radioactive)	Atomic mass	Percent abundance	Half-life and decay mode (if unstable)	
(Neutron)	*n*	1*	1.008665		10.4 m	β^-
Hydrogen	H	1	1.007825	99.985		
Deuterium	D	2	2.014102	0.015		
Tritium	T	3*	3.016049		12.33 y	β^-
Helium	He	3	3.016029	0.00014		
		4	4.002602	99.99986		
		6*	6.018886		0.81 s	β^-
		8*	8.033922		0.12 s	β^-
Lithium	Li	6	6.015121	7.5		
		7	7.016003	92.5		
		8*	8.022486		0.84 s	β^-
		9*	9.026789		0.18 s	β^-
		11*	11.043897		8.7 ms	β^-
Beryllium	Be	7*	7.016928		53.3 d	ec
		9	9.012174	100		
		10*	10.013534		1.5×10^6 y	β^-
		11*	11.021657		13.8 s	β^-
		12*	12.026921		23.6 ms	β^-
		14*	14.042866		4.3 ms	β^-
Boron	B	8*	8.024605		0.77 s	β^+
		10	10.012936	19.9		
		11	11.009305	80.1		
		12*	12.014352		0.0202 s	β^-
		13*	13.017780		17.4 ms	β^-
		14*	14.025404		13.8 ms	β^-
		15*	15.031100		10.3 ms	β^-
Carbon	C	9*	9.031030		0.13 s	β^+
		10*	10.016854		19.3 s	β^+
		11*	11.011433		20.4 m	β^+
		12	12.000000	98.90		
		13	13.003355	1.10		
		14*	14.003242		5730 y	β^-
		15*	15.010599		2.45 s	β^-
		16*	16.014701		0.75 s	β^-
		17*	17.022582		0.20 s	β^-

(Continued)

Element	Symbol	Mass number (*indicates radioactive)	Atomic mass	Percent abundance	Half-life and decay mode (if unstable)	
Nitrogen	N	12*	12.018613		0.0110 s	β^+
		13*	13.005738		9.96 m	β^+
		14	14.003074	99.63		
		15	15.000108	0.37		
		16*	16.006100		7.13 s	β^-
		17*	17.008450		4.17 s	β^-
		18*	18.014082		0.62 s	β^-
		19*	19.017038		0.24 s	β^-
Oxygen	O	13*	13.024813		8.6 ms	β^+
		14*	14.008595		70.6 s	β^+
		15*	15.003065		122 s	β^+
		16	15.994915	99.71		
		17	16.999132	0.039		
		18	17.999160	0.20		
		19*	19.003577		26.9 s	β^-
		20*	20.004076		13.6 s	β^-
		21*	21.008595		3.4 s	β^-
Fluorine	F	17*	17.002094		64.5 s	β^+
		18*	18.000937		109.8 m	β^+
		19	18.998404	100		
		20*	19.999982		11.0 s	β^-
		21*	20.999950		4.2 s	β^-
		22*	22.003036		4.2 s	β^-
		23*	23.003564		2.2 s	β^-
Neon	Ne	18*	18.005710		1.67 s	β^+
		19*	19.001880		17.2 s	β^+
		20	19.992435	90.48		
		21	20.993841	0.27		
		22	21.991383	9.25		
		23*	22.994465		37.2 s	β^-
		24*	23.993999		3.38 m	β^-
		25*	24.997789		0.60 s	β^-
Sodium	Na	21*	20.997650		22.5 s	β^+
		22*	21.994434		2.61 y	β^+
		23	22.989767	100		
		24*	23.990961		14.96 h	β^-
		25*	24.989951		59.1 s	β^-
		26*	25.992588		1.07 s	β^-
Magnesium	Mg	23*	22.994124		11.3 s	β^+
		24	23.985042	78.99		
		25	24.985838	10.00		
		26	25.982594	11.01		
		27*	26.984341		9.46 m	β^-
		28*	27.983876		20.9 h	β^-
		29*	28.375346		1.30 s	β^-

Element	Symbol	Mass number (*indicates radioactive)	Atomic mass	Percent abundance	Half-life and decay mode (if unstable)	
Aluminum	Al	25*	24.990429		7.18 s	β^+
		26*	25.986892		7.4×10^5 y	β^+
		27	26.981538	100		
		28*	27.981910		2.24 m	β^-
		29*	28.980445		6.56 m	β^-
		30*	29.982965		3.60 s	β^-
Silicon	Si	27*	26.986704		4.16 s	β^+
		28	27.976927	92.23		
		29	28.976495	4.67		
		30	28.973770	3.10		
		31*	30.975362		2.62 h	β^-
		32*	31.974148		172 y	β^-
		33*	32.977928		6.13 s	β^-
Phosphorus	P	30*	29.978307		2.50 m	β^+
		31	30.973762	100		
		32*	31.973762		14.26 d	β^-
		33*	32.971725		25.3 d	β^-
		34*	33.973636		12.43 s	β^-
Sulfur	S	31*	30.979554		2.57 s	β^+
		32	31.972071	95.02		
		33	32.971459	0.75		
		34	33.967867	4.21		
		35*	34.969033		87.5 d	β^-
		36	35.967081	0.02		
Chlorine	Cl	34*	33.973763		32.2 m	β^+
		35	34.968853	75.77		
		36*	35.968307		3.0×10^5 y	β^-
		37	36.965903	24.23		
		38*	37.968010		37.3 m	β^-
Argon	Ar	36	35.967547	0.337		
		37*	36.966776		35.04 d	ec
		38	37.962732	0.063		
		39*	38.964314		269 y	β^-
		40	39.962384	99.600		
		42*	41.963049		33 y	β^-
Potassium	K	39	38.963708	93.2581		
		40*	39.964000	0.0117	1.28×10^9 y	β^+, ec, β^-
		41	40.961827	6.7302		
		42*	41.962404		12.4 h	β^-
		43*	42.960716		22.3 h	β^-

(*Continued*)

Element	Symbol	Mass number (*indicates radioactive)	Atomic mass	Percent abundance	Half-life and decay mode (if unstable)
Calcium	Ca	40	39.962591	96.941	
		41*	40.962279		1.0×10^5 y ec
		42	41.958618	0.647	
		43	42.958767	0.135	
		44	43.955481	2.086	
		46	45.953687	0.004	
		48	47.952534	0.187	
Scandium	Sc	41*	40.969250		0.596 s β^+
		43*	42.961151		3.89 h β^+
		45	44.955911	100	
		46*	45.955170		83.8 d β^-
Titanium	Ti	44*	43.959691		49 y ec
		46	45.952630	8.0	
		47	46.951765	7.3	
		48	47.947947	73.8	
		49	48.947871	5.5	
		50	49.944792	5.4	
Vanadium	V	48*	47.952255		
		50*	49.947161	0.25	15.97 d β^+
		51	50.943962	99.75	1.5×10^{17} y β^+
Chromium	Cr	48*	47.954033		21.6 h ec
		50	49.946047	4.345	
		52	51.940511	83.79	
		53	52.940652	9.50	
		54	53.938883	2.365	
Manganese	Mn	53*	52.941292		3.74×10^6 y ec
		54*	53.940361		312.1 d ec
		55	54.938048	100	
		56*	55.938908		2.58 h β^-
Iron	Fe	54	53.939613	5.9	
		55*	54.938297		2.7 y ec
		56	55.934940	91.72	
		57	56.935396	2.1	
		58	57.933278	0.28	
		60*	59.934078		1.5×10^6 y β^-
Cobalt	Co	57*	56.936294		271.8 d ec
		58*	57.935755		70.9 h ec, β^+
		59	58.933198	100	
		60*	59.933820		5.27 y β^-
		61*	60.932478		1.65 h β^-
Nickel	Ni	58	57.935346	68.077	
		59*	58.934350		7.5×10^4 y ec, β^+
		60	59.930789	26.223	
		61	60.931058	1.140	
		62	61.928346	3.634	
		63*	62.929670		100 y β^-
		64	63.927967	0.926	

Element	Symbol	Mass number (*indicates radioactive)	Atomic mass	Percent abundance	Half-life and decay mode (if unstable)
Copper	Cu	63	62.929599	69.17	
		64*	63.929765		12.7 h ec
		65	64.927791	30.83	
		66*	65.928871		5.1 m β^-
Zinc	Zn	64	63.929144	48.6	
		66	65.926035	27.9	
		67	66.927129	4.1	
		68	67.924845	18.8	
		70	69.925323	0.6	
Gallium	Ga	69	68.925580	60.108	
		70*	69.926027		21.1 m β^-
		71	70.924703	39.892	
		72*	71.926367		14.1 h β^-
Germanium	Ge	69*	68.927969		39.1 h ec, β^+
		70	69.924250	21.23	
		72	71.922079	27.66	
		73	72.923462	7.73	
		74	73.921177	35.94	
		76	75.921402	7.44	
		77*	76.923547		11.3 h β^-
Arsenic	As	73*	72.923827		80.3 d ec
		74*	73.923928		17.8 d ec, β^+
		75	74.921594	100	
		76*	75.922393		1.1 d β^-
		77*	76.920645		38.8 h β^-
Selenium	Se	74	73.922474	0.89	
		76	75.919212	9.36	
		77	76.919913	7.63	
		78	77.917307	23.78	
		79*	78.918497		$\leq 6.5 \times 10^4$ y β^-
		80	79.916519	49.61	
		82*	81.916697	8.73	1.4×10^{20} y $2\beta^-$
Bromine	Br	79	78.918336	50.69	
		80*	79.918528		17.7 m β^+
		81	80.916287	49.31	
		82*	81.916802		35.3 h β^-
Krypton	Kr	78	77.920400	0.35	
		80	79.916377	2.25	
		81*	80.916589		2.11×10^5 y ec
		82	81.913481	11.6	
		83	82.914136	11.5	
		84	83.911508	57.0	
		85*	84.912531		10.76 y β^-
		86	85.910615	17.3	

(Continued)

Element	Symbol	Mass number (*indicates radioactive)	Atomic mass	Percent abundance	Half-life and decay mode (if unstable)
Rubidium	Rb	85	84.911793	72.17	
		86*	85.911171		18.6 d β^-
		87*	86.909186	27.83	4.75×10^{10} y β^-
		88*	87.911325		17.8 m β^-
Strontium	Sr	84	83.913428	0.56	
		86	85.909266	9.86	
		87	86.908883	7.00	
		88	87.905618	82.58	
		90*	89.907737		29.1 y β^-
Yttrium	Y	88*	87.909507		106.6 d ec, β^+
		89	88.905847	100	
		90*	89.914811		2.67 d β^-
Zirconium	Zr	90	89.904702	51.45	
		91	90.905643	11.22	
		92	91.905038	17.15	
		93*	92.906473		1.5×10^6 y β^-
		94	93.906314	17.38	
		96	95.908274	2.80	
Niobium	Nb	91*	90.906988		6.8×10^2 y ec
		92*	91.907191		3.5×10^7 y ec
		93	92.906376	100	
		94*	93.907280		2×10^4 y β^-
Molybdenum	Mo	92	91.906807	14.84	
		93*	92.906811		3.5×10^3 y ec
		94	93.905085	9.25	
		95	94.905841	15.92	
		96	95.904678	16.68	
		97	96.906020	9.55	
		98	97.905407	24.13	
		100	99.907476	9.63	
Technetium	Tc	97*	96.906363		2.6×10^6 y ec
		98*	97.907215		4.2×10^6 y β^-
		99*	98.906254		2.1×10^5 y β^-
Ruthenium	Ru	96	95.907597	5.54	
		98	97.905287	1.86	
		99	98.905939	12.7	
		100	99.904219	12.6	
		101	100.905558	17.1	
		102	101.904348	31.6	
		104	103.905428	18.6	
Rhodium	Rh	102*	101.906794		207 d ec
		103	102.905502	100	
		104*	103.906654		42 s β^-

Element	Symbol	Mass number (*indicates radioactive)	Atomic mass	Percent abundance	Half-life and decay mode (if unstable)
Palladium	Pd	102	101.905616	1.02	
		104	103.904033	11.14	
		105	104.905082	22.33	
		106	105.903481	27.33	
		107*	106.905126		6.5×10^6 y β^-
		108	107.903893	26.46	
		110	109.905158	11.72	
Silver	Ag	107	106.905091	51.84	
		108*	107.905953		2.39 m ec, β^+, β^-
		109	108.904754	48.16	
		110*	109.906110		24.6 s β^-
Cadmium	Cd	106	105.906457	1.25	
		108	107.904183	0.89	
		109*	108.904984		462 d ec
		110	109.903004	12.49	
		111	110.904182	12.80	
		112	111.902760	24.13	
		113*	112.904401	12.22	9.3×10^{15} y β^-
		114	113.903359	28.73	
		116	115.904755	7.49	
Indium	In	113	112.904060	4.3	
		114*	113.904916		1.2 m β^-
		115*	114.903876	95.7	4.4×10^{14} y β^-
		116*	115.905258		54.4 m β^-
Tin	Sn	112	111.904822	0.97	
		114	113.902780	0.65	
		115	114.903345	0.36	
		116	115.901743	14.53	
		117	116.902953	7.68	
		118	117.901605	24.22	
		119	118.903308	8.58	
		120	119.902197	32.59	
		121*	120.904237		55 y β^-
		122	121.903439	4.63	
		124	123.905274	5.79	
Antimony	Sb	121	120.903820	57.36	
		123	122.904215	42.64	
		125*	124.905251		2.7 y β^-
Tellurium	Te	120	119.904040	0.095	
		122	121.903052	2.59	
		123*	122.904271	0.905	1.3×10^{13} y ec
		124	123.902817	4.79	
		125	124.904429	7.12	
		126	125.903309	18.93	
		128*	127.904463	31.70	$> 8 \times 10^{24}$ y $2\beta^-$
		130*	129.906228	33.87	1.2×10^{21} y $2\beta^-$

(Continued)

Element	Symbol	Mass number (*indicates radioactive)	Atomic mass	Percent abundance	Half-life and decay mode (if unstable)
Iodine	I	126*	125.905619		13 d ec, β^+, β^-
		127	126.904474	100	
		128*	127.905812		25 m β^-, ec, β^+, β^-
		129*	128.904984		1.6×10^7 y
Xenon	Xe	124	123.905894	0.10	
		126	125.904268	0.09	
		128	127.903531	1.91	
		129	128.904779	26.4	
		130	129.903509	4.1	
		131	130.905069	21.2	
		132	131.904141	26.9	
		134	133.905394	10.4	
		136	135.907215	8.9	
Cesium	Cs	133	132.905436	100	
		134*	133.906703		2.1 y β^-
		135*	134.905891		2×10^6 y β^-
		137*	136.907078		30 y β^-
Barium	Ba	130	129.906289	0.106	
		132	131.905048	0.101	
		133*	132.905990		10.5 y ec
		134	133.904492	2.42	
		135	134.905671	6.593	
		136	135.904559	7.85	
		137	136.905816	11.23	
		138	137.905236	71.70	
Lanthanum	La	137*	136.906462		6×10^4 y ec
		138*	137.907105	0.0902	1.05×10^{11} y ec, β^+
		139	138.906346	99.9098	
Cerium	Ce	136	135.907139	0.19	
		138	137.905986	0.25	
		140	139.905434	88.43	
		142	141.909241	11.13	
Praseodymium	Pr	140*	139.909071		3.39 m ec, β^+
		141	140.907647	100	
		142*	141.910040		25.0 m β^-
Neodymium	Nd	142	141.907718	27.13	
		143	142.909809	12.18	
		144*	143.910082	23.80	2.3×10^{15} y α
		145	144.912568	8.30	
		146	145.913113	17.19	
		148	147.916888	5.76	
		150	149.920887	5.64	
Promethium	Pm	143*	142.910928		265 d ec
		145*	144.912745		17.7 y ec
		146*	145.914698		5.5 y ec
		147*	146.915134		2.623 y β^-

Element	Symbol	Mass number (*indicates radioactive)	Atomic mass	Percent abundance	Half-life and decay mode (if unstable)
Samarium	Sm	144	143.911996	3.1	
		146*	145.913043		1.0×10^8 y α
		147*	146.914894	15.0	1.06×10^{11} y α
		148*	147.914819	11.3	7×10^{15} y α
		149	148.917180	13.8	
		150	149.917273	7.4	
		151*	150.919928		90 y β^-
		152	151.919728	26.7	
		154	153.922206	22.7	
Europium	Eu	151	150.919846	47.8	
		152*	151.921740		13.5 y ec, β^+
		153	152.921226	52.2	
		154*	153.922975		8.59 y β^-
		155*	154.922888		4.7 y β^-
Gadolinium	Gd	148*	147.918112		75 y α
		150*	149.918657		1.8×10^6 y α
		152*	151.919787	0.20	1.1×10^{14} y α
		154	153.920862	2.18	
		155	154.922618	14.80	
		156	155.922119	20.47	
		157	156.923957	15.65	
		158	157.924099	24.84	
		160	159.927050	21.86	
Terbium	Tb	158*	157.925411		180 y ec, β^+, β^-
		159	158.925345	100	
		160*	159.927551		72.3 d β^-
Dysprosium	Dy	156	155.924277	0.06	
		158	157.924403	0.10	
		160	159.925193	2.34	
		161	160.926930	18.9	
		162	161.926796	25.5	
		163	162.928729	24.9	
		164	163.929172	28.2	
Holmium	Ho	165	164.930316	100	
		166*	165.932282		1.2×10^3 y β^-
Erbium	Er	162	161.928775	0.14	
		164	163.929198	1.61	
		166	165.930292	33.6	
		167	166.932047	22.95	
		168	167.932369	27.8	
		170	169.935462	14.9	
Thulium	Tm	169	168.934213	100	
		171*	170.936428		1.92 y β^-

(Continued)

Element	Symbol	Mass number (*indicates radioactive)	Atomic mass	Percent abundance	Half-life and decay mode (if unstable)
Ytterbium	Yb	168	167.933897	0.13	
		170	169.934761	3.05	
		171	170.936324	14.3	
		172	171.936380	21.9	
		173	172.938209	16.12	
		174	173.938861	31.8	
		176	175.942564	12.7	
Lutetium	Lu	173*	172.938930		1.37 y ec
		175	174.940772	97.41	
		176*	175.942679	2.59	3.8×10^{10} y β^-
Hafnium	Hf	174*	173.940042	0.162	2.0×10^{15} y α
		176	175.941404	5.206	
		177	176.943218	18.606	
		178	177.943697	27.297	
		179	178.945813	13.629	
		180	179.946547	35.100	
Tantalum	Ta	180	179.947542	0.012	
		181	180.947993	99.988	
Tungsten (Wolfram)	W	180	179.946702	0.12	
		182	181.948202	26.3	
		183	182.950221	14.28	
		184	183.950929	30.7	
		186	185.954358	28.6	
Rhenium	Re	185	184.952951	37.40	
		187*	186.955746	62.60	4.4×10^{10} y β^-
Osmium	Os	184	183.952486	0.02	
		186*	185.953834	1.58	2.0×10^{15} y α
		187	186.955744	1.6	
		188	187.955744	13.3	
		189	188.958139	16.1	
		190	189.958439	26.4	
		192	191.961468	41.0	
		194*	193.965172		6.0 y β^-
Iridium	Ir	191	190.960585	37.3	
		193	192.962916	62.7	
Platinum	Pt	190*	189.959926	0.01	6.5×10^{11} y α
		192	191.961027	0.79	
		194	193.962655	32.9	
		195	194.964765	33.8	
		196	195.964926	25.3	
		198	197.967867	7.2	
Gold	Au	197	196.966543	100	
		198*	197.968217		2.70 d β^-
		199*	198.968740		3.14 d β^-

Element	Symbol	Mass number (*indicates radioactive)	Atomic mass	Percent abundance	Half-life and decay mode (if unstable)
Mercury	Hg	196	195.965806	0.15	
		198	197.966743	9.97	
		199	198.968253	16.87	
		200	199.968299	23.10	
		201	200.970276	13.10	
		202	201.970617	29.86	
		204	203.973466	6.87	
Thallium	Tl	203	202.972320	29.524	
		204*	203.973839		3.78 y β^-
		205	204.974400	70.476	
	(Ra E″)	206*	205.976084		4.2 m β^-
	(Ac C″)	207*	206.977403		4.77 m β^-
	(Th C″)	208*	207.981992		3.053 m β^-
	(Ra C″)	210*	209.990057		1.30 m β^-
Lead	Pb	202*	201.972134		5×10^4 y ec
		204	203.973020	1.4	
		205*	204.974457		1.5×10^7 y ec
		206	205.974440	24.1	
		207	206.975871	22.1	
		208	207.976627	52.4	
	(Ra D)	210*	209.984163		22.3 y β^-
	(Ac B)	211*	210.988734		36.1 m β^-
	(Th B)	212*	211.991872		10.64 h β^-
	(Ra B)	214*	213.999798		26.8 m β^-
Bismuth	Bi	207*	206.978444		32.2 y ec, β^+
		208*	207.979717		3.7×10^5 y ec
		209	208.980374	100	
	(Ra E)	210*	209.984096		5.01 d α, β^-
	(Th C)	211*	210.987254		2.14 m α
	(Ra C)	212*	211.991259		60.6 m α, β^-
		214*	213.998692		19.9 m β^-
		215*	215.001836		7.4 m β^-
Polonium	Po	209*	208.982405		102 y α
	(Ra F)	210*	209.982848		138.38 d α
	(Ac C′)	211*	210.986627		0.52 s α
	(Th C′)	212*	211.988842		0.30 μs α
	(Ra C′)	214*	213.995177		164 μs α
	(Ac A)	215*	214.999418		0.0018 s α
	(Th A)	216*	216.001889		0.145 s α
	(Ra A)	218*	218.008965		3.10 m α
Astatine	At	215*	214.998638		$\approx$100 μs α
		218*	218.008685		1.6 s α
		219*	219.011297		0.9 m α
Radon	Rn				
	(An)	219*	219.009477		3.96 s α
	(Tn)	220*	220.011369		55.6 s α
	(Rn)	222*	222.017571		3.823 d α

(Continued)

Element	Symbol	Mass number (*indicates radioactive)	Atomic mass	Percent abundance	Half-life and decay mode (if unstable)
Francium		221*	221.01425		4.18 m α
	Fr	222*	222.017585		14.2 m β^-
	(Ac K)	223*	223.019733		22 m β^-
Radium	Ra	221*	221.01391		29 s α
	(Ac X)	223*	223.018499		11.43 d α
	(Th X)	224*	224.020187		3.66 d α
		225*			14.9 d β^-
	(Ra)	226*	226.025402		1600 y α
	($MsTh_1$)	228*	228.031064		5.75 y β^-
Actinium	Ac	225*			10 d α
	($Ms Th_2$)	227*	227.027749		21.77 y β^-
		228*	228.031015		6.15 h β^-
		229*			1.04 h β^-
Thorium	Th				
	(Rd Ac)	227*	227.027701		18.72 d α
	(Rd Th)	228*	228.028716		1.913 y α
		229*	229.031757		7300 y α
	(Io)	230*	230.033127		75,000 y α, sf
	(UY)	231*	231.036299	100	25.52 h β^-
	(Th)	232*	232.038051		1.40×10^{10} y α
	(UX_1)	234*	234.043593		24.1 d β^-
Protactinium	Pa	231*	231.035880		32,760 y α
	(UZ)	234*	234.043300		6.7 h β^-
Uranium	U	231*	231.036264		4.2 d β^+
		232*	232.037131		69 y α
		233*	233.039630		1.59×10^5 y α
	(UII)	234*	234.040946	0.0055	2.45×10^5 y α
	(Ac U)	235*	235.043924	0.720	7.04×10^8 y α
	(UI)	236*	236.045562		2.34×10^7 y α
		238*	238.050784	99.2745	4.47×10^9 y α
		239*	239.054290		23.5 m β^-
Neptunium	Np	235*	235.044057		396 d α
		236*	236.046559		1.54×10^5 y ec
		237*	237.048168		2.14×10^6 y α
Plutonium	Pu	236*	236.046033		2.87 y α, sf
		238*	238.049555		87.7 y α, sf
		239*	239.052157		24,120 y α, sf
		240*	240.053808		6560 y α, sf
		241*	241.056846		14.4 y β^-
		242*	242.058737		3.7×10^5 y α, sf
		244*	244.064200		8.1×10^7 y α, sf
Americium	Am	240*	240.055285		2.12 d ec
		241*	241.056824		432 y α, sf
Curium	Cm	247*	247.070347		1.56×10^7 y α
		248*	248.072344		3.4×10^5 y α, sf

Element	Symbol	Mass number (*indicates radioactive)	Atomic mass	Percent abundance	Half-life and decay mode (if unstable)
Berkelium	Bk	247*	247.070300		1380 y α
		249*	249.074979		327 d β^-
Californium	Cm	250*	250.076400		13.1 y α, sf
		251*	251.079580		898 y α
Einsteinium	Es	252*	252.082974		1.29 y α
		253*	253.084817		2.02 d α, sf
Fermium	Fm	253*	253.085173		3.00 d ec
		254*	254.086849		3.24 h α, sf
Mendelevium	Md	256*	256.093988		75.6 m ec, β^+
		258*	258.098594		55 d α
Nobelium	No	257*	257.096855		25 s α
		259*	259.100932		58 m α, sf
Lawrencium	Lr	259*	259.102888		6.14 s α, sf
		260*	260.105346		3.0 m α, sf
Rutherfordium	Rf	260*	260.160302		24 ms sf
		261*	261.108588		65 s α, sf
Dubnium	Db	261*	261.111830		1.8 s α
		262*	262.113763		35 s α
Seaborgium	Sg	263*	263.118310		0.78 s α, sf
Bohrium	Bh	262*	262.123081		0.10 s α, sf
Hassium	Hs	265*	265.129984		1.8 ms α
		267*	267.131770		60 ms α
Meitnerium	Mt	266*	266.137789		3.4 ms α, sf
		268*	268.138820		70 ms α
Darmstadtium	Ds	269*	269.145140		0.17 ms α
		271*	271.146080		1.1 ms α
		273*	272.153480		8.6 ms α
Roentgenium	Rg	272*	272.153480		1.5 ms α
Copernicium	Cn	277*	?		0.2 ms α
Nihonium	Nh	284*	?		? α
Flerovium	Fl	289*	?		? α
Moscovium	Mc	288*	?		? α
Livermorium	Lv	292*	?		? α
Tennessine	Ts	293*	?		? α
Oganesson	Og	294*	?		? α

GLOSSARY

absolute pressure Total pressure at a point in a fluid, equal to the sum of the gauge and the atmospheric pressures.

absolute zero The lowest temperature that is theoretically possible, at which the motion of particles is at a minimum; 0 on the Kelvin scale; $-273.15°C$ or $-459.67°F$.

absorption lines Dark lines in an otherwise continuous spectrum. These indicate certain wavelengths of light that are absorbed by the atoms of the intervening medium.

absorption spectrum A continuous spectrum, broken by a specific pattern of dark lines or bands, observed when light traverses a particular absorbing medium.

ac (alternating current) An electric current that reverses its direction many times a second at regular intervals.

acceleration due to gravity Acceleration of a body due to the pull of gravity; an object in free fall near Earth's surface has an acceleration of approximately 9.8 m/s^2.

acceleration vector (*or* instantaneous acceleration vector) The change in a velocity vector per unit over time.

acceleration The rate of change of velocity, due to changes in its direction or magnitude.

ac generator Alternating current generator; as its coil rotates in a magnetic field, an oscillating emf is generated in the turns of wire that make up the coil.

adiabatic process In thermodynamics, a process that occurs without transfer of heat or matter in or out of a system.

alpha decay Type of radioactive decay in which an atomic nucleus emits an alpha particle (helium nucleus) and thereby transforms into a nucleus with a mass number that is reduced by four and an atomic number that is reduced by two.

alpha particle A positively charged particle, indistinguishable from a helium nucleus and consisting of two protons and two neutrons.

alternating current (ac) *See* ac (alternating current).

ampere The SI unit of electric current, equal to a flow of one coulomb per second.

Ampère's law For any closed loop path, the circulation of a magnetic field created by an electric current is equal to the size of that electric current times the permeability of free space; discovered by the French physicist André-Marie Ampère.

Amperian loop An imaginary closed path in space around a current-carrying conductor.

amplitude (of a wave) The maximum displacement from equilibrium that occurs as a wave moves through its medium.

angular acceleration The rate of change of angular velocity.

angular displacement The angle through which an object has been rotated.

angular frequency Frequency of a periodic process (as electric oscillation or sound vibration) expressed in radians per second, equivalent to the frequency in cycles multiplied by 2π.

angular position On a rotating object, the angle of a line on the object from its original position.

angular resolution The angle that separates two point objects that are just barely resolved through a circular aperture.

angular speed The magnitude or absolute value of angular velocity.

angular velocity The rate of change of angular displacement.

angular wave number The reciprocal of wavelength multipled by 2π.

antimatter Particles with the same properties as ordinary particles, but with the opposite electric charge.

antineutrino The antiparticle of a neutrino.

antinode Positions along a standing wave at which the oscillation is maximal.

antiquark The antiparticle of a quark.

apparent weight Weight of an object submerged in a fluid. This is less than its true weight due to buoyant forces.

apparent weightlessness The perceived state of an object that is accelerating along with its surroundings.

Archimedes' principle The buoyant force on an object immersed in a fluid is equal to the weight of the fluid that the object displaces.

atmosphere (unit of pressure) The average value of atmospheric pressure at sea level; equal to 1.01325×10^5 Pa or about 14.7 pounds per square inch.

atomic number The number of protons in an atomic nucleus; determines the chemical properties of an element and its place in the periodic table.

average acceleration The change in velocity divided by the elapsed time.

average angular acceleration The change in angular velocity divided by the elapsed time.

average angular velocity The change in angular displacement divided by the elapsed time.

average speed The total distance an object travels divided by the time it takes for the object to travel that distance.

average velocity The total displacement of an object divided by the elapsed time.

baryon Any hadron that can be made up of three quarks.

battery An electrochemical cell that can set charges into motion.

beat frequency Rate of the periodic variations of amplitude when two waves of different frequencies interfere.

beats Periodic variations in amplitude when two waves of different frequency interfere.

becquerel The SI unit of radioactive decay rate, equal to one disintegration per second.

Bernoulli's equation A relationship among pressure, speed, and height in an ideal fluid in motion.

Bernoulli's principle In a moving fluid, the pressure is low where the fluid is moving rapidly.

beta decay A type of radioactive decay in which a beta particle (an electron or a positron) is emitted from an atomic nucleus.

beta-minus decay When a neutron (charge zero) changes into a proton, an electron, and an electron antineutrino.

beta-plus decay When a proton changes into a neutron (charge zero), a positron, and an electron neutrino.

Big Bang The rapid expansion of matter from a state of extremely high density and temperature that marked the origin of the universe.

binding energy The energy required to disassemble an atomic nucleus into its component protons and neutrons.

blackbody An object that does not reflect any light at all but absorbs all radiation falling on it.

blackbody radiation Light emitted by a perfect blackbody of emissivity = 1.

Bohr model Theory of atomic structure in which a small, positively charged nucleus is surrounded by electrons that travel in circular orbits around the nucleus.

Bohr orbit In the Bohr model, one of the orbits in which electrons in an atom travel around the nucleus.

Boltzmann constant Physical constant in the ideal gas law which has the same value for all gases and relates the average kinetic energy of the particles in a gas to the temperature of the gas; equal to 1.38065×10^{-23} J/K.

boundary layer In a fluid, the layer next to a solid surface within which the fluid speed increases from zero at the surface to full speed at the edge of the layer.

Brewster's angle An angle of incidence at which light with a particular polarization is perfectly transmitted through a transparent dielectric surface, with no reflection.

British thermal unit The quantity of heat required to increase the temperature of 1 lb of pure water from 63°F to 64°F.

bulk modulus A measure of how resistant to compression a substance is, defined as the ratio of the pressure increase to the resulting relative decrease of the volume.

buoyant force The upward force exerted by a fluid on an object placed in it.

calorie The quantity of heat required to increase the temperature of one gram (1 g) of pure water from 14.5°C to 15.5°C.

capacitance The ability of a system to store an electric charge.

capacitor A system or device that can store positive and negative charge, consisting of one or more pairs of conductors that may be separated by an insulator.

Carnot cycle An ideal, reversible thermodynamic cycle consisting of two isothermal processes and two adiabatic processes; the most efficient cycle in a heat engine; first proposed in the early 1800s by Sadi Carnot.

Cavendish experiment Experiment used to determine the value of the gravitational constant.

Celsius scale The most common temperature scale; based on the work of the eighteenth-century Swedish astronomer Anders Celsius. In this scale the freezing point of water is approximately 0°C, and the boiling point is approximately 100°C.

center of curvature (of a mirror) The center of the sphere defined by the surface of a concave mirror.

center of mass A point representing the average position of the matter in a body or system; moves as though all of the body's mass were concentrated at that point and all external forces act on it.

centripetal acceleration The rate of change of tangential velocity of an object in circular motion; points toward the center of the circle defined by the object's trajectory.

centripetal force The force that points toward the inside of an object's curving trajectory and produces the centripetal acceleration.

charged object A body with net electric charge.

circuit A complete loop in which a charge flows through a wire continuously from one terminal of a battery to the other.

circuit element Any component of a circuit, such as a battery, resistor, inductor, or capacitor.

circulation (of a magnetic field) For an Amperian loop, the sum of products (of the component of magnetic field parallel to each loop segment multiplied by the segment length) along the loop.

closed pipe A pipe which is open at one end and blocked at the other.

coefficient of kinetic friction The ratio of the force of friction acting on a sliding object to the object's normal force; depends on the properties of the surfaces in contact.

coefficient of linear expansion The fractional change in length of an object per unit change in temperature.

coefficient of performance (of a heat pump or refrigerator) The ratio of useful heating or cooling provided to work required.

coefficient of rolling friction The ratio of the force of friction acting on the point of a rolling object that is in contact with the surface to the object's normal force; depends on the properties of the surfaces in contact.

coefficient of static friction The ratio of the force of friction acting on an unmoving object to the object's normal force; depends on the properties of the surfaces in contact.

coefficient of volume expansion The fractional change in the volume of an object per unit change in temperature.

completely inelastic collision Encounter in which two objects stick together after they collide and the most mechanical energy is lost.

component (of a vector) The projection of a vector onto a coordinate axis.

compressible fluid A fluid that can be easily compacted by squeezing.

compression Pressure applied to all sides of a body, which may result in a reduction in volume.

compressive strain The length by which an object shrinks expressed as a fraction of its relaxed length.

compressive stress The force applied to an object being squeezed divided by the object's cross-sectional area.

Compton scattering Elastic scattering of a photon by a free charged particle, usually an electron.

concave Curved inward, such as a mirror.

condensation The phase change from gas to liquid.

conduction Transfer of heat by the direct collision of particles, with no net displacement of the particles.

conductor A substance in which charges can move freely.

conservation of angular momentum If there is no net external torque on a system, the angular momentum of the system is conserved.

conservative force A force that can be associated with a potential energy, such as the gravitational force or the force exerted by an ideal spring.

constant acceleration A situation in which velocity changes at a steady rate.

constant velocity Movement at a steady speed in the same direction.

constructive interference The mutual reinforcement of waves such that the amplitude of the total wave is the sum of the amplitudes of the individual waves.

contact force A force that arises only when two objects come in contact with one another.

contact time The amount of time colliding objects are in contact.

convection Energy transfer by the motion of a liquid or gas (such as air) caused by the tendency of hotter, less dense material to rise and colder, denser material to sink under the influence of gravity.

converging lens Lens that takes incoming parallel light rays and brings them to a focus on the principal axis.

convex Curved outward, such as a mirror.

coordinates Quantities indicating the position of an object in reference to the origin on a coordinate system.

coordinate system A system that can be used to denote the position of an object at a given time.

cosmic background radiation (cosmic microwave background) The thermal radiation left over from the Big Bang.

cosmological redshift A redshift caused by the expansion of the universe.

cosmology The science of the origin and evolution of the universe.

coulomb The unit of electric charge, equal to the amount of electricity conveyed in one second by a current of one ampere; named after the eighteenth-century French physicist Charles-Augustin de Coulomb, who uncovered the fundamental law that governs the interaction of charges.

Coulomb's constant A proportionality constant used to determine the magnitude of the electric forces between two point charges; equal to $8.99 \times 10^9\ \text{N} \cdot \text{m}^2/\text{C}^2$.

Coulomb's law The magnitude of the force of electrostatic attraction or repulsion acting between two electric charges is directly proportional to the product of the charges and inversely proportional to the square of the distance between them.

critical angle The angle of incidence beyond which total internal reflection occurs.

critically damped oscillations Case in which the minimum amount of damping is applied to result in a nonoscillatory response; when displaced from equilibrium, the system returns smoothly to equilibrium with no overshoot and hence no oscillation.

critical point A point on a phase diagram at which both the liquid and gas phases of a substance have the same density and are therefore indistinguishable.

cross product (vector product) The product of two vectors in three dimensions that is itself a vector at right angles to both the original vectors, with a direction given by the right-hand rule and a magnitude equal to the product of the magnitudes of the original vectors and the sine of the angle between their directions.

current The rate at which charge flows past any point in a circuit.

current loop A single loop that carries a current provided by a source of emf.

damped oscillations Oscillations that are diminished by a frictional force.

damping coefficient Proportionality constant related to the physical characteristics of a particular system and determining the degree to which oscillations are diminished.

dark energy A repulsive force that counteracts gravity and causes the universe to expand at an accelerating rate.

dark matter Nonluminous material that is postulated to exist in space and that is the dominant form of matter in the universe.

dc (direct current) An electric current that does not change direction in a circuit.

de Broglie wavelength The wavelength of a particle, given by Planck's constant divided by the momentum of the particle.

decay constant Proportionality between the size of a population of radioactive nuclei and the rate at which the population decreases because of radioactive decay.

decibel The units of sound intensity level.

degree of freedom In thermodynamics, a possible form of motion of an object.

density The mass of a substance divided by the volume that it occupies.

deposition The phase change from gas to solid.

destructive interference When two waves cancel each other out so that the amplitude of the total wave is zero.

diamagnetic Tending to become magnetized in a direction opposite to that of the applied magnetic field.

dielectric A material that is both an insulator and polarizable.

dielectric constant The greater the dielectric constant of a material, the more the material is polarized when it is placed in the electric field between the plates of a charged capacitor.

diffraction The bending of light around an obstacle or aperture.

diffraction maxima Locations in a diffraction pattern where wavelets interfere constructively, producing a bright fringe.

diffraction minima Locations in a diffraction pattern where wavelets interfere destructively, producing a dark fringe.

diffraction pattern The distinctive pattern of bright and dark fringes caused when light is diffracted through a slit or aperture.

diffuse light Light that reflects from an object's surface in many random directions.

dimensional analysis A method of checking the relations of physical quantities by identifying their dimensions.

diode A single-junction semiconductor device.

diopter A unit of power for a lens or mirror that is equal to the reciprocal of the focal length (in meters).

direct current (dc) *See* dc (direct current).

dispersion The separation of light according to wavelength due to differing propagation speeds of different wavelengths of light; a prism separates white light into its component colors because light of each color travels at a different speed through the glass.

displacement The difference between the positions of an object at two separate times.

displacement current A quantity appearing in Maxwell's equations that is defined in terms of the rate of change of the electric displacement field.

displacement vector Vector drawn from the starting point of an object's motion to its endpoint.

diverging lens Lens that takes incoming parallel light rays and causes them to spread away from the principal axis.

doping Adding small amounts of a different kind of atom to a substance.

Doppler effect The change in frequency of a wave caused by an observer moving relative to its source or its source moving relative to the observer.

drag force The force that resists the motion of an object through a liquid or a gas.

drift speed The average speed at which charges move through a conductor.

driving angular frequency The angular frequency of a driving force in an oscillation.

eccentricity Parameter determining the circularity of an ellipse; a perfect circle has zero eccentricity.

efficiency The useful work divided by the amount of energy taken in to do the work.

elapsed time The duration of a time interval.

elastic An elastic object returns to its original shape after being squeezed or stretched.

elastic collision A collision in which the forces between the colliding objects are conservative; both total momentum and total mechanical energy are conserved.

electric charge The physical property of matter that causes it to experience a force near other charged material.

electric dipole A combination of two point charges of the same magnitude but opposite signs.

electric energy density The energy per unit volume stored in an electric field, such as in a capacitor.

electric field lines Lines showing the direction of an electric field.

electric flux The area of a surface multiplied by the component of the electric field that's perpendicular to that surface.

electric force The force between electric charges.

electric potential The electric potential energy for a charge at a given position divided by the value of that charge.

electric potential difference The difference in electric potential between two locations.

electric potential energy Potential energy that results from conservative Coulomb forces; associated with the configuration of a particular set of point charges.

electromagnetic induction The process whereby a changing magnetic field induces an electric field.

electromagnetic spectrum The range of electromagnetic waves according to wavelength.

electromagnetic wave Waves that are propagated by simultaneous periodic variations of electric and magnetic fields. These include radio waves, infrared, visible light, ultraviolet, x rays, and gamma rays.

electromagnetism An umbrella term to cover both electricity and magnetism, since both involve interactions between charges.

electron volt A unit of energy equal to the work done on an electron in accelerating it through a potential difference of one volt.

emf (electromotive force) The voltage developed by any source of electrical energy such as a battery.

emission lines Bright lines in the emission spectrum of a gas.

emission spectrum A spectrum that consists only of specific emitted wavelengths.

emissivity How well or how poorly a surface radiates.

energy The capacity to do work.

energy level Quantized value of energy, for example of electrons in an atom.

entropy A measure of the amount of disorder present in a system.

equation of continuity In fluid dynamics, $A_1v_1 = A_2v_2$: the product of a pipe's cross-sectional area A and the flow speed v is conserved; it has the same value at point 1 as at point 2.

equation of hydrostatic equilibrium The equation $p = p_0 + \rho gd$, which must be satisfied for a fluid to remain at rest.

equation of state A relationship among the quantities of pressure, volume, and temperature.

equilibrium State in which the net external force on an object is zero.

equipartition theorem Principle stating that the energy of a molecule is shared equally among each degree of freedom.

equipotential A curve along which the electric potential has the same value at all points.

equipotential surface A surface on which the electric potential has the same value at all points.

equipotential volume Space inside a conductor in which the electric potential has the same value everywhere.

equivalent capacitance Effective capacitance of an arrangement of two or more connected capacitors.

equivalent resistance Effective resistance of an arrangement of two or more connected resistance.

escape speed The minimum speed at which an object must be launched from Earth's surface to escape to infinity.

event Something that happens at a certain point in time.

exchange particle A virtual particle that interacts with ordinary particles to mediate forces, producing the effects of attraction and repulsion.

exponent The superscript following the number 10 that denotes the number of zeros needed to write the long form of a number in scientific notation.

exponential function The irrational number *e* raised to a power.

external forces Forces exerted on an object by other objects.

Fahrenheit scale The official temperature scale used in the United States, the Cayman Islands, and Belize; originated by the German scientist Daniel Fahrenheit. On this scale water freezes at 32°F and boils at 212°F.

failure The point at which the structure of a material starts to lose its integrity, which eventually leads to the object breaking apart.

farad The SI unit of electrical capacitance, equal to the capacitance of a capacitor in which one coulomb of charge causes a potential difference of one volt; named after English physicist Michael Faraday.

Faraday's law (of induction) An emf is induced in a loop if the magnetic flux through that loop changes.

ferromagnetic A ferromagnetic material has a high susceptibility to magnetization, the strength of which depends on that of the applied magnetizing field and that may persist after removal of the applied field.

first law of thermodynamics In a thermodynamic process, the change in the internal energy of a system equals the heat that flows into the system during the process minus the work that the system does during the process.

fluid A substance (a gas or a liquid) that can flow because its molecules can move freely with respect to each other.

fluid resistance The resistance experienced by an object as it moves through a fluid.

focal length The distance from the focal point to the center of a mirror or lens.

focal point The point along the principal axis of a mirror or lens at which incident rays parallel to the principal axis converge and come to a common focus.

force A push or a pull.

forced oscillations Case in which a periodic driving force causes a system to oscillate at the frequency of that driving force.

force pair In an interaction between objects A and B, the forces of A on B and of B on A.

frame of reference (*or* reference frame) A coordinate system with respect to which we can make observations or measurements.

free-body diagram A graphical representation of all external forces acting on a body.

free fall The state of an object falling toward Earth without any effect from air resistance.

freezing The phase change from liquid to solid.

frequency The number of cycles of an oscillation per unit of time.

friction force Force resisting the sliding of an object across a surface, acting parallel to the surface and opposite to the motion of an object.

fringe The series of bright and dark patches in an interference or diffraction pattern.

front (of a wave) Wave crest.

fundamental frequency The lowest natural frequency of an oscillating object.

fundamental mode For an oscillating object, the standing wave mode corresponding to the fundamental frequency.

fusion The change from solid to liquid; melting.

galaxy A collection of a tremendous number of stars held together by gravity.

Galilean transformation Set of equations used to translate between the coordinates of two reference frames which differ only by constant relative motion.

Galilean velocity transformation Set of equations used to translate between the velocity of an object in two reference frames which differ only by constant relative motion.

gamma decay A radioactive process in which an atomic nucleus loses energy by emitting a gamma ray without a change in its atomic number or mass number.

gas A fluid that expands to fill whatever volume is available to it.

gauge pressure The amount by which the pressure exceeds atmospheric pressure.

Gaussian surface A closed surface used to enclose charge in order to apply Gauss's law.

Gauss's law (for the electric field) The net electric flux through a closed surface (called a Gaussian surface) equals the net charge enclosed by that surface divided by the permittivity. Charges outside the surface have no effect on the net electric flux through the surface; named after nineteenth-century German mathematician and physicist Carl Friedrich Gauss.

Gauss's law for the magnetic field For any closed Gaussian surface, there is zero net flux of the magnetic field through that surface.

general theory of relativity Albert Einstein's theory which provides a unified description of gravity as a geometric property of space and time, or spacetime.

geometrical optics The science of mirrors and lenses.

global warming An increase in the global average surface temperature caused by the greenhouse effect.

gluon A subatomic particle of a class that is thought to bind quarks together.

gravitational bending of light Effect in which a light beam passing near a massive object will curve under the influence of the object's gravity.

gravitational constant The constant involved in the calculation of gravitational force between two objects; equal to $6.67 \times 10^{-11}\ \mathrm{N \cdot m^2/kg^2}$.

gravitational force The force of attraction between all masses in the universe; especially the attraction of Earth's mass for bodies near its surface.

gravitational potential energy The ability to do work related to an object's vertical position in the presence of gravity.

gravitational slowing of time Effect in which gravity influences the rate of a ticking clock; clocks on the ground floor of a building tick more slowly than clocks on the top floor, which are farther from Earth's center.

gravitational waves Small variations in the curvature of spacetime that spread away from moving massive objects.

greenhouse effect Warming effect in which the atmosphere prevents some of the radiation emitted by Earth's surface from escaping into space.

greenhouse gas One of several gases in the atmosphere that are transparent to visible light but not to infrared radiation.

hadron A particle that can experience the strong force.

Hagen-Poiseuille equation Relates the pressure difference between the two ends of a pipe to the resulting flow rate of a viscous fluid.

half-life The time taken for the radioactivity of a specified isotope to fall to half its original value.

harmonic property Characteristic of simple harmonic motion in which the angular frequency, period, and frequency of an oscillation are independent of the amplitude if the restoring force obeys Hooke's law.

heat A form of energy arising from the random motion of the molecules of bodies, which may be transferred by conduction, convection, or radiation.

heat engine A system or device that converts heat to work.

Heisenberg uncertainty principle The shorter the duration of a phenomenon, the greater the uncertainty in the energy of that phenomenon.

hertz The SI unit of frequency, equal to one cycle per second.

Higgs field The theoretical field that gives fundamental particles their mass.

Higgs particle Subatomic particle associated with the Higgs field.

hole The lack of an electron at a position where one could exist in an atom.

Hooke's law The force needed to extend a spring by a certain distance is proportional to that distance.

Hubble constant The ratio of the speed of recession of a galaxy (due to the expansion of the universe) to its distance from the observer.

Hubble law The observation that the speed of recession of distant galaxies is proportional to their distance from the observer.

Huygens' principle The wave front at a later time is the superposition of all of the wavelets emitted at the starting time and is tangent to the leading edges of the wavelets.

hydrostatic equilibrium State in which a fluid is at rest.

ideal gas A theoretical gas composed of a set of randomly moving, noninteracting point particles.

ideal gas constant A constant in the ideal gas law, expressed in units of energy per temperature increment per mole; equal to $8.314 \text{ J}/(\text{mol} \cdot \text{K})$.

ideal gas law Equation of the state of a hypothetical ideal gas.

image An appearance of an object formed by light rays reflected by a mirror or focused by a lens.

image distance The distance from a mirror to an object's reflected image, or the distance from a lens to the image.

image height The height of an object's image reflected in a mirror or focused by a lens.

impedance The measure of the opposition that a circuit presents to an alternating current when an alternating voltage is applied.

impulse The product of the force acting on an object and the duration of the time interval over which that force acts.

incident light Incoming light that strikes a surface.

incompressible fluid A fluid whose volume and density change very little when squeezed.

index of refraction A measure of the speed of light traveling in a medium; the speed of light in a vacuum divided by the speed of light in the medium.

induced emf An emf induced around a loop by a changing magnetic field.

induced magnetic field A magnetic field produced by the current in a loop that is caused by an induced emf in that loop.

inductance For a current-carrying conducting coil, the ratio of the magnetic flux through the coil divided by the current that produces the flux.

inelastic collision A collision in which mechanical energy is not conserved.

inertia The tendency of an object to resist change in motion.

inertial frame of reference A frame of reference attached to an object that does not accelerate.

instantaneous acceleration Acceleration of an object at a specific instant.

instantaneous speed Speed of an object at a specific instant.

instantaneous velocity Velocity of an object at a specific instant.

insulator Substance in which charges are not able to move freely.

intensity Average wave power per unit area.

interference The combination of two or more electromagnetic waves to form a resultant wave in which the displacement is either reinforced or canceled.

interference maxima Locations in an interference pattern where wavelets interfere constructively, producing a bright fringe.

interference minima Locations in an interference pattern where wavelets interfere destructively, producing a dark fringe.

interference pattern Alternating bright and dark fringes produced when two or more light waves combine or cancel each other out.

internal energy The energy within an object due to the kinetic and potential energies associated with the individual molecules that comprise the object.

internal forces Forces exerted by one part of an object or system on another part.

internal resistance The resistance that mobile charges encounter as they pass through a battery.

inverse Lorentz velocity transformation Set of equations relating the velocity of an object in two reference frames moving relative to one another, consistent with special relativity.

inverse-square law for waves Intensity is inversely proportional to the square of the distance from the source.

inverted image Image (produced by a mirror or lens) that is flipped upside down relative to the object.

inviscid flow Flow of a fluid that is assumed to have no viscosity.

ionizing radiation Photons with enough energy to dislodge an electron from an atom.

irreversible process A process that cannot return both the system and the surroundings to their original conditions.

irrotational flow Flow in which the speed varies gradually from one part of the fluid to another, with no abrupt jumps.

isobaric process Process in which the pressure of the system remains constant.

isochoric process Process in which the volume of the system remains constant.

isotherm Contours of constant temperature.

isothermal process Process in which the temperature of the system remains constant.

isotope Variants of a particular chemical element that share the same number of protons in the nucleus of each atom but differ in neutron numbers.

joule The SI unit of work or energy, equal to the work done by a force of one newton when its point of application moves one meter in the direction of action of the force.

junction Points in a circuit where either the current breaks into two currents or two currents come together into one.

kelvin The SI base unit of thermodynamic temperature, equal in magnitude to the degree Celsius.

Kelvin scale Temperature scale based on the relationship between pressure and temperature of low-density gases; first proposed by the nineteenth-century Scottish physicist William Thomson (1st Baron Kelvin).

kilogram The SI unit of mass.

kinetic energy The energy that an object possesses by virtue of being in motion.

kinetic friction Force acting on a sliding object that opposes the object's motion.

Kirchhoff's junction rule The sum of the currents flowing into a junction equals the sum of the currents flowing out of it.

Kirchhoff's loop rule The sum of the changes in electric potential around a closed loop in a circuit must equal zero.

laminar flow Smooth fluid flow in which each object follows the object directly in front of it.

latent heat (of fusion or vaporization) The amount of heat per unit mass that must flow into or out of the substance to cause the phase change.

lateral magnification The ratio of the height of an image to the height of the corresponding object.

law of areas A line joining the Sun and a planet sweeps out equal areas in equal intervals of time, regardless of the position of the planet in the orbit.

law of conservation of angular momentum *See* conservation of angular momentum

law of conservation of energy One kind of energy can transform into another, but the total amount of energy of all forms remains the same.

law of conservation of momentum If the net external force on a system of objects is zero, then the total momentum of the system does not change.

law of orbits The orbit of each planet is an ellipse with the Sun located at one focus of the ellipse.

law of periods The square of the period of a planet's orbit is proportional to the cube of the semimajor axis of the orbit.

law of reflection The angle of the reflected light is the same as the angle of the incident light.

***LC* circuit** A circuit made up of an inductor of inductance L and a capacitor of capacitance C connected by ideal, zero-resistance wires.

length contraction The shortening of an object or distance moving at nearly the speed of light along the direction of motion; predicted by the theory of special relativity.

lens A piece of glass or other transparent material with a curved surface for concentrating or dispersing light rays.

lens equation Expression relating object distance and lens distance to focal length.

lensmaker's equation Expression relating the focal length of a lens in air to its index of refraction and curvature.

Lenz's law The direction of the magnetic field induced within a conducting loop opposes the change in magnetic flux that created it; named after Russian physicist Heinrich Lenz.

lepton A particle that is not affected by the strong force.

lever arm The perpendicular distance from the rotation axis of a body to the line of action of a force applied to that body.

light-emitting diode (LED) A semiconductor device that emits visible light when an electric current passes through it; produces much more light for a given power input than do incandescent light bulbs or fluorescent lamps.

linearly polarized light Light for which the electric field is oriented completely along one direction.

linear mass density Mass per unit length.

linear momentum (momentum) The product of an object's mass and its velocity vector.

linear motion Motion in a straight line.

line of action For a force applied to a body, an extension of the force vector through the point of application.

liquid A fluid that maintains the same volume regardless of the shape and size of its container.

longitudinal wave (pressure wave) A traveling disturbance or vibration in which the individual parts of the wave medium move in the direction parallel to the direction of wave propagation.

Lorentz transformation Set of equations used to translate between the coordinates of two reference frames moving relative to one another, consistent with special relativity.

Lorentz velocity transformation Set of equations relating the velocity of an object in two reference frames moving relative to one another, consistent with special relativity.

Mach angle For supersonic flow, the angle a shock wave makes with the direction of motion, determined by the velocity of the object and the velocity of shock propagation.

Mach cone The conical pressure wave front produced by a body moving at a speed greater than that of sound.

Mach number The ratio of an object's speed to the speed of sound.

magnet A material or object that produces a magnetic field.

magnetic dipole A pair of equal and opposite magnetic poles separated by a small distance; the magnetic field points away from the magnet's north pole and toward the magnet's south pole.

magnetic energy The energy required to set up a magnetic field in and around an inductor.

magnetic energy density The energy per unit volume stored in a magnetic field, such as in an inductor.

magnetic field A region around a magnetic material or a moving electric charge within which the force of magnetism acts.

magnetic flux The area of a surface multiplied by the component of the magnetic field that's perpendicular to that surface.

magnetic force Force of attraction or repulsion that arises between magnets, or between electrically charged particles because of their motion.

magnetic poles The two ends of a magnetic field.

magnetic resonance imaging (MRI) A form of medical imaging that measures the response of atomic nuclei in body tissues to radio waves when placed in a strong magnetic field, producing detailed images of internal organs.

magnetism The interaction between magnets or between electrically charged particles due to their motion.

magnitude (of a vector) The straight-line distance from the starting point of a vector to its endpoint.

mass The measure of the amount of material in an object.

mass number The total number of protons and neutrons in an atomic nucleus.

mass spectrometer A device used to determine the masses of individual atoms and molecules.

Maxwell–Ampère law Magnetic fields can be generated in two ways: by electrical current and by changing electric fields.

Maxwell's equations Four basic equations that describe all electromagnetic phenomena.

mean free path The average distance that a molecule travels from the time at which it collides with one molecule to when it collides with another molecule.

mechanical wave A propagating oscillation of matter that transfers energy through a medium.

medium The substance through which a mechanical wave propagates.

meson A subatomic particle that is intermediate in mass between an electron and a proton and that transmits the strong interaction that binds nucleons together in the atomic nucleus.

meter The SI unit of measurement for length.

Michelson–Morley experiment An experiment performed in 1887 attempting to detect the velocity of Earth with respect to the hypothetical luminiferous ether; discovered no evidence for its existence.

millimeters of mercury Unit of measurement for pressure; 760 mmHg equals 1 atm.

mirror equation Expression relating object distance and image distance to the focal length of a mirror.

mobile charge Charged particles that are free to move throughout a conducting material, as in a circuit.

molar specific heat The quantity of heat required to raise the temperature of one mole of the substance by one kelvin.

molar specific heat at constant pressure The quantity of heat required to make one mole of a substance undergo a temperature change of one kelvin if the pressure is held constant.

molar specific heat at constant volume The quantity of heat required to make one mole of a substance undergo a temperature change of one kelvin if the volume is held constant.

mole Unit measuring the quantity of a substance; one mole equals the number of atoms in exactly 12 grams of carbon-12, given by Avogadro's number: 6.022×10^{22}.

moment of inertia A property of a body that defines its resistance to a change in angular velocity about an axis of rotation.

momentum (linear momentum) *See* linear momentum.

monochromatic light Light that has a single definite wavelength.

motional emf A changing emf due to the motion of a conductor in a magnetic field.

motion diagram A diagram that visualizes the motion of an object using distance gained between equal time intervals.

motion in a plane Two-dimensional motion.

multiloop circuit A circuit with more than one pathway that a moving charge can take from the positive terminal of the battery through the circuit to the negative terminal.

mutual inductance An effect in a transformer in which a change in the current in one coil induces an emf and current in a second coil.

natural angular frequency The angular frequency at which a system oscillates when not subjected to an external force.

negative displacement The distance, in the negative direction along a defined coordinate axis, between the position of an object at one time and its position at an earlier time.

negative work Work done whenever the angle between the displacement of an object and the force acting on that object is greater than 90°.

net external force (net force) The vector sum of all external forces acting on an object.

neutral matter Matter that contains equal amounts of positive and negative charge.

neutrino A nearly massless, neutral particle.

neutron-induced fission The radioactive decay of an atomic nucleus initiated by the collision of a neutron.

neutron number The number of neutrons in the nucleus of an atom.

newton The SI unit of force; equal to the force that would give a mass of one kilogram an acceleration of one meter per second per second.

Newton's first law An object at rest tends to stay at rest, and an object in uniform motion tends to stay in motion with the same speed and in the same direction, unless acted upon by a net force.

Newton's law of universal gravitation Any two objects exert a gravitational force of attraction on each other in a direction along the line joining the objects, with a magnitude proportional to the product of the masses of the objects and inversely proportional to the square of the distance between them.

Newton's laws of motion Isaac Newton's three fundamental relationships between force and motion.

Newton's second law If a net external force acts on an object, the object accelerates. The net external force is equal to the product of the object's mass and the object's acceleration.

Newton's third law If object A exerts a force on object B, object B exerts a force on object A that has the same magnitude but is in the opposite direction. These two forces act on different objects.

node (of a wave) Any point where the displacement of a wave is always zero.

nonconservative force A dissipative force that does not have a defined potential energy, such as friction.

noninertial frame A frame of reference attached to an accelerated object.

normal A line that is perpendicular to a surface.

normal force The support force exerted upon an object that is in contact with another stable object, acting perpendicular to the surface of contact.

no-slip condition Requirement that the velocity of a fluid be zero next to a solid surface.

***n*-type semiconductor** A type of semiconductor doping provides extra electrons to the host material, creating an excess of negative electron charge carriers.

nuclear fission A nuclear reaction in which a heavy nucleus splits spontaneously or on impact with another particle, releasing energy.

nuclear fusion A nuclear reaction in which atomic nuclei of low atomic number fuse to form a heavier nucleus, releasing energy.

nuclear radiation The emission by a nucleus of either energy (in the form of a photon) or particles (such as an alpha particle).

nucleon A proton or neutron.

nuclide An atomic species characterized by the specific constitution of its nucleus, that is, by its number of protons and its number of neutrons.

object Anything that acts as a source of light rays for an optical device.

object distance An object's distance from a mirror or lens.

object height The vertical extent of an object that is reflected in a mirror or refracted by a lens.

ohm The SI unit of electrical resistance; the resistance in a circuit transmitting a current of one ampere when subjected to a potential difference of one volt.

one-dimensional wave A wave that propagates along a single dimension of space.

open pipe A pipe that is unblocked on both ends.

optical device An instrument that changes the direction of light rays in a regular way.

orbital period The time required to complete an orbit.

origin The location from which the points on a coordinate system are measured. Also called reference position.

oscillation The regular movement of an object back and forth around a point of equilibrium.

overdamped oscillations Case in which high damping is applied, resulting in a nonoscillatory response; when displaced from equilibrium, the system returns to equilibrium with no overshoot and hence no oscillation.

parabola A particular u-shaped curve; the shape of the path of a projectile under the influence of gravity.

parallel (capacitors) An arrangement of capacitors connected along multiple paths (not in series), resulting in a multiloop circuit; the total capacitance is equal to the sum of all the individual capacitances.

parallel (resistors) An arrangement of resistors connected along multiple paths (not in series), resulting in a multiloop circuit; the reciprocal of the total resistance is equal to the sum of the reciprocals of all the individual resistances.

parallel-axis theorem Expression determining the moment of inertia of an object about a given axis in terms of the moment of inertia about a parallel axis running through its center of mass and the distance between the two axes.

parallel-plate capacitor A capacitor formed using two parallel metal plates.

paramagnetic Tendency to be weakly attracted by the poles of a magnet but not retaining any permanent magnetism; if the magnetic field is turned off, random thermal motion will cause the atomic current loops to return to their original, nonaligned orientations.

partially polarized light Light in which the orientation of the electric field changes randomly but is more likely to be in one orientation than in other orientations.

particle physics The branch of physics that concerns the fundamental constituents of matter and how they interact.

pascal Unit of measurement for pressure; equal to one newton per square meter.

Pascal's principle Pressure applied to a confined, static fluid is transmitted undiminished to every part of the fluid as well as to the walls of the container.

path length difference The difference in the distance traversed by two waves traveling to the same point from different locations.

Pauli exclusion principle The quantum mechanical principle that no two electrons may occupy the same quantum state simultaneously.

pendulum A system that oscillates back and forth due to the restoring force of gravity.

period The time for one complete cycle of an oscillation.

permeability of free space A constant involved in the relationship between an electric current and the magnetic field that it produces in a vacuum.

permittivity of free space A constant involved in the relationship between an electric charge and the electric field that it produces in a vacuum.

phase A physically distinctive form of matter, such as a solid, liquid, gas, or plasma.

phase angle The amount that a wave is shifted, indicating where in the oscillation cycle the object is at $t = 0$.

phase change The transformation from one state of matter to another.

phase diagram A graph of pressure p versus temperature T for a substance, showing the values of p and T for each phase of the substance.

phase difference The mathematical difference between two phase angles, such as the phase angles of two different waves.

photoelectric effect The emission, or ejection, of electrons from the surface of a material in response to incident light.

photoelectrons Electrons emitted through the photoelectric effect.

photon The quantum of electromagnetic energy, regarded as a discrete particle having zero mass, no electric charge, and an indefinitely long lifetime.

photovoltaic solar cell An electrical device that converts the energy of light directly into electricity by the photovoltaic effect.

physical pendulum A pendulum whose mass is distributed throughout its volume.

Planck's constant A physical constant relating the ratio of the energy of a photon to its frequency; equal to $6.626 \times 10^{-34}\ \text{J} \cdot \text{s}$.

plane mirror A flat, reflecting surface.

plastic The state of an object when the tensile stress exceeds the yield strength and an object deforms permanently.

plates Two pieces of metal used in a capacitor to store charge.

***pn* junction** The region inside a semiconductor where *p*-type and *n*-type semiconductors meet.

point charges Very small charged objects whose size is much smaller than the separation between the charges.

polarization The orientation of the electric field in a light wave.

polarizing filter A transparent sheet which contains long-chain molecules that are all oriented in the same direction, used to polarize the light passing through it.

position The location of an object on a coordinate system.

position vector A vector that extends from an origin to a point where an object is located at a specific moment in time.

positive displacement The distance, in the positive direction along a defined coordinate axis, between the position of an object at one time and its position at an earlier time.

positron A subatomic particle with the same mass as an electron and a numerically equal but positive charge.

potential energy An ability to do work based on an object's position.

pound The English unit of force.

power The rate at which work is done or energy is transferred.

power (of a lens or mirror) The reciprocal of the focal length (in meters).

power of ten An alternative name for the exponent given in scientific notation, referring to how many tens must be multiplied together to give the desired number.

pressure The magnitude of the force per unit area on the surface of an object.

pressure amplitude In a sound wave, the maximum pressure variation above or below the pressure of the undisturbed air.

pressure wave (longitudinal wave) *See* longitudinal wave.

primary coil The winding of a transformer connected to the input voltage.

principal axis A line passing through the center of the surface of a lens or spherical mirror and through the centers of curvature of all segments of the lens or mirror.

principle of equivalence A gravitational field is equivalent to an accelerated frame of reference in the absence of gravity.

principle of Newtonian relativity The laws of motion are the same in all inertial frames of reference.

projectile Object undergoing free-fall motion under the influence of gravity.

projectile motion Free-fall motion under the influence of gravity, involving both vertical and horizontal motion.

propagation speed The rate at which a wave travels in a medium.

proper time The time interval between two events in a frame of reference in which the events occur at the same place.

***p*-type semiconductor** A type of semiconductor with an absence of negative charge and hence an abundance of positive charge carriers or holes.

***pV* diagram** A graph that plots the pressure of a system on the vertical axis versus the volume of the system on the horizontal axis.

quantized Restricted to only certain values, as energy.

quantum mechanics Branch of physics that deals with the motions and interactions of atoms and subatomic particles, incorporating the concepts of quantization of energy, wave-particle duality, and the uncertainty principle.

quantum number Number describing the value of a physical quantity in a quantum mechanical system, such as an atom.

quark Any of a number of subatomic particles carrying a fractional electric charge, postulated as building blocks of the hadrons.

quark confinement The phenomenon wherein quarks can never be removed from the hadrons they compose.

radiation Energy transfer by the emission (or absorption) of electromagnetic waves.

radioactive A radioactive nuclide is one that decays into another nuclide by emitting ionizing radiation or particles.

radius of curvature The distance from a curved surface (such as a mirror) to its center of curvature.

ratio of specific heats For a given substance, the ratio of molar specific heat at constant pressure to the molar specific heat at constant volume.

ray An arrow that points in the direction of light propagation.

ray diagram Drawing used to determine the position of an image made by a mirror or lens.

reaction force The force exerted by object A on object B in reaction to having a force exerted by object B on object A.

real image An image formed by light rays coming together.

redshift The displacement of spectral lines toward longer wavelengths (the red end of the spectrum) in radiation from distant galaxies and celestial objects.

reference frame *See* frame of reference.

reference position *See* origin.

refraction The change in direction of a beam of light that travels from one medium into another; the angle that the refracted light makes to the normal is not equal to the angle of the incident light to the normal.

refrigerator A device that takes in energy and uses it to transfer heat from an object at low temperature to an object at high temperature.

relativistic gamma A dimensionless quantity that is equal to 1 when an object is at rest and becomes infinitely large as the speed of the object approaches the speed of light.

relativistic speed A speed that is a significant proportion of the speed of light.

reservoir A part of a system large enough either to absorb or supply heat without a change in temperature.

resistance For an electrical conductor, the resistivity of the material of which the conductor is made multiplied by the length of the conductor and divided by its cross-sectional area.

resistivity A measure of how well or poorly a material inhibits the flow of electric charge.

resistor A circuit component intended to add resistance to the flow of current.

resolve To optically distinguish; as in telling two closely spaced objects apart.

resonance The condition in which an object or system is subjected to an oscillating force having a frequency close to its own natural frequency.

rest energy Energy of a particle that is not in motion.

restoring force A force that tends to bring an object back toward equilibrium.

reversible process An ideal process that can return both the system and the surroundings to their original conditions without increasing entropy; throughout the entire reversible process the system is in thermodynamic equilibrium with its surroundings.

Reynolds number The ratio of the forces due to pressure differences acting on a small piece of a fluid to the viscous forces acting on the same piece of fluid.

right-hand rule A rule that uses the right hand to determine the orientation of vector quantities normal to a plane; for example, used to find the direction of the angular momentum vector around an axis of rotation.

rigid object An object with a fixed shape; the distance between any two points in the object remains constant in time regardless of external forces exerted on it.

rolling friction Force acting on the point of contact between a rolling object and the surface opposite to the direction of motion.

rolling without slipping The state in which an object rolls uniformly across a surface without skidding.

root-mean-square speed (rms speed) (of molecules) A measure of how fast gas molecules move; the square root of the average value of the square of individual speeds.

root mean square value (rms value) (of an ac circuit) The square root of the average value of the ac voltage squared.

rotation Motion in which an object spins around an axis.

rotational kinematics The study of rotational motion, including angular velocities and angular acceleration, in the absence of forces.

rotational kinetic energy Energy possessed by an object by virtue of its rotational motion.

scalar A physical quantity that has only magnitude, not direction.

scientific notation A standard shorthand system used by physicists for extremely large or small numbers.

second The SI unit of measurement for time.

secondary coil The winding of a transformer that is the source of the output voltage.

second law of thermodynamics The amount of disorder in an isolated system either always increases or, if the system is in equilibrium, stays the same.

self-inductance The induction of a voltage in a current-carrying coil when the current in the coil itself is changing. The induced emf will oppose any change in the current.

semiconductors Substances with electrical properties that are intermediate between those of insulators and conductors.

semimajor axis Half of the distance of the longest diameter of an ellipse (the major axis).

series (*LRC* circuit) A circuit containing an inductor, resistor, and capacitor in series.

series (*RC* circuit) A circuit that contains both a resistor and capacitor in series.

series (capacitors) An arrangement of capacitors connected along a single path (not in parallel); the reciprocal of the total capacitance is equal to the sum of the reciprocals of the individual capacitances.

series (resistors) An arrangement of resistors connected along a single path (not in parallel); the total resistance is equal to the sum of all the individual resistances.

shear modulus A measure of the rigidity and resistance to deformation of a material.

shear strain The change in an object's shape caused by shear stress.

shear stress Forces applied parallel to the plane in which an object lies, deforming the object without making it expand or contract.

shock wave A sudden pressure increase in a narrow region of a medium (for example, air), such as that caused by a body moving faster than the speed of sound.

significant figures The number of digits in a figure that can be known with some degree of confidence.

simple harmonic motion (SHM) Oscillatory motion under a Hooke's law restoring force (which is proportional to the displacement from the equilibrium position).

simple pendulum A pendulum in which all of the mass is concentrated at a single point.

single-loop circuit A circuit with only a single path that moving charges can follow.

single-slit diffraction Experiment in which a wave passes through a narrow opening, producing a pattern of bright and dark fringes.

sinusoidal function A mathematical curve that describes a smooth repetitive oscillation, such as the sine and cosine functions.

sinusoidal wave A wave in which the wave pattern at any instant is a sinusoidal function.

Snell's law of refraction Expression describing the relationship between the angles of incidence and refraction when referring to light or other waves passing through a boundary between two different media, such as water, glass, and air.

solenoid A straight helical coil of wire.

solid A substance whose individual molecules cannot move freely but remain in essentially fixed positions relative to one another.

sonic boom A loud, explosive noise caused by the shock wave from an aircraft traveling faster than the speed of sound.

sound intensity level The power carried by sound waves per unit area.

sound wave A wave in air consisting of periodic variations in air pressure.

source of emf A device that originates voltage in a circuit, such as a battery.

special theory of relativity Theory developed by Albert Einstein that states: (1) All laws of physics are the same in all inertial frames, and (2) the speed of light in a vacuum is the same in all inertial frames, independent of both the speed of the source of the light and the speed of the observer.

specific gravity The density of a substance divided by the density of 4°C liquid water.

specific heat The amount of heat per unit mass required to raise the temperature by one kelvin.

specular reflection Type of surface reflection in which light rays moving in a single direction reflect from a smooth surface in a single outgoing direction.

speed The rate at which an object is moving; equal to the magnitude of the object's velocity.

speed of light The speed at which all electromagnetic waves—including radio waves, x rays, and others—travel in a vacuum.

spin An intrinsic characteristic of electrons, protons, and neutrons akin to the angular momentum of a rotating sphere.

spring constant Measure of the stiffness of a spring.

spring potential energy The energy stored in a spring, based on whether it is relaxed, stretched, or compressed.

Standard Model A mathematical description of the elementary particles of matter and the electromagnetic, weak, and strong forces by which they interact.

standing wave A wave in which each point in the medium has a constant amplitude, giving it the appearance of being stationary.

standing wave mode The conditions under which a standing wave is possible; for example, a whole number of half-wavelengths must fit onto a string.

state variables Quantities such as volume, temperature, pressure, and internal energy that depend on the state or condition of a substance.

static friction Force acting on a stationary object that opposes the object's sliding motion.

steady flow Type of fluid flow in which the flow pattern does not change with time.

Stefan–Boltzmann constant The constant of proportionality in the Stefan–Boltzmann law: The total energy radiated per unit surface area of a blackbody in unit time is proportional to the fourth power of the thermodynamic temperature; equal to $5.670 \times 10^{-8}\ \mathrm{W/(m^2 \cdot K^4)}$.

step-down transformer A transformer in which the number of windings in the secondary coil is less than the number of windings in the primary coil, so the output voltage is less than the input voltage.

step-up transformer A transformer in which the number of windings in the secondary coil is greater than the number of windings in the primary coil, so the output voltage is greater than the input voltage.

strain The amount of deformation that results from applied stress.

streamlines The paths followed by bits of fluid in laminar flow.

stress Force per area exerted on an object tending to cause the object to change in size or shape.

strong nuclear force An attractive force between protons and neutrons that is stronger than the repulsive electric force between protons.

sublimation The phase change from solid directly to gas.

surface charge density The amount of charge per unit area on a surface.

surface tension The attractive force exerted on molecules at the surface of a liquid by the molecules beneath, causing the liquid to assume the shape having the least surface area.

surface wave A wave that propagates along the interface between two media (for example, a seismic wave that travels along the surface of the Earth).

Système International (SI) The standard system of units based on the fundamental quantities.

temperature A measure of the kinetic energy associated with molecular motion.

tensile strain The distance an object stretches when pulled, expressed as a fraction of its relaxed length.

tensile stress Stretching force applied to an object divided by the object's cross-sectional area.

tension Stretching force applied at each end of an object; for example, the force exerted by a rope on an object it tows.

tesla The SI units of magnetic field strength.

test charge A point charge with such a small magnitude that it negligibly affects the field in which it is placed.

thermal conductivity A measure of how easily heat passes through a specified material.

thermal contact A state in which two or more systems can exchange thermal energy.

thermal energy Energy associated with the random motion of atoms and molecules.

thermal equilibrium The condition in which two objects in physical contact exchange no heat energy; in thermal equilibrium the objects are said to be at the same temperature.

thermal expansion The tendency of matter to increase in length, area, or volume in response to an increase in temperature.

thermodynamic process Any process that changes the state of a system.

thermodynamics The branch of physics that deals with relationships among properties of substances such as temperature, pressure, and volume, as well as the energy and flow of energy associated with these properties.

thermometer Instrument used to measure temperature changes.

thin film A very fine layer of a substance on a supporting material; for example, the thin lining behind the retina of the eyes of some animals.

thin lens A lens with a thickness that is negligible compared to the radii of curvature of the lens surfaces.

third law of thermodynamics It is possible for the temperature of a system to be arbitrarily close to absolute zero, but it can never reach absolute zero.

time constant The product of resistance multiplied by capacitance.

time dilation A difference of elapsed time between two events as measured by observers either moving relative to each other or located at different distances from large gravitational masses.

time interval A set length of time.

torque A force that causes rotation.

torr A unit measuring pressure; used especially in measuring partial vacuums; equal to 1/760 atm or 133.32 pascals.

total internal reflection The phenomenon occurring when a wave strikes a medium boundary at an angle larger than the critical angle with respect to the normal; 100% of the incident light is reflected back into the first medium.

total mechanical energy The sum of kinetic and potential energy.

total momentum The momentum of a system of objects.

trajectory The path followed by a projectile or an object.

transformer A device that can raise or lower an ac voltage to a desired value.

translation Motion in which an object as a whole moves through space.

translational kinetic energy The energy possessed by an object by virtue of its motion as a whole through space.

transverse wave A traveling disturbance or vibration in which the individual parts of the wave medium move in a direction perpendicular to the direction of wave propagation.

traveling wave A wave of the form $y(x,t) = A \cos (kx - vt)$ in which a disturbance propagates from one location to another.

triple point The particular combination of pressure and temperature of a material at which the solid, liquid, and vapor phases all coexist.

turbulent flow Type of fluid motion in which the velocity of the flow at any point is continuously undergoing changes in both magnitude and direction.

two-dimensional motion Vertical and horizontal movement of an object; motion confined to a plane.

ultimate strength The maximum tensile stress that a material can withstand before failure.

underdamped oscillation Case in which a small damping coefficient applied to an oscillating system causes the system to oscillate with ever decreasing amplitude.

uniform circular motion The motion of an object going around a circular path at a constant speed.

uniform density Constant mass density throughout a volume.

units The standard measurement for a specific quantity; for example, the second is a unit of measurement for time.

unpolarized light Natural light such as that emitted by the Sun or an ordinary light bulb, in which the orientation of the electric field changes randomly from one moment to the next.

unsteady flow Fluid motion in which the velocity changes with time.

upright image Image (produced by a lens or mirror) that has the same orientation as the object.

vaporization The phase change from solid to liquid.

vector A physical quantity that has both a magnitude and a direction.

vector addition The combination of two or more displacements to find the vector sum.

vector difference The result of subtracting one vector from another.

vector multiplication by a scalar An operation in which the product of a scalar and a vector pointing in a given direction is a new vector that has a magnitude equal to the product of the absolute value of the scalar and the magnitude of the original vector. The direction of the new vector is either the same as that of the original vector (if the scalar is positive) or opposite to it (if the scalar is negative).

vector product *See* cross product.

vector subtraction The process of taking a vector difference; the inverse of vector addition.

vector sum The result of vector addition.

velocity The magnitude and direction of the rate of change of an object's position.

velocity vector (or instantaneous velocity vector) The rate of change of the position of an object at a given point in time, expressed as a magnitude (speed) and direction.

virtual image An image from which rays of reflected or refracted light appear to diverge; for example, the image seen in a plane mirror.

virtual particle A particle whose existence is allowed by the uncertainty principle and that exhibits many of the characteristics of an ordinary particle, but that exists for a limited time.

viscosity A measure of the resistance to flow of a fluid.

visible light The range of wavelengths visible to the human eye.

volt The SI unit of electromotive force or electric potential; the emf required to drive one ampere of current against one ohm resistance.

voltage Electric potential difference.

volume flow rate Volume of fluid per unit time passing a given point.

volume strain The change in volume of an object under stress divided by the original volume.

volume stress Force applied perpendicularly to all faces of an object; pressure change.

v_x–t graph Chart depicting an object's velocity along the x axis versus time.

watt The SI unit of power; equal to one joule per second.

wave A disturbance or vibration that travels through space.

wave function A mathematical description of the properties of a wave, expressing the displacement of the wave medium at every position and at every time.

wavelength The distance between successive crests of a wave.

wavelet A tiny segment of a larger wave.

wave-particle duality The dual character of both light and matter, which have both wave and particle characteristics.

weak force An interaction between elementary particles, often involving neutrinos or antineutrinos, that is responsible for certain kinds of radioactive decay.

weight The magnitude of the gravitational force that acts on an object.

weighted average A mean calculated by giving values in a data set more influence according to some attribute of the data, such as how often a given value appears in the data set.

work The transfer of energy from one object to another.

work-energy theorem When an object undergoes a displacement, the work done on it by the net force equals the object's kinetic energy at the end of the displacement minus its kinetic energy at the beginning of the displacement.

work function The minimum amount of energy required to remove a single electron from a material.

***x* component** The component of a vector parallel to the x axis.

***x–t* graph** Chart depicting an object's position along the x axis versus time.

***x–y* plane** The coordinate plane formed by the x axis and the y axis.

***y* component** The component of a vector parallel to the y axis.

yield strength The tensile strength at which an object is permanently deformed and can no longer return to its normal strength.

Young's modulus A measure of the stiffness of a given material.

zeroth law of thermodynamics If two objects are each in thermal equilibrium with a third object, they are also in thermal equilibrium with each other.

Math Tutorial

In this tutorial, we review some of the basic results of algebra, geometry, trigonometry, and calculus. In many cases, we merely state results without proof. **Table M-1** lists some mathematical symbols.

M-1 Significant figures

Many numbers we work with in science are the result of measurement and are therefore known only within a degree of uncertainty. This uncertainty should be reflected in the number of digits used. For example, if you have a 1-meter-long rule with scale spacing of 1 cm, you know that you can measure the height of a box to within a fifth of a centimeter or so. Using this rule, you might find that the box height is 27.0 cm. If there is a scale with a spacing of 1 mm on your rule, you might perhaps measure the box height to be 27.03 cm. However, if there is a scale with a spacing of 1 mm on your rule, you might not be able to measure the height more accurately than 27.03 cm because the height might vary by 0.01 cm or so, depending on where you measure the height of the box. When you write down that the height of the box is 27.03 cm, you are stating that your best estimate of the height is 27.03 cm, but you are not claiming that it is exactly 27.030000 . . . cm high. The four digits in 27.03 cm are called **significant figures**. Your measured length, 27.03 cm, has four significant digits. Significant figures are also called significant digits.

The number of significant digits in an answer to a calculation will depend on the number of significant digits in the given data. When you work with numbers that have uncertainties, you should be careful not to include more digits than the certainty of measurement warrants. *Approximate* calculations (order-of-magnitude estimates) always result in answers that have only one significant digit or none. When you multiply, divide, add, or subtract numbers, you must consider the accuracy of the results. Listed below are some rules that will help you determine the number of significant digits of your results.

(1) When multiplying or dividing quantities, the number of significant digits in the final answer is no greater than that in the quantity with the fewest significant digits.

(2) When adding or subtracting quantities, the number of decimal places in the answer should match that of the term with the smallest number of decimal places.

(3) Exact values have an unlimited number of significant digits. For example, a value determined by counting, such as 2 tables, has no uncertainty and is an exact value. In addition, the conversion factor 0.0254000 . . . m/in. is an exact value because 1.000 . . . inches is exactly equal to 0.0254000 . . . meters. (The yard is, by definition, equal to exactly 0.9144 m, and 0.9144 divided by 36 is exactly equal to 0.0254.)

(4) Sometimes zeros are significant and sometimes they are not. If a zero is before a leading nonzero digit, then the zero is not significant. For example, the number 0.00890 has three significant digits. The first three zeroes are not significant digits but are merely markers to locate the decimal point. Note that the zero after the nine is significant.

TABLE M-1
Mathematical Symbols

Symbol	Meaning
$=$	is equal to
$\neq$	is not equal to
$\approx$	is approximately equal to
$\sim$	is of the order of
$\propto$	is proportional to
$>$	is greater than
$\geq$	is greater than or equal to
$\gg$	is much greater than
$<$	is less than
$\leq$	is less than or equal to
$\ll$	is much less than
Δx	change in x
$\lvert x \rvert$	absolute value of x
$n!$	$n(n-1)(n-2)\ldots 1$
Σ	sum

(5) Zeros that are between nonzero digits are significant. For example, 5603 has four significant digits.
(6) The number of significant digits in numbers with trailing zeros and no decimal point is ambiguous. For example, 31,000 could have as many as five significant digits or as few as two significant digits. To prevent ambiguity, you should report numbers by using scientific notation or by using a decimal point.

EXAMPLE M-1 Finding the Average of Three Numbers

Find the average of 19.90, −7.524, and −11.8179.

Set Up

You will be adding 3 numbers and then dividing the result by 3. The first number has four significant digits, the second number has four, and the third number has six.

Solve

Sum the three numbers.

$$19.90 + (-7.524) + (-11.8179) = 0.55\mathit{81}$$

If the problem only asked for the sum of the three numbers, we would round the answer to the least number of decimal places among all the numbers being added—the answer would be 0.56 (0.5581 rounds up to 0.56 to two significant digits). However, we must divide this intermediate result by 3, so we use the intermediate answer with the two extra digits (italicized and red).

$$\frac{0.5581}{3} = 0.18\mathit{60333}\ldots$$

Only two of the digits in the intermediate answer, 0.55*81* . . ., are significant digits, so we must round the final number to get our final answer. The number 3 in the denominator is a whole number and has an unlimited number of significant digits. Thus, the final answer has the same number of significant digits as the numerator, which is 2.

The final answer is 0.19.

Reflect

The sum in step 1 has two significant digits following the decimal point, the same as the number being summed with the least number of significant digits after the decimal point.

M-2 Equations

An **equation** is a statement written using numbers and symbols to indicate that two quantities, written on either side of an equal sign (=), are equal. The quantity on either side of the equal sign may consist of a single term, or of a sum or difference of two or more **terms**. For example, the equation $x = 1 - (ay + b)/(cx - d)$ contains three terms, x, 1, and $(ay + b)/(cx - d)$.

You can perform the following operations on equations:

(1) The same quantity can be added to or subtracted from each side of an equation.
(2) Each side of an equation can be multiplied or divided by the same quantity.
(3) Each side of an equation can be raised to the same power.

These operations are meant to be applied to each *side* of the equation rather than each term in the equation. (Because multiplication is distributive over addition, operation 2—and only operation 2—of the preceding operations also applies term by term.)

***Caution:** Division by zero is forbidden at* any *stage in solving an equation; results (if any) would be invalid.*

Adding or Subtracting Equal Amounts

To find x when $x - 3 = 7$, add 3 to both sides of the equation: $(x - 3) + 3 = 7 + 3$; thus, $x = 10$.

Multiplying or Dividing by Equal Amounts
If $3x = 17$, solve for x by dividing both sides of the equation by 3; thus, $x = \frac{17}{3}$, or 5.7.

EXAMPLE M-2 Simplifying Reciprocals in an Equation

Solve the following equation for x:

$$\frac{1}{x} + \frac{1}{4} = \frac{1}{3}$$

Equations containing reciprocals of unknowns occur in many circumstances in physics. Two instances of this are geometric optics and electric circuit analysis.

Set Up

In this equation, the term containing x is on the same side of the equation as a term not containing x. Furthermore, x is found in the denominator of a fraction. We'll start by isolating the $1/x$ term, find common denominators, and then multiply both sides of the equation by appropriate quantities.

Solve

Subtract $\frac{1}{4}$ from each side.

$$\frac{1}{x} = \frac{1}{3} - \frac{1}{4}$$

Simplify the right side of the equation by using the lowest common denominator.

Begin by multiplying both terms on the right-hand side by appropriate forms of 1.

$$\frac{1}{x} = \frac{1}{3}\frac{4}{4} - \frac{1}{4}\frac{3}{3} = \frac{4}{12} - \frac{3}{12}$$

$$= \frac{4-3}{12} = \frac{1}{12} \quad \text{so} \quad \frac{1}{x} = \frac{1}{12}$$

Multiply both sides of the equation by $12x$ to determine the value of x.

$$\cancel{12}x\frac{1}{\cancel{x}} = \cancel{12}x\frac{1}{\cancel{12}}$$

$$12 = x$$

Reflect

To check our answer, substitute 12 for x in the left side of original equation.

$$\frac{1}{x} + \frac{1}{4} = \frac{1}{12} + \frac{3}{12} = \frac{4}{12} = \frac{1}{3}$$

M-3 Direct and inverse proportions

When we say variable quantities x and y are **directly proportional**, we mean that as x and y change, the ratio x/y is constant. To say that two quantities are proportional is to say that they are directly proportional. When we say variable quantities x and y are **inversely proportional**, we mean that as x and y change, the ratio xy is constant.

Relationships of direct and inverse proportion are common in physics. Objects moving at the same velocity have momenta directly proportional to their masses. The ideal gas law ($PV = nRT$) states that pressure P is directly proportional to (absolute) temperature T, when volume V remains constant, and is inversely proportional to volume, when temperature remains constant. Ohm's law ($V = IR$) states that the voltage V across a resistor is directly proportional to the electric current in the resistor when the resistance remains constant.

Constant of Proportionality

When two quantities are directly proportional, the two quantities are related by a *constant of proportionality*. If you are paid for working at a regular rate R in dollars per day, for example, the money m you earn is directly proportional to the time t you work;

the rate R is the constant of proportionality that relates the money earned in dollars to the time worked t in days:

$$\frac{m}{t} = R \quad \text{or} \quad m = Rt$$

If you earn \$400 in 5 days, the value of R is \$400/(5 days) = \$80/day. To find the amount you earn in 8 days, you could perform the calculation

$$m = (\$80/\text{day})(8 \text{ days}) = \$640$$

Sometimes the constant of proportionality can be ignored in proportion problems. Because the amount you earn in 8 days is $\frac{8}{5}$ times what you earn in 5 days, this amount is

$$m_{8\text{ days}} = 8 \text{ days}\frac{\$400}{5 \text{ days}} = \$640$$

EXAMPLE M-3 Painting Cubes

You need 15.4 mL of paint to cover one side of a cube. The area of one side of the cube is 426 cm^2. What is the relation between the volume of paint needed and the area to be covered? How much paint do you need to paint one side of a cube on which the one side has an area of 503 cm^2?

Set Up

To determine the amount of paint for the side whose area is 503 cm^2 we will set up a proportion.

Solve

The volume V of paint needed increases in proportion to the area A to be covered.

V and A are directly proportional.

That is, $\frac{V}{A} = k$ or $V = kA$

where k is the proportionality constant

Determine the value of the proportionality constant using the given values $V_1 = 15.4$ mL and $A_1 = 426$ cm^2.

$$k = \frac{V_1}{A_1} = \frac{15.4 \text{ mL}}{426 \text{ cm}^2} = 0.0362 \text{ mL/cm}^2$$

Determine the volume of paint needed to paint a side of a cube whose area is 503 cm^2 using the proportionality constant in step 1.

$$V_2 = kA_2 = (0.0362 \text{ mL/cm}^2)(503 \text{ cm}^2) = 18.2 \text{ mL}$$

Reflect

Our value for V_2 is greater than the value for V_1, as expected. The amount of paint needed to cover an area equal to 503 cm^2 should be greater than the amount of paint needed to cover an area of 426 cm^2 because 503 cm^2 is larger than 426 cm^2.

M-4 Linear equations

A **linear equation** is an equation of the form $x + 2y - 4z = 3$. That is, an equation is linear if each term either is constant or is the product of a constant and a variable raised to the first power. Such equations are said to be linear because the plots of these equations form straight lines or planes. The equations of direct proportion between two variables are linear equations.

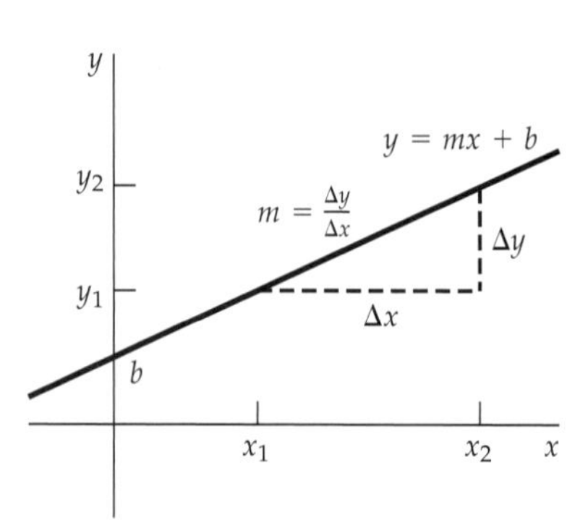

Figure M-1 Graph of the linear equation $y = mx + b$, where b is the y intercept and $m = \Delta y/\Delta x$ is the slope.

Graph of a Straight Line

A linear equation relating y and x can always be put into the standard form

$$y = mx + b \tag{M-1}$$

where m and b are constants that may be either positive or negative. **Figure M-1** shows a graph of the values of x and y that satisfy Equation M-1. The constant b, called the **y intercept**, is the value of y at $x = 0$. The constant m is the **slope** of the line, which equals the ratio of the change in y to the corresponding change in x. In the figure, we have indicated two points on the line, (x_1, y_1) and (x_2, y_2), and the changes $\Delta x = x_2 - x_1$ and $\Delta y = y_2 - y_1$. The slope m is then

$$m = \frac{y_2 - y_1}{x_2 - x_1} = \frac{\Delta y}{\Delta x}$$

If x and y are both unknown in the equation $y = mx + b$, there are no unique values of x and y that are solutions to the equation. Any pair of values (x_1, y_1) on the line in Figure M-1 will satisfy the equation. If we have two equations, each with the same two unknowns x and y, the equations can be solved simultaneously for the unknowns. Example M-4 shows two methods for simultaneously solving two linear equations.

EXAMPLE M-4 Using Two Equations to Solve for Two Unknowns

Find any and all values of x and y that simultaneously satisfy

$$3x - 2y = 8 \qquad \text{(M-2)}$$

and

$$y - x = 2 \qquad \text{(M-3)}$$

Set Up

Graph the two equations. At the point where the lines intersect, the values of x and y satisfy both equations.

We can solve two simultaneous equations by first solving either equation for one variable in terms of the other variable and then substituting the result into the second equation.

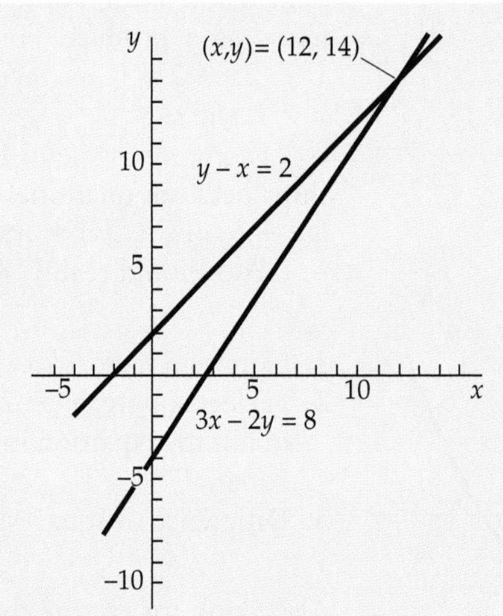

Figure M-2 Graph of Equations M-2 and M-3. At the point where the lines intersect, the values of *x* and *y* satisfy both equations.

Solve

Solve Equation M-3 for y.

$y = x + 2$

Substitute this value for y into Equation M-2.

$3x - 2(x + 2) = 8$

Simplify the equation and solve for x.

$3x - 2x - 4 = 8$
$x - 4 = 8$
$x = 12$

Use your solution for x and one of the given equations to find the value of y.

Return to Equation M-3 and substitute $x = 12$.

$y - x = 2$, where $x = 12$
$y - 12 = 2$
$y = 2 + 12 = 14$

Reflect

An alternative method is to multiply one equation by a constant such that one of the unknown terms is eliminated when the equations are added or subtracted.

We can multiply through Equation M-3 by 2

$2(y - x) = 2(2)$
$2y - 2x = 4$

Add the result to Equation M-2 and solve for x:

$\cancel{2y} - 2x = 4$
$3x - \cancel{2y} = 8$

$3x - 2x = 12 \Rightarrow x = 12$

Substitute into Equation M-3 and solve for y:

$y - 12 = 2 \Rightarrow y = 14$

M-5 Quadratic equations and factoring

A **quadratic equation** is an equation of the form $ax^2 + bxy + cy^2 + ex + fy + g = 0$, where x and y are variables and a, b, c, e, f, and g are constants. In each term of the equation the powers of the variables are integers that sum to 2, 1, or 0. The designation *quadratic equation* usually applies to a much simpler equation of one variable that can be written in the standard form

$$ax^2 + bx + c = 0 \qquad \text{(M-4)}$$

where a, b, and c are constants. The quadratic equation has two solutions or **roots**—values of x for which the equation is true.

Factoring

We can solve some quadratic equations by **factoring**. Very often terms of an equation can be grouped or organized into other terms. When we factor terms, we look for multipliers and multiplicands—which we now call **factors**—that will yield two or more new terms as a product. For example, we can find the roots of the quadratic equation $x^2 - 3x + 2 = 0$ by factoring the left side to get $(x - 2)(x - 1) = 0$. The roots are $x = 2$ and $x = 1$.

Factoring is useful for simplifying equations and for understanding the relationships between quantities. You should be familiar with the multiplication of the factors $(ax + by)(cx + dy) = acx^2 + (ad + bc)xy + bdy^2$.

You should readily recognize some typical factorable combinations:

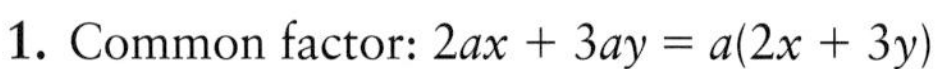

1. Common factor: $2ax + 3ay = a(2x + 3y)$
2. Perfect square: $x^2 - 2xy + y^2 = (x - y)^2$ (If the expression on the left side of a quadratic equation in standard form is a perfect square, the two roots will be equal.)
3. Difference of squares: $x^2 - y^2 = (x + y)(x - y)$

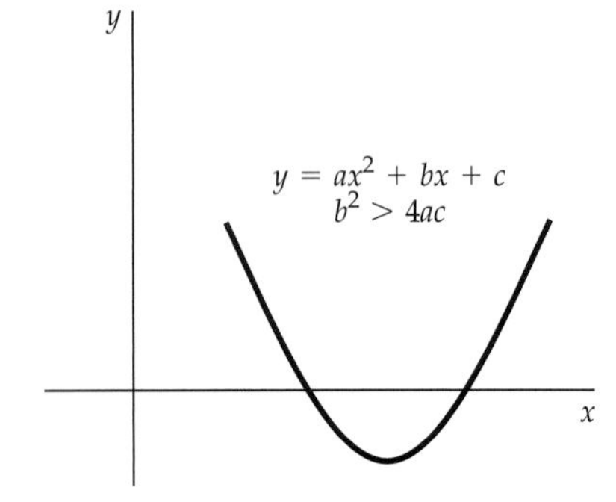

Figure M-3 Graph of y versus x when $y = ax^2 + bx + c$ for the case $b^2 > 4ac$. The two values of x for which $y = 0$ satisfy the quadratic equation (Equation M-4).

Also, look for factors that are prime numbers (2, 5, 7, etc.) because these factors can help you simplify terms quickly. For example, the equation $98x^2 - 140 = 0$ can be simplified because 98 and 140 share the common factor 2. That is, $98x^2 - 140 = 0$ becomes $2(49x^2 - 70) = 0$, so we have $49x^2 - 70 = 0$.

This result can be further simplified because 49 and 70 share the common factor 7. Thus, $49x^2 - 70 = 0$ becomes $7(7x^2 - 10) = 0$, so we have $7x^2 - 10 = 0$.

The Quadratic Formula

Not all quadratic equations can be solved by factoring. However, *any* quadratic equation in the standard form $ax^2 + bx + c = 0$ can be solved by the **quadratic formula**,

$$x = \frac{-b \pm \sqrt{b^2 - 4ac}}{2a} = -\frac{b}{2a} \pm \frac{1}{2a}\sqrt{b^2 - 4ac} \qquad \text{(M-5)}$$

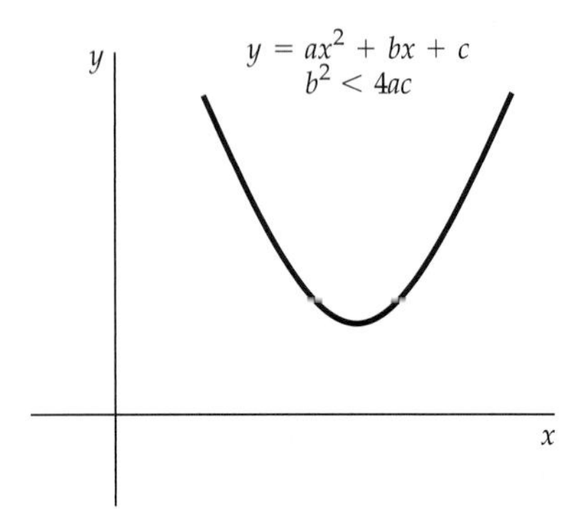

Figure M-4 Graph of y versus x when $y = ax^2 + bx + c$ for the case $b^2 < 4ac$. In this case, there are no real values of x for which $y = 0$.

When b^2 is greater than $4ac$, there are two solutions corresponding to the + and − signs, respectively. **Figure M-3** shows a graph of y versus x where $y = ax^2 + bx + c$. The curve, a **parabola**, crosses the x axis twice. (The simplest representation of a parabola in (x, y) coordinates is an equation of the form $y = ax^2 + bx + c$.) The two roots of this equation are the values for which $y = 0$; that is, they are the *x intercepts*.

When b^2 is less than $4ac$, the graph of y versus x does not intersect the x axis, as is shown in **Figure M-4**; there are still two roots, but they are not real numbers. When $b^2 = 4ac$, the graph of y versus x is tangent to the x axis at the point $x = -b/2a$; the two roots are each equal to $-b/2a$.

EXAMPLE M-5 Factoring a Second-Degree Polynomial

Factor the expression $6x^2 + 19xy + 10y^2$.

Set Up

We examine the coefficients of the terms to see whether the expression can be factored without resorting to more advanced methods. Remember that the multiplication $(ax + by)(cx + dy) = acx^2 + (ad + bc)xy + bdy^2$.

Solve

The coefficient of x^2 is 6, which can be factored two ways.

$ac = 6$
$3 \cdot 2 = 6 \quad \text{or} \quad 6 \cdot 1 = 6$

The coefficient of y^2 is 10, which can also be factored two ways.

$bd = 10$
$5 \cdot 2 = 10 \quad \text{or} \quad 10 \cdot 1 = 10$

List the possibilities for a, b, c, and d in a table. Include a column for $ad + bc$.

If $a = 3$, then $c = 2$, and vice versa. In addition, if $a = 6$, then $c = 1$, and vice versa. For each value of a there are four values for b.

a	b	c	d	$ad + bc$
3	5	2	2	16
3	2	2	5	19
3	10	2	1	23
3	1	2	10	32
2	5	3	2	19
2	2	3	5	16
2	10	3	1	32
2	1	3	10	23
6	5	1	2	17
6	2	1	5	32
6	10	1	1	16
6	1	1	10	61
1	5	6	2	32
1	2	6	5	17
1	10	6	1	61
1	1	6	10	16

Find a combination such that $ad + bc = 19$. As you can see from the table there are two such combinations.

$ad + bc = 19$
$3 \cdot 5 + 2 \cdot 2 = 19$ and
$2 \cdot 2 + 5 \cdot 3 = 19$

It doesn't matter which combination we choose. To finish this problem we will use the combination in the second row of the table to factor the expression in question:

$$6x^2 + 19xy + 10y^2 = (3x + 2y)(2x + 5y)$$

Reflect

As a check, expand $(3x + 2y)(2x + 5y)$ to see if we return to the original equation.

$$(3x + 2y)(2x + 5y) = 6x^2 + 15xy + 4xy + 10y^2$$
$$= 6x^2 + 19xy + 10y^2$$

You should be able to show that the combination in the fifth row is also an acceptable factoring.

M-6 Exponents and logarithms

Exponents

The notation x^n stands for the quantity obtained by multiplying x by itself n times. For example, $x^2 = x \cdot x$ and $x^3 = x \cdot x \cdot x$. The quantity n is called the **power**, or the **exponent**, of x (the **base**). Listed below are some rules that will help you simplify terms that have exponents.

(1) When two powers of x are multiplied, the exponents are added:

$$(x^m)(x^n) = x^{m+n} \qquad \text{(M-6)}$$

Example: $x^2x^3 = x^{2+3} = (x \cdot x)(x \cdot x \cdot x) = x^5$.

(2) Any number (except 0) raised to the 0 power is defined to be 1:

$$x^0 = 1 \tag{M-7}$$

(3) Based on rule 2,

$$x^n x^{-n} = x^0 = 1$$

$$x^{-n} = \frac{1}{x^n} \tag{M-8}$$

(4) When two powers are divided, the exponents are subtracted:

$$\frac{x^n}{x^m} = x^n x^{-m} = x^{n-m} \tag{M-9}$$

(5) When a power is raised to another power, the exponents are multiplied:

$$(x^n)^m = x^{nm} \tag{M-10}$$

(6) When exponents are written as fractions, they represent the roots of the base. For example,

$$x^{1/2} \cdot x^{1/2} = x$$

so

$$x^{1/2} = \sqrt{x} \quad (x > 0)$$

EXAMPLE M-6 Simplifying a Quantity That Has Exponents

Simplify $\dfrac{x^4 x^7}{x^8}$.

Set Up

According to rule 1, when two powers of x are multiplied, the exponents are added.

$$(x^m)(x^n) = x^{m+n} \tag{M-6}$$

Rule 4 states that when two powers are divided, the exponents are subtracted.

$$\frac{x^n}{x^m} = x^n x^{-m} = x^{n-m} \tag{M-9}$$

Solve

Simplify the numerator x^4x^7 using rule 1.

$$x^4x^7 = x^{4+7} = x^{11}$$

Simplify $\dfrac{x^{11}}{x^8}$ using rule 4.

$$\frac{x^{11}}{x^8} = x^{11}x^{-8} = x^{11-8} = x^3$$

Reflect

Use the value $x = 2$ to test our answer.

$$\frac{2^4 2^7}{2^8} = 2^3 = 8$$

$$\frac{2^4 2^7}{2^8} = \frac{(16)(128)}{256} = \frac{2048}{256} = 8$$

Logarithms

Any positive number can be expressed as some power of any other positive number except one. If y is related to x by $y = a^x$, then the number x is said to be the **logarithm** of y to the **base** a, and the relation is written

$$x = \log_a y$$

Thus, logarithms are *exponents*, and the rules for working with logarithms correspond to similar laws for exponents. Listed below are some rules that will help you simplify terms that have logarithms.

(1) If $y_1 = a^n$ and $y_2 = a^m$, then

$$y_1 y_2 = a^n a^m = a^{n+m}$$

Correspondingly,

$$\log_a y_1 y_2 = \log_a a^{n+m} = n + m = \log_a a^n + \log_a a^m = \log_a y_1 + \log_a y_2 \tag{M-11}$$

It then follows that

$$\log_a y^n = n \log_a y \qquad \text{(M-12)}$$

(2) Because $a^1 = a$ and $a^0 = 1$,

$$\log_a a = 1 \qquad \text{(M-13)}$$

and

$$\log_a 1 = 0 \qquad \text{(M-14)}$$

There are two bases in common use: logarithms to base 10 are called **common logarithms,** and logarithms to base e (where $e = 2.718\ldots$) are called **natural logarithms.**

In this text, the symbol ln is used for natural logarithms and the symbol log, without a subscript, is used for common logarithms. Thus,

$$\log_e x = \ln x \quad \text{and} \quad \log_{10} x = \log x \qquad \text{(M-15)}$$

and $y = \ln x$ implies

$$x = e^y \qquad \text{(M-16)}$$

Logarithms can be changed from one base to another. Suppose that

$$z = \log x \qquad \text{(M-17)}$$

Then

$$10^z = 10^{\log x} = x \qquad \text{(M-18)}$$

Taking the natural logarithm of both sides of Equation M-18, we obtain

$$z \ln 10 = \ln x$$

Substituting $\log x$ for z (see Equation M-17) gives

$$\ln x = (\ln 10)\log x \qquad \text{(M-19)}$$

EXAMPLE M-7 Converting between Common Logarithms and Natural Logarithms

The steps leading to Equation M-19 show that, in general, $\log_b x = (\log_b a)\log_a x$, and thus that conversion of logarithms from one base to another requires only multiplication by a constant. Describe the mathematical relation between the constant for converting common logarithms to natural logarithms and the constant for converting natural logarithms to common logarithms.

Set Up

We have a general mathematical formula for converting logarithms from one base to another. We look for the mathematical relation by exchanging a for b and vice versa in the formula.

Solve

We have a formula for converting logarithms from base a to base b.

$$\log_b x = (\log_b a)\log_a x$$

To convert from base b to base a, exchange all a for b and vice versa.

$$\log_a x = (\log_a b)\log_b x$$

Divide both sides of the equation in step 1 by $\log_a x$.

$$\frac{\log_b x}{\log_a x} = \log_b a$$

Divide both sides of the equation in step 2 by $(\log_a b)\log_a x$.

$$\frac{1}{\log_a b} = \frac{\log_b x}{\log_a x}$$

The results show that the conversion factors $\log_b a$ and $\log_a b$ are reciprocals of one another.

$$\frac{1}{\log_a b} = \log_b a$$

Reflect

For the value of $\log_{10} e$, your calculator will give 0.43429. For ln 10, your calculator will give 2.3026. Multiply 0.43429 by 2.3026; you will get 1.0000.

M-7 Geometry

The properties of the most common **geometric figures**—bounded shapes in two or three dimensions whose lengths, areas, or volumes are governed by specific ratios—are a basic analytical tool in physics. For example, the characteristic ratios within triangles give us the laws of *trigonometry* (see Section M-8), which in turn give us the theory of vectors, essential in analyzing motion in two or more dimensions. Circles and spheres are essential for understanding, among other concepts, angular momentum and the probability densities of quantum mechanics.

Basic Formulas in Geometry

Circle The ratio of the circumference of a circle to its diameter is a number π, which has the approximate value

$$\pi = 3.141\ 592$$

The circumference C of a circle is thus related to its diameter d and its radius r by

$$C = \pi d = 2\pi r \quad \text{circumference of circle} \tag{M-20}$$

The area of a circle is (**Figure M-5**)

$$A = \pi r^2 \quad \text{area of circle} \tag{M-21}$$

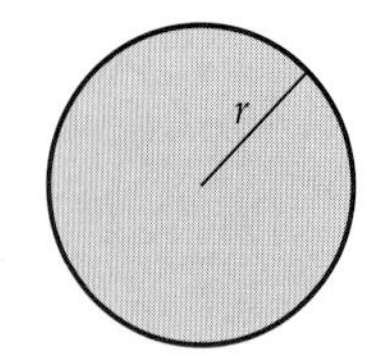

Figure M-5 Area of a circle.

Parallelogram The area of a parallelogram is the base b multiplied by the height h (**Figure M-6**):

$$A = bh$$

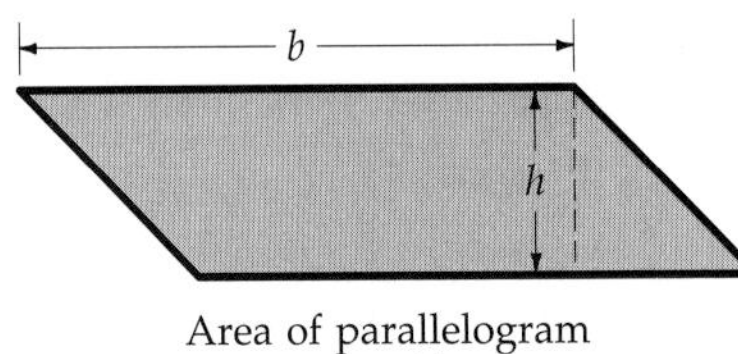

Figure M-6 Area of a parallelogram.

Triangle The area of a triangle is one-half the base multiplied by the height (**Figure M-7**):

$$A = \frac{1}{2}bh$$

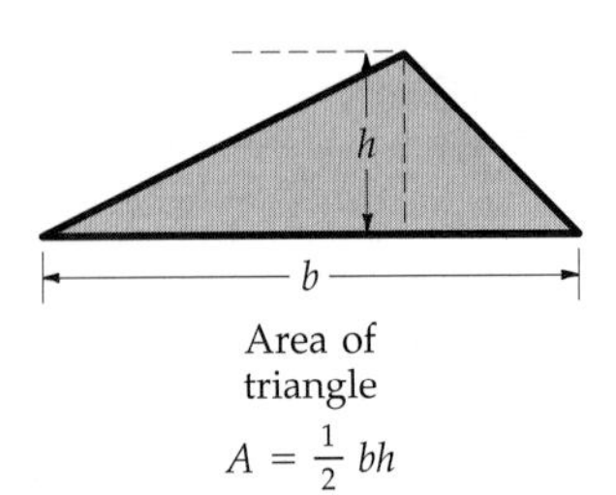

Figure M-7 Area of a triangle.

Sphere A sphere of radius r (**Figure M-8**) has a surface area given by

$$A = 4\pi r^2 \quad \text{surface area of sphere} \tag{M-22}$$

and a volume given by

$$V = \frac{4}{3}\pi r^3 \quad \text{volume of sphere} \tag{M-23}$$

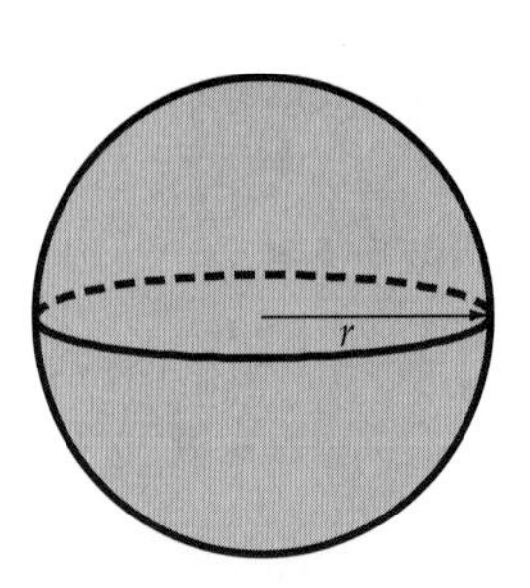

Figure M-8 Surface area and volume of a sphere.

Cylinder A cylinder of radius r and length L (**Figure M-9**) has a surface area (not including the end faces) of

$$A = 2\pi rL \quad \text{surface of cylinder} \tag{M-24}$$

and volume of

$$V = \pi r^2 L \quad \text{volume of cylinder} \tag{M-24}$$

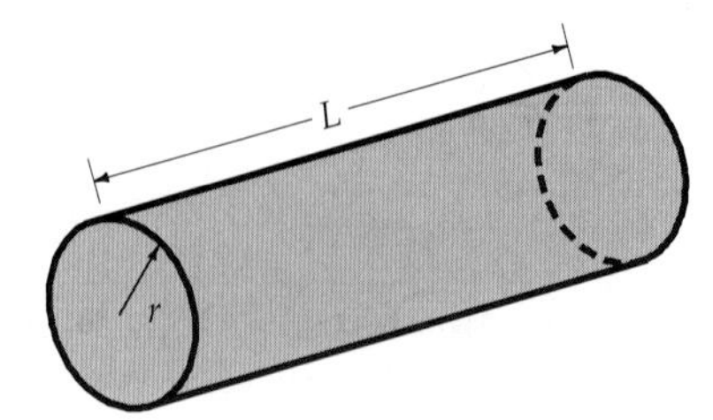

Figure M-9 Surface area (not including the end faces) and the volume of a cylinder.

EXAMPLE M-8 Calculating the Volume of a Spherical Shell

An aluminum spherical shell has an outer diameter of 40.0 cm and an inner diameter of 38.0 cm. What is the volume of the aluminum in this shell?

Set Up

The volume of the aluminum in the spherical shell is the volume that remains when we subtract the volume of the inner sphere having $d_i = 2r_i = 38.0$ cm from the volume of the outer sphere having $d_o = 2r_o = 40.0$ cm.

Spherical Volume:

$$V = \frac{4}{3}\pi r^3 \qquad \text{(M-23)}$$

Solve

Subtract the volume of the sphere of radius r_i from the volume of the sphere of radius r_o.

$$V = V_o - V_i = \frac{4}{3}\pi r_o^3 - \frac{4}{3}\pi r_i^3 = \frac{4}{3}\pi(r_o^3 - r_i^3)$$

Substitute 20.0 cm for r_o and 19.0 cm for r_i.

$$V = \frac{4}{3}\pi[(20.0\text{ cm})^3 - (19.0\text{ cm})^3]$$
$$= 4.78 \times 10^3\text{ cm}^3$$

Reflect

The volume calculated is less than the volume of the outer sphere.

$$V_o = \frac{4}{3}\pi r_o^3 = \frac{4}{3}\pi(20.0\text{ cm})^3$$
$$= 3.35 \times 10^4\text{ cm}^3$$

M-8 Trigonometry

Trigonometry, which gets its name from Greek roots meaning "triangle" and "measure," is the study of some important mathematical functions, called **trigonometric functions**. These functions are most simply defined as ratios of the sides of right triangles. However, these right-triangle definitions are of limited use because they are valid only for angles between zero and 90°. However, the validity of the right-triangle definitions can be extended by defining the trigonometric functions in terms of the ratio of the coordinates of points on a circle of unit radius drawn centered at the origin of the *xy* plane.

In physics, we first encounter trigonometric functions when we use vectors to analyze motion in two dimensions. Trigonometric functions are also essential in the analysis of any kind of periodic behavior, such as circular motion, oscillatory motion, and wave mechanics.

Angles and Their Measure: Degrees and Radians

The size of an angle formed by two intersecting straight lines is known as its **measure.** The standard way of finding the measure of an angle is to place the angle so that its **vertex**, or point of intersection of the two lines that form the angle, is at the center of a circle located at the origin of a graph that has Cartesian coordinates and one of the lines extends rightward on the positive *x* axis. The distance traveled *counterclockwise* on the circumference from the positive *x* axis to reach the intersection of the circumference with the other line defines the measure of the angle. (Traveling clockwise to the second line would simply give us a negative measure; to illustrate basic concepts, we position the angle so that the smaller rotation will be in the counterclockwise direction.)

One of the most familiar units for expressing the measure of an angle is the **degree,** which equals 1/360 of the full distance around the circumference of the circle. For greater precision, or for smaller angles, we either show degrees plus minutes (′) and seconds (″), with $1' = 1°/60$ and $1'' = 1'/60 = 1°/3600$; or show degrees as an ordinary decimal number.

For scientific work, a more useful measure of an angle is the **radian** (rad). Again, place the angle with its vertex at the center of a circle and measure counterclockwise rotation around the circumference. The measure of the angle in radians is then defined as the length of the circular arc from one line to the other divided by the radius of the

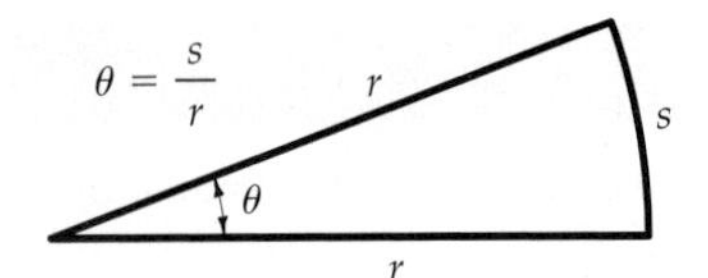

Figure M-10 The angle θ in radians is defined to be the ratio s/r, where s is the arc length intercepted on a circle of radius r.

circle (**Figure M-10**). If s is the arc length and r is the radius of the circle, the angle θ measured in radians is

$$\theta = \frac{s}{r} \tag{M-26}$$

Because the angle measured in radians is the ratio of two lengths, it is dimensionless. The relation between radians and degrees is

$$360° = 2\pi \text{ rad}$$

or

$$1 \text{ rad} = \frac{360°}{2\pi} = 57.3°$$

Figure M-11 shows some useful relations for angles.

The Trigonometric Functions

Figure M-12 shows a right triangle formed by drawing the line segment BC perpendicular to AC. The lengths of the sides are labeled a, b, and c. The right-triangle definitions of the trigonometric functions sin θ (the **sine**), cos θ (the **cosine**), and tan θ (the **tangent**) for an acute angle θ are

$$\sin\theta = \frac{a}{c} = \frac{\text{opposite side}}{\text{hypotenuse}} \tag{M-27}$$

$$\cos\theta = \frac{b}{c} = \frac{\text{adjacent side}}{\text{hypotenuse}} \tag{M-28}$$

$$\tan\theta = \frac{a}{b} = \frac{\text{opposite side}}{\text{adjacent side}} = \frac{\sin\theta}{\cos\theta} \tag{M-29}$$

(**Acute angles** are angles whose positive rotation around the circumference of a circle measures less than 90° or $\pi/2$.) Three other trigonometric functions—the **secant** (sec), the **cosecant** (csc), and the **cotangent** (cot), defined as the reciprocals of these functions—are

$$\csc\theta = \frac{c}{a} = \frac{1}{\sin\theta} \tag{M-30}$$

$$\sec\theta = \frac{c}{b} = \frac{1}{\cos\theta} \tag{M-31}$$

$$\cot\theta = \frac{b}{a} = \frac{1}{\tan\theta} = \frac{\cos\theta}{\sin\theta} \tag{M-32}$$

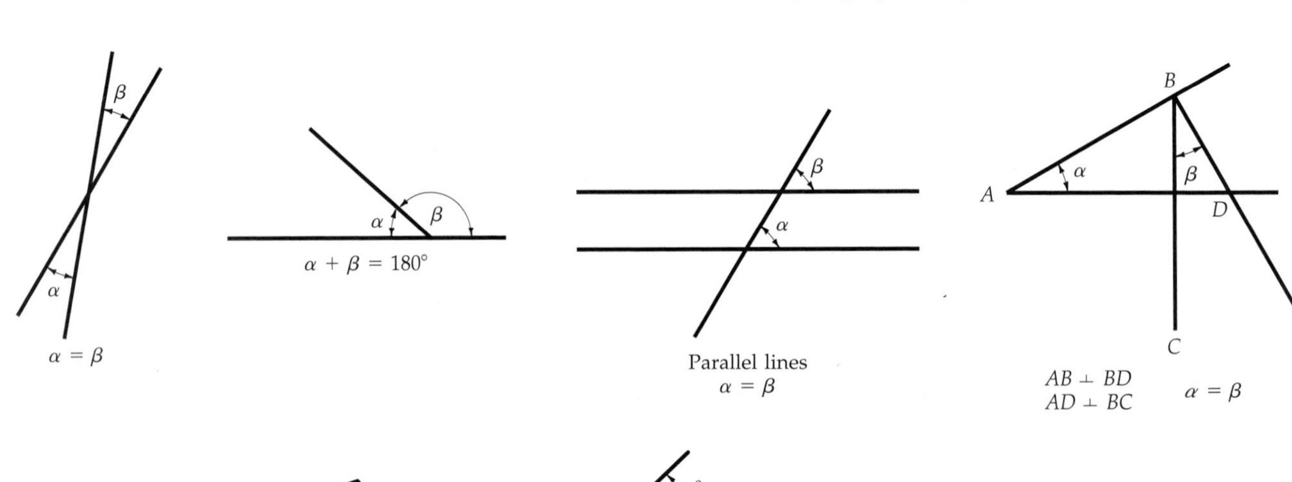

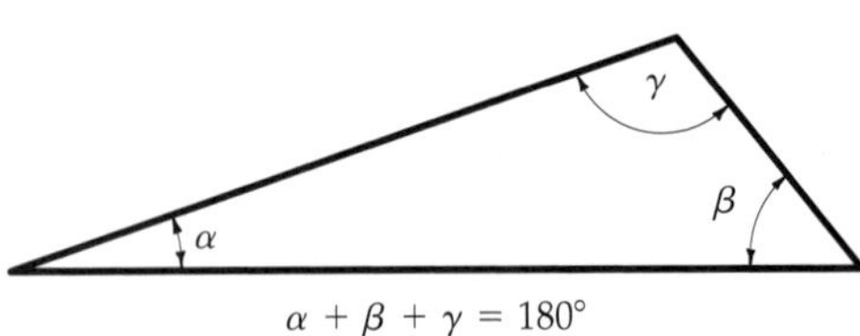

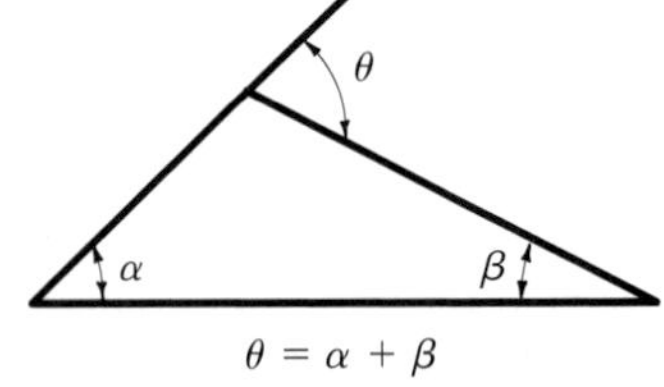

Figure M-11 Some useful relations for angles.

The angle θ, whose sine is x, is called the arcsine of x, and is written $\sin^{-1} x$. That is, if

$$\sin \theta = x$$

then

$$\theta = \arcsin x = \sin^{-1} x \tag{M-33}$$

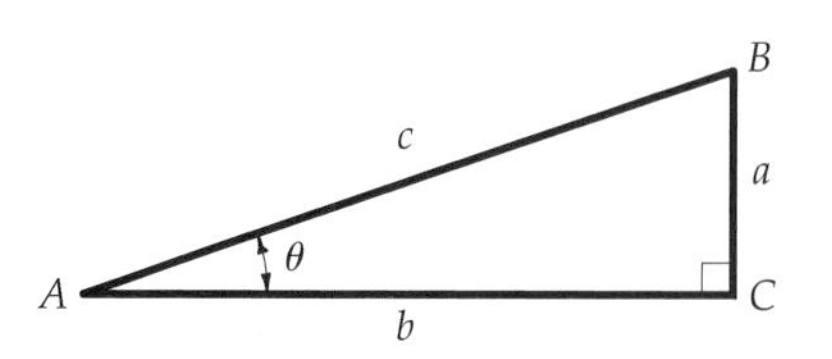

Figure M-12 A right triangle with sides of length a and b and a hypotenuse of length c.

The arcsine is the inverse of the sine. The inverse of the cosine and tangent are defined similarly. The angle whose cosine is y is the arccosine of y. That is, if

$$\cos \theta = y$$

then

$$\theta = \arccos y = \cos^{-1} y \tag{M-34}$$

The angle whose tangent is z is the arctangent of z. That is, if

$$\tan \theta = z$$

then

$$\theta = \arctan z = \tan^{-1} z \tag{M-35}$$

Trigonometric Identities

We can derive several useful formulas, called **trigonometric identities,** by examining relationships between the trigonometric functions. Equations M-30 through M-32 list three of the most obvious identities, formulas expressing some trigonometric functions as reciprocals of others. Almost as easy to discern are identities derived from the **Pythagorean theorem,**

$$a^2 + b^2 = c^2 \tag{M-36}$$

Simple algebraic manipulation of Equation M-36 gives us three more identities. First, if we divide each term in Equation M-36 by c^2, we obtain

$$\frac{a^2}{c^2} + \frac{b^2}{c^2} = 1$$

or, from the definitions of $\sin \theta$ (which is a/c) and $\cos \theta$ (which is b/c),

$$\sin^2 \theta + \cos^2 \theta = 1 \tag{M-37}$$

Similarly, we can divide each term in Equation M-36 by a^2 or b^2 and obtain

$$1 + \cot^2 \theta = \csc^2 \theta \tag{M-38}$$

and

$$1 + \tan^2 \theta = \sec^2 \theta \tag{M-39}$$

Table M-2 lists these last three and many more trigonometric identities. Notice that they fall into four categories: functions of sums or differences of angles, sums or differences of squared functions, functions of double angles (2θ), and functions of half angles ($\frac{1}{2}\theta$). Notice that some of the formulas contain paired alternatives, expressed with the signs $\pm$ and $\mp$; in such formulas, remember to always apply the formula with either all the upper or all the lower alternatives.

TABLE M-2 Trigonometric Identities

$$\sin(A \pm B) = \sin A \cos B \pm \cos A \sin B$$

$$\cos(A \pm B) = \cos A \cos B \mp \sin A \sin B$$

$$\tan(A \pm B) = \frac{\tan A \pm \tan B}{1 \mp \tan A \tan B}$$

$$\sin A \pm \sin B = 2 \sin\left[\frac{1}{2}(A \pm B)\right]\cos\left[\frac{1}{2}(A \mp B)\right]$$

$$\cos A + \cos B = 2 \cos\left[\frac{1}{2}(A + B)\right]\cos\left[\frac{1}{2}(A - B)\right]$$

$$\cos A - \cos B = 2 \sin\left[\frac{1}{2}(A + B)\right]\sin\left[\frac{1}{2}(B - A)\right]$$

$$\tan A \pm \tan B = \frac{\sin(A \pm B)}{\cos A \cos B}$$

$$\sin^2\theta + \cos^2\theta = 1; \sec^2\theta - \tan^2\theta = 1; \csc^2\theta - \cot^2\theta = 1$$

$$\sin 2\theta = 2 \sin \theta \cos \theta$$

$$\cos 2\theta = \cos^2\theta - \sin^2\theta = 2\cos^2\theta - 1 = 1 - 2\sin^2\theta$$

$$\tan 2\theta = \frac{2 \tan \theta}{1 - \tan^2\theta}$$

$$\sin \frac{1}{2}\theta = \pm\sqrt{\frac{1 - \cos\theta}{2}}; \cos \frac{1}{2}\theta = \pm\sqrt{\frac{1 + \cos\theta}{2}};$$

$$\tan \frac{1}{2}\theta = \pm\sqrt{\frac{1 - \cos\theta}{1 + \cos\theta}}$$

Some Important Values of the Functions

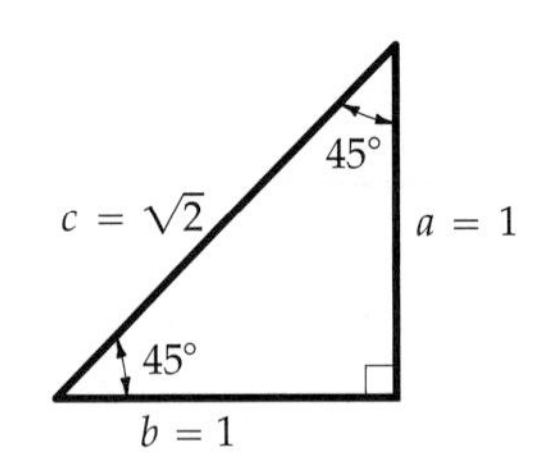

Figure M-13 An isosceles right triangle.

Figure M-13 is a diagram of an *isosceles* right triangle (an isosceles triangle is a triangle with two equal sides), from which we can find the sine, cosine, and tangent of 45°. The two acute angles of this triangle are equal. Because the sum of the three angles in a triangle must equal 180° and the right angle is 90°, each acute angle must be 45°. For convenience, let us assume that the equal sides each have a length of 1 unit. The Pythagorean theorem gives us a value for the hypotenuse of

$$c = \sqrt{a^2 + b^2} = \sqrt{1^2 + 1^2} = \sqrt{2} \text{ units}$$

We calculate the values of the functions as follows:

$$\sin 45° = \frac{a}{c} = \frac{1}{\sqrt{2}} = 0.707 \quad \cos 45° = \frac{b}{c} = \frac{1}{\sqrt{2}} = 0.707 \quad \tan 45° = \frac{a}{b} = \frac{1}{1} = 1$$

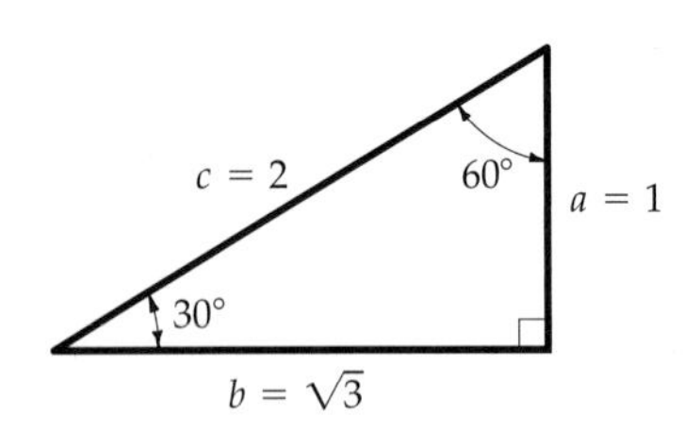

Figure M-14 A 30°–60°–90° right triangle.

Another common triangle, a 30°–60°–90° right triangle, is shown in **Figure M-14**. Because this particular right triangle is in effect half of an *equilateral triangle* (a 60°–60°–60° triangle or a triangle having three equal sides and three equal angles), we can see that the sine of 30° must be exactly 0.5 (**Figure M-15**). The equilateral triangle must have all sides equal to c, the hypotenuse of the 30°–60°–90° right triangle. Thus, side a is one-half the length of the hypotenuse, and so

$$\sin 30° = \frac{1}{2}$$

To find the other ratios within the 30°–60°–90° right triangle, let us assign a value of 1 to the side opposite the 30° angle. Then

$$c = \frac{1}{0.5} = 2 \qquad b = \sqrt{c^2 - a^2} = \sqrt{2^2 - 1^2} = \sqrt{3}$$

$$\cos 30° = \frac{b}{c} = \frac{\sqrt{3}}{2} = 0.866 \qquad \tan 30° = \frac{a}{b} = \frac{1}{\sqrt{3}} = 0.577$$

$$\sin 60° = \frac{b}{c} = \cos 30° = 0.866 \qquad \cos 60° = \frac{a}{c} = \sin 30° = \frac{1}{2}$$

$$\tan 60° = \frac{b}{a} = \frac{\sqrt{3}}{1} = 1.732$$

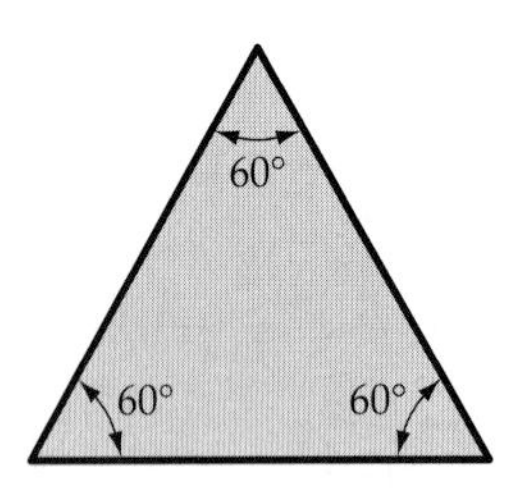

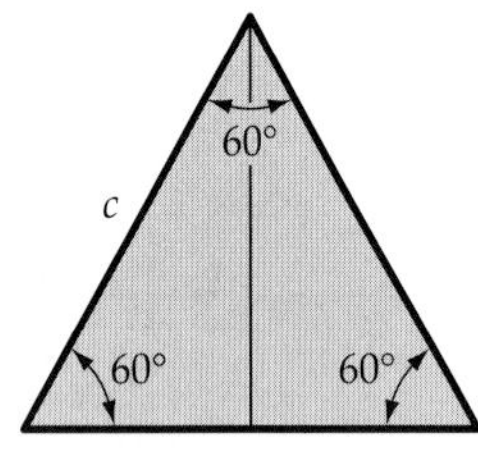

Figure M-15 (a) An equilateral triangle. (b) An equilateral triangle that has been bisected to form two 30°–60°–90° right triangles.

Small-Angle Approximation

For small angles, the length a is nearly equal to the arc length s, as can be seen in **Figure M-16**. The angle $\theta = s/c$ is therefore nearly equal to $\sin\theta = a/c$:

(M-40) $$\sin\theta \approx \theta \quad \text{for small values of } \theta$$

Similarly, the lengths c and b are nearly equal, so $\tan\theta = a/b$ is nearly equal to both θ and $\sin\theta$ for small values of θ:

(M-41) $$\tan\theta \approx \sin\theta \approx \theta \quad \text{for small values of } \theta$$

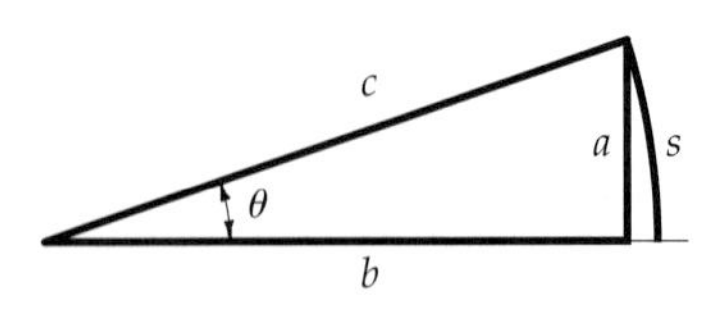

Figure M-16 For small angles, $\sin\theta = a/c$, $\tan\theta = a/b$, and the angle $\theta = s/c$ are all approximately equal.

Equations M-40 and M-41 hold only if θ is measured in radians. Because $\cos\theta = b/c$, and because these lengths are nearly equal for small values of θ, we have

(M-42) $$\cos\theta \approx 1 \quad \text{for small values of } \theta$$

Figure M-17 shows graphs of θ, $\sin\theta$, and $\tan\theta$ versus θ for small values of θ. If accuracy of a few percent is needed, small-angle approximations can be used only for angles of about a quarter of a radian (or about 15°) or less. Below this value, as the angle becomes smaller, the approximation $\theta \approx \sin\theta \approx \tan\theta$ is even more accurate.

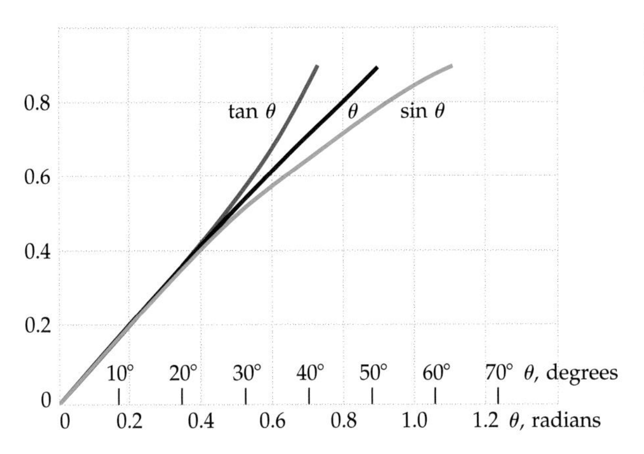

Figure M-17 Graphs of tan θ, θ, and sin θ versus θ for small values of θ.

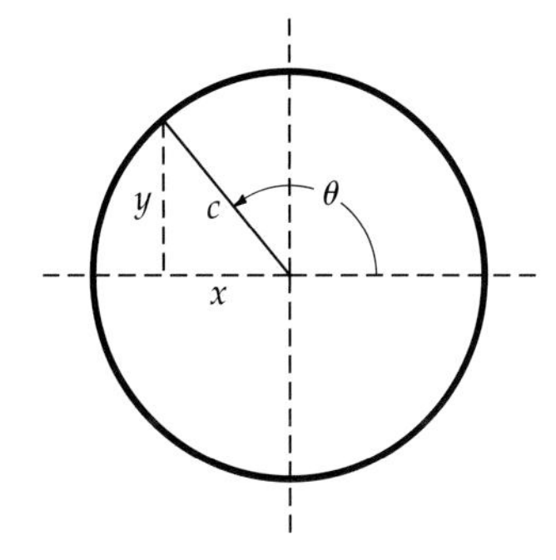

Figure M-18 Diagram for defining the trigonometric functions for an obtuse angle.

Trigonometric Functions as Functions of Real Numbers

So far we have illustrated the trigonometric functions as properties of angles. **Figure M-18** shows an *obtuse* angle with its vertex at the origin and one side along the x axis. The trigonometric functions for a "general" angle such as this are defined by

$$\sin\theta = \frac{y}{c} \qquad \text{(M-43)}$$

$$\cos\theta = \frac{x}{c} \qquad \text{(M-44)}$$

$$\tan\theta = \frac{y}{x} \qquad \text{(M-45)}$$

It is important to remember that values of x to the left of the vertical axis and values of y below the horizontal axis are negative; c in the figure is always regarded as positive. **Figure M-19** shows plots of the general sine, cosine, and tangent functions versus θ. The sine and cosine functions have a period of 2π rad. Thus, for any value of θ, $\sin(\theta + 2\pi) = \sin\theta$, and so forth. That is, when an angle changes by 2π rad, the function returns to its original value. The tangent function has a period of π rad. Thus, $\tan(\theta + \pi) = \tan\theta$, and so forth. Some other useful relations are

$$\sin(\pi - \theta) = \sin\theta \qquad \text{(M-46)}$$

$$\cos(\pi - \theta) = -\cos\theta \qquad \text{(M-47)}$$

$$\sin\left(\frac{1}{2}\pi - \theta\right) = \cos\theta \qquad \text{(M-48)}$$

$$\cos\left(\frac{1}{2}\pi - \theta\right) = \sin\theta \qquad \text{(M-49)}$$

Because the radian is dimensionless, it is not hard to see from the plots in Figure M-21 that the trigonometric functions are functions of all real numbers.

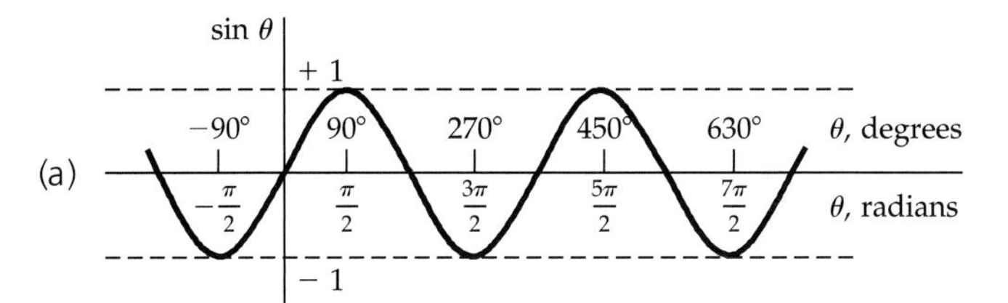

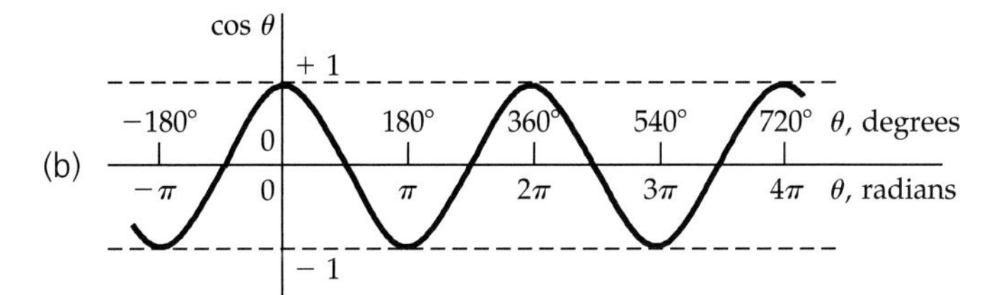

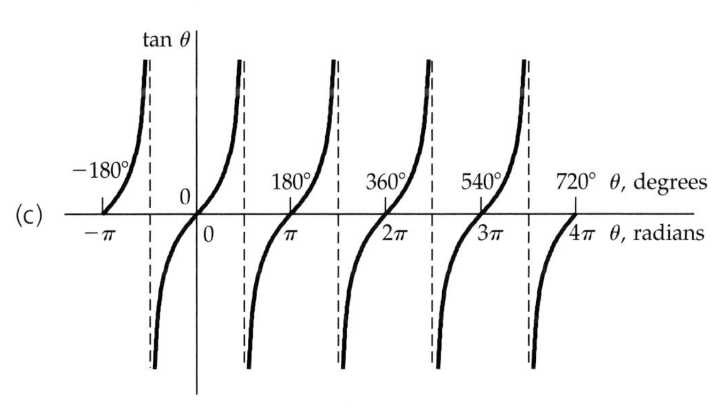

Figure M-19 The trigonometric functions sin θ, cos θ, and tan θ versus θ.

EXAMPLE M-9 Cosine of a Sum

Using the suitable trigonometric identity from Table M-2, find $\cos(135° + 22°)$. Give your answer with four significant figures.

Set Up

As long as all angles are given in degrees, there is no need to convert to radians, because all operations are numerical values of the functions. Be sure, however, that your calculator is in degree mode. The suitable identity is $\cos(A \pm B) = \cos A \cos B \mp \sin A \sin B$, where the upper signs are appropriate.

Solve

Write the trigonometric identity for the cosine of a sum, with $A = 135°$ and $B = 22°$:

$$\cos(135° + 22°) = (\cos 135°)(\cos 22°) - (\sin 135°)(\sin 22°)$$

Using a calculator, find cos 135°, sin 135°, cos 22°, and sin 22°:

$$\cos 135° = -0.7071$$
$$\cos 22° = 0.9272$$
$$\sin 135° = 0.7071$$
$$\sin 22° = 0.3746$$

Enter the values in the formula and calculate the answer:

$$\cos(135° + 22°) = (-0.7071)(0.9272) - (0.7071)(0.3746) = -0.9205$$

Reflect

The calculator shows that $\cos(135° + 22°) = \cos(157°) = -0.9205$.

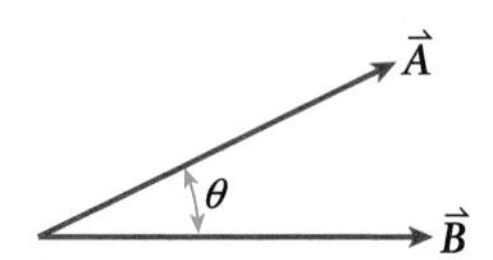

Figure M-20 Two vectors separated by an angle θ.

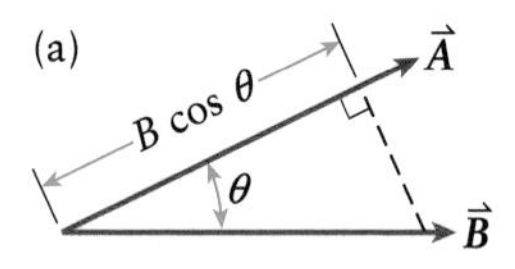

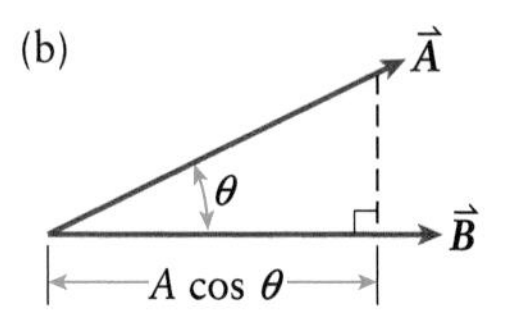

Figure M-21 The dot product is a measure of how parallel two vectors are. (a) $B\cos\theta$ is the component of $\vec{B}$ that is parallel to $\vec{A}$. (b) $A\cos\theta$ is the component of $\vec{A}$ that is parallel to $\vec{B}$.

M-9 The dot product

For two vectors $\vec{A}$ and $\vec{B}$ separated by angle θ, as shown in **Figure M-20**, their dot product C is defined as

$$C = \vec{A} \cdot \vec{B} = AB\cos\theta \qquad \text{(M-50)}$$

which you can read as, "C equals A dot B." In Equation M-50, A and B are the magnitudes of vectors $\vec{A}$ and $\vec{B}$, respectively. As a result, the dot product of two vectors is a scalar quantity. This is why $\vec{A} \cdot \vec{B}$ is also called the **scalar product** of $\vec{A}$ and $\vec{B}$.

Physically, the dot product $\vec{A} \cdot \vec{B}$ is a measure of how parallel the two vectors are. We can think of it as the magnitude of vector $\vec{A}$ multiplied by the component of vector $\vec{B}$ that is parallel to $\vec{A}$. Referring to **Figure M-21a**, we see that $B\cos\theta$ is the component of $\vec{B}$ that is parallel to $\vec{A}$. That is, $B\cos\theta$ tells us how much of $\vec{B}$ points in the direction of $\vec{A}$. Alternatively, the dot product $\vec{A} \cdot \vec{B}$ can be thought of as the magnitude of vector $\vec{B}$ multiplied by the component of vector $\vec{A}$ parallel to $\vec{B}$ (**Figure M-21b**).

The dot product is commutative; the order of the vectors in a dot product does not affect the result:

$$\vec{A} \cdot \vec{B} = \vec{B} \cdot \vec{A}$$

The dot product is also distributive, which means

$$\vec{A} \cdot (\vec{B} + \vec{C}) = \vec{A} \cdot \vec{B} + \vec{A} \cdot \vec{C}$$

Three special cases of the dot product are particularly important in physics. First, the dot product of two vectors $\vec{A}$ and $\vec{B}$ that point in the same direction (so $\theta = 0$ and $\cos\theta = \cos 0 = 1$) equals the product of their magnitudes:

$$\vec{A} \cdot \vec{B} = AB\cos 0 = AB \qquad \text{(M-51)}$$
(if $\vec{A}$ and $\vec{B}$ point in the same direction)

(As an example, the dot product of a vector $\vec{A}$ with itself is equal to the square of its magnitude: $\vec{A} \cdot \vec{A} = AA\cos 0 = A^2$.)

Second, the dot product of two perpendicular vectors $\vec{A}$ and $\vec{B}$ (so $\theta = 90°$ and $\cos\theta = \cos 90° = 0$) is zero:

$$\vec{A} \cdot \vec{B} = AB\cos 90° = 0 \qquad \text{(M-52)}$$
(if $\vec{A}$ and $\vec{B}$ are perpendicular)

Third, if two vectors $\vec{A}$ and $\vec{B}$ point in opposite directions (so $\theta = 180°$ and $\cos\theta = \cos 180° = -1$), their dot product equals the *negative* of the product of their magnitudes:

$$\vec{A} \cdot \vec{B}\cos 180° = -AB \qquad \text{(M-53)}$$
(if $\vec{A}$ and $\vec{B}$ point in opposite directions)

Finally, it's useful to know how to calculate the dot product of two vectors $\vec{A}$ and $\vec{B}$ that are expressed in terms of their components A_x, A_y, A_z, and B_x, B_y, and B_z:

$$\vec{A} \cdot \vec{B} = A_xB_x + A_yB_y + A_zB_z \qquad \text{(M-54)}$$

You can verify that Equation M-54 is correct by thinking of $\vec{A}$ as the sum of three vectors: $\vec{A}_1$, which has only an x-component A_x; $\vec{A}_2$, which has only a y-component A_y; and $\vec{A}_3$, which has only a z-component A_z. From the definition of the dot product, $\vec{A}_1 \cdot \vec{B}$ is equal to A_x multiplied by the component of $\vec{B}$ in the direction of $\vec{A}_1$, or $\vec{A}_1 \cdot \vec{B} = A_xB_x$. Similarly, $\vec{A}_2 \cdot \vec{B} = A_yB_y$ and $\vec{A}_3 \cdot \vec{B} = A_zB_z$. Since $\vec{A} = \vec{A}_1 + \vec{A}_2 + \vec{A}_3$ and the dot product is distributive, it follows that

$$\vec{A} \cdot \vec{B} = (\vec{A}_1 + \vec{A}_2 + \vec{A}_3) \cdot \vec{B} = \vec{A}_1 \cdot \vec{B} + \vec{A}_2 \cdot \vec{B} + \vec{A}_3 \cdot \vec{B} = A_xB_x + A_yB_y + A_zB_z$$

That's the same as Equation M-54. If the vectors have only x- and y-components, Equation M-54 simplifies to $\vec{A} \cdot \vec{B} = A_xB_x + A_yB_y$.

EXAMPLE M-10 The Dot Product

(a) Calculate the dot product of vector $\vec{A}$ with magnitude 5.00 pointed in a horizontal direction 36.9° north of east and vector $\vec{B}$ of magnitude 1.50 pointed in a horizontal direction 53.1° south of west. (b) What is the dot product of vector $\vec{C}$ with components $C_x = 4.00$, $C_y = 3.00$ and vector $\vec{D}$ with components $D_x = -0.900$, $D_y = -1.20$?

Set Up

In part (a) we know the magnitude and direction of the vectors, so we'll use Equation M-50. In part (b) the vectors are given in terms of components, so we'll evaluate the dot product using Equation M-54.

$$\vec{A} \cdot \vec{B} = AB\cos\theta \qquad \text{(M-50)}$$

Dot product of two vectors in terms of components:

$$\vec{A} \cdot \vec{B} = A_xB_x + A_yB_y + A_zB_z \qquad \text{(M-54)}$$

Solve

(a) The drawing shows that the angle between $\vec{A}$ and $\vec{B}$ is $\theta = 163.8°$. We use this in Equation M-50 to evaluate the dot product.

$$\begin{aligned}\vec{A} \cdot \vec{B} &= AB\cos\theta \\ &= (5.00)(1.50)\cos 163.8° \\ &= (5.00)(1.50)(-0.960) \\ &= -7.20\end{aligned}$$

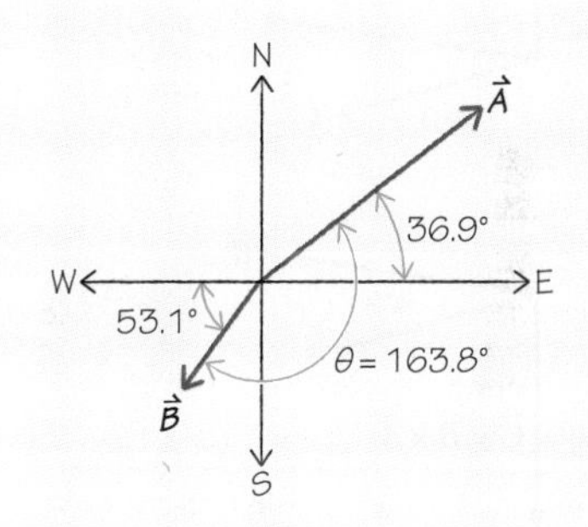

(b) Both $\vec{C}$ and $\vec{D}$ are in the x-y plane and have no z components, so we just need the first two terms in Equation M-54 to calculate their dot product.

$$\begin{aligned}\vec{C} \cdot \vec{D} &= C_xD_x + C_yD_y \\ &= (4.00)(-0.900) + (3.00)(-1.20) \\ &= -7.20\end{aligned}$$

Reflect

It's not a coincidence that we got the same result in part (b) as in part (a): Vectors $\vec{A}$ and $\vec{C}$ are the same, as are vectors $\vec{B}$ and $\vec{D}$. (You can verify this by using the techniques from Chapter 3 to calculate the components of the vectors $\vec{A}$ and $\vec{B}$ in part (a). You'll find that the components are the same as those of $\vec{C}$ and $\vec{D}$ in part (b).) This should give you confidence that the method of calculating the dot product using components gives you the same result as the method that involves the magnitudes and directions of the vectors.

Notice that the angle between vectors $\vec{A}$ and $\vec{B}$ is between 90° and 180°, and the dot product is negative.

M-10 The cross product

The dot product, described in section M-9, is only one way to multiply two vectors. We can also multiply two vectors $\vec{A}$ and $\vec{B}$ using the **cross product**

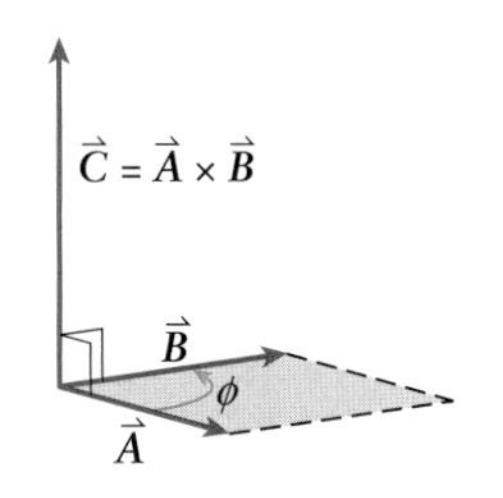

Figure M-22 The cross product is a vector $\vec{C}$ that is perpendicular to both $\vec{A}$ and $\vec{B}$, and has a magnitude $AB \sin \phi$, which equals the area of the parallelogram shown.

$$\vec{C} = \vec{A} \times \vec{B} \tag{M-55}$$

The symbol "×" represents the mathematical operation known as the cross product. As you can see from Equation M-55, the result of taking the cross product of two vectors is also a vector. The magnitude of the resulting vector is the product of the magnitudes of the two vectors and the sine of the angle between them. That is, the magnitude of the cross product of $\vec{A}$ and $\vec{B}$ is

$$C = |\vec{A} \times \vec{B}| = AB \sin \phi \tag{M-56}$$

where according to convention ϕ is defined as the angle that goes from $\vec{A}$ to $\vec{B}$. $\vec{C}$ points in the direction perpendicular to both $\vec{A}$ and $\vec{B}$ as shown in **Figure M-22**.

The magnitude of the cross product $\vec{A} \times \vec{B}$ can be interpreted as the magnitude of vector $\vec{A}$ multiplied by the component of vector $\vec{B}$ perpendicular to $\vec{A}$, or the magnitude of vector $\vec{B}$ multiplied by the component of vector $\vec{A}$ perpendicular to $\vec{B}$.

Note that the order of the two vectors in a cross product makes a difference. The cross product of $\vec{B}$ and $\vec{A}$ is the negative of the cross product of $\vec{A}$ and $\vec{B}$ or

$$\vec{A} \times \vec{B} = -\vec{B} \times \vec{A} \tag{M-57}$$

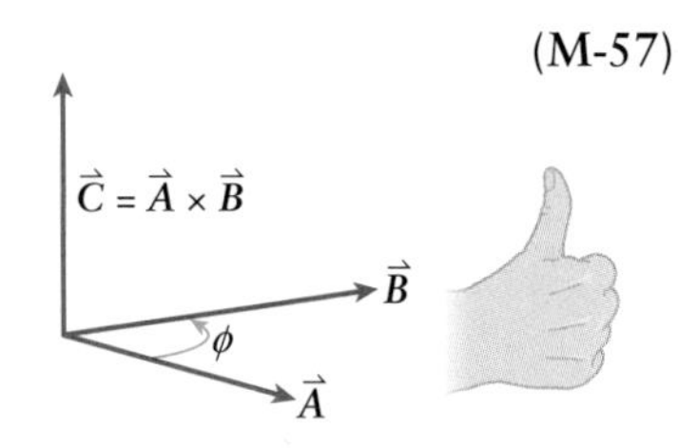

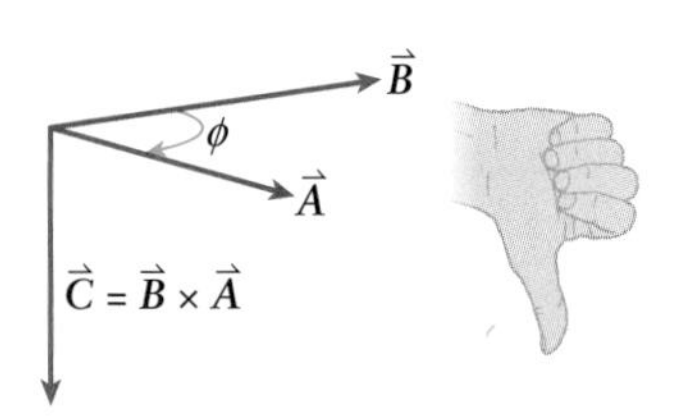

Figure M-23 (a) To find the direction of , point the fingers of your right hand in the direction of vector $\vec{A}$, then curl them toward vector $\vec{B}$. Your thumb points in the direction of the cross product. (b) The direction of $\vec{B} \times \vec{A}$ points in the opposite direction of.

This results from the definition of the angle ϕ in Equation M-56. Since ϕ is directed from the first vector to the second vector, if you travel the angle from the second vector to the first—in reverse direction—ϕ becomes negative. And the sine of a negative angle is also negative.

In addition, the cross product obeys the distributive law under addition:

$$\vec{A} \times (\vec{B} + \vec{C}) = \vec{A} \times \vec{B} + \vec{A} \times \vec{C} \tag{M-58}$$

To determine the direction of the cross product $\vec{C} = \vec{A} \times \vec{B}$, you can use the right-hand rule. To apply this rule, point the fingers of your right hand in the direction of the first vector of the cross product (in this case $\vec{A}$). Then curl your fingers toward the second vector, $\vec{B}$. If you stick your thumb straight out, it points in the direction of the cross product, vector $\vec{C}$ (**Figure M-23a**). If you instead want to find the direction of the cross product $\vec{B} \times \vec{A}$, begin by pointing the fingers of your right hand in the direction of vector $\vec{B}$. Then curl them toward vector $\vec{A}$. Your thumb again points in the direction of the cross product (**Figure M-23b**). Note that because you must curl your fingers in the opposite direction as for $\vec{C} = \vec{A} \times \vec{B}$, the cross product of $\vec{B} \times \vec{A}$ points in the opposite direction of $\vec{A} \times \vec{B}$, which is just what we stated in Equation M-58.

There are two special cases of the cross product that are worth pointing out. The first is the cross product for two perpendicular vectors, for which $\phi = 90°$, so $\sin \phi = 1$.

$$|\vec{A} \times \vec{B}| = AB \sin 90° = AB(1) = AB$$

(magnitude of the cross product of two perpendicular vectors)

The second special case is the cross product of two parallel vectors, for which $\phi = 0$, so $\sin \phi = 0$.

$$|\vec{A} \times \vec{B}| = AB \sin 0 = AB(0) = 0$$

(magnitude of the cross product for two parallel vectors)

One example of a cross product of two parallel vectors is the cross product of a vector with itself: $\vec{A} \times \vec{A} = 0$.

EXAMPLE M-11 The Cross Product

Evaluate $\vec{A} \times \vec{B}$, in which the components of vector $\vec{A}$ are $A_x = 5$, $A_y = 0$, and the components of vector $\vec{B}$ are $B_x = 9$, $B_y = 7$.

Set Up

We will use the definition of the magnitude of the cross product, Equation M-56, to find the magnitude of the cross product, and the right-hand rule to determine the direction of the cross product.

We will have to use the components of vector $\vec{B}$ to determine its magnitude and the angle it makes with the x axis and vector $\vec{A}$.

$$C = |\vec{A} \times \vec{B}| = AB \sin\phi \tag{M-56}$$

Finding vector magnitude and direction from vector components:

$$A = \sqrt{A_x^2 + A_y^2}$$

$$\tan\theta = \frac{A_y}{A_x} \tag{3-2}$$

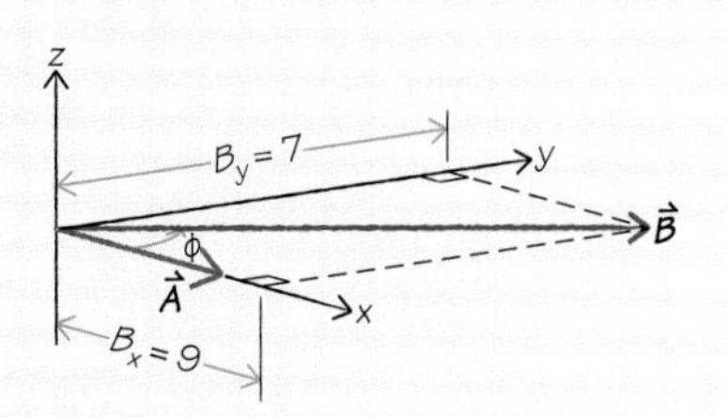

Solve

Begin by determining the magnitude and direction of vector $\vec{B}$ using its components and Equations 3-2.

Determine the magnitude of vector $\vec{B}$ from its components:

$$B = \sqrt{B_x^2 + B_y^2} = \sqrt{9^2 + 7^2} = 11.4$$

Determine the angle $\vec{B}$ makes with the x axis (and vector $\vec{A}$) from its components:

$$\tan\phi = \frac{B_y}{B_x} = \frac{7}{9} = 0.778, \text{ so}$$

$$\phi = \arctan 0.778 = 37.9°$$

Because vector $\vec{A}$ has only an x component, its magnitude is equal to its x component, and the angle it makes with the x axis is 0.

Determine the magnitude of vector $\vec{A}$ from its components:

$$A = \sqrt{A_x^2 + A_y^2} = \sqrt{5^2 + 0^2} = 5$$

Because $\vec{A}$ has only an x component, it makes an angle of zero degrees with the x axis.

Now that we know both the magnitude and direction of the vectors, we can use Equation M-56 to determine the magnitude of the cross product.

Apply Equation M-56 to the two vectors:

$$\begin{aligned} |\vec{A} \times \vec{B}| &= AB \sin\phi \\ &= (5.00)(11.4)\sin 37.9° \\ &= (5.00)(11.4)(0.614) \\ &= 35.0 \end{aligned}$$

Use the right-hand rule to determine the direction of the cross product.

From the figure, if we first point the fingers of our right hand in the direction of $\vec{A}$ (along the x axis), and then curl them toward $\vec{B}$, we see that the thumb points in the positive z direction. So the cross product $\vec{A} \times \vec{B}$ has a magnitude of 35 in the $+z$ direction.

Reflect

The vectors $\vec{A}$ and $\vec{B}$ lie in the xy plane, so the cross product, which must be perpendicular to both vectors, should point along the z axis, which is just what we found.

ANSWERS TO ODD PROBLEMS

Chapter 16

1. There would be no difference. The charges were originally arbitrarily defined as positive and negative.
3. The mass decreases because electrons are removed from the object to make it positively charged.
5. Our current understanding is that like charges repel and opposite charges attract. If a charged insulating object either repels or attracts both charged glass (positive) and charged rubber (negative), then you may have discovered a new kind of charge. However, you may also just be observing polarization (to be discussed in later chapters).
7. When the comb is run through your hair, electrons are transferred to the comb. The paper is polarized by the charged comb. The paper is then attracted to the comb. When they touch a small amount of charge is transferred to the paper so now the paper and comb are similarly charged and repel each other.
9. (a) No, the object could be attracted after being polarized. (b) Yes, to be repelled, the suspended object should be charged positively.
11. Similarities: The force varies as $1/r^2$; the force is directly proportional to the product of the masses or charges; the force is directed along the line that connects the two particles. Differences: Newton's law includes a minus sign, whereas Coulomb's law does not. Like charges repel; like masses always attract. The gravitational constant G is many orders of magnitude smaller than the Coulomb constant k; the gravitational force is many orders of magnitude weaker than the Coulomb force.
13. (a) If the charges creating the field move, the fact that the field propagates at the speed of light also allows us to understand the changes in the field and hence force on the other charged objects. With the electric field we see that charged particles take some time to experience the effects of other charges moving near them. (b) In electrostatics the field is just a computational device, and using it is merely a matter of convenience. However, in electrodynamics the field is necessary for energy and momentum to be conserved, so it is more than just convenience that leads us to the electric field.
15. The electric forces are equal in magnitude but opposite in direction, and the force on the proton is downward while the force on the electron is upward.
17. The electric field is the total electric field due to all charges both inside and outside the Gaussian surface. However, the charges outside the surface create a net zero electric flux through the surface.
19. B
21. C
23. A
25. C
27. B
29. Around 10^{-11} or 10^{-12} C
31. $q = 3 \times 10^{-8}$ C
33. About 0.05 $\text{N} \cdot \text{m}^2/\text{C}$
35. 4.6×10^{-18} C
37. 2.7×10^{7} C
39. 2.7×10^{13} electrons
41. (a) $2.33 \times 10^{5}\ \text{C/m}^3$; (b) Insulating. If it were conducting, one of the charges would be in the volume of the sphere. Charge would be only on the surface.
43. (a) $-1.25\ \mu\text{C}$; (c) $7.8 \times 10^{12}\ \text{e}^-$; (c) no more than three steps
45. 5.08 m
47. 0.0110 N in the $+x$ direction
49. $x = -27.3$ m
51. $F_A = 36.3$ N at the angle $\theta = 24.4°$ below the negative x axis
    $F_B = 43.5$ N at the angle $\theta = 11.0°$ below the positive x axis
    $F_C = 25.2$ N at the angle $\theta = 67.5°$ above the negative x axis
53. x component: -90.6 N; y component: 14.0 N
55. -3×10^{-4} C
57. -2.56×10^{-7} C
59. (a) x component: $54.0\ \dfrac{\text{N}}{\text{C}}$; y component: $27.6\ \dfrac{\text{N}}{\text{C}}$; (b) $F_x = 8.66 \times 10^{-18}$ N; $F_y = 4.42 \times 10^{-18}$ N
61. 0.38 nC
63. 0.075 $\text{N} \cdot \text{m}^2/\text{C}$
65. 6.64×10^{-5} C/m
67. $E = \sigma/\varepsilon_0$ pointing radially outward
69. (a) We're not told the direction of the electric field, so the charge could be positive or negative. We'll assume positive. $Q = 1.78 \times 10^{-11}$ C; (b) $3.54 \times 10^{-9}\ \text{C/m}^2$
71. $+3.20\ \mu\text{C}$ on the inner shell, $-9.00\ \mu\text{C}$ on the outer shell. The field points radially inward with magnitude 6.61×10^{7} N/C.
73. (a) With 94 protons confined to a small space, they are likely to be concentrated into an approximately uniform sphere of charge, which is equivalent to a point charge for points outside it. (b) $1.1 \times 10^{29}\ \text{m/s}^2$
75. (a) 1.6×10^{7} electrons; (b) $\sigma = -1.4 \times 10^{-2} \dfrac{\text{C}}{\text{m}^2} = 8.8 \times 10^{16}\ \text{electrons/m}^2$
77. 7.9×10^{-9} C
79. (a) 1.9×10^{-16} C; (b) 1200 electrons
81. 2.16×10^{10} N/C at the angle of 60° with respect to the $+x$ axis
83. 14.2 N/C in the direction of travel of the electron
85. (a) 1.8×10^{21} N/C, outward from the center of the nucleus; (b) 1.5×10^{13} N/C, outward from the center of the nucleus; (c) $2.6 \times 10^{24}\ \text{m/s}^2$, outward from the center of the nucleus
87. (a) The electric field is pointing downward. (b) Gravity pulls downward on the drop, so the electric force on it must be upward. Since the drop is negative, the electric field must point downward for the force to point upward. (c) 9070 N/C; (d) 7

89. (a) 0; (b) $\frac{\sigma_i R_i^2}{\varepsilon_0 r^2}$ pointing radially outward;

(c) $\frac{\sigma_i R_i^2 - \sigma_o R_o^2}{\varepsilon_0 r^2}$, pointing radially outward if $\sigma_i R_i^2 > \sigma_o R_o^2$ or inward if $\sigma_i R_i^2 < \sigma_o R_o^2$.

91. -2.20×10^{-6} C

Chapter 17

1. The electric potential is the electric potential energy per unit charge and is a scalar. The electric field is the electric force per unit charge and is a vector. The electric potential depends on both the electric field and the region over which the field extends.
3. Both electric field and potential result from the existence of a charge. If a charge exists, it will create both of these physical quantities. In order to have a nonzero potential energy, a force must be able to do work on an object as it is moved. If only one charge exists, then there is no electric force acting on the charge. If there is no force, then that force can do no work, so there is no potential energy.
5. This statement makes sense only if the zero point of the electric potential has been previously defined.
7. (a) Yes, a region of constant potential must have zero electric field. (b) No, if the electric field is zero, the potential need only be constant.
9. Zero. The electric field is perpendicular to an equipotential; therefore, the work done in moving along an equipotential is zero.
11. Increase plate area, decrease separation, and increase the dielectric constant.
13. The energy stored in the capacitor increases.
15. The total capacitance is the sum of the individual capacitances. This means that for a given voltage, capacitors in parallel store more charge than any one of the individual capacitors.
17. For a given potential across them, the capacitors in parallel have a greater total area and so can store a greater amount of charge, thus having a larger capacitance.
19. The larger dielectric strength increases the maximum voltage that can be applied across the electrodes. The larger dielectric constant decreases the voltage required to store a given amount of charge, thus increasing the capacitance. The dielectric material can also be used to maintain the separation of the electrodes.
21. D
23. C
25. C
27. B
29. A
31. ~10 V
33. 1×10^8 J
35. The car would need on the order of 30 million capacitors.
37. (a) -3.20×10^{-6} J; (b) 3.20×10^{-6} J; (c) The potential difference between the two points would remain the same. However, if a negative charge were placed at rest at point a, then it would never reach point b unless an external force acted upon it. The charge would accelerate in the $-x$ direction, away from point b.
39. -0.33 J
41. 242 V
43. 18 J
45. 28.8 V
47. -2.78×10^8 C
49. 7.14 cm and 16.7 cm
51. 2.13×10^6 V
53. -2.77×103 V
55. (a) 1 V/m to the left; (b) 1 V/m to the left.
57. (a)

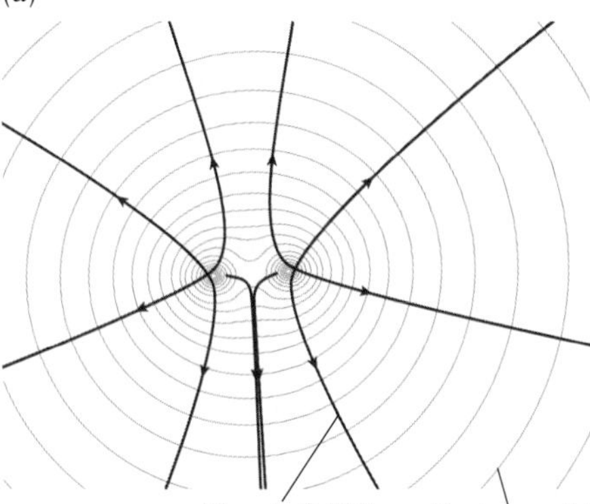

(b)

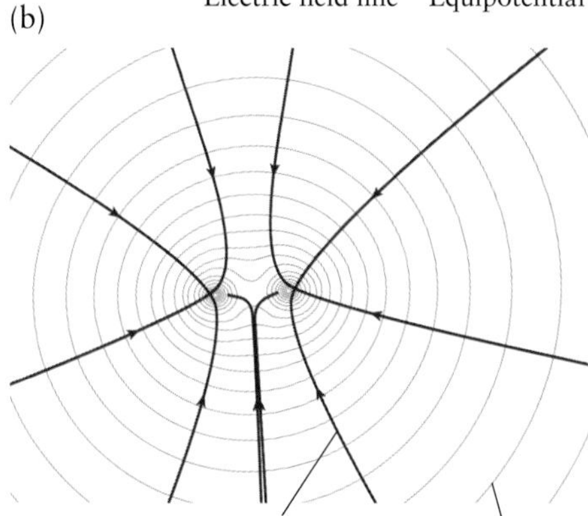

59.

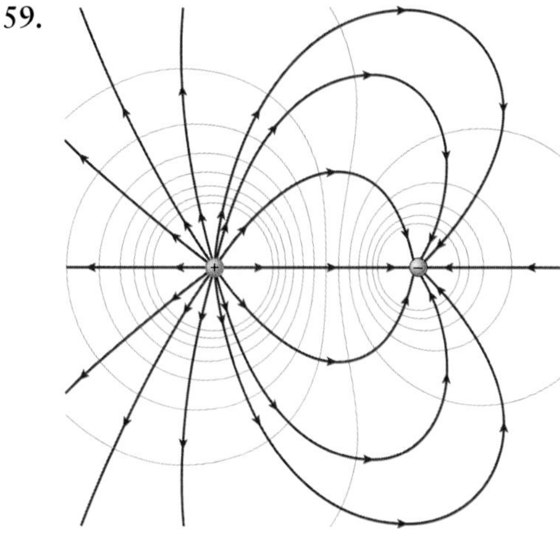

61. 24 μC
63. 8.85×10^{-9} F
65. 1.00 mm
67. 6.37×10^{-7} J
69. 2000 V
71. 1000 J
73. (a) 55.0 μF; (b) 5.00 μF
75. 4.00 μF and 4.00 μF

77. 11.0 μC on the 0.0500 μF capacitor and 22.0 μF on the 0.100 μF capacitor
79. 4.0 μF
81. (a) 7.41 μF; (b) 88.9 μC on each capacitor; (c) $V_{10} = 8.89$ V; $V_{40} = 2.22$, $V_{100} = 0.889$ V
83. 802
85. 2.15×10^{-7}C
87. (a) $C_{eq} = \frac{\varepsilon_0 A}{2d}\left(\frac{3k_2\, k_3}{k_2 + k_3} + \frac{k_1}{2}\right)$;

    (b) $k_3 = 1$ so $C_{eq} = \frac{\varepsilon_0 A}{2d}\left(\frac{3k_2}{k_2 + 1} + \frac{k_1}{2}\right)$
89. $1.45 \times 10^7 \frac{m}{s}$
91. (a) 6×10^8 J; (b) 2.65×10^2 kg
93. 1.12×10^{-20} J; increases
95. Combination 1: 2.00 μF; Combination 2: 12.5 μF; Combination 3: 8.33 μF; Combination 4: 5.00 μF; Combination 5: 5.71 μF; Combination 6: 3.75 μF; Combination 7: 2.86 μF; Combination 8: 4.29 μF; Combination 9: 12.0 μF; Combination 10: 8.00 μF; Combination 11: 50.0 μF.
97. 14 μF
99. 1×10^{-6} F
101. (a) $\frac{1}{2}$; (b) energy doubles, charge remains constant
103. (a) 6.3×10^{-10} J; (b) 1.4×10^{-7}
105. (a) 4.0 V; (b) $\frac{1}{9}$
107. $1 \times 10^5 \frac{\text{m}}{\text{s}}$

Chapter 18

1. Current is the amount of charge passing a point in the circuit per second. In this sense it is more like a flux than a vector. It has a direction and magnitude but is not really a vector.
3. There is no contradiction. If a conductor is in electrostatic equilibrium, the electric field within it must be zero. However, any conductor that is carrying a current is definitely not in electrostatic equilibrium.
5. The bird will not be electrocuted because there is no potential difference between the bird's feet. The bird grabs only one high-voltage wire, and it is not completing a circuit.
7. The cell membrane has the ability to alter its permeability properties. At rest the baseline membrane potential is lower on the inside compared to the outside of a cell (about 270 mV). If the membrane potential depolarizes (that is, becomes a little less negative on the inside surface), sodium ion channels open and sodium ions rush into the cell. The inner surface of the membrane becomes positively charged, and the potential difference quickly swings positive (about 140 mV). This is the peak of the action potential. Just as suddenly, as the potential difference across the cell membrane changes during the initial phase of an action potential, potassium ion channels begin to open. The membrane quickly becomes much more permeable to potassium ions than to sodium ions. Because the concentration of potassium ions is always higher inside cells, positive charge begins to flow out of the cell. This outward positive current drives the membrane potential to its resting level.
9. (a) Since the same current must flow through the ammeter and the circuit element, the two must be in series. (b) The equivalent resistance of resistors in series is the sum of the individual resistances, so the ammeter should have a very small resistance in order to have as little effect as possible on the resistance in the circuit.
11. This energy is dissipated as heat in the wire. The amount of energy dissipated is related to the amount of current running through the wire and the resistance of the wire.
13. It cannot be changed instantaneously because the resistor in the circuit limits the current (the rate of flow of charge). Because the current is finite, it requires time for the charge to flow on and off of the capacitor plates.
15. A
17. E
19. D
21. A
23. C
25. A laptop computer draws about 2 A of current. A hair dryer draws about 10 A of current. A compact fluorescent light bulb draws about 0.2 A of current.
27. In the United States a person uses about 1500 W; in developing countries a person uses about 100 W.
29. 1 kΩ, 100μF
31. 30 mΩ
33. 1.12×10^{22} electrons
35. 2.34×10^{-4} m/s
37. 0.3 mm/s
39. 12,000 K^+ ions
41. 0.0220 Ω
43. 16.0 Ω
45. 0.109 Ω
47. 0.56
49. 12 Ω
51. 4.0 V
53. 24.0 Ω
55. (a) 0.750 A in each resistor; (b) $V_9 = 6.75$ V, $V_3 = 2.25$ V
57. (a) 0.60 A in each of the 6-Ω resistors and 0.30 A in the 12-Ω resistor; (b) 1.5 A
59. 29 Ω
61. $R_A = 6.0\ \Omega$, $R_B = 3.0\ \Omega$
63. 12.5 A
65. 2.7 kW
67. 58 MW
69. $R_1 = 6.00\ \Omega$, $R_2 = R_3 = 24\ \Omega$
71. $P_b = 20$ W, $P_1 = 3.56$ W, $P_2 = 4.44$ W, $P_3 = 12$ W
73. 5×10^{-5} C
75. (a) (i) 0.00216 s, (ii) 0.00108 s; (b) (i) 0.0667 A, (ii) 0.211 A
77. 3.0 h/d
79. 5 Ω: 45 W; 8 Ω: 28 W; 10 Ω: 90 W; 16 Ω: 56 W
81. (a) 30 Ω; (b) 3 h and 45 min longer
83. (a) 16 Ω; (b) 20 Ω: 0.12 A, 12 Ω: 0.20 A, 8 Ω: 0.32 A; (c) 20 Ω: 0.29 W, 12 Ω: 0.49 W, 8 Ω: 0.84 W.
85. 300 $\Omega \cdot$ m
87. (a) $RC\ \ln(2) = 0.7\ RC$; (b) 5%
89. (a) Connect two 50-Ω resistors in parallel to make a 25-Ω equivalent resistance, and then connect that combination in series with another 50-Ω resistor to make a total equivalent resistance of 75 Ω. (b) Connect five 50-Ω resistors in parallel to make a 10-Ω equivalent resistance, and then connect that

combination in series with another 50-Ω resistor to make a total equivalent resistance of 60 Ω.

91. (a) 4.50×10^3 C; the "capacity" is a measure of the total charge that the battery can deliver as current. (b) 6.8×10^3 J (c) 3 h (d) 1×10^{-5} J; the battery stores 600 million times as much energy as the capacitor.

Chapter 19

1. Use both ends of one iron rod to approach the other iron rods. If both ends of the rod you are holding attract both ends of the other two rods, then the one you are holding is not magnetized iron.

3. Applying Newton's second law to a charged particle whose velocity is at right angles to a uniform magnetic field B results in the equation $qvB = mv^2/r$ (or $r = mv/qB$). As long as the particles have the same charge, the radius of their orbits depends on their momentum, not their velocity alone.

5. (a) The electric field points into the page. (b) The beam is deflected into the page. (c) The electron beam is not deflected.

7. Yes, because the width of the field determines the length of the wire in the magnetic field. The larger the length of wire in the magnetic field, the larger the force exerted on the wire will be.

9. Since the directions of the two currents are opposite, the magnetic field from one wire is opposite to the other. This means the magnetic fields will cancel each other to some extent.

11. There will be no net force on the wires, but there will be a torque. When two wires are perpendicular, each wire will experience a force on one end and another force, equal in magnitude but opposite in direction, on the other end. This results in zero net force. However, the forces will produce a nonzero torque that will make the wires want to align with each other such that the current in each wire is traveling in the same direction as the current in the neighboring wire.

13. C

15. D

17. A

19. A

21. E

23. about 4×10^{-6} T

25. 0.3 fN (3×10^{-16} N) vertically down

27.

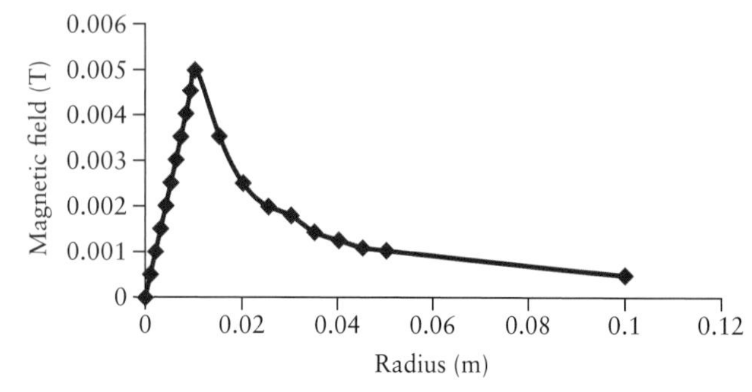

For $r < 0.01$ m: $B(r) = (0.501\ \text{T/m})r$
For $r > 0.01$ m: $B(r) = (0.000050\ \text{T}\cdot\text{m})/r$

29. (a) Out of the page; (b) down; (c) down; (d) to the left; (e) out of the page; (f) right

31. 0.7 N

33. 5.8×10^{-18} N to the left

35. 3.4×10^{-12} N in the $-z$ direction

37. 5×10^{-3} T in the $+x$ direction

39. 1.5×10^{-6} m

41. 0.511 T

43. 0 N

45. 2.5 T

47. 0.664 N·m at 30°
0.755 N·m at 10°
0.493 N·m at 50°

49. 1.44×10^{-5} N·m

51. B_O and B_P point in the $+z$ direction (out of the page). B_Q and B_R point in the $-z$ direction (into the page).

53. $B = \frac{\mu_0 I}{4}\left(\frac{R-r}{rR}\right)$ into the page

55. $B = 4 \times 10^{-6}\ \text{T} = 0.08\ B_{\text{Earth}}$

57. 120 m

59. 1.03×10^{-8} N away from the wire

61. 0.313 N

63. 0.0109 T

65. 9.12×10^{-31} kg

67. (a) $B = 1.0$ T at $r = 1.0$ m, $B = 0.001$ T at $r = 1.0$ km. Compared to Earth's magnetic field, the lightning bolt has a magnetic field 20,000 times as large at 1.0 m and 20 times as large at 1.0 km. (b) $B = 2.0 \times 10^{-6}$ T at $r = 1.0$ m, $B = 2.0 \times 10^{-9}$ T at $r = 1.0$ km. In each case the field due to the lightning is 500,000 times greater than the household current. (c) $r = 2.0$ mm

69. 0.40 N

71. (a) 298 A; (b) 0.62 T; (c) The easiest way to achieve the same field with less current is to increase the number of windings, N.

73. (a) 15.9 mA; (b) The loop will initially rotate to point its dipole moment in the direction of the magnetic field. At this instant there is no net torque on the loop, but because of its rotational momentum, it will continue to rotate past this point, slowing down. After it momentarily comes to rest, it will rotate back toward its initial position. If the system is frictionless, this harmonic oscillation will continue indefinitely.

75. 650 V

77. $B = \frac{\mu_0 i}{2\sqrt{2}\pi d}$ and makes an angle of 315° with the $+x$ axis.

79. $B = 1.5 \times 10^{-6}\ \text{T} = 0.031\ B_{\text{Earth}}$. Since the field is so much smaller than the Earth's magnetic field, there should be little or no cause for concern.

81. (a) $B = 0.7155\,\frac{\mu_0 N i}{R}$ in the $+x$ direction

(b) $B = 0.221\,\frac{\mu_0 N i}{R}$ in the $+x$ direction

83. (a) 2.64×10^{-6} T pointing westward; (b) 2.64×10^{-7} T pointing vertically upward; (c) 2.64×10^{-6} T pointing southward; (d) In (a) and (c), B is about 5% of B_{Earth}. This is fairly small, but it could possibly be enough to interfere with navigation. In part (b), B is about 0.5% of B_{Earth}, which seems too small to cause much of a problem.

85. (a) $i = 1.14$ A (b) clockwise

Chapter 20

1. A current is induced in loop b while the current in loop a is changing. If the current is in the direction shown and is increasing, the flux of its magnetic field through loop b is upward and increasing. In accordance with Lenz's law, the direction of the induced current in loop b is such that the flux of its magnetic field through loop b is downward, opposing the change in flux that produced it. This means that the current in loop b is in the opposite direction as the current in loop a, and the two loops repel. After the current in loop a stops changing, then the current in loop b becomes 0 and there is no force between the loops.
3. The induced emf will create an induced current in the clockwise direction, which will exert a force to the left on the sliding rod. This will gradually slow the rod until it comes to a stop.
5. As someone with a pacemaker or other electronic device walks through regions of spatially varying magnetic fields, the changing magnetic flux through the electronic circuitry will induce an emf in the device. This extra emf could cause unwanted, and possibly fatal, malfunctions of the device.
7. D
9. C
11. A
13. About 16,500 A
15. About 5 V/km or 15,000 V total. This is such a large variation that it would cause the cable to exceed the maximum current allowed and cease to function.

17.

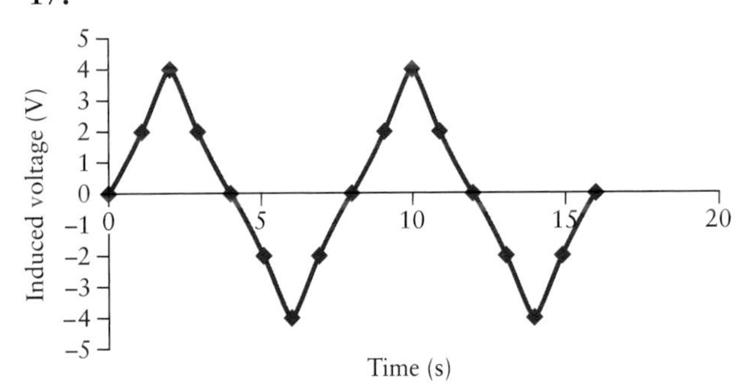

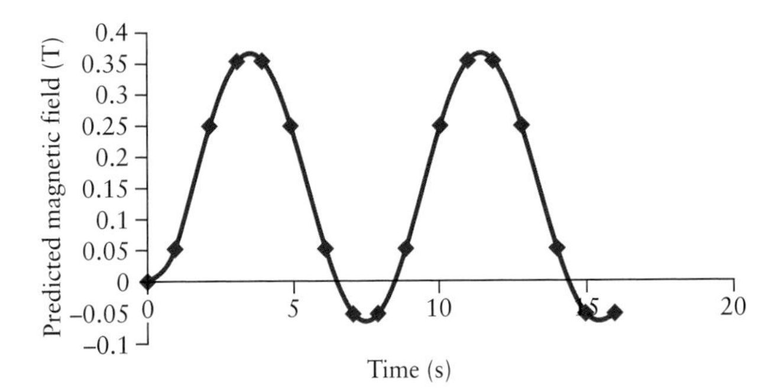

19. (a) 0.204 Wb; (b) 0.220 V
21. (a) Counterclockwise; (b) clockwise; (c) clockwise; (d) Counterclockwise; (e) clockwise; (f) no induced current
23.

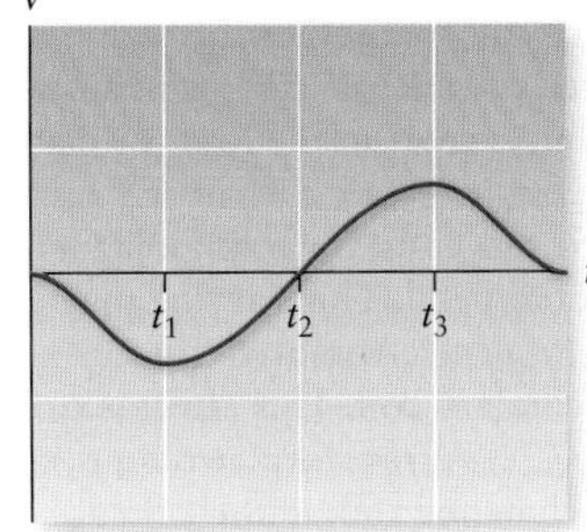

25. (a) For part (a) and part (c), $F = 0$ on all segments. For part (b), using x to represent the length of the given segment still in the magnetic field, $F = 0$ on the right-hand segment, $F = (0.0790\ \text{N/m})x$ in the upward direction for the top segment, $F = (0.0790\ \text{N/m})x$ in the downward direction for the bottom segment, and $F = 0.0790$ N to the left for the left-hand segment. (b) The current will be counterclockwise as viewed. The force on the left-hand segment will be zero. The force on the top and bottom segments will be $F = (0.0790\ \text{N/m})x$ downward and upward, respectively. The force on the right-hand segment will be $F = 0.0790$ N to the left.
27. 380 V
29. 1.9×10^6 rev/s. This is nearly 2 million rev/s, which does not seem feasible.
31. (a) 0.12 V; (b) 0.0047 A; (c) counterclockwise
33. 2.1 V
35. (a) 0.01 A; (b) up; (c) 0.001 N
37. 7.3 mW

Chapter 21

1. Many ac voltages are sinusoidal. A sine (or cosine) function has an average value of zero, since equal amounts of the function lie above and below the axis. If you calculate the root mean square value of the function, all of the values of the function—both negative and positive—are first squared before being averaged. Therefore, the negative values do not cancel out the positive values, and the general features of the waveform can be explained
3. For example:

Country	V_{rms}	Frequency (Hz)
Angola	220	50
Botswana	231	50
Ecuador	120–127	60
Mexico	127	60
Slovenia	220	50

5. No. These devices usually work over a range of voltages and will not function properly if a transformer is used. They contain "smart" circuitry that senses if the voltage is 120 or 240 V. (Many people needlessly buy transformers for their laptops. What may be needed, depending on destination, is an adapter plug.)

7. (a) A transformer is a device that transfers electrical energy from one circuit to another through inductively coupled conductors (the transformer's coils). (b) A varying current in the first (or primary) coil creates a varying magnetic flux in the transformer's core and, thus, a varying magnetic field through the secondary coil. This varying magnetic field induces a varying electromotive force (or voltage) in the secondary coil. This effect is called mutual inductance. (c) A transformer requires an ac input so that the changing flux in the primary coil will induce a varying electromotive force according to Faraday's law.
9. It saves copper. The primary power loss in transmission equals $i_{rms}^2 R$, where R is the resistance of the transmission lines, and the rate at which energy is transported equals $V_{rms} i_{rms}$. Thus, at high voltage, power can be transmitted with a lower current, and the lower the current, the less the transmission loss. The other way to reduce transmission loss would be to decrease the resistance of the transmission lines. The only practical way to do that is to use thicker wire, which requires more of the metal that the wires are made of. The voltages are stepped down for consumer safety.
11. The current drops suddenly toward zero when the switch is opened. This current drop induces a very large emf in the inductor, which produces a large potential difference across the switch gap. This potential is large enough to cause dielectric breakdown of the air in the gap, which causes the spark. The spark allows current to continue to flow for a brief period.
13. The electrical energy stored in the capacitor is $U_E = \dfrac{q^2}{2C}$, which varies as $\cos^2 \omega t$, and the magnetic energy stored in the inductor is $U_B = \dfrac{Li^2}{2}$, which varies as $\sin^2 \omega t$. Therefore, when a maximum amount of energy is stored in the capacitor, no energy is stored in the inductor, and vice versa.
15. No. This would be true only if there is negligible inductance and capacitance in the circuit or the circuit is at resonance.
17. Impedance describes the overall opposition to the flow of charge in a circuit driven by a time-varying voltage. The SI units of impedance are ohms (Ω).
19. D
21. A
23. C
25. C
27. Most household appliances draw between 1 and 20 amps. Most households have a main circuit breaker that "trips" at 150 A or so.
29. 56 nH
31. 75.0 V, 60 Hz
33. 0.5 A
35. 141 V
37. 1800 V; 8.49 A
39. 40 V
41. 9.00 A
43. 200 turns
45. 750 W
47. 21.4×10^{-6} V
49. (a) 9.65 mH; (b) The inductance doubles, assuming the number of turns and the length are held constant because the inductance is proportional to A: $\left(L = \dfrac{\mu N^2 A}{\ell}\right)$.
51. 2.6×10^{-11} F
53. 1×10^{-6} J
55. $q = Q_0/\sqrt{2}$
57. 2.25×10^3 Hz
59. 3.2 H
61. 2 A
63. (a) 1.2 A; (b) 1.7 A; (c) 140 W
65. (a) 0.5 A; (b) 2 A
67. 590 Ω, 65°
69. (a) 690 Ω; (b) 0.15 A; (c) $I(t) = (0.15\ \text{A}) \sin(1000\ \pi t + 1.56)$
71. 104.5 MHz, $i = 0.009$ A
73. 5.07 nF
75. 53 V
77. (a) 2.0×10^1 W; (b) United States: 0.042 A; Rome: 0.083 A; most likely will be damaged; (c) step-down transformer with twice as many windings in the primary coil as the secondary coil; (d) 2.9×10^3 V
79. (a) 1.2×10^3 W; (b) 12 W; (c) We should use a voltage of 120 V. A much smaller proportion of the total power is dissipated in the cord, which means more is delivered to the device. It is much more efficient than the 12.0-V delivery system.
81. (a) 2.29 V; (b) 52 A; (c) 74 A
85. (a) $i_{rms} = 1.414$ nA; (b) $\omega_0 = 785.4$ rad/s

Chapter 22

1. (a) e, c, b, d, a; (b) Gamma rays have the highest energy per photon. Microwaves have the lowest energy per photon.
3. The speed of light equals the product of the wavelength and the frequency ($c = \lambda f$).
5. Without the seminal work of Faraday, Gauss, Coulomb, Ampère, and others, it would not have been possible for Maxwell to complete his theory. Even though physics would have continued to progress with the work of Faraday, Gauss, and Ampère, it was Maxwell who really pushed the forefronts of science by understanding the connections among electricity, magnetism, and optics.
7. All EM waves move at the same speed ($c = 3.00 \times 10^8$ m/s). However, the wavelength multiplied by the frequency equals this constant ($c = \lambda f$). Therefore, the wavelength can vary (smaller/larger) in proportion to the frequency (larger/smaller).
9. The energy of the electromagnetic waves is proportional to the frequency of the oscillating electric and magnetic fields ($E \sim f$).
11. C
13. D
15. E
17. 10^{26} photons/s
19. 2.25 eV

21. About 10 nm

23. 0.24 sec (Note: The angular separation doesn't affect the estimate; that is, you get the same answer if you treat the satellite as being directly overhead and the sender/receiver at the same location. Even at the maximum separation of 10.8 time zones, the time is 0.28 s.)

25.

$t(s)$	E(N/C)	B($\times 10^{-9}$ T)
0.000	100	333
0.785	70. 7	236
1.571	0	0
2.356	−70.7	−236
3.142	−100	−333
3.927	−70.7	−236
4.12	0	0
5.498	70.7	236
6.283	100	333
7.069	70.7	236
7.854	0	0
8.639	−70.7	−236
9.425	−100	−333
10.210	−70.7	−236
10.996	0	0
11.781	70.7	236
12.566	100	333
13.352	70.7	236
14.137	0	0
14.923	−70.7	−236
15.708	−100	−333
16.493	−70.7	−236
17.279	0	0
18.064	70.7	236
18.850	100	333

The electric and magnetic fields oscillate in time in phase with one another.

27. A. 4.00×10^{-8} m
B. 5.00×10^{-7} m
C. 6.00×10^{-7} m
D. 6.99×10 m^{-7} m
E. 4.00×10^{-9} m
F. 1.13×10^{-8} m
G. 3.65×10^{-10} m
H. 5.00×10^{-11} m

29. A. 3.75×10^{14} Hz
B. 4.62×10^{14} Hz
C. 5.45×10^{14} Hz
D. 6.67×10^{14} Hz
E. 1.35×10^{17} Hz
F. 2.73×10^{16} Hz
G. 6.00×10^{12} Hz
H. 8.98×10^{9} Hz

31. 1.0×10^{6} Hz = 1.0 MHz

33. 1×10^{-3} s

35. $c = f\lambda = \left(\frac{\omega}{2\pi}\right)\left(\frac{2\pi}{k}\right) = \frac{\omega}{k}$

37. (a) 1.26×10^{11} m^{-1}; (b) $12\pi \times 10^{18}$ rad/s; (c) 4.99×10^{-11} m; (d) 6.02×10^{18} Hz

39. 2.8 A

41. 1.0×10^{-5} T

43. 2.39×10^{-6} N/C

45. (a) 6.67×10^{-7} T; (b) 53.1 W/m^2

47. (a) $\lambda = 853$ nm, $f = 3.52 \times 10^{14}$ Hz; (b) $\lambda = 442$ nm, $f = 6.79 \times 10^{14}$ Hz; (c) $\lambda = 621$ nm, $f = 4.83 \times 10^{14}$ Hz; (d) $\lambda = 232$ nm, $f = 1.29 \times 10^{15}$ Hz; (e) $\lambda = 19.6$ nm, $f = 1.53 \times 10^{16}$ Hz; (f) $\lambda = 142$ nm, $f = 2.11 \times 10^{15}$ Hz; (g) $\lambda = 627$ nm, $f = 4.78 \times 10^{14}$ Hz; (h) $\lambda = 273$ nm, $f = 1.10 \times 10^{15}$ Hz

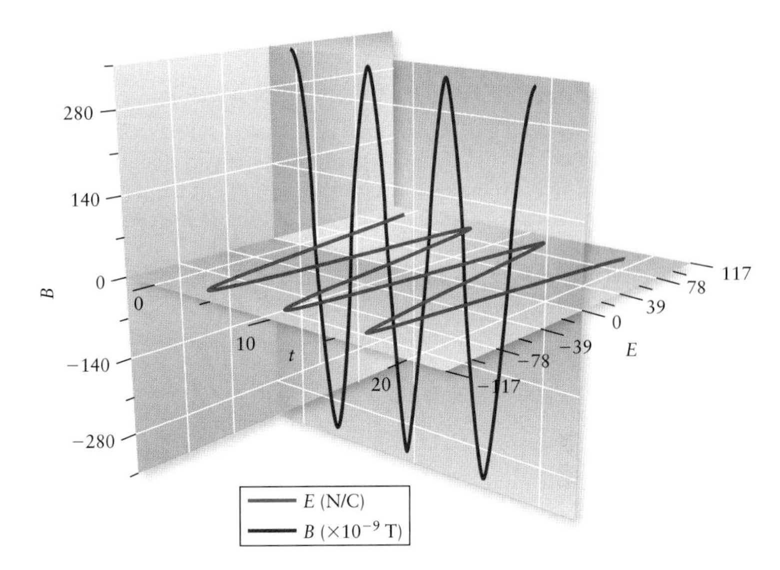

49. People 300 km away receive the news first. $t_{radio} = 0.001$ s and $t_{sound} = 0.0088$ s.

51. (a) 62.1 nm to 414 nm; (b) It extends from violet visible light into the UV.

53. (a) 0.45 m; (b) 101,000 J/m^3

Chapter 23

1. Christiaan Huygens suggested that every point along the front of a wave be treated as many separate sources of tiny "wavelets" that themselves move at the speed of the wave. This is important to see how light moves from one medium to another, different, medium and allows you to predict the bending that occurs in Snell's law.

3. No, the critical angle depends only on the index of refraction of the material that it enters and that it reflects off.

5. When light moves from a medium of larger index of refraction toward a medium with a smaller index of refraction, the angle of refraction will *increase*. If you make the incident angle larger and larger, the refracted angle will ultimately approach 90°. After the light moves past this critical value, it is reflected rather than refracted. In mathematical terms sinθ cannot be larger than 1. When Snell's law would require the sine of the angle of refraction to be greater than one, total internal reflection will occur.

7. When light enters a material with a negative index of refraction, the light refracts "back" away from the normal to the surface, as seen in the picture below:

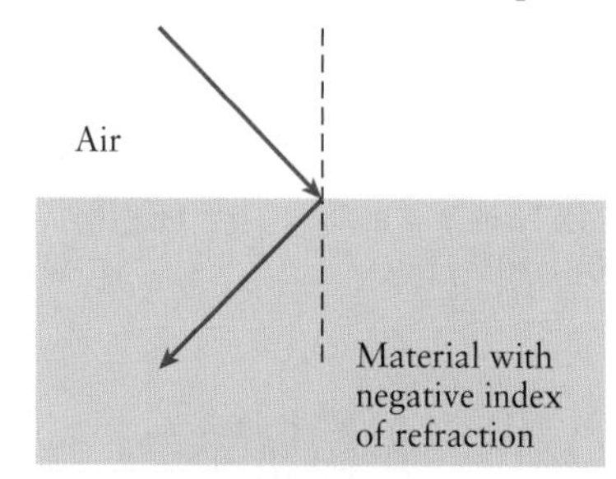

9. The light coming from the Sun is refracted as it passes through Earth's atmosphere. Because of the wavelength dependence of the index of refraction, red light bends less than orange than yellow than green than blue than violet. In addition, bluer light is scattered more than red light as it passes through the atmosphere. Because of these effects, more red/orange colored light can pass into the shadow of the Earth, making the Moon appear red.
11. Polarized sunglasses use polarizing filters to preferentially absorb light that harms your eyes. Since most glare is created by light that is reflected off of horizontal surfaces (for example, a lake, a ski slope), it is advantageous to block out the light that is horizontally polarized. As such, the lenses of polarized sunglasses allow only vertically polarized light to enter. Naturally, the intensity of the light that passes through such lenses is decreased in intensity.
13. 1. Soap bubbles reflect different colors. 2. Thin film coatings on photographic lenses ("nonreflective coatings"). 3. Oil floating in a puddle of water will show different colors.
15. They come from the sunlight, which contains all the wavelengths (colors) of the visible spectrum, as it is dispersed in the diamond.
19. D
21. B
23. B
25. A
27. A
29. air–diamond: ~(1.0 − 2.4)
31. 1.6%
33. $n_2/n_1 = 1.56$
35. 1.00029
37. 763 km
39. (a) 13.2°; (b) 22.1°; (c) 30.9°; (d) 20.1°
41. 452 nm
43. A. 1.34
    B. 1.37
    C. 1.48
    D. 1.74
    E. 1.21
    F. 1.61
    G. 2.65
    H. 1.04
45. light will always be totally internally reflected
47. 5.70 m
49. (a) 1.58 (rounded to 3 significant figures), 1.50; (b) 18.5°, 19.5°; (c) 33.3°, 35.3°
51. 1.50, 1.48
53. Vertically polarized and has an intensity of 39 W/m^2
55. 7.4×10^2 W/m^2
57. 49.4°
59. 36.9°, 48.8°
61. (a) Shifted by half a wavelength; (b) shifted by half a wavelength
63. 648 nm
65. $t = 93.8$ nm; $t = 282$ nm; $t = 469$ nm; $t = 657$ nm
67. $\lambda = 631$ nm and $\lambda = 473$ nm are not seen in the film.
69. 540 nm
71. (a) 5; (b) 8
73. $\theta_1 = 36.4°$, additional fringes are not visible.
75. 583 nm
77. 633 nm
79. 17 mm
81. 7.71×10^{-6} degree $= 1.34 \times 10^{-7}$ radians
83. 3.46×10^{-5} m
85.

Optical material with percentage of time required for light to pass through compared to an equal length of vacuum	Speed of light	Index of refraction
100%	3.00×10^8 m/s	1.00
125%	2.40×10^8 m/s	1.25
150%	2.00×10^8 m/s	1.50
200%	1.50×10^8 m/s	2.00
500%	6.00×10^7 m/s	5.00
1000%	3.00×10^7 m/s	10.00

87. $n_1 \sin(\theta_1) = n_n \sin(\theta_n)$ is true.
89. 40.5°
91. (a) 0.969 m; (b) You would have to change the index of refraction of water to be smaller than the index of refraction of air. Under this condition, total internal reflection cannot occur.
93. 9.14 m
95. (a) 9.38 W/m^2; (b) 45°
97. 902 nm
99. 188 μm
101. (a) 1.77; (b) 310 nm, 465 nm, and 620 nm
103. (a) 500 mi/h; (b) 1.2 h; (c) The formula $w \sin\theta = m\lambda$ applies only if the wavelength is smaller than the slit width. In this case $w = 100$ mi and $\lambda = 500$ mi, so the formula would not apply.
105. 9.35×10^{-4} rad. This is larger than the pupil, which makes sense because the pupil is larger in diameter than the hole.
107. (a) Ring 1 occurs at 3.05×10^{-4} rad $= 1.75 \times 10^{-2}$ degrees, ring 2 occurs at 5.58×10^{-4} rad $= 3.19 \times 10^{-2}$ degrees, ring 3 occurs at 8.10×10^{-4} rad $= 4.64 \times 10^{-2}$ degrees. (b) $x_1 = 7.6 \times 10^{-3}$ mm, $x_1 = 7.6 \times 10^{-3}$ mm, $x_3 = 2.0 \times 10^{-2}$ mm. (c) The dark rings, and hence the bright rings, are so close

together that the bright rings are essentially right next to each other and mask the dark rings, thus eliminating the diffraction effect and giving us only bright light.

109. (a) about 55 to 60 cm; (b) Telescopes are built larger than the diffraction-limited diameter for greater light-collecting power (it allows you to see dimmer objects).

111. (a) 13.1 cm; (b) 96.9 cm. Since the resolution with atmospheric turbulence is less than the diffraction limited resolution, the atmospheric resolution is the limiting factor, so the telescope can resolve only objects separated by about 1 m.

113. (a) 25; (b) 25

Chapter 24

1. Actually, a plane mirror does neither but instead inverts objects back to front. If the mirror inverted right and left, then the object's right hand that points east would appear on the image as a right hand pointing toward the west. Because the image is inverted back to front, the object facing north is transformed into an image that faces south. Also the object's right hand is transformed into a left-hand in the image.

3. A real image forms where light rays come together but no light rays actually meet where a virtual image forms.

5. This phrase refers to the fact that the curved mirrors produce images that are not located at the same point as an image in a plane mirror. In fact the mirrors are designed so that if you see a vehicle in the rearview mirror, you should not change lanes because it is too close.

7. Additional information is needed. In accordance with the lensmaker's equation, if the radius of curvature of the front surface is larger, it is a diverging lens; if the radius of curvature of the front surface is smaller, it is a converging lens.

9. The full image formed, but it is dimmer than before because it is formed with half as much light.

11. Since the brain can be "trained" to interpret all the nerve inputs that it receives, it is feasible that upright and inverted could be "redefined." It must be very disconcerting, however, for those several days when "up is down and down is up!"

13. The condition known as presbyopia (farsightedness) is associated with a weakening of the muscles around the eye and inflexibility in the crystalline lens system as a person ages. This loss in adaptive amplitude leads to the inability to focus on objects that are close up. Myopia, on the other hand, is a disorder that affects as many as 50% of the world's population, but it is not brought on by the aging process.

15. Since the focal length of the lens is fixed, as the object distance is decreased for a nearby object, the image distance must be increased by moving the lens away from the screen, which is the light detector.

17. B

19. B

21. D

23. A

25. B

27. The spherical mirror has a focal length of about -1 m.

29. A shiny doorknob (f about -1 cm), a shiny backside of a spoon (f about -1 cm), and a glass ornament (f about -2 cm)

31. About 22.86 cm

33. 60°

35. 0.900 m

37. The first five images to the left are 3, 17, 23, 37, and 43 m to the left of the left mirror. The first five images to the right are 7, 13, 27, 33, and 47 m to the right of the right mirror.

39.

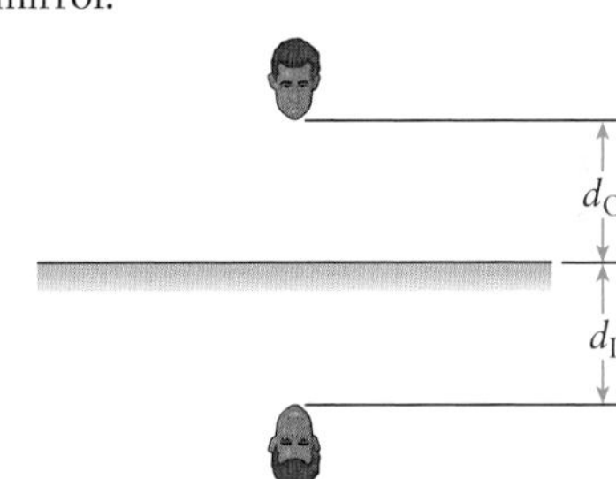

41. The camera will focus the lens on the surface of the mirror (2 m away from the virtual image). It will be out of focus a little bit.

43. As long as $d_O > f$, the image will be real in a concave, spherical mirror. In this case the focal point is located between the mirror and the object.

45. (a)

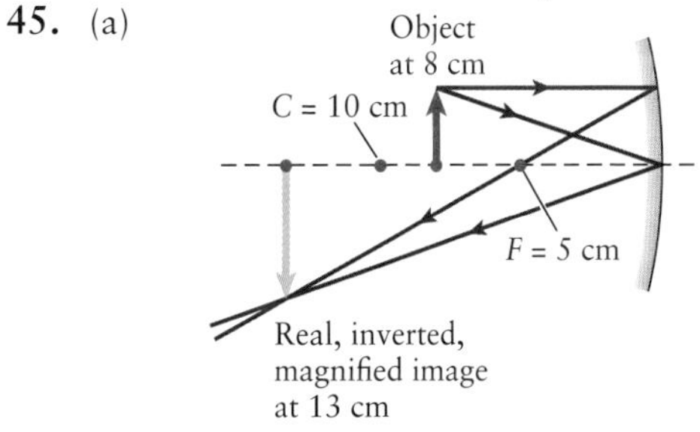

(b) $d_I = 13$ cm, $m = -1.7$; The image is real and inverted.

47. (a)

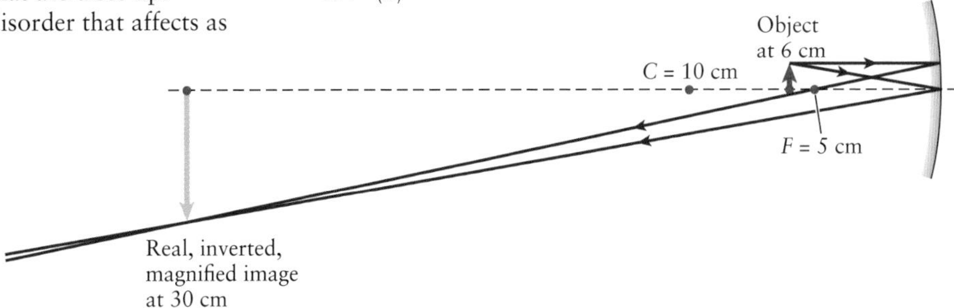

(b) $d_I = 30$ cm, $h = -5.0$ cm

49. (a) $d_I = 30$ cm, $h = 60$ cm; The image is real and inverted. (b) $d_I = 12$ cm, $h = 12$ cm; The image is real and inverted. (c) $d_I = 8.11$ cm, $h = 1.62$ cm; The image is real and inverted.

51.

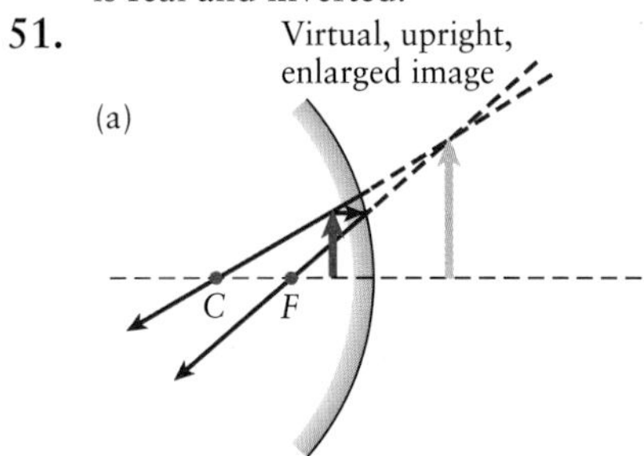

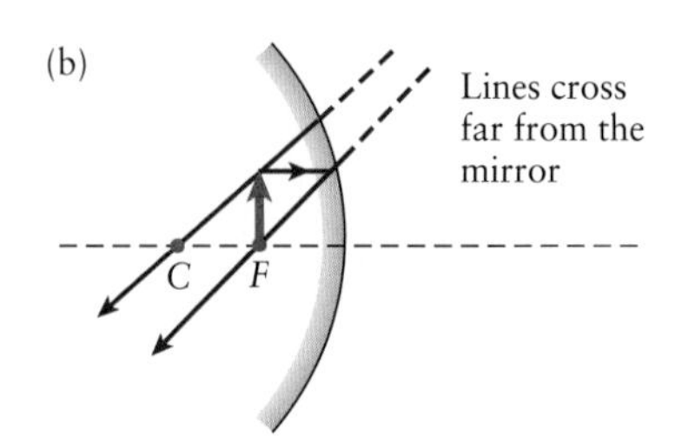

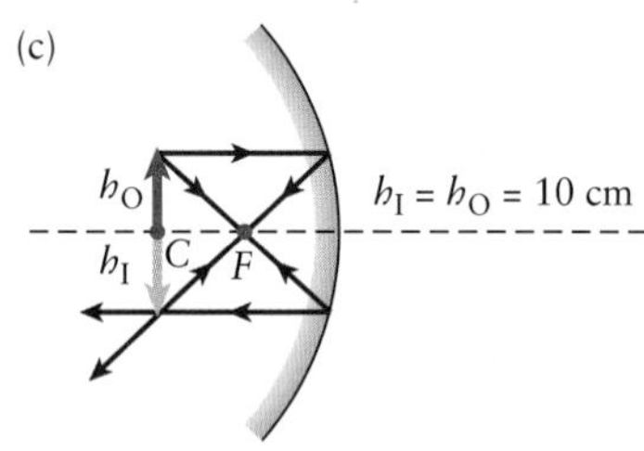

53. In both cases, the images are virtual. The image of an "up close" object is larger than the image of the same object "far away." In both cases the image is smaller than the object.

55. A convex mirror is the opposite of a concave mirror. Use a phrase such as "The cave goes in" to remember the shape of a concave mirror.

57. (a) $d_I = -6.67$ cm, $h = 3.33$ cm. The image is virtual and upright.
(b) $d_I = -8.33$ cm, $h = 1.87$ cm. The image is virtual and upright.
(c) $d_I = -9.09$ cm, $h = 0.909$ cm. The image is virtual and upright.

59. $d_I = -4.29$ cm, $h = 0.429$ m. The image is virtual and upright.

61. $d_I = -0.962$ cm, length $= 0.926$ cm. The image is virtual and upright.

63.

$$\frac{1}{d_O} + \frac{1}{d_I} = \frac{1}{f}$$

$$\frac{1}{d_I} = \frac{1}{f} - \frac{1}{d_O} = \frac{d_O - f}{fd_O}$$

$$d_I = \frac{fd_O}{d_O - f}$$

The focal length of a spherical convex mirror is always negative, and the distance of the object from the mirror, d_O, is positive for all mirrors, which means the term $d_O - f$ will always be positive for a real object. Therefore,

$$d_I = \frac{fd_O}{d_O - f} < 0$$

65. $d_I = -6.32$ cm, $h = 0.947$. The image is virtual and upright.

67. No, a real image in a converging lens occurs on the opposite side of the lens from the object.

69. Two. Images form where all the refracted rays (or their backward extensions) intersect. Therefore, wherever two rays intersect, the rest of the rays also intersect.

71. (a) $d_I = -6.7$ cm, $h = 13$ cm. The image is virtual and upright.

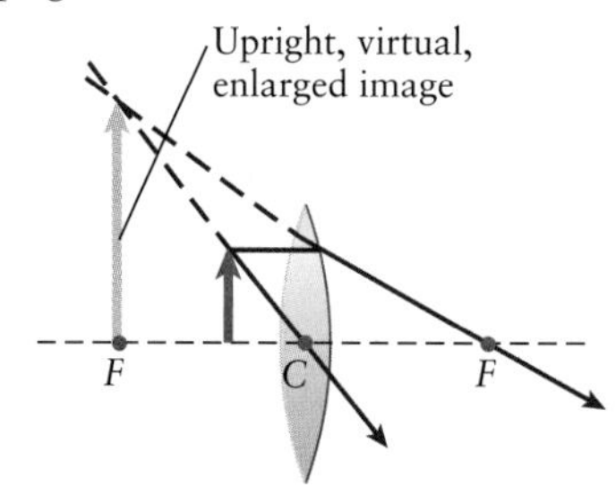

(b) $d_I = -20$ cm, $h = 20$ cm. The image is virtual and upright.

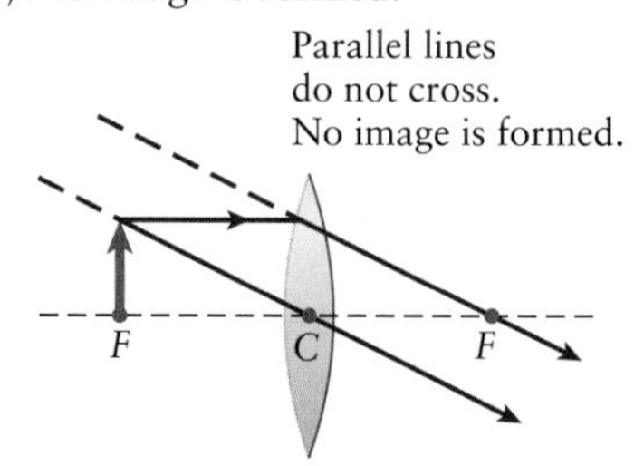

(c) No image is formed.

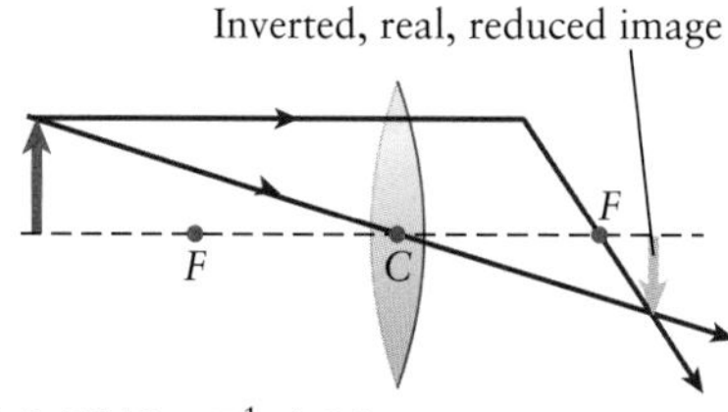

(d) $d_I = 33$ cm, $h = 6.7$ m. The image is real and inverted.

Inverted, real, reduced image

73. (a) 0.7543 m^{-1}; 1.33 m

75. (a) 45.0 cm, real, 3.00 cm tall and inverted;
(b)

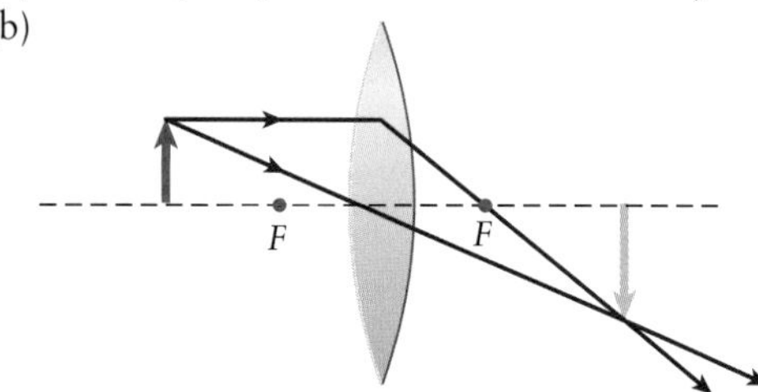

77. $f = 38.6$ cm, $P = 2.59$ diopters, converging lenses

79. (a) 3.3 mm × 5.8 mm;
(b) 1.2 μm^2/pixel; 1.1μm

81. (a) 9.00 m;
(b) 4.00 m

83. (a) 0.19 cm, 1.4 cm;
(b) As shown in the diagram below, the images are inverted because the ray from the top of the object forms the bottom of the image, while the ray from the bottom of the object forms the top of the image. The image is real in both cases because the rays travel through the lens and actually strike the retina.

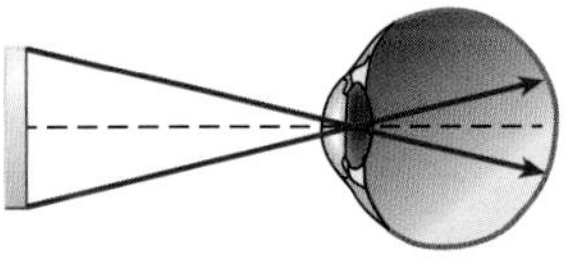

(not drawn to scale)

85. (a) 2.3 cm $\leq f \leq$ 2.5 cm; (b) converging lens

87. (a) 60.0 mm, −60.0 mm, −60.0 mm, 60.0 mm, 12.0 mm, −12.0 mm, 12.0 mm, −12.0 mm;
(b)

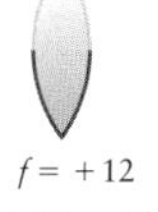

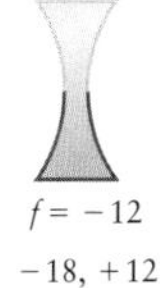

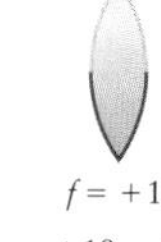

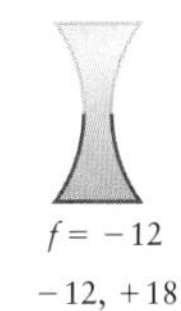

89. (a) 60 cm in front of the second lens; (b) virtual; (c) upright and three times as large as the object; (d) To see the final image, a viewer would need to look to the left through both lenses.

91. 1.52

93. (a) 3 cm (rounded to one significant figure); (b) 0.8 cm

95. −1.4 D

97. (a) $\frac{1}{f_{\text{combined}}} = \frac{1}{f_1} + \frac{1}{f_2}$; (b) Since the contact lens sits directly on top of the person's eye, the total combined power necessary to achieve corrected vision is equal to the power of the contact lens plus the power of the person's lens. (c) An eyeglass prescription cannot have the identical prescription as a contact lens because eyeglasses sit about an inch in front of the person's eye. The lenses are not pressed next to one another in this case.

99. −0.893 D; 2.80 D

101. (a) 67.1 mm; (b) −0.353

103. $|f_{\text{blue}} - f_{\text{red}}| = 0.3770$ mm

Chapter 25

1. If the frame were moving in both x and y directions, the given Galilean transformation equations would be as follows:

$$x' = x - V_x t$$
$$y' = y - V_y t$$
$$z' = z$$
$$t' = t$$

3. Earth's surface is a noninertial frame because Earth rotates around its axis and revolves around the Sun, both of which involve acceleration. We often approximate Earth's surface as an inertial frame because the accelerations associated with the rotational motions of Earth are small compared to the gravitational acceleration g.

5. The Michelson–Morley experiment was performed to determine our motion through the "luminiferous ether" (the hypothesized medium of light). In all situations, before this famous experiment, waves required some type of medium upon which to transmit their energy. So it was only natural to search for the medium upon which light would vibrate. When the null result was ascertained, it was difficult for physicists to abandon the long-held notion about waves and their media, so it was prudent to repeat the experiment again and again to make sure there was nothing wrong with the equipment. The experiment compared travel times for two beams of light that were moving at right angles to each other. As the beams were turned 90°, it was expected that a noticeable deviation would be perceived due to the "ether wind." The difference between the two travel times was related to the factor $1/\sqrt{1 - V^2/c^2}$.

7. If the object is moving relative to you, you measure its length by finding the difference between the coordinates of its endpoints at the same time.

9. It is impossible to accelerate an object to the speed of light. From the equation $E = \gamma mc^2$ as the speed of an object approaches the speed of light, the amount of energy required to increase the speed approaches infinity. It would take an infinite amount of work or an infinitely large force to accelerate the object to the speed of light.

11. The equivalence of gravity and acceleration is fundamental to the theory of general relativity. This leads to a context for Einsteinian physics that includes noninertial (that is, accelerating) frames of reference as well as inertial frames.

13. There is always a great deal of resistance to change, in any context, but especially in academia. Physics in this era (c. 1900) was predicated on Newtonian theories regarding motion, gravity, and cosmology. In addition, Maxwell's theory of electromagnetism was firmly in place. These two scientists were held in the highest esteem and were "above reproach." It was heretical for anyone, much less a young, upstart physicist who was not even part of a university, to challenge these time-tested, universally accepted underpinnings of physics. On a much more pragmatic level, it is always a "tough sell" when the new theory is complicated and not intuitive.

15. D

17. D

19. A

21. C

23. 1.01 or 1.04 for 5 kV and 20 kV

25. Yes. If $v/c = 0.6$, then $\gamma = 1.25$. If $v/c = 0.8$, then $\gamma = 1.67$.

27. About 5% of the world population

29.

v/c	γ	v/c	γ
0	1.0000	0.7	1.4003
0.1	1.0050	0.8	1.6667
0.2	1.0206	0.9	2.2942
0.3	1.0483	0.99	7.0888
0.4	1.0911	0.999	22.3663
0.5	1.1547	0.9999	70.7124
0.6	1.2500		

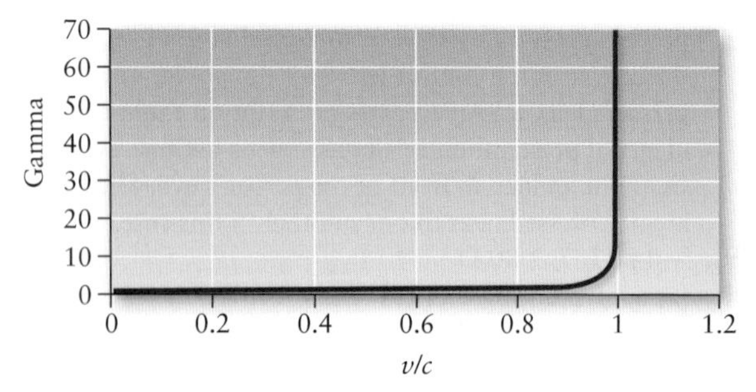

(a) 0.1; (b) $v/c = 0.1$, slope = 0.102; $v/c = 0.5$, slope = 0.770; $v/c = 0.9$, slope = 10.9; $v/c = 0.999$, slope = 11,200

31. (a) $x' = x - 4t, y' = y, z' = z$;
(b) $x' = -14$ m, $y' = 1$ m, $z' = 0$ m

33. (a) $x' = x - 15t, y' = y, z' = z$;
(b) $x'' = x - 4t, y'' = y, z'' = z$;
(c) $x' = x'' - 11t, y' = y'', z' = z''$

35. (a) $x = 12$ m, $y = 4$ m, $z = 6$ m, $t = 4 \times 10^{-3}$ s;
(b) $x = 8$ m, $y = 4$ y, $z = 6$ m, $t = 4 \times 10^{-3}$ s

37. 39.1 m/s in the direction of 39.8° north of the east direction

39. The massive slab of marble was used to keep any vibrations from disrupting the light as it moved through the interferometer. The slab needed to be rotated at different points in the experiment, so the mercury made it easier to move.

41. (a) 0.00667 μs; (b) 0.675 μs; (c) 23.8 μs; (d) 396 μs

43. 1.67

45. 2.19 μs

47. 0.9999999995c

49. 1.5 ns

51. 1.92 m

53. 0.980 m

55. $v = 2.9955$ m/ss; $v/c = 0.9985$

57. (a) 10^{-7} s, for both; (b) The backward photon registers first. (c) 1.5×10^{-7} s

59. The velocity of the truck relative to the car is $-0.110c$. The velocity of the car relative to the truck is 0.110c.

61. 0.548c

63. 0.99999999999989c

65. (a) 1.35×10^{-22} kg·m/s; (b) 9.51×10^{-15} J;
(c) 8.20×10^{-14} J; (d) 9.15×10^{-14} J

67. 0.866c

69. 0.209c

71. 1.8 m/s^2 in the downward direction

73. (a) 6×10^{-4} s; (b) 1.44×10^5 m

75. (a) 0.946c; (b) 0.946c; (c) Consider part (a). If the nucleus is "Earth," then the laboratory goes by at a speed of 0.8c, which corresponds to rocket B in part (b). The nucleus emits an electron in the forward direction, at a speed of 0.6c, and this electron corresponds to rocket A in part (b). In part (a), we determined the speed of the electron relative to the lab frame, which is then equivalent to the speed of rocket A relative to rocket B in this problem. Since two objects must always have the same speed relative to each other, our calculation of the speed of rocket B relative to rocket A better give the same answer.

77. (a) The muon does not make it to Earth's surface.
(b) 0.988c

79. 6×10^{15} m

81. 0.986c

83. (a) 8.4×10^{-2} kg = 84 g

85. (a) The speed of light is equal to c in all inertial reference frames, so it is the same for both the observer at rest in the rocket and the observer at rest on Earth.
(b) 1.07×10^{-6} s; 6.178×10^{-7} s; 1.853×10^{-6} s
(c) This is consistent with time dilation.

87. (a) (a) Both beams travel at the speed of light.
(b) c, $-c$ (c) 2c

Chapter 26

1. If light behaved simply as a wave, light of any frequency could produce a photoelectron if the intensity of that light was increased sufficiently to provide the required energy. The fact that light of a frequency lower than the cutoff frequency does not produce a photoelectron is consistent with the idea that light consists of particle-like packets called photons, each of which has an energy that depends on the frequency of light. Each photon can then produce a single photoelectron if the photon's frequency is sufficiently high.

3. Yes, it is possible for photoelectrons to be emitted from a metal at relativistic speeds, as long as the photon energy is sufficiently high.

5. Increasing the intensity means delivering photons to the surface of the metal at a greater rate; correspondingly, the photocurrent—the rate at which the electrons are emitted—increases.

7. Visible light can Compton scatter. However, because the Compton wavelength is so small compared to the wavelength of visible light (0.0024 nm compared to 400 to 750 nm), the change in the wavelength of a visible photon would be negligibly small.

9. The electron has the longer wavelength.

11. The de Broglie wavelength decreases as the kinetic energy increases.

13. The electron in the Bohr atom would be constantly accelerating, and so it would be constantly radiating. As the electron radiates, it would lose energy and slowly spiral into the nucleus.

15. Hydrogen is the first, most elemental, simplest element in the periodic table. Any cogent theory of atoms must describe hydrogen first and then work its way up to the larger, more complicated elements. In addition, because hydrogen only has one electron, it presents no complications to the orbit of an electron due to the interaction of other electrons.

17. The Compton wavelength for electrons is about 1800 times larger than the Compton wavelength for protons. So if you use electrons, you will not have to produce

electromagnetic radiation with such high frequency, and it will be easier to demonstrate the particle nature of light.

19. D

21. A

23. A

25. C

27. 5–25 eV; this is 10^5 times smaller than the nuclear binding energy.

29. Between 500 and 1000 W

31. (a) 1.4×10^{-34} m; (b) 1.7×10^{-34} m

33. 250 fm = 2.5×10^{-13} m

35.

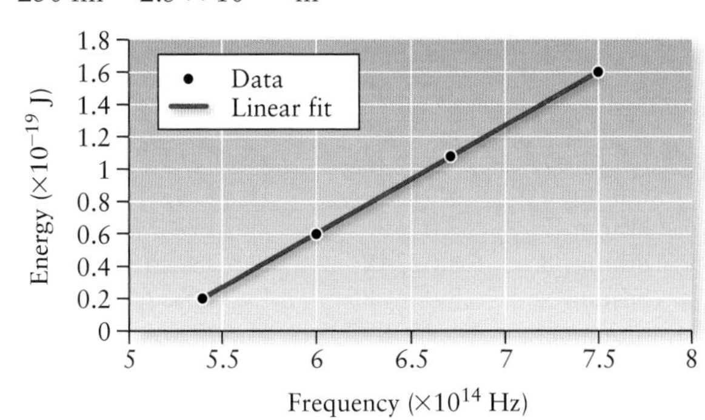

Using these data, we calculate $h = 6.8 \times 10^{-34}$ J·s = 4.12×10^{-15} eV·s

37. 1×10^{-30} J

39. 5.01×10^{-7} m = 501 nm; 5.99×10^{14} Hz

41. (a) 7.59×10^{-19} J; (b) 1.37×10^{-19} J

43. 9.85×10^{14} Hz

45. 4310 nm

47. 6110 K

49. 6 min

51. (a) 1.21×10^{-27} kg·m/s = 2.25 eV/c;
(b) 9.32×10^{-24} kg·m/s = 17400 eV/c

53. 25.8°

55. (a) $\lambda = 0.140$ nm, $K = 0$ eV
(b) $\lambda = 0.14033$ nm, $K = 20.8$ eV
(c) $\lambda = 0.14071$ nm, $K = 44.7$ eV
(d) $\lambda = 0.1412$ nm, $K = 75.3$ eV
(e) $\lambda = 0.14243$ nm, $K = 151$ eV
(f) $\lambda = 0.14486$ nm, $K = 297$ eV

57. 0.137 nm

59. (a) $f = 4.22 \times 10^{18}$ Hz, $E = 17440$ eV; (b) 0.0735 nm; (c) 16,860 eV; (d) 576 eV

61. 3.32×10^{-10} m

63. (a) 1.23×10^{-9} m
(b) 3.88×10^{-10} m
(c) 1.23×10^{-10} m
(d) 3.88×10^{-11} m
(e) 8.80×10^{-13} m
(f) 1.24×10^{-15} m

65. 1.43×10^{-10} m

67. 656.21 nm; 486.08 nm; 434.00 nm; 410.13 nm

69. Rydberg formula with $n = 2$:

$$\frac{1}{\lambda} = R_H\left(\frac{1}{2^2} - \frac{1}{m^2}\right) = R_H\left(\frac{1}{4} - \frac{1}{m^2}\right) = R_H\left(\frac{m^2-4}{4m^2}\right)$$

$$\lambda = \frac{1}{R_H}\left(\frac{4m^2}{m^2-4}\right) = \frac{4}{1.09737 \times 10^7\, m^{-1}}\left(\frac{m^2}{m^2-4}\right)$$

$$= (3.6451 \times 10^{-7}\, m)\left(\frac{m^2}{m^2-4}\right)$$

$$= (364.51\, nm)\left(\frac{m^2}{m^2-4}\right) \approx b\left(\frac{m^2}{m^2-4}\right)$$

71. 1.215×10^{-7} m = 121.5 nm; 1.026×10^{-7} m = 102.6 nm; 9.496×10^{-8} m = 94.96 nm; 9.376×10^{-8} m = 93.76 nm

73. 1.875×10^{-6} m = 1875 nm; 1.282×10^{-6} m = 1282 nm; 1.094×10^{-6} m = 1094 nm

75. $(8.2233 \times 10^{14}\text{ Hz})(\text{m}^2 - 4/\text{m}^2)$

77. (a) $L_n = nh$

(b) $\gamma_n = \dfrac{n^2k^2}{m_eke^2}$

(c) $K_n = \dfrac{m_e(ke^2)^2}{2n^2k^2}$

(d) $E_n = -K_n = -\dfrac{m_e(ke^2)^2}{2n^2k^2}$

(e) $v_n = \dfrac{ke^2}{nk}$

79. (a) $\boxed{L_n = nh}$

(b) $L_n = r_n m_e v_n$

$$v_n = \frac{L_n}{r_n m_e} = \frac{nh}{\left(\frac{n^2}{Z^{a_0}}\right)m_e} = \boxed{\frac{Zh}{na_0m_e}}$$

81. $\Delta f = 0.2$ MHz

83. 1.51 eV

85. (a) $n = 1$: State 1: $l = 0, m_l = 0, m_s = -1/2$,
State 2: $l = 0, m_l = 0, m_s = +1/2$
(b) $n = 2$: State 1: $l = 0, m_l = 0, m_s = -1/2$,
State 2: $l = 0, m_l = 0, m_s = +1/2$
State 3: $l = 1, m_l = -1, m_s = -1/2$,
State 4: $l = 0, m_l = -1, m_s = +1/2$
State 5: $l = 1, m_l = 0, m_s = -1/2$,
State 6: $l = 0, m_l = 0, m_s = +1/2$
State 7: $l = 1, m_l = +1, m_s = -1/2$,
State 8: $l = 0, m_l = +1, m_s = +1/2$

87. 7.37×10^{-19} J; 6.63×10^{-19} J

89. 2×10^{12}

91. (a) 1.21×10^{-14} J; (b) 38.0%

93. 8.3 eV

95. (a) 9.9×10^{-20} J; 6.0×10^{-26} J; (b) electron

97. $\lambda = \dfrac{h}{mv}\sqrt{1 - \dfrac{v^2}{c^2}}$

99. (a) 92.2 cm; (b) 352 MHz

101. (a) 1.11×10^{-7} nm; (b) 2.22×10^{-7} nm, ~22,000 times smaller than from an electron; (c) electron: 3.4%, carbon nucleus: 0.0003%; the wavelength shift from Compton scattering from carbon nuclei is extremely small and very difficult to measure; (d) 0.022 pm, 56 MeV

103. (a) 116 nm, 10.31 eV; (b) High-energy radiation from the star ionizes hydrogen in the nebula, which then produces an emission spectrum when electrons are recaptured. (c) 967 nm, 1.29 eV; no; no; not much emission is expected from this nebula; (d) That it is near a hot ($> 20{,}000$ K) star because there would be enough high-energy photons from the star to ionize hydrogen in the nebula and produce an emission spectrum.

Chapter 27

1. Isotopes are nuclides with the same atomic number Z but a different number of neutrons N. So they have different mass numbers $A = Z + N$.

3. (a) "Neutron rich" refers to the fact that heavier nuclei (larger than $A \sim 40$) possess a greater number of neutrons than protons. (b) Neutrons are important in order to balance out the repulsive Coulomb force between the positive protons. Since the neutrons are neutral, they do not repel one another. However, they do contribute to the attractive nuclear forces between the nucleons.

5. The whole nucleus weighs less than the sum of its parts because some of that mass is converted to energy when the parts are fused. In order to break apart the nucleus, you must add energy equivalent to the binding energy, which becomes mass after fission.

7. The nuclear force is charge independent, so it doesn't have attractive or repulsive components like the electrostatic Coulomb force. In addition, it is very short-ranged. Outside of a few femtometers, the nuclear force quickly goes to zero. Inside about 0.7 fm, the force is repulsive between nucleons, so the radius of the nucleus will never shrink to zero. The major competition inside the nucleus comes from the repulsive Coulomb force between the tightly packed, positively charged protons.

9. (a) Elements with a higher atomic number than iron are more likely to fission. (b) Elements with a lower atomic number than iron are more likely to undergo fusion.

11. Consider the typical beta decay, such as the decay of a neutron: $n \rightarrow p + \beta^- + \nu_e$. If the electron ($\beta^-$) were the only decay product, application of conservation of energy and momentum to the two-body decay would require that the β particle be ejected with a single unique energy. Instead we observe experimentally that β particles are produced with energies that range from zero to a maximum value. Further, because the original neutron had spin 1/2, conservation of angular momentum would be violated if the final decay products consisted of only the two particles p and β, each with spin 1/2.

13. As uranium begins to decay, the daughter nuclei are themselves often radioactive, decaying by various paths (which all lead down to a stable isotope in the periodic table). So, at any instant there are isotopes from all the many possible decay modes of all the radioactive progeny of the parent nuclei.

15. Because the atomic masses include Z electron masses in their values, it is already taken into account. There is 1 electron mass included in the atomic mass of the proton. There is no need to add in another one to account for the beta particle.

17. D

19. E

21. B

23. B

25. C

27. The nuclear force is about 100 times larger than the electrostatic force.

29. Energies associated with nuclear reactions are on the order of 10 to 100 MeV, whereas energies associated with chemical reactions are on the order of 1 to 10 eV.

31. After 3 half-lives, 12.5% of the initial sample remains. After 7 half-lives, 0.8% of the original sample remains.

33. Assuming the elemental composition of the body is 63% H, 26% O, 10% C, and 1% N, there would be about 2×10^{28} nuclei in a 50-kg body.

35. (a)

Isotope	Mass (u)	Binding Energy (MeV)	BE/ nucleon (MeV/ nucleon)
hydrogen-1	1.007825	0.000000	0.000000
hydrogen-2	2.014102	2.224408	1.112204
hydrogen-3	3.016049	8.482184	2.827395
helium-3	3.016029	7.718359	2.572786
helium-4	4.002602	28.296925	7.074231
helium-6	6.018886	29.271267	4.878545
helium-8	8.033922	31.408115	3.926014
lithium-6	6.015121	31.995887	5.332648
lithium-7	7.016003	39.245705	5.606529
lithium-8	8.022486	41.278225	5.159778
lithium-9	9.026789	45.341402	5.037934
lithium-11	11.043897	45.548194	4.140745
beryllium-7	7.016928	37.601618	5.371660
beryllium-9	9.012174	58.172732	6.463637
beryllium-10	10.013534	64.977295	6.497730
beryllium-11	11.021657	65.482165	5.952924
beryllium-12	12.026921	68.650176	5.720848
beryllium-14	14.024866	86.707187	6.193371
boron-8	8.024605	37.739479	4.717435
boron-10	10.012936	64.751874	6.475187
boron-11	11.009305	76.205524	6.927775
boron-12	12.014352	79.575669	6.631306
boron-13	13.017780	84.453904	6.496454
boron-14	14.025404	85.423589	6.101685
boron-15	15.031100	88.189194	5.879280

(b)

Isotope	Mass (u)	Binding Energy (MeV)	BE/nucleon (MeV/nucleon)
radium-221	221.01391	1701.965290	7.701200
radium-223	223.018499	1713.833455	7.685352
radium-224	224.020187	1720.332488	7.680056
radium-226	226.025402	1731.617538	7.662025
radium-228	228.031064	1742.486210	7.642483
actinium-227	227.027749	1736.720262	7.650750
actinium-228	228.031015	1741.749398	7.639252
thorium-227	227.027701	1735.982519	7.647500
thorium-228	228.028716	1743.108448	7.645212
thorium-229	229.031757	1748.347170	7.634704
thorium-230	230.033127	1755.142419	7.631054
thorium-231	231.036299	1760.259116	7.620169
thorium 232	232.038051	1766.698534	7.615080
thorium-234	234.043593	1777.678985	7.596919
protactinium-231	231.035880	1759.866957	7.618472
protactinium-234	234.043300	1777.169458	7.594741
uranium-231	231.036264	1758.726808	7.613536
uranium-232	232.037131	1765.990598	7.612028
uranium-233	233.039630	1771.734190	7.604009
uranium-234	234.040946	1778.579740	7.600768
uranium-235	235.043924	1783.877146	7.590967
uranium-236	236.045562	1790.422754	7.586537
uranium-238	238.050784	1801.701284	7.570173
uranium-239	239.054290	1806.506861	7.558606

(c) The lighter elements have E_B/nucleon values that vary radically and stay much less than the maximum of 8.79 MeV/nucleon at $^{56-26}$Fe. The heavier elements have E_B/nucleon values that are very stable and stay right near the maximum value of 8.79 MeV/nucleon.

37. (a) 1.7 fm; (b) 2.4 fm; (c) 3.6 fm; (d) 7.0 fm; (e) 5.6 fm; (f) 6.8 fm; (g) 6.9 fm; (h) 7.4 fm

39. 1.3×10^4 m

41. (a) 1.6735×10^{-24} g
(b) 6.6463×10^{-24} g
(c) 1.4965×10^{-23} g
(d) 1.9926×10^{-23} g
(e) 9.2880×10^{-23} g
(f) 1.4929×10^{-22} g
(g) 2.1405×10^{-22} g
(h) 3.9528×10^{-22} g

43. (a) 1.112 MeV/nucleon
(b) 7.0742 MeV/nucleon
(c) 5.3326 MeV/nucleon
(d) 7.6802 MeV/nucleon
(e) 8.79031 MeV/nucleon
(f) 8.69594 MeV/nucleon
(g) 8.43602 MeV/nucleon
(h) 7.59097 MeV/nucleon

45. 4.949 MeV

47. 165.949 MeV

49. (a) 7n
(b) ^{166}Pd
(c) ^{129}Ba
(d) ^{238}U

51. n; 177.992 MeV

53. 2.1×10^2 MeV

55. 1.39×10^6 g $\approx 1 \times 10^3$ kg (rounded to one significant figure)

57. (a) n
(b) n
(c) n
(d) ^{3}He
(e) ^{1}H

59. 5.494 MeV; 12.860 MeV; 26.73 MeV

61. 9×10^{20} reactions/s

63. (a) 1.58×10^3 y; (b) The radiation emission rate is slightly less than 1 Ci.

65. 0.12 h^{-1}; 6.0 h

67. $0.015625 \approx 1.56$; $0.00552 = 0.552$

69. 120 Bq

71. 108 Bq $\approx$ 100 Bq (rounded to one significant figure)

73. 1.54×10^9 y

75. (a) ^{14}N
(b) ^{239}Pu
(c) ^{60}Co
(d) ^{3}He
(e) υ_e, ^{13}C

77. (a) 63.929406 u
(b) 9.25×10^{-4} nm

79. 7.55 MeV/nucleon; 7.44854 MeV/nucleon

81. (a) Sodium-22 will undergo β^- decay:
$^{22}\text{Na} \rightarrow \text{e}^+ + \upsilon_e + {}^{22}\text{Ne}$
(b) Sodium-24 will undergo β^+ decay:
$^{24}\text{Na} \rightarrow \text{e}^- + \bar{\upsilon}_e + {}^{24}\text{Mg}$

83. 5.9×10^9 y

85. 1.15×10^4 y

87. (a) ^{237}Np; (b) 0.26%, 8%; (c) 5.4×10^{-9} g

89. 8.14978 MeV/nucleon; $^{26}\text{Al} \rightarrow \text{e}^+ + \upsilon_e + {}^{26}\text{Mg}$; 8.33159 MeV/nucleon; 8.30992 MeV/nucleon; $^{28}\text{Al} \rightarrow e^- + \bar{\upsilon}_e + {}^{28}\text{Si}$

91. (a) 7.40 μCi; (b) 197 d $\times$ 1 mo/30 d $\approx$ 6.6 mo

93. (a) 3.79×10^{-8} g; (b) 8.99×10^{24} cm $= 8.99$ μm

95. 4.599 MeV

97. -183.364 MeV

Chapter 28

1. (a) Baryons are hadrons made up of three quarks. (b) Mesons are hadrons made up of quark–antiquark pairs. (c) Quarks are fractionally charged particles that make up hadronic material (baryons and/or mesons). (d) Leptons are particles that are driven by the electroweak force and mediated with photons and W/Z bosons. (e) Antiparticles are the negative-energy counterpart to every particle. All particles possess antiparticles that are equal in mass but opposite in charge.

3. When two particles interact through one of the four fundamental forces (strong, weak, electromagnetic, or gravity), the interaction is mediated by the creation of an exchange particle that is transferred between the two interacting particles.

5. (a) The antiparticle, $\bar{\text{n}}$, is composed of one up-bar antiquark and two down-bar antiquarks: $\overline{\text{udd}}$.
(b) The antiparticle, $\bar{\text{p}}$, is composed of two up-bar antiquarks and one down-bar antiquark: $\overline{\text{uud}}$.
(c) The antiparticle, π^-, is composed of one down quark and one up-bar antiquark: $\text{d}\bar{\text{u}}$.

7. If only one photon were created, linear momentum could not be conserved because a zero momentum photon is not possible.

9. We know the universe is expanding because the spectrum of far-away galaxies appears red-shifted, meaning that they are receding away from us. This is known as the cosmological redshift. Furthermore, the farther away a galaxy is, the faster it is receding away from us. This means that the space between galaxies, and therefore the universe, is expanding.

11. B

13. D

15. B

17. C

19. 1.27×10^{-16} m $= 0.127$ fm

21. ~ 0.001 MeV/c^2, about 500 times less massive than the electron

23. No. It is impossible because neither baryon number nor lepton number is conserved.

25. (a) Λ or Σ^0 because these particles have the correct charge and quark configuration.
(b) Ξ^0 because this particle has the correct charge and quark configuration.
(c) Ξ^0 (uss) because the sum of the masses of its constituent quarks is greater.

27. (a) not possible; (b) possible

29. $E_\gamma = 273.5$ MeV, $\lambda = 4.53 \times 10^{-6}$ nm, and $p = 1.46 \times 10^{-19}$ kg·m/s for each photon.

31. (a) 1.02 MeV; (b) The nucleus is required to absorb the momentum so that conservation of momentum is not violated.

33. (a) $K_{\mathrm{p}} = 0.312$ GeV $= K_{\bar{\mathrm{p}}}$; (b) $K_{\mathrm{p}} = 0.347$ GeV and $K_{\bar{\mathrm{p}}} = 0.277$ GeV

35.

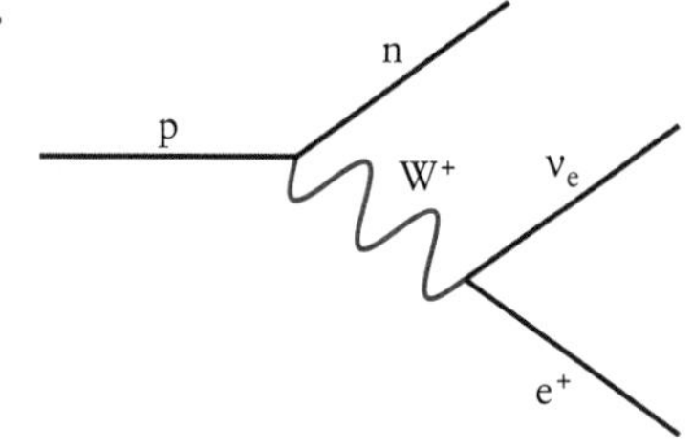

37.

e^+

γ

e^-

39. No. The Higgs particle has to have zero charge. This collection of particles will have a net negative charge.

41. 1500 km/s

43. 1.93×10^{-4} nm $= 193$ fm

45. 620 nm. These photons will appear reddish-orange in color to the human eye.

47. (a) 61.562 GeV; (b) 0.99989 c

49. (a) side length $= 0.91$ mm, volume $= 7.5 \times 10^{-10}$ m^3; (b) 3.2×10^{-18} kg/m^3

INDEX

Note: Page numbers preceded by A indicate appendices; those preceded by M indicate the Math Tutorial.

A

B

C

D

G

H

J

K

M

N

O

P

S

T

U

V

W

X

Y

Z